DISCOVERING STATISTICS USING SPSS

Student comments about *Discovering Statistics Using SPSS*, second edition

'This book is amazing, I love it! It is responsible for raising my degree classification to a 1st class. I want to thank you for making a down to earth, easy to follow no nonsense book about the stuff that really matters in statistics.' **Tim Kock**

'I wanted to tell you; I LOVE YOUR SENSE OF HUMOUR. Statistics make me cry usually but with your book I almost mastered it. All I can say is keep writing such books that make our life easier and make me love statistics.' **Naïlah Moussa**

'Just a quick note to say how fantastic your stats book is. I was very happy to find a stats book which had sensible and interesting (sex, drugs and rock and roll) worked examples.' **Josephine Booth**

'I am deeply in your debt for your having written *Discovering Statistics Using SPSS* (2nd edition). Thank you for a great contribution that has made life easier for so many of us.' **Bill Jervis Groton**

'I love the way that you write. You make this twoddle so lively. I am no longer crying, swearing, sweating or threatening to jack in this stupid degree just because I can't do the statistics. I am elated and smiling and jumping and grinning that I have come so far and managed to win over the interview panel using definitions and phrases that I read in your book!!! Bring it on you nasty exams. This candidate is Field Trained ...' **Sara Chamberlain**

'I just wanted to thank you for your book. I am working on my thesis and making sense of the statistics. Your book is wonderful!' **Katya Morgan**

'Sitting in front of a massive pile of books, in the midst of jamming up revision for exams, I cannot help distracting myself to tell you that your book keeps me smiling (or mostly laughing), although I am usually crying when I am studying. Thank you for your genius book. You have actually made a failing math student into a first class honors student, all with your amazing humor. Moreover, you have managed to convert me from absolutely detesting statistics to "actually" enjoying them. For this, I thank you immensely. At university we have a great laugh on your jokes ... till we finish our degrees your book will keep us going!' **Amber Atif Ghani**

'Your book has brought me out of the darkness to a place where I feel I might see the light and get through my exam. Stats is by far not my strong point but you make me feel like I could live with it! Thank you.' **Vicky Learmont**

'I just wanted to email you and thank you for writing your book, *Discovering Statistics Using SPSS*. I am a graduate student at the University of Victoria, Canada, and have found your book invaluable over the past few years. I hope that you will continue to write more in-depth stats books in the future! Thank you for making my life better!' **Leila Scannell**

'For a non-math book, this book is the best stat book that I have ever read.' **Dvir Kleper**

DISCOVERING STATISTICS USING SPSS THIRD EDITION

(and sex and drugs and rock 'n' roll)

ANDY FIELD

Los Angeles • London • New Delhi • Singapore • Washington DC

First edition published 2000
Second edition published 2005
This third Edition published in 2009
Reprinted 2009 (twice), 2010 and 2011 (twice)

SAGE Publications Ltd
1 Oliver's Yard
55 City Road
London EC1Y 1SP

SAGE Publications Inc.
2455 Teller Road
Thousand Oaks, California 91320

SAGE Publications India Pvt Ltd
B 1/I 1 Mohan Cooperative Industrial Area
Mathura Road
New Delhi 110 044

SAGE Publications Asia-Pacific Pte Ltd
3 Church Street
#10-04 Samsung Hub
Singapore 049483

Library of Congress Control Number: 2008930166

British Library Cataloguing in Publication data

A catalogue record for this book is available from the British Library

ISBN 978-1-84787-906-6
ISBN 978-1-84787-907-3

Typeset by C&M Digitals (P) Ltd, Chennai, India
Printed and bound in India by Replika Press Pvt. Ltd.
Printed on paper from sustainable resources

CONTENTS

PREFACE

Karma Police, arrest this man, he talks in maths, he buzzes like a fridge, he's like a detuned radio.

Radiohead *(1997)*

Introduction

Social science students despise statistics. For one thing, most have a non-mathematical background, which makes understanding complex statistical equations very difficult. The major advantage in being taught statistics in the early 1990s (as I was) compared to the 1960s was the development of computer software to do all of the hard work. The advantage of learning statistics now rather than 15 years ago is that Windows™/MacOS™ enable us to just click on stuff rather than typing in horribly confusing commands (although, as you will see, we can still type in horribly confusing commands if we want to). One of the most commonly used of these packages is SPSS; what on earth possessed me to write a book on it?

You know that you're a geek when you have favourite statistics textbooks; my favourites are Howell (2006), Stevens (2002) and Tabachnick and Fidell (2007). These three books are peerless as far as I am concerned and have taught me (and continue to teach me) more about statistics than you could possibly imagine. (I have an ambition to be cited in one of these books but I don't think that will ever happen.) So, why would I try to compete with these sacred tomes? Well, I wouldn't and I couldn't (intellectually these people are several leagues above me). However, these wonderful and clear books use computer examples as addenda to the theory. The advent of programs like SPSS provides the unique opportunity to teach statistics at a conceptual level without getting *too* bogged down in equations. However, many SPSS books concentrate on 'doing the test' at the expense of theory. Using SPSS without any statistical knowledge at all can be a dangerous thing (unfortunately, at the moment SPSS is a rather stupid tool, and it relies heavily on the users knowing what they are doing). As such, this book is an attempt to strike a good balance between theory and practice: I want to use SPSS as a tool for teaching statistical concepts in the hope that you will gain a better understanding of both theory and practice.

Primarily, I want to answer the kinds of questions that I found myself asking while learning statistics and using SPSS as an undergraduate (things like 'How can I understand how this statistical test works without knowing too much about the maths behind it?', 'What does that button do?', 'What the hell does this output mean?'). Like most academics I'm slightly high on the autistic spectrum, and I used to get fed up with people telling me to 'ignore' options or 'ignore that bit of the output'. I would lie awake for hours in my bed every night wondering 'Why is that bit of SPSS output there if we just ignore it?' So that no student has to suffer the mental anguish that I did, I aim to explain what different options do, what bits of the output mean, and if we ignore something, *why* we ignore it. Furthermore, I want to be non-prescriptive. Too many books tell the reader what to do ('click on this button', 'do this', 'do that', etc.) and this can create the impression that statistics and SPSS are inflexible. SPSS has many options designed to allow you to tailor a given test to your particular needs. Therefore, although I make recommendations, within the limits imposed by the senseless destruction of rainforests, I hope to give you enough background in theory to enable you to make your own decisions about which options are appropriate for the analysis you want to do.

A second, not in any way ridiculously ambitious, aim was to make this the only statistics textbook that anyone ever needs to buy. As such, it's a book that I hope will become your friend from first year right through to your professorship. I've tried, therefore, to write a book that can be read at several levels (see the next section for more guidance). There are chapters for first-year undergraduates (1, 2, 3, 4, 5, 6, 9 and 15), chapters for second-year undergraduates (5, 7, 10, 11, 12, 13 and 14) and chapters on more advanced topics that postgraduates might use (8, 16, 17, 18 and 19). All of these chapters should be accessible to everyone, and I hope to achieve this by flagging the level of each section (see the next section).

My third, final and most important aim is make the learning process fun. I have a sticky history with maths because I used to be terrible at it:

MATHEMATICS ADDL. MATHS.	43	59	27=	D	C	His work shows lack of discipline in thought and presentation. I hope it will mature next year.	[illegible]

Above is an extract of my school report at the age of 11. The '27' in the report is to say that I came equal 27th with another student out of a class of 29. That's almost bottom of the class. The 43 is my exam mark as a percentage! Oh dear. Four years later (at 15) this was my school report:

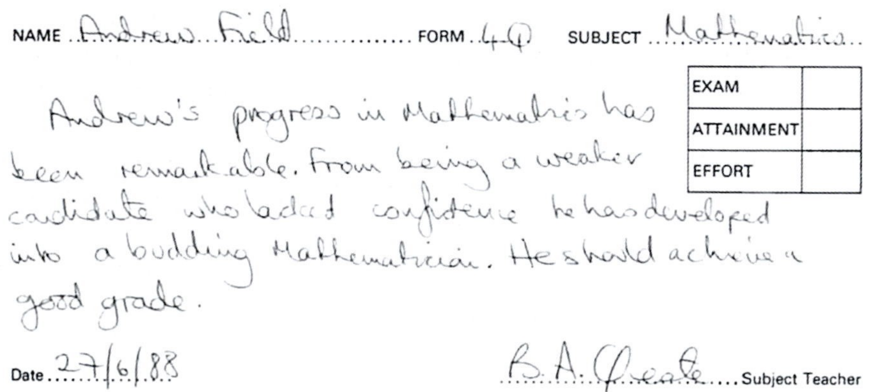
NAME Andrew Field FORM 4Q SUBJECT Mathematics

EXAM	
ATTAINMENT	
EFFORT	

Andrew's progress in Mathematics has been remarkable. From being a weaker candidate who lacked confidence he has developed into a budding Mathematician. He should achieve a good grade.

Date 27/6/88

[illegible] Subject Teacher

What led to this remarkable change? It was having a good teacher: my brother, Paul. In fact I owe my life as an academic to Paul's ability to do what my maths teachers couldn't: teach me stuff in an engaging way. To this day he still pops up in times of need to teach me things (a crash course in computer programming some Christmases ago springs to mind). Anyway, the reason he's a great teacher is because he's able to make things interesting and relevant to me. Sadly he seems to have got the 'good teaching' genes in the family (and he doesn't even work as a bloody teacher, so they're wasted!), but his approach inspires my lectures and books. One thing that I have learnt is that people appreciate the human touch, and so in previous editions I tried to inject a lot of my own personality and sense of humour (or lack of …). Many of the examples in this book, although inspired by some of the craziness that you find in the real world, are designed to reflect topics that play on the minds of the average student (i.e. sex, drugs, rock and roll, celebrity, people doing crazy stuff). There are also some examples that are there just because they made me laugh. So, the examples are light-hearted (some have said 'smutty' but I prefer 'light-hearted') and by the end, for better or worse, I think you will have some idea of what goes on in my head on a daily basis!

What's new?

Seeing as some people appreciated the style of the previous editions I've taken this as a green light to include even more stupid examples, more smut and more bad taste. I apologise to those who think it's crass, hate it, or think that I'm undermining the seriousness of science, but, come on, what's not funny about a man putting an eel up his anus?

Aside from adding more smut, I was forced reluctantly to expand the academic content! Most of the expansions have resulted from someone (often several people) emailing me to ask how to do something. So, in theory, this edition should answer any question anyone has asked me over the past four years! Mind you, I said that last time and still the questions come (will I never be free?). The general changes in the book are:

- **More introductory material**: The first chapter in the last edition was like sticking your brain into a food blender. I rushed chaotically through the entire theory of statistics in a single chapter at the pace of a cheetah on speed. I didn't really bother explaining any basic research methods, except when, out of the blue, I'd stick a section in some random chapter, alone and looking for friends. This time, I have written a brand-new Chapter 1, which eases you gently through the research process – why and how we do it. I also bring in some basic descriptive statistics at this point too.
- **More graphs**: Graphs are very important. In the previous edition information about plotting graphs was scattered about in different chapters making it hard to find. What on earth was I thinking? I've now written a self-contained chapter on how to use SPSS's Chart Builder. As such, everything you need to know about graphs (and I added a lot of material that wasn't in the previous edition) is now in Chapter 4.
- **More assumptions**: All chapters now have a section towards the end about what to do when assumptions are violated (although these usually tell you that SPSS can't do what needs to be done!).
- **More data sets**: You can never have too many examples, so I've added a lot of new data sets. There are 30 new data sets in the book at the last count (although I'm not very good at maths so it could be a few more or less).
- **More stupid faces**: I have added some more characters with stupid faces because I find stupid faces comforting, probably because I have one. You can find out more in the next section. Miraculously, the publishers stumped up some cash to get them designed by someone who can actually draw.
- **More reporting your analysis**: OK, I had these sections in the previous edition too, but then in some chapters I just seemed to forget about them for no good reason. This time every single chapter has one.
- **More glossary**: Writing the glossary last time nearly made me stick a vacuum cleaner into my ear to suck out my own brain. I thought I probably ought to expand it a bit. You can find my brain in the bottom of the vacuum cleaner in my house.
- **New! It's colour**: The publishers went full colour. This means that (1) I had to redo all of the diagrams to take advantage of the colour format, and (2) If you lick the orange bits they taste of orange (it amuses me that someone might try this to see whether I'm telling the truth).
- **New! Real-world data**: Lots of people said that they wanted more 'real data' to play with. The trouble is that real research can be quite boring. However, just for you, I trawled the world for examples of research on really fascinating topics (in my opinion). I then stalked the authors of the research until they gave me their data. Every chapter now has a real research example.
- **New! Self-test questions**: Everyone loves an exam, don't they? Well, everyone that is apart from people who breathe. Given how much everyone hates tests, I thought the

best way to commit commercial suicide was to liberally scatter tests throughout each chapter. These range from simple questions to test out what you have just learned to going back to a technique that you read about several chapters before and applying it in a new context. All of these questions have answers to them on the companion website. They are there so that you can check on your progress.

- **New! SPSS tips**: SPSS does weird things sometimes. In each chapter, I've included boxes containing tips, hints and pitfalls related to SPSS.
- **New! SPSS 17 compliant**: SPSS 17 looks different to earlier versions but in other respects is much the same. I updated the material to reflect the latest editions of SPSS.
- **New! Flash movies**: I've recorded some flash movies of using SPSS to accompany each chapter. They're on the companion website. They might help you if you get stuck.
- **New! Additional material**: Enough trees have died in the name of this book, but still it gets longer and still people want to know more. Therefore, I've written nearly 300 pages, yes, *three hundred*, of additional material for the book. So for some more technical topics and help with tasks in the book the material has been provided electronically so that (1) the planet suffers a little less, and (2) you can actually lift the book.
- **New! Multilevel modelling**: It's all the rage these days so I thought I should write a chapter on it. I didn't know anything about it, but I do now (sort of).
- **New! Multinomial logistic regression**: It doesn't get much more exciting than this; people wanted to know about logistic regression with several categorical outcomes and I always give people what they want (but only if they want smutty examples).

All of the chapters now have SPSS tips, self-test questions, additional material (Oliver Twisted boxes), real research examples (Labcoat Leni boxes), boxes on difficult topics (Jane Superbrain boxes) and flash movies. The specific changes in each chapter are:

- **Chapter 1 (Research methods)**: This is a completely new chapter. It basically talks about why and how to do research.
- **Chapter 2 (Statistics)**: I spent a lot of time rewriting this chapter but it was such a long time ago that I can't really remember what I changed. Trust me, though; it's much better than before.
- **Chapter 3 (SPSS)**: The old Chapter 2 is now SPSS 17 compliant. I restructured a lot of the material, and added some sections on other forms of variables (strings and dates).
- **Chapter 4 (Graphs)**: This chapter is completely new.
- **Chapter 5 (Assumptions)**: This retains some of the material from the old Chapter 4, but I've expanded the content to include P–P and Q–Q plots, a lot of new content on homogeneity of variance (including the variance ratio) and a new section on robust methods.
- **Chapter 6 (Correlation)**: The old Chapter 4; I redid one of the examples, added some material on confidence intervals for *r*, the biserial correlation, testing differences between dependent and independent *r*s and how certain eminent statisticians hate each other.
- **Chapter 7 (Regression)**: This chapter was already so long that the publishers banned me from extending it! Nevertheless I rewrote a few bits to make them clearer, but otherwise it's the same but with nicer diagrams and the bells and whistles that have been added to every chapter.
- **Chapter 8 (Logistic regression)**: I changed the main example from one about theory of mind (which is now an end of chapter task) to one about putting eels up your anus to cure constipation (based on a true story). Does this help you understand logistic regression? Probably not, but it really kept me entertained for days. I've extended the

chapter to include multinomial logistic regression, which was a pain because I didn't know how to do it.

- **Chapter 9 (*t*-tests)**: I stripped a lot of the methods content to go in Chapter 1, so this chapter is more purely about the *t*-test now. I added some discussion on median splits, and doing *t*-tests from only the means and standard deviations.
- **Chapter 10 (GLM 1)**: Is basically the same as the old Chapter 8.
- **Chapter 11 (GLM 2)**: Similar to the old Chapter 9, but I added a section on assumptions that now discusses the need for the covariate and treatment effect to be independent. I also added some discussion of eta-squared and partial eta-squared (SPSS produces partial eta-squared but I ignored it completely in the last edition). Consequently I restructured much of the material in this example (and I had to create a new data set when I realized that the old one violated the assumption that I had just spent several pages telling people not to violate).
- **Chapter 12 (GLM 3)**: This chapter is ostensibly the same as the old Chapter 10, but with nicer diagrams.
- **Chapter 13 (GLM 4)**: This chapter is more or less the same as the old Chapter 11. I edited it down quite a bit and restructured material so there was less repetition. I added an explanation of the between-participant sum of squares also. The first example (tutors marking essays) is now an end of chapter task, and the new example is one about celebrities eating kangaroo testicles on television. It needed to be done.
- **Chapter 14 (GLM 5)**: This chapter is very similar to the old Chapter 12 on mixed ANOVA.
- **Chapter 15 (Non-parametric statistics)**: This chapter is more or less the same as the old Chapter 13.
- **Chapter 16 (MANOVA)**: I rewrote a lot of the material on the interpretation of discriminant function analysis because I thought it pretty awful. It's better now.
- **Chapter 17 (Factor analysis)**: This chapter is very similar to the old Chapter 15. I wrote some material on interpretation of the determinant. I'm not sure why, but I did.
- **Chapter 18 (Categorical data)**: This is similar to Chapter 16 in the previous edition. I added some material on interpreting standardized residuals.
- **Chapter 19 (Multilevel linear models)**: This is a new chapter.

Goodbye

The first edition of this book was the result of two years (give or take a few weeks to write up my Ph.D.) of trying to write a statistics book that I would enjoy reading. The second edition was another two years of work and I was terrified that all of the changes would be the death of it. You'd think by now I'd have some faith in myself. Really, though, having spent an extremely intense six months in writing hell, I am still hugely anxious that I've just ruined the only useful thing that I've ever done with my life. I can hear the cries of lecturers around the world refusing to use the book because of cruelty to eels. This book has been part of my life now for over 10 years; it began and continues to be a labour of love. Despite this it isn't perfect, and I still love to have feedback (good or bad) from the people who matter most: you.

Andy
(My contact details are at www.statisticshell.com.)

HOW TO USE THIS BOOK

When the publishers asked me to write a section on 'How to use this book' it was obviously tempting to write 'Buy a large bottle of Olay anti-wrinkle cream (which you'll need to fend off the effects of ageing while you read), find a comfy chair, sit down, fold back the front cover, begin reading and stop when you reach the back cover.' However, I think they wanted something more useful.☺

What background knowledge do I need?

In essence, I assume you know nothing about statistics, but I do assume you have some very basic grasp of computers (I won't be telling you how to switch them on, for example) and maths (although I have included a quick revision of some very basic concepts so I really don't assume anything).

Do the chapters get more difficult as I go through the book?

In a sense they do (Chapter 16 on MANOVA is more difficult than Chapter 1), but in other ways they don't (Chapter 15 on non-parametric statistics is arguably less complex than Chapter 14, and Chapter 9 on the *t*-test is definitely less complex than Chapter 8 on logistic regression). Why have I done this? Well, I've ordered the chapters to make statistical sense (to me, at least). Many books teach different tests in isolation and never really give you a grip of the similarities between them; this, I think, creates an unnecessary mystery. Most of the tests in this book are the same thing expressed in slightly different ways. So, I wanted the book to tell this story. To do this I have to do certain things such as explain regression fairly early on because it's the foundation on which nearly everything else is built!

However, to help you through I've coded each section with an icon. These icons are designed to give you an idea of the difficulty of the section. It doesn't necessarily mean you can skip the sections (but see Smart Alex in the next section), but it will let you know whether a section is at about your level, or whether it's going to push you. I've based the icons on my own teaching so they may not be entirely accurate for everyone (especially as systems vary in different countries!):

① This means 'level 1' and I equate this to first-year undergraduate in the UK. These are sections that everyone should be able to understand.

② This is the next level and I equate this to second-year undergraduates in the UK. These are topics that I teach my second years and so anyone with a bit of background in statistics should be able to get to grips with them. However, some of these sections will be quite challenging even for second years. These are intermediate sections.

③ This is 'level 3' and represents difficult topics. I'd expect third-year (final-year) UK undergraduates and recent postgraduate students to be able to tackle these sections.

④ This is the highest level and represents very difficult topics. I would expect these sections to be very challenging to undergraduates and recent postgraduates, but postgraduates with a reasonable background in research methods shouldn't find them too much of a problem.

Why do I keep seeing stupid faces everywhere?

Brian Haemorrhage: Brian's job is to pop up to ask questions and look permanently confused. It's no surprise to note, therefore, that he doesn't look entirely different from the author. As the book progresses he becomes increasingly despondent. Read into that what you will.

Curious Cat: He also pops up and asks questions (because he's curious). Actually the only reason he's here is because I wanted a cat in the book … and preferably one that looks like mine. Of course the educational specialists think he needs a specific role, and so his role is to look cute and make bad cat-related jokes.

Cramming Sam: Samantha hates statistics. In fact, she thinks it's all a boring waste of time and she just wants to pass her exam and forget that she ever had to know anything about normal distributions. So, she appears and gives you a summary of the key points that you need to know. If, like Samantha, you're cramming for an exam, she will tell you the essential information to save you having to trawl through hundreds of pages of my drivel.

Jane Superbrain: Jane is the cleverest person in the whole universe (she makes Smart Alex look like a bit of an imbecile). The reason she is so clever is that she steals the brains of statisticians and eats them. Apparently they taste of sweaty tank tops, but nevertheless she likes them. As it happens, she is also able to absorb the contents of brains while she eats them. Having devoured some top statistics brains she knows all the really hard stuff and appears in boxes to tell you really advanced things that are a bit tangential to the main text. (Readers should note that Jane wasn't interested in eating my brain. That tells you all that you need to know about my statistics ability.)

Labcoat Leni: Leni is a budding young scientist and he's fascinated by real research. He says, 'Andy, man, I like an example about using an eel as a cure for constipation as much as the next man, but all of your examples are made up. Real data aren't like that, we need some real examples, dude!' So off Leni went; he walked the globe, a lone data warrior in a thankless quest for real data. He turned up at universities, cornered academics, kidnapped their families and threatened to put them in a bath of crayfish unless he was given real data. The generous ones relented, but others? Well, let's just say their families are sore. So, when you see Leni you know that you will get some real data, from a real research study to analyse. Keep it real.

Oliver Twisted: With apologies to Charles Dickens, Oliver, like his more famous fictional London urchin, is always asking, 'Please sir, can I have some more?' Unlike Master Twist, though, our young Master Twisted always wants more statistics information. Of course he does, who wouldn't? Let us not be the ones to disappoint a young, dirty, slightly smelly boy who dines on gruel, so when Oliver appears you can be certain of one thing: there is additional information to be found on the companion website. (Don't be shy; download it and bathe in the warm asp's milk of knowledge.)

Satan's Personal Statistics Slave: Satan is a busy boy – he has all of the lost souls to torture in hell; then there are the fires to keep fuelled, not to mention organizing enough carnage on the planet's surface to keep Norwegian black metal bands inspired. Like many of us, this leaves little time for him to analyse data, and this makes him very sad. So, he has his own personal slave, who, also like some of us, spends all day dressed in a gimp mask and tight leather pants in front of SPSS analysing Satan's data. Consequently, he knows a thing or two about SPSS, and when Satan's busy spanking a goat, he pops up in a box with SPSS tips.

Smart Alex: Alex is a very important character because he appears when things get particularly difficult. He's basically a bit of a smart alec and so whenever you see his face you know that something scary is about to be explained. When the hard stuff is over he reappears to let you know that it's safe to continue. Now, this is not to say that all of the rest of the material in the book is easy, he just let's you know the bits of the book that you can skip if you've got better things to do with your life than read all 800 pages! So, if you see Smart Alex then you can *skip the section* entirely and still understand what's going on. You'll also find that Alex pops up at the end of each chapter to give you some tasks to do to see whether you're as smart as he is.

What is on the companion website?

In this age of downloading, CD-ROMs are for losers (at least that's what the 'kids' tell me) so this time around I've put my cornucopia of additional funk on that worldwide interweb thing. This has two benefits: (1) The book is *slightly* lighter than it would have been, and (2) rather than being restricted to the size of a CD-ROM, there is no limit to the amount of fascinating extra material that I can give you (although Sage have had to purchase a new server to fit it all on). To enter my world of delights, go to **www.sagepub.co.uk/field3e** (see the image on the next page).

How will you know when there are extra goodies on this website? Easy-peasy, Oliver Twisted appears in the book to indicate that there's something you need (or something extra) on the website. The website contains resources for students and lecturers alike:

- **Data files**: You need data files to work through the examples in the book and they are all on the companion website. We did this so that you're forced to go there and once you're there you will never want to leave. There are data files here for a range of students, including those studying **psychology**, **business** and **health sciences.**
- **Flash movies**: Reading is a bit boring; it's much more amusing to listen to me explaining things in my camp English accent. Therefore, so that you can all have 'laugh at Andy' parties, I have created flash movies for each chapter that show you how to do the SPSS examples. I've also done extra ones that show you useful things that would otherwise have taken me pages of drivel to explain. Some of these movies are open access, but because the publishers want to sell some books, others are available only to lecturers. The idea is that they can put them on their virtual learning environments. If they don't, put insects under their office doors.
- **Podcast**: My publishers think that watching a film of me explaining what this book is all about is going to get people flocking to the bookshop. I think it will have people flocking to the medicine cabinet. Either way, if you want to see how truly uncharismatic I am, watch and cringe.

- **Self-assessment multiple-choice questions**: Organized by chapter, these will allow you to test whether wasting your life reading this book has paid off so that you can walk confidently into an examination much to the annoyance of your friends. If you fail said exam, you can employ a good lawyer and sue me.
- **Flashcard glossary**: As if a printed glossary wasn't enough, my publishers insisted that you'd like one in electronic format too. Have fun here flipping about between terms and definitions that are covered in the textbook, it's better than actually learning something.
- **Additional material**: Enough trees have died in the name of this book, but still it gets longer and still people want to know more. Therefore, I've written nearly 300 pages, yes, three hundred, of additional material for the book. So for some more technical topics and help with tasks in the book the material has been provided electronically so that (1) the planet suffers a little less, and (2) you can actually lift the book.
- **Answers**: each chapter ends with a set of tasks for you to test your newly acquired expertise. The chapters are also littered with self-test questions. How will you know if you get these correct? Well, the companion website contains around 300 hundred pages (that's a different three hundred pages to the three hundred above) of detailed answers. Will I ever stop writing?
- **Cyberworms of knowledge**: I have used nanotechnology to create cyberworms that crawl down your broadband connection, pop out of the USB port of your computer then fly through space into your brain. They re-arrange your neurons so that you understand statistics. You don't believe me? Well, you'll never know for sure unless you visit the companion website …

Happy reading, and don't get sidetracked by Facebook.

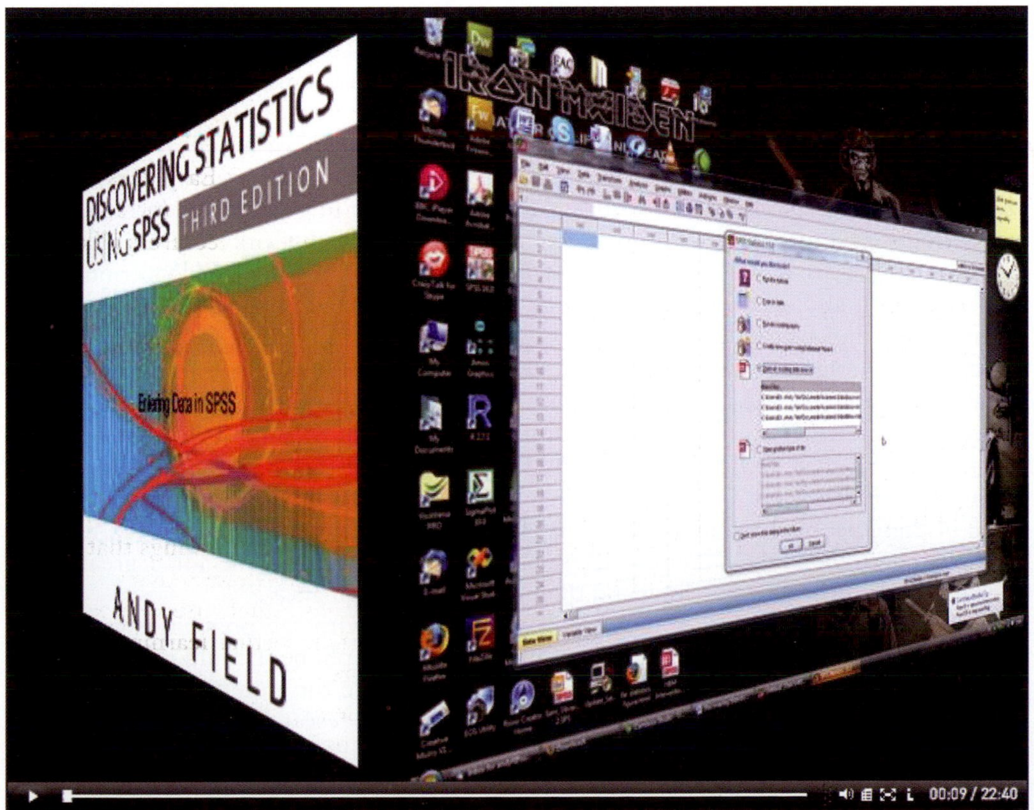

ACKNOWLEDGEMENTS

The first edition of this book wouldn't have happened if it hadn't been for Dan Wright, who not only had an unwarranted faith in a then-postgraduate to write the book, but also read and commented on draft chapters in all three editions. I'm really sad that he is leaving England to go back to the United States.

The last two editions have benefited from the following people emailing me with comments, and I really appreciate their contributions: John Alcock, Aliza Berger-Cooper, Sanne Bongers, Thomas Brügger, Woody Carter, Brittany Cornell, Peter de Heus, Edith de Leeuw, Sanne de Vrie, Jaap Dronkers, Anthony Fee, Andy Fugard, Massimo Garbuio, Ruben van Genderen, Daniel Hoppe, Tilly Houtmans, Joop Hox, Suh-Ing (Amy) Hsieh, Don Hunt, Laura Hutchins-Korte, Mike Kenfield, Ned Palmer, Jim Parkinson, Nick Perham, Thusha Rajendran, Paul Rogers, Alf Schabmann, Mischa Schirris, Mizanur Rashid Shuvra, Nick Smith, Craig Thorley, Paul Tinsley, Keith Tolfrey, Frederico Torracchi, Djuke Veldhuis, Jane Webster and Enrique Woll.

In this edition I have incorporated data sets from real research papers. All of these research papers are studies that I find fascinating and it's an honour for me to have these researchers' data in my book: Hakan Çetinkaya, Tomas Chamorro-Premuzic, Graham Davey, Mike Domjan, Gordon Gallup, Eric Lacourse, Sarah Marzillier, Geoffrey Miller, Peter Muris, Laura Nichols and Achim Schüetzwohl.

Jeremy Miles stopped me making a complete and utter fool of myself (in the book – sadly his powers don't extend to everyday life) by pointing out some glaring errors; he's also been a very nice person to know over the past few years (apart from when he's saying that draft sections of my books are, and I quote, 'bollocks'!). David Hitchin, Laura Murray, Gareth Williams and Lynne Slocombe made an enormous contribution to the last edition and all of their good work remains in this edition. In this edition, Zoë Nightingale's unwavering positivity and suggestions for many of the new chapters were invaluable. My biggest thanks go to Kate Lester who not only read every single chapter, but also kept my research laboratory ticking over while my mind was on this book. I literally could not have done it without her support and constant offers to take on extra work that she did not have to do so that I could be a bit less stressed. I am very lucky to have her in my research team.

All of these people have taken time out of their busy lives to help me out. I'm not sure what that says about their mental states, but they are all responsible for a great many improvements. May they live long and their data sets be normal.☺

Not all contributions are as tangible as those above. With the possible exception of them not understanding why sometimes I don't answer my phone, I could not have asked for more loving and proud parents – a fact that I often take for granted. Also, very early in my career Graham Hole made me realize that teaching research methods didn't have to be dull. My whole approach to teaching has been to steal all of his good ideas and I'm pleased that he has had the good grace not to ask for them back! He is also a rarity in being brilliant, funny *and* nice. I also thank my Ph.D. students Carina Ugland, Khanya Price-Evans and Saeid Rohani for their patience for the three months that I was physically away in Rotterdam, and for the three months that I was mentally away upon my return.

I appreciate everyone who has taken time to write nice reviews of this book on the various Amazon sites around the world (or any other website for that matter!). The success of this book has been in no small part due to these people being so positive and constructive in their reviews. I continue to be amazed and bowled over by the nice things that people write and if any of you are ever in Brighton, I owe you a pint!

The people at Sage are less hardened drinkers than they used to be, but I have been very fortunate to work with Michael Carmichael and Emily Jenner. Mike, despite his failings on the football field(!), has provided me with some truly memorable nights out and he also read some of my chapters this time around which, as an editor, made a pleasant change.☺ Both Emily and Mike took a lot of crap from me (especially when I was tired and stressed) and I'm grateful for their understanding. Emily I'm sure thinks I'm a grumpy sod, but she did a better job of managing me than she realizes. Also, Alex Lee did a fantastic job of turning the characters in my head into characters on the page. Thanks to Jill Rietema at SPSS Inc. who has been incredibly helpful over the past few years; it has been a pleasure working with her. The book (obviously) would not exist without SPSS Inc.'s kind permission to use screenshots of their software. Check out their web pages (http://www.SPSS.com) for support, contact information and training opportunities.

I wrote much of this edition while on sabbatical at the Department of Psychology at the Erasmus University, Rotterdam, The Netherlands. I'm grateful to the clinical research group (especially the white ape posse!) who so unreservedly made me part of the team. Part of me definitely stayed with you when I left – I hope it isn't annoying you too much.☺ Mostly, though, I thank Peter (Muris), Birgit (Mayer), Jip and Kiki who made me part of their family while in Rotterdam. They are all inspirational. I'm grateful for their kindness, hospitality, and for not getting annoyed when I was still in their kitchen having drunk all of their wine after the last tram home had gone. Mostly, I thank them for the wealth of happy memories that they gave me.

I always write listening to music. For the previous editions, I owed my sanity to: Abba, AC/DC, Arvo Pärt, Beck, The Beyond, Blondie, Busta Rhymes, Cardiacs, Cradle of Filth, DJ Shadow, Elliott Smith, Emperor, Frank Black and the Catholics, Fugazi, Genesis (Peter Gabriel era), Hefner, Iron Maiden, Janes Addiction, Love, Metallica, Massive Attack, Mercury Rev, Morrissey, Muse, Nevermore, Nick Cave, Nusrat Fateh Ali Khan, Peter Gabriel, Placebo, Quasi, Radiohead, Sevara Nazarkhan, Slipknot, Supergrass and The White Stripes. For this edition, I listened to the following, which I think tells you all that you need to know about my stress levels: 1349, Air, Angantyr, Audrey Horne, Cobalt, Cradle of Filth, Danzig, Dark Angel, Darkthrone, Death Angel, Deathspell Omega, Exodus, Fugazi, Genesis, High on Fire, Iron Maiden, The Mars Volta, Manowar, Mastodon, Megadeth, Meshuggah, Opeth, Porcupine Tree, Radiohead, Rush, Serj Tankian, She Said!, Slayer, Soundgarden, Taake, Tool and the Wedding Present.

Finally, all this book-writing nonsense requires many lonely hours (mainly late at night) of typing. Without some wonderful friends to drag me out of my dimly lit room from time to time I'd be even more of a gibbering cabbage than I already am. My eternal gratitude goes to Graham Davey, Benie MacDonald, Ben Dyson, Martin Watts, Paul Spreckley, Darren Hayman, Helen Liddle, Sam Cartwright-Hatton, Karina Knowles and Mark Franklin for reminding me that there is more to life than work. Also, my eternal gratitude to Gini Harrison, Sam Pehrson and Luke Anthony and especially my brothers of metal Doug Martin and Rob Mepham for letting me deafen them with my drumming on a regular basis. Finally, thanks to Leonora for her support while I was writing the last two editions of this book.

Dedication

Like the previous editions, this book is dedicated to my brother Paul and my cat Fuzzy, because one of them is a constant source of intellectual inspiration and the other wakes me up in the morning by sitting on me and purring in my face until I give him cat food: mornings will be considerably more pleasant when my brother gets over his love of cat food for breakfast.☺

SYMBOLS USED IN THIS BOOK

Mathematical operators

Σ	This symbol (called sigma) means 'add everything up'. So, if you see something like Σx_i it just means 'add up all of the scores you've collected'.
Π	This symbol means 'multiply everything'. So, if you see something like Πx_i it just means 'multiply all of the scores you've collected'.
$\sqrt{x}$	This means 'take the square root of x'.

Greek symbols

α	The probability of making a Type I error
β	The probability of making a Type II error
β_i	Standardized regression coefficient
χ^2	Chi-square test statistic
χ^2_F	Friedman's ANOVA test statistic
ε	Usually stands for 'error'
η^2	Eta-squared
μ	The mean of a population of scores
ρ	The correlation in the population
σ^2	The variance in a population of data
σ	The standard deviation in a population of data
$\sigma_{\bar{x}}$	The standard error of the mean
τ	Kendall's tau (non-parametric correlation coefficient)
ω^2	Omega squared (an effect size measure). This symbol also means 'expel the contents of your intestine immediately into your trousers'; you will understand why in due course

English symbols

b_i	The regression coefficient (unstandardized)
df	Degrees of freedom
e_i	The error associated with the *i*th person
F	*F*-ratio (test statistic used in ANOVA)
H	Kruskal–Wallis test statistic
k	The number of levels of a variable (i.e. the number of treatment conditions), or the number of predictors in a regression model
ln	Natural logarithm
MS	The mean squared error (Mean Square). The average variability in the data
N, n, n_i	The sample size. *N* usually denotes the total sample size, whereas *n* usually denotes the size of a particular group
P	Probability (the probability value, *p*-value or significance of a test are usually denoted by *p*)
r	Pearson's correlation coefficient
r_s	Spearman's rank correlation coefficient
r_{b}, r_{pb}	Biserial correlation coefficient and point–biserial correlation coefficient respectively
R	The multiple correlation coefficient
R^2	The coefficient of determination (i.e. the proportion of data explained by the model)
s^2	The variance of a sample of data
s	The standard deviation of a sample of data
SS	The sum of squares, or sum of squared errors to give it its full title
SS_A	The sum of squares for variable *A*
SS_M	The model sum of squares (i.e. the variability explained by the model fitted to the data)
SS_R	The residual sum of squares (i.e. the variability that the model can't explain – the error in the model)
SS_T	The total sum of squares (i.e. the total variability within the data)
t	Test statistic for Student's *t*-test
T	Test statistic for Wilcoxon's matched-pairs signed-rank test
U	Test statistic for the Mann–Whitney test
W_s	Test statistic for Wilcoxon's rank-sum test
$\bar{X}$ or $\bar{x}$	The mean of a sample of scores
z	A data point expressed in standard deviation units

SOME MATHS REVISION

1 **Two negatives make a positive**: Although in life two wrongs don't make a right, in mathematics they do! When we multiply a negative number by another negative number, the result is a positive number. For example, $-2 \times -4 = 8$.

2 **A negative number multiplied by a positive one make a negative number**: If you multiply a positive number by a negative number then the result is another negative number. For example, $2 \times -4 = -8$, or $-2 \times 6 = -12$.

3 **BODMAS**: This is an acronym for the order in which mathematical operations are performed. It stands for Brackets, Order, Division, Multiplication, Addition, Subtraction and this is the order in which you should carry out operations within an equation. Mostly these operations are self-explanatory (e.g. always calculate things within brackets first) except for order, which actually refers to power terms such as squares. Four squared, or 4^2, used to be called four raised to the order of 2, hence the reason why these terms are called 'order' in BODMAS (also, if we called it power, we'd end up with BPDMAS, which doesn't roll off the tongue quite so nicely). Let's look at an example of BODMAS: what would be the result of $1 + 3 \times 5^2$? The answer is 76 (not 100 as some of you might have thought). There are no brackets so the first thing is to deal with the order term: 5^2 is 25, so the equation becomes $1 + 3 \times 25$. There is no division, so we can move on to multiplication: 3×25, which gives us 75. BODMAS tells us to deal with addition next: $1 + 75$, which gives us 76 and the equation is solved. If I'd written the original equation as $(1 + 3) \times 5^2$, then the answer would have been 100 because we deal with the brackets first: $(1 + 3) = 4$, so the equation becomes 4×5^2. We then deal with the order term, so the equation becomes $4 \times 25 = 100$!

4 **http://www.easymaths.com** is a good site for revising basic maths.

1 Why is my evil lecturer forcing me to learn statistics?

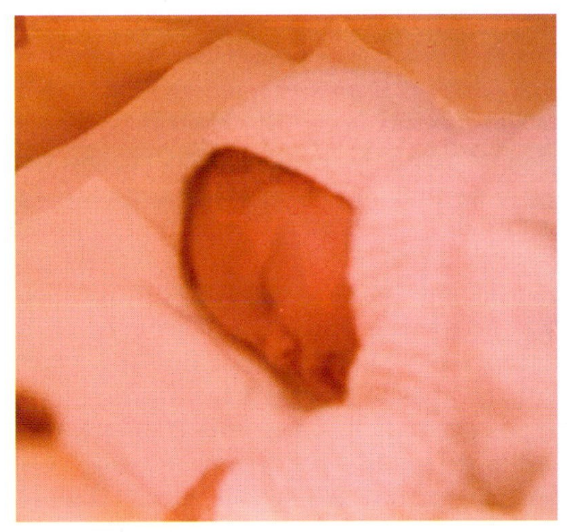

FIGURE 1.1 When I grow up, please don't let me be a statistics lecturer

1.1. What will this chapter tell me? ①

I was born on 21 June 1973. Like most people, I don't remember anything about the first few years of life and like most children I did go through a phase of driving my parents mad by asking 'Why?' every five seconds. 'Dad, why is the sky blue?', 'Dad, why doesn't mummy have a willy?' etc. Children are naturally curious about the world. I remember at the age of 3 being at a party of my friend Obe (this was just before he left England to return to Nigeria, much to my distress). It was a hot day, and there was an electric fan blowing cold air around the room. As I said, children are natural scientists and my little scientific brain was working through what seemed like a particularly pressing question: 'What happens when you stick your finger into a fan?' The answer, as it turned out, was that it hurts – a lot.[1] My point is this: my curiosity to explain the world never went away, and that's why

[1] In the 1970s fans didn't have helpful protective cages around them to prevent idiotic 3 year olds sticking their fingers into the blades.

I'm a scientist, and that's also why your evil lecturer is forcing you to learn statistics. It's because you have a curious mind too and you want to answer new and exciting questions. To answer these questions we need statistics. Statistics is a bit like sticking your finger into a revolving fan blade: sometimes it's very painful, but it does give you the power to answer interesting questions. This chapter is going to attempt to explain why statistics are an important part of doing research. We will overview the whole research process, from why we conduct research in the first place, through how theories are generated, to why we need data to test these theories. If that doesn't convince you to read on then maybe the fact that we discover whether Coca-Cola kills sperm will. Or perhaps not.

1.2. What the hell am I doing here? I don't belong here ①

You're probably wondering why you have bought this book. Maybe you liked the pictures, maybe you fancied doing some weight training (it *is* heavy), or perhaps you need to reach something in a high place (it *is* thick). The chances are, though, that given the choice of spending your hard-earned cash on a statistics book or something more entertaining (a nice novel, a trip to the cinema, etc.) you'd choose the latter. So, why have you bought the book (or downloaded an illegal pdf of it from someone who has way too much time on their hands if they can scan an 800-page textbook)? It's likely that you obtained it because you're doing a course on statistics, or you're doing some research, and you need to know how to analyse data. It's possible that you didn't realize when you started your course or research that you'd have to know this much about statistics but now find yourself inexplicably wading, neck high, through the Victorian sewer that is data analysis. The reason that you're in the mess that you find yourself in is because you have a curious mind. You might have asked yourself questions like why people behave the way they do (psychology) or why behaviours differ across cultures (anthropology), how businesses maximize their profit (business), how did the dinosaurs die (palaeontology), does eating tomatoes protect you against cancer (medicine, biology), is it possible to build a quantum computer (physics, chemistry), is the planet hotter than it used to be and in what regions (geography, environmental studies)? Whatever it is you're studying or researching, the reason you're studying it is probably because you're interested in answering questions. Scientists are curious people, and you probably are too. However, you might not have bargained on the fact that to answer interesting questions, you need two things: data and an explanation of those data.

The answer to 'what the hell are you doing here?' is, therefore, simple: to answer interesting questions you need data. Therefore, one of the reasons why your evil statistics lecturer is forcing you to learn about numbers is because they are a form of data and are vital to the research process. Of course there are forms of data other than numbers that can be used to test and generate theories. When numbers are involved the research involves **quantitative methods**, but you can also generate and test theories by analysing language (such as conversations, magazine articles, media broadcasts and so on). This involves **qualitative methods** and it is a topic for another book not written by me. People can get quite passionate about which of these methods is *best*, which is a bit silly because they are complementary, not competing, approaches and there are much more important issues in the world to get upset about. Having said that, all qualitative research is rubbish.[2]

[2] This is a joke. I thought long and hard about whether to include it because, like many of my jokes, there are people who won't find it remotely funny. Its inclusion is also making me fear being hunted down and forced to eat my own entrails by a hoard of rabid qualitative researchers. However, it made me laugh, a lot, and despite being vegetarian I'm sure my entrails will taste lovely.

1.2.1. The research process ①

How do you go about answering an interesting question? The research process is broadly summarized in Figure 1.2. You begin with an observation that you want to understand, and this observation could be anecdotal (you've noticed that your cat watches birds when they're on TV but not when jellyfish are on[3]) or could be based on some data (you've got several cat owners to keep diaries of their cat's TV habits and have noticed that lots of them watch birds on TV). From your initial observation you generate explanations, or theories, of those observations, from which you can make predictions (hypotheses). Here's where the data come into the process because to test your predictions you need data. First you collect some relevant data (and to do that you need to identify things that can be measured) and then you analyse those data. The analysis of the data may support your theory or give you cause to modify the theory. As such, the processes of data collection and analysis and generating theories are intrinsically linked: theories lead to data collection/analysis and data collection/analysis informs theories! This chapter explains this research process in more detail.

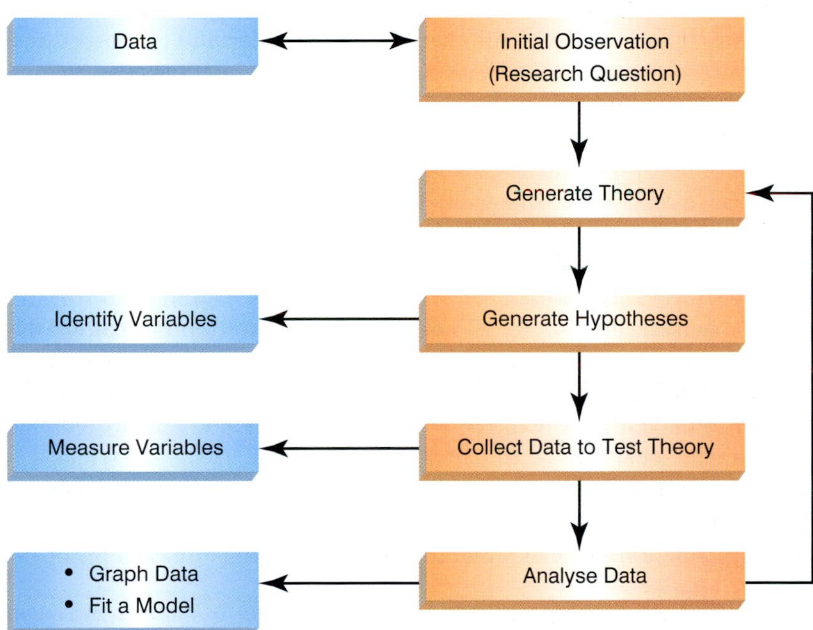

FIGURE 1.2 The research process

1.3. Initial observation: finding something that needs explaining ①

The first step in Figure 1.2 was to come up with a question that needs an answer. I spend rather more time than I should watching reality TV. Every year I swear that I won't get hooked on *Big Brother*, and yet every year I find myself glued to the TV screen waiting for

[3] My cat does actually climb up and stare at the TV when it's showing birds flying about.

the next contestant's meltdown (I am a psychologist, so really this is just research – honestly). One question I am constantly perplexed by is why every year there are so many contestants with really unpleasant personalities (my money is on narcissistic personality disorder[4]) on the show. A lot of scientific endeavour starts this way: not by watching *Big Brother*, but by observing something in the world and wondering why it happens.

Having made a casual observation about the world (*Big Brother* contestants on the whole have profound personality defects), I need to collect some data to see whether this observation is true (and not just a biased observation). To do this, I need to define one or more **variables** that I would like to measure. There's one variable in this example: the personality of the contestant. I could measure this variable by giving them one of the many well-established questionnaires that measure personality characteristics. Let's say that I did this and I found that 75% of contestants did have narcissistic personality disorder. These data support my observation: a lot of *Big Brother* contestants have extreme personalities.

1.4. Generating theories and testing them ①

The next logical thing to do is to explain these data (Figure 1.2). One explanation could be that people with narcissistic personality disorder are more likely to audition for *Big Brother* than those without. This is a **theory**. Another possibility is that the producers of *Big Brother* are more likely to select people who have narcissistic personality disorder to be contestants than those with less extreme personalities. This is another theory. We verified our original observation by collecting data, and we can collect more data to test our theories. We can make two predictions from these two theories. The first is that the number of people turning up for an audition that have narcissistic personality disorder will be higher than the general level in the population (which is about 1%). A prediction from a theory, like this one, is known as a **hypothesis** (see Jane Superbrain Box 1.1). We could test this hypothesis by getting a team of clinical psychologists to interview each person at the *Big Brother* audition and diagnose them as having narcissistic personality disorder or not. The prediction from our second theory is that if the *Big Brother* selection panel are more likely to choose people with narcissistic personality disorder then the rate of this disorder in the final contestants will be even higher than the rate in the group of people going for auditions. This is another hypothesis. Imagine we collected these data; they are in Table 1.1.

In total, 7662 people turned up for the audition. Our first hypothesis is that the percentage of people with narcissistic personality disorder will be higher at the audition than the general level in the population. We can see in the table that of the 7662 people at the audition,

TABLE 1.1 A table of the number of people at the *Big Brother* audition split by whether they had narcissistic personality disorder and whether they were selected as contestants by the producers

	No Disorder	Disorder	Total
Selected	3	9	12
Rejected	6805	845	7650
Total	6808	854	7662

[4] This disorder is characterized by (among other things) a grandiose sense of self-importance, arrogance, lack of empathy for others, envy of others and belief that others envy them, excessive fantasies of brilliance or beauty, the need for excessive admiration and exploitation of others.

854 were diagnosed with the disorder, this is about 11% (854/7662 × 100) which is much higher than the 1% we'd expect. Therefore, hypothesis 1 is supported by the data. The second hypothesis was that the *Big Brother* selection panel have a bias to choose people with narcissistic personality disorder. If we look at the 12 contestants that they selected, 9 of them had the disorder (a massive 75%). If the producers did not have a bias we would have expected only 11% of the contestants to have the disorder. The data again support our hypothesis. Therefore, my initial observation that contestants have personality disorders was verified by data, then my theory was tested using specific hypotheses that were also verified using data. Data are *very* important!

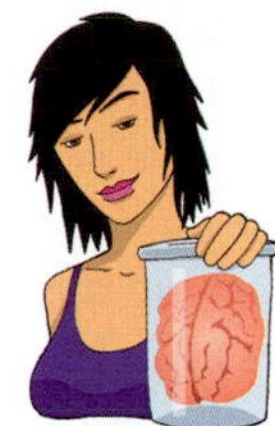

JANE SUPERBRAIN 1.1

When is a hypothesis not a hypothesis? ①

A good theory should allow us to make statements about the state of the world. Statements about the world are good things: they allow us to make sense of our world, and to make decisions that affect our future. One current example is global warming. Being able to make a definitive statement that global warming is happening, and that it is caused by certain practices in society, allows us to change these practices and, hopefully, avert catastrophe. However, not all statements are ones that can be tested using science. Scientific statements are ones that can be verified with reference to empirical evidence, whereas non-scientific statements are ones that cannot be empirically tested. So, statements such as 'The Led Zeppelin reunion concert in London in 2007 was the best gig ever',[5] 'Lindt chocolate is the best food', and 'This is the worst statistics book in the world' are all non-scientific; they cannot be proved or disproved. Scientific statements can be confirmed or disconfirmed empirically. 'Watching *Curb Your Enthusiasm* makes you happy', 'having sex increases levels of the neurotransmitter dopamine' and 'Velociraptors ate meat' are all things that can be tested empirically (provided you can quantify and measure the variables concerned). Non-scientific statements can sometimes be altered to become scientific statements, so 'The Beatles were the most influential band ever' is non-scientific (because it is probably impossible to quantify 'influence' in any meaningful way) but by changing the statement to 'The Beatles were the best-selling band ever' it becomes testable (we can collect data about worldwide record sales and establish whether The Beatles have, in fact, sold more records than any other music artist). Karl Popper, the famous philosopher of science, believed that non-scientific statements were nonsense, and had no place in science. Good theories should, therefore, produce hypotheses that are scientific statements.

I would now be smugly sitting in my office with a contented grin on my face about how my theories and observations were well supported by the data. Perhaps I would quit while I'm ahead and retire. It's more likely, though, that having solved one great mystery, my excited mind would turn to another. After another few hours (well, days probably) locked up at home watching *Big Brother* I would emerge triumphant with another profound observation, which is that these personality-disordered contestants, despite their obvious character flaws, enter the house convinced that the public will love them and that they will win.[6] My hypothesis would, therefore, be that if I asked the contestants if they thought that they would win, the people with a personality disorder would say yes.

[5] It was pretty awesome actually.

[6] One of the things I like about *Big Brother* in the UK is that year upon year the winner tends to be a nice person, which does give me faith that humanity favours the nice.

Are *Big Brother* contestants odd?
GET ANDY OUT

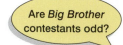

Let's imagine I tested my hypothesis by measuring their expectations of success in the show, by just asking them, 'Do you think you will win *Big Brother*?'. Let's say that 7 of 9 contestants with personality disorders said that they thought that they will win, which confirms my observation. Next, I would come up with another theory: these contestants think that they will win because they don't realize that they have a personality disorder. My hypothesis would be that if I asked these people about whether their personalities were different from other people they would say 'no'. As before, I would collect some more data and perhaps ask those who thought that they would win whether they thought that their personalities were different from the norm. All 7 contestants said that they thought their personalities were different from the norm. These data seem to contradict my theory. This is known as **falsification**, which is the act of disproving a hypothesis or theory.

It's unlikely that we would be the only people interested in why individuals who go on *Big Brother* have extreme personalities and think that they will win. Imagine these researchers discovered that: (1) people with narcissistic personality disorder think that they are more interesting than others; (2) they also think that they deserve success more than others; and (3) they also think that others like them because they have 'special' personalities.

This additional research is even worse news for my theory: if they didn't realize that they had a personality different from the norm then you wouldn't expect them to think that they were more interesting than others, and you certainly wouldn't expect them to think that others will like their unusual personalities. In general, this means that my theory sucks: it cannot explain all of the data, predictions from the theory are not supported by subsequent data, and it cannot explain other research findings. At this point I would start to feel intellectually inadequate and people would find me curled up on my desk in floods of tears wailing and moaning about my failing career (no change there then).

At this point, a rival scientist, Fester Ingpant-Stain, appears on the scene with a rival theory to mine. In his new theory, he suggests that the problem is not that personality-disordered contestants don't realize that they have a personality disorder (or at least a personality that is unusual), but that they falsely believe that this special personality is perceived positively by other people (put another way, they believe that their personality makes them likeable, not dislikeable). One hypothesis from this model is that if personality-disordered contestants are asked to evaluate what other people think of them, then they will overestimate other people's positive perceptions. To test this hypothesis, Fester Ingpant-Stain collected yet more data. When each contestant came to the diary room they had to fill out a questionnaire evaluating all of the other contestants' personalities, and also answer each question as if they were each of the contestants responding about them. (So, for every contestant there is a measure of what they thought of every other contestant, and also a measure of what they believed every other contestant thought of them.) He found out that the contestants with personality disorders did overestimate their housemate's view of them; in comparison the contestants without personality disorders had relatively accurate impressions of what others thought of them. These data, irritating as it would be for me, support the rival theory that the contestants with personality disorders know they have unusual personalities but believe that these characteristics are ones that others would feel positive about. Fester Ingpant-Stain's theory is quite good: it explains the initial observations and brings together a range of research findings. The end result of this whole process (and my career) is that we should be able to make a general statement about the state of the world. In this case we could state: '*Big Brother* contestants who have personality disorders overestimate how much other people like their personality characteristics'.

SELF-TEST Based on what you have read in this section, what qualities do you think a scientific theory should have?

1.5. Data collection 1: what to measure ①

We have seen already that data collection is vital for testing theories. When we collect data we need to decide on two things: (1) what to measure, (2) how to measure it. This section looks at the first of these issues.

1.5.1. Variables ①

1.5.1.1. Independent and dependent variables ①

To test hypotheses we need to measure variables. Variables are just things that can change (or vary); they might vary between people (e.g. IQ, behaviour) or locations (e.g. unemployment) or even time (e.g. mood, profit, number of cancerous cells). Most hypotheses can be expressed in terms of two variables: a proposed cause and a proposed outcome. For example, if we take the scientific statement 'Coca-Cola is an effective spermicide'[7] then proposed cause is 'Coca-Cola' and the proposed effect is dead sperm. Both the cause and the outcome are variables: for the cause we could vary the type of drink, and for the outcome, these drinks will kill different amounts of sperm. The key to testing such statements is to measure these two variables.

A variable that we think is a cause is known as an **independent variable** (because its value does not depend on any other variables). A variable that we think is an effect is called a **dependent variable** because the value of this variable depends on the cause (independent variable). These terms are very closely tied to experimental methods in which the cause is actually manipulated by the experimenter (as we will see in section 1.6.2). In cross-sectional research we don't manipulate any variables, and we cannot make causal statements about the relationships between variables, so it doesn't make sense to talk of dependent and independent variables because all variables are dependent variables in a sense. One possibility is to abandon the terms dependent and independent variable and use the terms **predictor variable** and **outcome variable**. In experimental work the cause, or independent variable, is a predictor, and the effect, or dependent variable, is simply an outcome. This terminology also suits cross-sectional work where, statistically at least, we can use one or more variables to make predictions about the other(s) without needing to imply causality.

CRAMMING SAM'S TIPS **Some important terms**

When doing research there are some important generic terms for variables that you will encounter:

Independent variable: A variable thought to be the cause of some effect. This term is usually used in experimental research to denote a variable that the experimenter has manipulated.

Dependent variable: A variable thought to be affected by changes in an independent variable. You can think of this variable as an outcome.

Predictor variable: A variable thought to predict an outcome variable. This is basically another term for independent variable (although some people won't like me saying that; I think life would be easier if we talked only about predictors and outcomes).

Outcome variable: A variable thought to change as a function of changes in a predictor variable. This term could be synonymous with 'dependent variable' for the sake of an easy life.

[7] Actually, there is a long-standing urban myth that a post-coital douche with the contents of a bottle of Coke is an effective contraceptive. Unbelievably, this hypothesis has been tested and Coke does affect sperm motility, and different types of Coke are more or less effective – Diet Coke is best apparently (Umpierre, Hill, & Anderson, 1985). Nevertheless, a Coke douche is ineffective at preventing pregnancy.

1.5.1.2. Levels of measurement ①

As we have seen in the examples so far, variables can take on many different forms and levels of sophistication. The relationship between what is being measured and the numbers that represent what is being measured is known as the **level of measurement**. Broadly speaking, variables can be categorical or continuous, and can have different levels of measurement.

A **categorical variable** is made up of categories. A categorical variable that you should be familiar with already is your species (e.g. human, domestic cat, fruit bat, etc.). You are a human or a cat or a fruit bat: you cannot be a bit of a cat and a bit of a bat, and neither a batman nor (despite many fantasies to the contrary) a catwoman (not even one in a nice PVC suit) exist. A categorical variable is one that names distinct entities. In its simplest form it names just two distinct types of things, for example male or female. This is known as a **binary variable**. Other examples of binary variables are being alive or dead, pregnant or not, and responding 'yes' or 'no' to a question. In all cases there are just two categories and an entity can be placed into only one of the two categories.

When two things that are equivalent in some sense are given the same name (or number), but there are more than two possibilities, the variable is said to be a **nominal variable**. It should be obvious that if the variable is made up of names it is pointless to do arithmetic on them (if you multiply a human by a cat, you do not get a hat). However, sometimes numbers are used to denote categories. For example, the numbers worn by players in a rugby or football (soccer) team. In rugby, the numbers of shirts denote specific field positions, so the number 10 is always worn by the fly-half (e.g. England's Jonny Wilkinson),[8] and the number 2 is always the hooker (the ugly-looking player at the front of the scrum). These numbers do not tell us anything other than what position the player plays. We could equally have shirts with FH and H instead of 10 and 1. A number 10 player is not necessarily better than a number 1 (most managers would not want their fly-half stuck in the front of the scrum!). It is equally as daft to try to do arithmetic with nominal scales where the categories are denoted by numbers: the number 10 takes penalty kicks, and if the England coach found that Jonny Wilkinson (his number 10) was injured he would not get his number 4 to give number 6 a piggyback and then take the kick. The only way that nominal data can be used is to consider frequencies. For example, we could look at how frequently number 10s score tries compared to number 4s.

JANE SUPERBRAIN 1.2

Self-report data ①

A lot of self-report data are ordinal. Imagine if two judges at our beauty pageant were asked to rate Billie's beauty on a 10-point scale. We might be confident that a judge who gives a rating of 10 found Billie more beautiful than one who gave a rating of 2, but can we be certain that the first judge found her five times more beautiful than the second? What about if both judges gave a rating of 8, could we be sure they found her equally beautiful? Probably not: their ratings will depend on their subjective feelings about what constitutes beauty. For these reasons, in any situation in which we ask people to rate something subjective (e.g. rate their preference for a product, their confidence about an answer, how much they have understood some medical instructions) we should probably regard these data as ordinal although many scientists do not.

[8] Unlike, for example, NFL American football where a quarterback could wear any number from 1 to 19.

So far the categorical variables we have considered have been unordered (e.g. different brands of Coke with which you're trying to kill sperm) but they can be ordered too (e.g. increasing concentrations of Coke with which you're trying to kill sperm). When categories are ordered, the variable is known as an **ordinal variable**. Ordinal data tell us not only that things have occurred, but also the order in which they occurred. However, these data tell us nothing about the differences between values. Imagine we went to a beauty pageant in which the three winners were Billie, Freema and Elizabeth. The names of the winners don't provide any information about where they came in the contest; however labelling them according to their performance does – first, second and third. These categories are ordered. In using ordered categories we now know that the woman who won was better than the women who came second and third. We still know nothing about the differences between categories, though. We don't, for example, know how much better the winner was than the runners-up: Billie might have been an easy victor, getting much higher ratings from the judges than Freema and Elizabeth, or it might have been a very close contest that she won by only a point. Ordinal data, therefore, tell us more than nominal data (they tell us the order in which things happened) but they still do not tell us about the differences between points on a scale.

The next level of measurement moves us away from categorical variables and into continuous variables. A **continuous variable** is one that gives us a score for each person and can take on any value on the measurement scale that we are using. The first type of continuous variable that you might encounter is an **interval variable**. Interval data are considerably more useful than ordinal data and most of the statistical tests in this book rely on having data measured at this level. To say that data are interval, we must be certain that equal intervals on the scale represent equal differences in the property being measured. For example, on www.ratemyprofessors.com students are encouraged to rate their lecturers on several dimensions (some of the lecturers' rebuttals of their negative evaluations are worth a look). Each dimension (i.e. helpfulness, clarity, etc.) is evaluated using a 5-point scale. For this scale to be interval it must be the case that the difference between helpfulness ratings of 1 and 2 is the same as the difference between say 3 and 4, or 4 and 5. Similarly, the difference in helpfulness between ratings of 1 and 3 should be identical to the difference between ratings of 3 and 5. Variables like this that look interval (and are treated as interval) are often ordinal – see Jane Superbrain Box 1.2.

Ratio variables go a step further than interval data by requiring that in addition to the measurement scale meeting the requirements of an interval variable, the ratios of values along the scale should be meaningful. For this to be true, the scale must have a true and meaningful zero point. In our lecturer ratings this would mean that a lecturer rated as 4 would be twice as helpful as a lecturer rated with a 2 (who would also be twice as helpful as a lecturer rated as 1!). The time to respond to something is a good example of a ratio variable. When we measure a reaction time, not only is it true that, say, the difference between 300 and 350 ms (a difference of 50 ms) is the same as the difference between 210 and 260 ms or 422 and 472 ms, but also it is true that distances along the scale are divisible: a reaction time of 200ms is twice as long as a reaction time of 100 ms and twice as short as a reaction time of 400 ms.

Continuous variables can be, well, continuous (obviously) but also discrete. This is quite a tricky distinction (Jane Superbrain Box 1.3). A truly continuous variable can be measured to any level of precision, whereas a **discrete variable** can take on only certain values (usually whole numbers) on the scale. What does this actually mean? Well, our example in the text of rating lecturers on a 5-point scale is an example of a discrete variable. The range of the scale is 1–5, but you can enter only values of 1, 2, 3, 4 or 5; you cannot enter a value of 4.32 or 2.18. Although a continuum exists underneath the scale (i.e. a rating of 3.24 makes sense), the actual values that the variable takes on are limited. A continuous variable would be something like age, which can be measured at an infinite level of precision (you could be 34 years, 7 months, 21 days, 10 hours, 55 minutes, 10 seconds, 100 milliseconds, 63 microseconds, 1 nanosecond old).

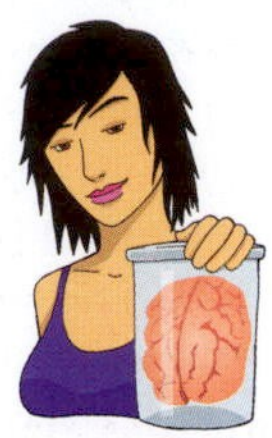

JANE SUPERBRAIN 1.3

Continuous and discrete variables ①

The distinction between discrete and continuous variables can be very blurred. For one thing, continuous variables can be measured in discrete terms; for example, when we measure age we rarely use nanoseconds but use years (or possibly years and months). In doing so we turn a continuous variable into a discrete one (the only acceptable values are years). Also, we often treat discrete variables as if they were continuous. For example, the number of boyfriends/girlfriends that you have had is a discrete variable (it will be, in all but the very weird cases, a whole number). However, you might read a magazine that says 'the average number of boyfriends that women in their 20s have has increased from 4.6 to 8.9'. This assumes that the variable is continuous, and of course these averages are meaningless: no one in their sample actually had 8.9 boyfriends.

CRAMMING SAM'S TIPS Levels of measurement

Variables can be split into categorical and continuous, and within these types there are different levels of measurement:

Categorical (entities are divided into distinct categories):

- **Binary variable:** There are only two categories (e.g. dead or alive).
- **Nominal variable:** There are more than two categories (e.g. whether someone is an omnivore, vegetarian, vegan, or fruitarian).
- **Ordinal variable:** The same as a nominal variable but the categories have a logical order (e.g. whether people got a fail, a pass, a merit or a distinction in their exam).

Continuous (entities get a distinct score):

- **Interval variable:** Equal intervals on the variable represent equal differences in the property being measured (e.g. the difference between 6 and 8 is equivalent to the difference between 13 and 15).
- **Ratio variable:** The same as an interval variable, but the ratios of scores on the scale must also make sense (e.g. a score of 16 on an anxiety scale means that the person is, in reality, twice as anxious as someone scoring 8).

1.5.2. Measurement error ①

We have seen that to test hypotheses we need to measure variables. Obviously, it's also important that we measure these variables accurately. Ideally we want our measure to be calibrated such that values have the same meaning over time and across situations. Weight is one example: we would expect to weigh the same amount regardless of who weighs us, or where we take the measurement (assuming it's on Earth and not in an anti-gravity chamber). Sometimes variables can be directly measured (profit, weight, height) but in other cases we are forced to use indirect measures such as self-report, questionnaires and computerized tasks (to name a few).

Let's go back to our Coke as a spermicide example. Imagine we took some Coke and some water and added them to two test tubes of sperm. After several minutes, we measured the motility (movement) of the sperm in the two samples and discovered no difference. A few years passed and another scientist, Dr Jack Q. Late, replicated the study but found that sperm motility was worse in the Coke sample. There are two measurement-related issues that could explain his success and our failure: (1) Dr Late might have used more Coke in the test tubes (sperm might need a critical mass of Coke before they are affected); (2) Dr Late measured the outcome (motility) differently to us.

The former point explains why chemists and physicists have devoted many hours to developing standard units of measurement. If you had reported that you'd used 100 ml of Coke and 5 ml of sperm, then Dr Late could have ensured that he had used the same amount – because millilitres are a standard unit of measurement we would know that Dr Late used exactly the same amount of Coke that we used. Direct measurements such as the millilitre provide an objective standard: 100 ml of a liquid is known to be twice as much as only 50 ml.

The second reason for the difference in results between the studies could have been to do with how sperm motility was measured. Perhaps in our original study we measured motility using absorption spectrophotometry, whereas Dr Late used laser light-scattering techniques.[9] Perhaps his measure is more sensitive than ours.

There will often be a discrepancy between the numbers we use to represent the thing we're measuring and the actual value of the thing we're measuring (i.e. the value we would get if we could measure it directly). This discrepancy is known as **measurement error**. For example, imagine that you know as an absolute truth that you weight 83 kg. One day you step on the bathroom scales and it says 80 kg. There is a difference of 3 kg between your actual weight and the weight given by your measurement tool (the scales): there is a measurement error of 3 kg. Although properly calibrated bathroom scales should produce only very small measurement errors (despite what we might want to believe when it says we have gained 3 kg), self-report measures do produce measurement error because factors other than the one you're trying to measure will influence how people respond to our measures. Imagine you were completing a questionnaire that asked you whether you had stolen from a shop. If you had, would you admit it, or might you be tempted to conceal this fact?

1.5.3. Validity and reliability ①

One way to try to ensure that measurement error is kept to a minimum is to determine properties of the measure that give us confidence that it is doing its job properly. The first property is **validity**, which is whether an instrument actually measures what it sets out to measure. The second is **reliability**, which is whether an instrument can be interpreted consistently across different situations.

Validity refers to whether an instrument measures what it was designed to measure; a device for measuring sperm motility that actually measures sperm count is not valid. Things like reaction times and physiological measures are valid in the sense that a reaction time does in fact measure the time taken to react and skin conductance does measure the conductivity of your skin. However, if we're using these things to infer other things (e.g. using skin conductance to measure anxiety) then they will be valid only if there are no other factors other than the one we're interested in that can influence them.

Criterion validity is whether the instrument is measuring what it claims to measure (does your lecturers' helpfulness rating scale actually measure lecturers' helpfulness?). In an ideal world, you could assess this by relating scores on your measure to real-world observations.

[9] In the course of writing this chapter I have discovered more than I think is healthy about the measurement of sperm.

For example, we could take an objective measure of how helpful lecturers were and compare these observations to student's ratings on ratemyprofessor.com. This is often impractical and, of course, with attitudes you might not be interested in the reality so much as the person's perception of reality (you might not care whether they are a psychopath but whether they think they are a psychopath). With self-report measures/questionnaires we can also assess the degree to which individual items represent the construct being measured, and cover the full range of the construct (**content validity**).

Validity is a necessary but not sufficient condition of a measure. A second consideration is reliability, which is the ability of the measure to produce the same results under the same conditions. To be valid the instrument must first be reliable. The easiest way to assess reliability is to test the same group of people twice: a reliable instrument will produce similar scores at both points in time (**test–retest reliability**). Sometimes, however, you will want to measure something that does vary over time (e.g. moods, blood-sugar levels, productivity). Statistical methods can also be used to determine reliability (we will discover these in Chapter 17).

SELF-TEST What is the difference between reliability and validity?

1.6. Data collection 2: how to measure ①

1.6.1. Correlational research methods ①

So far we've learnt that scientists want to answer questions, and that to do this they have to generate data (be they numbers or words), and to generate good data they need to use accurate measures. We move on now to look briefly at how the data are collected. If we simplify things quite a lot then there are two ways to test a hypothesis: either by observing what naturally happens, or by manipulating some aspect of the environment and observing the effect it has on the variable that interests us.

The main distinction between what we could call **correlational** or **cross-sectional research** (where we observe what naturally goes on in the world without directly interfering with it) and **experimental research** (where we manipulate one variable to see its effect on another) is that experimentation involves the direct manipulation of variables. In correlational research we do things like observe natural events or we take a snapshot of many variables at a single point in time. As some examples, we might measure pollution levels in a stream and the numbers of certain types of fish living there; lifestyle variables (smoking, exercise, food intake) and disease (cancer, diabetes); workers' job satisfaction under different managers; or children's school performance across regions with different demographics. Correlational research provides a very natural view of the question we're researching because we are not influencing what happens and the measures of the variables should not be biased by the researcher being there (this is an important aspect of **ecological validity**).

At the risk of sounding like I'm absolutely obsessed with using Coke as a contraceptive (I'm not, but my discovery that people in the 1950s and 1960s actually tried this has, I admit, intrigued me), let's return to that example. If we wanted to answer the question 'is Coke an effective contraceptive?' we could administer questionnaires about sexual practices (quantity of sexual activity, use of contraceptives, use of fizzy drinks as contraceptives, pregnancy, etc.). By looking at these variables we could see which variables predict pregnancy, and in particular whether those reliant on Coke as a form of contraceptive were more likely to end up pregnant than those using other contraceptives, and less likely than those using no contraceptives at all. This is the only way to answer a question like this because we cannot manipulate any of these variables particularly easily. Even if we could, it would be totally unethical to insist on some people using Coke as a contraceptive (or indeed to do anything that would make a person likely to produce a child that they didn't intend to produce). However, there is a price to pay, which relates to causality.

1.6.2. Experimental research methods ①

Most scientific questions imply a causal link between variables; we have seen already that dependent and independent variables are named such that a causal connection is implied (the dependent variable *depends* on the independent variable). Sometimes the causal link is very obvious in the research question: 'Does low self-esteem cause dating anxiety?' Sometimes the implication might be subtler, such as 'Is dating anxiety all in the mind?' The implication is that a person's mental outlook causes them to be anxious when dating. Even when the cause–effect relationship is not explicitly stated, most research questions can be broken down into a proposed cause (in this case mental outlook) and a proposed outcome (dating anxiety). Both the cause and the outcome are variables: for the cause some people will perceive themselves in a negative way (so it is something that varies); and for the outcome, some people will get anxious on dates and others won't (again, this is something that varies). The key to answering the research question is to uncover how the proposed cause and the proposed outcome relate to each other; is it the case that the people who have a low opinion of themselves are the same people that get anxious on dates?

David Hume (1748; see Hume (1739–40) for more detail),[10] an influential philosopher, said that to infer cause and effect: (1) cause and effect must occur close together in time (contiguity); (2) the cause must occur before an effect does; and (3) the effect should never occur without the presence of the cause. These conditions imply that causality can be inferred through corroborating evidence: cause is equated to high degrees of correlation between contiguous events. In our dating example, to infer that low self-esteem caused dating anxiety, it would be sufficient to find that whenever someone had low self-esteem they would feel anxious when on a date, that the low self-esteem emerged before the dating anxiety did, and that the person should never have dating anxiety if they haven't been suffering from low self-esteem.

In the previous section on correlational research, we saw that variables are often measured simultaneously. The first problem with doing this is that it provides no information about the contiguity between different variables: we might find from a questionnaire study that people with low self-esteem also have dating anxiety but we wouldn't know whether the low self-esteem or the dating anxiety came first.

Let's imagine that we find that there are people who have low self-esteem but do not get dating anxiety. This finding doesn't violate Hume's rules: he doesn't say anything about the cause happening without the effect. It could be that both low self-esteem and dating anxiety are caused by a third variable (e.g. poor social skills which might make you feel generally worthless but also puts pressure on you in dating situations). This illustrates a second

[10] Both of these can be read online at http://www.utilitarian.net/hume/ or by doing a Google search for David Hume.

problem with correlational evidence: the **tertium quid** ('a third person or thing of indeterminate character'). For example, a correlation has been found between having breast implants and suicide (Koot, Peeters, Granath, Grobbee, & Nyren, 2003). However, it is unlikely that having breast implants causes you to commit suicide – presumably, there is an external factor (or factors) that causes both; for example, low self-esteem might lead you to have breast implants and also attempt suicide. These extraneous factors are sometimes called **confounding variables** or confounds for short.

What's the difference between experimental and correlational research?

The shortcomings of Hume's criteria led John Stuart Mill (1865) to add a further criterion: that all other explanations of the cause–effect relationship be ruled out. Put simply, Mill proposed that, to rule out confounding variables, an effect should be present when the cause is present and that when the cause is absent the effect should be absent also. Mill's ideas can be summed up by saying that the only way to infer causality is through comparison of two controlled situations: one in which the cause is present and one in which the cause is absent. This is what *experimental methods* strive to do: to provide a comparison of situations (usually called *treatments* or *conditions*) in which the proposed cause is present or absent.

As a simple case, we might want to see what the effect of motivators has on learning about statistics. I might, therefore, randomly split some students into three different groups in which I change my style of teaching in the seminars on the course:

- **Group 1 (positive reinforcement)**: During seminars I congratulate all students in this group on their hard work and success. Even when they get things wrong, I am supportive and say things like 'that was very nearly the right answer, you're coming along really well' and then give them a nice piece of chocolate.
- **Group 2 (punishment)**: This group receives seminars in which I give relentless verbal abuse to all of the students even when they give the correct answer. I demean their contributions and am patronizing and dismissive of everything they say. I tell students that they are stupid, worthless and shouldn't be doing the course at all.
- **Group 3 (no motivator)**: This group receives normal university-style seminars (some might argue that this is the same as group 2!). Students are not praised or punished and instead I give them no feedback at all.

The thing that I have manipulated is the teaching method (positive reinforcement, punishment or no motivator). As we have seen earlier in this chapter, this variable is known as the independent variable and in this situation it is said to have three *levels*, because it has been manipulated in three ways (i.e. motivator has been split into three types: positive reinforcement, punishment and none). Once I have carried out this manipulation I must have some kind of outcome that I am interested in measuring. In this case it is statistical ability, and I could measure this variable using a statistics exam after the last seminar. We have also already discovered that this outcome variable is known as the dependent variable because we assume that these scores will depend upon the type of teaching method used (the independent variable). The critical thing here is the inclusion of the 'no motivator' group because this is a group where our proposed cause (motivator) is absent, and we can compare the outcome in this group against the two situations where the proposed cause is present. If the statistics scores are different in each of the motivation groups (cause is present) compared to the group for which no motivator was given (cause is absent) then this difference can be attributed to the types of motivator used. In other words, the type of motivator used caused a difference in statistics scores (Jane Superbrain Box 1.4).

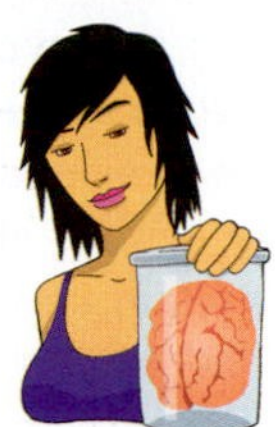

JANE SUPERBRAIN 1.4

Causality and statistics ①

People sometimes get confused and think that certain statistical procedures allow causal inferences and others don't. This isn't true, it's the fact that in experiments we manipulate the causal variable systematically to see its effect on an outcome (the effect). In correlational research we observe the co-occurrence of variables; we do not manipulate the causal variable first and then measure the effect, therefore we cannot compare the effect when the causal variable is present against when it is absent. In short, we cannot say which variable causes a change in the other; we can merely say that the variables co-occur in a certain way. The reason why some people think that certain statistical tests allow causal inferences is because historically certain tests (e.g. ANOVA, *t*-tests, etc.) have been used to analyse experimental research, whereas others (e.g. regression, correlation) have been used to analyse correlational research (Cronbach, 1957). As you'll discover, these statistical procedures are, in fact, mathematically identical.

1.6.2.1. Two methods of data collection ①

When we collect data in an experiment, we can choose between two methods of data collection. The first is to manipulate the independent variable using different participants. This method is the one described above, in which different groups of people take part in each experimental condition (a **between-groups**, **between-subjects** or **independent design**). The second method is to manipulate the independent variable using the same participants. Simplistically, this method means that we give a group of students positive reinforcement for a few weeks and test their statistical abilities and then begin to give this same group punishment for a few weeks before testing them again, and then finally giving them no motivator and testing them for a third time (a **within-subject** or **repeated-measures design**). As you will discover, the way in which the data are collected determines the type of test that is used to analyse the data.

1.6.2.2. Two types of variation ①

Imagine we were trying to see whether you could train chimpanzees to run the economy. In one training phase they are sat in front of a chimp-friendly computer and press buttons which change various parameters of the economy; once these parameters have been changed a figure appears on the screen indicating the economic growth resulting from those parameters. Now, chimps can't read (I don't think) so this feedback is meaningless. A second training phase is the same except that if the economic growth is good, they get a banana (if growth is bad they do not) – this feedback is valuable to the average chimp. This is a repeated-measures design with two conditions: the same chimps participate in condition 1 *and* in condition 2.

Let's take a step back and think what would happen if we did *not* introduce an experimental manipulation (i.e. there were no bananas in the second training phase so condition 1 and condition 2 were identical). If there is no experimental manipulation then we expect a chimp's behaviour to be the same in both conditions. We expect this because external factors such as age, gender, IQ, motivation and arousal will be similar for both conditions (a chimp's gender etc. will not change from when they are tested in condition 1 to when they are tested in condition 2). If the performance measure is reliable (i.e. our test of how well they run the economy), and the variable or characteristic that we are measuring (in this case ability to run an economy) remains stable over time, then a participant's performance in condition 1 should be very highly related to their performance in condition 2. So, chimps who score highly in condition 1 will also score highly in condition 2, and those who have low scores for condition 1 will have low scores in condition 2. However, performance won't be *identical*, there will be small differences in performance created by unknown factors. This variation in performance is known as **unsystematic variation**.

If we introduce an experimental manipulation (i.e. provide bananas as feedback in one of the training sessions), then we do something different to participants in condition 1 to what we do to them in condition 2. So, the *only* difference between conditions 1 and 2 is the manipulation that the experimenter has made (in this case that the chimps get bananas as a positive reward in one condition but not in the other). Therefore, any differences between the means of the two conditions is probably due to the experimental manipulation. So, if the chimps perform better in one training phase than the other then this *has* to be due to the fact that bananas were used to provide feedback in one training phase but not the other. Differences in performance created by a specific experimental manipulation are known as **systematic variation**.

Now let's think about what happens when we use different participants – an independent design. In this design we still have two conditions, but this time different participants participate in each condition. Going back to our example, one group of chimps receives training without feedback, whereas a second group of different chimps does receive feedback on their performance via bananas.[11] Imagine again that we didn't have an experimental manipulation. If we did nothing to the groups, then we would still find some variation in behaviour between the groups because they contain different chimps who will vary in their ability, motivation, IQ and other factors. In short, the type of factors that were held constant in the repeated-measures design are free to vary in the independent-measures design. So, the unsystematic variation will be bigger than for a repeated-measures design. As before, if we introduce a manipulation (i.e. bananas) then we will see additional variation created by this manipulation. As such, in both the repeated-measures design and the independent-measures design there are always two sources of variation:

- **Systematic variation**: This variation is due to the experimenter doing something to all of the participants in one condition but not in the other condition.
- **Unsystematic variation**: This variation results from random factors that exist between the experimental conditions (such as natural differences in ability, the time of day, etc.).

The role of statistics is to discover how much variation there is in performance, and then to work out how much of this is systematic and how much is unsystematic.

In a repeated-measures design, differences between two conditions can be caused by only two things: (1) the manipulation that was carried out on the participants, or (2) any other factor that might affect the way in which a person performs from one time to the next. The latter factor is likely to be fairly minor compared to the influence of the experimental manipulation. In an independent design, differences between the two conditions can also be caused by one of two things: (1) the manipulation that was carried out on the participants, or (2) differences between the characteristics of the people allocated to each of the groups. The latter factor in this instance is likely to create considerable random variation both within each condition and between them. Therefore, the effect of our experimental manipulation is likely to be more apparent in a repeated-measures design than in a between-group design,

[11] When I say 'via' I don't mean that the bananas developed little banana mouths that opened up and said 'well done old chap, the economy grew that time' in chimp language. I mean that when they got something right they received a banana as a reward for their correct response.

because in the former unsystematic variation can be caused only by differences in the way in which someone behaves at different times. In independent designs we have differences in innate ability contributing to the unsystematic variation. Therefore, this error variation will almost always be much larger than if the same participants had been used. When we look at the effect of our experimental manipulation, it is always against a background of 'noise' caused by random, uncontrollable differences between our conditions. In a repeated-measures design this 'noise' is kept to a minimum and so the effect of the experiment is more likely to show up. This means that, other things being equal, repeated-measures designs have more power to detect effects than independent designs.

1.6.3. Randomization ①

In both repeated-measures and independent-measures designs it is important to try to keep the unsystematic variation to a minimum. By keeping the unsystematic variation as small as possible we get a more sensitive measure of the experimental manipulation. Generally, scientists use the **randomization** of participants to treatment conditions to achieve this goal. Many statistical tests work by identifying the systematic and unsystematic sources of variation and then comparing them. This comparison allows us to see whether the experiment has generated considerably more variation than we would have got had we just tested participants without the experimental manipulation. Randomization is important because it eliminates most other sources of systematic variation, which allows us to be sure that any systematic variation between experimental conditions is due to the manipulation of the independent variable. We can use randomization in two different ways depending on whether we have an independent- or repeated-measures design.

Let's look at a repeated-measures design first. When the same people participate in more than one experimental condition they are naive during the first experimental condition but they come to the second experimental condition with prior experience of what is expected of them. At the very least they will be familiar with the dependent measure (e.g. the task they're performing). The two most important sources of systematic variation in this type of design are:

- **Practice effects**: Participants may perform differently in the second condition because of familiarity with the experimental situation and/or the measures being used.
- **Boredom effects**: Participants may perform differently in the second condition because they are tired or bored from having completed the first condition.

Although these effects are impossible to eliminate completely, we can ensure that they produce no systematic variation between our conditions by **counterbalancing** the order in which a person participates in a condition. We can use randomization to determine in which order the conditions are completed. That is, we randomly determine whether a participant completes condition 1 before condition 2, or condition 2 before condition 1. Let's look at the teaching method example and imagine that there were just two conditions: no motivator and punishment. If the same participants were used in all conditions, then we might find that statistical ability was higher after the punishment condition. However, if every student experienced the punishment after the no motivator seminars then they would enter the punishment condition already having a better knowledge of statistics than when they began the no motivator condition. So, the apparent improvement after punishment would not be due to the experimental manipulation (i.e. it's not because punishment works), but because participants had attended more statistics seminars by the end of the punishment condition compared to the no motivator one. We can use randomization to ensure that the number of statistics seminars does not introduce a systematic bias by randomly assigning students to have the punishment seminars first or the no motivator seminars first.

If we turn our attention to independent designs, a similar argument can be applied. We know that different participants participate in different experimental conditions and that these participants will differ in many respects (their IQ, attention span, etc.). Although we know that these confounding variables contribute to the variation between conditions, we need to make sure that these variables contribute to the unsystematic variation and *not* the systematic variation. The way to ensure that confounding variables are unlikely to contribute systematically to the variation between experimental conditions is to randomly allocate participants to a particular experimental condition. This should ensure that these confounding variables are evenly distributed across conditions.

A good example is the effects of alcohol on personality. You might give one group of people 5 pints of beer, and keep a second group sober, and then count how many fights each person gets into. The effect that alcohol has on people can be very variable because of different tolerance levels: teetotal people can become very drunk on a small amount, while alcoholics need to consume vast quantities before the alcohol affects them. Now, if you allocated a bunch of teetotal participants to the condition that consumed alcohol, then you might find no difference between them and the sober group (because the teetotal participants are all unconscious after the first glass and so can't become involved in any fights). As such, the person's prior experiences with alcohol will create systematic variation that cannot be dissociated from the effect of the experimental manipulation. The best way to reduce this eventuality is to randomly allocate participants to conditions.

SELF-TEST Why is randomization important?

1.7. Analysing data ①

The final stage of the research process is to analyse the data you have collected. When the data are quantitative this involves both looking at your data graphically to see what the general trends in the data are, and also fitting statistical models to the data.

1.7.1. Frequency distributions ①

Once you've collected some data a very useful thing to do is to plot a graph of how many times each score occurs. This is known as a **frequency distribution**, or **histogram**, which is a graph plotting values of observations on the horizontal axis, with a bar showing how many times each value occurred in the data set. Frequency distributions can be very useful for assessing properties of the distribution of scores. We will find out how to create these types of charts in Chapter 4.

Frequency distributions come in many different shapes and sizes. It is quite important, therefore, to have some general descriptions for common types of distributions. In an ideal world our data would be distributed symmetrically around the centre of all scores. As such, if we drew a vertical line through the centre of the distribution then it should look the same on both sides. This is known as a **normal distribution** and is characterized by the bell-shaped curve with which you might already be familiar. This shape basically implies that the majority of scores lie around the centre of the distribution (so the largest bars on the histogram are all around the central value).

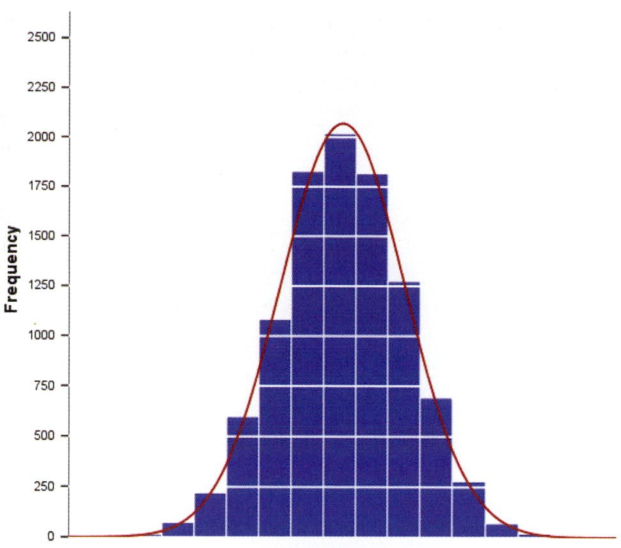

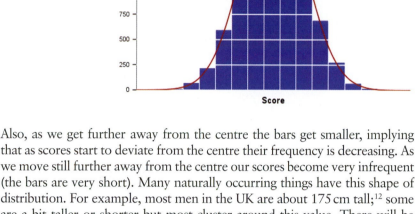

FIGURE 1.3 A 'normal' distribution (the curve shows the idealized shape)

What is a frequency distribution and when is it normal?

Also, as we get further away from the centre the bars get smaller, implying that as scores start to deviate from the centre their frequency is decreasing. As we move still further away from the centre our scores become very infrequent (the bars are very short). Many naturally occurring things have this shape of distribution. For example, most men in the UK are about 175 cm tall;[12] some are a bit taller or shorter but most cluster around this value. There will be very few men who are really tall (i.e. above 205 cm) or really short (i.e. under 145 cm). An example of a normal distribution is shown in Figure 1.3.

There are two main ways in which a distribution can deviate from normal: (1) lack of symmetry (called **skew**) and (2) pointyness (called **kurtosis**). Skewed distributions are not symmetrical and instead the most frequent scores (the tall bars on the graph) are clustered at one end of the scale. So, the typical pattern is a cluster of frequent scores at one end of the scale and the frequency of scores tailing off towards the other end of the scale. A skewed distribution can be either *positively skewed* (the frequent scores are clustered at the lower end and the tail points towards the higher or more positive scores) or *negatively skewed* (the frequent scores are clustered at the higher end and the tail points towards the lower or more negative scores). Figure 1.4 shows examples of these distributions.

Distributions also vary in their kurtosis. Kurtosis, despite sounding like some kind of exotic disease, refers to the degree to which scores cluster at the ends of the distribution (known as the *tails*) and how pointy a distribution is (but there are other factors that can affect how pointy the distribution looks – see Jane Superbrain Box 2.2). A distribution with *positive kurtosis* has many scores in the tails (a so-called heavy-tailed distribution) and is pointy. This is known as a **leptokurtic** distribution. In contrast, a distribution with *negative kurtosis* is relatively thin in the tails (has light tails) and tends to be flatter than normal. This distribution is called **platykurtic**. Ideally, we want our data to be normally distributed (i.e. not too skewed, and not too many or too few scores at the extremes!). For everything there is to know about kurtosis read DeCarlo (1997).

In a normal distribution the values of skew and kurtosis are 0 (i.e. the tails of the distribution are as they should be). If a distribution has values of skew or kurtosis above or below 0 then this indicates a deviation from normal: Figure 1.5 shows distributions with kurtosis values of +1 (left panel) and –4 (right panel).

[12] I am exactly 180 cm tall. In my home country this makes me smugly above average. However, I'm writing this in The Netherlands where the average male height is 185 cm (a massive 10 cm higher than the UK), and where I feel like a bit of a dwarf.

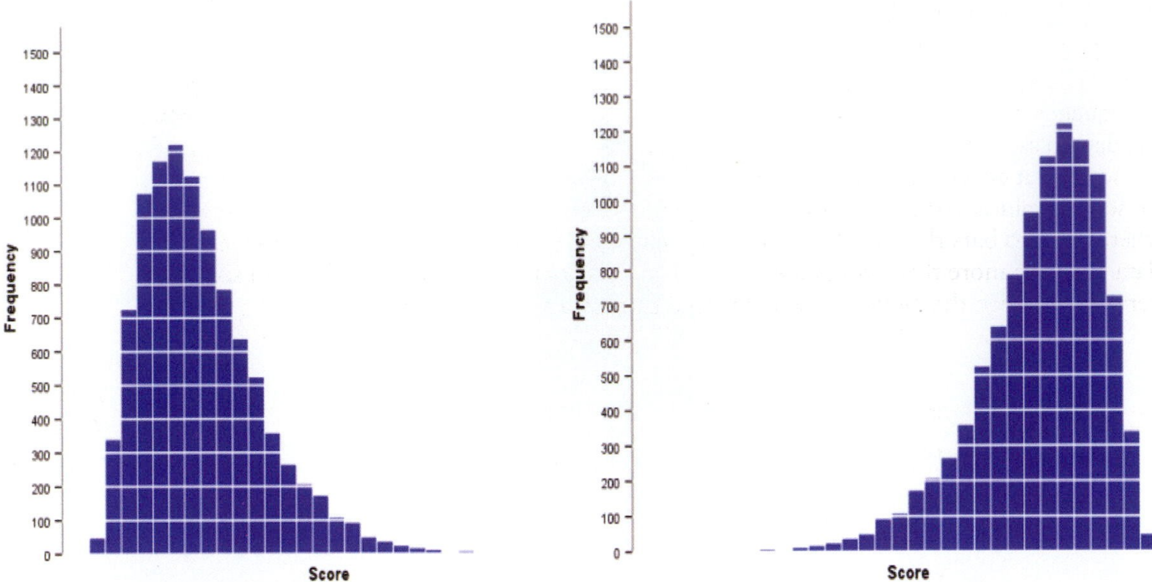

FIGURE 1.4 A positively (left figure) and negatively (right figure) skewed distribution

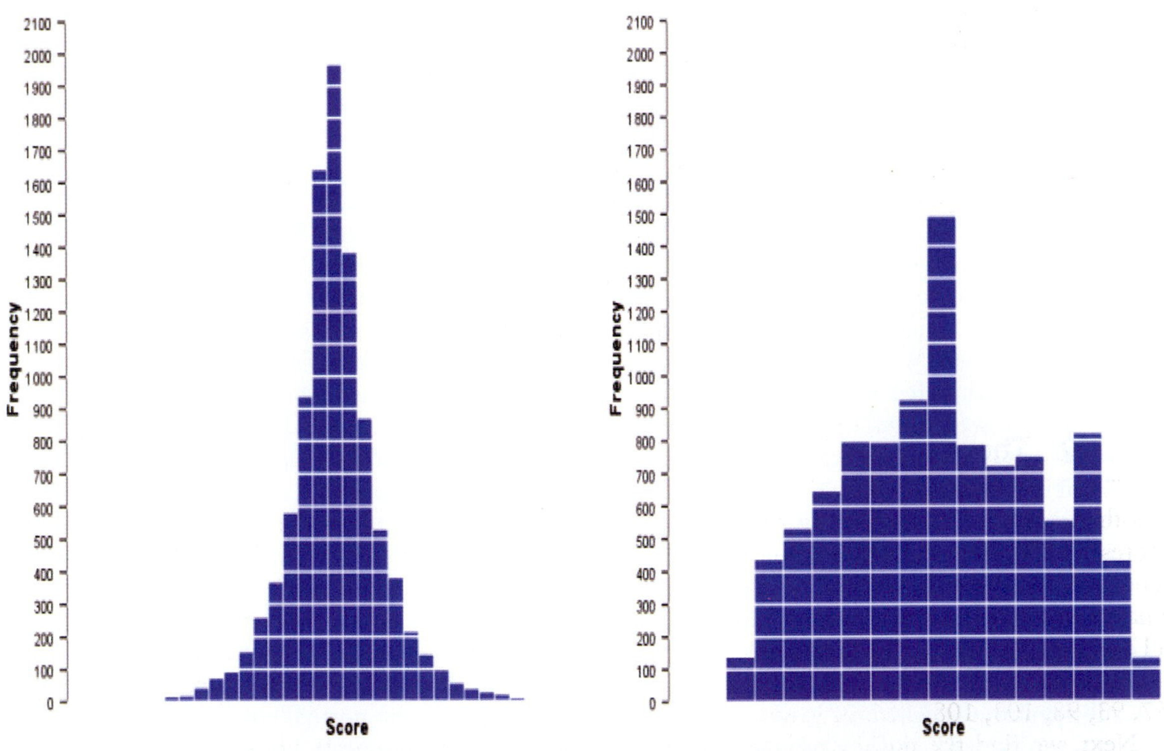

FIGURE 1.5 Distributions with positive kurtosis (leptokurtic, left figure) and negative kurtosis (platykurtic, right figure)

1.7.2. The centre of a distribution ①

We can also calculate where the centre of a frequency distribution lies (known as the **central tendency**). There are three measures commonly used: the mean, the mode and the median.

1.7.2.1. The mode ①

The **mode** is simply the score that occurs most frequently in the data set. This is easy to spot in a frequency distribution because it will be the tallest bar! To calculate the mode, simply place the data in ascending order (to make life easier), count how many times each score occurs, and the score that occurs the most is the mode! One problem with the mode is that it can often take on several values. For example, Figure 1.6 shows an example of a distribution with two modes (there are two bars that are the highest), which is said to be **bimodal**. It's also possible to find data sets with more than two modes (**multimodal**). Also, if the frequencies of certain scores are very similar, then the mode can be influenced by only a small number of cases.

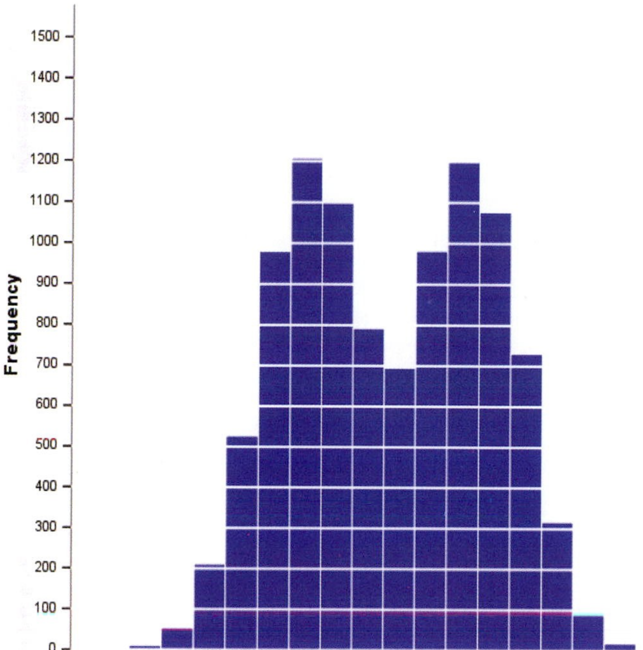

FIGURE 1.6 A bimodal distribution

1.7.2.2. The median ①

Another way to quantify the centre of a distribution is to look for the middle score when scores are ranked in order of magnitude. This is called the **median**. For example, Facebook is a popular social networking website, in which users can sign up to be 'friends' of other users. Imagine we looked at the number of friends that a selection (actually, some of my friends) of 11 Facebook users had. Number of friends: 108, 103, 252, 121, 93, 57, 40, 53, 22, 116, 98.

To calculate the median, we first arrange these scores into ascending order: 22, 40, 53, 57, 93, 98, 103, 108, 116, 121, 252.

Next, we find the position of the middle score by counting the number of scores we have collected (n), adding 1 to this value, and then dividing by 2. With 11 scores, this gives us $(n + 1)/2 = (11 + 1)/2 = 12/2 = 6$. Then, we find the score that is positioned at the location we have just calculated. So, in this example we find the sixth score:

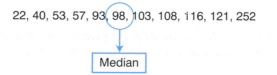

22, 40, 53, 57, 93, 98, 103, 108, 116, 121, 252

Median

What are the mode, median and mean?

This works very nicely when we have an odd number of scores (as in this example) but when we have an even number of scores there won't be a middle value. Let's imagine that we decided that because the highest score was so big (more than twice as large as the next biggest number), we would ignore it. (For one thing, this person is far too popular and we hate them.) We have only 10 scores now. As before, we should rank-order these scores: 22, 40, 53, 57, 93, 98, 103, 108, 116, 121. We then calculate the position of the middle score, but this time it is $(n + 1)/2 = 11/2 = 5.5$. This means that the median is halfway between the fifth and sixth scores. To get the median we add these two scores and divide by 2. In this example, the fifth score in the ordered list was 93 and the sixth score was 98. We add these together (93 + 98 = 191) and then divide this value by 2 (191/2 = 95.5). The median number of friends was, therefore, 95.5.

The median is relatively unaffected by extreme scores at either end of the distribution: the median changed only from 98 to 95.5 when we removed the extreme score of 252. The median is also relatively unaffected by skewed distributions and can be used with ordinal, interval and ratio data (it cannot, however, be used with nominal data because these data have no numerical order).

1.7.2.3. The mean ①

The **mean** is the measure of central tendency that you are most likely to have heard of because it is simply the average score and the media are full of average scores.[13] To calculate the mean we simply add up all of the scores and then divide by the total number of scores we have. We can write this in equation form as:

$$\overline{X} = \frac{\sum_{i=1}^{n} x_i}{n} \tag{1.1}$$

This may look complicated, but the top half of the equation simply means 'add up all of the scores' (the x_i just means 'the score of a particular person'; we could replace the letter i with each person's name instead), and the bottom bit means divide this total by the number of scores you have got (n). Let's calculate the mean for the Facebook data. First, we first add up all of the scores:

$$\sum_{i=1}^{n} x_i = 22 + 40 + 53 + 57 + 93 + 98 + 103 + 108 + 116 + 121 + 252$$
$$= 1063$$

We then divide by the number of scores (in this case 11):

$$\overline{X} = \frac{\sum_{i=1}^{n} x_i}{n} = \frac{1063}{11} = 96.64$$

The mean is 96.64 friends, which is not a value we observed in our actual data (it would be ridiculous to talk of having 0.64 of a friend). In this sense the mean is a statistical model – more on this in the next chapter.

[13] I'm writing this on 15 February 2008, and to prove my point the BBC website is running a headline about how PayPal estimates that Britons will spend an average of £71.25 each on Valentine's Day gifts, but uSwitch.com said that the average spend would be £22.69!

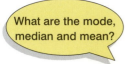

SELF-TEST Compute the mean but excluding the score of 252.

If you calculate the mean without our extremely popular person (i.e. excluding the value 252), the mean drops to 81.1 friends. One disadvantage of the mean is that it can be influenced by extreme scores. In this case, the person with 252 friends on Facebook increased the mean by about 15 friends! Compare this difference with that of the median. Remember that the median hardly changed if we included or excluded 252, which illustrates how the median is less affected by extreme scores than the mean. While we're being negative about the mean, it is also affected by skewed distributions and can be used only with interval or ratio data.

If the mean is so lousy then why do we use it all of the time? One very important reason is that it uses every score (the mode and median ignore most of the scores in a data set). Also, the mean tends to be stable in different samples.

1.7.3. The dispersion in a distribution ①

It can also be interesting to try to quantify the spread, or dispersion, of scores in the data. The easiest way to look at dispersion is to take the largest score and subtract from it the smallest score. This is known as the **range** of scores. For our Facebook friends data, if we order these scores we get 22, 40, 53, 57, 93, 98, 103, 108, 116, 121, 252. The highest score is 252 and the lowest is 22; therefore, the range is 252 – 22 = 230. One problem with the range is that because it uses only the highest and lowest score it is affected dramatically by extreme scores.

SELF-TEST Compute the range but excluding the score of 252.

If you have done the self-test task you'll see that without the extreme score the range drops dramatically from 230 to 99 – less than half the size!

One way around this problem is to calculate the range when we exclude values at the extremes of the distribution. One convention is to cut off the top and bottom 25% of scores and calculate the range of the middle 50% of scores – known as the **interquartile range**. Let's do this with the Facebook data. First we need to calculate what are called **quartiles**. Quartiles are the three values that split the sorted data into four equal parts. First we calculate the median, which is also called the **second quartile**, which splits our data into two equal parts. We already know that the median for these data is 98. The **lower quartile** is the median of the lower half of the data and the **upper quartile** is the median of the upper half of the data. One rule of thumb is that the median is not included in the two halves when they are split (this is convenient if you have an odd number of values), but you can include it (although which half you put it in is another question). Figure 1.7 shows how we would calculate these values for the Facebook data. Like the median, the upper and lower quartile need not be values that actually appear in the data (like the median, if each half of the data had an even number of values in it then the upper and lower quartiles would be the average

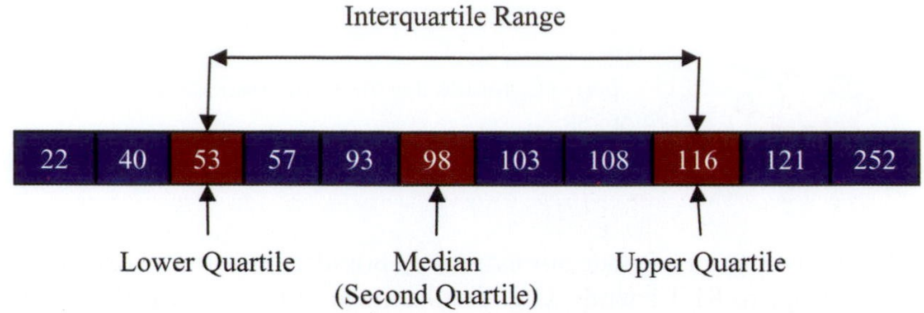

FIGURE 1.7 Calculating quartiles and the interquartile range

of two values in the data set). Once we have worked out the values of the quartiles, we can calculate the interquartile range, which is the difference between the upper and lower quartile. For the Facebook data this value would be 116 – 53 = 63. The advantage of the interquartile range is that it isn't affected by extreme scores at either end of the distribution. However, the problem with it is that you lose a lot of data (half of it in fact!).

SELF-TEST Twenty-one heavy smokers were put on a treadmill at the fastest setting. The time in seconds was measured until they fell off from exhaustion: 18, 16, 18, 24, 23, 22, 22, 23, 26, 29, 32, 34, 34, 36, 36, 43, 42, 49, 46, 46, 57

Compute the mode, median, mean, upper and lower quartiles, range and interquartile range.

1.7.4. Using a frequency distribution to go beyond the data ①

Another way to think about frequency distributions is not in terms of how often scores actually occurred, but how likely it is that a score would occur (i.e. probability). The word 'probability' induces suicidal ideation in most people (myself included) so it seems fitting that we use an example about throwing ourselves off a cliff. Beachy Head is a large, windy cliff on the Sussex coast (not far from where I live) that has something of a reputation for attracting suicidal people, who seem to like throwing themselves off it (and after several months of rewriting this book I find my thoughts drawn towards that peaceful chalky cliff top more and more often). Figure 1.8 shows a frequency distribution of some completely made up data of the number of suicides at Beachy Head in a year by people of different ages (although I made these data up, they are roughly based on general suicide statistics such as those in Williams, 2001). There were 172 suicides in total and you can see that the suicides were most frequently aged between about 30 and 35 (the highest bar). The graph also tells us that, for example, very few people aged above 70 committed suicide at Beachy Head.

I said earlier that we could think of frequency distributions in terms of probability. To explain this, imagine that someone asked you 'how likely is it that a person who committed suicide at Beachy Head is a 70 year old?' What would your answer be? The chances are that if you looked at the frequency distribution you might respond 'not very likely' because you can see that only 3 people out of the 172 suicides were aged around 70. What about if someone asked you 'how likely is it that a 30 year old committed suicide?' Again, by looking at the graph, you might say 'it's actually quite likely' because 33 out of the 172 suicides were by people aged around 30 (that's more than 1 in every 5 people that committed suicide). So

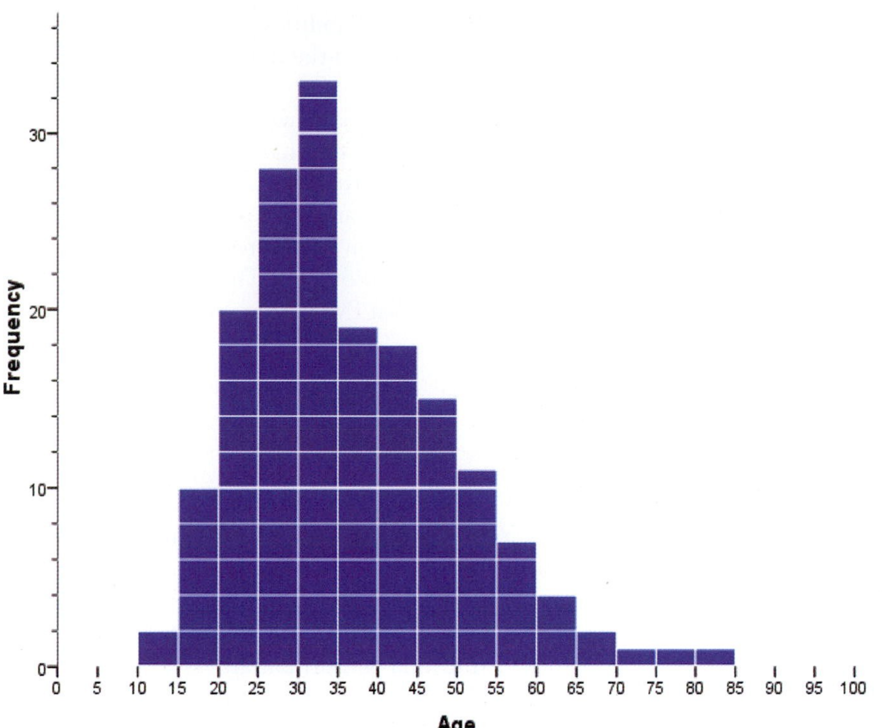

FIGURE 1.8 Frequency distribution showing the number of suicides at Beachy Head in a year by age

on the frequencies of different scores it should start to become clear that we could use this information to estimate the probability that a particular score will occur. We could ask, based on our data, 'what's the probability of a suicide victim being aged 16–20?' A probability value can range from 0 (there's no chance whatsoever of the event happening) to 1 (the event will definitely happen). So, for example, when I talk to my publishers I tell them there's a probability of 1 that I will have completed the revisions to this book by April 2008. However, when I talk to anyone else, I might, more realistically, tell them that there's a .10 probability of me finishing the revisions on time (or put another way, a 10% chance, or 1 in 10 chance that I'll complete the book in time). In reality, the probability of my meeting the deadline is 0 (not a chance in hell) because I never manage to meet publisher's deadlines! If probabilities don't make sense to you then just ignore the decimal point and think of them as percentages instead (i.e. .10 probability that something will happen = 10% chance that something will happen).

I've talked in vague terms about how frequency distributions can be used to get a rough idea of the probability of a score occurring. However, we can be precise. For any distribution of scores we could, in theory, calculate the probability of obtaining a score of a certain size – it would be incredibly tedious and complex to do it, but we could. To spare our sanity, statisticians have identified several common distributions. For each one they have worked out mathematical formulae that specify idealized versions of these distributions (they are specified in terms of a curved line). These idealized distributions are known as **probability distributions** and from these distributions it is possible to calculate the probability of getting particular scores based on the frequencies with which a particular score occurs in a distribution with these common shapes. One of these 'common' distributions is the normal distribution, which I've already mentioned in section 1.7.1. Statisticians have calculated the probability of certain scores occurring in a normal distribution with a mean of 0 and a standard deviation of 1. Therefore, if we have any data that are shaped like a normal distribution, then if the mean and standard deviation

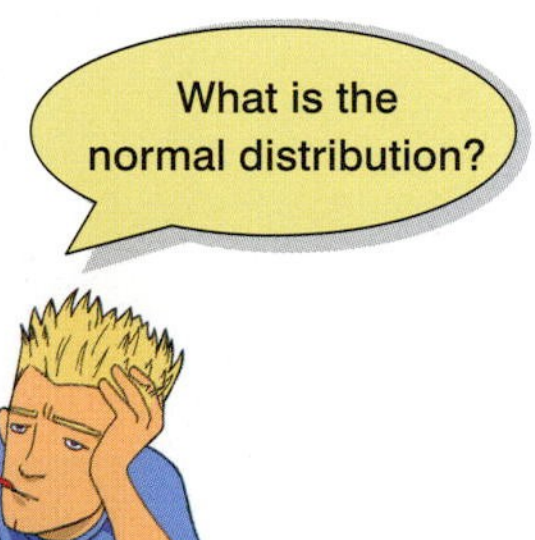

are 0 and 1 respectively we can use the tables of probabilities for the normal distribution to see how likely it is that a particular score will occur in the data (I've produced such a table in the Appendix to this book).

The obvious problem is that not all of the data we collect will have a mean of 0 and standard deviation of 1. For example, we might have a data set that has a mean of 567 and a standard deviation of 52.98. Luckily any data set can be converted into a data set that has a mean of 0 and a standard deviation of 1. First, to centre the data around zero, we take each score (x) and subtract from it the mean ($\bar{x}$)of all scores. Then, we divide the resulting score by the standard deviation (s) to ensure the data have a standard deviation of 1. The resulting scores are known as **z-scores** and in equation form, the conversion that I've just described is:

$$z = \frac{X - \overline{X}}{s} \tag{1.2}$$

The table of probability values that have been calculated for the standard normal distribution is shown in the Appendix. Why is this table important? Well, if we look at our suicide data, we can answer the question 'what's the probability that someone who threw themselves off Beachy Head was 70 or older?' First we convert 70 into a z-score. Say, the mean of the suicide scores was 36, and the standard deviation 13; then 70 will become (70 – 36)/13 = 2.62. We then look up this value in the column labelled 'Smaller Portion' (i.e. the area above the value 2.62). You should find that the probability is .0044, or put another way, only a 0.44% chance that a suicide victim would be 70 years old or more. By looking at the column labelled 'Bigger Portion' we can also see the probability that a suicide victim was aged 70 or less! This probability is .9956, or put another way, there's a 99.56% chance that a suicide victim was less than 70 years old!

Hopefully you can see from these examples that the normal distribution and z-scores allow us to go a first step beyond our data in that from a set of scores we can calculate the probability that a particular score will occur. So, we can see whether scores of a certain size are likely or unlikely to occur in a distribution of a particular kind. You'll see just how useful this is in due course, but it is worth mentioning at this stage that certain z-scores are particularly important. This is because their value cuts off certain important percentages of the distribution. The first important value of z is 1.96 because this cuts off the top 2.5% of the distribution, and its counterpart at the opposite end (–1.96) cuts off the bottom 2.5% of the distribution. As such, taken together, this value cuts off 5% of scores, or put another way, 95% of z-scores lie between –1.96 and 1.96. The other two important benchmarks are ±2.58 and ±3.29, which cut off 1% and 0.1% of scores respectively. Put another way, 99% of z-scores lie between –2.58 and 2.58, and 99.9% of them lie between –3.29 and 3.29. Remember these values because they'll crop up time and time again.

SELF-TEST Assuming the same mean and standard deviation for the Beachy Head example above, what's the probability that someone who threw themselves off Beachy Head was 30 or younger?

1.7.5. Fitting statistical models to the data ①

Having looked at your data (and there is a lot more information on different ways to do this in Chapter 4), the next step is to fit a statistical model to the data. I should really just write 'insert the rest of the book here', because most of the remaining chapters discuss the various models that you can fit to the data. However, I do want to talk here briefly about two very important types of hypotheses that are used when analysing the data. Scientific statements, as we have seen, can be split into testable hypotheses. The hypothesis or prediction that comes from your theory is usually saying that an effect will be present. This hypothesis is called the *alternative hypothesis* and is denoted by H_1. (It is sometimes also called the *experimental hypothesis* but because this term relates to a specific type of methodology it's probably best to use 'alternative hypothesis'.) There is another type of hypothesis, though, and this is called the **null hypothesis** and is denoted by H_0. This hypothesis is the opposite of the alternative hypothesis and so would usually state that an effect is absent. Taking our *Big Brother* example from earlier in the chapter we might generate the following hypotheses:

- **Alternative hypothesis**: *Big Brother* contestants will score higher on personality disorder questionnaires than members of the public.
- **Null hypothesis**: *Big Brother* contestants and members of the public will not differ in their scores on personality disorder questionnaires.

The reason that we need the null hypothesis is because we cannot prove the experimental hypothesis using statistics, but we can reject the null hypothesis. If our data give us confidence to reject the null hypothesis then this provides support for our experimental hypothesis. However, be aware that even if we can reject the null hypothesis, this doesn't prove the experimental hypothesis – it merely supports it. So, rather than talking about accepting or rejecting a hypothesis (which some textbooks tell you to do) we should be talking about 'the chances of obtaining the data we've collected assuming that the null hypothesis is true'.

Using our *Big Brother* example, when we collected data from the auditions about the contestants' personalities we found that 75% of them had a disorder. When we analyse our data, we are really asking, 'Assuming that contestants are no more likely to have personality disorders than members of the public, is it likely that 75% or more of the contestants would have personality disorders?' Intuitively the answer is that the chances are very low: if the null hypothesis is true, then most contestants would not have personality disorders because they are relatively rare. Therefore, we are very unlikely to have got the data that we did if the null hypothesis were true.

What if we found that only 1 contestant reported having a personality disorder (about 8%)? If the null hypothesis is true, and contestants are no different in personality to the general population, then only a small number of contestants would be expected to have a personality disorder. The chances of getting these data if the null hypothesis is true are, therefore, higher than before.

When we collect data to test theories we have to work in these terms: we cannot talk about the null hypothesis being true or the experimental hypothesis being true, we can only talk in terms of the probability of obtaining a particular set of data if, hypothetically speaking, the null hypothesis was true. We will elaborate on this idea in the next chapter.

Finally, hypotheses can also be directional or non-directional. A directional hypothesis states that an effect will occur, but it also states the direction of the effect. For example, 'readers will know more about research methods after reading this chapter' is a one-tailed hypothesis because it states the direction of the effect (readers will know more). A non-directional hypothesis states that an effect will occur, but it doesn't state the direction of the effect. For example, 'readers' knowledge of research methods will change after they have read this chapter' does not tell us whether their knowledge will improve or get worse.

What have I discovered about statistics? ①

Actually, not a lot because we haven't really got to the statistics bit yet. However, we have discovered some stuff about the process of doing research. We began by looking at how research questions are formulated through observing phenomena or collecting data about a 'hunch'. Once the observation has been confirmed, theories can be generated about why something happens. From these theories we formulate hypotheses that we can test. To test hypotheses we need to measure things and this leads us to think about the variables that we need to measure and how to measure them. Then we can collect some data. The final stage is to analyse these data. In this chapter we saw that we can begin by just looking at the shape of the data but that ultimately we should end up fitting some kind of statistical model to the data (more on that in the rest of the book). In short, the reason that your evil statistics lecturer is forcing you to learn statistics is because it is an intrinsic part of the research process and it gives you enormous power to answer questions that are interesting; or it could be that they are a sadist who spends their spare time spanking politicians while wearing knee-high PVC boots, a diamond-encrusted leather thong and a gimp mask (that'll be a nice mental image to keep with you throughout your course). We also discovered that I was a curious child (you can interpret that either way). As I got older I became more curious, but you will have to read on to discover what I was curious about.

Key terms that I've discovered

Alternative hypothesis
Between-group design
Between-subject design
Bimodal
Binary variable
Boredom effect
Categorical variable
Central tendency
Confounding variable
Content validity
Continuous variable
Correlational research
Counterbalancing
Criterion validity
Cross-Sectional research
Dependent variable
Discrete variable
Ecological validity
Experimental hypothesis
Experimental research
Falsification
Frequency distribution
Histogram
Hypothesis
Independent design
Independent variable
Interquartile range
Interval variable
Kurtosis
Leptokurtic
Level of measurement
Lower quartile
Mean
Measurement error
Median
Mode
Multimodal
Negative skew
Nominal variable
Normal distribution
Null hypothesis
Ordinal variable
Outcome variable
Platykurtic
Positive skew
Practice effect

Predictor variable
Probability distribution
Qualitative methods
Quantitative methods
Quartile
Randomization
Range
Ratio variable
Reliability
Repeated-measures design
Second quartile
Skew
Systematic variation
Tertium quid
Test–retest reliability
Theory
Unsystematic variation
Upper quartile
Validity
Variables
Within-subject design
z-scores

Smart Alex's stats quiz

Smart Alex knows everything there is to know about statistics and SPSS. He also likes nothing more than to ask people stats questions just so that he can be smug about how much he knows. So, why not really annoy him and get all of the answers right!

1 What are (broadly speaking) the five stages of the research process? ①

2 What is the fundamental difference between experimental and correlational research? ①

3 What is the level of measurement of the following variables? ①
   a. The number of downloads of different bands' songs on iTunes
   b. The names of the bands that were downloaded.
   c. The position in the iTunes download chart.
   d. The money earned by the bands from the downloads.
   e. The weight of drugs bought by the bands with their royalties.
   f. The type of drugs bought by the bands with their royalties.
   g. The phone numbers that the bands obtained because of their fame.
   h. The gender of the people giving the bands their phone numbers.
   i. The instruments played by the band members.
   j. The time they had spent learning to play their instruments.

4 Say I own 857 CDs. My friend has written a computer program that uses a webcam to scan the shelves in my house where I keep my CDs and measure how many I have. His program says that I have 863 CDs. Define measurement error. What is the measurement error in my friend's CD-counting device? ①

5 Sketch the shape of a normal distribution, a positively skewed distribution and a negatively skewed distribution. ①

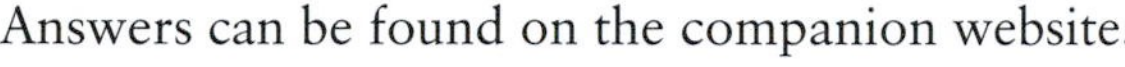

Answers can be found on the companion website.

Further reading

Field, A. P., & Hole, G. J. (2003). *How to design and report experiments*. London: Sage. (I am rather biased, but I think this is a good overview of basic statistical theory and research methods.)

Miles, J. N. V., & Banyard, P. (2007). *Understanding and using statistics in psychology: a practical introduction*. London: Sage. (A fantastic and amusing introduction to statistical theory.)
Wright, D. B., & London, K. (2009). *First steps in statistics* (2nd ed.). London: Sage. (This book is a very gentle introduction to statistical theory.)

Interesting real research

Umpierre, S. A., Hill, J. A., & Anderson, D. J. (1985). Effect of Coke on sperm motility. *New England Journal of Medicine*, *313*(21), 1351–1351.

Everything you ever wanted to know about statistics (well, sort of)

2

FIGURE 2.1
The face of innocence … but what are the hands doing?

2.1. What will this chapter tell me? ①

As a child grows, it becomes important for them to fit models to the world: to be able to reliably predict what will happen in certain situations. This need to build models that accurately reflect reality is an essential part of survival. According to my parents (conveniently I have no memory of this at all), while at nursery school one model of the world that I was particularly enthusiastic to try out was 'If I get my penis out, it will be really funny'. No doubt to my considerable disappointment, this model turned out to be a poor predictor of positive outcomes. Thankfully for all concerned, I soon learnt that the model 'If I get my penis out at nursery school the teachers and mummy and daddy are going to be quite annoyed' was a better 'fit' of the observed data. Fitting models that accurately reflect the observed data is important to establish whether a theory is true. You'll be delighted to know that this chapter is all about fitting statistical models (and not about my penis). We edge sneakily away from the frying pan of research methods and trip accidentally into the fires of statistics hell. We begin by discovering what a statistical model is by using the mean as a straightforward example. We then see how we can use the properties of data to go beyond the data we have collected and to draw inferences about the world at large. In a nutshell then, this chapter lays the foundation for the whole of the rest of the book, so it's quite important that you read it or nothing that comes later will make any sense. Actually, a lot of what comes later probably won't make much sense anyway because *I've* written it, but there you go.

2.2. Building statistical models ①

Why do we build statistical models?

We saw in the previous chapter that scientists are interested in discovering something about a phenomenon that we assume actually exists (a 'real-world' phenomenon). These real-world phenomena can be anything from the behaviour of interest rates in the economic market to the behaviour of undergraduates at the end-of-exam party. Whatever the phenomenon we desire to explain, we collect data from the real world to test our hypotheses about the phenomenon. Testing these hypotheses involves building statistical models of the phenomenon of interest.

The reason for building statistical models of real-world data is best explained by analogy. Imagine an engineer wishes to build a bridge across a river. That engineer would be pretty daft if she just built any old bridge, because the chances are that it would fall down. Instead, an engineer collects data from the real world: she looks at bridges in the real world and sees what materials they are made from, what structures they use and so on (she might even collect data about whether these bridges are damaged). She then uses this information to construct a model. She builds a scaled-down version of the real-world bridge because it is impractical, not to mention expensive, to build the actual bridge itself. The model may differ from reality in several ways – it will be smaller for a start – but the engineer will try to build a model that best fits the situation of interest based on the data available. Once the model has been built, it can be used to predict things about the real world: for example, the engineer might test whether the bridge can withstand strong winds by placing the model in a wind tunnel. It seems obvious that it is important that the model is an accurate representation of the real world. Social scientists do much the same thing as engineers: they build models of real-world processes in an attempt to predict how these processes operate under certain conditions (see Jane Superbrain Box 2.1 below). We don't have direct access to the processes, so we collect data that represent the processes and then use these data to build statistical models (we reduce the process to a statistical model). We then use this statistical model to make predictions about the real-world phenomenon. Just like the engineer, we want our models to be as accurate as possible so that we can be confident that the predictions we make are also accurate. However, unlike engineers we don't have access to the real-world situation and so we can only ever *infer* things about psychological, societal, biological or economic processes based upon the models we build. If we want our inferences to be accurate then the statistical model we build must represent the data collected (the *observed data*) as closely as possible. The degree to which a statistical model represents the data collected is known as the **fit** of the model.

Figure 2.2 illustrates the kinds of models that an engineer might build to represent the real-world bridge that she wants to create. The first model (a) is an excellent representation of the real-world situation and is said to be a *good fit* (i.e. there are a few small differences but the model is basically a very good replica of reality). If this model is used to make predictions about the real world, then the engineer can be confident that these predictions will be very accurate, because the model so closely resembles reality. So, if the model collapses in a strong

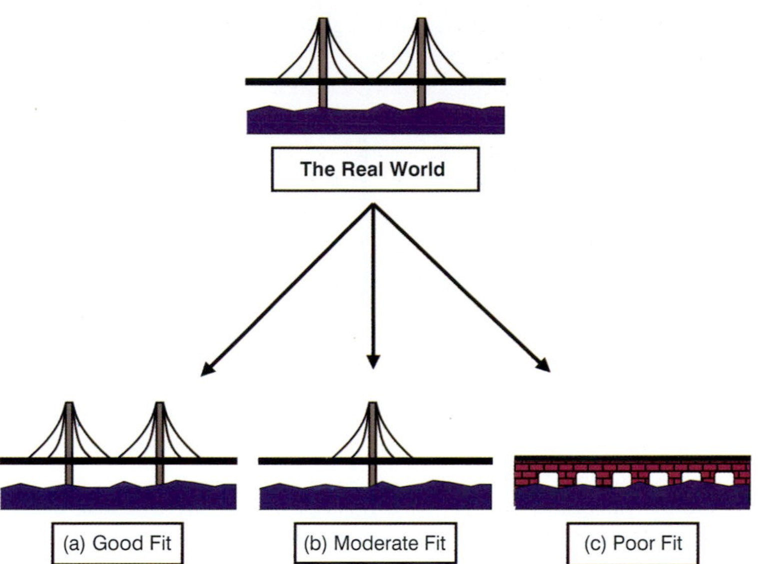

FIGURE 2.2 Fitting models to real-world data (see text for details)

wind, then there is a good chance that the real bridge would collapse also. The second model (b) has some similarities to the real world: the model includes some of the basic structural features, but there are some big differences from the real-world bridge (namely the absence of one of the supporting towers). This is what we might term a *moderate fit* (i.e. there are some differences between the model and the data but there are also some great similarities). If the engineer uses this model to make predictions about the real world then these predictions may be inaccurate and possibly catastrophic (e.g. the model predicts that the bridge will collapse in a strong wind, causing the real bridge to be closed down, creating 100-mile tailbacks with everyone stranded in the snow; all of which was unnecessary because the real bridge was perfectly safe – the model was a bad representation of reality). We can have some confidence, but not complete confidence, in predictions from this model. The final model (c) is completely different to the real-world situation; it bears no structural similarities to the real bridge and is a poor fit (in fact, it might more accurately be described as an abysmal fit). As such, any predictions based on this model are likely to be completely inaccurate. Extending this analogy to the social sciences we can say that it is important when we fit a statistical model to a set of data that this model fits the data well. If our model is a poor fit of the observed data then the predictions we make from it will be equally poor.

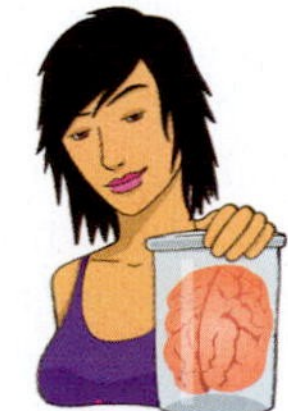

JANE SUPERBRAIN 2.1

Types of statistical models ①

As behavioural and social scientists, most of the models that we use to describe data tend to be **linear models**. For example, analysis of variance (ANOVA) and regression are identical systems based on linear models (Cohen, 1968), yet they have different names and, in psychology at least, are used largely in different contexts due to historical divisions in methodology (Cronbach, 1957).

A linear model is simply a model that is based upon a straight line; this means that we are usually trying to summarize our observed data in terms of a straight line. Suppose we measured how many chapters of this book a person had read, and then measured their spiritual enrichment. We could represent these hypothetical data in the form of a scatterplot in which each dot represents an individual's score on both variables (see section 4.8). Figure 2.3 shows two versions of such a graph that summarizes the pattern of these data with either a straight

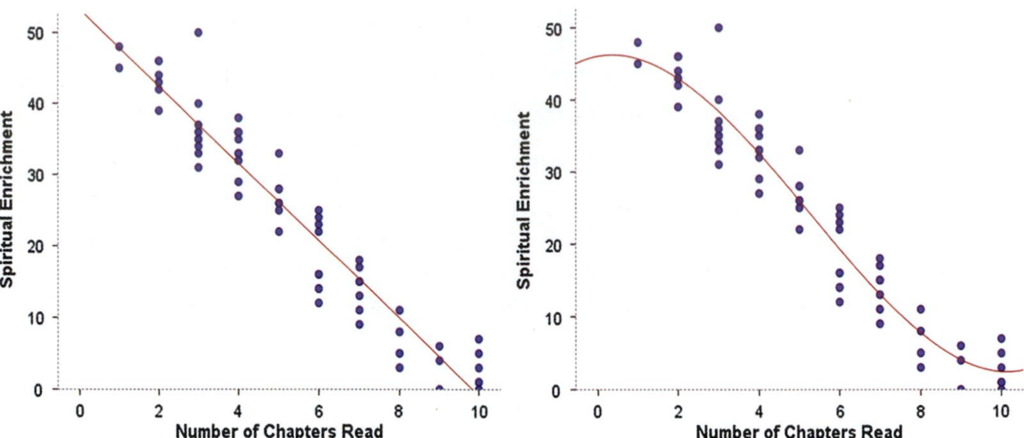

FIGURE 2.3 A scatterplot of the same data with a linear model fitted (left), and with a non-linear model fitted (right)

(left) or curved (right) line. These graphs illustrate how we can fit different types of models to the same data. In this case we can use a straight line to represent our data and it shows that the more chapters a person reads, the less their spiritual enrichment. However, we can also use a curved line to summarize the data and this shows that when most, or all, of the chapters have been read, spiritual enrichment seems to increase slightly (presumably because once the book is read everything suddenly makes sense – yeah, as if!). Neither of the two types of model is necessarily correct, but it will be the case that one model fits the data better than another and this is why when we use statistical models it is important for us to assess how well a given model fits the data.

It's possible that many scientific disciplines are progressing in a biased way because most of the models that we tend to fit are linear (mainly because books like this tend to ignore more complex curvilinear models). This could create a bias because most published scientific studies are ones with statistically significant results and there may be cases where a linear model has been a poor fit of the data (and hence the paper was not published), yet a non-linear model would have fitted the data well. This is why it is useful to plot your data first: plots tell you a great deal about what models should be applied to data. If your plot seems to suggest a non-linear model then investigate this possibility (which is easy for me to say when I don't include such techniques in this book!).

2.3. Populations and samples ①

As researchers, we are interested in finding results that apply to an entire population of people or things. For example, psychologists want to discover processes that occur in all humans, biologists might be interested in processes that occur in all cells, economists want to build models that apply to all salaries, and so on. A population can be very general (all human beings) or very narrow (all male ginger cats called Bob). Usually, scientists strive to infer things about general populations rather than narrow ones. For example, it's not very interesting to conclude that psychology students with brown hair who own a pet hamster named George recover more quickly from sports injuries if the injury is massaged (unless, like René Koning,[1] you happen to be a psychology student with brown hair who has a pet hamster named George). However, if we can conclude that *everyone's* sports injuries are aided by massage this finding has a much wider impact.

Scientists rarely, if ever, have access to every member of a population. Psychologists cannot collect data from every human being and ecologists cannot observe every male ginger cat called Bob. Therefore, we collect data from a small subset of the population (known as a **sample**) and use these data to infer things about the population as a whole. The bridge-building

[1] A brown-haired psychology student with a hamster called Sjors (Dutch for George, apparently), who, after reading one of my web resources, emailed me to weaken my foolish belief that this is an obscure combination of possibilities.

engineer cannot make a full-size model of the bridge she wants to build and so she builds a small-scale model and tests this model under various conditions. From the results obtained from the small-scale model the engineer infers things about how the full-sized bridge will respond. The small-scale model may respond differently to a full-sized version of the bridge, but the larger the model, the more likely it is to behave in the same way as the full-size bridge. This metaphor can be extended to scientists. We never have access to the entire population (the real-size bridge) and so we collect smaller samples (the scaled-down bridge) and use the behaviour within the sample to infer things about the behaviour in the population. The bigger the sample, the more likely it is to reflect the whole population. If we take several random samples from the population, each of these samples will give us slightly different results. However, on average, large samples should be fairly similar.

2.4. Simple statistical models ①

2.4.1. The mean: a very simple statistical model ①

One of the simplest models used in statistics is the mean, which we encountered in section 1.7.2.3. In Chapter 1 we briefly mentioned that the mean was a statistical model of the data because it is a hypothetical value that doesn't have to be a value that is actually observed in the data. For example, if we took five statistics lecturers and measured the number of friends that they had, we might find the following data: 1, 2, 3, 3 and 4. If we take the mean number of friends, this can be calculated by adding the values we obtained, and dividing by the number of values measured: $(1 + 2 + 3 + 3 + 4)/5 = 2.6$. Now, we know that it is impossible to have 2.6 friends (unless you chop someone up with a chainsaw and befriend their arm, which, frankly is probably not beyond your average statistics lecturer) so the mean value is a *hypothetical* value. As such, the mean is a model created to summarize our data.

2.4.2. Assessing the fit of the mean: sums of squares, variance and standard deviations ①

With any statistical model we have to assess the fit (to return to our bridge analogy we need to know how closely our model bridge resembles the real bridge that we want to build). With most statistical models we can determine whether the model is accurate by looking at how different our real data are from the model that we have created. The easiest way to do this is to look at the difference between the data we observed and the model fitted. Figure 2.4 shows the number of friends that each statistics lecturer had, and also the mean number that we calculated earlier on. The line representing the mean can be thought of as our model, and the circles are the observed data. The diagram also has a series of vertical lines that connect each observed value to the mean value. These lines represent the **deviance** between the observed data and our model and can be thought of as the error in the model. We can calculate the magnitude of these deviances by simply subtracting the mean value ($\bar{x}$) from each of the observed values (x_i).[2] For example, lecturer 1 had only 1 friend (a glove puppet of an ostrich called Kevin) and so the difference is $x_1 - \bar{x} = 1 - 2.6 = -1.6$. You might notice that the deviance is a negative number, and this represents the fact that our model *overestimates* this lecturer's popularity: it

[2] The x_i simply refers to the observed score for the ith person (so, the i can be replaced with a number that represents a particular individual). For these data: for lecturer 1, $x_i = x_1 = 1$; for lecturer 3, $x_i = x_3 = 3$; for lecturer 5, $x_i = x_5 = 4$.

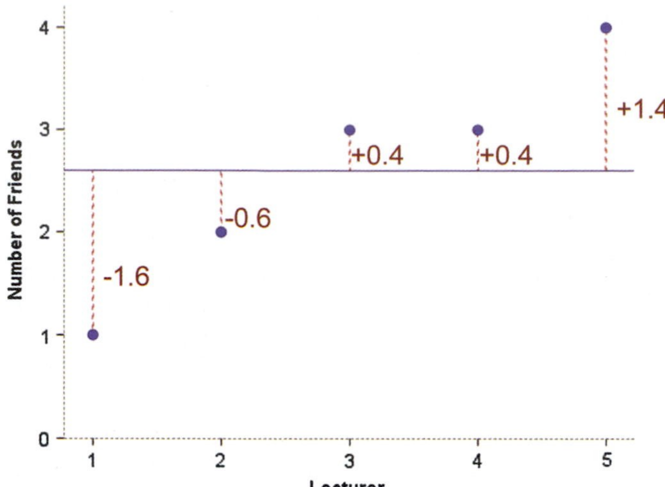

FIGURE 2.4 Graph showing the difference between the observed number of friends that each statistics lecturer had, and the mean number of friends

predicts that he will have 2.6 friends yet in reality he has only 1 friend (bless him!). Now, how can we use these deviances to estimate the accuracy of the model? One possibility is to add up the deviances (this would give us an estimate of the total error). If we were to do this we would find that (don't be scared of the equations, we will work through them step by step – if you need reminding of what the symbols mean there is a guide at the beginning of the book):

$$\begin{aligned}\text{total error} &= \text{sum of deviances}\\ &= \sum(x_i - \bar{x}) = (-1.6) + (-0.6) + (0.4) + (0.4) + (1.4) = 0\end{aligned}$$

So, in effect the result tells us that there is no total error between our model and the observed data, so the mean is a perfect representation of the data. Now, this clearly isn't true: there were errors but some of them were positive, some were negative and they have simply cancelled each other out. It is clear that we need to avoid the problem of which direction the error is in and one mathematical way to do this is to square each error,[3] that is multiply each error by itself. So, rather than calculating the sum of errors, we calculate the sum of squared errors. In this example:

$$\begin{aligned}\text{sum of squared errors (SS)} &= \sum(x_i - \bar{x})(x_i - \bar{x})\\ &= (-1.6)^2 + (-0.6)^2 + (0.4)^2 + (0.4)^2 + (1.4)^2\\ &= 2.56 + 0.36 + 0.16 + 0.16 + 1.96\\ &= 5.20\end{aligned}$$

The **sum of squared errors (SS)** is a good measure of the accuracy of our model. However, it is fairly obvious that the sum of squared errors is dependent upon the amount of data that has been collected – the more data points, the higher the SS. To overcome this problem we calculate the average error by dividing the SS by the number of observations (*N*). If we are interested only in the average error for the sample, then we can divide by *N* alone. However, we are generally interested in using the error in the sample to estimate the error in the population and so we divide the SS by the number of observations minus 1 (the reason why is explained in Jane Superbrain Box 2.2 below). This measure is known as the **variance** and is a measure that we will come across a great deal:

[3] When you multiply a negative number by itself it becomes positive.

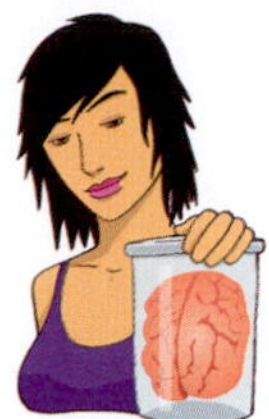

JANE SUPERBRAIN 2.2

Degrees of freedom ②

Degrees of freedom (*df*) is a very difficult concept to explain. I'll begin with an analogy. Imagine you're the manager of a rugby team and you have a team sheet with 15 empty slots relating to the positions on the playing field. There is a standard formation in rugby and so each team has 15 specific positions that must be held constant for the game to be played. When the first player arrives, you have the choice of 15 positions in which to place this player. You place his name in one of the slots and allocate him to a position (e.g. scrum-half) and, therefore, one position on the pitch is now occupied. When the next player arrives, you have the choice of 14 positions but you still have the freedom to choose which position this player is allocated. However, as more players arrive, you will reach the point at which 14 positions have been filled and the final player arrives. With this player you have no freedom to choose where they play – there is only one position left. Therefore there are 14 degrees of freedom; that is, for 14 players you have some degree of choice over where they play, but for 1 player you have no choice. The degrees of freedom is one less than the number of players.

In statistical terms the degrees of freedom relate to the number of observations that are free to vary. If we take a sample of four observations from a population, then these four scores are free to vary in any way (they can be any value). However, if we then use this sample of four observations to calculate the standard deviation of the population, we have to use the mean of the sample as an estimate of the population's mean. Thus we hold one parameter constant. Say that the mean of the sample was 10; then we assume that the population mean is 10 also and we keep this value constant. With this parameter fixed, can all four scores from our sample vary? The answer is no, because to keep the mean constant only three values are free to vary. For example, if the values in the sample were 8, 9, 11, 12 (mean = 10) and we changed three of these values to 7, 15 and 8, then the final value *must* be 10 to keep the mean constant. Therefore, if we hold one parameter constant then the degrees of freedom must be one less than the sample size. This fact explains why when we use a sample to estimate the standard deviation of a population, we have to divide the sums of squares by $N-1$ rather than N alone.

$$\text{variance } (s^2) = \frac{\text{SS}}{N-1} = \frac{\sum(x_i - \bar{x})^2}{N-1} = \frac{5.20}{4} = 1.3 \qquad (2.1)$$

The variance is, therefore, the average error between the mean and the observations made (and so is a measure of how well the model fits the actual data). There is one problem with the variance as a measure: it gives us a measure in units squared (because we squared each error in the calculation). In our example we would have to say that the average error in our data (the variance) was 1.3 friends squared. It makes little enough sense to talk about 1.3 friends, but it makes even less to talk about friends squared! For this reason, we often take the square root of the variance (which ensures that the measure of average error is in the same units as the original measure). This measure is known as the standard deviation and is simply the square root of the variance. In this example the **standard deviation** is:

$$\begin{aligned} s &= \sqrt{\frac{\sum(x_i - \bar{x})^2}{N-1}} \\ &= \sqrt{1.3} \\ &= 1.14 \end{aligned} \qquad (2.2)$$

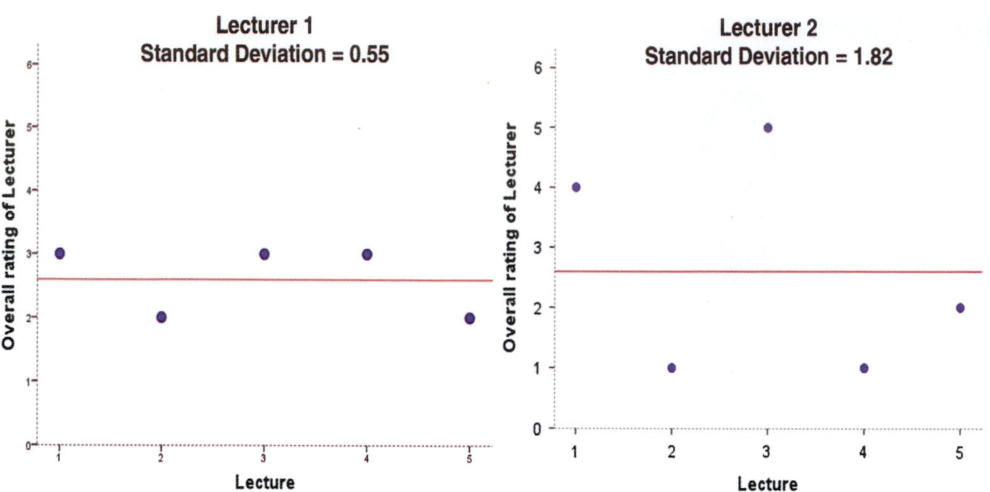

FIGURE 2.5 Graphs illustrating data that have the same mean but different standard deviations

The sum of squares, variance and standard deviation are all, therefore, measures of the 'fit' (i.e. how well the mean represents the data). Small standard deviations (relative to the value of the mean itself) indicate that data points are close to the mean. A large standard deviation (relative to the mean) indicates that the data points are distant from the mean (i.e. the mean is not an accurate representation of the data). A standard deviation of 0 would mean that all of the scores were the same. Figure 2.5 shows the overall ratings (on a 5-point scale) of two lecturers after each of five different lectures. Both lecturers had an average rating of 2.6 out of 5 across the lectures. However, the first lecturer had a standard deviation of 0.55 (relatively small compared to the mean). It should be clear from the graph that ratings for this lecturer were consistently close to the mean rating. There was a small fluctuation, but generally his lectures did not vary in popularity. As such, the mean is an accurate representation of his ratings. The mean is a good fit of the data. The second lecturer, however, had a standard deviation of 1.82 (relatively high compared to the mean). The ratings for this lecturer are clearly more spread from the mean; that is, for some lectures he received very high ratings, and for others his ratings were appalling. Therefore, the mean is not such an accurate representation of his performance because there was a lot of variability in the popularity of his lectures. The mean is a poor fit of the data. This illustration should hopefully make clear why the standard deviation is a measure of how well the mean represents the data.

SELF-TEST In section 1.7.2.2 we came across some data about the number of friends that 11 people had on Facebook (22, 40, 53, 57, 93, 98, 103, 108, 116, 121, 252). We calculated the mean for these data as 96.64. Now calculate the sums of squares, variance, and standard deviation.

SELF-TEST Calculate these values again but excluding the extreme score (252).

2.4.3. Expressing the mean as a model ②

The discussion of means, sums of squares and variance may seem a side track from the initial point about fitting statistical models, but it's not: the mean is a simple statistical model

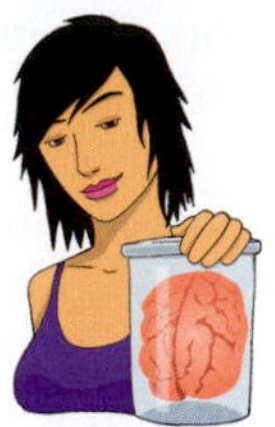

JANE SUPERBRAIN 2.3

The standard deviation and the shape of the distribution ①

As well as telling us about the accuracy of the mean as a model of our data set, the variance and standard deviation also tell us about the shape of the distribution of scores. As such, they are measures of dispersion like those we encountered in section 1.7.3. If the mean represents the data well then most of the scores will cluster close to the mean and the resulting standard deviation is small relative to the mean. When the mean is a worse representation of the data, the scores cluster more widely around the mean (think back to Figure 2.5) and the standard deviation is larger. Figure 2.6 shows two distributions that have the same mean (50) but different standard deviations. One has a large standard deviation relative to the mean (*SD* = 25) and this results in a flatter distribution that is more spread out, whereas the other has a small standard deviation relative to the mean (*SD* = 15) resulting in a more pointy distribution in which scores close to the mean are very frequent but scores further from the mean become increasingly infrequent. The main message is that as the standard deviation gets larger, the distribution gets fatter. This can make distributions look platykurtic or leptokurtic when, in fact, they are not.

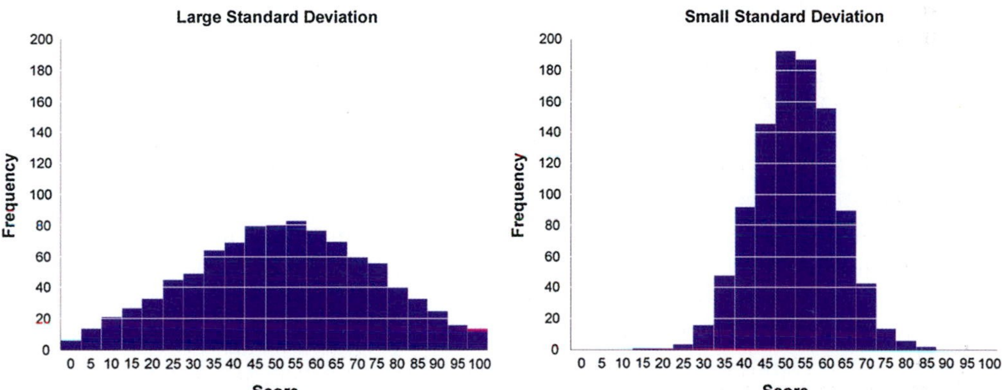

FIGURE 2.6 Two distributions with the same mean, but large and small standard deviations

that can be fitted to data. What do I mean by this? Well, everything in statistics essentially boils down to one equation:

$$\text{outcome}_i = (\text{model}) + \text{error}_i \tag{2.3}$$

This just means that the data we observe can be predicted from the model we choose to fit to the data plus some amount of error. When I say that the mean is a simple statistical model, then all I mean is that we can replace the word 'model' with the word 'mean' in that equation. If we return to our example involving the number of friends that statistics lecturers have and look at lecturer 1, for example, we observed that they had one friend and the mean of all lecturers was 2.6. So, the equation becomes:

$$\begin{aligned}\text{outcome}_{\text{lecturer1}} &= \overline{X} + \varepsilon_{\text{lecturer1}}\\ 1 &= 2.6 + \varepsilon_{\text{lecturer1}}\end{aligned}$$

From this we can work out that the error is 1 – 2.6, or −1.6. If we replace this value in the equation we get 1 = 2.6 – 1.6 or 1 = 1. Although it probably seems like I'm stating the obvious, it is worth bearing this general equation in mind throughout this book because if you do you'll discover that most things ultimately boil down to this one simple idea!

Likewise, the variance and standard deviation illustrate another fundamental concept: how the goodness of fit of a model can be measured. If we're looking at how well a model fits the data (in this case our model is the mean) then we generally look at deviation from the model, we look at the sum of squared error, and in general terms we can write this as:

$$\text{deviation} = \sum(\text{observed} - \text{model})^2 \tag{2.4}$$

Put another way, we assess models by comparing the data we observe to the model we've fitted to the data, and then square these differences. Again, you'll come across this fundamental idea time and time again throughout this book.

2.5. Going beyond the data ①

Using the example of the mean, we have looked at how we can fit a statistical model to a set of observations to summarize those data. It's one thing to summarize the data that you have actually collected but usually we want to go beyond our data and say something general about the world (remember in Chapter 1 that I talked about how good theories should say something about the world). It is one thing to be able to say that people in our sample responded well to medication, or that a sample of high-street stores in Brighton had increased profits leading up to Christmas, but it's more useful to be able to say, based on our sample, that all people will respond to medication, or that all high-street stores in the UK will show increased profits. To begin to understand how we can make these general inferences from a sample of data we can first look not at whether our model is a good fit of the sample from which it came, but whether it is a good fit of the **population** from which the sample came.

2.5.1. The standard error ①

We've seen that the standard deviation tells us something about how well the mean represents the sample data, but I mentioned earlier on that usually we collect data from samples because we don't have access to the entire population. If you take several samples from a population, then these samples will differ slightly; therefore, it's also important to know how well a particular sample represents the population. This is where we use the **standard error**. Many students get confused about the difference between the standard deviation and the standard error (usually because the difference is never explained clearly). However, the standard error is an important concept to grasp, so I'll do my best to explain it to you.

We have already learnt that social scientists use samples as a way of estimating the behaviour in a population. Imagine that we were interested in the ratings of all lecturers (so, lecturers in general were the population). We could take a sample from this population. When someone takes a sample from a population, they are taking one of many

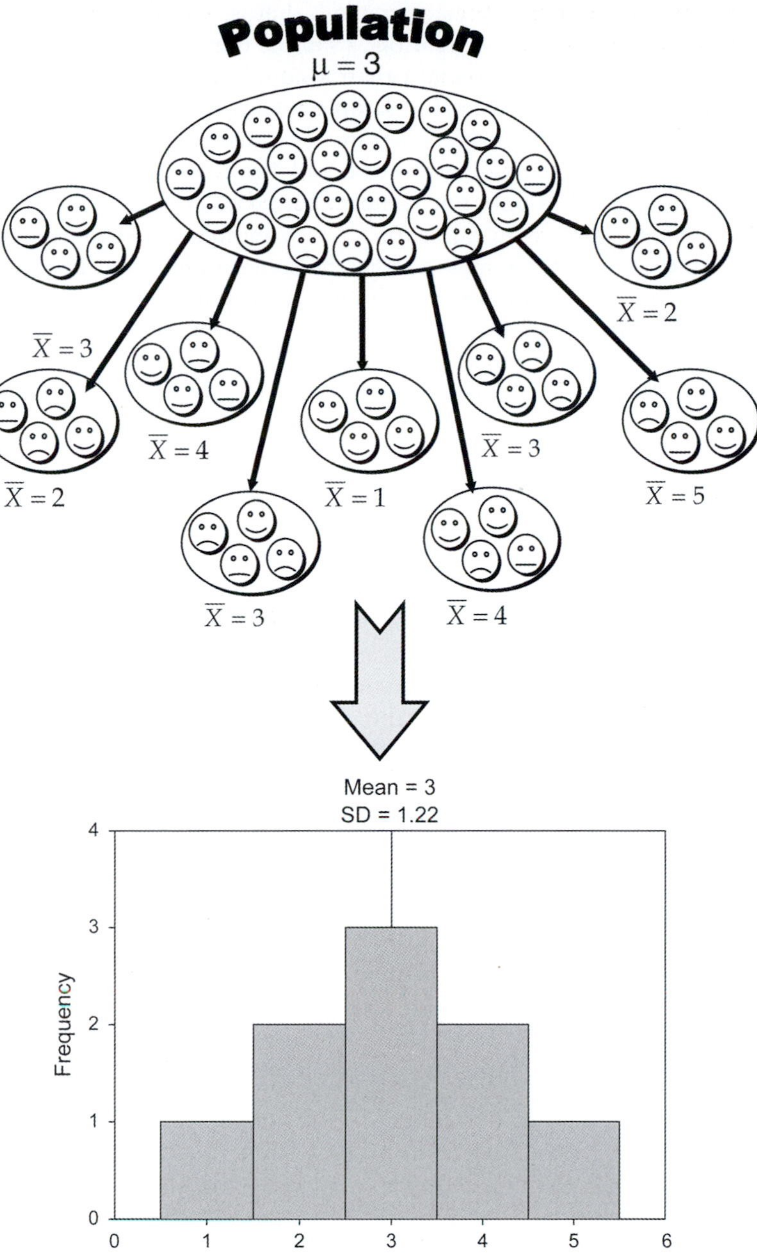

FIGURE 2.7 Illustration of the standard error (see text for details)

possible samples. If we were to take several samples from the same population, then each sample has its own mean, and some of these sample means will be different.

Figure 2.7 illustrates the process of taking samples from a population. Imagine that we could get ratings of all lecturers on the planet and that, on average, the rating is 3 (this is the *population mean*, μ). Of course, we can't collect ratings of all lecturers, so we use a sample. For each of these samples we can calculate the average, or *sample mean*. Let's imagine we took nine different samples (as in the diagram); you can see that some of the samples have the same mean as the population but some have different means: the first sample of lecturers were rated, on average, as 3, but the second sample were, on average, rated as only 2. This illustrates **sampling variation**: that is, samples will vary because they contain different members of the population; a sample that by chance includes some very good lecturers will have a higher average than a sample that, by chance, includes some awful lecturers. We can actually plot the sample means as a frequency distribution, or histogram,[4] just like I have done in the diagram. This distribution shows that there were three samples that had a mean of 3, means of 2 and 4 occurred in two samples each, and means of 1 and 5 occurred in only one sample each. The end result is a nice symmetrical distribution known as a **sampling distribution**. A sampling distribution is simply the frequency distribution of sample means from the same population. In theory you need to imagine that we're taking hundreds or thousands of samples to construct a sampling distribution, but I'm just using nine to keep the diagram simple.[5] The sampling distribution tells us about the behaviour of samples from the population, and you'll notice that it is centred at the same value as the mean of the population (i.e. 3). This means that if we took the average of all sample means we'd get the value of the population mean. Now, if the average of the sample means is the same value as the population mean, then if we know the accuracy of that average we'd know something about how likely it is that a given sample is representative of the population. So how do we determine the accuracy of the population mean?

Think back to the discussion of the standard deviation. We used the standard deviation as a measure of how representative the mean was of the observed data. Small standard deviations represented a scenario in which most data points were close to the mean, a large standard deviation represented a situation in which data points were widely spread from the mean. If you were to calculate the standard deviation between *sample means* then this too would give you a measure of how much variability there was between the means of different samples. The standard deviation of sample means is known as the **standard error of the mean (SE)**. Therefore, the standard error could be calculated by taking the difference between each sample mean and the overall mean, squaring these differences, adding them up, and then dividing by the number of samples. Finally, the square root of this value would need to be taken to get the standard deviation of sample means, the standard error.

Of course, in reality we cannot collect hundreds of samples and so we rely on approximations of the standard error. Luckily for us some exceptionally clever statisticians have demonstrated that as samples get large (usually defined as greater than 30), the sampling distribution has a normal distribution with a mean equal to the population mean, and a standard deviation of:

$$\sigma_{\overline{X}} = \frac{s}{\sqrt{N}} \tag{2.5}$$

This is known as the **central limit theorem** and it is useful in this context because it means that if our sample is large we can use the above equation to approximate the standard error (because, remember, it is the standard deviation of the sampling distribution).[6] When the sample is relatively small (fewer than 30) the sampling distribution has a different shape, known as a *t*-distribution, which we'll come back to later.

[4] This is just a graph of each sample mean plotted against the number of samples that has that mean – see section 1.7.1 for more details.

[5] It's worth pointing out that I'm talking hypothetically. We don't need to *actually* collect these samples because clever statisticians have worked out what these sampling distributions would look like and how they behave.

[6] In fact it should be the *population* standard deviation (σ) that is divided by the square root of the sample size; however, for large samples this is a reasonable approximation.

CRAMMING SAM'S TIPS **The standard error**

The standard error is the standard deviation of sample means. As such, it is a measure of how representative a sample is likely to be of the population. A large standard error (relative to the sample mean) means that there is a lot of variability between the means of different samples and so the sample we have might not be representative of the population. A small standard error indicates that most sample means are similar to the population mean and so our sample is likely to be an accurate reflection of the population.

2.5.2. Confidence intervals ②

2.5.2.1. Calculating confidence intervals ②

Remember that usually we're interested in using the sample mean as an estimate of the value in the population. We've just seen that different samples will give rise to different values of the mean, and we can use the standard error to get some idea of the extent to which sample means differ. A different approach to assessing the accuracy of the sample mean as an estimate of the mean in the population is to calculate boundaries within which we believe the true value of the mean will fall. Such boundaries are called **confidence intervals**. The basic idea behind confidence intervals is to construct a range of values within which we think the population value falls.

Let's imagine an example: Domjan, Blesbois, and Williams (1998) examined the learnt release of sperm in Japanese quail. The basic idea is that if a quail is allowed to copulate with a female quail in a certain context (an experimental chamber) then this context will serve as a cue to copulation and this in turn will affect semen release (although during the test phase the poor quail were tricked into copulating with a terry cloth with an embalmed female quail head stuck on top)[7]. Anyway, if we look at the mean amount of sperm released in the experimental chamber, there is a true mean (the mean in the population); let's imagine it's 15 million sperm. Now, in our actual sample, we might find the mean amount of sperm released was 17 million. Because we don't know the true mean, we don't really know whether our sample value of 17 million is a good or bad estimate of this value. What we can do instead is use an interval estimate: we use our sample value as the mid-point, but set a lower and upper limit as well. So, we might say, we think the true value of the mean sperm release is somewhere between 12 million and 22 million spermatozoa (note that 17 million falls exactly between these values). Of course, in this case the true value (15 million) does fall within these limits. However, what if we'd set smaller limits, what if we'd said we think the true value falls between 16 and 18 million (again, note that 17 million is in the middle)? In this case the interval does not contain the true value of the mean. Let's now imagine that you were particularly fixated with Japanese quail sperm, and you repeated the experiment 50 times using different samples. Each time you did the experiment again you constructed an interval around the sample mean as I've just described. Figure 2.8 shows this scenario: the circles represent the mean for each sample with the lines sticking out from them representing the intervals for these means. The true value of the mean (the mean in the population) is 15 million and is shown by a vertical line. The first thing to note is that the sample means are different from

What is a confidence interval?

[7] This may seem a bit sick, but the male quails didn't appear to mind too much, which probably tells us all we need to know about male mating behaviour.

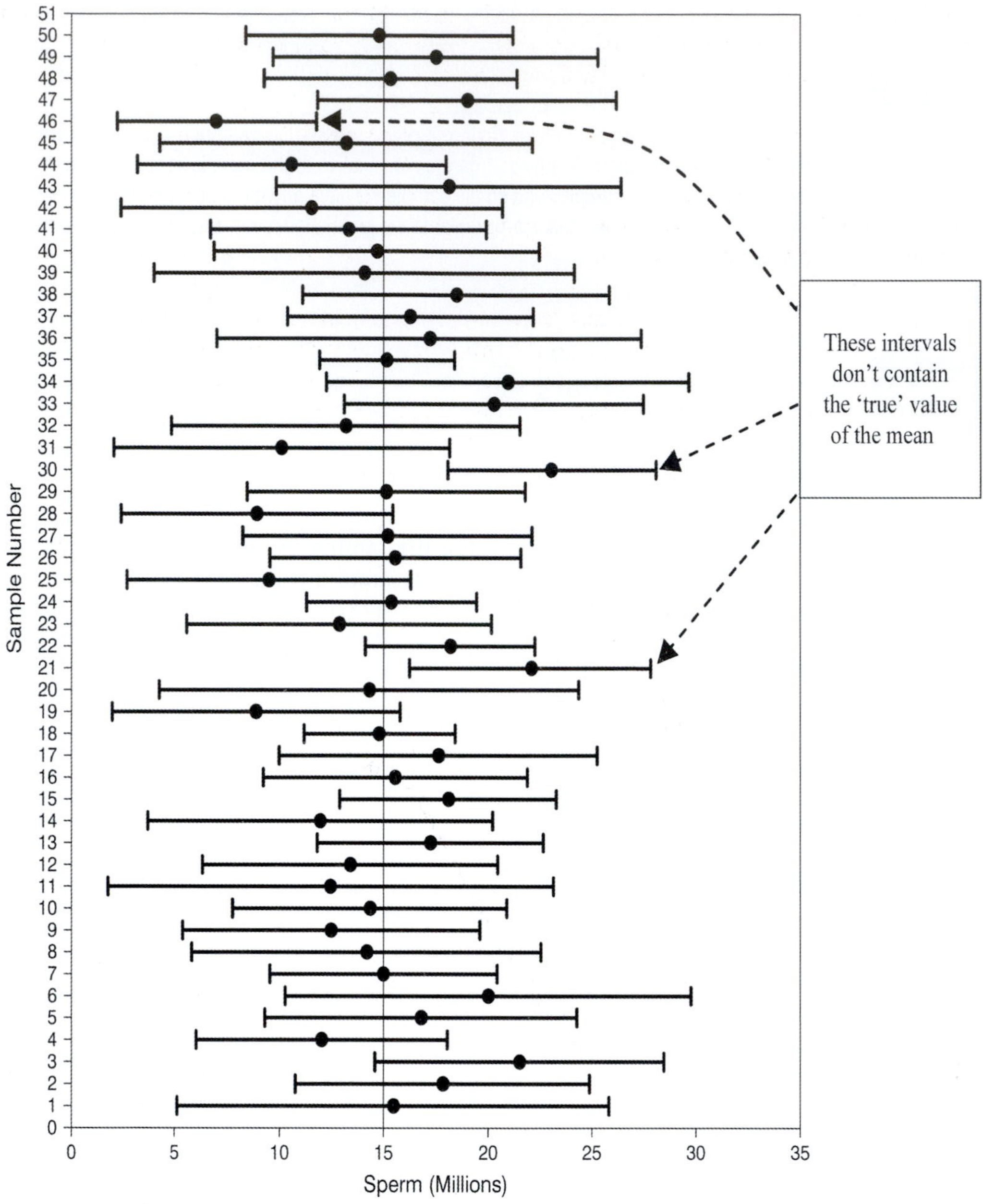
Sample Number
Sperm (Millions)
These intervals don't contain the 'true' value of the mean

FIGURE 2.8 The confidence intervals of the sperm counts of Japanese quail (horizontal axis) for 50 different samples (vertical axis)

the true mean (this is because of sampling variation as described in the previous section). Second, although most of the intervals do contain the true mean (they cross the vertical line, meaning that the value of 15 million spermatozoa falls somewhere between the lower and upper boundaries), a few do not.

Up until now I've avoided the issue of how we might calculate the intervals. The crucial thing with confidence intervals is to construct them in such a way that they tell us something useful. Therefore, we calculate them so that they have certain properties: in particular they tell us the likelihood that they contain the true value of the thing we're trying to estimate (in this case, the mean).

Typically we look at 95% confidence intervals, and sometimes 99% confidence intervals, but they all have a similar interpretation: they are limits constructed such that for a certain percentage of the time (be that 95% or 99%) the true value of the population mean will fall within these limits. So, when you see a 95% confidence interval for a mean, think of it like this: if we'd collected 100 samples, calculated the mean and then calculated a confidence interval for that mean (a bit like in Figure 2.8) then for 95 of these samples, the confidence intervals we constructed would contain the true value of the mean in the population.

To calculate the confidence interval, we need to know the limits within which 95% of means will fall. How do we calculate these limits? Remember back in section 1.7.4 that I said that 1.96 was an important value of z (a score from a normal distribution with a mean of 0 and standard deviation of 1) because 95% of z-scores fall between –1.96 and 1.96. This means that if our sample means were normally distributed with a mean of 0 and a standard error of 1, then the limits of our confidence interval would be –1.96 and +1.96. Luckily we know from the central limit theorem that in large samples (above about 30) the sampling distribution will be normally distributed (see section 2.5.1). It's a pity then that our mean and standard deviation are unlikely to be 0 and 1; except not really because, as you might remember, we can convert scores so that they do have a mean of 0 and standard deviation of 1 (z-scores) using equation (1.2):

$$z = \frac{X - \overline{X}}{s}$$

If we know that our limits are –1.96 and 1.96 in z-scores, then to find out the corresponding scores in our raw data we can replace z in the equation (because there are two values, we get two equations):

$$1.96 = \frac{X - \overline{X}}{s} \qquad -1.96 = \frac{X - \overline{X}}{s}$$

We rearrange these equations to discover the value of X:

$$1.96 \times s = X - \overline{X} \qquad -1.96 \times s = X - \overline{X}$$

$$(1.96 \times s) + \overline{X} = X \qquad (-1.96 \times s) + \overline{X} = X$$

Therefore, the confidence interval can easily be calculated once the standard deviation (s in the equation above) and mean ($\overline{X}$ in the equation) are known. However, in fact we use the standard error and not the standard deviation because we're interested in the variability of sample means, not the variability in observations within the sample. The lower boundary of the confidence interval is, therefore, the mean minus 1.96 times the standard error, and the upper boundary is the mean plus 1.96 standard errors.

$$\text{lower boundary of confidence interval} = \overline{X} - (1.96 \times SE)$$

$$\text{upper boundary of confidence interval} = \overline{X} + (1.96 \times SE)$$

As such, the mean is always in the centre of the confidence interval. If the mean represents the true mean well, then the confidence interval of that mean should be small. We know that 95% of confidence intervals contain the true mean, so we can assume this confidence interval contains the true mean; therefore, if the interval is small, the sample mean must be very close to the true mean. Conversely, if the confidence interval is very wide then the sample mean could be very different from the true mean, indicating that it is a bad representation of the population You'll find that confidence intervals will come up time and time again throughout this book.

2.5.2.2. Calculating other confidence intervals ②

The example above shows how to compute a 95% confidence interval (the most common type). However, we sometimes want to calculate other types of confidence interval such as a 99% or 90% interval. The 1.96 and −1.96 in the equations above are the limits within which 95% of *z*-scores occur. Therefore, if we wanted a 99% confidence interval we could use the values within which 99% of *z*-scores occur (−2.58 and 2.58). In general then, we could say that confidence intervals are calculated as:

$$\text{lower boundary of confidence interval} = \overline{X} - \left(z_{\frac{1-p}{2}} \times \text{SE}\right)$$

$$\text{upper boundary of confidence interval} = \overline{X} + \left(z_{\frac{1-p}{2}} \times \text{SE}\right)$$

in which *p* is the probability value for the confidence interval. So, if you want a 95% confidence interval, then you want the value of *z* for (1 − 0.95)/2 = 0.025. Look this up in the 'smaller portion' column of the table of the standard normal distribution (see the Appendix) and you'll find that *z* is 1.96. For a 99% confidence interval we want *z* for (1 − 0.99)/2 = 0.005, which from the table is 2.58. For a 90% confidence interval we want *z* for (1 − 0.90)/2 = 0.05, which from the table is 1.65. These values of *z* are multiplied by the standard error (as above) to calculate the confidence interval. Using these general principles we could work out a confidence interval for any level of probability that takes our fancy.

2.5.2.3. Calculating confidence intervals in small samples ②

The procedure that I have just described is fine when samples are large, but for small samples, as I have mentioned before, the sampling distribution is not normal, it has a *t*-distribution. The *t*-distribution is a family of probability distributions that change shape as the sample size gets bigger (when the sample is very big, it has the shape of a normal distribution). To construct a confidence interval in a small sample we use the same principle as before but instead of using the value for *z* we use the value for *t*:

$$\text{lower boundary of confidence interval} = \overline{X} - (t_{n-1} \times \text{SE})$$

$$\text{upper boundary of confidence interval} = \overline{X} + (t_{n-1} \times \text{SE})$$

The $n - 1$ in the equations is the degrees of freedom (see Jane Superbrain Box 2.3) and tells us which of the *t*-distributions to use. For a 95% confidence interval we find the value of *t* for a two-tailed test with probability of .05, for the appropriate degrees of freedom.

SELF-TEST In section 1.7.2.2 we came across some data about the number of friends that 11 people had on Facebook. We calculated the mean for these data as 96.64 and standard deviation as 61.27. Calculate a 95% confidence interval for this mean.

SELF-TEST Recalculate the confidence interval assuming that the sample size was 56

2.5.2.4. Showing confidence intervals visually ②

Confidence intervals provide us with very important information about the mean, and, therefore, you often see them displayed on graphs. (We will discover more about how to create these graphs in Chapter 4.) The confidence interval is usually displayed using something called an error bar, which just looks like the letter 'I'. An error bar can represent the standard deviation, or the standard error, but more often than not it shows the 95% confidence interval of the mean. So, often when you see a graph showing the mean, perhaps displayed as a bar (section 4.6) or a symbol (section 4.7), it is often accompanied by this funny I-shaped bar. Why is it useful to see the confidence interval visually?

What's an error bar?

We have seen that the 95% confidence interval is an interval constructed such that in 95% of samples the true value of the population mean will fall within its limits. We know that it is possible that any two samples could have slightly different means (and the standard error tells us a little about how different we can expect sample means to be). Now, the confidence interval tells us the limits within which the population mean is likely to fall (the size of the confidence interval will depend on the size of the standard error). By comparing the confidence intervals of different means we can start to get some idea about whether the means came from the same population or different populations.

Taking our previous example of quail sperm, imagine we had a sample of quail and the mean sperm release had been 9 million sperm with a confidence interval of 2 to 16. Therefore, we know that the population mean is probably between 2 and 16 million sperm. What if we now took a second sample of quail and found the confidence interval ranged from 4 to 15? This interval overlaps a lot with our first sample:

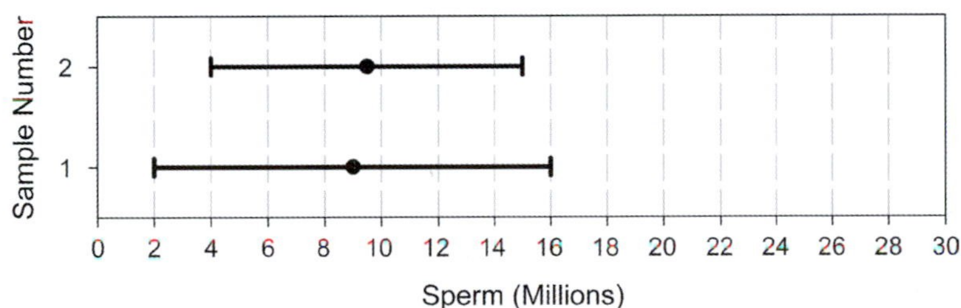

The fact that the confidence intervals overlap in this way tells us that these means could plausibly come from the same population: in both cases the intervals are likely to contain the true value of the mean (because they are constructed such that in 95% of studies they will), and both intervals overlap considerably, so they contain many similar values. What if the confidence interval for our second sample ranges from 18 to 28? If we compared this to our first sample we'd get:

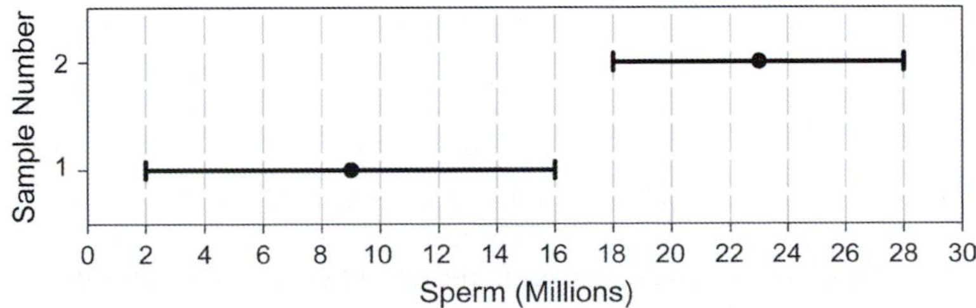

Now, these confidence intervals don't overlap at all. So, one confidence interval, which is likely to contain the population mean, tells us that the population mean is somewhere between 2 and 16 million, whereas the other confidence interval, which is also likely to contain the population mean, tells us that the population mean is somewhere between 18 and 28. This suggests that either our confidence intervals both do contain the population mean, but they come from different populations (and, therefore, so do our samples), or both samples come from the same population but one of the confidence intervals doesn't contain the population mean. If we've used 95% confidence intervals then we know that the second possibility is unlikely (this happens only 5 times in 100 or 5% of the time), so the first explanation is more plausible.

OK, I can hear you all thinking 'so what if the samples come from a different population?' Well, it has a very important implication in experimental research. When we do an experiment, we introduce some form of manipulation between two or more conditions (see section 1.6.2). If we have taken two random samples of people, and we have tested them on some measure (e.g. fear of statistics textbooks), then we expect these people to belong to the same population. If their sample means are so different as to suggest that, in fact, they come from different populations, why might this be? The answer is that our experimental manipulation has induced a difference between the samples.

To reiterate, when an experimental manipulation is successful, we expect to find that our samples have come from different populations. If the manipulation is unsuccessful, then we expect to find that the samples came from the same population (e.g. the sample means should be fairly similar). Now, the 95% confidence interval tells us something about the likely value of the population mean. If we take samples from two populations, then we expect the confidence intervals to be different (in fact, to be sure that the samples were from different populations we would not expect the two confidence intervals to overlap). If we take two samples from the same population, then we expect, if our measure is reliable, the confidence intervals to be very similar (i.e. they should overlap completely with each other).

This is why error bars showing 95% confidence intervals are so useful on graphs, because if the bars of any two means do not overlap then we can infer that these means are from different populations – they are significantly different.

CRAMMING SAM'S TIPS **Confidence intervals**

A confidence interval for the mean is a range of scores constructed such that the population mean will fall within this range in 95% of samples.

The confidence interval is not an interval within which we are 95% confident that the population mean will fall.

2.6. Using statistical models to test research questions ①

In Chapter 1 we saw that research was a five-stage process:

1 Generate a research question through an initial observation (hopefully backed up by some data).
2 Generate a theory to explain your initial observation.
3 Generate hypotheses: break your theory down into a set of testable predictions.

4 Collect data to test the theory: decide on what variables you need to measure to test your predictions and how best to measure or manipulate those variables.

5 Analyse the data: fit a statistical model to the data – this model will test your original predictions. Assess this model to see whether or not it supports your initial predictions.

This chapter has shown us that we can use a sample of data to estimate what's happening in a larger population to which we don't have access. We have also seen (using the mean as an example) that we can fit a statistical model to a sample of data and assess how well it fits. However, we have yet to see how fitting models like these can help us to test our research predictions. How do statistical models help us to test complex hypotheses such as 'is there a relationship between the amount of gibberish that people speak and the amount of vodka jelly they've eaten?', or 'is the mean amount of chocolate I eat higher when I'm writing statistics books than when I'm not?' We've seen in section 1.7.5 that hypotheses can be broken down into a null hypothesis and an alternative hypothesis.

SELF-TEST What are the null and alternative hypotheses for the following questions:

- ✓ 'Is there a relationship between the amount of gibberish that people speak and the amount of vodka jelly they've eaten?'
- ✓ 'Is the mean amount of chocolate eaten higher when writing statistics books than when not?'

Most of this book deals with *inferential statistics*, which tell us whether the alternative hypothesis is likely to be true – they help us to confirm or reject our predictions. Crudely put, we fit a statistical model to our data that represents the alternative hypothesis and see how well it fits (in terms of the variance it explains). If it fits the data well (i.e. explains a lot of the variation in scores) then we assume our initial prediction is true: we gain confidence in the alternative hypothesis. Of course, we can never be completely sure that either hypothesis is correct, and so we calculate the probability that our model would fit if there were no effect in the population (i.e. the null hypothesis is true). As this probability decreases, we gain greater confidence that the alternative hypothesis is actually correct and that the null hypothesis can be rejected. This works provided we make our predictions before we collect the data (see Jane Superbrain Box 2.4).

To illustrate this idea of whether a hypothesis is likely, Fisher (1925/1991) (Figure 2.9) describes an experiment designed to test a claim by a woman that she could determine, by tasting a cup of tea, whether the milk or the tea was added first to the cup. Fisher thought that he should give the woman some cups of tea, some of which had the milk added first and some of which had the milk added last, and see whether she could correctly identify them. The woman would know that there are an equal number of cups in which milk was added first or last but wouldn't know in which order the cups were placed. If we take the simplest situation in which there are only two cups then the woman has 50% chance of guessing correctly. If she did guess correctly we wouldn't be that confident in concluding that she can tell the difference between cups in which the milk was added first from those in which it was added last, because even by guessing she would be correct half of the time. However, what about if we complicated things by having six cups? There are 20 orders in which these cups can be arranged and the woman would guess the correct order only 1 time in 20 (or 5% of the time). If she got the correct order we would be much more

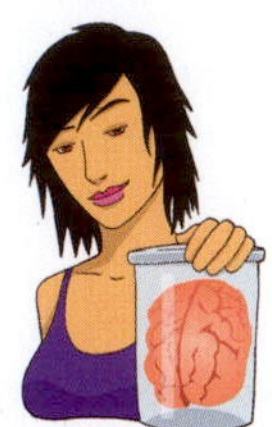

JANE SUPERBRAIN 2.4

Cheating in research ①

The process I describe in this chapter works only if you generate your hypotheses and decide on your criteria for whether an effect is significant before collecting the data. Imagine I wanted to place a bet on who would win the Rugby World Cup. Being an Englishman, I might want to bet on England to win the tournament. To do this I'd: (1) place my bet, choosing my team (England) and odds available at the betting shop (e.g. 6/4); (2) see which team wins the tournament; (3) collect my winnings (if England do the decent thing and actually win).

To keep everyone happy, this process needs to be equitable: the betting shops set their odds such that they're not paying out too much money (which keeps them happy), but so that they do pay out sometimes (to keep the customers happy). The betting shop can offer any odds before the tournament has ended, but it can't change them once the tournament is over (or the last game has started). Similarly, I can choose any team before the tournament, but I can't then change my mind halfway through, or after the final game!

The situation in research is similar: we can choose any hypothesis (rugby team) we like before the data are collected, but we can't change our minds halfway through data collection (or after data collection). Likewise we have to decide on our probability level (or betting odds) before we collect data. *If* we do this, the process works. However, researchers sometimes cheat. They don't write down their hypotheses before they conduct their experiments, sometimes they change them when the data are collected (like me changing my team after the World Cup is over), or worse still decide on them after the data are collected! With the exception of some complicated procedures called *post hoc* tests, this is cheating. Similarly, researchers can be guilty of choosing which significance level to use after the data are collected and analysed, like a betting shop changing the odds after the tournament.

Every time that you change your hypothesis or the details of your analysis you appear to increase the chance of finding a significant result, but in fact you are making it more and more likely that you will publish results that other researchers can't reproduce (which is very embarrassing!). If, however, you follow the rules carefully and do your significance testing at the 5% level you at least know that in the long run at most only 1 result out of every 20 will risk this public humiliation.

(With thanks to David Hitchin for this box, and with apologies to him for turning it into a rugby example.)

confident that she could genuinely tell the difference (and bow down in awe of her finely tuned palette). If you'd like to know more about Fisher and his tea-tasting antics see David Salsburg's excellent book *The lady tasting tea* (Salsburg, 2002). For our purposes the take-home point is that only when there was a very small probability that the woman could complete the tea-task by luck alone would we conclude that she had genuine skill in detecting whether milk was poured into a cup before or after the tea.

It's no coincidence that I chose the example of six cups above (where the tea-taster had a 5% chance of getting the task right by guessing), because Fisher suggested that 95% is a useful threshold for confidence: only when we are 95% certain that a result is genuine (i.e. not a chance finding) should we accept it as being true.[8] The opposite way to look at this is to say that if there is only a 5% chance (a probability of .05) of something occurring by chance then we can accept that it is a genuine effect: we say it is a *statistically significant* finding (see Jane Superbrain Box 2.5 to find out how the criterion of .05 became popular!).

[8] Of course, in reality, it might not be true – we're just prepared to believe that it is.

FIGURE 2.9 Sir Ronald A. Fisher, probably the cleverest person ever ($p < .0001$)

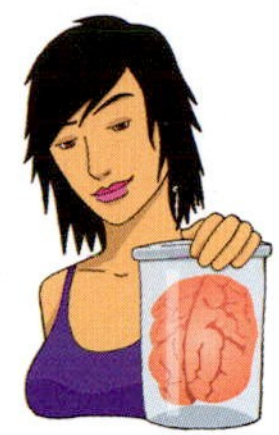

JANE SUPERBRAIN 2.5

Why do we use .05? ①

This criterion of 95% confidence, or a .05 probability, forms the basis of modern statistics and yet there is very little justification for it. How it arose is a complicated mystery to unravel. The significance testing that we use today is a blend of Fisher's idea of using the probability value p as an index of the weight of evidence against a null hypothesis, and Jerzy Neyman and Egron Pearson's idea of testing a null hypothesis *against* an alternative hypothesis. Fisher objected to Neyman's use of an alternative hypothesis (among other things), and Neyman objected to Fisher's exact probability approach (Berger, 2003; Lehmann, 1993). The confusion arising from both parties' hostility to each other's ideas led scientists to create a sort of bastard child of both approaches.

This doesn't answer the question of why we use .05. Well, it probably comes down to the fact that back in the days before computers, scientists had to compare their test statistics against published tables of 'critical values' (they did not have SPSS to calculate exact probabilities for them). These critical values had to be calculated by exceptionally clever people like Fisher. In his incredibly influential textbook *Statistical methods for research workers* (Fisher, 1925)[9] Fisher produced tables of these critical values, but to save space produced tables for particular probability values (.05, .02 and .01). The impact of this book should not be underestimated (to get some idea of its influence 25 years after publication see Mather, 1951; Yates, 1951) and these tables were very frequently used – even Neyman and Pearson admitted the influence that these tables had on them (Lehmann, 1993). This disastrous combination of researchers confused about the Fisher and Neyman–Pearson approaches and the availability of critical values for only certain levels of probability led to a trend to report test statistics as being significant at the now infamous $p < .05$ and $p < .01$ (because critical values were readily available at these probabilities).

However, Fisher acknowledged that the dogmatic use of a fixed level of significance was silly: 'no scientific worker has a fixed level of significance at which from year to year, and in all circumstances, he rejects hypotheses; he rather gives his mind to each particular case in the light of his evidence and his ideas' (Fisher, 1956).

The use of effect sizes (section 2.6.4) strikes a balance between using arbitrary cut-off points such as $p < .05$ and assessing whether an effect is meaningful within the research context. The fact that we still worship at the shrine of $p < .05$ and that research papers are more likely to be published if they contain significant results does make me wonder about a parallel universe where Fisher had woken up in a $p < .10$ kind of mood. My filing cabinet full of research with p just bigger than .05 are published and I am Vice-Chancellor of my university (although, if this were true, the parallel universe version of my university would be in utter chaos, but it would have a campus full of cats).

[9] You can read this online at http://psychclassics.yorku.ca/Fisher/Methods/.

2.6.1. Test statistics ①

We have seen that we can fit statistical models to data that represent the hypotheses that we want to test. Also, we have discovered that we can use probability to see whether scores are likely to have happened by chance (section 1.7.4). If we combine these two ideas then we can test whether our statistical models (and therefore our hypotheses) are significant fits of the data we collected. To do this we need to return to the concepts of systematic and unsystematic variation that we encountered in section 1.6.2.2. Systematic variation is variation that can be explained by the model that we've fitted to the data (and, therefore, due to the hypothesis that we're testing). Unsystematic variation is variation that cannot be explained by the model that we've fitted. In other words, it is error, or variation not attributable to the effect we're investigating. The simplest way, therefore, to test whether the model fits the data, or whether our hypothesis is a good explanation of the data we have observed, is to compare the systematic variation against the unsystematic variation. In doing so we compare how good the model/hypothesis is at explaining the data against how bad it is (the error):

$$\text{test statistic} = \frac{\text{variance explained by the model}}{\text{variance not explained by the model}} = \frac{\text{effect}}{\text{error}}$$

This ratio of systematic to unsystematic variance or effect to error is a **test statistic**, and you'll discover later in the book there are lots of them: t, F and χ^2 to name only three. The exact form of this equation changes depending on which test statistic you're calculating, but the important thing to remember is that they all, crudely speaking, represent the same thing: the amount of variance explained by the model we've fitted to the data compared to the variance that can't be explained by the model (see Chapters 7 and 9 in particular for a more detailed explanation). The reason why this ratio is so useful is intuitive really: if our model is good then we'd expect it to be able to explain more variance than it can't explain. In this case, the test statistic will be greater than 1 (but not necessarily significant).

A test statistic is a statistic that has known properties; specifically we know how frequently different values of this statistic occur. By knowing this, we can calculate the probability of obtaining a particular value (just as we could estimate the probability of getting a score of a certain size from a frequency distribution in section 1.7.4). This allows us to establish how likely it would be that we would get a test statistic of a certain size if there were no effect (i.e. the null hypothesis were true). Field and Hole (2003) use the analogy of the age at which people die. Past data have told us the distribution of the age of death. For example, we know that on average men die at about 75 years old, and that this distribution is top heavy; that is, most people die above the age of about 50 and it's fairly unusual to die in your twenties. So, the frequencies of the age of demise at older ages are very high but are lower at younger ages. From these data, it would be possible to calculate the probability of someone dying at a certain age. If we randomly picked someone and asked them their age, and it was 53, we could tell them how likely it is that they will die before their next birthday (at which point they'd probably punch us). Also, if we met a man of 110, we could calculate how probable it was that he would have lived that long (it would be a very small probability because most people die before they reach that age). The way we use test statistics is rather similar: we know their distributions and this allows us, once we've calculated the test statistic, to discover the probability of having found a value as big as we have. So, if we calculated a test statistic and its value was 110 (rather like our old man) we can then calculate the probability of obtaining a value that large. The more variation our model explains (compared to the variance it can't explain), the bigger the test statistic will be, and the more unlikely it is to occur by chance (like our 110 year old man). So, as test statistics get bigger, the probability of them occurring becomes smaller. When this probability falls below .05 (Fisher's criterion), we accept this as giving us enough confidence to assume that the test statistic is as large as it is because our model explains a sufficient amount of variation to reflect what's genuinely happening in the real world (the population). The test statistic is said to be *significant* (see Jane Superbrain Box 2.6 for a discussion of what statistically significant actually means). Given that the statistical model that we fit to the data reflects the hypothesis that we set out to test, then a significant test statistic tells us that the model would be unlikely to fit this well if the there was no effect in the population (i.e. the null hypothesis was true). Therefore, we can reject our null hypothesis and gain confidence that the alternative hypothesis is true (but, remember, we don't accept it – see section 1.7.5).

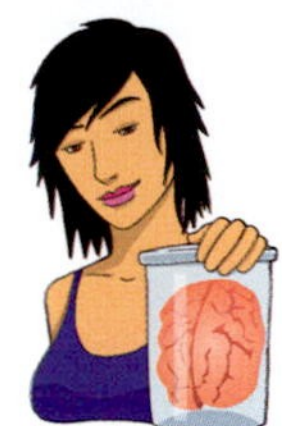

JANE SUPERBRAIN 2.6

What we can and can't conclude from a significant test statistic ②

- **The importance of an effect:** We've seen already that the basic idea behind hypothesis testing involves us generating an experimental hypothesis and a null hypothesis, fitting a statistical model to the data, and assessing that model with a test statistic. If the probability of obtaining the value of our test statistic by chance is less than .05 then we generally accept the experimental hypothesis as true: there is an effect in the population. Normally we say 'there is a *significant* effect of …'. However, don't be fooled by that word 'significant', because even if the probability of our effect being a chance result is small (less than .05) it doesn't necessarily follow that the effect is important. Very small and unimportant effects can turn out to be statistically significant just because huge numbers of people have been used in the experiment (see Field & Hole, 2003: 74).
- **Non-significant results:** Once you've calculated your test statistic, you calculate the probability of that test statistic occurring by chance; if this probability is greater than .05 you reject your alternative hypothesis. However, this does *not* mean that the null hypothesis is true. Remember that the null hypothesis is that there is no effect in the population. All that a non-significant result tells us is that the effect is not big enough to be anything other than a chance finding – it doesn't tell us that the effect is zero. As Cohen (1990) points out, a non-significant result should never be interpreted (despite the fact that it often is) as 'no difference between means' or 'no relationship between variables'. Cohen also points out that the null hypothesis is *never* true because we know from sampling distributions (see section 2.5.1) that two random samples will have slightly different means, and even though these differences can be very small (e.g. one mean might be 10 and another might be 10.00001) they are nevertheless different. In fact, even such a small difference would be deemed as statistically significant if a big enough sample were used. So, significance testing can never tell us that the null hypothesis is true, because it never is.
- **Significant results:** OK, we may not be able to accept the null hypothesis as being true, but we can at least conclude that it is false when our results are significant, right? Wrong! A significant test statistic is based on probabilistic reasoning, which severely limits what we can conclude. Again, Cohen (1994), who was an incredibly lucid writer on statistics, points out that formal reasoning relies on an initial statement of fact followed by a statement about the current state of affairs, and an inferred conclusion. This syllogism illustrates what I mean:
  - If a man has no arms then he can't play guitar:
  - This man plays guitar.
  - Therefore, this man has arms.

  The syllogism starts with a statement of fact that allows the end conclusion to be reached because you can deny the man has no arms (the antecedent) by denying that he can't play guitar (the consequent).[10] A comparable version of the null hypothesis is:

- If the null hypothesis is correct, then this test statistic cannot occur:
- This test statistic has occurred.
- Therefore, the null hypothesis is false.

This is all very nice except that the null hypothesis is not represented in this way because it is based on probabilities. Instead it should be stated as follows:

- If the null hypothesis is correct, then this test statistic is highly unlikely:
- This test statistic has occurred.
- Therefore, the null hypothesis is highly unlikely.

If we go back to the guitar example we could get a similar statement:

- If a man plays guitar then he probably doesn't play for Fugazi (this is true because there are thousands of people who play guitar but only two who play guitar in the band Fugazi):
- Guy Picciotto plays for Fugazi:
- Therefore, Guy Picciotto probably doesn't play guitar.

This should hopefully seem completely ridiculous – the conclusion is wrong because Guy Picciotto does play guitar. This illustrates a common fallacy in hypothesis testing. In fact significance testing allows us to say very little about the null hypothesis.

2.6.2. One- and two-tailed tests ①

We saw in section 1.7.5 that hypotheses can be directional (e.g. 'the more someone reads this book, the more they want to kill its author') or non-directional (i.e. 'reading more of this book could increase or decrease the reader's desire to kill its author'). A statistical model that tests a directional hypothesis is called a **one-tailed test**, whereas one testing a non-directional hypothesis is known as a **two-tailed test**.

Imagine we wanted to discover whether reading this book increased or decreased the desire to kill me. We could do this either (experimentally) by taking two groups, one who had read this book and one who hadn't, or (correlationally) by measuring the amount of this book that had been read and the corresponding desire to kill me. If we have no directional hypothesis then there are three possibilities. (1) People who read this book want to kill me more than those who don't so the difference (the mean for those reading the book minus the mean for non-readers) is positive. Correlationally, the more of the book you read, the more you want to kill me – a positive relationship. (2) People who read this book want to kill me less than those who don't so the difference (the mean for those reading the book minus the mean for non-readers) is negative. Correlationally, the more of the book you read, the less you want to kill me – a negative relationship. (3) There is no difference between readers and non-readers in their desire to kill me – the mean for readers minus the mean for non-readers is exactly zero. Correlationally, there is no relationship between reading this book and wanting to kill me. This final option is the null hypothesis. The direction of the test statistic (i.e. whether it is positive or negative) depends on whether the difference is positive or negative. Assuming there is a positive difference or relationship (reading this book makes you want to kill me), then to detect this difference we have to take account of the fact that the mean for readers is bigger than for non-readers (and so derive a positive test statistic). However, if we've predicted incorrectly and actually reading this book makes readers want to kill me less then the test statistic will actually be negative.

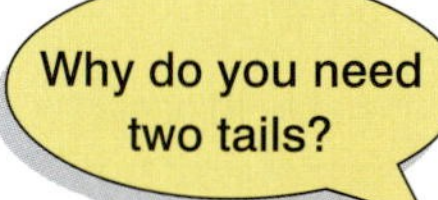
Why do you need two tails?

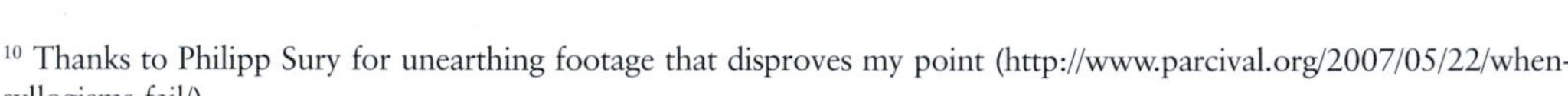
[10] Thanks to Philipp Sury for unearthing footage that disproves my point (http://www.parcival.org/2007/05/22/when-syllogisms-fail/).

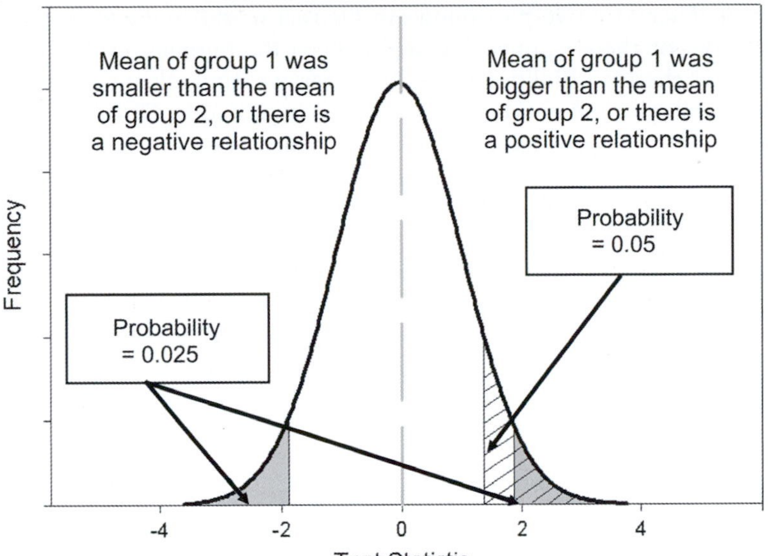

FIGURE 2.10 Diagram to show the difference between one- and two-tailed tests

What are the consequences of this? Well, if at the .05 level we needed to get a test statistic bigger than say 10 and the one we get is actually –12, then we would reject the hypothesis even though a difference does exist. To avoid this we can look at both ends (or tails) of the distribution of possible test statistics. This means we will catch both positive and negative test statistics. However, doing this has a price because to keep our criterion probability of .05 we have to split this probability across the two tails: so we have .025 at the positive end of the distribution and .025 at the negative end. Figure 2.10 shows this situation – the tinted areas are the areas above the test statistic needed at a .025 level of significance. Combine the probabilities (i.e. add the two tinted areas together) at both ends and we get .05, our criterion value. Now if we have made a prediction, then we put all our eggs in one basket and look only at one end of the distribution (either the positive or the negative end depending on the direction of the prediction we make). So, in Figure 2.10, rather than having two small tinted areas at either end of the distribution that show the significant values, we have a bigger area (the lined area) at only one end of the distribution that shows significant values. Consequently, we can just look for the value of the test statistic that would occur by chance with a probability of .05. In Figure 2.10, the lined area is the area above the positive test statistic needed at a .05 level of significance. Note on the graph that the value that begins the area for the .05 level of significance (the lined area) is smaller than the value that begins the area for the .025 level of significance (the tinted area). This means that if we make a specific prediction then we need a smaller test statistic to find a significant result (because we are looking in only one tail of the distribution), but if our prediction happens to be in the wrong direction then we'll miss out on detecting the effect that does exist. In this context it's important to remember what I said in Jane Superbrain Box 2.4: you can't place a bet or change your bet when the tournament is over. If you didn't make a prediction of direction before you collected the data, you are too late to predict the direction and claim the advantages of a one-tailed test.

2.6.3. Type I and Type II errors ①

We have seen that we use test statistics to tell us about the true state of the world (to a certain degree of confidence). Specifically, we're trying to see whether there is an effect in our population. There are two possibilities in the real world: there is, in reality, an effect in the population, or there is, in reality, no effect in the population. We have no way of knowing which of these possibilities is true; however, we can look at test statistics and their associated probability to tell us which of the two is more likely. Obviously, it is important that we're as accurate as possible, which is why Fisher originally said that we should be very conservative and only believe that a result is genuine when we are 95% confident that it is – or when there is only a 5% chance that the results could occur if there was not an effect (the null hypothesis is true). However, even if we're 95% confident there is still a small chance that we get it wrong. In fact there are two mistakes we can make: a Type I and a Type II error. A **Type I error** occurs when we believe that there is a genuine effect in our population, when in fact there isn't. If we use Fisher's criterion then the probability of this error is .05 (or 5%) when there is no effect in the population – this value is known as the **α-level**. Assuming there is no effect in our population, if we replicated our data collection 100 times we could expect that on five occasions we would obtain a test statistic large enough to make us think that there was a genuine effect in the population even though there isn't. The opposite is a **Type II error**, which occurs when we believe that there is no effect in the population when, in reality, there is. This would occur when we obtain a small test statistic (perhaps because there is a lot of natural variation between our samples). In an ideal world, we want the probability of this error to be very small (if there is an effect in the population then it's important that we can detect it). Cohen (1992) suggests that the maximum acceptable probability of a Type II error would be .2 (or 20%) – this is called the **β-level**. That would mean that if we took 100 samples of data from a population in which an effect exists, we would fail to detect that effect in 20 of those samples (so we'd miss 1 in 5 genuine effects).

There is obviously a trade-off between these two errors: if we lower the probability of accepting an effect as genuine (i.e. make α smaller) then we increase the probability that we'll reject an effect that does genuinely exist (because we've been so strict about the level at which we'll accept that an effect is genuine). The exact relationship between the Type I and Type II error is not straightforward because they are based on different assumptions: to make a Type I error there has to be no effect in the population, whereas to make a Type II error the opposite is true (there has to be an effect that we've missed). So, although we know that as the probability of making a Type I error decreases, the probability of making a Type II error increases, the exact nature of the relationship is usually left for the researcher to make an educated guess (Howell, 2006, gives a great explanation of the trade-off between errors).

2.6.4. Effect sizes ②

The framework for testing whether effects are genuine that I've just presented has a few problems, most of which have been briefly explained in Jane Superbrain Box 2.6. The first problem we encountered was knowing how important an effect is: just because a test statistic is significant doesn't mean that the effect it measures is meaningful or important. The solution to this criticism is to measure the size of the effect that we're testing in a standardized way. When we measure the size of an effect (be that an experimental manipulation or the strength of a relationship between variables) it is known as an **effect size**. An effect size is simply an objective and (usually) standardized measure of the magnitude of observed effect. The fact that the measure is standardized just means that we can compare effect sizes across different studies that have measured different variables, or have used different scales of measurement (so an effect size based on speed in milliseconds could be compared to an effect size based on heart rates). Such is the utility of effect size estimates that the American Psychological Association is now recommending that all psychologists report these effect sizes in the results of any published work. So, it's a habit well worth getting into.

Many measures of effect size have been proposed, the most common of which are Cohen's *d*, Pearson's correlation coefficient *r* (Chapter 6) and the odds ratio (Chapter 18). Many of you will be familiar with the correlation coefficient as a measure of the strength of relationship between two variables (see Chapter 6 if you're not); however, it is also a very versatile measure of the strength of an experimental effect. It's a bit difficult to reconcile how the humble correlation coefficient can also be used in this way; however, this is only because students are typically taught about it within the context of non-experimental research. I don't want to get into it now, but as you read through Chapters 6, 9 and 10 it will (I hope) become clear what I mean. Personally, I prefer Pearson's correlation coefficient, *r*, as an effect size measure because it is constrained to lie between 0 (no effect) and 1 (a perfect effect).[11] However, there are situations in which *d* may be favoured; for example, when group sizes are very discrepant *r* can be quite biased compared to *d* (McGrath & Meyer, 2006).

Can we measure how important an effect is?

Effect sizes are useful because they provide an objective measure of the importance of an effect. So, it doesn't matter what effect you're looking for, what variables have been measured, or how those variables have been measured – we know that a correlation coefficient of 0 means there is no effect, and a value of 1 means that there is a perfect effect. Cohen (1988, 1992) has also made some widely used suggestions about what constitutes a large or small effect:

- $r = .10$ **(small effect)**: In this case the effect explains 1% of the total variance.
- $r = .30$ **(medium effect)**: The effect accounts for 9% of the total variance.
- $r = .50$ **(large effect)**: The effect accounts for 25% of the variance.

It's worth bearing in mind that *r* is not measured on a linear scale so an effect with $r = .6$ isn't twice as big as one with $r = .3$. Although these guidelines can be a useful rule of thumb to assess the importance of an effect (regardless of the significance of the test statistic), it is worth remembering that these 'canned' effect sizes are no substitute for evaluating an effect size within the context of the research domain that it is being used (Baguley, 2004; Lenth, 2001).

A final thing to mention is that when we calculate effect sizes we calculate them for a given sample. When we looked at means in a sample we saw that we used them to draw inferences about the mean of the entire population (which is the value in which we're actually interested). The same is true of effect sizes: the size of the effect in the population is the value in which we're interested, but because we don't have access to this value, we use the effect size in the sample to estimate the likely size of the effect in the population. We can also combine effect sizes from different studies researching the same question to get better estimates of the population effect sizes. This is called **meta-analysis** – see Field (2001, 2005b).

[11] The correlation coefficient can also be negative (but not below −1), which is useful when we're measuring a relationship between two variables because the sign of *r* tells us about the direction of the relationship, but in experimental research the sign of *r* merely reflects the way in which the experimenter coded their groups (see Chapter 6).

2.6.5. Statistical power ②

Effect sizes are an invaluable way to express the importance of a research finding. The effect size in a population is intrinsically linked to three other statistical properties: (1) the sample size on which the sample effect size is based; (2) the probability level at which we will accept an effect as being statistically significant (the α-level); and (3) the ability of a test to detect an effect of that size (known as the statistical **power**, not to be confused with statistical powder, which is an illegal substance that makes you understand statistics better). As such, once we know three of these properties, then we can always calculate the remaining one. It will also depend on whether the test is a one- or two-tailed test (see section 2.6.2). Typically, in psychology we use an α-level of .05 (see earlier) so we know this value already. The power of a test is the probability that a given test will find an effect assuming that one exists in the population. If you think back you might recall that we've already come across the probability of failing to detect an effect when one genuinely exists (β, the probability of a Type II error). It follows that the probability of detecting an effect if one exists must be the opposite of the probability of not detecting that effect (i.e. $1 - \beta$). I've also mentioned that Cohen (1988, 1992) suggests that we would hope to have a .2 probability of failing to detect a genuine effect, and so the corresponding level of power that he recommended was 1 – .2, or .8. We should aim to achieve a power of .8, or an 80% chance of detecting an effect if one genuinely exists. The effect size in the population can be estimated from the effect size in the sample, and the sample size is determined by the experimenter anyway so that value is easy to calculate. Now, there are two useful things we can do knowing that these four variables are related:

1. **Calculate the power of a test**: Given that we've conducted our experiment, we will have already selected a value of α, we can estimate the effect size based on our sample, and we will know how many participants we used. Therefore, we can use these values to calculate β, the power of our test. If this value turns out to be .8 or more we can be confident that we achieved sufficient power to detect any effects that might have existed, but if the resulting value is less, then we might want to replicate the experiment using more participants to increase the power.
2. **Calculate the sample size necessary to achieve a given level of power**: Given that we know the value of α and β, we can use past research to estimate the size of effect that we would hope to detect in an experiment. Even if no one had previously done the exact experiment that we intend to do, we can still estimate the likely effect size based on similar experiments. We can use this estimated effect size to calculate how many participants we would need to detect that effect (based on the values of α and β that we've chosen).

The latter use is the more common: to determine how many participants should be used to achieve the desired level of power. The actual computations are very cumbersome, but fortunately there are now computer programs available that will do them for you (one example is G*Power, which is free and can be downloaded from a link on the companion website; another is nQuery Adviser but this has to be bought!). Also, Cohen (1988) provides extensive tables for calculating the number of participants for a given level of power (and vice versa). Based on Cohen (1992) we can use the following guidelines: if we take the standard α-level of .05 and require the recommended power of .8, then we need 783 participants to detect a small effect size (r = .1), 85 participants to detect a medium effect size (r = .3) and 28 participants to detect a large effect size (r = .5).

What have I discovered about statistics? ①

OK, that has been your crash course in statistical theory – hopefully your brain is still relatively intact. The key point I want you to understand is that when you carry out research you're trying to see whether some effect genuinely exists in your population (the effect you're interested in will depend on your research interests and your specific predictions). You won't be able to collect data from the entire population (unless you want to spend your entire life, and probably several after-lives, collecting data) so you use a sample instead. Using the data from this sample, you fit a statistical model to test your predictions, or, put another way, detect the effect you're looking for. Statistics boil down to one simple idea: observed data can be predicted from some kind of model and an error associated with that model. You use that model (and usually the error associated with it) to calculate a test statistic. If that model can explain a lot of the variation in the data collected (the probability of obtaining that test statistic is less than .05) then you infer that the effect you're looking for genuinely exists in the population. If the probability of obtaining that test statistic is more than .05, then you conclude that the effect was too small to be detected. Rather than rely on significance, you can also quantify the effect in your sample in a standard way as an *effect size* and this can be helpful in gauging the importance of that effect. We also discovered that I managed to get myself into trouble at nursery school. It was soon time to move on to primary school and to new and scary challenges. It was a bit like using SPSS for the first time.

Key terms that I've discovered

α-level
β-level
Central limit theorem
Confidence interval
Degrees of freedom
Deviance
Effect size
Fit
Linear model
Meta-analysis
One-tailed test
Population
Power
Sample
Sampling distribution
Sampling variation
Standard deviation
Standard error
Standard error of the mean (SE)
Sum of squared errors (SS)
Test statistic
Two-tailed test
Type I error
Type II error
Variance

Smart Alex's stats quiz

1 Why do we use samples? ①
2 What is the mean and how do we tell if it's representative of our data? ①

3 What's the difference between the standard deviation and the standard error? ①

4 In Chapter 1 we used an example of the time taken for 21 heavy smokers to fall off a treadmill at the fastest setting (18, 16, 18, 24, 23, 22, 22, 23, 26, 29, 32, 34, 34, 36, 36, 43, 42, 49, 46, 46, 57). Calculate the sums of squares, variance, standard deviation, standard error and 95% confidence interval of these data. ①

5 What do the sum of squares, variance and standard deviation represent? How do they differ? ①

6 What is a test statistic and what does it tell us? ①

7 What are Type I and Type II errors? ①

8 What is an effect size and how is it measured? ②

9 What is statistical power? ②

Answers can be found on the companion website.

Further reading

Cohen, J. (1990). Things I have learned (so far). *American Psychologist*, *45*(12), 1304–1312.

Cohen, J. (1994). The earth is round ($p < .05$). *American Psychologist*, *49*(12), 997–1003. (A couple of beautiful articles by the best modern writer of statistics that we've had.)

Field, A. P., & Hole, G. J. (2003). *How to design and report experiments*. London: Sage. (I am rather biased, but I think this is a good overview of basic statistical theory.)

Miles, J. N. V., & Banyard, P. (2007). *Understanding and using statistics in psychology: a practical introduction*. London: Sage. (A fantastic and amusing introduction to statistical theory.)

Wright, D. B., & London, K. (2009). *First steps in statistics* (2nd ed.). London: Sage. (This book has very clear introductions to sampling, confidence intervals and other important statistical ideas.)

Interesting real research

Domjan, M., Blesbois, E., & Williams, J. (1998). The adaptive significance of sexual conditioning: Pavlovian control of sperm release. *Psychological Science*, *9*(5), 411–415.

The SPSS environment 3

FIGURE 3.1 All I want for Christmas is … some tasteful wallpaper

3.1. What will this chapter tell me? ①

At about 5 years old I moved from nursery (note that I moved, I was not 'kicked out' for showing my …) to primary school. Even though my older brother was already there, I remember being really scared about going. None of my nursery school friends were going to the same school and I was terrified about meeting all of these new children. I arrived in my classroom and, as I'd feared, it was full of scary children. In a fairly transparent ploy to make me think that I'd be spending the next 6 years building sand castles, the teacher told me to play in the sand pit. While I was nervously trying to discover whether I could build a pile of sand high enough to bury my head in it, a boy came and joined me. He was Jonathan Land,

and he was really nice. Within an hour he was my new best friend (5 year olds are fickle …) and I loved school. Sometimes new environments seem more scary than they really are. This chapter introduces you to a scary new environment: SPSS. The SPSS environment is a generally more unpleasant environment in which to spend time than your normal environment; nevertheless, we have to spend time there if we are to analyse our data. The purpose of this chapter is, therefore, to put you in a sand pit with a 5 year old called Jonathan. I will orient you in your new home and everything will be fine. We will explore the key windows in SPSS (the **data editor**, the **viewer** and the **syntax editor**) and also look at how to create variables, enter data and adjust the properties of your variables. We finish off by looking at how to load files and save them.

3.2. Versions of SPSS ①

Which version of SPSS do I needed to use this book?

This book is based primarily on version 17 of SPSS (at least in terms of the diagrams); however, don't be fooled too much by version numbers because SPSS has a habit of releasing 'new' versions fairly regularly. Although this makes SPSS a lot of money and creates a nice market for people writing books on SPSS, there are few differences in these new releases that most of us would actually notice. Occasionally they have a major overhaul (version 16 actually looks quite different to version 15, and the output viewer changed quite a bit too), but most of the time you can get by with a book that doesn't explicitly cover the version you're using (the last edition of this book was based on version 13, but could be used easily with all subsequent versions up to this revision!). So, this third edition, although dealing with version 17, will happily cater for earlier versions (certainly back to version 10). I also suspect it'll be useful with versions 18, 19 and 20 when they appear (although it's always a possibility that SPSS may decide to change everything just to annoy me). In case you're very old school there is a file called **Field2000(Chapter1).pdf** on the book's website that covers data entry in versions of SPSS before version 10, but most of you can ignore this file.

3.3. Getting started ①

SPSS mainly uses two windows: the data editor (this is where you input your data and carry out statistical functions) and the viewer (this is where the results of any analysis appear).[1] There are several additional windows that can be activated such as the SPSS Syntax Editor (see section 3.7), which allows you to enter SPSS commands manually (rather than using the window-based menus). For many people, the syntax window is redundant because you can carry out most analyses by clicking merrily with your mouse. However, there are various additional functions that can be accessed using syntax and in many situations it can save time. This is why sick individuals who enjoy statistics find numerous uses for syntax and start dribbling excitedly when discussing it. Because I wish to drown in a pool of my own excited dribble, there are some sections of the book where I'll force you to use it.

Once SPSS has been activated, a start-up window will appear (see Figure 3.2), which allows you to select various options. If you already have a data file on disk that you would like to open then select *Open an existing data source* by clicking on the ○ so that it looks like ◉: this is the default option. In the space underneath this option there will be a list of recently used data files that you can select with the mouse. To open a selected file click on OK. If you want to open a data file that isn't in the list then simply select *More Files …*

[1] There is also the SmartViewer window, which actually isn't very smart, but more on that later.

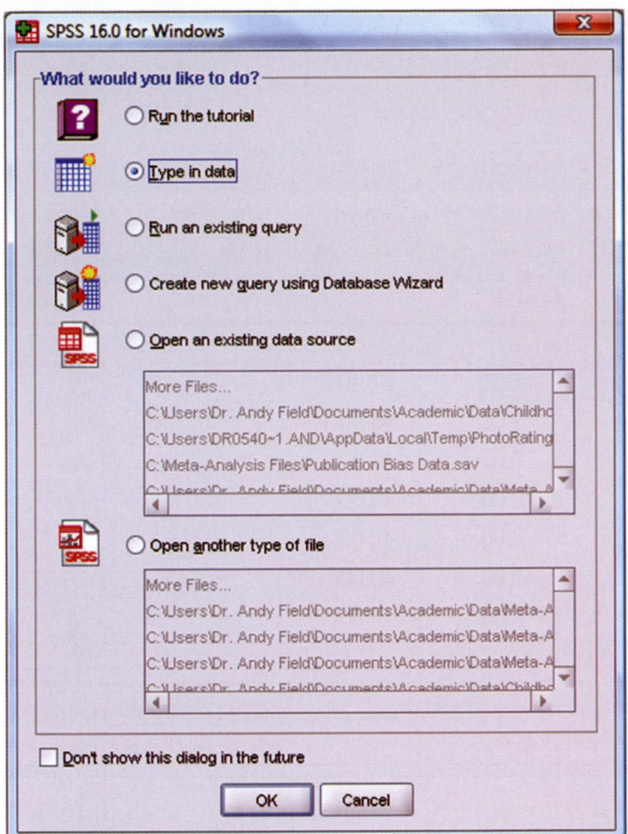

FIGURE 3.2
The start-up window of SPSS

with the mouse and click on OK. This will open a standard explorer window that allows you to browse your computer and find the file you want (see section 3.9). Now it might be the case that you want to open something other than a data file, for example a *viewer* document containing the results of your last analysis. You can do this by selecting *Open another type of file* by clicking on the ○ (so that it looks like ◉) and either selecting a file from the list or selecting *More Files …* and browsing your computer. If you're starting a new analysis (as we are here) then we want to type our data into a new data editor. Therefore, we need to select *Type in data* (by again clicking on the appropriate ○) and then clicking on OK. This will load a blank data editor window.

3.4. The data editor ①

The main SPSS window includes a data editor for entering data. This window is where most of the action happens. At the top of this screen is a menu bar similar to the ones you might have seen in other programs. Figure 3.3 shows this menu bar and the data editor. There are several menus at the top of the screen (e.g. File Edit View) that can be activated by using the computer mouse to move the on-screen arrow onto the desired menu and then pressing the left mouse button once (I'll refer to pressing this button as *clicking*). When you have clicked on a menu, a menu box will appear that displays a list of options that can be activated by moving the on-screen arrow so that it is pointing at the desired option and then clicking with the mouse. Often, selecting an option from a menu makes a window appear; these windows are referred to as *dialog boxes*. When referring to selecting options in a menu I will use images to notate the menu paths; for example, if I were to say that you should select the *Save As …* option in the *File* menu, you will see File Save As....

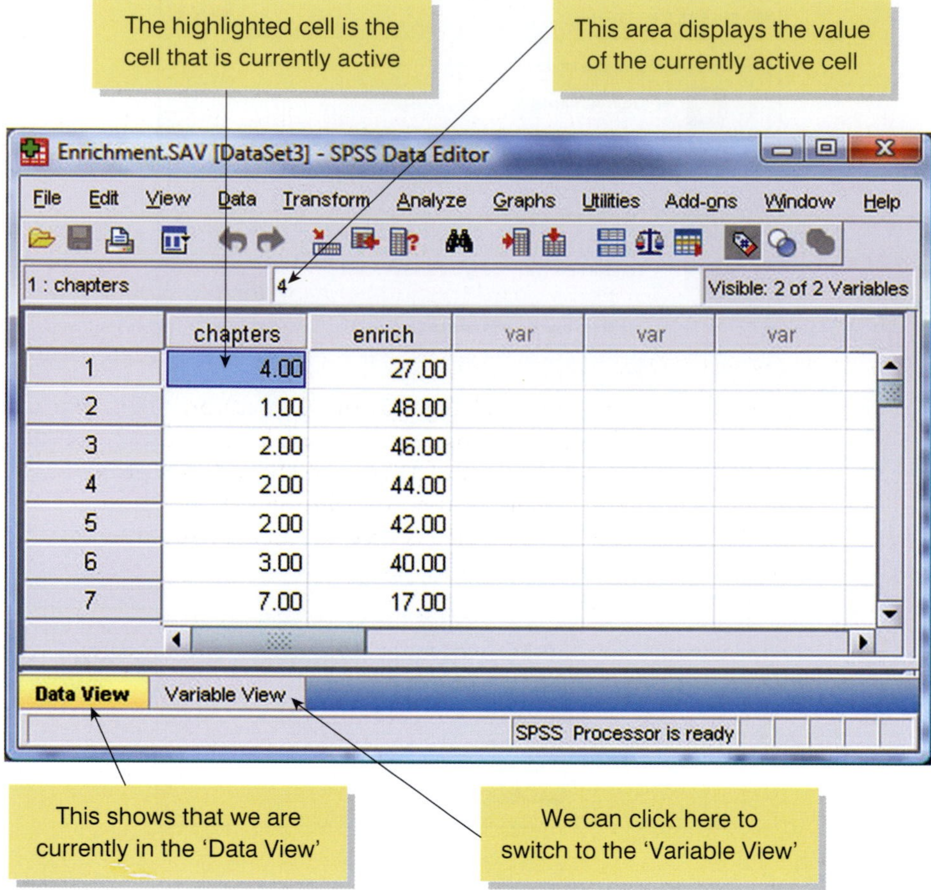

FIGURE 3.3 The SPSS Data Editor

The data editor has two views: the **data view** and the **variable view**. The data view is for entering data into the data editor, and the variable view allows us to define various characteristics of the variables within the data editor. At the bottom of the data editor, you should notice that there are two tabs labelled 'Data View' and 'Variable View' (**Data View** | Variable View) and all we do to switch between these two views is click on these tabs (the highlighted tab tells you which view you're in, although it will be obvious). Let's look at some general features of the data editor, features that don't change when we switch between the data view and the variable view. First off, let's look at the menus.

In many computer packages you'll find that within the menus some letters are underlined: these underlined letters represent the *keyboard shortcut* for accessing that function. It is possible to select many functions without using the mouse, and the experienced keyboard user may find these shortcuts faster than manoeuvring the mouse arrow to the appropriate place on the screen. The letters underlined in the menus indicate that the option can be obtained by simultaneously pressing *Alt* on the keyboard and the underlined letter. So, to access the *Save As…* option, using only the keyboard, you should press *Alt* and F on the keyboard simultaneously (which activates the *File* menu), then, keeping your finger on the *Alt* key, press A (which is the underlined letter).[2]

Below is a brief reference guide to each of the menus and some of the options that they contain. This is merely a summary and we will discover the wonders of each menu as we progress through the book:

[2] If these underlined letters are not visible (in Windows XP they seemed to disappear, but in Vista they appear to have come back) then try pressing *Alt* and then the underlined letters should become visible.

- File This menu allows you to do general things such as saving data, graphs or output. Likewise, you can open previously saved files and print graphs data or output. In essence, it contains all of the options that are customarily found in *File* menus.
- Edit This menu contains edit functions for the data editor. In SPSS it is possible to *cut* and *paste* blocks of numbers from one part of the data editor to another (which can be very handy when you realize that you've entered lots of numbers in the wrong place). You can also use the Options... to select various preferences such as the font that is used for the output. The default preferences are fine for most purposes.
- View This menu deals with system specifications such as whether you have grid lines on the data editor, or whether you display value labels (exactly what value labels are will become clear later).
- Data This menu allows you to make changes to the data editor. The important features are Insert Variable, which is used to insert a new variable into the data editor (i.e. add a column); Insert Cases, which is used to add a new row of data between two existing rows of data; Split File..., which is used to split the file by a grouping variable (see section 5.4.3); and Select Cases..., which is used to run analyses on only a selected sample of cases.
- Transform You should use this menu if you want to manipulate one of your variables in some way. For example, you can use *recode* to change the values of certain variables (e.g. if you wanted to adopt a slightly different coding scheme for some reason) – see SPSS Tip 7.1. The *compute* function is also useful for transforming data (e.g. you can create a new variable that is the average of two existing variables). This function allows you to carry out any number of calculations on your variables (see section 5.7.3).
- Analyze The fun begins here, because the statistical procedures lurk in this menu. Below is a brief guide to the options in the statistics menu that will be used during the course of this book (this is only a small portion of what is available):
  - Descriptive Statistics ▸ This menu is for conducting descriptive statistics (mean, mode, median, etc.), frequencies and general data exploration. There is also a command called *crosstabs* that is useful for exploring frequency data and performing tests such as chi-square, Fisher's exact test and Cohen's kappa.
  - Compare Means ▸ This is where you can find *t*-tests (related and unrelated – Chapter 9) and one-way independent ANOVA (Chapter 10).
  - General Linear Model ▸ This menu is for complex ANOVA such as two-way (unrelated, related or mixed), one-way ANOVA with repeated measures and multivariate analysis of variance (MANOVA) – see Chapters 11, 12, 13, 14, and 16.
  - Mixed Models ▸ This menu can be used for running multilevel linear models (MLMs). At the time of writing I know absolutely nothing about these, but seeing as I've promised to write a chapter on them I'd better go and do some reading. With luck you'll find a chapter on it later in the book, or 30 blank sheets of paper. It could go either way.
  - Correlate ▸ It doesn't take a genius to work out that this is where the correlation techniques are kept! You can do bivariate correlations such as Pearson's *R*, Spearman's rho (ρ) and Kendall's tau (τ) as well as partial correlations (see Chapter 6).
  - Regression ▸ There are a variety of regression techniques available in SPSS. You can do simple linear regression, multiple linear regression (Chapter 7) and more advanced techniques such as logistic regression (Chapter 8).
  - Loglinear ▸ Loglinear analysis is hiding in this menu, waiting for you, and ready to pounce like a tarantula from its burrow (Chapter 13).

- Data Reduction You'll find factor analysis here (Chapter 17).
- Scale Here you'll find reliability analysis (Chapter 17).
- Nonparametric Tests There are a variety of non-parametric statistics available such as the chi-square goodness-of-fit statistic, the binomial test, the Mann–Whitney test, the Kruskal–Wallis test, Wilcoxon's test and Friedman's ANOVA (Chapter 15).

- Graphs SPSS has some graphing facilities and this menu is used to access the Chart Builder (see Chapter 4). The types of graphs you can do include: bar charts, histograms, scatterplots, box–whisker plots, pie charts and error bar graphs to name but a few.
- Window This menu allows you to switch from window to window. So, if you're looking at the output and you wish to switch back to your data sheet, you can do so using this menu. There are icons to shortcut most of the options in this menu so it isn't particularly useful.
- Utilities In this menu there is an option, Data File Comments..., that allows you to comment on your data set. This can be quite useful because you can write yourself notes about from where the data come, or the date they were collected and so on.
- Add-ons SPSS sells several add-ons that can be accessed through this menu. For example, SPSS has a program called *Sample Power* that computes the sample size required for studies, and power statistics (see section 2.6.5). However, because most people won't have these add-ons (including me) I'm not going to discuss them in the book.
- Help This is an invaluable menu because it offers you online help on both the system itself and the statistical tests. The statistics help files are fairly incomprehensible at times (the program is not designed to teach you statistics) and are certainly no substitute for acquiring a good book like this – erm, no, I mean acquiring a good knowledge of your own. However, they can get you out of a sticky situation.

SPSS TIP 3.1 Save time and avoid RSI ①

By default, when you try to open a file from SPSS it will go to the directory in which the program is stored on your computer. This is fine if you happen to store all of your data and output in that folder, but if not then you will find yourself spending time navigating around your computer trying to find your data. If you use SPSS as much as I do then this has two consequences: (1) all those seconds have added up and I have probably spent weeks navigating my computer when I could have been doing something useful like playing my drum kit; (2) I have increased my chances of getting RSI in my wrists, and if I'm going to get RSI in my wrists I can think of more enjoyable ways to achieve it than navigating my computer (drumming again, obviously). Well, we can get around this by telling SPSS where we'd like it to start looking for data files. Select Edit Options... to open the *Options* dialog box below and select the *File Locations* tab.

This dialog box allows you to select a folder in which SPSS will initially look for data files and other files. For example, I keep all of my data files in a single folder called, rather unimaginatively, 'Data'. In the dialog box here I have clicked on Browse... and then navigated to my data folder. SPSS will use this as the default location when I try to open files and my wrists are spared the indignity of RSI. You can also select the option for SPSS to use the *Last folder used,* in which case SPSS remembers where you were last time it was loaded and uses that folder when you try to open up data files.

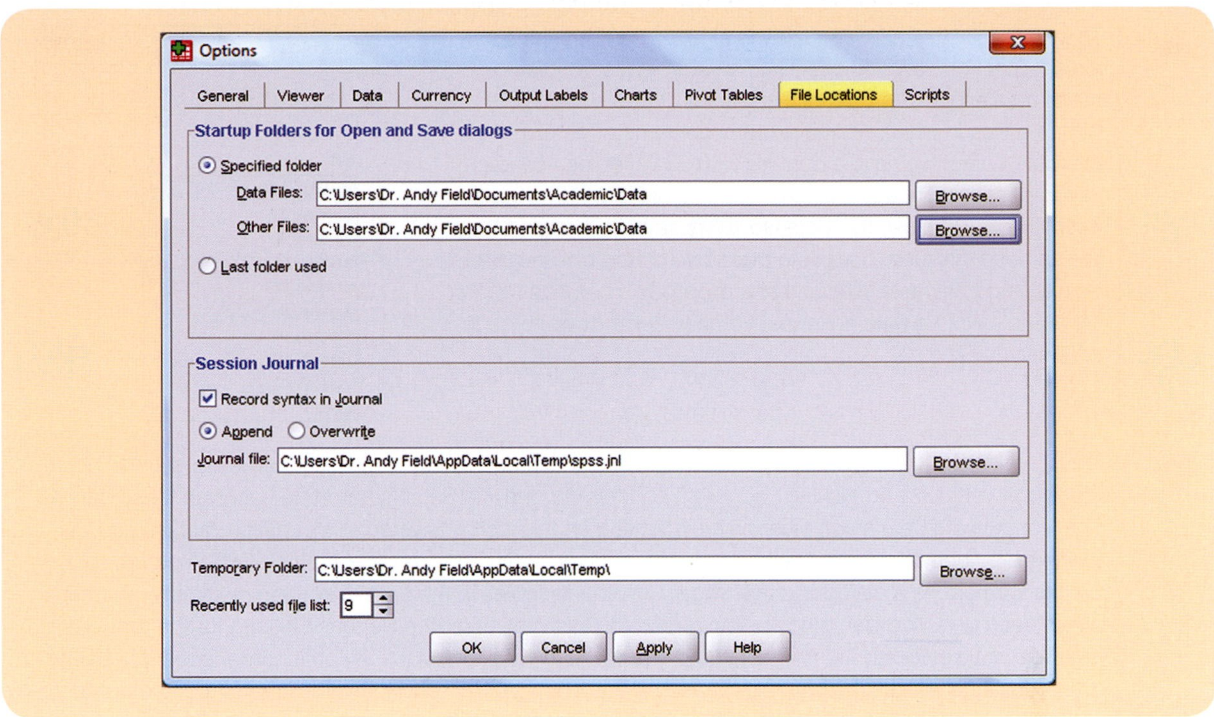

As well as the menus there is also a set of *icons* at the top of the data editor window (see Figure 3.3) that are shortcuts to specific, frequently used, facilities. All of these facilities can be accessed via the menu system but using the icons will save you time. Below is a brief list of these icons and their functions:

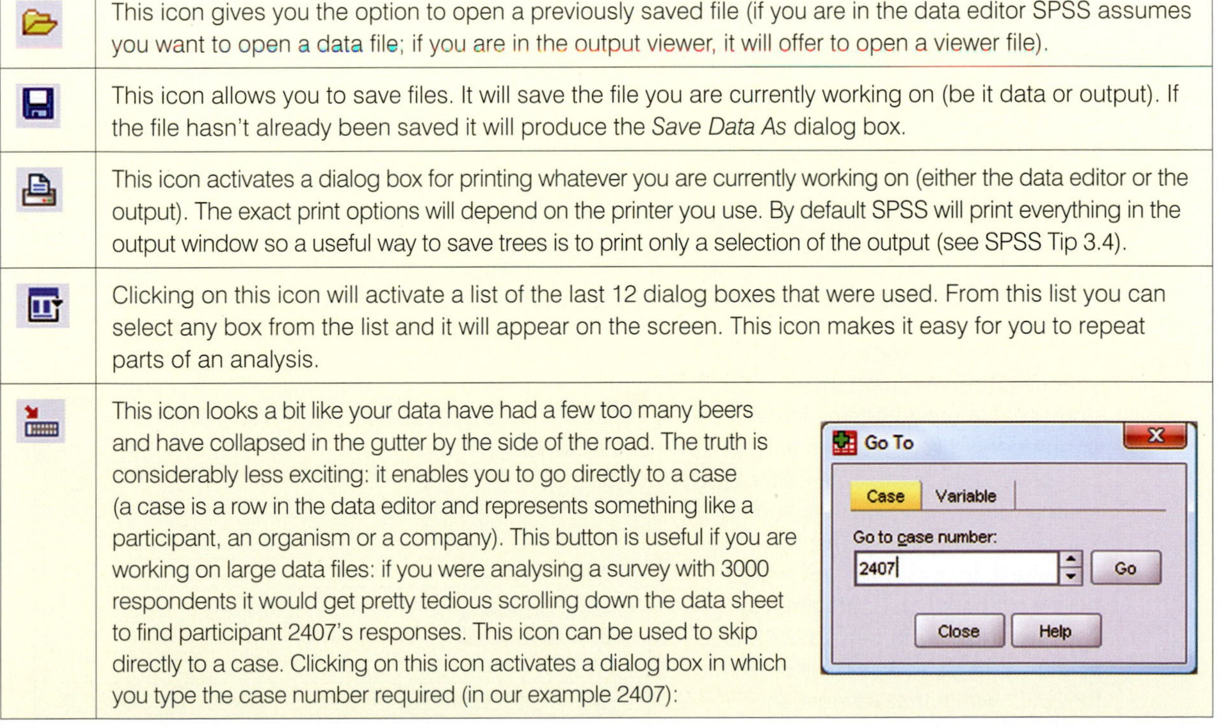

	This icon gives you the option to open a previously saved file (if you are in the data editor SPSS assumes you want to open a data file; if you are in the output viewer, it will offer to open a viewer file).
	This icon allows you to save files. It will save the file you are currently working on (be it data or output). If the file hasn't already been saved it will produce the *Save Data As* dialog box.
	This icon activates a dialog box for printing whatever you are currently working on (either the data editor or the output). The exact print options will depend on the printer you use. By default SPSS will print everything in the output window so a useful way to save trees is to print only a selection of the output (see SPSS Tip 3.4).
	Clicking on this icon will activate a list of the last 12 dialog boxes that were used. From this list you can select any box from the list and it will appear on the screen. This icon makes it easy for you to repeat parts of an analysis.
	This icon looks a bit like your data have had a few too many beers and have collapsed in the gutter by the side of the road. The truth is considerably less exciting: it enables you to go directly to a case (a case is a row in the data editor and represents something like a participant, an organism or a company). This button is useful if you are working on large data files: if you were analysing a survey with 3000 respondents it would get pretty tedious scrolling down the data sheet to find participant 2407's responses. This icon can be used to skip directly to a case. Clicking on this icon activates a dialog box in which you type the case number required (in our example 2407):

This icon activates a function that is similar to the previous one except that you can skip directly to a variable (i.e. a column in the data editor). As before, this is useful when working with big data files in which you have many columns of data. In the example below, we have a data file with 23 variables and each variable represents a question on a questionnaire and is named accordingly (we'll use this data file, **SAQ.sav**, in Chapter 17). We can use this icon to activate the *Go To* dialog box, but this time to find a variable. Notice that a drop-down box lists the first 10 variables in the data editor but you can scroll down to go to others.

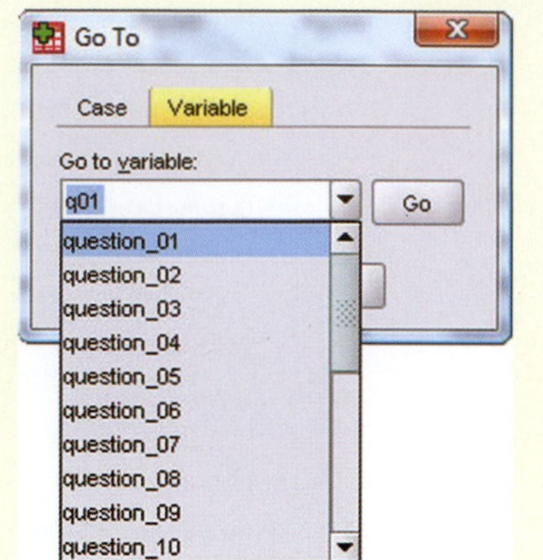

This icon implies to me (what with the big question mark and everything) that you click on it when you're looking at your data scratching your head and thinking 'how in Satan's name do I analyse these data?' The world would be a much nicer place if clicking on this icon answered this question, but instead the shining diamond of hope is snatched cruelly from you by the cloaken thief that is SPSS. Instead, clicking on this icon opens a dialog box that shows you the variables in the data editor and summary information about each one. The dialog box below shows the information for the file that we used for the previous icon. We have selected the first variable in this file, and we can see the variable name (question_01), the label (Statistics makes me cry), the measurement level (ordinal), and the value labels (e.g. the number 1 represents the response of 'strongly agree').

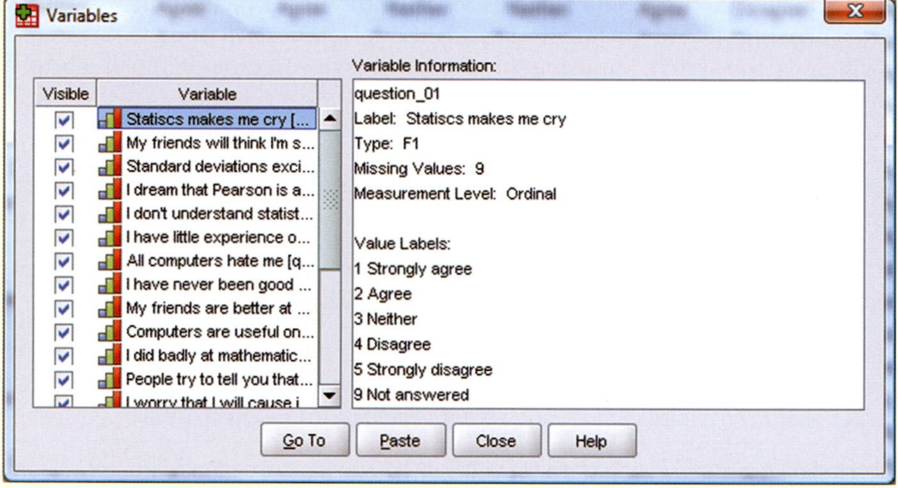

This icon doesn't allow you to spy on your neighbours (unfortunately), but it does enable you to search for words or numbers in your data file and output window. In the data editor it will search within the variable (column) that is currently active. This option is useful if, for example, you realize from a graph of your data that you have typed 20.02 instead of 2.02 (see section 4.4): you can simply search for 20.02 within that variable and replace that value with 2.02:

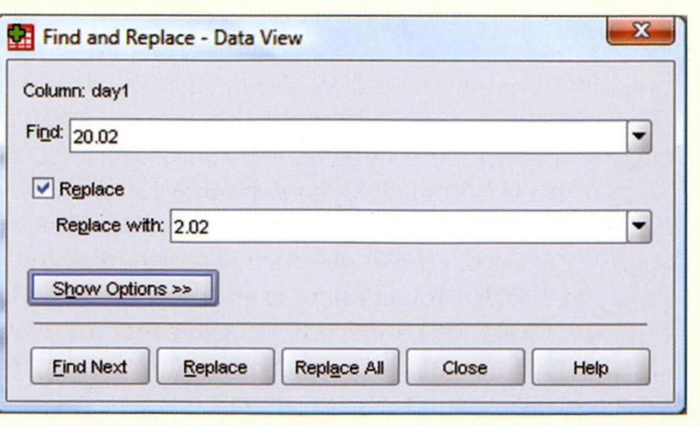

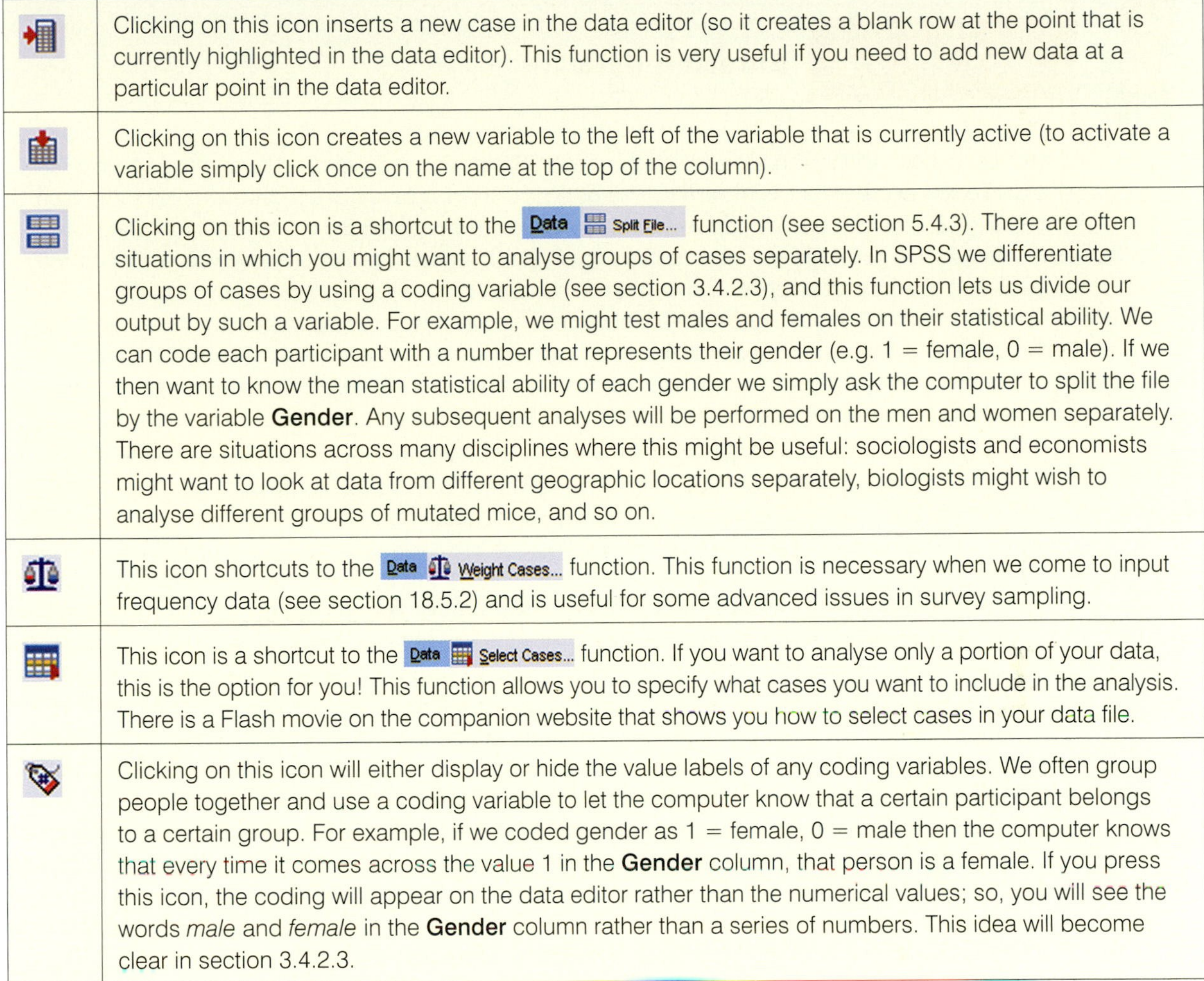

Icon	Description
	Clicking on this icon inserts a new case in the data editor (so it creates a blank row at the point that is currently highlighted in the data editor). This function is very useful if you need to add new data at a particular point in the data editor.
	Clicking on this icon creates a new variable to the left of the variable that is currently active (to activate a variable simply click once on the name at the top of the column).
	Clicking on this icon is a shortcut to the Data ▸ Split File... function (see section 5.4.3). There are often situations in which you might want to analyse groups of cases separately. In SPSS we differentiate groups of cases by using a coding variable (see section 3.4.2.3), and this function lets us divide our output by such a variable. For example, we might test males and females on their statistical ability. We can code each participant with a number that represents their gender (e.g. 1 = female, 0 = male). If we then want to know the mean statistical ability of each gender we simply ask the computer to split the file by the variable **Gender**. Any subsequent analyses will be performed on the men and women separately. There are situations across many disciplines where this might be useful: sociologists and economists might want to look at data from different geographic locations separately, biologists might wish to analyse different groups of mutated mice, and so on.
	This icon shortcuts to the Data ▸ Weight Cases... function. This function is necessary when we come to input frequency data (see section 18.5.2) and is useful for some advanced issues in survey sampling.
	This icon is a shortcut to the Data ▸ Select Cases... function. If you want to analyse only a portion of your data, this is the option for you! This function allows you to specify what cases you want to include in the analysis. There is a Flash movie on the companion website that shows you how to select cases in your data file.
	Clicking on this icon will either display or hide the value labels of any coding variables. We often group people together and use a coding variable to let the computer know that a certain participant belongs to a certain group. For example, if we coded gender as 1 = female, 0 = male then the computer knows that every time it comes across the value 1 in the **Gender** column, that person is a female. If you press this icon, the coding will appear on the data editor rather than the numerical values; so, you will see the words *male* and *female* in the **Gender** column rather than a series of numbers. This idea will become clear in section 3.4.2.3.

3.4.1. Entering data into the data editor ①

When you first load SPSS it will provide a blank data editor with the title *Untitled1* (this of course is daft because once it has been given the title 'untitled' it ceases to be untitled!). When inputting a new set of data, you must input your data in a logical way. The SPSS Data Editor is arranged such that *each row represents data from one entity while each column represents a variable*. There is no discrimination between independent and dependent variables: both types should be placed in a separate column. The key point is that each row represents one entity's data (be that entity a human, mouse, tulip, business, or water sample). Therefore, any information about that case should be entered across the data editor. For example, imagine you were interested in sex differences in perceptions of pain created by hot and cold stimuli. You could place some people's hands in a bucket of very cold water for a minute and ask them to rate how painful they thought the experience was on a scale of 1 to 10. You could then ask them to hold a hot potato and again measure their perception of pain. Imagine I was a participant. You would have a single row representing my data, so there would be a different column for my name, my gender, my pain perception for cold water and my pain perception for a hot potato: Andy, male, 7, 10.

SPSS TIP 3.2 Entering data ①

There is a simple rule for how variables should be placed in the SPSS Data Editor: data from different things go in different rows of the data editor, whereas data from the same things go in different columns of the data editor. As such, each person (or mollusc, goat, organization, or whatever you have measured) is represented in a different row. Data within each person (or mollusc etc.) go in different columns. So, if you've prodded your mollusc, or human, several times with a pencil and measured how much it twitches as an outcome, then each prod will be represented by a column.

In experimental research this means that any variable measured with the same participants (a repeated measure) should be represented by several columns (each column representing one level of the repeated-measures variable). However, any variable that defines different groups of things (such as when a between-group design is used and different participants are assigned to different levels of the independent variable) is defined using a single column. This idea will become clearer as you learn about how to carry out specific procedures. This golden rule is broken in mixed models but until Chapter 19 we can overlook this annoying anomaly.

The column with the information about my gender is a grouping variable: I can belong to either the group of males or the group of females, but not both. As such, this variable is a between-group variable (different people belong to different groups). Rather than representing groups with words, in SPSS we have to use numbers. This involves assigning each group a number, and then telling SPSS which number represents which group. Therefore, between-group variables are represented by a single column in which the group to which the person belonged is defined using a number (see section 3.4.2.3). For example, we might decide that if a person is male then we give them the number 0, and if they're female we give them the number 1. We then have to tell SPSS that every time it sees a 1 in a particular column the person is a female, and every time it sees a 0 the person is a male. Variables that specify to which of several groups a person belongs can be used to split up data files (so in the pain example you could run an analysis on the male and female participants separately – see section 5.4.3).

Finally, the two measures of pain are a repeated measure (all participants were subjected to hot and cold stimuli). Therefore, levels of this variable (see SPSS Tip 3.2) can be entered in separate columns (one for pain to a hot stimulus and one for pain to a cold stimulus).

The data editor is made up of lots of *cells*, which are just boxes in which data values can be placed. When a cell is active it becomes highlighted in blue (as in Figure 3.3). You can move around the data editor, from cell to cell, using the arrow keys ← ↑ ↓ → (found on the right of the keyboard) or by clicking the mouse on the cell that you wish to activate. To enter a number into the data editor simply move to the cell in which you want to place the data value, type the value, then press the appropriate arrow button for the direction in which you wish to move. So, to enter a row of data, move to the far left of the row, type the value and then press → (this process inputs the value and then moves you into the next cell on the right).

The first step in entering your data is to create some variables using the 'Variable View' of the data editor, and then to input your data using the 'Data View' of the data editor. We'll go through these two steps by working through an example.

3.4.2. The 'Variable View' ①

Before we input any data into the data editor, we need to create the variables. To create variables we use the 'Variable View' of the data editor. To access this view click on the 'Variable View' tab at the bottom of the data editor (Data View | Variable View); the contents of the window will change (see Figure 3.4).

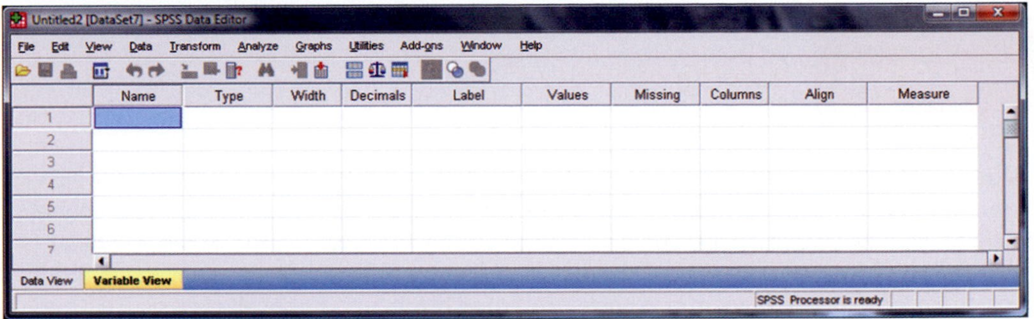

FIGURE 3.4 The 'Variable View' of the SPSS Data Editor

Every row of the variable view represents a variable, and you set characteristics of a particular variable by entering information into the labelled columns. You can change various characteristics of the variable by entering information into the following columns (play around and you'll get the hang of it):

Name	You can enter a name in this column for each variable. This name will appear at the top of the corresponding column in the data view, and helps you to identify variables in the data view. In current versions of SPSS you can more or less write what you like, but there are certain symbols you can't use (mainly symbols that have other uses in SPSS such as +, −, $, &) , and you can't use spaces. (Many people use a 'hard' space in variable names, which replaces the space with an underscore, for example, i.e. Andy_Field instead of Andy Field.) If you use a character that SPSS doesn't like you'll get an error message saying that the variable name is invalid when you click on a different cell, or try to move off the cell using the arrow keys.
Type	You can have different types of data. Mostly you will use **numeric variables** (which just means that the variable contains numbers – SPSS assumes this data type). You will come across **string variables**, which consist of strings of letters. If you wanted to type in people's names, for example, you would need to change the variable type to be string rather than numeric. You can also have **currency variables** (i.e. £s, $s, euro) and **date variables** (e.g. 21-06-1973)
Width	By default, when a new variable is created, SPSS sets it up to be *numeric* and to store 8 digits, but you can change this value by typing a new number in this column in the dialog box. Normally 8 digits is fine, but if you are doing calculations that need to be particularly precise you could make this value bigger.
Decimals	Another default setting is to have 2 decimal places displayed. (You'll notice that if you don't change this option then when you type in whole numbers to the data editor SPSS adds a decimal place with two zeros after it – this can be disconcerting initially!) If you want to change the number of decimal places for a given variable then replace the 2 with a new value or increase or decrease the values using ⬍, when in the cell that you want to change.
Label	The name of the variable (see above) has some restrictions on characters, and you also wouldn't want to use huge long names at the top of your columns (they become hard to read). Therefore, you can write a longer variable description in this column. This may seem pointless, but is actually one of the best habits you can get into (see SPSS Tip 3.3).
Values	This column is for assigning numbers to represent groups of people (see Section 3.4.2.3 below).
Missing	This column is for assigning numbers to missing data (see Section 3.4.3 below).
Columns	Enter a number into this column to determine the width of the column that is how many characters are displayed in the column. (This differs from Width, which determines the width of the variable itself – you could have a variable of 10 characters but by setting the column width to 8 you would only see 8 of the 10 characters of the variable in the data editor.) It can be useful to increase the column width if you have a string variable (section 3.4.2.1) that exceeds 8 characters, or a coding variable (section 3.4.2.3) with value labels that exceed 8 characters.
Align	You can use this column to select the alignment of the data in the corresponding column of the data editor. You can choose to align the data to the Left, Right or Center.
Measure	This is where you define the level at which a variable was measured (*Nominal*, *Ordinal* or *Scale* – section 1.5.1.2).

SPSS TIP 3.3 Naming variables ①

Why is it a good idea to take the time and effort to type in long variable names for your variables? Surely it's a waste of my time? In a way, I can understand why it would seem to be so, but as you go through your course accumulating data files, you will be grateful that you did. Imagine you had a variable called 'number of times I wanted to shoot myself during Andy Field's statistics lecture'; then you might have called the column in SPSS 'shoot'. If you don't add a label, SPSS will use this variable name in all of the output from an analysis. That's all well and good, but what happens in three weeks' time when you look at your data and output again? The chances are that you'll probably think 'What did shoot stand for? Number of shots at goal? Number of shots I drank?' Imagine the chaos you could get into if you had used an acronym for the variable 'workers attending new kiosk'. I have many data sets with variables called things like 'sftg45c', and if I didn't give them proper labels I would be in all sorts of trouble. Get into a good habit and label all of your variables!

Let's use the variable view to create some variables. Imagine we were interested in looking at the differences between lecturers and students. We took a random sample of five psychology lecturers from the University of Sussex and five psychology students and then measured how many friends they had, their weekly alcohol consumption (in units), their yearly income and how neurotic they were (higher score is more neurotic). These data are in Table 3.1.

TABLE 3.1 Some data with which to play

Name	Birth Date	Job	No. of Friends	Alcohol (Units)	Income	Neuroticism
Leo	17-Feb-1977	Lecturer	5	10	20,000	10
Martin	24-May-1969	Lecturer	2	15	40,000	17
Andy	21-Jun-1973	Lecturer	0	20	35,000	14
Paul	16-Jul-1970	Lecturer	4	5	22,000	13
Graham	10-Oct-1949	Lecturer	1	30	50,000	21
Carina	05-Nov-1983	Student	10	25	5,000	7
Karina	08-Oct-1987	Student	12	20	100	13
Doug	16-Sep-1989	Student	15	16	3,000	9
Mark	20-May-1973	Student	12	17	10,000	14
Mark	28-Mar-1990	Student	17	18	10	13

3.4.2.1. Creating a string variable ①

The first variable in our data set is the name of the lecturer/student. This variable consists of names; therefore, it is a *string variable*. To create this variable follow these steps:

1 Move the on-screen arrow (using the mouse) to the first white cell in the column labelled *Name*.

2 Type the word *Name*.

3 Move off this cell using the arrow keys on the keyboard (you can also just click on a different cell, but this is a very slow way of doing it).

You've just created your first variable! Notice that once you've typed a name, SPSS creates default settings for the variable (such as assuming it's numeric and assigning 2 decimal places). The problem is that although SPSS has assumed that we want a numeric variable (i.e. numbers), we don't; we want to enter people's names, namely a *string* variable. Therefore, we have to change the variable type. Move into the column labelled Type using the arrow keys on the keyboard. The cell will now look like this Numeric .... Click on ... to activate the dialog box in Figure 3.5. By default, SPSS selects the numeric variable type (Numeric) – see the left panel of Figure 3.5. To change the variable to a string variable, click on String and the dialog box will change to look like the right panel of Figure 3.5. You can choose how many characters you want in your string variable (i.e. the maximum number of characters you will type for a given case of data). The default is 8, which is fine for us because our longest name is only six letters; however, if we were entering surnames as well, we would need to increase this value. When you have finished, click on OK to return to the variable view.

Now because I want you to get into good habits, move to the cell in the Label column and type a description of the variable, such as 'Participant's First Name'. Finally, we can specify the level at which a variable was measured (see section 1.5.1.2) by going to the column labelled *Measure* and selecting either *Nominal*, *Ordinal* or *Scale* from the drop-down list. In this case, we have a string variable, so they represent only names of cases and provide no information about the order of cases, or the magnitude of one case compared to another. Therefore, we need to select Nominal.

Once the variable has been created, you can return to the data view by clicking on the 'Data View' tab at the bottom of the data editor (Data View | Variable View). The contents of the window will change, and you'll notice that the first column now has the label *Name*. To enter the data, click on the white cell at the top of the column labelled *Name* and type the first name, 'Leo'. To register this value in this cell, we have to move to a different cell and because we are entering data down a column, the most sensible way to do this is to press the ↓ key on the keyboard. This action moves you down to the next cell, and the word 'Leo' should appear in the cell above. Enter the next name, 'Martin', and then press ↓ to move down to the next cell, and so on.

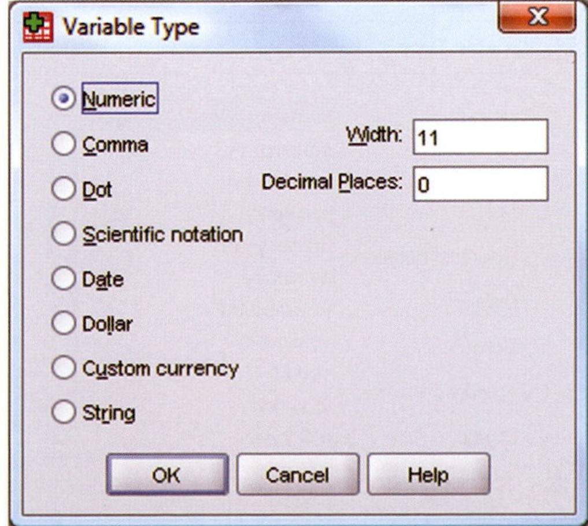

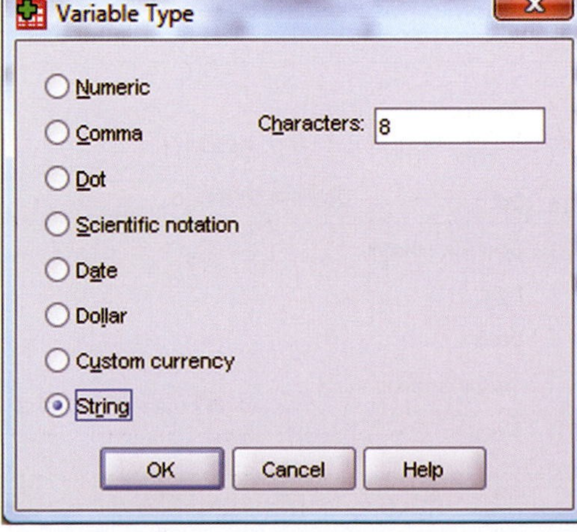

FIGURE 3.5
Defining a string variable in SPSS

3.4.2.2. Creating a date variable ①

Notice that the second column in our table contains dates (birth dates to be exact). To enter date variables into SPSS we use the same procedure as with the previous variable, except that we need to change the variable type. First, move back to the 'Variable View' using the tab at the bottom of the data editor (Data View | Variable View). As with the previous variable, move to the cell in row 2 of the column labelled *Name* (under the previous variable you created). Type the word 'Birth_Date' (note that we have used a hard space to separate the words). Move into the column labelled Type using the → key on the keyboard (SPSS will create default settings in the other columns). The cell will now look like this Numeric ···. Click on ··· to activate the dialog box in Figure 3.6. By default, SPSS selects the numeric variable type (Numeric) – see the left panel of Figure 3.6. To change the variable to a date, click on Date and the dialog box will change to look like the right panel of Figure 3.6. You can then choose your preferred date format; being British, I am used to the days coming before the month and I have stuck with the default option of dd-mmm-yyyy (i.e. 21-Jun-1973), but Americans, for example, will be used to the month and date being the other way around and could select mm/dd/yyyy (06/21/1973). When you have selected a format for your dates, click on OK to return to the variable view. Finally, move to the cell in the column labelled *Label* and type 'Date of Birth'.

Now that the variable has been created, you can return to the data view by clicking on the 'Data View' tab (Data View | Variable View) and input the dates of birth. The second column now has the label *Birth_Date*; click on the white cell at the top of this column and type the first value, 17-Feb-1977. To register this value in this cell, move down to the next cell by pressing the ↓ key on the keyboard. Now enter the next date, and so on.

3.4.2.3. Creating coding variables ①

A coding variable (also known as a grouping variable) is a variable that uses numbers to represent different groups of data. As such, it is a *numeric variable*, but these numbers represent names (i.e. it is a nominal variable). These groups of data could be levels of a treatment variable in an experiment, different groups of people (men or women, an experimental group, or a control group, ethnic groups, etc.), different geographic locations, different organizations, etc.

In experiments, coding variables represent independent variables that have been measured between groups (i.e. different participants were assigned to different groups). If you were to run an experiment with one group of participants in an experimental condition and a different group of participants in a control group, you might assign the experimental group a code

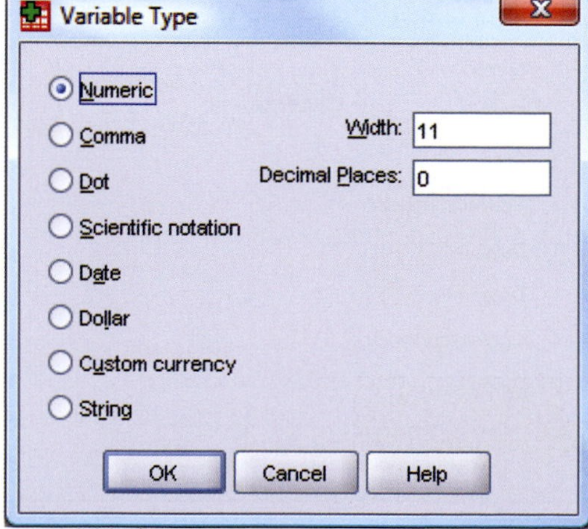

FIGURE 3.6 Defining variable types in SPSS

of 1 and the control group a code of 0. When you come to put the data into the data editor, you would create a variable (which you might call **group**) and type in the value 1 for any participants in the experimental group, and 0 for any participant in the control group. These codes tell SPSS that all of the cases that have been assigned the value 1 should be treated as belonging to the same group, and likewise for the cases assigned the value 0. In situations other than experiments, you might simply use codes to distinguish naturally occurring groups of people (e.g. you might give students a code of 1 and lecturers a code of 0).

We have a coding variable in our data: the one describing whether a person was a lecturer or student. To create this coding variable, we follow the steps for creating a normal variable, but we also have to tell SPSS which numeric codes have been assigned to which groups. So, first of all, return to the variable view (Data View | Variable View) if you're not already in it and then move to the cell in the third row of the data editor and in the column labelled *Name* type a name (let's call it **Group**). I'm still trying to instil good habits, so move along the third row to the column called *Label* and give the variable a full description such as 'Is the person a lecturer or a student?' Then to define the group codes, move along the row to the column labelled Values and into this cell: None ... . Click on ... to access the *Value Labels* dialog box (see Figure 3.7).

The *Value Labels* dialog box is used to specify group codes. This can be done in three easy steps. First, click with the mouse in the white space next to where it says *Value* (or press *Alt* and *u* at the same time) and type in a code (e.g. 1). These codes are completely arbitrary; for the sake of convention people typically use 0, 1, 2, 3, etc., but in practice you could have a code of 495 if you were feeling particularly arbitrary. The second step is to click the mouse in the white space below, next to where it says *Value Label* (or press *Tab*, or *Alt* and *e* at the same time) and type in an appropriate label for that group. In Figure 3.7 I have already defined a code of 1 for the lecturer group, and then I have typed in 2 as my code and given this a label of *Student*. The third step is to add this coding to the list by clicking on Add . When you have defined all of your coding values you can click on Spelling... and SPSS will check your variable labels for spelling errors (which can be very handy if you are as bad at spelling as I am). To finish, click on OK ; if you click on OK and have forgotten to add your final coding to the list, SPSS will display a message warning you that any pending changes will be lost. In plain English

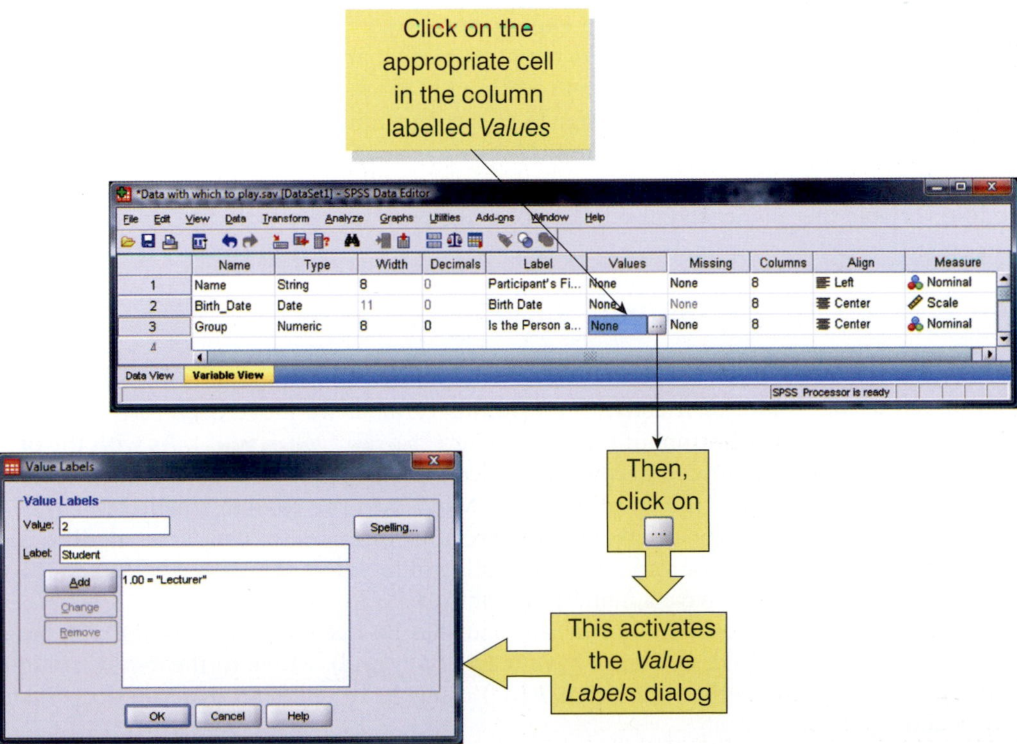

FIGURE 3.7 Defining coding variables and their values in SPSS

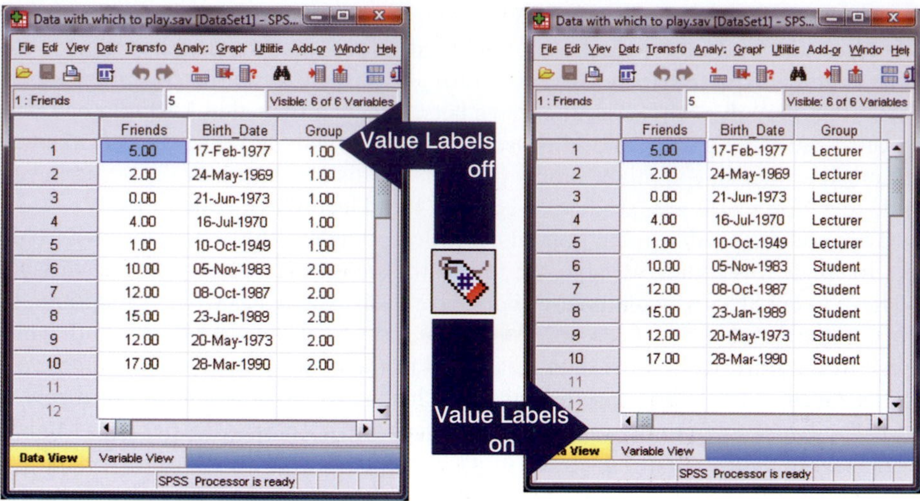

FIGURE 3.8 Coding values in the data editor with the value labels switched off and on

this simply tells you to go back and click on Add before continuing. Finally, coding variables always represent categories and so the level at which they are measured is nominal (or ordinal if the categories have a meaningful order). Therefore, you should specify the level at which the variable was measured by going to the column labelled *Measure* and selecting Nominal (or Ordinal if the groups have a meaningful order) from the drop-down list (see earlier).

Having defined your codes, switch to the data view and type these numerical values into the appropriate column (so if a person was a lecturer, type 1, but if they were a student then type 2). You can get SPSS to display the numeric codes, or the value labels that you assigned to them by clicking on (see Figure 3.8), which is pretty groovy. Figure 3.8 shows how the data should be arranged for a coding variable. Now remember that each row of the data editor represents data from one entity and in this example our entities were people (well, arguably in the case of the lecturers). The first five participants were lecturers whereas participants 6–10 were students.

This example should clarify why in experimental research grouping variables are used for variables that have been measured between participants: because by using a coding variable it is impossible for a participant to belong to more than one group. This situation should occur in a between-group design (i.e. a participant should not be tested in both the experimental and the control group). However, in repeated-measures designs (within subjects) each participant is tested in every condition and so we would not use this sort of coding variable (because each participant does take part in every experimental condition).

3.4.2.4. Creating a numeric variable ①

Numeric variables are the easiest ones to create because SPSS assumes this format for data. Our next variable is **No. of friends**; to create this variable we move back to the variable view using the tab at the bottom of the data editor (Data View **Variable View**). As with the previous variables, move to the cell in row 4 of the column labelled *Name* (under the previous variable you created). Type the word 'Friends'. Move into the column labelled Type using the → key on the keyboard. As with the previous variables we have created, SPSS has assumed that this is a numeric variable, so the cell will look like this Numeric .... We can leave this as it is, because we do have a numeric variable.

Notice that our data for the number of friends has no decimal places (unless you are a very strange person indeed, you can't have 0.23 of a friend). Move to the Decimals column and type '0' (or decrease the value from 2 to 0 using) to tell SPSS that you don't want any decimal places.

Next, let's continue our good habit of naming variables and move to the cell in the column labelled *Label* and type 'Number of Friends'. Finally, we can specify the level at which a variable was measured (see section 1.5.1.2) by going to the column labelled *Measure* and selecting Scale from the drop-down list (this will have been done automatically actually, but it's worth checking).

SELF-TEST Why is the 'Number of Friends' variable a 'scale' variable?

Once the variable has been created, you can return to the data view by clicking on the 'Data View' tab at the bottom of the data editor (Data View | Variable View). The contents of the window will change, and you'll notice that the first column now has the label *Friends*. To enter the data, click on the white cell at the top of the column labelled *Friends* and type the first value, 5. To register this value in this cell, we have to move to a different cell and because we are entering data down a column, the most sensible way to do this is to press the ↓ key on the keyboard. This action moves you down to the next cell, and the number 5 should appear in the cell above. Enter the next number, 2, and then press ↓ to move down to the next cell, and so on.

SELF-TEST Having created the first four variables with a bit of guidance, try to enter the rest of the variables in Table 3.1 yourself.

3.4.3. Missing values ①

Although as researchers we strive to collect complete sets of data, it is often the case that we have missing data. Missing data can occur for a variety of reasons: in long questionnaires participants accidentally (or, depending on how paranoid you're feeling, deliberately just to piss you off) miss out questions; in experimental procedures mechanical faults can lead to a datum not being recorded; and in research on delicate topics (e.g. sexual behaviour) participants may exert their right not to answer a question. However, just because we have missed out on some data for a participant doesn't mean that we have to ignore the data we do have (although it sometimes creates statistical difficulties). Nevertheless, we do need to tell SPSS that a value is missing for a particular case. The principle behind missing values is quite similar to that of coding variables in that we choose a numeric value to represent the missing data point. This value tells SPSS that there is no recorded value for a participant for a certain variable. The computer then ignores that cell of the data editor (it does not use the value you select in the analysis). You need to be careful that the chosen code doesn't correspond to any naturally occurring data value. For example, if we tell the computer to regard the value 9 as a missing value and several participants genuinely scored 9, then the computer will treat their data as missing when, in reality, they are not.

To specify missing values you simply click in the column labelled Missing in the variable view and then click on ... to activate the *Missing Values* dialog box in Figure 3.9. By default SPSS assumes that no missing values exist, but if you do have data with missing

FIGURE 3.9 Defining missing values

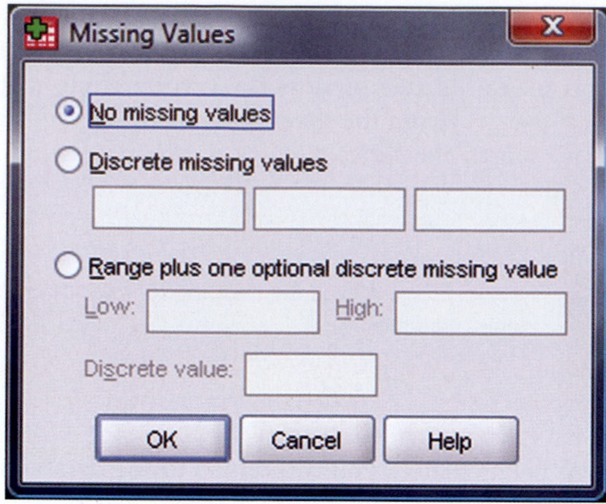

values you can choose to define them in one of three ways. The first is to select discrete values (by clicking on the circle next to where it says *Discrete missing values*) which are single values that represent missing data. SPSS allows you to specify up to three discrete values to represent missing data. The reason why you might choose to have several numbers to represent missing values is that you can assign a different meaning to each discrete value. For example, you could have the number 8 representing a response of 'not applicable', a code of 9 representing a 'don't know' response, and a code of 99 meaning that the participant failed to give any response. As far as the computer is concerned it will ignore any data cell containing these values; however, using different codes may be a useful way to remind you of why a particular score is missing. Usually, one discrete value is enough and in an experiment in which attitudes are measured on a 100-point scale (so scores vary from 1 to 100) you might choose 666 to represent missing values because (1) this value cannot occur in the data that have been collected and (2) missing data create statistical problems, and you will regard the people who haven't given you responses as children of Satan! The second option is to select a range of values to represent missing data and this is useful in situations in which it is necessary to exclude data falling between two points. So, we could exclude all scores between 5 and 10. The final option is to have a range of values and one discrete value.

3.5. The SPSS Viewer ①

Alongside the SPSS Data Editor window, there is a second window known as the *SPSS Viewer*. The viewer window has come a long way; not that you care, I'm sure, but in days of old, this window displayed statistical results in a rather bland and drab font, graphs appeared in a window called the carousel (not nearly as exciting as it sounds), and all of the computations were done by little nerdy dinosaurs that lived inside your computer. These days, the SPSS Viewer displays everything you could want (well, OK, it doesn't display photos of your cat) and it's generally prettier to look at than it used to be. Sadly, however, my prediction in previous editions of this book that future versions of SPSS will include a tea-making facility in the viewer have not come to fruition (SPSS Inc. take note!).

Figure 3.10 shows the basic layout of the output viewer. On the right-hand side there is a large space in which the output is displayed. SPSS displays both graphs and the results of statistical analyses in this part of the viewer. It is also possible to edit graphs and to do this

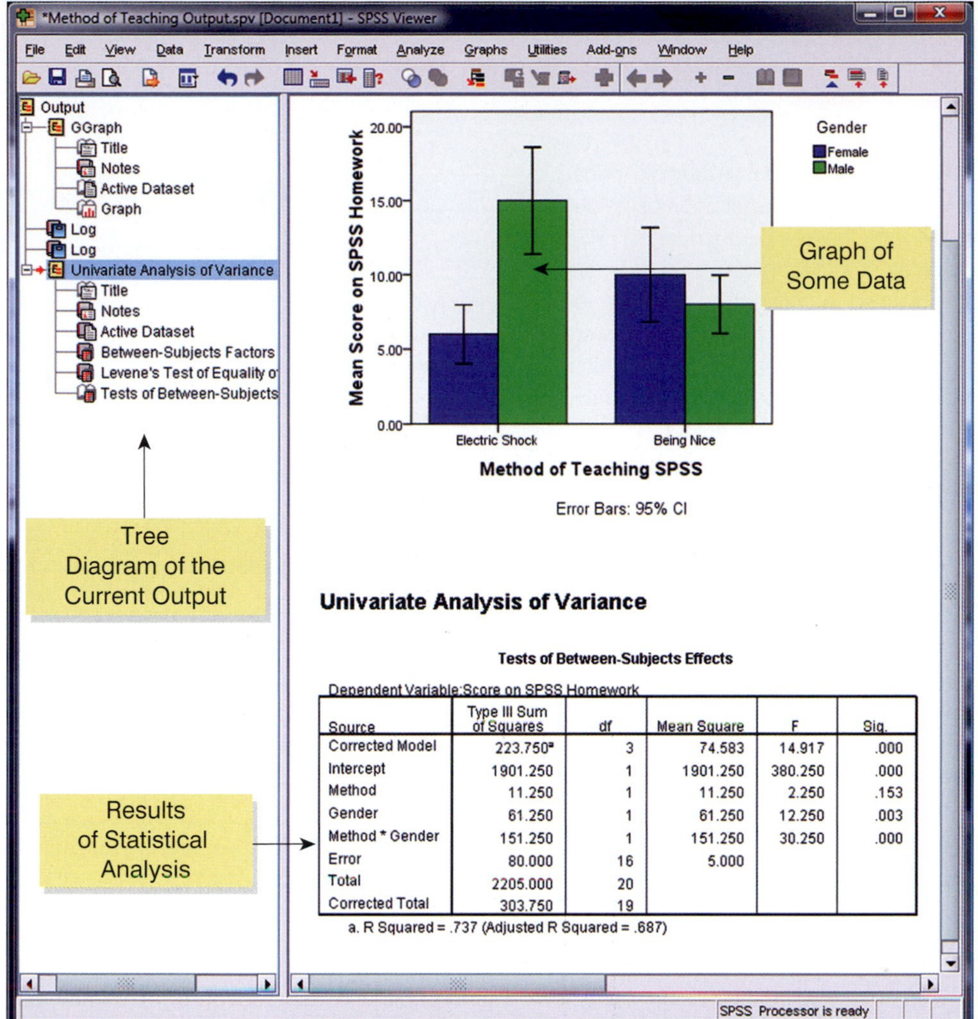

FIGURE 3.10 The SPSS Viewer

you simply double-click on the graph you wish to edit (this creates a new window in which the graph can be edited – see section 4.9). On the left-hand side of the output viewer there is a tree diagram illustrating the structure of the output. This tree diagram is useful when you have conducted several analyses because it provides an easy way of accessing specific parts of the output. The tree structure is fairly self-explanatory in that every time you do something in SPSS (such as drawing a graph or running a statistical procedure), it lists this procedure as a main heading.

In Figure 3.10 I conducted a graphing procedure followed by a univariate analysis of variance (ANOVA) and so these names appear as main headings. For each procedure there are a series of sub-headings that represent different parts of the analysis. For example, in the ANOVA procedure, which you'll learn more about later in the book, there are several sections to the output such as Levene's test (see section 5.6.1) and a table of the between-group effects (i.e. the *F*-test of whether the means are significantly different). You can skip to any one of these sub-components of the ANOVA output by clicking on the appropriate branch of the tree diagram. So, if you wanted to skip straight to the between-group effects you should move the on-screen arrow to the left-hand portion of the window and click where it says *Tests of Between-Subjects Effects*. This action will highlight this part of the output in the main part of the viewer (see SPSS Tip 3.4).

SPSS TIP 3.4 Printing and saving the planet ①

Rather than printing all of your SPSS output on reams of paper, you can help the planet by printing only a selection of the output. You can do this by using the tree diagram in the SPSS Viewer to select parts of the output for printing. For example, if you decided that you wanted to print out a graph but you didn't want to print the whole output, you can click on the word *GGraph* in the tree structure and that graph will become highlighted in the output. It is then possible through the print menu to select to print only the selected part of the output.

It is worth noting that if you click on a main heading (such as *Univariate Analysis of Variance*) then SPSS will highlight not only that main heading but all of the sub-components as well. This is useful for printing the results of a single statistical procedure.

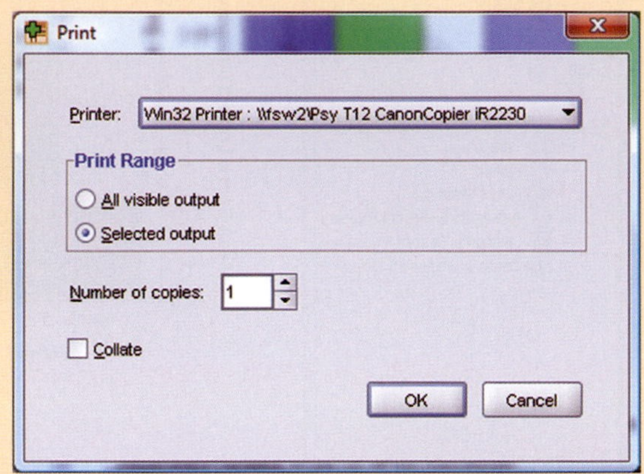

There are several icons in the output viewer window that help you to do things quickly without using the drop-down menus. Some of these icons are the same as those described for the data editor window so I will concentrate mainly on the icons that are unique to the viewer window:

	As with the data editor window, this icon activates the print menu. However, when this icon is pressed in the viewer window it activates a menu for printing the output (see SPSS Tip 3.4).
	This icon returns you to the data editor in a flash!
	This icon takes you to the last output in the viewer (so it returns you to the last procedure you conducted).
	This icon *promotes* the currently active part of the tree structure to a higher branch of the tree. For example, in Figure 3.10 the *Tests of Between-Subjects Effects* are a sub-component under the heading of *Univariate Analysis of Variance*. If we wanted to promote this part of the output to a higher level (i.e. to make it a main heading) then this is done using this icon.
	This icon is the opposite of the above in that it *demotes* parts of the tree structure. For example, in Figure 3.10 if we didn't want the *Univariate Analysis of Variance* to be a unique section we could select this heading and demote it so that it becomes part of the previous heading (the *Graph* heading). This button is useful for combining parts of the output relating to a specific research question.
	This icon collapses parts of the tree structure, which simply means that it hides the sub-components under a particular heading. For example, in Figure 3.10 if we selected the heading *Univariate Analysis of Variance* and pressed this icon, all of the sub-headings would disappear. The sections that disappear from the tree structure don't disappear from the output itself; the tree structure is merely condensed. This can be useful when you have been conducting lots of analyses and the tree diagram is becoming very complex.
	This icon expands any collapsed sections. By default all of the main headings are displayed in the tree diagram in their expanded form. If, however, you have opted to collapse part of the tree diagram (using the icon above) then you can use this icon to undo your dirty work.
	This icon and the following one allow you to show and hide parts of the output itself. So you can select part of the output in the tree diagram and click on this icon and that part of the output will disappear. It isn't erased, but it is hidden from view. This icon is similar to the collapse icon listed above except that it affects the output rather than the tree structure. This is useful for hiding less relevant parts of the output.

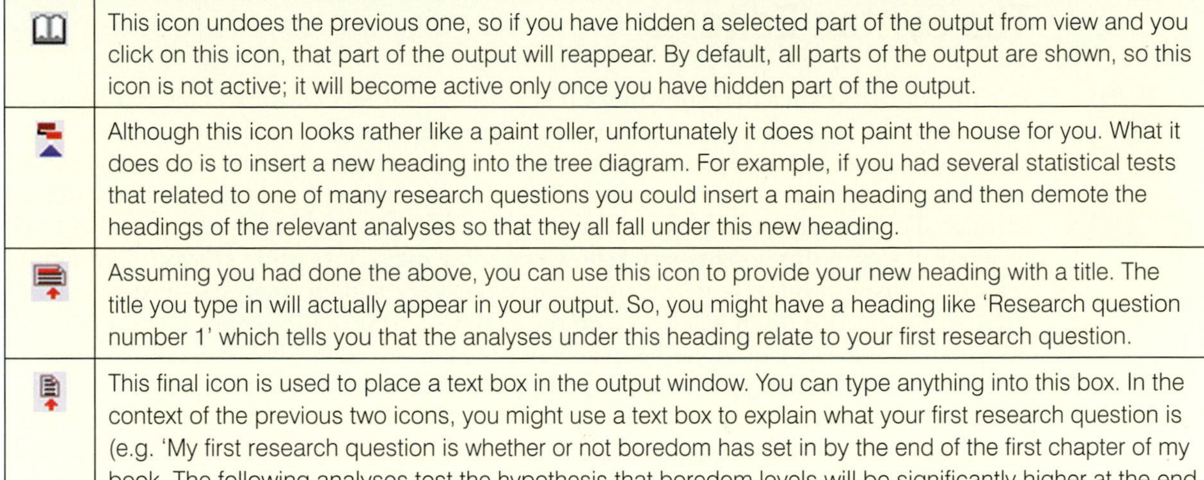

Icon	Description
	This icon undoes the previous one, so if you have hidden a selected part of the output from view and you click on this icon, that part of the output will reappear. By default, all parts of the output are shown, so this icon is not active; it will become active only once you have hidden part of the output.
	Although this icon looks rather like a paint roller, unfortunately it does not paint the house for you. What it does do is to insert a new heading into the tree diagram. For example, if you had several statistical tests that related to one of many research questions you could insert a main heading and then demote the headings of the relevant analyses so that they all fall under this new heading.
	Assuming you had done the above, you can use this icon to provide your new heading with a title. The title you type in will actually appear in your output. So, you might have a heading like 'Research question number 1' which tells you that the analyses under this heading relate to your first research question.
	This final icon is used to place a text box in the output window. You can type anything into this box. In the context of the previous two icons, you might use a text box to explain what your first research question is (e.g. 'My first research question is whether or not boredom has set in by the end of the first chapter of my book. The following analyses test the hypothesis that boredom levels will be significantly higher at the end of the first chapter than at the beginning').

SPSS TIP 3.5 **Funny numbers** ①

You might notice that SPSS sometimes reports numbers with the letter 'E' placed in the mix just to confuse you. For example, you might see a value such as 9.612 E–02 and many students find this notation confusing. Well, this notation means 9.61×10^{-2} (which might be a more familiar notation, or could be even more confusing). OK, some of you are still confused. Well think of E–02 as meaning 'move the decimal place 2 places to the left', so 9.612 E–02 becomes 0.09612. If the notation read 9.612 E–01, then that would be 0.9612, and if it read 9.612 E–03, that would be 0.009612. Likewise, think of E+02 (notice the minus sign has changed) as meaning 'move the decimal place 2 places to the right'. So 9.612 E+02 becomes 961.2.

3.6. The SPSS SmartViewer ①

Progress is all well and good, but it usually comes at a price and with version 16 of SPSS the output viewer changed quite dramatically (not really in terms of what you see, but behind the scenes). When you save an output file in version 17, SPSS uses the file extension **.spv**, and it calls it an SPSS Viewer file. In versions of SPSS before version 16, the viewer documents were saved as SPSS output files (**.spo**). Why does this matter? Well, it matters mainly because if you try to open an SPSS Viewer file in a version earlier than 16, SPSS won't know what the hell it is and will scream at you (well, it won't open the file). Similarly in versions after 16, if you try to open an output file (**.spo**) from an earlier version of SPSS you won't be able to do it. I know what you're thinking: 'this is bloody SPSS madness – statistics is hard enough as it is without versions of SPSS not being compatible with each other; is it trying to drive me insane?' Well, yes, it probably is.

This is why when you install SPSS on your computer you can install the SPSS **SmartViewer**. The SPSS SmartViewer looks suspiciously like the normal SPSS Viewer but without the icons. In fact, it looks so similar that I'm not even going to put a screenshot of it in this book – we can live on the edge, just this once. So what's so smart about the SmartViewer? Actually bugger all, that's what; it's just a way for you to open and read your old pre-version 17 SPSS output files. So, if you're not completely new to SPSS, this is a useful program to have.

The Syntax Editor ③

I've mentioned earlier that sometimes it's useful to use SPSS syntax. This is a language of commands for carrying out statistical analyses and data manipulations. Most of the time you'll do the things you need to using SPSS dialog boxes, but SPSS syntax can be useful. For one thing there are certain things you can do with syntax that you can't do through dialog boxes (admittedly most of these things are fairly advanced, but there will be a few places in this book where I show you some nice tricks using syntax). The second reason for using syntax is if you often carry out very similar analyses on data sets. In these situations it is often quicker to do the analysis and save the syntax as you go along. Fortunately this is easily done because many dialog boxes in SPSS have a Paste button. When you've specified your analysis using the dialog box, if you click on this button it will paste the syntax into a syntax editor window for you. To open a syntax editor window simply use the menus File New ▸ Syntax and a blank syntax editor will appear as in Figure 3.11. In this window you can type your syntax commands into the command area. The rules of SPSS syntax can be quite frustrating; for example, each line has to end with a full stop and if you forget this full stop you'll get an error message. In fact, if you have made a syntax error, SPSS will produce an error message in the Viewer window that identifies the line in the syntax window in which the error occurred; notice that in the syntax window SPSS helpfully numbers each line so that you can find the line in which the error occurred easily.

As we go through the book I'll show you a few things that will give you a flavour of how syntax can be used. Most of you won't have to use it, but for those that do this flavour will hopefully be enough to start you on your way. The window also has a navigation area (rather like the Viewer window). When you have a large file of syntax commands this navigation area can be helpful for negotiating your way to the bit of syntax that you actually need. Once you've typed in your syntax you have to run it using the Run menu. Run All will run all of the syntax in the window (clicking on ▸ will also do this), or you can highlight a selection of your syntax using the mouse and use Run Selection to process the selected syntax. You can also run the current command by using Run Current (or press *Ctrl* and *R* on the keyboard), or run all the syntax from the cursor to the end of the syntax window using Run To End. Another thing to note is that in SPSS you can have several data files open at once. Rather than have a syntax window for each data file, which could get confusing, you can use the same syntax window, but select the data set that you want to run the syntax commands on before you run them using the drop-down list DataSet3 ▾.

OLIVER TWISTED

Please, Sir, can I have some more ... syntax?

'I want to drown in a puddle of my own saliva' froths Oliver, 'tell me more about the syntax Window'. Really? Well, OK, there is a Flash movie on the companion website that guides you through how to use the syntax window.

3.8. Saving files ①

Although most of you should be familiar with how to save files, it is a vital thing to know and so I will briefly describe what to do. To save files simply use the 💾 icon (or use the

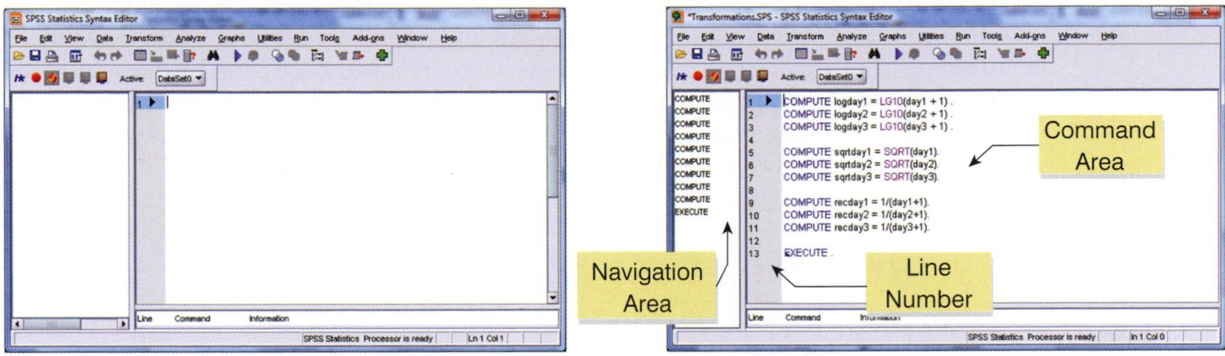

FIGURE 3.11 A new syntax window (top) and a syntax window with some syntax in it (bottom)

menus File Save or File Save As...). If the file is a new file, then clicking on this icon will activate the *Save As* ... dialog box (see Figure 3.12). If you are in the data editor when you select *Save As* ... then SPSS will save the data file you are currently working on, but if you are in the viewer window then it will save the current output.

There are several features of the dialog box in Figure 3.12. First, you need to select a location at which to store the file. There are many types of locations where you can save data: the hard drive (or drives), a USB drive, a CD or DVD, etc. (you could have many other choices of location on your particular computer). The first thing to do is select a main location by double-clicking on it: your hard drive (), a CD or DVD (), or a USB stick or other external drive (). Once you have chosen a main location the dialog box will display all of the available folders on that particular device. Once you have selected a folder in which to save your file, you need to give your file a name. If you click in the space next to where it says *File name*, a cursor will appear and you can type a name. By default, the file will be saved in an SPSS format, so if it is a data file it will have the file extension **.sav**, if it is a viewer document it will have the file extension **.spv**, and if it is a syntax file it will have the file extension

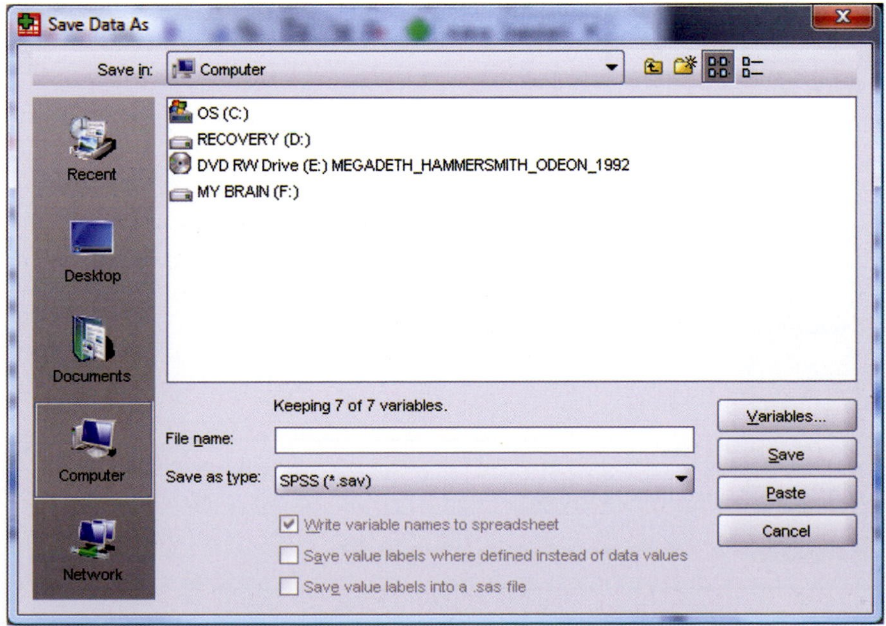

FIGURE 3.12 The *Save Data As* dialog box

.sps. However, you can save data in different formats such as Microsoft Excel files and tab-delimited text. To do this just click on SPSS (*.sav) and a list of possible file formats will be displayed. Click on the file type you require. Once a file has previously been saved, it can be saved again (updated) by clicking on the save icon. This icon appears in both the data editor and the viewer, and the file saved depends on the window that is currently active. The file will be saved in the location at which it is currently stored.

3.9. Retrieving a file ①

Throughout this book you will work with data files that you need to download from the companion website. It is, therefore, important that you know how to load these data files into SPSS. The procedure is very simple. To open a file, simply use the open-folder icon (or use the menus File Open ▸ Data...) to activate the dialog box in Figure 3.13. First, you need to find the location at which the file is stored. Navigate to wherever you downloaded the files from the website (a USB stick, or a folder on your hard drive). You should see a list of files and folders that can be opened. As with saving a file, if you are currently in the data editor then SPSS will display only SPSS data files to be opened (if you are in the viewer window then only output files will be displayed). If you use the menus and used the path File Open ▸ Data... then data files will be displayed, but if you used the path File Open ▸ Output... then viewer files will be displayed, and if you used File Open ▸ Syntax... then syntax files will be displayed (you get the general idea). You can open a folder by double-clicking on the folder icon. Once you have tracked down the required file you can open it either by selecting it with the mouse and then clicking on Open, or by double-clicking on the icon next to the file you want (e.g. double-clicking on the data-file icon). The data/output will then appear in the appropriate window. If you are in the data editor and you want to open a viewer file, then click on SPSS (*.sav) and a list of alternative file formats will be displayed. Click on the appropriate file type (viewer document (*.**spv**), syntax file (*.**sps**), Excel file (*.**xls**), text file (*.**dat**, *.**txt**)) and any files of that type will be displayed for you to open.

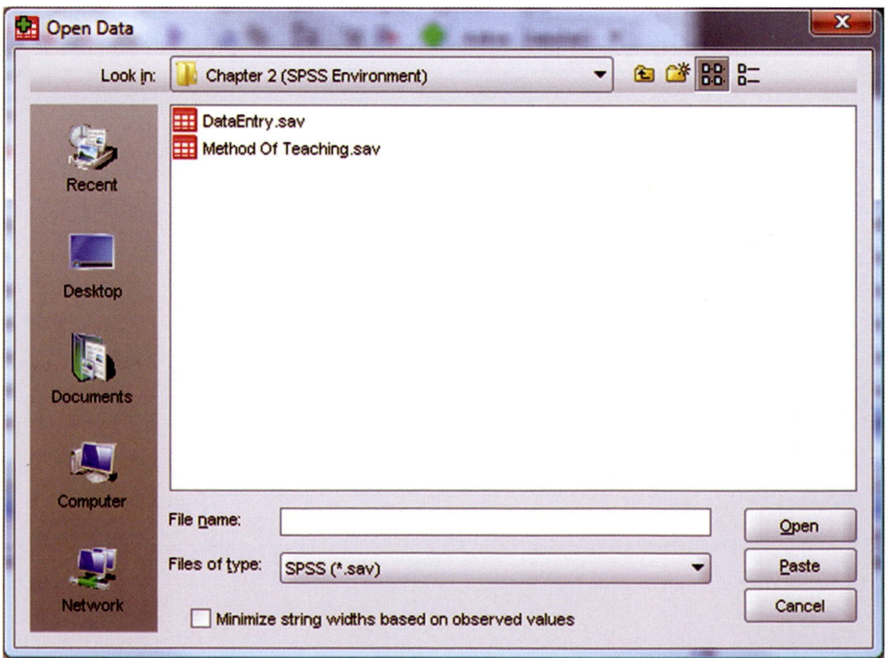

FIGURE 3.13
Dialog box to open a file

What have I discovered about statistics? ①

This chapter has provided a basic introduction to the SPSS environment. We've seen that SPSS uses two main windows: the data editor and the viewer. The data editor has both a data view (where you input the raw scores) and a variable view (where you define variables and their properties). The viewer is a window in which any output appears, such as tables, statistics and graphs. You also created your first data set by creating some variables and inputting some data. In doing so you discovered that we can code groups of people using numbers (coding variables) and discovered that rows in the data editor represent people (or cases of data) and columns represent different variables. Finally, we had a look at the syntax window and were told how to open and save files.

We also discovered that I was scared of my new school. However, with the help of Jonathan Land my confidence grew. With this new confidence I began to feel comfortable not just at school but in the world at large. It was time to explore.

Key terms that I've discovered

Currency variable
Data editor
Data view
Date variable
Numeric variable
SmartViewer
String variable
Syntax editor
Variable view
Viewer

Smart Alex's tasks

- **Task 1**: Smart Alex's first task for this chapter is to save the data that you've entered in this chapter. Save it somewhere on the hard drive of your computer (or a USB stick if you're not working on your own computer). Give it a sensible title and save it somewhere easy to find (perhaps create a folder called 'My Data Files' where you can save all of your files when working through this book).
- **Task 2**: Your second task is to enter the data that I used to create Figure 3.10. These data show the score (out of 20) for 20 different students, some of whom are male and some female, and some of whom were taught using positive reinforcement (being nice) and others who were taught using punishment (electric shock). Just to make it hard, the data should not be entered in the same way that they are laid out below:

Male		Female	
Electric Shock	Being Nice	Electric Shock	Being Nice
15	12	6	10
14	10	7	9
20	7	5	8
13	8	4	8
13	13	8	7

- **Task 3**: Research has looked at emotional reactions to infidelity and found that men get homicidal and suicidal and women feel undesirable and insecure (Shackelford, LeBlanc, & Drass, 2000). Let's imagine we did some similar research: we took some men and women and got their partners to tell them they had slept with someone else. We then took each person to two shooting galleries and each time gave them a gun and 100 bullets. In one gallery was a human-shaped target with a picture of their own face on it, and in the other was a target with their partner's face on it. They were left alone with each target for 5 minutes and the number of bullets used was measured. The data are below; enter them into SPSS and save them as **Infidelity.sav** (clue: they are not entered in the format in the table!).

Male		Female	
Partner's Face	Own Face	Partner's Face	Own Face
69	33	70	97
76	26	74	80
70	10	64	88
76	51	43	100
72	34	51	100
65	28	93	58
82	27	48	95
71	9	51	83
71	33	74	97
75	11	73	89
52	14	41	69
34	46	84	82

Answers can be found on the companion website.

Further reading

There are many good introductory SPSS books on the market that go through similar material to this chapter. Pallant's *SPSS survival manual* and Kinnear and Gray's *SPSS XX made simple* (insert a version number where I've typed XX because they update it regularly) are both excellent guides for people new to SPSS. There are many others on the market as well, so have a hunt around.

Online tutorials

The companion website contains the following Flash movie tutorials to accompany this chapter:

- Entering data
- Exporting SPSS output to Word
- Importing text data to SPSS
- Selecting cases
- The Syntax window
- The Viewer window
- Using Excel with SPSS

Exploring data with graphs 4

FIGURE 4.1
Explorer Field borrows a bike and gets ready to ride it recklessly around a caravan site

4.1. What will this chapter tell me? ①

As I got a bit older I used to love exploring. At school they would teach you about maps and how important it was to know where you were going and what you were doing. I used to have a more relaxed view of exploration and there is a little bit of a theme of me wandering off to whatever looked most exciting at the time. I got lost at a holiday camp once when I was about 3 or 4. I remember nothing about this but apparently my parents were frantically running around trying to find me while I was happily entertaining myself (probably by throwing myself head first out of a tree or something). My older brother, who was supposed to be watching me, got a bit of flak for that but he was probably working out equations to bend time and space at the time. He did that a lot when he was 7. The careless explorer in me hasn't really gone away: in new cities I tend to just wander off and hope for the best, and usually get lost and fortunately usually don't die (although I tested my luck once by wandering through part of New Orleans where apparently tourists get mugged a lot – it seemed fine to me). When exploring data you can't afford not to have a map; to explore data in the way that the 6 year old me used to explore the world is to spin around 8000 times while drunk and then run along the edge of a cliff. Wright (2003) quotes Rosenthal who said that researchers should 'make friends with their data'. This wasn't meant to imply that people who use statistics may as well befriend their data because the data are the only friend they'll have; instead Rosenthal meant that researchers often rush their analysis. Wright makes the analogy of a fine wine: you should savour the bouquet and delicate flavours to truly enjoy the experience. That's perhaps overstating the joys of data analysis, but rushing your analysis is, I suppose, a bit like gulping down a bottle of wine: the outcome is messy and incoherent! To negotiate your way around your data you need a map, maps of data are called graphs, and it is into this tranquil and tropical ocean that we now dive (with a compass and ample supply of oxygen, obviously).

4.2. The art of presenting data ①

4.2.1. What makes a good graph? ①

Before we get down to the nitty-gritty of how to draw graphs in SPSS, I want to begin by talking about some general issues when presenting data. SPSS (and other packages) make it very easy to produce very snazzy-looking graphs (see section 4.9), and you may find yourself losing consciousness at the excitement of colouring your graph bright pink (really, it's amazing how excited my undergraduate psychology students get at the prospect of bright pink graphs – personally I'm not a fan of pink). Much as pink graphs might send a twinge of delight down your spine, I want to urge you to remember why you're doing the graph – it's not to make yourself (or others) purr with delight at the pinkness of your graph, it's to present information (dull, perhaps, but true).

Tufte (2001) wrote an excellent book about how data should be presented. He points out that graphs should, among other things:

- ✓ Show the data.
- ✓ Induce the reader to think about the data being presented (rather than some other aspect of the graph, like how pink it is).
- ✓ Avoid distorting the data.
- ✓ Present many numbers with minimum ink.
- ✓ Make large data sets (assuming you have one) coherent.
- ✓ Encourage the reader to compare different pieces of data.
- ✓ Reveal data.

However, graphs often don't do these things (see Wainer, 1984, for some examples).

Let's look at an example of a bad graph. When searching around for the worst example of a graph that I have ever seen, it turned out that I didn't need to look any further than myself – it's in the first edition of this book (Field, 2000). Overexcited by SPSS's ability to put all sorts of useless crap on graphs (like 3-D effects, fill effects and so on – Tufte calls these **chartjunk**) I literally went into some weird orgasmic state and produced an absolute abomination (I'm surprised Tufte didn't kill himself just so he could turn in his grave at the sight of it). The only consolation was that because the book was published in black and white, it's not bloody pink! The graph is reproduced in Figure 4.2 (you should compare this to the more sober version in this edition, Figure 16.11). What's wrong with this graph?

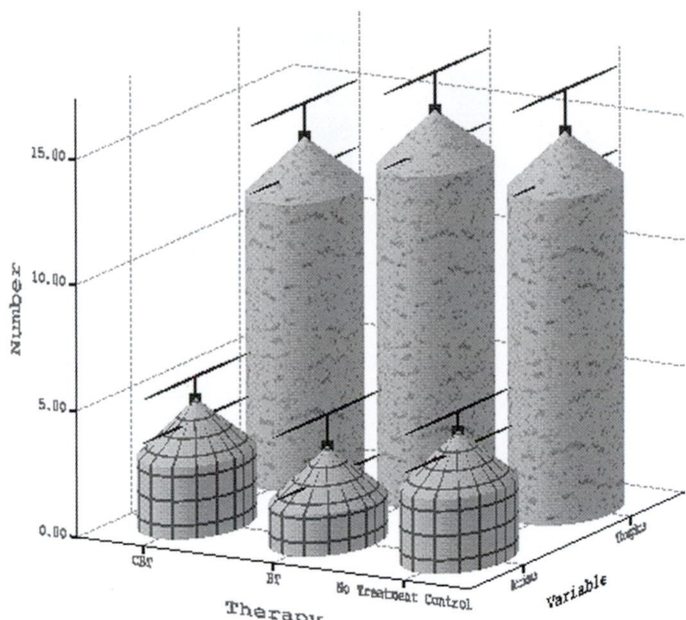

FIGURE 4.2
A cringingly bad example of a graph from the first edition of this book

- × The bars have a 3-D effect: Never use 3-D plots for a graph plotting two variables because it obscures the data.[1] In particular it makes it hard to see the values of the bars because of the 3-D effect. This graph is a great example because the 3-D effect makes the error bars almost impossible to read.
- × Patterns: The bars also have patterns, which, although very pretty, merely distract the eye from what matters (namely the data). These are completely unnecessary!
- × Cylindrical bars: What's that all about eh? Again, they muddy the data and distract the eye from what is important.
- × Badly labelled y-axis: 'Number' of what? Delusions? Fish? Cabbage-eating sea lizards from the eighth dimension? Idiots who don't know how to draw graphs?

Now, take a look at the alternative version of this graph (Figure 4.3). Can you see what improvements have been made?

- ✓ A 2-D plot: The completely unnecessary third dimension is gone making it much easier to compare the values across therapies and thoughts/behaviours.
- ✓ The *y*-axis has a more informative label: We now know that it was the number of obsessive thoughts or actions per day that was being measured.
- ✓ Distractions: There are fewer distractions like patterns, cylindrical bars and the like!
- ✓ Minimum ink: I've got rid of superfluous ink by getting rid of the axis lines and by using lines on the bars rather than grid lines to indicate values on the *y*-axis. Tufte would be pleased.

You have to do a fair bit of editing to get your graphs to look like this in SPSS but section 4.9 explains how.

[1] If you do 3-D plots when you're plotting only two variables then a bearded statistician will come to your house, lock you in a room and make you write I *μυστ νοτ δο* 3-*Δ γραπησ* 75,172 times on the blackboard. Really, they will.

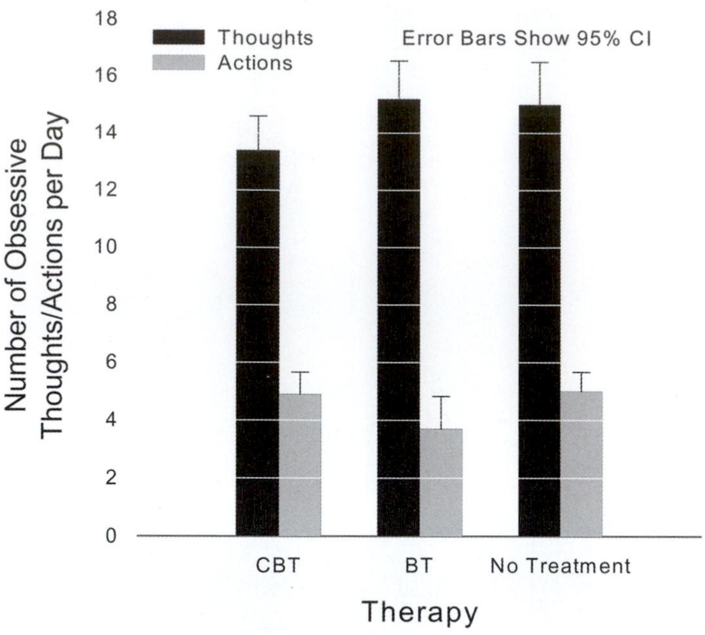

FIGURE 4.3 Figure 4.2 drawn properly

4.2.2. Lies, damned lies, and ... erm ... graphs ①

Governments lie with statistics, but scientists shouldn't. How you present your data makes a huge difference to the message conveyed to the audience. As a big fan of cheese, I'm often curious about whether the urban myth that it gives you nightmares is true. Shee (1964) reported the case of a man who had nightmares about his workmates: 'He dreamt of one, terribly mutilated, hanging from a meat-hook.[2] Another he dreamt of falling into a bottomless abyss. When cheese was withdrawn from his diet the nightmares ceased.' This would not be good news if you were the minister for cheese in your country.

Figure 4.4 shows two graphs that, believe it or not, display exactly the same data: the number of nightmares had after eating cheese. The first panel shows how the graph should probably be scaled. The *y*-axis reflects the maximum of the scale, and this creates the correct impression: that people have more nightmares about colleagues hanging from meat hooks if

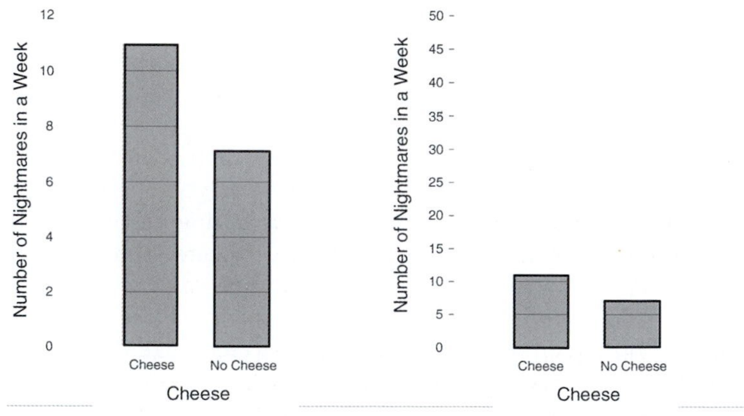

FIGURE 4.4 Two graphs about cheese

[2] I have similar dreams but that has more to do with some of my workmates than cheese!

CRAMMING SAM'S TIPS Graphs

- ✓ The vertical axis of a graph is known as the *y*-axis (or *ordinate*) of the graph.
- ✓ The horizontal axis of a graph is known as the *x*-axis (or *abscissa*) of the graph.

If you want to draw a good graph follow the cult of Tufte:

- ✓ Don't create false impressions of what the data actually show (likewise, don't hide effects!) by scaling the *y*-axis in some weird way.
- ✓ Abolish chartjunk: Don't use patterns, 3-D effects, shadows, pictures of spleens, photos of your Uncle Fred or anything else.
- ✓ Avoid excess ink: This is a bit radical, and difficult to achieve on SPSS, but if you don't need the axes, then get rid of them.

they eat cheese before bed. However, as minister for cheese, you want people to think the opposite; all you have to do is rescale the graph (by extending the *y*-axis way beyond the average number of nightmares) and there suddenly seems to be a little difference. Tempting as it is, don't do this (unless, of course, you plan to be a politician at some point in your life).

4.3. The SPSS Chart Builder ①

Graphs are a really useful way to look at your data before you get to the nitty-gritty of actually analysing them. You might wonder why you should bother drawing graphs – after all, you are probably drooling like a rabid dog to get into the statistics and to discover the answer to your really interesting research question. Graphs are just a waste of your precious time, right? Data analysis is a bit like Internet dating (actually it's not, but bear with me), you can scan through the vital statistics and find a perfect match (good IQ, tall, physically fit, likes arty French films, etc.) and you'll think you have found the perfect answer to your question. However, if you haven't looked at a picture, then you don't really know how to interpret this information – your perfect match might turn out to be Rimibald the Poisonous, King of the Colorado River Toads, who has genetically combined himself with a human to further his plan to start up a lucrative rodent farm (they like to eat small rodents).[3] Data analysis is much the same: inspect your data with a picture, see how it looks and only then can you interpret the more vital statistics.

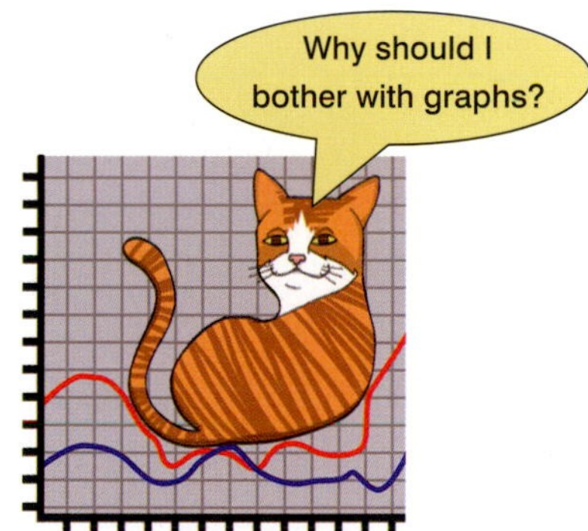
Why should I bother with graphs?

Without delving into the long and boring history of graphs in SPSS, in versions of SPSS after 16 most of the old-school menus for doing graphs were eliminated in favour of the all-singing and all-dancing **Chart Builder**.[4] In general, SPSS's graphing facilities have got better (you can certainly edit graphs more than you used to be able to – see section 4.9) but they are still quite limited for repeated-measures data (for this reason some of the graphs in later chapters are done in SigmaPlot in case you're wondering why you can't replicate them in SPSS).

[3] On the plus side, he would have a long sticky tongue and if you smoke his venom (which, incidentally, can kill a dog) you'll hallucinate (if you're lucky, you'd hallucinate that he wasn't a Colorado River Toad-Human hybrid).

[4] Unfortunately it's dancing like an academic at a conference disco and singing 'I will always love you' in the wrong key after 34 pints of beer.

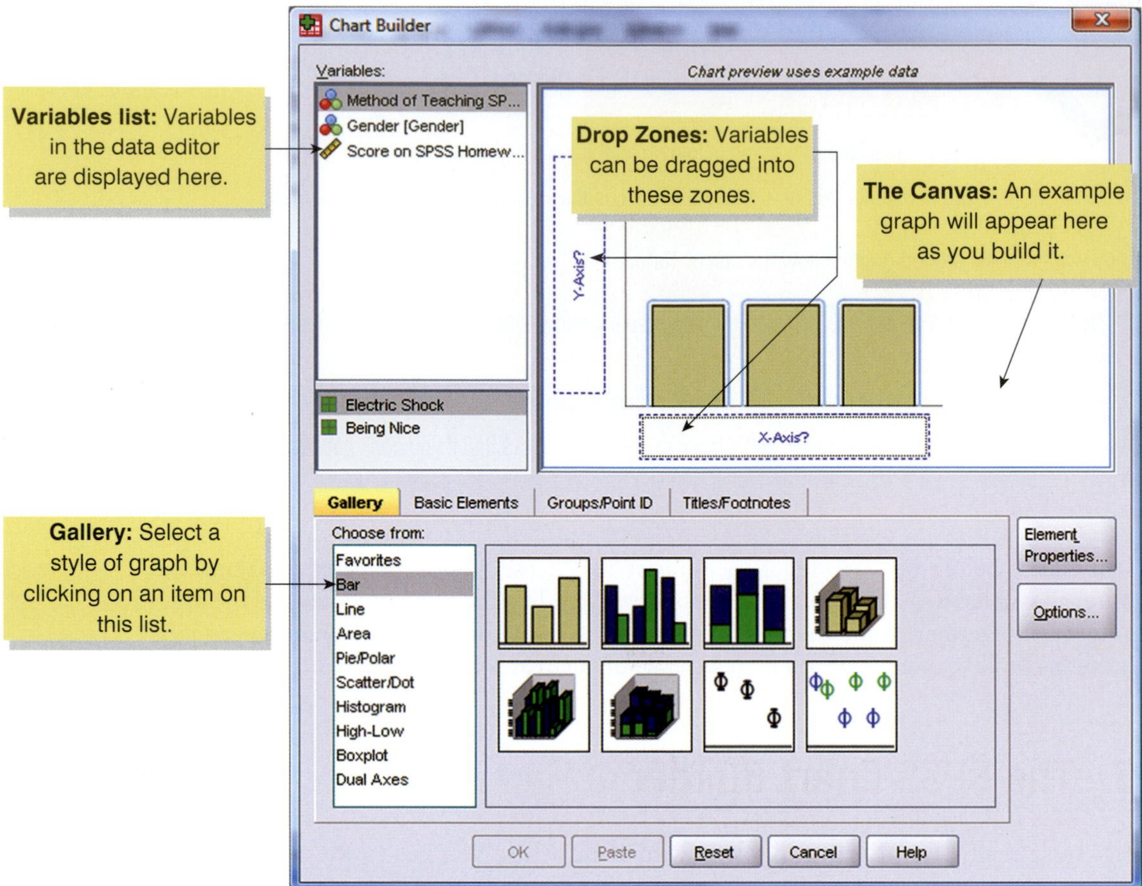

FIGURE 4.5 The SPSS Chart Builder

Figure 4.5 shows the basic *Chart Builder* dialog box, which is accessed through the Graphs Chart Builder... menu. There are some important parts of this dialog box:

- *Gallery*: For each type of graph, a gallery of possible variants is shown. Double-click on an icon to select a particular type of graph.
- *Variables list*: The variables in the data editor are listed here. These can be dragged into drop zones to specify what is shown in a given graph.
- *The canvas*: This is the main area in the dialog box and is where a preview of the graph is displayed as you build it.
- *Drop zones*: These zones are designated with blue dotted lines. You can drag variables from the variable list into these zones.

There are two ways to build a graph: the first is by using the gallery of predefined graphs and the second is by building a graph on an element-by-element basis. The gallery is the default option and this tab (Gallery | Basic Elements | Groups/Point ID | Titles/Footnotes) is automatically selected; however, if you want to build your graph from basic elements then click on the *Basic Elements* tab (Gallery | Basic Elements | Groups/Point ID | Titles/Footnotes). This changes the bottom of the dialog box in Figure 4.5 to look like Figure 4.6.

We will have a look at building various graphs throughout this chapter rather than trying to explain everything in this introductory section (see also SPSS Tip 4.1). Most graphs that you are likely to need can be obtained using the gallery view, so I will tend to stick with this method.

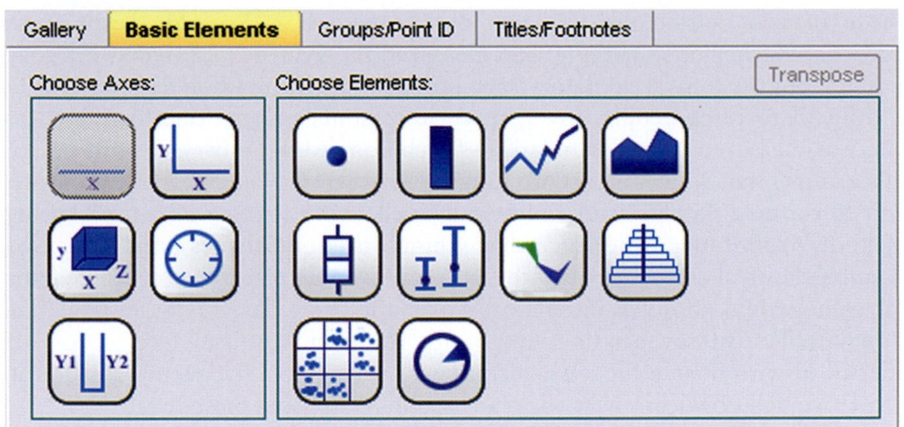

FIGURE 4.6 Building a graph from basic elements

SPSS TIP 4.1 Strange dialog boxes ①

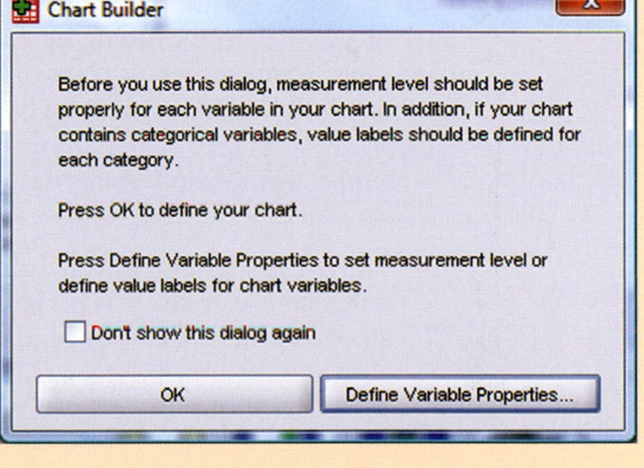

When you first use the Chart Builder to draw a graph you will see a dialog box that seems to signal an impending apocalypse. In fact, SPSS is just helpfully(?!) reminding you that for the Chart Builder to work, you need to have set the level of measurement correctly for each variable. That is, when you defined each variable you must have set them correctly to be *Scale, Ordinal* or *Nominal* (see section 3.4.2). This is because SPSS needs to know whether variables are categorical (nominal) or continuous (scale) when it creates the graphs. If you have been diligent and set these properties when you entered the data then simply click on OK to make the dialog disappear. If you forgot to set the level of measurement for any variables then click on Define Variable Properties... to go to a new dialog box in which you can change the properties of the variables in the data editor.

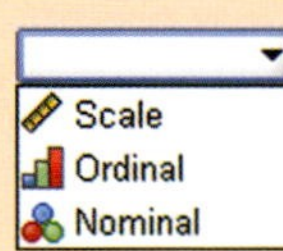

4.4. Histograms: a good way to spot obvious problems ①

In this section we'll look at how we can use frequency distributions to screen our data.[5] We'll use an example to illustrate what to do. A biologist was worried about the potential health effects of music festivals. So, one year she went to the Download Music Festival[6]

[5] An alternative way to graph the distribution is a density plot, which we'll discuss in section 4.8.5.

[6] http://www.downloadfestival.co.uk.

(for those of you outside the UK, you can pretend it is Roskilde Festival, Ozzfest, Lollopalooza, Wacken or something) and measured the hygiene of 810 concert-goers over the three days of the festival. In theory each person was measured on each day but because it was difficult to track people down, there were some missing data on days 2 and 3. Hygiene was measured using a standardized technique (don't worry, it wasn't licking the person's armpit) that results in a score ranging between 0 (you smell like a corpse that's been left to rot up a skunk's arse) and 4 (you smell of sweet roses on a fresh spring day). Now I know from bitter experience that sanitation is not always great at these places (the Reading Festival seems particularly bad) and so this researcher predicted that personal hygiene would go down dramatically over the three days of the festival. The data file, **DownloadFestival.sav**, can be found on the companion website (see section 3.9 for a reminder of how to open a file). We encountered histograms (frequency distributions) in Chapter 1; we will now learn how to create one in SPSS using these data.

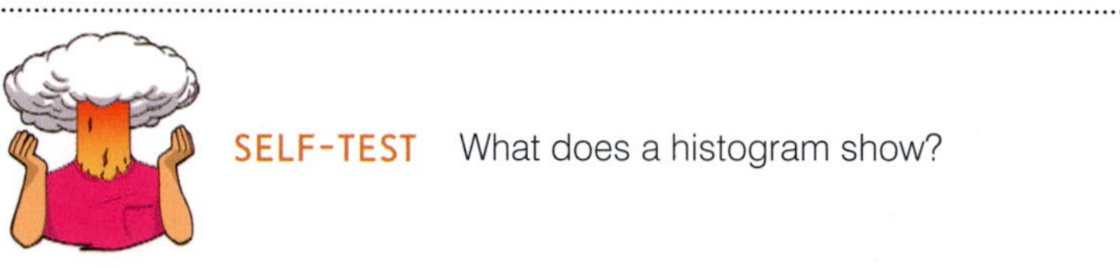

First, access the chart builder as in Figure 4.5 and then select *Histogram* in the list labelled *Choose from* to bring up the gallery shown in Figure 4.7. This gallery has four icons representing different types of histogram, and you should select the appropriate one either by double-clicking on it, or by dragging it onto the canvas in the Chart Builder:

- *Simple histogram*: Use this option when you just want to see the frequencies of scores for a single variable.
- *Stacked histogram*: If you had a grouping variable (e.g. whether men or women attended the festival) you could produce a histogram in which each bar is split by group. In the example of gender, each bar would have two colours, one representing men and the other women. This is a good way to compare the relative frequency of scores across groups (e.g. were there more smelly women than men?).

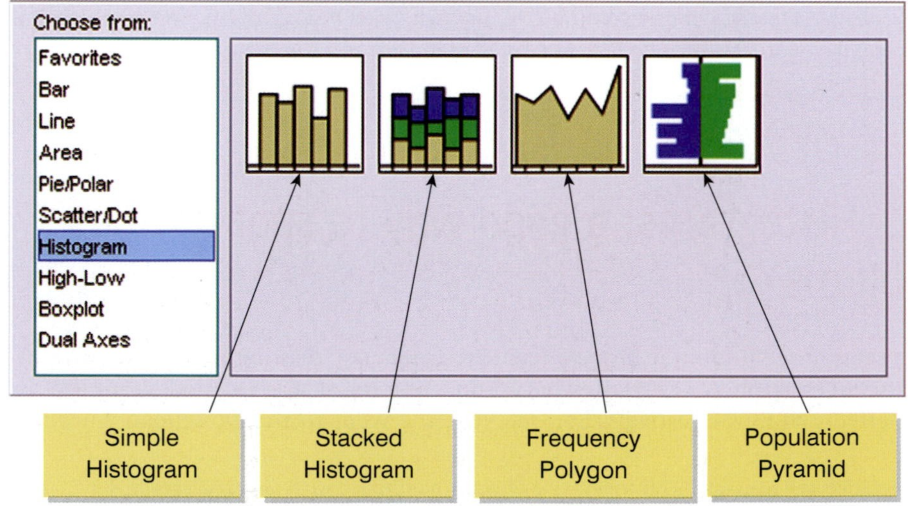

FIGURE 4.7 The histogram gallery

- *Frequency polygon*: This option displays the same data as the simple histogram except that it uses a line instead of bars to show the frequency, and the area below the line is shaded.
- *Population pyramid*: Like a stacked histogram this shows the relative frequency of scores in two populations. It plots the variable (in this case hygiene) on the vertical axis and the frequencies for each population on the horizontal: the populations appear back to back on the graph. If the bars either side of the dividing line are equally long then the distributions have equal frequencies.

We are going to do a simple histogram so double-click on the icon for a simple histogram (Figure 4.7). The *Chart Builder* dialog box will now show a preview of the graph in the canvas area. At the moment it's not very exciting (top of Figure 4.8) because we haven't told SPSS which variables we want to plot. Note that the variables in the data editor are listed on the left-hand side of the Chart Builder, and any of these variables can be dragged into any of the spaces surrounded by blue dotted lines (called *drop zones*).

A histogram plots a single variable (x-axis) against the frequency of scores (y-axis), so all we need to do is select a variable from the list and drag it into X-Axis?. Let's do this for the hygiene scores on day 1 of the festival. Click on this variable in the list and drag it to X-Axis? as shown in Figure 4.8; you will now find the histogram previewed on the canvas. (Although SPSS calls the resulting graph a preview it's not really because it does not use your data to generate this image – it is a preview only of the general form of the graph, and not what your specific graph will actually look like.) To draw the histogram click on OK (see also SPSS Tip 4.2).

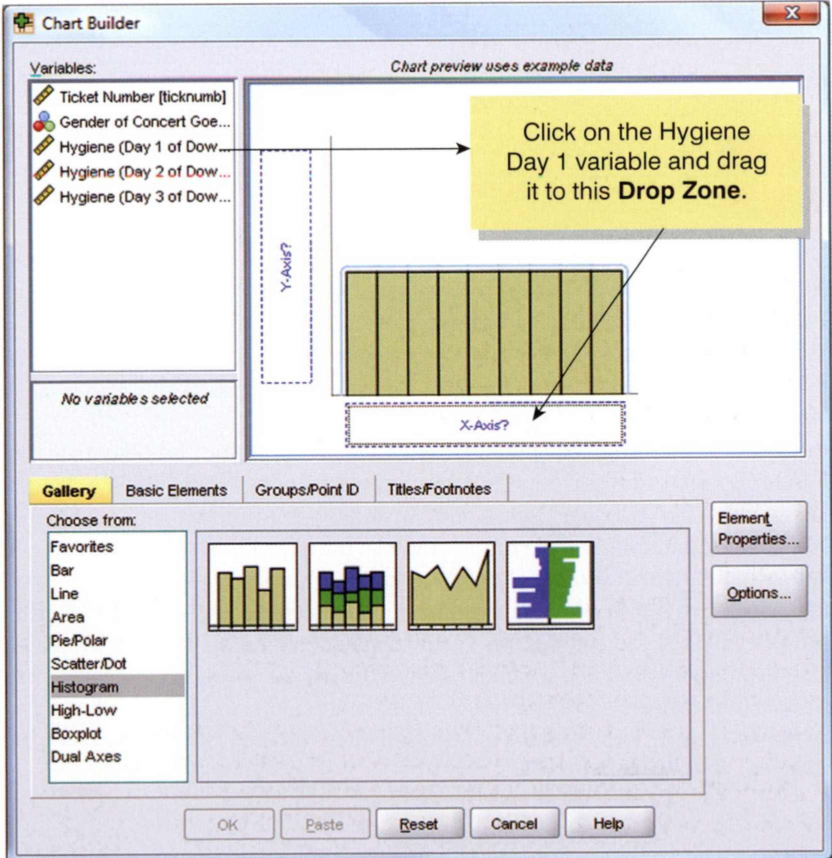

FIGURE 4.8 Defining a histogram in the Chart Builder

FIGURE 4.8
(continued)

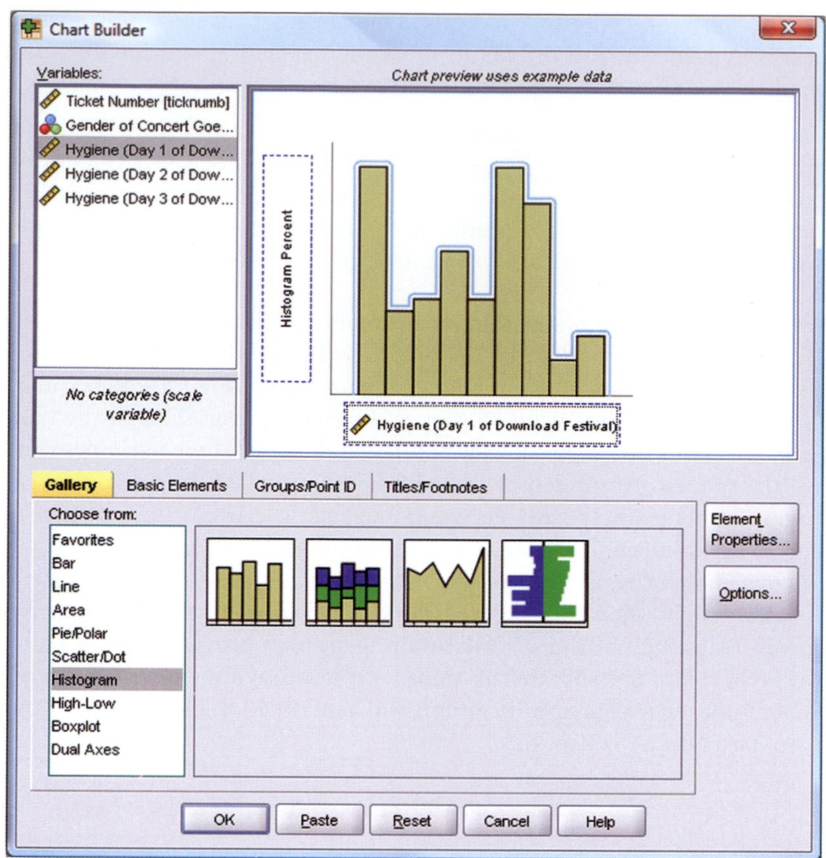

SPSS TIP 4.2 Further histogram options ①

You might notice another dialog box floating about making a nuisance of itself (if not, then consider yourself lucky, or click on Element Properties...). This dialog box allows you to edit various features of a histogram (Figure 4.9). For example, you can change the statistic displayed: the default is *Histogram* but if you wanted to express values as a percentage rather than a frequency, you could select *Histogram Percent*. You can also decide manually how you want to divide up your data to compute frequencies.

If you click on Set Parameters... then another dialog box appears (Figure 4.9), in which you can determine properties of the 'bins' used to make the histogram. You can think of a bin as, well, a rubbish bin (this is a pleasing analogy as you will see): on each rubbish bin you write a score (e.g. 3), or a range of scores (e.g. 1–3), then you go through each score in your data set and throw it into the rubbish bin with the appropriate label on it (so, a score of 2 gets thrown into the bin labelled 1–3). When you have finished throwing your data into these rubbish bins, you count how many scores are in each bin. A histogram is created in much the same way; either SPSS can decide how the bins are labelled (the default), or you can decide. Our hygiene scores range from 0 to 4, therefore we might decide that our bins should begin with 0 and we could set the ⦿ Custom value for anchor: property to 0. We might also decide that we want each bin to contain scores between whole numbers (i.e. 0–1, 1–2, 2–3, etc.), in which case we could set the ⦿ Interval width: to be 1. This is what I've done in Figure 4.9, but for the time being leave the default settings (i.e. everything set to ⦿ Automatic).

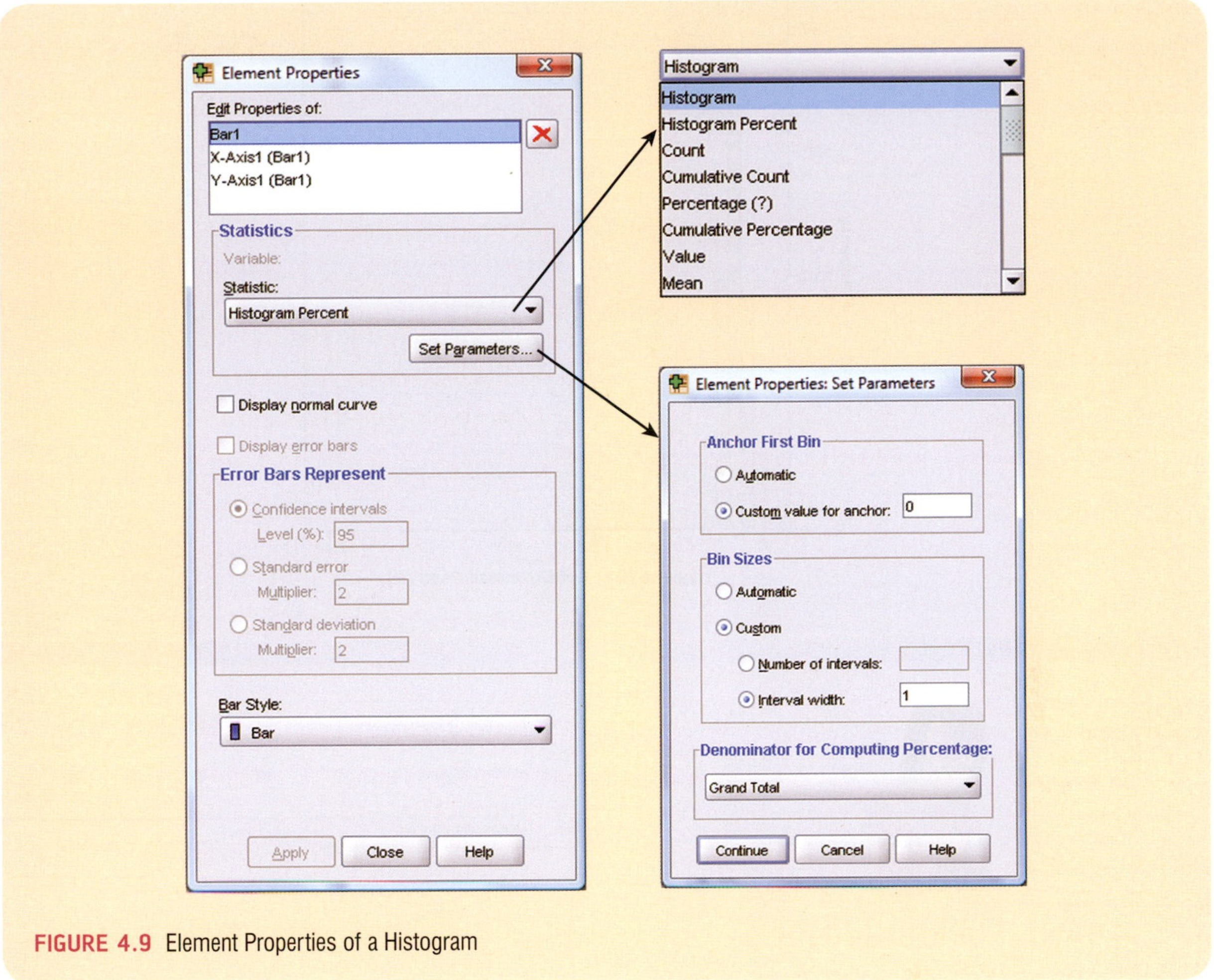

FIGURE 4.9 Element Properties of a Histogram

The resulting histogram is shown in Figure 4.10. The first thing that should leap out at you is that there appears to be one case that is very different to the others. All of the scores appear to be squashed up at one end of the distribution because they are all less than 5 (yielding a very pointy distribution!) except for one, which has a value of 20! This is an **outlier**: a score very different to the rest (Jane Superbrain Box 4.1). Outliers bias the mean and inflate the standard deviation (you should have discovered this from the self-test tasks in Chapters 1 and 2) and screening data is an important way to detect them. You can look for outliers in two ways: (1) graph the data with a histogram (as we have done here) or a boxplot (as we will do in the next section), or (2) look at *z*-scores (this is quite complicated but if you want to know see Jane Superbrain Box 4.2).

The outlier shown on the histogram is particularly odd because it has a score of 20, which is above the top of our scale (remember, our hygiene scale ranged only from 0 to 4), and so it must be a mistake (or the person had obsessive compulsive disorder and had washed themselves into a state of extreme cleanliness). However, with 810 cases, how on earth do we find out which case it was? You could just look through the data, but that would certainly give you a headache and so instead we can use a **boxplot** which is another very useful graph for spotting outliers.

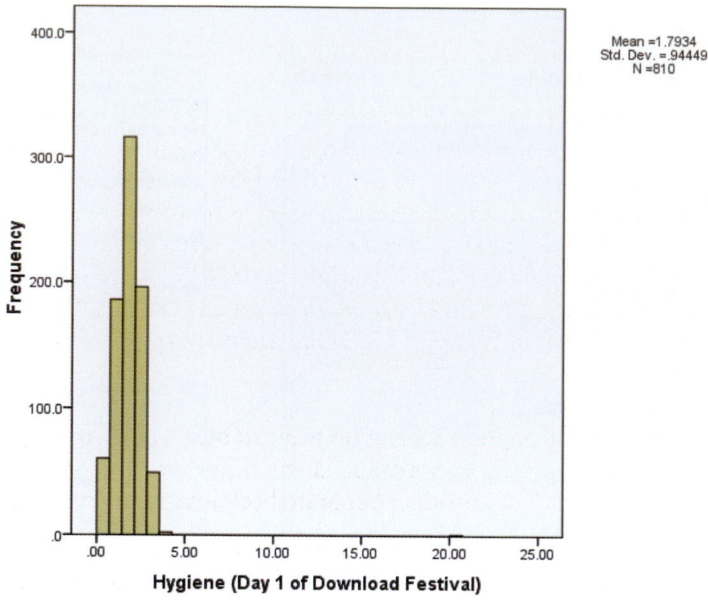

FIGURE 4.10 Histogram of the Day 1 Download Festival hygiene scores

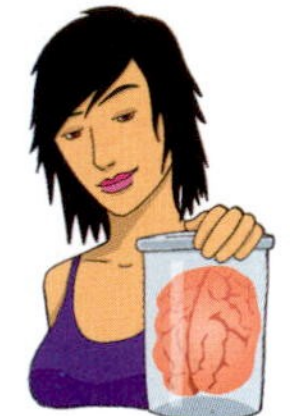

JANE SUPERBRAIN 4.1

What is an outlier? ①

An outlier is a score very different from the rest of the data. When we analyse data we have to be aware of such values because they bias the model we fit to the data. A good example of this bias can be seen by looking at the mean. When the first edition of this book came out in 2000, I was quite young and became very excited about obsessively checking the book's ratings on Amazon.co.uk. These ratings can range from 1 to 5 stars. Back in 2002, the first edition of this book had seven ratings (in the order given) of 2, 5, 4, 5, 5, 5, 5. All but one of these ratings are fairly similar (mainly 5 and 4) but the first rating was quite different from the rest – it was a rating of 2 (a mean and horrible rating). The graph plots seven reviewers on the horizontal axis and their ratings on the vertical axis and there is also a horizontal line that represents the mean rating (4.43 as it happens). It should be clear that all of the scores except one lie close to this line. The score of 2 is very different and lies some way below the mean. This score is an example of an outlier – a weird and unusual person (sorry, I mean score) that deviates from the rest of humanity (I mean, data set). The dashed

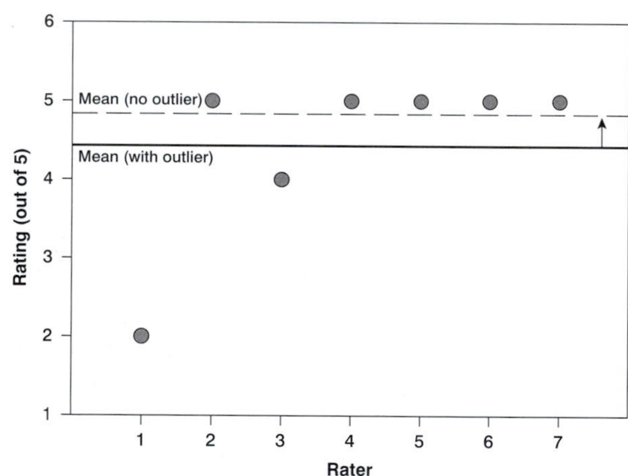

horizontal line represents the mean of the scores when the outlier is not included (4.83). This line is higher than the original mean indicating that by ignoring this score the mean increases (it increases by 0.4). This example shows how a single score, from some mean-spirited badger turd, can bias the mean; in this case the first rating (of 2) drags the average down. In practical terms this had a bigger implication because Amazon rounded off to half numbers, so that single score made a difference between the average rating reported by Amazon as a generally glowing 5 stars and the less impressive 4.5 stars. (Nowadays Amazon sensibly produces histograms of the ratings and has a better rounding system.) Although I am consumed with bitterness about this whole affair, it has at least given me a great example of an outlier! (Data for this example were taken from http://www.amazon.co.uk/ in about 2002.)

4.5. Boxplots (box–whisker diagrams) ①

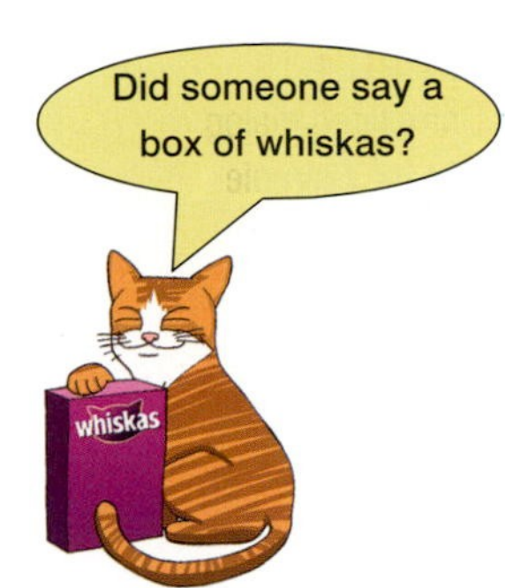

Boxplots or box–whisker diagrams are really useful ways to display your data. At the centre of the plot is the *median*, which is surrounded by a box the top and bottom of which are the limits within which the middle 50% of observations fall (the interquartile range). Sticking out of the top and bottom of the box are two whiskers which extend to the most and least extreme scores respectively. First, we will plot some using the Chart Builder and then we'll look at what they tell us in more detail.

In the Chart Builder (Figure 4.5) select *Boxplot* in the list labelled *Choose from* to bring up the gallery shown in Figure 4.11. There are three types of boxplot you can choose:

- *Simple boxplot*: Use this option when you want to plot a boxplot of a single variable, but you want different boxplots produced for different categories in the data (for these hygiene data we could produce separate boxplots for men and women).
- *Clustered boxplot*: This option is the same as the simple boxplot except that you can select a second categorical variable on which to split the data. Boxplots for this second variable are produced in different colours. For example, we might have measured whether our festival-goer was staying in a tent or a nearby hotel during the festival. We could produce boxplots not just for men and women, but within men and women we could have different coloured boxplots for those who stayed in tents and those who stayed in hotels.
- *1-D boxplot*: Use this option when you just want to see a boxplot for a single variable. (This differs from the simple boxplot only in that no categorical variable is selected for the *x*-axis.)

In the data file of hygiene scores we also have information about the gender of the concert-goer. Let's plot this information as well. To make our boxplot of the day 1 hygiene scores for males and females, double-click on the simple boxplot icon (Figure 4.11), then from the variable list select the hygiene day 1 score variable and drag it into [Y-Axis?] and select the variable gender and drag it to [X-Axis?]. The dialog should now look like Figure 4.12 – note that the variable names are displayed in the drop zones, and the canvas now displays a preview of our graph (e.g. there are two boxplots representing each gender). Click on [OK] to produce the graph.

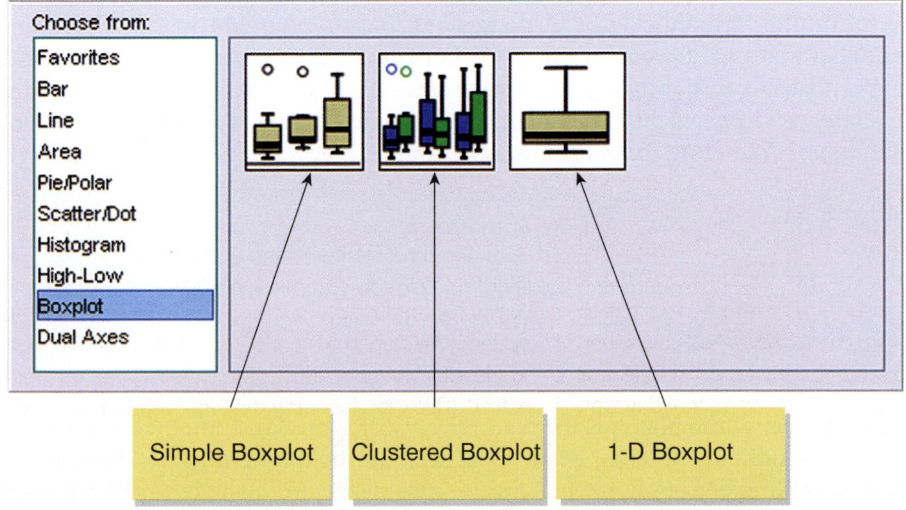

FIGURE 4.11 The boxplot gallery

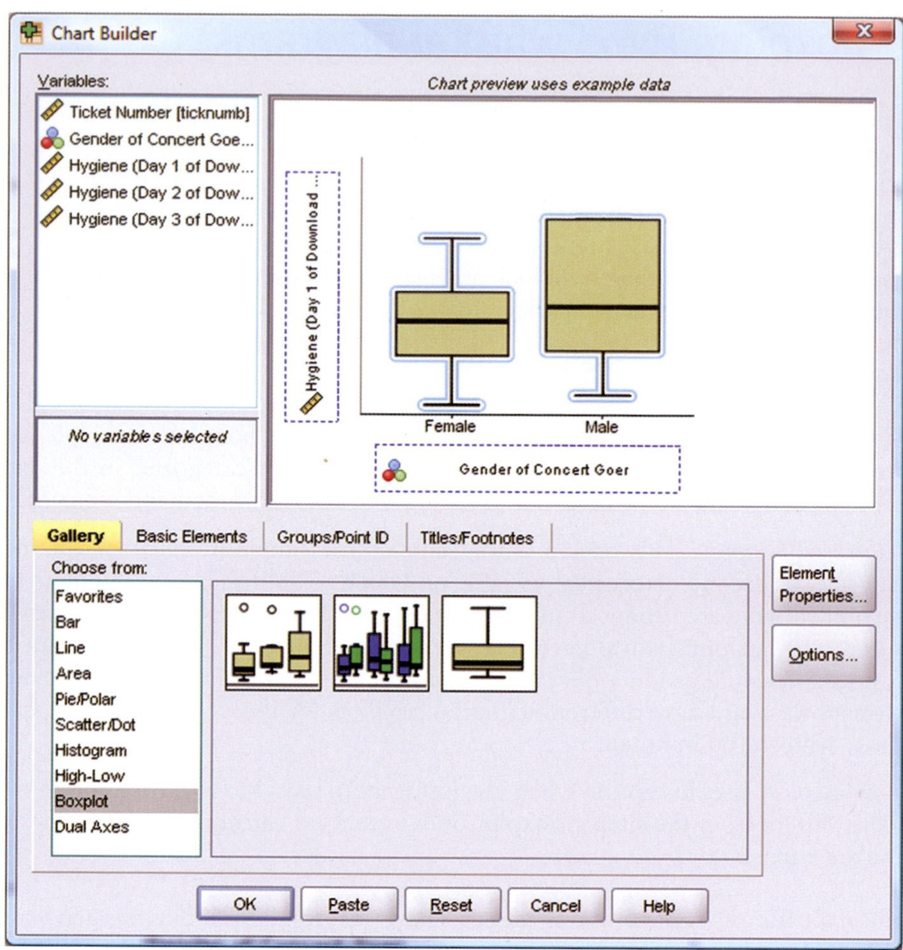

FIGURE 4.12 Completed dialog box for a simple boxplot

The resulting boxplot is shown in Figure 4.13. It shows a separate boxplot for the men and women in the data. You may remember that the whole reason that we got into this boxplot malarkey was to help us to identify an outlier from our histogram (if you have skipped straight to this section then you might want to backtrack a bit). The important thing to note is that the outlier that we detected in the histogram is shown up as an asterisk (*) on the boxplot and next to it is the number of the case (611) that's producing this outlier. (We can also tell that this case was a female.) If we go to the data editor (data view), we can locate this case quickly by clicking on and typing 611 in the dialog box that appears. That takes us straight to case 611. Looking at this case reveals a score of 20.02, which is probably a mistyping of 2.02. We'd have to go back to the raw data and check. We'll assume we've checked the raw data and it should be 2.02, so replace the value 20.02 with the value 2.02 before we continue this example.

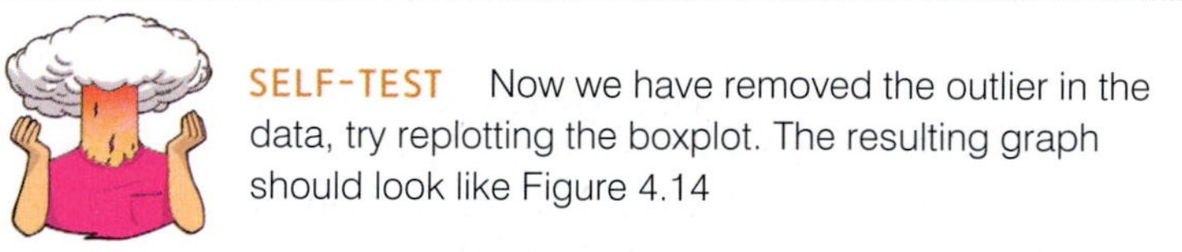

SELF-TEST Now we have removed the outlier in the data, try replotting the boxplot. The resulting graph should look like Figure 4.14

Figure 4.14 shows the boxplots for the hygiene scores on day 1 after the outlier has been corrected. Let's look now in more detail about what the boxplot represents. First, it shows us the lowest score (the bottom horizontal line on each plot) and the highest (the top horizontal

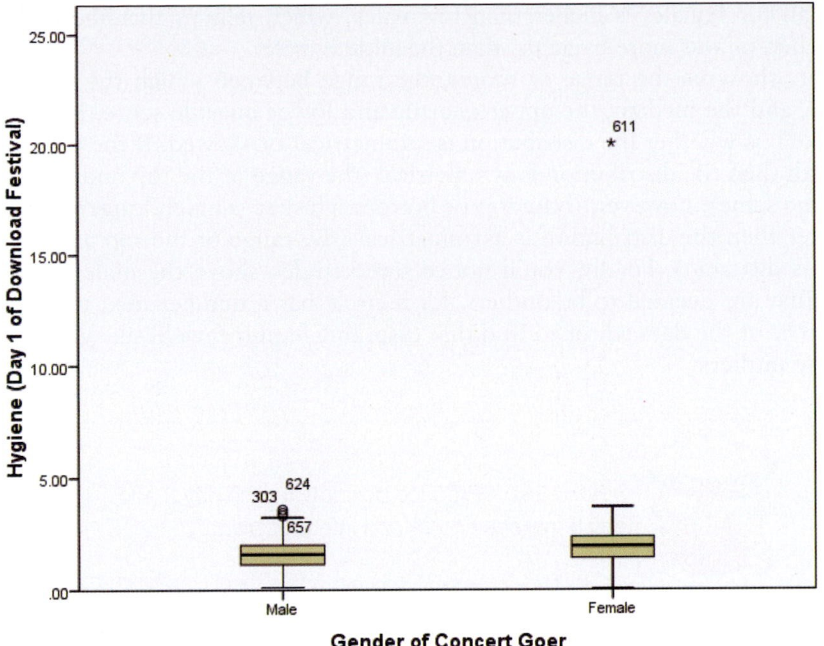

FIGURE 4.13 Boxplot of hygiene scores on day 1 of the Download Festival split by gender

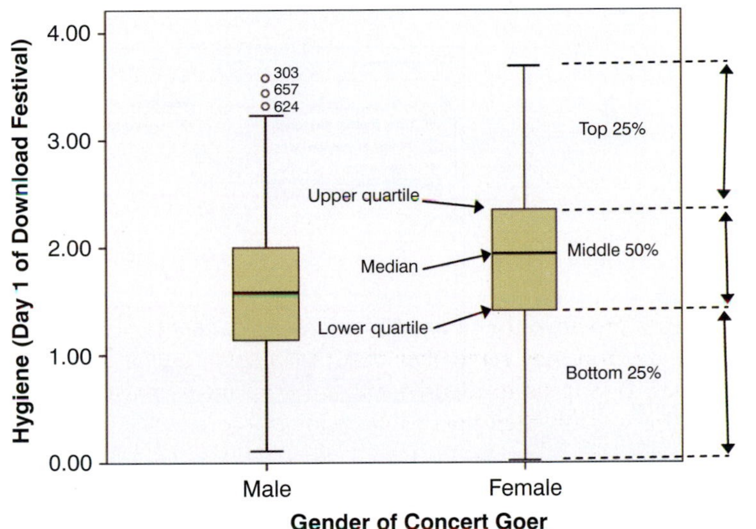

FIGURE 4.14 Boxplot of hygiene scores on day 1 of the Download Festival split by gender

line of each plot). Comparing the males and females we can see they both had similar low scores (0, or very smelly) but the women had a slightly higher top score (i.e. the most fragrant female was more hygienic than the cleanest male). The lowest edge of the tinted box is the lower quartile (see section 1.7.3); therefore, the distance between the lowest horizontal line and the lowest edge of the tinted box is the range between which the lowest 25% of scores fall. This range is slightly larger for women than for men, which means that if we take the most unhygienic 25% females then there is more variability in their hygiene scores than the lowest 25% of males. The box (the tinted area) shows the interquartile range (see section 1.7.3): that is, 50% of the scores are bigger than the lowest part of the tinted area but smaller than the top part of the tinted area. These boxes are of similar size in the males and females.

The top edge of the tinted box shows the value of the upper quartile (see section 1.7.3); therefore, the distance between the top edge of the shaded box and the top horizontal line shows the range between which the top 25% of scores fall. In the middle of the tinted box is a slightly thicker horizontal line. This represents the value of the median (see section 1.7.2).

The median for females is higher than for males, which tells us that the middle female scored higher, or was more hygienic, than the middle male.

Boxplots show us the range of scores, the range between which the middle 50% of scores fall, and the median, the upper quartile and lower quartile score. Like histograms, they also tell us whether the distribution is symmetrical or skewed. If the whiskers are the same length then the distribution is symmetrical (the range of the top and bottom 25% of scores is the same); however, if the top or bottom whisker is much longer than the opposite whisker then the distribution is asymmetrical (the range of the top and bottom 25% of scores is different). Finally, you'll notice some circles above the male boxplot. These are cases that are deemed to be outliers. Each circle has a number next to it that tells us in which row of the data editor to find that case. In Chapter 5 we'll see what can be done about these outliers.

SELF-TEST Produce boxplots for the day 2 and day 3 hygiene scores and interpret them.

JANE SUPERBRAIN 4.2

Using z-scores to find outliers ③

To check for outliers we can look at z-scores. We saw in section 1.7.4 that z-scores are simply a way of *standardizing* a data set by expressing the scores in terms of a distribution with a mean of 0 and a standard deviation of 1. In doing so we can use benchmarks that we can apply to any data set (regardless of what its original mean and standard deviation were). We also saw in this section that to convert a score into z-scores we simply take each score (X) and convert it to a z-score by subtracting the mean of all scores ($\bar{X}$) from it and dividing by the standard deviation of all scores (s).

We can get SPSS to do this conversion for us using the Analyze Descriptive Statistics ▸ Descriptives... dialog box, selecting a variable (such as day 2 of the hygiene data as in the diagram), or several variables, and selecting the *Save standardized values as variables* before we click on OK. SPSS creates a new variable in the data editor (with the same name prefixed with the letter z). To look for outliers we could use these z-scores and count how many fall within certain important limits. If we take the absolute value

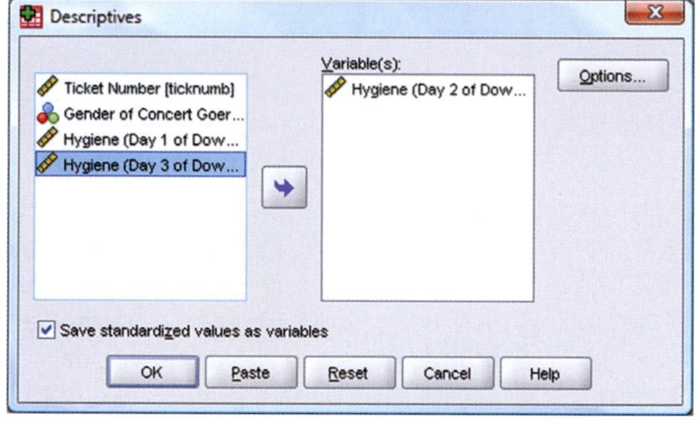

(i.e. we ignore whether the z-score is positive or negative) then in a normal distribution we'd expect about 5% to have absolute values greater than 1.96 (we often use 2 for convenience), and 1% to have absolute values greater than 2.58, and none to be greater than about 3.29.

Alternatively, you could use some SPSS syntax in a syntax window to create the z-scores and count them for you. I've written the file **Outliers (Percentage of Z-scores).sps** (on the companion website) to produce a table for day 2 of the Download Festival hygiene data. Load this file and run the syntax, or open a syntax window (see section 3.7) and type the following (remembering all of the full stops – the explanations of the code are surrounded by *s and don't need to be typed).

```
DESCRIPTIVES
VARIABLES= day2/SAVE.
```

*This uses SPSS's descriptives function on the variable **day2** (instead of using the dialog box) to save the z-scores in the data editor (these will be saved as a variable called **zday2**).*

```
COMPUTE outlier1=abs(zday2).
EXECUTE.
```

*This creates a new variable in the data editor called **outlier1**, which contains the absolute values of the z-scores that we just created.*

```
RECODE
outlier1 (3.29 thru Highest=4) (2.58 thru Highest=3)
(1.96 thru Highest=2) (Lowest thru 2=1).
EXECUTE.
```

*This recodes the variable called **outlier1** according to the benchmarks I've described. So, if a value is greater than 3.29 it's assigned a code of 4, if it's between 2.58 and 3.29 then it's assigned a code of 3, if it's between 1.96 and 2.58 it's assigned a code of 2, and if it's less than 1.96 it gets a code of 1.*

```
VALUE LABELS outlier1
1 'Absolute z-score less than 2' 2 'Absolute z-score
greater than 1.96' 3 'Absolute z-score greater than 2.58'
4 'Absolute z-score greater than 3.29'.
```

This assigns appropriate labels to the codes we defined above.

```
FREQUENCIES
VARIABLES=outlier1
/ORDER=ANALYSIS.
```

*Finally, this syntax uses the *frequencies* facility of SPSS to produce a table telling us the percentage of 1s, 2s, 3s and 4s found in the variable **outlier1**.*

The table produced by this syntax is shown below. Look at the column labelled 'Valid Percent'. We would expect to see 95% of cases with absolute value less than 1.96, 5% (or less) with an absolute value greater than 1.96, and 1% (or less) with an absolute value greater than 2.58. Finally, we'd expect no cases above 3.29 (well, these cases are significant outliers). For hygiene scores on day 2 of the festival, 93.2% of values had z-scores less than 1.96; put another way, 6.8% were above (looking at the table we get this figure by adding 4.5% + 1.5% + 0.8%). This is slightly more than the 5% we would expect in a normal distribution. Looking at values above 2.58, we would expect to find only 1%, but again here we have a higher value of 2.3% (1.5% + 0.8%). Finally, we find that 0.8% of cases were above 3.29 (so, 0.8% are significant outliers). This suggests that there may be too many outliers in this data set and we might want to do something about them!

OUTLIER1

		Frequency	Percent	Valid Percent	Cumulative Percent
Valid	Absolute z-score less than 2	246	30.4	93.2	93.2
	Absolute z-score greater than 1.96	12	1.5	4.5	97.7
	Absolute z-score greater than 2.58	4	.5	1.5	99.2
	Absolute z-score greater than 3.29	2	.2	.8	100.0
	Total	264	32.6	100.0	
Missing	System	546	67.4		
Total		810	100.0		

4.6. Graphing means: bar charts and error bars ①

Bar charts are the usual way for people to display means. How you create these graphs in SPSS depends largely on how you collected your data (whether the means come from independent cases and are, therefore, independent, or came from the same cases and so are related). For this reason we will look at a variety of situations.

In all of these situations, our starting point is the Chart Builder (Figure 4.5). In this dialog box select *Bar* in the list labelled *Choose from* to bring up the gallery shown in Figure 4.15. This gallery has eight icons representing different types of bar chart, and you should select the appropriate one either by double-clicking on it, or by dragging it onto the canvas.

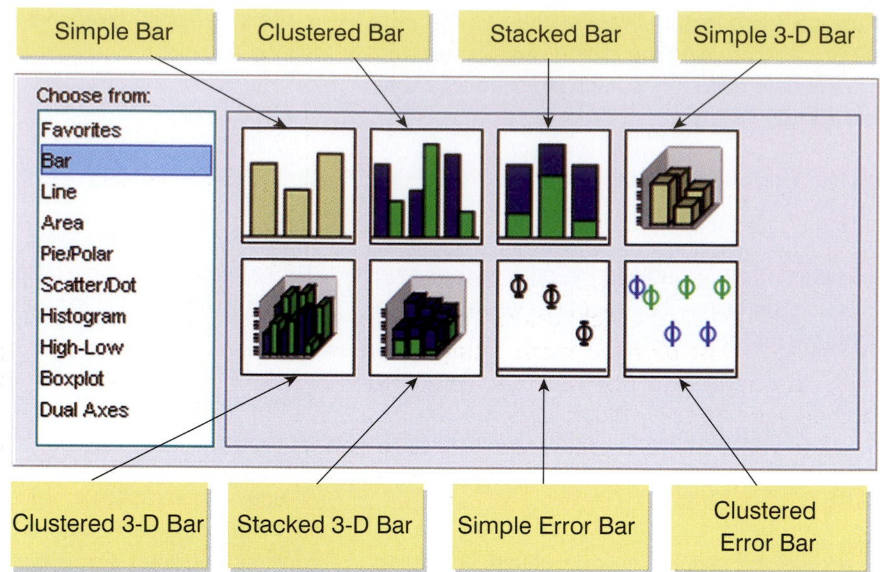

FIGURE 4.15 The bar chart gallery

- *Simple bar*: Use this option when you just want to see the means of scores across different groups of cases. For example, you might want to plot the mean ratings of two films.
- *Clustered bar*: If you had a second grouping variable you could produce a simple bar chart (as above) but with bars produced in different colours for levels of a second grouping variable. For example, you could have ratings of the two films, but for each film have a bar representing ratings of 'excitement' and another bar showing ratings of 'enjoyment'.
- *Stacked bar*: This is really the same as the clustered bar except that the different coloured bars are stacked on top of each other rather than being side by side.
- *Simple 3-D bar*: This is also the same as the clustered bar except that the second grouping variable is displayed not by different coloured bars, but by an additional axis. Given what I said in section 4.2 about 3-D effects obscuring the data, my advice is not to use this type of graph, but to stick to a clustered bar chart.
- *Clustered 3-D bar*: This is like the clustered bar chart above except that you can add a third categorical variable on an extra axis. The means will almost certainly be impossible for anyone to read on this type of graph so don't use it.
- *Stacked 3-D bar*: This graph is the same as the clustered 3-D graph except the different coloured bars are stacked on top of each other instead of standing side by side. Again, this is not a good type of graph for presenting data clearly.
- *Simple error bar*: This is the same as the simple bar chart except that instead of bars, the mean is represented by a dot, and a line represents the precision of the estimate of the mean (usually the 95% confidence interval is plotted, but you can plot the standard deviation or standard error of the mean also). You can add these error bars to a bar chart anyway, so really the choice between this type of graph and a bar chart with error bars is largely down to personal preference.
- *Clustered error bar*: This is the same as the clustered bar chart except that the mean is displayed as a dot with an error bar around it. These error bars can also be added to a clustered bar chart.

4.6.1. Simple bar charts for independent means ①

To begin with, imagine that a film company director was interested in whether there was really such a thing as a 'chick flick' (a film that typically appeals to women more than men). He took 20 men and 20 women and showed half of each sample a film that was supposed to be a 'chick flick' (*Bridget Jones' Diary*), and the other half of each sample a film that didn't fall into the category of 'chick flick' (*Memento*, a brilliant film by the way). In all cases he measured their physiological arousal as an indicator of how much they enjoyed the film. The data are in a file called **ChickFlick.sav** on the companion website. Load this file now.

First of all, let's just plot the mean rating of the two films. We have just one grouping variable (the film) and one outcome (the arousal); therefore, we want a simple bar chart. In the Chart Builder double-click on the icon for a simple bar chart (Figure 4.15). On the canvas you will see a graph and two drop zones: one for the *y*-axis and one for the *x*-axis. The *y*-axis needs to be the dependent variable, or the thing you've measured, or more simply the thing for which you want to display the mean. In this case it would be **arousal**, so select arousal from the variable list and drag it into the *y*-axis drop zone (Y-Axis?). The *x*-axis should be the variable by which we want to split the arousal data. To plot the means for the two films, select the variable **film** from the variable list and drag it into the drop zone for the *x*-axis (X-Axis?).

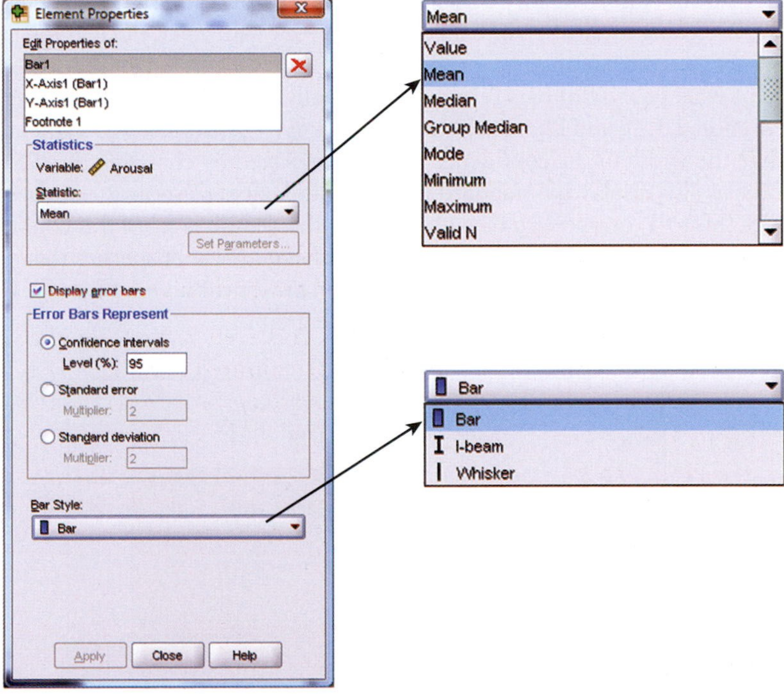

FIGURE 4.16 Element Properties of a bar chart

Figure 4.16 shows some other options for the bar chart. The main dialog box should appear when you select the type of graph you want, but if it doesn't click on Element Properties... in the Chart Builder. There are three important features of this dialog box. The first is that, by default, the bars will display the mean value. This is fine, but just note that you can plot other summary statistics such as the median or mode. Second, just because you've selected a simple bar chart doesn't mean that you have to have a bar chart. You can select to show an I-bar (the bar is reduced to a line with bars showing the top and bottom), or just a whisker (the bar is reduced to a vertical line).

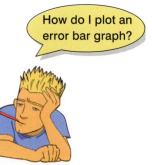

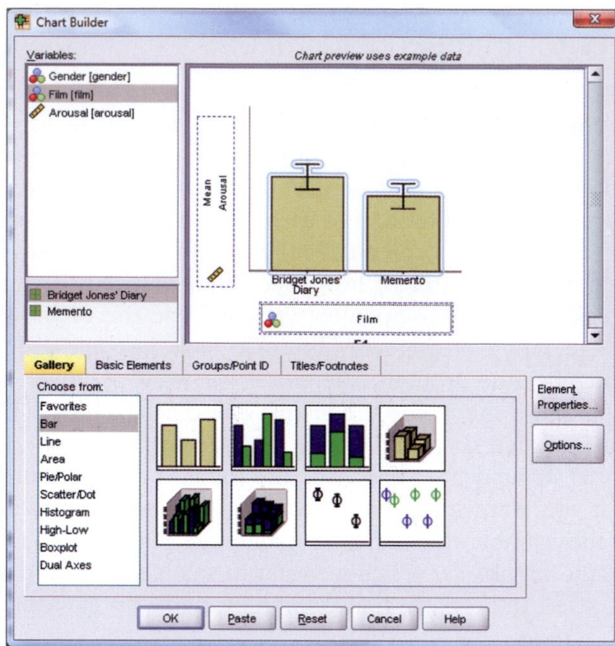

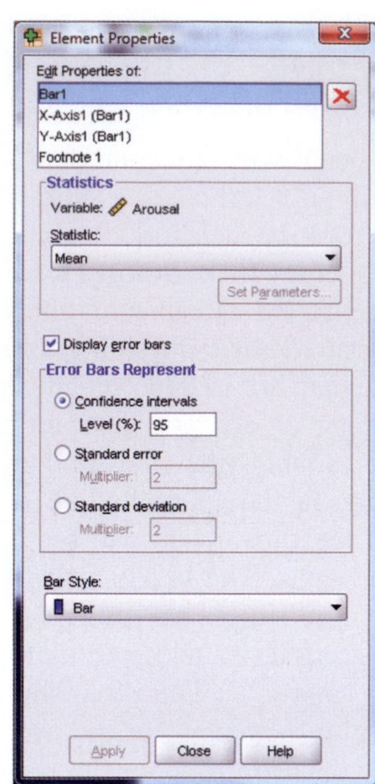

FIGURE 4.17 Dialog boxes for a simple bar chart with error bar

The I-bar and whisker options might be useful when you're not planning on showing error bars, but because we are going to show error bars we should stick with a bar. Finally, you can ask SPSS to add error bars to your bar chart to create an **error bar chart** by selecting Display error bars. You have a choice of what your error bars represent. Normally, error bars show the 95% confidence interval (see section 2.5.2), and I have selected this option (Confidence intervals).[7] Note, though, that you can change the width of the confidence interval displayed by changing the '95' to a different value. You can also display the standard error (the default is to show 2 standard errors, but you can change this to 1) or standard deviation (again, the default is 2 but this could be changed to 1 or another value). It's important that when you change these properties that you click on Apply: if you don't then the changes will not be applied to Chart Builder. Figure 4.17 shows the completed Chart Builder. Click on OK to produce the graph.

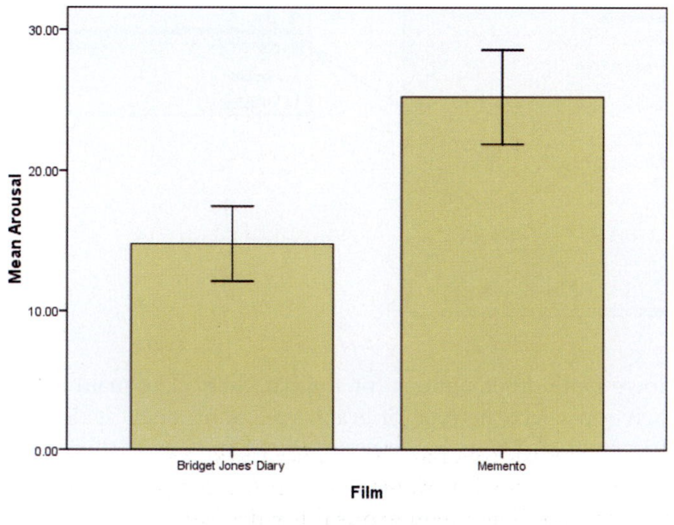

FIGURE 4.18 Bar chart of the mean arousal for each of the two films

[7] It's also worth mentioning at this point that error bars from SPSS are suitable only for normally distributed data (see section 5.4).

Figure 4.18 shows the resulting bar chart. This graph displays the mean (and the confidence interval of those means) and shows us that on average, people were more aroused by *Memento* than they were by *Bridget Jones' Diary*. However, we originally wanted to look for gender effects, so this graph isn't really telling us what we need to know. The graph we need is a *clustered graph*.[8]

4.6.2. Clustered bar charts for independent means ①

To do a clustered bar chart for means that are independent (i.e. have come from different groups) we need to double-click on the clustered bar chart icon in the Chart Builder (Figure 4.15). On the canvas you will see a graph as with the simple bar chart but there is now an extra drop zone: Cluster on X: set color. All we need to do is to drag our second grouping variable into this drop zone. As with the previous example, select **arousal** from the variable list and drag it into Y-Axis?, then select **film** from the variable list and drag it into X-Axis?. In addition, though, select the **Gender** variable and drag it into Cluster on X: set color. This will mean that bars representing males and females will be displayed in different colours (but see SPSS Tip 4.3). As in the previous section, select error bars in the properties dialog box and click on Apply to apply them to the Chart Builder. Figure 4.19 shows the completed Chart Builder. Click on OK to produce the graph.

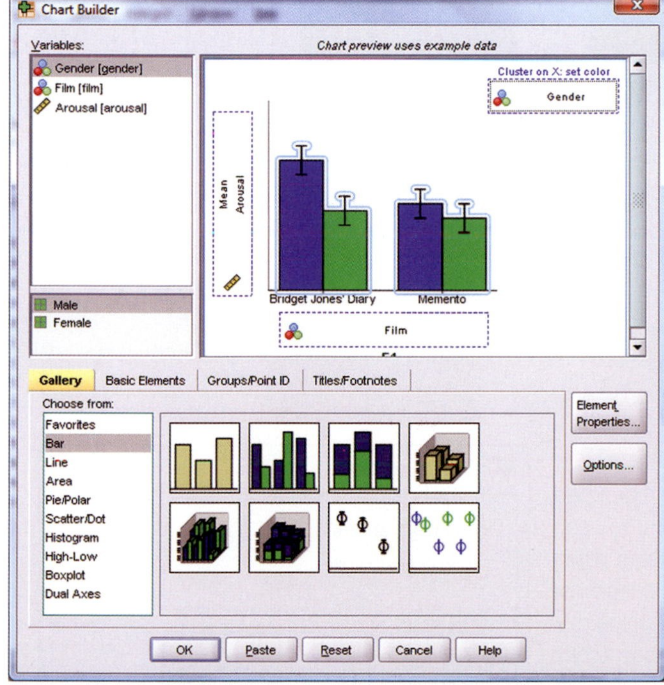

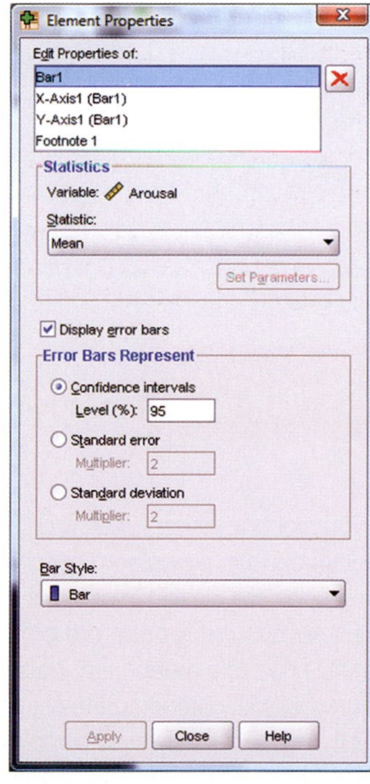

FIGURE 4.19 Dialog boxes for a clustered bar chart with error bar

Figure 4.20 shows the resulting bar chart. Like the simple bar chart this graph tells us that arousal was overall higher for *Memento* than *Bridget Jones' Diary*, but it also splits this information by gender. The mean arousal for *Bridget Jones' Diary* shows that males

[8] You can also use a drop-line graph, which is described in section 4.8.6.

were actually more aroused during this film than females. This indicates they enjoyed the film more than the women did. Contrast this with *Memento*, for which arousal levels are comparable in males and females. On the face of it, this contradicts the idea of a 'chick flick': it actually seems that men enjoy chick flicks more than the chicks (probably because it's the only help we get to understand the complex workings of the female mind!).

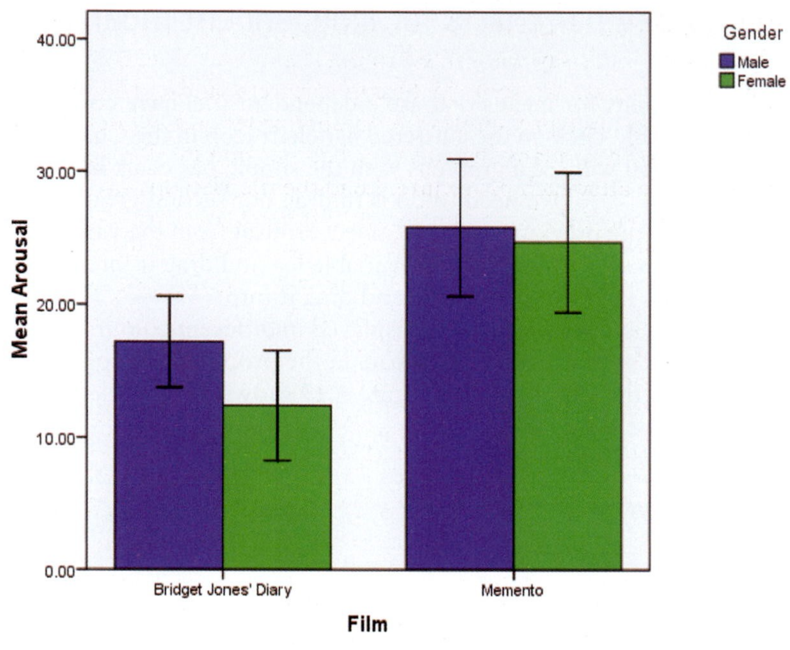

Error Bars: 95% CI

FIGURE 4.20 Bar chart of the mean arousal for each of the two films

SPSS TIP 4.3 Colours or patterns? ①

By default, when you plot graphs on which you group the data by some categorical variable (e.g. a clustered bar chart, or a grouped scatterplot) these groups are plotted in different colours. You can change this default so that the groups are plotted using different patterns. In a bar chart this means that bars will be filled not with different colours, but with different patterns. With a scatterplot (see section 4.8.2) it means that different symbols are used to plot data from different groups. To make this change, double-click in the Cluster on X: set color drop zone (bar chart) or Set color (scatterplot) to bring up a new dialog box. Within this dialog box there is a drop-down list labelled *Distinguish Groups by* and in this list you can select *Color* or *Pattern*. To change the default select *Pattern* and then click on OK. Obviously you can switch back to displaying different groups in different colours in the same way

Grouping Zone
Grouping Variable:
Gender
Distinguish Groups by:
Color
Color
Pattern
OK Cancel Help

4.6.3. Simple bar charts for related means ①

Hiccups can be a serious problem: Charles Osborne apparently got a case of hiccups while slaughtering a hog (well, who wouldn't?) that lasted 67 years. People have many methods for stopping hiccups (a surprise; holding your breath), but actually medical science has put its collective mind to the task too. The official treatment methods include tongue-pulling manoeuvres, massage of the carotid artery, and, believe it or not, digital rectal massage (Fesmire, 1988). I don't know the details of what the digital rectal massage involved, but I can probably imagine. Let's say we wanted to put digital rectal massage to the test (erm, as a cure of hiccups I mean). We took 15 hiccup sufferers, and during a bout of hiccups administered each of the three procedures (in random order and at intervals of 5 minutes) after taking a baseline of how many hiccups they had per minute. We counted the number of hiccups in the minute after each procedure. Load the file **Hiccups.sav** from the companion website. Note that these data are laid out in different columns; there is no grouping variable that specifies the interventions because each patient experienced all interventions. In the previous two examples we have used grouping variables to specify aspects of the graph (e.g. we used the grouping variable **film** to specify the *x*-axis). For repeated-measures data we will not have these grouping variables and so the process of building a graph is a little more complicated (but not a lot more).

How do I plot a bar graph of repeated measures data?

To plot the mean number of hiccups go to the Chart Builder and double-click on the icon for a simple bar chart (Figure 4.15). As before, you will see a graph on the canvas with drop zones for the *x*- and *y*-axis. Previously we specified the column in our data that contained data from our outcome measure on the *y*-axis, but for these data we have four columns containing data on the number of hiccups (the outcome variable). What we have to do then is to drag all four of these variables from the variable list into the *y-axis* drop zone. We have to do this simultaneously. First, we need to select multiple items in the variable list: to do this select the first variable by clicking on it with the mouse. The variable will be highlighted in blue. Now, hold down the *Ctrl* key on the keyboard and click on a second variable. Both variables are now highlighted in blue. Again, hold down the *Ctrl* key and click on a third variable in the variable list and so on for the fourth. In cases in which you want to select a list of consecutive variables, you can do this very quickly by simply clicking on the first variable that you want to select (in this case **baseline**), hold down the *Shift* key on the keyboard and then click on the last variable that you want to select (in this case **digital rectal massage**); notice that all of the variables in between have been selected too. Once the four variables are selected you can drag them by clicking on any one of the variables and then dragging them into Y-Axis? as shown in Figure 4.21.

Once you have dragged the four variables onto the *y*-axis drop zones a new dialog box appears (Figure 4.22). This box tells us that SPSS is creating two temporary variables. One is called **Summary**, which is going to be the outcome variable (i.e. what we measured – in this case the number of hiccups per minute). The other is called **index** and this variable will represent our independent variable (i.e. what we manipulated – in this case the type of intervention). SPSS uses these temporary names because it doesn't know what our particular variables represent, but we should change them to something more helpful. Just click on OK to get rid of this dialog box.

We need to edit some of the properties of the graph. Figure 4.23 shows the options that need to be set: if you can't see this dialog box then click on Element Properties in the Chart Builder. In the left panel of Figure 4.23 just note that I have selected to display error bars (see the previous two sections for more information). The middle panel is accessed by clicking on *X-Axis1 (Bar1)* in the list labelled *Edit Properties of* which allows us to edit properties of the horizontal axis. The first

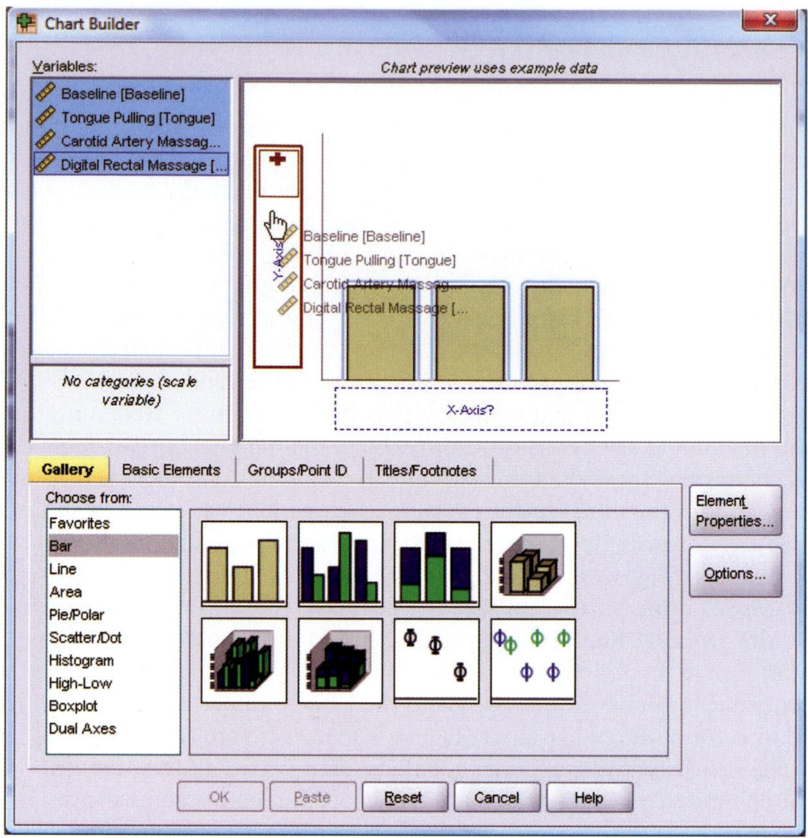

FIGURE 4.21 Specifying a simple bar chart for repeated-measures data

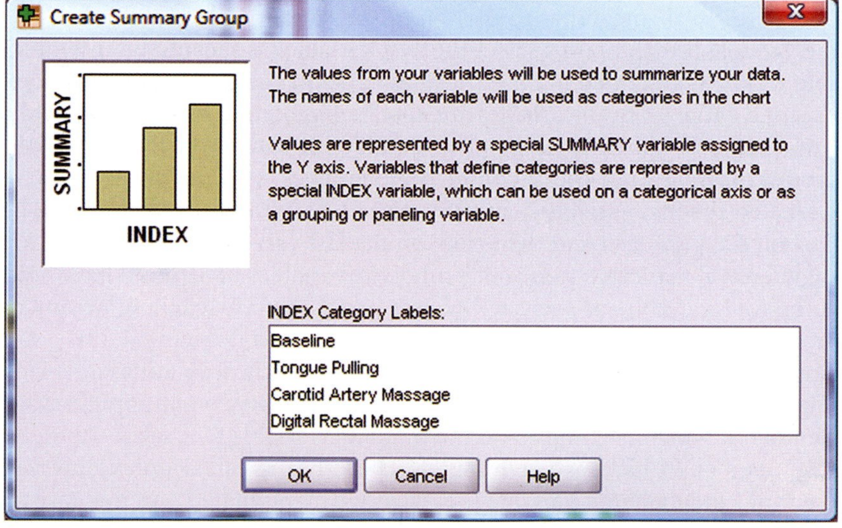

FIGURE 4.22 The *Create Summary Group* dialog box

thing we need to do is give the axis a title and I have typed *Intervention* in the space labelled *Axis Label*. This label will appear on the graph. Also, we can change the order of our variables if we want to by selecting a variable in the list labelled *Order* and moving it up or down using ↑ and ↓. If we change our mind about displaying one of our variables then we can also remove it from the list by selecting it and clicking on ✕. Click on Apply for these changes to take effect. The right panel of Figure 4.23 is accessed by clicking on *Y-Axis1 (Bar1)* in the list labelled *Edit Properties of* which allows us to edit properties of the vertical axis. The main change that

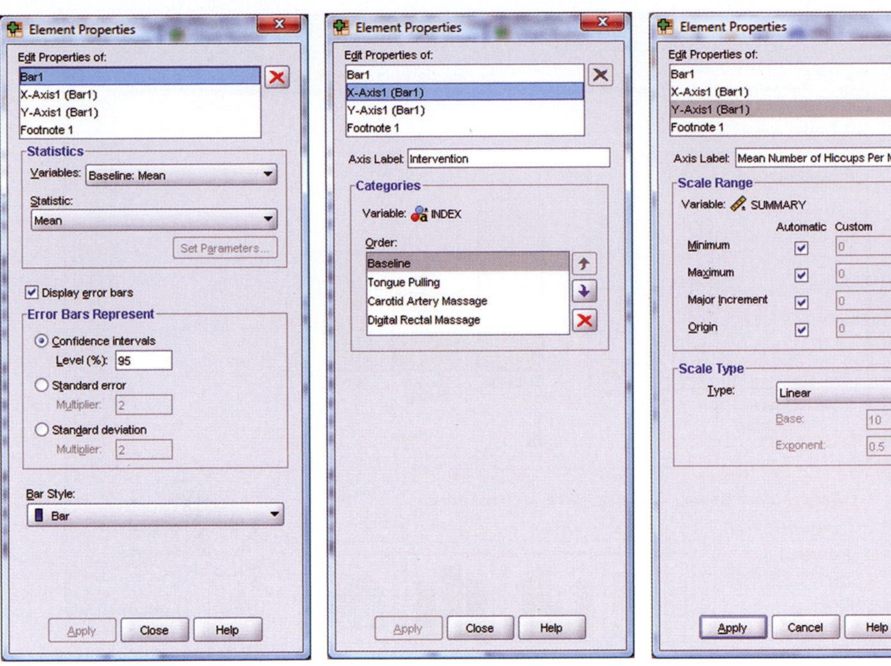

FIGURE 4.23 Setting Element Properties for a repeated-measures graph

I have made here is to give the axis a label so that the final graph has a useful description on the axis (by default it will just say Mean, which isn't very helpful). I have typed 'Mean Number of Hiccups Per Minute' in the box labelled *Axis Label*. Also note that you can use this dialog box to set the scale of the vertical axis (the minimum value, maximum value and the major increment, which is how often a mark is made on the axis). Mostly you can let SPSS construct the scale automatically and it will be fairly sensible – and even if it's not you can edit it later. Click on Apply to apply the changes.

Figure 4.24 shows the completed Chart Builder. Click on OK to produce the graph. The resulting bar chart in Figure 4.25 displays the mean (and the confidence interval of those means) number of hiccups at baseline and after the three interventions. Note that the axis labels that I typed in have appeared on the graph. The error bars on graphs of repeated-measures designs aren't actually correct as we will see in Chapter 9; I don't want to get into the reasons why here because I want to keep things simple, but if you're doing a graph of your own data then I would read section 9.2 before you do.

We can conclude that the amount of hiccups after tongue pulling was about the same as at baseline; however, carotid artery massage reduced hiccups, but not by as much as a good old-fashioned digital rectal massage. The moral here is: if you have hiccups, find something digital and go amuse yourself for a few minutes.

4.6.4. Clustered bar charts for related means ①

Now we have seen how to plot means that are related (i.e. show different conditions applied to the same group of cases), you might well wonder what you do if you have a second independent variable that had been measured in the same sample. You'd do a clustered bar chart, right? Wrong? Actually, the SPSS Chart Builder doesn't appear to be able to cope with this situation at all – at least not that I can work out from playing about with it. (Cue a deluge of emails along the general theme of 'Dear Dr Field, I was recently looking through my FEI Titan 80-300 monochromated scanning transmission electron microscope and I think I may have found your brain. I have enclosed it for you – good luck finding it in the

FIGURE 4.24 Completed Chart Builder for a repeated-measures graph

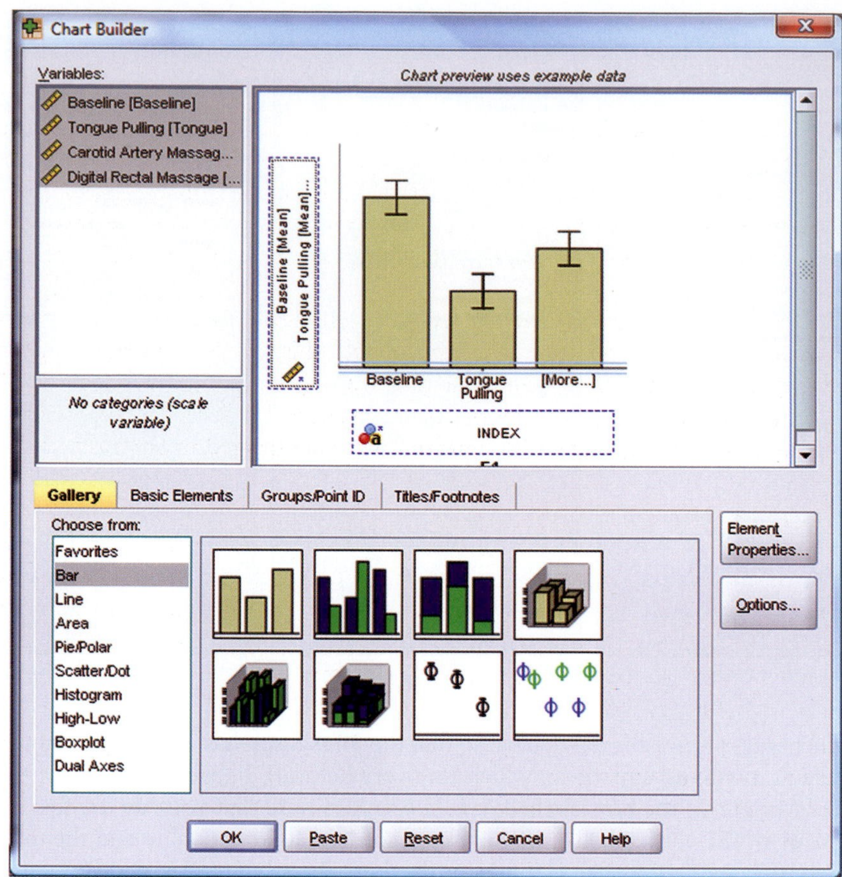

FIGURE 4.25 Bar chart of the mean number of hiccups at baseline and after various interventions

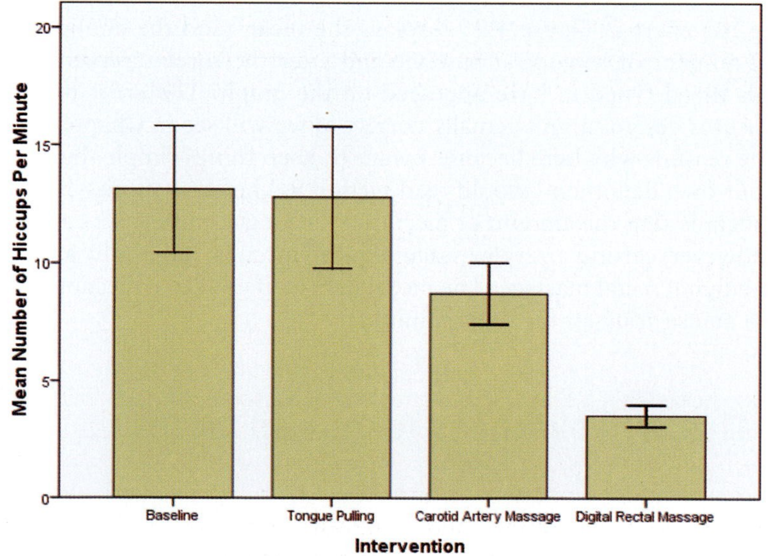

envelope. May I suggest that you take better care next time there is a slight gust of wind or else, I fear, it might blow out of your head again. Yours, Professor Enormobrain. PS Doing clustered charts for related means in SPSS is simple for anyone whose mental acumen can raise itself above that of a louse.')

4.6.5. Clustered bar charts for 'mixed' designs ①

The Chart Builder might not be able to do charts for multiple repeated-measures variables, but it can graph what is known as a mixed design (see Chapter 14). This is a design in which you have one or more independent variables measured using different groups, and one or more independent variables measured using the same sample. Basically, the Chart Builder can produce a graph provided you have only one variable that was a repeated measure.

We all like to text-message (especially students in my lectures who feel the need to text-message the person next to them to say 'Bloody hell, this guy is so boring I need to poke out my own eyes.'). What will happen to the children, though? Not only will they develop super-sized thumbs, they might not learn correct written English. Imagine we conducted an experiment in which a group of 25 children was encouraged to send text messages on their mobile phones over a six-month period. A second group of 25 children was forbidden from sending text messages for the same period. To ensure that kids in this latter group didn't use their phones, this group was given armbands that administered painful shocks in the presence of microwaves (like those emitted from phones).[9] The outcome was a score on a grammatical test (as a percentage) that was measured both before and after the intervention. The first independent variable was, therefore, text message use (text messagers versus controls) and the second independent variable was the time at which grammatical ability was assessed (baseline or after six months). The data are in the file **Text Messages.sav**.

To graph these data we need to follow the procedure for graphing related means in section 4.6.3. Our repeated-measures variable is time (whether grammatical ability was measured at baseline or six months) and is represented in the data file by two columns, one for the baseline data and the other for the follow-up data. In the Chart Builder select these two variables simultaneously by clicking on one and then holding down the *Ctrl* key on the keyboard and clicking on the other. When they are both highlighted click on either one and drag it into [Y-Axis?] as shown in Figure 4.26. The second variable (whether children text messaged or not) was measured using different children and so is represented in the data file by a grouping variable (**group**). This variable can be selected in the variable list and dragged into [Cluster on X: set color]. The two groups will be displayed as different-coloured bars. The finished Chart Builder is in Figure 4.27. Click on [OK] to produce the graph.

SELF-TEST Use what you learnt in section 4.6.3 to add error bars to this graph and to label both the *x*- (I suggest 'Time') and *y*-axis (I suggest 'Mean Grammar Score (%)').

Figure 4.28 shows the resulting bar chart. It shows that at baseline (before the intervention) the grammar scores were comparable in our two groups; however, after the intervention, the grammar scores were lower in the text messagers than in the controls. Also, if you compare the two blue bars you can see that text messagers' grammar scores have fallen over the six months; compare this to the controls (green bars) whose grammar scores are fairly similar over time. We could, therefore, conclude that text messaging has a detrimental effect on children's understanding of English grammar and civilization will crumble, with Abaddon rising cackling from his bottomless pit to claim our wretched souls. Maybe.

[9] Although this punished them for any attempts to use a mobile phone, because other people's phones also emit microwaves, an unfortunate side effect was that these children acquired a pathological fear of anyone talking on a mobile phone.

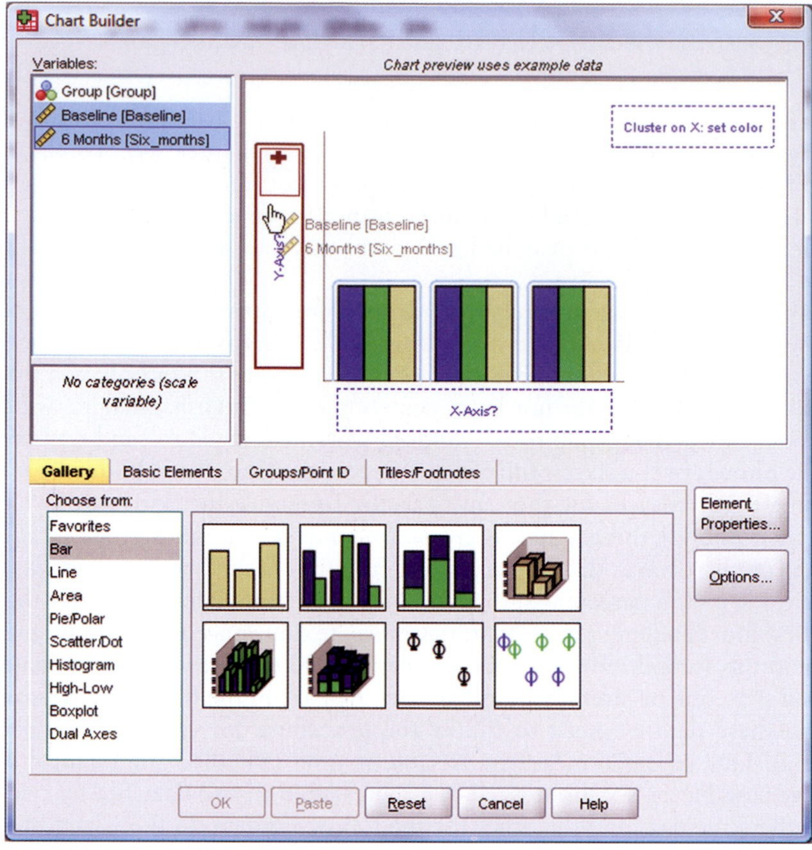

FIGURE 4.26 Selecting the repeated-measures variable in the Chart Builder

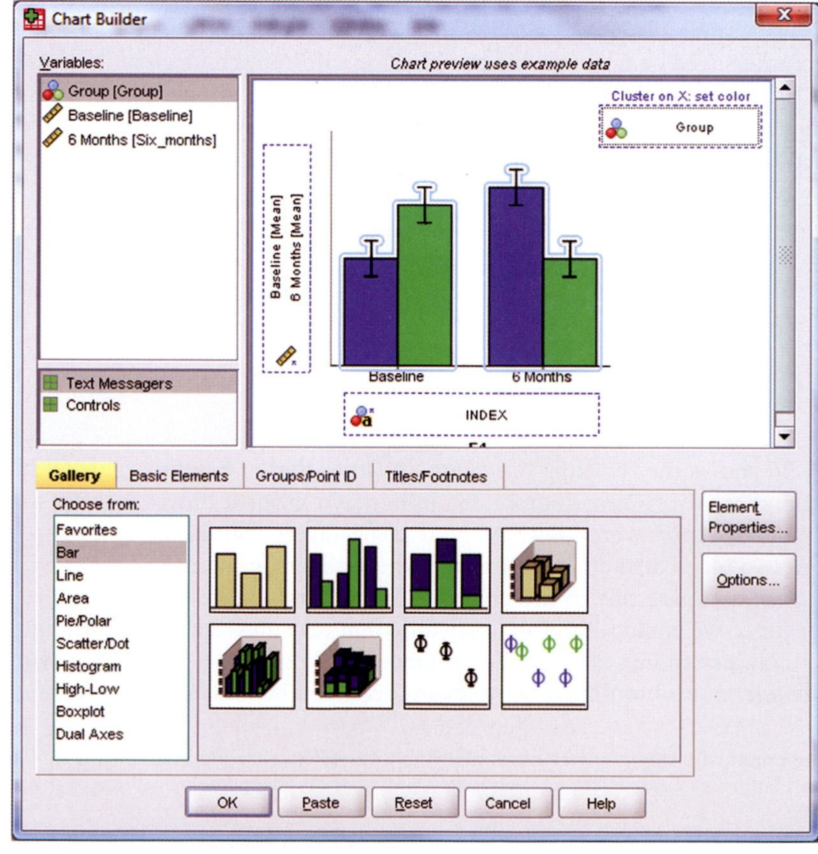

FIGURE 4.27 Completed dialog box for an error bar graph of a mixed design

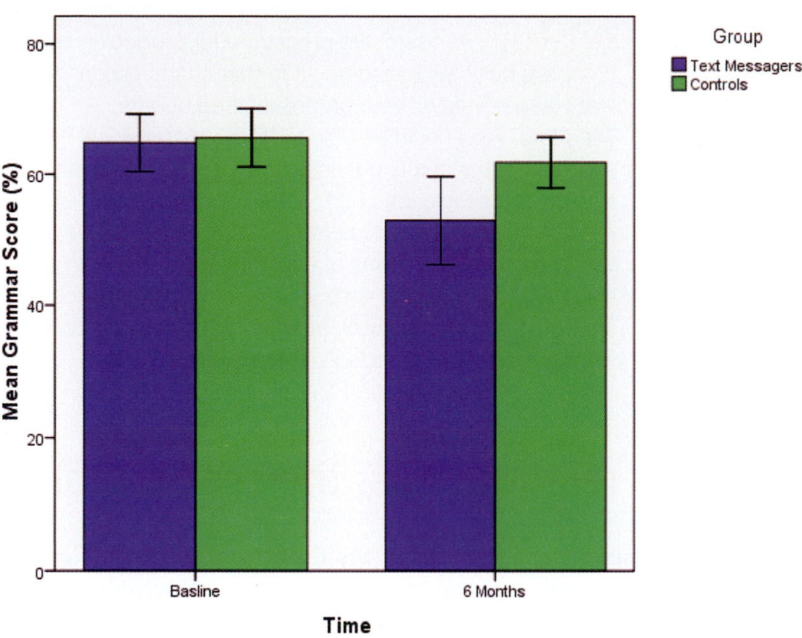

FIGURE 4.28 Error bar graph of the mean grammar score over six months in children who were allowed to text-message versus those who were forbidden

4.7. Line charts ①

Line charts are bar charts but with lines instead of bars. Therefore, everything we have just done with bar charts we can display as a line chart instead. As ever, our starting point is the Chart Builder (Figure 4.5). In this dialog box select *Line* in the list labelled *Choose from* to bring up the gallery shown in Figure 4.29. This gallery has two icons and you should select the appropriate one either by double-clicking on it, or by dragging it onto the canvas.

- *Simple line*: Use this option when you just want to see the means of scores across different groups of cases.
- *Multiple line*: This is equivalent to the clustered bar chart in the previous section in that you can plot means of a particular variable but produce different-coloured lines for each level of a second variable.

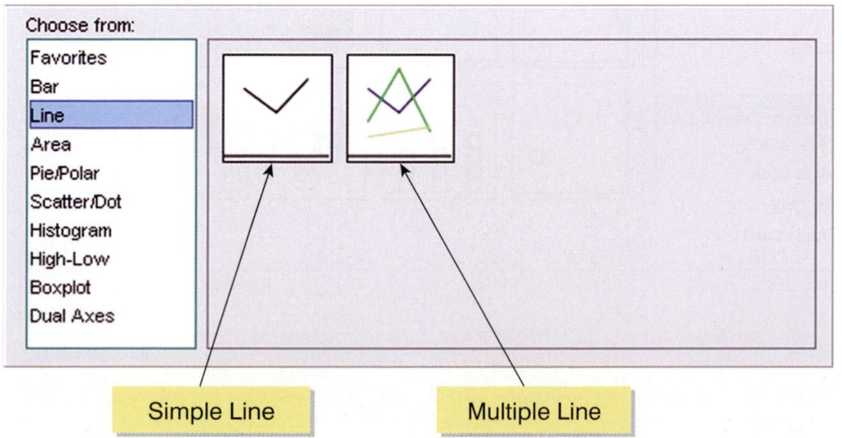

FIGURE 4.29 The line chart gallery

SELF-TEST As I said, the procedure for producing line graphs is basically the same as for bar charts except that you get lines on your graphs instead of bars. Therefore, you should be able to follow the previous sections for bar charts but selecting a simple line chart instead of a simple bar chart, and selecting a multiple line chart instead of a clustered bar chart. I would like you to produce line charts of each of the bar charts in the previous section. In case you get stuck, the self-test answers that can be downloaded from the companion website will take you through it step by step.

4.8. Graphing relationships: the scatterplot ①

How do I draw a graph of the relationship between two variables?

Sometimes we need to look at the relationships between variables (rather than their means, or frequencies). A **scatterplot** is simply a graph that plots each person's score on one variable against their score on another. A scatterplot tells us several things about the data, such as whether there seems to be a relationship between the variables, what kind of relationship it is and whether any cases are markedly different from the others. We saw earlier that a case that differs substantially from the general trend of the data is known as an outlier and such cases can severely bias statistical procedures (see Jane Superbrain Box 4.1 and section 7.6.1.1 for more detail). We can use a scatterplot to show us if any cases look like outliers.

Drawing a scatterplot using SPSS is dead easy using the Chart Builder. As with all of the graphs in this chapter, our starting point is the Chart Builder (Figure 4.5). In this dialog box select *Scatter/Dot* in the list labelled *Choose from* to bring up the gallery shown in Figure 4.30. This gallery has eight icons representing different types of scatterplot, and you should select the appropriate one either by double-clicking on it, or by dragging it onto the canvas.

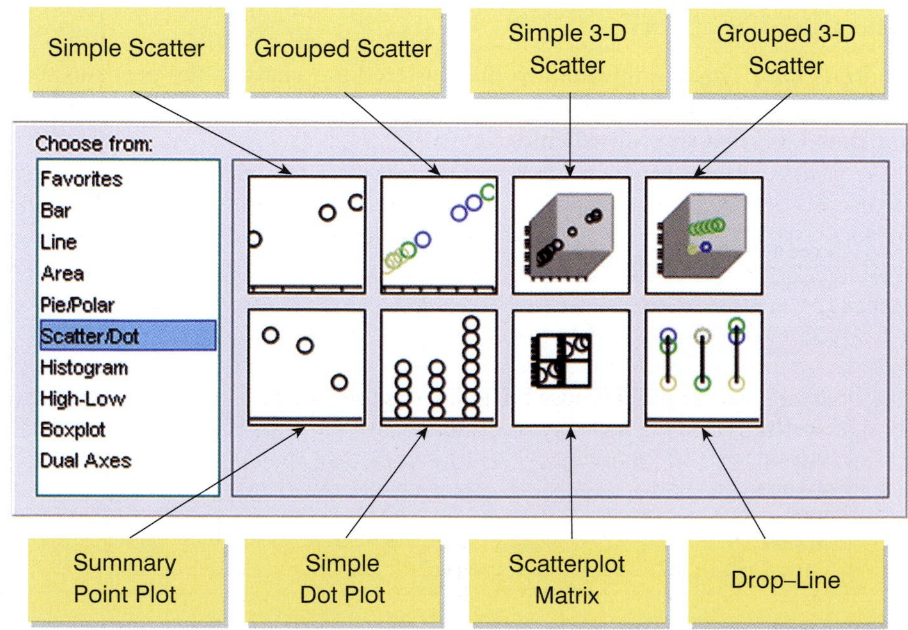

FIGURE 4.30 The scatter/dot gallery

- *Simple scatter*: Use this option when you want to plot values of one continuous variable against another.
- *Grouped scatter*: This is like a simple scatterplot except that you can display points belonging to different groups in different colours (or symbols).
- *Simple 3-D scatter*: Use this option to plot values of one continuous variable against values of two others.
- *Grouped 3-D scatter*: Use this option if you want to plot values of one continuous variable against two others but differentiating groups of cases with different-coloured dots.
- *Summary point plot*: This graph is the same as a bar chart (see section 4.6) except that a dot is used instead of a bar.
- *Simple dot plot*: Otherwise known as **density plot**, this graph is similar to a histogram (see section 4.4) except that rather than having a summary bar representing the frequency of scores, a density plot shows each individual score as a dot. They can be useful, like a histogram, for looking at the shape of a distribution.
- *Scatterplot matrix*: This option produces a grid of scatterplots showing the relationships between multiple pairs of variables.
- *Drop-line*: This option produces a graph that is similar to a clustered bar chart (see, for example, section 4.6.2) but with a dot representing a summary statistic (e.g. the mean) instead of a bar, and with a line connecting means of different groups. These graphs can be useful for comparing statistics, such as the mean, across different groups.

4.8.1. Simple scatterplot ①

This type of scatterplot is for looking at just two variables. For example, a psychologist was interested in the effects of exam stress on exam performance. So, she devised and validated a questionnaire to assess state anxiety relating to exams (called the Exam Anxiety Questionnaire, or EAQ). This scale produced a measure of anxiety scored out of 100. Anxiety was measured before an exam, and the percentage mark of each student on the exam was used to assess the exam performance. The first thing that the psychologist should do is draw a scatterplot of the two variables (her data are in the file **Exam Anxiety.sav** and you should load this file into SPSS).

In the Chart Builder double-click on the icon for a simple scatterplot (Figure 4.31). On the canvas you will see a graph and two drop zones: one for the *y*-axis and one for the *x*-axis. The *y*-axis needs to be the dependent variable (the outcome that was measured).[10] In this case the outcome is **Exam Performance (%)**, so select it from the variable list and drag it into the *y*-axis drop zone (Y-Axis?). The horizontal axis should display the independent variable (the variable that predicts the outcome variable). In this case it is **Exam Anxiety**, so click on this variable in the variable list and drag it into the drop zone for the *x*-axis (X-Axis?). Figure 4.17 shows the completed Chart Builder. Click on OK to produce the graph.

Figure 4.32 shows the resulting scatterplot; yours won't have a funky line on it yet, but don't get too depressed about it because I'm going to show you how to add this line very soon. The scatterplot tells us that the majority of students suffered from high levels of anxiety (there are very few cases that had anxiety levels below 60). Also, there are no obvious outliers in

[10] In experimental research the independent variable is usually plotted on the horizontal axis and the dependent variable on the vertical axis because changes in the independent variable (the variable that the experimenter has manipulated) cause changes in the dependent variable. In correlational research, variables are measured simultaneously and so no cause-and-effect relationship can be established. As such, these terms are used loosely.

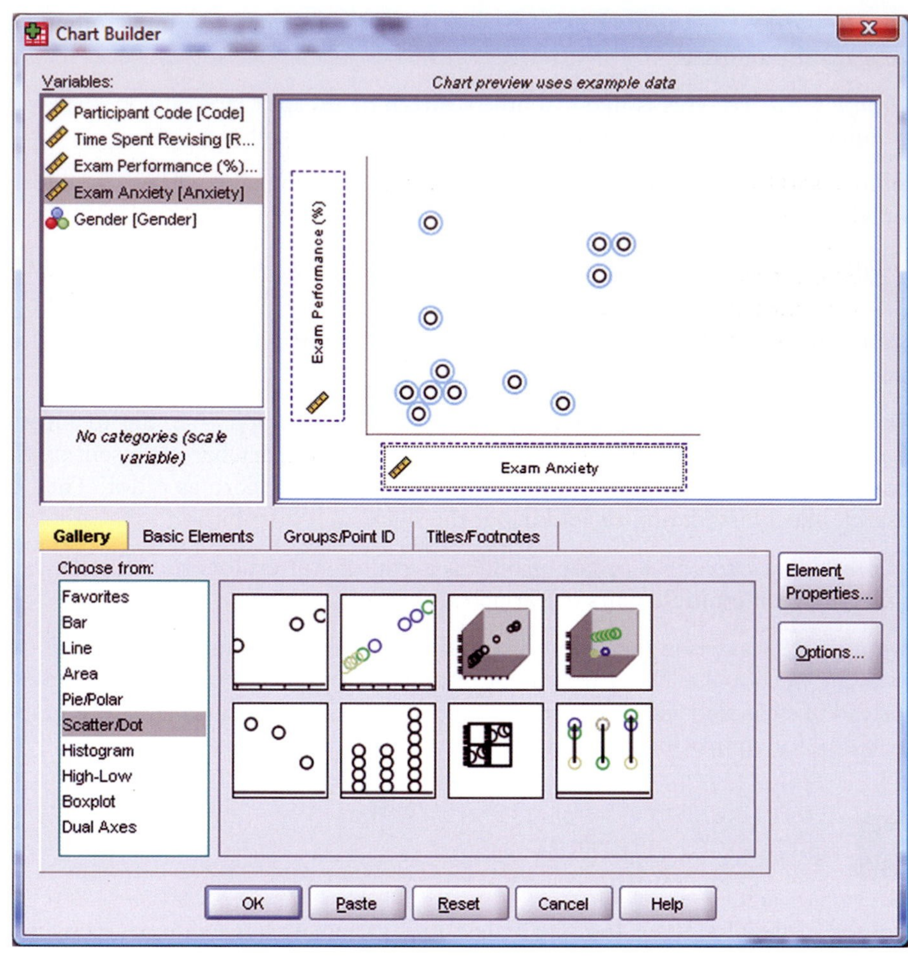

FIGURE 4.31 Completed *Chart Builder* dialog box for a simple scatterplot

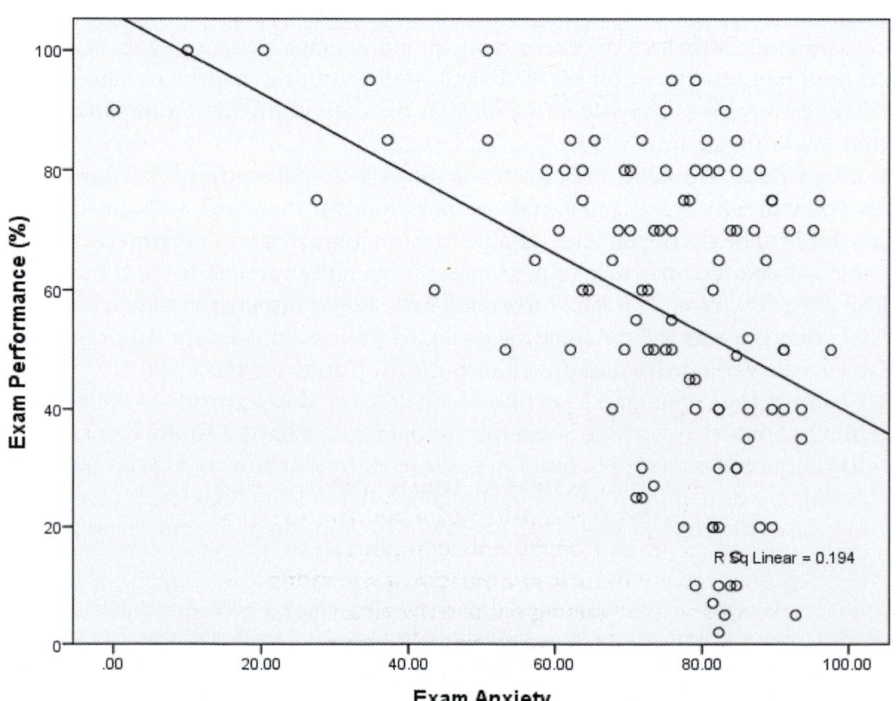

FIGURE 4.32 Scatterplot of exam anxiety and exam performance

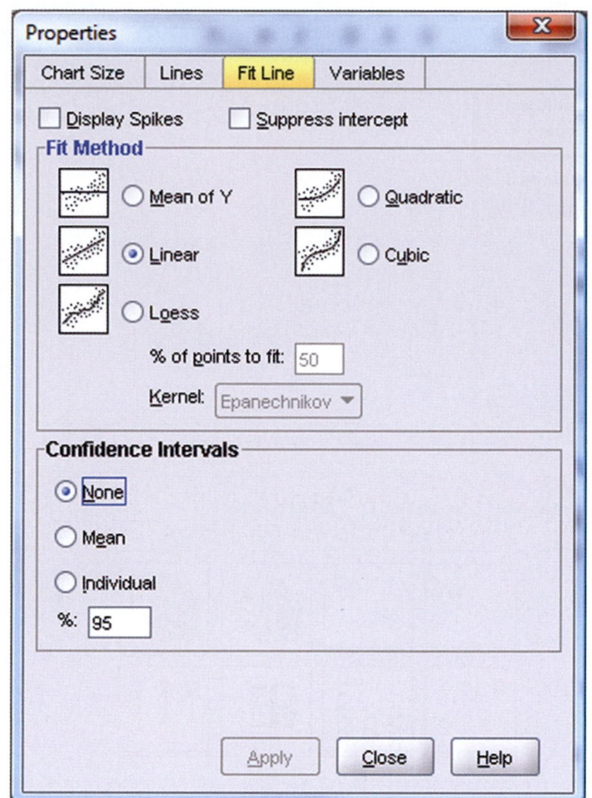

FIGURE 4.33 *Properties* dialog box for a simple scatterplot

that most points seem to fall within the vicinity of other points. There also seems to be some general trend in the data, shown by the line, such that higher levels of anxiety are associated with lower exam scores and low levels of anxiety are almost always associated with high examination marks. Another noticeable trend in these data is that there were no cases having low anxiety and low exam performance – in fact, most of the data are clustered in the upper region of the anxiety scale.

How do I fit a regression line to a scatterplot?

Often when you plot a scatterplot it is useful to plot a line that summarizes the relationship between variables (this is called a **regression line** and we will discover more about it in Chapter 7). All graphs in SPSS can be edited by double-clicking on them in the SPSS Viewer to open them in the SPSS Chart Editor (see Figure 4.40). For more detail on editing graphs see section 4.9; for now, just click on in the Chart Editor to open the *Properties* dialog box (Figure 4.33). Using this dialog box we can add a line to the graph that represents the overall mean of all data, a linear (straight line) model, a quadratic model, a cubic model and so on (these trends are described in section 10.2.11.5). Let's look at the linear regression line; select this option and then click on Apply to apply the changes to the graph. It should now look like Figure 4.32. What if we want to see whether male and female students had different reactions to exam anxiety? To do this, we need a grouped scatterplot.

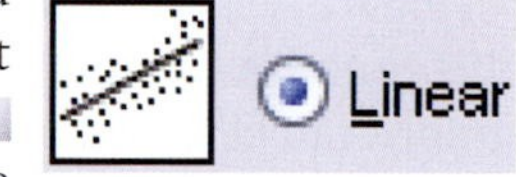

4.8.2. Grouped scatterplot ①

This type of scatterplot is for looking at two continuous variables, but when you want to colour data points by a third categorical variable. Sticking with our previous example, we

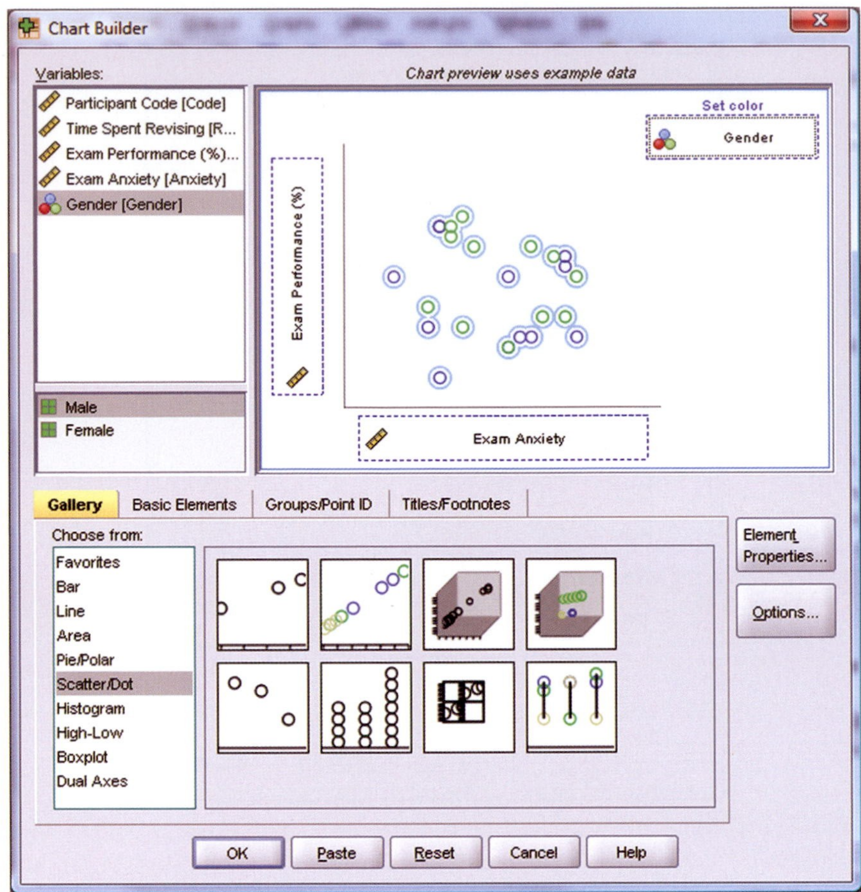

FIGURE 4.34 Completed *Chart Builder* dialog box for a grouped scatterplot

could look at the relationship between exam anxiety and exam performance in males and females (our grouping variable). To do this we double-click on the grouped scatter icon in the Chart Builder (Figure 4.30). As in the previous example, we select **Exam Performance (%)** from the variable list, and drag it into the Y-Axis? drop zone, and select **Exam Anxiety** and drag it into X-Axis? drop zone. There is an additional drop zone (Set color) into which we can drop any categorical variable. In this case, **Gender** is the only categorical variable in our variable list, so select it and drag it into this drop zone. (If you want to display the different genders using different-shaped symbols rather than different-coloured symbols then read SPSS Tip 4.3). Figure 4.34 shows the completed Chart Builder. Click on OK to produce the graph.

Figure 4.35 shows the resulting scatterplot; as before I have added regression lines, but this time I have added different lines for each group. We saw in the previous section that graphs can be edited by double-clicking on them in the SPSS Viewer to open them in the SPSS Chart Editor (Figure 4.40). We also saw that we could fit a regression line that summarized the whole data set by clicking on [icon]. We could do this again, if we wished. However, having split the data by gender it might be more interesting to fit separate lines for our two groups. This is easily achieved by clicking on [icon] in the Chart Editor. As before, this action opens the *Properties* dialog box (Figure 4.33) and we can ask for a linear model to be fitted to the data (see the previous section); however, this time when we click on Apply SPSS will fit a separate line for the men and women. These lines (Figure 4.35) tell us that the relationship between exam anxiety and exam performance was slightly stronger in males (the line is steeper) indicating that men's exam performance was more adversely affected by anxiety than women's exam anxiety. (Whether this difference is significant is another issue – see section 6.7.1.)

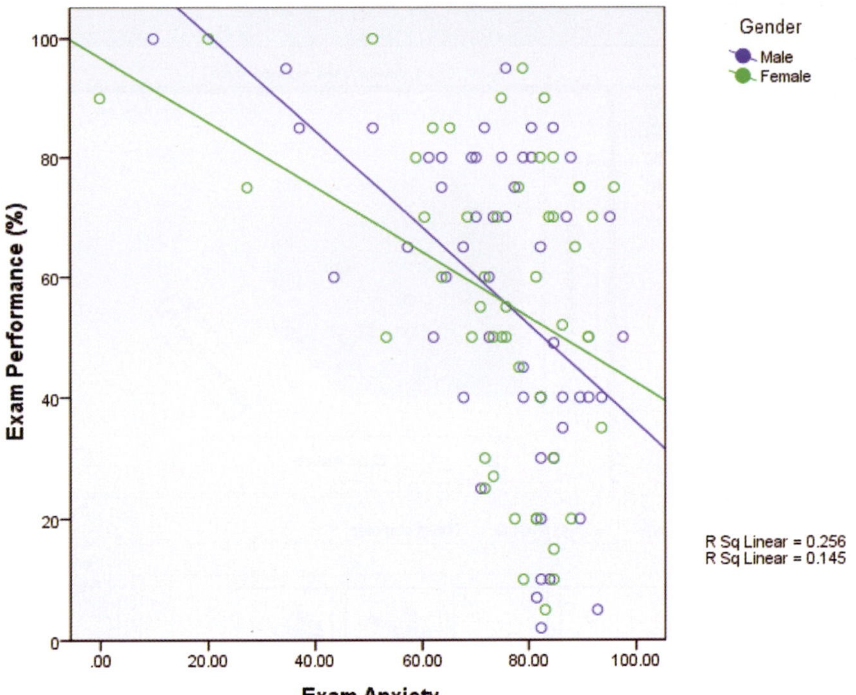

FIGURE 4.35 Scatterplot of exam anxiety and exam performance split by gender

4.8.3. Simple and grouped 3-D scatterplots ①

I'm now going to show you one of the few times when you can use a 3-D graph without a bearded statistician locking you in a room. Having said that, even in this situation it is, arguably, still not a clear way to present data. A 3-D scatterplot is used to display the relationship between three variables. The reason why it's alright to use a 3-D graph here is because the third dimension is actually telling us something useful (and isn't just there to look pretty). As an example, imagine our researcher decided that exam anxiety might not be the only factor contributing to exam performance. So, she also asked participants to keep a revision diary from which she calculated the number of hours spent revising for the exam. She wanted to look at the relationships between these variables simultaneously.

We can use the same data set as in the previous example, but this time in the Chart Builder double-click on the simple 3-D scatter icon (Figure 4.30). The graph preview on the canvas differs from ones that we have seen before in that there is a third axis and a new drop zone (Z-Axis?). It doesn't take a genius to work out that we simply repeat what we have done for previous scatterplots by dragging another continuous variable into this new drop zone. So, select **Exam Performance (%)** from the variable list and drag it into the Y-Axis? drop zone, and select **Exam Anxiety** and drag it into the X-Axis? drop zone. Now select **Time Spent Revising** in the variable list and drag it into the Z-Axis? drop zone. Figure 4.36 shows the completed Chart Builder. Click on OK to produce the graph.

A 3-D scatterplot is great for displaying data concisely; however, as the resulting scatterplot in Figure 4.37 shows, it can be quite difficult to interpret (can you really see what the relationship between exam revision and exam performance is?). As such, its usefulness in exploring data can be limited. However, you can try to improve the interpretability by changing the angle of rotation of the graph (it sometimes helps, and sometimes doesn't). To do this double-click on the scatterplot in the SPSS Viewer to open it in the SPSS Chart Editor (Figure 4.40). Then click on [icon] in the Chart Editor to open the *3-D Rotation* dialog box. You can change the values in this dialog box to achieve different views of data cloud.

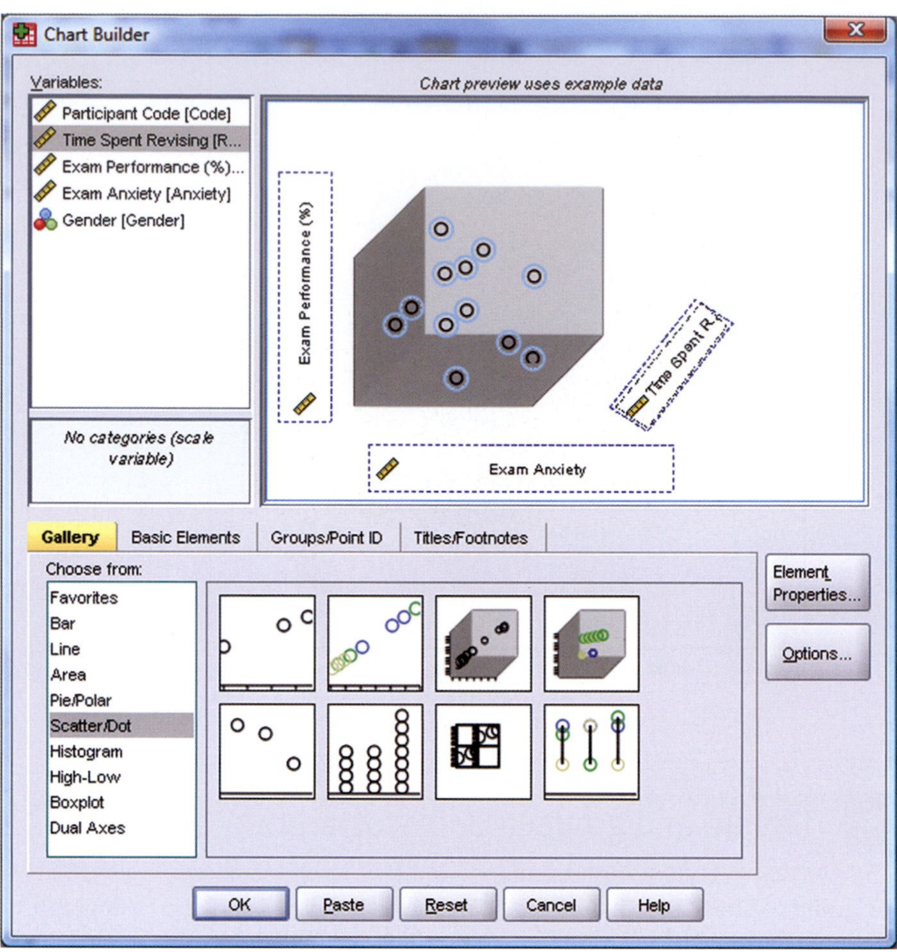

FIGURE 4.36 Completed *Chart Builder* for a 3-D scatterplot

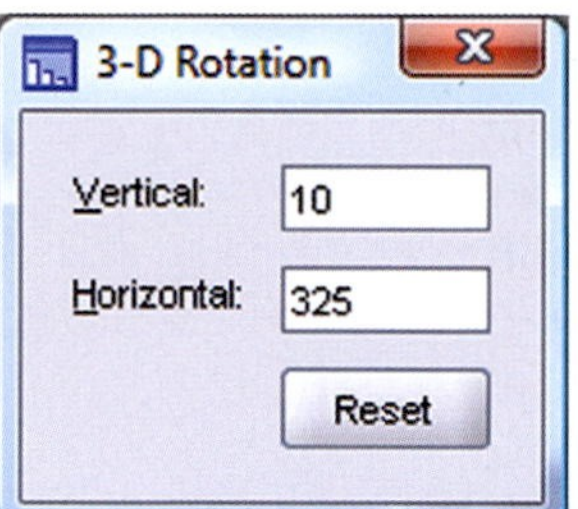

Have a play with the different values and see what happens; ultimately you'll have to decide which angle best represents the relationship for when you want to put the graph on a 2-D piece of paper! If you end up in some horrid pickle by putting random numbers in the *3-D Rotation* dialog box then simply click on Reset to restore the original view.

What about if we wanted to split the data cloud on our 3-D plot into different groups? This is very simple (so simple that I'm sure you can work it out for yourself); in the Chart Builder double-click on the grouped 3-D scatter icon (Figure 4.15). The graph preview on the canvas will look the same as for the simple 3-D scatterplot except that our old friend the Set color drop zone is back. We can simply drag our categorical variable (in this case **Gender**) into this drop zone.

SELF-TEST Based on my minimal (and no doubt unhelpful) summary, produce a 3-D scatterplot of the data in Figure 4.37 but with the data split by gender. To make things a bit more tricky see if you can get SPSS to display different symbols for the two groups rather than two colours (see SPSS Tip 4.3). A full guided answer can be downloaded from the companion website

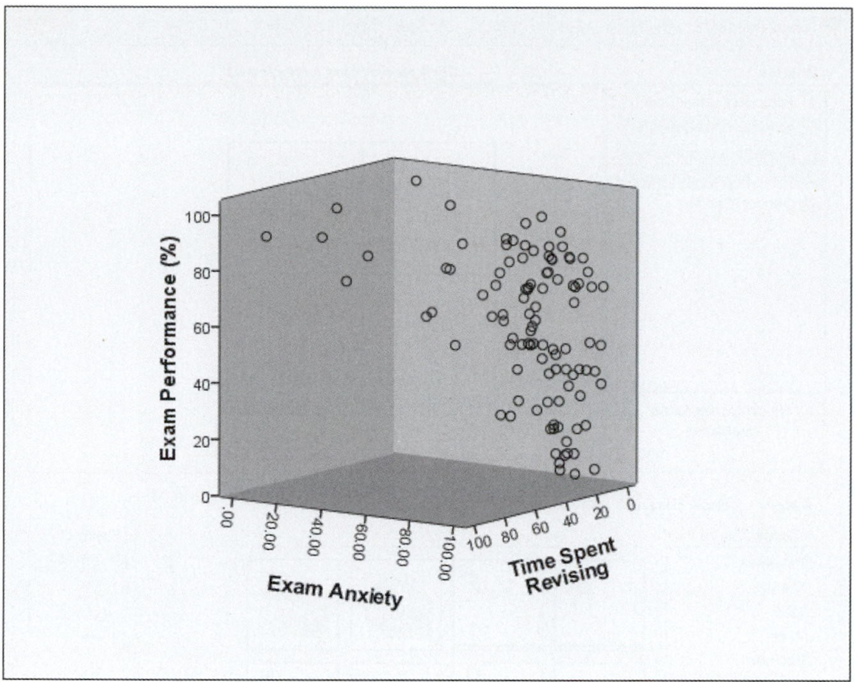

FIGURE 4.37 A 3-D scatterplot of exam performance plotted against exam anxiety and the amount of time spent revising for the exam

4.8.4. Matrix scatterplot ①

Instead of plotting several variables on the same axes on a 3-D scatterplot (which, as we have seen, can be difficult to interpret), it is possible to plot a matrix of 2-D scatterplots. This type of plot allows you to see the relationship between all combinations of many different pairs of variables. We'll use the same data set as with the other scatterplots in this chapter. First, access the Chart Builder and double-click on the icon for a scatterplot matrix (Figure 4.30). A different type of graph to what you have seen before will appear on the canvas, and it has only one drop zone (Scattermatrix?). We need to drag all of the variables that we would like to see plotted against each other into this single drop zone. We have dragged multiple variables into a drop zone in previous sections, but, to recap, we first need to select multiple items in the variable list: to do this select the first variable (**Time Spent Revising**) by clicking on it with the mouse. The variable will be highlighted in blue. Now, hold down the *Ctrl* key on the keyboard and click on a second variable (**Exam Performance %**). Both variables are now highlighted in blue. Again, hold down the *Ctrl* key and click on a third variable (**Exam Anxiety**). (We could also have simply clicked on **Time Spent Revising**, then held down the *Shift* key on the keyboard and then clicked on **Exam Anxiety**.) Once the three variables are selected, click on any one of them and then drag them into Scattermatrix? as shown in Figure 4.38. Click on OK to produce the graph in Figure 4.39.

The six scatterplots in Figure 4.39 represent the various combinations of each variable plotted against each other variable. So, the grid references represent the following plots:

- **B1**: revision time (*Y*) vs. exam performance (*X*)
- **C1**: revision time (*Y*) vs. anxiety (*X*)
- **C2**: exam performance (*Y*) vs. anxiety (*X*)
- **A2**: exam performance (*Y*) vs. revision time (*X*)
- **B3**: anxiety (*Y*) vs. exam performance (*X*)
- **A3**: anxiety (*Y*) vs. revision time (*X*)

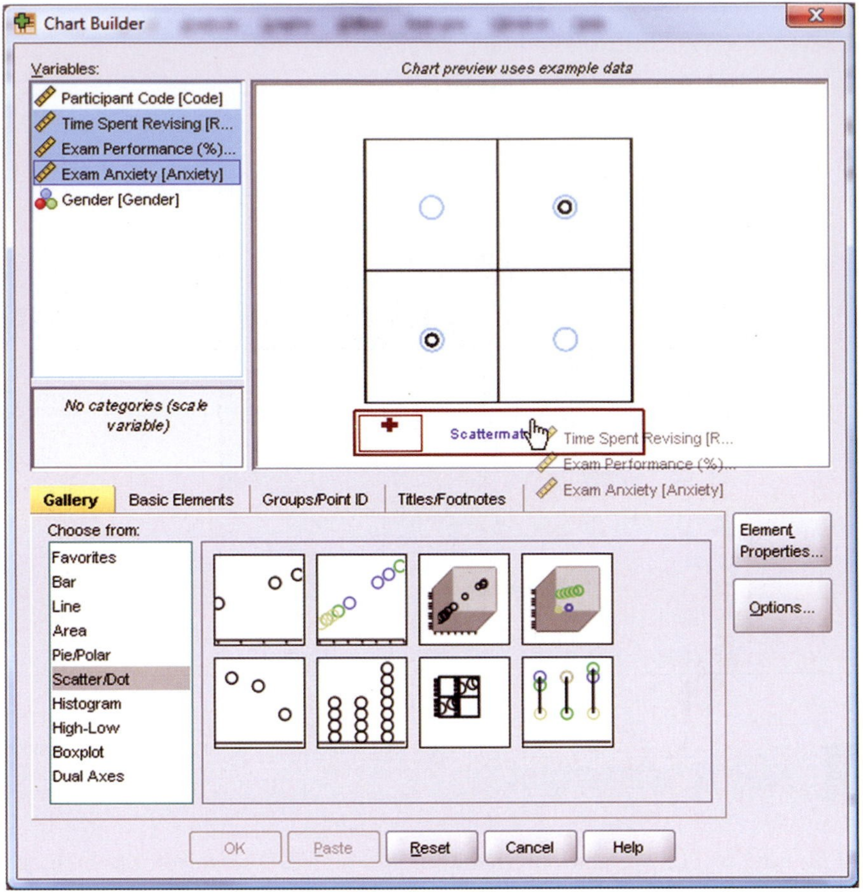

FIGURE 4.38 *Chart Builder* dialog box for a matrix scatterplot

Thus, the three scatterplots below the diagonal of the matrix are the same plots as the ones above the diagonal but with the axes reversed. From this matrix we can see that revision time and anxiety are inversely related (so, the more time spent revising the less anxiety the participant had about the exam). Also, in the scatterplot of revision time against anxiety (grids C1 and A3) there looks as though there is one possible outlier – there is a single participant who spent very little time revising yet suffered very little anxiety about the exam. As all participants who had low anxiety scored highly on the exam, we can deduce that this person also did well on the exam (don't you just hate a smart alec!). We could choose to examine this case more closely if we believed that their behaviour was caused by some external factor (such as taking brain-pills!). Matrix scatterplots are very convenient for examining pairs of relationships between variables (see SPSS Tip 4.4). However, I don't recommend plotting them for more than three or four variables because they become very confusing indeed!

SPSS TIP 4.4 Regression lines on a scatterplot matrix ①

You can add regression lines to each scatterplot in the matrix in exactly the same way as for a simple scatterplot. First, double-click on the scatterplot matrix in the SPSS Viewer to open it in the SPSS Chart Editor, then click on to open the *Properties* dialog box. Using this dialog box add a line to the graph that represents the linear model (this should be set by default). Click on Apply to apply the changes. Each panel of the matrix should now show a regression line.

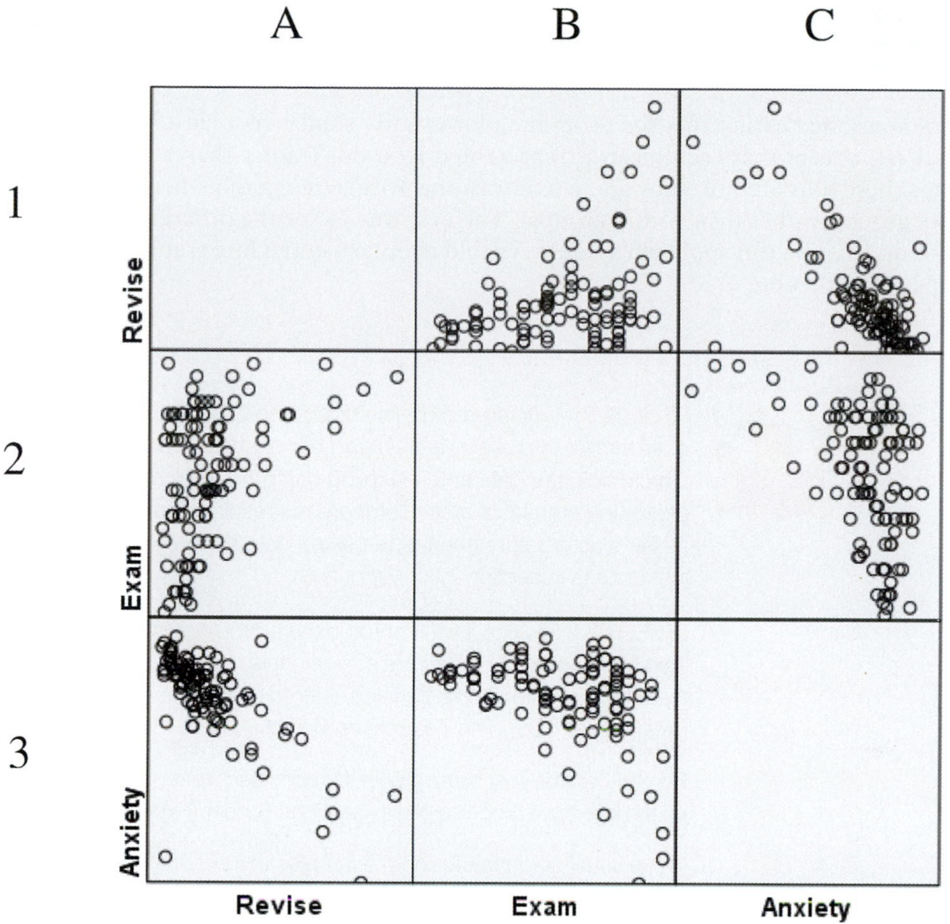

FIGURE 4.39 Matrix scatterplot of exam performance, exam anxiety and revision time. Grid references have been added for clarity

4.8.5. Simple dot plot or density plot ①

I mentioned earlier that the simple dot plot or density plot as it is also known is a histogram except that each data point is plotted (rather than using a single summary bar to show each frequency). Like a histogram, the data are still placed into bins (SPSS Tip 4.2) but a dot is used to represent each data point. As such, you should be able to follow the instructions for a histogram to draw one.

SELF-TEST Doing a simple dot plot in the Chart Builder is quite similar to drawing a histogram. Reload the **DownloadFestival.sav** data and see if you can produce a simple dot plot of the Download Festival day 1 hygiene scores. Compare the resulting graph to the earlier histogram of the same data (Figure 4.10). Remember that your starting point is to double-click on the icon for a simple dot plot in the Chart Builder (Figure 4.30). The instructions for drawing a histogram (section 4.4) might then help – if not there is full guidance in the additional material on the companion website.

4.8.6. Drop-line graph ①

I also mentioned earlier that the drop-line plot is fairly similar to a clustered bar chart (or line chart) except that each mean is represented by a dot (rather than a bar), and within groups these dots are linked by a line (contrast this with a line graph where dots are joined across groups, rather than within groups). The best way to see the difference is to plot one and to do this you can apply what you were told about clustered line graphs (section 4.6.2) to this new situation.

SELF-TEST Doing a drop-line plot in the Chart Builder is quite similar to drawing a clustered bar chart. Reload the **ChickFlick.sav** data and see if you can produce a drop-line plot of the arousal scores. Compare the resulting graph to the earlier clustered bar chart of the same data (Figure 4.20). The instructions in section 4.6.2 might help.

SELF-TEST Now see if you can produce a drop-line plot of the **Text Messages.sav** data from earlier in this chapter. Compare the resulting graph to the earlier clustered bar chart of the same data (Figure 4.28). The instructions in section 4.6.5 might help.

Remember that your starting point for both tasks is to double-click on the icon for a drop-line plot in the Chart Builder (Figure 4.30).

There is full guidance for both examples in the additional material on the companion website.

4.9. Editing graphs ①

We have already seen how to add regression lines to scatterplots (section 4.8.1). You can edit almost every aspect of the graph by double-clicking on the graph in the SPSS Viewer to open it in a new window called the **Chart Editor** (Figure 4.40). Once in the Chart Editor you can click on virtually anything that you want to change and change it. There are also many buttons that you can click on to add elements to the graph (such as grid lines, regression lines, data labels). You can change the bar colours, the axes titles, the scale of each axis and so on. You can also do things like make the bars three-dimensional. However, tempting as these tools may be (it can look quite pretty) try to remember the advice I gave at the start of this chapter when editing your graphs.

Once in the Chart Editor (Figure 4.41) there are several icons that you can click on to change aspects of the graph. Whether a particular icon is active depends on the type of chart that you are editing (e.g. the icon to fit a regression line will not work on a bar chart). The figure tells you what most of the icons do, and to be honest most of them are fairly self-explanatory (you don't need me to explain what the icon for adding a title does). I would suggest playing around with these features.

You can also edit parts of the graph by selecting them and then changing their properties. To select part of the graph simply click on it, it will become highlighted in blue and a new dialog box will appear (Figure 4.42). This *Properties* dialog box enables you to change virtually anything about the item that you have selected. Rather than spend a lot of time here showing you the various properties (there are lots) there is a tutorial in the additional website material (see Oliver Twisted).

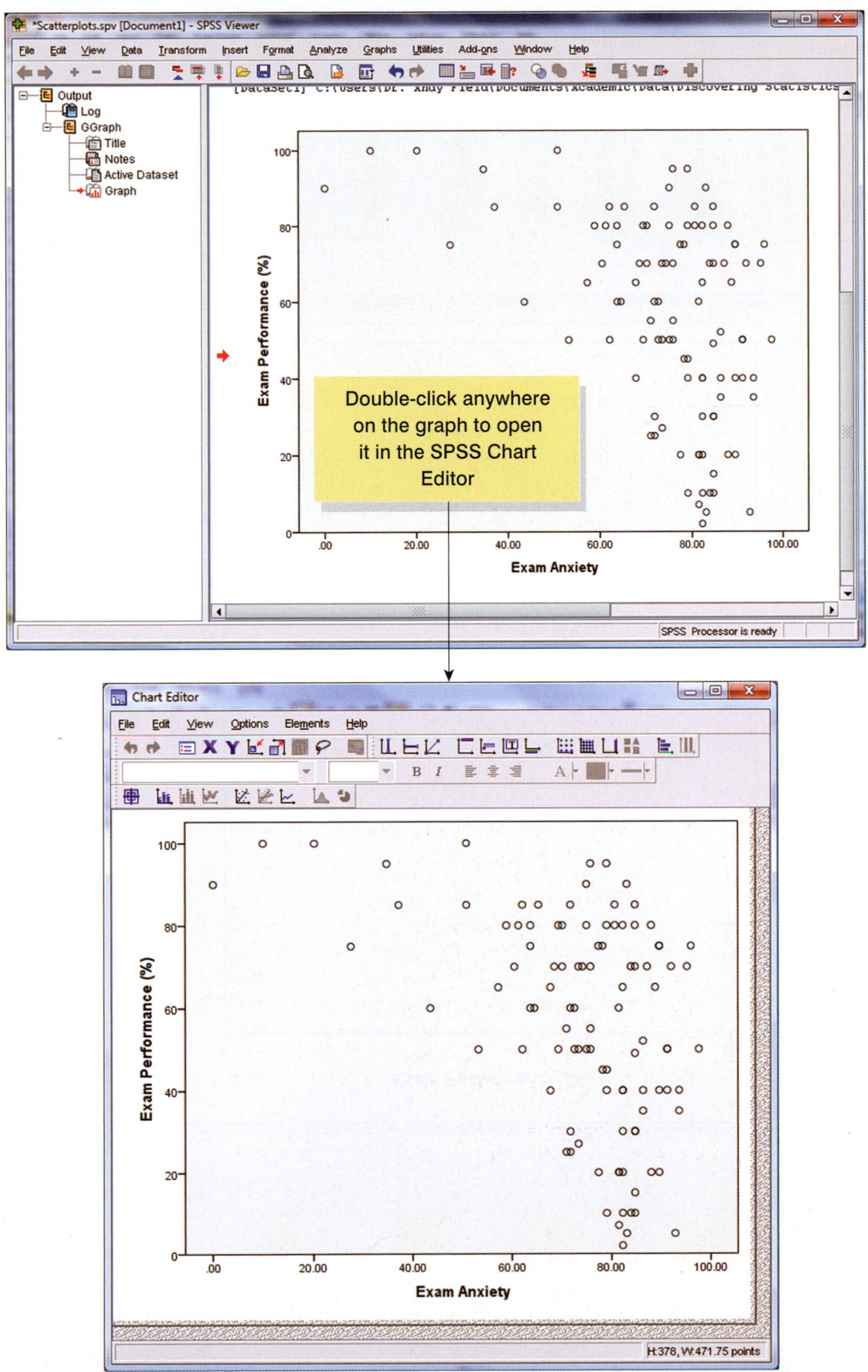

FIGURE 4.40 Opening a graph for editing in the SPSS Chart Editor

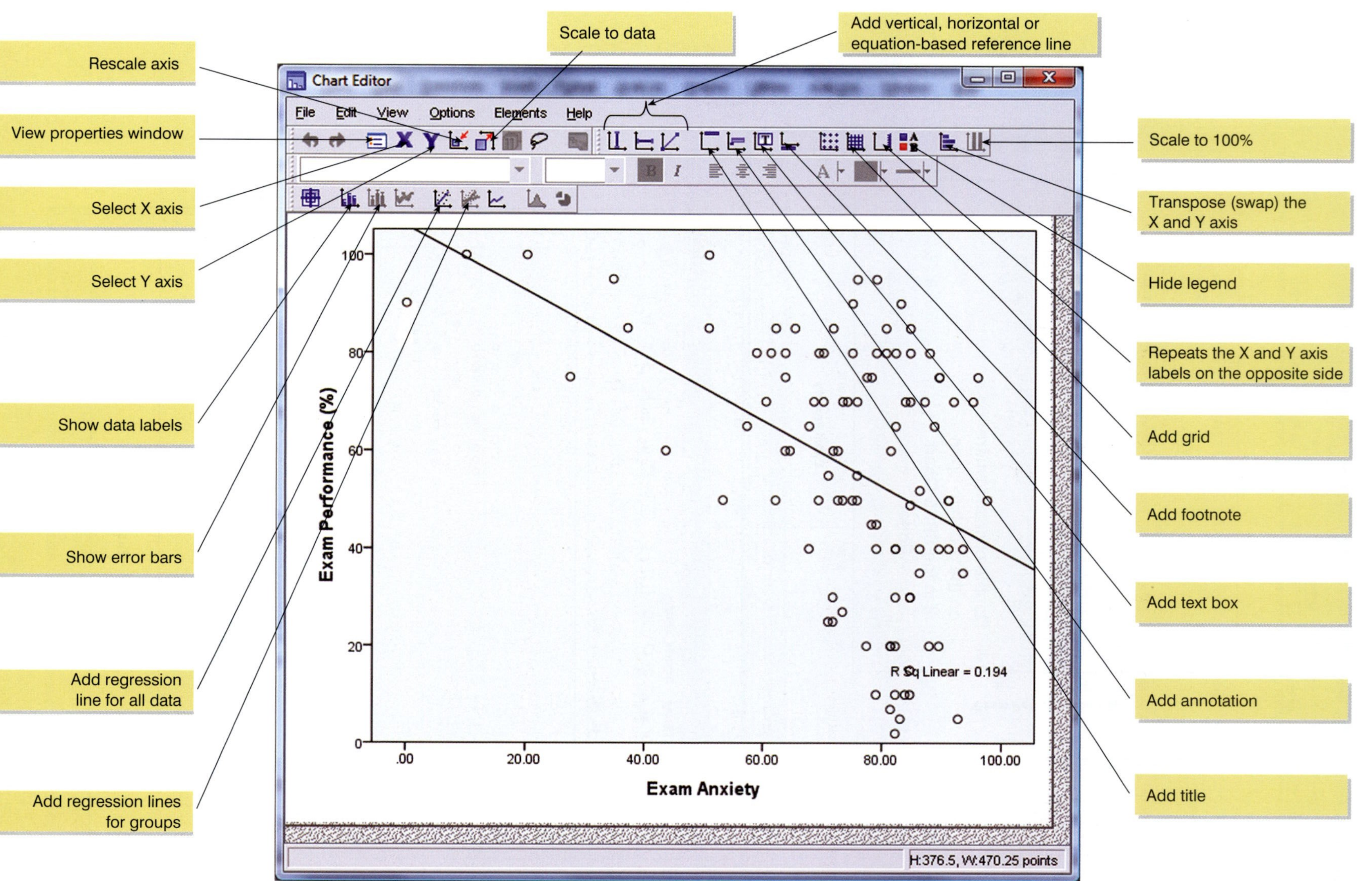

FIGURE 4.41 The Chart Editor

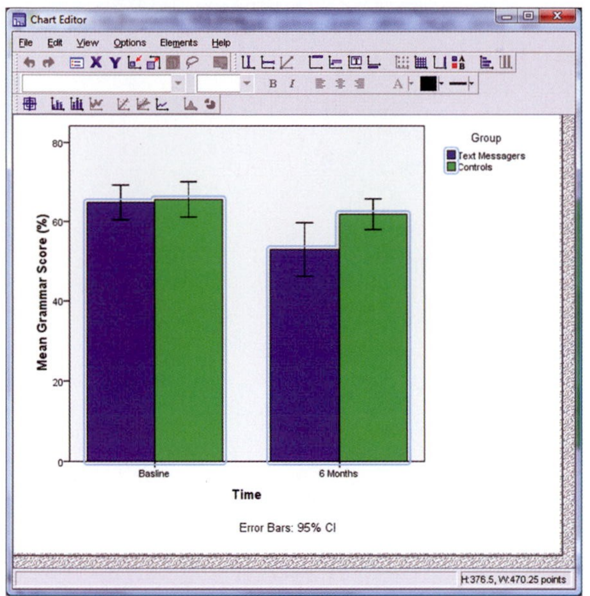

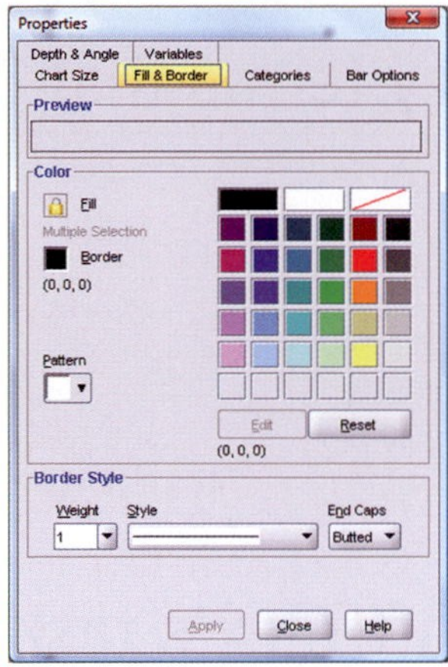

FIGURE 4.42 To select an element in the graph simply click on it and its *Properties* dialog box will appear

OLIVER TWISTED

Please, Sir, can I have some more … graphs?

'Blue and green should never be seen!', shrieks Oliver with so much force that his throat starts to hurt. 'This graph offends my delicate artistic sensibilities. It must be changed immediately!' Never fear Oliver, the editing functions on SPSS are quite a lot better than they used to be and it's possible to create some very tasteful graphs. However, these facilities are so extensive that I could probably write a whole book on them. In the interests of saving trees, I have prepared a tutorial and flash movie that is available in the additional material that can be downloaded from the companion website. We look at an example of how to edit an error bar chart to make it conform to some of the guidelines that I talked about at the beginning of this chapter. In doing so we will look at how to edit the axes, add grid lines, change the bar colours, change the background and borders. It's a very extensive tutorial!

What have I discovered about statistics? ①

This chapter has looked at how to inspect your data using graphs. We've covered a lot of different graphs. We began by covering some general advice on how to draw graphs and we can sum that up as minimal is best: no pink, no 3-D effects, no pictures of Errol your pet ferret superimposed on the graph – oh, and did I mention no pink? We have looked at graphs that tell you about the distribution of your data (histograms, boxplots and density plots), that show summary statistics about your data (bar charts, error bar charts, line charts, drop-line charts) and that show relationships between variables (scatterplots). We ended the chapter by looking at how we can edit graphs in SPSS to make them look minimal (and of course to colour them pink, but we know better than to do that, don't we?).

We also discovered that I liked to explore as a child. I was constantly dragging my dad (or was it the other way around?) over piles of rocks along any beach we happened to be on. However, at this time I also started to explore great literature, although unlike my cleverer older brother who was reading Albert Einstein's papers (well, Isaac Asimov) as an embryo, my literary preferences were more in keeping with my intellect as we shall see.

Key terms that I've discovered

Bar chart
Boxplot (box–whisker plot)
Chart Builder
Chart Editor
Chartjunk
Density plot
Error bar chart
Line chart
Outlier
Regression line
Scatterplot

Smart Alex's tasks

- **Task 1**: Using the data from Chapter 2 (which you should have saved, but if you didn't re-enter it from Table 3.1) plot and interpret the following graphs: ①
  - An error bar chart showing the mean number of friends for students and lecturers.
  - An error bar chart showing the mean alcohol consumption for students and lecturers.
  - An error line chart showing the mean income for students and lecturers.
  - An error line chart showing the mean neuroticism for students and lecturers.
  - A scatterplot with regression lines of alcohol consumption and neuroticism grouped by lecturer/student.
  - A scatterplot matrix with regression lines of alcohol consumption, neuroticism and number of friends.

- **Task 2**: Using the **Infidelity.sav** data from Chapter 3 (see Smart Alex's task) plot a clustered error bar chart of the mean number of bullets used against the self and the partner for males and females. ①

Answers can be found on the companion website.

Further reading

Tufte, E. R. (2001). *The visual display of quantitative information* (2nd ed.). Cheshire, CT: Graphics Press.

Wainer, H. (1984). How to display data badly. *American Statistician*, *38*(2), 137–147.

Wright, D. B., & Williams, S. (2003). Producing bad results sections. *The Psychologist*, *16*, 646–648. (This is a very accessible article on how to present data. It is currently available from http://www.sussex.ac.uk/Users/danw/nc/rm1graphlinks.htm or Google Dan Wright.)

http.//junkcharts.typepad.com/is an amusing look at bad graphs.

Online tutorial

The companion website contains the following Flash movie tutorial to accompany this chapter:

- Editing graphs

Interesting real research

Fesmire, F. M. (1988). Termination of intractable hiccups with digital rectal massage. *Annals of Emergency Medicine*, *17*(8), 872.

Exploring assumptions 5

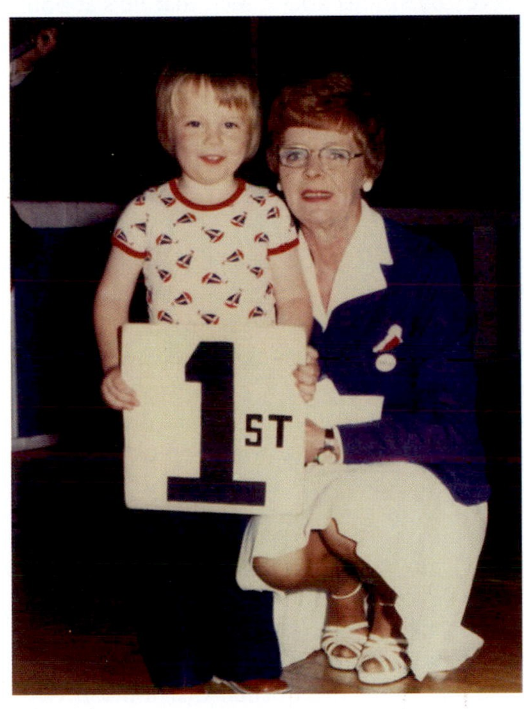

FIGURE 5.1
I came first in the competition for who has the smallest brain

5.1. What will this chapter tell me? ①

When we were learning to read at primary school, we used to read versions of stories by the famous storyteller Hans Christian Andersen. One of my favourites was the story of the ugly duckling. This duckling was a big ugly grey bird, so ugly that even a dog would not bite him. The poor duckling was ridiculed, ostracized and pecked by the other ducks. Eventually, it became too much for him and he flew to the swans, the royal birds, hoping that they would end his misery by killing him because he was so ugly. As he stared into the water, though, he saw not an ugly grey bird but a beautiful swan. Data are much the same. Sometimes they're just big, grey and ugly and don't do any of the things that they're supposed to do. When we get data like these, we swear at them, curse them, peck them and hope that they'll fly away and be killed by the swans. Alternatively, we can try to force our data into becoming beautiful swans. That's what this chapter is all about: assessing how much of an ugly duckling of a data set you have, and discovering how to turn it into a swan. Remember, though, a swan can break your arm.[1]

5.2. What are assumptions? ①

Some academics tend to regard assumptions as rather tedious things about which no one really need worry. When I mention statistical assumptions to my fellow psychologists they tend to give me that raised eyebrow, 'good grief, get a life' look and then ignore me. However, there are good reasons for taking assumptions seriously. Imagine that I go over to a friend's house, the lights are on and it's obvious that someone is at home. I ring the doorbell and no one answers. From that experience, I conclude that my friend hates me and that I am a terrible, unlovable, person. How tenable is this conclusion? Well, there is a reality that I am trying to tap (i.e. whether my friend likes or hates me), and I have collected data about that reality (I've gone to his house, seen that he's at home, rang the doorbell and got no response). Imagine that in reality my friend likes me (he never was a good judge of character!); in this scenario, my conclusion is false. Why have my data led me to the wrong conclusion? The answer is simple: I had assumed that my friend's doorbell was working and under this assumption the conclusion that I made from my data was accurate (my friend heard the bell but chose to ignore it because he hates me). However, this assumption was not true – his doorbell was not working, which is why he didn't answer the door – and as a consequence the conclusion I drew about reality was completely false.

Enough about doorbells, friends and my social life: the point to remember is that when assumptions are broken we stop being able to draw accurate conclusions about reality. Different statistical models assume different things, and if these models are going to reflect reality accurately then these assumptions need to be true. This chapter is going to deal with some particularly ubiquitous assumptions so that you know how to slay these particular beasts as we battle our way through the rest of the book. However, be warned: some tests have their own unique two-headed, fire-breathing, green-scaled assumptions and these will jump out from behind a mound of blood-soaked moss and try to eat us alive when we least expect them to. Onward into battle …

5.3. Assumptions of parametric data ①

What are the assumptions of parametric data?

Many of the statistical procedures described in this book are **parametric tests** based on the normal distribution (which is described in section 1.7.4). A parametric test is one that requires data from one of the large catalogue of distributions that statisticians have described and for data to be parametric certain assumptions must be true. If you use a parametric test when your data are not parametric then the results are likely to be inaccurate. Therefore, it is very important that you check the assumptions before deciding which statistical test is appropriate. Throughout this book you will become aware of my obsession with assumptions and checking them. Most parametric tests based on the normal distribution have four basic assumptions that must be met for the test to be accurate. Many students find checking assumptions a pretty tedious affair, and often get confused about how to tell whether or not an assumption has been met. Therefore, this chapter is designed to take you on a step-by-step tour of the world of parametric assumptions (wow, how exciting!). Now, you may think that

[1] Although it is theoretically possible, apparently you'd have to be weak boned, and swans are nice and wouldn't do that sort of thing.

assumptions are not very exciting, but they can have great benefits: for one thing you can impress your supervisor/lecturer by spotting all of the test assumptions that they have violated throughout their careers. You can then rubbish, on statistical grounds, the theories they have spent their lifetime developing – and they can't argue with you[2] – but they can poke your eyes out! The assumptions of parametric tests are:

1. **Normally distributed data**: This a tricky and misunderstood assumption because it means different things in different contexts. For this reason I will spend most of the chapter discussing this assumption! In short, the rationale behind hypothesis testing relies on having something that is normally distributed (in some cases it's the sampling distribution, in others the errors in the model) and so if this assumption is not met then the logic behind hypothesis testing is flawed (we came across these principles in Chapters 1 and 2).
2. **Homogeneity of variance**: This assumption means that the variances should be the same throughout the data. In designs in which you test several groups of participants this assumption means that each of these samples comes from populations with the same variance. In correlational designs, this assumption means that the variance of one variable should be stable at all levels of the other variable (see section 5.6).
3. **Interval data**: Data should be measured at least at the interval level. This assumption is tested by common sense and so won't be discussed further (but do reread section 1.5.1.2 to remind yourself of what we mean by interval data).
4. **Independence**: This assumption, like that of normality, is different depending on the test you're using. In some cases it means that data from different participants are independent, which means that the behaviour of one participant does not influence the behaviour of another. In repeated-measures designs (in which participants are measured in more than one experimental condition), we expect scores in the experimental conditions to be non-independent for a given participant, but behaviour between different participants should be independent. As an example, imagine two people, Paul and Julie, were participants in an experiment where they had to indicate whether they remembered having seen particular photos earlier on in the experiment. If Paul and Julie were to confer about whether they'd seen certain pictures then their answers would *not* be independent: Julie's response to a given question would depend on Paul's answer, and this would violate the assumption of independence. If Paul and Julie were unable to confer (if they were locked in different rooms) then their responses should be independent (unless they're telepathic): Paul's responses should not be influenced by Julie's. In regression, however, this assumption also relates to the errors in the regression model being uncorrelated, but we'll discuss that more in Chapter 7.

We will, therefore, focus in this chapter on the assumptions of normality and homogeneity of variance.

5.4. The assumption of normality ①

We encountered the normal distribution back in Chapter 1; we know what it looks like and we (hopefully) understand it. You'd think then that this assumption would be easy to

[2] When I was doing my Ph.D., we were set a task by our statistics lecturer in which we had to find some published papers and criticize the statistical methods in them. I chose one of my supervisor's papers and proceeded to slag off every aspect of the data analysis (and I was being *very* pedantic about it all). Imagine my horror when my supervisor came bounding down the corridor with a big grin on his face and declared that, unbeknownst to me, he was the second marker of my essay. Luckily, he had a sense of humour and I got a good mark.☺

understand – it just means that our data are normally distributed, right? Actually, no. In many statistical tests (e.g. the *t*-test) we assume that the sampling distribution is normally distributed. This is a problem because we don't have access to this distribution – we can't simply look at its shape and see whether it is normally distributed. However, we know from the central limit theorem (section 2.5.1) that if the sample data are approximately normal then the sampling distribution will be also. Therefore, people tend to look at their sample data to see if they are normally distributed. If so, then they have a little party to celebrate and assume that the sampling distribution (which is what actually matters) is also. We also know from the central limit theorem that in big samples the sampling distribution tends to be normal anyway – regardless of the shape of the data we actually collected (and remember that the sampling distribution will tend to be normal regardless of the population distribution in samples of 30 or more). As our sample gets bigger then, we can be more confident that the sampling distribution is normally distributed (but see Jane Superbrain Box 5.1).

The assumption of normality is also important in research using regression (or general linear models). General linear models, as we will see in Chapter 7, assume that errors in the model (basically, the deviations we encountered in section 2.4.2) are normally distributed.

In both cases it might be useful to test for normality and that's what this section is dedicated to explaining. Essentially, we can look for normality visually, look at values that quantify aspects of a distribution (i.e. skew and kurtosis) and compare the distribution we have to a normal distribution to see if it is different.

5.4.1. Oh no, it's that pesky frequency distribution again: checking normality visually ①

We discovered in section 1.7.1 that frequency distributions are a useful way to look at the shape of a distribution. In addition, we discovered how to plot these graphs in section 4.4. Therefore, we are already equipped to look for normality in our sample using a graph. Let's return to the Download Festival data from Chapter 4. Remember that a biologist had visited the Download Festival (a rock and heavy metal festival in the UK) and assessed people's hygiene over the three days of the festival using a standardized technique that results in a score ranging between 0 (you smell like a rotting corpse that's hiding up a skunk's anus) and 5 (you smell of sweet roses on a fresh spring day). The data file can be downloaded from the companion website (**DownloadFestival.sav**) – remember to use the version of the data for which the outlier has been corrected (if you haven't a clue what I mean then read section 4.4 or your graphs will look very different to mine!).

SELF-TEST Using what you learnt in section 4.4 plot histograms for the hygiene scores for the three days of the Download Festival.

There is another useful graph that we can inspect to see if a distribution is normal called a **P–P plot** (probability–probability plot). This graph plots the cumulative probability of a variable against the cumulative probability of a particular distribution (in this case we would specify a normal distribution). What this means is that the data are ranked and sorted. Then for each rank the corresponding *z*-score is calculated. This is the expected value that the score should have in a normal distribution. Next the score itself is converted

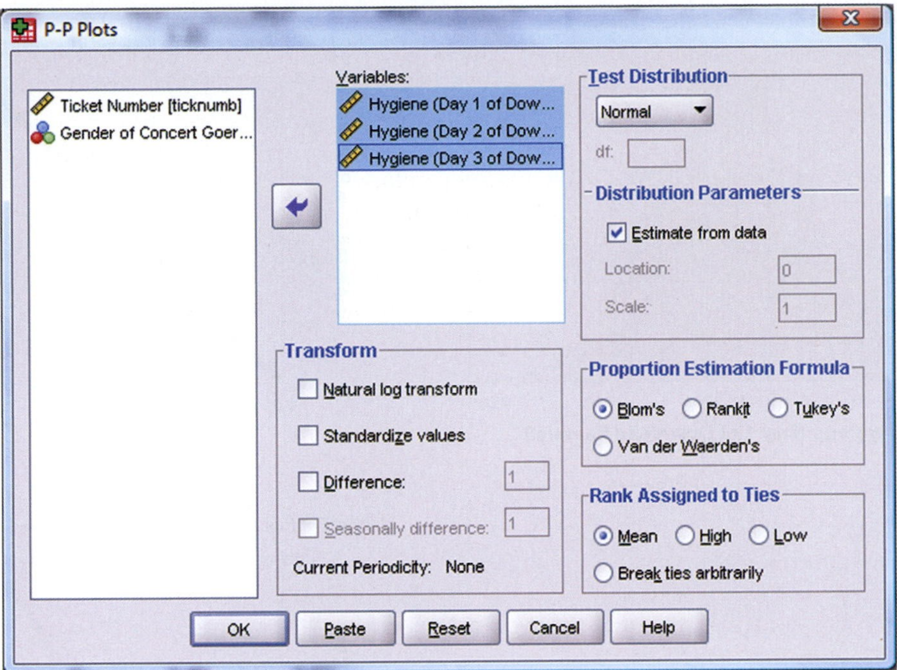

FIGURE 5.2 Dialog box for obtaining P–P plots

to a *z*-score (see section 1.7.4). The actual *z*-score is plotted against the expected *z*-score. If the data are normally distributed then the actual *z*-score will be the same as the expected *z*-score and you'll get a lovely straight diagonal line. This ideal scenario is helpfully plotted on the graph and your job is to compare the data points to this line. If values fall on the diagonal of the plot then the variable is normally distributed, but deviations from the diagonal show deviations from normality.

To get a P–P plot use Analyze ▸ Descriptive Statistics ▸ P-P Plots... to access the dialog box in Figure 5.2. There's not a lot to say about this dialog box really because the default options will compare any variables selected to a normal distribution, which is what we want (although note that there is a drop-down list of different distributions against which you could compare your data). Select the three hygiene score variables in the variable list (click on the day 1 variable, then hold down *Shift* and select the day 3 variable and the day 2 scores will be selected as well). Transfer the selected variables to the box labelled *Variables* by clicking on ➜. Click on OK to draw the graphs.

Figure 5.3 shows the histograms (from the self-test task) and the corresponding P–P plots. The first thing to note is that the data from day 1 look a lot more healthy since we've removed the data point that was mis-typed back in section 4.5. In fact the distribution is amazingly normal looking: it is nicely symmetrical and doesn't seem too pointy or flat – these are good things! This is echoed by the P–P plot: note that the data points all fall very close to the 'ideal' diagonal line.

However, the distributions for days 2 and 3 are not nearly as symmetrical. In fact, they both look positively skewed. Again, this can be seen in the P–P plots by the data values deviating away from the diagonal. In general, what this seems to suggest is that by days 2 and 3, hygiene scores were much more clustered around the low end of the scale. Remember that the lower the score, the less hygienic the person is, so this suggests that generally people became smellier as the festival progressed. The skew occurs because a substantial minority insisted on upholding their levels of hygiene (against all odds!) over the course of the festival (baby wet-wipes are indispensable I find). However, these skewed distributions might cause us a problem if we want to use parametric tests. In the next section we'll look at ways to try to quantify the skewness and kurtosis of these distributions.

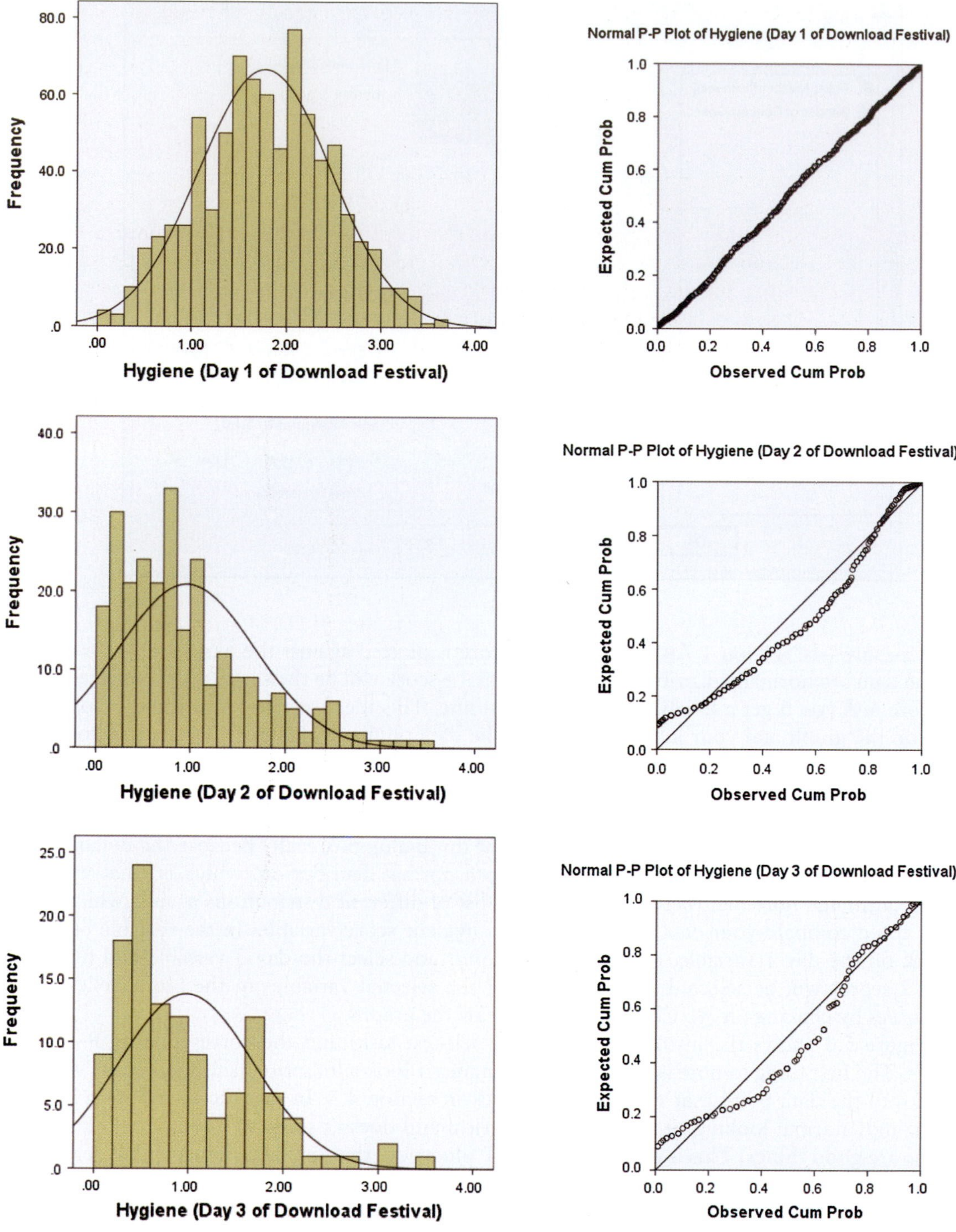

FIGURE 5.3 Histograms (left) and P–P plots (right) of the hygiene scores over the three days of the Download Festival

5.4.2. Quantifying normality with numbers ①

It is all very well to look at histograms, but they are subjective and open to abuse (I can imagine researchers sitting looking at a completely distorted distribution and saying 'yep, well Bob, that looks normal to me', and Bob replying 'yep, sure does'). Therefore, having inspected the distribution of hygiene scores visually, we can move on to look at ways to quantify the shape of the distributions and to look for outliers. To further explore the distribution of the variables, we can use the *frequencies* command (Analyze Descriptive Statistics ▸ 123 Frequencies...). The main dialog box is shown in Figure 5.4. The variables in the data editor are listed on the left-hand side, and they can be transferred to the box labelled *Variable(s)* by clicking on a variable (or highlighting several with the mouse) and then clicking on ➔. If a variable listed in the *Variable(s)* box is selected using the mouse, it can be transferred back to the variable list by clicking on the arrow button (which should now be pointing in the opposite direction). By default, SPSS produces a frequency distribution of all scores in table form. However, there are two other dialog boxes that can be selected that provide other options. The statistics dialog box is accessed by clicking on Statistics..., and the charts dialog box is accessed by clicking on Charts... .

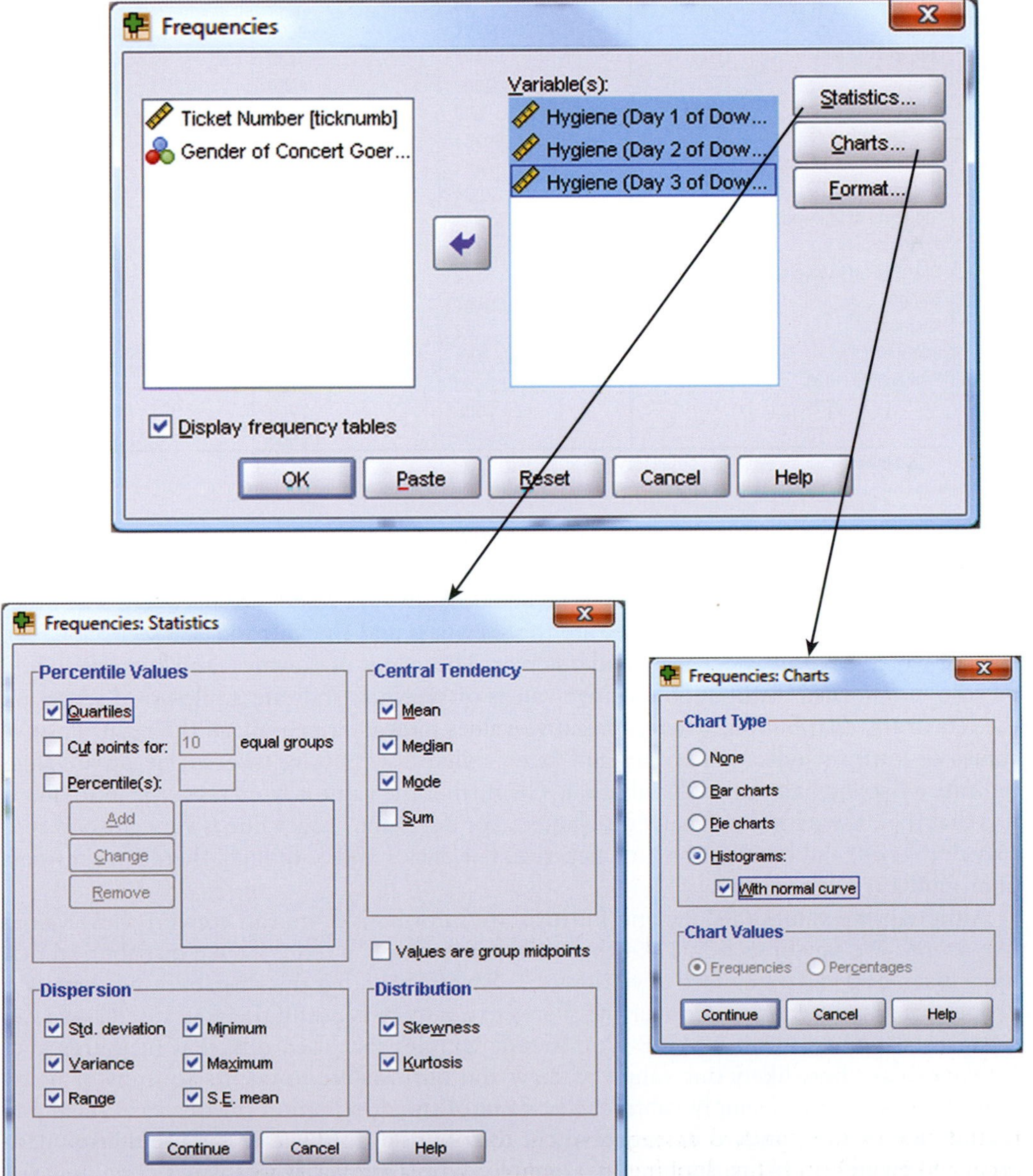

FIGURE 5.4 Dialog boxes for the *frequencies* command

The statistics dialog box allows you to select several ways in which a distribution of scores can be described, such as measures of central tendency (mean, mode, median), measures of variability (range, standard deviation, variance, quartile splits), measures of shape (kurtosis and skewness). To describe the characteristics of the data we should select the mean, mode, median, standard deviation, variance and range. To check that a distribution of scores is normal, we need to look at the values of kurtosis and skewness (see section 1.7.1). The *charts* option provides a simple way to plot the frequency distribution of scores (as a bar chart, a pie chart or a histogram). We've already plotted histograms of our data so we don't need to select these options, but you could use these options in future analyses. When you have selected the appropriate options, return to the main dialog box by clicking on Continue. Once in the main dialog box, click on OK to run the analysis.

SPSS OUTPUT 5.1

Statistics

		Hygiene (Day 1 of Download Festival)	Hygiene (Day 2 of Download Festival)	Hygiene (Day 3 of Download Festival)
N	Valid	810	264	123
	Missing	0	546	687
Mean		1.7711	.9609	.9765
Std. Error of Mean		.02437	.04436	.06404
Median		1.7900	.7900	.7600
Mode		2.00	.23	.44[a]
Std. Deviation		.69354	.72078	.71028
Variance		.481	.520	.504
Skewness		-.004	1.095	1.033
Std. Error of Skewness		.086	.150	.218
Kurtosis		-.410	.822	.732
Std. Error of Kurtosis		.172	.299	.433
Range		3.67	3.44	3.39
Minimum		.02	.00	.02
Maximum		3.69	3.44	3.41
Percentiles	25	1.3050	.4100	.4400
	50	1.7900	.7900	.7600
	75	2.2300	1.3500	1.5500

a. Multiple modes exist. The smallest value is shown

SPSS Output 5.1 shows the table of descriptive statistics for the three variables in this example. From this table, we can see that, on average, hygiene scores were 1.77 (out of 5) on day 1 of the festival, but went down to 0.96 and 0.98 on days 2 and 3 respectively. The other important measures for our purposes are the skewness and the kurtosis (see section 1.7.1), both of which have an associated standard error. The values of skewness and kurtosis should be zero in a normal distribution. Positive values of skewness indicate a pile-up of scores on the left of the distribution, whereas negative values indicate a pile-up on the right. Positive values of kurtosis indicate a pointy and heavy-tailed distribution, whereas negative values indicate a flat and light-tailed distribution. The further the value is from zero, the more likely it is that the data are not normally distributed. For day 1 the skew value is very close to zero (which is good) and kurtosis is a little negative. For days 2 and 3, though, there is a skewness of around 1 (positive skew).

Although the values of skew and kurtosis are informative, we can convert these values to *z*-scores. We saw in section 1.7.4 that a *z*-score is simply a score from a distribution that has a mean of 0 and a standard deviation of 1. We also saw that this distribution has known properties that we can use. Converting scores to a *z*-score is useful then because (1) we can compare skew and kurtosis values in different samples that used different measures, and (2) we can see how likely our values of skew and kurtosis are to occur. To transform any score to a *z*-score you simply subtract the mean of the distribution (in this case zero) and then divide by the standard deviation of the distribution (in this case we use the standard error). Skewness and kurtosis are converted to *z*-scores in exactly this way.

$$Z_{\text{skewness}} = \frac{S - 0}{SE_{\text{skewness}}} \quad Z_{\text{kurtosis}} = \frac{K - 0}{SE_{\text{kurtosis}}}$$

In the above equations, the values of S (skewness) and K (kurtosis) and their respective standard errors are produced by SPSS. These z-scores can be compared against values that you would expect to get by chance alone (i.e. known values for the normal distribution shown in the Appendix). So, an absolute value greater than 1.96 is significant at $p < .05$, above 2.58 is significant at $p < .01$ and absolute values above about 3.29 are significant at $p < .001$. Large samples will give rise to small standard errors and so when sample sizes are big, significant values arise from even small deviations from normality. In smallish samples it's OK to look for values above 1.96; however, in large samples this criterion should be increased to the 2.58 one and in very large samples, because of the problem of small standard errors that I've described, no criterion should be applied! If you have a large sample (200 or more) it is more important to look at the shape of the distribution visually and to look at the value of the skewness and kurtosis statistics rather than calculate their significance.

For the hygiene scores, the z-score of skewness is $-0.004/0.086 = 0.047$ on day 1, $1.095/0.150 = 7.300$ on day 2 and $1.033/0.218 = 4.739$ on day 3. It is pretty clear then that although on day 1 scores are not at all skewed, on days 2 and 3 there is a very significant positive skew (as was evident from the histogram) – however, bear in mind what I just said about large samples! The kurtosis z-scores are: $-0.410/0.172 = -2.38$ on day 1, $0.822/0.299 = 2.75$ on day 2 and $0.732/0.433 = 1.69$ on day 3. These values indicate significant kurtosis (at $p < .05$) for all three days; however, because of the large sample, this isn't surprising and so we can take comfort in the fact that all values of kurtosis are below our upper threshold of 3.29.

CRAMMING SAM'S TIPS Skewness and kurtosis

- To check that the distribution of scores is approximately normal, we need to look at the values of *skewness* and *kurtosis* in the SPSS output.
- Positive values of skewness indicate too many low scores in the distribution, whereas negative values indicate a build-up of high scores.
- Positive values of kurtosis indicate a pointy and heavy-tailed distribution, whereas negative values indicate a flat and light-tailed distribution.
- The further the value is from zero, the more likely it is that the data are not normally distributed.
- You can convert these scores to z-scores by dividing by their standard error. If the resulting score (when you ignore the minus sign) is greater than 1.96 then it is significant ($p < .05$).
- Significance tests of skew and kurtosis should not be used in large samples (because they are likely to be significant even when skew and kurtosis are not too different from normal).

OLIVER TWISTED

Please, Sir, can I have some more … frequencies?

In your SPSS output you will also see tabulated frequency distributions of each variable. This table is reproduced in the additional online material along with a description.

5.4.3. Exploring groups of data ①

Can I analyse groups of data?

Sometimes we have data in which there are different groups of people (men and women, different universities, people with depression and people without, for example). There are several ways to produce basic descriptive statistics for separate groups of people (and we will come across some of these methods in section 5.5.1). However, I intend to use this opportunity to introduce you to the *split file* function. This function allows you to specify a grouping variable (remember, these variables are used to specify categories of cases). Any subsequent procedure in SPSS is then carried out on each category of cases separately.

You're probably getting sick of the hygiene data from the Download Festival so let's use the data in the file **SPSSExam.sav**. This file contains data regarding students' performance on an SPSS exam. Four variables were measured: **exam** (first-year SPSS exam scores as a percentage), **computer** (measure of computer literacy in percent), **lecture** (percentage of SPSS lectures attended) and **numeracy** (a measure of numerical ability out of 15). There is a variable called **uni** indicating whether the student attended Sussex University (where I work) or Duncetown University. To begin with, open the file **SPSSExam.sav** (see section 3.9). Let's begin by looking at the data as a whole.

5.4.3.1. Running the analysis for all data ①

To see the distribution of the variables, we can use the *frequencies* command, which we came across in the previous section (see Figure 5.4). Use this dialog box and place all four variables (**exam**, **computer**, **lecture** and **numeracy**) in the *Variable(s)* box. Then click on Statistics... to select the statistics dialog box and select some measures of central tendency (mean, mode, median), measures of variability (range, standard deviation, variance, quartile splits) and measures of shape (kurtosis and skewness). Also click on Charts... to access the charts dialog box and select a frequency distribution of scores with a normal curve (see Figure 5.4 if you need any help with any of these options). Return to the main dialog box by clicking on Continue and once in the main dialog box, click on OK to run the analysis.

SPSS Output 5.2 shows the table of descriptive statistics for the four variables in this example. From this table, we can see that, on average, students attended nearly 60% of lectures, obtained 58% in their SPSS exam, scored only 51% on the computer literacy test, and only 5 out of 15 on the numeracy test. In addition, the standard deviation for computer literacy was relatively small compared to that of the percentage of lectures attended and exam scores. These latter two variables had several modes (multimodal). The other important measures are the skewness and the kurtosis, both of which have an associated standard error. We came across these measures earlier on and found that we can convert these values to *z*-scores by dividing by their standard errors. For the SPSS exam scores, the *z*-score of skewness is $-0.107/0.241 = -0.44$. For numeracy, the *z*-score of skewness is $0.961/0.241 = 3.99$. It is pretty clear then that the numeracy scores are significantly positively skewed ($p < .05$) because the *z*-score is greater than 1.96, indicating a pile-up of scores on the left of the distribution (so, most students got low scores).

SELF-TEST Calculate and interpret the *z*-scores for skewness of the other variables (computer literacy and percentage of lectures attended).

SELF-TEST Calculate and interpret the *z*-scores for kurtosis of all of the variables.

The output provides tabulated frequency distributions of each variable (not reproduced here). These tables list each score and the number of times that it is found within the data set. In addition, each frequency value is expressed as a percentage of the sample (in this case the frequencies and percentages are the same because the sample size was 100). Also, the cumulative percentage is given, which tells us how many cases (as a percentage) fell below a certain score. So, for example, we can see that 66% of numeracy scores were 5 or less, 74% were 6 or less, and so on. Looking in the other direction, we can work out that only 8% (100 – 92%) got scores greater than 8.

SPSS OUTPUT 5.2

Statistics

		Percentage on SPSS exam	Computer literacy	Percentage of lectures attended	Numeracy
N	Valid	100	100	100	100
	Missing	0	0	0	0
Mean		58.10	50.71	59.765	4.85
Std. Error of Mean		2.132	.826	2.1685	.271
Median		60.00	51.50	62.000	4.00
Mode		72[a]	54	48.5[a]	4
Std. Deviation		21.316	8.260	21.6848	2.706
Variance		454.354	68.228	470.230	7.321
Skewness		-.107	-.174	-.422	.961
Std. Error of Skewness		.241	.241	.241	.241
Kurtosis		-1.105	.364	-.179	.946
Std. Error of Kurtosis		.478	.478	.478	.478
Range		84	46	92.0	13
Minimum		15	27	8.0	1
Maximum		99	73	100.0	14

a. Multiple modes exist. The smallest value is shown

Finally, we are given histograms of each variable with the normal distribution overlaid. These graphs are displayed in Figure 5.5 and show us several things. The exam scores are very interesting because this distribution is quite clearly not normal; in fact, it looks suspiciously bimodal (there are two peaks indicative of two modes). This observation corresponds with the earlier information from the table of descriptive statistics. It looks as though computer literacy is fairly normally distributed (a few people are very good with computers and a few are very bad, but the majority of people have a similar degree of knowledge) as is the lecture attendance. Finally, the numeracy test has produced very positively skewed data (i.e. the majority of people did very badly on this test and only a few did well). This corresponds to what the skewness statistic indicated.

Descriptive statistics and histograms are a good way of getting an instant picture of the distribution of your data. This snapshot can be very useful: for example, the bimodal distribution of SPSS exam scores instantly indicates a trend that students are typically either very good at statistics or struggle with it (there are relatively few who fall in between these extremes). Intuitively, this finding fits with the nature of the subject: statistics is very easy once everything falls into place, but before that enlightenment occurs it all seems hopelessly difficult!

5.4.3.2. Running the analysis for different groups ①

If we want to obtain separate descriptive statistics for each of the universities, we can split the file, and then proceed using the *frequencies* command described in the previous section. To split the file, select Data Split File... or click on . In the resulting dialog box (Figure 5.6) select the option *Organize output by groups*. Once this option is selected, the *Groups Based on* box will activate. Select the variable containing the group codes by which you wish to repeat the analysis (in this example select **Uni**), and drag it to the box or click on . By default, SPSS will sort the file by these groups (i.e. it will list one category followed by the other in the data editor window). Once you have split the file, use the *frequencies* command (see the previous section). Let's request statistics for only **numeracy** and **exam** scores for the time being.

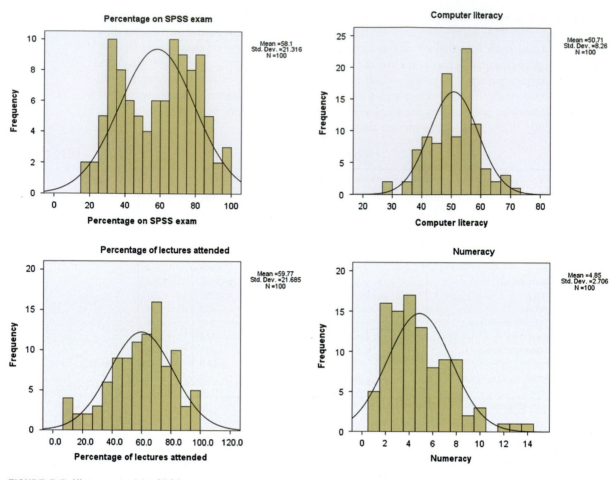

FIGURE 5.5 Histograms of the SPSS exam data

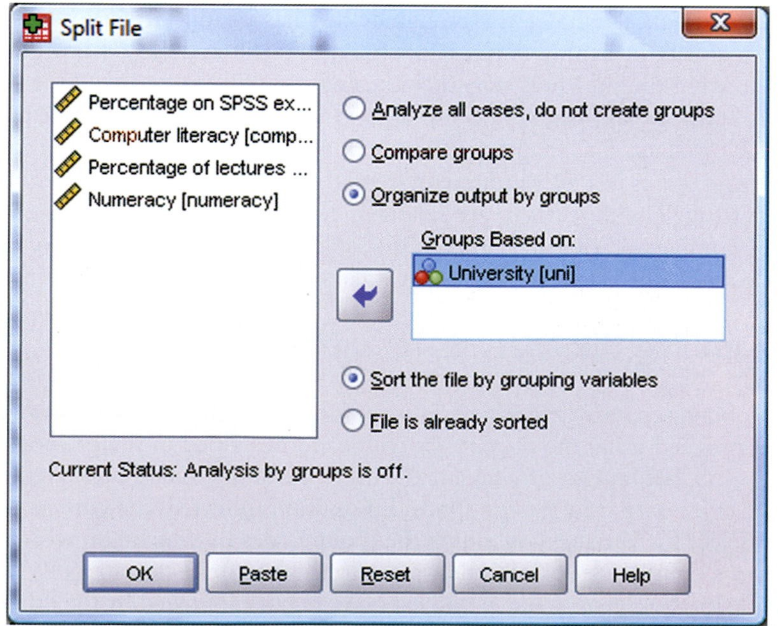

FIGURE 5.6 Dialog box for the *split file* command

SPSS OUTPUT 5.3

Duncetown University

Statistics[b]

		Percentage on SPSS exam	Numeracy
N	Valid	50	50
	Missing	0	0
Mean		40.18	4.12
Std. Error of Mean		1.780	.292
Median		38.00	4.00
Mode		34[a]	4
Std. Deviation		12.589	2.067
Variance		158.477	4.271
Skewness		.309	.512
Std. Error of Skewness		.337	.337
Kurtosis		-.567	-.484
Std. Error of Kurtosis		.662	.662
Range		51	8
Minimum		15	1
Maximum		66	9

a. Multiple modes exist. The smallest value is shown

b. University = Duncetown University

Sussex University

Statistics[b]

		Percentage on SPSS exam	Numeracy
N	Valid	50	50
	Missing	0	0
Mean		76.02	5.58
Std. Error of Mean		1.443	.434
Median		75.00	5.00
Mode		72[a]	5
Std. Deviation		10.205	3.071
Variance		104.142	9.432
Skewness		.272	.793
Std. Error of Skewness		.337	.337
Kurtosis		-.264	.260
Std. Error of Kurtosis		.662	.662
Range		43	13
Minimum		56	1
Maximum		99	14

a. Multiple modes exist. The smallest value is shown

b. University = Sussex University

The SPSS output is split into two sections: first the results for students at Duncetown University, then the results for those attending Sussex University. SPSS Output 5.3 shows the two main summary tables. From these tables it is clear that Sussex students scored higher on both their SPSS exam and the numeracy test than their Duncetown counterparts. In fact, looking at the means reveals that, on average, Sussex students scored an amazing 36% more on the SPSS exam than Duncetown students, and had higher numeracy scores too (what can I say, my students are the best).

Figure 5.7 shows the histograms of these variables split according to the university attended. The first interesting thing to note is that for exam marks, the distributions are both fairly normal. This seems odd because the overall distribution was bimodal. However, it starts to make sense when you consider that for Duncetown the distribution is centred around a mark of about 40%, but for Sussex the distribution is centred around a mark of about 76%. This illustrates how important it is to look at distributions within groups. If we were interested in comparing Duncetown to Sussex it wouldn't matter that overall the distribution of scores was bimodal; all that's important is that each group comes from a normal distribution, and in this case it appears to be true. When the two samples are combined, these two normal distributions create a bimodal one (one of the modes being around the centre of the Duncetown distribution, and the other being around the centre of the Sussex data!). For numeracy scores, the distribution is slightly positively skewed (there is a larger concentration at the lower end of scores) in both the Duncetown and Sussex groups. Therefore, the overall positive skew observed before is due to the mixture of universities (the Duncetown students contaminate Sussex's normally distributed scores!). When you have finished with the *split file* command, remember to *switch it off* (otherwise SPSS will carry on doing every analysis on each group separately). To switch this function off, return to the *Split File* dialog box (Figure 5.6) and select *Analyze all cases, do not create groups*.

SELF-TEST Repeat these analyses for the computer literacy and percentage of lectures attended and interpret the results.

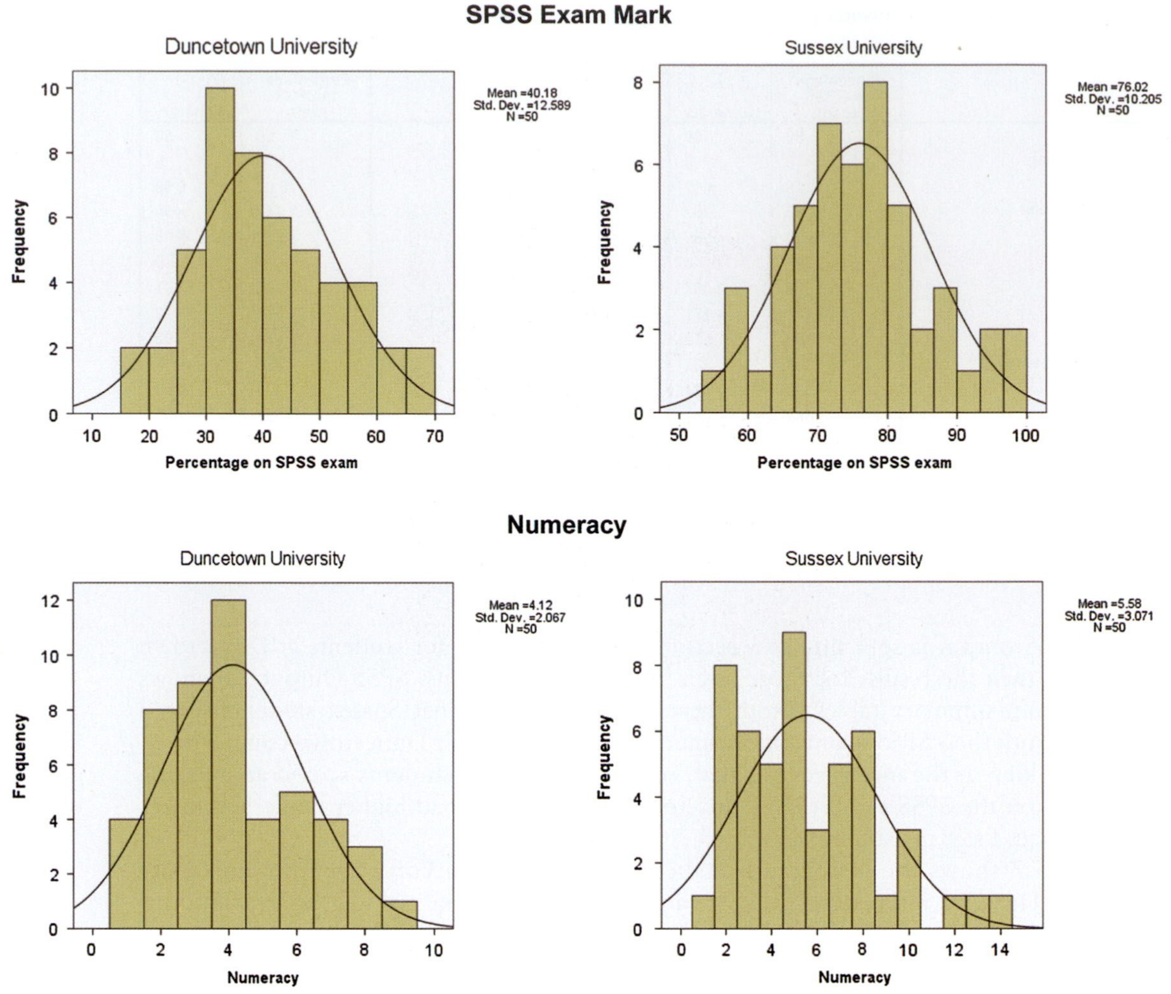

FIGURE 5.7 Distributions of exam and numeracy scores for Duncetown University and Sussex University students

5.5. Testing whether a distribution is normal ①

Did someone say Smirnov? Great, I need a drink after all this data analysis!

Another way of looking at the problem is to see whether the distribution as a whole deviates from a comparable normal distribution. The **Kolmogorov–Smirnov test** and **Shapiro–Wilk test** do just this: they compare the scores in the sample to a normally distributed set of scores with the same mean and standard deviation. If the test is non-significant ($p > .05$) it tells us that the distribution of the sample is not significantly different from a normal distribution (i.e. it is probably normal). If, however, the test is significant ($p < .05$) then the distribution in question is significantly different from a normal distribution (i.e. it is non-normal). These tests seem great: in one easy procedure they tell us whether our scores are normally distributed (nice!). However, they have their limitations because with large sample sizes it is very easy to get significant results from small deviations from normality, and so a significant test doesn't necessarily tell us whether the deviation from normality is enough to bias any statistical procedures that we apply to the data. I guess the take-home message is: by all means use these tests, but plot your data as well and try to make an informed decision about the extent of non-normality.

FIGURE 5.8 Andrei Kolmogorov, wishing he had a Smirnov

5.5.1. Doing the Kolmogorov–Smirnov test on SPSS ①

The Kolmogorov–Smirnov (K–S from now on; Figure 5.8) test can be accessed through the *explore* command (Analyze Descriptive Statistics ▸ Explore...). Figure 5.9 shows the dialog boxes for the *explore* command. First, enter any variables of interest in the box labelled *Dependent List* by highlighting them on the left-hand side and transferring them by clicking on ➜. For this example, just select the exam scores and numeracy scores. It is also possible to select a factor (or grouping variable) by which to split the output (so, if you select **Uni** and transfer it to the box labelled *Factor List*, SPSS will produce exploratory analysis for each group – a bit like the *split file* command). If you click on Statistics... a dialog box appears, but the default option is fine (it will produce means, standard deviations and so on). The more interesting option for our current purposes is accessed by clicking on Plots... . In this dialog box select the option ☑ Normality plots with tests, and this will produce both the K–S test and some graphs called *normal Q–Q plots*. A **Q–Q plot** is very similar to the P-P plot that we encountered in section 5.4.1 except that it plots the **quantiles** of the data set instead of every individual score in the data. Quantiles are just values that split a data set into equal portions. We have already used quantiles without knowing it because quartiles (as in the interquartile range in section 1.7.3) are a special case of quantiles that split the data into four equal parts. However, you can have other quantiles such as **percentiles** (points that split the data into 100 equal parts), **noniles** (points that split the data into nine equal parts) and so on. In short, then, the Q–Q plot can be interpreted in the same way as a P–P plot but it will have less points on it because rather than plotting every single data point it plots only values that divide the data into equal parts (so, they can be easier to interpret if you have a lot of scores). By default, SPSS will produce boxplots (split according to group if a factor has been specified) and stem and leaf diagrams as well. Click on Continue to return to the main dialog box and then click on OK to run the analysis.

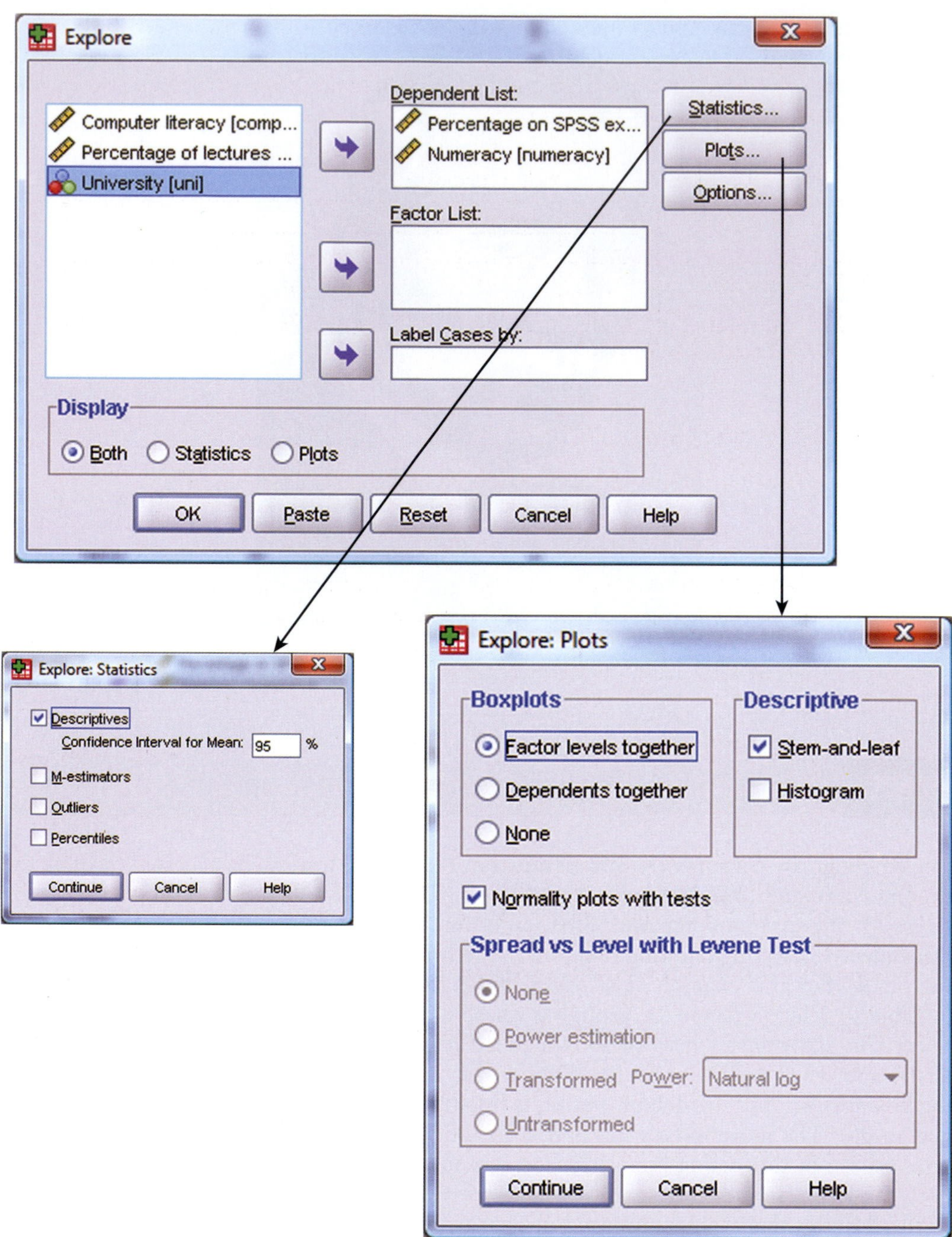

FIGURE 5.9 Dialog boxes for the *explore* command

5.5.2. Output from the explore procedure ①

The first table produced by SPSS contains descriptive statistics (mean etc.) and should have the same values as the tables obtained using the frequencies procedure. The important table is that of the K–S test (SPSS Output 5.4). This table includes the test statistic itself, the degrees of freedom (which should equal the sample size) and the significance value of this test. Remember that a significant value (*Sig.* less than .05) indicates a deviation from normality. For both numeracy and SPSS exam scores, the K–S test is highly significant, indicating that both distributions are not normal. This result is likely to reflect the bimodal distribution found for exam scores, and the positively skewed distribution observed in the numeracy scores. However, these tests confirm that these deviations were *significant*. (But bear in mind that the sample is fairly big.)

SPSS OUTPUT 5.4

Tests of Normality

	Kolmogorov-Smirnov[a]			Shapiro-Wilk		
	Statistic	df	Sig.	Statistic	df	Sig.
Percentage on SPSS exam	.102	100	.012	.961	100	.005
Numeracy	.153	100	.000	.924	100	.000

a. Lilliefors Significance Correction

OLIVER TWISTED

Please, Sir, can I have some more … normality tests?

'There is another test reported in the table (the Shapiro–Wilk test)', whispers Oliver as he creeps up behind you, knife in hand, 'and a footnote saying that "Lilliefors' significance correction" has been applied. What the hell is going on?'. (If you do the K–S test through the non-parametric test menu rather than the explore menu this correction is not applied.) Well, Oliver, all will be revealed in the additional material for this chapter on the companion website: you can find out more about the K–S test, and information about the Lilliefor correction and Shapiro–Wilk test. What are you waiting for?

As a final point, bear in mind that when we looked at the exam scores for separate groups, the distributions seemed quite normal; now if we'd asked for separate tests for the two universities (by placing **Uni** in the box labelled *Factor List* as in Figure 5.9) the K–S test might not have been significant. In fact if you try this out, you'll get the table in SPSS Output 5.5, which shows that the percentages on the SPSS exam are indeed normal within the two groups (the values in the *Sig.* column are greater than .05). This is important because if our analysis involves comparing groups, then what's important is not the overall distribution but the distribution in each group.

SPSS OUTPUT 5.5

Tests of Normality

		Kolmogorov-Smirnov[a]			Shapiro-Wilk		
	University	Statistic	df	Sig.	Statistic	df	Sig.
Percentage on SPSS exam	Duncetown University	.106	50	.200*	.972	50	.283
	Sussex University	.073	50	.200*	.984	50	.715
Numeracy	Duncetown University	.183	50	.000	.941	50	.015
	Sussex University	.155	50	.004	.932	50	.007

*. This is a lower bound of the true significance.

a. Lilliefors Significance Correction

SPSS also produces a normal Q–Q plot for any variables specified (see Figure 5.10). The normal Q–Q chart plots the values you would expect to get if the distribution were normal (expected values) against the values actually seen in the data set (observed values). The expected values are a straight diagonal line, whereas the observed values are plotted as individual points. If the data are normally distributed, then the observed values (the dots on the chart) should fall exactly along the straight line (meaning that the observed values are the same as you would expect to get from a normally distributed data set). Any

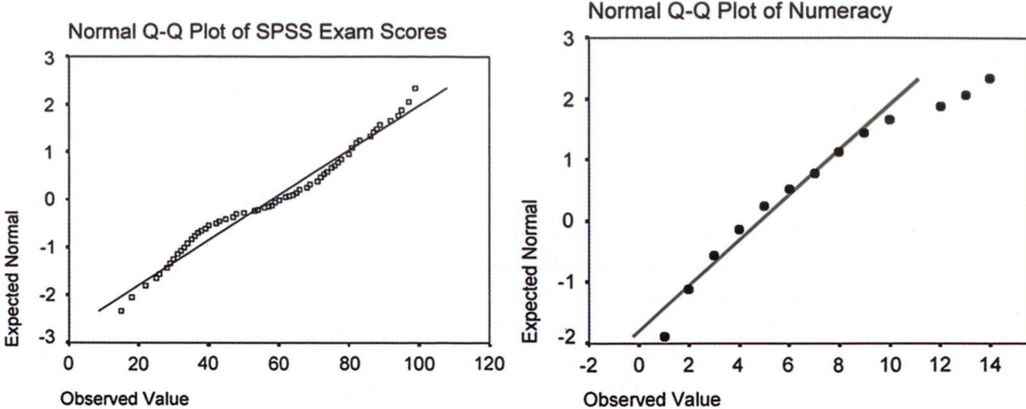

FIGURE 5.10 Normal Q–Q plots of numeracy and SPSS exam scores

deviation of the dots from the line represents a deviation from normality. So, if the Q–Q plot looks like a straight line with a wiggly snake wrapped around it then you have some deviation from normality! Specifically, when the line sags consistently below the diagonal, or consistently rises above it, then this shows that the kurtosis differs from a normal distribution, and when the curve is S-shaped, the problem is skewness.

In both of the variables analysed we already know that the data are not normal, and these plots confirm this observation because the dots deviate substantially from the line. It is noteworthy that the deviation is greater for the numeracy scores, and this is consistent with the higher significance value of this variable on the K–S test.

5.5.3. Reporting the K–S test ①

The test statistic for the K–S test is denoted by D and we must also report the degrees of freedom (df) from the table in brackets after the D. We can report the results in SPSS Output 5.4 in the following way:

✓ The percentage on the SPSS exam, $D(100) = 0.10$, $p < .05$, and the numeracy scores, $D(100) = 0.15$, $p < .001$, were both significantly non-normal.

CRAMMING SAM'S TIPS Normality tests

- The K–S test can be used to see if a distribution of scores significantly differs from a normal distribution.
- If the K–S test is significant (*Sig.* in the SPSS table is less than .05) then the scores are significantly different from a normal distribution.
- Otherwise, scores are approximately normally distributed.
- The Shapiro–Wilk test does much the same thing, but it has more power to detect differences from normality (so, you might find this test is significant when the K–S test is not).
- **Warning**: In large samples these tests can be significant even when the scores are only slightly different from a normal distribution. Therefore, they should always be interpreted in conjunction with histograms, P–P or Q–Q plots, and the values of skew and kurtosis.

5.6. Testing for homogeneity of variance ①

So far I've concentrated on the assumption of normally distributed data; however, at the beginning of this chapter I mentioned another assumption: homogeneity of variance. This assumption means that as you go through levels of one variable, the variance of the other should not change. If you've collected groups of data then this means that the variance of your outcome variable or variables should be the same in each of these groups. If you've collected continuous data (such as in correlational designs), this assumption means that the variance of one variable should be stable at all levels of the other variable. Let's illustrate this with an example. An audiologist was interested in the effects of loud concerts on people's hearing. So, she decided to send 10 people on tour with the loudest band she could find, Motörhead. These people went to concerts in Brixton (London), Brighton, Bristol, Edinburgh, Newcastle, Cardiff and Dublin and after each concert the audiologist measured the number of hours after the concert that these people had ringing in their ears.

Figure 5.11 shows the number of hours that each person had ringing in their ears after each concert (each person is represented by a circle). The horizontal lines represent the average number of hours that there was ringing in the ears after each concert and these means are connected by a line so that we can see the general trend of the data. Remember that for each concert, the circles are the scores from which the mean is calculated. Now, we can see in both graphs that the means increase as the people go to more concerts. So, after the first concert their ears ring for about 12 hours, but after the second they ring for about 15–20 hours, and by the final night of the tour, they ring for about 45–50 hours (2 days). So, there is a cumulative effect of the concerts on ringing in the ears. This pattern is found in both graphs; the difference between the graphs is not in terms of the means (which are roughly the same), but in terms of the spread of scores around the mean. If you look at the left-hand graph the spread of scores around the mean stays the same after each concert (the scores are fairly tightly packed around the mean). Put another way, if you measured the vertical distance between the lowest score and the highest score after the Brixton concert, and then did the same after the other concerts, all of these distances would be fairly similar. Although the means increase, the spread of scores for hearing loss is the same at each level of the concert variable (the spread of scores is the same after Brixton, Brighton, Bristol, Edinburgh, Newcastle, Cardiff and Dublin). This is what we mean by *homogeneity of variance*. The right-hand graph shows a different picture: if you look at the spread of scores after the Brixton concert, they are quite tightly packed around the mean (the vertical distance from

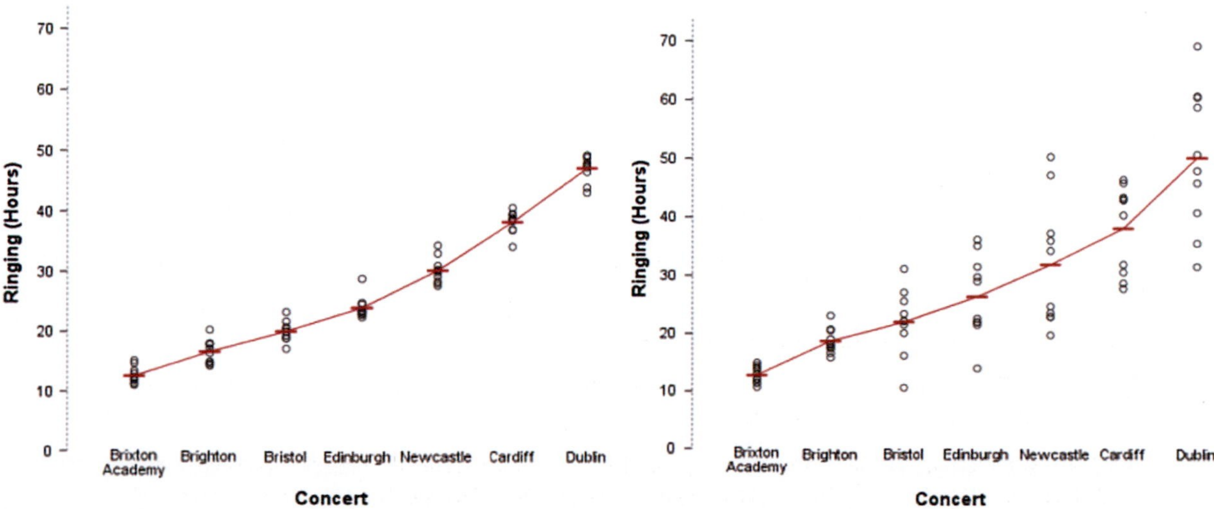

FIGURE 5.11 Graphs illustrating data with homogeneous (left) and heterogeneous (right) variances

the lowest score to the highest score is small), but after the Dublin show (for example) the scores are very spread out around the mean (the vertical distance from the lowest score to the highest score is large). This is an example of *heterogeneity of variance*: that is, at some levels of the concert variable the variance of scores is different to other levels (graphically, the vertical distance from the lowest to highest score is different after different concerts).

5.6.1. Levene's test ①

Hopefully you've got a grip of what homogeneity of variance actually means. Now, how do we test for it? Well, we could just look at the values of the variances and see whether they are similar. However, this approach would be very subjective and probably prone to academics thinking 'Ooh look, the variance in one group is only 3000 times larger than the variance in the other: that's roughly equal'. Instead, in correlational analysis such as regression we tend to use graphs (see section 7.8.7) and for groups of data we tend to use a test called **Levene's test** (Levene, 1960). Levene's test tests the null hypothesis that the variances in different groups are equal (i.e. the difference between the variances is zero). It's a very simple and elegant test that works by doing a one-way ANOVA (see Chapter 10) conducted on the deviation scores; that is, the absolute difference between each score and the mean of the group from which it came (see Glass, 1966, for a very readable explanation).[3] For now, all we need to know is that if Levene's test is significant at $p \leq .05$ then we can conclude that the null hypothesis is incorrect and that the variances are significantly different – therefore, the assumption of homogeneity of variances has been violated. If, however, Levene's test is non-significant (i.e. $p > .05$) then the variances are roughly equal and the assumption is tenable. Although Levene's test can be selected as an option in many of the statistical tests that require it, it can also be examined when you're exploring data (and strictly speaking it's better to examine Levene's test now than wait until your main analysis).

As with the K–S test (and other tests of normality), when the sample size is large, small differences in group variances can produce a Levene's test that is significant (because, as we saw in Chapter 1, the power of the test is improved). A useful double check, therefore, is to look at **Hartley's F_{Max}**, also known as the **variance ratio** (Pearson & Hartley, 1954). This is the ratio of the variances between the group with the biggest variance and the group with the smallest variance. This ratio was compared to critical values in a table published by Hartley. Some of the critical values (for a .05 level of significance) are shown in Figure 5.12 (see Oliver Twisted); as you can see the critical values depend on the number of cases per group (well, $n - 1$ actually), and the number of variances being compared. From this graph you can see that with sample sizes (n) of 10 per group, an F_{Max} of less than 10 is more or less always going to be non-significant, with 15–20 per group the ratio needs to be less than about 5, and with samples of 30–60 the ratio should be below about 2 or 3.

OLIVER TWISTED

Please, Sir, can I have some more … Hartley's F_{Max}?

Oliver thinks that my graph of critical values is stupid. 'Look at that graph,' he laughed, 'it's the most stupid thing I've ever seen since I was at Sussex Uni and I saw my statistics lecturer, Andy Fie…'. Well, go choke on your gruel you Dickensian bubo because the full table of critical values is in the additional material for this chapter on the companion website.

[3] We haven't covered ANOVA yet so this explanation won't make much sense to you now, but in Chapter 10 we will look in more detail at how Levene's test works.

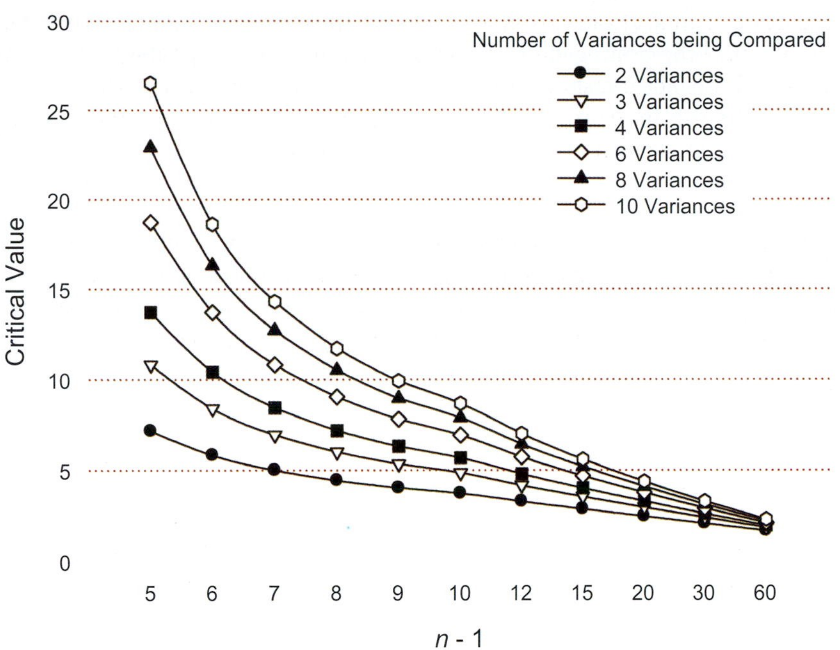

FIGURE 5.12 Selected critical values for Hartley's F_{Max} test

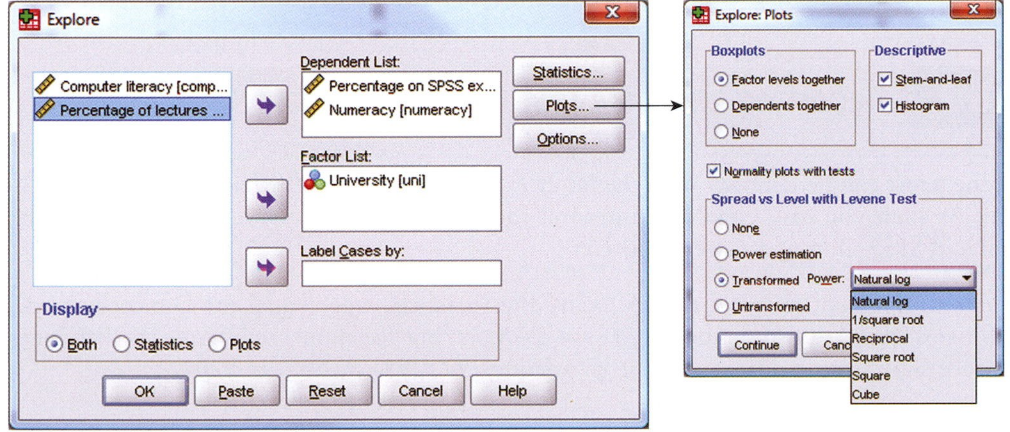

FIGURE 5.13 Exploring groups of data and obtaining Levene's test

We can get Levene's test using the *explore* menu that we used in the previous section. For this example, we'll use the SPSS exam data that we used in the previous section (in the file **SPSSExam.sav**). Once the data are loaded, use Analyze Descriptive Statistics ▸ Explore... to open the dialog box in Figure 5.13. To keep things simple we'll just look at the SPSS exam scores and the numeracy scores from this file, so transfer these two variables from the list on the left-hand side to the box labelled *Dependent List* by clicking on the ➔ next to this box, and because we want to split the output by the grouping variable to compare the variances, select the variable **Uni** and transfer it to the box labelled *Factor List* by clicking on the appropriate ➔. Then click on Plots... to open the other dialog box in Figure 5.13. To get Levene's test we need to select one of the options where it says *Spread vs. level with Levene's test*. If you select Untransformed Levene's test is carried out on the raw data (a good place to start). When you've finished with this dialog box click on Continue to return to the main *Explore* dialog box and then click on OK to run the analysis.

SPSS Output 5.6 shows the table for Levene's test. You should read the statistics based on the mean. Levene's test is non-significant for the SPSS exam scores (values in the *Sig.*

column are more than .05) indicating that the variances are not significantly different (i.e. they are similar and the homogeneity of variance assumption is tenable). However, for the numeracy scores, Levene's test is significant (values in the *Sig.* column are less than .05) indicating that the variances are significantly different (i.e. they are not the same and the homogeneity of variance assumption has been violated). We can also calculate the variance ratio. To do this we need to divide the largest variance by the smallest. You should find the variances in your output, but if not we obtained these values in SPSS Output 5.3. For SPSS exam scores the variance ratio is 158.48/104.14 = 1.52 and for numeracy scores the value is 9.43/4.27 = 2.21. Our group sizes are 50 and we're comparing 2 variances so the critical value is (from the table in the additional material) approximately 1.67. These ratios concur with Levene's test: variances are significantly different for numeracy scores (2.21 is bigger than 1.67) but not for SPSS exam scores (1.52 is smaller than 1.67).

SPSS OUTPUT 5.6

Test of Homogeneity of Variance

		Levene Statistic	df1	df2	Sig.
Percentage on SPSS exam	Based on Mean	2.584	1	98	.111
	Based on Median	2.089	1	98	.152
	Based on Median and with adjusted df	2.089	1	94.024	.152
	Based on trimmed mean	2.523	1	98	.115
Numeracy	Based on Mean	7.368	1	98	.008
	Based on Median	5.366	1	98	.023
	Based on Median and with adjusted df	5.366	1	83.920	.023
	Based on trimmed mean	6.766	1	98	.011

5.6.2. Reporting Levene's test ①

Levene's test can be denoted with the letter F and there are two different degrees of freedom. As such you can report it, in general form, as F(df1, df2) = value, *sig*. So, for the results in SPSS Output 5.6 we could say:

✓ For the percentage on the SPSS exam, the variances were equal for Duncetown and Sussex University students, $F(1, 98) = 2.58$, *ns*, but for numeracy scores the variances were significantly different in the two groups, $F(1, 98) = 7.37, p < .01$.

CRAMMING SAM'S TIPS Homogeneity of variance

- Homogeneity of variance is the assumption that the spread of scores is roughly equal in different groups of cases, or more generally that the spread of scores is roughly equal at different points on the predictor variable.
- When comparing groups, this assumption can be tested with Levene's test and the variance ratio (Hartley's F_{Max}).
- If Levene's test is significant (*Sig.* in the SPSS table is less than .05) then the variances are significantly different in different groups.
- Otherwise, homogeneity of variance can be assumed.
- The variance ratio is the largest group variance divided by the smallest. This value needs to be smaller than the critical values in Figure 5.12.
- **Warning:** In large samples Levene's test can be significant even when group variances are not very different. Therefore, it should be interpreted in conjunction with the variance ratio.

5.7. Correcting problems in the data ②

The previous section showed us various ways to explore our data; we saw how to look for problems with our distribution of scores and how to detect heterogeneity of variance. In Chapter 4 we also discovered how to spot outliers in the data. The next question is what to do about these problems?

5.7.1. Dealing with outliers ②

If you detect outliers in the data there are several options for reducing the impact of these values. However, before you do any of these things, it's worth checking that the data have been entered correctly for the problem cases. If the data are correct then the three main options you have are:

1. *Remove the case*: This entails deleting the data from the person who contributed the outlier. However, this should be done only if you have good reason to believe that this case is not from the population that you intended to sample. For example, if you were investigating factors that affected how much cats purr and one cat didn't purr at all, this would likely be an outlier (all cats purr). Upon inspection, if you discovered that this cat was actually a dog wearing a cat costume (hence why it didn't purr), then you'd have grounds to exclude this case because it comes from a different population (dogs who like to dress as cats) than your target population (cats).
2. *Transform the data*: Outliers tend to skew the distribution and, as we will see in the next section, this skew (and, therefore, the impact of the outliers) can sometimes be reduced by applying **transformations** to the data.
3. *Change the score*: If transformation fails, then you can consider replacing the score. This on the face of it may seem like cheating (you're changing the data from what was actually corrected); however, if the score you're changing is very unrepresentative and biases your statistical model anyway then changing the score is the lesser of two evils! There are several options for how to change the score:
   (a) *The next highest score plus one*: Change the score to be one unit above the next highest score in the data set.
   (b) *Convert back from a z-score*: A z-score of 3.29 constitutes an outlier (see Jane Superbrain Box 4.1) so we can calculate what score would give rise to a z-score of 3.29 (or perhaps 3) by rearranging the z-score equation in section 1.7.4, which gives us $X = (z \times s) + \bar{X}$. All this means is that we calculate the mean ($\bar{X}$) and standard deviation (s) of the data; we know that z is 3 (or 3.29 if you want to be exact) so we just add three times the standard deviation to the mean, and replace our outliers with that score.
   (c) *The mean plus two standard deviations*: A variation on the above method is to use the mean plus two times the standard deviation (rather than three times the standard deviation).

5.7.2. Dealing with non-normality and unequal variances ②

5.7.2.1. Transforming data ②

The next section is quite hair raising so don't worry if it doesn't make much sense – many undergraduate courses won't cover transforming data so feel free to ignore this section if you want to!

What do I do if my data are not normal?

We saw in the previous section that you can deal with outliers by transforming the data and these transformations are also useful for correcting problems with normality and the assumption of homogeneity of variance. The idea behind transformations is that you do something to every score to correct for distributional problems, outliers or unequal variances. Although some students often (understandably) think that transforming data sounds dodgy (the phrase 'fudging your results' springs to some people's minds!), in fact it isn't because you do the same thing to all of your scores.[4] As such, transforming the data won't change the relationships between variables (the relative differences between people for a given variable stay the same), but it does change the differences between different variables (because it changes the units of measurement). Therefore, if you are looking at relationships between variables (e.g., regression) it is alright just to transform the problematic variable, but if you are looking at differences between variables (e.g., change in a variable over time) then you need to tranform all of those variables.

Let's return to our Download Festival data (**DownloadFestival.sav**) from earlier in the chapter. These data were not normal on days 2 and 3 of the festival (section 5.4). Now, we might want to look at how hygiene levels changed across the three days (i.e. compare the mean on day 1 to the means on days 2 and 3 to see if people got smellier). The data for days 2 and 3 were skewed and need to be transformed, but because we might later compare the data to scores on day 1, we would also have to transform the day 1 data (even though scores were not skewed). If we don't change the day 1 data as well, then any differences in hygiene scores we find from day 1 to days 2 or 3 will be due to us transforming one variable and not the others.

There are various transformations that you can do to the data that are helpful in correcting various problems.[5] However, whether these transformations are necessary or useful is quite a complex issue (see Jane Superbrain Box 5.1). Nevertheless, because they *are* used by researchers Table 5.1 shows some common transformations and their uses.

5.7.2.2. Choosing a transformation ②

Given that there are many transformations that you can do, how can you decide which one is best? The simple answer is trial and error: try one out and see if it helps and if it doesn't then try a different one. If you are looking at differences between variables you *must apply the same transformation to all variables* (you cannot, for example, apply a log transformation to one variable and a square root transformation to another). This can be quite time consuming. However, for homogeneity of variance we can see the effect of a transformation quite quickly. In section 5.6.1 we saw how to use the *explore* function to get Levene's test. In that section we ran the analysis selecting the raw scores (◉ Untransformed). However, if the variances turn out to be unequal, as they did in our example, you can use the same dialog box (Figure 5.13) but select ◉ Transformed. When you do this you should notice a drop-down list that becomes active and if you click on this you'll notice that it lists several transformations including the ones that I have just described. If you select a transformation from this list (*Natural log* perhaps or *Square root*) then SPSS will calculate what Levene's test would be if you were to transform the data using this method. This can save you a lot of time trying out different transformations.

[4] Although there aren't statistical consequences of transforming data, there may be empirical or scientific implications that outweigh the statistical benefits (see Jane Superbrain Box 5.1).

[5] You'll notice in this section that I keep writing X_i. We saw in Chapter 1 that this refers to the observed score for the *i*th person (so, the *i* could be replaced with the name of a particular person, thus for Graham, $X_i = X_{Graham}$ = Graham's score, and for Carol, $X_i = X_{Carol}$ = Carol's score).

TABLE 5.1 Data transformations and their uses

Data Transformation	*Can Correct For*
Log transformation ($\log(X_i)$): Taking the logarithm of a set of numbers squashes the right tail of the distribution. As such it's a good way to reduce positive skew. However, you can't get a log value of zero or negative numbers, so if your data tend to zero or produce negative numbers you need to add a constant to all of the data before you do the transformation. For example, if you have zeros in the data then do $\log(X_i + 1)$, or if you have negative numbers add whatever value makes the smallest number in the data set positive.	Positive skew, unequal variances
Square root transformation ($\sqrt{X_i}$): Taking the square root of large values has more of an effect than taking the square root of small values. Consequently, taking the square root of each of your scores will bring any large scores closer to the centre – rather like the log transformation. As such, this can be a useful way to reduce positive skew; however, you still have the same problem with negative numbers (negative numbers don't have a square root).	Positive skew, unequal variances
Reciprocal transformation ($1/X_i$): Dividing 1 by each score also reduces the impact of large scores. The transformed variable will have a lower limit of 0 (very large numbers will become close to 0). One thing to bear in mind with this transformation is that it reverses the scores: scores that were originally large in the data set become small (close to zero) after the transformation, but scores that were originally small become big after the transformation. For example, imagine two scores of 1 and 10; after the transformation they become 1/1 = 1, and 1/10 = 0.1: the small score becomes bigger than the large score after the transformation. However, you can avoid this by reversing the scores before the transformation, by finding the highest score and changing each score to the highest score minus the score you're looking at. So, you do a transformation $1/(X_{Highest} - X_i)$.	Positive skew, unequal variances
Reverse score transformations: Any one of the above transformations can be used to correct negatively skewed data, but first you have to reverse the scores. To do this, subtract each score from the highest score obtained, or the highest score + 1 (depending on whether you want your lowest score to be 0 or 1). If you do this, don't forget to reverse the scores back afterwards, or to remember that the interpretation of the variable is reversed: big scores have become small and small scores have become big!	Negative skew

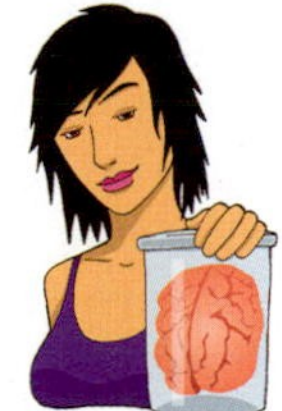

JANE SUPERBRAIN 5.1

To transform or not to transform, that is the question ③

Not everyone agrees that transforming data is a good idea; for example, Glass, Peckham, and Sanders (1972) in a very extensive review commented that 'the payoff of normalizing transformations in terms of more valid probability statements is low, and they are seldom considered to be worth the effort' (p. 241). In which case, should we bother?

The issue is quite complicated (especially for this early in the book), but essentially we need to know whether the statistical models we apply perform better on transformed data than they do when applied to data that violate the assumption that the transformation corrects. If a statistical model is still accurate even when its assumptions are broken it is said to be a **robust test** (section 5.7.4). I'm not going to discuss whether particular tests are robust here, but I will discuss the issue for particular tests in their respective chapters. The question of whether to transform is linked to this issue of robustness (which in turn is linked to what test you are performing on your data).

A good case in point is the *F*-test in ANOVA (see Chapter 10), which is often claimed to be robust (Glass et al., 1972). Early findings suggested that *F* performed as it should in skewed distributions and that transforming the data helped as often as it hindered the accuracy of *F* (Games & Lucas, 1966). However, in a lively but informative exchange Levine and Dunlap (1982) showed that transformations of skew did improve the performance of *F*; however, in a response Games (1983) argued that their conclusion was incorrect, which Levine and Dunlap (1983) contested in a response to the response. Finally, in a response to the response of the response, Games (1984) pointed out several important questions to consider:

1 The central limit theorem (section 2.5.1) tells us that in big samples the sampling distribution will be normal regardless, and this is what's actually important so the debate is academic in anything other than small samples. Lots of early research did indeed show that with samples of 40 the normality of the sampling distribution was, as predicted, normal. However, this research focused on distributions with light tails and subsequent work has shown that with heavy-tailed distributions larger samples would be necessary to invoke the central limit theorem (Wilcox, 2005). This research suggests that transformations might be useful for such distributions.

2 By transforming the data you change the hypothesis being tested (when using a log transformation and comparing means you change from comparing arithmetic means to comparing geometric means). Transformation also means that you're now addressing a different construct to the one originally measured, and this has obvious implications for interpreting that data (Grayson, 2004).

3 In small samples it is tricky to determine normality one way or another (tests such as K–S will have low power to detect deviations from normality and graphs will be hard to interpret with so few data points).

4 The consequences for the statistical model of applying the 'wrong' transformation could be worse than the consequences of analysing the untransformed scores.

As we will see later in the book, there is an extensive library of robust tests that can be used and which have considerable benefits over transforming data. The definitive guide to these is Wilcox's (2005) outstanding book.

5.7.3. Transforming the data using SPSS ②

5.7.3.1. The *Compute* function ②

To do transformations on SPSS we use the *compute* command, which enables us to carry out various functions on columns of data in the data editor. Some typical functions are adding scores across several columns, taking the square root of the scores in a column or calculating the mean of several variables. To access the *Compute Variable* dialog box, use the mouse to specify Transform ▸ Compute Variable.... The resulting dialog box is shown in Figure 5.14; it has a list of functions on the right-hand side, a calculator-like keyboard in the centre and a blank space that I've labelled the command area. The basic idea is that you type a name for a new variable in the area labelled *Target Variable* and then you write some kind of command in the command area to tell SPSS how to create this new variable. You use a combination of existing variables selected from the list on the left, and numeric expressions. So, for example, you could use it like a calculator to add variables (i.e. add two columns in the data editor to make a third). However, you can also use it to generate data without using existing variables too. There are hundreds of built-in functions that SPSS has grouped together. In the dialog box it lists these groups in the area labelled *Function group*; upon selecting a function group, a list of available functions within that group will appear in the box labelled *Functions and Special Variables*. If you select a function, then a description of that function appears in the grey box indicated in Figure 5.14. You can enter variable names into the command area by selecting the variable required from the variables list and then clicking on ➔. Likewise, you can select a certain function from the list of available functions and enter it into the command area by clicking on ⬆.

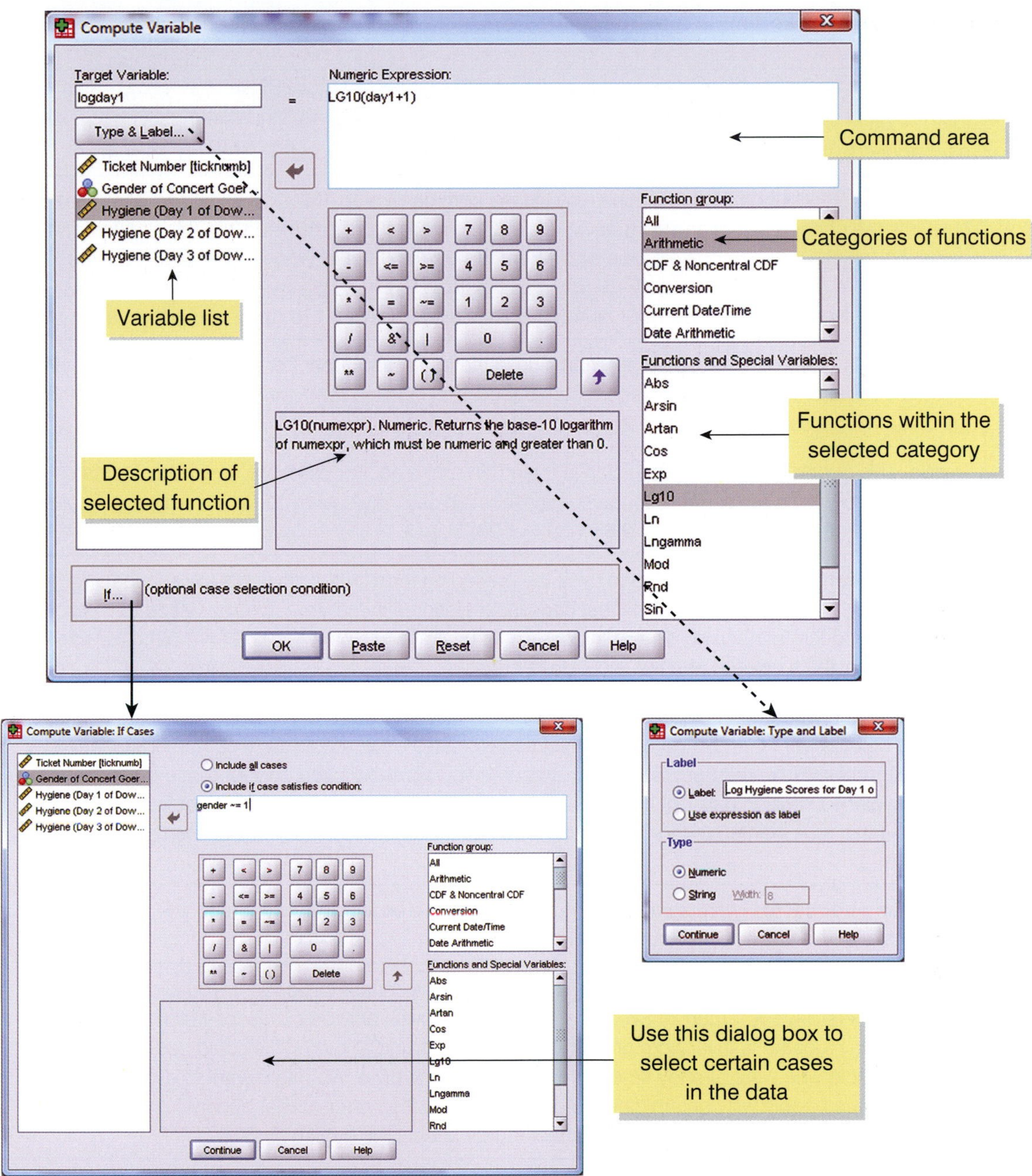

FIGURE 5.14 Dialog box for the *compute* function

The basic procedure is to first type a variable name in the box labelled *Target Variable*. You can then click on Type & Label... and another dialog box appears, where you can give the variable a descriptive label, and where you can specify whether it is a numeric or string variable (see section 3.4.2). Then when you have written your command for SPSS to execute, click on OK to run the command and create the new variable. If you type in a variable name that already exists in the data editor then SPSS will tell you and ask you whether you want to replace this existing variable. If you respond with *Yes* then SPSS will replace the data in the existing column with the result of the *compute* function; if you respond with *No* then nothing will happen and you will need to rename the target variable. If you're computing a lot of new variables it can be quicker to use syntax (see SPSS Tip 5.1).

Let's first look at some of the simple functions:

Button	Description
+	**Addition**: This button places a plus sign in the command area. For example, with our hygiene data, 'day1 + day2' creates a column in which each row contains the hygiene score from the column labelled *day1* added to the score from the column labelled *day2* (e.g. for participant 1: 2.65 + 1.35 = 4).
-	**Subtraction**: This button places a minus sign in the command area. For example, if we wanted to calculate the change in hygiene from day 1 to day 2 we could type 'day2 – day1'. This creates a column in which each row contains the score from the column labelled *day1* subtracted from the score from the column labelled *day2* (e.g. for participant 1: 2.65 – 1.35 = –1.30). Therefore, this person's hygiene went down by 1.30 (on our 5-point scale) from day 1 to day 2 of the festival.
*	**Multiply**: This button places a multiplication sign in the command area. For example, 'day1 * day2' creates a column that contains the score from the column labelled *day1* multiplied by the score from the column labelled *day2* (e.g. for participant 1: 2.65 × 1.35 = 3.58).
/	**Divide**: This button places a division sign in the command area. For example, 'day1/day2' creates a column that contains the score from the column labelled *day1* divided by the score from the column labelled *day2* (e.g. for participant 1: 2.65/1.35 = 1.96).
**	**Exponentiation**: This button is used to raise the preceding term by the power of the succeeding term. So, 'day1**2' creates a column that contains the scores in the *day1* column raised to the power of 2 (i.e. the square of each number in the *day1* column: for participant 1, $(2.65)^2 = 7.02$). Likewise, 'day1**3' creates a column with values of **day1** cubed.
<	**Less than**: This operation is usually used for 'include case' functions. If you click on the If... button, a dialog box appears that allows you to select certain cases on which to carry out the operation. So, if you typed 'day1 < 1', then SPSS would carry out the *compute* function only for those participants whose hygiene score on day 1 of the festival was less than 1 (i.e. if **day1** was 0.99 or less). So, we might use this if we wanted to look only at the people who were already smelly on the first day of the festival!
<=	**Less than or equal to**: This operation is the same as above except that in the example above, cases that are exactly 1 would be included as well.
>	**More than**: This operation is used to include cases above a certain value. So, if you clicked on If... and then typed 'day1 > 1' then SPSS will carry out any analysis only on cases for which hygiene scores on day 1 of the festival were greater than 1 (i.e. 1.01 and above). This could be used to exclude people who were already smelly at the start of the festival. We might want to exclude them because these people will contaminate the data (not to mention our nostrils) because they reek of putrefaction to begin with so the festival cannot further affect their hygiene!
>=	**More than or equal to**: This operation is the same as above but will include cases that are exactly 1 as well.
=	**Equal to**: You can use this operation to include cases for which participants have a specific value. So, if you clicked on If... and typed 'day1 = 1' then only cases that have a value of exactly 1 for the **day1** variable are included. This is most useful when you have a coding variable and you want to look at only one of the groups. For example, if we wanted to look only at females at the festival we could type 'gender = 1', then the analysis would be carried out on only females (who are coded as 1 in the data).
~=	**Not equal to**: This operation will include all cases except those with a specific value. So, 'gender ~= 1' (as in Figure 5.14) will include all cases except those that were female (have a 1 in the gender column). In other words, it will carry out the *compute* command only on the males.

TABLE 5.2 Some useful compute functions

Function	Name	Example Input	Output
MEAN(?,?, ..)	Mean	Mean(day1, day2, day3)	For each row, SPSS calculates the average hygiene score across the three days of the festival
SD(?,?, ..)	Standard deviation	SD(day1, day2, day3)	Across each row, SPSS calculates the standard deviation of the values in the columns labelled *day1*, *day2* and *day3*
SUM(?,?, ..)	Sum	SUM(day1, day2)	For each row, SPSS adds the values in the columns labelled *day1* and *day2*
SQRT(?)	Square root	SQRT(day2)	Produces a column containing the square root of each value in the column labelled *day2*
ABS(?)	Absolute value	ABS(day1)	Produces a variable that contains the absolute value of the values in the column labelled *day1* (absolute values are ones where the signs are ignored: so –5 becomes +5 and +5 stays as +5)
LG10(?)	Base 10 logarithm	LG10(day1)	Produces a variable that contains the logarithmic (to base 10) values of the variable *day1*
RV.NORMAL (mean, stddev)	Normal random numbers	Normal(20, 5)	Produces a variable of pseudo-random numbers from a normal distribution with a mean of 20 and a standard deviation of 5

Some of the most useful functions are listed in Table 5.2, which shows the standard form of the function, the name of the function, an example of how the function can be used and what SPSS would output if that example were used. There are several basic functions for calculating means, standard deviations and sums of columns. There are also functions such as the square root and logarithm that are useful for transforming data that are skewed and we will use these functions now. For the interested reader, the SPSS help files have details of all of the functions available through the *Compute Variable* dialog box (click on Help when you're in the dialog box).

5.7.3.2. The log transformation on SPSS ②

Now we've found out some basic information about the *compute* function, let's use it to transform our data. First open the main Compute dialog box by selecting Transform Compute Variable.... Enter the name **logday1** into the box labelled *Target Variable* and then click on Type & Label... and give the variable a more descriptive name such as *Log transformed hygiene scores for day 1 of Download festival*. In the list box labelled *Function group* click on *Arithmetic* and then in the box labelled *Functions and Special Variables* click on *Lg10* (this is the log transformation to base 10, *Ln* is the natural log) and transfer it to the command area by clicking on ↑. When the command is transferred, it appears in the command area as 'LG10(?)' and the question mark should be replaced with a variable name (which can be typed manually or transferred from the variables list). So replace the question mark with the variable **day1** by either selecting the variable in the list and dragging it across, clicking on ➜, or just typing 'day1' where the question mark is.

For the day 2 hygiene scores there is a value of 0 in the original data, and there is no logarithm of the value 0. To overcome this we should add a constant to our original scores before we take the log of those scores. Any constant will do, provided that it makes all of the scores greater than 0. In this case our lowest score is 0 in the data set so we can simply add 1 to all of the scores and that will ensure that all scores are greater than zero. To do this, make sure the cursor is still inside the brackets and click on + and then 1. The final dialog box should look like Figure 5.14. Note that the expression reads LG10(day1 + 1); that is, SPSS will add one to each of the day1 scores and then take the log of the resulting values. Click on OK to create a new variable **logday1** containing the transformed values.

SELF-TEST Have a go at creating similar variables **logday2** and **logday3** for the day 2 and day 3 data. Plot histograms of the transformed scores for all three days.

5.7.3.3. The square root transformation on SPSS ②

To do a square root transformation, we run through the same process, by using a name such as **sqrtday1** in the box labelled *Target Variable* (and click on Type & Label... to give the variable a more descriptive name). In the list box labelled *Function group* click on *Arithmetic* and then in the box labelled *Functions and Special Variables* click on *Sqrt* and drag it to the command area or click on ↑. When the command is transferred, it appears in the command area as SQRT(?). Replace the question mark with the variable **day1** by selecting the variable in the list and dragging it, clicking on ➜, or just typing 'day1' where the question mark is. The final expression will read *SQRT(day1)*. Click on OK to create the variable.

SELF-TEST Repeat this process for **day2** and **day3** to create variables called **sqrtday2** and **sqrtday3**. Plot histograms of the transformed scores for all three days.

5.7.3.4. The reciprocal transformation on SPSS ②

To do a reciprocal transformation on the data from day 1, we could use a name such as **recday1** in the box labelled *Target Variable*. Then we can simply click on 1 and then /. Ordinarily you would select the variable name that you want to transform from the list and drag it across, click on ➜ or just type the name of the variable. However, the day 2 data contain a zero value and if we try to divide 1 by 0 then we'll get an error message (you can't divide by 0). As such we need to add a constant to our variable just as we did for the log transformation. Any constant will do, but 1 is a convenient number for these data. So, instead of selecting the variable we want to transform, click on (). This places a pair of brackets into the box labelled *Numeric Expression*; then make sure the cursor is between these two brackets and select the variable you want to transform from the list and transfer it across by clicking on ➜ (or type the name of the variable manually). Now click on + and then 1 (or type + *1* using your keyboard). The box labelled *Numeric Expression* should now contain the text *1/(day1 + 1)*. Click on OK to create a new variable containing the transformed values.

SELF-TEST Repeat this process for **day2** and **day3**. Plot histograms of the transformed scores for all three days..

SPSS TIP 5.1 Using syntax to compute new variables ③

If you're computing a lot of new variables it can be quicker to use syntax. For example, to create the transformed data in the example in this chapter. I've written the file **Transformations.sps** to do all nine of the transformations we've discussed in this section. Download this file from the companion website and open it. Alternatively, open a syntax window (see section 3.7) and type the following:

```
COMPUTE logday1 = LG10(day1 + 1) .
COMPUTE logday2 = LG10(day2 + 1) .
COMPUTE logday3 = LG10(day3 + 1) .
COMPUTE sqrtday1 = SQRT(day1).
COMPUTE sqrtday2 = SQRT(day2).
COMPUTE sqrtday3 = SQRT(day3).
COMPUTE recday1 = 1/(day1+1).
COMPUTE recday2 = 1/(day2+1).
COMPUTE recday3 = 1/(day3+1).
EXECUTE .
```

Each COMPUTE command above is doing the equivalent of what you'd do using the *Compute Variable* dialog box in Figure 5.14. So, the first three lines just ask SPSS to create three new variables (**logday1, logday2** and **logday3**), which are just the log transformations of the variables **day1, day2** and **day3** plus 1. The next three lines do much the same but use the SQRT function, and so take the square root of **day1**, **day2** and **day3** to create new variables called **sqrtday1**, **sqrtday2** and **sqrtday3** respectively. The next three lines do the reciprocal transformations in much the same way. The final line has the command EXECUTE without which none of the COMPUTE commands beforehand will be executed! Note also that every line ends with a full stop.

5.7.3.5. The effect of transformations ②

Figure 5.15 shows the distributions for days 1 and 2 of the festival after the three different transformations. Compare these to the untransformed distributions in Figure 5.3. Now, you can see that all three transformations have cleaned up the hygiene scores for day 2: the positive skew is reduced (the square root transformation in particular has been useful). However, because our hygiene scores on day 1 were more or less symmetrical to begin with, they have now become slightly negatively skewed for the log and square root transformation, and positively skewed for the reciprocal transformation![6] If we're using scores from day 2 alone then we could use the transformed scores; however, if we wanted to look at the change in scores then we'd have to weigh up whether the benefits of the transformation for the day 2 scores outweigh the problems it creates in the day 1 scores – data analysis can be frustrating sometimes!

[6] The reversal of the skew for the reciprocal transformation is because, as I mentioned earlier, the reciprocal has the effect of reversing the scores.

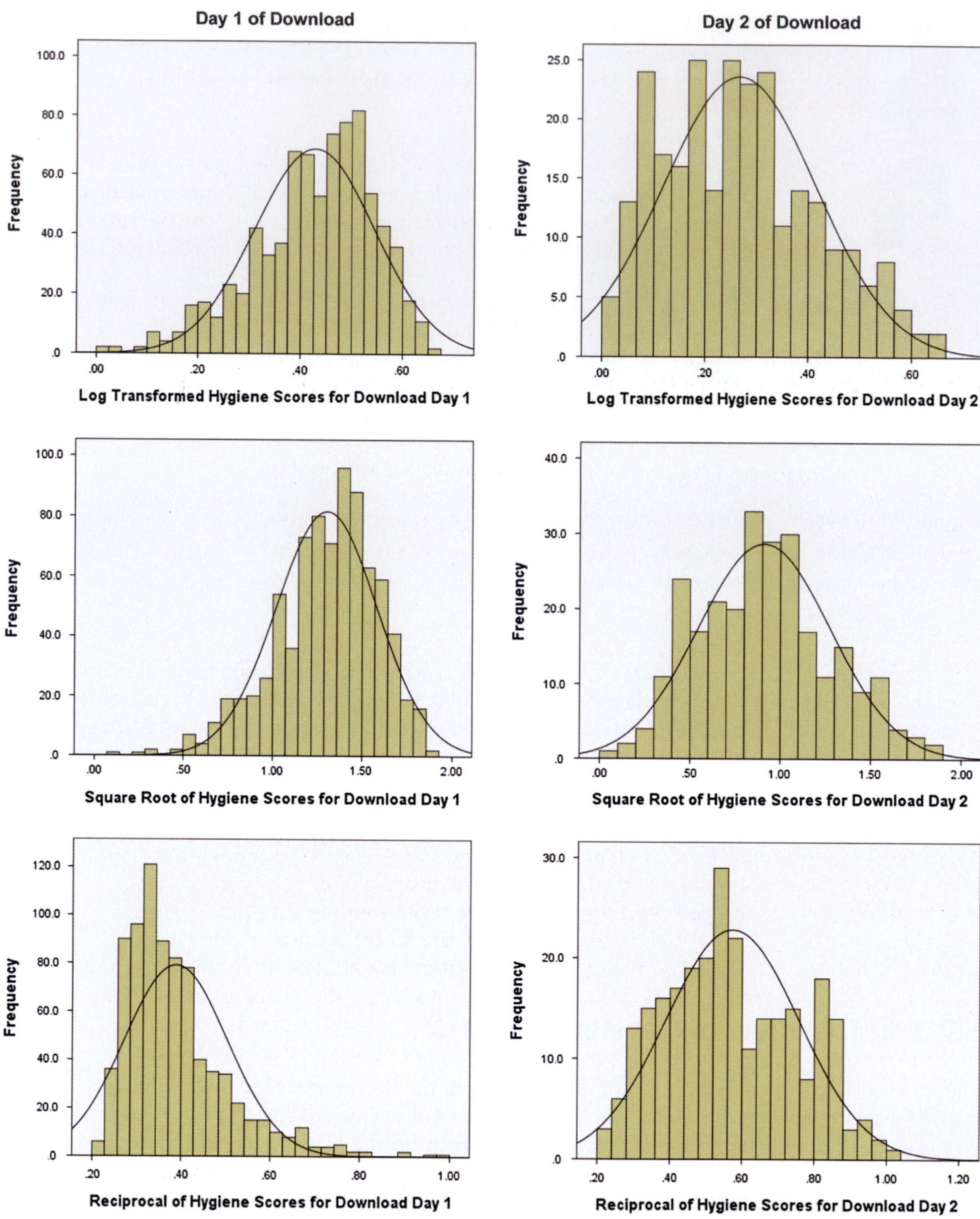

FIGURE 5.15 Distributions of the hygiene data on day 1 and day 2 after various transformations

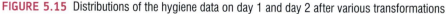

5.7.4. When it all goes horribly wrong ③

It's very easy to think that transformations are the answers to all of your broken assumption prayers. However, as we have seen, there are reasons to think that transformations are not necessarily a good idea (see Jane Superbrain Box 5.1) and even if you think that they are they do not always solve the problem, and even when they do solve the problem they often create different problems in the process. This happens more frequently than you might imagine (messy data are the norm).

If you find yourself in the unenviable position of having irksome data then there are some other options available to you (other than sticking a big samurai sword through your head). The first is to use a test that does not rely on the assumption of normally distributed data and as you go through the various chapters of this book I'll point out these tests – there is also a whole chapter dedicated to them later on.[7] One thing that you will quickly discover about non-parametric tests is that they have been developed for only a fairly limited range of situations. So, happy days if you want to compare two means, but sad lonely days listening to Joy Division if you have a complex experimental design.

A much more promising approach is to use robust methods (which I mentioned in Jane Superbrain Box 5.1). These tests have developed as computers have got more sophisticated (doing these tests without computers would be only marginally less painful than ripping off your skin and diving into a bath of salt). How these tests work is beyond the scope of this book (and my brain) but two simple concepts will give you the general idea. Some of these procedures use a **trimmed mean**. A trimmed mean is simply a mean based on the distribution of scores after some percentage of scores has been removed from each extreme of the distribution. So, a 10% trimmed mean will remove 10% of scores from the top and bottom before the mean is calculated. We saw in Chapter 2 that the accuracy of the mean depends on a symmetrical distribution, but a trimmed mean produces accurate results even when the distribution is not symmetrical, because by trimming the ends of the distribution we remove outliers and skew that bias the mean. Some robust methods work by taking advantage of the properties of the trimmed mean.

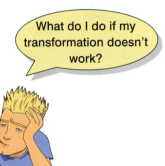

What do I do if my transformation doesn't work?

The second general procedure is the **bootstrap** (Efron & Tibshirani, 1993). The idea of the bootstrap is really very simple and elegant. The problem that we have is that we don't know the shape of the sampling distribution, but normality in our data allows us to infer that the sampling distribution is normal (and hence we can know the probability of a particular test statistic occurring). Lack of normality prevents us from knowing the shape of the sampling distribution unless we have big samples (but see Jane Superbrain Box 5.1). Bootstrapping gets around this problem by estimating the properties of the sampling distribution from the sample data. In effect, the sample data are treated as a population from which smaller samples (called bootstrap samples) are taken (putting the data back before a new sample is drawn). The statistic of interest (e.g. the mean) is calculated in each sample, and by taking many samples the sampling distribution can be estimated (rather like in Figure 2.7). The standard error of the statistic is estimated from the standard deviation of this sampling distribution created from the bootstrap samples. From this standard error, confidence intervals and significance tests can be computed. This is a very neat way of getting around the problem of not knowing the shape of the sampling distribution.

These techniques sound pretty good don't they? It might seem a little strange then that I haven't written a chapter on them. The reason why is that SPSS does not do most of them, which is something that I hope it will correct sooner rather than later. However, thanks to Rand Wilcox you can do them using a free statistics program called R (www.r-project.org) and a non-free program called S-Plus. Wilcox provides a very comprehensive review of robust methods in his excellent book *Introduction to robust estimation and hypothesis testing* (2005) and has written programs to run these methods using R. Among many other things, he has files to run robust versions of many tests discussed in this book: ANOVA, ANCOVA, correlation and multiple regression. If there is a robust method, it is likely to be in his book, and he will have written a macro procedure to run it! You can also download these macros from his website.[8] There are

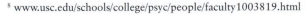

[7] For convenience a lot of textbooks refer to these tests as *non-parametric tests* or *assumption-free* tests and stick them in a separate chapter. Actually neither of these terms are particularly accurate (e.g. none of these tests is assumption-free) but in keeping with tradition I've put them in a chapter (15) on their own ostracized from their 'parametric' counterparts and feeling lonely.

[8] www.usc.edu/schools/college/psyc/people/faculty1003819.html

several good introductory books to tell you how to use R also (e.g. Dalgaard, 2002). There is also an SPSS plugin that you can download (http://www.spss.com/devcentral/index.cfm?pg=plugins) that allows you to use R commands from the SPSS syntax window. With this plugin you can analyse an open SPSS data file using the robust methods of R. These analyses are quite technical so I don't discuss them in the book, but if you'd like to know more see Oliver Twisted.

OLIVER TWISTED

Please, Sir, can I have some more … R?

'Why is it called R?', cackles Oliver 'Is it because using it makes you shout Arrghhh!!?' No, Oliver, it's not. The R plugin is a useful tool but it's very advanced. However, I've prepared a flash movie on the companion website that shows you how to use it. Look out for some other R-related demos on there too.

What have I discovered about statistics? ①

'You promised us swans,' I hear you cry, 'and all we got was normality this, homosomethingorother that, transform this – it's all a waste of time that. Where were the bloody swans?!' Well, the Queen owns them all so I wasn't allowed to have them. Nevertheless, this chapter did negotiate Dante's eighth circle of hell (Malebolge), where data of deliberate and knowing evil dwell. That is, data that don't conform to all of those pesky assumptions that make statistical tests work properly. We began by seeing what assumptions need to be met for parametric tests to work, but we mainly focused on the assumptions of normality and homogeneity of variance. To look for normality we rediscovered the joys of frequency distributions, but also encountered some other graphs that tell us about deviations from normality (P–P and Q–Q plots). We saw how we can use skew and kurtosis values to assess normality and that there are statistical tests that we can use (the Kolmogorov–Smirnov test). While negotiating these evildoers, we discovered what homogeneity of variance is, and how to test it with Levene's test and Hartley's F_{Max}. Finally, we discovered redemption for our data. We saw we can cure their sins, make them good, with transformations (and on the way we discovered some of the uses of the *transform* function of SPSS and the *split-file* command). Sadly, we also saw that some data are destined to always be evil.

We also discovered that I had started to read. However, reading was not my true passion; it was music. One of my earliest memories is of listening to my dad's rock and soul records (back in the days of vinyl) while waiting for my older brother to come home from school, so I must have been about 3 at the time. The first record I asked my parents to buy me was 'Take on the world' by Judas Priest which I'd heard on *Top of the Pops* (a now defunct UK TV show) and liked. This record came out in 1978 when I was 5. Some people think that this sort of music corrupts young minds. Let's see if it did …

Key terms that I've discovered

Bootstrap
Hartley's F_{Max}
Homogeneity of variance
Independence
Kolmogorov–Smirnov test
Levene's test
Noniles
Normally distributed data
Parametric test
Percentiles
P–P plot
Q–Q plot
Quantiles
Robust test
Shapiro–Wilk test
Transformation
Trimmed mean
Variance ratio

Smart Alex's tasks

- **Task 1**: Using the **ChickFlick.sav** data from Chapter 4, check the assumptions of normality and homogeneity of variance for the two films (ignore gender): are the assumptions met? ①
- **Task 2**: Remember that the numeracy scores were positively skewed in the **SPSSExam.sav** data (see Figure 5.5)? Transform these data using one of the transformations described in this chapter: do the data become normal? ②

Answers can be found on the companion website.

Online tutorial

The companion website contains the following Flash movie tutorial to accompany this chapter:

- Using the compute command to transform variables

Further reading

Tabachnick, B. G., & Fidell, L. S. (2006). *Using multivariate statistics* (5th ed.). Boston: Allyn & Bacon. (Chapter 4 is the definitive guide to screening data!)
Wilcox, R. R. (2005). *Introduction to robust estimation and hypothesis testing* (2nd ed.). Burlington, MA: Elsevier. (Quite technical, but this is the definitive book on robust methods.)

6 Correlation

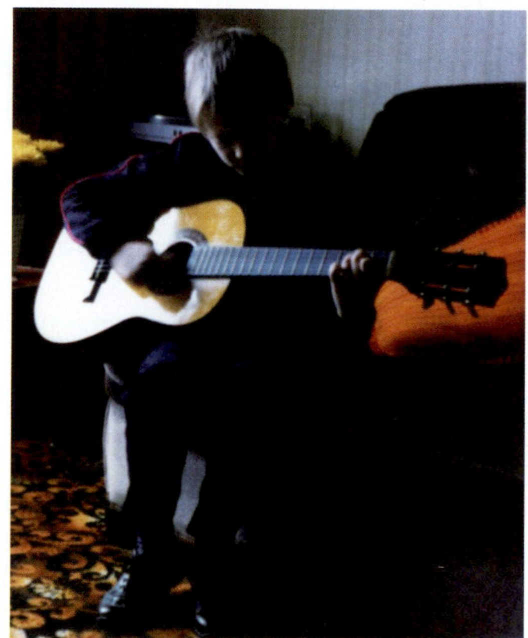

FIGURE 6.1 I don't have a photo from Christmas 1981, but this was taken about that time at my grandparents' house. I'm trying to play an 'E' by the looks of it, no doubt because it's in 'Take on the world'

6.1. What will this chapter tell me? ①

When I was 8 years old, my parents bought me a guitar for Christmas. Even then, I'd desperately wanted to play the guitar for years. I could not contain my excitement at getting this gift (had it been an *electric* guitar I think I would have actually exploded with excitement). The guitar came with a 'learn to play' book and after a little while of trying to play what was on page 1 of this book, I readied myself to unleash a riff of universe-crushing power onto the world (well, 'skip to my Lou' actually). But, I couldn't do it. I burst into tears and ran upstairs to hide.[1] My dad sat with me and said, 'Don't worry, Andy, everything is hard to begin with, but the more you practise the easier it gets.'

[1] This is not a dissimilar reaction to the one I have when publishers ask me for new editions of statistics textbooks.

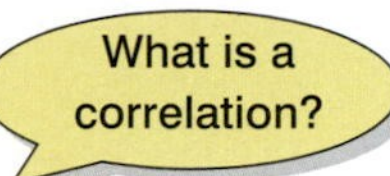

In his comforting words, my dad was inadvertently teaching me about the relationship, or correlation, between two variables. These two variables could be related in three ways: (1) *positively related*, meaning that the more I practised my guitar, the better guitar player I would become (i.e. my dad was telling me the truth); (2) *not related* at all, meaning that as I practise the guitar my playing ability remains completely constant (i.e. my dad has fathered a cretin); or (3) *negatively related*, which would mean that the more I practised my guitar the worse a guitar player I became (i.e. my dad has fathered an indescribably strange child). This chapter looks first at how we can express the relationships between variables statistically by looking at two measures: *covariance* and the *correlation coefficient*. We then discover how to carry out and interpret correlations in SPSS. The chapter ends by looking at more complex measures of relationships; in doing so it acts as a precursor to the chapter on multiple regression.

6.2. Looking at relationships ①

In Chapter 4 I stressed the importance of looking at your data graphically before running any other analysis on them. I just want to begin by reminding you that our first starting point with a correlation analysis should be to look at some scatterplots of the variables we have measured. I am not going to repeat how to get SPSS to produce these graphs, but I am going to urge you (if you haven't done so already) to read section 4.8 before embarking on the rest of this chapter.

6.3. How do we measure relationships? ①

6.3.1. A detour into the murky world of covariance ①

The simplest way to look at whether two variables are associated is to look at whether they *covary*. To understand what **covariance** is, we first need to think back to the concept of variance that we met in Chapter 2. Remember that the variance of a single variable represents the average amount that the data vary from the mean. Numerically, it is described by:

$$\text{variance}(s^2) = \frac{\sum(x_i - \bar{x})^2}{N-1} = \frac{\sum(x_i - \bar{x})(x_i - \bar{x})}{N-1} \tag{6.1}$$

The mean of the sample is represented by $\bar{x}$, x_i is the data point in question and N is the number of observations (see section 2.4.1). If we are interested in whether two variables are related, then we are interested in whether changes in one variable are met with similar changes in the other variable. Therefore, when one variable deviates from its mean we would expect the other variable to deviate from its mean in a similar way. To illustrate what I mean, imagine we took five people and subjected them to a certain number of advertisements promoting toffee sweets, and then measured how many packets of those sweets each person bought during the next week. The data are in Table 6.1 as well as the mean and standard deviation (s) of each variable.

TABLE 6.1

Subject	1	2	3	4	5	Mean	S
Adverts Watched	5	4	4	6	8	**5.4**	**1.67**
Packets Bought	8	9	10	13	15	**11.0**	**2.92**

If there were a relationship between these two variables, then as one variable deviates from its mean, the other variable should deviate from its mean in the same or the directly opposite way. Figure 6.2 shows the data for each participant (green circles represent the number of packets bought and blue circles represent the number of adverts watched); the green line is the average number of packets bought and the blue line is the average number of adverts watched. The vertical lines represent the differences (remember that these differences are called *deviations*) between the observed values and the mean of the relevant variable. The first thing to notice about Figure 6.2 is that there is a very similar pattern of deviations for both variables. For the first three participants the observed values are below the mean for both variables, for the last two people the observed values are above the mean for both variables. This pattern is indicative of a potential relationship between the two variables (because it seems that if a person's score is below the mean for one variable then their score for the other will also be below the mean).

So, how do we calculate the exact similarity between the pattern of differences of the two variables displayed in Figure 6.2? One possibility is to calculate the total amount of deviation but we would have the same problem as in the single variable case: the positive and negative deviations would cancel out (see section 2.4.1). Also, by simply adding the deviations, we would gain little insight into the *relationship* between the variables. Now, in the single variable case, we squared the deviations to eliminate the problem of positive and negative deviations cancelling out each other. When there are two variables, rather than squaring each deviation, we can multiply the deviation for one variable by the corresponding deviation for the second variable. If both deviations are positive or negative then this will give us a positive value (indicative of the deviations being in the same direction), but if one deviation is positive and one negative then the resulting product will be negative

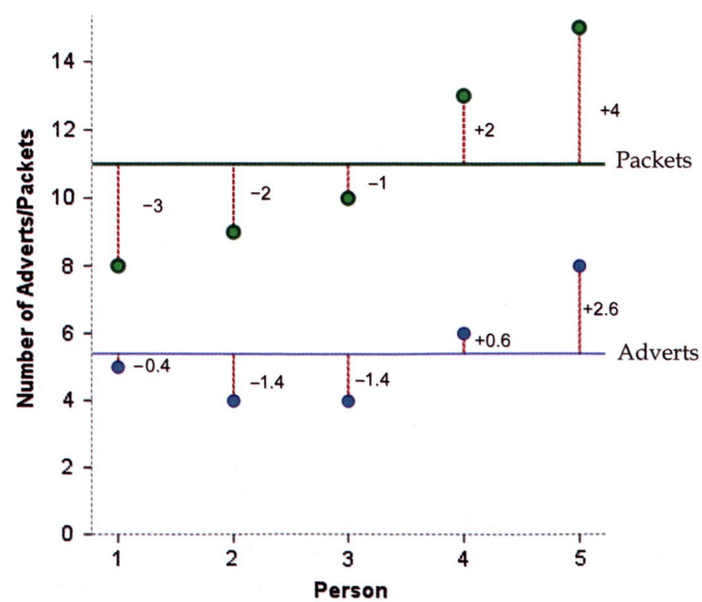

FIGURE 6.2 Graphical display of the differences between the observed data and the means of two variables

(indicative of the deviations being opposite in direction). When we multiply the deviations of one variable by the corresponding deviations of a second variable, we get what is known as the **cross-product deviations**. As with the variance, if we want an average value of the combined deviations for the two variables, we must divide by the number of observations (we actually divide by $N - 1$ for reasons explained in Jane Superbrain Box 2.3). This averaged sum of combined deviations is known as the covariance. We can write the covariance in equation form as in equation (6.2) – you will notice that the equation is the same as the equation for variance, except that instead of squaring the differences, we multiply them by the corresponding difference of the second variable:

$$\text{cov}(x, y) = \frac{\sum(x_i - \bar{x})(y_i - \bar{y})}{N - 1} \tag{6.2}$$

For the data in Table 6.1 and Figure 6.2 we reach the following value:

$$\begin{aligned}
\text{cov}(x, y) &= \frac{\sum(x_i - \bar{x})(y_i - \bar{y})}{N - 1} \\
&= \frac{(-0.4)(-3) + (-1.4)(-2) + (-1.4)(-1) + (0.6)(2) + (2.6)(4)}{4} \\
&= \frac{1.2 + 2.8 + 1.4 + 1.2 + 10.4}{4} \\
&= \frac{17}{4} \\
&= 4.25
\end{aligned}$$

Calculating the covariance is a good way to assess whether two variables are related to each other. A positive covariance indicates that as one variable deviates from the mean, the other variable deviates in the same direction. On the other hand, a negative covariance indicates that as one variable deviates from the mean (e.g. increases), the other deviates from the mean in the opposite direction (e.g. decreases).

There is, however, one problem with covariance as a measure of the relationship between variables and that is that it depends upon the scales of measurement used. So, covariance is not a standardized measure. For example, if we use the data above and assume that they represented two variables measured in miles then the covariance is 4.25 (as calculated above). If we then convert these data into kilometres (by multiplying all values by 1.609) and calculate the covariance again then we should find that it increases to 11. This dependence on the scale of measurement is a problem because it means that we cannot compare covariances in an objective way – so, we cannot say whether a covariance is particularly large or small relative to another data set unless both data sets were measured in the same units.

6.3.2. Standardization and the correlation coefficient ①

To overcome the problem of dependence on the measurement scale, we need to convert the covariance into a standard set of units. This process is known as **standardization**. A very basic form of standardization would be to insist that all experiments use the same units of measurement, say metres – that way, all results could be easily compared. However, what happens if you want to measure attitudes – you'd be hard pushed to measure them in metres! Therefore, we need a unit of measurement into which any scale of measurement can be converted. The unit of measurement we use is the *standard deviation*. We came across this measure in section 2.4.1 and saw that, like the variance, it is a measure of the average deviation from the mean. If we divide any distance from the mean by the standard deviation, it gives us that distance in standard deviation units. For example, for the data in Table 6.1, the standard deviation for the number of packets bought is approximately 3.0 (the exact value is 2.91). In Figure 6.2 we can see that the observed value for participant 1 was 3 packets less than the mean (so, there was an error of −3 packets of sweets). If we divide this deviation, −3, by the standard deviation, which is approximately 3, then we get a value of −1. This tells us that the difference between participant 1's score and the mean was −1 standard deviation. So, we can express the deviation from the mean for a participant in standard units by dividing the observed deviation by the standard deviation.

It follows from this logic that if we want to express the covariance in a standard unit of measurement we can simply divide by the standard deviation. However, there are two variables and, hence, two standard deviations. Now, when we calculate the covariance we actually calculate two deviations (one for each variable) and then multiply them. Therefore, we do the same for the standard deviations: we multiply them and divide by the product of this multiplication. The standardized covariance is known as a *correlation coefficient* and is defined by equation (6.3) in which s_x is the standard deviation of the first variable and s_y is the standard deviation of the second variable (all other letters are the same as in the equation defining covariance):

$$r = \frac{\text{cov}_{xy}}{s_x s_y} = \frac{\sum(x_i - \bar{x})(y_i - \bar{y})}{(N-1)s_x s_y} \tag{6.3}$$

The coefficient in equation (6.3) is known as the Pearson product-moment correlation coefficient or **Pearson correlation coefficient** (for a really nice explanation of why it was originally called the 'product-moment' correlation see Miles & Banyard, 2007) and was invented by Karl Pearson (see Jane Superbrain Box 6.1).[2] If we look back at Table 6.1 we see that the standard deviation for the number of adverts watched (s_x) was 1.67, and for the number of packets of crisps bought (s_y) was 2.92. If we multiply these together we get 1.67 × 2.92 = 4.88. Now, all we need to do is take the covariance, which we calculated a few pages ago as being 4.25, and divide by these multiplied standard deviations. This gives us $r = 4.25/4.88 = .87$.

By standardizing the covariance we end up with a value that has to lie between −1 and +1 (if you find a correlation coefficient less than −1 or more than +1 you can be sure that something has gone hideously wrong!). A coefficient of +1 indicates that the two variables are perfectly positively correlated, so as one variable increases, the other increases by a proportionate amount. Conversely, a coefficient of −1 indicates a perfect negative relationship: if one variable increases, the other decreases by a proportionate amount. A coefficient of zero indicates no linear relationship at all and so if one variable changes, the other stays the same. We also saw in section 2.6.4 that because the correlation coefficient is a standardized measure of an observed effect, it is a commonly used measure of the size of an effect and that values of ±.1 represent a small effect, ±.3 is a medium effect and ±.5 is a large effect (although I re-emphasize my caveat that these canned effect sizes are no substitute for interpreting the effect size within the context of the research literature).

[2] You will find Pearson's product-moment correlation coefficient denoted by both r and R. Typically, the upper-case form is used in the context of regression because it represents the multiple correlation coefficient; however, for some reason, when we square r (as in section 6.5.2.3) an upper case R is used. Don't ask me why – it's just to confuse me I suspect.

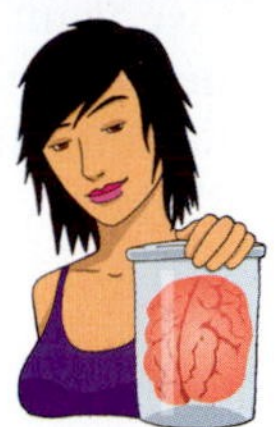

JANE SUPERBRAIN 6.1

Who said statistics was dull? ①

Students often think that statistics is dull, but back in the early 1900s it was anything but dull with various prominent figures entering into feuds on a soap opera scale. One of the most famous was between Karl Pearson and Ronald Fisher (whom we met in Chapter 2). It began when Pearson published a paper of Fisher's in his journal but made comments in his editorial that, to the casual reader, belittled Fisher's work. Two years later Pearson's group published work following on from Fisher's paper without consulting him. The antagonism persisted with Fisher turning down a job to work in Pearson's group and publishing 'improvements' on Pearson's ideas. Pearson for his part wrote in his own journal about apparent errors made by Fisher.

Another prominent statistician, Jerzy Neyman, criticized some of Fisher's most important work in a paper delivered to the Royal Statistical Society on 28 March 1935 at which Fisher was present. Fisher's discussion of the paper at that meeting directly attacked Neyman. Fisher more or less said that Neyman didn't know what he was talking about and didn't understand the background material on which his work was based. Relations soured so much that while they both worked at University College London, Neyman openly attacked many of Fisher's ideas in lectures to his students. The two feuding groups even took afternoon tea (a common practice in the British academic community of the time) in the same room but at different times! The truth behind who fuelled these feuds is, perhaps, lost in the mists of time, but Zabell (1992) makes a sterling effort to unearth it.

Basically then, the founders of modern statistical methods were a bunch of squabbling children. Nevertheless, these three men were astonishingly gifted individuals. Fisher, in particular, was a world leader in genetics, biology and medicine as well as possibly the most original mathematical thinker ever (Barnard, 1963; Field, 2005d; Savage, 1976).

6.3.3. The significance of the correlation coefficient ③

Although we can directly interpret the size of a correlation coefficient, we have seen in Chapter 2 that scientists like to test hypotheses using probabilities. In the case of a correlation coefficient we can test the hypothesis that the correlation is different from zero (i.e. different from 'no relationship'). If we find that our observed coefficient was very unlikely to happen if there was no effect in the population then we can gain confidence that the relationship that we have observed is statistically meaningful.

There are two ways that we can go about testing this hypothesis. The first is to use our trusty z-scores that keep cropping up in this book. As we have seen, z-scores are useful because we know the probability of a given value of z occurring, if the distribution from which it comes is normal. There is one problem with Pearson's r, which is that it is known to have a sampling distribution that is not normally distributed. This is a bit of a nuisance, but luckily thanks to our friend Fisher we can adjust r so that its sampling distribution *is* normal as follows (Fisher, 1921):

$$z_r = \frac{1}{2}\log_e\left(\frac{1+r}{1-r}\right) \tag{6.4}$$

The resulting z_r has a standard error of:

$$\text{SE}_{z_r} = \frac{1}{\sqrt{N-3}} \tag{6.5}$$

For our advert example, our $r = .87$ becomes 1.33 with a standard error of .71.

We can then transform this adjusted r into a z-score just as we have done for raw scores, and for skewness and kurtosis values in previous chapters. If we want a z-score that represents the size of the correlation relative to a particular value, then we simply compute a z-score using the value that we want to test against and the standard error. Normally we want to see whether the correlation is different from 0, in which case we can subtract 0 from the observed value of r and divide by the standard error (in other words, we just divide z_r by its standard error):

$$z = \frac{z_r}{\text{SE}z_r} \tag{6.6}$$

For our advert data this gives us 1.33/.71 = 1.87. We can look up this value of z (1.87) in the table for the normal distribution in the Appendix and get the one-tailed probability from the column labelled 'Smaller Portion'. In this case the value is .0307. To get the two-tailed probability we simply multiply the one-tailed probability value by 2, which gives us .0614. As such the correlation is significant, $p < .05$ one-tailed, but not two-tailed.

In fact, the hypothesis that the correlation coefficient is different from 0 is usually (SPSS, for example, does this) tested not using a z-score, but using a t-statistic with $N - 2$ degrees of freedom, which can be directly obtained from r:

$$t_r = \frac{r\sqrt{N-2}}{\sqrt{1-r^2}} \tag{6.7}$$

You might wonder then why I told you about z-scores. Partly it was to keep the discussion framed in concepts with which you are already familiar (we don't encounter the t-test properly for a few chapters), but also it is useful background information for the next section.

6.3.4. Confidence intervals for *r* ③

I was moaning earlier on about how SPSS doesn't make tea for you. Another thing that it doesn't do is compute confidence intervals for r. This is a shame because as we have seen in Chapter 2 these intervals tell us something about the likely value (in this case of the correlation) in the population. However, we can calculate these manually. To do this we need to take advantage of what we learnt in the previous section about converting r to z_r (to make the sampling distribution normal), and using the associated standard errors. We can then construct a confidence interval in the usual way. For a 95% confidence interval we have (see section 2.5.2.1):

$$\text{lower boundary of confidence interval} = \overline{X} - (1.96 \times \text{SE})$$

$$\text{upper boundary of confidence interval} = \overline{X} + (1.96 \times \text{SE})$$

In the case of our transformed correlation coefficients these equations become:

$$\text{lower boundary of confidence interval} = z_r - (1.96 \times \text{SE}_{z_r})$$

$$\text{upper boundary of confidence interval} = z_r + (1.96 \times \text{SE}_{z_r})$$

For our advert data this gives us 1.33 – (1.96 × .71) = –0.062, and 1.33 + (1.96 × .71) = 2.72. Remember that these values are in the z_r metric and so we have to convert back to correlation coefficients using:

$$r = \frac{e^{(2z_r)} - 1}{e^{(2z_r)} + 1} \tag{6.8}$$

This gives us an upper bound of $r = .991$ and a lower bound of −0.062 (because this value is so close to zero the transformation to z has no impact).

OLIVER TWISTED

Please, Sir, can I have some more … confidence intervals?

'These confidence intervals are rubbish,' says Oliver, 'they're too confusing and I hate equations, and the values we get will only be approximate. Can't we get SPSS to do it for us while we check Facebook?' Well, no you can't. Except you sort of can with some syntax. I've written some SPSS syntax which will compute confidence intervals for r for you. To find out more, read the additional material for this chapter on the companion website. Or check Facebook, the choice is yours.

CRAMMING SAM'S TIPS

- A crude measure of the relationship between variables is the *covariance*.
- If we standardize this value we get *Pearson's correlation coefficient, r*.
- The correlation coefficient has to lie between −1 and +1.
- A coefficient of +1 indicates a perfect positive relationship, a coefficient of −1 indicates a perfect negative relationship, a coefficient of 0 indicates no linear relationship at all.
- The correlation coefficient is a commonly used measure of the size of an effect: values of ±.1 represent a small effect, ±.3 is a medium effect and ±.5 is a large effect.

6.3.5. A word of warning about interpretation: causality ①

Considerable caution must be taken when interpreting correlation coefficients because they give no indication of the direction of *causality*. So, in our example, although we can conclude that as the number of adverts watched increases, the number of packets of toffees bought increases also, we cannot say that watching adverts *causes* us to buy packets of toffees. This caution is for two reasons:

- **The third-variable problem:** We came across this problem in section 1.6.2. To recap, in any correlation, causality between two variables cannot be assumed because there may be other measured or unmeasured variables affecting the results. This is known as the *third-variable* problem or the *tertium quid* (see section 1.6.2 and Jane Superbrain Box 1.1).

- **Direction of causality**: Correlation coefficients say nothing about which variable causes the other to change. Even if we could ignore the third-variable problem described above, and we could assume that the two correlated variables were the only important ones, the correlation coefficient doesn't indicate in which direction causality operates. So, although it is intuitively appealing to conclude that watching adverts causes us to buy packets of toffees, there is no *statistical* reason why buying packets of toffees cannot cause us to watch more adverts. Although the latter conclusion makes less intuitive sense, the correlation coefficient does not tell us that it isn't true.

6.4. Data entry for correlation analysis using SPSS ①

Data entry for correlation, regression and multiple regression is straightforward because each variable is entered in a separate column. So, for each variable you have measured, create a variable in the data editor with an appropriate name, and enter a participant's scores across one row of the data editor. There may be occasions on which you have one or more categorical variables (such as gender) and these variables can also be entered in a column (but remember to define appropriate value labels). As an example, if we wanted to calculate the correlation between the two variables in Table 6.1 we would enter these data as in Figure 6.3. You can see that each variable is entered in a separate column, and each row represents a single individual's data (so the first consumer saw 5 adverts and bought 8 packets).

SELF-TEST Enter the advert data and use the Chart Editor to produce a scatterplot (number of packets bought on the *y*-axis, and adverts watched on the *x*-axis) of the data.

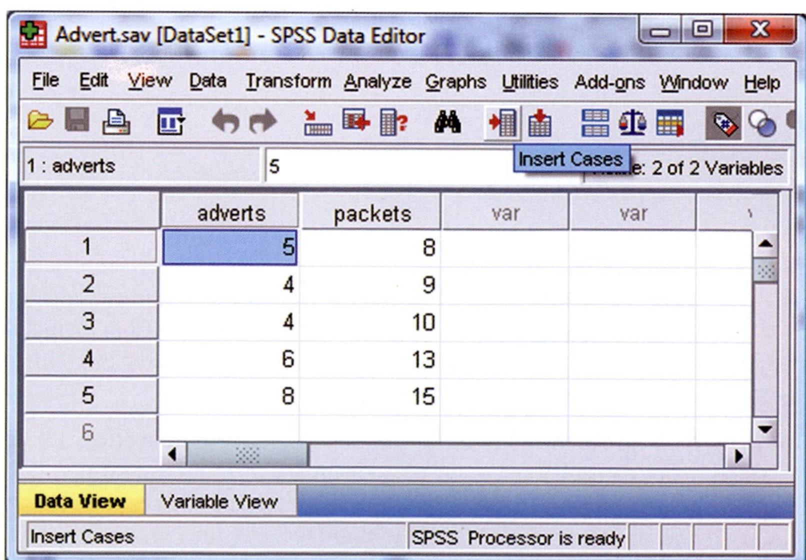

FIGURE 6.3 Data entry for correlation

6.5. Bivariate correlation ①

6.5.1. General procedure for running correlations on SPSS ①

There are two types of correlation: *bivariate* and *partial*. A **bivariate correlation** is a correlation between two variables (as described at the beginning of this chapter) whereas a **partial correlation** looks at the relationship between two variables while 'controlling' the effect of one or more additional variables. Pearson's product-moment correlation coefficient (described earlier) and Spearman's rho (see section 6.5.3) are examples of bivariate correlation coefficients.

Let's return to the example from Chapter 4 about exam scores. Remember that a psychologist was interested in the effects of exam stress and revision on exam performance. She had devised and validated a questionnaire to assess state anxiety relating to exams (called the Exam Anxiety Questionnaire, or EAQ). This scale produced a measure of anxiety scored out of 100. Anxiety was measured before an exam, and the percentage mark of each student on the exam was used to assess the exam performance. She also measured the number of hours spent revising. These data are in **Exam Anxiety.sav** on the companion website. We have already created scatterplots for these data (section 4.8) so we don't need to do that again.

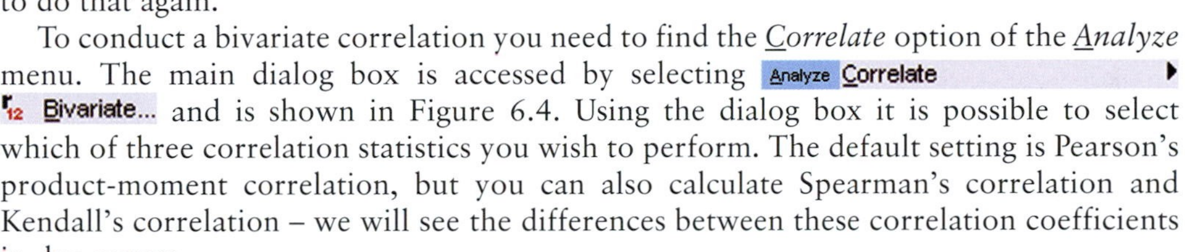

To conduct a bivariate correlation you need to find the *Correlate* option of the *Analyze* menu. The main dialog box is accessed by selecting Analyze Correlate ▸ Bivariate... and is shown in Figure 6.4. Using the dialog box it is possible to select which of three correlation statistics you wish to perform. The default setting is Pearson's product-moment correlation, but you can also calculate Spearman's correlation and Kendall's correlation – we will see the differences between these correlation coefficients in due course.

Having accessed the main dialog box, you should find that the variables in the data editor are listed on the left-hand side of the dialog box (Figure 6.4). There is an empty box

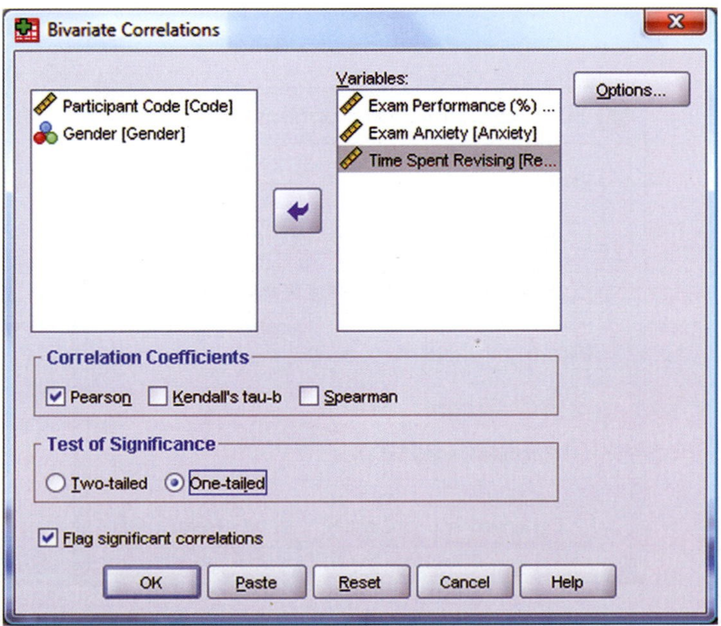

FIGURE 6.4 Dialog box for conducting a bivariate correlation

labelled *Variables* on the right-hand side. You can select any variables from the list using the mouse and transfer them to the *Variables* box by dragging them there or clicking on ➜. SPSS will create a table of correlation coefficients for all of the combinations of variables. This table is called a correlation matrix. For our current example, select the variables **Exam performance, Exam anxiety** and **Time spent revising** and transfer them to the *Variables* box by clicking on ➜. Having selected the variables of interest you can choose between three correlation coefficients: Pearson's product-moment correlation coefficient (Pearson) Spearman's rho (☑ Spearman) and Kendall's tau (☑ Kendall's tau-b). Any of these can be selected by clicking on the appropriate tick-box with a mouse.

In addition, it is possible to specify whether or not the test is one- or two-tailed (see section 2.6.2). To recap, a one-tailed test should be selected when you have a directional hypothesis (e.g. 'the more anxious someone is about an exam, the worse their mark will be'). A two-tailed test (the default) should be used when you cannot predict the nature of the relationship (i.e. 'I'm not sure whether exam anxiety will improve or reduce exam marks'). Therefore, if you have a directional hypothesis click on ⦿ One-tailed, whereas if you have a non-directional hypothesis click on ⦿ Two-tailed.

If you click on Options... then another dialog box appears with two *Statistics* options and two options for missing values (Figure 6.5). The *Statistics* options are enabled only when Pearson's correlation is selected; if Pearson's correlation is not selected then these options are disabled (they appear in a light grey rather than black and you can't activate them). This deactivation occurs because these two options are meaningful only for parametric data and the Pearson correlation is used with those kinds of data. If you select the tick-box labelled *Means and standard deviations* then SPSS will produce the mean and standard deviation of all of the variables selected for analysis. If you activate the tick-box labelled *Cross-product deviations and covariances* then SPSS will give you the values of these statistics for each of the variables in the analysis. The cross-product deviations tell us the sum of the products of mean corrected variables, which is simply the numerator (top half) of equation (6.2). The covariances option gives us values of the covariance between variables, which could be calculated manually using equation (6.2). In other words, these covariance values are the cross-product deviations divided by (N–1) and represent the unstandardized correlation coefficient. In most instances, you will not need to use these options but they occasionally come in handy (see Oliver Twisted). We can also decide how to deal with missing values (see SPSS Tip 6.1).

FIGURE 6.5 Dialog box for bivariate correlation options

OLIVER TWISTED

Please, Sir, can I have some more ... options?

Oliver is so excited to get onto analysing his data that he doesn't want me to spend pages waffling on about options that you will probably never use. 'Stop writing, you waffling fool,' he says. 'I want to analyse my data.' Well, he's got a point. If you want to find out more about what the Options... do in correlation, then the additional material for this chapter on the companion website will tell you.

SPSS TIP 6.1 Pairwise or listwise? ①

As we run through the various analyses in this book, many of them have additional options that can be accessed by clicking on Options... . Often part of the resulting *Options* dialog box will ask you if you want to exclude cases 'pairwise', 'analysis by analysis' or 'listwise'. First, we can exclude cases listwise, which means that if a case has a missing value for any variable, then they are excluded from the whole analysis. So, for example, in our exam anxiety data if one of our students had reported their anxiety and we knew their exam performance but we didn't have data about their revision time, then their data would not be used to calculate any of the correlations: *they would be completely excluded from the analysis*. Another option is to exclude cases on a pairwise or analysis-by-analysis basis, which means that if a participant has a score missing for a particular variable or analysis, then their data are excluded only from calculations involving the variable for which they have no score. For our student about whom we don't have any revision data, this means that their data would be excluded when calculating the correlation between exam scores and revision time, and when calculating the correlation between exam anxiety and revision time; however, the student's scores would be *included* when calculating the correlation between exam anxiety and exam performance because for this pair of variables we have both of their scores.

6.5.2. Pearson's correlation coefficient ①

6.5.2.1. Assumptions of Pearson's *r* ①

Pearson's correlation coefficient was described in full at the beginning of this chapter. Pearson's correlation requires only that data are interval (see section 1.5.1.2) for it to be an accurate measure of the linear relationship between two variables. However, if you want to establish whether the correlation coefficient is significant, then more assumptions are required: for the test statistic to be valid the sampling distribution has to be normally distributed and as we saw in Chapter 5 we assume that it is if our sample data are normally distributed (or if we have a large sample). Although typically, to assume that the sampling distribution is normal, we would want both variables to be normally distributed, there is one exception to this rule: one of the variables can be a categorical variable provided there are only two categories (in fact, if you look at section 6.5.5 you'll see that this is the same as doing a *t*-test, but I'm jumping the gun a bit). In any case, if your data are non-normal (see Chapter 5) or are not measured at the interval level then you should deselect the Pearson tick-box.

FIGURE 6.6
Karl Pearson

6.5.2.2. Running Pearson's *r* on SPSS ①

We have already seen how to access the main dialog box and select the variables for analysis earlier in this section (Figure 6.4). To obtain Pearson's correlation coefficient simply select the appropriate box (☑ Pearson) – SPSS selects this option by default. Our researcher predicted that (1) as anxiety increases, exam performance will decrease, and (2) as the time spent revising increases, exam performance will increase. Both of these are directional hypotheses, so both tests are one-tailed. To ensure that the output displays the one-tailed significance values select ⦿ One-tailed and then click on OK to run the analysis.

SPSS OUTPUT 6.1
Output for a Pearson's correlation

Correlations

		Exam performance (%)	Exam Anxiety	Time spent revising
Exam performance (%)	Pearson Correlation	1.000	-.441**	.397**
	Sig. (1-tailed)	.	.000	.000
	N	103	103	103
Exam Anxiety	Pearson Correlation	-.441**	1.000	-.709**
	Sig. (1-tailed)	.000	.	.000
	N	103	103	103
Time spent revising	Pearson Correlation	.397**	-.709**	1.000
	Sig. (1-tailed)	.000	.000	.
	N	103	103	103

**. Correlation is significant at the 0.01 level (1-tailed).

SPSS Output 6.1 provides a matrix of the correlation coefficients for the three variables. Underneath each correlation coefficient both the significance value of the correlation and the sample size (*N*) on which it is based are displayed. Each variable is perfectly correlated with itself (obviously) and so $r = 1$ along the diagonal of the table. Exam performance is negatively related to exam anxiety with a Pearson correlation coefficient of $r = -.441$ and the significance value is less than .001 (as indicated by the double asterisk after the coefficient). This significance value tells us that the probability of getting a correlation coefficient this big in a sample of 103 people if the null hypothesis were true (there was no relationship between these variables) is very low (close to zero in fact). Hence, we can gain confidence that there is a genuine relationship between exam performance and anxiety. Our criterion for significance is usually .05 (see section 2.6.1) so SPSS marks any correlation coefficient significant at this level with an asterisk. The output also shows that exam performance is positively related to the amount of time spent revising, with a coefficient of $r = .397$, which is also significant at $p < .001$. Finally, exam anxiety appears to be negatively related to the time spent revising, $r = -.709$, $p < .001$.

In psychological terms, this all means that as anxiety about an exam increases, the percentage mark obtained in that exam decreases. Conversely, as the amount of time revising increases, the percentage obtained in the exam increases. Finally, as revision time increases, the student's anxiety about the exam decreases. So there is a complex interrelationship between the three variables.

6.5.2.3. Using R^2 for interpretation ①

Although we cannot make direct conclusions about causality from a correlation, we can take the correlation coefficient a step further by squaring it. The correlation coefficient squared (known as the **coefficient of determination**, R^2) is a measure of the amount of variability in one variable that is shared by the other. For example, we may look at the relationship between exam anxiety and exam performance. Exam performances vary from person to person because of any number of factors (different ability, different levels of preparation and so on). If we add up all of this variability (rather like when we calculated the sum of squares in section 2.4.1) then we would have an estimate of how much variability exists in exam performances. We can then use R^2 to tell us how much of this variability is shared by exam anxiety. These two variables had a correlation of −0.4410 and so the value of R^2 will be $(-0.4410)^2 = 0.194$. This value tells us how much of the variability in exam performance is shared by exam anxiety.

If we convert this value into a percentage (multiply by 100) we can say that exam anxiety shares 19.4% of the variability in exam performance. So, although exam anxiety was highly correlated with exam performance, it can account for only 19.4% of variation in exam scores. To put this value into perspective, this leaves 80.6% of the variability still to be accounted for by other variables. I should note at this point that although R^2 is an extremely useful measure of the substantive importance of an effect, it cannot be used to infer causal relationships. Although we usually talk in terms of 'the variance in *y accounted for* by *x*', or even the variation in one variable *explained* by the other, this still says nothing about which way causality runs. So, although exam anxiety can account for 19.4% of the variation in exam scores, it does not necessarily cause this variation.

6.5.3. Spearman's correlation coefficient ①

Spearman's correlation coefficient (Spearman, 1910; Figure 6.7), r_s, is a non-parametric statistic and so can be used when the data have violated parametric assumptions such as non-normally distributed data (see Chapter 5). You'll sometimes hear the test referred to as Spearman's rho (pronounced 'row', as in 'row your boat gently down the stream').

FIGURE 6.7 Charles Spearman, ranking furiously

Spearman's test works by first ranking the data (see section 15.3.1), and then applying Pearson's equation (equation (6.3)) to those ranks.

I was born in England, which has some bizarre traditions. One such oddity is The World's Biggest Liar Competition held annually at the Santon Bridge Inn in Wasdale (in the Lake District). The contest honours a local publican, 'Auld Will Ritson', who in the nineteenth century was famous in the area for his far-fetched stories (one such tale being that Wasdale turnips were big enough to be hollowed out and used as garden sheds). Each year locals are encouraged to attempt to tell the biggest lie in the world (lawyers and politicians are apparently banned from the competition). Over the years there have been tales of mermaid farms, giant moles, and farting sheep blowing holes in the ozone layer. (I am thinking of entering next year and reading out some sections of this book.)

What if my data are not parametric?

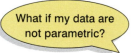

Imagine I wanted to test a theory that more creative people will be able to create taller tales. I gathered together 68 past contestants from this competition and asked them where they were placed in the competition (first, second, third, etc.) and also gave them a creativity questionnaire (maximum score 60). The position in the competition is an ordinal variable (see section 1.5.1.2) because the places are categories but have a meaningful order (first place is better than second place and so on). Therefore, Spearman's correlation coefficient should be used (Pearson's *r* requires interval or ratio data). The data for this study are in the file **The Biggest Liar.sav**. The data are in two columns: one labelled **Creativity** and one labelled **Position** (there's actually a third variable in there but we will ignore it for the time being). For the **Position** variable, each of the categories described above has been coded with a numerical value. First place has been coded with the value 1, with positions being labelled 2, 3 and so on. Note that for each numeric code I have provided a value label (just like we did for coding variables). I have also set the *Measure* property of this variable to Ordinal.

The procedure for doing a Spearman correlation is the same as for a Pearson correlation except that in the *Bivariate Correlations* dialog box (Figure 6.4), we need to select Spearman and deselect the option for a Pearson correlation. At this stage, you should also specify whether you require a one- or two-tailed test. I predicted that more creative people would tell better lies. This hypothesis is directional and so a one-tailed test should be selected.

SPSS OUTPUT 6.2

Correlations

			Creativity	Position in Best Liar Competition
Spearman's rho	Creativity	Correlation Coefficient	1.000**	-.373**
		Sig. (1-tailed)	.	.001
		N	68	68
	Position in Best Liar Competition	Correlation Coefficient	-.373**	1.000**
		Sig. (1-tailed)	.001	.
		N	68	68

**. Correlation is significant at the 0.01 level (1-tailed).

SPSS Output 6.2 shows the output for a Spearman correlation on the variables **Creativity** and **Position**. The output is very similar to that of the Pearson correlation: a matrix is displayed giving the correlation coefficient between the two variables (–.373), underneath is the significance value of this coefficient (.001) and finally the sample size (68).[3] The significance value for this correlation coefficient is less than .05; therefore, it can be concluded that there is a significant relationship between creativity scores and how well someone did in the World's Biggest Liar Competition. Note that the relationship is negative: as creativity increased, position decreased. This might seem contrary to what we predicted until you remember that a low number means that you did well in the competition (a low number such as 1 means you came first, and a high number like 4 means you came fourth). Therefore, our hypothesis is supported: as creativity increased, so did success in the competition.

SELF-TEST Did creativity cause success in the World's Biggest Liar Competition?

6.5.4. Kendall's tau (non-parametric) ①

Kendall's tau, τ, is another non-parametric correlation and it should be used rather than Spearman's coefficient when you have a small data set with a large number of tied ranks. This means that if you rank all of the scores and many scores have the same rank, then Kendall's tau should be used. Although Spearman's statistic is the more popular of the two coefficients, there is much to suggest that Kendall's statistic is actually a better estimate of the correlation in the population (see Howell, 1997: 293). As such, we can draw more accurate generalizations from Kendall's statistic than from Spearman's. To carry out Kendall's correlation on the world's biggest liar data simply follow the same steps as for the Pearson and Spearman correlations but select ☑ Kendall's tau-b and deselect the Pearson and Spearman options. The output is much the same as for Spearman's correlation.

[3] It is good to check that the value of *N* corresponds to the number of observations that were made. If it doesn't then data may have been excluded for some reason.

SPSS OUTPUT 6.3

Correlations

			Creativity	Position in Best Liar Competition
Kendall's tau_b	Creativity	Correlation Coefficient	1.000**	-.300**
		Sig. (1-tailed)	.	.001
		N	68	68
	Position in Best Liar Competition	Correlation Coefficient	-.300**	1.000**
		Sig. (1-tailed)	.001	.
		N	68	68

**. Correlation is significant at the 0.01 level (1-tailed).

You'll notice from SPSS Output 6.3 that the actual value of the correlation coefficient is closer to zero than the Spearman correlation (it has increased from −.373 to −.300). Despite the difference in the correlation coefficients we can still interpret this result as being a highly significant relationship (because the significance value of .001 is less than .05). However, Kendall's value is a more accurate gauge of what the correlation in the population would be. As with the Pearson correlation, we cannot assume that creativity caused success in the World's Biggest Liar Competition.

SELF-TEST Conduct a Pearson correlation analysis of the advert data from the beginning of the chapter.

6.5.5. Biserial and point–biserial correlations ③

SMART ALEX ONLY

The biserial and point–biserial correlation coefficients are distinguished by only a conceptual difference yet their statistical calculation is quite different. These correlation coefficients are used when one of the two variables is dichotomous (i.e. it is categorical with only two categories). An example of a dichotomous variable is being pregnant, because a woman can be either pregnant or not (she cannot be 'a bit pregnant'). Often it is necessary to investigate relationships between two variables when one of the variables is dichotomous. The difference between the use of biserial and point–biserial correlations depends on whether the dichotomous variable is discrete or continuous. This difference is very subtle. A discrete, or true, dichotomy is one for which there is no underlying continuum between the categories. An example of this is whether someone is dead or alive: a person can be only dead or alive, they can't be 'a bit dead'. Although you might describe a person as being 'half-dead' – especially after a heavy drinking session – they are clearly still alive if they are still breathing! Therefore, there is no continuum between the two categories. However, it is possible to have a dichotomy for which a continuum does exist. An example is passing or failing a statistics test: some people will only just fail while others will fail by a large margin; likewise some people will scrape a pass while others will clearly excel. So although participants fall into only two categories there is clearly an underlying continuum along which people lie. Hopefully, it is clear that in this case there is some kind of continuum underlying the dichotomy, because some people passed or failed more dramatically than others. The **point–biserial correlation** coefficient (r_{pb}) is used when one variable is a discrete dichotomy (e.g. pregnancy), whereas the **biserial correlation** coefficient (r_b) is used when one variable is a continuous dichotomy (e.g. passing or failing an exam). The biserial correlation coefficient cannot be calculated directly in SPSS: first you must calculate the point–biserial correlation coefficient and then use an equation to adjust that figure.

Imagine that I was interested in the relationship between the gender of a cat and how much time it spent away from home (what can I say? I love cats so these things interest me). I had heard that male cats disappeared for substantial amounts of time on long-distance roams around the neighbourhood (something about hormones driving them to find mates) whereas female cats tended to be more homebound. So, I used this as a purr-fect (sorry!) excuse to go and visit lots of my friends and their cats. I took a note of the gender of the cat and then asked the owners to note down the number of hours that their cat was absent from home over a week. Clearly the time spent away from home is measured at an interval level – and let's assume it meets the other assumptions of parametric data – while the gender of the cat is discrete dichotomy. A point–biserial correlation has to be calculated and this is simply a Pearson correlation when the dichotomous variable is coded with 0 for one category and 1 for the other (actually you can use any values and SPSS will change the lower one to 0 and the higher one to 1 when it does the calculations). So, to conduct these correlations in SPSS assign the **Gender** variable a coding scheme as described in section 3.4.2.3 (in the saved data the coding is 1 for a male and 0 for a female). The **time** variable simply has time in hours recorded as normal. These data are in the file **pbcorr.sav**.

SELF-TEST Carry out a Pearson correlation on these data (as in 6.5.2.2).

Congratulations: if you did the self-test task then you have just conducted your first point–biserial correlation. See, despite the horrible name, it's really quite easy to do. You should find that you have the same output as SPSS Output 6.4, which shows the correlation matrix of **time** and **Gender**. The point–biserial correlation coefficient is r_{pb} = .378, which has a one-tailed significance value of .001. The significance test for this correlation is actually the same as performing an independent-samples *t*-test on the data (see Chapter 9). The sign of the correlation (i.e. whether the relationship was positive or negative) will depend entirely on which way round the coding of the dichotomous variable was made. To prove that this is the case, the data file **pbcorr.sav** has an extra variable called **recode** which is the same as the variable **Gender** except that the coding is reversed (1 = female, 0 = male). If you repeat the Pearson correlation using **recode** instead of **Gender** you will find that the correlation coefficient becomes −0.378. The sign of the coefficient is completely dependent on which category you assign to which code and so we must ignore all information about the direction of the relationship. However, we can still interpret R^2 as before. So in this example, $R^2 = (0.378)^2 = .143$. Hence, we can conclude that gender accounts for 14.3% of the variability in time spent away from home.

SPSS OUTPUT 6.4

Correlations

		Time away from home (hours)	Gender of cat
Time Away from Home (Hours)	Pearson Correlation	1.000	.378**
	Sig. (1-tailed)	.	.001
	N	60	60
Gender of Cat	Pearson Correlation	.378**	1.000
	Sig. (1-tailed)	.001	.
	N	60	60

**. Correlation is significant at the 0.01 level (1-tailed).

Imagine now that we wanted to convert the point–biserial correlation into the biserial correlation coefficient (r_b) (because some of the male cats were neutered and so there might be a continuum of maleness that underlies the gender variable). We must use equation (6.9) in which p is the proportion of cases that fell into the largest category and q is the proportion of cases that fell into the smallest category. Therefore, p and q are simply the number of male and female cats. In this equation y is the ordinate of the normal distribution at the point where there is p% of the area on one side and q% on the other (this will become clearer as we do an example):

$$r_{\mathrm{b}} = \frac{r_{\mathrm{pb}}\sqrt{pq}}{y} \tag{6.9}$$

To calculate p and q access the *Frequencies* dialog box using Analyze Descriptive Statistics ▸ 123 Frequencies... and select the variable **Gender**. There is no need to click on any further options as the defaults will give you what you need to know (namely the percentage of male and female cats). It turns out that 53.3% (0.533 as a proportion) of the sample were female (this is p because it is the largest portion) while the remaining 46.7% (0.467 as a proportion) were male (this is q because it is the smallest portion). To calculate y, we use these values and the values of the normal distribution displayed in the Appendix. Figure 6.8

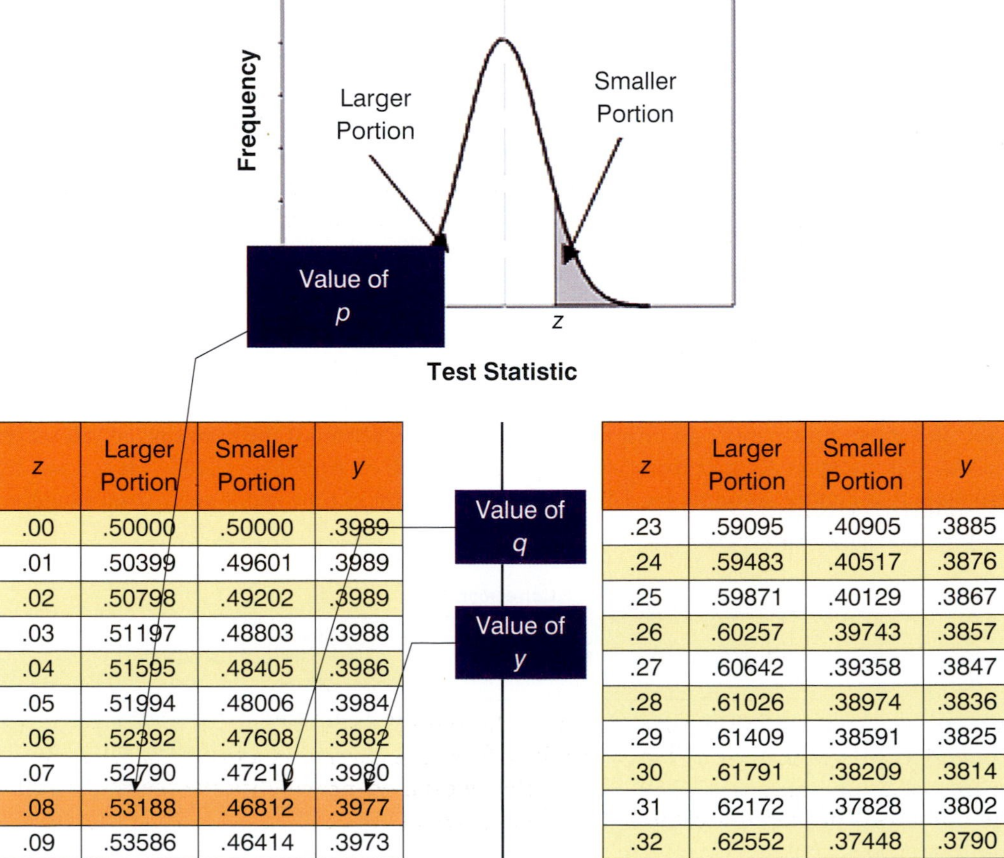

z	Larger Portion	Smaller Portion	y
.00	.50000	.50000	.3989
.01	.50399	.49601	.3989
.02	.50798	.49202	.3989
.03	.51197	.48803	.3988
.04	.51595	.48405	.3986
.05	.51994	.48006	.3984
.06	.52392	.47608	.3982
.07	.52790	.47210	.3980
.08	.53188	.46812	.3977
.09	.53586	.46414	.3973

z	Larger Portion	Smaller Portion	y
.23	.59095	.40905	.3885
.24	.59483	.40517	.3876
.25	.59871	.40129	.3867
.26	.60257	.39743	.3857
.27	.60642	.39358	.3847
.28	.61026	.38974	.3836
.29	.61409	.38591	.3825
.30	.61791	.38209	.3814
.31	.62172	.37828	.3802
.32	.62552	.37448	.3790

FIGURE 6.8 Getting the 'ordinate' of the normal distribution

shows how to find the ordinate (the value in the column labelled y) when the normal curve is split with .467 as the smaller portion and .533 as the larger portion. The figure shows which columns represent p and q and we look for our values in these columns (the exact values of 0.533 and 0.467 are not in the table so instead we use the nearest values that we can find, which are .5319 and .4681 respectively). The ordinate value is in the column y and is .3977.

If we replace these values in equation (6.9) we get .475 (see below), which is quite a lot higher than the value of the point–biserial correlation (0.378). This finding just shows you that whether you assume an underlying continuum or not can make a big difference to the size of effect that you get:

$$r_b = \frac{r_{pb}\sqrt{pq}}{y} = \frac{(0.378)\sqrt{(0.533 \times 0.467)}}{0.3977} = 0.475$$

If this process freaks you out, you can also convert the point–biserial r to the biserial r using a table published by Terrell (1982b) in which you can use the value of the point–biserial correlation (i.e. Pearson's r) and p, which is just the proportion of people in the largest group (in the above example, .53). This spares you the trouble of having to work out y in the above equation (which you're also spared from using). Using Terrell's table we get a value in this example of .48 which is the same as we calculated to 2 decimal places.

To get the significance of the biserial correlation we need to first work out its standard error. If we assume the null hypothesis (that the biserial correlation in the population is zero) then the standard error is given by (Terrell, 1982a):

$$SE_{r_b} = \frac{\sqrt{pq}}{y\sqrt{N}} \qquad (6.10)$$

This equation is fairly straightforward because it uses the values of p, q and y that we already used to calculate the biserial r. The only additional value is the sample size (N), which in this example was 60. So, our standard error is:

$$SE_{r_b} = \frac{\sqrt{(.533 \times .467)}}{.3977 \times \sqrt{60}} = .162$$

The standard error helps us because we can create a z-score (see section 1.7.4). To get a z-score we take the biserial correlation, subtract the mean in the population and divide by the standard error. We have assumed that the mean in the population is 0 (the null hypothesis), so we can simply divide the biserial correlation by its standard error:

$$Z_{r_b} = \frac{r_b - \bar{r}_b}{SE_{r_b}} = \frac{r_b - 0}{SE_{r_b}} = \frac{r_b}{SE_{r_b}} = \frac{.475}{.162} = 2.93$$

EVERYBODY

We can look up this value of z (2.93) in the table for the normal distribution in the Appendix and get the one-tailed probability from the column labelled 'Smaller Portion'. In this case the value is .00169. To get the two-tailed probability we simply multiply the one-tailed probability value by 2, which gives us .00338. As such the correlation is significant, $p < .01$.

CRAMMING SAM'S TIPS

- We can measure the relationship between two variables using *correlation coefficients*.
- These coefficients lie between −1 and +1.
- *Pearson's correlation coefficient, r,* is a parametric statistic and requires interval data for both variables. To test its significance we assume normality too.
- *Spearman's correlation coefficient,* r_s, is a non-parametric statistic and requires only ordinal data for both variables.
- *Kendall's correlation coefficient,* τ, is like Spearman's r_s but probably better for small samples.
- *The point–biserial correlation coefficient,* r_{pb}, quantifies the relationship between a continuous variable and a variable that is a discrete dichotomy (e.g. there is no continuum underlying the two categories, such as dead or alive).
- *The biserial correlation coefficient,* r_b, quantifies the relationship between a continuous variable and a variable that is a continuous dichotomy (e.g. there is a continuum underlying the two categories, such as passing or failing an exam).

6.6. Partial correlation ②

6.6.1. The theory behind part and partial correlation ②

SMART ALEX ONLY

I mentioned earlier that there is a type of correlation that can be done that allows you to look at the relationship between two variables when the effects of a third variable are held constant. For example, analyses of the exam anxiety data (in the file **ExamAnxiety.sav**) showed that exam performance was negatively related to exam anxiety, but positively related to revision time, and revision time itself was negatively related to exam anxiety. This scenario is complex, but given that we know that revision time is related to both exam anxiety and exam performance, then if we want a pure measure of the relationship between exam anxiety and exam performance we need to take account of the influence of revision time. Using the values of R^2 for these relationships, we know that exam anxiety accounts for 19.4% of the variance in exam performance, that revision time accounts for 15.7% of the variance in exam performance, and that revision time accounts for 50.2% of the variance in exam anxiety. If revision time accounts for half of the variance in exam anxiety, then it seems feasible that at least some of the 19.4% of variance in exam performance that is accounted for by anxiety is the same variance that is accounted for by revision time. As such, some of the variance in exam performance explained by exam anxiety is not *unique* and can be accounted for by revision time. A correlation between two variables in which the effects of other variables are held constant is known as a partial correlation.

Let's return to our example of exam scores, revision time and exam anxiety to illustrate the principle behind partial correlation (Figure 6.9). In part 1 of the diagram there is a box for exam performance that represents the total variation in exam scores (this value would be the variance of exam performance). There is also a box that represents the variation in exam anxiety (again, this is the variance of that variable). We know already that exam anxiety and exam performance share 19.4% of their variation (this value is the correlation coefficient squared). Therefore, the variations of these two variables overlap (because they share variance) creating a third box (the one with diagonal lines). The overlap of the boxes representing exam performance and exam anxiety is the common variance. Likewise, in

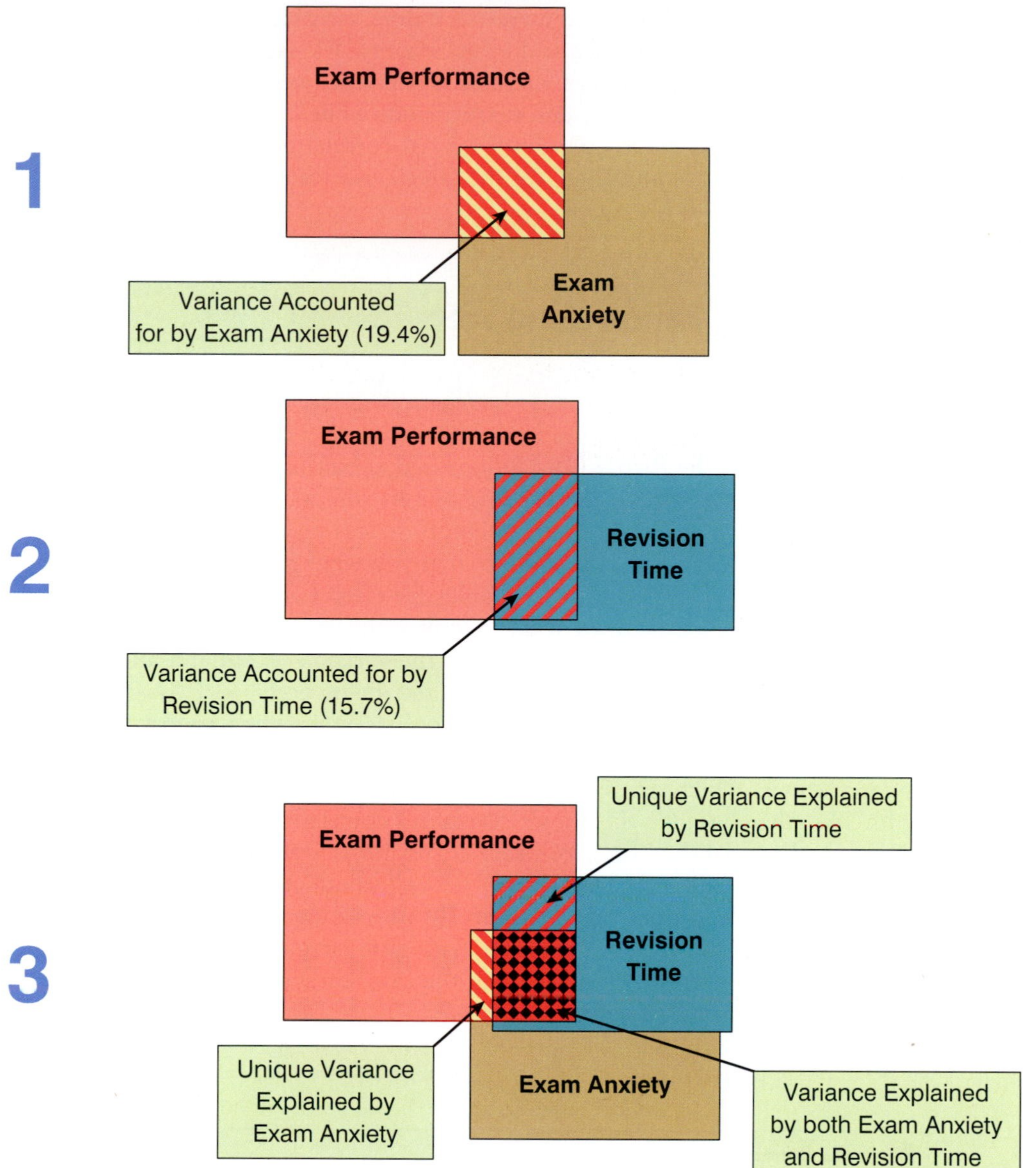

FIGURE 6.9 Diagram showing the principle of partial correlation

part 2 of the diagram the shared variation between exam performance and revision time is illustrated. Revision time shares 15.7% of the variation in exam scores. This shared variation is represented by the area of overlap (filled with diagonal lines). We know that revision time and exam anxiety also share 50% of their variation; therefore, it is very probable that some of the variation in exam performance shared by exam anxiety is the same as the variance shared by revision time.

Part 3 of the diagram shows the complete picture. The first thing to note is that the boxes representing exam anxiety and revision time have a large overlap (this is because they share 50% of their variation). More important, when we look at how revision time and anxiety contribute to exam performance we see that there is a portion of exam performance that is shared by both anxiety and revision time (the dotted area). However, there are still small chunks of the variance in exam performance that are unique to the other two variables. So, although in part 1 exam anxiety shared a large chunk of variation in exam performance, some of this overlap is also shared by revision time. If we remove the portion of variation that is also shared by revision time, we get a measure of the unique relationship between exam performance and exam anxiety. We use partial correlations to find out the size of the unique portion of variance. Therefore, we could conduct a partial correlation between exam anxiety and exam performance while 'controlling' for the effect of revision time. Likewise, we could carry out a partial correlation between revision time and exam performance while 'controlling' for the effects of exam anxiety.

6.6.2. Partial correlation using SPSS ②

Reload the **Exam Anxiety.sav** file so that, as I suggested above, we can conduct a partial correlation between exam anxiety and exam performance while 'controlling' for the effect of revision time.. To access the *Partial Correlations* dialog box (Figure 6.10) select Analyze Correlate ▸ Partial.... This dialog box lists all of the variables in the data editor on the left-hand side and there are two empty spaces on the right-hand side. The space labelled *Variables* is for listing the variables that you want to correlate and the space labelled *Controlling for* is for declaring any variables the effects of which you want to control. In the example I have described, we want to look at the unique effect of exam anxiety on exam performance and so we want to correlate the variables **exam** and **anxiety** while controlling for **revise**. Figure 6.10 shows the completed dialog box.

If you click on Options... then another dialog box appears as shown in Figure 6.11. These options are similar to those in bivariate correlation except that you can choose to compute zero-order correlations, which are simply the bivariate correlation coefficients without controlling for any other variables (i.e. they're Pearson's correlation coefficient). So, in our example, if we select the tick-box for zero-order correlations SPSS will produce a correlation matrix of **anxiety**, **exam** and **revise**. If you haven't conducted bivariate correlations before the partial correlation then this is a useful way to compare the correlations that haven't been controlled against those that have. This comparison gives you some insight into the contribution of different variables. Tick the box for zero-order correlations but leave the rest of the options as they are.

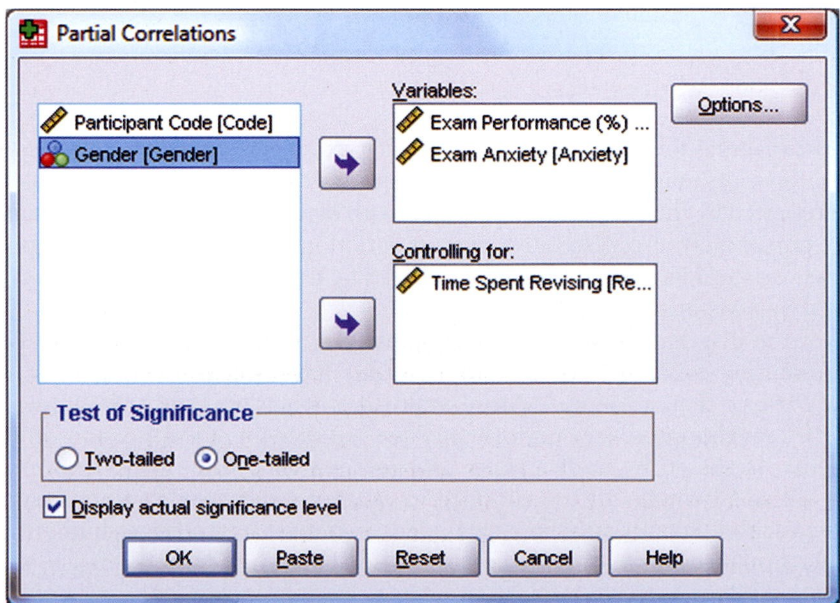

FIGURE 6.10 Main dialog box for conducting a partial correlation

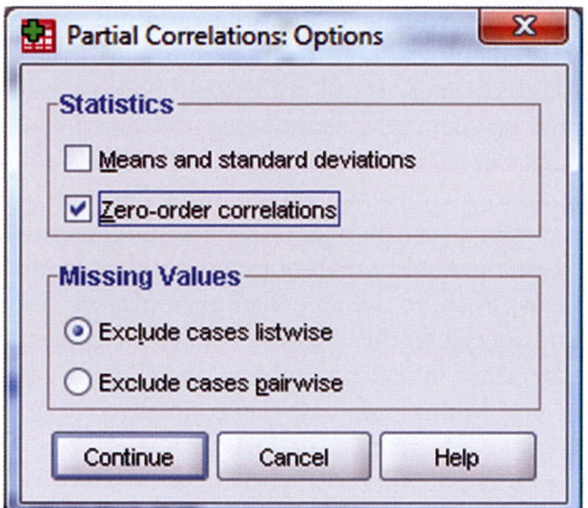

FIGURE 6.11
Options for partial correlation

Correlations

Control Variables			Exam Performance (%)	Exam Anxiety	Time Spent Revising
-none-[a]	Exam Performance (%)	Correlation	1.000	-.441	.397
		Significance (1-tailed)	.	.000	.000
		df	0	101	101
	Exam Anxiety	Correlation	-.441	1.000	-.709
		Significance (1-tailed)	.000	.	.000
		df	101	0	101
	Time Spent Revising	Correlation	.397	-.709	1.000
		Significance (1-tailed)	.000	.000	.
		df	101	101	0
Time Spent Revising	Exam Performance (%)	Correlation	1.000	-.247	
		Significance (1-tailed)	.	.006	
		df	0	100	
	Exam Anxiety	Correlation	-.247	1.000	
		Significance (1-tailed)	.006	.	
		df	100	0	

a. Cells contain zero-order (Pearson) correlations.

SPSS OUTPUT 6.5
Output from a partial correlation

SPSS Output 6.5 shows the output for the partial correlation of exam anxiety and exam performance controlling for revision time. The first thing to notice is the matrix of zero-order correlations, which we asked for using the options dialog box. The correlations displayed here are identical to those obtained from the Pearson correlation procedure (compare this matrix to the one in SPSS Output 6.1). Underneath the zero-order correlations is a matrix of correlations for the variables **anxiety** and **exam** but controlling for the effect of revision. In this instance we have controlled for one variable and so this is known as a first-order partial correlation. It is possible to control for the effects of two variables at the same time (a second-order partial correlation) or control three variables (a third-order partial correlation) and so on. First, notice that the partial correlation between exam performance and exam anxiety is −.247, which is considerably less than the correlation when the effect of revision time is not controlled for ($r = -.441$). In fact, the correlation coefficient is nearly half what it was before. Although this correlation is still statistically significant (its p-value is still below .05), the relationship is diminished. In terms of variance, the value of R^2 for the partial correlation is .06, which means that exam anxiety can now account for only 6% of the variance in exam performance. When the effects of revision time were not controlled for, exam anxiety shared 19.4% of the variation in exam scores and so the inclusion of revision time has severely diminished the amount of variation in exam scores shared by anxiety. As such, a truer measure of the role of exam anxiety has been obtained. Running this analysis has shown us that exam anxiety alone does explain some of the variation in exam scores, but there is a complex relationship between anxiety, revision and exam performance that might otherwise have been ignored. Although causality is still not certain, because relevant variables are being included, the third variable problem is, at least, being addressed in some form.

These partial correlations can be done when variables are dichotomous (including the 'third' variable). So, for example, we could look at the relationship between bladder relaxation (did the person wet themselves or not?) and the number of large tarantulas crawling up your leg controlling for fear of spiders (the first variable is dichotomous, but the second variable and 'controlled for' variables are continuous). Also, to use an earlier example, we could examine the relationship between creativity and success in the world's greatest liar contest controlling for whether someone had previous experience in the competition (and therefore had some idea of the type of tale that would win) or not. In this latter case the 'controlled for' variable is dichotomous.[4]

6.6.3. Semi-partial (or part) correlations ②

In the next chapter, we will come across another form of correlation known as a **semi-partial correlation** (also referred to as a part correlation). While I'm babbling on about partial correlations it is worth my explaining the difference between this type of correlation and a semi-partial correlation. When we do a partial correlation between two variables, we control for the effects of a third variable. Specifically, the effect that the third variable has on *both* variables in the correlation is controlled. In a semi-partial correlation we control for the effect that the third variable has on only one of the variables in the correlation. Figure 6.12 illustrates this principle for the exam performance data. The partial correlation that we calculated took account not only of the effect of revision on exam performance, but also of the effect of revision on anxiety. If we were to calculate the semi-partial correlation for the same data, then this would control for only the effect of revision on exam performance (the effect of revision on exam anxiety is ignored). Partial correlations are most useful for looking at the unique relationship between two variables when other variables are ruled out. Semi-partial correlations are, therefore, useful when trying to explain the variance in one particular variable (an outcome) from a set of predictor variables. (Bear this in mind when you read Chapter 7.)

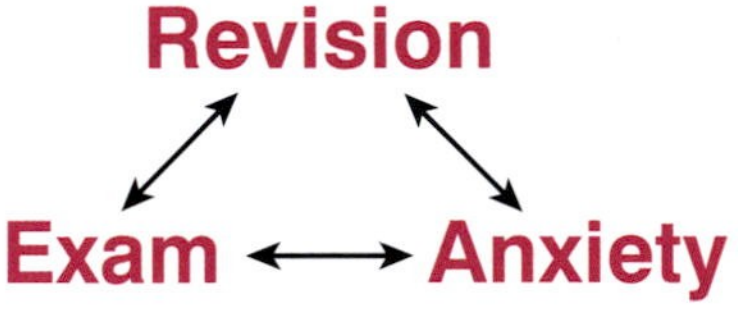

Partial Correlation

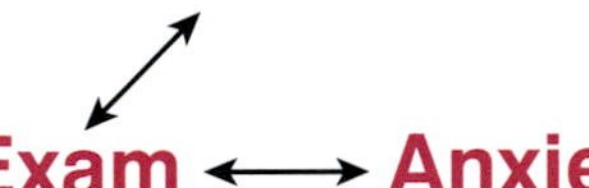

Semi-Partial Correlation

FIGURE 6.12 The difference between a partial and a semi-partial correlation

CRAMMING SAM'S TIPS

- A *partial correlation* quantifies the relationship between two variables while controlling for the effects of a third variable on *both* variables in the original correlation.
- A *semi-partial correlation* quantifies the relationship between two variables while controlling for the effects of a third variable on only *one* of the variables in the original correlation.

[4] Both these examples are, in fact, simple cases of hierarchical regression (see the next chapter) and the first example is also an example of analysis of covariance. This may be confusing now, but as we progress through the book I hope it'll become clearer that virtually all of the statistics that you use are actually the same things dressed up in different names.

6.7. Comparing correlations ③

6.7.1. Comparing independent *r*s ③

Sometimes we want to know whether one correlation coefficient is bigger than another. For example, when we looked at the effect of exam anxiety on exam performance, we might have been interested to know whether this correlation was different in men and women. We could compute the correlation in these two samples, but then how would we assess whether the difference was meaningful?

SELF-TEST Use the *split file* command to compute the correlation coefficient between exam anxiety and exam performance in men and women.

If we did this, we would find that the correlations were $r_{Male} = -.506$ and $r_{Female} = -.381$. These two samples are independent; that is, they contain different entities. To compare these correlations we can again use what we discovered in section 6.3.3 to convert these coefficients to z_r (just to remind you, we do this because it makes the sampling distribution normal and we know the standard error). If you do the conversion, then we get z_r (males) = $-.557$ and z_r (females) = $-.401$. We can calculate a z-score of the differences between these correlations as:

$$Z_{Difference} = \frac{z_{r_1} - z_{r_2}}{\sqrt{\frac{1}{N_1 - 3} + \frac{1}{N_2 + 3}}} \tag{6.11}$$

We had 52 men and 51 women so we would get:

$$Z_{Difference} = \frac{-.557 - (-.401)}{\sqrt{\frac{1}{49} + \frac{1}{48}}} = \frac{-.156}{0.203} = -0.768$$

We can look up this value of z (0.768, we can ignore the minus sign) in the table for the normal distribution in the Appendix and get the one-tailed probability from the column labelled 'Smaller Portion'. In this case the value is .221. To get the two-tailed probability we simply multiply the one-tailed probability value by 2, which gives us .442. As such the correlation between exam anxiety and exam performance is not significantly different in men and women. (see Oliver Twisted for how to do this on SPSS).

6.7.2. Comparing dependent *r*s ③

If you want to compare correlation coefficients that come from the same entities then things are a little more complicated. You can use a t-statistic to test whether a difference between two dependent correlations from the same sample is significant. For example, in our exam anxiety data we might want to see whether the relationship between exam anxiety (x) and exam performance (y) is stronger than the relationship between revision (z) and exam performance. To calculate this, all we need are the three rs that quantify the relationships between these variables: r_{xy}, the relationship between exam anxiety and exam performance (–.441); r_{zy}, the relationship between revision and exam performance (.397); and r_{xz}, the relationship between exam anxiety and revision (–.709). The t-statistic is computed as (Chen & Popovich, 2002):

$$t_{Difference} = \left(r_{xy} - r_{zy}\right)\sqrt{\frac{(n-3)(1+r_{xz})}{2\left(1 - r_{xy}^2 - r_{xz}^2 - r_{zy}^2 + 2r_{xy}r_{xz}r_{zy}\right)}} \tag{6.12}$$

Admittedly that equation looks hideous, but really it's not too bad: it just uses the three correlation coefficients and the sample size N. Place the numbers from the exam anxiety example in it (N was 103) and you should end up with:

$$t_{Difference} = (-.838)\sqrt{\frac{29.1}{2(1 - .194 - .503 - .158 + 0.248}} = -5.09$$

EVERYBODY

This value can be checked against the appropriate critical value in the Appendix with $N - 3$ degrees of freedom (in this case 100). The critical values in the table are 1.98 ($p < .05$) and 2.63 ($p < .01$), two-tailed. As such we can say that the correlation between exam anxiety and exam performance was significantly higher than the correlation between revision time and exam performance (this isn't a massive surprise given that these relationships went in the opposite directions to each other).

OLIVER TWISTED

Please, Sir, can I have some more … comparing of correlations?

'Are you having a bloody laugh with that equation?' yelps Oliver. 'I'd rather smother myself with cheese sauce and lock myself in a room of hungry mice.' Yes, yes, Oliver, enough of your sexual habits. To spare the poor mice I have written some SPSS syntax to run the comparisons mentioned in this section. For a guide on how to use them read the additional material for this chapter on the companion website. Go on, be kind to the mice!

6.8. Calculating the effect size ①

Can I use r^2 for non-parametric correlations?

Calculating effect sizes for correlation coefficients couldn't be easier because, as we saw earlier in the book, correlation coefficients *are* effect sizes! So, no calculations (other than those you have already done) necessary! However, I do want to point out one caveat when using non-parametric correlation coefficients as effect sizes. Although the Spearman and Kendall correlations are comparable in many respects (their power, for example, is similar under parametric conditions), there are two important differences (Strahan, 1982).

First, we saw for Pearson's r that we can square this value to get the proportion of shared variance, R^2. For Spearman's r_s we can do this too because it uses the same equation as Pearson's r.

However, the resulting R_s^2 needs to be interpreted slightly differently: it is the proportion of variance in the *ranks* that two variables share. Having said this, R_s^2 is usually a good approximation of R^2 (especially in conditions of near-normal distributions). Kendall's τ, however, is not numerically similar to either r or r_s and so τ^2 does not tell us about the proportion of variance shared by two variables (or the ranks of those two variables).

Second, Kendall's τ is 66–75% smaller than both Spearman's r_s and Pearson's r, but r and r_s are generally similar sizes (Strahan, 1982). As such, if τ is used as an effect size it should be borne in mind that it is not comparable to r and r_s and should not be squared. A related issue is that the point–biserial and biserial correlations differ in size too (as we saw in this chapter, the biserial correlation was bigger than the point–biserial). In this instance you should be careful to decide whether your dichotomous variable has an underlying continuum, or whether it is a truly discrete variable. More generally, when using correlations as effect sizes you should remember (both when reporting your own analysis and when interpreting others) that the choice of correlation coefficient can make a substantial difference to the apparent size of the effect.

6.9. How to report correlation coefficents ①

Reporting correlation coefficients is pretty easy: you just have to say how big they are and what their significance value was (although the significance value isn't *that* important because the correlation coefficient is an effect size in its own right!). Five things to note are that: (1) there should be no zero before the decimal point for the correlation coefficient or the probability value (because neither can exceed 1); (2) coefficients are reported to 2 decimal places; (3) if you are quoting a one-tailed probability, you should say so; (4) each correlation coefficient is represented by a different letter (and some of them are Greek!); and (5) there are standard criteria of probabilities that we use (.05, .01 and .001). Let's take a few examples from this chapter:

- ✓ There was a significant relationship between the number of adverts watched and the number of packets of sweets purchased, $r = .87$, p (one-tailed) $< .05$.
- ✓ Exam performance was significantly correlated with exam anxiety, $r = -.44$, and time spent revising, $r = .40$; the time spent revising was also correlated with exam anxiety, $r = -.71$ (all $ps < .001$).
- ✓ Creativity was significantly related to how well people did in the World's Biggest Liar Competition, $r_s = -.37$, $p < .001$.
- ✓ Creativity was significantly related to how well people did in the World's Biggest Liar Competition, $\tau = -.30$, $p < .001$. (Note that I've quoted Kendall's τ here.)
- ✓ The gender of the cat was significantly related to the time the cat spent away from home, $r_{pb} = .38$, $p < .01$.
- ✓ The gender of the cat was significantly related to the time the cat spent away from home, $r_b = .48$, $p < .01$.

Scientists, rightly or wrongly, tend to use several *standard* levels of statistical significance. Primarily, the most important criterion is that the significance value is less than .05; however, if the exact significance value is much lower then we can be much more confident about the strength of the experimental effect. In these circumstances we like to make a big song and dance about the fact that our result isn't just significant at .05, but is significant at a much lower level as well (hooray!). The values we use are .05, .01, .001 and .0001. You are rarely ever going to be in the fortunate position of being able to report an effect that is significant at a level less than .0001!

TABLE 6.2 An example of reporting a table of correlations

	Exam Performance	Exam Anxiety	Revision Time
Exam Performance	1	−.44***	.40***
Exam Anxiety	101	1	−.71***
Revision Time	101	101	1

Ns = not significant ($p > .05$), *$p < .05$, ** $p < .01$, *** $p < .001$

When we have lots of correlations we sometimes put them into a table. For example, our exam anxiety correlations could be reported as in Table 6.2. Note that above the diagonal I have reported the correlation coefficients and used symbols to represent different levels of significance. Under the table there is a legend to tell readers what symbols represent. (Actually, none of the correlations were non-significant or had p bigger than .001 so most of these are here simply to give you a reference point – you would normally include symbols that you had actually used in the table in your legend.) Finally, in the lower part of the table I have reported the sample sizes. These are all the same (101) but sometimes when you have missing data it is useful to report the sample sizes in this way because different values of the correlation will be based on different sample sizes. For some more ideas on how to report correlations have a look at Labcoat Leni's Real Research 6.1.

LABCOAT LENI'S REAL RESEARCH 6.1

Why do you like your lecturers? ①

As students you probably have to rate your lecturers at the end of the course. There will be some lecturers you like and others that you hate. As a lecturer I find this process horribly depressing (although this has a lot to do with the fact that I tend focus on negative feedback and ignore the good stuff). There is some evidence that students tend to pick courses of lecturers who they perceive to be enthusastic and good communicators. In a fascinating study, Tomas Chamorro-Premuzic and his colleagues (Chamorro-Premuzic, Furnham, Christopher, Garwood, & Martin, 2008) tested a slightly different hypothesis, which was that students tend to like lecturers who are like themselves. (This hypothesis will have the students on my course who like my lectures screaming in horror.)

First of all the authors measured students' own personalities using a very well-established measure (the NEO-FFI) which gives rise to scores on five fundamental personality traits: Neuroticism, Extroversion, Openness to experience, Agreeableness and Conscientiousness. They also gave students a questionnaire that asked them to rate how much they wanted their lecturer to have each of a list of characteristics. For example, they would be given the description 'warm: friendly, warm, sociable, cheerful, affectionate, outgoing' and asked to rate how much they wanted to see this in a lecturer from −5 (they don't want this characteristic at all) through 0 (the characteristic is not important) to +5 (I really want this characteristic in my lecturer). The characteristics on the questionnaire all related to personality characteristics measured by the NEO-FFI. As such, the authors had a measure of how much a student had each of the five core personality characteristics, but also a measure of how much they wanted to see those same characteristics in their lecturer.

In doing so, Tomas and his colleagues could test whether, for instance, extroverted students want extrovert lecturers. The data from this study (well, for the variables that I've mentioned) are in the file **Chamorro-Premuzic.sav**. Run some Pearson correlations on these variables to see if students with certain personality characteristics want to see those characteristics in their lecturers. What conclusions can you draw?

Answers are in the additional material on the companion website (or look at Table 3 in the original article, which will also show you how to report a large number of correlations).

CHAMORRO-PREMUZIC, T., ET AL. (2008). PERSONALITY AND INDIVIDUAL DIFFERENCES, 44, 965–976.

What have I discovered about statistics? ①

This chapter has looked at ways to study relationships between variables. We began by looking at how we might measure relationships statistically by developing what we already know about variance (from Chapter 1) to look at variance shared between variables. This shared variance is known as *covariance*. We then discovered that when data are parametric we can measure the strength of a relationship using Pearson's correlation coefficient, *r*. When data violate the assumptions of parametric tests we can use Spearman's r_s, or for small data sets Kendall's τ may be more accurate. We also saw that correlations can be calculated between two variables when one of those variables is a dichotomy (i.e. composed of two categories); when the categories have no underlying continuum then we use the point–biserial correlation, r_{pb}, but when the categories do have an underlying continuum we use the biserial correlation, r_b. Finally, we looked at the difference between *partial correlations*, in which the relationship between two variables is measured controlling for the effect that one or more variables has on both of those variables, and *semi-partial correlations*, in which the relationship between two variables is measured controlling for the effect that one or more variables has on only one of those variables. We also discovered that I had a guitar and, like my favourite record of the time, I was ready to 'Take on the world'. Well, Wales at any rate …

Key terms that I've discovered

Biserial correlation
Bivariate correlation
Coefficient of determination
Covariance
Cross-product deviations
Kendall's tau
Partial correlation
Pearson correlation coefficient
Point–biserial correlation
Semi-partial correlation
Spearman's correlation coefficient
Standardization

Smart Alex's tasks

- **Task 1**: A student was interested in whether there was a positive relationship between the time spent doing an essay and the mark received. He got 45 of his friends and timed how long they spent writing an essay (**hours**) and the percentage they got in the essay (**essay**). He also translated these grades into their degree classifications (**grade**): first, upper second, lower second and third class. Using the data in the file **EssayMarks.sav** find out what the relationship was between the time spent doing an essay and the eventual mark in terms of percentage and degree class (draw a scatterplot too!). ①
- **Task 2**: Using the **ChickFlick.sav** data from Chapter 3, is there a relationship between gender and arousal? Using the same data, is there a relationship between the film watched and arousal? ①

- **Task 3**: As a statistics lecturer I am always interested in the factors that determine whether a student will do well on a statistics course. One potentially important factor is their previous expertise with mathematics. Imagine I took 25 students and looked at their degree grades for my statistics course at the end of their first year at university. In the UK, a student can get a first-class mark (the best), an upper-second-class mark, a lower second, a third, a pass or a fail (the worst). I also asked these students what grade they got in their GCSE maths exams. In the UK GCSEs are school exams taken at age 16 that are graded A, B, C, D, E or F (an A grade is better than all of the lower grades). The data for this study are in the file **grades.sav**. Carry out the appropriate analysis to see if GCSE maths grades correlate with first-year statistics grades. ①

Answers can be found on the companion website.

Further reading

Chen, P. Y., & Popovich, P. M. (2002). *Correlation: Parametric and nonparametric measures*. Thousand Oaks, CA: Sage.

Howell, D. C. (2006). *Statistical methods for psychology* (6th ed.). Belmont, CA: Duxbury. (Or you might prefer his *Fundamental Statistics for the Behavioral Sciences*, also in its 6th edition, 2007. Both are excellent texts that are a bit more technical than this book so they are a useful next step.)

Miles, J. N. V., & Banyard, P. (2007). *Understanding and using statistics in psychology: a practical introduction*. London: Sage. (A fantastic and amusing introduction to statistical theory.)

Wright, D. B., & London, K. (2009). *First steps in statistics* (2nd ed.). London: Sage. (This book is a very gentle introduction to statistical theory.)

Online tutorial

The companion website contains the following Flash movie tutorial to accompany this chapter:

- Correlations using SPSS

Interesting real research

Chamorro-Premuzic, T., Furnham, A., Christopher, A. N., Garwood, J., & Martin, N. (2008). Birds of a feather: Students' preferences for lecturers' personalities as predicted by their own personality and learning approaches. *Personality and Individual Differences*, *44*, 965–976.

Regression 7

FIGURE 7.1 Me playing with my ding-a-ling in the Holimarine Talent Show. Note the groupies queuing up at the front

7.1. What will this chapter tell me? ①

Although none of us can know the future, predicting it is so important that organisms are hard wired to learn about predictable events in their environment. We saw in the previous chapter that I received a guitar for Christmas when I was 8. My first foray into public performance was a weekly talent show at a holiday camp called 'Holimarine' in Wales (it doesn't exist anymore because I am old and this was 1981). I sang a Chuck Berry song called 'My ding-a-ling'[1] and to my absolute amazement I won the competition.[2] Suddenly other 8 year olds across the land (well, a ballroom in Wales) worshipped me (I made lots of friends after the competition). I had tasted success, it tasted like praline chocolate, and so I wanted to enter the competition in the second week of our holiday. To ensure success, I needed to know why I had won in the first week. One way to do this would have been to collect data and to use these data to predict people's evaluations of children's performances in the contest

[1] It appears that even then I had a passion for lowering the tone of things that should be taken seriously.

[2] I have a very grainy video of this performance recorded by my dad's friend on a video camera the size of a medium-sized dog that had to be accompanied at all times by a 'battery pack' the size and weight of a tank. Maybe I'll put it up on the companion website ...

from certain variables: the age of the performer, what type of performance they gave (singing, telling a joke, magic tricks), and maybe how cute they looked. A regression analysis on these data would enable us to predict future evaluations (success in next week's competition) based on values of the predictor variables. If, for example, singing was an important factor in getting a good audience evaluation, then I could sing again the following week; however, if jokers tended to do better then I could switch to a comedy routine. When I was 8 I wasn't the sad geek that I am today, so I didn't know about regression analysis (nor did I wish to know); however, my dad thought that success was due to the winning combination of a cherub-looking 8 year old singing songs that can be interpreted in a filthy way. He wrote me a song to sing in the competition about the keyboard player in the Holimarine Band 'messing about with his organ', and first place was mine again. There's no accounting for taste.

7.2. An introduction to regression ①

In the previous chapter we looked at how to measure relationships between two variables. These correlations can be very useful but we can take this process a step further and predict one variable from another. A simple example might be to try to predict levels of stress from the amount of time until you have to give a talk. You'd expect this to be a negative relationship (the smaller the amount of time until the talk, the larger the anxiety). We could then extend this basic relationship to answer a question such as 'if there's 10 minutes to go until someone has to give a talk, how anxious will they be?' This is the essence of regression analysis: we fit a model to our data and use it to predict values of the dependent variable (DV) from one or more independent variables (IVs).[3] Regression analysis is a way of predicting an **outcome variable** from one **predictor variable** (**simple regression**) or several predictor variables (**multiple regression**). This tool is incredibly useful because it allows us to go a step beyond the data that we collected.

In section 2.4.3 I introduced you to the idea that we can predict any data using the following general equation:

$$\text{outcome}_i = (\text{model}) + \text{error}_i \tag{7.1}$$

This just means that the outcome we're trying to predict for a particular person can be predicted by whatever model we fit to the data plus some kind of error. In regression, the model we fit is linear, which means that we summarize a data set with a straight line (think back to Jane Superbrain Box 2.1). As such, the word 'model' in the equation above simply gets replaced by 'things' that define the line that we fit to the data (see the next section).

How do I fit a straight line to my data?

With any data set there are several lines that could be used to summarize the general trend and so we need a way to decide which of many possible lines to choose. For the sake of making accurate predictions we want to fit a model that *best* describes the data. The simplest way to do this would be to use your eye to gauge a line that looks as though it summarizes the data well. You don't need to be a genius to realize that the 'eyeball' method is very subjective and so offers no assurance that the model is the best one that could have been chosen. Instead, we use a mathematical technique called the *method of least squares* to establish the line that best describes the data collected.

[3] I want to remind you here of something I discussed in Chapter 1: SPSS refers to regression variables as dependent and independent variables (as in controlled experiments). However, correlational research by its nature seldom controls the independent variables to measure the effect on a dependent variable and so I will talk about 'independent variables' as *predictors*, and the 'dependent variable' as the *outcome*.

7.2.1. Some important information about straight lines ①

I mentioned above that in our general equation the word 'model' gets replaced by 'things that define the line that we fit to the data'. In fact, any straight line can be defined by two things: (1) the slope (or gradient) of the line (usually denoted by b_1); and (2) the point at which the line crosses the vertical axis of the graph (known as the *intercept* of the line, b_0). In fact, our general model becomes equation (7.2) below in which Y_i is the outcome that we want to predict and X_i is the *i*th participant's score on the predictor variable.[4] Here b_1 is the gradient of the straight line fitted to the data and b_0 is the intercept of that line. These parameters b_1 and b_0 are known as the *regression coefficients* and will crop up time and time again in this book, where you may see them referred to generally as *b* (without any subscript) or b_i (meaning the *b* associated with variable *i*). There is a residual term, ε_i, which represents the difference between the score predicted by the line for participant *i* and the score that participant *i* actually obtained. The equation is often conceptualized without this residual term (so, ignore it if it's upsetting you); however, it is worth knowing that this term represents the fact our model will not fit the data collected perfectly:

$$Y_i = (b_0 + b_1X_i) + \varepsilon_i \tag{7.2}$$

A particular line has a specific intercept and gradient. Figure 7.2 shows a set of lines that have the same intercept but different gradients, and a set of lines that have the same gradient but different intercepts. Figure 7.2 also illustrates another useful point: the gradient of the line tells us something about the nature of the relationship being described. In Chapter 6 we saw how relationships can be either positive or negative (and I don't mean the difference between getting on well with your girlfriend and arguing all the time!). A line that has a gradient with a positive value describes a positive relationship, whereas a line with a negative gradient describes a negative relationship. So, if you look at the graph in Figure 7.2

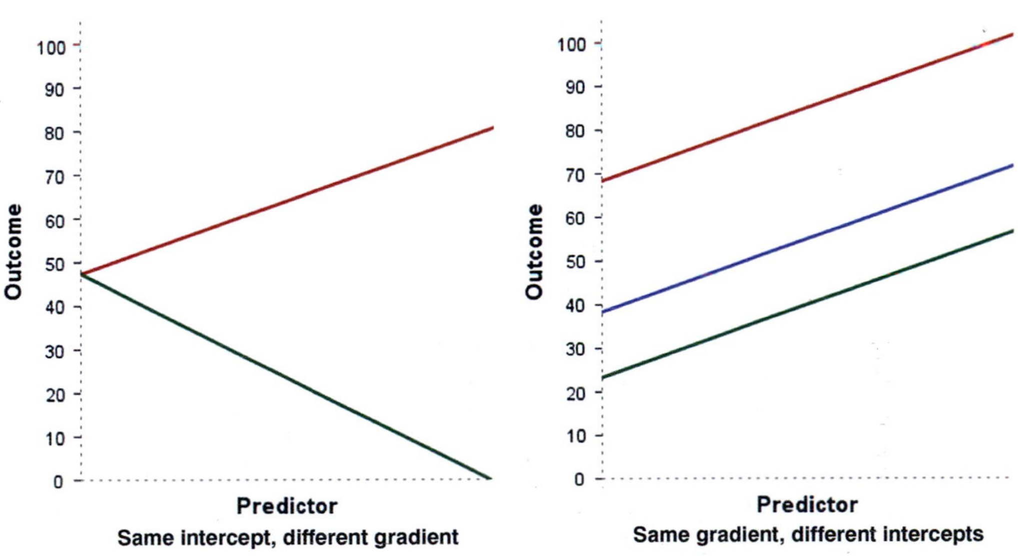

FIGURE 7.2 Lines with the same gradients but different intercepts, and lines that share the same intercept but have different gradients

[4] You'll sometimes see this equation written as:

$$Y_i = (\beta_0 + \beta_1X_i) + \varepsilon_i$$

The only difference is that this equation has got *β*s in it instead of *b*s and in fact both versions are the same thing, they just use different letters to represent the coefficients. When SPSS estimates the coefficients in this equation it labels them *b* and to be consistent with the SPSS output that you'll end up looking at, I've used *b*s instead of *β*s.

in which the gradients differ but the intercepts are the same, then the red line describes a positive relationship whereas the green line describes a negative relationship. Basically then, the gradient (b) tells us what the model looks like (its shape) and the intercept (b_0) tells us where the model is (its location in geometric space).

If it is possible to describe a line knowing only the gradient and the intercept of that line, then we can use these values to describe our model (because in linear regression the model we use is a straight line). So, the model that we fit to our data in linear regression can be conceptualized as a straight line that can be described mathematically by equation (7.2). With regression we strive to find the line that best describes the data collected, then estimate the gradient and intercept of that line. Having defined these values, we can insert different values of our predictor variable into the model to estimate the value of the outcome variable.

7.2.2. The method of least squares ①

I have already mentioned that the method of least squares is a way of finding the line that best fits the data (i.e. finding a line that goes through, or as close to, as many of the data points as possible). This 'line of best fit' is found by ascertaining which line, of all of the possible lines that could be drawn, results in the least amount of difference between the observed data points and the line. Figure 7.3 shows that when any line is fitted to a set of data, there will be small differences between the values predicted by the line and the data that were actually observed.

Back in Chapter 2 we saw that we could assess the fit of a model (the example we used was the mean) by looking at the deviations between the model and the actual data collected. These deviations were the vertical distances between what the model predicted and each data point that was actually observed. We can do exactly the same to assess the fit of a regression line (which, like the mean, is a statistical model). So, again we are interested in the vertical differences between the line and the actual data because the line is our model: we use it to predict values of Y from values of the X variable. In regression these differences are usually called **residuals** rather than deviations, but they are the same thing. As with the mean, data points fall both above (the model underestimates their value) and below (the

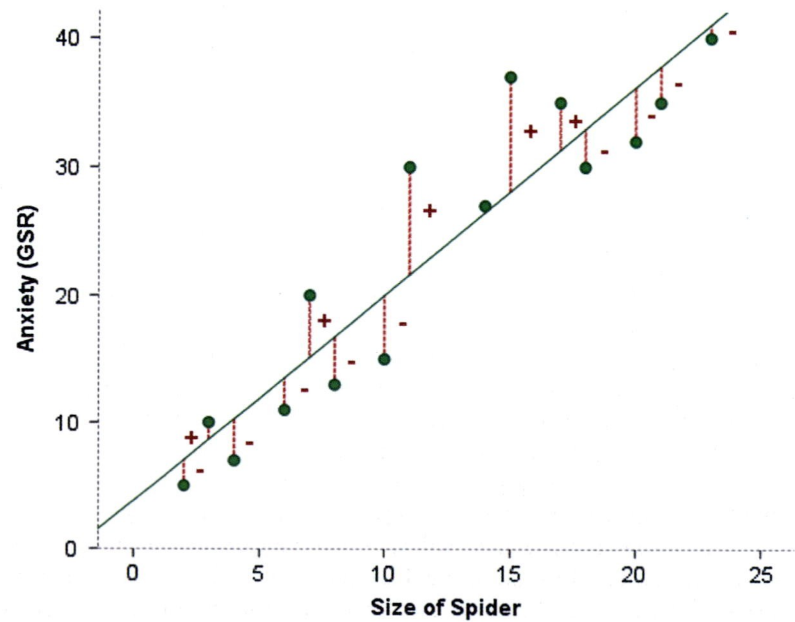

FIGURE 7.3 This graph shows a scatterplot of some data with a line representing the general trend. The vertical lines (dotted) represent the differences (or residuals) between the line and the actual data

model overestimates their value) the line yielding both positive and negative differences. In the discussion of variance in section 2.4.2 I explained that if we sum positive and negative differences then they tend to cancel each other out and that to circumvent this problem we square the differences before adding them up. We do the same thing here. The resulting squared differences provide a gauge of how well a particular line fits the data: if the squared differences are large, the line is not representative of the data; if the squared differences are small, the line is representative.

You could, if you were particularly bored, calculate the sum of squared differences (or SS for short) for every possible line that is fitted to your data and then compare these 'goodness-of-fit' measures. The one with the lowest SS is the line of best fit. Fortunately we don't have to do this because the method of least squares does it for us: it selects the line that has the lowest sum of squared differences (i.e. the line that best represents the observed data). How exactly it does this is by using a mathematical technique for finding maxima and minima and this technique is used to find the line that minimizes the sum of squared differences. I don't really know much more about it than that, to be honest, so I tend to think of the process as a little bearded wizard called Nephwick the Line Finder who just magically finds lines of best fit. Yes, he lives inside your computer. The end result is that Nephwick estimates the value of the slope and intercept of the 'line of best fit' for you. We tend to call this line of best fit a regression line.

7.2.3. Assessing the goodness of fit: sums of squares, R and R^2 ①

Once Nephwick the Line Finder has found the line of best fit it is important that we assess how well this line fits the actual data (we assess the **goodness of fit** of the model). We do this because even though this line is the best one available, it can still be a lousy fit to the data! In section 2.4.2 we saw that one measure of the adequacy of a model is the sum of squared differences (or more generally we assess models using equation (7.3) below). If we want to assess the line of best fit, we need to compare it against something, and the thing we choose is the most basic model we can find. So we use equation (7.3) to calculate the fit of the most basic model, and then the fit of the best model (the line of best fit), and basically if the best model is any good then it should fit the data significantly better than our basic model:

$$\text{deviation} = \sum(\text{observed} - \text{model})^2 \tag{7.3}$$

How do I tell if my model is good?

This is all quite abstract so let's look at an example. Imagine that I was interested in predicting record sales (Y) from the amount of money spent advertising that record (X). One day my boss came in to my office and said 'Andy, I know you wanted to be a rock star and you've ended up working as my stats-monkey, but how many records will we sell if we spend £100,000 on advertising?' If I didn't have an accurate model of the relationship between record sales and advertising, what would my best guess be? Well, probably the best answer I could give would be the mean number of record sales (say, 200,000) because on average that's how many records we expect to sell. This response might well satisfy a brainless record company executive (who didn't offer my band a record contract). However, what if he had asked 'How many records will we sell if we spend £1 on advertising?' Again, in the absence of any accurate information, my best guess would be to give the average number of sales (200,000). There is a problem: whatever amount of money is spent on advertising I always predict the same levels of sales. As such, the mean is a model of 'no relationship' at all between the variables. It should be pretty clear then that the mean is fairly useless as a model of a relationship between two variables – but it is the simplest model available.

So, as a basic strategy for predicting the outcome, we might choose to use the mean, because on average it will be a fairly good guess of an outcome. Using the mean as a model, we can calculate the difference between the observed values, and the values predicted by the mean (equation (7.3)). We saw in section 2.4.2 that we square all of these differences to give us the sum of squared differences. This sum of squared differences is known as the **total sum of squares** (denoted SS_T) because it is the total amount of differences present when the most basic model is applied to the data. This value represents how good the mean is as a model of the observed data. Now, if we fit the more sophisticated model to the data, such as a line of best fit, we can again work out the differences between this new model and the observed data (again using equation (7.3)). In the previous section we saw that the method of least squares finds the best possible line to describe a set of data by minimizing the difference between the model fitted to the data and the data themselves. However, even with this optimal model there is still some inaccuracy, which is represented by the differences between each observed data point and the value predicted by the regression line. As before, these differences are squared before they are added up so that the directions of the differences do not cancel out. The result is known as the *sum of squared residuals* or **residual sum of squares** (SS_R). This value represents the degree of inaccuracy when the best model is fitted to the data. We can use these two values to calculate how much better the regression line (the line of best fit) is than just using the mean as a model (i.e. how much better is the best possible model than the worst model?). The improvement in prediction resulting from using the regression model rather than the mean is calculated by calculating the difference between SS_T and SS_R. This difference shows us the reduction in the inaccuracy of the model resulting from fitting the regression model to the data. This improvement is the **model sum of squares** (SS_M). Figure 7.4 shows each sum of squares graphically.

If the value of SS_M is large then the regression model is very different from using the mean to predict the outcome variable. This implies that the regression model has made a big improvement to how well the outcome variable can be predicted. However, if SS_M is small then using the regression model is little better than using the mean (i.e. the regression model is no better than taking our 'best guess'). A useful measure arising from these sums of squares is the proportion of improvement due to the model. This is easily calculated by dividing the sum of squares for the model by the total sum of squares. The resulting value is called R^2 and to express this value as a percentage you should multiply it by 100. R^2 represents the amount of variance in the outcome explained by the model (SS_M) relative to how much variation there was to explain in the first place (SS_T). Therefore, as a percentage, it represents the percentage of the variation in the outcome that can be explained by the model:

$$R^2 = \frac{SS_M}{SS_T} \tag{7.4}$$

This R^2 is the same as the one we met in Chapter 6 (section 6.5.2.3) and you might have noticed that it is interpreted in the same way. Therefore, in simple regression we can take the square root of this value to obtain Pearson's correlation coefficient. As such, the correlation coefficient provides us with a good estimate of the overall fit of the regression model, and R^2 provides us with a good gauge of the substantive size of the relationship.

A second use of the sums of squares in assessing the model is through the *F*-test. I mentioned way back in Chapter 2 that test statistics (like *F*) are usually the amount of systematic variance divided by the amount of unsystematic variance, or, put another way, the model compared against the error in the model. This is true here: *F* is based upon the ratio of the improvement due to the model (SS_M) and the difference between the model and the observed data (SS_R). Actually, because the sums of squares depend on the number

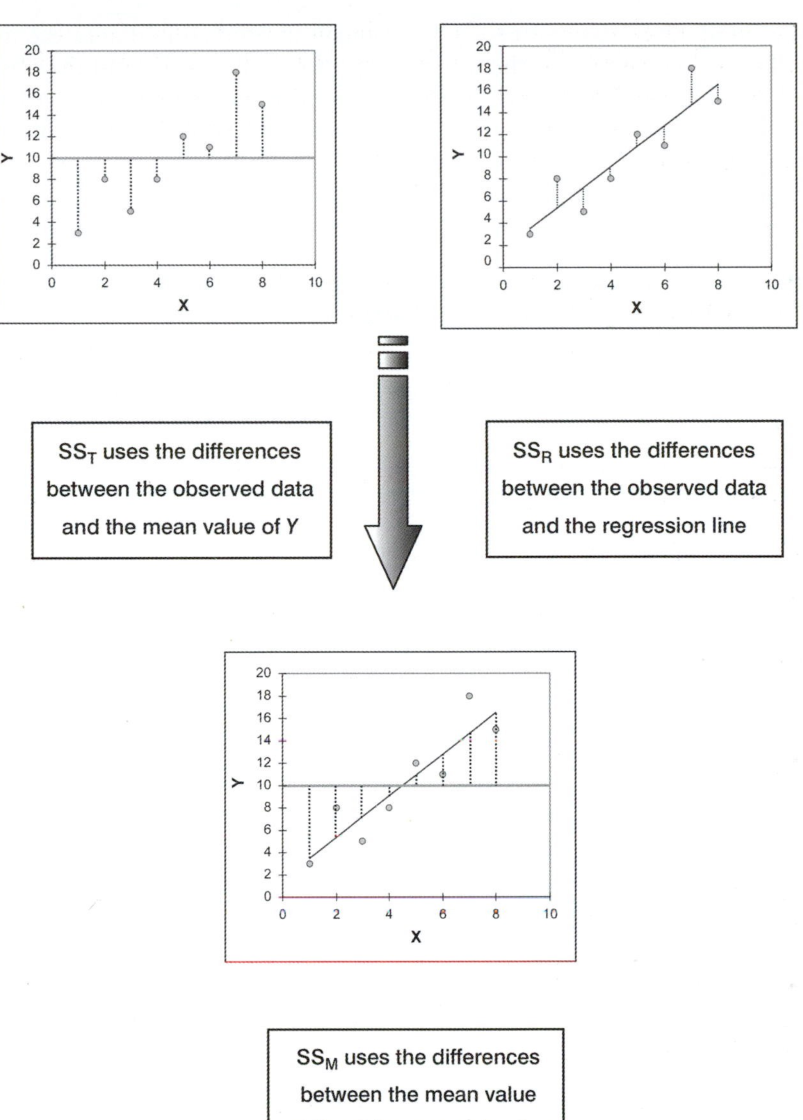
SS$_T$ uses the differences between the observed data and the mean value of Y
SS$_R$ uses the differences between the observed data and the regression line
SS$_M$ uses the differences between the mean value of Y and the regression line

FIGURE 7.4 Diagram showing from where the regression sums of squares derive

of differences that we have added up, we use the average sums of squares (referred to as the **mean squares** or MS). To work out the mean sums of squares we divide by the degrees of freedom (this is comparable to calculating the variance from the sums of squares – see section 2.4.2). For SS_M the degrees of freedom are simply the number of variables in the model, and for SS_R they are the number of observations minus the number of parameters being estimated (i.e. the number of beta coefficients including the constant). The result is the mean squares for the model (MS_M) and the residual mean squares (MS_R). At this stage it isn't essential that you understand how the mean squares are derived (it is explained in Chapter 10). However, it is important that you understand that the **F-ratio** (equation (7.5)) is a measure of how much the model has improved the prediction of the outcome compared to the level of inaccuracy of the model:

$$F = \frac{MS_M}{MS_R} \tag{7.5}$$

If a model is good, then we expect the improvement in prediction due to the model to be large (so, MS_M will be large) and the difference between the model and the observed data to be small (so, MS_R will be small). In short, a good model should have a large F-ratio (greater than 1 at least) because the top of equation (7.5) will be bigger than the bottom. The exact magnitude of this F-ratio can be assessed using critical values for the corresponding degrees of freedom (as in the Appendix).

7.2.4. Assessing individual predictors ①

We've seen that the predictor in a regression model has a coefficient (b_1), which in simple regression represents the gradient of the regression line. The value of b represents the change in the outcome resulting from a unit change in the predictor. If the model was useless at predicting the outcome, then if the value of the predictor changes, what might we expect the change in the outcome to be? Well, if the model is very bad then we would expect the change in the outcome to be zero. Think back to Figure 7.4 (see the panel representing SS_T) in which we saw that using the mean was a very bad way of predicting the outcome. In fact, the line representing the mean is flat, which means that as the predictor variable changes, the value of the outcome does *not* change (because for each level of the predictor variable, we predict that the outcome will equal the mean value). The important point here is that a bad model (such as the mean) will have regression coefficients of 0 for the predictors. A regression coefficient of 0 means: (1) a unit change in the predictor variable results in no change in the predicted value of the outcome (the predicted value of the outcome does not change at all); and (2) the gradient of the regression line is 0, meaning that the regression line is flat. Hopefully, you'll see that it logically follows that if a variable significantly predicts an outcome, then it should have a b-value significantly different from zero. This hypothesis is tested using a t-test (see Chapter 9). The **t-statistic** tests the null hypothesis that the value of b is 0: therefore, if it is significant we gain confidence in the hypothesis that the b-value is significantly different from 0 and that the predictor variable contributes significantly to our ability to estimate values of the outcome.

Like F, the t-statistic is also based on the ratio of explained variance against unexplained variance or error. Well, actually, what we're interested in here is not so much variance but whether the b we have is big compared to the amount of error in that estimate. To estimate how much error we could expect to find in b we use the standard error. The standard error tells us something about how different b-values would be across different samples. We could take lots and lots of samples of data regarding record sales and advertising budgets and calculate the b-values for each sample. We could plot a frequency distribution of these samples to discover whether the b-values from all samples would be relatively similar, or whether they would be very different (think back to section 2.5.1). We can use the standard deviation of this distribution (known as the *standard error*) as a measure of the similarity of b-values across samples. If the standard error is very small, then it means that most samples are likely to have a b-value similar to the one in our sample (because there is little variation across samples). The t-test tells us whether the b-value is different from 0 relative to the variation in b-values across samples. When the standard error is small even a small deviation from zero can reflect a meaningful difference because b is representative of the majority of possible samples.

Equation (7.6) shows how the t-test is calculated and you'll find a general version of this equation in Chapter 9 (equation (9.1)). The b_{expected} is simply the value of b that we would expect to obtain if the null hypothesis were true. I mentioned earlier that the null hypothesis is that b is 0 and so this value can be replaced by 0. The equation simplifies to become the observed value of b divided by the standard error with which it is associated:

$$t = \frac{b_{\text{observed}} - b_{\text{expected}}}{\text{SE}_b} = \frac{b_{\text{observed}}}{\text{SE}_b} \quad (7.6)$$

The values of t have a special distribution that differs according to the degrees of freedom for the test. In regression, the degrees of freedom are $N - p - 1$, where N is the total sample size and p is the number of predictors. In simple regression when we have only one predictor, this reduces down to $N - 2$. Having established which t-distribution needs to be used, the observed value of t can then be compared to the values that we would expect to find if there was no effect (i.e. $b = 0$): if t is very large then it is unlikely to have occurred when there is no effect (these values can be found in the Appendix). SPSS provides the exact probability that the observed value or a larger value of t would occur if the value of b was, in fact, 0. As a general rule, if this observed significance is less than .05, then scientists assume that b is significantly different from 0; put another way, the predictor makes a significant contribution to predicting the outcome.

7.3. Doing simple regression on SPSS ①

So far, we have seen a little of the theory behind regression, albeit restricted to the situation in which there is only one predictor. To help clarify what we have learnt so far, we will go through an example of a simple regression on SPSS. Earlier on I asked you to imagine that I worked for a record company and that my boss was interested in predicting record sales from advertising. There are some data for this example in the file **Record1.sav**. This data file has 200 rows, each one representing a different record. There are also two columns, one representing the sales of each record in the week after release and the other representing the amount (in pounds) spent promoting the record before release. This is the format for entering regression data: the outcome variable and any predictors should be entered in different columns, and each row should represent independent values of those variables.

The pattern of the data is shown in Figure 7.5 and it should be clear that a positive relationship exists: so, the more money spent advertising the record, the more it is likely to sell. Of course there are some records that sell well regardless of advertising (top left of scatterplot), but there are none that sell badly when advertising levels are high (bottom right of scatterplot). The scatterplot also shows the line of best fit for these data: bearing in mind that the mean would be represented by a flat line at around the 200,000 sales mark, the regression line is noticeably different.

To find out the parameters that describe the regression line, and to see whether this line is a useful model, we need to run a regression analysis. To do the analysis you need to access the main dialog box by selecting Analyze Regression ▸Linear ▾. Figure 7.6 shows the resulting dialog box. There is a space labelled *Dependent* in which you should place the outcome variable (in this example **sales**). So, select **sales** from the list on the left-hand side, and transfer it by dragging it or clicking on ➜. There is another space labelled *Independent(s)* in which any predictor variable should be placed. In simple regression we use only one predictor (in this example, **adverts**) and so you should select **adverts** from the list and click on ➜ to transfer it to the list of predictors. There are a variety of options available, but these will be explored within the context of multiple regression. For the time being just click on OK to run the basic analysis.

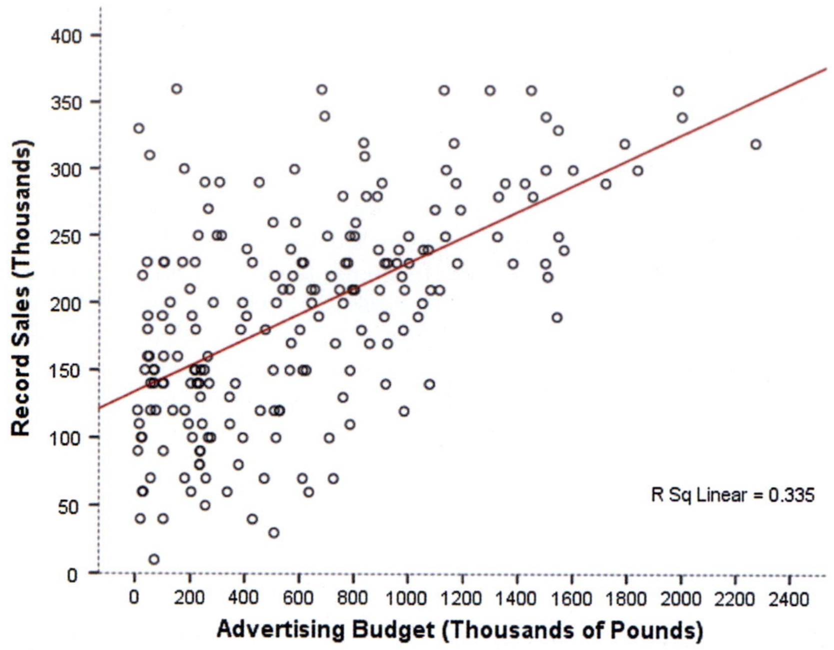

FIGURE 7.5 Scatterplot showing the relationship between record sales and the amount spent promoting the record

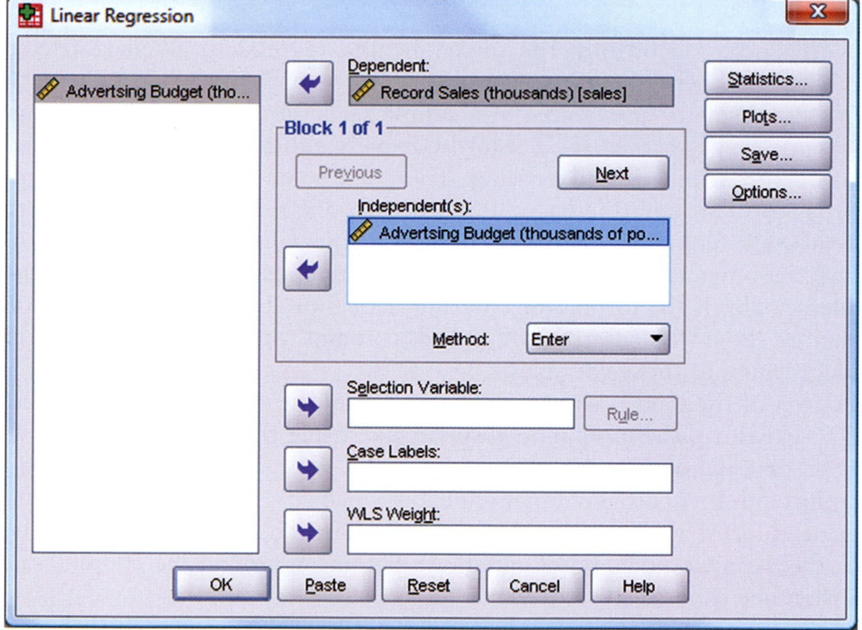

FIGURE 7.6 Main dialog box for regression

7.4. Interpreting a simple regression ①

7.4.1. Overall fit of the model ①

The first table provided by SPSS is a summary of the model (SPSS Output 7.1). This summary table provides the value of R and R^2 for the model that has been derived. For these data, R has a value of .578 and because there is only one predictor, this value represents the simple correlation between advertising and record sales (you can confirm this by running a correlation using what you were taught in Chapter 6). The value of R^2 is .335, which tells us that advertising expenditure can account for 33.5% of the variation in record sales. In other words, if we are trying to explain why some records sell more than others, we can look at the variation in sales of different records. There might be many factors that can explain this variation, but our model, which includes only advertising expenditure, can explain approximately 33% of it. This means that 66% of the variation in record sales cannot be explained by advertising alone. Therefore, there must be other variables that have an influence also.

SPSS OUTPUT 7.1

Model Summary

Model	R	R Square	Adjusted R Square	Std. Error of the Estimate
1	.578[a]	.335	.331	65.991

a. Predictors: (Constant), Advertising Budget (thousands of pounds)

The next part of the output (SPSS Output 7.2) reports an analysis of variance (ANOVA – see Chapter 10). The summary table shows the various sums of squares described in Figure 7.4 and the degrees of freedom associated with each. From these two values, the average sums of squares (the mean squares) can be calculated by dividing the sums of squares by the associated degrees of freedom. The most important part of the table is the *F*-ratio, which is calculated using equation (7.5), and the associated significance value of that *F*-ratio. For these data, *F* is 99.59, which is significant at $p < .001$ (because the value in the column labelled *Sig.* is less than .001). This result tells us that there is less than a 0.1% chance that an *F*-ratio this large would happen if the null hypothesis were true. Therefore, we can conclude that our regression model results in significantly better prediction of record sales than if we used the mean value of record sales. In short, the regression model overall predicts record sales significantly well.

SPSS OUTPUT 7.2

ANOVA[b]

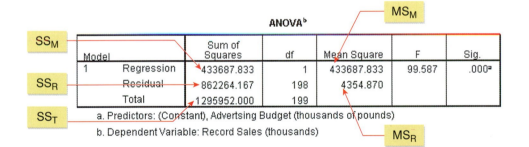

Model		Sum of Squares	df	Mean Square	F	Sig.
1	Regression	433687.833	1	433687.833	99.587	.000[a]
	Residual	862264.167	198	4354.870		
	Total	1295952.000	199			

a. Predictors: (Constant), Advertising Budget (thousands of pounds)

b. Dependent Variable: Record Sales (thousands)

SS_M SS_R SS_T MS_M MS_R

7.4.2. Model parameters ①

The ANOVA tells us whether the model, overall, results in a significantly good degree of prediction of the outcome variable. However, the ANOVA doesn't tell us about the individual contribution of variables in the model (although in this simple case there is only one variable in the model and so we can infer that this variable is a good predictor). The table in SPSS Output 7.3 provides details of the model parameters (the beta values) and the significance of these values. We saw in equation (7.2) that b_0 was the *Y* intercept and this value is the value *B* (in the SPSS output) for the constant. So, from the table, we can say that b_0 is 134.14, and this can be interpreted as meaning that when no money is spent on

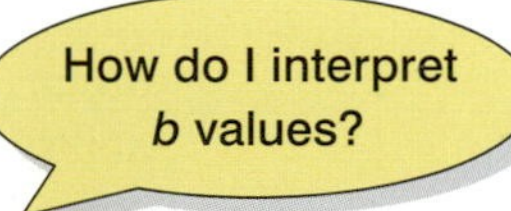

How do I interpret *b* values?

advertising (when $X = 0$), the model predicts that 134,140 records will be sold (remember that our unit of measurement was thousands of records). We can also read off the value of b_1 from the table and this value represents the gradient of the regression line. It is 0.096. Although this value is the slope of the regression line, it is more useful to think of this value as representing *the change in the outcome associated with a unit change in the predictor*. Therefore, if our predictor variable is increased by one unit (if the advertising budget is increased by 1), then our model predicts that 0.096 extra records will be sold. Our units of measurement were thousands of pounds and thousands of records sold, so we can say that for an increase in advertising of £1000 the model predicts 96 ($0.096 \times 1000 = 96$) extra record sales. As you might imagine, this investment is pretty bad for the record company: it invests £1000 and gets only 96 extra sales. Fortunately, as we already know, advertising accounts for only one-third of record sales.

SPSS OUTPUT 7.3

Coefficients[a]

Model		Unstandardized Coefficients		Standardized Coefficients	t	Sig.
		B	Std. Error	Beta		
1	(Constant)	134.140	7.537		17.799	.000
	Advertsing Budget (thousands of pounds)	.096	.010	.578	9.979	.000

a. Dependent Variable: Record Sales (thousands)

We saw earlier that, in general, values of the regression coefficient b represent the change in the outcome resulting from a unit change in the predictor and that if a predictor is having a significant impact on our ability to predict the outcome then this b should be different from 0 (and big relative to its standard error). We also saw that the t-test tells us whether the b-value is different from 0. SPSS provides the exact probability that the observed value of t would occur if the value of b in the population were 0. If this observed significance is less than .05, then scientists agree that the result reflects a genuine effect (see Chapter 2). For these two values, the probabilities are .000 (zero to 3 decimal places) and so we can say that the probability of these t-values or larger occurring if the values of b in the population were 0 is less than .001. Therefore, the bs are different from 0 and we can conclude that the advertising budget makes a significant contribution ($p < .001$) to predicting record sales.

SELF-TEST How is the t in SPSS Output 7.3 calculated? Use the values in the table to see if you can get the same value as SPSS.

7.4.3. Using the model ①

So far, we have discovered that we have a useful model, one that significantly improves our ability to predict record sales. However, the next stage is often to use that model to make some predictions. The first stage is to define the model by replacing the b-values in equation (7.2) with the values from SPSS Output 7.3. In addition, we can replace the X and Y with the variable names so that the model becomes:

$$\begin{aligned} \text{record sales}_i &= b_0 + b_1\text{advertising budget}_i \\ &= 134.14 + (0.096 \times \text{advertising budget}_i) \end{aligned} \tag{7.7}$$

It is now possible to make a prediction about record sales, by replacing the advertising budget with a value of interest. For example, imagine a record executive wanted to spend £100,000 on advertising a new record. Remembering that our units are already in thousands of pounds, we can simply replace the advertising budget with 100. He would discover that record sales should be around 144,000 for the first week of sales:

$$\begin{aligned}\text{record sales}_i &= 134.14 + (0.096 \times \text{advertising budget}_i)\\ &= 134.14 + (0.096 \times 100)\\ &= 143.74\end{aligned} \tag{7.8}$$

SELF-TEST How many records would be sold if we spent £666,000 on advertising the latest CD by black metal band Abgott?

CRAMMING SAM'S TIPS **Simple regression**

- Simple regression is a way of predicting values of one variable from another.
- We do this by fitting a statistical model to the data in the form of a straight line.
- This line is the line that best summarizes the pattern of the data.
- We have to assess how well the line fits the data using:
  - R^2 which tells us how much variance is explained by the model compared to how much variance there is to explain in the first place. It is the proportion of variance in the outcome variable that is shared by the predictor variable.
  - *F*, which tells us how much variability the model can explain relative to how much it can't explain (i.e. it's the ratio of how good the model is compared to how bad it is).
- The *b*-value tells us the gradient of the regression line and the strength of the relationship between a predictor and the outcome variable. If it is significant (*Sig.* < .05 in the SPSS table) then the predictor variable significantly predicts the outcome variable.

7.5. Multiple regression: the basics ②

What is the difference between simple and multiple regression?

To summarize what we have learnt so far, in simple linear regression the outcome variable *Y* is predicted using the equation of a straight line (equation (7.2)). Given that we have collected several values of *Y* and *X*, the unknown parameters in the equation can be calculated. They are calculated by fitting a model to the data (in this case a straight line) for which the sum of the squared differences between the line and the actual data points is minimized. This method is called the method of least squares. Multiple regression is a logical extension of these principles to situations in which there are several predictors. Again, we still use our basic equation of:

$$\text{outcome}_i = (\text{model}) + \text{error}_i$$

but this time the model is slightly more complex. It is basically the same as for simple regression except that for every extra predictor you include, you have to add a coefficient; so, each predictor variable has its own coefficient, and the outcome variable is predicted from a combination of all the variables multiplied by their respective coefficients plus a residual term (see equation (7.9) – the brackets aren't necessary, they're just to make the connection to the general equation above):

$$Y_i = (b_0 + b_1X_{1i} + b_2X_{2i} + \ldots + b_nX_{ni}) + \varepsilon_i \tag{7.9}$$

Y is the outcome variable, b_1 is the coefficient of the first predictor (X_1), b_2 is the coefficient of the second predictor (X_2), b_n is the coefficient of the nth predictor (X_n), and ε_i is the difference between the predicted and the observed value of Y for the ith participant. In this case, the model fitted is more complicated, but the basic principle is the same as simple regression. That is, we seek to find the linear combination of predictors that correlate maximally with the outcome variable. Therefore, when we refer to the regression model in multiple regression, we are talking about a model in the form of equation (7.9).

7.5.1. An example of a multiple regression model ②

Imagine that our record company executive was interested in extending his model of record sales to incorporate another variable. We know already that advertising accounts for 33% of variation in record sales, but a much larger 67% remains unexplained. The record executive could measure a new predictor in an attempt to explain some of the unexplained variation in record sales. He decides to measure the number of times the record is played on Radio 1 (the UK's biggest national radio station) during the week prior to release. The existing model that we derived using SPSS (see equation (7.7)) can now be extended to include this new variable (**airplay**):

$$\text{record sales}_i = b_0 + b_1\text{advertising budget}_i + b_2\text{airplay}_i + \varepsilon_i \tag{7.10}$$

The new model is based on equation (7.9) and includes a *b*-value for both predictors (and, of course, the constant). If we calculate the *b*-values, we could make predictions about record sales based not only on the amount spent on advertising but also in terms of radio play. There are only two predictors in this model and so we could display this model graphically in three dimensions (Figure 7.7).

Equation (7.9) describes the tinted trapezium in the diagram (this is known as the regression *plane*) and the dots represent the observed data points. Like simple regression, the plane fitted to the data aims to best-predict the observed data. However, there are invariably some differences between the model and the real-life data (this fact is evident because some of the dots do not lie exactly on the tinted area of the graph). The *b*-value for advertising describes the slope of the left and right sides of the regression plane, whereas the *b*-value for airplay describes the slope of the top and bottom of the regression plane. Just like simple regression, knowledge of these two slopes tells us about the shape of the model (what it looks like) and the intercept locates the regression plane in space.

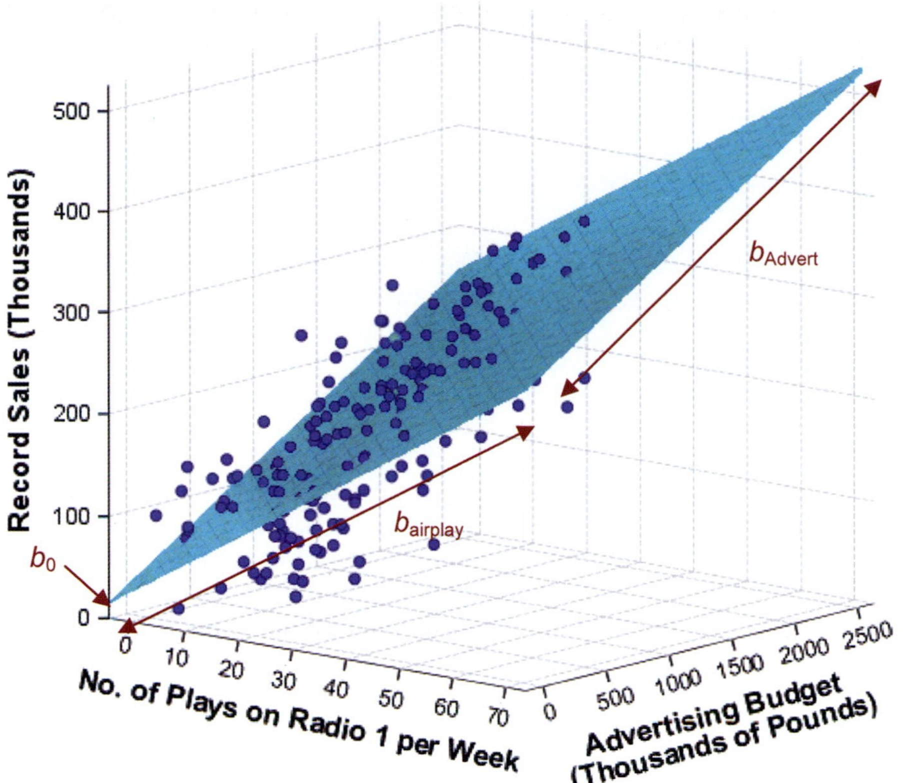

FIGURE 7.7 Scatterplot of the relationship between record sales, advertising budget and radio play

It is fairly easy to visualize a regression model with two predictors, because it is possible to plot the regression plane using a 3-D scatterplot. However, multiple regression can be used with three, four or even ten or more predictors. Although you can't immediately visualize what such complex models look like, or visualize what the *b*-values represent, you should be able to apply the principles of these basic models to more complex scenarios.

7.5.2. Sums of squares, *R* and R^2 ②

When we have several predictors, the partitioning of sums of squares is the same as in the single variable case except that the model we refer to takes the form of equation (7.9) rather than simply being a 2-D straight line. Therefore, SS_T can be calculated that represents the difference between the observed values and the mean value of the outcome variable. SS_R still represents the difference between the values of *Y* predicted by the model and the observed values. Finally, SS_M can still be calculated and represents the difference between the values of *Y* predicted by the model and the mean value. Although the computation of these values is much more complex than in simple regression, conceptually these values are the same.

When there are several predictors it does not make sense to look at the simple correlation coefficient and instead SPSS produces a multiple correlation coefficient (labelled **Multiple *R***). Multiple *R* is the correlation between the observed values of *Y* and the values of *Y* predicted by the multiple regression model. Therefore, large values of the multiple *R* represent a large correlation between the predicted and observed values of the outcome. A multiple *R* of 1 represents a situation in which the model perfectly predicts the observed data. As such, multiple R is a gauge of how well the model predicts the observed data. It follows that the resulting R^2 can be interpreted in the same way as simple regression: it is the amount of variation in the outcome variable that is accounted for by the model.

7.5.3. Methods of regression ②

If we are interested in constructing a complex model with several predictors, how do we decide which predictors to use? A great deal of care should be taken in selecting predictors for a model because the values of the regression coefficients depend upon the variables in the model. Therefore, the predictors included and the way in which they are entered into the model can have a great impact. In an ideal world, predictors should be selected based on past research.[5] If new predictors are being added to existing models then select these new variables based on the substantive *theoretical* importance of these variables. One thing *not* to do is select hundreds of random predictors, bung them all into a regression analysis and hope for the best. In addition to the problem of selecting predictors, there are several ways in which variables can be entered into a model. When predictors are all completely uncorrelated the order of variable entry has very little effect on the parameters calculated; however, we rarely have uncorrelated predictors and so the method of predictor selection is crucial.

7.5.3.1. Hierarchical (blockwise entry) ②

In **hierarchical regression** predictors are selected based on past work and the experimenter decides in which order to enter the predictors into the model. As a general rule, known predictors (from other research) should be entered into the model first in order of their importance in predicting the outcome. After known predictors have been entered, the experimenter can add any new predictors into the model. New predictors can be entered either all in one go, in a stepwise manner, or hierarchically (such that the new predictor suspected to be the most important is entered first).

7.5.3.2. Forced entry ②

Forced entry (or *Enter* as it is known in SPSS) is a method in which all predictors are forced into the model simultaneously. Like hierarchical, this method relies on good theoretical reasons for including the chosen predictors, but unlike hierarchical the experimenter makes no decision about the order in which variables are entered. Some researchers believe that this method is the only appropriate method for theory testing (Studenmund & Cassidy, 1987) because stepwise techniques are influenced by random variation in the data and so seldom give replicable results if the model is retested.

7.5.3.3. Stepwise methods ②

In **stepwise regressions** decisions about the order in which predictors are entered into the model are based on a purely mathematical criterion. In the *forward* method, an initial model is defined that contains only the constant (b_0). The computer then searches for the predictor (out of the ones available) that best predicts the outcome variable – it does this

[5] I might cynically qualify this suggestion by proposing that predictors be chosen based on past research that has utilized good methodology. If basing such decisions on regression analyses, select predictors based only on past research that has used regression appropriately and yielded reliable, generalizable models!

by selecting the predictor that has the highest simple correlation with the outcome. If this predictor significantly improves the ability of the model to predict the outcome, then this predictor is retained in the model and the computer searches for a second predictor. The criterion used for selecting this second predictor is that it is the variable that has the largest semi-partial correlation with the outcome. Let me explain this in plain English. Imagine that the first predictor can explain 40% of the variation in the outcome variable; then there is still 60% left unexplained. The computer searches for the predictor that can explain the biggest part of the remaining 60% (so, it is not interested in the 40% that is already explained). As such, this semi-partial correlation gives a measure of how much 'new variance' in the outcome can be explained by each remaining predictor (see section 6.6). The predictor that accounts for the most new variance is added to the model and, if it makes a significant contribution to the predictive power of the model, it is retained and another predictor is considered.

The *stepwise* method in SPSS is the same as the forward method, except that each time a predictor is added to the equation, a removal test is made of the least useful predictor. As such the regression equation is constantly being reassessed to see whether any redundant predictors can be removed. The *backward* method is the opposite of the forward method in that the computer begins by placing all predictors in the model and then calculating the contribution of each one by looking at the significance value of the *t*-test for each predictor. This significance value is compared against a removal criterion (which can be either an absolute value of the test statistic or a probability value for that test statistic). If a predictor meets the removal criterion (i.e. if it is not making a statistically significant contribution to how well the model predicts the outcome variable) it is removed from the model and the model is re-estimated for the remaining predictors. The contribution of the remaining predictors is then reassessed.

If you do decide to use a stepwise method then the backward method is preferable to the forward method. This is because of **suppressor effects**, which occur when a predictor has a significant effect but only when another variable is held constant. Forward selection is more likely than backward elimination to exclude predictors involved in suppressor effects. As such, the forward method runs a higher risk of making a Type II error (i.e. missing a predictor that does in fact predict the outcome).

7.5.3.4. Choosing a method ②

SPSS allows you to opt for any one of these methods and it is important to select an appropriate one. The forward, backward and stepwise methods all come under the general heading of *stepwise methods* because they all rely on the computer selecting variables based upon mathematical criteria. Many writers argue that this takes many important methodological decisions out of the hands of the researcher. What's more, the models derived by computer often take advantage of random sampling variation and so decisions about which variables should be included will be based upon slight differences in their semi-partial correlation. However, these slight statistical differences may contrast dramatically with the theoretical importance of a predictor to the model. There is also the danger of over-fitting (having too many variables in the model that essentially make little contribution to predicting the outcome) and under-fitting (leaving out important predictors) the model. For this reason stepwise methods are best avoided except for exploratory model building. If you must do a stepwise regression then it is advisable to **cross-validate** your model by splitting the data (see section 7.6.2.2).

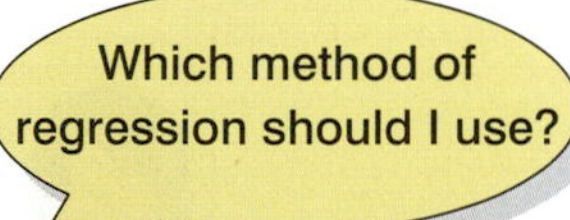
Which method of
regression should I use?

When there is a sound theoretical literature available, then base your model upon what past research tells you. Include any meaningful variables in the model in their order of importance. After this initial analysis, repeat the regression but exclude any variables that were statistically redundant the first time around. There are important considerations in deciding which predictors should be included. First, it is important not to include too many predictors. As a general rule, the fewer predictors the better, and certainly include only predictors for which you have a good theoretical grounding (it is meaningless to measure hundreds of variables and then put them all into a regression model). So, be selective and remember that you should have a decent sample size – see section 7.6.2.3.

7.6. How accurate is my regression model? ②

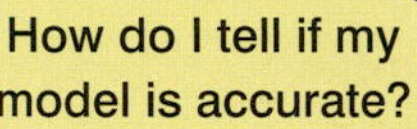

How do I tell if my model is accurate?

When we have produced a model based on a sample of data there are two important questions to ask: (1) does the model fit the observed data well, or is it influenced by a small number of cases; and (2) can my model generalize to other samples? These questions are vital to ask because they affect how we use the model that has been constructed. These questions are also, in some sense, hierarchical because we wouldn't want to generalize a bad model. However, it is a mistake to think that because a model fits the observed data well we can draw conclusions beyond our sample. **Generalization** is a critical additional step and if we find that our model is not generalizable, then we must restrict any conclusions based on the model to the sample used. First, we will look at how we establish whether a model is an accurate representation of the actual data, and in section 7.6.2 we move on to look at how we assess whether a model can be used to make inferences beyond the sample of data that has been collected.

7.6.1. Assessing the regression model I: diagnostics ②

To answer the question of whether the model fits the observed data well, or if it is influenced by a small number of cases, we can look for outliers and influential cases (the difference is explained in Jane Superbrain Box 7.1). We will look at these in turn.

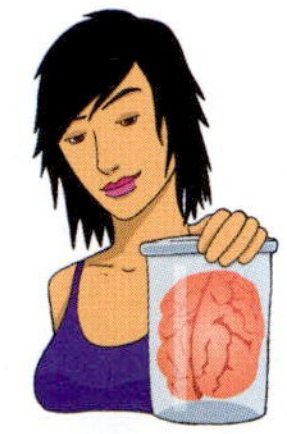

JANE SUPERBRAIN 7.1

The difference between residuals and influence statistics ③

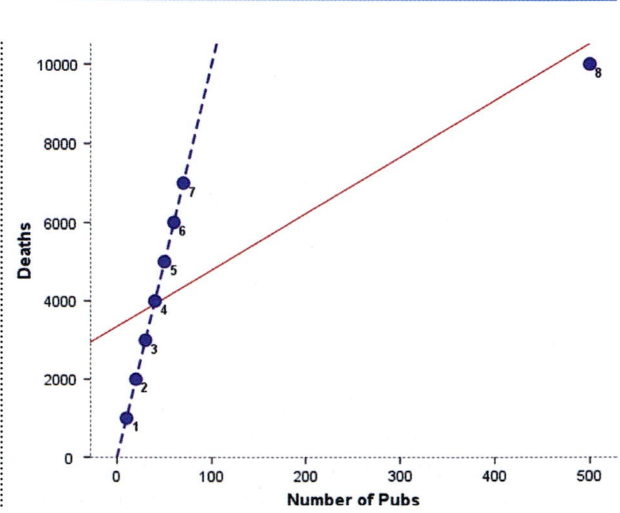

In this section I've described two ways to look for cases that might bias the model: residual and influence statistics. To illustrate how these measures differ, imagine that the Mayor of London at the turn of the last century was interested in how drinking affected mortality. London is divided up into different regions called boroughs, and so he might measure the number of pubs and the number of deaths over a period of time in eight of his boroughs. The data are in a file called **pubs.sav**.

The scatterplot of these data reveals that without the last case there is a perfect linear relationship (the dashed straight line). However, the presence of the last case (case 8) changes the line of best fit dramatically (although this line is still a significant fit of the data – do the regression analysis and see for yourself).

What's interesting about these data is when we look at the residuals and influence statistics. The standardized residual for case 8 is the second *smallest*: this outlier produces a very small residual (most of the non-outliers have larger residuals) because it sits very close to the line that has been fitted to the data. How can this be? Look at the influence statistics below and you'll see that they're massive for case 8: it exerts a huge influence over the model.

Case Summaries[a]

	Standardized Residual	Mahalanobis Distance	Cook's Distance	Centered Leverage Value	DFFIT	DFBETA Intercept	DFBETA pubs
1	-1.33839	.28515	.21328	.04074	-495.72692	-509.65184	1.39249
2	-.87895	.22370	.08530	.03196	-305.09716	-321.12768	.80153
3	-.41950	.16969	.01814	.02424	-137.20167	-147.10661	.33016
4	.03995	.12314	.00015	.01759	12.38769	13.45081	-.02658
5	.49940	.08403	.02294	.01200	147.81622	161.44976	-.27267
6	.95885	.05237	.08092	.00748	273.00807	297.67748	-.41116
7	1.41830	.02817	.17107	.00402	391.72124	422.81664	-.44422
8	-.27966	6.03375	227.14286	.86196	-39478.58473	3351.95531	-85.66108
Total N	8	8	8	8	8	8	8

a. Limited to first 100 cases.

As always when you see a statistical oddity you should ask what was happening in the real world. The last data point represents the City of London, a tiny area of only 1 square mile in the centre of London where very few people lived but where thousands of commuters (even then) came to work and had lunch in the pubs. Hence the pubs didn't rely on the resident population for their business and the residents didn't consume all of their beer! Therefore, there was a massive number of pubs.

This illustrates that a case exerting a massive influence can produce a small residual – so look at both! (I'm very grateful to David Hitchin for this example, and he in turn got it from Dr Richard Roberts.)

7.6.1.1. Outliers and residuals ②

An outlier is a case that differs substantially from the main trend of the data (see Jane Superbrain Box 4.1). Figure 7.8 shows an example of such a case in regression. Outliers can cause your model to be biased because they affect the values of the estimated regression coefficients. For example, Figure 7.8 uses the same data as Figure 7.3 except that the score of one participant has been changed to be an outlier (in this case a person who was very calm in the presence of a very big spider). The change in this one point has had a dramatic effect on the regression model chosen to fit the data. With the outlier present, the regression model changes: its gradient is reduced (the line becomes flatter) and the intercept increases (the new line will cross the Y-axis at a higher point). It should be clear from this diagram that it is important to try to detect outliers to see whether the model is biased in this way.

How do you think that you might detect an outlier? Well, we know that an outlier, by its nature, is very different from all of the other scores. This being true, do you think that the model will predict that person's score very accurately? The answer is *no*: looking at Figure 7.8 it is evident that even though the outlier has biased the model, the model still predicts that one value very badly (the regression line is long way from the outlier). Therefore, if we were to work out the differences between the data values that were

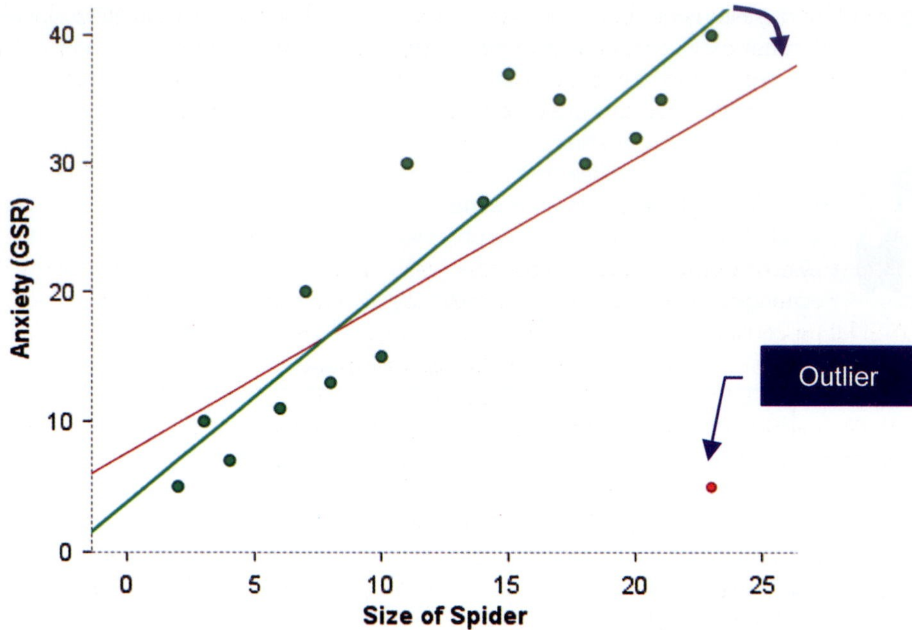

FIGURE 7.8 Graph demonstrating the effect of an outlier. The green line represents the original regression line for these data (see Figure 7.3), whereas the red line represents the regression line when an outlier is present

collected, and the values predicted by the model, we could detect an outlier by looking for large differences. This process is the same as looking for cases that the model predicts inaccurately. The differences between the values of the outcome predicted by the model and the values of the outcome observed in the sample are known as *residuals*. These residuals represent the error present in the model. If a model fits the sample data well then all residuals will be small (if the model was a perfect fit of the sample data – all data points fall on the regression line – then all residuals would be zero). If a model is a poor fit of the sample data then the residuals will be large. Also, if any cases stand out as having a large residual, then they could be outliers.

The *normal* or **unstandardized residuals** described above are measured in the same units as the outcome variable and so are difficult to interpret across different models. What we can do is to look for residuals that stand out as being particularly large. However, we cannot define a universal cut-off point for what constitutes a large residual. To overcome this problem, we use **standardized residuals**, which are the residuals divided by an estimate of their standard deviation. We came across standardization in section 6.3.2 as a means of converting variables into a standard unit of measurement (the standard deviation); we also came across *z*-scores (see section 1.7.4) in which variables are converted into standard deviation units (i.e. they're converted into scores that are distributed around a mean of 0 with a standard deviation of 1). By converting residuals into *z*-scores (standardized residuals) we can compare residuals from different models and use what we know about the properties of *z*-scores to devise universal guidelines for what constitutes an acceptable (or unacceptable) value. For example, we know from Chapter 1 that in a normally distributed sample, 95% of *z*-scores should lie between −1.96 and +1.96, 99% should lie between −2.58 and +2.58, and 99.9% (i.e. nearly all of them) should lie between −3.29 and +3.29. Some general rules for standardized residuals are derived from these facts: (1) standardized residuals with an absolute value greater than 3.29 (we can use 3 as an approximation) are cause for concern because in an average sample case a value this high is unlikely to happen by chance; (2) if more than 1% of our sample cases have standardized residuals with an absolute value greater than 2.58 (we usually just say 2.5) there is evidence that the level of error within our model is unacceptable (the model is a fairly poor fit of the sample data); and (3) if more than 5% of cases have standardized residuals with an absolute value greater than 1.96 (we can use 2 for convenience) then there is also evidence that the model is a poor representation of the actual data.

A third form of residual is the **Studentized residual**, which is the unstandardized residual divided by an estimate of its standard deviation that varies point by point. These residuals have the same properties as the standardized residuals but usually provide a more precise estimate of the error variance of a specific case.

7.6.1.2. Influential cases ③

As well as testing for outliers by looking at the error in the model, it is also possible to look at whether certain cases exert undue influence over the parameters of the model. So, if we were to delete a certain case, would we obtain different regression coefficients? This type of analysis can help to determine whether the regression model is stable across the sample, or whether it is biased by a few influential cases. Again, this process will unveil outliers.

There are several residual statistics that can be used to assess the influence of a particular case. One statistic is the **adjusted predicted value** for a case when that case is excluded from the analysis. In effect, the computer calculates a new model without a particular case and then uses this new model to predict the value of the outcome variable for the case that was excluded. If a case does not exert a large influence over the model then we would expect the adjusted predicted value to be very similar to the predicted value when the case is included. Put simply, if the model is stable then the predicted value of a case should be the same regardless of whether or not that case was used to calculate the model. The difference between the adjusted predicted value and the original predicted value is known as **DFFit** (see below). We can also look at the residual based on the adjusted predicted value: that is, the difference between the adjusted predicted value and the original observed value. This is the **deleted residual**. The deleted residual can be divided by the standard deviation to give a standardized value known as the **Studentized deleted residual**. This residual can be compared across different regression analyses because it is measured in standard units.

SMART ALEX ONLY

The deleted residuals are very useful to assess the influence of a case on the ability of the model to predict that case. However, they do not provide any information about how a case influences the model as a whole (i.e. the impact that a case has on the model's ability to predict *all* cases). One statistic that does consider the effect of a single case on the model as a whole is **Cook's distance**. Cook's distance is a measure of the overall influence of a case on the model and Cook and Weisberg (1982) have suggested that values greater than 1 may be cause for concern.

A second measure of influence is **leverage** (sometimes called **hat values**), which gauges the influence of the observed value of the outcome variable over the predicted values. The average leverage value is defined as $(k + 1)/n$ in which k is the number of predictors in the model and n is the number of participants.[6] The maximum value for leverage is (N–1)/N; however, SPSS calculates a version of the leverage that takes a maximum value of 1 (indicating that the case has complete influence over prediction). If no cases exert undue influence over the model then we would expect all of the leverage values to be close to the average value ($(k + 1)/n$). Hoaglin and Welsch (1978) recommend investigating cases with values greater than twice the average ($2(k + 1)/n$) and Stevens (2002) recommends using three times the average ($3(k + 1)/n$) as a cut-off point for identifying cases having undue influence. We will see how to use these cut-off points later. However, cases with large leverage values will not necessarily have a large influence on the regression coefficients because they are measured on the outcome variables rather than the predictors.

Related to the leverage values are the **Mahalanobis distances** (Figure 7.9), which measure the distance of cases from the mean(s) of the predictor variable(s). You need to look for the cases with the highest values. It is not easy to establish a cut-off point at which to worry,

[6] You may come across the average leverage denoted as p/n in which p is the number of parameters being estimated. In multiple regression, we estimate parameters for each predictor and also for a constant and so p is equivalent to the number of predictors plus one, $(k + 1)$.

FIGURE 7.9 Prasanta Chandra Mahalanobis staring into his distances

although Barnett and Lewis (1978) have produced a table of critical values dependent on the number of predictors and the sample size. From their work it is clear that even with large samples ($N = 500$) and 5 predictors, values above 25 are cause for concern. In smaller samples ($N = 100$) and with fewer predictors (namely 3) values greater than 15 are problematic, and in very small samples ($N = 30$) with only 2 predictors values greater than 11 should be examined. However, for more specific advice, refer to Barnett and Lewis's table.

It is possible to run the regression analysis with a case included and then rerun the analysis with that same case excluded. If we did this, undoubtedly there would be some difference between the b coefficients in the two regression equations. This difference would tell us how much influence a particular case has on the parameters of the regression model. To take a hypothetical example, imagine two variables that had a perfect negative relationship except for a single case (case 30). If a regression analysis was done on the 29 cases that were perfectly linearly related then we would get a model in which the predictor variable X perfectly predicts the outcome variable Y, and there are no errors. If we then ran the analysis but this time include the case that didn't conform (case 30), then the resulting model has different parameters. Some data are stored in the file **dfbeta.sav** which illustrate such a situation. Try running a simple regression first with all the cases included and then with case 30 deleted. The results are summarized in Table 7.1, which shows: (1) the parameters for the regression model when the extreme case is included or excluded; (2) the resulting regression equations; and (3) the value of Y predicted from participant 30's score on the X variable (which is obtained by replacing the X in the regression equation with participant 30's score for X, which was 1).

When case 30 is excluded, these data have a perfect negative relationship; hence the coefficient for the predictor (b_1) is −1 (remember that in simple regression this term is the same as Pearson's correlation coefficient), and the coefficient for the constant (the intercept, b_0) is 31. However, when case 30 is included, both parameters are reduced[7] and the difference between the parameters is also displayed. The difference between a parameter estimated using all cases and estimated when one case is excluded is known as the **DFBeta** in SPSS. DFBeta is calculated for every case and for each of the parameters in the model. So, in our hypothetical example, the DFBeta for the constant is −2, and the DFBeta for the predictor variable is 0.1. By looking at the values of DFBeta, it is possible to identify cases that have a large influence on the parameters of the regression model. Again, the units of measurement used will affect these values and so SPSS produces a **standardized DFBeta**. These standardized

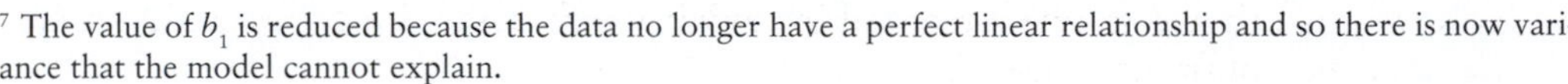

[7] The value of b_1 is reduced because the data no longer have a perfect linear relationship and so there is now variance that the model cannot explain.

TABLE 7.1 The difference in the parameters of the regression model when one case is excluded

Parameter (b)	Case 30 Included	Case 30 Excluded	Difference
Constant (intercept)	29.00	31.00	−2.00
Predictor (gradient)	−0.90	−1.00	0.10
Model (regression line):	$Y = (-0.9)X + 29$	$Y = (-1)X + 31$	
Predicted Y	28.10	30.00	−1.90

values are easier to use because universal cut-off points can be applied. In this case absolute values above 1 indicate cases that substantially influence the model parameters (although Stevens, 2002, suggests looking at cases with absolute values greater than 2).

A related statistic is the **DFFit**, which is the difference between the predicted value for a case when the model is calculated including that case and when the model is calculated excluding that case: in this example the value is −1.90 (see Table 7.1). If a case is not influential then its DFFit should be zero – hence, we expect non-influential cases to have small DFFit values. However, we have the problem that this statistic depends on the units of measurement of the outcome and so a DFFit of 0.5 will be very small if the outcome ranges from 1 to 100, but very large if the outcome varies from 0 to 1. Therefore, SPSS also produces standardized versions of the DFFit values (**Standardized DFFit**). A final measure is that of the **covariance ratio (CVR)**, which is a measure of whether a case influences the variance of the regression parameters. A description of the computation of this statistic would leave most readers dazed and confused, so suffice to say that when this ratio is close to 1 the case is having very little influence on the variances of the model parameters. Belsey, Kuh and Welsch (1980) recommend the following:

- If $CVR_i > 1 + [3(k + 1)/n]$ then deleting the ith case will damage the precision of some of the model's parameters.
- If $CVR_i < 1 - [3(k + 1)/n]$ then deleting the ith case will improve the precision of some of the model's parameters.

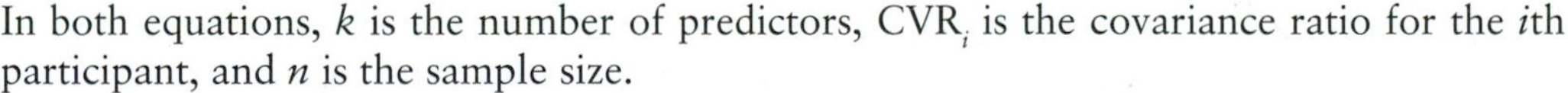

In both equations, k is the number of predictors, CVR_i is the covariance ratio for the ith participant, and n is the sample size.

7.6.1.3. A final comment on diagnostic statistics ②

There are a lot of diagnostic statistics that should be examined after a regression analysis, and it is difficult to summarize this wealth of material into a concise conclusion. However, one thing I would like to stress is a point made by Belsey et al. (1980) who noted the dangers inherent in these procedures. The point is that diagnostics are tools that enable you to see how good or bad your model is in terms of fitting the sampled data. They are a way of assessing your model. They are *not*, however, a way of justifying the removal of data points to effect some desirable change in the regression parameters (e.g. deleting a case that changes a non-significant b-value into a significant one). Stevens (2002), as ever, offers excellent advice:

> If a point is a significant outlier on Y, but its Cook's distance is < 1, there is no real need to delete that point since it does not have a large effect on the regression analysis. However, one should still be interested in studying such points further to understand why they did not fit the model. (p. 135)

7.6.2. Assessing the regression model II: generalization ②

When a regression analysis is done, an equation can be produced that is correct for the sample of observed values. However, in the social sciences we are usually interested in generalizing our findings outside of the sample. So, although it can be useful to draw conclusions about a particular sample of people, it is usually more interesting if we can then assume that our conclusions are true for a wider population. For a regression model to generalize we must be sure that underlying assumptions have been met, and to test whether the model does generalize we can look at cross-validating it.

7.6.2.1. Checking assumptions ②

To draw conclusions about a population based on a regression analysis done on a sample, several assumptions must be true (see Berry, 1993):

- **Variable types**: All predictor variables must be quantitative or categorical (with two categories), and the outcome variable must be quantitative, continuous and unbounded. By quantitative I mean that they should be measured at the interval level and by unbounded I mean that there should be no constraints on the variability of the outcome. If the outcome is a measure ranging from 1 to 10 yet the data collected vary between 3 and 7, then these data are constrained.
- **Non-zero variance**: The predictors should have some variation in value (i.e. they do not have variances of 0).
- **No perfect multicollinearity**: There should be no perfect linear relationship between two or more of the predictors. So, the predictor variables should not correlate too highly (see section 7.6.2.4).
- **Predictors are uncorrelated with 'external variables'**: *External variables* are variables that haven't been included in the regression model which influence the outcome variable.[8] These variables can be thought of as similar to the 'third variable' that was discussed with reference to correlation. This assumption means that there should be no external variables that correlate with any of the variables included in the regression model. Obviously, if external variables do correlate with the predictors, then the conclusions we draw from the model become unreliable (because other variables exist that can predict the outcome just as well).
- **Homoscedasticity**: At each level of the predictor variable(s), the variance of the residual terms should be constant. This just means that the residuals at each level of the predictor(s) should have the same variance (**homoscedasticity**); when the variances are very unequal there is said to be **heteroscedasticity** (see section 5.6 as well).
- **Independent errors**: For any two observations the residual terms should be uncorrelated (or independent). This eventuality is sometimes described as a lack of **autocorrelation**. This assumption can be tested with the **Durbin–Watson test**, which tests for serial correlations between errors. Specifically, it tests whether adjacent residuals are correlated. The test statistic can vary between 0 and 4 with a value of 2 meaning that the residuals are uncorrelated. A value greater than 2 indicates a negative correlation between

[8] Some authors choose to refer to these external variables as part of an error term that includes any random factor in the way in which the outcome varies. However, to avoid confusion with the residual terms in the regression equations I have chosen the label 'external variables'. Although this term implicitly washes over any random factors, I acknowledge their presence here!

adjacent residuals, whereas a value below 2 indicates a positive correlation. The size of the Durbin–Watson statistic depends upon the number of predictors in the model and the number of observations. For accuracy, you should look up the exact acceptable values in Durbin and Watson's (1951) original paper. As a very conservative rule of thumb, values less than 1 or greater than 3 are definitely cause for concern; however, values closer to 2 may still be problematic depending on your sample and model.

- **Normally distributed errors:** It is assumed that the residuals in the model are random, normally distributed variables with a mean of 0. This assumption simply means that the differences between the model and the observed data are most frequently zero or very close to zero, and that differences much greater than zero happen only occasionally. Some people confuse this assumption with the idea that predictors have to be normally distributed. In fact, predictors do not need to be normally distributed (see section 7.11).
- **Independence:** It is assumed that all of the values of the outcome variable are independent (in other words, each value of the outcome variable comes from a separate entity).
- **Linearity:** The mean values of the outcome variable for each increment of the predictor(s) lie along a straight line. In plain English this means that it is assumed that the relationship we are modelling is a linear one. If we model a non-linear relationship using a linear model then this obviously limits the generalizability of the findings.

This list of assumptions probably seems pretty daunting but as we saw in Chapter 5, assumptions are important. When the assumptions of regression are met, the model that we get for a sample can be accurately applied to the population of interest (the coefficients and parameters of the regression equation are said to be *unbiased*). Some people assume that this means that when the assumptions are met the regression model from a sample is always identical to the model that would have been obtained had we been able to test the entire population. Unfortunately, this belief isn't true. What an unbiased model does tell us is that *on average* the regression model from the sample is the same as the population model. However, you should be clear that even when the assumptions are met, it is possible that a model obtained from a sample may not be the same as the population model – but the likelihood of them being the same is increased.

7.6.2.2. Cross-validation of the model ③

Even if we can't be confident that the model derived from our sample accurately represents the entire population, there are ways in which we can assess how well our model can predict the outcome in a different sample. Assessing the accuracy of a model across different samples is known as cross-validation. If a model can be generalized, then it must be capable of accurately predicting the same outcome variable from the same set of predictors in a different group of people. If the model is applied to a different sample and there is a severe drop in its predictive power, then the model clearly does *not* generalize. As a first rule of thumb, we should aim to collect enough data to obtain a reliable regression model (see the next section). Once we have a regression model there are two main methods of cross-validation:

- **Adjusted R^2:** In SPSS, not only are the values of R and R^2 calculated, but also an **adjusted R^2**. This adjusted value indicates the loss of predictive power or **shrinkage**. Whereas R^2 tells us how much of the variance in Y is accounted for by the regression model from our sample, the adjusted value tells us how much variance in Y would be accounted for if the model had been derived from the population from which the sample was taken. SPSS derives the adjusted R^2 using Wherry's equation. However, this equation has been criticized because it tells us nothing about how well the regression model would predict an entirely different set of data (how well can the model predict scores of a different sample of data from the same population?). One version of R^2 that does tell us how well the model cross-validates uses Stein's formula which is shown in equation (7.11) (see Stevens, 2002):

$$\text{adjusted } R^2 = 1 - \left[\left(\frac{n-1}{n-k-1}\right)\left(\frac{n-2}{n-k-2}\right)\left(\frac{n+1}{n}\right)\right](1-R^2) \tag{7.11}$$

In Stein's equation, R^2 is the unadjusted value, n is the number of participants and k is the number of predictors in the model. For the more mathematically minded of you, it is worth using this equation to cross-validate a regression model.

- **Data splitting:** This approach involves randomly splitting your data set, computing a regression equation on both halves of the data and then comparing the resulting models. When using stepwise methods, cross-validation is a good idea; you should run the stepwise regression on a random selection of about 80% of your cases. Then force this model on the remaining 20% of the data. By comparing the values of R^2 and b in the two samples you can tell how well the original model generalizes (see Tabachnick & Fidell, 2007, for more detail).

7.6.2.3. Sample size in regression ③

How much data should I collect?

In the previous section I said that it's important to collect enough data to obtain a reliable regression model. Well, how much is enough? You'll find a lot of rules of thumb floating about, the two most common being that you should have 10 cases of data for each predictor in the model, or 15 cases of data per predictor. So, with five predictors, you'd need 50 or 75 cases respectively (depending on the rule you use). These rules are very pervasive (even I used the 15 cases per predictor rule in the first edition of this book) but they over simplify the issue considerably. In fact, the sample size required will depend on the size of effect that we're trying to detect (i.e. how strong the relationship is that we're trying to measure) and how much power we want to detect these effects. The simplest rule of thumb is that the bigger the sample size, the better! The reason is that the estimate of R that we get from regression is dependent on the number of predictors, k, and the sample size, N. In fact the expected R for random data is $k/(N-1)$ and so with small sample sizes random data can appear to show a strong effect: for example, with six predictors and 21 cases of data, $R = 6/(21-1) = .3$ (a medium effect size by Cohen's criteria described in section 6.3.2). Obviously for random data we'd want the expected R to be 0 (no effect) and for this to be true we need large samples (to take the previous example, if we had 100 cases not 21, then the expected R would be a more acceptable .06).

It's all very well knowing that larger is better, but researchers usually need some more concrete guidelines (much as we'd all love to collect 1000 cases of data it isn't always practical!). Green (1991) makes two rules of thumb for the *minimum* acceptable sample size, the first based on whether you want to test the overall fit of your regression model (i.e. test the R^2), and the second based on whether you want to test the individual predictors within the model (i.e. test b-values of the model). If you want to test the model overall, then he recommends a minimum sample size of $50 + 8k$, where k is the number of predictors. So, with five predictors, you'd need a sample size of $50 + 40 = 90$. If you want to test the individual predictors then he suggests a minimum sample size of $104 + k$, so again taking the example of 5 predictors you'd need a sample size of $104 + 5 = 109$. Of course, in most cases we're interested both in the overall fit and in the contribution of individual predictors, and in this situation Green recommends you calculate both of the minimum sample sizes I've just described, and use the one that has the largest value (so, in the five-predictor example, we'd use 109 because it is bigger than 90).

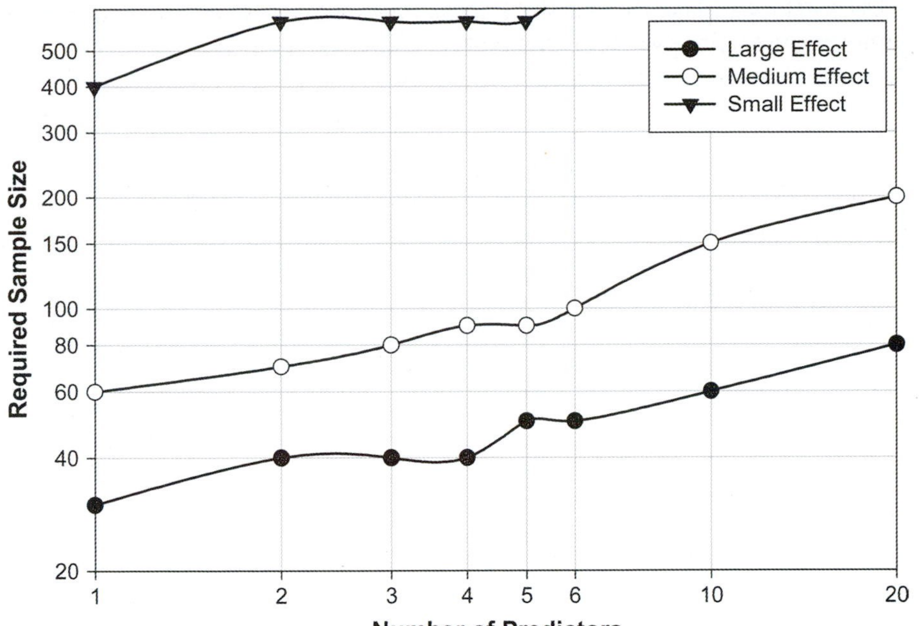

FIGURE 7.10 Graph to show the sample size required in regression depending on the number of predictors and the size of expected effect

Now, these guidelines are all right as a rough and ready guide, but they still oversimplify the problem. As I've mentioned, the sample size required actually depends on the size of the effect (i.e. how well our predictors predict the outcome) and how much statistical power we want to detect these effects. Miles and Shevlin (2001) produce some extremely useful graphs that illustrate the sample sizes needed to achieve different levels of power, for different effect sizes, as the number of predictors vary. For precise estimates of the sample size you should be using, I recommend using these graphs. I've summarized some of the general findings in Figure 7.10. This diagram shows the sample size required to achieve a high level of power (I've taken Cohen's, 1988, benchmark of .8) depending on the number of predictors and the size of expected effect. To summarize the graph very broadly: (1) if you expect to find a large effect then a sample size of 80 will always suffice (with up to 20 predictors) and if there are fewer predictors then you can afford to have a smaller sample; (2) if you're expecting a medium effect, then a sample size of 200 will always suffice (up to 20 predictors), you should always have a sample size above 60, and with six or fewer predictors you'll be fine with a sample of 100; and (3) if you're expecting a small effect size then just don't bother unless you have the time and resources to collect at least 600 cases of data (and many more if you have six or more predictors).

7.6.2.4. Multicollinearity ②

Multicollinearity exists when there is a strong correlation between two or more predictors in a regression model. Multicollinearity poses a problem only for multiple regression because (without wishing to state the obvious) simple regression requires only one predictor. **Perfect collinearity** exists when at least one predictor is a perfect linear combination of the others (the simplest example being two predictors that are perfectly correlated – they have a correlation coefficient of 1). If there is perfect collinearity between predictors it becomes impossible to obtain unique estimates of the regression coefficients because there are an infinite number of combinations of coefficients that would work equally well. Put simply, if we have two predictors that are perfectly correlated, then the values of b for each variable are interchangeable. The good news is that perfect collinearity is rare in real-life data. The bad news is that less than perfect collinearity is virtually unavoidable. Low levels of collinearity pose little threat to the models generated by SPSS, but as collinearity increases there are three problems that arise:

- **Untrustworthy *b*s**: As collinearity increases so do the standard errors of the *b* coefficients. If you think back to what the standard error represents, then big standard errors for *b* coefficients means that these *b*s are more variable across samples. Therefore, it means that the *b* coefficient in our sample is less likely to represent the population. Crudely put, multicollinearity means that the *b*-values are less trustworthy. Don't lend them money and don't let them go for dinner with your boy- or girlfriend. Of course if the *b*s are variable from sample to sample then the resulting predictor equations will be unstable across samples too.
- **It limits the size of *R***: Remember that *R* is a measure of the multiple correlation between the predictors and the outcome and that R^2 indicates the variance in the outcome for which the predictors account. Imagine a situation in which a single variable predicts the outcome variable fairly successfully (e.g. $R = .80$) and a second predictor variable is then added to the model. This second variable might account for a lot of the variance in the outcome (which is why it is included in the model), but the variance it accounts for is the same variance accounted for by the first variable. In other words, once the variance accounted for by the first predictor has been removed, the second predictor accounts for very little of the remaining variance (the second variable accounts for very little *unique variance*). Hence, the overall variance in the outcome accounted for by the two predictors is little more than when only one predictor is used (so *R* might increase from .80 to .82). This idea is connected to the notion of partial correlation that was explained in Chapter 6. If, however, the two predictors are completely uncorrelated, then the second predictor is likely to account for different variance in the outcome to that accounted for by the first predictor. So, although in itself the second predictor might account for only a little of the variance in the outcome, the variance it does account for is different to that of the other predictor (and so when both predictors are included, *R* is substantially larger, say .95). Therefore, having uncorrelated predictors is beneficial.
- **Importance of predictors**: Multicollinearity between predictors makes it difficult to assess the individual importance of a predictor. If the predictors are highly correlated, and each accounts for similar variance in the outcome, then how can we know which of the two variables is important? Quite simply we can't tell which variable is important – the model could include either one, interchangeably.

One way of identifying multicollinearity is to scan a correlation matrix of all of the predictor variables and see if any correlate very highly (by very highly I mean correlations of above .80 or .90). This is a good 'ball park' method but misses more subtle forms of multicollinearity. Luckily, SPSS produces various collinearity diagnostics, one of which is the **variance inflation factor (VIF)**. The VIF indicates whether a predictor has a strong linear relationship with the other predictor(s). Although there are no hard and fast rules about what value of the VIF should cause concern, Myers (1990) suggests that a value of 10 is a good value at which to worry. What's more, if the average VIF is greater than 1, then multicollinearity may be biasing the regression model (Bowerman & O'Connell, 1990). Related to the VIF is the **tolerance** statistic, which is its reciprocal (1/VIF). As such, values below 0.1 indicate serious problems although Menard (1995) suggests that values below 0.2 are worthy of concern.

Other measures that are useful in discovering whether predictors are dependent are the *eigenvalues of the scaled, uncentred cross-products matrix*, the *condition indexes* and the *variance proportions*. These statistics are extremely complex and will be covered as part of the interpretation of SPSS output (see section 7.8.5). If none of this has made any sense then have a look at Hutcheson and Sofroniou (1999: 78–85) who give a really clear explanation of multicollinearity.

7.7. How to do multiple regression using SPSS ②

7.7.1. Some things to think about before the analysis ②

A good strategy to adopt with regression is to measure predictor variables for which there are sound theoretical reasons for expecting them to predict the outcome. Run a regression analysis in which all predictors are entered into the model and examine the output to see which predictors contribute substantially to the model's ability to predict the outcome. Once you have established which variables are important, rerun the analysis including only the important predictors and use the resulting parameter estimates to define your regression model. If the initial analysis reveals that there are two or more significant predictors then you could consider running a forward stepwise analysis (rather than forced entry) to find out the individual contribution of each predictor.

I have spent a lot of time explaining the theory behind regression and some of the diagnostic tools necessary to gauge the accuracy of a regression model. It is important to remember that SPSS may appear to be very clever, but in fact it is not. Admittedly, it can do lots of complex calculations in a matter of seconds, but what it can't do is control the quality of the model that is generated – to do this requires a human brain (and preferably a trained one). SPSS will happily generate output based on any garbage you decide to feed into the data editor and SPSS will not judge the results or give any indication of whether the model can be generalized or if it is valid. However, SPSS provides the statistics necessary to judge these things, and at this point our brains must take over the job – which is slightly worrying (especially if your brain is as small as mine)!

7.7.2. Main options ②

Imagine that the record company executive was now interested in extending the model of record sales to incorporate other variables. He decides to measure two new variables: (1) the number of times the record is played on Radio 1 during the week prior to release (**airplay**); and (2) the attractiveness of the band (**attract**). Before a record is released, the executive notes the amount spent on advertising, the number of times the record is played on radio the week before release, and the attractiveness of the band. He does this for 200 different records (each made by a different band). Attractiveness was measured by asking a random sample of the target audience to rate the attractiveness of each band on a scale from 0 (hideous potato-heads) to 10 (gorgeous sex objects). The mode attractiveness given by the sample was used in the regression (because he was interested in what the majority of people thought, rather than the average of people's opinions).

These data are in the file **Record2.sav** and you should note that each variable has its own column (the same layout as for correlation) and each row represents a different record. So, the first record had £10,260 spent advertising it, sold 330,000 copies, received 43 plays on Radio 1 the week before release, and was made by a band that the majority of people rated as gorgeous sex objects (Figure 7.11).

The executive has past research indicating that advertising budget is a significant predictor of record sales, and so he should include this variable in the model first. His new variables (**airplay** and **attract**) should, therefore, be entered into the model *after* advertising budget. This method is hierarchical (the researcher decides in which order to enter variables into the model based on past research). To do a hierarchical regression in SPSS we have to enter the variables in blocks (each block representing one step in the hierarchy). To get to the main regression dialog box (Figure 7.12) select Analyze Regression ▸ Linear...; this dialog box is the same as when we encountered it for simple regression.

The main dialog box is fairly self-explanatory in that there is a space to specify the dependent variable (outcome) and a space to place one or more independent variables

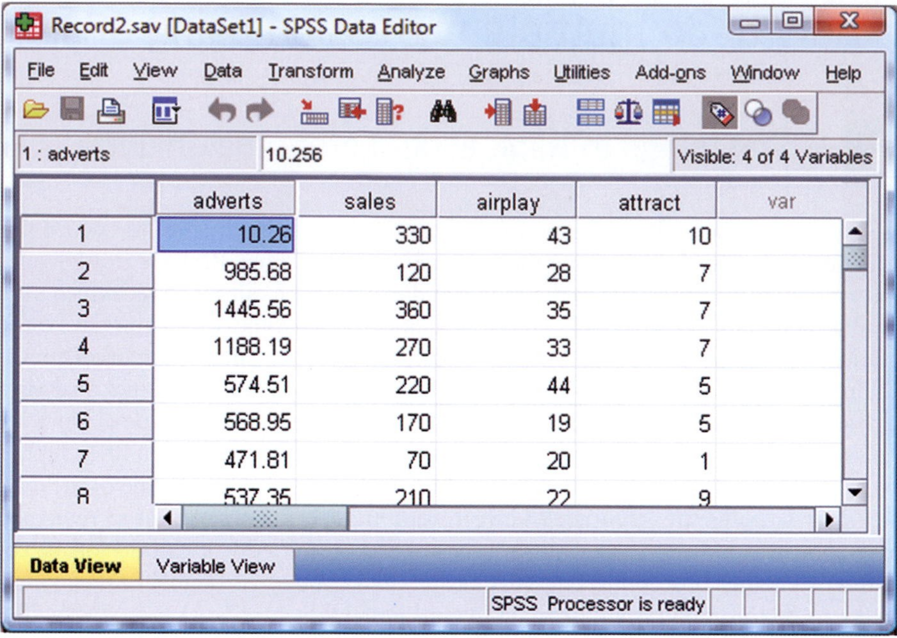

FIGURE 7.11 Data layout for multiple regression

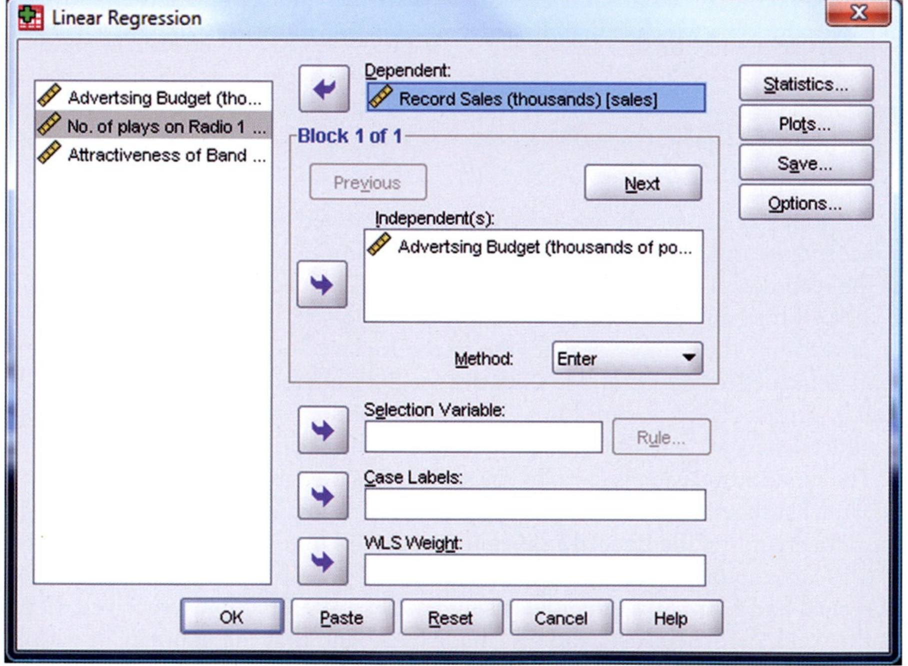

FIGURE 7.12 Main dialog box for block 1 of the multiple regression

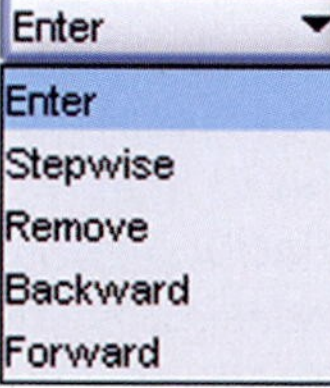

(predictor variables). As usual, the variables in the data editor are listed on the left-hand side of the box. Highlight the outcome variable (record sales) in this list by clicking on it and then drag it to the box labelled *Dependent* or click on ➜. We also need to specify the predictor variable for the first block. We decided that advertising budget should be entered into the model first (because past research indicates that it is an important predictor), so highlight this variable in the list and drag it to the box labelled *Independent(s)* or click on ➜. Underneath the *Independent(s)* box, there

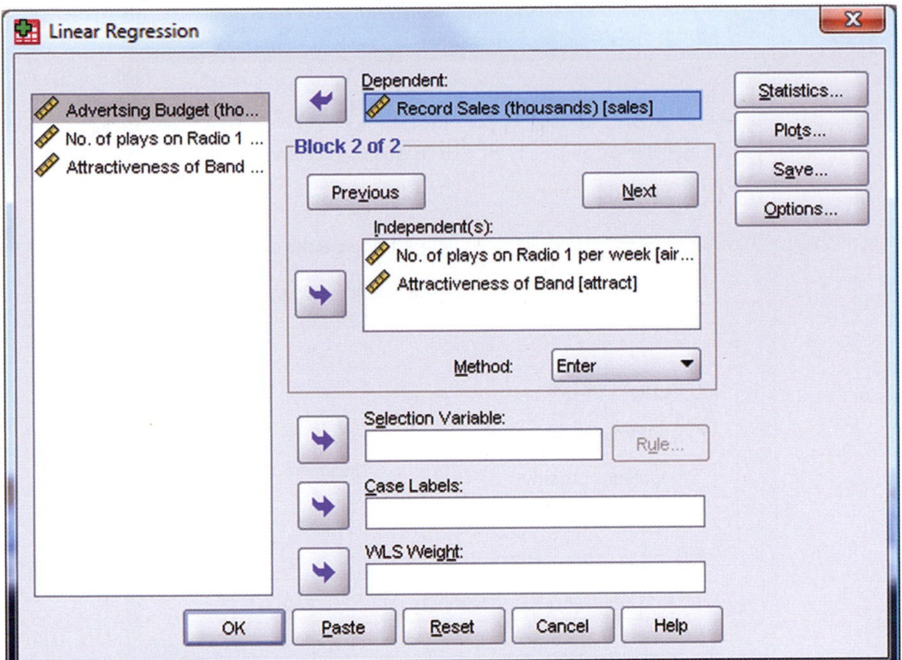

FIGURE 7.13 Main dialog box for block 2 of the multiple regression

is a drop-down menu for specifying the *Method* of regression (see section 7.5.3). You can select a different method of variable entry for each block by clicking on Enter, next to where it says *Method*. The default option is forced entry, and this is the option we want, but if you were carrying out more exploratory work, you might decide to use one of the stepwise methods (forward, backward, stepwise or remove).

Having specified the first block in the hierarchy, we need to move on to the second. To tell the computer that you want to specify a new block of predictors you must click on Next. This process clears the *Independent(s)* box so that you can enter the new predictors (you should also note that above this box it now reads *Block 2 of 2* indicating that you are in the second block of the two that you have so far specified). We decided that the second block would contain both of the new predictors and so you should click on **airplay** and **attract** (while holding down *Ctrl*) in the variables list and drag them to the *Independent(s)* box or click on ➜. The dialog box should now look like Figure 7.13. To move between blocks use the Previous and Next buttons (so, for example, to move back to block 1, click on Previous).

It is possible to select different methods of variable entry for different blocks in a hierarchy. So although we specified forced entry for the first block, we could now specify a stepwise method for the second. Given that we have no previous research regarding the effects of attractiveness and airplay on record sales, we might be justified in requesting a stepwise method for this block. However, because of the problems with stepwise methods, I am going to stick with forced entry for both blocks in this example.

7.7.3. Statistics ②

In the main regression dialog box click on Statistics... to open a dialog box for selecting various important options relating to the model (Figure 7.14). Most of these options relate to the parameters of the model; however, there are procedures available for checking the assumptions of no multicollinearity (collinearity diagnostics) and serial independence of errors

FIGURE 7.14
Statistics dialog box for regression analysis

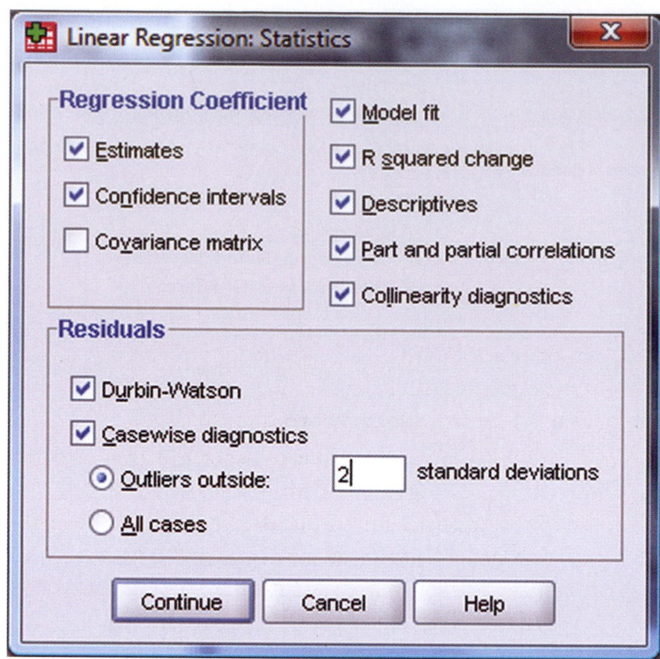

(Durbin–Watson). When you have selected the statistics you require (I recommend all but the covariance matrix as a general rule) click on Continue to return to the main dialog box:

- *Estimates*: This option is selected by default because it gives us the estimated coefficients of the regression model (i.e. the estimated *b*-values). Test statistics and their significance are produced for each regression coefficient: a *t*-test is used to see whether each *b* differs significantly from zero (see section 7.2.4).
- *Confidence intervals*: This option produces confidence intervals for each of the unstandardized regression coefficients. Confidence intervals can be a very useful tool in assessing the likely value of the regression coefficients in the population – I will describe their exact interpretation later.
- *Covariance matrix*: This option will display a matrix of the covariances, correlation coefficients and variances between the regression coefficients of each variable in the model. A variance–covariance matrix is produced with variances displayed along the diagonal and covariances displayed as off-diagonal elements. The correlations are produced in a separate matrix.
- *Model fit*: This option is vital and so is selected by default. It provides not only a statistical test of the model's ability to predict the outcome variable (the *F*-test, see section 7.2.3), but also the value of *R* (or multiple *R*), the corresponding R^2 and the adjusted R^2.
- *R squared change*: This option displays the change in R^2 resulting from the inclusion of a new predictor (or block of predictors). This measure is a useful way to assess the contribution of new predictors (or blocks) to explaining variance in the outcome.
- *Descriptives*: If selected, this option displays a table of the mean, standard deviation and number of observations of all of the variables included in the analysis. A correlation matrix is also displayed showing the correlation between all of the variables and the one-tailed probability for each correlation coefficient. This option is extremely useful because the correlation matrix can be used to assess whether predictors are interrelated (which can be used to establish whether there is multicollinearity).
- *Part and partial correlations*: This option produces the zero-order correlation (the Pearson correlation) between each predictor and the outcome variable. It also

produces the partial correlation between each predictor and the outcome, controlling for all other predictors in the model. Finally, it produces the part correlation (or semi-partial correlation) between each predictor and the outcome. This correlation represents the relationship between each predictor and the part of the outcome that is not explained by the other predictors in the model. As such, it measures the unique relationship between a predictor and the outcome (see section 6.6).

- *Collinearity diagnostics*: This option is for obtaining collinearity statistics such as the VIF, tolerance, eigenvalues of the scaled, uncentred cross-products matrix, condition indexes and variance proportions (see section 7.6.2.3).
- *Durbin-Watson*: This option produces the **Durbin–Watson test** statistic, which tests the assumption of independent errors. Unfortunately, SPSS does not provide the significance value of this test, so you must decide for yourself whether the value is different enough from 2 to be cause for concern (see section 7.6.2.1).
- *Casewise diagnostics*: This option, if selected, lists the observed value of the outcome, the predicted value of the outcome, the difference between these values (the residual) and this difference standardized. Furthermore, it will list these values either for all cases, or just for cases for which the standardized residual is greater than 3 (when the ± sign is ignored). This criterion value of 3 can be changed, and I recommend changing it to 2 for reasons that will become apparent. A summary table of residual statistics indicating the minimum, maximum, mean and standard deviation of both the values predicted by the model and the residuals (see section 7.7.5) is also produced.

7.7.4. Regression plots ②

Once you are back in the main dialog box, click on Plots... to activate the regression plots dialog box shown in Figure 7.15. This dialog box provides the means to specify several graphs, which can help to establish the validity of some regression assumptions. Most of these plots involve various *residual* values, which will be described in more detail in section 7.7.5.

On the left-hand side of the dialog box is a list of several variables.

- **DEPENDNT** (*the outcome variable*).
- ***ZPRED** (*the standardized predicted values* of the dependent variable based on the model). These values are standardized forms of the values predicted by the model.
- ***ZRESID** (*the standardized residuals*, or errors). These values are the standardized differences between the observed data and the values that the model predicts).
- ***DRESID** (*the deleted residuals*). See section 7.6.1.1 for details.
- ***ADJPRED** (*the adjusted predicted values*). See section 7.6.1.1 for details.
- ***SRESID** (*the Studentized residual*). See section 7.6.1.1 for details.
- ***SDRESID** (*the Studentized deleted residual*). This value is the deleted residual divided by its standard deviation.

The variables listed in this dialog box all come under the general heading of residuals, and were discussed in detail in section 7.6.1.1. For a basic analysis it is worth plotting *ZRESID (*Y*-axis) against *ZPRED (*X*-axis), because this plot is useful to determine whether the assumptions of random errors and homoscedasticity have been met. A plot of *SRESID (*y*-axis) against *ZPRED (*x*-axis) will show up any heteroscedasticity also. Although often these two plots are virtually identical, the latter is more sensitive on a case-by-case basis. To create these plots simply select a variable from the list, and transfer it to the space labelled

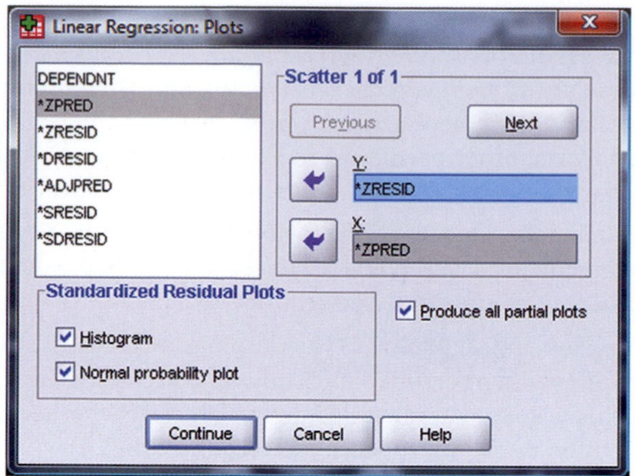

FIGURE 7.15 *Linear Regression: Plots* dialog box

either *x* or *y* (which refer to the axes) by clicking on ➜. When you have selected two variables for the first plot (as is the case in Figure 7.15) you can specify a new plot by clicking on Next. This process clears the spaces in which variables are specified. If you click on Next and would like to return to the plot that you last specified, then simply click on Previous. You can specify up to nine plots.

You can also select the tick-box labelled *Produce all partial plots* which will produce scatterplots of the residuals of the outcome variable and each of the predictors when both variables are regressed separately on the remaining predictors. Regardless of whether the previous sentence made any sense to you, these plots have several important characteristics that make them worth inspecting. First, the gradient of the regression line between the two residual variables is equivalent to the coefficient of the predictor in the regression equation. As such, any obvious outliers on a partial plot represent cases that might have undue influence on a predictor's regression coefficient. Second, non-linear relationships between a predictor and the outcome variable are much more detectable using these plots. Finally, they are a useful way of detecting collinearity. For these reasons, I recommend requesting them.

There are several options for plots of the standardized residuals. First, you can select a histogram of the standardized residuals (this is extremely useful for checking the assumption of normality of errors). Second, you can ask for a normal probability plot, which also provides information about whether the residuals in the model are normally distributed. When you have selected the options you require, click on Continue to take you back to the main regression dialog box.

7.7.5. Saving regression diagnostics ②

In section 7.6 we met two types of regression diagnostics: those that help us assess how well our model fits our sample and those that help us detect cases that have a large influence on the model generated. In SPSS we can choose to save these diagnostic variables in the data editor (so, SPSS will calculate them and then create new columns in the data editor in which the values are placed).

To save regression diagnostics you need to click on Save... in the main regression dialog box. This process activates the *save* new variables dialog box (see Figure 7.16). Once this dialog box is active, it is a simple matter to tick the boxes next to the required statistics. Most of the available options were explained in section 7.6 and Figure 7.16 shows, what I consider to be, a fairly basic set of diagnostic statistics. Standardized (and Studentized) versions of these diagnostics are generally easier to interpret and so I suggest selecting them in preference to the unstandardized versions. Once the regression has been run, SPSS creates a column in your data editor for each statistic requested and it has a standard set of variable names to describe each one. After the name, there will be a number that refers to the analysis that has been run. So, for the first regression run on a data set the variable names will be followed by a 1, if you carry out a second regression it will create a new set of variables with names followed by a 2, and so on. The names of the variables that will be created are below. When you have selected the diagnostics you require (by clicking in the appropriate boxes), click on Continue to return to the main regression dialog box:

- **pre_1**: unstandardized predicted value.
- **zpr_1**: standardized predicted value.
- **adj_1**: adjusted predicted value.
- **sep_1**: standard error of predicted value.
- **res_1**: unstandardized residual.
- **zre_1**: standardized residual.
- **sre_1**: Studentized residual.
- **dre_1**: deleted residual.
- **sdr_1**: Studentized deleted residual.
- **mah_1**: Mahalanobis distance.
- **coo_1**: Cook's distance.
- **lev_1**: centred leverage value.
- **sdb0_1**: standardized DFBETA (intercept).
- **sdb1_1**: standardized DFBETA (predictor 1).
- **sdb2_1**: standardized DFBETA (predictor 2).
- **sdf_1**: standardized DFFIT.
- **cov_1**: covariance ratio.

7.7.6. Further options ②

As a final step in the analysis, you can click on Options... to take you to the options dialog box (Figure 7.17). The first set of options allows you to change the criteria used for entering variables in a stepwise regression. If you insist on doing stepwise regression, then it's probably best that you leave the default criterion of .05 probability for entry alone. However, you can make this criterion more stringent (.01). There is also the option to build a model that doesn't include a constant (i.e. has no *Y* intercept). This option should also be left alone (almost always). Finally, you can select a method for dealing with missing data points (see SPSS Tip 6.1). By default, SPSS excludes cases listwise, which in regression means that if a person has a missing value for any variable, then they are excluded from the whole analysis. So, for example, if our record company executive didn't have an attractiveness score for one of his bands, their data would not be used in the regression model. Another option is to excluded cases on a pairwise basis, which means that if a participant has a score missing for a particular variable, then their data are excluded only from calculations involving the variable for which they have no score. So, data for the band for which there was no attractiveness rating would still be used to calculate the relationships between advertising budget, airplay and record sales. However, if you do this many of your variables may not make sense, and you can end up with absurdities such as R^2 either negative or greater than 1.0. So, it's not a good option.

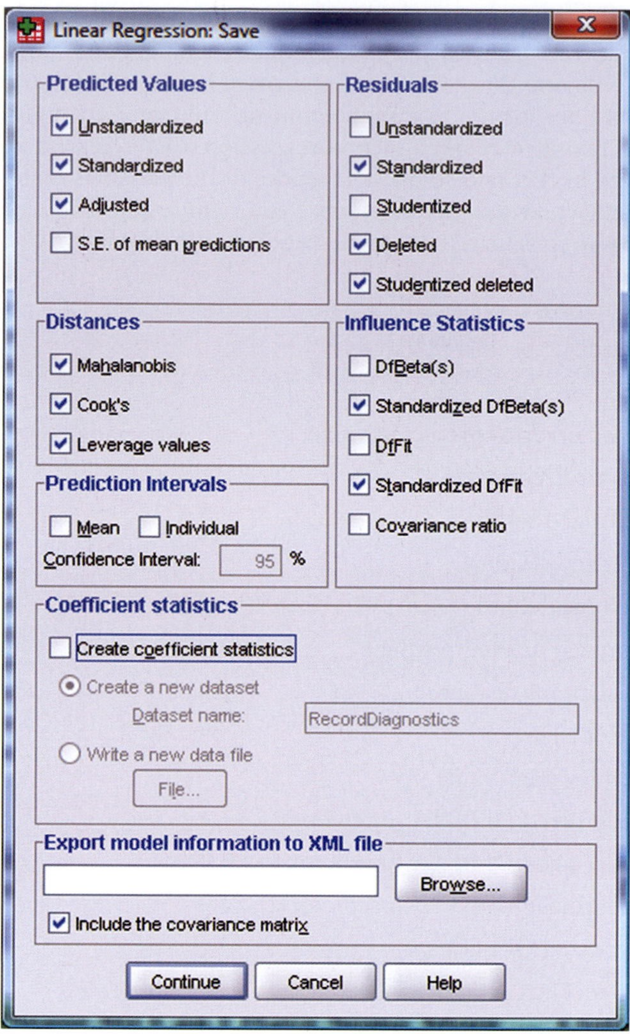

FIGURE 7.16 Dialog box for regression diagnostics

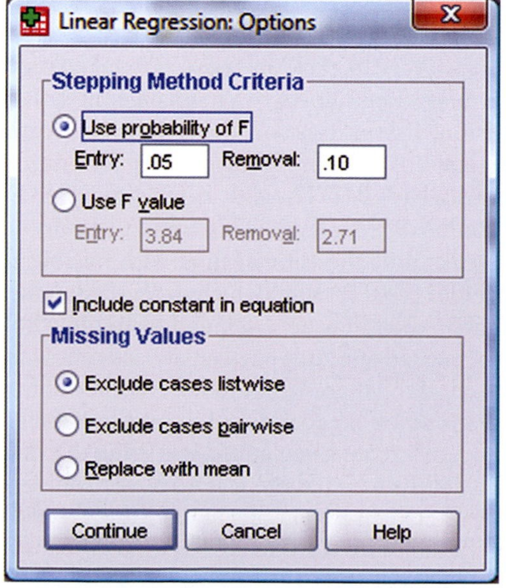

FIGURE 7.17 Options for linear regression

Another possibility is to replace the missing score with the average score for this variable and then include that case in the analysis (so, our example band would be given an attractiveness rating equal to the average attractiveness of all bands). The problem with this final choice is that it is likely to suppress the true value of the standard deviation (and more importantly the standard error). The standard deviation will be suppressed because for any replaced case there will be no difference between the mean and the score, whereas if data had been collected for that case there would, almost certainly, have been some difference between the score and the mean. Obviously, if the sample is large and the number of missing values small then this is not a serious consideration. However, if there are many missing values this choice is potentially dangerous because smaller standard errors are more likely to lead to significant results that are a product of the data replacement rather than a genuine effect. The final option is to use the Missing Value Analysis routine in SPSS. This is for experts. It makes use of the fact that if two or more variables are present and correlated for most cases in the file, and an occasional value is missing, you can replace the missing values with estimates far better than the mean (Tabachnick & Fidell, 2007: Chapter 4, describe some of these procedures).

7.8. Interpreting multiple regression ②

Having selected all of the relevant options and returned to the main dialog box, we need to click on OK to run the analysis. SPSS will spew out copious amounts of output in the viewer window, and we now turn to look at how to make sense of this information.

7.8.1. Descriptives ②

The output described in this section is produced using the options in the *Linear Regression: Statistics* dialog box (see Figure 7.14). To begin with, if you selected the *Descriptives* option, SPSS will produce the table seen in SPSS Output 7.4. This table tells us the mean and standard deviation of each variable in our data set, so we now know that the average number of record sales was 193,000. This table isn't necessary for interpreting the regression model, but it is a useful summary of the data. In addition to the descriptive statistics, selecting this option produces a correlation matrix too. This table shows three things. First, the table shows the value of Pearson's correlation coefficient between every pair of variables (e.g. we can see that the advertising budget had a large positive correlation with record sales, $r = .578$). Second, the one-tailed significance of each correlation is displayed (e.g. the correlation above is significant, $p < .001$). Finally, the number of cases contributing to each correlation ($N = 200$) is shown.

You might notice that along the diagonal of the matrix the values for the correlation coefficients are all 1.00 (i.e. a perfect positive correlation). The reason for this is that these values represent the correlation of each variable with itself, so obviously the resulting values are 1. The correlation matrix is extremely useful for getting a rough idea of the relationships between predictors and the outcome, and for a preliminary look for multicollinearity. If there is no multicollinearity in the data then there should be no substantial correlations ($r > .9$) between predictors.

If we look only at the predictors (ignore record sales) then the highest correlation is between the attractiveness of the band and the amount of airplay which is significant at a .01 level ($r = .182$, $p = .005$). Despite the significance of this correlation, the coefficient is small and so it looks as though our predictors are measuring different things (there is little collinearity). We can see also that of all of the predictors the number of plays on Radio 1 correlates best with the outcome ($r = .599$, $p < .001$) and so it is likely that this variable will best predict record sales.

CRAMMING SAM'S TIPS

Use the descriptive statistics to check the correlation matrix for multicollinearity; that is, predictors that correlate too highly with each other, $R > .9$.

SPSS OUTPUT 7.4 Descriptive statistics for regression analysis

Descriptive Statistics

	Mean	Std. Deviation	N
Record Sales (thousands)	193.2000	80.6990	200
Advertising Budget (thousands of pounds)	614.4123	485.6552	200
No. of plays on Radio 1 per week	27.5000	12.2696	200
Attractiveness of Band	6.7700	1.3953	200

Correlations

		Record Sales (thousands)	Advertising Budget (thousands of pounds)	No. of plays on Radio 1 per week	Attractiveness of Band
Pearson Correlation	Record Sales (thousands)	1.000	.578	.599	.326
	Advertising Budget (thousands of pounds)	.578	1.000	.102	.081
	No. of plays on Radio 1 per week	.599	.102	1.000	.182
	Attractiveness of Band	.326	.081	.182	1.000
Sig. (1-tailed)	Record Sales (thousands)	.	.000	.000	.000
	Advertising Budget (thousands of pounds)	.000	.	.076	.128
	No. of plays on Radio 1 per week	.000	.076	.	.005
	Attractiveness of Band	.000	.128	.005	.
N	Record Sales (thousands)	200	200	200	200
	Advertising Budget (thousands of pounds)	200	200	200	200
	No. of plays on Radio 1 per week	200	200	200	200
	Attractiveness of Band	200	200	200	200

7.8.2. Summary of model ②

The next section of output describes the overall model (so it tells us whether the model is successful in predicting record sales). Remember that we chose a hierarchical method and so each set of summary statistics is repeated for each stage in the hierarchy. In SPSS Output 7.5 you should note that there are two models. Model 1 refers to the first stage in the hierarchy when only advertising budget is used as a predictor. Model 2 refers to when all three predictors are used. SPSS Output 7.5 is the *model summary* and this table was produced using the Model fit option. This option is selected by default in SPSS because it provides us with some very important information about the model: the values of R, R^2 and the adjusted R^2. If the *R squared change* and *Durbin-Watson* options were selected, then these values are included also (if they weren't selected you'll find that you have a smaller table).

The model summary table is shown in SPSS Output 7.5 and you should notice that under this table SPSS tells us what the dependent variable (outcome) was and what the predictors were in each of the two models. In the column labelled *R* are the values of the multiple correlation coefficient between the predictors and the outcome. When only advertising budget is used as a predictor, this is the simple correlation between advertising and record sales (0.578). In fact all of the statistics for model 1 are the same as the simple regression model earlier (see section 7.4). The next column gives us a value of R^2, which we already know is a measure of how much of the variability in the outcome is accounted for by the predictors. For the first model its value is .335, which means that advertising budget accounts for 33.5% of the variation in record sales. However, when the other two predictors are included as well (model 2), this value increases to .665 or 66.5% of the variance in record sales. Therefore, if advertising accounts for 33.5%, we can tell that attractiveness and radio play account for an additional 33%.[9] So, the inclusion of the two new predictors has explained quite a large amount of the variation in record sales.

SPSS OUTPUT 7.5 Regression model summary

Model Summary[c]

Model	R	R Square	Adjusted R Square	Std. Error of the Estimate	Change Statistics: R Square Change	F Change	df1	df2	Sig. F Change	Durbin-Watson
1	.578[a]	.335	.331	65.9914	.335	99.587	1	198	.000	
2	.815[b]	.665	.660	47.0873	.330	96.447	2	196	.000	1.950

a. Predictors: (Constant), Advertising Budget (thousands of pounds)

b. Predictors: (Constant), Advertising Budget (thousands of pounds), Attractiveness of Band, No. of plays on Radio 1 per week

c. Dependent Variable: Record Sales (thousands)

The adjusted R^2 gives us some idea of how well our model generalizes and ideally we would like its value to be the same, or very close to, the value of R^2. In this example the difference for the final model is small (in fact the difference between the values is .665 – .660 = .005 (about 0.5%). This shrinkage means that if the model were derived from the population rather than a sample it would account for approximately 0.5% less variance in the outcome. Advanced students might like to apply Stein's formula to the R^2 to get some idea of its likely value in different samples. Stein's formula was given in equation (7.11) and can be applied by replacing n with the sample size (200) and k with the number of predictors (3):

$$\begin{aligned}\text{adjusted } R^2 &= 1 - \left[\left(\frac{200-1}{200-3-1}\right)\left(\frac{200-2}{200-3-2}\right)\left(\frac{200+1}{200}\right)\right](1-0.665)\\ &= 1 - [(1.015)(1.015)(1.005)](0.335)\\ &= 1 - 0.347\\ &= 0.653\end{aligned}$$

This value is very similar to the observed value of R^2 (.665) indicating that the cross-validity of this model is very good.

The change statistics are provided only if requested and these tell us whether the change in R^2 is significant. The significance of R^2 can actually be tested using an F-ratio, and this F is calculated from the following equation (in which N is the number of cases or participants, and k is the number of predictors in the model):

$$F = \frac{(N-k-1)R^2}{k(1-R^2)}$$

[9] That is, 33% = 66.5% – 33.5% (this value is the *R Square Change* in the table).

In SPSS Output 7.5, the change in this F is reported for each block of the hierarchy. So, model 1 causes R^2 to change from 0 to .335, and this change in the amount of variance explained gives rise to an F-ratio of 99.587, which is significant with a probability less than .001. Bearing in mind for this first model that we have only one predictor (so $k = 1$) and 200 cases ($N = 200$), this F comes from the equation above:[10]

$$F_{Model1} = \frac{(200 - 1 - 1)0.334648}{1(1 - 0.334648)} = 99.587$$

The addition of the new predictors (model 2) causes R^2 to increase by .330 (see above). We can calculate the F-ratio for this change using the same equation, but because we're looking at the change in models we use the change in R^2, R^2_{Change}, and the R^2 in the new model (model 2 in this case so I've called it R^2_2) and we also use the change in the number of predictors, k_{Change} (model 1 had one predictor and model 2 had three predictors, so the change in the number of predictors is 3–1 = 2), and the number of predictors in the new model, k_2 (in this case because we're looking at model 2). Again, if we use a few more decimal places than in the SPSS table, we get approximately the same answer as SPSS:

$$\begin{aligned} F_{Change} &= \frac{(N - k_2 - 1)R^2_{Change}}{k_{Change}(1 - R^2_2)} \\ &= \frac{(200 - 3 - 1) \times 0.330}{2(1 - 0.664668)} \\ &= 96.44 \end{aligned}$$

As such, the change in the amount of variance that can be explained gives rise to an F-ratio of 96.44, which is again significant ($p < .001$). The change statistics therefore tell us about the difference made by adding new predictors to the model.

Finally, if you requested the Durbin–Watson statistic it will be found in the last column of the table in SPSS Output 7.5. This statistic informs us about whether the assumption of independent errors is tenable (see section 7.6.2.1). As a conservative rule I suggested that values less than 1 or greater than 3 should definitely raise alarm bells (although I urge you to look up precise values for the situation of interest). The closer to 2 that the value is, the better, and for these data the value is 1.950, which is so close to 2 that the assumption has almost certainly been met.

SPSS Output 7.6 shows the next part of the output, which contains an ANOVA that tests whether the model is significantly better at predicting the outcome than using the mean as a 'best guess'. Specifically, the F-ratio represents the ratio of the improvement in prediction that results from fitting the model, relative to the inaccuracy that still exists in the model (see section 7.2.3). This table is again split into two sections: one for each model. We are told the value of the sum of squares for the model (this value is SS_M in section 7.2.3 and represents the improvement in prediction resulting from fitting a regression line to the data rather than using the mean as an estimate of the outcome). We are also told the residual sum of squares (this value is SS_R in section 7.2.3 and represents the total difference between the model and the observed data). We are also told the degrees of freedom (df) for each term. In the case of the improvement due to the model, this value is equal to the number of predictors (1 for the first model and 3 for the second), and for SS_R it is the number of observations (200) minus the number of coefficients in the regression model. The first

[10] To get the same values as SPSS we have to use the exact value of R^2, which is 0.3346480676231 (if you don't believe me double-click on the table in the SPSS output that reports this value, then double-click on the cell of the table containing the value of R^2 and you'll see that .335 becomes the value that I've just typed!).

model has two coefficients (one for the predictor and one for the constant) whereas the second has four (one for each of the three predictors and one for the constant). Therefore, model 1 has 198 degrees of freedom whereas model 2 has 196. The average sum of squares (MS) is then calculated for each term by dividing the SS by the *df*. The *F*-ratio is calculated by dividing the average improvement in prediction by the model (MS_M) by the average difference between the model and the observed data (MS_R). If the improvement due to fitting the regression model is much greater than the inaccuracy within the model then the value of *F* will be greater than 1 and SPSS calculates the exact probability of obtaining the value of *F* by chance. For the initial model the *F*-ratio is 99.587, which is very unlikely to have happened by chance ($p < .001$). For the second model the value of *F* is even higher (129.498), which is also highly significant ($p < .001$). We can interpret these results as meaning that the initial model significantly improved our ability to predict the outcome variable, but that the new model (with the extra predictors) was even better (because the *F*-ratio is more significant).

SPSS OUTPUT 7.6

ANOVA[c]

Model		Sum of Squares	df	Mean Square	F	Sig.
1	Regression	433687.833	1	433687.833	99.587	.000[a]
	Residual	862264.167	198	4354.870		
	Total	1295952.0	199			
2	Regression	861377.418	3	287125.806	129.498	.000[b]
	Residual	434574.582	196	2217.217		
	Total	1295952.0	199			

a. Predictors: (Constant), Advertising Budget (thousands of pounds)

b. Predictors: (Constant), Advertising Budget (thousands of pounds), Attractiveness of Band, No. of Plays on Radio 1 per Week

c. Dependent Variable: Record Sales (thousands)

CRAMMING SAM'S TIPS

The fit of the regression model can be assessed using the **Model Summary** and **ANOVA** tables from SPSS. Look for the R^2 to tell you the proportion of variance explained by the model. If you have done a hierarchical regression then you can assess the improvement of the model at each stage of the analysis by looking at the change in R^2 and whether this change is significant (look for values less than .05 in the column labelled *Sig F Change.*). The ANOVA also tells us whether the model is a significant fit of the data overall (look for values less than .05 in the column labelled *Sig.*). Finally, there is an assumption that errors in regression are independent; this assumption is likely to be met if the Durbin–Watson statistic is close to 2 (and between 1 and 3).

7.8.3. Model parameters ②

So far we have looked at several summary statistics telling us whether or not the model has improved our ability to predict the outcome variable. The next part of the output is concerned with the parameters of the model. SPSS Output 7.7 shows the model parameters for both steps in the hierarchy. Now, the first step in our hierarchy was to include advertising budget (as we did for the simple regression earlier in this chapter) and so the parameters for the first model are identical to the parameters obtained in SPSS Output 7.3. Therefore, we will be concerned only with the parameters for the final model (in which all predictors were included). The format of the table of coefficients will depend on the options selected. The confidence interval for the *b*-values, collinearity diagnostics and the part and partial correlations will be present only if selected in the dialog box in Figure 7.14.

SPSS OUTPUT 7.7
Coefficients of the regression model[11]

Coefficients[a]

Model		Unstandardized Coefficients		Standardized Coefficients			95% Confidence Interval for B	
		B	Std. Error	Beta	t	Sig.	Lower Bound	Upper Bound
1	(Constant)	134.140	7.537		17.799	.000	119.278	149.002
	Advertising Budget (thousands of pounds)	.096	.010	.578	9.979	.000	.077	.115
2	(Constant)	-26.613	17.350		-1.534	.127	-60.830	7.604
	Advertising Budget (thousands of pounds)	.085	.007	.511	12.261	.000	.071	.099
	No. of plays on Radio 1 per week	3.367	.278	.512	12.123	.000	2.820	3.915
	Attractiveness of Band	11.086	2.438	.192	4.548	.000	6.279	15.894

a. Dependent Variable: Record Sales (thousands)

Coefficients[a]

Model		Correlations			Collinearity Statistics	
		Zero-order	Partial	Part	Tolerance	VIF
1	Advertising Budget (thousands of pounds)	.578	.578	.578	1.000	1.000
2	Advertising Budget (thousands of pounds)	.578	.659	.507	.986	1.015
	No. of plays on Radio 1 per week	.599	.655	.501	.959	1.043
	Attractiveness of Band	.326	.309	.188	.963	1.038

a. Dependent Variable: Record Sales (thousands)

Remember that in multiple regression the model takes the form of equation (7.9) and in that equation there are several unknown quantities (the *b*-values). The first part of the table gives us estimates for these *b*-values and these values indicate the individual contribution of each predictor to the model. If we replace the *b*-values in equation (7.9) we find that we can define the model as follows:

$$\begin{aligned}\text{sales}_i &= b_0 + b_1\text{advertising}_i + b_2\text{airplay}_i + b_3\text{attractiveness}_i \\ &= -26.61 + (0.08\text{advertising}_i) + (3.37\text{airplay}_i) \\ &\quad + (11.09\text{attractiveness}_i)\end{aligned} \tag{7.12}$$

The *b*-values tell us about the relationship between record sales and each predictor. If the value is positive we can tell that there is a positive relationship between the predictor and the outcome, whereas a negative coefficient represents a negative relationship. For these data all three predictors have positive *b*-values indicating positive relationships. So, as advertising budget increases, record sales increase; as plays on the radio increase, so do record sales; and finally more attractive bands will sell more records. The *b*-values tell us more than this, though. They tell us to what degree each predictor affects the outcome *if the effects of all other predictors are held constant*:

[11] To spare your eyesight I have split this part of the output into two tables; however, it should appear as one long table in the SPSS Viewer.

- **Advertising budget** (b = 0.085): This value indicates that as advertising budget increases by one unit, record sales increase by 0.085 units. Both variables were measured in thousands; therefore, for every £1000 more spent on advertising, an extra 0.085 thousand records (85 records) are sold. This interpretation is true only if the effects of attractiveness of the band and airplay are held constant.
- **Airplay** (b = 3.367): This value indicates that as the number of plays on radio in the week before release increases by one, record sales increase by 3.367 units. Therefore, every additional play of a song on radio (in the week before release) is associated with an extra 3.367 thousand records (3367 records) being sold. This interpretation is true only if the effects of attractiveness of the band and advertising are held constant.
- **Attractiveness** (b = 11.086): This value indicates that a band rated one unit higher on the attractiveness scale can expect additional record sales of 11.086 units. Therefore, every unit increase in the attractiveness of the band is associated with an extra 11.086 thousand records (11,086 records) being sold. This interpretation is true only if the effects of radio airplay and advertising are held constant.

Each of these beta values has an associated standard error indicating to what extent these values would vary across different samples, and these standard errors are used to determine whether or not the b-value differs significantly from zero. As we saw in section 7.4.2, a t-statistic can be derived that tests whether a b-value is significantly different from 0. In simple regression, a significant value of t indicates that the slope of the regression line is significantly different from horizontal, but in multiple regression, it is not so easy to visualize what the value tells us. Well, it is easiest to conceptualize the t-tests as measures of whether the predictor is making a significant contribution to the model. Therefore, if the t-test associated with a b-value is significant (if the value in the column labelled *Sig.* is less than .05) then the predictor is making a significant contribution to the model. The smaller the value of *Sig.* (and the larger the value of t), the greater the contribution of that predictor. For this model, the advertising budget ($t(196) = 12.26$, $p < .001$), the amount of radio play prior to release ($t(196) = 12.12$, $p < .001$) and attractiveness of the band ($t(196) = 4.55$, $p < .001$) are all significant predictors of record sales.[12] From the magnitude of the t-statistics we can see that the advertising budget and radio play had a similar impact, whereas the attractiveness of the band had less impact.

The b-values and their significance are important statistics to look at; however, the standardized versions of the b-values are in many ways easier to interpret (because they are not dependent on the units of measurement of the variables). The standardized beta values are provided by SPSS (labelled as Beta, β_1) and they tell us the number of standard deviations that the outcome will change as a result of one standard deviation change in the predictor. The standardized beta values are all measured in standard deviation units and so are directly comparable: therefore, they provide a better insight into the 'importance' of a predictor in the model. The standardized beta values for airplay and advertising budget are virtually identical (.512 and .511 respectively) indicating that both variables have a comparable degree of importance in the model (this concurs with what the magnitude of the t-statistics told us). To interpret these values literally, we need to know the standard deviations of all of the variables and these values can be found in SPSS Output 7.4.

[12] For all of these predictors I wrote $t(196)$. The number in brackets is the degrees of freedom. We saw in section 7.2.4 that in regression the degrees of freedom are $N - p - 1$, where N is the total sample size (in this case 200) and p is the number of predictors (in this case 3). For these data we get 200 − 3 − 1 = 196.

- **Advertising budget** (*standardized* β = .511): This value indicates that as advertising budget increases by one standard deviation (£485,655), record sales increase by 0.511 standard deviations. The standard deviation for record sales is 80,699 and so this constitutes a change of 41,240 sales (0.511 × 80,699). Therefore, for every £485,655 more spent on advertising, an extra 41,240 records are sold. This interpretation is true only if the effects of attractiveness of the band and airplay are held constant.
- **Airplay** (*standardized* β = .512): This value indicates that as the number of plays on radio in the week before release increases by 1 standard deviation (12.27), record sales increase by 0.512 standard deviations. The standard deviation for record sales is 80,699 and so this constitutes a change of 41,320 sales (0.512 × 80,699). Therefore, if Radio 1 plays the song an extra 12.27 times in the week before release, 41,320 extra record sales can be expected. This interpretation is true only if the effects of attractiveness of the band and advertising are held constant.
- **Attractiveness** (*standardized* β = .192): This value indicates that a band rated one standard deviation (1.40 units) higher on the attractiveness scale can expect additional record sales of 0.192 standard deviations units. This constitutes a change of 15,490 sales (0.192 × 80,699). Therefore, a band with an attractiveness rating 1.40 higher than another band can expect 15,490 additional sales. This interpretation is true only if the effects of radio airplay and advertising are held constant.

SELF-TEST Think back to what the confidence interval of the mean represented (section 2.5.2). Can you guess what the confidence interval for *b* represents?

Imagine that we collected 100 samples of data measuring the same variables as our current model. For each sample we could create a regression model to represent the data. If the model is reliable then we hope to find very similar parameters in all samples. Therefore, each sample should produce approximately the same *b*-values. The confidence intervals of the unstandardized beta values are boundaries constructed such that in 95% of these samples these boundaries will contain the true value of *b* (see section 2.5.2). Therefore, if we'd collected 100 samples, and calculated the confidence intervals for *b*, we are saying that 95% of these confidence intervals would contain the true value of *b*. Therefore, we can be fairly confident that the confidence interval we have constructed for this sample will contain the true value of *b* in the population. This being so, a good model will have a small confidence interval, indicating that the value of *b* in this sample is close to the true value of *b* in the population. The sign (positive or negative) of the *b*-values tells us about the direction of the relationship between the predictor and the outcome. Therefore, we would expect a very bad model to have confidence intervals that cross zero, indicating that in some samples the predictor has a negative relationship to the outcome whereas in others it has a positive relationship. In this model, the two best predictors (advertising and airplay) have very tight confidence intervals indicating that the estimates for the current model are likely to be representative of the true population values. The interval for attractiveness is wider (but still does not cross zero) indicating that the parameter for this variable is less representative, but nevertheless significant.

If you asked for part and partial correlations, then they will appear in the output in separate columns of the table. The zero-order correlations are the simple Pearson's correlation coefficients (and so correspond to the values in SPSS Output 7.4). The partial correlations represent the relationships between each predictor and the outcome variable, controlling for the effects of the other two predictors. The part correlations represent the relationship between each predictor and the outcome, controlling for the effect that the other two variables have on the outcome. In effect, these part correlations represent the unique relationship that each predictor has with the outcome. If you opt to do a stepwise regression, you would find that variable entry is based initially on the variable with the largest zero-order correlation and then on the part correlations of the remaining variables. Therefore, airplay would be entered first (because it has the largest zero-order correlation), then advertising budget (because its part correlation is bigger than attractiveness) and then finally attractiveness. Try running a forward stepwise regression on these data to see if I'm right! Finally, we are given details of the collinearity statistics, but these will be discussed in section 7.8.5.

CRAMMING SAM'S TIPS

The individual contribution of variables to the regression model can be found in the **Coefficients** table from SPSS. If you have done a hierarchical regression then look at the values for the final model. For each predictor variable, you can see if it has made a significant contribution to predicting the outcome by looking at the column labelled *Sig.* (values less than .05 are significant). You should also look at the standardized beta values because these tell you the importance of each predictor (bigger absolute value = more important). The Tolerance and VIF values will also come in handy later on, so make a note of them!

7.8.4. Excluded variables ②

At each stage of a regression analysis SPSS provides a summary of any variables that have not yet been entered into the model. In a hierarchical model, this summary has details of the variables that have been specified to be entered in subsequent steps, and in stepwise regression this table contains summaries of the variables that SPSS is considering entering into the model. For this example, there is a summary of the excluded variables (SPSS Output 7.8) for the first stage of the hierarchy (there is no summary for the second stage because all predictors are in the model). The summary gives an estimate of each predictor's beta value if it was entered into the equation at this point and calculates a *t*-test for this value. In a stepwise regression, SPSS should enter the predictor with the highest *t*-statistic and will continue entering predictors until there are none left with *t*-statistics that have significance values less than .05. The partial correlation also provides some indication as to what contribution (if any) an excluded predictor would make if it were entered into the model.

SPSS OUTPUT 7.8

Excluded Variables[b]

Model		Beta In	t	Sig.	Partial Correlation	Collinearity Statistics		
						Tolerance	VIF	Minimum Tolerance
1	No. of plays on Radio 1 per week	.546[a]	12.513	.000	.665	.990	1.010	.990
	Attractiveness of Band	.281[a]	5.136	.000	.344	.993	1.007	.993

a. Predictors in the Model: (Constant), Advertising Budget (thousands of pounds)

b. Dependent Variable: Record Sales (thousands)

7.8.5. Assessing the assumption of no multicollinearity ②

SPSS Output 7.7 provided some measures of whether there is collinearity in the data. Specifically, it provides the VIF and tolerance statistics (with tolerance being 1 divided by the VIF). There are a few guidelines from section 7.6.2.3 that can be applied here:

- If the largest VIF is greater than 10 then there is cause for concern (Bowerman & O'Connell, 1990; Myers, 1990).
- If the average VIF is substantially greater than 1 then the regression may be biased (Bowerman & O'Connell, 1990).
- Tolerance below 0.1 indicates a serious problem.
- Tolerance below 0.2 indicates a potential problem (Menard, 1995).

For our current model the VIF values are all well below 10 and the tolerance statistics all well above 0.2; therefore, we can safely conclude that there is no collinearity within our data. To calculate the average VIF we simply add the VIF values for each predictor and divide by the number of predictors (*k*):

$$\overline{\text{VIF}} = \frac{\sum_{i=1}^{k} \text{VIF}_i}{k} = \frac{1.015 + 1.043 + 1.038}{3} = 1.032$$

The average VIF is very close to 1 and this confirms that collinearity is not a problem for this model. SPSS also produces a table of eigenvalues of the scaled, uncentred cross-products matrix, condition indexes and variance proportions (see Jane Superbrain Box 7.2). There is a lengthy discussion, and example, of collinearity in section 8.8.1 and how to detect it using variance proportions, so I will limit myself now to saying that we are looking for large variance proportions on the same *small* eigenvalues. Therefore, in SPSS Output 7.9 we look at the bottom few rows of the table (these are the small eigenvalues) and look for any variables that both have high variance proportions for that eigenvalue. The variance proportions vary between 0 and 1, and for each predictor should be distributed across different dimensions (or eigenvalues). For this model, you can see that each predictor has most of its variance loading onto a different dimension (advertising has 96% of variance on dimension 2, airplay has 93% of variance on dimension 3 and attractiveness has 92% of variance on dimension 4). These data represent a classic example of no multicollinearity. For an example of when collinearity exists in the data and some suggestions about what can be done, see Chapters 8 (section 8.8.1) and 17 (section 17.3.3.3).

SPSS OUTPUT 7.9

Collinearity Diagnostics[a]

Model	Dimension	Eigenvalue	Condition Index	Variance Proportions (Constant)	Advertising Budget (thousands of pounds)	No. of plays on Radio 1 per week	Attractiveness of Band
1	1	1.785	1.000	.11	.11		
	2	.215	2.883	.89	.89		
2	1	3.562	1.000	.00	.02	.01	.00
	2	.308	3.401	.01	**.96**	.05	.01
	3	.109	5.704	.05	.02	**.93**	.07
	4	2.039E-02	13.219	.94	.00	.00	**.92**

a. Dependent Variable: Record Sales (thousands)

CRAMMING SAM'S TIPS

To check the assumption of no multicollinearity, use the VIF values from the table **Labelled** coefficients in the SPSS output. If these values are less than 10 then that indicates there probably isn't cause for concern. If you take the average of VIF values, and this average is not substantially greater than 1, then that also indicates that there's no cause for concern.

JANE SUPERBRAIN 7.2

What are eigenvectors and eigenvalues? ④

The definitions and mathematics of eigenvalues and eigenvectors are very complicated and most of us need not worry about them (although they do crop up again in Chapters 16 and 17). However, although the mathematics of them is hard, they are quite easy to visualize! Imagine we have two variables: the salary a supermodel earns in a year, and how attractive she is. Also imagine these two variables are normally distributed and so can be considered together as a bivariate normal distribution. If these variables are correlated, then their scatterplot forms an ellipse. This is shown in the scatterplots below: if we draw a dashed line around the outer values of the scatterplot we get something oval shaped. Now, we can draw two lines to measure the length and height of this ellipse. These lines are the *eigenvectors* of the original correlation matrix for these two variables (a vector is just a set of numbers that tells us the location of a line in geometric space). Note that the two lines we've drawn (one for height and one for width of the oval) are perpendicular; that is, they are at 90 degrees, which means that they are independent of one another). So, with two variables, eigenvectors are just lines measuring the length and height of the ellipse that surrounds the scatterplot of data for those variables. If we add a third variable (e.g. experience of the supermodel) then all that happens is our scatterplot gets a third dimension, the ellipse turns into something shaped like a rugby ball (or American football), and because we now have a third dimension (height, width and depth) we get an extra eigenvector to measure this extra dimension. If we add a fourth variable, a similar logic applies (although it's harder to visualize): we get an extra dimension, and an eigenvector to measure that dimension. Now, each eigenvector has an *eigenvalue* that tells us its length (i.e. the distance from one end of the eigenvector to the other). So, by looking at all of the eigenvalues for a data set, we know the dimensions of the ellipse or rugby ball: put more generally, we know the dimensions of the data. Therefore, the eigenvalues show how evenly (or otherwise) the variances of the matrix are distributed.

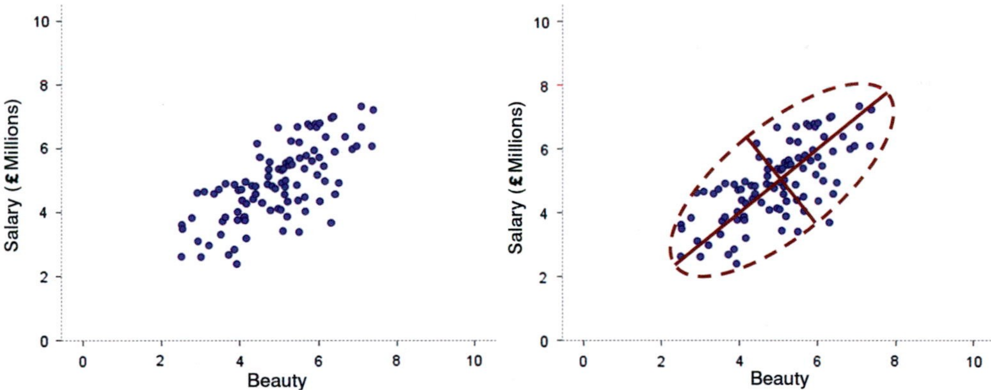

In the case of two variables, the *condition* of the data is related to the ratio of the larger eigenvalue to the smaller. Let's look at the two extremes: when there is no relationship at all between variables, and when there is a perfect relationship. When there is no relationship, the scatterplot will, more or less, be contained within a circle (or a sphere if we had three variables). If we again draw lines that measure the height and width of this circle we'll find that these lines are the same length. The eigenvalues measure the length, therefore the eigenvalues will also be the same. So, when we divide the largest eigenvalue by the smallest we'll get a value of 1 (because the eigenvalues are the same). When the variables are perfectly correlated (i.e. there is perfect collinearity) then the scatterplot forms a straight line and the ellipse surrounding it will also collapse to a straight line. Therefore, the height of the ellipse will be very small indeed (it will approach zero). Therefore, when we divide the largest eigenvalue by the smallest we'll get a value that tends to infinity (because the smallest eigenvalue is close to zero). Therefore, an infinite condition index is a sign of deep trouble.

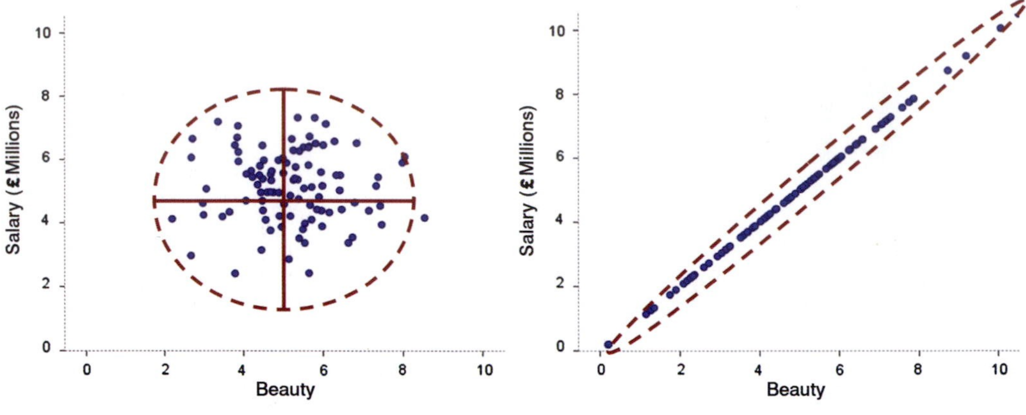

7.8.6. Casewise diagnostics ②

SPSS produces a summary table of the residual statistics and these should be examined for extreme cases. SPSS Output 7.10 shows any cases that have a standardized residual less than −2 or greater than 2 (remember that we changed the default criterion from 3 to 2 in Figure 7.14). I mentioned in section 7.6.1.1 that in an ordinary sample we would expect 95% of cases to have standardized residuals within about ±2. We have a sample of 200, therefore it is reasonable to expect about 10 cases (5%) to have standardized residuals outside of these limits. From SPSS Output 7.10 we can see that we have 12 cases (6%) that are outside of the limits: therefore, our sample is within 1% of what we would expect. In addition, 99% of cases should lie within ±2.5 and so we would expect only 1% of cases to lie outside of these limits. From the cases listed here, it is clear that two cases (1%) lie outside of the limits (cases 164 and 169). Therefore, our sample appears to conform to what we would expect for a fairly accurate model. These diagnostics give us no real cause for concern except that case 169 has a standardized residual greater than 3, which is probably large enough for us to investigate this case further.

SPSS OUTPUT 7.10

Casewise Diagnostics[a]

Case Number	Std. Residual	Record Sales (thousands)	Predicted Value	Residual
1	2.125	330.00	229.9203	100.0797
2	-2.314	120.00	228.9490	-108.9490
10	2.114	300.00	200.4662	99.5338
47	-2.442	40.00	154.9698	-114.9698
52	2.069	190.00	92.5973	97.4027
55	-2.424	190.00	304.1231	-114.1231
61	2.098	300.00	201.1897	98.8103
68	-2.345	70.00	180.4156	-110.4156
100	2.066	250.00	152.7133	97.2867
164	-2.577	120.00	241.3240	-121.3240
169	3.061	360.00	215.8675	144.1325
200	-2.064	110.00	207.2061	-97.2061

a. Dependent Variable: Record Sales (thousands)

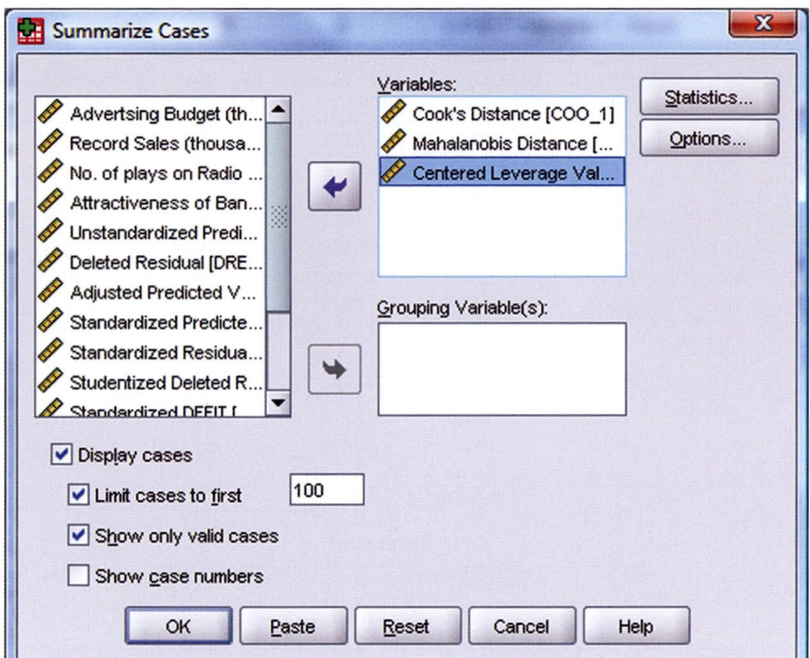

FIGURE 7.18 The summarize cases dialog box

You may remember that in section 7.7.5 we asked SPSS to save various diagnostic statistics. You should find that the data editor now contains columns for these variables. It is perfectly acceptable to check these values in the data editor, but you can also get SPSS to list the values in your viewer window too. To list variables you need to use the *Case Summaries* command, which can be found by selecting Analyze Reports ▸ Case Summaries.... Figure 7.18 shows the dialog box for this function. Simply select the variables that you want to list and transfer them to the box labelled *Variables* by clicking on ➜. By default, SPSS will limit the output to the first 100 cases, but if you want to list all of your cases then simply deselect this option. It is also very important to select the *Show case numbers* option because otherwise you might not be able to identify a problem case.

One useful strategy is to use the casewise diagnostics to identify cases that you want to investigate further. So, to save space, I created a coding variable (1 = include, 0 = exclude) so that I could specify the 12 cases listed in SPSS Output 7.11 in one group, and all other cases in the other. By using this coding variable and specifying it as a grouping variable in the *Summarize Cases* dialog box, I could look at those 12 cases together and discard all others.

SPSS Output 7.11 shows the influence statistics for the 12 cases that I selected. None of them have a Cook's distance greater than 1 (even case 169 is well below this criterion) and so none of the cases is having an undue influence on the model. The average leverage can be calculated as 0.02 ($k + 1/n = 4/200$) and so we are looking for values either twice as large as this (0.04) or three times as large (0.06) depending on which statistician you trust most (see section 7.6.1.2)! All cases are within the boundary of three times the average and only case 1 is close to two times the average. Finally, from our guidelines for the Mahalanobis distance we saw that with a sample of 100 and three predictors, values greater than 15 were problematic. We have three predictors and a larger sample size, so this value will be a conservative cut-off, yet none of our cases comes close to exceeding this criterion. The evidence suggests that there are no influential cases within our data (although all cases would need to be examined to confirm this fact).

We can look also at the DFBeta statistics to see whether any case would have a large influence on the regression parameters. An absolute value greater than 1 is a problem and

SPSS OUTPUT 7.11

Case Summaries

	Case Number	Standardized DFBETA Intercept	Standardized DFBETA ADVERTS	Standardized DFBETA AIRPLAY	Standardized DFBETA ATTRACT	Standardized DFFIT	COVRATIO
1	1	-.31554	-.24235	.15774	.35329	.48929	.97127
2	2	.01259	-.12637	.00942	-.01868	-.21110	.92018
3	10	-.01256	-.15612	.16772	.00672	.26896	.94392
4	47	.06645	.19602	.04829	-.17857	-.31469	.91458
5	52	.35291	-.02881	-.13667	-.26965	.36742	.95995
6	55	.17427	-.32649	-.02307	-.12435	-.40736	.92486
7	61	.00082	-.01539	.02793	.02054	.15562	.93654
8	68	-.00281	.21146	-.14766	-.01760	-.30216	.92370
9	100	.06113	.14523	-.29984	.06766	.35732	.95888
10	164	.17983	.28988	-.40088	-.11706	-.54029	.92037
11	169	-.16819	-.25765	.25739	.16968	.46132	.85325
12	200	.16633	-.04639	.14213	-.25907	-.31985	.95435
Total N		12	12	12	12	12	12

Case Summaries

	Case Number	Cook's Distance	Mahalanobis Distance	Centered Leverage Value
1	1	.05870	8.39591	.04219
2	2	.01089	.59830	.00301
3	10	.01776	2.07154	.01041
4	47	.02412	2.12475	.01068
5	52	.03316	4.81841	.02421
6	55	.04042	4.19960	.02110
7	61	.00595	.06880	.00035
8	68	.02229	2.13106	.01071
9	100	.03136	4.53310	.02278
10	164	.07077	6.83538	.03435
11	169	.05087	3.14841	.01582
12	200	.02513	3.49043	.01754
Total N		12	12	12

in all cases the values lie within ±1, which shows that these cases have no undue influence over the regression parameters. There is also a column for the covariance ratio. We saw in section 7.6.1.2 that we need to use the following criteria:

- $CVR_i > 1 + [3(k + 1)/n] = 1 + [3(3 + 1)/200] = 1.06$.
- $CVR_i < 1 - [3(k + 1)/n] = 1 - [3(3 + 1)/200] = 0.94$.

Therefore, we are looking for any cases that deviate substantially from these boundaries. Most of our 12 potential outliers have CVR values within or just outside these boundaries. The only case that causes concern is case 169 (again!) whose CVR is some way below the bottom limit. However, given the Cook's distance for this case, there is probably little cause for alarm.

You would have requested other diagnostic statistics and from what you know from the earlier discussion of them you would be well advised to glance over them in case of any unusual cases in the data. However, from this minimal set of diagnostics we appear to have a fairly reliable model that has not been unduly influenced by any subset of cases.

CRAMMING SAM'S TIPS

You need to look for cases that might be influencing the regression model:

- Look at standardized residuals and check that no more than 5% of cases have absolute values above 2, and that no more than about 1% have absolute values above 2.5. Any case with the value above about 3, could be an outlier.
- Look in the data editor for the values of Cook's distance: any value above 1 indicates a case that might be influencing the model.
- Calculate the average leverage (the number of predictors plus 1, divided by the sample size) and then look for values greater than twice or three times this average value.
- For Mahalanobis distance, a crude check is to look for values above 25 in large samples (500) and values above 15 in smaller samples (100). However, Barnett and Lewis (1978) should be consulted for more detailed analysis.
- Look for absolute values of DFBeta greater than 1.
- Calculate the upper and lower limit of acceptable values for the covariance ratio, CVR. The upper limit is 1 plus three times the average leverage, whereas the lower limit is 1 minus three times the average leverage. Cases that have a CVR that fall outside of these limits may be problematic.

7.8.7. Checking assumptions ②

As a final stage in the analysis, you should check the assumptions of the model. We have already looked for collinearity within the data and used Durbin–Watson to check whether the residuals in the model are independent. In section 7.7.4 we asked for a plot of *ZRESID against *ZPRED and for a histogram and normal probability plot of the residuals. The graph of *ZRESID and *ZPRED should look like a random array of dots evenly dispersed around zero. If this graph funnels out, then the chances are that there is heteroscedasticity in the data. If there is any sort of curve in this graph then the chances are that the data have broken the assumption of linearity. Figure 7.19 shows several examples of the plot of standardized residuals against standardized predicted values. Panel (a) shows the graph for the data in our record sales example. Note how the points are randomly and evenly dispersed throughout the plot. This pattern is indicative of a situation in which the assumptions of linearity and homoscedasticity have been met. Panel (b) shows a similar plot for a data set that violates the assumption of homoscedasticity. Note that the points form the shape of a funnel so they become more spread out across the graph. This funnel shape is typical of heteroscedasticity and indicates increasing variance across the residuals. Panel (c) shows a plot of some data in which there is a non-linear relationship between the outcome and the predictor. This pattern is shown up by the residuals. A line illustrating the curvilinear relationship has been drawn over the top of the graph to illustrate the trend in the data. Finally, panel (d) represents a situation in which the data not only represent a non-linear relationship, but also show heteroscedasticity. Note first the curved trend in the data, and then also note that at one end of the plot the points are very close together whereas at the other end they are widely dispersed. When these assumptions have been violated you will not see these exact patterns, but hopefully these plots will help you to understand the types of anomalies you should look out for.

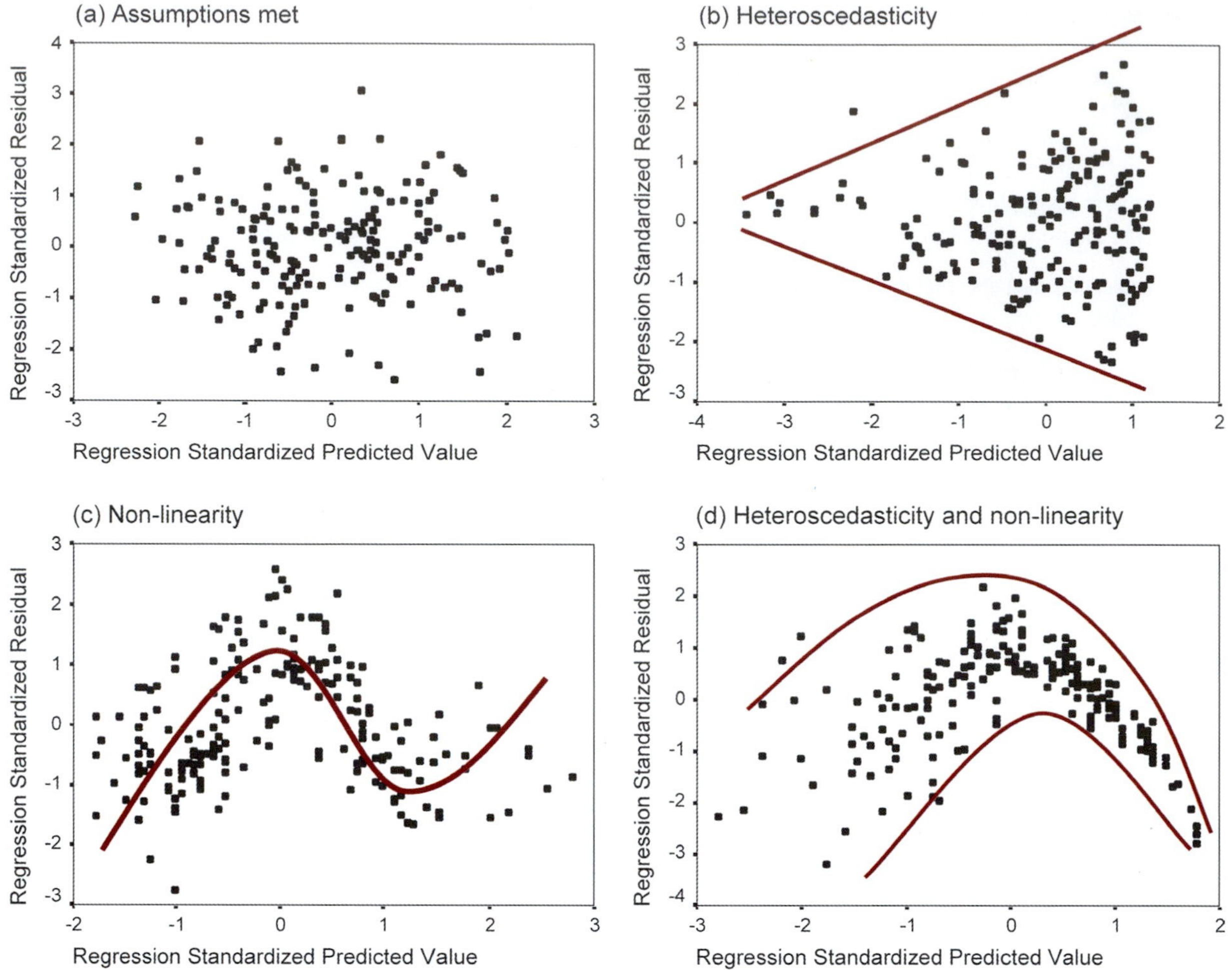

FIGURE 7.19 Plots of *ZRESID against *ZPRED

To test the normality of residuals, we must look at the histogram and normal probability plot selected in Figure 7.15. Figure 7.20 shows the histogram and normal probability plot of the data for the current example (left-hand side). The histogram should look like a normal distribution (a bell-shaped curve). SPSS draws a curve on the histogram to show the shape of the distribution. For the record company data, the distribution is roughly normal (although there is a slight deficiency of residuals exactly on zero). Compare this histogram to the extremely non-normal histogram next to it and it should be clear that the non-normal distribution is extremely skewed (unsymmetrical). So, you should look for a curve that has the same shape as the one for the record sales data: any deviation from this curve is a sign of non-normality and the greater the deviation, the more non-normally distributed the residuals. The normal probability plot also shows up deviations from normality (see Chapter 5). The straight line in this plot represents a normal distribution, and the points represent the observed residuals. Therefore, in a perfectly normally distributed data set, all points will lie on the line. This is pretty much what we see for the record sales data. However, next to the normal probability plot of the record sales data is an example of an extreme deviation from normality. In this plot, the dots

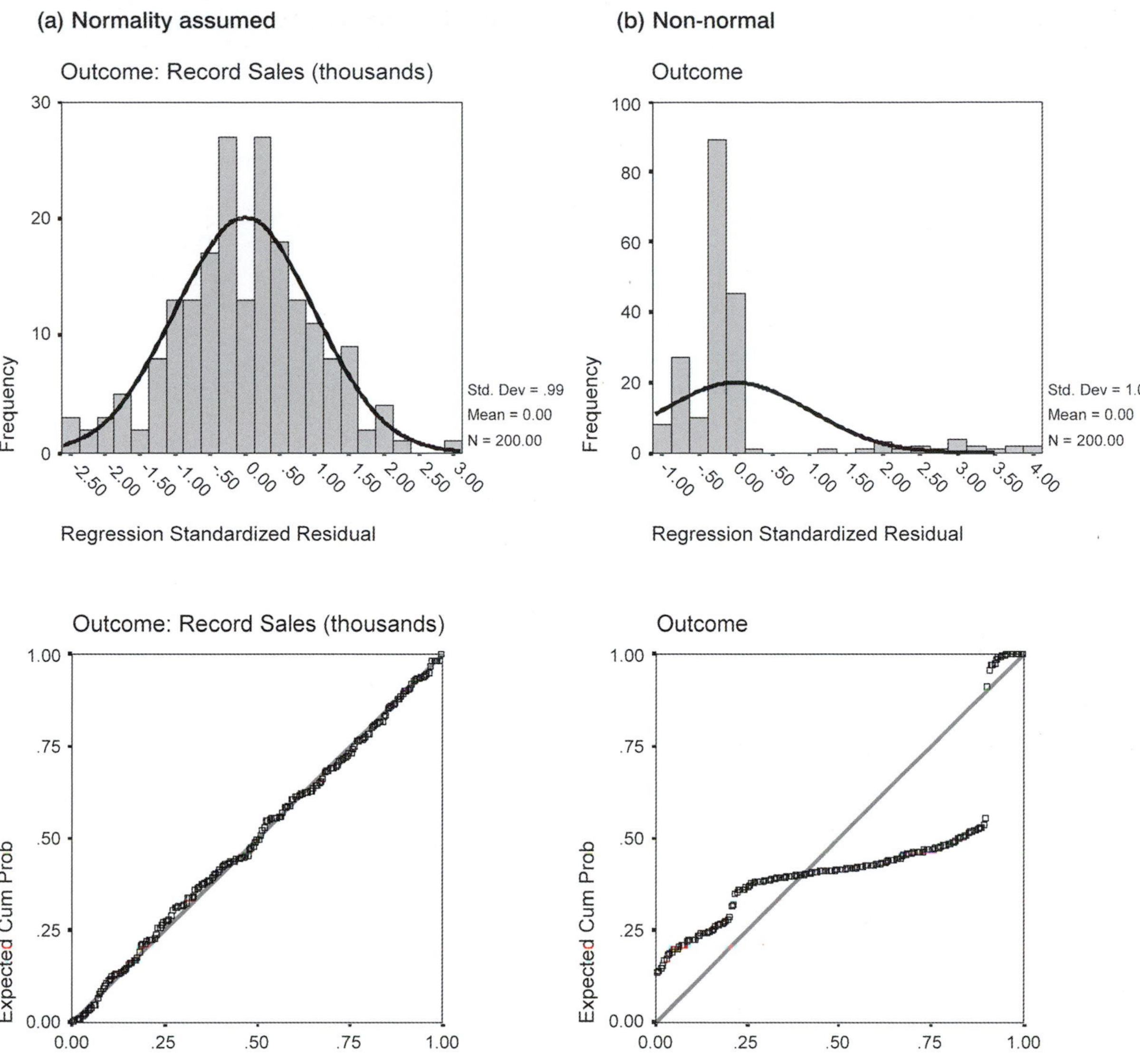

FIGURE 7.20 Histograms and normal P–P plots of normally distributed residuals (left-hand side) and non-normally distributed residuals (right-hand side)

are very distant from the line, which indicates a large deviation from normality. For both plots, the non-normal data are extreme cases and you should be aware that the deviations from normality are likely to be subtler. Of course, you can use what you learnt in Chapter 5 to do a K–S test on the standardized residuals to see whether they deviate significantly from normality.

A final set of plots specified in Figure 7.15 was the partial plots. These plots are scatterplots of the residuals of the outcome variable and each of the predictors when both variables are regressed separately on the remaining predictors. I mentioned earlier that obvious outliers on a partial plot represent cases that might have undue influence on a predictor's regression coefficient and that non-linear relationships and heteroscedasticity can be detected using these plots as well:

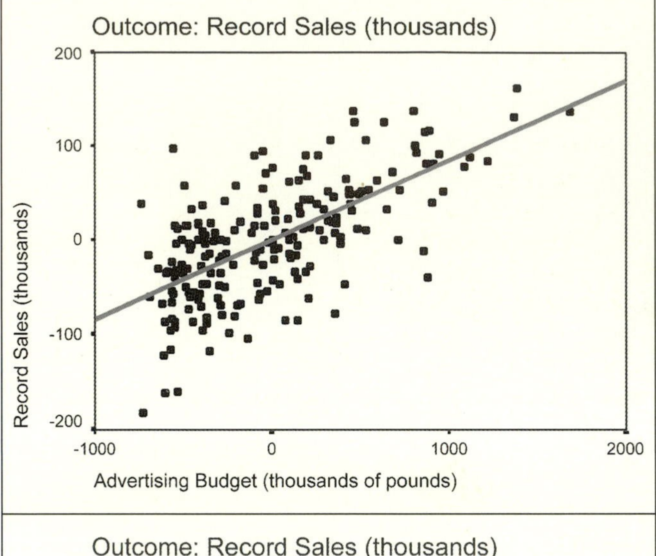

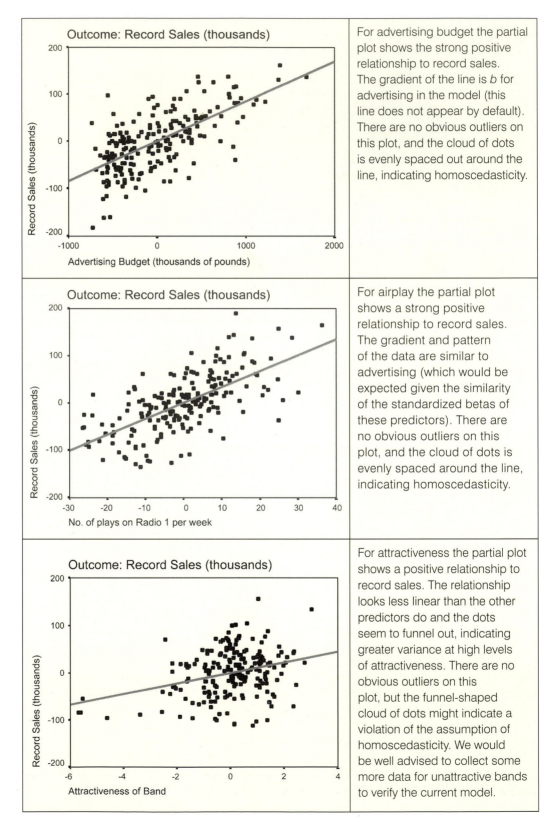

For advertising budget the partial plot shows the strong positive relationship to record sales. The gradient of the line is *b* for advertising in the model (this line does not appear by default). There are no obvious outliers on this plot, and the cloud of dots is evenly spaced out around the line, indicating homoscedasticity.

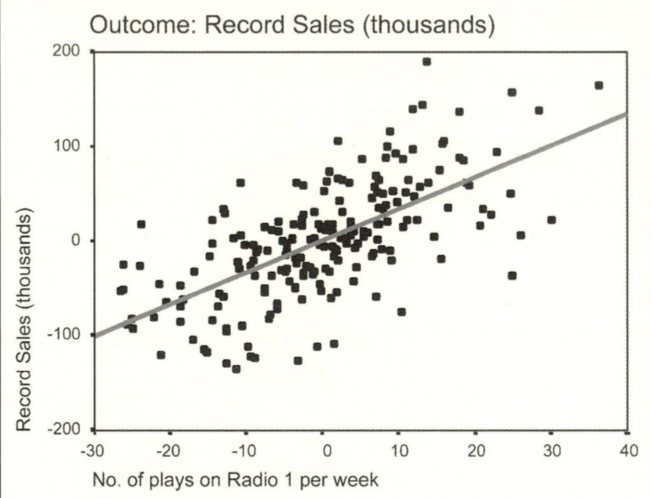

For airplay the partial plot shows a strong positive relationship to record sales. The gradient and pattern of the data are similar to advertising (which would be expected given the similarity of the standardized betas of these predictors). There are no obvious outliers on this plot, and the cloud of dots is evenly spaced around the line, indicating homoscedasticity.

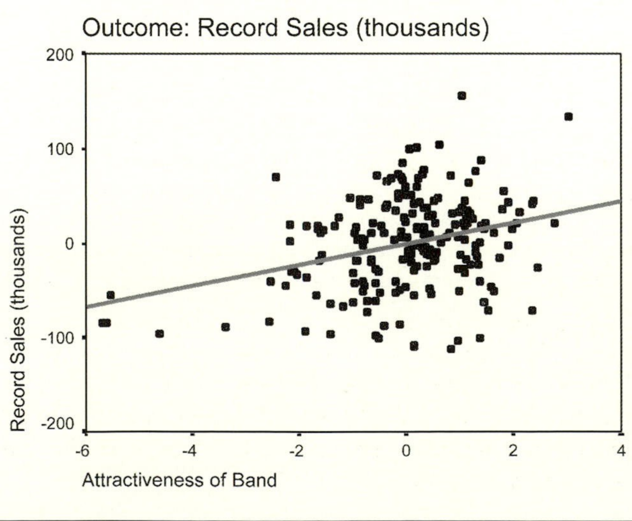

For attractiveness the partial plot shows a positive relationship to record sales. The relationship looks less linear than the other predictors do and the dots seem to funnel out, indicating greater variance at high levels of attractiveness. There are no obvious outliers on this plot, but the funnel-shaped cloud of dots might indicate a violation of the assumption of homoscedasticity. We would be well advised to collect some more data for unattractive bands to verify the current model.

CRAMMING SAM'S TIPS

You need to check some of the assumptions of regression to make sure your model generalizes beyond your sample:

- Look at the graph of* ZRESID plotted against* ZPRED. If it looks like a random array of dots then this is good. If the dots seem to get more or less spread out over the graph (look like a funnel) then this is probably a violation of the assumption of homogeneity of variance. If the dots have a pattern to them (i.e. a curved shape) then this is probably a violation of the assumption of linearity. If the dots seem to have a pattern and are more spread out at some points on the plot than others then this probably reflects violations of both homogeneity of variance *and* linearity. Any of these scenarios puts the validity of your model into question. Repeat the above for all partial plots too.
- Look at histograms and P–P plots. If the histograms look like normal distributions (and the P–P plot looks like a diagonal line), then all is well. If the histogram looks non-normal and the P–P plot looks like a wiggly snake curving around a diagonal line then things are less good! Be warned, though: distributions can look very non-normal in small samples even when they are!

We could summarize by saying that the model appears, in most senses, to be both accurate for the sample and generalizable to the population. The only slight glitch is some concern over whether attractiveness ratings had violated the assumption of homoscedasticity. Therefore, we could conclude that in our sample advertising budget and airplay are fairly equally important in predicting record sales. Attractiveness of the band is a significant predictor of record sales but is less important than the other two predictors (and probably needs verification because of possible heteroscedasticity). The assumptions seem to have been met and so we can probably assume that this model would generalize to any record being released.

7.9. What if I violate an assumption? ②

It's worth remembering that you can have a perfectly good model for your data (no outliers, influential cases, etc.) and you can use that model to draw conclusions about your sample, even if your assumptions are violated. However, it's much more interesting to generalize your regression model and this is where assumptions become important. If they have been violated then you cannot generalize your findings beyond your sample. The options for correcting for violated assumptions are a bit limited. If residuals show problems with heteroscedasticity or non-normality you could try transforming the raw data – but this won't necessarily affect the residuals! If you have a violation of the linearity assumption then you could see whether you can do logistic regression instead (described in the next chapter). Finally, there are a series of robust regression techniques (see section 5.7.4), which are described extremely well by Rand Wilcox in Chapter 10 of his book. SPSS can't do these methods directly (well, technically it can do robust parameter estimates but it's not easy), but you can attempt a robust regression using some of Wilcox's files for the software *R* (Wilcox, 2005) using SPSS's *R* plugin.

OLIVER TWISTED

Please, Sir, can I have some more … robust regression?

'R you going to show us how to do robust regression?,' mumbles Oliver as he dribbles down his shirt 'I want to do one.' I've prepared a flash movie on the companion website that shows you how to use the *R* plugin to do robust regression. You have no excuses now for not using it.

7.10. How to report multiple regression ②

If you follow the American Psychological Association guidelines for reporting multiple regression then the implication seems to be that tabulated results are the way forward. The APA also seem in favour of reporting, as a bare minimum, the standardized betas, their significance value and some general statistics about the model (such as the R^2). If you do decide to do a table then the beta values and their standard errors are also very useful. Personally I'd like to see the constant as well because then readers of your work can construct the full regression model if they need to. Also, if you've done a hierarchical regression you should report these values at each stage of the hierarchy. So, basically, you want to reproduce the table labelled **Coefficients** from the SPSS output and omit some of the non-essential information. For the example in this chapter we might produce a table like that in Table 7.2.

See if you can look back through the SPSS output in this chapter and work out from where the values came. Things to note are: (1) I've rounded off to 2 decimal places throughout; (2) for the standardized betas there is no zero before the decimal point (because these values cannot exceed 1) but for all other values less than 1 the zero is present; (3) the significance of the variable is denoted by an asterisk with a footnote to indicate the significance level being used (if there's more than one level of significance being used you can denote this with multiple asterisks, such as $*p < .05$, $**p < .01$, and $***p < .001$); and (4) the R^2 for the initial model and the change in R^2 (denoted as ΔR^2) for each subsequent step of the model are reported below the table.

TABLE 7.2 How to report multiple regression

	B	*SE B*	β
Step 1			
Constant	134.14	7.54	
Advertising Budget	0.10	0.01	.58*
Step 2			
Constant	–26.61	17.35	
Advertising Budget	0.09	0.01	.51*
Plays on BBC Radio 1	3.37	0.28	.51*
Attractiveness	11.09	2.44	.19*

Note: $R^2 = .34$ for Step 1, $\Delta R^2 = .33$ for Step 2 ($p < .001$). * $p < .001$.

CHAMORRO-PREMUZIC, T., ET AL. (2008). *PERSONALITY AND INDIVIDUAL DIFFERENCES*, 44, 965–976.

LABCOAT LENI'S REAL RESEARCH 7.1

Why do you like your lecturers? ①

In the previous chapter we encountered a study by Chamorro-Premuzic et al. in which they measured students' personality characteristics and asked them to rate how much they wanted these same characteristics in their lecturers (see Labcoat Leni's Real Research 6.1 for a full description). In that chapter we correlated these scores; however, we could go a step further and see whether students' personality characteristics predict the characteristics that they would like to see in their lecturers.

The data from this study are in the file **Chamorro-Premuzic.sav**. Labcoat Leni wants you to carry out five multiple regression analyses: the outcome variable in each of the five analyses is the ratings of how much students want to see Neuroticism, Extroversion, Openness to experience, Agreeableness and Conscientiousness. For each of these outcomes, force Age and Gender into the analysis in the first step of the hierarchy, then in the second block force in the five student personality traits (Neuroticism, Extroversion, Openness to experience, Agreeableness and Conscientiousness). For each analysis create a table of the results.

Answers are in the additional material on the companion website (or look at Table 4 in the original article).

7.11. Categorical predictors and multiple regression ③

SMART ALEX ONLY

Often in regression analysis you'll collect data about groups of people (e.g. ethnic group, gender, socio-economic status, diagnostic category). You might want to include these groups as predictors in the regression model; however, we saw from our assumptions that variables need to be continuous or categorical with only two categories. We saw in section 6.5.5 that a point–biserial correlation is Pearson's *r* between two variables when one is continuous and the other has two categories coded as 0 and 1. We've also learnt that simple regression is based on Pearson's *r*, so it shouldn't take a great deal of imagination to see that, like the point–biserial correlation, we could construct a regression model with a predictor that has two categories (e.g. gender). Likewise, it shouldn't be too inconceivable that we could then extend this model to incorporate several predictors that had two categories. All that is important is that we code the two categories with the values of 0 and 1. Why is it important that there are only two categories and that they're coded 0 and 1? Actually, I don't want to get into this here because this chapter is already too long, the publishers are going to break my legs if it gets any longer, and I explain it anyway later in the book (sections 9.7 and 10.2.3) so, for the time being, just trust me!

7.11.1. Dummy coding ③

The obvious problem with wanting to use categorical variables as predictors is that often you'll have more than two categories. For example, if you'd measured religiosity you might have categories of Muslim, Jewish, Hindu, Catholic, Buddhist, Protestant, Jedi (for those of you not in the UK, we had a census here a few years back in which a significant portion of people put down Jedi as their religion). Clearly these groups cannot be distinguished using a single variable coded with zeros and ones. In these cases we have to use what's called **dummy variables**. Dummy coding is a way of representing groups of people using only zeros and ones. To do it, we have to create several variables; in fact, the number of variables we need is one less than the number of groups we're recoding. There are eight basic steps:

1 Count the number of groups you want to recode and subtract 1.

2 Create as many new variables as the value you calculated in step 1. These are your dummy variables.

3 Choose one of your groups as a baseline (i.e. a group against which all other groups should be compared). This should usually be a control group, or, if you don't have a specific hypothesis, it should be the group that represents the majority of people (because it might be interesting to compare other groups against the majority).

4 Having chosen a baseline group, assign that group values of 0 for all of your dummy variables.

5 For your first dummy variable, assign the value 1 to the first group that you want to compare against the baseline group. Assign all other groups 0 for this variable.

6 For the second dummy variable assign the value 1 to the second group that you want to compare against the baseline group. Assign all other groups 0 for this variable.

7 Repeat this until you run out of dummy variables.

8 Place all of your dummy variables into the regression analysis!

Let's try this out using an example. In Chapter 4 we came across an example in which a biologist was worried about the potential health effects of music festivals. She collected some data at the Download Festival, which is a music festival specializing in heavy metal. The biologist was worried that the findings that she had were a function of the fact that she had tested only one type of person: metal fans. Perhaps it's not the festival that makes people smelly, maybe it's only metal fans at festivals who get smellier (as a metal fan, I would at this point sacrifice the biologist to Satan for being so prejudiced). Anyway, to answer this question she went to another festival that had a more eclectic clientele. So, she went to the Glastonbury Music Festival which attracts all sorts of people because it has all styles of music there. Again, she measured the hygiene of concert-goers over the three days of the festival using a technique that results in a score ranging between 0 (you smell like you've bathed in sewage) and 5 (you smell of freshly baked bread). Now, in Chapters 4 and 5, we just looked at the distribution of scores for the three days of the festival, but now the biologist wanted to look at whether the type of music you like (your cultural group) predicts whether hygiene decreases over the festival. The data are in the file called **GlastonburyFestivalRegression.sav**. This file contains the hygiene scores for each of three days of the festival, but it also contains a variable called **change** which is the change in hygiene over the three days of the festival (so it's the change from day 1 to day 3).[13] Finally, the biologist categorized people according to their musical affiliation: if they mainly liked alternative music she called them 'indie kid', if they mainly liked heavy metal she called them a 'metaller' and if they mainly liked sort of hippy/folky/ambient type of stuff then she labelled them a 'crusty'. Anyone not falling into these categories was labelled 'no musical affiliation'. In the data file she coded these groups 1, 2, 3 and 4 respectively.

The first thing we should do is calculate the number of dummy variables. We have four groups, so there will be three dummy variables (one less than the number of groups). Next we need to choose a baseline group. We're interested in comparing those that have different musical affiliations against those that don't, so our baseline category will be 'no musical

[13] Not everyone could be measured on day 3, so there is a change score only for a subset of the original sample.

TABLE 7.3 Dummy coding for the Glastonbury Festival data

	Dummy Variable 1	Dummy Variable 2	Dummy Variable 3
No Affiliation	0	0	0
Indie Kid	0	0	1
Metaller	0	1	0
Crusty	1	0	0

affiliation'. We give this group a code of 0 for all of our dummy variables. For our first dummy variable, we could look at the 'crusty' group, and to do this we give anyone that was a crusty a code of 1, and everyone else a code of 0. For our second dummy variable, we could look at the 'metaller' group, and to do this we give anyone that was a metaller a code of 1, and everyone else a code of 0. We have one dummy variable left and this will have to look at our final category: 'indie kid'. To do this we give anyone that was an indie kid a code of 1, and everyone else a code of 0. The resulting coding scheme is shown in Table 7.3. The thing to note is that each group has a code of 1 on only one of the dummy variables (except the base category that is always coded as 0).

As I said, we'll look at why dummy coding works in sections 9.7 and 10.2.3, but for the time being let's look at how to recode our grouping variable into these dummy variables using SPSS. To recode variables you need to use the *Recode* function. Select Transform x→y Recode into Different Variables... to access the dialog box in Figure 7.21.

The recode dialog box lists all of the variables in the data editor, and you need to select the one you want to recode (in this case **music**) and transfer it to the box labelled *Numeric Variable → Output Variable* by clicking on ➜. You then need to name the new variable (the output variable as SPSS calls it), so go to the part that says *Output Variable* and in the box below where it says *Name* write a name for your first dummy variable (you might call it **Crusty**). You can also give this variable a more descriptive name by typing something in the box labelled *Label* (for this first dummy variable I've called it No Affiliation vs. Crusty). When you've done this click on Change to transfer this new variable to the box labelled *Numeric Variable → Output Variable* (this box should now say *music → Crusty*).

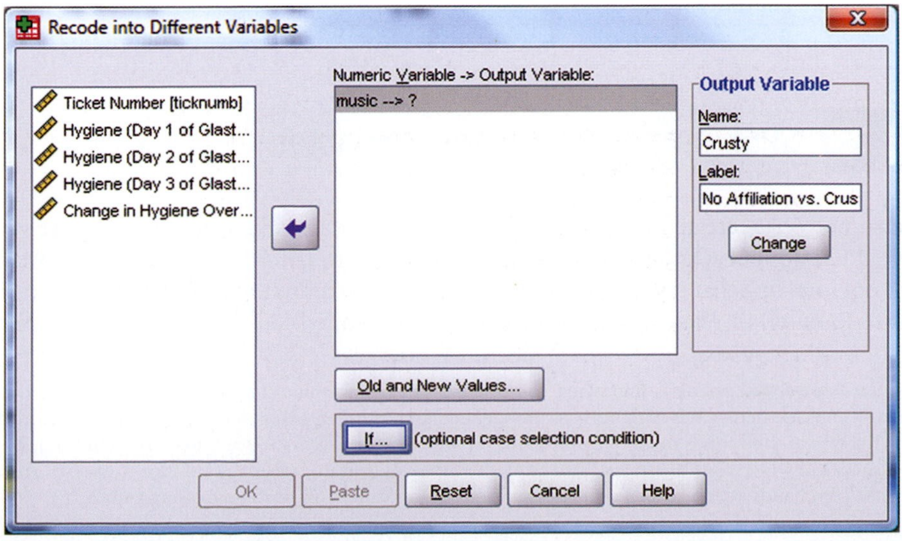

FIGURE 7.21 The recode dialog box

OLIVER TWISTED

Please, Sir, can I have some more ... recode?

'Our data set has missing values,' worries Oliver. 'What do we do if we only want to recode cases for which we have data?' Well, we can set some other options at this point, that's what we can do. This is getting a little bit more involved so if you want to know more, the additional material for this chapter on the companion website will tell you. Stop worrying Oliver, everything will be OK.

Having defined the first dummy variable, we need to tell SPSS how to recode the values of the variable **music** into the values that we want for the new variable, **music1**. To do this click on Old and New Values... to access the dialog box in Figure 7.22. This dialog box is used to change values of the original variable into different values for the new variable. For our first dummy variable, we want anyone who was a crusty to get a code of 1 and everyone else to get a code of 0. Now, crusty was coded with the value 3 in the original variable, so you need to type the value 3 in the section labelled *Old Value* in the box labelled *Value*. The new value we want is 1, so we need to type the value 1 in the section labelled *New Value* in the box labelled *Value*. When you've done this, click on Add to add this change to the list of changes (the list is displayed in the box labelled *Old → New*, which should now say *3 → 1* as in the diagram). The next thing we need to do is to change the remaining groups to have a value of 0 for the first dummy variable. To do this just select All other values and type the value 0 in the section labelled *New Value* in the box labelled *Value*.[14] When you've done this, click on Add to add this change to the list of changes (this list will now also say *ELSE → 0*). When you've done this click on Continue to return to the main dialog box, and then click on OK to create the first dummy variable. This variable will appear as a new column in the data editor, and you should notice that it will have a value of 1 for anyone originally classified as a crusty and a value of 0 for everyone else.

SELF-TEST Try creating the remaining two dummy variables (call them **Metaller** and **Indie_Kid**) using the same principles.

7.11.2. SPSS output for dummy variables ③

Let's assume you've created the three dummy coding variables (if you're stuck there is a data file called **GlastonburyDummy.sav** (the 'Dummy' refers to the fact it has dummy variables in it – I'm not implying that if you need to use this file you're a dummy!)). With dummy variables, you have to enter all related dummy variables in the same block (so use the *Enter* method).

[14] Using this All other values option is fine when you don't have missing values in the data, but just note that when you do (as is the case here) cases with both system-defined and user-defined missing values will be included in the recode. One way around this is to recode only cases for which there is a value (see Oliver Twisted Box). The alternative is specifically to recode missing values using the Range: option. It is also a good idea to use the *frequencies* or *crosstabs* commands after a recode and check that you have caught all of these missing values.

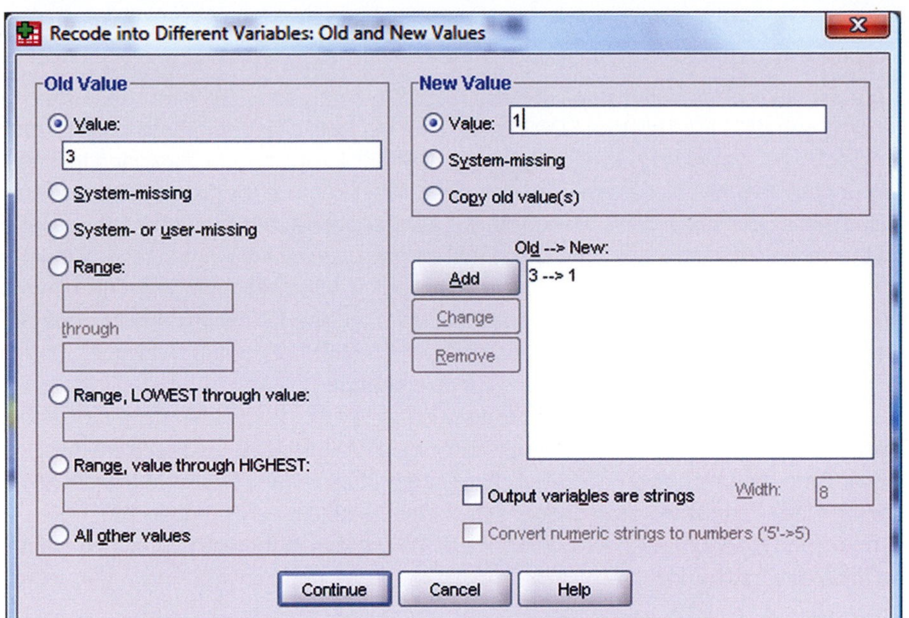

FIGURE 7.22 Dialog boxes for the *Recode* function (see also SPSS Tip 7.1)

SPSS TIP 7.1 Using syntax to recode ③

If you're doing a lot of recoding it soon becomes pretty tedious using the dialog boxes all of the time. I've written the syntax file, **RecodeGlastonburyData.sps**, to create all of the dummy variables we've discussed. Load this file and run the syntax, or open a syntax window (see Section 3.7) and type the following:

```
DO IF (1-MISSING(change)).
RECODE music (3=1)(ELSE = 0) INTO Crusty.
RECODE music (2=1)(ELSE = 0) INTO Metaller.
RECODE music (1=1)(ELSE = 0) INTO Indie_Kid.
END IF.
VARIABLE LABELS Crusty 'No Affiliation vs. Crusty'.
VARIABLE LABELS Metaller 'No Affiliation vs. Metaller'.
VARIABLE LABELS Indie_Kid 'No Affiliation vs. Indie Kid'.
VARIABLE LEVEL Crusty Metaller Indie_Kid (Nominal).
FORMATS Crusty Metaller Indie_Kid (F1.0).
EXECUTE.
```

Each RECODE command is doing the equivalent of what you'd do using the compute dialog box in Figure 7.22. So, the three lines beginning RECODE ask SPSS to create three new variables (**Crusty, Metaller** and **Indie_Kid**), which are based on the original variable **music**. For the first variable, if **music** is 3 then it becomes 1, and every other value becomes 0. For the second, if **music** is 2 then it becomes 1, and every other value becomes

(Continued)

(Continued)

0, and so on for the third dummy variable. Note that all of these RECODE commands are within an IF statement (beginning DO IF and ending with END IF). This tells SPSS to carry out the RECODE commands only if a certain condition is met. The condition we have set is (1- MISSING(change)). MISSING is a built-in command that returns 'true' (i.e. the value 1) for a case that has a system- or user-defined missing value for the specified variable; it returns 'false' (i.e. the value 0) if a case has a value. Hence, MISSING(change) returns a value of 1 for cases that have a missing value for the variable 'change' and 0 for cases that do have values. We want to recode the cases that *do* have a value for the variable change, therefore I have specified '1- MISSING(change)'. This command reverses MISSING(change) so that it returns 1 (true) for cases that have a value for the variable **change** and 0 (false) for system- or user-defined missing values. To sum up, the DO IF (1- MISSING(change)) tells SPSS 'Do the following RECODE commands if the case has a value for the variable **change**.'

The lines beginning VARIABLE LABELS just tells SPSS to assign the text in the quotations as labels for the variables **Crusty**, **Metaller** and **Indie_Kid** respectively. The line beginning VARIABLE LEVEL then sets these three variables to be 'nominal', and the line beginning FORMATS changes the three variables to have a width of 1 and 0 decimal places (hence the 1.0) – in other words, it changes the format to be a binary number.

The final line has the command EXECUTE without which none of the commands beforehand will be executed! Note also that every line ends with a full stop.

So, in this case we have to enter our dummy variables in the same block; however, if we'd had another variable (e.g. socio-economic status) that had been transformed into dummy variables, we could enter these dummy variables in a different block (so, it's only dummy variables that have recoded the same variable that need to be entered in the same block).

SELF-TEST Use what you've learnt in this chapter to run a multiple regression using the change scores as the outcome, and the three dummy variables (entered in the same block) as predictors.

Let's have a look at the output.

SPSS OUTPUT 7.12

Model Summary[b]

Model	R	R Square	Adjusted R Square	Std. Error of the Estimate	Change Statistics: R Square Change	F Change	df1	df2	Sig. F Change	Durbin-Watson
1	.276[a]	.076	.053	.68818	.076	3.270	3	119	.024	1.893

a. Predictors: (Constant), No Affiliation vs. Indie Kid, No Affiliation vs. Crusty, No Affiliation vs. Metaller

b. Dependent Variable: Change in Hygiene Over The Festival

ANOVA[b]

Model		Sum of Squares	df	Mean Square	F	Sig.
1	Regression	4.646	3	1.549	3.270	.024[a]
	Residual	56.358	119	.474		
	Total	61.004	122			

a. Predictors: (Constant), No Affiliation vs. Indie Kid, No Affiliation vs. Crusty, No Affiliation vs. Metaller

b. Dependent Variable: Change in Hygiene Over The Festival

SPSS Output 7.12 shows the model statistics. This shows that by entering the three dummy variables we can explain 7.6% of the variance in the change in hygiene scores (the R^2-value × 100). In other words, 7.6% of the variance in the change in hygiene can be explained by the musical affiliation of the person. The ANOVA (which shows the same thing as the R^2 change statistic because there is only one step in this regression) tells us that the model is significantly better at predicting the change in hygiene scores than having no model (or, put another way, the 7.6% of variance that can be explained is a significant amount). Most of this should be clear from what you've read in this chapter already; what's more interesting is how we interpret the individual dummy variables.

SPSS OUTPUT 7.13

Coefficients[a]

Model		Unstandardized Coefficients		Standardized Coefficients	t	Sig.
		B	Std. Error	Beta		
1	(Constant)	-.554	.090		-6.134	.000
	No Affiliation vs. Crusty	-.412	.167	-.232	-2.464	.015
	No Affiliation vs. Metaller	.028	.160	.017	.177	.860
	No Affiliation vs. Indie Kid	-.410	.205	-.185	-2.001	.048

a. Dependent Variable: Change in Hygiene Over The Festival

SPSS Output 7.13 shows a basic **Coefficients** table for the dummy variables (I've excluded the confidence intervals and collinearity diagnostics). The first thing to notice is that each dummy variable appears in the table with a useful label (such as No Affiliation vs. Crusty). This has happened because when we recoded our variables we gave each variable a label; if we hadn't done this then the table would contain the rather less helpful variable names (crusty, metaller and Indie_Kid). The labels that I suggested giving to each variable give us a hint about what each dummy variable represents. The first dummy variable (No Affiliation vs. Crusty) shows the difference between the change in hygiene scores for the no affiliation group and the crusty group. Remember that the beta value tells us the change in the outcome due to a unit change in the predictor. In this case, a unit change in the predictor is the change from 0 to 1. As such it shows the shift in the change in hygiene scores that results from the dummy variable changing from 0 to 1 (Crusty). By including all three dummy variables at the same time, our baseline category is always zero, so this actually represents the difference in the change in hygiene scores if a person has no musical affiliation, compared to someone who is a crusty. This difference is the difference between the two group means.

To illustrate this fact, I've produced a table (SPSS Output 7.14) of the group means for each of the four groups and also the difference between the means for each group and the no affiliation group. These means represent the average change in hygiene scores for the three groups (i.e. the mean of each group on our outcome variable). If we calculate the difference in these means for the No Affiliation group and the crusty group we get, Crusty – no affiliation = (–0.966) – (–0.554) = –0.412. In other words, the change in hygiene scores is greater for the crusty group than it is for the no affiliation group (crusties' hygiene decreases more over the festival than those with no musical affiliation). This value is the same as the *unstandardized* beta value in SPSS Output 7.13! So, the beta values tell us the relative difference between each group and the group that we chose as a baseline category. This beta value is converted to a t-statistic and the significance of this t reported. This t-statistic is testing, as we've seen before, whether the beta value is 0 and when we have two categories coded with 0 and 1, that means it's testing whether the difference between group means is 0. If it is significant then it means that the group coded with 1 is significantly different from the baseline category – so, it's testing the difference between two means, which is the context in which students are most familiar with the *t*-statistic (see Chapter 9). For our first dummy variable, the *t*-test is significant, and the beta value has a negative value so we could say that the change in hygiene scores goes down as a person changes from having no affiliation to being a crusty. Bear in mind that a decrease in hygiene scores represents more change (you're becoming smellier) so what this actually means is that hygiene decreased significantly more in crusties compared to those with no musical affiliation!

SPSS OUTPUT 7.14

OLAP Cubes

Variables=Change in Hygiene Over The Festival

Musical Affiliation	Mean	Std. Deviation	N
Indie Kid	-0.964	0.670	14
Metaller	-0.526	0.576	27
Crusty	-0.966	0.760	24
No Musical Affiliation	-0.554	0.708	58
Crusty - No Musical Affiliation	-0.412	0.052	-34
Metaller - No Musical Affiliation	0.028	-0.133	-31
Indie Kid - No Musical Affiliation	-0.410	-0.038	-44
Total	-0.675	0.707	123

Moving on to our next dummy variable, this compares metallers to those that have no musical affiliation. The beta value again represents the shift in the change in hygiene scores if a person has no musical affiliation, compared to someone who is a metaller. If we calculate the difference in the group means for the no affiliation group and the metaller group we get, metaller – no affiliation = (−0.526) – (−0.554) = 0.028. This value is again the same as the unstandardized beta value in SPSS Output 7.13! For this second dummy variable, the *t*-test is not significant, so we could say that the change in hygiene scores is the same if a person changes from having no affiliation to being a metaller. In other words, the change in hygiene scores is not predicted by whether someone is a metaller compared to if they have no musical affiliation.

For the final dummy variable, we're comparing indie kids to those that have no musical affiliation. The beta value again represents the shift in the change in hygiene scores if a person has no musical affiliation, compared to someone who is an indie kid. If we calculate the difference in the group means for the no affiliation group and the indie kid group we get, indie kid – no affiliation = (−0.964) – (−0.554) = −0.410. It should be no surprise to you by now that this is the unstandardized beta value in SPSS Output 7.13! The *t*-test is significant, and the beta value has a negative value so, as with the first dummy variable, we could say that the change in hygiene scores goes down as a person changes from having no affiliation to being an indie kid. Bear in mind that a decrease in hygiene scores represents more change (you're becoming smellier) so what this actually means is that hygiene decreased significantly more in indie kids compared to those with no musical affiliation!

EVERYBODY

So, overall this analysis has shown that compared to having no musical affiliation, crusties and indie kids get significantly smellier across the three days of the festival, but metallers don't. This section has introduced some really complex ideas that I expand upon in Chapters 9 and 10. It might all be a bit much to take in, and so if you're confused or want to know more about why dummy coding works in this way I suggest reading sections 9.7 and 10.2.3 and then coming back here. Alternatively, read Hardy's (1993) excellent monograph!

What have I discovered about statistics? ①

This chapter is possibly the longest book chapter ever written, and if you feel like you aged several years while reading it then, well, you probably have (look around, there are cobwebs in the room, you have a long beard, and when you go outside you'll discover a second ice age has been and gone leaving only you and a few woolly mammoths to populate the planet). However, on the plus side, you now know more or less everything you ever need to know about statistics. Really, it's true; you'll discover in the coming chapters that everything else we discuss is basically a variation on the theme of regression. So, although you may be near death having spent your life reading this chapter (and I'm certainly near death having written it) you are a stats genius – it's official!

We started the chapter by discovering that at 8 years old I could have really done with regression analysis to tell me which variables are important in predicting talent competition success. Unfortunately I didn't have regression, but fortunately I had my dad instead (and he's better than regression). We then looked at how we could use statistical models to make similar predictions by looking at the case of when you have one predictor and one outcome. This allowed us to look at some basic principles such as the equation of a straight line, the method of least squares, and how to assess how well our model fits the data using some important quantities that you'll come across in future chapters: the model sum of squares, SS_M, the residual sum of squares, SS_R, and the total sum of squares, SS_T. We used these values to calculate several important statistics such as R^2 and the *F*-ratio. We also learnt how to do a regression on SPSS, and how we can plug the resulting beta values into the equation of a straight line to make predictions about our outcome.

Next, we saw that the question of a straight line can be extended to include several predictors and looked at different methods of placing these predictors in the model (hierarchical, forced entry, stepwise). Next, we looked at factors that can affect the accuracy of a model (outliers and influential cases) and ways to identify these factors. We then moved on to look at the assumptions necessary to generalize our model beyond the sample of data we've collected before discovering how to do the analysis on SPSS, and how to interpret the output, create our multiple regression model and test its reliability and generalizability. I finished the chapter by looking at how we can use categorical predictors in regression (and in passing we discovered the *recode* function). In general, multiple regression is a long process and should be done with care and attention to detail. There are a lot of important things to consider and you should approach the analysis in a systematic fashion. I hope this chapter helps you to do that!

So, I was starting to get a taste for the rock-idol lifestyle: I had friends, a fortune (well, two gold-plated winner's medals), fast cars (a bike) and dodgy-looking 8 year olds were giving me suitcases full of lemon sherbet to lick off mirrors. However, my parents and teachers were about to impress reality upon my young mind …

Key terms that I've discovered

Adjusted predicted value
Adjusted R^2
Autocorrelation
b_i

β_i
Cook's distance
Covariance ratio (CVR)
Cross-validation
Deleted residual
DFBeta
DFFit
Dummy variables
Durbin–Watson test
F-ratio
Generalization
Goodness of fit
Hat values
Heteroscedasticity
Hierarchical regression
Homoscedasticity
Independent errors
Leverage
Mahalanobis distances
Mean squares
Model sum of squares
Multicollinearity
Multiple R
Multiple regression
Outcome variable
Perfect collinearity
Predictor variable
Residual
Residual sum of squares
Shrinkage
Simple regression
Standardized DFBeta
Standardized DFFit
Standardized residuals
Stepwise regression
Studentized deleted residuals
Studentized residuals
Suppressor effects
t-statistic
Tolerance
Total sum of squares
Unstandardized residuals
Variance inflation factor (VIF)

Smart Alex's tasks

- **Task 1**: A fashion student was interested in factors that predicted the salaries of catwalk models. She collected data from 231 models. For each model she asked them their salary per day on days when they were working (**salary**), their age (**age**), how many years they had worked as a model (**years**), and then got a panel of experts from modelling agencies to rate the attractiveness of each model as a percentage with 100% being perfectly attractive (**beauty**). The data are in the file **Supermodel.sav**. Unfortunately, this fashion student bought some substandard statistics text and so doesn't know how to analyse her data.☺ Can you help her out by conducting a multiple regression to see which variables predict a model's salary? How valid is the regression model? ②
- **Task 2**: Using the Glastonbury data from this chapter (with the dummy coding in **GlastonburyDummy.sav**), which you should've already analysed, comment on whether you think the model is reliable and generalizable. ③
- **Task 3**: A study was carried out to explore the relationship between **Aggression** and several potential predicting factors in 666 children that had an older sibling. Variables measured were **Parenting_Style** (high score = bad parenting practices), **Computer_Games** (high score = more time spent playing computer games), **Television** (high score = more time spent watching television), **Diet** (high score = the child has a good diet low in E-numbers), and **Sibling_Aggression** (high score = more aggression seen in their older sibling). Past research indicated that parenting style and sibling aggression were good predictors of the level of aggression in the younger child. All other variables were treated in an exploratory fashion. The data are in the file **Child Aggression.sav**. Analyse them with multiple regression. ②

Answers can be found on the companion website.

Further reading

Bowerman, B. L., & O'Connell, R. T. (1990). *Linear statistical models: An applied approach* (2nd ed.). Belmont, CA: Duxbury. (This text is only for the mathematically minded or postgraduate students but provides an extremely thorough exposition of regression analysis.)

Hardy, M. A. (1993). *Regression with dummy variables*. Sage university paper series on quantitative applications in the social sciences, 07–093. Newbury Park, CA: Sage.

Howell, D. C. (2006). *Statistical methods for psychology* (6th ed.). Belmont, CA: Duxbury. (Or you might prefer his *Fundamental Statistics for the Behavioral Sciences,* also in its 6th edition, 2007. Both are excellent introductions to the mathematics behind regression analysis.)

Miles, J. N. V., & Shevlin, M. (2001). *Applying regression and correlation: a guide for students and researchers*. London: Sage. (This is an extremely readable text that covers regression in loads of detail but with minimum pain – highly recommended.)

Stevens, J. (2002). *Applied multivariate statistics for the social sciences* (4th ed.). Hillsdale, NJ: Erlbaum. Chapter 3.

Online tutorial

The companion website contains the following Flash movie tutorials to accompany this chapter:

- Regression using SPSS
- Robust Regression

Interesting real research

Chamorro-Premuzic, T., Furnham, A., Christopher, A. N., Garwood, J., & Martin, N. (2008). Birds of a feather: Students' preferences for lecturers' personalities as predicted by their own personality and learning approaches. *Personality and Individual Differences*, 44, 965–976.

8 Logistic regression

FIGURE 8.1 Practising for my career as a rock star by slaying the baying throng of Grove Primary School at the age of 10. (Note the girl with her hands covering her ears)

8.1. What will this chapter tell me? ①

We saw in the previous chapter that I had successfully conquered the holiday camps of Wales with my singing and guitar playing (and the Welsh know a thing or two about good singing). I had jumped on a snowboard called oblivion and thrown myself down the black run known as world domination. About 10 metres after starting this slippery descent I hit the lumpy patch of ice called 'adults'. I was 9, life was fun, and yet every adult that I seemed to encounter was obsessed with my future. 'What do you want to be when you grow up?' they would ask. I was 9 and 'grown-up' was a lifetime away; all I knew was that I was going to marry Clair Sparks (more on her in the next chapter) and that I was a rock legend who didn't need to worry about such adult matters as having a job. It was a difficult question, but adults require answers and I wasn't going to let them know that I didn't care about 'grown-up' matters. We saw in the previous chapter that we can use regression to predict future outcomes based on past data, when the outcome is a continuous variable, but this question had a categorical outcome (e.g. would I be a fireman, a doctor, an evil dictator?). Luckily, though, we can use an extension of regression called logistic regression to deal with these situations. What a result; bring on the rabid wolves of categorical data. To make a prediction about a categorical outcome then, as with regression, I needed to draw on past data: I hadn't tried conducting brain surgery, neither had I experience of sentencing psychopaths to prison sentences for eating their husbands, nor had I taught anyone. I had, however, had a go at singing and playing guitar; 'I'm going to be a rock star' was my prediction. A prediction can be accurate (which would mean that I *am* a rock star) or it can be inaccurate (which would mean that I'm writing a statistics textbook). This chapter looks at the theory and application of **logistic regression**, an extension of regression that allows us to predict categorical outcomes based on predictor variables.

8.2. Background to logistic regression ①

In a nutshell, logistic regression is multiple regression but with an outcome variable that is a categorical variable and predictor variables that are continuous or categorical. In its simplest form, this means that we can predict which of two categories a person is likely to belong to given certain other information. A trivial example is to look at which variables predict whether a person is male or female. We might measure laziness, pig-headedness, alcohol consumption and number of burps that a person does in a day. Using logistic regression, we might find that all of these variables predict the gender of the person, but the technique will also allow us to predict whether a new person is likely to be male or female. So, if we picked a random person and discovered they scored highly on laziness, pig-headedness, alcohol consumption and the number of burps, then the regression model might tell us that, based on this information, this person is likely to be male. Admittedly, it is unlikely that a researcher would ever be interested in the relationship between flatulence and gender (it is probably too well established by experience to warrant research!), but logistic regression can have life-saving applications. In medical research logistic regression is used to generate models from which predictions can be made about the likelihood that a tumour is cancerous or benign (for example). A database of patients can be used to establish which variables are influential in predicting the malignancy of a tumour. These variables can then be measured for a new patient and their values placed in a logistic regression model, from which a probability of malignancy could be estimated. If the probability value of the tumour being malignant is suitably low then the doctor may decide not to carry out expensive and painful surgery that in all likelihood is unnecessary. We might not face such life-threatening decisions but logistic regression can nevertheless be a very useful tool. When we are trying to predict membership of only two categorical outcomes the analysis is known as **binary logistic regression**, but when we want to predict membership of more than two categories we use **multinomial (or polychotomous) logistic regression**.

8.3. What are the principles behind logistic regression? ③

I don't wish to dwell on the underlying principles of logistic regression because they aren't necessary to understand the test (I am living proof of this fact). However, I do wish to draw a few parallels to normal regression so that you can get the gist of what's going on using a framework that will be familiar to you already (what do you mean you haven't read the regression chapter yet?!). To keep things simple I'm going to explain binary logistic regression, but most of the principles extend easily to when there are more than two outcome categories. Now would be a good time for the equation-phobes to look away. In simple linear regression, we saw that the outcome variable Y is predicted from the equation of a straight line:

$$Y_i = b_0 + b_1 X_{1i} + \varepsilon_i \tag{8.1}$$

in which b_0 is the Y intercept, b_1 is the gradient of the straight line, X_1 is the value of the predictor variable and ε is a residual term. Given the values of Y and X_1, the unknown parameters in the equation can be estimated by finding a solution for which the squared distance between the observed and predicted values of the dependent variable is minimized (the method of least squares).

This stuff should all be pretty familiar by now. In multiple regression, in which there are several predictors, a similar equation is derived in which each predictor has its own coefficient. As such, Y is predicted from a combination of each predictor variable multiplied by its respective regression coefficient.

$$Y_i = b_0 + b_1 X_{1i} + b_2 X_{2i} + \ldots + b_n X_{ni} + \varepsilon_i \tag{8.2}$$

in which b_n is the regression coefficient of the corresponding variable X_n. In logistic regression, instead of predicting the value of a variable Y from a predictor variable X_1 or several predictor variables (Xs), we predict the *probability* of Y occurring given known values of X_1 (or Xs). The logistic regression equation bears many similarities to the regression equations just described. In its simplest form, when there is only one predictor variable X_1, the logistic regression equation from which the probability of Y is predicted is given by:

$$P(Y) = \frac{1}{1 + e^{-(b_0 + b_1 X_{1i})}} \tag{8.3}$$

in which $P(Y)$ is the probability of Y occurring, e is the base of natural logarithms, and the other coefficients form a linear combination much the same as in simple regression. In fact, you might notice that the bracketed portion of the equation is identical to the linear regression equation in that there is a constant (b_0), a predictor variable (X_1) and a coefficient (or weight) attached to that predictor (b_1). Just like linear regression, it is possible to extend this equation so as to include several predictors. When there are several predictors the equation becomes:

$$P(Y) = \frac{1}{1 + e^{-(b_0 + b_1 X_{1i} + b_2 X_{2i} + \ldots + b_n X_{ni})}} \tag{8.4}$$

Equation (8.4) is the same as the equation used when there is only one predictor except that the linear combination has been extended to include any number of predictors. So, whereas the one-predictor version of the logistic regression equation contained the simple linear regression equation within it, the multiple-predictor version contains the multiple regression equation.

Despite the similarities between linear regression and logistic regression, there is a good reason why we cannot apply linear regression directly to a situation in which the outcome variable is categorical. The reason is that one of the assumptions of linear regression is that the

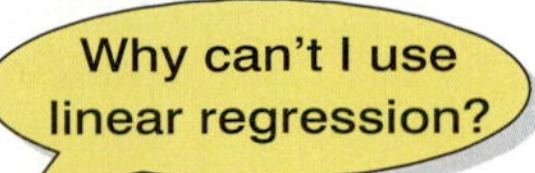
Why can't I use
linear regression?

relationship between variables is linear (see section 7.6.2.1). In that section we saw how important it is that the assumptions of a model are met for it to be accurate. Therefore, for linear regression to be a valid model, the observed data should contain a linear relationship. When the outcome variable is categorical, this assumption is violated (Berry, 1993). One way around this problem is to transform the data using the logarithmic transformation (see Berry & Feldman, 1985, and Chapter 5). This transformation is a way of expressing a non-linear relationship in a linear way. The logistic regression equation described above is based on this principle: it expresses the multiple linear regression equation in logarithmic terms (called the *logit*) and thus overcomes the problem of violating the assumption of linearity.

The exact form of the equation can be arranged in several ways but the version I have chosen expresses the equation in terms of the probability of Y occurring (i.e. the probability that a case belongs in a certain category). The resulting value from the equation, therefore, varies between 0 and 1. A value close to 0 means that Y is very unlikely to have occurred, and a value close to 1 means that Y is very likely to have occurred. Also, just like linear regression, each predictor variable in the logistic regression equation has its own coefficient. When we run the analysis we need to estimate the value of these coefficients so that we can solve the equation. These parameters are estimated by fitting models, based on the available predictors, to the observed data. The chosen model will be the one that, when values of the predictor variables are placed in it, results in values of Y closest to the observed values. Specifically, the values of the parameters are estimated using **maximum-likelihood estimation**, which selects coefficients that make the observed values most likely to have occurred. So, as with multiple regression, we try to fit a model to our data that allows us to estimate values of the outcome variable from known values of the predictor variable or variables.

8.3.1. Assessing the model: the log-likelihood statistic ③

We've seen that the logistic regression model predicts the probability of an event occurring for a given person (we would denote this as $P(Y_i)$ the probability that Y occurs for the ith person), based on observations of whether or not the event did occur for that person (we could denote this as Y_i, the actual outcome for the ith person). So, for a given person, Y will be either 0 (the outcome didn't occur) or 1 (the outcome did occur), and the predicted value, $P(Y)$, will be a value between 0 (there is no chance that the outcome will occur) and 1 (the outcome will certainly occur). We saw in multiple regression that if we want to assess whether a model fits the data we can compare the observed and predicted values of the outcome (if you remember, we use R^2, which is the Pearson correlation between observed values of the outcome and the values predicted by the regression model). Likewise, in logistic regression, we can use the observed and predicted values to assess the fit of the model. The measure we use is the **log-likelihood**:

$$\text{log-likelihood} = \sum_{i=1}^{N} [Y_i \ln(P(Y_i)) + (1 - Y_i) \ln(1 - P(Y_i))] \qquad (8.5)$$

The log-likelihood is based on summing the probabilities associated with the predicted and actual outcomes (Tabachnick & Fidell, 2007). The log-likelihood statistic is analogous to the residual sum of squares in multiple regression in the sense that it is an indicator of how much unexplained information there is after the model has been fitted. It, therefore, follows that large values of the log-likelihood statistic indicate poorly fitting statistical models, because the larger the value of the log-likelihood, the more unexplained observations there are.

Now, it's possible to calculate a log-likelihood for different models and to compare these models by looking at the difference between their log-likelihoods. One use of this is to compare the state of a logistic regression model against some kind of baseline state. The baseline state that's usually used is the model when only the constant is included. In multiple regression, the baseline model we use is the mean of all scores (this is our best guess of the outcome when we have no other information). In logistic regression, if we want to predict the outcome, what would our best guess be? Well, we can't use the mean score because our outcome is made of zeros and ones and so the mean is meaningless! However, if we know the frequency of zeros and ones, then the best guess will be the category with the largest number of cases. So, if the outcome occurs 107 times, and doesn't occur only 72 times, then our best guess of the outcome will be that it occurs (because it occurs 107 times compared to only 72 times when it doesn't occur). As such, like multiple regression, our baseline model is the model that gives us the best prediction when we know nothing other than the values of the outcome: in logistic regression this will be to predict the outcome that occurs most often. This is, the logistic regression model when only the constant is included. If we then add one or more predictors to the model, we can compute the improvement of the model as follows:

$$\chi^2 = 2[\text{LL(new)} - \text{LL(baseline)}] \quad (df = k_{\text{new}} - k_{\text{baseline}}) \tag{8.6}$$

So, we merely take the new model and subtract from it the baseline model (the model when only the constant is included). You'll notice that we multiply this value by 2; this is because it gives the result a chi-square distribution (see Chapter 18 and the Appendix) and so makes it easy to calculate the significance of the value. The chi-square distribution we use has degrees of freedom equal to the number of parameters, k in the new model minus the number of parameters in the baseline model. The number of parameters in the baseline model will always be 1 (the constant is the only parameter to be estimated); any subsequent model will have degrees of freedom equal to the number of predictors plus 1 (i.e. the number of predictors plus one parameter representing the constant).

8.3.2. Assessing the model: R and R^2 ③

When we talked about linear regression, we saw that the multiple correlation coefficient R and the corresponding R^2-value were useful measures of how well the model fits the data.

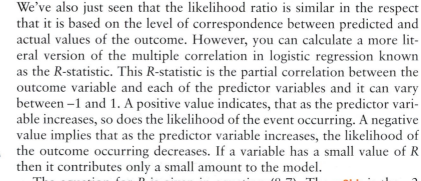

We've also just seen that the likelihood ratio is similar in the respect that it is based on the level of correspondence between predicted and actual values of the outcome. However, you can calculate a more literal version of the multiple correlation in logistic regression known as the R-statistic. This R-statistic is the partial correlation between the outcome variable and each of the predictor variables and it can vary between −1 and 1. A positive value indicates, that as the predictor variable increases, so does the likelihood of the event occurring. A negative value implies that as the predictor variable increases, the likelihood of the outcome occurring decreases. If a variable has a small value of R then it contributes only a small amount to the model.

The equation for R is given in equation (8.7). The **−2LL** is the −2 log-likelihood for the original model, the Wald statistic is calculated as described in the next section, and the degrees of freedom can be read from the summary table for the variables in the equation. However, because this value of R is dependent upon the Wald statistic it is by no means an accurate measure (we'll see in the next section that the Wald statistic can be inaccurate under certain circumstances). For this reason the value of R should be treated with some caution, and it is invalid to square this value and interpret it as you would in linear regression:

$$R = \pm\sqrt{\left(\frac{\text{Wald} - (2 \times df)}{-2\text{LL(original)}}\right)} \tag{8.7}$$

There is some controversy over what would make a good analogue to the R^2-value in linear regression, but one measure described by Hosmer and Lemeshow (1989) can be easily calculated. In SPSS terminology, **Hosmer and Lemeshow's R^2_L** measure is calculated as:

$$R^2_L = \frac{-2\text{LL(model)}}{-2\text{LL(original)}} \tag{8.8}$$

As such, R^2_L is calculated by dividing the model chi-square (based on the log-likelihood) by the *original* −2LL (the log-likelihood of the model before any predictors were entered). R^2_L is the proportional reduction in the absolute value of the log-likelihood measure and as such it is a measure of how much the badness of fit improves as a result of the inclusion of the predictor variables. It can vary between 0 (indicating that the predictors are useless at predicting the outcome variable) and 1 (indicating that the model predicts the outcome variable perfectly).

However, this is not the measure used by SPSS. SPSS uses **Cox and Snell's R^2_{CS}** (1989), which is based on the log-likelihood of the model (LL(*new*)) and the log-likelihood of the original model (LL(*baseline*)), and the sample size, n:

$$R^2_{CS} = 1 - e^{\left[-\frac{2}{n}(\text{LL(new)}) - (\text{LL(baseline)})\right]} \tag{8.9}$$

However, this statistic never reaches its theoretical maximum of 1. Therefore, Nagelkerke (1991) suggested the following amendment (**Nagelkerke's R^2_N**):

$$R^2_N = \frac{R^2_{CS}}{1 - e^{\left[\frac{2(\text{LL(baseline)})}{n}\right]}} \tag{8.10}$$

Although all of these measures differ in their computation (and the answers you get), conceptually they are somewhat the same. So, in terms of interpretation they can be seen as similar to the R^2 in linear regression in that they provide a gauge of the substantive significance of the model.

8.3.3. Assessing the contribution of predictors: the Wald statistic ②

As in linear regression, we want to know not only how well the model overall fits the data, but also the individual contribution of predictors. In linear regression, we used the estimated regression coefficients (b) and their standard errors to compute a t-statistic. In logistic regression there is an analogous statistic known as the **Wald statistic**, which has a special distribution known as the **chi-square distribution**. Like the t-test in linear regression, the Wald statistic tells us whether the b coefficient for that predictor is significantly different from zero. If the coefficient is significantly different from zero then we can assume that the predictor is making a significant contribution to the prediction of the outcome (Y):

$$\text{Wald} = \frac{b}{\text{SE}_b} \quad (8.11)$$

Equation (8.11) shows how the Wald statistic is calculated and you can see it's basically identical to the t-statistic in linear regression (see equation (7.6)): it is the value of the regression coefficient divided by its associated standard error. The Wald statistic (Figure 8.2) is usually used to ascertain whether a variable is a significant predictor of the outcome; however, it is probably more accurate to examine the likelihood ratio statistics. The reason why the Wald statistic should be used cautiously is because, when the regression coefficient (b) is large, the standard error tends to become inflated, resulting in the Wald statistic being underestimated (see Menard, 1995). The inflation of the standard error increases the probability of rejecting a predictor as being significant when in reality it is making a significant contribution to the model (i.e. you are more likely to make a Type II error).

FIGURE 8.2 Abraham Wald writing 'I must not devise test statistics prone to having inflated standard errors' on the blackboard 100 times

8.3.4. The odds ratio: *Exp(B)* ③

More crucial to the *interpretation* of logistic regression is the value of the odds ratio (**Exp(B)** in the SPSS output), which is an indicator of the change in odds resulting from a unit change in the predictor. As such, it is similar to the b coefficient in logistic regression but easier to understand (because it doesn't require a logarithmic transformation). When the predictor variable is categorical the odds ratio is easier to explain, so imagine we had a simple example in which we were trying to predict whether or not someone got pregnant from whether or not they used a condom last time they made love. The **odds** of an event occurring are defined as the probability of an event occurring divided by the probability of that event not occurring (see equation (8.12)) and should not be confused with the more colloquial usage of the word to refer to probability. So, for example, the odds of becoming pregnant are the probability of becoming pregnant divided by the probability of not becoming pregnant:

$$\text{odds} = \frac{P(\text{event})}{P(\text{no event})}$$

$$P(\text{event } Y) = \frac{1}{1+e^{-(b_0+b_1X_1)}} \tag{8.12}$$

$$P(\text{no event } Y) = 1 - P(\text{event } Y)$$

To calculate the change in odds that results from a unit change in the predictor, we must first calculate the odds of becoming pregnant given that a condom *wasn't* used. We then calculate the odds of becoming pregnant given that a condom *was* used. Finally, we calculate the proportionate change in these two odds.

To calculate the first set of odds, we need to use equation (8.3) to calculate the probability of becoming pregnant given that a condom wasn't used. If we had more than one predictor we would use equation (8.4). There are three unknown quantities in this equation: the coefficient of the constant (b_0), the coefficient for the predictor (b_1) and the value of the predictor itself (*X*). We'll know the value of *X* from how we coded the condom use variable (chances are we would've used 0 = condom wasn't used and 1 = condom was used). The values of b_1 and b_0 will be estimated for us. We can calculate the odds as in equation (8.12).

Next, we calculate the same thing after the predictor variable has changed by one unit. In this case, because the predictor variable is dichotomous, we need to calculate the odds of getting pregnant, given that a condom *was* used. So, the value of *X* is now 1 (rather than 0).

We now know the odds before and after a unit change in the predictor variable. It is a simple matter to calculate the proportionate change in odds by dividing the odds after a unit change in the predictor by the odds before that change:

$$\Delta\text{odds} = \frac{\text{odds after a unit change in the predictor}}{\text{original odds}} \tag{8.13}$$

This proportionate change in odds is the odds ratio, and we can interpret it in terms of the change in odds: if the value is greater than 1 then it indicates that as the predictor increases, the odds of the outcome occurring increase. Conversely, a value less than 1 indicates that as the predictor increases, the odds of the outcome occurring decrease. We'll see how this works with a real example shortly.

8.3.5. Methods of logistic regression ②

As with multiple regression (section 7.5.3), there are several different methods that can be used in logistic regression.

8.3.5.1. The forced entry method ②

The default method of conducting the regression is 'enter'. This is the same as forced entry in multiple regression in that all of the predictors are placed into the regression model in one block, and parameter estimates are calculated for each block.

8.3.5.2. Stepwise methods ②

If you are undeterred by the criticisms of stepwise methods in the previous chapter, then you can select either a forward or a backward stepwise method. When the forward method is employed the computer begins with a model that includes only a constant and then adds single predictors to the model based on a specific criterion. This criterion is the value of the *score* statistic: the variable with the most significant score statistic is added to the model. The computer proceeds until none of the remaining predictors have a significant score statistic (the cut-off point for significance being .05). At each step, the computer also examines the variables in the model to see whether any should be removed. It does this in one of three ways. The first way is to use the likelihood ratio statistic described in section 18.3.3 (the *Forward:LR* method) in which case the current model is compared to the model when that predictor is removed. If the removal of that predictor makes a significant difference to how well the model fits the observed data, then the computer retains that predictor (because the model is better if the predictor is included). If, however, the removal of the predictor makes little difference to the model then the computer rejects that predictor. Rather than using the likelihood ratio statistic, which estimates how well the model fits the observed data, the computer could use the conditional statistic as a removal criterion (*Forward:Conditional*). This statistic is an arithmetically less intense version of the likelihood ratio statistic and so there is little to recommend it over the likelihood ratio method. The final criterion is the Wald statistic, in which case any predictors in the model that have significance values of the Wald statistic (above the default removal criterion of .1) will be removed. Of these methods the likelihood ratio method is the best removal criterion because the Wald statistic can, at times, be unreliable (see section 8.3.3).

The opposite of the forward method is the backward method. This method uses the same three removal criteria, but instead of starting the model with only a constant, it begins the model with all predictors included. The computer then tests whether any of these predictors can be removed from the model without having a substantial effect on how well the model fits the observed data. The first predictor to be removed will be the one that has the least impact on how the model fits the data.

8.3.5.3. How do I select a method? ②

As with ordinary regression (previous chapter), the method of regression chosen will depend on several things. The main consideration is whether you are testing a theory or merely carrying out exploratory work. As noted earlier, some people believe that stepwise methods have no value for theory testing. However, stepwise methods are defensible when used in situations in which no previous research exists on which to base hypotheses for testing, and in situations where causality is not of interest and you merely wish to find a model to fit your data (Agresti & Finlay, 1986; Menard, 1995). Also, as I mentioned for ordinary regression, if you do decide to use a stepwise method then the backward method is preferable to the forward method. This is because of **suppressor effects**, which occur when a predictor has a significant effect but only when another variable is held constant. Forward selection is more likely than backward elimination to exclude predictors involved in suppressor effects. As such, the forward method runs a higher risk of making a Type II error. In terms of the test statistic used in stepwise methods, the Wald statistic, as we have seen, has a tendency to be inaccurate in certain circumstances and so the likelihood ratio method is best.

Which method should I use?

8.4. Assumptions and things that can go wrong ④

8.4.1. Assumptions ②

Logistic regression shares some of the assumptions of normal regression:

1 **Linearity**: In ordinary regression we assumed that the outcome had linear relationships with the predictors. In logistic regression the outcome is categorical and so this assumption is violated. As I explained before, this is why we use the log (or *logit*) of the data. The assumption of linearity in logistic regression, therefore, assumes that there is a linear relationship between any continuous predictors and the logit of the outcome variable. This assumption can be tested by looking at whether the interaction term between the predictor and its log transformation is significant (Hosmer & Lemeshow, 1989). We will go through an example in section 8.8.1.

2 **Independence of errors**: This assumption is the same as for ordinary regression (see section 7.6.2.1). Basically it means that cases of data should not be related; for example, you cannot measure the same people at different points in time. Violating this assumption produces overdispersion (see section 8.4.4).

3 **Multicollinearity**: Although not really an assumption as such, multicollinearity is a problem as it was for ordinary regression (see section 7.6.2.1). In essence, predictors should not be too highly correlated. As with ordinary regression, this assumption can be checked with tolerance and VIF statistics, the eigenvalues of the scaled, uncentred cross-products matrix, the condition indexes and the variance proportions. We go through an example in section 8.8.1.

Logistic regression also has some unique problems of its own (not assumptions, but things that can go wrong). SPSS solves logistic regression problems by an iterative procedure (SPSS Tip 8.1). Sometimes, instead of pouncing on the correct solution quickly, you'll notice nothing happening: SPSS begins to move infinitely slowly, or appears to have just got fed up with you asking it to do stuff and has gone on strike. If it can't find a correct solution, then sometimes it actually does give up, quietly offering you (without any apology) a result which is completely incorrect. Usually this is revealed by implausibly large standard errors. Two situations can provoke this situation, both of which are related to the ratio of cases to variables: incomplete information and complete separation.

8.4.2. Incomplete information from the predictors ④

Imagine you're trying to predict lung cancer from smoking (a foul habit believed to increase the risk of cancer) and whether or not you eat tomatoes (which are believed to reduce the risk of cancer). You collect data from people who do and don't smoke, and from people who do and don't eat tomatoes; however, this isn't sufficient unless you collect data from all combinations of smoking and tomato eating. Imagine you ended up with the following data:

Do you smoke?	Do you eat tomatoes?	Do you have cancer?
Yes	No	Yes
Yes	Yes	Yes
No	No	Yes
No	Yes	??????

SPSS TIP 8.1 **Error messages about 'failure to coverage'** ③

Many statistical procedures use an *iterative process*, which means that SPSS attempts to estimate the parameters of the model by finding successive approximations of those parameters. Essentially, it starts by estimating the parameters with a 'best guess'. It then attempts to approximate them more accurately (known as an *iteration*). It then tries again, and then again, and so on through many iterations. It stops either when the approximations of parameters converge (i.e. at each new attempt the 'approximations' of parameters are the same or very similar to the previous attempt), or when it reaches the maximum number of attempts (iterations).

Sometimes you will get an error message in the output that says something like '*Maximum number of iterations were exceeded, and the log-likelihood value and/or the parameter estimates cannot converge*'. What this means is that SPSS has attempted to estimate the parameters the maximum number of times (as specified in the options) but they are not converging (i.e. at each iteration SPSS is getting quite different estimates). This certainly means that you should ignore any output that SPSS has produced, and it might mean that your data are beyond help. You can try increasing the number of iterations that SPSS attempts, or make the criteria that SPSS uses to assess 'convergence' less strict.

Observing only the first three possibilities does not prepare you for the outcome of the fourth. You have no way of knowing whether this last person will have cancer or not based on the other data you've collected. Therefore, SPSS will have problems unless you've collected data from all combinations of your variables. This should be checked before you run the analysis using a crosstabulation table, and I describe how to do this in Chapter 18. While you're checking these tables, you should also look at the expected frequencies in each cell of the table to make sure that they are greater than 1 and no more than 20% are less than 5 (see section 18.4). This is because the goodness-of-fit tests in logistic regression make this assumption.

This point applies not only to categorical variables, but also to continuous ones. Suppose that you wanted to investigate factors related to human happiness. These might include age, gender, sexual orientation, religious beliefs, levels of anxiety and even whether a person is right-handed. You interview 1000 people, record their characteristics, and whether they are happy ('yes' or 'no'). Although a sample of 1000 seems quite large, is it likely to include an 80 year old, highly anxious, Buddhist left-handed lesbian? If you found one such person and she was happy, should you conclude that everyone else in the same category is happy? It would, obviously, be better to have several more people in this category to confirm that this combination of characteristics causes happiness. One solution is to collect more data.

As a general point, whenever samples are broken down into categories and one or more combinations are empty it creates problems. These will probably be signalled by coefficients that have unreasonably large standard errors. Conscientious researchers produce and check multiway crosstabulations of all categorical independent variables. Lazy but cautious ones don't bother with crosstabulations, but look carefully at the standard errors. Those who don't bother with either should expect trouble.

8.4.3. Complete separation ④

A second situation in which logistic regression collapses might surprise you: it's when the outcome variable can be perfectly predicted by one variable or a combination of variables! This is known as **complete separation**.

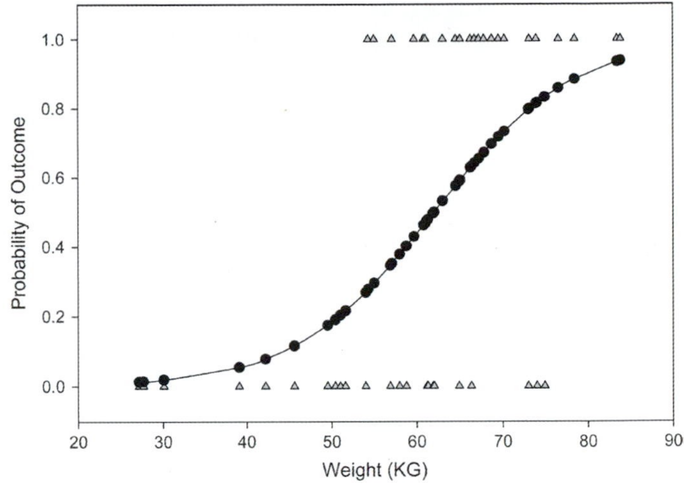

Let's look at an example: imagine you placed a pressure pad under your door mat and connected it to your security system so that you could detect burglars when they creep in at night. However, because your teenage children (which you would have if you're old enough and rich enough to have security systems and pressure pads) and their friends are often coming home in the middle of the night, when they tread on the pad you want it to work out the probability that the person is a burglar and not your teenager. Therefore, you could measure the weight of some burglars and some teenagers and use logistic regression to predict the outcome (teenager or burglar) from the weight. The graph would show a line of triangles at zero (the data points for all of the teenagers you weighed) and a line of triangles at 1 (the data points for burglars you weighed). Note that these lines of triangles overlap (some teenagers are as heavy as burglars). We've seen that in logistic regression, SPSS tries to predict the probability of the outcome given a value of the predictor. In this case, at low weights the fitted probability follows the bottom line of the plot, and at high weights it follows the top line. At intermediate values it tries to follow the probability as it changes.

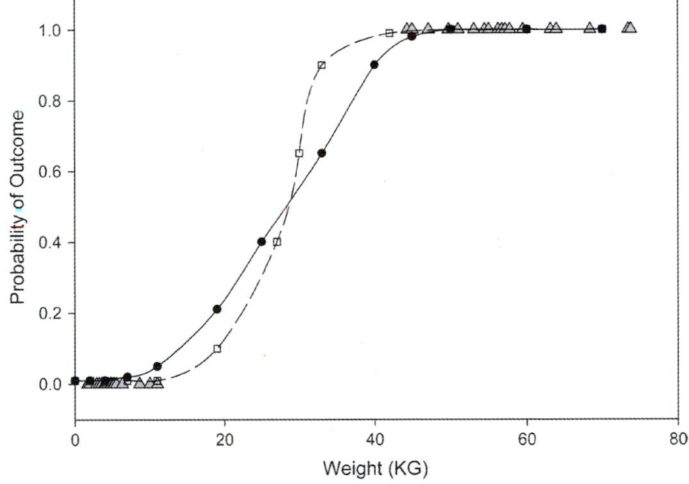

Imagine that we had the same pressure pad, but our teenage children had left home to go to university. We're now interested in distinguishing burglars from our pet cat based on weight. Again, we can weigh some cats and weigh some burglars. This time the graph still has a row of triangles at zero (the cats we weighed) and a row at 1 (the burglars) but this time the rows of triangles do not overlap: there is no burglar who weighs the same as a cat – obviously there were no cat burglars in the sample (groan now at that sorry excuse for a joke!). This is known as perfect separation: the outcome (cats and burglars) can be perfectly predicted from weight (anything less than 15 kg is a cat, anything more than 40 kg is a burglar). If we try to calculate the probabilities of the outcome given a certain weight then we run into trouble. When the weight is low, the probability is 0, and when the weight is high, the probability is 1, but what happens in between? We have no data in between 15 and 40 kg on which to base these probabilities. The figure shows two possible probability curves that we could fit to these data: one much steeper than the other. Either one of these curves is valid based on the data we have available. The lack of data means that SPSS will be uncertain about how steep it should make the intermediate slope and it will try to bring the centre as close to vertical as possible, but its estimates veer unsteadily towards infinity (hence large standard errors).

This problem often arises when too many variables are fitted to too few cases. Often the only satisfactory solution is to collect more data, but sometimes a neat answer is found by adopting a simpler model.

8.4.4. Overdispersion ④

I'm not a statistician, and most of what I've read on overdispersion doesn't make an awful lot of sense to me. From what I can gather, it is when the observed variance is bigger than expected from the logistic regression model. This can happen for two reasons. The first is correlated observations (i.e. when the assumption of independence is broken) and the second is due to variability in success probabilities. For example, imagine our outcome was whether a puppy in a litter survived or died. Genetic factors mean that within a given litter the chances of success (living) depend on the litter from which the puppy came. As such success probabilities vary across litters (Halekoh & Højsgaard, 2007), this example of dead puppies is particularly good – not because I'm a cat lover, but because it shows how variability in success probabilities can create correlation between observations (the survival rates of puppies from the same litter are not independent).

Overdispersion creates a problem because it tends to limit standard errors and result in narrower confidence intervals for test statistics of predictors in the logistic regression model. Given that the test statistics are computed by dividing by the standard error, if the standard error is too small then the test statistic will be bigger than it should be, and more likely to be deemed significant. Similarly, narrow confidence intervals will give us overconfidence in the effect of our predictors on the outcome. In short, there is more chance of Type I errors. The parameters themselves (i.e. the *b*-values) are unaffected.

SPSS produces a chi-square goodness-of-fit statistic, and overdispersion is present if the ratio of this statistic to its degrees of freedom is greater than 1 (this ratio is called the *dispersion parameter,* ϕ). Overdispersion is likely to be problematic if the dispersion parameter approaches or is greater than 2. (Incidentally, underdispersion is shown by values less than 1, but this problem is much less common in practice.) There is also the *deviance* goodness-of-fit statistic, and the dispersion parameter can be based on this statistic instead (again by dividing by the degrees of freedom). When the chi-square and deviance statistics are very discrepant, then overdispersion is likely.

The effects of overdispersion can be reduced by using the dispersion parameter to rescale the standard errors and confidence intervals. For example, the standard errors are multiplied by $\sqrt{\phi}$ to make them bigger (as a function of how big the overdispersion is). You can base these corrections on the deviance statistic too, and whether you rescale using this statistic or the Pearson chi-square statistic depends on which one is bigger. The bigger statistic will have the bigger dispersion parameter (because their degrees of freedom are the same), and will make the bigger correction; therefore, correct by the bigger of the two.

CRAMMING SAM'S TIPS **Issues in logistic regression**

- In logistic regression, like ordinary regression, we assume linearity, no multicollinearity and independence of errors.
- The linearity assumption is that each predictor has a linear relationship with the log of the outcome variable.
- If we created a table that combined all possible values of all variables then we should ideally have some data in every cell of this table. If you don't then watch out for big standard errors.
- If the outcome variable can be predicted perfectly from one predictor variable (or a combination of predictor variables) then we have *complete separation*. This problem creates large standard errors too.
- *Overdispersion* is where the variance is larger than expected from the model. This can be caused by violating the assumption of independence. This problem makes the standard errors too small!

8.5. Binary logistic regression: an example that will make you feel eel ②

It's amazing what you find in academic journals sometimes. It's a bit of a hobby of mine trying to unearth bizarre academic papers (really, if you find any email them to me). I believe that science should be fun, and so I like finding research that makes me laugh. A research paper by Lo and colleagues is the one that (so far) has made me laugh the most (Lo, Wong, Leung, Law, & Yip, 2004). Lo and colleagues report the case of a 50 year old man who reported to the Accident and Emergency Department (ED for the Americans) with abdominal pain. A physical examination revealed peritonitis so they took an X-ray of the man's abdomen. Although it somehow slipped the patient's mind to mention this to the receptionist upon arrival at the hospital, the X-ray revealed the shadow of an eel. The authors don't directly quote the man's response to this news, but I like to imagine it was something to the effect of 'Oh, that! Erm, yes, well I didn't think it was terribly relevant to my abdominal pain so I didn't mention it, but I did insert an eel into my anus. Do you think that's the problem?' Whatever he *did* say, the authors report that he admitted inserting an eel up his anus to 'relieve constipation'.

Can an eel cure constipation?

I can have a lively imagination at times, and when I read this article I couldn't help thinking about the poor eel. There it was, minding it's own business swimming about in a river (or fish tank possibly), thinking to itself 'Well, today seems like a nice day, there are no eel-eating sharks about, the sun is out, the water is nice, what could possibly go wrong?' The next thing it knows, it's being shoved up the anus of a man from Hong Kong. 'Well, I didn't see that coming,' thinks the eel. Putting myself in the mindset of an eel for a moment, he has found himself in a tight dark tunnel, there's no light, there's a distinct lack of water compared to his usual habitat, and he's probably fearing for his life. His day has gone *very* wrong. How can he escape this horrible fate? Well, doing what any self-respecting eel would do, he notices that his prison cell is fairly soft and decides 'bugger this,[1] I'll *eat* my way out of here'. Unfortunately he didn't make it, but he went out with a fight (there's a fairly unpleasant photograph in the article of the eel biting the splenic flexure). The authors conclude that 'Insertion of a live animal into the rectum causing rectal perforation has never been reported. This may be related to a bizarre healthcare belief, inadvertent sexual behaviour, or criminal assault. However, the true reason may never be known.' Quite.

OK, so this is a really grim tale.[2] It's not really very funny for the man or the eel, but it was so unbelievably bizarre that I did laugh. Of course my instant reaction was that sticking an eel up your anus to 'relieve constipation' is the poorest excuse for bizarre sexual behaviour I have ever heard. But upon reflection I wondered if I was being harsh on the man – maybe an eel up the anus really can cure constipation. If we wanted to test this, we could collect some data. Our outcome might be 'constipated' vs. 'not constipated', which is a dichotomous variable that we're trying to predict. One predictor variable would be

[1] Literally.

[2] As it happens, it isn't an isolated grim tale. Through this article I found myself hurtling down a road of morbid curiosity that was best left untravelled. Although the eel was my favourite example, I could have chosen from a very large stone (Sachdev, 1967), a test tube (Hughes, Marice, & Gathright, 1976), a baseball (McDonald & Rosenthal, 1977), an aerosol deodorant can, hose pipe, iron bar, broomstick, penknife, marijuana, bank notes, blue plastic tumbler, vibrator and primus stove (Clarke, Buccimazza, Anderson, & Thomson, 2005), or (a close second place to the eel) a toy pirate ship, with or without pirates I'm not sure (Bemelman & Hammacher, 2005). So, although I encourage you to send me bizarre research, if it involves objects in the rectum then probably don't, unless someone has managed to put Buckingham Palace up there.

intervention (eel up the anus) vs. waiting list (no treatment). We might also want to factor how many days the patient had been constipated before treatment. This scenario is perfect for logistic regression (but not for eels). The data are in **Eel.sav**.

I'm quite aware that many statistics lecturers would rather not be discussing eel-created rectal perforations with their students, so I have named the variables in the file more generally:

- **Outcome (dependent variable): Cured** (cured or not cured).
- **Predictor (independent variable): Intervention** (intervention or no treatment).
- **Predictor (independent variable): Duration** (the number of days before treatment that the patient had the problem).

In doing so, your lecturer can adapt the example to something more palatable if they wish to, but you will secretly know that it's all about having eels up your bum!

8.5.1. The main analysis ②

To carry out logistic regression, the data must be entered as for normal regression: they are arranged in the data editor in three columns (one representing each variable). Looking at the data editor you should notice that both of the categorical variables have been entered as coding variables (see section 3.4.2.3); that is, numbers have been specified to represent categories. For ease of interpretation, the outcome variable should be coded 1 (event occurred) and 0 (event did not occur); in this case, 1 represents being cured and 0 represents not being cured. For the intervention a similar coding has been used (1 = intervention, 0 = no treatment). To open the main *Logistic Regression* dialog box (Figure 8.3) select Analyze Regression ▸ Binary Logistic... .

The main dialog box is very similar to the standard *regression* dialog box. There is a space to place a dependent variable (or outcome variable). In this example, the outcome was whether or not the patient was cured, so we can simply click on **Cured** and drag it to the *Dependent* box or click on ➜. There is also a box for specifying the covariates (the predictor variables). It is possible to specify the **main effect** of a predictor variable, which

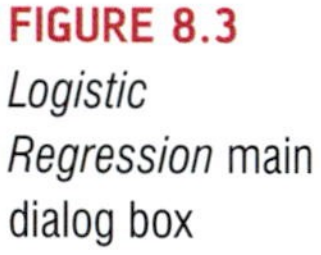

FIGURE 8.3 *Logistic Regression* main dialog box

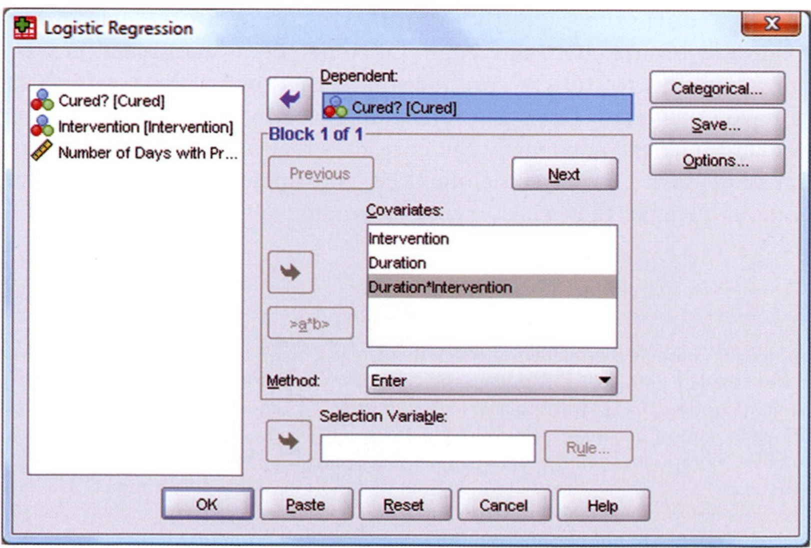

is simply the effect (on an outcome variable) of a variable *on its own*. You can also specify an **interaction effect**, which is the effect (on an outcome variable) of two or more variables *in combination*. We will discover more about main effects and interactions in Chapter 12. To specify a main effect, simply select one predictor (e.g. **Duration**) and then drag it to the *Covariates* box or click on . To input an interaction, click on more than one variable on the left-hand side of the dialog box (i.e. click on several variables while holding down the *Ctrl* key) and then click on >a*b> to move them to the *Covariates* box. In this example there are only two predictors and therefore there is only one possible interaction (the **Duration** × **Intervention** interaction), but if you have three predictors then you can select several interactions using two predictors, and an interaction involving all three. In the figure I have selected the two main effects of **Duration**, **Intervention** and the **Duration** × **Intervention** interaction. Select these variables too.

8.5.2. Method of regression ②

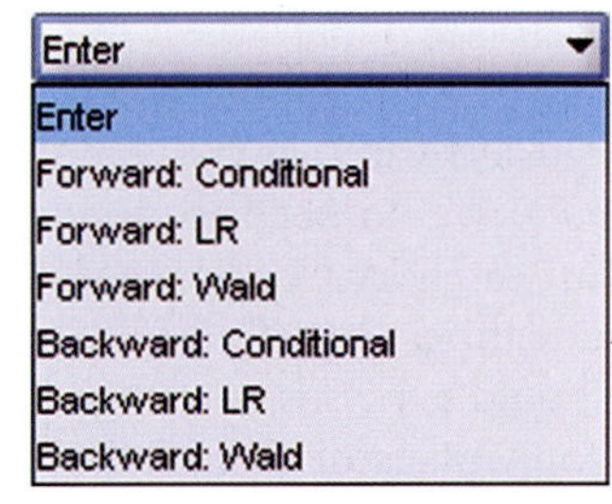

As with multiple regression, there are different ways of doing logistic regression (see section 8.3.5). You can select a particular method of regression by clicking on Enter ▾ and then clicking on a method in the resulting drop-down menu. For this analysis select a *Forward:LR* method of regression. Having spent a vast amount of time telling you never to do stepwise analyses, it's probably a bit strange to hear me suggest doing forward regression here. Well, for one thing this study is the first in the field and so we have no past research to tell us which variables to expect to be reliable predictors. Second, I didn't show you a stepwise example in the regression chapter and so this will be a useful way to demonstrate how a stepwise procedure operates!

8.5.3. Categorical predictors ②

It is necessary to 'tell' SPSS which predictor variables, if any, are categorical by clicking on Categorical... in the main *Logistic Regression* dialog box to activate the dialog box in Figure 8.4. In this dialog box, the covariates are listed on the left-hand side, and there is a space on the right-hand side in which categorical covariates can be placed. Highlight any categorical variables you have (in this example we have only one, so click on **Intervention**) and drag them to the *Categorical Covariates* box or click on .

There are many ways in which you can treat categorical predictors. In section 7.11 we saw that categorical predictors could be incorporated into regression by recoding them using zeros and ones (known as dummy coding). Actually, there are different ways that you can code categorical variables depending on what you want to compare, and SPSS has several 'standard' ways built into it that you can select. By default SPSS uses *Indicator* coding, which is the standard dummy variable coding that I explained in section 7.11 (and you can choose to have either the first or last category as your baseline). To change to a different kind of contrast click on Indicator ▾ to access a drop-down list of possible contrasts: it is possible to select simple contrasts; difference contrasts, Helmert contrasts, repeated contrasts, polynomial contrasts and deviation contrasts. These techniques will be discussed in detail in Chapter 10 and I'll explain what these contrasts actually do

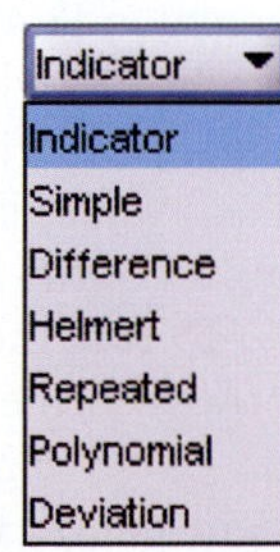

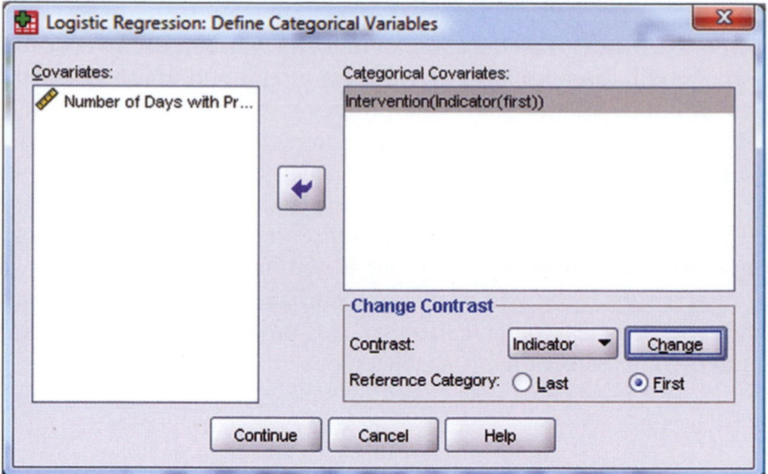

FIGURE 8.4 Defining categorical variables in logistic regression

then (see Table 10.6). However, you won't come across indicator contrasts in that chapter and so we'll use them here. To reiterate, when an indicator contrast is used, levels of the categorical variable are recoded using standard dummy variable coding (see sections 7.11 and 9.7). We do need to decide whether to use the first category as our baseline (First) or the last category (Last). In this case it doesn't make much difference because we have only two categories, but if you had a categorical predictor with more than two categories, then you should either use the highest number to code your control category and then select Last for your indicator contrast, or use the lowest number to code your control category and then change the indicator contrast to compare against First. In our data, I coded 'cured' as 1 and 'not cured' (our control category) as 0; therefore, select the contrast, then click on First and then Change so that the completed dialog box looks like Figure 8.4 .

8.5.4. Obtaining residuals ②

As with linear regression, it is possible to save a set of residuals (see section 7.6.1.1) as new variables in the data editor. These residual variables can then be examined to see how well the model fits the observed data. To save residuals click on Save in the main *Logistic Regression* dialog box (Figure 8.3). SPSS saves each of the selected variables into the data editor but they can be listed in the output viewer by using the *Case Summaries* command (see section 7.8.6) and selecting the residual variables of interest. The residuals dialog box in Figure 8.5 gives us several options and most of these are the same as those in multiple regression (refer to section 7.7.5). Two residuals that are unique to logistic regression are the *predicted probabilities* and the *predicted group memberships*. The predicted probabilities are the probabilities of *Y* occurring given the values of each predictor for a given participant. As such, they are derived from equation (8.4) for a given case. The predicted group membership is self-explanatory in that it predicts to which of the two outcome categories a participant is most likely to belong based on the model. The group memberships are based on the predicted probabilities and I will explain these values in more detail when we consider how to interpret the residuals. It is worth selecting all of the available options, or as a bare minimum select the same options as in Figure 8.5.

To reiterate a point from the previous chapter, running a regression without checking how well the model fits the data is like buying a new pair of trousers without trying them on – they might look fine on the hanger but get them home and you find you're Johnny-tight-pants. The trousers do their job (they cover your legs and keep you warm) but they

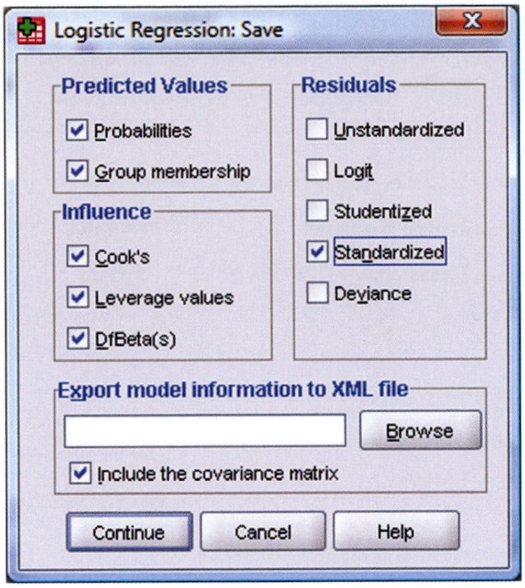

FIGURE 8.5 Dialog box for obtaining residuals for logistic regression

have no real-life value (because they cut off the blood circulation to your legs, which then have to be amputated). Likewise, regression does its job regardless of the data – it will create a model – but the real-life value of the model may be limited (see section 7.6).

8.5.5. Further options ②

There is a final dialog box that offers further options. This box is shown in Figure 8.6 and is accessed by clicking on Options... in the main *Logistic Regression* dialog box. For the most part, the default settings in this dialog box are fine. I mentioned in section 8.5.2 that when a stepwise method is used there are default criteria for selecting and removing predictors from the model. These default settings are displayed in the options dialog box under *Probability for Stepwise*. The probability thresholds can be changed, but there is really no need unless you have a good reason for wanting harsher criteria for variable selection. Another default is to arrive at a model after a maximum of 20 iterations (SPSS Tip 8.1). Unless you have a very complex model, 20 iterations will be more than adequate. We saw in Chapter 7 that regression equations contain a constant that represents the *Y* intercept (i.e. the value of *Y* when the value of the predictors is 0). By default SPSS includes this constant in the model, but it is possible to run the analysis without this constant and this has the effect of making the model pass through the origin (i.e. *Y* is 0 when *X* is 0). Given that we are usually interested in producing a model that best fits the data we have collected, there is little point in running the analysis without the constant included.

A classification plot is a histogram of the actual and predicted values of the outcome variable. This plot is useful for assessing the fit of the model to the observed data. It is also possible to do a *Casewise listing of residuals* either for any cases for which the standardized residual is greater than 2 standard deviations (this value can be changed but the default is sensible), or for all cases. I recommend a more thorough examination of residuals but this option can be useful for a quick inspection. You can ask SPSS to display a confidence interval (see section 2.5.2) for the odds ratio (see section 8.3.4) *Exp(B)*, and by default a 95% confidence interval is used, which is appropriate (if it says anything else then change it to 95%) and a useful statistic to have. More important, you can request the *Hosmer-Lemeshow goodness-of-fit* statistic, which can be used to assess how well the chosen model fits the data.

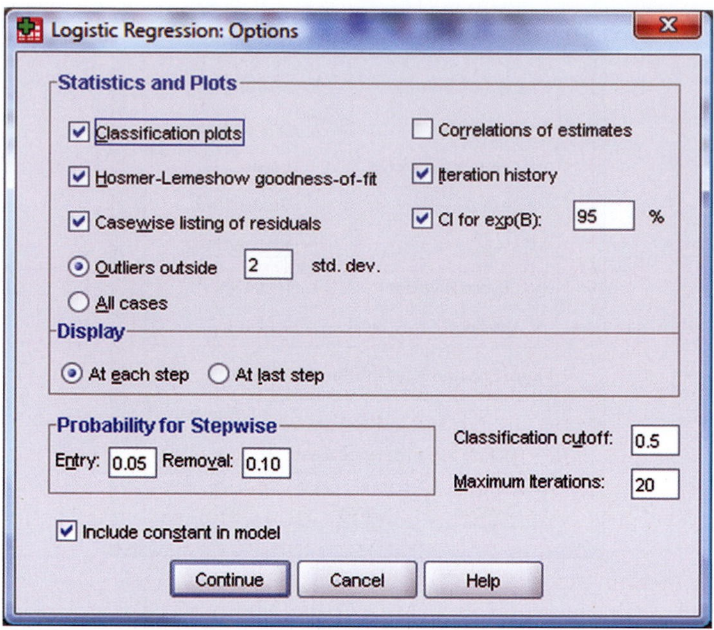

FIGURE 8.6 Dialog box for logistic regression options

The remaining options are fairly unimportant: you can choose to display all statistics and graphs at each stage of an analysis (the default), or only after the final model has been fitted. Finally, you can display a correlation matrix of parameter estimates for the terms in the model, and you can display coefficients and log-likelihood values at each iteration of the parameter estimation process – the practical function of doing this is lost on most of us mere mortals!

8.6. Interpreting logistic regression ②

When you have selected all of the options that I've just described, click on OK and watch the output spew out in the viewer window.

8.6.1. The initial model ②

SPSS Output 8.1 tells us two things. First it tells us how we coded our outcome variable, and this table merely reminds us that 0 = not cured, and 1 = cured.[3] The second is that it tell us how SPSS has coded the categorical predictors. The parameter codings are also given for the categorical predictor variable (**Intervention**). Indicator coding was chosen with two categories, and so the coding is the same as the values in the data editor (0 = no treatment, 1 = treatment). If *deviation* coding had been chosen then the coding would have been −1 (Treatment) and 1 (No Treatment). With a *simple* contrast the codings would have been −0.5 (**Intervention** = No Treatment) and 0.5 (**Intervention** = Treatment) if First was selected as the reference category or vice versa if Last was selected as the reference category. The parameter codings are important for calculating the probability of the outcome variable ($P(Y)$), but we will come to that later.

[3] These values are the same as the data editor so this table might seem pointless; however, had we used codes other than 0 and 1 (e.g. 1 = not cured, 2 = cured) then SPSS changes these to zeros and ones and this table informs you of which category is represented by 0 and which by 1. This is important when it comes to interpretation.

SPSS OUTPUT 8.1

Dependent Variable Encoding

Original Value	Internal Value
Not Cured	0
Cured	1

Categorical Variables Codings

		Frequency	Parameter coding (1)
Intervention	No Treatment	56	.000
	Intervention	57	1.000

For this first analysis we requested a forward stepwise method[4] and so the initial model is derived using only the constant in the regression equation. SPSS Output 8.2 tells us about the model when only the constant is included (i.e. all predictor variables are omitted). The table labelled **Iteration History** tells us that the log-likelihood of this baseline model (see section 8.3.1) is 154.08. This represents the fit of the most basic model to the data. When including only the constant, the computer bases the model on assigning every participant to a single category of the outcome variable. In this example, SPSS can decide either to predict that the patient was cured, or that every patient was not cured. It could make this decision arbitrarily, but because it is crucial to try to maximize how well the model predicts the observed data, SPSS will predict that every patient belongs to the category in which most observed cases fell. In this example there were 65 patients who were cured, and only 48 who were not cured. Therefore, if SPSS predicts that every patient was cured then this prediction will be correct 65 times out of 113 (i.e. 58% approx.). However, if SPSS predicted that every patient was not cured, then this prediction would be correct only 48 times out of 113 (42% approx.). As such, of the two available options it is better to predict that all patients were cured because this results in a greater number of correct predictions. The output shows a contingency table for the model in this basic state. You can see that SPSS has predicted that all patients are cured, which results in 0% accuracy for the patients who were not cured, and 100% accuracy for those observed to be cured. Overall, the model correctly classifies 57.5% of patients.

SPSS OUTPUT 8.2

Iteration History[a,b,c]

Iteration		-2 Log likelihood	Coefficients: Constant
Step 0	1	154.084	.301
	2	154.084	.303
	3	154.084	.303

a. Constant is included in the model.

b. Initial -2 Log Likelihood: 154.084

c. Estimation terminated at iteration number 3 because parameter estimates changed by less than .001.

Classification Table[a,b]

Observed			Predicted: Cured? Not Cured	Cured	Percentage Correct
Step 0	Cured?	Not Cured	0	48	.0
		Cured	0	65	100.0
	Overall Percentage				57.5

a. Constant is included in the model.

b. The cut value is .500

[4] Actually, this is a *really* bad idea when you have an interaction term because to look at an interaction you need to include the main effects of the variables in the interaction term. I chose this method *only* to illustrate how stepwise methods work.

SPSS Output 8.3 summarizes the model (**Variables in the Equation**), and at this stage this entails quoting the value of the constant (b_0), which is equal to 0.30. The table labelled **Variables not in the Equation** tells us that the residual chi-square statistic is 9.83 which is significant at $p < .05$ (it labels this statistic *Overall Statistics*). This statistic tells us that the coefficients for the variables not in the model are significantly different from zero – in other words, that the addition of one or more of these variables to the model will significantly affect its predictive power. If the probability for the residual chi-square had been greater than .05 it would have meant that forcing all of the variables excluded from the model into the model would not have made a significant contribution to its predictive power.

The remainder of this table lists each of the predictors in turn with a value of **Roa's efficient score statistic** for each one (column labelled *Score*). In large samples when the null hypothesis is true, the score statistic is identical to the Wald statistic and the likelihood ratio statistic. It is used at this stage of the analysis because it is computationally less intensive than the Wald statistic and so can still be calculated in situations when the Wald statistic would prove prohibitive. Like any test statistic, Roa's score statistic has a specific distribution from which statistical significance can be obtained. In this example, **Intervention** and the **Intervention × Duration** interaction both have significant score statistics at $p < .01$ and could potentially make a contribution to the model, but **Duration** alone does not look likely to be a good predictor because its score statistic is non-significant $p > .05$. As mentioned in section 8.5.2, the stepwise calculations are relative and so the variable that will be selected for inclusion is the one with the highest value for the score statistic that has a significance below .05. In this example, that variable will be **Intervention** because its score statistic (9.77) is the biggest.

SPSS OUTPUT 8.3

Variables in the Equation

		B	S.E.	Wald	df	Sig.	Exp(B)
Step 0	Constant	.303	.190	2.538	1	.111	1.354

Variables not in the Equation

			Score	df	Sig.
Step 0	Variables	Intervention(1)	9.771	1	.002
		Duration	.609	1	.435
		Duration by Intervention (1)	9.052	1	.003
	Overall Statistics		9.827	3	.020

8.6.2. Step 1: intervention ③

As I predicted in the previous section, whether or not an intervention was given to the patient (**Intervention**) is added to the model in the first step. As such, a patient is now classified as being cured or not based on whether they had an intervention or not (waiting list). This can be explained easily if we look at the crosstabulation for the variables **Intervention** and **Cured**.[5] The model will use whether a patient had an intervention or not to predict whether they were cured or not by applying the crosstabulation table shown in Table 8.1.

The model predicts that all of the patients who had an intervention were cured. There were 57 patients who had an intervention, so the model predicts that these 57 patients were cured; it is correct for 41 of these patients, but misclassifies 16 people as 'cured' who were not cured – see Table 8.1. In addition, this new model predicts that all of the 56 patients who received no treatment were not cured; for these patients the model is correct 32 times but misclassifies as 'not cured' 24 people who were.

[5] The dialog box to produce this table can be obtained by selecting Analyze Descriptive Statistics ▸ 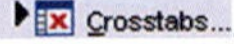 Crosstabs....

TABLE 8.1 Crosstabulation of intervention with outcome status (cured or not)

		Intervention or Not (Intervention)	
		No Treatment	Intervention
Cured? (Cured)	Not Cured	32	16
	Cured	24	41
	Total	56	57

SPSS OUTPUT 8.4

Omnibus Tests of Model Coefficients

		Chi-square	df	Sig.
Step 1	Step	9.926	1	.002
	Block	9.926	1	.002
	Model	9.926	1	.002

Model Summary

Step	-2 Log likelihood	Cox & Snell R Square	Nagelkerke R Square
1	144.158[a]	.084	.113

a. Estimation terminated at iteration number 3 because parameter estimates changed by less than .001.

Classification Table[a]

	Observed		Predicted		
			Cured?		Percentage Correct
			Not Cured	Cured	
Step 1	Cured?	Not Cured	32	16	66.7
		Cured	24	41	63.1
	Overall Percentage				64.6

a. The cut value is .500

SPSS Output 8.4 shows summary statistics about the new model (which we've already seen contains **Intervention**). The overall fit of the new model is assessed using the log-likelihood statistic (see section 8.3.1). In SPSS, rather than reporting the log-likelihood itself, the value is multiplied by −2 (and sometimes referred to as −2LL): this multiplication is done because −2LL has an approximately chi-square distribution and so it makes it possible to compare values against those that we might expect to get by chance alone. Remember that large values of the log-likelihood statistic indicate poorly fitting statistical models.

At this stage of the analysis the value of −2LL should be less than the value when only the constant was included in the model (because lower values of −2LL indicate that the model is predicting the outcome variable more accurately). When only the constant was included, −2LL = 154.08, but now **Intervention** has been included this value has been reduced to 144.16. This reduction tells us that the model is better at predicting whether someone was cured than it was before **Intervention** was added. The question of how much better the model predicts the outcome variable can be assessed using the *model chi-square statistic*, which measures the difference between the model as it currently stands and the model when only the constant was included. We saw in section 8.3.1 that we could assess the significance of the change in a model by taking the log-likelihood of the new model and subtracting the log-likelihood of the baseline model from it. The value of the model chi-square statistic works on this principle and is, therefore, equal to −2LL with **Intervention** included minus the value of −2LL when only the constant was in the model (154.08

– 144.16 = 9.92). This value has a chi-square distribution and so its statistical significance can be calculated easily.[6] In this example, the value is significant at a .05 level and so we can say that overall the model is predicting whether a patient is cured or not significantly better than it was with only the constant included. The model chi-square is an analogue of the *F*-test for the linear regression (see Chapter 7). In an ideal world we would like to see a non-significant overall −2LL (indicating that the amount of unexplained data is minimal) and a highly significant model chi-square statistic (indicating that the model including the predictors is significantly better than without those predictors). However, in reality it is possible for both statistics to be highly significant.

There is a second statistic called the *step* statistic that indicates the improvement in the predictive power of the model since the last stage. At this stage there has been only one step in the analysis and so the value of the improvement statistic is the same as the model chi-square. However, in more complex models in which there are three or four stages, this statistic gives a measure of the improvement of the predictive power of the model since the last step. Its value is equal to −2LL at the current step minus −2LL at the previous step. If the improvement statistic is significant then it indicates that the model now predicts the outcome significantly better than it did at the last step, and in a forward regression this can be taken as an indication of the contribution of a predictor to the predictive power of the model. Similarly, the *block* statistic provides the change in −2LL since the last block (for use in hierarchical or blockwise analyses).

SPSS Output 8.4 also tells us the values of Cox and Snell's and Nagelkerke's R^2, but we will discuss these a little later. There is also a classification table that indicates how well the model predicts group membership; because the model is using **Intervention** to predict the outcome variable, this classification table is the same as Table 8.1. The current model correctly classifies 32 patients who were not cured but misclassifies 16 others (it correctly classifies 66.7% of cases). The model also correctly classifies 41 patients who were cured but misclassifies 24 others (it correctly classifies 63.1% of cases). The overall accuracy of classification is, therefore, the weighted average of these two values (64.6%). So, when only the constant was included, the model correctly classified 57.5% of patients, but now, with the inclusion of **Intervention** as a predictor, this has risen to 64.6%.

SPSS OUTPUT 8.5

Variables in the Equation

		B	S.E.	Wald	df	Sig.	Exp(B)	95.0% C.I.for EXP(B)	
								Lower	Upper
Step 1[a]	Intervention(1)	1.229	.400	9.447	1	.002	3.417	1.561	7.480
	Constant	-.288	.270	1.135	1	.287	.750		

a. Variable(s) entered on step 1: Intervention.

The next part of the output (SPSS Output 8.5) is crucial because it tells us the estimates for the coefficients for the predictors included in the model. This section of the output gives us the coefficients and statistics for the variables that have been included in the model at this point (namely **Intervention** and the constant). The *b*-value is the same as the *b*-value in linear regression: they are the values that we need to replace in equation (8.4) to establish the probability that a case falls into a certain category. We saw in linear regression that the value of *b* represents the change in the outcome resulting from a unit change in the predictor variable. The interpretation of this coefficient in logistic regression is very similar in that it represents the change in the logit of the outcome variable associated with a one-unit change in the predictor variable. The logit of the outcome is simply the natural logarithm of the odds of *Y* occurring.

[6] The degrees of freedom will be the number of parameters in the new model (the number of predictors plus 1, which in this case with one predictor means 2) minus the number of parameters in the baseline model (which is 1, the constant). So, in this case, $df = 2 - 1 = 1$.

The crucial statistic is the Wald statistic[7] which has a chi-square distribution and tells us whether the *b* coefficient for that predictor is significantly different from zero. If the coefficient is significantly different from zero then we can assume that the predictor is making a significant contribution to the prediction of the outcome (*Y*). We came across the Wald statistic in section 8.3.3 and saw that it should be used cautiously because when the regression coefficient (*b*) is large, the standard error tends to become inflated, resulting in the Wald statistic being underestimated (see Menard, 1995). However, for these data it seems to indicate that having the intervention (or not) is a significant predictor of whether the patient is cured (note that the significance of the Wald statistic is less than .05).

In section 8.3.2 we saw that we could calculate an analogue of *R* using equation (8.7). For these data, the Wald statistic and its *df* can be read from SPSS Output 8.5 (9.45 and 1 respectively), and the original –2LL was 154.08. Therefore, *R* can be calculated as:

$$\begin{aligned} R &= \pm\sqrt{\frac{9.45-(2\times 1)}{154.08}} \\ &= .22 \end{aligned} \tag{8.14}$$

In the same section we saw that Hosmer and Lemeshow's measure (R^2_L) is calculated by dividing the model chi-square by the *original* –2LL. In this example the model chi-square after **Intervention** has been entered into the model is 9.93, and the original –2LL (before any variables were entered) was 154.08. So, $R^2_L = 9.93/154.08 = .06$, which is different to the value we would get by squaring the value of *R* given above ($R^2 = .22^2 = 0.05$). Earlier on in SPSS Output 8.4, SPSS gave us two other measures of R^2 that were described in section 8.3.2. The first is Cox and Snell's measure, which SPSS reports as .084, and the second is Nagelkerke's adjusted value, which SPSS reports as .113. As you can see, all of these values differ, but they can be used as effect size measures for the model.

SELF-TEST Using equations (8.9) and (8.10) calculate the values of Cox and Snell's and Nagelkerke's R^2 reported by SPSS. [Hint: These equations use the log-likelihood, whereas SPSS reports –2 × log-likelihood. LL(*new*) is, therefore, 144.16/–2 = –72.08, and LL(*baseline*) = 154.08/–2 = –77.04. The sample size, *n*, is 113.]

The final thing we need to look at is the odds ratio (*Exp(B)* in the SPSS output), which was described in section 8.3.4. To calculate the change in odds that results from a unit change in the predictor for this example, we must first calculate the odds of a patient being cured given that they *didn't* have the intervention. We then calculate the odds of a patient being cured given that they *did* have the intervention. Finally, we calculate the proportionate change in these two odds.

To calculate the first set of odds, we need to use equation (8.12) to calculate the probability of a patient being cured given that they *didn't* have the intervention. The parameter coding at the beginning of the output told us that patients who did not have the intervention were coded with a 0, so we can use this value in place of X. The value of b_1 has been estimated for us as 1.229 (see **Variables in the Equation** in SPSS Output 8.5), and the

[7] As we have seen, this is simply *b* divided by its standard error (1.229/.40 = 3.0725); however, SPSS actually quotes the Wald statistic squared. For these data $3.0725^2 = 9.44$ as reported (within rounding error) in the table.

coefficient for the constant can be taken from the same table and is –0.288. We can calculate the odds as:

$$\begin{aligned}
P(\text{Cured}) &= \frac{1}{1+e^{-(b_0+b_1X_1)}} \\
&= \frac{1}{1+e^{-[-0.288+(1.299\times 0)]}} \\
&= 0.428 \\
P(\text{Not Cured}) &= 1 - P(\text{Cured}) \\
&= 1 - 0.428 \\
&= 0.572 \\
\text{odds} &= \frac{0.428}{0.572} \\
&= 0.748
\end{aligned} \tag{8.15}$$

Now, we calculate the same thing after the predictor variable has changed by one unit. In this case, because the predictor variable is dichotomous, we need to calculate the odds of a patient being cured, given that they have had the intervention. So, the value of the intervention variable, X, is now 1 (rather than 0). The resulting calculations are as follows:

$$\begin{aligned}
P(\text{Cured}) &= \frac{1}{1+e^{-(b_0+b_1X_1)}} \\
&= \frac{1}{1+e^{-[-0.288+(1.299\times 1)]}} \\
&= 0.719 \\
P(\text{Not Cured}) &= 1 - P(\text{Cured}) \\
&= 1 - 0.719 \\
&= 0.281 \\
\text{odds} &= \frac{0.719}{0.281} \\
&= 2.559
\end{aligned} \tag{8.16}$$

We now know the odds before and after a unit change in the predictor variable. It is now a simple matter to calculate the proportionate change in odds by dividing the odds after a unit change in the predictor by the odds before that change:

$$\begin{aligned}
\Delta\text{odds} &= \frac{\text{odds after a unit change in the predictor}}{\text{original odds}} \\
&= \frac{2.56}{0.75} \\
&= 3.41
\end{aligned} \tag{8.17}$$

You should notice that the value of the proportionate change in odds is the same as the value that SPSS reports for *Exp(B)* (allowing for differences in rounding). We can interpret the odds ratio in terms of the change in odds. If the value is greater than 1 then it indicates that as the predictor increases, the odds of the outcome occurring increase. Conversely, a value less than 1 indicates that as the predictor increases, the odds of the outcome occurring decrease. In this example, we can say that the odds of a patient who is treated being cured are 3.41 times higher than those of a patient who is not treated.

In the options (see section 8.5.5), we requested a confidence interval for the odds ratio and it can also be found in the output. The way to interpret this confidence interval is the same as any other confidence interval (section 2.5.2): if we calculated confidence intervals for the value of the odds ratio in 100 different samples, then these intervals would encompass the actual value of the odds ratio in the population (rather than the sample) in 95 of those samples. In this case, we can be fairly confident that the population value of the odds ratio lies between 1.56 and 7.48. However, our sample could be one of the 5% that produces a confidence interval that 'misses' the population value.

The important thing about this confidence interval is that it doesn't cross 1 (both values are greater than 1). This is important because values greater than 1 mean that as the predictor variable increases, so do the odds of (in this case) being cured. Values less than 1 mean the opposite: as the predictor variable increases, the odds of being cured decrease. The fact that both limits of our confidence interval are above 1 gives us confidence that the direction of the relationship that we have observed is true in the population (i.e. it's likely that having an intervention compared to not increases the odds of being cured). If the lower limit had been below 1 then it would tell us that there is a chance that in the population the direction of the relationship is the opposite to what we have observed. This would mean that we could not trust that our intervention increases the odds of being cured.

SPSS OUTPUT 8.6

Model if Term Removed

Variable		Model Log Likelihood	Change in -2 Log Likelihood	df	Sig. of the Change
Step 1	Intervention	-77.042	9.926	1	.002

Variables not in the Equation

			Score	df	Sig.
Step 1	Variables	Duration	.002	1	.964
		Duration by Intervention (1)	.043	1	.835
	Overall Statistics		.063	2	.969

The test statistics for **Intervention** if it were removed from the model are in SPSS Output 8.6. Now, remember that earlier on I said how the regression would place variables into the equation and then test whether they then met a removal criterion. Well, the **Model if Term Removed** part of the output tells us the effects of removal. The important thing to note is the significance value of the log-likelihood ratio (log LR). The log LR for this model is significant ($p < .01$) which tells us that removing **Intervention** from the model would have a significant effect on the predictive ability of the model – in other words, it would be a very bad idea to remove it!

Finally, we are told about the variables currently not in the model. First of all, the residual chi-square (labelled *Overall Statistics* in the output), which is non-significant, tells us that none of the remaining variables have coefficients significantly different from zero. Furthermore, each variable is listed with its score statistic and significance value, and for both variables their coefficients are not significantly different from zero (as can be seen from the significance values of .964 for **Duration** and .835 for the **Duration×Intervention** interaction). Therefore, no further variables will be added to the model.

SPSS Output 8.7 displays the classification plot that we requested in the options dialog box. This plot is a histogram of the predicted probabilities of a patient being cured. If the model perfectly fits the data, then this histogram should show all of the cases for which the event has occurred on the right-hand side, and all the cases for which the event hasn't occurred on the left-hand side. In other words, all of the patients who were cured should appear on the right and all those who were not cured should appear on the left. In this example, the only significant predictor is dichotomous and so there are only two columns of cases on the plot. If the

SPSS OUTPUT 8.7

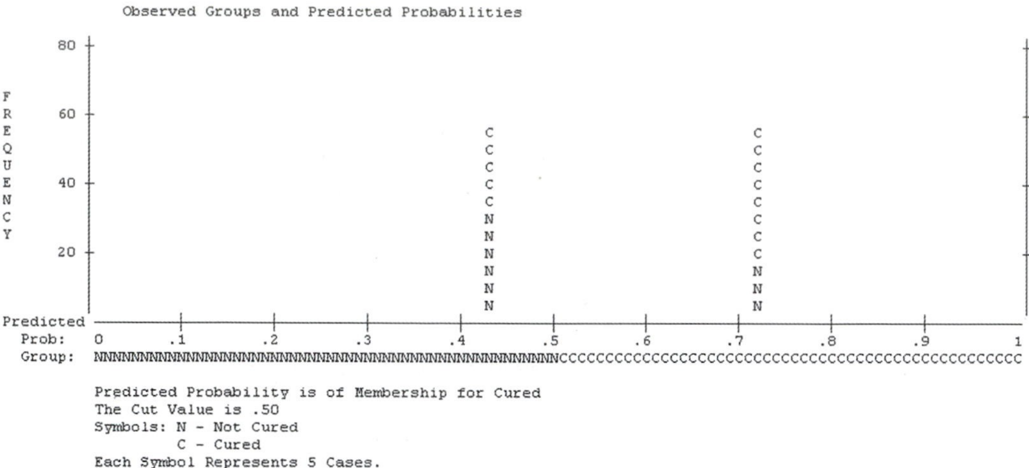

predictor is a continuous variable, the cases are spread out across many columns. As a rule of thumb, the more that the cases cluster at each end of the graph, the better; such a plot would show that when the outcome did actually occur (i.e. the patient was cured) the predicted probability of the event occurring is also high (i.e. close to 1). Likewise, at the other end of the plot it would show that when the event didn't occur (i.e. the patient still had a problem) the predicted probability of the event occurring is also low (i.e. close to 0). This situation represents a model that is correctly predicting the observed outcome data. If, however, there are a lot of points clustered in the centre of the plot then it shows that for many cases the model is predicting a probability of .5 that the event will occur. In other words, for these cases there is little more than a 50:50 chance that the data are correctly predicted by the model – the model could predict these cases just as accurately by tossing a coin! In SPSS Output 8.7 it's clear that cured cases are predicted relatively well by the model (the probabilities are not that close to .5), but for not cured cases the model is less good (the probability of classification is only slightly lower than .5 or chance!). Also, a good model will ensure that few cases are misclassified; in this example there are a few Ns (not cureds) appearing on the cured side, but more worryingly there are quite a few Cs (cureds) appearing on the N side.

CRAMMING SAM'S TIPS

- The overall fit of the final model is shown by the −2 log-likelihood statistic and its associated chi-square statistic. If the significance of the chi-square statistic is less than .05, then the model is a significant fit of the data.
- Check the table labelled **Variables in the equation** to see which variables significantly predict the outcome.
- For each variable in the model, look at the Wald statistic and its significance (which again should be below .05). More important though, use the odds ratio, *Exp(B)*, for interpretation. If the value is greater than 1 then as the predictor increases, the odds of the outcome occurring increase. Conversely, a value less than 1 indicates that as the predictor increases, the odds of the outcome occurring decrease. For the aforementioned interpretation to be reliable the confidence interval of *Exp(B)* should not cross 1!
- Check the table labelled **Variables not in the equation** to see which variables did not significantly predict the outcome.

8.6.3. Listing predicted probabilities ②

It is possible to list the expected probability of the outcome variable occurring based on the final model. In section 8.5.4 we saw that SPSS could save residuals and also predicted probabilities. SPSS saves these predicted probabilities and predicted group memberships as variables in the data editor and names them PRE_1 and PGR_1 respectively. These probabilities can be listed using the Analyze Reports ▸ Case Summaries... dialog box (see section 7.8.6).

SELF-TEST Use the *case summaries* function in SPSS to create a table for the first 15 cases in the file **eel.sav** showing the values of **Cured, Intervention, Duration**, the predicted probability (**PRE_1**) and the predicted group membership (**PGR_1**) for each case.

SPSS Output 8.8 shows a selection of the predicted probabilities (because the only significant predictor was a dichotomous variable, there will be only two different probability values). It is also worth listing the predictor variables as well to clarify from where the predicted probabilities come.

SPSS OUTPUT 8.8

Case Summaries[a]

	Case Number	Cured?	Intervention	Number of Days with Problem before Treatment	Predicted probability	Predicted group
1	1	Not Cured	No Treatment	7	.42857	Not Cured
2	2	Not Cured	No Treatment	7	.42857	Not Cured
3	3	Not Cured	No Treatment	6	.42857	Not Cured
4	4	Cured	No Treatment	8	.42857	Not Cured
5	5	Cured	Intervention	7	.71930	Cured
6	6	Cured	No Treatment	6	.42857	Not Cured
7	7	Not Cured	Intervention	7	.71930	Cured
8	8	Cured	Intervention	7	.71930	Cured
9	9	Cured	No Treatment	8	.42857	Not Cured
10	10	Not Cured	No Treatment	7	.42857	Not Cured
11	11	Cured	Intervention	7	.71930	Cured
12	12	Cured	No Treatment	7	.42857	Not Cured
13	13	Cured	No Treatment	5	.42857	Not Cured
14	14	Not Cured	Intervention	9	.71930	Cured
15	15	Not Cured	No Treatment	6	.42857	Not Cured
Total N		15	15	15	15	15

a. Limited to first 15 cases.

We found from the model that the only significant predictor of being cured was having the intervention. This could have a value of either 1 (have the intervention) or 0 (no intervention). If these two values are placed into equation (8.4) with the respective regression coefficients, then the two probability values are derived. In fact, we calculated these values as part of equation (8.15) and equation (8.16) and you should note that the calculated probabilities – *P*(Cured) in these equations – correspond to the values in PRE_1. These values tells us that when a patient is not treated (**Intervention** = 0, No Treatment), there is a probability of .429 that they will be cured – basically, about 43% of people get better without any treatment. However, if the patient does have the intervention (**Intervention** = 1, yes), there is a probability of .719 that they will get better – about 72% of people treated get better. When you consider that a probability of 0 indicates no chance of getting better, and a probability of 1 indicates that the patient will definitely get better, the values obtained provide strong evidence that having the intervention increases your chances of getting better (although the probability of recovery without the intervention is still not bad).

Assuming we are content that the model is accurate and that the intervention has some substantive significance, then we could conclude that our intervention (which, to remind you, was putting an eel up the anus) is the single best predictor of getting better (not being constipated). Furthermore, the duration of the constipation pre-intervention and its interaction with the intervention did not significantly predict whether a person got better.

SELF-TEST Rerun this analysis using the forced entry method of analysis – how do your conclusions differ?

8.6.4. Interpreting residuals ②

Our conclusions so far are fine in themselves, but to be sure that the model is a good one, it is important to examine the residuals. In section 8.5.4 we saw how to get SPSS to save various residuals in the data editor. We can now interpret them.

We saw in the previous chapter that the main purpose of examining residuals in any regression is to (1) isolate points for which the model fits poorly, and (2) isolate points that exert an undue influence on the model. To assess the former we examine the residuals, especially the Studentized residual, standardized residual and deviance statistics. To assess the latter we use influence statistics such as Cook's distance, DFBeta and leverage statistics. These statistics were explained in detail in section 7.6 and their interpretation in logistic regression is the same; therefore, Table 8.2 summarizes the main statistics that you should look at and what to look for, but for more detail consult the previous chapter.

If you request these residual statistics, SPSS saves them as new columns in the data editor. You can look at the values in the data editor, or produce a table using the Analyze Reports ▸ Case Summaries... dialog box.

The basic residual statistics for this example (Cook's distance, leverage, standardized residuals and DFBeta values) are pretty good: note that all cases have DFBetas less than 1, and leverage statistics (LEV_1) are very close to the calculated expected value of 0.018. There are

OLIVER TWISTED

Please, Sir, can I have some more … diagnostics?

'What about the trees?' protests eco-warrior Oliver. 'These SPSS outputs take up so much room, why don't you put them on the website instead?' It's a valid point so I have produced a table of the diagnostic statistics for this example, but it's in the additional material for this chapter on the companion website

TABLE 8.2 Summary of residual statistics saved by SPSS

Label	Name	Comment
PRE_1	Predicted Value	
PGR_1	Predicted Group	
COO_1	Cook's Distance	Should be less than 1
LEV_1	Leverage	Lies between 0 (no influence) and 1 (complete influence). The expected leverage is $(k+1)/N$, where k is the number of predictors and N is the sample size. In this case it would be 2/113 = .018
SRE_1	Studentized Residual	Only 5% should lie outside ±1.96, and about 1% should lie outside ±2.58. Cases above 3 are cause for concern and cases close to 3 warrant inspection
ZRE_1	Standardized Residual	
DEV_1	Deviance	
DFB0_1	DFBeta for the Constant	Should be less than 1
DFB1_1	DFBeta for the First Predictor (**Intervention**)	

also no unusually high values of Cook's distance (COO_1) which, all in all, means that there are no influential cases having an effect on the model. The standardized residuals all have values of less than ± 2 and so there seems to be very little here to concern us.

You should note that these residuals are slightly unusual because they are based on a single predictor that is categorical. This is why there isn't a lot of variability in the values of the residuals. Also, if substantial outliers or influential cases had been isolated, you are not justified in eliminating these cases to make the model fit better. Instead these cases should be inspected closely to try to isolate a good reason why they were unusual. It might simply be an error in inputting data, or it could be that the case was one which had a special reason for being unusual: for example, there were other medical complications that might contribute to the constipation that were noted during the patient's assessment. In such a case, you may have good reason to exclude the case and duly note the reasons why.

CRAMMING SAM'S TIPS Diagnostic statistics

You need to look for cases that might be influencing the logistic regression model:

- Look at standardized residuals and check that no more than 5% of cases have absolute values above 2, and that no more than about 1% have absolute values above 2.5. Any case with a value above about 3 could be an outlier.
- Look in the data editor for the values of Cook's distance: any value above 1 indicates a case that might be influencing the model.
- Calculate the average leverage (the number of predictors plus 1, divided by the sample size) and then look for values greater than twice or three times this average value.
- Look for absolute values of DFBeta greater than 1.

8.6.5. Calculating the effect size ②

We've already seen (section 8.3.2) that SPSS produces a value of *r* for each predictor, based on the Wald statistic. This can be used as your effect size measure for a predictor; however, it's worth bearing in mind what I've already said about it: it won't be accurate when the Wald statistic is inaccurate. The better effect size to use is the odds ratio (see section 18.5.5).

8.7. How to report logistic regression ②

Logistic regression is fairly rarely used in my discipline of psychology, so it's difficult to find any concrete guidelines about how to report one. My personal view is that you should report it much the same as linear regression (see section 7.9). I'd be inclined to tabulate the results, unless it's a very simple model. As a bare minimum, report the beta values and their standard errors and significance value and some general statistics about the model (such as the R^2 and goodness-of-fit statistics). I'd also highly recommend reporting the odds ratio and its confidence interval. I'd also include the constant so that readers of your work can construct the full regression model if they need to. You might also consider reporting the variables that were not significant predictors because this can be as valuable as knowing about which predictors were significant.

For the example in this chapter we might produce a table like that in Table 8.3. Hopefully you can work out from where the values came by looking back through the chapter so far. As with multiple regression, I've rounded off to 2 decimal places throughout; for the R^2 there is no zero before the decimal point (because these values cannot exceed 1) but for all other values less than 1 the zero is present; the significance of the variable is denoted by an asterisk with a footnote to indicate the significance level being used.

TABLE 8.3 How to report logistic regression

		95% CI for Odds Ratio		
	B (SE)	Lower	Odds Ratio	Upper
Included				
Constant	−0.29 (0.27)			
Intervention	1.23* (0.40)	1.56	3.42	7.48

Note: R^2 = .06 (Hosmer & Lemeshow), .08 (Cox & Snell), .11 (Nagelkerke). Model $\chi^2(1) = 9.93$, $p < .01$. * $p < .01$.

8.8. Testing assumptions: another example ②

This example was originally inspired by events in the soccer World Cup of 1998 (a long time ago now, but such crushing disappointments are not easily forgotten). Unfortunately for me (being an Englishman), I was subjected to watching England get knocked out of the competition by losing a penalty shootout. Reassuringly, six years later I watched England get knocked out of the European Championship in another penalty shootout. Even more reassuring, a few months ago I saw them fail to even qualify for the European Championship (not a penalty shootout this time, just playing like cretins).

Now, if I were the England coach, I'd probably shoot the spoilt overpaid prima donnas, or I might be interested in finding out what factors predict whether or not a player will score a penalty. Those of you who hate football can read this example as being factors that predict success in a free-throw in basketball or netball, a penalty in hockey or a penalty kick in rugby[8] or field goal in American football. Now, this research question is perfect for logistic regression because our outcome variable is a dichotomy: a penalty can be either scored or missed. Imagine that past research (Eriksson, Beckham, & Vassell, 2004; Hoddle, Batty, & Ince, 1998) had shown that there are two factors that reliably predict whether a penalty kick will be missed or scored. The first factor is whether the player taking the kick is a worrier (this factor can be measured using a measure such as the Penn State Worry Questionnaire, PSWQ). The second factor is the player's past success rate at scoring (so, whether the player has a good track record of scoring penalty kicks). It is fairly well accepted that anxiety has detrimental effects on the performance of a variety of tasks and so it was also predicted that state anxiety might be able to account for some of the unexplained variance in penalty success.

Why do the England football team always miss penalties?

This example is a classic case of building on a well-established model, because two predictors are already known and we want to test the effect of a new one. So, 75 football players were selected at random and before taking a penalty kick in a competition they were given a state anxiety questionnaire to complete (to assess anxiety before the kick was taken). These players were also asked to complete the PSWQ to give a measure of how much they worried about things generally, and their past success rate was obtained from a database. Finally, a note was made of whether the penalty was scored or missed. The data can be found in the file **penalty.sav**, which contains four variables – each in a separate column:

- **Scored**: This variable is our outcome and it is coded such that 0 = penalty missed and 1 = penalty scored.
- **PSWQ**: This variable is the first predictor variable and it gives us a measure of the degree to which a player worries.
- **Previous**: This variable is the percentage of penalties scored by a particular player in their career. As such, it represents previous success at scoring penalties.
- **Anxious**: This variable is our third predictor and it is a variable that has not previously been used to predict penalty success. **Anxious** is a measure of state anxiety before taking the penalty.

SELF-TEST We learnt how to do hierarchical regression in the previous chapter. Try to conduct a hierarchical logistic regression analysis on these data. Enter **Previous** and **PSWQ** in the first block and **Anxious** in the second (forced entry). There is a full guide on how to do the analysis and its interpretation in the additional material on the companion website.

[8] Although this would be an unrealistic example because our rugby team, unlike their football counterparts, have Jonny Wilkinson who is the lord of penalty kicks and we bow at his great left foot in wonderment (well, I do).

8.8.1. Testing for linearity of the logit ③

In this example we have three continuous variables, therefore we have to check that each one is linearly related to the log of the outcome variable (**Scored**). I mentioned earlier in this chapter that to test this assumption we need to run the logistic regression but include predictors that are the interaction between each predictor and the log of itself (Hosmer & Lemeshow, 1989). To create these interaction terms, we need to use Transform Compute Variable... (see section 5.7.3). For each variable create a new variable that is the log of the original variable. For example, for **PSWQ**, create a new variable called **LnPSWQ** by entering this name into the box labelled *Target Variable* and then click on Type & Label... and give the variable a more descriptive label such as *Ln(PSWQ)*. In the list box labelled *Function group* click on *Arithmetic* and then in the box labelled *Functions and Special Variables* click on *Ln* (this is the natural log transformation) and transfer it to the command area by clicking on ⬆. When the command is transferred, it appears in the command area as 'LN(?)' and the question mark should be replaced with a variable name (which can be typed manually or transferred from the variables list). So replace the question mark with the variable **PSWQ** by either selecting the variable in the list and clicking on ➡ or just typing 'PSWQ' where the question mark is. Click on OK to create the variable.

SELF-TEST Try creating two new variables that are the natural logs of **Anxious** and **Previous**.

To test the assumption we need to redo the analysis exactly the same as before except that we should force all variables in a single block (i.e. we don't need to do it hierarchically), and we also need to put in three new interaction terms of each predictor and their logs. Select Analyze Regression ▶ Binary Logistic..., then in the main dialog box click on **Scored** and drag it to the *Dependent* box or click on ➡. Specify the main effects by clicking on **PSWQ**, **Anxious** and **Previous** while holding down the *Ctrl* key and then drag them to the *Covariates* box or click on ➡. To input the interactions, click on the two variables in the interaction while holding down the *Ctrl* key (e.g. click on **PSWQ** then, while holding down *Ctrl*, click on **Ln(PSWQ)**) and then click on >a*b> to move them to the *Covariates* box. This specifies the **PSWQ×Ln(PSWQ)** interaction; specify the **Anxious×Ln(Anxious)** and **Previous×Ln(Previous)** interactions in the same way. The completed dialog box is in Figure 8.7 (note that the final **Previous×Ln(Previous)** interaction isn't visible but it is there!).

SPSS Output 8.9 shows the part of the output that tests the assumption. We're interested only in whether the interaction terms are significant. Any interaction that is significant indicates that the main effect has violated the assumption of linearity of the logit. All three interactions have significance values greater than .05 indicating that the assumption of linearity of the logit has been met for **PSWQ**, **Anxious** and **Previous**.

SPSS OUTPUT 8.9

Variables in the Equation

		B	S.E.	Wald	df	Sig.	Exp(B)
Step 1[a]	PSWQ	-.422	1.103	.147	1	.702	.656
	Anxious	-2.645	2.797	.894	1	.344	.071
	Previous	1.666	1.482	1.264	1	.261	5.291
	LnPSWQ by PSWQ	.044	.297	.022	1	.882	1.045
	Anxious by LnAnxious	.681	.653	1.088	1	.297	1.975
	LnPrevious by Previous	-.319	.317	1.008	1	.315	.727
	Constant	-3.879	14.924	.068	1	.795	.021

a. Variable(s) entered on step 1: PSWQ, Anxious, Previous, LnPSWQ * PSWQ , Anxious * LnAnxious , LnPrevious * Previous .

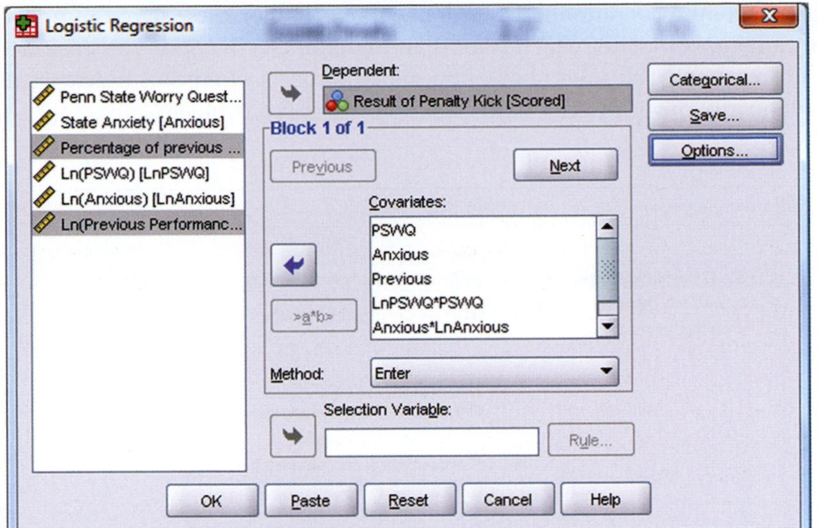

FIGURE 8.7 Dialog box for testing the assumption of linearity in logistic regression

8.8.2. Testing for multicollinearity ③

In section 7.6.2.4 we saw how multicollinearity can affect the parameters of a regression model. Logistic regression is just as prone to the biasing effect of collinearity and it is essential to test for collinearity following a logistic regression analysis. Unfortunately, SPSS does not have an option for producing collinearity diagnostics in logistic regression (which can create the illusion that it is unnecessary to test for it!). However, you can obtain statistics such as the tolerance and VIF by simply running a linear regression analysis using the same outcome and predictors. For the penalty example in the previous section, access the *Linear Regression* dialog box by selecting Analyze Regression ▸ Linear... . The completed dialog box is shown in Figure 8.8. It is unnecessary to specify lots of options (we are using this technique only to obtain tests of collinearity) but it is essential that you click on Statistics... and then select *Collinearity diagnostics* in the dialog box. Once you have selected ☑ Collinearity diagnostics, switch off all of the default options, click on Continue to return to the *Linear Regression* dialog box, and then click on OK to run the analysis.

The results of the linear regression analysis are shown in SPSS Output 8.10. From the first table we can see that the tolerance values are 0.014 for **Previous** and **Anxious** and 0.575 for **PSWQ**. In Chapter 7 we saw various criteria for assessing collinearity. To recap, Menard (1995) suggests that a tolerance value less than 0.1 almost certainly indicates a serious collinearity problem. Myers (1990) also suggests that a VIF value greater than 10 is cause for concern and in these data the values are over 70 for both **Anxious** and **Previous**. It seems from these values that there is an issue of collinearity between the predictor variables. We can investigate this issue further by examining the collinearity diagnostics.

SPSS Output 8.10 also shows a table labelled **Collinearity Diagnostics**. In this table, we are given the eigenvalues of the scaled, uncentred cross-products matrix, the condition index and the variance proportions for each predictor. If any of the eigenvalues in this table are much larger than others then the uncentred cross-products matrix is said to be ill-conditioned, which means that the solutions of the regression parameters can be greatly affected by small changes in the predictors or outcome. In plain English, these values give us some idea as to how accurate our regression model is: if the eigenvalues are fairly similar then the derived model is likely to be unchanged by small changes in the measured variables. The *condition indexes* are another way of expressing these eigenvalues and represent the square root of the ratio of the largest eigenvalue to the eigenvalue of interest (so, for the dimension with the largest eigenvalue, the condition index will always be 1). For these data the final dimension has a condition index of 81.3, which is massive compared to the other dimensions. Although there are no hard and fast rules about how much larger a condition index needs to be to indicate collinearity problems, this case clearly shows that a problem exists.

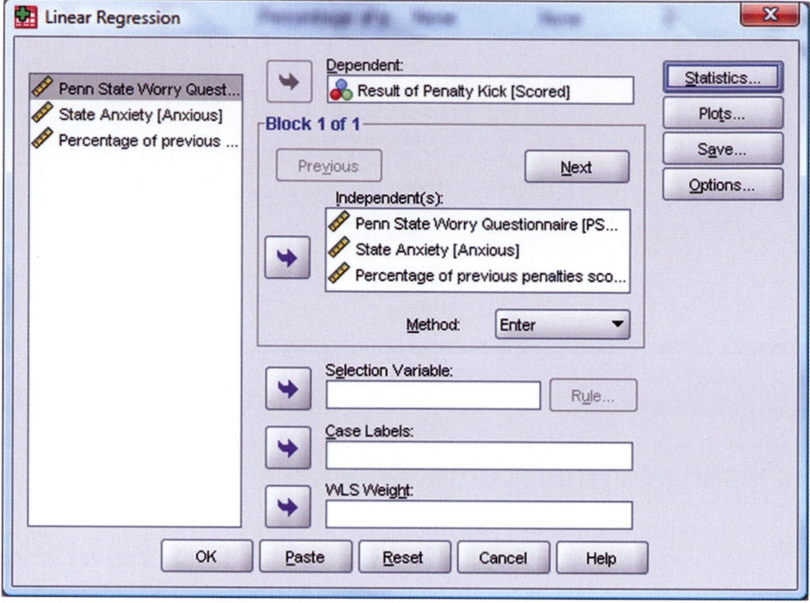

FIGURE 8.8 *Linear Regression* dialog box for the penalty data

SPSS OUTPUT 8.10 Collinearity diagnostics for the penalty data

Coefficients[a]

Model		Collinearity Statistics	
		Tolerance	VIF
1	Penn State Worry Questionnaire	.575	1.741
	State Anxiety	.014	71.764
	Percentage of previous penalties scored	.014	70.479

a. Dependent Variable: Result of Penalty Kick

Collinearity Diagnostics[a]

Model	Dimension	Eigenvalue	Condition Index	Variance Proportions			
				(Constant)	Penn State Worry Questionnaire	State Anxiety	Percentage of previous penalties scored
1	1	3.434	1.000	.00	.01	.00	.00
	2	.492	2.641	.00	.04	.00	.00
	3	.073	6.871	.00	.95	.01	.00
	4	.001	81.303	1.00	.00	.99	.99

a. Dependent Variable: Result of Penalty Kick

The final step in analysing this table is to look at the variance proportions. The variance of each regression coefficient can be broken down across the eigenvalues and the variance proportions tell us the proportion of the variance of each predictor's regression coefficient that is attributed to each eigenvalue. These proportions can be converted to percentages by multiplying them by 100 (to make them more easily understood). So, for example, for **PSWQ** 95% of the variance of the regression coefficient is associated with eigenvalue number 3, 4% is associated with eigenvalue number 2 and 1% is associated with eigenvalue number 1. In terms of collinearity, we are looking for predictors that have high proportions on the same *small* eigenvalue, because this would indicate that the variances of their regression coefficients are dependent. So we are interested mainly in the bottom few rows of the table (which represent small eigenvalues). In this example, 99% of the variance in the regression coefficients of both **Anxiety** and **Previous** is associated with eigenvalue number 4 (the smallest eigenvalue), which clearly indicates dependency between these variables.

The result of this analysis is pretty clear cut: there is collinearity between state anxiety and previous experience of taking penalties and this dependency results in the model becoming biased.

SELF-TEST Using what you learned in Chapter 6, carry out a Pearson correlation between all of the variables in this analysis. Can you work out why we have a problem with collinearity?

If you have identified collinearity then, unfortunately, there's not much that you can do about it. One obvious solution is to omit one of the variables (so, for example, we might stick with the model from block 1 that ignored state anxiety). The problem with this should be obvious: there is no way of knowing which variable to omit. The resulting theoretical

LABCOAT LENI'S REAL RESEARCH 8.1

Mandatory suicide? ②

Although I have fairly eclectic tastes in music, my favourite kind of music is heavy metal. One thing that is mildly irritating about liking heavy music is that everyone assumes that you're a miserable or aggressive bastard. When not listening (and often while listening) to heavy metal, I spend most of my time researching clinical psychology: I research how anxiety develops in children. Therefore, I was literally beside myself with excitement when a few years back I stumbled on a paper that combined these two interests: Lacourse, Claes, and Villeneuve (2001) carried out a study to see whether a love of heavy metal could predict suicide risk. Fabulous stuff!

Eric Lacourse and his colleagues used questionnaires to measure several variables: suicide risk (yes or no), marital status of parents (together or divorced/separated), the extent to which the person's mother and father were neglectful, self-estrangement/powerlessness (adolescents who have negative self-perceptions, bored with life, etc.), social isolation (feelings of a lack of support), normlessness (beliefs that socially disapproved behaviours can be used to achieve certain goals), meaninglessness (doubting that school is relevant to gain employment) and drug use. In addition, the author measured liking of heavy metal; they included the sub-genres of classic (Black Sabbath, Iron Maiden), thrash metal (Slayer, Metallica), death/black metal (Obituary, Burzum) and gothic (Marilyn Manson). As well as liking they measured behavioural manifestations of worshipping these bands (e.g. hanging posters, hanging out with other metal fans) and vicarious music listening (whether music was used when angry or to bring out aggressive moods). They used logistic regression to predict suicide risk from these predictors for males and females separately.

The data for the female sample are in the file **Lacourse et al. (2001) Females.sav**. Labcoat Leni wants you to carry out a logistic regression predicting **Suicide_Risk** from all of the predictors (forced entry). (To make your results easier to compare to the published results, enter the predictors in the same order as Table 3 in the paper: **Age**, **Marital_Status**, **Mother_Negligence**, **Father_Negligence**, **Self_Estrangement**, **Isolation**, **Normlessness**, **Meaninglessness**, **Drug_Use**, **Metal**, **Worshipping**, **Vicarious**). Create a table of the results. Does listening to heavy metal make girls suicidal? If not, what does?

Answers are in the additional material on the companion website (or look at Table 3 in the original article).

LACOURSE, E. ET AL. (2001). *JOURNAL OF YOUTH AND ADOLESCENCE, 30*, 321–332.

conclusions are meaningless because, statistically speaking, any of the collinear variables could be omitted. There are no statistical grounds for omitting one variable over another. Even if a predictor is removed, Bowerman and O'Connell (1990) recommend that another equally important predictor that does not have such strong multicollinearity replaces it. They also suggest collecting more data to see whether the multicollinearity can be lessened. Another possibility when there are several predictors involved in the multicollinearity is to run a factor analysis on these predictors and to use the resulting factor scores as a predictor (see Chapter 17). The safest (although unsatisfactory) remedy is to acknowledge the unreliability of the model. So, if we were to report the analysis of which factors predict penalty success, we might acknowledge that previous experience significantly predicted penalty success in the first model, but propose that this experience might affect penalty taking by increasing state anxiety. This statement would be highly speculative because the correlation between **Anxious** and **Previous** tells us nothing of the direction of causality, but it would acknowledge the inexplicable link between the two predictors. I'm sure that many of you may find the lack of remedy for collinearity grossly unsatisfying – unfortunately statistics is frustrating sometimes!

8.9. Predicting several categories: multinomial logistic regression ③

I mentioned earlier that it is possible to use logistic regression to predict membership of more than two categories and that this is called multinomial logistic regression. Essentially, this form of logistic regression works in the same way as binary logistic regression so there's no need for any additional equations to explain what is going on (hooray!). The analysis breaks the outcome variable down into a series of comparisons between two categories (which helps explain why no extra equations are really necessary). For example, if you have three outcome categories (A, B and C), then the analysis will consist of two comparisons. The form that these comparisons take depends on how you specify the analysis: you can compare everything against your first category (e.g. A vs. B and A vs. C), or your last category (e.g. A vs. C and B vs. C), or a custom category, for example category B (e.g. B vs. A and B vs. C). In practice, this means that you have to select a baseline category. The important parts of the analysis and output are much the same as we have just seen for binary logistic regression.

What do I do when I have more than two outcome categories?

Let's look at an example. There has been some recent work looking at how men and women evaluate chat-up lines (Bale, Morrison, & Caryl, 2006; Cooper, O'Donnell, Caryl, Morrison, & Bale, 2007). This research has looked at how the content (e.g. whether the chat-up line is funny, has sexual content or reveals desirable personality characteristics) affects how favourably the chat-up line is viewed. To sum up this research, it has found that men and women like different things in chat-up lines: men prefer chat-up lines with a high sexual content and women prefer chat-up lines that are funny and show good moral fibre!

Imagine that we wanted to assess how *successful* these chat-up lines were. We did a study in which we recorded the chat-up lines used by 348 men and 672 women in a night-club. Our outcome was whether the chat-up line resulted in one of the following three events: the person got no response or the recipient walked away, the person obtained the recipient's phone number, or the person left the night-club with the recipient. Afterwards, the chat-up lines used in each case were rated by a panel of judges for how funny they were

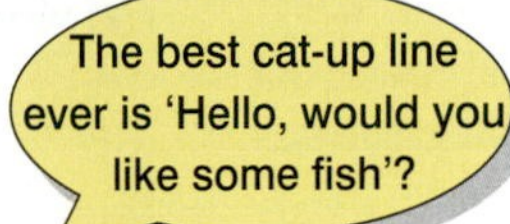
The best cat-up line ever is 'Hello, would you like some fish'?

(0 = not funny at all, 10 = the funniest thing that I have ever heard), sexuality (0 = no sexual content at all, 10 = very sexually direct) and whether the chat-up line reflected good moral vales (0 = the chat-up line does not reflect good characteristics, 10 = the chat-up line is very indicative of good characteristics). For example, 'I may not be Fred Flintstone, but I bet I could make your bed rock' would score high on sexual content, low on good characteristics and medium on humour; 'I've been looking all over for YOU, the woman of my dreams' would score high on good characteristics, low on sexual content and low on humour (but high on cheese had it been measured). We predict based on past research that the success of different types of chat-up line will interact with gender.

This situation is perfect for multinomial regression. The data are in the file **Chat-Up Lines.sav**. There is one outcome variable (**Success**) with three categories (no response, phone number, go home with recipient) and four predictors: funniness of the chat-up line (**Funny**), sexual content of the chat-up line (**Sex**), degree to which the chat-up line reflects good characteristics (**Good_Mate**) and the gender of the person being chatted up (**Gender**).

8.9.1. Running multinomial logistic regression in SPSS ③

To run multinomial logistic regression in SPSS, first select the main dialog box by selecting Analyze Regression ▸ Multinomial Logistic.... In this dialog box there are spaces to place the outcome variable (*Dependent*), any categorical predictors (*Factor(s)*) and any continuous predictors (*Covariate(s)*). In this example, the outcome variable is **Success** so select this variable from the list and transfer it to the box labelled *Dependent* by dragging it there or clicking on ➜. We also have to tell SPSS whether we want to compare categories against the first category or the last and we do this by clicking on Reference Category... .

SELF-TEST Think about the three categories that we have as an outcome variable. Which of these categories do you think makes most sense to use as a baseline category?

By default SPSS uses the last category, but in our case it makes most sense to use the first category (No response/walk off) because this category represents failure (the chat-up line did not have the desired effect) whereas the other two categories represent some form of success (getting a phone number or leaving the club together). To change the reference category to be the first category, click on ⦿ First Category and then click on Continue to return to the main dialog box (Figure 8.9).

Next we have to specify the predictor variables. We have one categorical predictor variable, which is **Gender**, so select this variable next and transfer it to the box labelled *Factor(s)* by dragging it there or clicking on ➜. Finally, we have three continuous predictors or covariates (**Funny**, **Sex** and **Good_Mate**). You can select all of these variables simultaneously by holding down the *Ctrl* key as you click on each one. Drag all three to the box labelled *Covariate(s)* or click on ➜. For a basic analysis in which all of these predictors are forced into the model, this is all we really need to do. However, as we saw in the regression chapter you will often want to do a hierarchical regression and so for this analysis we'll look at how to do this in SPSS.

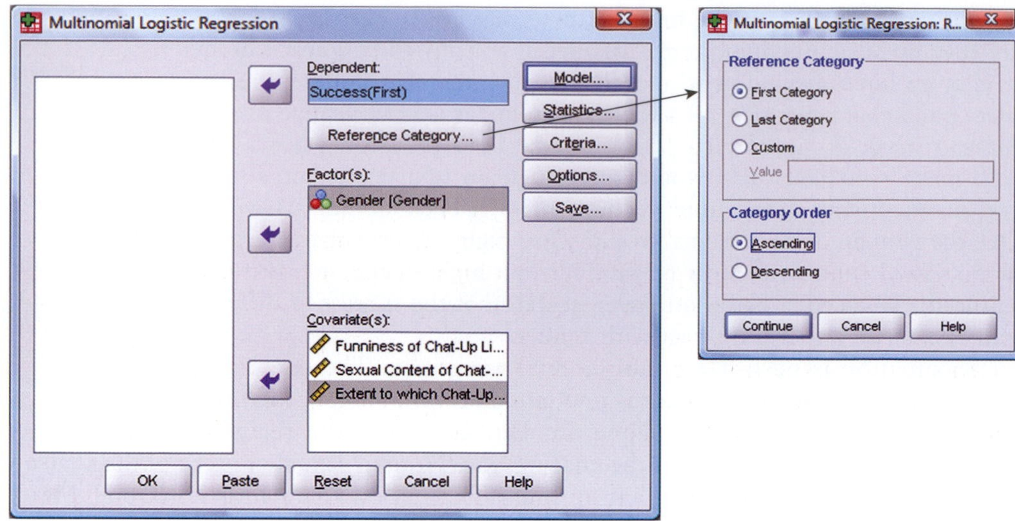

FIGURE 8.9 Main dialog box for multinomial logistic regression

8.9.1.1. Customizing the model ③

In binary logistic regression, SPSS allowed us to specify interactions between predictor variables in the main dialog box, but with multinomial logistic regression we cannot do this. Instead we have to specify what SPSS calls a 'custom model', and this is done by clicking on Model... to open the dialog box in Figure 8.10. You'll see that, by default, SPSS just looks at the main effects of the predictor variables. In this example, however, the main effects are not particularly interesting: based on past research we don't necessarily expect funny chat-up lines to be successful, but we do expect them to be more successful when

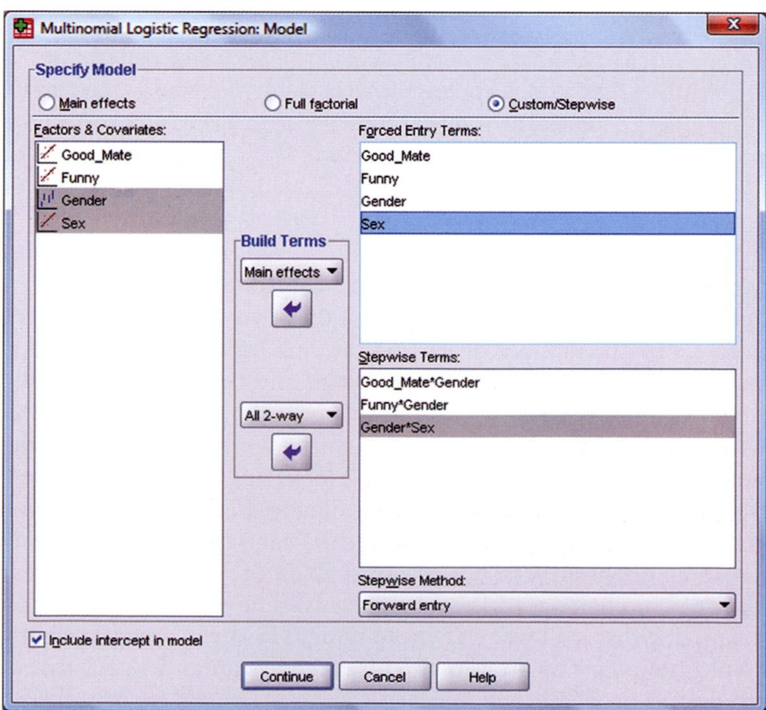

FIGURE 8.10 Specifying a custom model

used on women than on men. What this prediction implies is that the *interaction* of **Gender** and **Funny** will be significant. Similarly, chat-up lines with a high sexual content might not be successful overall, but expect them to be relatively successful when used on men. Again, this means that we might not expect the **Sex** main effect to be significant, but we do expect the **Sex×Gender** interaction to be significant. As such, we need to enter some interaction terms into the model.

To customize the model we first have to select ⦿ Custom/Stepwise to activate the rest of the dialog box. There are two main ways that we can specify terms: we can force them in (by moving them to the box labelled *Forced Entry Terms*) or we can put them into the model using a stepwise procedure (by moving them into the box labelled *Stepwise terms*). If we want to look at interaction terms, we must force the main effects into the model. If we look at interactions without the corresponding main effects being in the model then we allow the interaction term to explain variance that might otherwise be attributed to the main effect (in other words, we're not really looking at the interaction any more). So, select all of the variables in the box labelled *Factors and covariates* by clicking on them while holding down *Ctrl* (or selecting the first variable and then clicking on the last variable while holding down *Shift*). There is a drop-down list that determines whether you transfer these effects as main effects or interactions. We want to transfer them as main effects so set this box to Main effects ▾ and click on ➔.

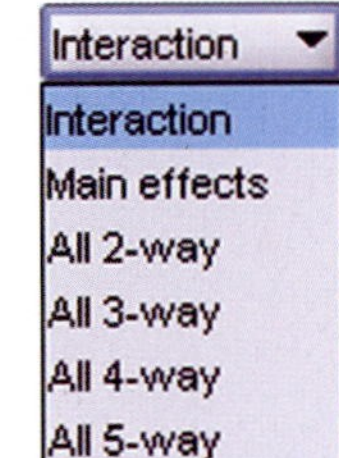

To specify interactions we can do much the same: we can select two or more variables and then set the drop-down box to be Interaction ▾ and then click on ➔. If, for example, we selected **Funny** and **Sex**, then doing this would specify the **Funny×Sex** interaction. We can also specify multiple interactions at once. For example, if we selected **Funny**, **Sex** and **Gender** and then set the drop-down box to All 2-way ▾, it would transfer *all* of the interactions involving two variables (i.e. **Funny×Sex, Funny×Gender** and **Sex×Gender**). You get the general idea. We could also select ⦿ Full factorial which would automatically enter all main effects (**Funny, Sex, Good_Mate, Gender**), all interactions with two variables (**Funny×Sex, Funny×Gender, Funny×Good_Mate, Sex×Gender, Sex×Good_Mate, Gender×Good_Mate**), all interactions with three variables (**Funny×Sex×Gender, Funny×Sex×Good_Mate, Good_Mate×Sex×Gender, Funny×Good_Mate×Gender**) and the interaction of all four variables (**Funny×Sex×Gender×Good_Mate**).

In this scenario, we want to specify interactions between the ratings of the chat-up lines and gender only (we're not interested in any interactions involving three variables, or all four variables). We can either force these interaction terms into the model by putting them in the box labelled *Forced Entry Terms* or we can put them into the model using a stepwise procedure (by moving them into the box labelled *Stepwise terms*). We're going to do the latter, so interactions will be entered into the model only if they are significant predictors of the success of the chat-up line. Let's first enter the **Funny×Gender** interaction first. Click on **Funny** and then **Gender** in the *Factors and covariates* box while holding down the *Ctrl* key. Then next to the box labelled *Stepwise terms* change the drop-down menu to be Interaction ▾ and then click on ➔. You should now see **Gender×Funny** (SPSS orders the variables in reverse alphabetical order for some reason) listed in the *Stepwise terms* box. Specify the **Sex×Gender** and **Good_Mate×Gender** in the same way. Once the three interaction terms have been specified we can decide how we want to carry out the stepwise analysis. There is a drop-down list of methods under the heading *Stepwise Method* and this list enables you to choose between forward and stepwise entry, or backward and stepwise elimination (i.e. terms are removed from the model if they do not make a significant contribution). I've described these methods elsewhere, so select forward entry for this analysis.

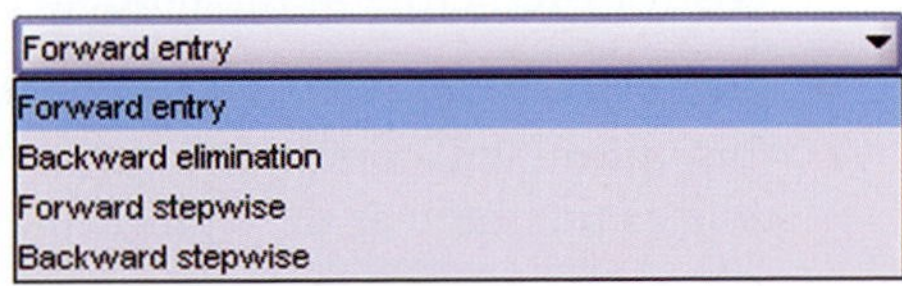

8.9.2. Statistics ③

If you click on Statistics... you will see the dialog box in Figure 8.11, in which you can specify certain statistics:

- *Pseudo R-square*: This produces the Cox and Snell and Nagelkerke R^2 statistics. These can be used as effect sizes so this is a useful option to select.
- *Step summary*: This option should be selected for the current analysis because we have a stepwise component to the model; this option produces a table that summarizes the predictors entered or removed at each step.
- *Model fitting information*: This option produces a table that compares the model (or models in a stepwise analysis) to the baseline (the model with only the intercept term in it and no predictor variables). This table can be useful to compare whether the model had improved (from the baseline) as a result of entering the predictors that you have.
- *Information criteria*: This option produces Akaike's information criterion (AIC) and Schwarz's Bayesian information criterion (BIC). Both of these statistics are a useful way to compare models. Therefore, if you're using stepwise methods, or if you want to compare different models containing different combinations of predictors, then select this option. Low values of the AIC and BIC indicate good fit; therefore, models with lower values of the AIC or BIC fit the data relatively better than models with higher values.
- *Cell probabilities*: This option produces a table of the observed and expected frequencies. This is basically the same as the classification table produced in binary logistic regression and is probably worth inspecting.
- *Classification table*: This option produces a contingency table of observed versus predicted responses for all combinations of predictor variables. I wouldn't select this option, unless you're running a relatively small analysis (i.e. a small number of predictors made up of a small number of possible values). In this example, we have three covariates with 11 possible values and one predictor (gender) with 2 possible values. Tabulating all combinations of these variables will create a very big table indeed.
- *Goodness-of-fit*: This option is important because it produces Pearson and likelihood-ratio chi-square statistics for the model.
- *Monotonicity measures*: This option is worth selecting only if your outcome variable has two outcomes (which in our case it doesn't). It will produce measures of monotonic association such as the concordance index, which measures the probability that, using a previous example, a person who scored a penalty kick is classified by the model as having scored and can range from .5 (guessing) to 1 (perfect prediction).
- *Estimates*: This option produces the beta values, test statistics and confidence intervals for predictors in the model. This option is very important.
- *Likelihood ratio tests*: The model overall is tested using likelihood ratio statistics, but this option will compute the same test for individual effects in the model. (Basically it tells us the same as the significance values for individual predictors.)
- *Asymptotic correlations and covariances*: These produce a table of correlations (or covariances) between the betas in the model.

Set the options as in Figure 8.11 and click on Continue to return to the main dialog box.

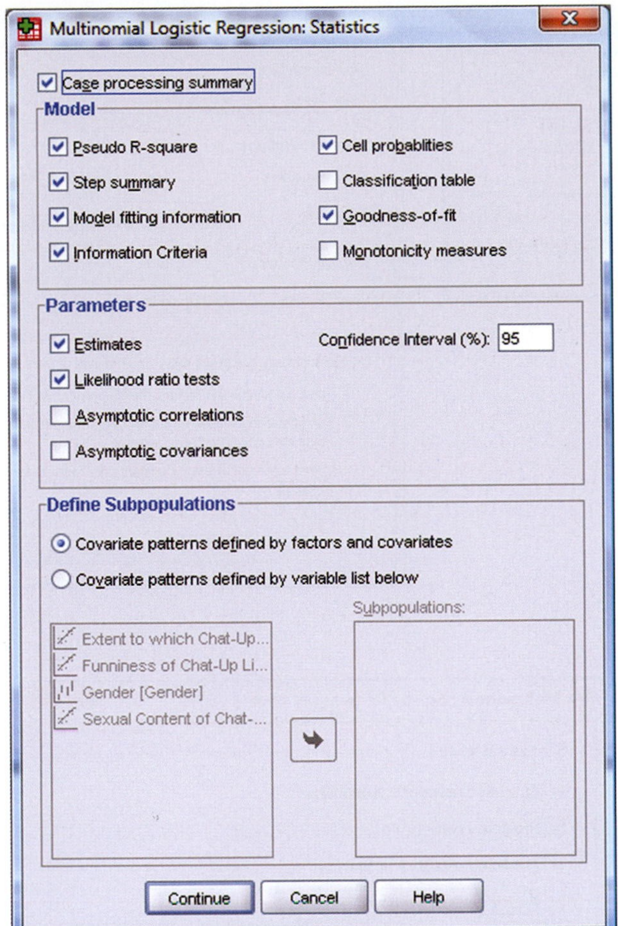

FIGURE 8.11 Statistics options for multinomial logistic regression

8.9.3. Other options ③

If you click on Criteria... you'll access the dialog box in Figure 8.12 (left). Logistic regression works through an iterative process (SPSS Tip 8.1). The options available here relate to this process. For example, by default, SPSS will make 100 attempts (iterations) and the threshold for how similar parameter estimates have to be to 'converge' can be made more or less strict (the default is .0000001). You should leave these options alone unless when you run the analysis you get an error message saying something about 'failing to converge', in which case you could try increasing the *Maximum iterations* (to 150 or 200), the *Parameter convergence* (to .00001) or *Log-likelihood convergence* (to greater than 0). However, bear in mind that a failure to converge can reflect messy data and forcing the model to converge does not necessarily mean that parameters are accurate or stable across samples.

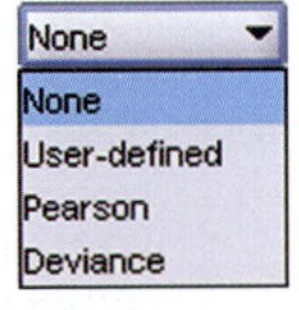

You can also click on Options... in the main dialog box to access the dialog box in Figure 8.12 (right). The *Scale* option here can be quite useful; I mentioned in section 8.4.4 that overdispersion can be a problem in logistic regression because it reduces the standard errors that are used to test the significance and construct the confidence intervals of the parameter estimates for individual predictors in the model. I also mentioned that this problem could be counteracted by rescaling the standard errors. Should you be in a situation where you need to do this (i.e. you have run the analysis and found evidence of overdispersion)

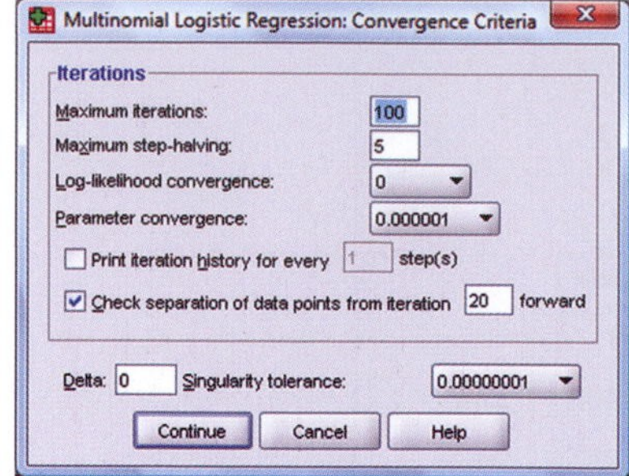

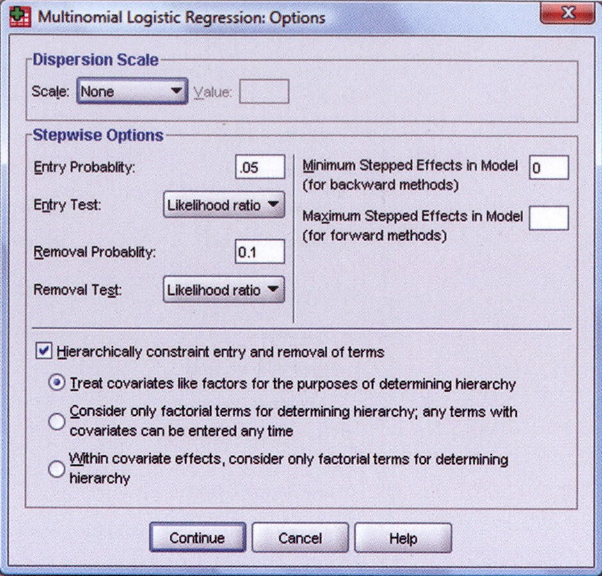

FIGURE 8.12 Criteria and options for multinomial logistic regression

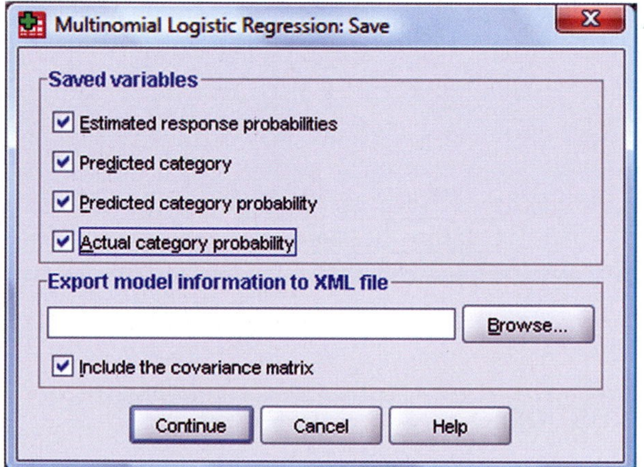

FIGURE 8.13 *Save* options for multinomial logistic regression

then you need to come to this dialog box and use the drop-down list to select to correct the standard errors by the dispersion parameter based on either the Deviance or Pearson statistic. You should select whichever of these two statistics was bigger in the original analysis (because this will produce the bigger correction).

Finally, if you click on Save... you can opt to save predicted probabilities and predicted group membership (the same as for binary logistic regression except that they are called *Estimated response probabilities* and *Predicted category* (Figure 8.13)).

8.9.4. Interpreting the multinomial logistic regression output ③

Our SPSS output from this analysis begins with a warning (SPSS Tip 8.2). It's always nice after months of preparation, weeks entering data, years reading chapters of stupid statistics

SPSS TIP 8.2 Warning! Zero frequencies ③

Warnings

There are 504 (53.5%) cells (i.e., dependent variable levels by subpopulations) with zero frequencies.

Sometimes in logistic regression you get a warning about zero frequencies. This relates to the problem that I discussed in section 8.4.2 of 80 year old, highly anxious, Buddhist left-handed lesbians (well, incomplete information). Imagine we had just looked at gender as a predictor of chat-up line success. We have three outcome categories and two gender categories. There are six possible combinations of these two variables and ideally we would like a large number of observations in each of these combinations. However, in this case, we have three variables (**Funny**, **Sex** and **Good_Mate**) with 11 possible outcomes, and **Gender** with 2 possible outcomes and an outcome variable with 3 outcomes. It should be clear that by including the three covariates, the number of combinations of these variables has escalated considerably. This error message tells us that there are some combinations of these variables for which there are no observations. So, we really have a situation where we didn't find an 80 year old, highly anxious, Buddhist left-handed lesbian; well, we didn't find (for example) a chat-up line that was the most funny, showed the most good characteristics, had the most sexual content and was used on both a man and woman! In fact 53.5% of our possible combinations of variables had no data!

Whenever you have covariates it is inevitable that you will have empty cells, so you will get this kind of error message. To some extent, given its inevitability, we can just ignore it (in this study, for example, we have 1020 cases of data and half of our cells are empty, so we would need to at least double the sample size to stand any chance of filling those cells). However, it is worth reiterating what I said earlier that empty cells create problems and that when you get a warning like this you should look for coefficients that have unreasonably large standard errors and if you find them be wary of them.

textbooks, and sleepless nights with equations slicing at your brain with little pick-axes, to see at the start of your analysis: 'Warning! Warning! abandon ship! flee for your life! bad data alert! bad data alert!' Still, such is life.

Once, like all good researchers, we have ignored the warnings then the first part of the output tells us about our model overall (SPSS Output 8.11). First, because we requested a stepwise analysis for our interaction terms, we get a table summarizing the steps in the analysis. You can see here that after the main effects were entered (Model 0), the **Gender×Funny** interaction term was entered (Model 1) followed by the **Sex×Gender** interaction (Model 2). The chi-square statistics for each of these steps are highly significant, indicating that these interactions have a significant effect on predicting whether a chat-up line was significant (this fact is also self-evident because these terms wouldn't have been entered into the model had they not been significant). Also note that the AIC gets smaller as these terms are added to the model, indicating that the fit of the model is getting better as these terms are added (the BIC changes less, but still shows that having the interaction terms in the model results in a better fit than when just the main effects are present). Underneath the step summary, we see the statistics for the final model, which replicates the model-fitting criteria from the last line of the step summary table. This table also produces a likelihood ratio test of the overall model.

SELF-TEST What does the log-likelihood measure?

Remember that the log-likelihood is a measure of how much unexplained variability there is in the data; therefore, the difference or change in log-likelihood indicates how much new variance has been explained by the model. The chi-square test tests the decrease in unexplained variance from the baseline model (1149.53) to the final model (871.00), which is a difference of 1149.53 − 871 = 278.53. This change is significant, which means that our final model explains a significant amount of the original variability (in other words, it's a better fit than the original model).

SPSS OUTPUT 8.11

Step Summary

Model	Action	Effect(s)	Model Fitting Criteria: AIC	BIC	-2 Log Likelihood	Effect Selection Tests: Chi-Square[a]	df	Sig.
0	Entered	Intercept, Good_Mate, Funny, Gender, Sex	937.572	986.848	917.572	.		
1	Entered	Gender * Funny	908.451	967.582	884.451	33.121	2	.000
2	Entered	Gender * Sex	899.002	967.987	871.002	13.450	2	.001

Stepwise Method: Forward Entry

a. The chi-square for entry is based on the likelihood ratio test.

Model Fitting Information

Model	Model Fitting Criteria: AIC	BIC	-2 Log Likelihood	Likelihood Ratio Tests: Chi-Square	df	Sig.
Intercept Only	1.154E3	1.163E3	1149.526			
Final	899.002	967.987	871.002	278.525	12	.000

SPSS OUTPUT 8.12

Goodness-of-Fit

	Chi-Square	df	Sig.
Pearson	886.616	614	.000
Deviance	617.481	614	.453

Pseudo R-Square

Cox and Snell	.239
Nagelkerke	.277
McFadden	.138

The next part of the output (SPSS Output 8.12) relates to the fit of the model to the data. We know that the model is significantly better than no model, but is it a good fit of the data? The Pearson and deviance statistics test the same thing, which is whether the predicted values from the model differ significantly from the observed values. If these statistics are not significant then the predicted values are not significantly different from the observed values; in other words, the model is a good fit. Here we have contrasting results: the deviance statistic says that the model is a good fit of the data ($p = .45$, which is much higher than .05), but the Pearson test indicates the opposite, namely that predicted values are significantly different from the observed values ($p < .001$). Oh dear.

SELF-TEST Why might the Pearson and deviance statistics be different? What could this be telling us?

One answer is that differences between these statistics can be caused by overdispersion. This is a possibility that we need to look into. However, there are other reasons for this conflict: for example, the Pearson statistic can be very inflated by low expected frequencies (which could happen because we have so many empty cells as indicated by our warning). One thing that is certain is that conflicting deviance and Pearson chi-square statistics are not good news!

Let's look into the possibility of overdispersion. We can compute the dispersion parameters from both statistics:

$$\phi_{\text{Pearson}} = \frac{\chi^2_{\text{Pearson}}}{df} = \frac{886.62}{614} = 1.44$$

$$\phi_{\text{Deviance}} = \frac{\chi^2_{\text{Deviance}}}{df} = \frac{617.48}{614} = 1.01$$

Neither of these is particularly high, and the one based on the deviance statistic is close to the ideal value of 1. The value based on Pearson is greater than 1, but not close to 2, so again does not give us an enormous cause for concern that the data are overdispersed.[9]

The output also shows us the two other measures of R^2 that were described in section 8.3.2. The first is Cox and Snell's measure, which SPSS reports as .24, and the second is Nagelkerke's adjusted value, which SPSS reports as .28. As you can see, they are reasonably similar values and represent relatively decent-sized effects.

SPSS Output 8.13 shows the results of the likelihood ratio tests and these can be used to ascertain the significance of predictors to the model. The first thing to note is that no significance values are produced for covariates that are involved in higher-order interactions (this is why there are blank spaces in the *Sig.* column for the effects of **Funny** and **Sex**). This table tells us, though, that gender had a significant main effect on success rates of chat-up lines, $\chi^2(2) = 18.54$, $p < .001$, as did whether the chat-up lined showed evidence of being a good partner, $\chi^2(2) = 6.32$, $p < .042$. Most interesting are the interactions which showed that the humour in the chat-up line interacted with gender to predict success at getting a date, $\chi^2(2) = 35.81$, $p < .001$; also the sexual content of the chat-up line interacted with the gender of the person being chatted up in predicting their reaction, $\chi^2(2) = 13.45$, $p < .001$. These likelihood statistics can be seen as sorts of overall statistics that tell us which predictors significantly enable us to predict the outcome category, but they don't really tell us specifically what the effect is. To see this we have to look at the individual parameter estimates.

SPSS OUTPUT 8.13

Likelihood Ratio Tests

	Model Fitting Criteria			Likelihood Ratio Tests		
Effect	AIC of Reduced Model	BIC of Reduced Model	-2 Log Likelihood of Reduced Model	Chi-Square	df	Sig.
Intercept	899.002	967.987	8.710E2ᵃ	.000	0	.
Good_Mate	901.324	960.454	877.324	6.322	2	.042
Funny	899.002	967.987	8.710E2ᵃ	.000	0	.
Gender	913.540	972.671	889.540	18.538	2	.000
Sex	899.002	967.987	8.710E2ᵃ	.000	0	.
Gender * Funny	930.810	989.941	906.810	35.808	2	.000
Gender * Sex	908.451	967.582	884.451	13.450	2	.001

The chi-square statistic is the difference in -2 log-likelihoods between the final model and a reduced model. The reduced model is formed by omitting an effect from the final model. The null hypothesis is that all parameters of that effect are 0.

a. This reduced model is equivalent to the final model because omitting the effect does not increase the degrees of freedom.

[9] Incidentally, large dispersion parameters can occur for reasons other than overdispersion, for example omitted variables or interactions (in this example there were several interaction terms that we could have entered but chose not to), and predictors that violate the linearity of the logit assumption.

SPSS Output 8.14 shows the individual parameter estimates. Note that the table is split into two halves. This is because these parameters compare pairs of outcome categories. We specified the first category as our reference category; therefore, the part of the table labelled *Get Phone Number* is comparing this category against the 'No response/walked away' category. Let's look at the effects one by one; because we are just comparing two categories the interpretation is the same as for binary logistic regression (so if you don't understand my conclusions reread the start of this chapter):

- **Good_Mate**: Whether the chat-up line showed signs of good moral fibre significantly predicted whether you got a phone number or no response/walked away, $b = 0.13$, Wald $\chi^2(1) = 6.02$, $p < .05$. The odds ratio tells us that as this variable increases, so as chat-up lines show one more unit of moral fibre, the change in the odds of getting a phone number (rather than no response/walked away) is 1.14. In short, you're more likely to get a phone number than not if you use a chat-up line that demonstrates good moral fibre. (Note that this effect is superseded by the interaction with gender below.)
- **Funny**: Whether the chat-up line was funny did not significantly predict whether you got a phone number or no response, $b = 0.14$, Wald $\chi^2(1) = 1.60$, $p > .05$. Note that although this predictor is not significant, the odds ratio is approximately the same as for the previous predictor (which was significant). So, the effect size is comparable, but the non-significance stems from a relatively higher standard error.
- **Gender**: The gender of the person being chatted up significantly predicted whether they gave out their phone number or gave no response, $b = -1.65$, Wald $\chi^2(1) = 4.27$, $p < .05$. Remember that 0 = female and 1 = male, so this is the effect of females compared to males. The odds ratio tells us that as gender changes from female (0) to male (1) the change in the odds of giving out a phone number compared to not responding is 0.19. In other words, the odds of a man giving out his phone number compared to not responding are 1/0.19 = 5.26 times more than for a woman. Men are cheap.
- **Sex**: The sexual content of the chat-up line significantly predicted whether you got a phone number or no response/walked away, $b = 0.28$, Wald $\chi^2(1) = 9.59$, $p < .01$. The odds ratio tells us that as the sexual content increased by a unit, the change in the odds of getting a phone number (rather than no response) is 1.32. In short, you're more likely to get a phone number than not if you use a chat-up line with high sexual content. (But this effect is superseded by the interaction with gender.)
- **Funny×Gender**: The success of funny chat-up lines depended on whether they were delivered to a man or a woman because in interaction these variables predicted whether or not you got a phone number, $b = 0.49$, Wald $\chi^2(1) = 12.37$, $p < .001$. Bearing in mind how we interpreted the effect of gender above, the odds ratio tells us that as gender changes from female (0) to male (1) in combination with funniness increasing, the change in the odds of giving out a phone number compared to not responding was 1.64. In other words, as funniness increases, women become more likely to hand out their phone number than men. Funny chat-up lines are more successful when used on women than men.
- **Sex×Gender**: The success of chat-up lines with sexual content depended on whether they were delivered to a man or a woman because in interaction these variables predicted whether or not you got a phone number, $b = -0.35$, Wald $\chi^2(1) = 10.82$, $p < .01$. Bearing in mind how we interpreted the interaction above (note that b is negative here but positive above), the odds ratio tells us that as gender changes from female (0) to male (1) in combination with the sexual content increasing, the change in the odds of giving out a phone number compared to not responding is 0.71. In other words, as sexual content increases, women become *less* likely than men to hand out their phone number. Chat-up lines with a high sexual content are more successful when used on men than women.

SPSS OUTPUT 8.14

Parameter Estimates

Success of Chat-Up Line[a]		B	Std. Error	Wald	df	Sig.	Exp(B)	95% Confidence Interval for Exp (B) Lower Bound	Upper Bound
Get Phone Number	Intercept	-1.783	.670	7.087	1	.008			
	Good_Mate	.132	.054	6.022	1	.014	1.141	1.027	1.268
	Funny	.139	.110	1.602	1	.206	1.150	.926	1.427
	[Gender=0]	-1.646	.796	4.274	1	.039	.193	.040	.918
	[Gender=1]	0[b]	.	.	0	.	.	.	.
	Sex	.276	.089	9.589	1	.002	1.318	1.107	1.570
	[Gender=0] * Funny	.492	.140	12.374	1	.000	1.636	1.244	2.153
	[Gender=1] * Funny	0[b]	.	.	0	.	.	.	.
	[Gender=0] * Sex	-.348	.106	10.824	1	.001	.706	.574	.869
	[Gender=1] * Sex	0[b]	.	.	0	.	.	.	.
Go Home with Person	Intercept	-4.286	.941	20.731	1	.000			
	Good_Mate	.130	.084	2.423	1	.120	1.139	.967	1.341
	Funny	.318	.125	6.459	1	.011	1.375	1.076	1.758
	[Gender=0]	-5.626	1.329	17.934	1	.000	.004	.000	.049
	[Gender=1]	0[b]	.	.	0	.	.	.	.
	Sex	.417	.122	11.683	1	.001	1.518	1.195	1.928
	[Gender=0] * Funny	1.172	.199	34.627	1	.000	3.230	2.186	4.773
	[Gender=1] * Funny	0[b]	.	.	0	.	.	.	.
	[Gender=0] * Sex	-.477	.163	8.505	1	.004	.621	.451	.855
	[Gender=1] * Sex	0[b]	.	.	0	.	.	.	.

a. The reference category is: No response/Walk Off.

b. This parameter is set to zero because it is redundant.

The bottom half of SPSS Output 8.14 shows the individual parameter estimates for the *Go Home with Person* category compared to the 'No response/walked away' category. We can interpret these effects as follows:

- **Good_Mate**: Whether the chat-up line showed signs of good moral fibre did not significantly predict whether you went home with the date or got a slap in the face, $b = 0.13$, Wald $\chi^2(1) = 2.42$, $p > .05$. In short, you're not significantly more likely to go home with the person if you use a chat-up line that demonstrates good moral fibre.
- **Funny**: Whether the chat-up line was funny significantly predicted whether you went home with the date or no response, $b = 0.32$, Wald $\chi^2(1) = 6.46$, $p < .05$. The odds ratio tells us that as chat-up lines are one more unit funnier, the change in the odds of going home with the person (rather than no response) is 1.38. In short, you're more likely to go home with the person than get no response if you use a chat-up line that is funny. (This effect, though, is superseded by the interaction with gender below.)
- **Gender**: The gender of the person being chatted up significantly predicted whether they went home with the person or gave no response, $b = -5.63$, Wald $\chi^2(1) = 17.93$, $p < .001$. The odds ratio tells us that as gender changes from female (0) to male (1), the change in the odds of going home with the person compared to not responding is 0.004. In other words, the odds of a man going home with someone compared to not responding are 1/0.004 = 250 times more likely than for a woman. Men are *really* cheap.
- **Sex**: The sexual content of the chat-up line significantly predicted whether you went home with the date or got a slap in the face, $b = 0.42$, Wald $\chi^2(1) = 11.68$, $p < .01$. The odds ratio tells us that as the sexual content increased by a unit, the change in the odds of going home with the person (rather than no response) is 1.52: you're more likely to go home with the person than not if you use a chat-up line with high sexual content.
- **Funny×Gender**: The success of funny chat-up lines depended on whether they were delivered to a man or a woman because in interaction these variables predicted whether or not you went home with the date, $b = 1.17$, Wald $\chi^2(1) = 34.63$, $p < .001$. The odds ratio tells us that as gender changes from female (0) to male (1) in combination with funniness increasing, the change in the odds of going home with the person compared to not responding is 3.23. As funniness increases, women become more likely to go home with the person than men. Funny chat-up lines are more successful when used on women compared to men.

- **Sex×Gender**: The success of chat-up lines with sexual content depended on whether they were delivered to a man or a woman because in interaction these variables predicted whether or not you went home with the date, $b = -0.48$, Wald $\chi^2(1) = 8.51$, $p < .01$. The odds ratio tells us that as gender changes from female (0) to male (1) in combination with the sexual content increasing, the change in the odds of going home with the date compared to not responding is 0.62. As sexual content increases, women become less likely than men to go home with the person. Chat-up lines with sexual content are more successful when used on men than women.

SELF-TEST Use what you learnt earlier in this chapter to check the assumptions of multicollinearity and linearity of the logit

8.9.5. Reporting the results

We can report the results as with binary logistic regression using a table (see Table 8.4). Note that I have split the table by the outcome categories being compared, but otherwise it is the same as before. These effects are interpreted as in the previous section.

TABLE 8.4 How to report multinomial logistic regression

		95% CI for Odds Ratio		
	***B* (SE)**	***Lower***	***Odds Ratio***	***Upper***
Phone Number vs. No Response				
Intercept	−1.78 (0.67)**			
Good Mate	.13 (0.05)*	1.03	1.14	1.27
Funny	.14 (0.11)	.93	1.15	1.43
Gender	−1.65 (0.80)*	.04	.19	.92
Sexual Content	.28 (0.09)**	1.11	1.32	1.57
Gender × Funny	.49 (0.14)***	1.24	1.64	2.15
Gender × Sex	−.35 (0.11)**	.57	.71	.87
Going Home vs. No Response				
Intercept	−4.29 (0.94)***			
Good Mate	.13 (0.08)	.97	1.14	1.34
Funny	.32 (0.13)*	1.08	1.38	1.76
Gender	−5.63 (1.33)***	.00	.00	.05
Sexual Content	.42 (0.12)**	1.20	1.52	1.93
Gender × Funny	1.17 (0.20)***	2.19	3.23	4.77
Gender × Sex	−.48 (0.16)**	.45	.62	.86

Note: $R^2 = .24$ (Cox & Snell), .28 (Nagelkerke). Model $\chi^2(12) = 278.53$, $p < .001$. * $p < .05$, ** $p < .01$, *** $p < .001$.

What have I discovered about statistics? ①

At the age of 10 I thought I was going to be a rock star. Such was my conviction about this that even today (many years on) I'm still not entirely sure how I ended up *not* being a rock star (lack of talent, not being a very cool person, inability to write songs that don't make people want to throw rotting vegetables at you, are all possible explanations). Instead of the glitzy and fun life that I anticipated I am instead reduced to writing chapters about things that I don't even remotely understand.

We began the chapter by looking at why we can't use linear regression when we have a categorical outcome, but instead have to use binary logistic regression (two outcome categories) or multinomial logistic regression (several outcome categories). We then looked into some of the theory of logistic regression by looking at the regression equation and what it means. Then we moved onto assessing the model and talked about the log-likelihood statistic and the associated chi-square test. I talked about different methods of obtaining equivalents to R^2 in regression (Hosmer & Lemeshow, Cox & Snell and Nagelkerke). We also discovered the Wald statistic and *Exp(B)*. The rest of the chapter looked at three examples using SPSS to carry out various logistic regression. So, hopefully, you should have a pretty good idea of how to conduct and interpret a logistic regression by now.

Having decided that I was going to be a rock star I put on my little denim jacket with Iron Maiden patches sewn onto it and headed off down the rocky road of stardom. The first stop was … my school.

Key terms that I've discovered

–2LL
Binary logistic regression
Chi-square distribution
Complete separation
Cox and Snell's R^2_{CS}
Exp(B)
Hosmer and Lemeshow's R^2_L
Interaction effect
Likelihood
Logistic regression
Log-likelihood
Main effect
Maximum-likelihood estimation
Multinomial logistic regression
Nagelkerke's R^2_N
Odds
Polychotomous logistic regression
Roa's efficient score statistic
Suppressor effects
Wald statistic

Smart Alex's tasks

- **Task 1**: A psychologist was interested in whether children's understanding of display rules can be predicted from their age, and whether the child possesses a theory of mind. A display rule is a convention of displaying an appropriate emotion in a given situation. For example, if you receive a Christmas present that you don't like, the appropriate emotional display is to smile politely and say 'Thank you Auntie Kate,

I've always wanted a rotting cabbage.' The inappropriate emotional display is to start crying and scream 'Why did you buy me a rotting cabbage you selfish old bag?' Using appropriate display rules has been linked to having a theory of mind (the ability to understand what another person might be thinking). To test this theory, children were given a false belief task (a task used to measure whether someone has a theory of mind), a display rule task (which they could either pass or fail) and their age in months was measured. The data are in **Display.sav**. Run a logistic regression to see whether possession of display rule understanding (did the child pass the test: Yes/No?) can be predicted from possession of a theory of mind (did the child pass the false belief task: Yes/No?), age in months and their interaction. ③

- **Task 2**: Recent research has shown that lecturers are among the most stressed workers. A researcher wanted to know exactly what it was about being a lecturer that created this stress and subsequent burnout. She took 467 lecturers and administered several questionnaires to them that measured: **Burnout** (burnt out or not), **Perceived Control** (high score = low perceived control), **Coping Style** (high score = high ability to cope with stress), **Stress from Teaching** (high score = teaching creates a lot of stress for the person), **Stress from Research** (high score = research creates a lot of stress for the person) and **Stress from Providing Pastoral Care** (high score = providing pastoral care creates a lot of stress for the person). The outcome of interest was burnout, and Cooper, Sloan, and Williams' (1988) model of stress indicates that perceived control and coping style are important predictors of this variable. The remaining predictors were measured to see the unique contribution of different aspects of a lecturer's work to their burnout. Can you help her out by conducting a logistic regression to see which factors predict burnout? The data are in **Burnout.sav**. ③
- **Task 3**: A health psychologist interested in research into HIV wanted to know the factors that influenced condom use with a new partner (relationship less than 1 month old). The outcome measure was whether a condom was used (Use: condom used = 1, not used = 0). The predictor variables were mainly scales from the Condom Attitude Scale (CAS) by Sacco, Levine, Reed, and Thompson (*Psychological Assessment: A Journal of Consulting and Clinical Psychology*, 1991): **Gender** (gender of the person); **Safety** (relationship safety, measured out of 5, indicates the degree to which the person views this relationship as 'safe' from sexually transmitted disease); **Sexexp** (sexual experience, measured out of 10, indicates the degree to which previous experience influences attitudes towards condom use); **Previous** (a measure not from the CAS, this variable measures whether or not the couple used a condom in their previous encounter, 1 = condom used, 0 = not used, 2 = no previous encounter with this partner); **selfcon** (self-control, measured out of 9, indicates the degree of self-control that a person has when it comes to condom use, i.e. do they get carried away with the heat of the moment, or do they exert control?); **Perceive** (perceived risk, measured out of 6, indicates the degree to which the person feels at risk from unprotected sex). Previous research (Sacco, Rickman, Thompson, Levine, and Reed, 1993) has shown that **Gender**, **relationship safety** and **perceived risk** predict condom use. Carry out an appropriate analysis to verify these previous findings, and to test whether self-control, previous usage and sexual experience can predict any of the remaining variance in condom use. (1) Interpret all important parts of the SPSS output. (2) How reliable is the final model? (3) What are the probabilities that participants 12, 53 and 75 will use a condom? (4) A female who used a condom in her previous encounter with her new partner scores 2 on all variables except perceived risk (for which she scores 6). Use the model to estimate the probability that she will use a condom in her next encounter. Data are in the file **condom.sav**. ③

Answers can be found on the companion website.

Further reading

Hutcheson, G., & Sofroniou, N. (1999). *The multivariate social scientist*. London: Sage. Chapter 4.

Menard, S. (1995). *Applied logistic regression analysis*. Sage university paper series on quantitative applications in the social sciences, 07–106. Thousand Oaks, CA: Sage. (This is a fairly advanced text, but great nevertheless. Unfortunately, few basic-level texts include logistic regression so you'll have to rely on what I've written!)

Miles, J. N. V., & Shevlin, M. (2001). *Applying regression and correlation: a guide for students and researchers*. London: Sage. (Chapter 6 is a nice introduction to logistic regression.)

Online tutorial

The companion website contains the following Flash movie tutorial to accompany this chapter:

- Logistic regression using SPSS

Interesting real research

Bale, C., Morrison, R., & Caryl, P. G. (2006). Chat-up lines as male sexual displays. *Personality and Individual Differences*, *40*(4), 655–664.

Bemelman, M., & Hammacher, E. R. (2005). Rectal impalement by pirate ship: A case report. *Injury Extra*, *36*, 508–510.

Cooper, M., O'Donnell, D., Caryl, P. G., Morrison, R., & Bale, C. (2007). Chat-up lines as male displays: Effects of content, sex, and personality. *Personality and Individual Differences*, *43*(5), 1075–1085.

Lacourse, E., Claes, M., & Villeneuve, M. (2001). Heavy metal music and adolescent suicidal risk. *Journal of Youth and Adolescence*, *30*(3), 321–332.

Lo, S. F., Wong, S. H., Leung, L. S., Law, I. C., & Yip, A. W. C. (2004). Traumatic rectal perforation by an eel. *Surgery*, *135*(1), 110–111.

9 Comparing two means

FIGURE 9.1 My (probably) 8th birthday. L–R: My brother Paul (who still hides behind cakes rather than have his photo taken), Paul Spreckley, Alan Palsey, Clair Sparks and me

9.1. What will this chapter tell me? ①

Having successfully slayed audiences at holiday camps around the country, my next step towards global domination was my primary school. I had learnt another Chuck Berry song ('Johnny B. Goode'), but also broadened my repertoire to include songs by other artists (I have a feeling 'Over the edge' by Status Quo was one of them).[1] Needless to say, when the opportunity came to play at a school assembly I jumped at it. The headmaster tried to have me banned,[2] but the show went on. It was a huge success (I want to reiterate my

[1] This would have been about 1982, so just before they became the most laughably bad band on the planet. Some would argue that they were *always* the most laughably bad band on the planet, but they were the first band that I called my favourite band.

[2] Seriously! Can you imagine, a headmaster banning a 10 year old from assembly? By this time I had an electric guitar and he used to play hymns on an acoustic guitar; I can assume only that he somehow lost all perspective on the situation and decided that a 10 year old blasting out some Quo in a squeaky little voice was subversive or something.

earlier point that 10 year olds are very easily impressed). My classmates carried me around the playground on their shoulders. I was a hero. Around this time I had a childhood sweetheart called Clair Sparks. Actually, we had been sweethearts since before my new-found rock legend status. I don't think the guitar playing and singing impressed her much, but she rode a motorbike (really, a little child's one) which impressed *me* quite a lot; I was utterly convinced that we would one day get married and live happily ever after. I was utterly convinced, that is, until she ran off with Simon Hudson. Being 10, she probably literally did run off with him – across the playground. To make this important decision of which boyfriend to have, Clair had needed to compare two things (Andy and Simon) to see which one was better; sometimes in science we want to do the same thing, to compare two things to see if there is evidence that one is different to the other. This chapter is about the process of comparing two means using a *t*-test.

9.2. Looking at differences ①

Rather than looking at relationships between variables, researchers are sometimes interested in looking at differences between groups of people. In particular, in experimental research we often want to manipulate what happens to people so that we can make causal inferences. For example, if we take two groups of people and randomly assign one group a programme of dieting pills and the other group a programme of sugar pills (which they think will help them lose weight) then if the people who take the dieting pills lose more weight than those on the sugar pills we can infer that the diet pills caused the weight loss. This is a powerful research tool because it goes one step beyond merely observing variables and looking for relationships (as in correlation and regression).[3] This chapter is the first of many that looks at this kind of research scenario, and we start with the simplest scenario: when we have two groups, or, to be more specific, when we want to compare two means. As we have seen (Chapter 1), there are two different ways of collecting data: we can either expose different people to different experimental manipulations (*between-group* or *independent* design), or take a single group of people and expose them to different experimental manipulations at different points in time (a *repeated-measures* design).

9.2.1. A problem with error bar graphs of repeated-measures designs ①

We also saw in Chapter 4 that it is important to visualize group differences using error bars. We're now going to look at a problem that occurs when we graph repeated-measures error bars. To do this, we're going to look at an example that I use throughout this chapter (not because I am too lazy to think up different data sets, but because it allows me to illustrate various things). The example relates to whether arachnophobia (fear of spiders) is specific to real spiders or whether pictures of spiders can evoke similar levels of anxiety. Twenty-four arachnophobes were used in all. Twelve were asked to play with a big hairy tarantula spider with big fangs and an evil look in its eight eyes. Their subsequent anxiety was measured. The remaining twelve were shown only pictures of the same big hairy tarantula and again their anxiety was measured. The data are in Table 9.1 (and **spiderBG.sav** if you're having difficulty entering them into SPSS yourself). Remember that each row in the data editor represents a

[3] People sometimes get confused and think that certain statistical procedures allow causal inferences and others don't (see Jane Superbrain Box 1.4).

TABLE 9.1 Data from **spiderBG.sav**

Participant	*Group*	*Anxiety*
1	0	30
2	0	35
3	0	45
4	0	40
5	0	50
6	0	35
7	0	55
8	0	25
9	0	30
10	0	45
11	0	40
12	0	50
13	1	40
14	1	35
15	1	50
16	1	55
17	1	65
18	1	55
19	1	50
20	1	35
21	1	30
22	1	50
23	1	60
24	1	39

different participant's data. Therefore, you need a column representing the group to which they belonged and a second column representing their anxiety. The data in Table 9.1 show only the group codes and not the corresponding label. When you enter the data into SPSS, remember to tell the computer that a code of 0 represents the group that were shown the picture, and that a code of 1 represents the group that saw the real spider (see section 3.4.2.3).

SELF-TEST Enter these data into SPSS. Using what you learnt in Chapter 4, plot an error bar graph of the spider data.

OK, now let's imagine that we'd collected these data using the *same* participants; that is, all participants had their anxiety rated after seeing the real spider, but also after seeing

TABLE 9.2 Data from **spiderRM.sav**

Subject	Picture (Anxiety score)	Real (Anxiety Score)
1	30	40
2	35	35
3	45	50
4	40	55
5	50	65
6	35	55
7	55	50
8	25	35
9	30	30
10	45	50
11	40	60
12	50	39

the picture (in counterbalanced order obviously!). The data would now be arranged differently in SPSS. Instead of having a coding variable, and a single column with anxiety scores in, we would arrange the data in two columns (one representing the **picture** condition and one representing the **real** condition). The data are displayed in Table 9.2 (and **spiderRM.sav** if you're having difficulty entering them into SPSS yourself). Note that the anxiety scores are identical to the between-group data (Table 9.1) – it's just that we're pretending that they came from the same people rather than different people.

SELF-TEST Enter these data into SPSS. Using what you learnt in Chapter 4, plot an error bar graph of the spider data.

Figure 9.2 shows the error bar graphs from the two different designs. Remember that the data are exactly the same, all that has changed is whether the design used the same participants (repeated measures) or different (independent). Now, we discovered in Chapter 1 that repeated-measures designs eliminate some extraneous variables (such as age, IQ and so on) and so can give us more sensitivity in the data. Therefore, we would expect our graphs to be different: the repeated-measures graph should reflect the increased sensitivity in the design. Looking at the two error bar graphs, can you spot this difference between the graphs?

Hopefully you're answer was 'no' because, of course, the graphs are identical! This similarity reflects the fact that when you create an error bar graph of repeated-measures data, SPSS treats the data as though different groups of participants were used! In other words, the error bars do not reflect the 'true' error around the means for repeated-measures designs. Fortunately we can correct this problem manually, and that's what we will discover now.

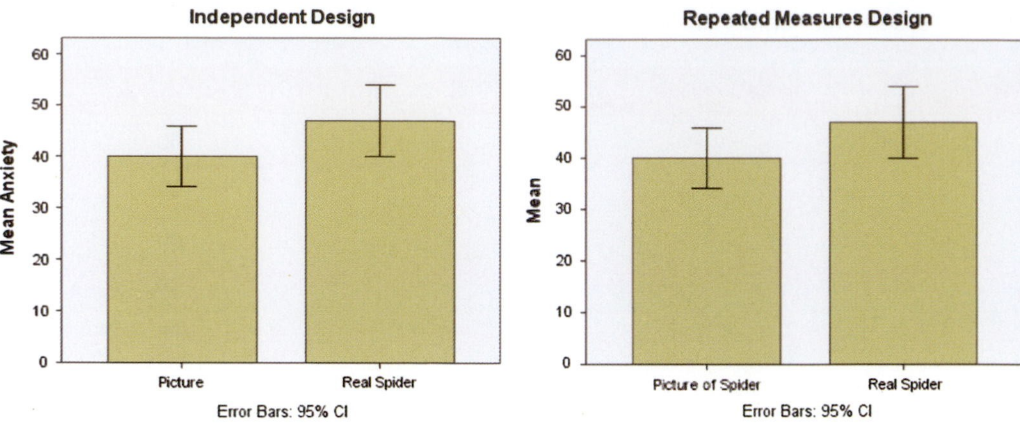

FIGURE 9.2 Two error bar graphs of anxiety data in the presence of a real spider or a photograph. The data on the left are treated as though they are different participants, whereas those on the right are treated as though they are from the same participants

9.2.2. Step 1: calculate the mean for each participant ②

To correct the repeated-measures error bars, we need to use the *compute* command that we encountered in Chapter 5. To begin with, we need to calculate the average anxiety for each participant and so we use the *mean* function. Access the main compute dialog box by selecting Transform Compute Variable.... Enter the name **Mean** into the box labelled *Target Variable* and then in the list labelled *Function group* select *Statistical* and then in the list labelled *Functions and Special Variables* select *Mean*. Transfer this command to the command area by clicking on . When the command is transferred, it appears in the command area as *MEAN(?,?)*; the question marks should be replaced with variable names (which can be typed manually or transferred from the variables list). So replace the first question mark with the variable **picture** and the second one with the variable **real**. The completed dialog box should look like the one in Figure 9.3. Click on OK to create this new variable, which will appear as a new column in the data editor.

9.2.3. Step 2: calculate the grand mean ②

The **grand mean** is the mean of all scores (regardless of which condition the score comes from) and so for the current data this value will be the mean of all 24 scores. One way to calculate this is by hand (i.e. add up all of the scores and divide by 24); however, an easier way is to use the means that we have just calculated. The means we have just calculated represent the average score for each participant and so if we take the average of those mean scores, we will have the mean of all participants (i.e. the grand mean) – phew, there were a lot of means in that sentence! OK, to do this we can use a useful little gadget called the *descriptives* command (you could also use the *Explore* or *Frequencies* functions that we came across in Chapter 5, but as I've already covered those we'll try something different). Access the *descriptives* command by selecting Analyze Descriptive Statistics ▸ Descriptives.... The dialog box in Figure 9.4 should appear. The *descriptives* command is used to get basic descriptive statistics for variables and by clicking on Options... a second dialog box is activated. Select the variable **Mean** from the list and transfer it to the box labelled *Variable(s)* by clicking on . Then, use the options dialog box to specify only the mean (you can leave the default settings as they are, but it is only the mean in which we are interested). If you run this analysis the output should provide you with

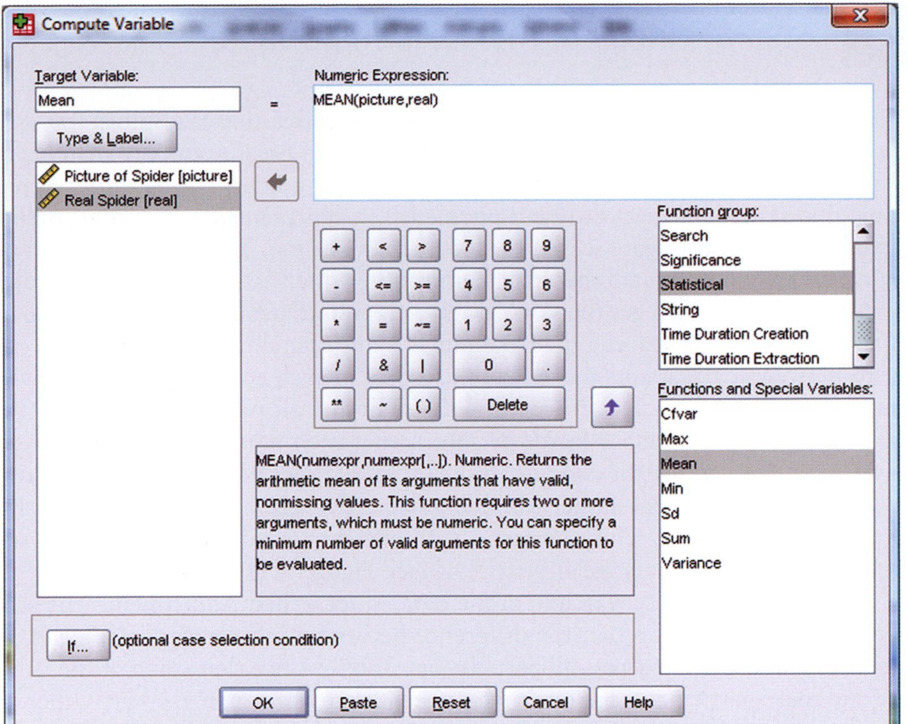

FIGURE 9.3 Using the compute function to calculate the mean of two columns

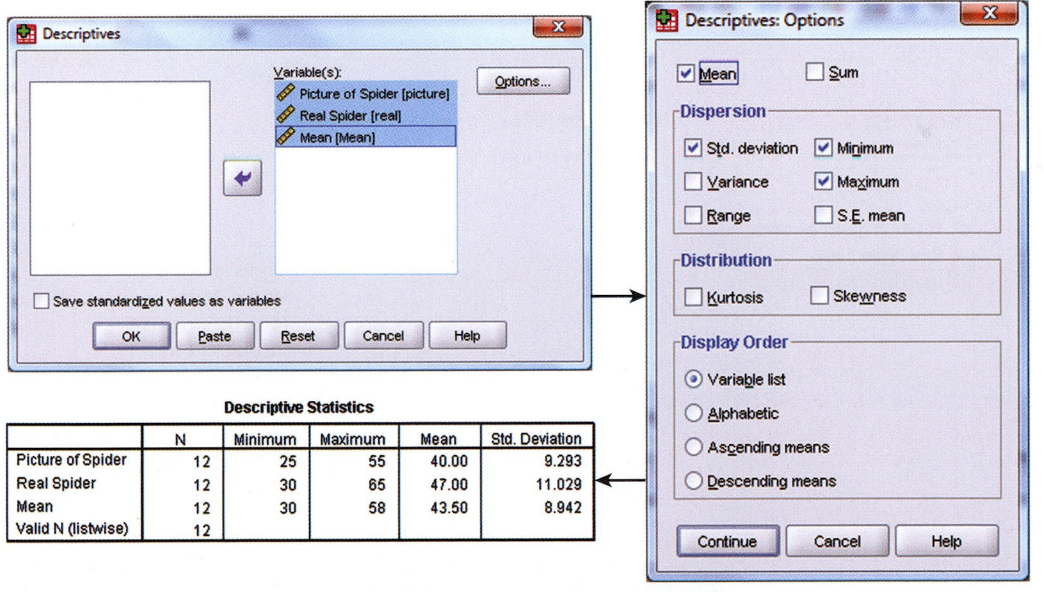

Descriptive Statistics

	N	Minimum	Maximum	Mean	Std. Deviation
Picture of Spider	12	25	55	40.00	9.293
Real Spider	12	30	65	47.00	11.029
Mean	12	30	58	43.50	8.942
Valid N (listwise)	12				

FIGURE 9.4 Dialog boxes and output for descriptive statistics

some self-explanatory descriptive statistics for each of the three variables (assuming you selected all three). You should see that we get the mean of the picture condition, and the mean of the real spider condition, but it's actually the final variable we're interested in: the mean of the picture and spider condition. The mean of this variable is the grand mean, and you can see from the summary table that its value is 43.50. We will use this grand mean in the following calculations.

9.2.4. Step 3: calculate the adjustment factor ②

If you look at the variable labelled **Mean**, you should notice that the values for each participant are different, which tells us that some people had greater anxiety than others did across the conditions. The fact that participants' mean anxiety scores differ represents individual differences between different people (so, it represents the fact that some of the participants are generally more scared of spiders than others). These differences in natural anxiety to spiders contaminate the error bar graphs, which is why if we don't adjust the values that we plot, we will get the same graph as if an independent design had been used. Loftus and Masson (1994) argue that to eliminate this contamination we should equalize the means between participants (i.e. adjust the scores in each condition such that when we take the mean score across conditions, it is the same for all participants). To do this, we need to calculate an adjustment factor by subtracting each participant's mean score from the grand mean. We can use the *compute* function to do this calculation for us. Activate the compute dialog box, give the target variable a name (I suggest **Adjustment**) and then use the command '43.5-mean'. This command will take the grand mean (43.5) and subtract from it each participant's average anxiety level (see Figure 9.5).

This process creates a new variable in the data editor called **Adjustment**. The scores in the column **Adjustment** represent the difference between each participant's mean anxiety and the mean anxiety level across all participants. You'll notice that some of the values are positive and these participants are one's who were less anxious than average. Other participants were more anxious than average and they have negative adjustment scores. We can now use these adjustment values to eliminate the between-subject differences in anxiety.

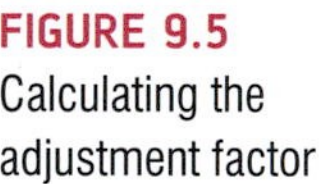

FIGURE 9.5 Calculating the adjustment factor

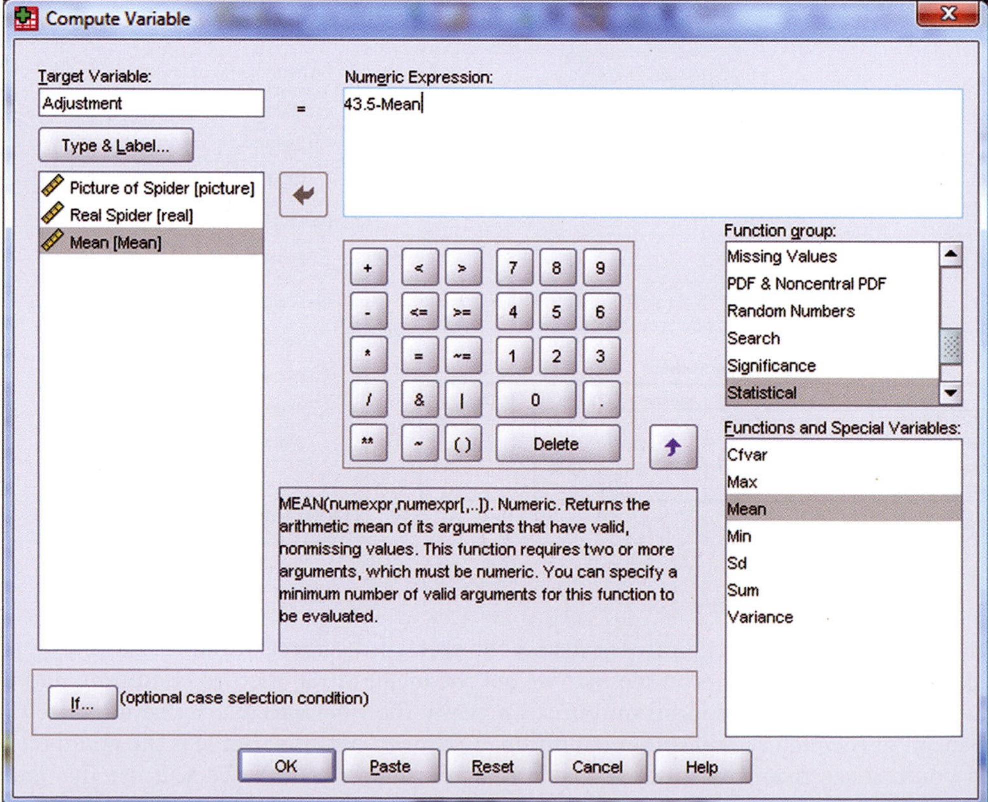

9.2.5. Step 4: create adjusted values for each variable ②

So far, we have calculated the difference between each participant's mean score and the mean score of all participants (the grand mean). This difference can be used to adjust the existing scores for each participant. First we need to adjust the scores in the **picture** condition. Once again, we can use the *compute* command to make the adjustment. Activate the compute dialog box in the same way as before, and then title our new variable **Picture_Adjusted** (you can then click on Type & Label... and give this variable a label such as 'Picture Condition: Adjusted Values'). All we are going to do is to add each participant's score in the **picture** condition to their adjustment value. Select the variable **picture** and transfer it to the command area by clicking on , then click on + and select the variable **Adjustment** and transfer it to the command area by clicking on . The completed dialog box is shown in Figure 9.6. Now do the same thing for the variable **real**: create a variable called **Real_Adjusted** that contains the values of **real** added to the value in the **Adjustment** column.

Now, the variables **Real_Adjusted** and **Picture_Adjusted** represent the anxiety experienced in each condition, adjusted so as to eliminate any between-subject differences. If you don't believe me, then use the *compute* command to create a variable **Mean2** that is the average of **Real_Adjusted** and **Picture_Adjusted** (just like we did in section 9.2.2). You should find that the value in this column is the same for every participant, thus proving that the between-subject variability in means is gone: the value will be 43.50 (the grand mean).

SELF-TEST Create an error bar chart of the mean of the adjusted values that you have just made (**Real_Adjusted** and **Picture_Adjusted**).

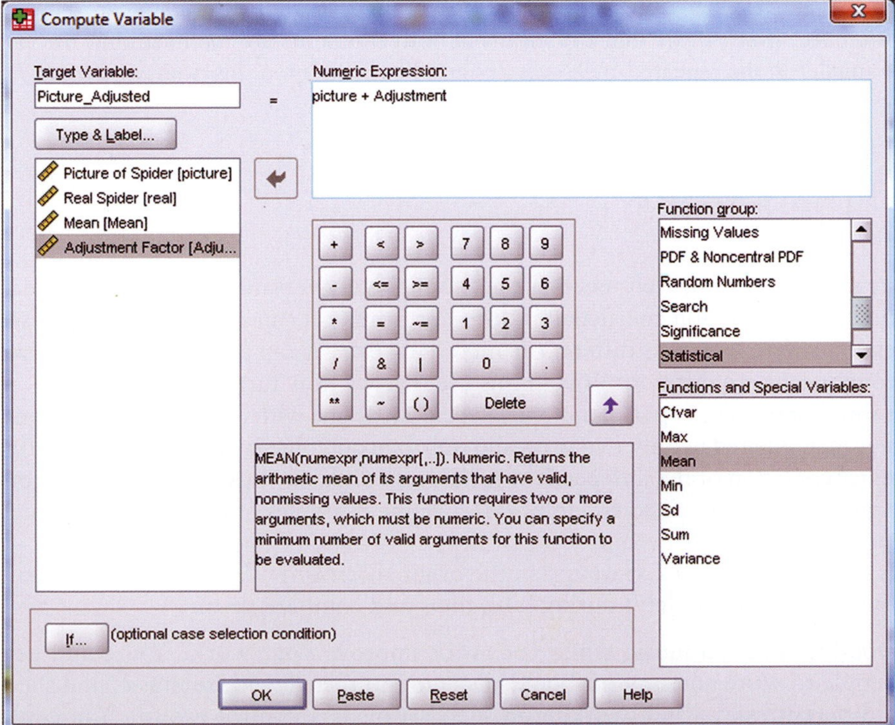

FIGURE 9.6 Adjusting the values of **picture**

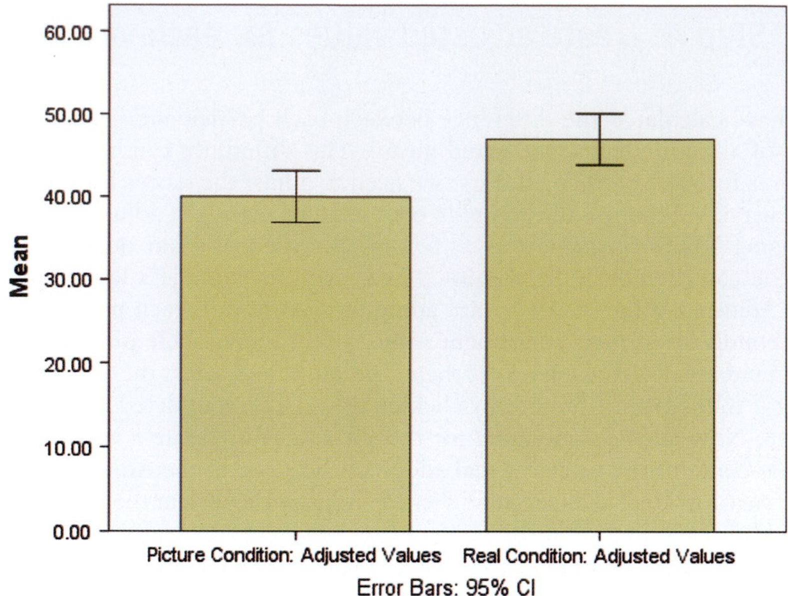

FIGURE 9.7 Error bar graph of the adjusted values of **spiderRM.sav**

The resulting error bar graph is shown in Figure 9.7. Compare this graph to the graphs in Figure 9.2 – what differences do you see? The first thing to notice is that the means in the two conditions have not changed. However, the error bars have changed: they have got smaller. Also, whereas in Figure 9.2 the error bars overlap, in this new graph they do not. In Chapter 2 we discovered that when error bars do not overlap we can be fairly confident that our samples have not come from the same population (and so our experimental manipulation has been successful). Therefore, when we plot the proper error bars for the repeated-measures data it shows the extra sensitivity that this design has: the differences between conditions appear to be significant, whereas when different participants are used, there does not appear to be a significant difference. (Remember that the means in both situations are identical, but the sampling error is smaller in the repeated-measures design.) I expand upon this point in section 9.6.

9.3. The *t*-test ①

We have seen in previous chapters that the *t*-test is a very versatile statistic: it can be used to test whether a correlation coefficient is different from 0; it can also be used to test whether a regression coefficient, *b*, is different from 0. However, it can also be used to test whether two group means are different. It is to this use that we now turn.

The simplest form of experiment that can be done is one with only one independent variable that is manipulated in only two ways and only one outcome is measured. More often than not the manipulation of the independent variable involves having an experimental condition and a control group (see Field & Hole, 2003). Some examples of this kind of design are:

- Is the movie *Scream 2* scarier than the original *Scream*? We could measure heart rates (which indicate anxiety) during both films and compare them.
- Does listening to music while you work improve your work? You could get some people to write an essay (or book!) listening to their favourite music, and then write a different essay when working in silence (this is a control group). You could then compare the essay grades!

- Does listening to Andy's favourite music improve your work? You could repeat the above but rather than letting people work with their favourite music, you could play them some of my favourite music (as listed in the acknowledgements) and watch the quality of their work plummet!

What's the difference between the independent and dependent *t*-test?

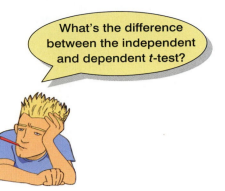

The *t*-test can analyse these sorts of scenarios. Of course, there are more complex experimental designs and we will look at these in subsequent chapters. There are, in fact, two different *t*-tests and the one you use depends on whether the independent variable was manipulated using the same participants or different:

- **Independent-means *t*-test:** This test is used when there are two experimental conditions and different participants were assigned to each condition (this is sometimes called the *independent-measures* or *independent-samples t*-test).
- **Dependent-means *t*-test:** This test is used when there are two experimental conditions and the same participants took part in both conditions of the experiment (this test is sometimes referred to as the *matched-pairs* or *paired-samples t*-test).

9.3.1. Rationale for the *t*-test ①

Both *t*-tests have a similar rationale, which is based on what we learnt in Chapter 2 about hypothesis testing:

- Two samples of data are collected and the sample means calculated. These means might differ by either a little or a lot.
- If the samples come from the same population, then we expect their means to be roughly equal (see section 2.5.1). Although it is possible for their means to differ by chance alone, we would expect large differences between sample means to occur very infrequently. Under the null hypothesis we assume that the experimental manipulation has no effect on the participants: therefore, we expect the sample means to be very similar.
- We compare the difference between the sample means that we collected to the difference between the sample means that we would expect to obtain if there were no effect (i.e. if the null hypothesis were true). We use the standard error (see section 2.5.1) as a gauge of the variability between sample means. If the standard error is small, then we expect most samples to have very similar means. When the standard error is large, large differences in sample means are more likely. If the difference between the samples we have collected is larger than what we would expect based on the standard error then we can assume one of two things:
    - There is no effect and sample means in our population fluctuate a lot and we have, by chance, collected two samples that are atypical of the population from which they came.
    - The two samples come from different populations but are typical of their respective parent population. In this scenario, the difference between samples represents a genuine difference between the samples (and so the null hypothesis is incorrect).
- As the observed difference between the sample means gets larger, the more confident we become that the second explanation is correct (i.e. that the null hypothesis should be rejected). If the null hypothesis is incorrect, then we gain confidence that the two sample means differ because of the different experimental manipulation imposed on each sample.

I mentioned in section 2.6.1 that most test statistics can be thought of as the 'variance explained by the model' divided by the 'variance that the model can't explain'. In other words, effect/error. When comparing two means the 'model' that we fit to the data (the effect) is the difference between the two group means. We saw also in Chapter 2 that means vary from sample to sample (sampling variation) and that we can use the standard error as a measure of how much means fluctuate (in other words, the error in the estimate of the mean). Therefore, we can also use the standard error of the differences between the two means as an estimate of the error in our model (or the error in the difference between means). Therefore, we calculate the *t*-test using equation (9.1) below. The top half of the equation is the 'model' (our model being the difference between means is bigger than the expected difference, which in most cases will be 0 – we expect the difference between means to be different to zero). The bottom half is the 'error'. So, just as I said in Chapter 2, we're basically getting the test statistic by dividing the model (or effect) by the error in the model. The exact form that this equation takes depends on whether the same or different participants were used in each experimental condition:

$$t = \frac{\begin{array}{c}\text{observed difference}\\ \text{between sample means}\end{array} - \begin{array}{c}\text{expected difference}\\ \text{between population means}\\ \text{(if null hypothesis is true)}\end{array}}{\begin{array}{c}\text{estimate of the standard error of the}\\ \text{difference between two sample means}\end{array}} \tag{9.1}$$

9.3.2. Assumptions of the *t*-test ①

Both the **independent *t*-test** and the **dependent *t*-test** are *parametric tests* based on the normal distribution (see Chapter 5). Therefore, they assume:

- The sampling distribution is normally distributed. In the dependent *t*-test this means that the sampling distribution of the *differences* between scores should be normal, not the scores themselves (see section 9.4.3).
- Data are measured at least at the interval level.

The independent *t*-test, because it is used to test different groups of people, also assumes:

- Variances in these populations are roughly equal (*homogeneity of variance*).
- Scores are independent (because they come from different people).

These assumptions were explained in detail in Chapter 5 and, in that chapter, I emphasized the need to check these assumptions before you reach the point of carrying out your statistical test. As such, I won't go into them again, but it does mean that if you have ignored my advice and haven't checked these assumptions then you need to do it now! SPSS also incorporates some procedures into the *t*-test (e.g. Levene's test, see section 5.6.1, can be done at the same time as the *t*-test). Let's now look at each of the two *t*-tests in turn.

9.4. The dependent *t*-test ①

If we stay with our repeated-measures data for the time being we can look at the dependent *t*-test, or paired-samples *t*-test. The dependent *t*-test is easy to calculate. In effect, we use a numeric version of equation (9.1):

$$t = \frac{\overline{D} - \mu_D}{s_D/\sqrt{N}} \tag{9.2}$$

Equation (9.2) compares the mean difference between our samples ($\bar{D}$) to the difference that we would expect to find between population means (μ_D), and then takes into account the standard error of the differences ($s_D/\sqrt{N}$). If the null hypothesis is true, then we expect there to be no difference between the population means (hence $\mu_D = 0$).

9.4.1. Sampling distributions and the standard error ①

In equation (9.1) I referred to the lower half of the equation as the standard error of differences. The standard error was introduced in section 2.5.1 and is simply the standard deviation of the sampling distribution. Have a look back at this section now to refresh your memory about sampling distributions and the standard error. Sampling distributions have several properties that are important. For one thing, if the population is normally distributed then so is the sampling distribution; in fact, if the samples contain more than about 50 scores the sampling distribution should be normally distributed. The mean of the sampling distribution is equal to the mean of the population, so the average of all possible sample means should be the same as the population mean. This property makes sense because if a sample is representative of the population then you would expect its mean to be equal to that of the population. However, sometimes samples are unrepresentative and their means differ from the population mean. On average, though, a sample mean will be very close to the population mean and only rarely will the sample mean be substantially different from that of the population. A final property of a sampling distribution is that its standard deviation is equal to the standard deviation of the population divided by the square root of the number of observations in the sample. As I mentioned before, this standard deviation is known as the standard error.

We can extend this idea to look at the *differences* between sample means. If you were to take several pairs of samples from a population and calculate their means, then you could also calculate the difference between their means. I mentioned earlier that *on average* sample means will be very similar to the population mean: as such, most samples will have very similar means. Therefore, most of the time the difference between sample means from the same population will be zero, or close to zero. However, sometimes one or both of the samples could have a mean very deviant from the population mean and so it is possible to obtain large differences between sample means by chance alone. However, this would happen less frequently.

In fact, if you plotted these differences between sample means as a histogram, you would again have a sampling distribution with all of the properties previously described. The standard deviation of this sampling distribution is called the **standard error of differences**. A small standard error tells us that most pairs of samples from a population will have very similar means (i.e. the difference between sample means should normally be very small). A large standard error tells us that sample means can deviate quite a lot from the population mean and so differences between pairs of samples can be quite large by chance alone.

9.4.2. The dependent *t*-test equation explained ①

In an experiment, a person's score in condition 1 will be different to their score in condition 2, and this difference could be very large or very small. If we calculate the differences between each person's score in each condition and add up these differences we would get the total amount of difference. If we then divide this total by the number of participants we get the average difference (thus how much, on average, a person's score differed in condition 1 compared to condition 2). This average difference is $\bar{D}$ in equation (9.2) and it is an indicator

How does the
t-test actually work?

of the systematic variation in the data (i.e. it represents the experimental effect). We need to compare this systematic variation against some kind of measure of the 'systematic variation that we could naturally expect to find'. In Chapter 2 we saw that the standard deviation was a measure of the 'fit' of the mean to the observed data (i.e. it measures the error in the model when the model is the mean), but it does not measure the fit of the mean to the population. To do this we need the standard error (see the previous section, where we revised this idea).

The standard error is a measure of the error in the mean as a model of the population. In this context, we know that if we had taken two random samples from a population (and not done anything to these samples) then the means could be different just by chance. The standard error tells us by how much these samples could differ. A small standard error means that sample means should be quite similar, so a big difference between two sample means is unlikely. In contrast, a large standard error tells us that big differences between the means of two random samples are more likely. Therefore it makes sense to compare the average difference between means against the standard error of these differences. This gives us a test statistic that, as I've said numerous times in previous chapters, represents model/error. Our model is the average difference between condition means, and we divide by the standard error which represents the error associated with this model (i.e. how similar two random samples are likely to be from this population).

To clarify, imagine that an alien came down and cloned me millions of times. This population is known as Landy of the Andys (this would be possibly the most dreary and strangely terrifying place I could imagine). Imagine the aliens were interested in spider phobia in this population (because I am petrified of spiders). Everyone in this population (my clones) will be the same as me, and would behave in an identical way to me. If you took two samples from this population and measured their spider fear, then the means of these samples would be the same (we are clones), so the difference between sample means would be zero. Also, because we are all identical, then all samples from the population will be perfect reflections of the population (the standard error would be zero also). Therefore, if we were to get two samples that differed even very slightly then this would be very unlikely indeed (because our population is full of cloned Andys). Therefore, a difference between samples must mean that they have come from different populations. Of course, in reality we don't have samples that perfectly reflect the population, but the standard error gives an idea of how well samples reflect the population from which they came.

Therefore, by dividing by the standard error we are doing two things: (1) standardizing the average difference between conditions (this just means that we can compare values of t without having to worry about the scale of measurement used to measure the outcome variable); and (2) contrasting the difference between means that we have against the difference that we could *expect* to get based on how well the samples represent the populations from which they came. If the standard error is large, then large differences between samples are more common (because the distribution of differences is more spread out). Conversely, if the standard error is small, then large differences between sample means are uncommon (because the distribution is very narrow and centred around zero). Therefore, if the average difference between our samples is large, and the standard error of differences is small, then we can be confident that the difference we observed in our sample is not a chance result. If the difference is not a chance result then it must have been caused by the experimental manipulation.

In a perfect world, we could calculate the standard error by taking all possible pairs of samples from a population, calculating the differences between their means, and then working out the standard deviation of these differences. However, in reality this is impossible. Therefore, we estimate the standard error from the standard deviation of differences obtained within the sample (s_D) and the sample size (N). Think back to section 2.5.1 where we saw that the standard error is simply the standard deviation divided by the square root of the sample size; likewise the standard error of differences ($\sigma_{\bar{D}}$) is simply the standard deviation of differences divided by the square root of the sample size:

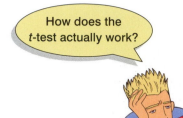

$$(\sigma_{\overline{D}}) = \frac{s_D}{\sqrt{N}}$$

If the standard error of differences is a measure of the unsystematic variation within the data, and the sum of difference scores represents the systematic variation, then it should be clear that the *t*-statistic is simply the ratio of the systematic variation in the experiment to the unsystematic variation. If the experimental manipulation creates any kind of effect, then we would expect the systematic variation to be much greater than the unsystematic variation (so at the very least, *t* should be greater than 1). If the experimental manipulation is unsuccessful then we might expect the variation caused by individual differences to be much greater than that caused by the experiment (so *t* will be less than 1). We can compare the obtained value of *t* against the maximum value we would expect to get by chance alone in a *t*-distribution with the same degrees of freedom (these values can be found in the Appendix); if the value we obtain exceeds this critical value we can be confident that this reflects an effect of our independent variable.

9.4.3. The dependent *t*-test and the assumption of normality ①

We talked about the assumption of normality in Chapter 5 and discovered that parametric tests (like the dependent *t*-test) assume that the sampling distribution is normal. This should be true in large samples, but in small samples people often check the normality of their data because if the data themselves are normal then the sampling distribution is likley to be also. With the dependent *t*-test we analyse the *differences* between scores because we're interested in the sampling distribution of these differences (not the raw data). Therefore, if you want to test for normality before a dependent *t*-test then what you should do is compute the differences between scores, and then check if this new variable is normally distributed (or use a big sample and not worry about normality!). It is possible to have two measures that are highly non-normal that produce beautifully distributed differences!

SELF-TEST Using the **spiderRM.sav** data, compute the differences between the picture and real condition and check the assumption of normality for these differences.

9.4.4. Dependent *t*-tests using SPSS ①

Using our spider data (**spiderRM.sav**), we have 12 spider-phobes who were exposed to a picture of a spider (**picture**) and on a separate occasion a real live tarantula (**real**). Their anxiety was measured in each condition (half of the participants were exposed to the picture before the real spider while the other half were exposed to the real spider first). I have already described how the data are arranged, and so we can move straight on to doing the test itself. First, we need to access the main dialog box by selecting Analyze Compare Means ▸ Paired-Samples T Test... (Figure 9.8). Once the dialog box is activated, you need to select pairs of variables to be analysed. In this case we have only one pair (**Real** vs. **Picture**). To select a pair you should click on the first variable that you want to select (in this case **Picture**), then hold down the *Ctrl* key on the keyboard and select the second (in this case **Real**). To transfer these

FIGURE 9.8 Main dialog box for paired-samples *t*-test

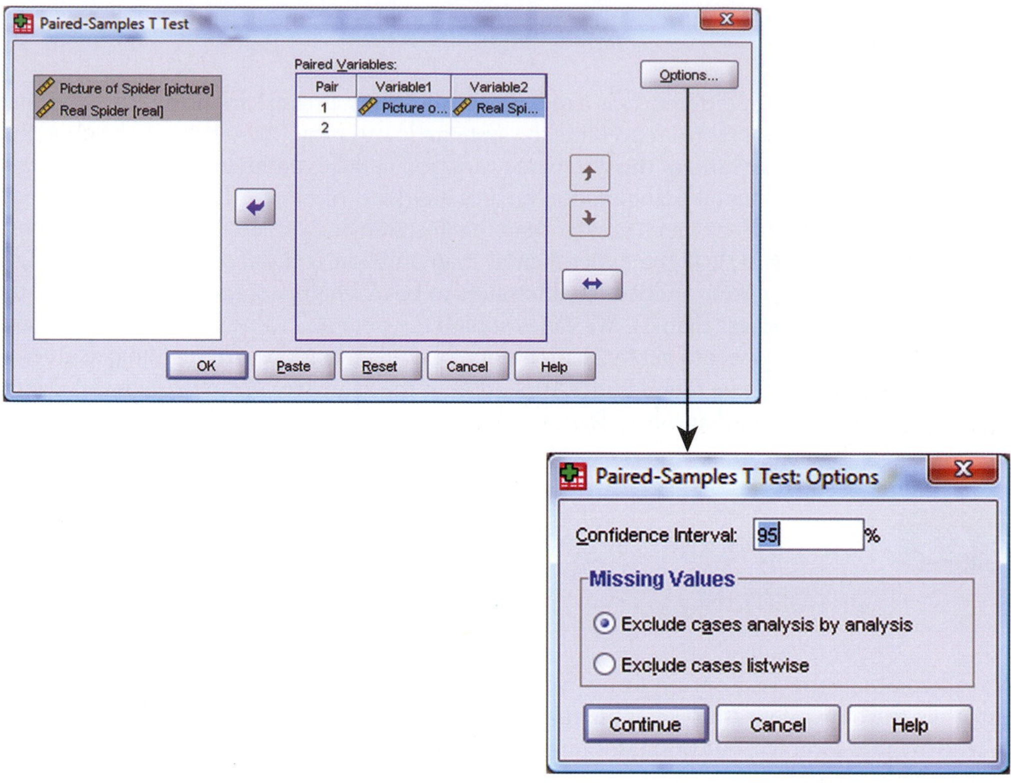

two variables to the box labelled *Paired Variables* click on ➡. (You can also select each variable individually and transfer it by clicking on ➡, but the method using the *Ctrl* key to select both variables is quicker.) If you want to carry out several *t*-tests then you can select another pair of variables, transfer them to the variables list, then select another pair and so on. In this case, we want only one test. If you click on Options... then another dialog box appears that gives you the chance to change the width of the confidence interval that is calculated. The default setting is for a 95% confidence interval and this is fine; however, if you want to be stricter about your analysis you could choose a 99% confidence interval but you run a higher risk of failing to detect a genuine effect (a Type II error). You can also select how to deal with missing values (see SPSS Tip 6.1). To run the analysis click on OK.

9.4.5. Output from the dependent *t*-test ①

The resulting output produces three tables. SPSS Output 9.1 shows a table of summary statistics for the two experimental conditions. For each condition we are told the mean, the number of participants (*N*) and the standard deviation of the sample. In the final column we are told the standard error (see section 9.4.1), which is the sample standard deviation divided by the square root of the sample size ($SE = s/\sqrt{N}$), so for the picture condition $SE = 9.2932/\sqrt{12} = 9.2932/3.4641 = 2.68$.

SPSS Output 9.1 also shows the Pearson correlation between the two conditions. When repeated measures are used it is possible that the experimental conditions will correlate (because the data in each condition come from the same people and so there could be some constancy in their responses). SPSS provides the value of Pearson's *r* and the two-tailed significance value (see Chapter 6). For these data the experimental conditions yield a fairly large correlation coefficient ($r = .545$) but are not significantly correlated because $p > .05$.

SPSS OUTPUT 9.1

Paired Samples Statistics

		Mean	N	Std. Deviation	Std. Error Mean
Pair 1	Picture of Spider	40.00	12	9.293	2.683
	Real Spider	47.00	12	11.029	3.184

Paired Samples Correlations

		N	Correlation	Sig.
Pair 1	Picture of Spider & Real Spider	12	.545	.067

SPSS Output 9.2 shows the most important of the tables: the one that tells us whether the difference between the means of the two conditions was large enough to *not* be a chance result. First, the table tells us the mean difference between scores (this value, i.e. $\bar{D}$ in equation (9.2), is the difference between the mean scores of each condition: 40 – 47 = –7). The table also reports the standard deviation of the differences between the means and more important the standard error of the differences between participants' scores in each condition (see section 9.4.1). The test statistic, *t*, is calculated by dividing the mean of differences by the standard error of differences (see equation (9.2): $t = -7/2.8311 = -2.47$). The size of *t* is compared against known values based on the degrees of freedom. When the same participants have been used, the degrees of freedom are simply the sample size minus 1 ($df = N - 1 = 11$). SPSS uses the degrees of freedom to calculate the exact probability that a value of *t* as big as the one obtained could occur if the null hypothesis were true (i.e. there was no difference between these means). This probability value is in the column labelled *Sig.* By default, SPSS provides only the two-tailed probability, which is the probability when no prediction was made about the direction of group differences. If a specific prediction was made (e.g. we might predict that anxiety will be higher when a real spider is used) then the one-tailed probability should be reported and this value is obtained by dividing the two-tailed probability by 2 (see SPSS Tip 9.1). The two-tailed probability for the spider data is very low ($p = .031$) and in fact it tells us that there is only a 3.1% chance that a value of *t* this big could happen if the null hypothesis were true. We saw in Chapter 2 that we generally accept a $p < .05$ as statistically meaningful; therefore, this *t* is significant because .031 is smaller than .05. The fact that the *t*-value is a negative number tells us that the first condition (the **picture** condition) had a smaller mean than the second (the **real** condition) and so the real spider led to greater anxiety than the picture. Therefore, we can conclude that exposure to a real spider caused significantly more reported anxiety in spider-phobes than exposure to a picture, $t(11) = -2.47$, $p < .05$. This result was predicted by the error bar chart in Figure 9.7.

SPSS OUTPUT 9.2

Paired Samples Test

		Paired Differences					t	df	Sig. (2-tailed)
					95% Confidence Interval of the Difference				
		Mean	Std. Deviation	Std. Error Mean	Lower	Upper			
Pair 1	Picture of Spider - Real Spider	-7.000	9.807	2.831	-13.231	-.769	-2.473	11	.031

Finally, this output provides a 95% confidence interval for the mean difference. Imagine we took 100 samples from a population of difference scores and calculated their means ($\bar{D}$) and a confidence interval for that mean. In 95 of those samples the constructed confidence intervals contains the true value of the mean difference. The confidence interval tells us the boundaries within which the true mean difference is likely to lie.[4] So, assuming this sample's confidence interval is one of the 95 out of 100 that contains the population value,

[4] We saw in section 2.5.2 that these intervals represent the value of two (well, 1.96 to be precise) standard errors either side of the mean of the sampling distribution. For these data, in which the mean difference was –7 and the standard error was 2.8311, these limits will be $-7 \pm (1.96 \times 2.8311)$. However, because we're using the *t*-distribution, not the normal distribution, we use the critical value of *t* to compute the confidence intervals. This value is (with $df = 11$ as in this example) 2.201 (two-tailed), which gives us $-7 \pm (2.201 \times 2.8311)$.

SPSS TIP 9.1 One- and two-tailed significance in SPSS ①

Some students get a bit upset by the fact that SPSS produces only the two-tailed significance much of the time and are confused by why there isn't an option that can be selected to produce the one-tailed significance. The answer is simple: there is no need for an option because the one-tailed probability can be ascertained by dividing the two-tailed significance value by 2. For example, if the two-tailed probability is .107, then the one-tailed probability is .107/2 = .054.

we can say that the true mean difference lies between −13.23 and −0.77. The importance of this interval is that it does not contain zero (i.e. both limits are negative) because this tells us that the true value of the mean difference is unlikely to be zero. Crucially, if we were to compare pairs of random samples from a population we would expect most of the differences between sample means to be zero. This interval tells us that, based on our two samples, the true value of the difference between means is unlikely to be zero. Therefore, we can be confident that our two samples do not represent random samples from the same population. Instead they represent samples from different populations induced by the experimental manipulation.

9.4.6. Calculating the effect size ②

Even though our t-statistic is statistically significant, this doesn't mean our effect is important in practical terms. To discover whether the effect is substantive we need to use what we know about effect sizes (see section 2.6.4). I'm going to stick with the effect size r because it's widely understood, frequently used, and yes, I'll admit it, I actually like it! Converting a t-value into an r-value is actually really easy; we can use the following equation (e.g. Rosenthal, 1991; Rosnow & Rosenthal, 2005).[5]

$$r = \sqrt{\frac{t^2}{t^2 + df}}$$

We know the value of t and the df from the SPSS output and so we can compute r as follows:

$$r = \sqrt{\frac{-2.473^2}{-2.473^2 + 11}} = \sqrt{\frac{6.116}{17.116}} = .60$$

If you think back to our benchmarks for effect sizes this represents a very large effect (it is above .5, the threshold for a large effect). Therefore, as well as being statistically significant, this effect is large and so represents a substantive finding.

[5] Actually, this will overestimate the effect size because of the correlation between the two conditions. This is quite a technical issue and I'm trying to keep things simple here, but bear this in mind and if you're interested read Dunlap, Cortina, Vaslow, and Burke (1996).

9.4.7. Reporting the dependent *t*-test ①

There is a fairly standard way to report any test statistic: you usually state the finding to which the test relates and then report the test statistic, its degrees of freedom and the probability value of that test statistic. There has also been a recent move (by the American Psychological Association among others) to recommend that an estimate of the effect size is routinely reported. Although effect sizes are still rather sporadically used, I want to get you into good habits so we'll start thinking about effect sizes now. In this example the SPSS output tells us that the value of t was −2.47, that the degrees of freedom on which this was based was 11, and that it was significant at $p = .031$. We can also see the means for each group. We could write this as:

✓ On average, participants experienced significantly greater anxiety to real spiders ($M = 47.00$, $SE = 3.18$) than to pictures of spiders ($M = 40.00$, $SE = 2.68$), $t(11) = -2.47$, $p < .05$, $r = .60$.

Note how we've reported the means in each group (and standard errors) in the standard format. For the test statistic, note that we've used an italic t to denote the fact that we've calculated a t-statistic, then in brackets we've put the degrees of freedom and then stated the value of the test statistic. The probability can be expressed in several ways: often people report things to a standard level of significance (such as .05) as I have done here, but sometimes people will report the exact significance. Finally, note that I've reported the effect size at the end – you won't always see this in published papers but that's no excuse for you not to report it!

Try to avoid writing vague, unsubstantiated things like this:

× People were more scared of real spiders ($t = -2.47$).

More scared than what? Where are the *df*? Was the result statistically significant? Was the effect important (what was the effect size)?

CRAMMING SAM'S TIPS

- The dependent *t*-test compares two means, when those means have come from the same entities; for example, if you have used the same participants in each of two experimental conditions.
- Look at the column labelled *Sig.* If the value is less than .05 then the means of the two conditions are significantly different.
- Look at the values of the means to tell you how the conditions differ.
- SPSS provides only the two-tailed significance value; if you want the one-tailed significance just divide the value by 2.
- Report the *t*-statistic, the degrees of freedom and the significance value. Also report the means and their corresponding standard errors (or draw an error bar chart).
- If you're feeling brave, calculate and report the effect size too!

9.5. The independent *t*-test ①

9.5.1. The independent *t*-test equation explained ①

The independent *t*-test is used in situations in which there are two experimental conditions and different participants have been used in each condition. There are two different equations that can be used to calculate the *t*-statistic depending on whether the samples contain an equal number of people. As with the dependent *t*-test we can calculate the *t*-statistic by using a numerical version of equation (9.1); in other words, we are comparing the model or effect against the error. With the dependent *t*-test we could look at differences between pairs of scores, because the scores came from the same participants and so individual differences between conditions were eliminated. Hence, the difference in scores should reflect only the effect of the experimental manipulation. Now, when different participants participate in different conditions then pairs of scores will differ not just because of the experimental manipulation, but also because of other sources of variance (such as individual differences between participants' motivation, IQ, etc.). If we cannot investigate differences between conditions on a *per participant* basis (by comparing pairs of scores as we did for the dependent *t*-test) then we must make comparisons on a *per condition* basis (by looking at the overall effect in a condition – see equation (9.3)):

$$t = \frac{(\overline{X}_1 - \overline{X}_2) - (\mu_1 - \mu_2)}{\text{estimate of the standard error}} \tag{9.3}$$

Instead of looking at differences between pairs of scores, we now look at differences between the overall means of the two samples and compare them to the differences we would expect to get between the means of the two populations from which the samples come. If the null hypothesis is true then the samples have been drawn from the same population. Therefore, under the null hypothesis $\mu_1 = \mu_2$ and therefore $\mu_1 - \mu_2 = 0$. Therefore, under the null hypothesis the equation becomes:

$$t = \frac{\overline{X}_1 - \overline{X}_2}{\text{estimate of the standard error}} \tag{9.4}$$

In the dependent *t*-test we divided the mean difference between pairs of scores by the standard error of these differences. For the independent *t*-test we are looking at differences between groups and so we need to divide by the standard deviation of differences between groups. We can still apply the logic of sampling distributions to this situation. Now, imagine we took several pairs of samples – each pair containing one sample from the two different populations – and compared the means of these samples. From what we have learnt about sampling distributions, we know that the majority of samples from a population will have fairly similar means. Therefore, if we took several pairs of samples (from different populations), the differences between the sample means will be similar across pairs. However, often the difference between a pair of sample means will deviate by a small amount and very occasionally it will deviate by a large amount. If we could plot a sampling distribution of the differences between every pair of sample means that could be taken from two populations, then we would find that it had a normal distribution with a mean equal to the difference between population means ($\mu_1 - \mu_2$). The sampling distribution would tell us by how much we can expect the means of two (or more) samples to differ. As before, the standard deviation of the sampling distribution (the standard error) tells us how variable the differences between sample means are by chance alone. If the standard deviation is high then large differences between sample means can occur by chance; if it is small then only small differences between sample means are expected. It, therefore, makes sense that we use the standard error of the sampling distribution to assess whether the difference between two sample means is statistically meaningful or simply a chance result. Specifically, we divide the difference between sample means by the standard deviation of the sampling distribution.

So, how do we obtain the standard deviation of the sampling distribution of differences between sample means? Well, we use the **variance sum law**, which states that the variance of a difference between two independent variables is equal to the sum of their variances (see, for example, Howell, 2006). This statement means that the variance of the sampling distribution is equal to the sum of the variances of the two populations from which the samples were taken. We saw earlier that the standard error is the standard deviation of the sampling distribution of a population. We can use the sample standard deviations to calculate the standard error of each population's sampling distribution:

$$\text{SE of sampling distribution of population 1} = \frac{s_1}{\sqrt{N_1}}$$

$$\text{SE of sampling distribution of population 2} = \frac{s_2}{\sqrt{N_2}}$$

Therefore, remembering that the variance is simply the standard deviation squared, we can calculate the variance of each sampling distribution:

$$\text{variance of sampling distribution of population 1} = \left(\frac{s_1}{\sqrt{N_1}}\right)^2 = \frac{s_1^2}{N_1}$$

$$\text{variance of sampling distribution of population 2} = \left(\frac{s_2}{\sqrt{N_2}}\right)^2 = \frac{s_2^2}{N_2}$$

The variance sum law means that to find the variance of the sampling distribution of differences we merely add together the variances of the sampling distributions of the two populations:

$$\text{variance of sampling distribution of differences} = \frac{s_1^2}{N_1} + \frac{s_2^2}{N_2}$$

To find out the standard error of the sampling distribution of differences we merely take the square root of the variance (because variance is the standard deviation squared):

$$\text{SE of the sampling distribution of differences} = \sqrt{\left(\frac{s_1^2}{N_1} + \frac{s_2^2}{N_2}\right)}$$

Therefore, equation (9.4) becomes:

$$t = \frac{\bar{X}_1 - \bar{X}_2}{\sqrt{\left(\frac{s_1^2}{N_1} + \frac{s_2^2}{N_2}\right)}} \quad (9.5)$$

Equation (9.5) is true only when the sample sizes are equal. Often in the social sciences it is not possible to collect samples of equal size (because, for example, people may not complete an experiment). When we want to compare two groups that contain different numbers of participants then equation (9.5) is not appropriate. Instead the pooled variance estimate *t*-test is used which takes account of the difference in sample size by *weighting* the variance of each sample. We saw in Chapter 1 that large samples are better than small ones because they more closely approximate the population; therefore, we weight the variance by the size of sample on which it's based (we actually weight by the number of degrees of freedom, which is the sample size minus 1). Therefore, the pooled variance estimate is:

$$s_p^2 = \frac{(n_1 - 1)s_1^2 + (n_2 - 1)s_2^2}{n_1 + n_2 - 2}$$

This is simply a weighted average in which each variance is multiplied (weighted) by its degrees of freedom, and then we divide by the sum of weights (or sum of the two degrees of freedom). The resulting weighted average variance is then just replaced in the *t*-test equation:

$$t = \frac{\overline{X}_1 - \overline{X}_2}{\sqrt{\frac{s_p^2}{n_1} + \frac{s_p^2}{n_2}}}$$

As with the dependent *t*-test we can compare the obtained value of *t* against the maximum value we would expect to get by chance alone in a *t*-distribution with the same degrees of freedom (these values can be found in the Appendix); if the value we obtain exceeds this critical value we can be confident that this reflects an effect of our independent variable. One thing that should be apparent from the equation for *t* is that to compute it you don't actually need any data! All you need are the means, standard deviations and sample sizes (see SPSS Tip 9.2).

The derivation of the *t*-statistic is merely to provide a conceptual grasp of what we are doing when we carry out a *t*-test on SPSS. Therefore, if you don't know what on earth I'm babbling on about then don't worry about it (just spare a thought for my cat: he has to listen to this rubbish all the time!) because SPSS knows how to do it and that's all that matters!

SPSS TIP 9.2 Computing *t* from means, *SDs* and *Ns* ③

Using syntax, you can compute a *t*-test in SPSS from only the two group means, the two group standard deviations and the two group sizes. Open a data editor window and set up six new variables: **x1** (mean of group 1), **x2** (mean of group 2), **sd1** (standard deviation of group 1), **sd2** (standard deviation of group 2), **n1** (sample size of group 1) and **n2** (sample size of group 2). Type the values of each of these in the first row of the data editor. Open a syntax window and type the following:

```
COMPUTE df = n1+n2-2.
COMPUTE poolvar = (((n1-1)*(sd1 ** 2))+((n2-1)*(sd2 ** 2)))/df.
COMPUTE t = (x1-x2)/sqrt(poolvar*((1/n1)+(1/n2))).
COMPUTE sig = 2*(1-(CDF.T(abs(t),df))) .
Variable labels sig 'Significance (2-tailed)'.
EXECUTE .
```

The first line computes the degrees of freedom, the second computes the pooled variance, s_p^2, the third computes *t* and the fourth its two-tailed significance. All of these values will be created in a new column in the data editor. The line beginning 'Variable labels' simply labels the significance variable so that we know that it is two-tailed. If you want to display the results in the SPSS Viewer you could type:

```
SUMMARIZE
 /TABLES= x1 x2 df t sig
 /FORMAT=VALIDLIST NOCASENUM TOTAL LIMIT=100
 /TITLE='T-test'
 /MISSING=VARIABLE
 /CELLS=NONE.
```

These commands will produce a table of the variables **x1**, **x2**, **df**, **t** and **sig** so you'll see the means of the two groups, the degrees of freedom, the value of *t* and its two-tailed significance.

You can run lots of *t*-tests at the same time by putting different values for the means, SDs and sample sizes in different rows. If you do this, though, I suggest having a string variable called **Outcome** in the file in which you type what was being measured (or some other information so that you can identify to what the *t*-test relates).

I have put these commands in a syntax file called **Independent t from means.sps**. My file is actually a bit more complicated because it calculates an effect size measure (Cohen's *d*). For an example of how to use this file see Labcoat Leni's Real Research 9.1.

LABCOAT LENI'S REAL RESEARCH 9.1

You don't have to be mad here, but it helps ③

BOARD, B. J., & FRITZON, K. (2005). *PSYCHOLOGY, CRIME & LAW*, *11*, 17–32.

In the UK you often see the 'humorous' slogan 'You don't have to be mad to work here, but it helps' stuck up in work places. Well, Board and Fritzon (2005) took this a step further by measuring whether 39 senior business managers and chief executives from leading UK companies were mad (well, had personality disorders, PDs). They gave them The Minnesota Multiphasic Personality Inventory Scales for DSM III Personality Disorders (MMPI-PD), which is a well-validated measure of 11 personality disorders: Histrionic, Narcissistic, Antisocial, Borderline, Dependent, Compulsive, Passive–aggressive, Paranoid, Schizotypal, Schizoid and Avoidant. They needed a comparison group, and what better one to choose than 317 legally classified psychopaths at Broadmoor Hospital (a famous high-security psychiatric hospital in the UK).

The authors report the means and SDs for these two groups in Table 2 of their paper. Using these values and the syntax file **Independent t from means.sps** we can run *t*-tests on these means. The data from Board and Fritzon's Table 2 are in the file **Board and Fritzon 2005.sav**. Use this file and the syntax file to run *t*-tests to see whether managers score higher on personality disorder questionnaires than legally classified psychopaths. Report these results. What do you conclude?

Answers are in the additional material on the companion website (or look at Table 2 in the original article).

9.5.2. The independent *t*-test using SPSS ①

I have probably bored most of you to the point of wanting to eat your own legs by now. Equations are boring and that is why SPSS was invented to help us minimize our contact with them. Using our spider data again (**spiderBG.sav**), we have 12 spider-phobes who were exposed to a picture of a spider and 12 different spider-phobes who were exposed to a real-life tarantula (the groups are coded using the variable **group**). Their anxiety was measured in each condition (**anxiety**). I have already described how the data are arranged (see section 9.2), so we can move straight on to doing the test itself. First, we need to access the main dialog box by selecting Analyze Compare Means ▸ Independent-Samples T Test... (see Figure 9.9). Once the dialog box is activated, select the dependent variable from the list (click on **anxiety**) and transfer it to the box labelled *Test Variable(s)* by clicking on ➜. If you want to carry out *t*-tests on several dependent variables then you can select other dependent variables and transfer them to the variables list. However, there are good reasons why it is not a good idea to carry out lots of tests (see Chapter 10).

Next, we need to select an independent variable (the grouping variable). In this case, we need to select **group** and then transfer it to the box labelled *Grouping Variable*. When your grouping variable has been selected the Define Groups... button will become active and you should click on it to activate the *Define Groups* dialog box. SPSS needs to know what numeric codes you assigned to your two groups, and there is a space for you to type the codes. In this example, we coded our picture group as 0 and our real group as 1, and so these are the codes that we type. Alternatively you can specify a *Cut point* in which case SPSS will assign all cases greater than or equal to that value to one group and all the values below the cut point to the second group. This facility is useful if you are testing different groups of participants based on something like a median split (see Jane Superbrain Box 9.1) – you would simply type the median value in the box labelled *Cut point*. When you have defined the groups, click on Continue to return to the main dialog box. If you click on Options... then another dialog box appears that gives you the same options as for the dependent *t*-test. To run the analysis click on OK.

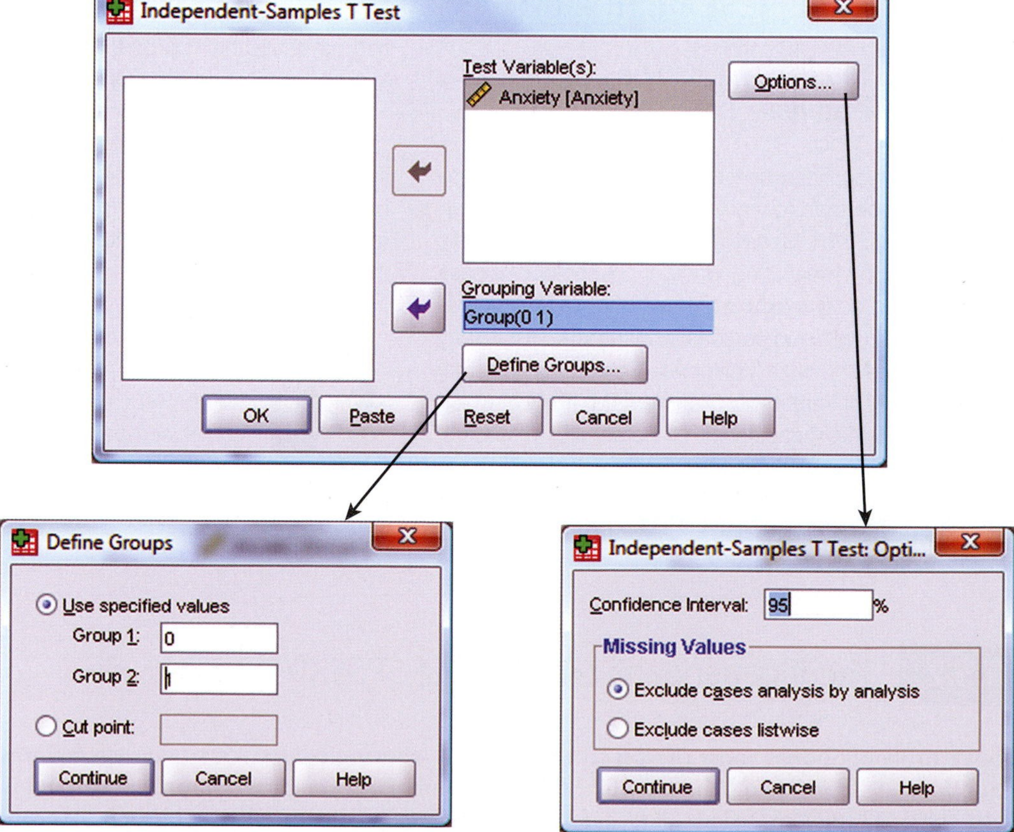

FIGURE 9.9 Dialog boxes for the independent-samples *t*-test

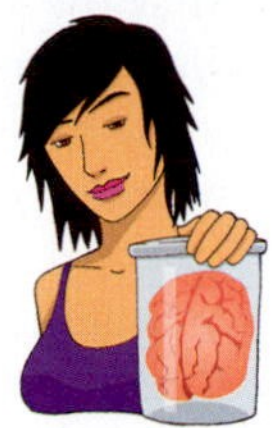

JANE SUPERBRAIN 9.1

Are median splits the devil's work? ②

Often in research papers you see that people have analysed their data using a 'median split'. In our spider phobia example, this means that you measure scores on a spider phobia questionnaire and calculate the median. You then classify anyone with a score above the median as a 'phobic', and those below the median as 'non-phobic'. In doing this you 'dichotomize' a continuous variable. This practice is quite common, but is it sensible?

MacCallum, Zhang, Preacher, and Rucker (2002) wrote a splendid paper pointing out various problems on turning a perfectly decent continuous variable into a categorical variable:

1 Imagine there are four people: Peter, Birgit, Jip and Kiki. We measure how scared of spiders they are as a percentage and get Jip (100%), Kiki (60%), Peter (40%) and Birgit (0%). If we split these four people at the median (50%) then we're saying that Jip and Kiki are the same (they get a score of 1 = phobic) and Peter and Birgit are the same (they both get a score of 0 = not phobic). In reality, Kiki and Peter are the most similar of the four people, but they have been put in different groups. So, median splits change the original information quite dramatically (Peter and Kiki are originally very similar but become very different after the split, Jip and Kiki are relatively dissimilar originally but become identical after the split).

2 Effect sizes get smaller: if you correlate two continuous variables then the effect size will be larger than if you correlate the same variables after one of them has been dichotomized. Effect sizes also get smaller in ANOVA and regression.

3 There is an increased chance of finding spurious effects.

So, if your supervisor has just told you to do a median split, have a good think about whether it is the right thing to do (and read MacCallum et al.'s paper). One of the rare situations in which dichotomizing a continuous variable is justified, according to MacCallum et al., is when there is a clear theoretical rationale for distinct categories of people based on a meaningful break point (i.e. not the median); for example, phobic versus not phobic based on diagnosis by a trained clinician would be a legitimate dichotomization of anxiety.

9.5.3. Output from the independent *t*-test ①

The output from the independent *t*-test contains only two tables. The first table (SPSS Output 9.3) provides summary statistics for the two experimental conditions. From this table, we can see that both groups had 12 participants (column labelled *N*). The group who saw the picture of the spider had a mean anxiety of 40, with a standard deviation of 9.29. What's more, the standard error of that group (the standard deviation of the sampling distribution) is 2.68 ($SE = 9.293/\sqrt{12} = 9.293/3.464 = 2.68$). In addition, the table tells us that the average anxiety level in participants who were shown a real spider was 47, with a standard deviation of 11.03 and a standard error of 3.18 ($SE = 11.029/\sqrt{12} = 11.029/3.464 = 3.18$).

SPSS OUTPUT 9.3

Group Statistics

	Spider or Picture?	N	Mean	Std. Deviation	Std. Error Mean
Anxiety	Picture	12	40.00	9.293	2.683
	Real Spider	12	47.00	11.029	3.184

The second table of output (SPSS Output 9.4) contains the main test statistics. The first thing to notice is that there are two rows containing values for the test statistics: one row is labelled *Equal variances assumed*, while the other is labelled *Equal variances not assumed*. In Chapter 5, we saw that parametric tests assume that the variances in experimental groups are roughly equal. Well, in reality there are adjustments that can be made in situations in which the variances are not equal. The rows of the table relate to whether or not this assumption has been broken. How do we know whether this assumption has been broken?

We saw in section 5.6.1 that we can use Levene's test to see whether variances are different in different groups, and SPSS produces this test for us. Remember that Levene's test is similar to a *t*-test in that it tests the hypothesis that the variances in the two groups are equal (i.e. the difference between the variances is zero). Therefore, if Levene's test is significant at $p \leq .05$, we can gain confidence in the hypothesis that the variances are significantly different and that the assumption of homogeneity of variances has been violated. If, however, Levene's test is non-significant (i.e. $p > .05$) then we do not have sufficient evidence to reject the null hypothesis that the difference between the variances is zero – in other words, we can assume that the variances are roughly equal and the assumption is tenable. For these data, Levene's test is non-significant (because $p = .386$, which is greater than .05) and so we should read the test statistics in the row labelled *Equal variances assumed*. Had Levene's test been significant, then we would have read the test statistics from the row labelled *Equal variances not assumed*.

SPSS OUTPUT 9.4

Independent Samples Test

		Levene's Test for Equality of Variances		t-test for Equality of Means						
									95% Confidence Interval of the Difference	
		F	Sig.	t	df	Sig. (2-tailed)	Mean Difference	Std. Error Difference	Lower	Upper
Anxiety	Equal variances assumed	.782	.386	-1.681	22	.107	-7.000	4.163	-15.634	1.634
	Equal variances not assumed			-1.681	21.385	.107	-7.000	4.163	-15.649	1.649

Having established that the assumption of homogeneity of variances is met, we can move on to look at the *t*-test itself. We are told the mean difference $(\bar{X}_1 - \bar{X}_2 = 40 - 47 = -7)$ and the standard error of the sampling distribution of differences, which is calculated using the lower half of equation (9.5):

$$\begin{aligned}\sqrt{\left(\frac{s_1^2}{N_1} + \frac{s_2^2}{N_2}\right)} &= \sqrt{\left(\frac{9.29^2}{12} + \frac{11.03^2}{12}\right)} \\ &= \sqrt{(7.19 + 10.14)} \\ &= \sqrt{17.33} \\ &= 4.16\end{aligned}$$

The *t*-statistic is calculated by dividing the mean difference by the standard error of the sampling distribution of differences ($t = -7/4.16 = -1.68$). The value of t is then assessed against the value of t you might expect to get by chance when you have certain degrees of freedom. For the independent *t*-test, degrees of freedom are calculated by adding the two sample sizes and then subtracting the number of samples ($df = N_1 + N_2 - 2 = 12 + 12 - 2 = 22$). SPSS produces the exact significance value of t, and we are interested in whether this value is less than or greater than .05. In this case the two-tailed value of p is .107, which is greater than .05, and so we would have to conclude that there was no significant difference between the means of these two samples. In terms of the experiment, we can infer that spider-phobes are made equally anxious by pictures of spiders as they are by the real thing.

Now, we use the two-tailed probability when we have made no specific prediction about the direction of our effect (see section 2.6.2). For example, if we were unsure whether a real spider would induce more or less anxiety, then we would have to use a two-tailed test. However, often in research we can make specific predictions about which group has the highest mean. In this example, it is likely that we would have predicted that a real spider would induce greater anxiety than a picture and so we predict that the mean of the real group would be greater than the mean of the picture group. In this case, we can use a one-tailed test (for more discussion of this issue see section 2.6.2). The one-tailed probability is .107/2 = .054 (see SPSS Tip 9.1). The one-tailed probability is still greater than .05 (albeit by a small margin) and so we would still have to conclude that spider-phobes' anxiety when presented with a real spider was not significantly different to spider-phobes who were presented with a picture of the same spider. This result was predicted by the error bar chart in Figure 9.2.

9.5.4. Calculating the effect size ②

To discover whether our effect is substantive we can use the same equation as in section 9.4.6 to convert the *t*-statistics into a value of *r*. We know the value of *t* and the *df* from the SPSS output and so we can compute *r* as follows:

$$\begin{aligned} r &= \sqrt{\frac{-1.681^2}{-1.681^2 + 22}} \\ &= \sqrt{\frac{2.826}{24.826}} \\ &= .34 \end{aligned}$$

If you think back to our benchmarks for effect sizes this represents a medium effect (it is around .3, the threshold for a medium effect). Therefore, even though the effect was non-significant, it still represented a fairly substantial effect. You may also notice that the effect has shrunk, which may seem slightly odd given that we used exactly the same data (but see section 9.6)!

9.5.5. Reporting the independent *t*-test ①

The rules that I made up, erm, I mean, reported, for the dependent *t*-test pretty much apply for the independent *t*-test. The SPSS output tells us that the value of *t* was −1.68, that the number of degrees of freedom on which this was based was 22, and that it was not significant at $p < .05$. We can also see the means for each group. We could write this as:

✓ On average, participants experienced greater anxiety to real spiders (M = 47.00, SE = 3.18) than to pictures of spiders (M = 40.00, SE = 2.68). This difference was not significant $t(22) = -1.68, p > .05$; however, it did represent a medium-sized effect $r = .34$.

Note how we've reported the means in each group (and standard errors) as before. For the test statistic everything is much the same as before except that I've had to report that *p* was greater than (>) .05 rather than less than (<). Finally, note that I've commented on the effect size at the end.

CRAMMING SAM'S TIPS

- The independent *t*-test compares two means, when those means have come from different groups of entities; for example, if you have used different participants in each of two experimental conditions.
- Look at the column labelled *Levene's Test for Equality of Variance*. If the *Sig.* value is less than .05 then the assumption of homogeneity of variance has been broken and you should look at the row in the table labelled *Equal variances not assumed*. If the *Sig.* value of Levene's test is bigger than .05 then you should look at the row in the table labelled *Equal variances assumed*.
- Look at the column labelled *Sig.* If the value is less than .05 then the means of the two groups are significantly different.
- Look at the values of the means to tell you how the groups differ.
- SPSS provides only the two-tailed significance value; if you want the one-tailed significance just divide the value by 2.
- Report the *t*-statistic, the degrees of freedom and the significance value. Also report the means and their corresponding standard errors (or draw an error bar chart).
- Calculate and report the effect size. Go on, you can do it!

9.6. Between groups or repeated measures? ①

The two examples in this chapter are interesting (honestly!) because they illustrate the difference between data collected using the same participants and data collected using different participants. The two examples in this chapter use the same scores in each condition. When analysed as though the data came from the same participants the result was a significant difference between means, but when analysed as though the data came from different participants there was no significant difference between group means. This may seem like a puzzling finding – after all the numbers were identical in both examples. What this illustrates is the relative *power* of repeated-measures designs. When the same participants are used across conditions the unsystematic variance (often called the error variance) is reduced dramatically, making it easier to detect any systematic variance. It is often assumed that the way in which you collect data is irrelevant, but I hope to have illustrated that it can make the difference between detecting a difference and not detecting one. In fact, researchers have carried out studies using the same participants in experimental conditions, then repeated the study using different participants in experimental conditions, then used the method of data collection as an independent variable in the analysis. Typically, they have found that the method of data collection interacts significantly with the results found (see Erlebacher, 1977).

9.7. The *t*-test as a general linear model ②

A lot of you might think it's odd that I've chosen to represent the effect size for my *t*-tests using *r*, the correlation coefficient. In fact you might well be thinking 'but correlations show relationships, not differences between means'. I used to think this too until I read a fantastic paper by Cohen (1968), which made me realize what I'd been missing; the complex, thorny, weed-infested and large Andy-eating tarantula-inhabited world of statistics suddenly turned into a beautiful meadow filled with tulips and little bleating lambs all jumping for joy at the wonder of life. Actually, I'm still a bumbling fool trying desperately to avoid having the blood sucked from my flaccid corpse by the tarantulas of statistics, but it was a good paper! What I'm about to say will either make no sense at all, or might help you to appreciate what I've said in most of the chapters so far: all statistical procedures are basically the same, they're just more or less elaborate versions of the correlation coefficient!

In Chapter 7 we saw that the *t*-test was used to test whether the regression coefficient of a predictor was equal to zero. The experimental design for which the independent *t*-test is used can be conceptualized as a regression equation (after all, there is one independent variable (predictor) and one dependent variable (outcome)). If we want to predict our outcome, then we can use the general equation that I've mentioned at various points:

$$\text{outcome}_i = (\text{model}) + \text{error}_\text{i}$$

If we want to use a linear model, then we saw that this general equation becomes equation (7.2) in which the model is defined by the slope and intercept of a straight line. Equation (9.6) shows a very similar equation in which A_i is the dependent variable (outcome), b_0 is the intercept, b_1 is the weighting of the predictor and G_i is the independent variable (predictor). Now, I've also included the same equation but with some of the letters replaced with what they represent in the spider experiment (so, A = **anxiety**, G = **group**). When we run an experiment with two conditions, the independent variable has only two values (group 1 or group 2). There are several ways in which these groups can be coded (in the spider example we coded group 1 with the value 0 and group 2 with the value 1). This coding variable is known as a *dummy variable* and values of this variable represent groups of entities. We have come across this coding in section 7.11:

$$\begin{aligned} A_i &= b_0 + b_1 G_i + \varepsilon_i \\ \text{Anxiety}_i &= b_0 + b_1 \text{group}_i + \varepsilon_i \end{aligned} \tag{9.6}$$

Using the spider example, we know that the mean **anxiety** of the picture group was 40, and that the **group** variable is equal to 0 for this condition. Look at what happens when the **group** variable is equal to 0 (the picture condition): equation (9.6) becomes (if we ignore the residual term):

$$\begin{aligned} \overline{X}_\text{Picture} &= b_0 + (b_1 \times 0) \\ b_0 &= \overline{X}_\text{Picture} \\ b_0 &= 40 \end{aligned}$$

Therefore, b_0 (the intercept) is equal to the mean of the picture group (i.e. it is the mean of the group coded as 0). Now let's look at what happens when the **group** variable is equal to 1. This condition is the one in which a real spider was used, therefore the mean **anxiety** ($\overline{X}_\text{Real}$) of this condition was 47. Remembering that we have just found out that b_0 is equal to the mean of the picture group ($\overline{X}_\text{Picture}$), equation (9.6) becomes:

$$\begin{aligned} \overline{X}_\text{Real} &= b_0 + (b_1 \times 1) \\ \overline{X}_\text{Real} &= \overline{X}_\text{Picture} + b_1 \\ b_1 &= \overline{X}_\text{Real} - \overline{X}_\text{Picture} \\ &= 47 - 40 \\ &= 7 \end{aligned}$$

b_1, therefore, represents the difference between the group means. As such, we can represent a two-group experiment as a regression equation in which the coefficient of the independent variable (b_1) is equal to the difference between group means, and the intercept (b_0) is equal to the mean of the group coded as 0. In regression, the *t*-test is used to ascertain whether the regression coefficient (b_1) is equal to 0, and when we carry out a *t*-test on grouped data we, therefore, test whether the difference between group means is equal to 0.

SELF-TEST To prove that I'm not making it up as I go along, run a regression on the data in **spiderBG.sav** with **group** as the predictor and **anxiety** as the outcome. **Group** is coded using zeros and ones and represents the dummy variable described above.

The resulting SPSS output should contain the regression summary table shown in SPSS Output 9.5. The first thing to notice is the value of the constant (b_0): its value is 40, the same as the mean of the base category (the picture group). The second thing to notice is that the value of the regression coefficient b_1 is 7, which is the difference between the two group means (47 – 40 = 7). Finally, the *t*-statistic, which tests whether b_1 is significantly different from zero, is the same as for the independent *t*-test (see SPSS Output 9.4) and so is the significance value.[6]

SPSS OUTPUT 9.5 Regression analysis of between-group spider data

Coefficients[a]

Model		Unstandardized Coefficients		Standardized Coefficients	t	Sig.
		B	Std. Error	Beta		
1	(Constant)	40.000	2.944		13.587	.000
	Condition	7.000	4.163	.337	1.681	.107

a. Dependent Variable: Anxiety

This section has demonstrated that differences between means can be represented in terms of linear models and this concept is essential in understanding the following chapters on the general linear model.

9.8. What if my data are not normally distributed? ②

We've seen in this chapter that there are adjustments that can be made to the *t*-test when the assumption of homogeneity of variance is broken, but what about when you have non-normally distributed data? The first thing to note is that although a lot of early evidence suggested that *t* was accurate when distributions were skewed, the *t*-test *can be* biased when

[6] In fact, the value of the *t*-statistic is the same but has a positive sign rather than negative. You'll remember from the discussion of the point–biserial correlation in section 6.5.5 that when you correlate a dichotomous variable the direction of the correlation coefficient depends entirely upon which cases are assigned to which groups. Therefore, the direction of the *t*-statistic here is similarly influenced by which group we select to be the base category (the category coded as 0).

the assumption of normality is not met (Wilcox, 2005). Second, we need to remember that it's the shape of the sampling distribution that matters, not the sample data. One option then is to use a big sample and rely on the central limit theorem (section 2.5.1) which says that the sampling distribution should be normal when samples are big. You could also try to correct the distribution using a transformation (but see Jane Superbrain Box 5.1). Another useful solution is to use one of a group of tests commonly referred to as *non-parametric tests*. These tests have fewer assumptions than their parametric counterparts and so are useful when your data violate the assumptions of parametric data described in Chapter 5. Some of these tests are described in Chapter 15. The non-parametric counterpart of the *dependent t-test* is called the *Wilcoxon signed-rank Test* (section 15.4), and the independent *t*-test has two non-parametric counterparts (both extremely similar) called the *Wilcoxon rank-sum test* and the *Mann–Whitney test* (section 15.3). I'd recommend reading these sections before moving on.

A final option is to use robust methods (see section 5.7.4). There are various robust ways to test differences between means that involve using trimmed means or a bootstrap. However, SPSS doesn't do any of these directly. Should you wish to do these then plugin for SPSS. Look at the companion website for some demos of how to use the R plugin.

What have I discovered about statistics? ①

We started this chapter by looking at my relative failures as a human being compared to Simon Hudson before investigating some problems with the way SPSS produces error bars for repeated-measures designs. We then had a look at some general conceptual features of the *t*-test, a parametric test that's used to test differences between two means. After this general taster, we moved on to look specifically at the dependent *t*-test (used when your conditions involve the same entities). I explained how it was calculated, how to do it on SPSS and how to interpret the results. We then discovered much the same for the independent *t*-test (used when your conditions involve different entities). After this I droned on excitedly about how a situation with two conditions can be conceptualized as a general linear model, by which point those of you who have a life had gone to the pub for a stiff drink. My excitement about things like general linear models could explain why Clair Sparks chose Simon Hudson all those years ago. Perhaps she could see the writing on the wall! Fortunately, I was a ruthless pragmatist at the age of 10, and the Clair Sparks episode didn't seem to concern me unduly; I just set my sights elsewhere during the obligatory lunchtime game of kiss chase. These games were the last I would see of women for quite some time …

Key terms that I've discovered

Dependent *t*-test
Grand mean
Independent *t*-test
Standard error of differences
Variance sum law

Smart Alex's tasks

These scenarios are taken from Field and Hole (2003). In each case analyse the data on SPSS:

- **Task 1**: One of my pet hates is 'pop psychology' books. Along with banishing Freud from all bookshops, it is my avowed ambition to rid the world of these rancid putrefaction-ridden wastes of trees. Not only do they give psychology a very bad name by stating the bloody obvious and charging people for the privilege, but they are also considerably less enjoyable to look at than the trees killed to produce them (admittedly the same could be said for the turgid tripe that I produce in the name of education but let's not go there just for now!). Anyway, as part of my plan to rid the world of popular psychology I did a little experiment. I took two groups of people who were in relationships and randomly assigned them to one of two conditions. One group read the famous popular psychology book *Women are from Bras and men are from Penis*, whereas another group read *Marie Claire*. I tested only 10 people in each of these groups, and the dependent variable was an objective measure of their happiness with their relationship after reading the book. I didn't make any specific prediction about which reading material would improve relationship happiness. The data are in the file **Penis.sav.** Analyse them with the appropriate *t*-test. ①
- **Task 2**: Imagine Twaddle and Sons, the publishers of *Women are from Bras and men are from Penis*, were upset about my claims that their book was about as useful as a paper umbrella. They decided to take me to task and design their own experiment in which participants read their book and one of my books (Field and Hole) at different times. Relationship happiness was measured after reading each book. To maximize their chances of finding a difference they used a sample of 500 participants, but got each participant to take part in both conditions (they read both books). The order in which books were read was counterbalanced and there was a delay of six months between reading the books. They predicted that reading their wonderful contribution to popular psychology would lead to greater relationship happiness than reading some dull and tedious book about experiments. The data are in **Field&Hole.sav.** Analyse them using the appropriate *t*-test. ①

Answers can be found on the companion website (or for more detail see Field and Hole, 2003).

Further reading

Field, A. P., & Hole, G. (2003). *How to design and report experiments*. London: Sage. (In my completely unbiased opinion this is a useful book to get some more background on experimental methods.)

Miles, J. N. V., & Banyard, P. (2007). *Understanding and using statistics in psychology: a practical introduction*. London: Sage. (A fantastic and amusing introduction to statistical theory.)

Rosnow, R. L., & Rosenthal, R. (2005). *Beginning behavioural research: a conceptual primer* (5th ed.). Englewood Cliffs, NJ: Pearson/Prentice Hall.

Wright, D. B., & London, K. (2009). *First steps in statistics* (2nd ed.) London: Sage. (This book has very clear introductions to the *t*-test.)

Online tutorial

The companion website contains the following Flash movie tutorial to accompany this chapter:

- *t*-tests using SPSS

Interesting real research

Board, B. J., & Fritzon, K. (2005). Disordered personalities at work. *Psychology, Crime & Law*, *11*(1), 17–32.

Comparing several means: ANOVA (GLM 1)

10

FIGURE 10.1 My brother Paul (left) and I (right) in our very fetching school uniforms

10.1. What will this chapter tell me? ①

There are pivotal moments in everyone's life, and one of mine was at the age of 11. Where I grew up in England there were three choices when leaving primary school and moving on to secondary school: (1) state school (where most people go); (2) grammar school (where clever people who pass an exam called the 11+ go); and (3) private school (where rich people go). My parents were not rich and I am not clever and consequently I failed my 11+, so private school and grammar school (where my clever older brother had gone) were out. This left me to join all of my friends at the local state school. I could not have been happier. Imagine everyone's shock when my parents received a letter saying that some extra spaces had become available at the grammar school; although the local authority could scarcely believe it and had re-marked the 11+ papers several million times to confirm their findings, I was next on their list. I could not have been unhappier. So, I waved goodbye to all of my friends and trundled off to join my brother at Ilford County High School For Boys (a school that still hit students with a cane if they were particularly bad and that, for some considerable time and with good reason, had

'H.M. Prison' painted in huge white letters on its roof). It was goodbye to normality, and hello to six years of learning how not to function in society. I often wonder how my life would have turned out had I not gone to this school; in the parallel universes where the letter didn't arrive and the Andy went to state school, or where my parents were rich and the Andy went to private school, what became of him? If we wanted to compare these three situations we couldn't use a *t*-test because there are more than two conditions.[1] However, this chapter tells us all about the statistical models that we use to analyse situations in which we want to compare more than two conditions: **analysis of variance** (or ANOVA to its friends). This chapter will begin by explaining the theory of ANOVA when different participants are used (**independent ANOVA**). We'll then look at how to carry out the analysis on SPSS and interpret the results.

10.2. The theory behind ANOVA ②

10.2.1. Inflated error rates ②

Before explaining how ANOVA works, it is worth mentioning why we don't simply carry out several *t*-tests to compare all combinations of groups that have been tested. Imagine a situation in which there were three experimental conditions and we were interested in differences between these three groups. If we were to carry out *t*-tests on every pair of groups, then we would have to carry out three separate tests: one to compare groups 1 and 2, one to compare groups 1 and 3, and one to compare groups 2 and 3. If each of these *t*-tests uses a .05 level of significance then for each test the probability of falsely rejecting the null hypothesis (known as a Type I error) is only 5%. Therefore, the probability of no Type I errors is .95 (95%) for each test. If we assume that each test is independent (hence, we can multiply the probabilities) then the overall probability of no Type I errors is $(.95)^3 = .95 \times .95 \times .95 = .857$, because the probability of no Type I errors is .95 for each test and there are three tests. Given that the probability of no Type I errors is .857, then we can calculate the probability of making at least one Type I error by subtracting this number from 1 (remember that the maximum probability of any event occurring is 1). So, the probability of at least one Type I error is $1 - .857 = .143$, or 14.3%. Therefore, across this group of tests, the probability of making a Type I error has increased from 5% to 14.3%, a value greater than the criterion accepted by social scientists. This error rate across statistical tests conducted on the same experimental data is known as the **familywise** or **experimentwise error rate**. An experiment with three conditions is a relatively simple design, and so the effect of carrying out several tests is not severe. If you imagine that we now increase the number of experimental conditions from three to five (which is only two more groups) then the number of *t*-tests that would need to done increases to 10.[2] The familywise error rate can be calculated using the

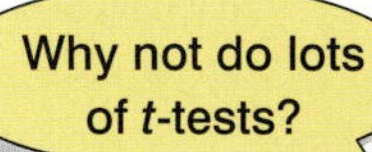
Why not do lots of t-tests?

[1] Really, this is the least of our problems: there's the small issue of needing access to parallel universes.

[2] These comparisons are group 1 vs. 2, 1 vs. 3, 1 vs. 4, 1 vs. 5, 2 vs. 3, 2 vs. 4, 2 vs. 5, 3 vs. 4, 3 vs. 5 and 4 vs. 5. The number of tests required is calculated using this equation:

$$\text{number of comparisons, C} = \frac{k!}{2(k-2)!}$$

in which k is the number of experimental conditions. The ! symbol stands for *factorial*, which means that you multiply the value preceding the symbol by all of the whole numbers between zero and that value (so $5! = 5 \times 4 \times 3 \times 2 \times 1 = 120$). Thus, with five conditions we find that:

$$C = \frac{5!}{2(5-2)!} = \frac{120}{2 \times 6} = 10$$

general equation (10.1) below, in which n is the number of tests carried out on the data. With 10 tests carried out, the familywise error rate is .40 ($1 - .95^{10} = .40$), which means that there is a 40% chance of having made at least one Type I error. For this reason we use ANOVA rather than conducting lots of t-tests:

$$\text{familywise error} = 1 - (.95)^n \qquad (10.1)$$

10.2.2. Interpreting *F* ②

When we perform a t-test, we test the hypothesis that the two samples have the same mean. Similarly, ANOVA tells us whether three or more means are the same, so it tests the null hypothesis that all group means are equal. An ANOVA produces an *F-statistic* or *F-ratio*, which is similar to the t-statistic in that it compares the amount of systematic variance in the data to the amount of unsystematic variance. In other words, F is the ratio of the model to its error.

What does an ANOVA tell me?

ANOVA is an *omnibus* test, which means that it tests for an overall experimental effect: so, there are things that ANOVA cannot tell us. Although ANOVA tells us whether the experimental manipulation was generally successful, it does not provide specific information about which groups were affected. Assuming an experiment was conducted with three different groups, the F-ratio tells us that the means of these three samples are not equal (i.e. that $\bar{X}_1 = \bar{X}_2 = \bar{X}_3$ is *not* true). However, there are several ways in which the means can differ. The first possibility is that all three sample means are significantly different ($\bar{X}_1 \neq \bar{X}_2 \neq \bar{X}_3$). A second possibility is that the means of group 1 and 2 are the same but group 3 has a significantly different mean from both of the other groups ($\bar{X}_1 = \bar{X}_2 \neq \bar{X}_3$). Another possibility is that groups 2 and 3 have similar means but group 1 has a significantly different mean ($\bar{X}_1 \neq \bar{X}_2 = \bar{X}_3$). Finally, groups 1 and 3 could have similar means but group 2 has a significantly different mean from both ($\bar{X}_1 = \bar{X}_3 \neq \bar{X}_2$). So, in an experiment, the F-ratio tells us only that the experimental manipulation has had some effect, but it doesn't tell us specifically what the effect was.

10.2.3. ANOVA as regression ②

I've hinted several times that all statistical tests boil down to variants on regression. In fact, ANOVA is just a special case of regression. This surprises many scientists because ANOVA and regression are usually used in different situations. The reason is largely historical in that two distinct branches of methodology developed in the social sciences: correlational research and experimental research. Researchers interested in controlled experiments adopted ANOVA as their statistic of choice whereas those looking for real-world relationships adopted multiple regression. As we all know, scientists are intelligent, mature and rational people and so neither group was tempted to slag off the other and claim that their own choice of methodology was far superior to the other (yeah right!). With the divide in methodologies came a chasm between the statistical methods adopted by the two opposing camps (Cronbach, 1957, documents this divide in a lovely article). This divide has lasted many decades to the extent that now students are generally taught regression and ANOVA in very different contexts and many textbooks teach ANOVA in an entirely different way to regression.

Although many considerably more intelligent people than me have attempted to redress the balance (notably the great Jacob Cohen, 1968), I am passionate about making my own small, feeble-minded attempt to enlighten you (and I set the ball rolling in sections 7.11 and 9.7). There are several good reasons why I think ANOVA should be taught within the context of regression. First, it provides a familiar context: I wasted many trees trying to explain regression, so why not use this base of knowledge to explain a new concept (it should make it easier to understand)? Second, the traditional method of teaching ANOVA (known as the variance-ratio method) is fine for simple designs, but becomes impossibly cumbersome in more complex situations (such as analysis of covariance). The regression model extends very logically to these more complex designs without anyone needing to get bogged down in mathematics. Finally, the variance-ratio method becomes extremely unmanageable in unusual circumstances such as when you have unequal sample sizes.[3] The regression method makes these situations considerably simpler. Although these reasons are good enough, it is also the case that SPSS has moved away from the variance-ratio method of ANOVA and progressed towards solely using the regression model (known as the general linear model, or GLM).

I have mentioned that ANOVA is a way of comparing the ratio of systematic variance to unsystematic variance in an experimental study. The ratio of these variances is known as the *F*-ratio. However, any of you who have read Chapter 7 should recognize the *F*-ratio (see section 7.2.3) as a way to assess how well a regression model can predict an outcome compared to the error within that model. If you haven't read Chapter 7 (surely not!), have a look before you carry on (it should only take you a couple of weeks to read). How can the *F*-ratio be used to test differences between means *and* whether a regression model fits the data? The answer is that when we test differences between means we *are* fitting a regression model and using *F* to see how well it fits the data, but the regression model contains only categorical predictors (i.e. grouping variables). So, just as the *t*-test could be represented by the linear regression equation (see section 9.7), ANOVA can be represented by the multiple regression equation in which the number of predictors is one less than the number of categories of the independent variable.

Let's take an example. There was a lot of controversy, when I wrote the first edition of this book, surrounding the drug Viagra. Admittedly there's less controversy now, but the controversy has been replaced by an alarming number of spam emails on the subject (for which I'll no doubt be grateful in 20 years' time), so I'm going to stick with the example. Viagra is a sexual stimulant (used to treat impotence) that broke into the black market under the belief that it will make someone a better lover (oddly enough, there was a glut of journalists taking the stuff at the time in the name of 'investigative journalism' … hmmm!). Suppose we tested this belief by taking three groups of participants and administering one group with a placebo (such as a sugar pill), one group with a low dose of Viagra and one with a high dose. The dependent variable was an objective measure of libido (I will tell you only that it was measured over the course of a week – the rest I will leave to your own imagination). The data can be found in the file **Viagra.sav** (which is described in detail later in this chapter) and are in Table 10.1.

TABLE 10.1 Data in **Viagra.sav**

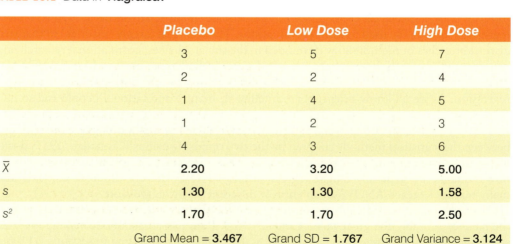

	Placebo	*Low Dose*	*High Dose*
	3	5	7
	2	2	4
	1	4	5
	1	2	3
	4	3	6
$\bar{X}$	**2.20**	**3.20**	**5.00**
s	**1.30**	**1.30**	**1.58**
s^2	**1.70**	**1.70**	**2.50**
	Grand Mean = **3.467**	Grand SD = **1.767**	Grand Variance = **3.124**

[3] Having said this, it is well worth the effort in trying to obtain equal sample sizes in your different conditions because unbalanced designs do cause statistical complications (see section 10.2.10).

If we want to predict levels of libido from the different levels of Viagra then we can use the general equation that keeps popping up:

$$\text{outcome}_i = (\text{model}) + \text{error}_i$$

If we want to use a linear model, then we saw in section 9.7 that when there are only two groups we could replace the 'model' in this equation with a linear regression equation with one dummy variable to describe two experimental groups. This dummy variable was a categorical variable with two numeric codes (0 for one group and 1 for the other). With three groups, however, we can extend this idea and use a multiple regression model with two dummy variables. In fact, as a general rule we can extend the model to any number of groups and the number of dummy variables needed will be one less than the number of categories of the independent variable. In the two-group case, we assigned one category as a base category (remember that in section 9.7 we chose the picture condition to act as a base) and this category was coded with 0. When there are three categories we also need a base category and you should choose the condition to which you intend to compare the other groups. Usually this category will be the control group. In most well-designed social science experiments there will be a group of participants who act as a baseline for other categories. This baseline group should act as the reference or base category, although the group you choose will depend upon the particular hypotheses that you want to test. In unbalanced designs (in which the group sizes are unequal) it is important that the base category contains a fairly large number of cases to ensure that the estimates of the regression coefficients are reliable. In the Viagra example, we can take the placebo group as the base category because this group was a placebo control. We are interested in comparing both the high- and low-dose groups to the group which received no Viagra at all. If the placebo group is the base category then the two dummy variables that we have to create represent the other two conditions: so, we should have one dummy variable called High and the other one called Low). The resulting equation is described as:

$$\text{Libido}_i = b_0 + b_2\text{High}_i + b_1\text{Low}_i + \varepsilon_i \tag{10.2}$$

In equation (10.2), a person's libido can be predicted from knowing their group code (i.e. the code for the High and Low dummy variables) and the intercept (b_0) of the model. The dummy variables in equation (10.2) can be coded in several ways, but the simplest way is to use a similar technique to that of the *t*-test. The base category is always coded as 0. If a participant was given a high dose of Viagra then they are coded with a 1 for the High dummy variable and 0 for all other variables. If a participant was given a low dose of Viagra then they are coded with the value 1 for the Low dummy variable and coded with 0 for all other variables (this is the same type of scheme we used in section 7.11). Using this coding scheme we can express each group by combining the codes of the two dummy variables (see Table 10.2).

TABLE 10.2 Dummy coding for the three-group experimental design

Group	Dummy Variable 1 (High)	Dummy Variable 2 (Low)
Placebo	0	0
Low-Dose Viagra	0	1
High-Dose Viagra	1	0

Placebo group: Let's examine the model for the placebo group. In the placebo group both the High and Low dummy variables are coded as 0. Therefore, if we ignore the error term (ε_i), the regression equation becomes:

$$\text{Libido}_i = b_0 + (b_2 \times 0) + (b_1 \times 0)$$
$$\text{Libido}_i = b_0$$
$$\overline{X}_{\text{Placebo}} = b_0$$

This is a situation in which the high- and low-dose groups have both been excluded (because they are coded with 0). We are looking at predicting the level of libido when both doses of Viagra are ignored, and so the predicted value will be the mean of the placebo group (because this group is the only one included in the model). Hence, the intercept of the regression model, b_0, is always the mean of the base category (in this case the mean of the placebo group).

High-dose group: If we examine the high-dose group, the dummy variable High will be coded as 1 and the dummy variable Low will be coded as 0. If we replace the values of these codes into equation (10.2) the model becomes:

$$\text{Libido}_i = b_0 + (b_2 \times 1) + (b_1 \times 0)$$
$$\text{Libido}_i = b_0 + b_2$$

We know already that b_0 is the mean of the placebo group. If we are interested in only the high-dose group then the model should predict that the value of Libido for a given participant equals the mean of the high-dose group. Given this information, the equation becomes:

$$\text{Libido}_i = b_0 + b_2$$
$$\overline{X}_{\text{High}} = \overline{X}_{\text{Placebo}} + b_2$$
$$b_2 = \overline{X}_{\text{High}} - \overline{X}_{\text{Placebo}}$$

Hence, b_2 represents the difference between the means of the high-dose group and the placebo group.

Low-dose group: Finally, if we look at the model when a low dose of Viagra has been taken, the dummy variable Low is coded as 1 (and hence High is coded as 0). Therefore, the regression equation becomes:

$$\text{Libido}_i = b_0 + (b_2 \times 0) + (b_1 \times 1)$$
$$\text{Libido}_i = b_0 + b_1$$

We know that the intercept is equal to the mean of the base category and that for the low-dose group the predicted value should be the mean libido for a low dose. Therefore the model can be reduced down to:

$$\text{Libido}_i = b_0 + b_1$$
$$\overline{X}_{\text{Low}} = \overline{X}_{\text{Placebo}} + b_1$$
$$b_1 = \overline{X}_{\text{Low}} - \overline{X}_{\text{Placebo}}$$

Hence, b_1 represents the difference between the means of the low-dose group and the placebo group. This form of dummy variable coding is the simplest form, but as we will see later, there are other ways in which variables can be coded to test specific hypotheses. These alternative coding schemes are known as *contrasts* (see section 10.2.11.2). The idea behind contrasts is that you code the dummy variables in such a way that the b-values represent differences between groups that you are interested in testing.

SELF-TEST To illustrate exactly what is going on I have created a file called **dummy.sav**. This file contains the Viagra data but with two additional variables (**dummy1** and **dummy2**) that specify to which group a data point belongs (as in Table 10.2). Access this file and run multiple regression analysis using libido as the outcome and **dummy1** and **dummy2** as the predictors. If you're stuck on how to run the regression then read Chapter 7 again (see, these chapters are ordered for a reason)!

The resulting analysis is shown in SPSS Output 10.1. It might be a good idea to remind yourself of the group means from Table 10.1. The first thing to notice is that just as in the regression chapter, an ANOVA has been used to test the overall fit of the model. This test is significant, $F(2, 12) = 5.12$, $p < .05$. Given that our model represents the group differences, this ANOVA tells us that using group means to predict scores is significantly better than using the overall mean: in other words, the group means are significantly different.

In terms of the regression coefficients, *b*s, the constant is equal to the mean of the base category (the placebo group). The regression coefficient for the first dummy variable (b_2) is equal to the difference between the means of the high-dose group and the placebo group (5.0 – 2.2 = 2.8). Finally, the regression coefficient for the second dummy variable (b_1) is equal to the difference between the means of the low-dose group and the placebo group (3.2 – 2.2 = 1). This analysis demonstrates how the regression model represents the three-group situation. We can see from the significance values of the *t*-tests that the difference between the high-dose group and the placebo group (b_2) is significant because $p < .05$. The difference between the low-dose and the placebo group is not, however, significant ($p = .282$).

SPSS OUTPUT 10.1

ANOVA[b]

Model		Sum of Squares	df	Mean Square	F	Sig.
1	Regression	20.133	2	10.067	5.119	.025[a]
	Residual	23.600	12	1.967		
	Total	43.733	14			

a. Predictors: (Constant), Dummy Variable 2, Dummy Variable 1

b. Dependent Variable: Libido

Coefficients[a]

Model		Unstandardized Coefficients		Standardized Coefficients	t	Sig.
		B	Std. Error	Beta		
1	(Constant)	2.200	.627		3.508	.004
	Dummy Variable 1	2.800	.887	.773	3.157	.008
	Dummy Variable 2	1.000	.887	.276	1.127	.282

a. Dependent Variable: Libido

A four-group experiment can be described by extending the three-group scenario. I mentioned earlier that you will always need one less dummy variable than the number of groups in the experiment: therefore, this model requires three dummy variables. As before, we need to specify one category that is a base category (a control group). This base category should have a code of 0 for all three dummy variables. The remaining three conditions will have a code of 1 for the dummy variable that described that condition and a code of 0 for the other two dummy variables. Table 10.3 illustrates how the coding scheme would work.

TABLE 10.3 Dummy coding for the four-group experimental design

	Dummy Variable 1	Dummy Variable 2	Dummy Variable 3
Group 1	1	0	0
Group 2	0	1	0
Group 3	0	0	1
Group 4 (base)	0	0	0

10.2.4. Logic of the *F*-ratio ②

In Chapter 7 we learnt a little about the *F*-ratio and its calculation. To recap, we learnt that the *F*-ratio is used to test the overall fit of a regression model to a set of observed data. In other words, it is the ratio of how good the model is compared to how bad it is (its error). I have just explained how ANOVA can be represented as a regression equation, and this should help you to understand what the *F*-ratio tells you about your data. Figure 10.2 shows the Viagra data in graphical form (including the group means, the overall mean and the difference between each case and the group mean). In this example, there were three groups; therefore, we want to test the hypothesis that the means of three groups are different (so, the null hypothesis is that the group means are the same). If the group means were all the same, then we would not expect the placebo group to differ from the low-dose group or the high-dose group, and we would not expect the low-dose group to differ from the high-dose group. Therefore, on the diagram, the three coloured lines would be in the same vertical position (the exact position would be the grand mean – the dashed line in the figure). We can see from the diagram that the group means are actually different because the coloured lines (the group means) are in different vertical positions. We have just found out that in the regression model, b_2 represents the difference between the means of the placebo and the high-dose group, and b_1 represents the difference in means between the low-dose and placebo groups. These two distances are represented in Figure 10.2 by the vertical arrows. If the null hypothesis is true

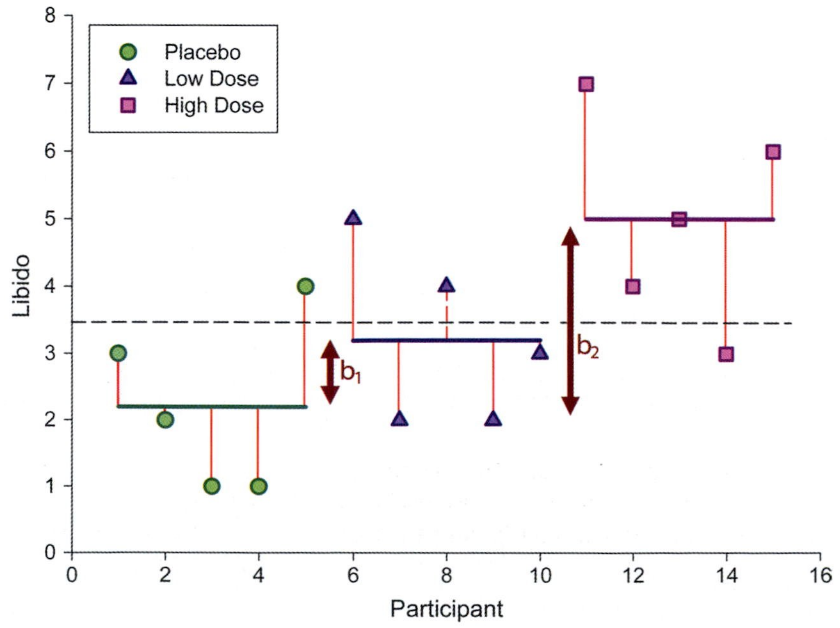

FIGURE 10.2 The Viagra data in graphical form. The coloured horizontal lines represent the mean libido of each group. The shapes represent the libido of individual participants (different shapes indicate different experimental groups). The dashed horizontal line is the average libido of all participants

and all the groups have the same means, then these b coefficients should be zero (because if the group means are equal then the difference between them will be zero).

The logic of ANOVA follows from what we understand about regression:

- The simplest model we can fit to a set of data is the grand mean (the mean of the outcome variable). This basic model represents 'no effect' or 'no relationship between the predictor variable and the outcome'.
- We can fit a different model to the data collected that represents our hypotheses. If this model fits the data well then it must be better than using the grand mean. Sometimes we fit a linear model (the line of best fit) but in experimental research we often fit a model based on the means of different conditions.
- The intercept and one or more regression coefficients can describe the chosen model.
- The regression coefficients determine the shape of the model that we have fitted; therefore, the bigger the coefficients, the greater the deviation between the line and the grand mean.
- In correlational research, the regression coefficients represent the slope of the line, but in experimental research they represent the differences between group means.
- The bigger the differences between group means, the greater the difference between the model and the grand mean.
- If the differences between group means are large enough, then the resulting model will be a better fit of the data than the grand mean.
- If this is the case we can infer that our model (i.e. predicting scores from the group means) is better than not using a model (i.e. predicting scores from the grand mean). Put another way, our group means are significantly different.

Just like when we used ANOVA to test a regression model, we can compare the improvement in fit due to using the model (rather than the grand mean) to the error that still remains. Another way of saying this is that when the grand mean is used as a model, there will be a certain amount of variation between the data and the grand mean. When a model is fitted it will explain some of this variation but some will be left unexplained. The F-ratio is the ratio of the explained to the unexplained variation. Look back at section 7.2.3 to refresh you memory on these concepts before reading on. This may all sound quite complicated, but actually most of it boils down to variations on one simple equation (see Jane Superbrain Box 10.1).

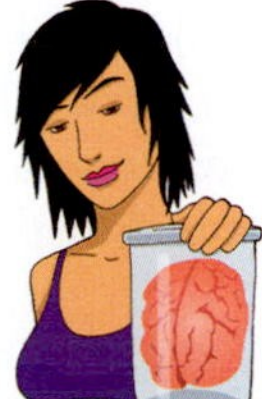

JANE SUPERBRAIN 10.1

You might be surprised to know that ANOVA boils down to one equation (well, sort of) ②

At every stage of the ANOVA we're assessing variation (or deviance) from a particular model (be that the most basic model, or the most sophisticated model). We saw back in section 2.4.1 that the extent to which a model deviates from the observed data can be expressed, in general, in the form of equation (10.3). So, in ANOVA, as in regression, we use equation (10.3) to calculate the fit of the most basic model, and then the fit of the best model (the line of best fit). If the best model is any good then it should fit the data significantly better than our basic model:

$$\text{deviation} = \sum(\text{observed} - \text{model})^2 \qquad (10.3)$$

The interesting point is that all of the sums of squares in ANOVA are variations on this one basic equation. All that changes is what we use as the model, and what the corresponding observed data are. Look through the various sections on the sums of squares and compare the resulting equations to equation (10.3); hopefully, you can see that they are all basically variations on this general form of the equation!

10.2.5. Total sum of squares (SS_T) ②

To find the total amount of variation within our data we calculate the difference between each observed data point and the grand mean. We then square these differences and add them together to give us the total sum of squares (SS_T):

$$SS_T = \sum (x_i - \bar{x}_{grand})^2 \tag{10.4}$$

We also saw in section 2.4.1 that the variance and the sums of squares are related such that variance, $s^2 = SS/(N - 1)$, where N is the number of observations. Therefore, we can calculate the total sums of squares from the variance of all observations (the **grand variance**) by rearranging the relationship ($SS = s^2(N - 1)$). The grand variance is the variation between all scores, regardless of the experimental condition from which the scores come. Therefore, in Figure 10.2 it would be the sum of the squared distances between each point and the dashed horizontal line. The grand variance for the Viagra data is given in Table 10.1, and if we count the number of observations we find that there were 15 in all. Therefore, SS_T is calculated as follows:

$$\begin{aligned} SS_T &= s^2_{grand}(n - 1) \\ &= 3.124(15 - 1) \\ &= 3.124 \times 14 \\ &= 43.74 \end{aligned}$$

Before we move on, it is important to understand degrees of freedom, so have a look back at Jane Superbrain Box 2.2 to refresh your memory. We saw before that when we estimate population values, the degrees of freedom are typically one less than the number of scores used to calculate the population value. This is because to get these estimates we have to hold something constant in the population (in this case the mean), which leaves all but one of the scores free to vary (see Jane Superbrain Box 2.2). For SS_T, we used the entire sample (i.e. 15 scores) to calculate the sums of squares and so the total degrees of freedom (df_T) are one less than the total sample size ($N - 1$). For the Viagra data, this value is 14.

10.2.6. Model sum of squares (SS_M) ②

So far, we know that the total amount of variation within the data is 43.74 units. We now need to know how much of this variation the regression model can explain. In the ANOVA scenario, the model is based upon differences between group means and so the model sums of squares tell us how much of the total variation can be explained by the fact that different data points come from different groups.

In section 7.2.3 we saw that the model sum of squares is calculated by taking the difference between the values predicted by the model and the grand mean (see Figure 7.4). In ANOVA, the values predicted by the model are the group means (therefore, in Figure 10.2 the coloured horizontal lines represented the values of libido predicted by the model). For each participant the value predicted by the model is the mean for the group to which the participant belongs. In the Viagra example, the predicted value for the five participants in the placebo group will be 2.2, for the five participants in the low-dose condition it will be 3.2, and for the five participants in the high-dose condition it will be 5. The model sum of squares requires us to calculate the differences between each participant's predicted value and the grand mean. These differences are then squared and added together (for reasons that should be clear in your mind by now). We know that the predicted value for participants in a particular group is the mean of that group. Therefore, the easiest way to calculate SS_M is to:

1. Calculate the difference between the mean of each group and the grand mean.
2. Square each of these differences.
3. Multiply each result by the number of participants within that group (n_k).
4. Add the values for each group together.

The mathematical expression of this process is:

$$SS_M = \sum n_k (\bar{x}_k - \bar{x}_{grand})^2 \quad (10.5)$$

Using the means from the Viagra data, we can calculate SS_M as follows:

$$\begin{aligned} SS_M &= 5(2.200 - 3.467)^2 + 5(3.200 - 3.467)^2 + 5(5.000 - 3.467)^2 \\ &= 5(-1.267)^2 + 5(-0.267)^2 + 5(1.533)^2 \\ &= 8.025 + 0.355 + 11.755 \\ &= 20.135 \end{aligned}$$

For SS_M, the degrees of freedom (df_M) will always be one less than the number of parameters estimated. In short, this value will be the number of groups minus one (which you'll see denoted as $k - 1$). So, in the three-group case the degrees of freedom will always be 2 (because the calculation of the sums of squares is based on the group means, two of which will be free to vary in the population if the third is held constant).

10.2.7. Residual sum of squares (SS_R) ②

We now know that there are 43.74 units of variation to be explained in our data, and that our model can explain 20.14 of these units (nearly half). The final sum of squares is the residual sum of squares (SS_R), which tells us how much of the variation cannot be explained by the model. This value is the amount of variation caused by extraneous factors such as individual differences in weight, testosterone or whatever. Knowing SS_T and SS_M already, the simplest way to calculate SS_R is to subtract SS_M from SS_T ($SS_R = SS_T - SS_M$); however, telling you to do this provides little insight into what is being calculated and, of course, if you've messed up the calculations of either SS_M or SS_T (or indeed both!) then SS_R will be incorrect also. We saw in section 7.2.3 that the residual sum of squares is the difference between what the model predicts and what was actually observed. We already know that for a given participant, the model predicts the mean of the group to which that person belongs. Therefore, SS_R is calculated by looking at the difference between the score obtained by a person and the mean of the group to which the person belongs. In graphical terms the vertical lines in Figure 10.2 represent this sum of squares. These distances between each data point and the group mean are squared and then added together to give the residual sum of squares, SS_R, thus:

$$SS_R = \sum (x_{ik} - \bar{x}_k)^2 \quad (10.6)$$

Now, the sum of squares for each group represents the sum of squared differences between each participant's score in that group and the group mean. Therefore, we can express SS_R as $SS_R = SS_{group1} + SS_{group2} + SS_{group3} \ldots$ and so on. Given that we know the relationship between the variance and the sums of squares, we can use the variances for each group of the Viagra data to create an equation like we did for the total sum of squares. As such, SS_R can be expressed as:

$$SS_R = \sum s_k^2 (n_k - 1) \quad (10.7)$$

This just means take the variance from each group (s_k^2) and multiply it by one less than the number of people in that group ($n_k - 1$). When you've done this for each group, add them all up. For the Viagra data, this gives us:

$$\begin{aligned} SS_R &= s^2_{group1}(n_1 - 1) + s^2_{group2}(n_2 - 1) + s^2_{group3}(n_3 - 1) \\ &= (1.70)(5-1) + (1.70)(5-1) + (2.50)(5-1) \\ &= (1.70 \times 4) + (1.70 \times 4) + (2.50 \times 4) \\ &= 6.8 + 6.8 + 10 \\ &= 23.60 \end{aligned}$$

The degrees of freedom for SS_R (df_R) are the total degrees of freedom minus the degrees of freedom for the model ($df_R = df_T - df_M = 14 - 2 = 12$). Put another way, it's $N - k$: the total sample size, N, minus the number of groups, k.

10.2.8. Mean squares ②

SS_M tells us the *total* variation that the regression model (e.g. the experimental manipulation) explains and SS_R tells us the *total* variation that is due to extraneous factors. However, because both of these values are summed values they will be influenced by the number of scores that were summed; for example, SS_M used the sum of only 3 different values (the group means) compared to SS_R and SS_T, which used the sum of 12 and 14 values respectively. To eliminate this bias we can calculate the average sum of squares (known as the *mean squares*, MS), which is simply the sum of squares divided by the degrees of freedom. The reason why we divide by the degrees of freedom rather than the number of parameters used to calculate the SS is because we are trying to extrapolate to a population and so some parameters within that populations will be held constant (this is the same reason that we divide by $N - 1$ when calculating the variance, see Jane Superbrain Box 2.2). So, for the Viagra data we find the following mean squares:

$$MS_M = \frac{SS_M}{df_M} = \frac{20.135}{2} = 10.067$$

$$MS_R = \frac{SS_R}{df_R} = \frac{23.60}{12} = 1.967$$

MS_M represents the average amount of variation explained by the model (e.g. the systematic variation), whereas MS_R is a gauge of the average amount of variation explained by extraneous variables (the unsystematic variation).

10.2.9. The *F*-ratio ②

The *F*-ratio is a measure of the ratio of the variation explained by the model and the variation explained by unsystematic factors. In other words, it is the ratio of how good the model is against how bad it is (how much error there is). It can be calculated by dividing the model mean squares by the residual mean squares.

$$F = \frac{MS_M}{MS_R} \tag{10.8}$$

As with the independent *t*-test, the *F*-ratio is, therefore, a measure of the ratio of systematic variation to unsystematic variation. In experimental research, it is the ratio of the experimental effect to the individual differences in performance. An interesting point about the *F*-ratio is that because it is the ratio of systematic variance to unsystematic variance, if its value is less than 1 then it must, by definition, represent a non-significant effect. The reason why this statement is true is because if the *F*-ratio is less than 1 it means that MS_R is greater than MS_M, which in real terms means that there is more unsystematic than systematic variance. You can think of this in terms of the effect of natural differences in ability being greater than differences brought about by the experiment. In this scenario, we can, therefore, be sure that our experimental manipulation has been unsuccessful (because it has brought about less change than if we left our participants alone!). For the Viagra data, the *F*-ratio is:

$$F = \frac{MS_M}{MS_R} = \frac{10.067}{1.967} = 5.12$$

This value is greater than 1, which indicates that the experimental manipulation had some effect above and beyond the effect of individual differences in performance. However, it doesn't yet tell us whether the *F*-ratio is large enough to not be a chance result. To discover this we can compare the obtained value of *F* against the maximum value we would expect to get by chance if the group means were equal in an *F*-distribution with the same degrees of freedom (these values can be found in the Appendix); if the value we obtain exceeds this critical value we can be confident that this reflects an effect of our independent variable (because this value would be very unlikely if there were no effect in the population). In this case, with 2 and 12 degrees of freedom the critical values are 3.89 (p = .05) and 6.93 (p = .01). The observed value, 5.12, is, therefore, significant at a .05 level of significance but not significant at a .01 level. The exact significance produced by SPSS should, therefore, fall somewhere between .05 and .01 (which, incidentally, it does).

10.2.10. Assumptions of ANOVA ③

The assumptions under which the *F* statistic is reliable are the same as for all parametric tests based on the normal distribution (see section 5.2). That is, the variances in each experimental condition need to be fairly similar, observations should be independent and the dependent variable should be measured on at least an interval scale. In terms of normality, what matters is that distributions *within groups* are normally distributed.

You often hear people say 'ANOVA is a robust test', which means that it doesn't matter much if we break the assumptions of the test: the *F* will still be accurate. There is some truth to this statement, but it is also an oversimplification of the situation. For one thing, the term *ANOVA* covers many different situations and the performance of *F* has been investigated in only some of those situations. There are two issues to consider: (1) does the *F* control the Type I error rate or is it significant even when there are no differences between means; and (2) does the *F* have enough power (i.e. is it able to detect differences when they are there)? Let's have a look at the evidence.

Is the *F* statistic robust?

Looking at normality first, Glass et al. (1972) reviewed a lot of evidence that suggests that *F* controls the Type I error rate well under conditions of skew, kurtosis and non-normality. Skewed distributions seem to have little effect on the error rate and power for two-tailed tests (but can have serious consequences for one-tailed tests). However, some of this evidence has been questioned (see Jane

Superbrain Box 5.1). In terms of kurtosis, leptokurtic distributions make the Type I error rate too low (too many null effects are significant) and consequently the power is too high; platykurtic distributions have the opposite effect. The effects of kurtosis seem unaffected by whether sample sizes are equal or not. One study that is worth mentioning in a bit of detail is by Lunney (1970) who investigated the use of ANOVA in about the most non-normal situation you could imagine: when the dependent variable is binary (it could have values of only 0 or 1). The results showed that when the group sizes were equal, ANOVA was accurate when there were at least 20 degrees of freedom and the smallest response category contained at least 20% of all responses. If the smaller response category contained less than 20% of all responses then ANOVA performed accurately only when there were 40 or more degrees of freedom. The power of *F* also appears to be relatively unaffected by non-normality (Donaldson, 1968). This evidence suggests that *when group sizes are equal* the *F*-statistic can be quite robust to violations of normality.

However, when group sizes are not equal the accuracy of *F* is affected by skew, and non-normality also affects the power of *F* in quite unpredictable ways (Wilcox, 2005). One situation that Wilcox describes shows that when means are equal the error rate (which should be 5%) can be as high as 18%. If you make the differences between means bigger you should find that power increases, but actually he found that initially power *decreased* (although it increased when he made the group differences bigger still). As such *F* can be biased when normality is violated.

In terms of violations of the assumption of homogeneity of variance, ANOVA is fairly robust in terms of the error rate when sample sizes are equal. However, when sample sizes are unequal, ANOVA is not robust to violations of homogeneity of variance (this is why earlier on I said it's worth trying to collect equal-sized samples of data across conditions!). When groups with larger sample sizes have larger variances than the groups with smaller sample sizes, the resulting *F*-ratio tends to be conservative. That is, it's more likely to produce a non-significant result when a genuine difference does exist in the population. Conversely, when the groups with larger sample sizes have smaller variances than the groups with smaller samples sizes, the resulting *F*-ratio tends to be liberal. That is, it is more likely to produce a significant result when there is no difference between groups in the population (put another way, the Type I error rate is not controlled) – see Glass et al. (1972) for a review. When variances are proportional to the means then the power of *F* seems to be unaffected by the heterogeneity of variance and trying to stabilize variances does not substantially improve power (Budescu, 1982; Budescu & Appelbaum, 1981). Problems resulting from violations of homogeneity of variance assumption can be corrected (see Jane Superbrain Box 10.2).

Violations of the assumption of independence are very serious indeed. Scariano and Davenport (1987) showed that when this assumption is violated (i.e. observations across groups are correlated) then the Type I error rate is substantially inflated. For example, using the conventional .05 Type I error rate when observations are independent, if these observations are made to correlate moderately (say, with a Pearson coefficient of .5), when comparing three groups to 10 observations per group the actual Type I error rate is .74 (a substantial inflation!). Therefore, if observations are correlated you might think that you are working with the accepted .05 error rate (i.e. you'll incorrectly find a significant result only 5% of the time) when in fact your error rate is closer to .75 (i.e. you'll find a significant result on 75% of occasions when, in reality, there is no effect in the population)!

10.2.11. Planned contrasts ②

The *F*-ratio tells us only whether the model fitted to the data accounts for more variation than extraneous factors, but it doesn't tell us where the differences between groups lie.

So, if the *F*-ratio is large enough to be statistically significant, then we know only that one or more of the differences between means is statistically significant (e.g. either b_2 or b_1 is statistically significant). It is, therefore, necessary after conducting an ANOVA to carry out further analysis to find out which groups differ. In multiple regression, each *b* coefficient is tested individually using a *t*-test and we could do the same for ANOVA. However, we would need to carry out two *t*-tests, which would inflate the familywise error rate (see section 10.2). Therefore, we need a way to contrast the different groups without inflating the Type I error rate. There are two ways in which to achieve this goal. The first is to break down the variance accounted for by the model into component parts, the second is to compare every group (as if conducting several *t*-tests) but to use a stricter acceptance criterion such that the familywise error rate does not rise above .05. The first option can be done using planned comparisons (also known as **planned contrasts**)[4] whereas the latter option is done using *post hoc* comparisons (see next section). The difference between planned comparisons and ***post hoc* tests** can be likened to the difference between one- and two-tailed tests in that planned comparisons are done when you have specific hypotheses that you want to test, whereas *post hoc* tests are done when you have no specific hypotheses. Let's first look at planned contrasts.

10.2.11.1. Choosing which contrasts to do ②

In the Viagra example we could have had very specific hypotheses. For one thing, we would expect any dose of Viagra to change libido compared to the placebo group. As a second hypothesis we might believe that a high dose should increase libido more than a low dose. To do planned comparisons, these hypotheses must be derived *before* the data are collected. It is fairly standard in social sciences to want to compare experimental conditions to the control conditions as the first contrast, and then to see where the differences lie between the experimental groups. ANOVA is based upon splitting the total variation into two component parts: the variation due to the experimental manipulation (SS_M) and the variation due to unsystematic factors (SS_R) (see Figure 10.3).

Planned comparisons take this logic a step further by breaking down the variation due to the experiment into component parts (see Figure 10.4). The exact comparisons that are carried out depend upon the hypotheses you want to test. Figure 10.4 shows a situation in which the experimental variance is broken down to look at how much variation is created by the two drug conditions compared to the placebo condition (*contrast 1*). Then the variation explained by taking Viagra is broken down to see how much is explained by taking a high dose relative to a low dose (*contrast 2*).

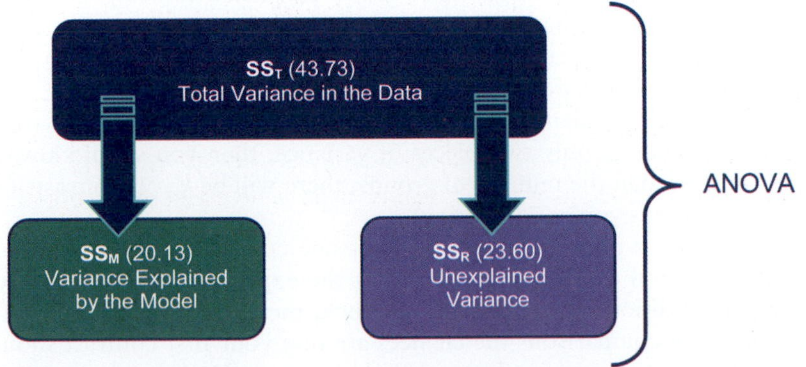

FIGURE 10.3 Partitioning variance for ANOVA

[4] The terms *comparison* and *contrast* are used interchangeably.

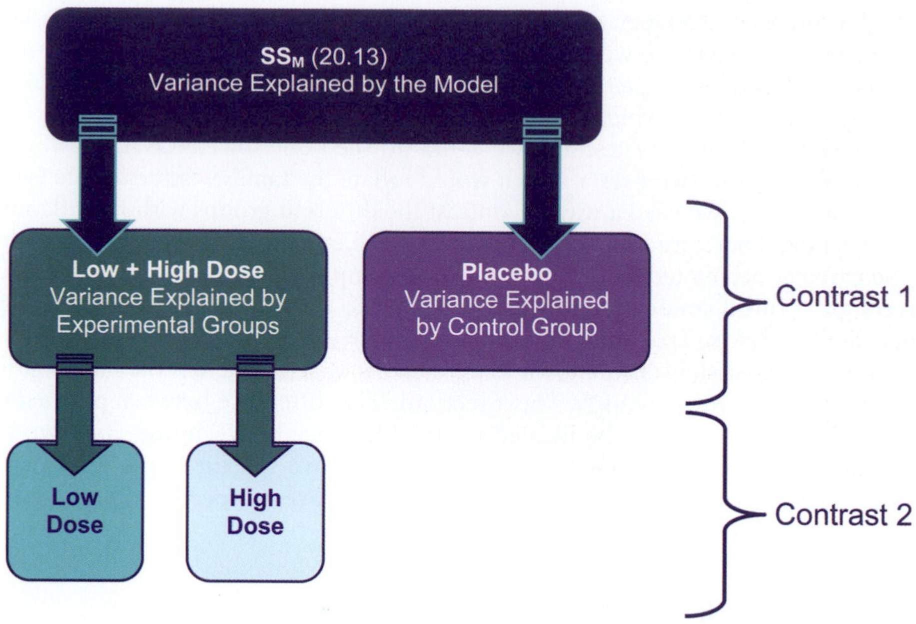

FIGURE 10.4 Partitioning of experimental variance into component comparisons

Typically, students struggle with the notion of planned comparisons, but there are several rules that can help you to work out what to do. The important thing to remember is that we are breaking down one chunk of variation into smaller independent chunks. This means several things. First, if a group is singled out in one comparison, then it should not reappear in another comparison. So, in Figure 10.4 contrast 1 involved comparing the placebo group to the experimental groups; because the placebo group is singled out, it should not be incorporated into any other contrasts. You can think of partitioning variance as being similar to slicing up a cake. You begin with a cake (the total sum of squares) and you then cut this cake into two pieces (SS_M and SS_R). You then take the piece of cake that represents SS_M and divide this up into smaller pieces. Once you have cut off a piece of cake you cannot stick that piece back onto the original slice, and you cannot stick it onto other pieces of cake, but you can divide it into smaller pieces of cake. Likewise, once a slice of variance has been split from a larger chunk, it cannot be attached to any other pieces of variance, it can only be subdivided into smaller chunks of variance. Now, all of this talk of cake is making me hungry, but hopefully it illustrates a point.

Each contrast must compare only two chunks of variance. This rule is so that we can draw firm conclusions about what the contrast tells us. The *F*-ratio tells us that some of our means differ, but not which ones, and if we were to perform a contrast on more than two chunks of variance we would have the same problem. By comparing only two chunks of variance we can be sure that a significant result represents a difference between these two portions of experimental variation.

If you follow the independence of contrasts rule that I've just explained (the cake slicing!), and always compare only two pieces of variance, then you should always end up with one less contrast than the number of groups; there will be $k - 1$ contrasts (where k is the number of conditions you're comparing).

In most social science research we use at least one control condition, and in the vast majority of experimental designs we predict that the experimental conditions will differ from the control condition (or conditions). As such, the biggest hint that I can give you is that when planning comparisons the chances are that your first contrast should be one that compares all of the experimental groups with the control group (or groups). Once you have done this first comparison, any remaining comparisons will depend upon which of the experimental groups you predict will differ.

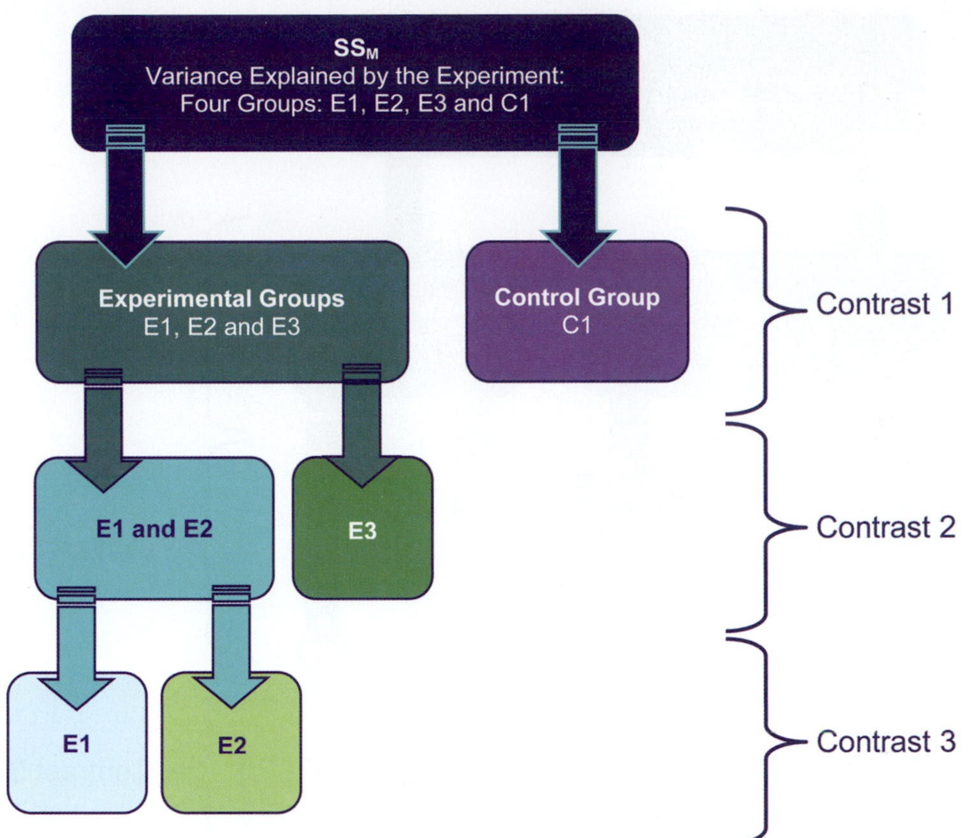

FIGURE 10.5 Partitioning variance for planned comparisons in a four-group experiment using one control group

To illustrate these principles, Figure 10.5 and Figure 10.6 show the contrasts that might be done in a four-group experiment. The first thing to notice is that in both scenarios there are three possible comparisons (one less than the number of groups). Also, every contrast compares only two chunks of variance. What's more, in both scenarios the first contrast is the same: the experimental groups are compared against the control group or groups. In Figure 10.5 there was only one control condition and so this portion of variance is used only in the first contrast (because it cannot be broken down any further). In Figure 10.6 there were two control groups, and so the portion of variance due to the control conditions (contrast 1) can be broken down again so as to see whether or not the scores in the control groups differ from each other (contrast 3).

In Figure 10.5, the first contrast contains a chunk of variance that is due to the three experimental groups and this chunk of variance is broken down by first looking at whether groups E1 and E2 differ from E3 (contrast 2). It is equally valid to use contrast 2 to compare groups E1 and E3 to E2, or to compare groups E2 and E3 to E1. The exact comparison that you choose depends upon your hypotheses. For contrast 2 in Figure 10.5 to be valid we need to have a good reason to expect group E3 to be different from the other two groups. The third comparison in Figure 10.5 depends on the comparison chosen for contrast 2. Contrast 2 necessarily had to involve comparing two experimental groups against a third, and the experimental groups chosen to be combined must be separated in the final comparison. As a final point, you'll notice that in Figure 10.5 and Figure 10.6, once a group has been singled out in a comparison, it is never used in any subsequent contrasts.

When we carry out a planned contrast, we compare 'chunks' of variance and these chunks often consist of several groups. It is

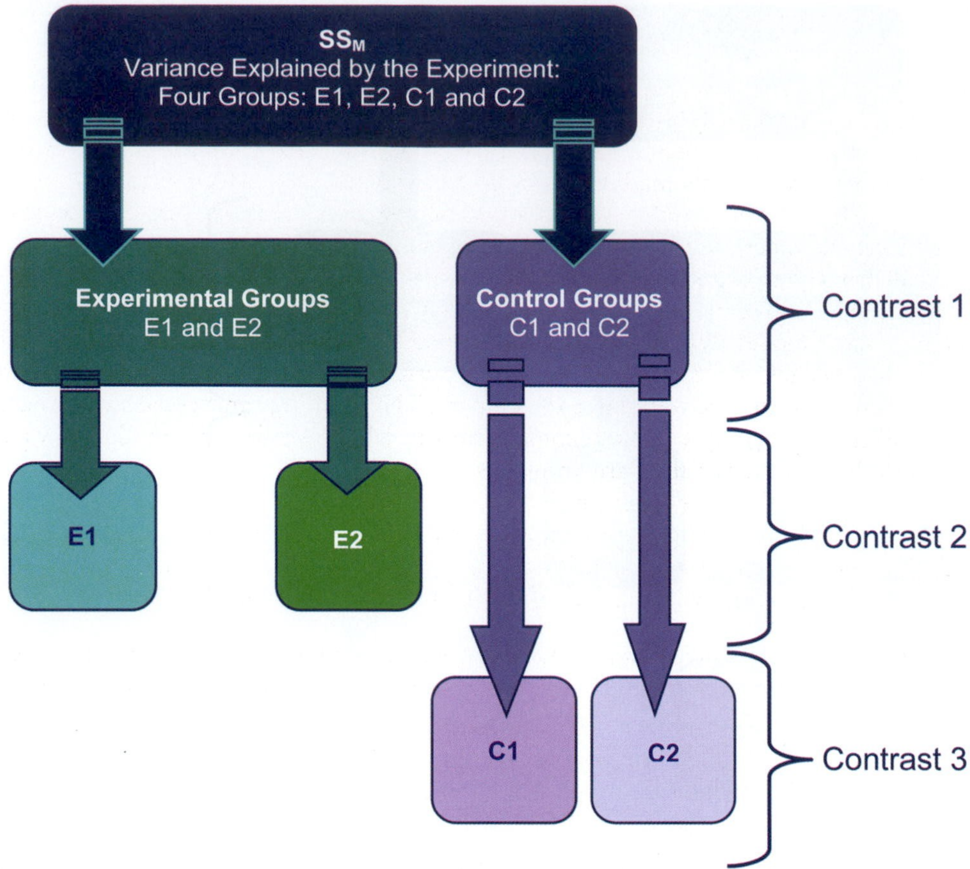

FIGURE 10.6 Partitioning variance for planned comparisons in a four-group experiment using two control groups

perhaps confusing to understand exactly what these contrasts tell us. Well, when you design a contrast that compares several groups to one other group, you are comparing the means of the groups in one chunk with the mean of the group in the other chunk. As an example, for the Viagra data I suggested that an appropriate first contrast would be to compare the two dose groups with the placebo group. The means of the groups are 2.20 (placebo), 3.20 (low dose) and 5.00 (high dose) and so the first comparison, which compared the two experimental groups to the placebo, is comparing 2.20 (the mean of the placebo group) to the average of the other two groups ((3.20 + 5.00)/2 = 4.10). If this first contrast turns out to be significant, then we can conclude that 4.10 is significantly greater than 2.20, which in terms of the experiment tells us that the average of the experimental groups is significantly different to the average of the controls. You can probably see that logically this means that, if the standard errors are the same, the experimental group with the highest mean (the high-dose group) will be significantly different from the mean of the placebo group. However, the experimental group with the lower mean (the low-dose group) might not necessarily differ from the placebo group; we have to use the final comparison to make sense of the experimental conditions. For the Viagra data the final comparison looked at whether the two experimental groups differ (i.e. is the mean of the high-dose group significantly different from the mean of the low-dose group?). If this comparison turns out to be significant then we can conclude that having a high dose of Viagra significantly affected libido compared to having a low dose. If the comparison is non-significant then we have to conclude that the dosage of Viagra made no significant difference to libido. In this latter scenario it is likely that both doses affect libido more than placebo, whereas the former case implies that having a low dose may be no different to having a placebo. However, the word *implies* is important here: it is possible that the low-dose group might not differ from the placebo. To be completely sure we must carry out *post hoc* tests.

10.2.11.2. Defining contrasts using weights ②

Hopefully by now you have got some idea of how to plan which comparisons to do (i.e. if your brain hasn't exploded by now). Much as I'd love to tell you that all of the hard work is now over and SPSS will magically carry out the comparisons that you've selected, it won't. To get SPSS to carry out planned comparisons we need to tell it which groups we would like to compare and doing this can be quite complex. In fact, when we carry out contrasts we assign values to certain variables in the regression model (sorry, I'm afraid that I have to start talking about regression again) – just as we did when we used dummy coding for the main ANOVA. To carry out contrasts we assign certain values to the dummy variables in the regression model. Whereas before we defined the experimental groups by assigning the dummy variables values of 1 or 0, when we perform contrasts we use different values to specify which groups we would like to compare. The resulting coefficients in the regression model (b_2 and b_1) represent the comparisons in which we are interested. The values assigned to the dummy variables are known as **weights**.

This procedure is horribly confusing, but there are a few basic rules for assigning values to the dummy variables to obtain the comparisons you want. I will explain these simple rules before showing how the process actually works. Remember the previous section when you read through these rules, and remind yourself of what I mean by a 'chunk' of variation!

- **Rule 1**: Choose sensible comparisons. Remember that you want to compare only two chunks of variation and that if a group is singled out in one comparison, that group should be excluded from any subsequent contrasts.
- **Rule 2**: Groups coded with positive weights will be compared against groups coded with negative weights. So, assign one chunk of variation positive weights and the opposite chunk negative weights.
- **Rule 3**: The sum of weights for a comparison should be zero. If you add up the weights for a given contrast the result should be zero.
- **Rule 4**: If a group is not involved in a comparison, automatically assign it a weight of 0. If we give a group a weight of 0 then this eliminates that group from all calculations.
- **Rule 5**: For a given contrast, the weights assigned to the group(s) in one chunk of variation should be equal to the number of groups in the opposite chunk of variation.

OK, let's follow some of these rules to derive the weights for the Viagra data. The first comparison we chose was to compare the two experimental groups against the control:

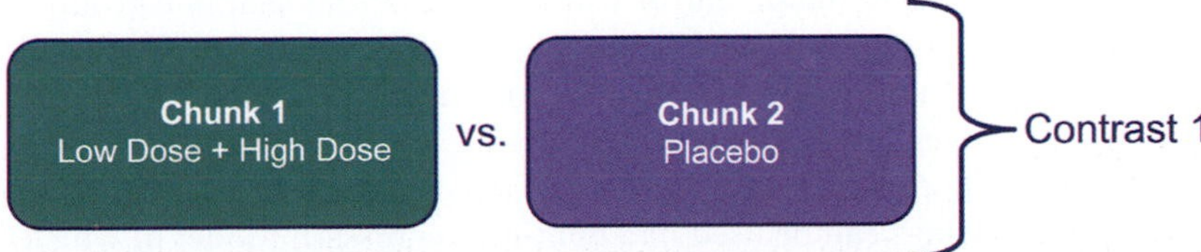

Therefore, the first chunk of variation contains the two experimental groups, and the second chunk contains only the placebo group. Rule 2 states that we should assign one chunk positive weights, and the other negative. It doesn't matter which way round we do this, but for convenience let's assign chunk 1 positive weights, and chunk 2 negative weights:

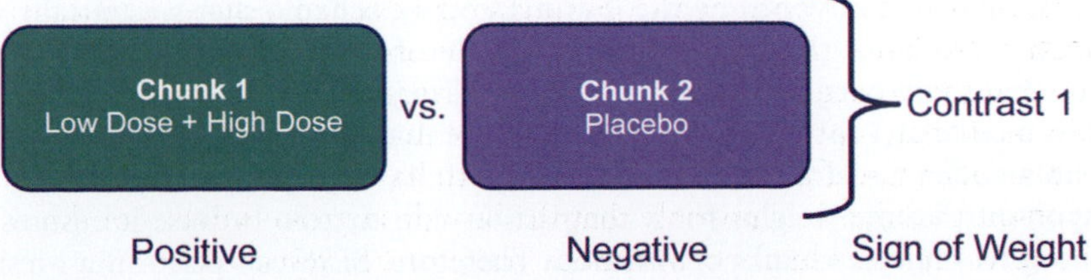

Using rule 5, the weight we assign to the groups in chunk 1 should be equivalent to the number of groups in chunk 2. There is only one group in chunk 2 and so we assign each group in chunk 1 a weight of 1. Likewise, we assign a weight to the group in chunk 2 that is equal to the number of groups in chunk 1. There are two groups in chunk 1 so we give the placebo group a weight of 2. Then we combine the sign of the weights with the magnitude to give us weights of –2 (placebo), 1 (low dose) and 1 (high dose):

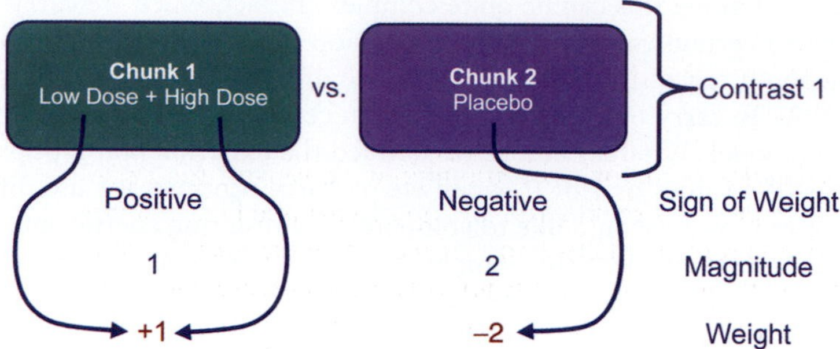

Rule 3 states that for a given contrast, the weights should add up to zero, and by following rules 2 and 5 this rule will always be followed (if you haven't followed these rules properly then this will become clear when you add the weights). So, let's check by adding the weights: sum of weights = 1 + 1 – 2 = 0:

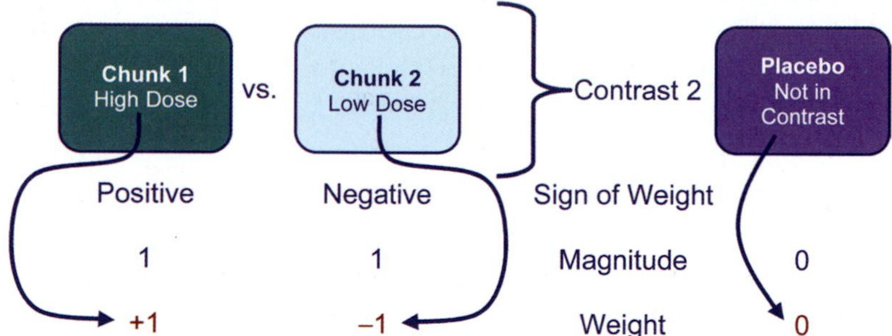

The second contrast was to compare the two experimental groups and so we want to ignore the placebo group. Rule 4 tells us that we should automatically assign this group a weight of 0 (because this will eliminate this group from any calculations). We are left with two chunks of variation: chunk 1 contains the low-dose group and chunk 2 contains the high-dose group. By following rules 2 and 5 it should be obvious that one group is assigned a weight of +1 while the other is assigned a weight of –1. The control group is ignored (and so given a weight of 0). If we add the weights for contrast 2 we should find that they again add up to zero: sum of weights = 1 – 1 + 0 = 0.

The weights for each contrast are codings for the two dummy variables in equation (10.2). Hence, these codings can be used in a multiple regression model in which b_2 represents contrast 1 (comparing the experimental groups to the control), b_1 represents contrast 2 (comparing the high-dose group to the low-dose group), and b_0 is the grand mean:

$$\text{Libido}_i = b_0 + b_1\text{Contrast}_{1i} + b_2\text{Contrast}_{2i} \tag{10.9}$$

Each group is specified now not by the 0 and 1 coding scheme that we initially used, but by the coding scheme for the two contrasts. A code of –2 for contrast 1 and a code of 0 for contrast 2 identify participants in the placebo group. Likewise, the high-dose group is identified by a code of 1 for both variables, and the low-dose group has a code of 1 for one contrast and a code of –1 for the other (see Table 10.4).

It is important that the weights for a comparison sum to zero because it ensures that you are comparing two unique chunks of variation. Therefore, SPSS can perform a *t*-test. A more

TABLE 10.4 Orthogonal contrasts for the Viagra data

Group	Dummy Variable 1 ($Contrast_1$)	Dummy Variable 2 ($Contrast_2$)	Product $Contrast_1 \times Contrast_2$
Placebo	–2	0	0
Low Dose	1	–1	–1
High Dose	1	1	1
Total	0	0	0

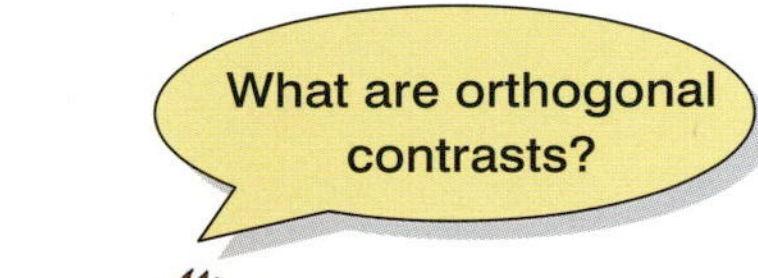
What are orthogonal contrasts?

important consideration is that when you multiply the weights for a particular group, these products should also add up to zero (see final column of Table 10.4). If the products add to zero then we can be sure that the contrasts are *independent* or **orthogonal**. It is important for interpretation that contrasts are orthogonal. When we used dummy variable coding and ran a regression on the Viagra data, I commented that we couldn't look at the individual *t*-tests done on the regression coefficients because the familywise error rate is inflated (see section 10.2.11 and SPSS Output 10.1). However, if the contrasts are independent then the *t*-tests done on the *b* coefficients are independent also and so the resulting *p*-values are uncorrelated. You might think that it is very difficult to ensure that the weights you choose for your contrasts conform to the requirements for independence but, provided you follow the rules I have laid out, you should always derive a set of *orthogonal* comparisons. You should double-check by looking at the sum of the multiplied weights and if this total is not zero then go back to the rules and see where you have gone wrong (see last column of Table 10.4).

SMART ALEX ONLY

Earlier on, I mentioned that when you used contrast codings in dummy variables in a regression model the *b*-values represented the differences between the means that the contrasts were designed to test. Although it is reasonable for you to trust me on this issue, for the more advanced students I'd like to take the trouble to show you how the regression model works (this next part is not for the faint-hearted and so equation-phobes should move on to the next section!). When we do planned contrasts, the intercept b_0 is equal to the grand mean (i.e. the value predicted by the model when group membership is not known), which when group sizes are equal is:

$$b_0 = \text{grand mean} = \frac{\overline{X}_{\text{High}} + \overline{X}_{\text{Low}} + \overline{X}_{\text{Placebo}}}{3}$$

Placebo group: If we use the contrast codings for the placebo group (see Table 10.4), the predicted value of libido equals the mean of the placebo group. The regression equation can, therefore, be expressed as:

$$\text{Libido}_i = b_0 + b_1\text{Contrast}_{1i} + b_2\text{Contrast}_{2i}$$

$$\overline{X}_{\text{Placebo}} = \left(\frac{\overline{X}_{\text{High}} + \overline{X}_{\text{Low}} + \overline{X}_{\text{Placebo}}}{3}\right) + (-2b_1) + (b_2 \times 0)$$

Now, if we rearrange this equation and then multiply everything by 3 (to get rid of the fraction) we get:

$$2b_1 = \left(\frac{\overline{X}_{\text{High}} + \overline{X}_{\text{Low}} + \overline{X}_{\text{Placebo}}}{3}\right) - \overline{X}_{\text{Placebo}}$$

$$6b_1 = \overline{X}_{\text{High}} + \overline{X}_{\text{Low}} + \overline{X}_{\text{Placebo}} - 3\overline{X}_{\text{Placebo}}$$

$$6b_1 = \overline{X}_{\text{High}} + \overline{X}_{\text{Low}} - 2\overline{X}_{\text{Placebo}}$$

We can then divide everything by 2 to reduce the equation to its simplest form:

$$3b_1 = \left(\frac{\overline{X}_{\text{High}} + \overline{X}_{\text{Low}}}{2}\right) - \overline{X}_{\text{Placebo}}$$

$$b_1 = \frac{1}{3}\left[\left(\frac{\overline{X}_{\text{High}} + \overline{X}_{\text{Low}}}{2}\right) - \overline{X}_{\text{Placebo}}\right]$$

This equation shows that b_1 represents the difference between the average of the two experimental groups and the control group:

$$\begin{aligned} 3b_1 &= \left(\frac{\overline{X}_{\text{High}} + \overline{X}_{\text{Low}}}{2}\right) - \overline{X}_{\text{Placebo}} \\ &= \frac{5 + 3.2}{2} - 2.2 \\ &= 1.9 \end{aligned}$$

We planned contrast 1 to look at the difference between the average of the experimental groups and the control and so it should now be clear how b_1 represents this difference. The observant among you will notice that rather than being the true value of the difference between experimental and control groups, b_1 is actually a third of this difference ($b_1 = 1.9/3 = 0.633$). The reason for this division is that the familywise error is controlled by making the regression coefficient equal to the actual difference divided by the number of groups in the contrast (in this case 3).

High-dose group: For the situation in which the codings for the high-dose group (see Table 10.4) are used, the predicted value of libido is the mean for the high-dose group, and so the regression equation becomes:

$$\begin{aligned} \text{Libido}_i &= b_0 + b_1\text{Contrast}_{1i} + b_2\text{Contrast}_{2i} \\ \overline{X}_{\text{High}} &= b_0 + (b_1 \times 1) + (b_2 \times 1) \\ b_2 &= \overline{X}_{\text{High}} - b_1 - b_0 \end{aligned}$$

We know already what b_1 and b_0 represent, so we place these values into the equation and then multiply by 3 to get rid of some of the fractions:

$$b_2 = \overline{X}_{\text{High}} - b_1 - b_0$$

$$b_2 = \overline{X}_{\text{High}} - \left\{\frac{1}{3}\left[\left(\frac{\overline{X}_{\text{High}} + \overline{X}_{\text{Low}}}{2}\right) - \overline{X}_{\text{Placebo}}\right]\right\} - \left(\frac{\overline{X}_{\text{High}} + \overline{X}_{\text{Low}} + \overline{X}_{\text{Placebo}}}{3}\right)$$

$$3b_2 = 3\overline{X}_{\text{High}} - \left[\left(\frac{\overline{X}_{\text{High}} + \overline{X}_{\text{Low}}}{2}\right) - \overline{X}_{\text{Placebo}}\right] - \left(\overline{X}_{\text{High}} + \overline{X}_{\text{Low}} + \overline{X}_{\text{Placebo}}\right)$$

If we multiply everything by 2 to get rid of the other fraction, expand all of the brackets and then simplify the equation we get:

$$6b_2 = 6\overline{X}_{\text{High}} - \left(\overline{X}_{\text{High}} + \overline{X}_{\text{Low}} - 2\overline{X}_{\text{Placebo}}\right) - 2\left(\overline{X}_{\text{High}} + \overline{X}_{\text{Low}} + \overline{X}_{\text{Placebo}}\right)$$

$$6b_2 = 6\overline{X}_{\text{High}} - \overline{X}_{\text{High}} - \overline{X}_{\text{Low}} + 2\overline{X}_{\text{Placebo}} - 2\overline{X}_{\text{High}} - 2\overline{X}_{\text{Low}} - 2\overline{X}_{\text{Placebo}}$$

$$6b_2 = 3\overline{X}_{\text{High}} - 3\overline{X}_{\text{Low}}$$

Finally, we can divide the equation by 6 to find out what b_2 represents (remember that $3/6 = 1/2$):

$$b_2 = \frac{1}{2}(\overline{X}_{\text{High}} - \overline{X}_{\text{Low}})$$

We planned contrast 2 to look at the difference between the experimental groups:

$$\overline{X}_{\text{High}} - \overline{X}_{\text{Low}} = 5 - 3.2 = 1.8$$

It should now be clear how b_2 represents this difference. Again, rather than being the absolute value of the difference between the experimental groups, b_2 is actually half of this difference ($1.8/2 = 0.9$). The familywise error is again controlled, by making the regression coefficient equal to the actual difference divided by the number of groups in the contrast (in this case 2).

SELF-TEST To illustrate these principles, I have created a file called **Contrast.sav** in which the Viagra data are coded using the contrast coding scheme used in this section. Run multiple regression analyses on these data using libido as the outcome and using **dummy1** and **dummy2** as the predictor variables (leave all default options).

EVERYBODY

SPSS Output 10.2 shows the result of this regression. The main ANOVA for the model is the same as when dummy coding was used (compare it to SPSS Output 10.1) showing that the model fit is the same (it should be because the model represents the group means and these have not changed); however, the regression coefficients have now changed. The first thing to notice is that the intercept is the grand mean, 3.467 (see, I wasn't telling lies). Second, the regression coefficient for contrast 1 is one-third of the difference between the average of the experimental conditions and the control condition (see above). Finally, the regression coefficient for contrast 2 is half of the difference between the experimental groups (see above). So, when a planned comparison is done in ANOVA a t-test is conducted comparing the mean of one chunk of variation with the mean of a different chunk. From the significance values of the t-tests we can see that our experimental groups were significantly different from the control ($p < .05$) but that the experimental groups were not significantly different ($p > .05$).

Coefficients[a]

Model		Unstandardized Coefficients B	Unstandardized Coefficients Std. Error	Standardized Coefficients Beta	t	Sig.
1	(Constant)	3.467	.362		9.574	.000
	Dummy Variable 1	.633	.256	.525	2.474	.029
	Dummy Variable 2	.900	.443	.430	2.029	.065

a. Dependent Variable: Libido

SPSS OUTPUT 10.2

10.2.11.3. Non-orthogonal comparisons ②

I have spent a lot of time labouring how to design appropriate orthogonal comparisons without mentioning the possibilities that non-orthogonal contrasts provide. Non-orthogonal contrasts

CRAMMING SAM'S TIPS Planned contrasts

- After an ANOVA you need more analysis to find out which groups differ.
- When you have generated specific hypotheses before the experiment use *planned contrasts*.
- Each contrast compares two 'chunks' of variance. (A chunk can contain one or more groups.)
- The first contrast will usually be experimental groups vs. control groups.
- The next contrast will be to take one of the chunks that contained more than one group (if there were any) and divide it into two chunks.
- You then repeat this process: if there are any chunks in previous contrasts that contained more than one group that haven't already been broken down into smaller chunks, then create a new contrast that breaks it down into smaller chunks.
- Carry on creating contrasts until each group has appeared in a chunk on its own in one of your contrasts.
- You should end up with one less contrast than the number of experimental conditions. If not, you've done it wrong.
- In each contrast assign a 'weight' to each group that is the value of the number of groups in the opposite chunk in that contrast.
- For a given contrast, randomly select one chunk, and for the groups in that chunk change their weights to be negative numbers.
- Breathe a sigh of relief.

are comparisons that are in some way related and the best way to get them is to disobey rule 1 in the previous section. Using my cake analogy again, non-orthogonal comparisons are where you slice up your cake and then try to stick slices of cake together again! So, for the Viagra data a set of non-orthogonal contrasts might be to have the same initial contrast (comparing experimental groups against the placebo), but then to compare the high-dose group to the placebo. This disobeys rule 1 because the placebo group is singled out in the first contrast but used again in the second contrast. The coding for this set of contrasts is shown in Table 10.5 and by looking at the last column it is clear that when you multiply and add the codings from the two contrasts the sum is not zero. This tells us that the contrasts are not orthogonal.

Are non-orthogonal contrasts legitimate?

There is nothing intrinsically wrong with performing non-orthogonal contrasts. However, if you choose to perform this type of contrast you must be very careful in how you interpret the results. With non-orthogonal contrasts, the comparisons you do are related and so the resulting test statistics and *p*-values will be correlated to some extent. For this reason you should use a more conservative probability level to accept that a given contrast is statistically meaningful (see section 10.2.12).

TABLE 10.5 Non-orthogonal contrasts for the Viagra data

Group	Dummy Variable 1 (Contrast$_1$)	Dummy Variable 2 (Contrast$_2$)	Product Contrast$_1$ × Contrast$_2$
Placebo	−2	−1	2
Low Dose	1	0	0
High Dose	1	1	1
Total	0	0	3

10.2.11.4. Standard contrasts ②

Although under most circumstances you will design your own contrasts, there are special contrasts that have been designed to compare certain situations. Some of these contrasts are orthogonal whereas others are non-orthogonal. Many procedures in SPSS allow you to choose to carry out the contrasts mentioned in this section.

Table 10.6 shows the contrasts that are available in SPSS for procedures such as logistic regression (see section 8.5.3), factorial ANOVA and repeated-measures ANOVA (see Chapters 12 and 13). Although the exact codings are not provided in Table 10.6, examples of the comparisons done in a three- and four-group situation are given (where the groups are labelled 1, 2, 3 and 1, 2, 3, 4 respectively). When you code variables in the data editor, SPSS will treat the lowest-value code as group 1, the next highest code as group 2, and so on. Therefore, depending on which comparisons you want to make you should code your grouping variable appropriately (and then use Table 10.6 as a guide to which comparisons SPSS will carry out). One thing that clever readers might notice about the contrasts in Table 10.6 is that some are orthogonal (i.e. Helmert and difference contrasts) while others are non-orthogonal (deviation, simple and repeated). You might also notice that the comparisons calculated using simple contrasts are the same as those given by using the dummy variable coding described in Table 10.2.

TABLE 10.6 Standard contrasts available in SPSS

Name	*Definition*	*Contrast*	*Three Groups*			*Four Groups*		
Deviation (first)	Compares the effect of each category (except first) to the overall experimental effect	1	2	vs.	(1,2,3)	2	vs.	(1,2,3,4)
		2	3	vs.	(1,2,3)	3	vs.	(1,2,3,4)
		3				4	vs.	(1,2,3,4)
Deviation (last)	Compares the effect of each category (except last) to the overall experimental effect	1	1	vs.	(1,2,3)	1	vs.	(1,2,3,4)
		2	2	vs.	(1,2,3)	2	vs.	(1,2,3,4)
		3				3	vs.	(1,2,3,4)
Simple (first)	Each category is compared to the first category	1	1	vs.	2	1	vs.	2
		2	1	vs.	3	1	vs.	3
		3				1	vs.	4
Simple (last)	Each category is compared to the last category	1	1	vs.	3	1	vs.	4
		2	2	vs.	3	2	vs.	4
		3				3	vs.	4
Repeated	Each category (except the first) is compared to the previous category	1	1	vs.	2	1	vs.	2
		2	2	vs.	3	2	vs.	3
		3				3	vs.	4
Helmert	Each category (except the last) is compared to the mean effect of all subsequent categories	1	1	vs.	(2, 3)	1	vs.	(2, 3, 4)
		2	2	vs.	3	2	vs.	(3, 4)
		3				3	vs.	4
Difference (reverse Helmert)	Each category (except the first) is compared to the mean effect of all previous categories	1	3	vs.	(2, 1)	4	vs.	(3, 2, 1)
		2	2	vs.	1	3	vs.	(2, 1)
		3				2	vs.	1

10.2.11.5. Polynomial contrasts: trend analysis ②

One type of contrast deliberately omitted from Table 10.6 is the **polynomial contrast**. This contrast tests for trends in the data and in its most basic form it looks for a linear trend (i.e. that the group means increase proportionately). However, there are more complex trends such as quadratic, cubic and quartic trends that can be examined. Figure 10.7 shows examples of the types of trend that can exist in data sets. The *linear* trend should be familiar to you all by now and represents a simply proportionate change in the value of the dependent variable across ordered categories (the diagram shows a positive linear trend but of course it could be negative). A **quadratic trend** is where there is one change in the direction of the line (e.g. the line is curved in one place). An example of this might be a situation in which a drug enhances performance on a task at first but then as the dose increases the performance drops again. To find a quadratic trend you need at least three groups (because in the two-group situation there are not enough categories of the independent variable for the means of the dependent variable to change one way and then another). A **cubic trend** is where there are two changes in the direction of the trend. So, for example, the mean of the dependent variable at first goes up across the first couple of categories of the independent variable, then across the succeeding categories the means go down, but then across the last few categories the means rise again. To have two changes in the direction of the mean you must have at least four categories of the independent variable. The final trend that you are likely to come across is the **quartic trend**, and this trend has three changes of direction (so you need at least five categories of the independent variable).

Polynomial trends should be examined in data sets in which it makes sense to order the categories of the independent variable (so, for example, if you have administered five doses of a drug it makes sense to examine the five doses in order of magnitude). For the Viagra data there are only three groups and so we can expect to find only a linear or quadratic trend (and it would be pointless to test for any higher-order trends).

Each of these trends has a set of codes for the dummy variables in the regression model, so we are doing the same thing that we did for planned contrasts except that the codings have already been devised to represent the type of trend of interest. In fact, the graphs in Figure 10.7 have been constructed by plotting the coding values for the five groups. Also, if you add the codes for a given trend the sum will equal zero and if you multiply the codes you will find that the sum of the products also equals zero. Hence, these contrasts are orthogonal. The great thing about these contrasts is that you don't need to construct your own coding values to do them, because the codings already exist.

10.2.12. *Post hoc* procedures ②

Often it is the case that you have no specific a priori predictions about the data you have collected and instead you are interested in exploring the data for any between-group differences between means that exist. This procedure is sometimes called *data mining* or *exploring data*. Now, personally I have always thought that these two terms have certain 'rigging the data' connotations to them and so I prefer to think of these procedures as 'finding the differences that I should have predicted if only I'd been clever enough'.

Post hoc tests consist of **pairwise comparisons** that are designed to compare all different combinations of the treatment groups. So, it is rather like taking every pair of groups and then performing a t-test on each pair of groups. Now, this might seem like a particularly stupid thing to say (but then again, I am particularly stupid) in the light of what I have already told you about the problems of inflated familywise error rates. However, pairwise comparisons control the familywise error by correcting the level of significance for each test such that the overall Type I error rate (α) across all comparisons remains at .05. There are several ways in which the familywise error rate can be controlled. The most popular (and easiest) way is to divide α by the number of

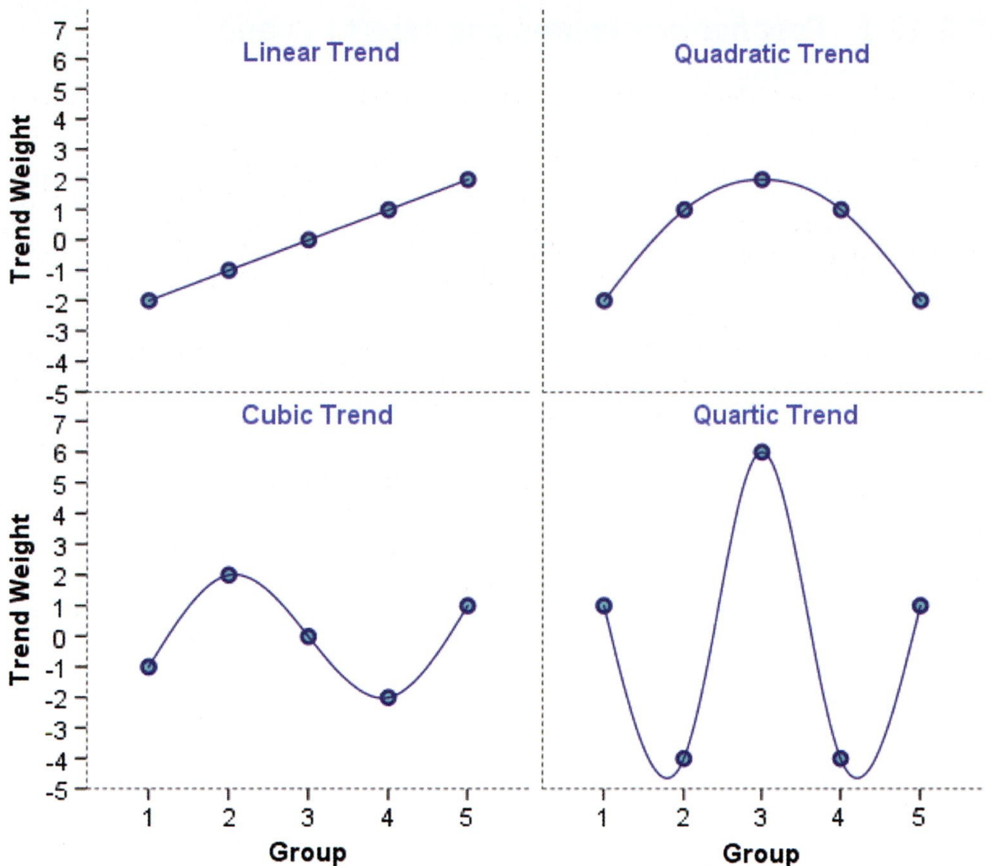

FIGURE 10.7 Linear, quadratic, cubic and quartic trends across five groups

comparisons, thus ensuring that the cumulative Type I error is below .05. Therefore, if we conduct 10 tests, we use .005 as our criterion for significance. This method is known as the **Bonferroni correction** (Figure 10.8). There is a trade-off for controlling the familywise error rate and that is a loss of statistical power. This means that the probability of rejecting an effect that does actually exist is increased (this is called a Type II error). By being more conservative in the Type I error rate for each comparison, we increase the chance that we will miss a genuine difference in the data.

Therefore, when considering which *post hoc* procedure to use we need to consider three things: (1) does the test control the Type I error rate; (2) does the test control the Type II error rate (i.e. does the test have good statistical power); and (3) is the test reliable when the test assumptions of ANOVA have been violated?

Although I would love to go into tedious details about how all of the various *post hoc* tests work, there is really very little point. For one thing, there are some excellent texts already available for those who wish to know (Klockars & Sax, 1986; Toothaker, 1993), for another, SPSS provides no less than 18 *post hoc* procedures and so it would use up several square miles of rainforest to explain them. By far the best reason, though, is that to explain them I would have to learn about them first, and I may be a nerd but even I draw the line at reading up on 18 different *post hoc* tests. However, it *is* important that you know which *post hoc* tests perform best according to the aforementioned criteria.

FIGURE 10.8 Carlo Bonferroni before the celebrity of his correction led to drink, drugs and statistics groupies

10.2.12.1. *Post hoc* procedures and Type I (α) and Type II error rates ②

SMART ALEX ONLY

The Type I error rate and the statistical power of a test are linked. Therefore, there is always a trade-off: if a test is conservative (the probability of a Type I error is small) then it is likely to lack statistical power (the probability of a Type II error will be high). Therefore, it is important that multiple comparison procedures control the Type I error rate but without a substantial loss in power. If a test is too conservative then we are likely to reject differences between means that are, in reality, meaningful.

The least-significant difference (LSD) pairwise comparison makes no attempt to control the Type I error and is equivalent to performing multiple *t*-tests on the data. The only difference is that the LSD requires the overall ANOVA to be significant. The Studentized Newman–Keuls (SNK) procedure is also a very liberal test and lacks control over the familywise error rate. *Bonferroni's* and *Tukey's* tests both control the Type I error rate very well but are conservative tests (they lack statistical power). Of the two, Bonferroni has more power when the number of comparisons is small, whereas Tukey is more powerful when testing large numbers of means. Tukey generally has greater power than *Dunn* and *Scheffé*. The Ryan, Einot, Gabriel and Welsch *Q* procedure (REGWQ) has good power and tight control of the Type I error rate. In fact, when you want to test all pairs of means this procedure is probably the best. However, when group sizes are different this procedure should not be used.

10.2.12.2. *Post hoc* procedures and violations of test assumptions ②

Most research on *post hoc* tests has looked at whether the test performs well when the group sizes are different (an unbalanced design), when the population variances are very different, and when data are not normally distributed. The good news is that most multiple comparison procedures perform relatively well under small deviations from normality. The bad news is that they perform badly when group sizes are unequal and when population variances are different.

Hochberg's GT2 and *Gabriel's* pairwise test procedure were designed to cope with situations in which sample sizes are different. Gabriel's procedure is generally more powerful but can become too liberal when the sample sizes are very different. Also, Hochberg's GT2 is very unreliable when the population variances are different and so should be used only when you are sure that this is not the case. There are several multiple comparison procedures that have been specially designed for situations in which population variances differ. SPSS provides four options for this situation: *Tamhane's T2*, *Dunnett's T3*, *Games–Howell* and *Dunnett's* C. Tamhane's T2 is conservative and Dunnett's T3 and C keep very tight Type I error control. The Games–Howell procedure is the most powerful but can be liberal when sample sizes are small. However, Games–Howell is also accurate when sample sizes are unequal.

EVERYBODY

10.2.12.3. Summary of *post hoc* procedures ②

The choice of comparison procedure will depend on the exact situation you have and whether it is more important for you to keep strict control over the familywise error rate or to have greater statistical power. However, some general guidelines can be drawn (Toothaker, 1993). When you have equal sample sizes and you are confident that your population variances are similar then use REGWQ or Tukey as both have good power and tight control over the Type I error rate. Bonferroni is generally conservative, but if you want guaranteed control over the Type I error rate then this is the test to use. If sample sizes are slightly different then use

Gabriel's procedure because it has greater power, but if sample sizes are very different use Hochberg's GT2. If there is any doubt that the population variances are equal then use the Games–Howell procedure because this generally seems to offer the best performance. I recommend running the Games–Howell procedure in addition to any other tests you might select because of the uncertainty of knowing whether the population variances are equivalent.

Although these general guidelines provide a convention to follow, be aware of the other procedures available and when they might be useful to use (e.g. Dunnett's test is the only multiple comparison that allows you to test means against a control mean).

CRAMMING SAM'S TIPS ***Post hoc* tests**

- After an ANOVA you need a further analysis to find out which groups differ.
- When you have no specific hypotheses before the experiment use *post hoc* tests.
- When you have equal sample sizes and group variances are similar use REGWQ or Tukey.
- If you want guaranteed control over the Type I error rate then use Bonferroni.
- If sample sizes are slightly different then use Gabriel's, but if sample sizes are very different use Hochberg's GT2.
- If there is any doubt that group variances are equal then use the Games–Howell procedure.

10.3. Running one-way ANOVA on SPSS ②

Hopefully you should all have some appreciation for the theory behind ANOVA, so let's put that theory into practice by conducting an ANOVA test on the Viagra data. As with the independent *t*-test we need to enter the data into the data editor using a coding variable to specify to which of the three groups the data belong. So, the data must be entered in two columns (one called **dose** which specifies how much Viagra the participant was given and one called **libido** which indicates the person's libido over the following week). The data are in the file **Viagra.sav** but I recommend entering them by hand to gain practice in data entry. I have coded the grouping variable so that 1 = placebo, 2 = low dose and 3 = high dose (see section 3.4.2.3).

To conduct one-way ANOVA we have to access the main dialog box by selecting Analyze Compare Means One-Way ANOVA... (Figure 10.9). This dialog box has a space in which you can list one or more dependent variables and a second space to specify a grouping variable, or *factor*. Factor is another term for independent variable and should not be confused with the factors that we will come across when we learn about factor analysis. For the Viagra data we need select only **libido** from the variables list and drag it to the box labelled *Dependent List* (or click on ➔). Then select the grouping variable **dose** and drag it to the box labelled *Factor (or* click on ➔).

One thing that I dislike about SPSS is that in various procedures, such as one-way ANOVA, the program encourages the user to carry out multiple tests, which as we have seen is not a good thing. For example, in this procedure you are allowed to specify several dependent variables on which to conduct several ANOVAs. In reality, if you had measured several dependent variables (say you had measured not just libido but physiological arousal and anxiety too) it would be preferable to analyse these data using MANOVA rather than treating each dependent measure separately (see Chapter 16).

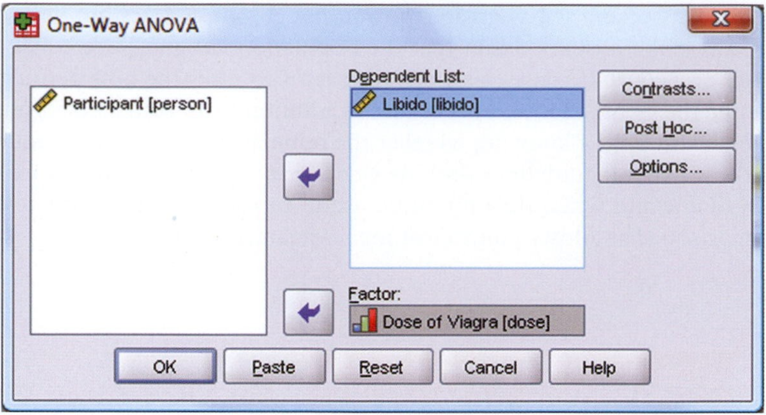

FIGURE 10.9 Main dialog box for one-way ANOVA

10.3.1. Planned comparisons using SPSS ②

If you click on Contrasts... you access the dialog box that allows you to conduct the planned comparisons described in section 10.2.11.

The dialog box is shown in Figure 10.10 and has two sections. The first section is for specifying trend analyses. If you want to test for trends in the data then tick the box labelled *Polynomial*. Once this box is ticked, you can select the degree of polynomial you would like. The Viagra data have only three groups and so the highest degree of trend there can be is a quadratic trend (see section 10.2.11.3). Now, it is important from the point of view of trend analysis that we have coded the grouping variable in a meaningful order. Also, we expect libido to be smallest in the placebo group, to increase in the low-dose group and then to increase again in the high-dose group. To detect a meaningful trend, we need to have coded these groups in ascending order. We have done this by coding the placebo group with the lowest value 1, the low-dose group with the middle value 2 and the high-dose group with the highest coding value of 3. If we coded the groups differently, this would influence both whether a trend is detected and, if a trend is detected, whether it is statistically meaningful.

For the Viagra data there are only three groups and so we should select the polynomial option ☑ Polynomial, then select a quadratic degree by clicking on Linear ▾ and then select *Quadratic* (the drop-down list should now say Quadratic ▾). If a quadratic trend is selected SPSS will test for both linear and quadratic trends.

The lower part of the dialog box in Figure 10.10 is for specifying any planned comparisons. To conduct planned comparisons we need to tell SPSS what weights to assign to each group. The first step is to decide which comparisons you want to do and then what weights must be assigned to each group for each of the contrasts. We have already gone through this process in section 10.2.11.2, so we know that the weights for contrast 1 were −2 (placebo group), +1 (low-dose group) and +1 (high-dose group). We will specify this contrast first. It is important to make sure that you enter the correct weight for each group, so you should remember that the first weight that you enter should be the weight for the *first* group (i.e. the group coded with the lowest value in the data editor). For the Viagra data, the group coded with the lowest value was the placebo group (which had a code of 1) so we should enter the weighting for this group first. Click in the box labelled *Coefficients* with the mouse and then type '-2' in this box and click on Add. Next, we need to input the weight for the second group, which for the Viagra data is the low-dose group (because this group was coded in the data editor with the second-highest value). Click in the box labelled *Coefficients* with the mouse and then type '1' in this box and click on Add. Finally, we need to input the weight for the last group, which for the Viagra data is the high-dose

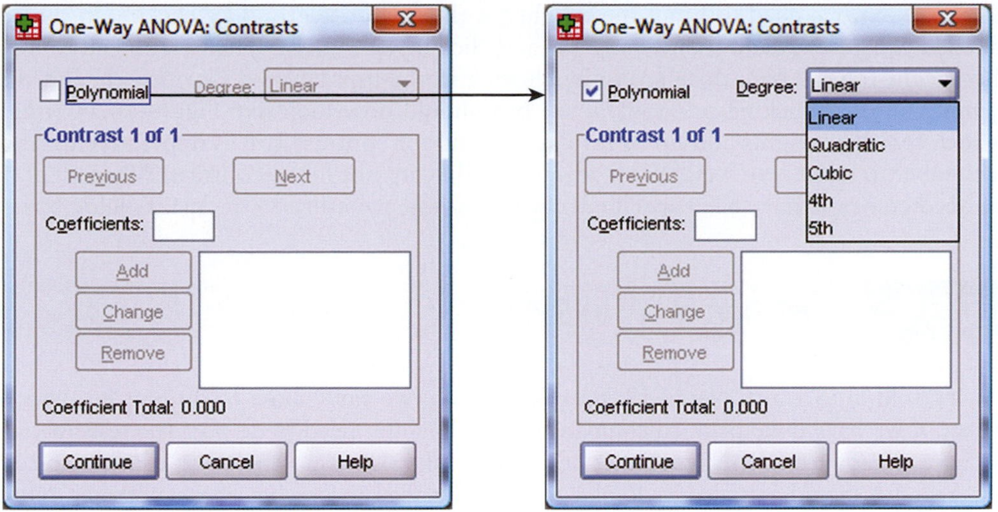

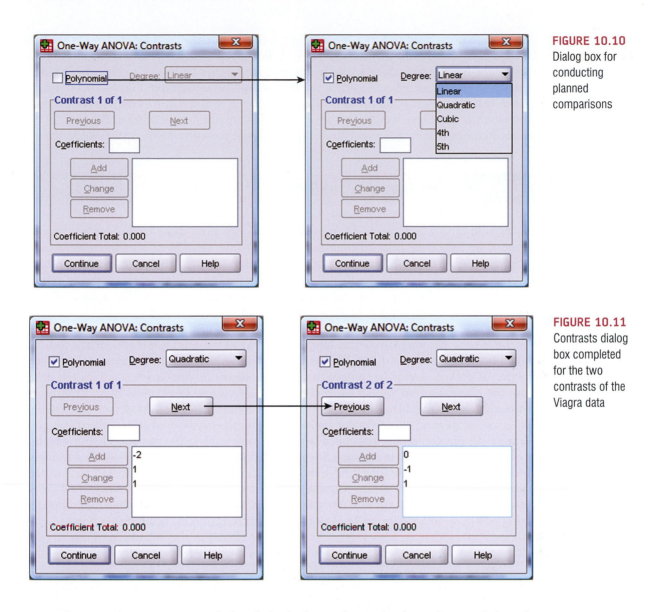

FIGURE 10.10 Dialog box for conducting planned comparisons

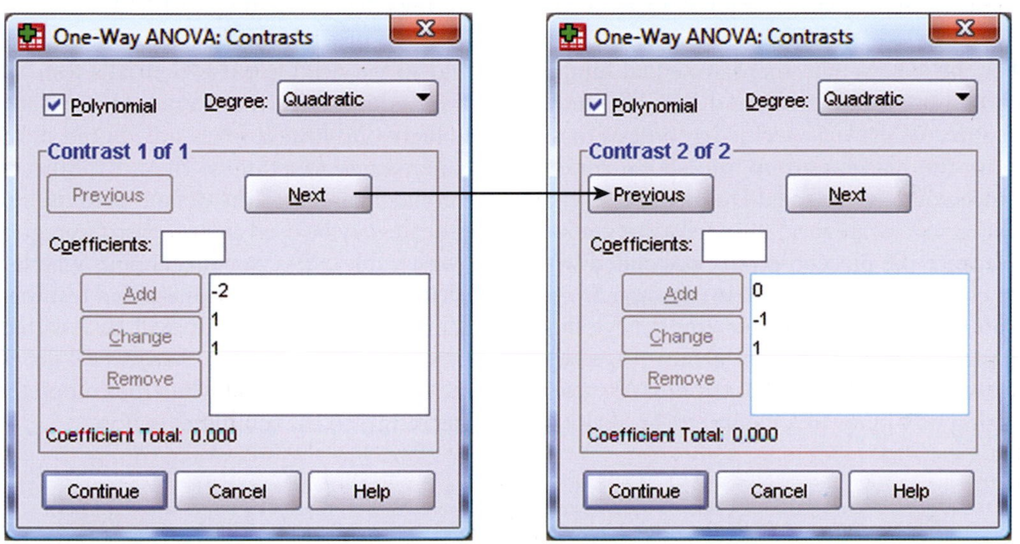

FIGURE 10.11 Contrasts dialog box completed for the two contrasts of the Viagra data

group (because this group was coded with the highest value in the data editor). Click in the box labelled *Coefficients* with the mouse and then type '1' in this box and click on Add. The box should now look like Figure 10.11 (left).

Once you have inputted the weights you can change or remove any one of them by using the mouse to select the weight that you want to change. The weight will then appear in the box labelled *Coefficients* where you can type in a new weight and then click on Change. Alternatively, you can click on any of the weights and remove it completely by clicking Remove. Underneath the weights SPSS calculates the coefficient total which, as we saw in section 10.2.11.2, should equal zero. If the coefficient number is anything other than zero you should go back and check that the contrasts you have planned make sense and that you have followed the appropriate rules for assigning weights.

Once you have specified the first contrast, click on Next. The weights that you have just entered will disappear and the dialog box will now read *Contrast 2 of 2*. We know from section 10.2.11.2 that the weights for contrast 2 were: 0 (placebo group), −1 (low-dose group) and +1 (high-dose group). We can specify this contrast as before. Remembering that the first weight we enter will be for the placebo group, we must enter the value 0 as the first weight. Click in the box labelled *Coefficients* with the mouse and then type '0' and click on

Add. Next, we need to input the weight for the low-dose group by clicking in the box labelled *Coefficients* and then typing '-1' and clicking on Add. Finally, we need to input the weight for the high-dose group by clicking in the box labelled *Coefficients* and then typing '+1' and clicking on Add. The box should now look like Figure 10.11 (right). Notice that the weights add up to zero as they did for contrast 1. It is imperative that you remember to input zero weights for any groups that are not in the contrast. When all of the planned contrasts have been specified, click on Continue to return to the main dialog box.

10.3.2. *Post hoc* tests in SPSS ②

Having told SPSS which planned comparisons to do, we can choose to do *post hoc* tests. In theory, if we have done planned comparisons we shouldn't need to do *post hoc* tests (because we have already tested the hypotheses of interest). Likewise, if we choose to conduct *post hoc* tests then we should not need to do planned contrasts (because we have no hypotheses to test). However, for the sake of space we will conduct some *post hoc* tests on the Viagra data. Click on Post Hoc... in the main dialog box to access the *post hoc* tests dialog box (Figure 10.12).

In section 10.2.12.3, I recommended various *post hoc* procedures for various situations. For the Viagra data there are equal sample sizes and so we need not use Gabriel's test. We should use Tukey's test and REGWQ and check the findings with the Games–Howell procedure. We have a specific hypothesis that both the high- and low-dose groups should differ from the placebo group and so we could use Dunnett's test to examine these hypotheses. Once you have selected Dunnett's test, change the control category (the default is to use the Last category) to specify that the First category be used as the control category (because the placebo group was coded with the lowest value). You can also choose whether to conduct a two-tailed test (2-sided) or a one-tailed test. If you choose a one-tailed test then you must predict whether you believe that the mean of the control group will be less than a particular experimental group (> Control) or greater than a particular experimental group (< Control). These are all of the *post hoc* tests that need to be specified and when the completed dialog box looks like Figure 10.12 click on Continue to return to the main dialog box.

FIGURE 10.12 Dialog box for specifying *post hoc* tests

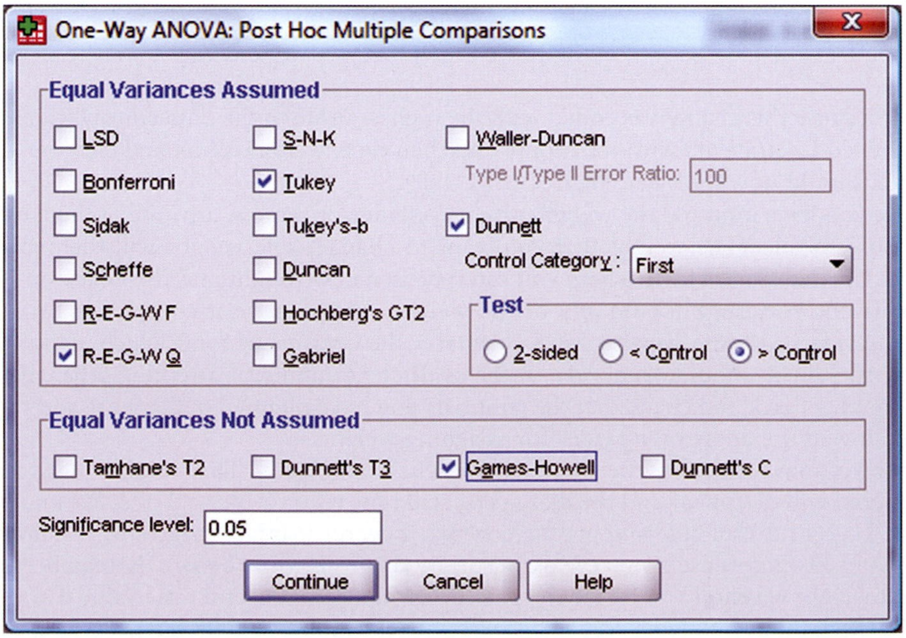

10.3.3. Options ②

The options for one-way ANOVA are fairly straightforward (Figure 10.13). First you can ask for some descriptive statistics which will produce a table of the means, standard deviations, standard errors, ranges and confidence intervals for the means of each group. This option is useful to select because it assists in interpreting the final results. A vital option to select is the homogeneity of variance tests. As with the *t*-test, there is an assumption that the variances of the groups are equal and selecting this option tests this assumption. SPSS uses Levene's test, which tests the hypothesis that the variances of each group are equal (see section 5.6.1). If the homogeneity of variance assumption is broken, then SPSS offers us two alternative versions of the *F*-ratio: the **Brown–Forsythe *F*** (1974), and **Welch's *F*** (1951). If you're really bored, these two statistics are discussed in Jane Superbrain Box 10.2, but suffice it to say they're worth selecting just in case the assumption is broken.

There is also an option to have a *Means plot* and if this option is selected then a line graph of the group means will be produced in the output. However, the resulting graph is a leprotic tramp compared to what we can create using the Chart Builder and, as I have said before, it's best to graph your data *before* the analysis. Finally, the options let us specify whether we want to exclude cases on a listwise basis or on a per analysis basis (SPSS Tip 6.1). This option is useful only if you are conducting several ANOVAs on different dependent variables. The first option (*Exclude cases analysis by analysis*) excludes any case that has a missing value for either the independent or the dependent variable used in that particular analysis. *Exclude cases listwise* will exclude from *all analyses* any case that has a missing value for the independent variable or any of the dependent variables specified. If you stick to good practice and don't conduct hundreds of ANOVAs on different dependent variables (see Chapter 16 on MANOVA) the default settings are fine. When you have selected the appropriate options, click on Continue to return to the main dialog box and then click on OK to run the analysis.

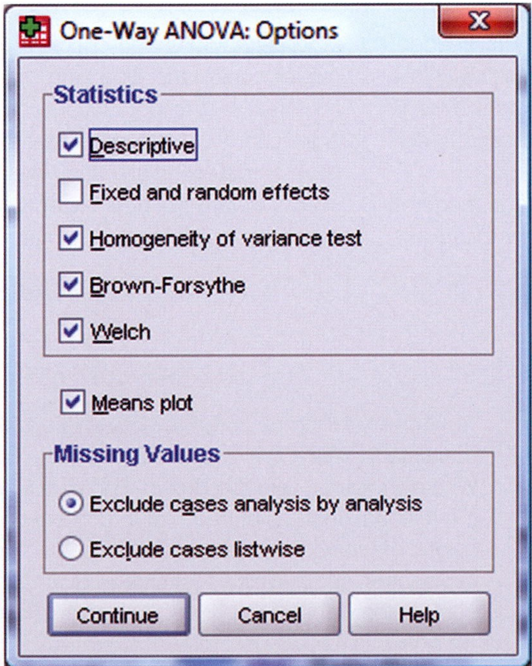

FIGURE 10.13 Options for one-way ANOVA

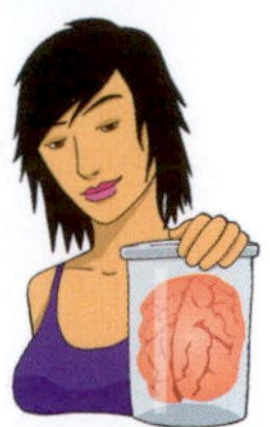

JANE SUPERBRAIN 10.2

What do I do in ANOVA when the homogeneity of variance assumption is broken? ③

In section 10.2.10 I mentioned that when group sizes are unequal, violations of the assumption of homogeneity of variance can have quite serious consequences. SPSS incorporates options for two alternative *F*-ratios, which have been derived to be robust when homogeneity of variance has been violated. The first is the Brown and Forsythe (1974) *F*-ratio, which is fairly easy to explain. I mentioned earlier that when group sizes are unequal and the large groups have the biggest variance, then this biases the *F*-ratio to be conservative. If you think back to equation (10.9) this makes perfect sense because to calculate SS_R variances are multiplied by their sample size (minus one), so in this situation you get a large sample size cross-multiplied with a large variance, which will inflate the value of SS_R. What effect does this have on the *F*-ratio? Well, the *F*-ratio is proportionate to SS_M/SS_R, so if SS_R is big, then the *F*-ratio gets smaller (which is why it would be more conservative: its value is being overly reduced!). Brown and Forsythe get around this problem by weighting the group variances, not by their sample size, but by the inverse of their sample sizes (actually they use *n*/*N* so it's the sample size as a proportion of the total sample size). This means that the impact of large sample sizes with large variance is reduced:

$$F_{BF} = \frac{SS_M}{SS_{BF}} = \frac{SS_M}{\sum s_k^2\left(1 - \frac{n_k}{N}\right)}$$

So, for the Viagra data, SS_M is the same as before (20.135), so the equation becomes:

$$\begin{aligned} F_{BF} &= \frac{20.135}{s^2_{group1}\left(1 - \frac{n_{group1}}{N}\right) + s^2_{group2}\left(1 - \frac{n_{group2}}{N}\right) + s^2_{group3}\left(1 - \frac{n_{group3}}{N}\right)} \\ &= \frac{20.135}{1.70\left(1 - \frac{5}{15}\right) + 1.70\left(1 - \frac{5}{15}\right) + 2.50\left(1 - \frac{5}{15}\right)} \\ &= \frac{20.135}{3.933} \\ &= 5.119 \end{aligned}$$

This statistic is evaluated using degrees of freedom for the model and error terms. For the model, df_M is the same as before (i.e. $k - 1 = 2$), but an adjustment is made to the residual degrees of freedom, df_R. The second correction is Welch's (1951) *F* – see Oliver Twisted.

The obvious question is which of the two procedures is best? Tomarken and Serlin (1986) review these and other techniques and seem to conclude that both techniques control the Type I error rate well (i.e. when there's no effect in the population you do indeed get a non-significant *F*). However, in terms of power (i.e. which test is best a detecting an effect when it exists), the Welch test seems to fare the best except when there is an extreme mean that has a large variance.

OLIVER TWISTED

Please, Sir, can I have some more … Welch's F?

'You're only telling us about the Brown–Forsythe *F* because you don't understand Welch's *F*,' taunts Oliver, 'Andy, Andy, brains all sandy ….' Whatever, Oliver. Like the Brown–Forsythe *F*, Welch's *F* adjusts *F* and the residual degrees of freedom to combat problems arising from violations of the homogeneity of variance assumption. There is a lengthy explanation about Welch's *F* in the additional material available on the companion website. Oh, and Oliver, microchips are made of sand.

10.4. Output from one-way ANOVA ②

If you load up the Viagra data (or enter it in by hand) and select all of the options I have suggested, you should find that the output looks the same as what follows. If your output is different we should panic because one of us has done it wrong – hopefully not me or a lot of trees have died for nothing.

In Chapter 4 we saw that it is always a good idea to look at a graph of your data. In this case we should produce an error bar graph or a line graph with error bars.

SELF-TEST Produce a line chart with error bars for the Viagra data.

Figure 10.14 shows a line chart with error bars of the Viagra data (I have edited my graph; see if you can use the SPSS Chart Editor to make yours look like mine). It's clear from this chart that all of the error bars overlap, indicating that, on face value, there are no between-group differences (although this measure is only approximate). The line that joins the means seems to indicate a linear trend in that, as the dose of Viagra increases, so does the mean level of libido.

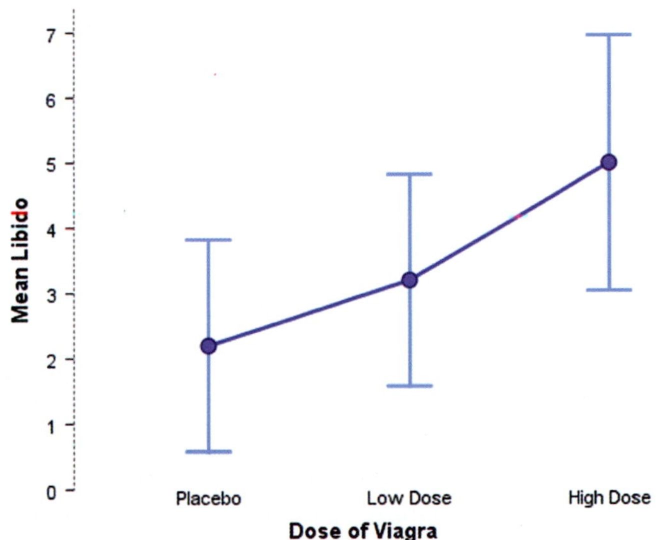

FIGURE 10.14 Error bar chart of the Viagra data

10.4.1. Output for the main analysis ②

SPSS Output 10.3 shows the table of descriptive statistics from the one-way procedure for the Viagra data. The first thing to notice is that the means and standard deviations correspond to those shown in Table 10.1. In addition we are told the standard error. You should remember that the standard error is the standard deviation of the sampling distribution of these data (so, for the placebo group, if you took lots of samples from the population from which these data come, the means of these samples would have a standard deviation of 0.5831). We are also given confidence intervals for the mean. By now, you should be familiar with what a confidence interval tells us, and that is that if we took 100 samples from the population from which the placebo group came and constructed confidence intervals for the mean, then 95 of these intervals would contain the true value of the mean: in other words, the true value of the mean is likely to be between 0.5811 and 3.8189. Although these diagnostics are not immediately important, we will refer back to them throughout the analysis.

SPSS OUTPUT 10.3

Descriptives

Libido

	N	Mean	Std. Deviation	Std. Error	95% Confidence Interval for Mean		Minimum	Maximum
					Lower Bound	Upper Bound		
Placebo	5	2.20	1.304	.583	.58	3.82	1	4
Low Dose	5	3.20	1.304	.583	1.58	4.82	2	5
High Dose	5	5.00	1.581	.707	3.04	6.96	3	7
Total	15	3.47	1.767	.456	2.49	4.45	1	7

The next part of the output is a summary table of Levene's test (see section 5.6.1). This test is designed to test the null hypothesis that the variances of the groups are the same. It is an ANOVA test conducted on the absolute differences between the observed data and the mean from which the data came (see Oliver Twisted). In this case, Levene's test is, therefore, testing whether the variances of the three groups are significantly different. If Levene's test is significant (i.e. the value of *Sig.* is less than .05) then we can say that the variances are significantly different. This would mean that we had violated one of the assumptions of ANOVA and we would have to take steps to rectify this matter. As we saw in Chapter 5, one common way to rectify differences between group variances is to transform all of the data and then reanalyse these transformed values (see Chapter 5). However, given the apparent utility of Welch's *F*, or the Brown–Forsythe *F*, and the fact that transformations often don't help at all, you can instead report Welch's *F* (and I'd probably suggest reporting this instead of the Brown–Forsythe *F* unless you have an extreme mean that is also causing the problem with the variances). Luckily, for these data the variances are very similar (hence the high probability value); in fact, if you look at SPSS Output 10.3 you'll see that the variances of the placebo and low-dose groups are identical.

SPSS OUTPUT 10.4

Test of Homogeneity of Variances

	Levene Statistic	df1	df2	Sig.
Libido	.092	2	12	.913

SPSS Output 10.5 shows the main ANOVA summary table. The table is divided into between-group effects (effects due to the model – the experimental effect) and within-group

OLIVER TWISTED

Please, Sir, can I have some more … Levene's test?

'Liar! Liar! Pants on fire,' screams Oliver his cheeks red and eyes about to explode. 'You promised in Chapter 5 to explain Levene's test properly and you haven't, you spatula head.' True enough, Oliver, I do have a spatula for a head. I also have a very nifty little demonstration of Levene's test in the additional material for this chapter on the companion website. It will tell you more than you could possibly want to know. Let's go fry an egg …

effects (this is the unsystematic variation in the data). The between-group effect is further broken down into a linear and quadratic component and these components are the trend analyses described in section 10.2.11.5. The between-group effect labelled *Combined* is the overall experimental effect. In this row we are told the sums of squares for the model (SS_M = 20.13) and this value corresponds to the value calculated in section 10.2.6. The degrees of freedom are equal to 2 and the mean squares for the model corresponds to the value calculated in section 10.2.8 (10.067). The sum of squares and mean squares represent the experimental effect. This overall effect is then broken down because we asked SPSS to conduct trend analyses of these data (we will return to these trends in due course). Had we not specified this in section 10.3.1, then these two rows of the summary table would not be produced. The row labelled *Within groups* gives details of the unsystematic variation within the data (the variation due to natural individual differences in libido and different reactions to Viagra). The table tells us how much unsystematic variation exists (the residual sum of squares, SS_R) and this value (23.60) corresponds to the value calculated in section 10.2.7. The table then gives the average amount of unsystematic variation, the mean squares (MS_R), which corresponds to the value (1.967) calculated in section 10.2.8. The test of whether the group means are the same is represented by the *F*-ratio for the combined between-group effect. The value of this ratio is 5.12, which is the same as was calculated in section 10.2.9. Finally SPSS tells us whether this value is likely to have happened by chance. The final column labelled *Sig.* indicates the likelihood of an *F*-ratio the size of the one obtained occurring if there was no effect in the population (see also SPSS Tip 10.1). In this case, there is a probability of .025 that an *F*-ratio of this size would occur if in reality there was no effect (that's only a 2.5% chance!). We have seen in previous chapters that we use a cut-off point of .05 as a criterion for statistical significance. Hence, because the observed significance value is less than .05 we can say that there was a significant effect of Viagra. However, at this stage we still do not know exactly what the effect of Viagra was (we don't know which groups differed). One thing that is interesting here is that we obtained a significant experimental effect yet our error bar plot indicated that no significant difference would be found. This contradiction illustrates how the error bar chart can act only as a rough guide to the data.

SPSS OUTPUT 10.5

ANOVA

Libido

			Sum of Squares	df	Mean Square	F	Sig.
Between Groups	(Combined)		20.133	2	10.067	5.119	.025
	Linear Term	Contrast	19.600	1	19.600	9.966	.008
		Deviation	.533	1	.533	.271	.612
	Quadratic Term	Contrast	.533	1	.533	.271	.612
Within Groups			23.600	12	1.967		
Total			43.733	14			

Knowing that the overall effect of Viagra was significant, we can now look at the trend analysis. The trend analysis breaks down the experimental effect to see whether it can be explained by either a linear or a quadratic relationship in the data. First, let's look at the linear component. This comparison tests whether the means increase across groups in a linear way. Again the sum of squares and mean squares are given, but the most important things to note are the value of the *F*-ratio and the corresponding significance value. For the linear trend the *F*-ratio is 9.97 and this value is significant at a .008 level. Therefore, we can say that as the dose of Viagra increased from nothing to a low dose to a high dose, libido increased proportionately. Moving on to the quadratic trend, this comparison is testing whether the pattern of means is curvilinear (i.e. is represented by a curve that has one bend). The error bar graph of the data suggests that the means cannot be represented by a curve and the results for the quadratic trend bear this out. The *F*-ratio for the quadratic trend is non-significant (in fact, the value of *F* is less than 1, which immediately indicates that this contrast will not be significant).

SPSS TIP 10.1 One- and two-tailed tests in ANOVA ②

A question I get asked a lot by students is 'is the significance of the ANOVA one- or two-tailed, and if it's two-tailed can I divide by 2 to get the one-tailed value?' The answer is that to do a one-tailed test you have to be making a directional hypothesis (i.e. the mean for cats is greater than for dogs). ANOVA is a non-specific test, so it just tells us generally whether there is a difference or not and because there are several means you can't possibly make a directional hypothesis. As such, it's invalid to halve the significance value.

Finally, SPSS Output 10.6 shows Welch's and the Brown–Forsythe *F*-ratios. As it turned out we didn't need these because our Levene's test was not significant, indicating that our variances were equal. However, when homogeneity of variance has been violated you should look at these *F*-ratios *instead* of the ones in the main table. If you're interested in how these values are calculated then look at Jane Superbrain Box 10.2, but to be honest it's just bloody confusing if you ask me; you're much better off just looking at the values in SPSS Output 10.6 and trusting that they do what they're supposed to do (you should also note that the error degrees of freedom have been adjusted and you should remember this when you report the values!).

SPSS OUTPUT 10.6

Robust Tests of Equality of Means

Libido

	Statistic[a]	df1	df2	Sig.
Welch	4.320	2	7.943	.054
Brown-Forsythe	5.119	2	11.574	.026

a. Asymptotically F distributed.

10.4.2. Output for planned comparisons ②

In section 10.3.1 we told SPSS to conduct two planned comparisons: one to test whether the control group was different to the two groups which received Viagra, and one to see whether the two doses of Viagra made a difference to libido. SPSS Output 10.7 shows the results of the planned comparisons that we requested for the Viagra data. The first table displays the contrast coefficients; these values are the ones that we entered in section 10.3.1 and it is well worth looking at this table to double-check that the contrasts are comparing what they are supposed to! As a quick rule of thumb, remember that when we do planned comparisons we arrange the weights such that we compare any group with a positive weight against any group with a negative weight. Therefore, the table of weights shows that contrast 1 compares the placebo group against the two experimental groups, and contrast 2 compares the low-dose group to the high-dose group. It is useful to check this table to make sure that the weights that we entered into SPSS correspond to the weights we intended to enter into SPSS!

SPSS OUTPUT 10.7

Contrast Coefficients

Contrast	Dose of Viagra		
	Placebo	Low Dose	High Dose
1	-2	1	1
2	0	-1	1

Contrast Tests

		Contrast	Value of Contrast	Std. Error	t	df	Sig. (2-tailed)
Libido	Assume equal variances	1	3.80	1.536	2.474	12	.029
		2	1.80	.887	2.029	12	.065
	Does not assume equal variances	1	3.80	1.483	2.562	8.740	.031
		2	1.80	.917	1.964	7.720	.086

The second table gives the statistics for each contrast. The first thing to notice is that statistics are produced for situations in which the group variances are equal, and when they are unequal. If Levene's test was significant then you should read the part of the table labelled *Does not assume equal variances*. However, for these data Levene's test was not significant and we can, therefore, use the part of the table labelled *Assume equal variances*. The table tells us the value of the contrast itself, which is the weighted sum of the group means. This value is obtained by taking each group mean, multiplying it by the weight for the contrast of interest, and then adding these values together.[5] The table also gives the standard error of each contrast and a t-statistic. The t-statistic is derived by dividing the contrast value by the standard error ($t = 3.8/1.5362 = 2.47$) and is compared against critical values of the t-distribution. The significance value of the contrast is given in the final column and this value is two-tailed. Using the first contrast as an example, if we had used this contrast to test the general hypothesis that the experimental groups would differ from the placebo group, then we should use this two-tailed value. However, in reality we tested the hypothesis that the experimental groups would increase libido above the levels seen in the placebo group: this hypothesis is one-tailed. Provided the means for the groups bear out the hypothesis we can divide the significance values by 2 to obtain the one-tailed probability. Hence, for contrast 1, we can say that taking Viagra significantly increased libido compared to the control group ($p = .0145$). For contrast 2 we also had a one-tailed hypothesis (that a high dose of Viagra would increase libido significantly more than a low dose) and the means bear this hypothesis out. The significance of contrast 2 tells us that a high dose of Viagra increased libido significantly more than a low dose (p(one-tailed) $= .065/2 = .0325$). Notice that had we not had a specific hypothesis regarding which group would have the highest mean, then we would have had to conclude that the dose of Viagra had no significant effect on libido. For this reason it can be important as scientists that we generate hypotheses before collecting any data because this method of scientific discovery is more powerful.

In summary, there is an overall effect of Viagra on libido. Furthermore, the planned contrasts revealed that having Viagra significantly increased libido compared to a control group, $t(12) = 2.47$, $p < .05$, and having a high dose significantly increased libido compared to a low dose, $t(12) = 2.03$, $p < .05$ (one-tailed).

10.4.3. Output for *post hoc* tests ②

If we had no specific hypotheses about the effect that Viagra might have on libido then we could carry out *post hoc* tests to compare all groups of participants with each other. In fact, we asked SPSS to do this (see section 10.3.2) and the results of this analysis are shown in SPSS Output 10.8. This table shows the results of Tukey's test (known as Tukey's HSD)[6], the Games–Howell procedure and Dunnett's test, which were all specified earlier on. If we look at Tukey's test first (because we have no reason to doubt that the population variances are unequal) it is clear from the table that each group of participants is compared to all of the

[5] For the first contrast this value is:

$$\Sigma(\bar{X}W) = [(2.2\times-2)+(3.2\times1)+(5.0\times1)] = 3.8$$

[6] The HSD stands for 'honestly significant difference', which has a slightly dodgy ring to it if you ask me!

remaining groups. For each pair of groups the difference between group means is displayed, the standard error of that difference, the significance level of that difference and a 95% confidence interval. First of all, the placebo group is compared to the low-dose group and reveals a non-significant difference (*Sig.* is greater than .05), but when compared to the high-dose group there is a significant difference (*Sig.* is less than .05).

SELF-TEST Our planned comparison showed that any dose of Viagra produced a significant increase in libido, yet the *post hoc* tests indicate that a low dose does not. Why is there this contradiction?

In section 10.2.11.2, I explained that the first planned comparison would compare the experimental groups to the placebo group. Specifically, it would compare the average of the two group means of the experimental groups ((3.2 + 5.0)/2 = 4.1) to the mean of the placebo group (2.2). So, it was assessing whether the difference between these values (4.1 − 2.2 = 1.9) was significant. In the *post hoc* tests, when the low dose is compared to the placebo, the contrast is testing whether the difference between the means of these two groups is significant. The difference in this case is only 1, compared to a difference of 1.9 for the planned comparison. This explanation illustrates how it is possible to have apparently contradictory results from planned contrasts and *post hoc* comparisons. More important, it illustrates how careful we must be in interpreting planned contrasts.

The low-dose group is then compared to both the placebo group and the high-dose group. The first thing to note is that the contrast involving the low-dose and placebo groups is identical to the one just described. The only new information is the comparison between the two experimental conditions. The group means differ by 1.8 which is not significant. This result contradicts the planned comparisons (remember that contrast 2 compared these groups and found a significant difference).

SELF-TEST Why does the *post hoc* test show a non-significant difference between high and low dose, when the planned comparison showed a significant difference?

This contradiction occurs for two possible reasons. First, *post hoc* tests by their nature are two-tailed (you use them when you have made no specific hypotheses and you cannot predict the direction of hypotheses that don't exist!) and contrast 2 was significant only when considered as a one-tailed hypothesis. However, even at the two-tailed level the planned comparison was closer to significance than the *post hoc* test and this fact illustrates that *post hoc* procedures are more conservative (i.e. have less power to detect true effects) than planned comparisons.

The rest of the table describes the Games–Howell test and a quick inspection reveals the same pattern of results: the only groups that differed significantly were the high-dose and placebo groups. These results give us confidence in our conclusions from Tukey's test because even if the populations variances are not equal (which seems unlikely given that the sample variances are very similar), then the profile of results still holds true. Finally, Dunnett's test is described and you'll hopefully remember that we asked the computer to compare both experimental groups against the control using a one-tailed hypothesis that the mean of the control group would be smaller than both experimental groups. Even as a one-tailed hypothesis, levels of libido in the low-dose group are equivalent to the placebo group. However, the high-dose group has a significantly higher libido than the placebo group.

SPSS OUTPUT 10.8

Multiple Comparisons

Dependent Variable:Libido

	(I) Dose of Viagra	(J) Dose of Viagra	Mean Difference (I-J)	Std. Error	Sig.	95% Confidence Interval Lower Bound	Upper Bound
Tukey HSD	Placebo	Low Dose	-1.000	.887	.516	-3.37	1.37
		High Dose	-2.800*	.887	.021	-5.17	-.43
	Low Dose	Placebo	1.000	.887	.516	-1.37	3.37
		High Dose	-1.800	.887	.147	-4.17	.57
	High Dose	Placebo	2.800*	.887	.021	.43	5.17
		Low Dose	1.800	.887	.147	-.57	4.17
Games-Howell	Placebo	Low Dose	-1.000	.825	.479	-3.36	1.36
		High Dose	-2.800*	.917	.039	-5.44	-.16
	Low Dose	Placebo	1.000	.825	.479	-1.36	3.36
		High Dose	-1.800	.917	.185	-4.44	.84
	High Dose	Placebo	2.800*	.917	.039	.16	5.44
		Low Dose	1.800	.917	.185	-.84	4.44
Dunnett t (>control) a	Low Dose	Placebo	1.000	.887	.227	-.87	
	High Dose	Placebo	2.800*	.887	.008	.93	

*. The mean difference is significant at the 0.05 level.

a. Dunnett t-tests treat one group as a control, and compare all other groups against it.

The table in SPSS Output 10.9 shows the results of Tukey's test and the REGWQ test. These tests display subsets of groups that have the same means. Therefore, Tukey's test creates two subsets of groups with statistically similar means. The first subset contains the placebo and low-dose groups (indicating that these two groups have the similar means) whereas the second subset contains the high- and low-dose groups. These results demonstrate that the placebo group has a similar mean to the low-dose group but not the high-dose group, and that the low-dose group has a similar mean to both the placebo and high-dose groups. In other words, the only groups that have significantly different means are the high-dose and placebo groups. The tests provide a significance value for each subset and it's clear from these significance values that the groups in subsets have non-significant means (as indicated by values of *Sig.* that are greater than .05).

These calculations use the harmonic mean sample size. The **harmonic mean** is a weighted version of the mean that takes account of the relationship between variance and sample size. Although you don't need to know the intricacies of the harmonic mean, it is useful that the harmonic sample size is used because it reduces bias that might be introduced through having unequal sample sizes. However, as we have seen, these tests are still biased when sample sizes are unequal.

SPSS OUTPUT 10.9

Libido

	Dose of Viagra	N	Subset for alpha = 0.05: 1	2
Tukey HSD[a]	Placebo	5	2.20	
	Low Dose	5	3.20	3.20
	High Dose	5		5.00
	Sig.		.516	.147
Ryan-Einot-Gabriel-Welsch Range	Placebo	5	2.20	
	Low Dose	5	3.20	3.20
	High Dose	5		5.00
	Sig.		.282	.065

Means for groups in homogeneous subsets are displayed.

a. Uses Harmonic Mean Sample Size = 5.000.

CRAMMING SAM'S TIPS One-way ANOVA

- The one-way independent ANOVA compares several means, when those means have come from different groups of people. For example; if you have several experimental conditions and have used different participants in each condition.
- When you have generated specific hypotheses before the experiment use *planned comparisons*, but if you don't have specific hypotheses use *post hoc* tests.
- There are lots of different *post hoc* tests: when you have equal sample sizes and homogeneity of variance is met use *REGWQ* or *Tukey's* HSD. If sample sizes are slightly different then use *Gabriel's* procedure, but if sample sizes are very different use *Hochberg's GT2*. If there is any doubt about homogeneity of variance use the *Games–Howell* procedure.
- Test for homogeneity of variance using *Levene's test*. Find the table with this label: if the value in the column labelled *Sig.* is less than .05 then the assumption is violated. If this is the case go to the table labelled **Robust Tests of Equality of Means**. If homogeneity of variance has been met (the significance of Levene's test is greater than *.05)* go to the table labelled **ANOVA**.
- In the table labelled **ANOVA** *(*or **Robust Tests of Equality of Means** – see above), look at the column labelled *Sig.* if the value is less than .05 then the means of the groups are significantly different.
- For contrasts and *post hoc* tests, again look to the columns labelled *Sig.* to discover if your comparisons are significant (they will be if the significance value is less than .05).

LABCOAT LENI'S REAL RESEARCH 10.1

Scraping the barrel? ①

Evolution has endowed us with many beautiful things (cats, dolphins, the Great Barrier Reef, etc.), all selected to fit their ecological niche. Given evolution's seemingly limitless capacity to produce beauty, it's something of a wonder how it managed to produce such a monstrostity as the human penis. One theory is that the penis evolved into the shape that it is because of sperm competition. Specifically, the human penis has an unusually large glans (the 'bell-end' as it's affectionately known) compared to other primates, and this may have evolved so that the penis can displace seminal fluid from other males by 'scooping it out' during intercourse. To put this idea to the test, Gordon Gallup and his colleagues came up with an ingenious study (Gallup, Burch, Zappieri, Parvez, Stockwell, & Davis, 2003). Armed with various female masturbatory devices from Hollywood Exotic Novelties, an artificial vagina from California Exotic Novelties, and some water and cornstarch to make fake sperm, they loaded the artificial vagina with 2.6 ml of fake sperm and inserted one of three female sex toys into it before withdrawing it. Over several trials, three different female sex toys were used: a control phallus that had no coronal ridge (i.e. no bell-end), a phallus with a minimal coronal ridge (small bell-end) and a phallus with a coronal ridge.

They measured sperm displacement as a percentage using the following equation (included here because it is more interesting than all of the other equations in this book):

$$\frac{\text{weight of vagina with semen} - \text{weight of vagina following insertion and removal of phallus}}{\text{weight of vagina with semen} - \text{weight of empty vagina}} \times 100$$

As such, 100% means that all of the sperm was displaced by the phallus, and 0% means that none of the sperm was displaced. If the human penis evolved as a sperm displacement device then Gallup et al. predicted: (1) that having a bell-end would displace more sperm than not; and (2) the phallus with the larger coronal ridge would displace more sperm than the phallus with the minimal coronal ridge. The conditions are ordered (no ridge, minimal ridge, normal ridge) so we might also predict a linear trend. The data can be found in the file **Gallup et al.sav**. Draw an error bar graph of the means of the three conditions. Conduct a one-way ANOVA with planned comparisons to test the two hypotheses above. What did Gallup et al. find?

Answers are in the additional material on the companion website (or look at pages 280–281 in the original article).

GALLUP, G.G.J. ET AL. (2003). *EVOLUTION AND HUMAN BEHAVIOR*, 24, 277–289.

10.5. Calculating the effect size ②

One thing you will notice is that SPSS doesn't routinely provide an effect size for one-way independent ANOVA. However, we saw in equation (7.4) that:

$$R^2 = \frac{SS_M}{SS_T}$$

Of course we know these values from the SPSS output. So we can simply calculate r^2 using the between-group effect (SS_M) and the total amount of variance in the data (SS_T) – although for some bizarre reason it's usually called **eta squared,** η^2. It is then a simple matter to take the square root of this value to give us the effect size r:

$$\begin{aligned} r^2 &= \eta^2 \\ &= \frac{SS_M}{SS_T} \\ &= \frac{20.13}{43.73} \\ &= .46 \\ r &= \sqrt{.46} \\ &= .68 \end{aligned}$$

Using the benchmarks for effect sizes this represents a large effect (it is above the .5 threshold for a large effect). Therefore, the effect of Viagra on libido is a substantive finding.

However, this measure of effect size is slightly biased because it is based purely on sums of squares from the sample and no adjustment is made for the fact that we're trying to estimate the effect size in the population. Therefore, we often use a slightly more complex measure called **omega squared** (ω^2). This effect size estimate is still based on the sums of squares that we've met in this chapter, but like the F-ratio it uses the variance explained by the model, and the error variance (in both cases the average variance, or mean squared error, is used):

$$\omega^2 = \frac{SS_M - (df_M)MS_R}{SS_T + MS_R}$$

The df_M in the equation is the degrees of freedom for the effect, which you can get from the SPSS output (in the case of the main effect this is the number of experimental conditions minus one). So, in this example we'd get:

$$\begin{aligned} \omega^2 &= \frac{20.13 - (2)1.97}{43.73 + 1.97} \\ &= \frac{16.19}{45.70} \\ &= .35 \\ \omega &= .60 \end{aligned}$$

As you can see, this has led to a slightly lower estimate to using r, and in general ω is a more accurate measure. Although in the sections on ANOVA I will use ω as my effect size measure, think of it as you would r (because it's basically an unbiased estimate of r anyway). People normally report ω^2 and it has been suggested that values of .01, .06 and .14 represent small, medium and large effects respectively (Kirk, 1996). Remember, though, that these are rough guidelines and that effect sizes need to be interpreted within the context of the research literature.

Most of the time it isn't that interesting to have effect sizes for the overall ANOVA because it's testing a general hypothesis. Instead, we really want effect sizes for the contrasts (because these compare only two things, so the effect size is considerably easier to interpret). Planned comparisons are tested with the t-statistic and, therefore, we can use the same equation as in section 9.4.6:

$$r_{\text{contrast}} = \sqrt{\frac{t^2}{t^2 + df}}$$

We know the value of t and the df from SPSS Output 10.7 and so we can compute r as follows:

$$\begin{aligned} r_{\text{contrast1}} &= \sqrt{\frac{2.474^2}{2.474^2 + 12}} \\ &= \sqrt{\frac{6.12}{18.12}} \\ &= 0.58 \end{aligned}$$

If you think back to our benchmarks for effect sizes this represents a large effect (it is above .5, the threshold for a large effect). Therefore, as well as being statistically significant, this effect is large and so represents a substantive finding. For contrast 2 we get:

$$\begin{aligned} r_{\text{contrast2}} &= \sqrt{\frac{2.029^2}{2.029^2 + 12}} \\ &= \sqrt{\frac{4.12}{16.12}} \\ &= 0.51 \end{aligned}$$

This too is a substantive finding and represents a large effect size.

10.6. Reporting results from one-way independent ANOVA ②

When we report an ANOVA, we have to give details of the F-ratio and the degrees of freedom from which it was calculated. For the experimental effect in these data the F-ratio was derived by dividing the mean squares for the effect by the mean squares for the residual.

Therefore, the degrees of freedom used to assess the *F*-ratio are the degrees of freedom for the effect of the model ($df_M = 2$) and the degrees of freedom for the residuals of the model ($df_R = 12$). Therefore, the correct way to report the main finding would be:

✓ There was a significant effect of Viagra on levels of libido, $F(2, 12) = 5.12$, $p < .05$, $\omega = .60$.

Notice that the value of the *F*-ratio is preceded by the values of the degrees of freedom for that effect. Also, we rarely state the exact significance value of the *F*-ratio: instead we report that the significance value, *p*, was less than the criterion value of .05 and include an effect size measure. The linear contrast can be reported in much the same way:

✓ There was a significant linear trend, $F(1, 12) = 9.97$, $p < .01$, $\omega = .62$, indicating that as the dose of Viagra increased, libido increased proportionately.

Notice that the degrees of freedom have changed to reflect how the *F*-ratio was calculated. I've also included an effect size measure (have a go at calculating this as we did for the main *F*-ratio and see if you get the same value). Also, we have now reported that the *F*-value was significant at a value less than the criterion value of .01. We can also report our planned contrasts:

✓ Planned contrasts revealed that having any dose of Viagra significantly increased libido compared to having a placebo, $t(12) = 2.47$, $p < .05$ (1-tailed), $r = .58$, and that having a high dose significantly increased libido compared to having a low dose, $t(12) = 2.03$, $p < .05$ (1-tailed), $r = .51$.

Note that in both cases I've stated that we used a one-tailed probability.

10.7. Violations of assumptions in one-way independent ANOVA ②

I've mentioned several times in this chapter that ANOVA can be robust to violations of its assumptions, but not always. We also saw that there are measures that can be taken when you have heterogeneity of variance (Jane Superbrain Box 10.2). However, there is another alternative. There are a group of tests (often called assumption-free, distribution-free and non-parametric tests, none of which are particularly accurate names!). Well, the one-way independent ANOVA also has a non-parametric counterpart called the Kruskal–Wallis test. If you have non-normally distributed data, or have violated some other assumption, then this test can be a useful way around the problem. This test is described in Chapter 15.

There are also robust methods available (see section 5.7.4) to compare independent means (and even medians) that involve, for example, using 20% trimmed means or a bootstrap. SPSS doesn't do any of them so I advise investigating Wilcox's Chapter 7 and the associated files for the software R (Wilcox, 2005). You can use the R plugin for SPSS to conduct these tests. See the companion website for some training demos about using R in SPSS.

What have I discovered about statistics? ①

This chapter has introduced you to analysis of variance (ANOVA), which is the topic of the next few chapters also. One-way independent ANOVA is used in situations when you want to compare several means, and you've collected your data using different participants in each condition. I started off explaining that if we just do lots of *t*-tests on the same data then our Type I error rate becomes inflated. Hence we use ANOVA instead. I looked at how ANOVA can be conceptualized as a general linear model (GLM) and so is in fact the same as multiple regression. Like multiple regression, there are three important measures that we use in ANOVA: the total sum of squares, SS_T (a measure of the variability in our data), the model sum of squares, SS_M (a measure of how much of that variability can be explained by our experimental manipulation), and SS_R (a measure of how much variability can't be explained by our experimental manipulation). We discovered that, crudely speaking, the *F*-ratio is just the ratio of variance that we can explain against the variance that we can't. We also discovered that a significant *F*-ratio tells us only that our groups differ, not how they differ. To find out where the differences lie we have two options: specify specific contrasts to test hypotheses (*planned contrasts*), or test every group against every other group (*post hoc tests*). The former are used when we have generated hypotheses before the experiment, whereas the latter are for exploring data when no hypotheses have been made. Finally we discovered how to implement these procedures on SPSS.

We also saw that my life was changed by a letter that popped through the letterbox one day saying only that I could go to the local grammar school if I wanted to. When my parents told me, rather than being in celebratory mood, they were very down beat; they knew how much it meant to me to be with my friends and how I had got used to my apparent failure. Sure enough, my initial reaction was to say that I wanted to go to the local school. I was unwavering in this view. Unwavering, that is, until my brother convinced me that being at the same school as him would be really cool. It's hard to measure how much I looked up to him, and still do, but the fact that I willingly subjected myself to a lifetime of social dysfunction just to be with him is a measure of sorts. As it turned out, being at school with him was not always cool – he was bullied for being a boffin (in a school of boffins) and being the younger brother of a boffin made me a target. Luckily, unlike my brother, I was stupid and played football, which seemed to be good enough reasons for them to leave me alone. Most of the time.

Key terms that I've discovered

Analysis of variance (ANOVA)
Bonferroni correction
Brown–Forsythe *F*
Cubic trend
Deviation contrast
Difference contrast (reverse Helmert contrast)
Eta squared, η^2
Experimentwise error rate
Familywise error rate
Grand variance
Harmonic mean
Helmert contrast
Independent ANOVA
Omega squared, ω^2
Orthogonal
Pairwise comparisons
Planned contrasts
Polynomial contrast
Post hoc tests
Quadratic trend
Quartic trend
Repeated contrast
Simple contrast
Weights
Welch's *F*

Smart Alex's tasks

- **Task 1**: Imagine that I was interested in how different teaching methods affected students' knowledge. I noticed that some lecturers were aloof and arrogant in their teaching style and humiliated anyone who asked them a question, while others were encouraging and supporting of questions and comments. I took three statistics courses where I taught the same material. For one group of students I wandered around with a large cane and beat anyone who asked daft questions or got questions wrong (*punish*). In the second group I used my normal teaching style, which is to encourage students to discuss things that they find difficult and to give anyone working hard a nice sweet (*reward*). The final group I remained indifferent to and neither punished nor rewarded students' efforts (*indifferent*). As the dependent measure I took the students' exam marks (percentage). Based on theories of operant conditioning, we expect punishment to be a very unsuccessful way of reinforcing learning, but we expect reward to be very successful. Therefore, one prediction is that reward will produce the best learning. A second hypothesis is that punishment should actually retard learning such that it is worse than an indifferent approach to learning. The data are in the file **Teach.sav.** Carry out a one-way ANOVA and use planned comparisons to test the hypotheses that (1) reward results in better exam results than either punishment or indifference; and (2) indifference will lead to significantly better exam results than punishment. ②
- **Task 2**: In Chapter 15 (section 15.5) there are some data looking at whether eating Soya meals reduces your sperm count. Have a look at this section, access the data for that example, but analyse them with ANOVA. What's the difference between what you find and what is found in section 15.5.4? Why do you think this difference has arisen? ②
- **Task 3**: Students (and lecturers for that matter) love their mobile phones, which is rather worrying given some recent controversy about links between mobile phone use and brain tumours. The basic idea is that mobile phones emit microwaves, and so holding one next to your brain for large parts of the day is a bit like sticking your brain in a microwave oven and selecting the 'cook until well done' button. If we wanted to test this experimentally, we could get six groups of people and strap a mobile phone on their heads (that they can't remove). Then, by remote control, we turn the phones on for a certain amount of time each day. After six months, we measure the size of any tumour (in mm^3) close to the site of the phone antenna (just behind the ear). The six groups experienced 0, 1, 2, 3, 4 or 5 hours per day of phone microwaves for six months. The data are in **Tumour.sav**. (From Field & Hole, 2003, so there is a very detailed answer in there.) ②
- **Task 4**: Using the Glastonbury data from Chapter 7 (**GlastonburyFestival.sav**), carry out a one-way ANOVA on the data to see if the change in hygiene (**change**) is significant across people with different musical tastes (**music**). Do a simple contrast to compare each group against 'No Affiliation'. Compare the results to those described in section 7.11. ②
- **Task 5**: Labcoat Leni's Real Research 15.2 describes an experiment (Çetinkaya & Domjan, 2006) on quails with fetishes for terrycloth objects (really, it does). In this example, you are asked to analyse two of the variables that they measured with a Kruskal–Wallis test. However, there were two other outcome variables (time spent near the terrycloth object and copulatory efficiency). These data can be analysed with one-way ANOVA. Read Labcoat Leni's Real Research 15.2 to get the full story, then carry out two one-way ANOVAs and Bonferroni *post hoc* tests on the aforementioned outcome variables. ②

Answers can be found on the companion website.

Further reading

Howell, D. C. (2006). *Statistical methods for psychology* (6th ed.). Belmont, CA: Duxbury. (Or you might prefer his *Fundamental Statistics for the Behavioral Sciences,* also in its 6th edition, 2007. Both are excellent texts that provide very detailed coverage of the standard variance approach to ANOVA but also the GLM approach that I have discussed.)

Iversen, G. R., & Norpoth, H. (1987). *ANOVA* (2nd ed.). Sage university paper series on quantitative applications in the social sciences, 07-001. Newbury Park, CA: Sage. (Quite high-level, but a good read for those with a mathematical brain.)

Klockars, A. J., & Sax, G. (1986). *Multiple comparisons.* Sage university paper series on quantitative applications in the social sciences, 07-061. Newbury Park, CA: Sage. (High-level but thorough coverage of multiple comparisons – in my view this book is better than Toothaker for planned comparisons.)

Rosenthal, R., Rosnow, R. L., & Rubin, D. B. (2000). *Contrasts and effect sizes in behavioural research: a correlational approach.* Cambridge: Cambridge University Press. (Fantastic book on planned comparisons by three of the great writers on statistics.)

Rosnow, R. L., & Rosenthal, R. (2005). *Beginning behavioural research: a conceptual primer* (5th ed.). Englewood Cliffs, NJ: Pearson/Prentice Hall. (Look, they wrote another great book!)

Toothaker, L. E. (1993). *Multiple comparison procedures.* Sage university paper series on quantitative applications in the social sciences, 07-089. Newbury Park, CA: Sage. (Also high-level, but gives an excellent precis of *post hoc* procedures.)

Wright, D. B., & London, K. (2009). *First steps in statistics* (2nd ed.). London: Sage. (If this chapter is too complex then this book is a very readable basic introduction to ANOVA.)

Online tutorials

The companion website contains the following Flash movie tutorials to accompany this chapter:

- One-Way Independent ANOVA using SPSS
- The *R* plugin

Interesting real research

Gallup, G. G. J., Burch, R. L., Zappieri, M. L., Parvez, R., Stockwell, M., & Davis, J. A. (2003). The human penis as a semen displacement device. *Evolution and Human Behavior, 24,* 277–289.

Analysis of covariance, ANCOVA (GLM 2)

11

FIGURE 11.1 Davey Murray (guitarist from Iron Maiden) and me backstage in London in 1986; my grimace reflects the utter terror I was feeling at meeting my hero

11.1. What will this chapter tell me? ②

My road to rock stardom had taken a bit of a knock with my unexpected entry to an all-boys grammar school (rock bands and grammar schools really didn't go together). I needed to be inspired and I turned to the masters: Iron Maiden. I first heard Iron Maiden at the age of 11 when a friend of mine lent me 'Piece of Mind' and told me to listen to 'The Trooper'. It was, to put it mildly, an epiphany. I became their smallest (I was 11) biggest fan and started to obsess about them in the most unhealthy way possible. I started stalking the man who ran their fan club with letters, and, bless him, he replied. Eventually this stalking paid off and he arranged for me to go backstage when they played Hammersmith Odeon in London (now the Carling Apollo Hammersmith) on 5 November 1986 (*Somewhere on Tour* in case you're interested). Not only was it the first time that I had seen them live, but I got to meet them. It's hard to put into words how bladder-splittingly exciting this was. I was so utterly awe-struck that I managed to say precisely no words to them. As usual, then, a social situation provoked me to make an utter tit of myself.[1] When it was over I was in no doubt that this was the best day of my life. In fact, I thought, I should just kill myself there and then because nothing would ever be as good as that again. This may be true, but I have subsequently had many other very nice experiences, so who is to say that they were not better? I could compare experiences to see which one is the best, but there is an important confound: my age. At the age of 13, meeting Iron Maiden was bowel-weakeningly exciting, but adulthood (sadly) dulls your capacity for this kind of unqualified joy of life. Therefore, to really see which experience was best, I would have to take account of the variance in enjoyment that is attributable to my age at the time. This will give me a purer measure of how much variance in my enjoyment is attributable to the event itself. This chapter describes **analysis of covariance**, which extends the basic idea of ANOVA from the previous chapter to situations when we want to factor in other variables that influence the outcome variable.

11.2. What is ANCOVA? ②

In the previous chapter we saw how one-way ANOVA could be characterized in terms of a multiple regression equation that used dummy variables to code group membership. In addition, in Chapter 7 we saw how multiple regression could incorporate several continuous predictor variables. It should, therefore, be no surprise that the regression equation for ANOVA can be extended to include one or more continuous variables that predict the outcome (or dependent variable). Continuous variables such as these, that are not part of the main experimental manipulation but have an influence on the dependent variable, are known as *covariates* and they can be included in an ANOVA analysis. When we measure covariates and include them in an analysis of variance we call it analysis of covariance (or ANCOVA for short). This chapter focuses on this technique.

In the previous chapter, we used an example about looking at the effects of Viagra on libido. Let's think about things other than Viagra that might influence libido: well, the obvious one is the libido of the participant's sexual partner (after all 'it takes two to tango'!), but there are other things too such as other medication that suppresses libido (such as antidepressants or the contraceptive pill) and fatigue. If these variables (the **covariates**) are measured, then it is possible to control for the influence they have on the dependent variable by including them in the regression model. From what we know of hierarchical regression (see Chapter 7) it should be clear that if we enter the covariate into the regression model first, and then enter the dummy variables representing the experimental manipulation, we can see what effect an independent variable has *after* the effect of the covariate. As such, we **partial out** the effect of the covariate. There are two reasons for including covariates in ANOVA:

- **To reduce within-group error variance**: In the discussion of ANOVA and *t*-tests we got used to the idea that we assess the effect of an experiment by comparing the amount of variability in the data that the experiment can explain against the variability that it cannot explain. If we can explain some of this 'unexplained' variance (SS_R) in terms of other variables (covariates), then we reduce the error variance, allowing us to more accurately assess the effect of the independent variable (SS_M).

[1] In my teens I stalked many bands and Iron Maiden are by far the nicest of the bands I've met.

- **Elimination of confounds**: In any experiment, there may be unmeasured variables that confound the results (i.e. variables other than the experimental manipulation that affect the outcome variable). If any variables are known to influence the dependent variable being measured, then ANCOVA is ideally suited to remove the bias of these variables. Once a possible confounding variable has been identified, it can be measured and entered into the analysis as a covariate.

There are other reasons for including covariates in ANOVA but because I do not intend to describe the computation of ANCOVA in any detail I recommend that the interested reader consult my favourite sources on the topic (Stevens, 2002; Wildt & Ahtola, 1978).

Imagine that the researcher who conducted the Viagra study in the previous chapter suddenly realized that the libido of the participants' sexual partners would affect the participants' own libido (especially because the measure of libido was behavioural). Therefore, they repeated the study on a different set of participants, but this time took a measure of the partner's libido. The partner's libido was measured in terms of how often they tried to initiate sexual contact. In the previous chapter, we saw that this experimental scenario could be characterized in terms of equation (10.2). Think back to what we know about multiple regression (Chapter 7) and you can hopefully see that this equation can be extended to include this covariate as follows:

$$
\begin{aligned}
\text{Libido}_i &= b_0 + b_3\text{Covariate}_i + b_2\text{High}_i + b_1\text{Low}_i + \varepsilon_i \\
\text{Libido}_i &= b_0 + b_3\text{Partner's Libido}_i + b_2\text{High}_i + b_1\text{Low}_i + \varepsilon_i
\end{aligned}
\tag{11.1}
$$

11.3. Assumptions and issues in ANCOVA ③

ANCOVA has the same assumptions as ANOVA except that there are two important additional considerations: (1) independence of the covariate and treatment effect, and (2) homogeneity of regression slopes.

11.3.1. Independence of the covariate and treatment effect ③

I said in the previous section that one use of ANCOVA is to reduce within-group error variance by allowing the covariate to explain some of this error variance. However, for this to be true the covariate must be independent from the experimental effect.

Figure 11.2 shows three different scenarios. Part A shows a basic ANOVA and is similar to Figure 10.3; it shows that the experimental effect (in our example libido) can be partitioned into two parts that represent the experimental or treatment effect (in this case the administration of Viagra) and the error or unexplained variance (i.e. factors that affect libido that we haven't measured). Part B shows the ideal scenario for ANCOVA in which the covariate shares its variance only with the bit of libido that is currently unexplained. In other words, it is completely independent from the treatment effect (it does not overlap with the effect of Viagra at all). This scenario is the only one in which ANCOVA is appropriate. Part C shows a situation in which people often use ANCOVA when they should not. In this situation the effect of the covariate overlaps with the experimental effect. In other words, the experimental effect is confounded with the effect of the covariate. In this situation, the covariate will reduce (statistically speaking) the experimental effect because it explains some of the variance that would otherwise be attributable to the experiment. When the covariate and the experimental effect (independent variable) are not independent, the

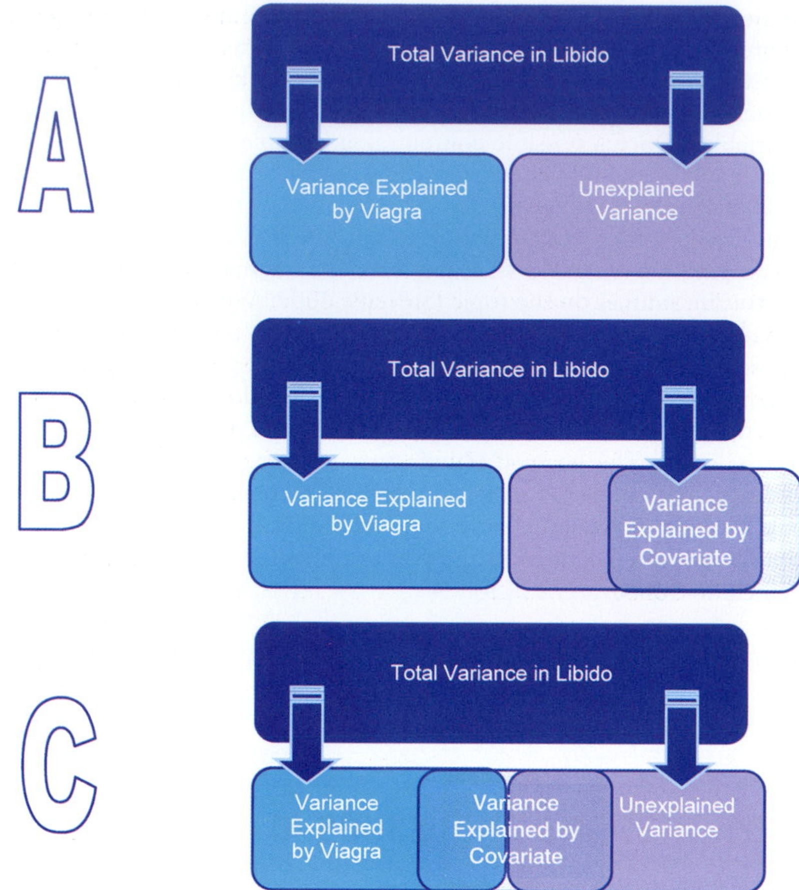

FIGURE 11.2 The role of the covariate in ANCOVA (see text for details)

treatment effect is obscured, spurious treatment effects can arise and at the very least the interpretation of the ANCOVA is seriously compromised (Wildt & Ahtola, 1978).

The problem of the covariate and treatment sharing variance is common and is ignored or misunderstood by many people (Miller & Chapman, 2001). Miller and Chapman, in a very readable review, cite many situations in which people misapply ANCOVA and I recommend reading this paper. To summarize the main issue, when treatment groups differ on the covariate then putting the covariate into the analysis will not 'control for' or 'balance out' those differences (Lord, 1967, 1969). This situation arises mostly when participants are not randomly assigned to experimental treatment conditions. For example, anxiety and depression are closely correlated (anxious people tend to be depressed) so if you wanted to compare an anxious group of people against a non-anxious group on some task, the chances are that the anxious group would also be more depressed than the non-anxious group. You might think that by adding depression as a covariate into the analysis you can look at the 'pure' effect of anxiety but you can't. This would be the situation in part C of Figure 11.2; the effect of the covariate (depression) would contain some of the variance from the effect of anxiety. Statistically speaking all that we know is that anxiety and depression share variance; we cannot separate this shared variance into 'anxiety variance' and 'depression variance', it will always just be 'shared'. Another common example is if you happen to find that your experimental groups differ in their ages. Placing age into the analysis as a covariate will not solve this problem – it is still confounded with the experimental manipulation. ANCOVA is not a magic solution to this problem.

This problem can be avoided by randomizing participants to experimental groups, or by matching experimental groups on the covariate (in our anxiety example, you could try to find participants for the low anxious group who score high on depression). We can check whether this problem is likely to be an issue by checking whether experimental groups differ on the covariate before we run the ANCOVA. To use our anxiety example again, we could test whether our high and low anxious groups differ on levels of depression (with a *t*-test or ANOVA). If the groups do not significantly differ then we can use depression as a covariate.

11.3.2. Homogeneity of regression slopes ③

When an ANCOVA is conducted we look at the overall relationship between the outcome (dependent variable) and the covariate: we fit a regression line to the entire data set, ignoring to which group a person belongs. In fitting this overall model we, therefore, assume that this overall relationship is true for all groups of participants. For example, if there's a positive relationship between the covariate and the outcome in one group, we assume that there is a positive relationship in all of the other groups too. If, however, the relationship between the outcome (dependent variable) and covariate differs across the groups then the overall regression model is inaccurate (it does not represent all of the groups). This assumption is very important and is called the assumption of **homogeneity of regression slopes.** The best way to think of this assumption is to imagine plotting a scatterplot for each experimental condition with the covariate on one axis and the outcome on the other. If you then calculated, and drew, the regression line for each of these scatterplots you should find that the regression lines look more or less the same (i.e. the values of *b* in each group should be equal). We will have a look at an example of this assumption and how to test it in section 11.7.

11.4. Conducting ANCOVA on SPSS ②

11.4.1. Inputting data ①

The data for this example are in Table 11.1 and can be found in the file **ViagraCovariate.sav**. Table 11.1 shows the participant's libido and their partner's libido and Table 11.2 shows the means and standard deviations of these data. For practice, let's enter these data into the data editor by hand. This can be done in much the same way as the Viagra data from the previous chapter except that an extra variable must be created in which to place the values of the covariate.

In essence, the data should be laid out in the data editor as they are in Table 11.1. So, create a coding variable called **Dose** and use the *Labels* option to define value labels (as in Chapter 10 let's use 1 = placebo, 2 = low dose, 3 = high dose). There were different numbers of participants in each condition, so you need to enter 9 values of 1 into this column (so that the first 9 rows contain the value 1), followed by 8 rows containing the value 2, followed by 13 rows containing the value of 3. At this point, you should have one column with 30 rows of data entered. Next, create a second variable called **Libido** and enter the 30 scores that correspond to the person's libido. Finally, create a third variable called **Partner_Libido** and use the *Labels* option to give this variable a title of 'Partner's libido'. Then, enter the 30 scores that correspond to the partner's libido.

SELF-TEST Use SPSS to find out the means and standard deviations of both the participant's libido and that of their partner in the three groups. (Answers are in Table 11.2.)

TABLE 11.1 Data from **ViagraCovariate.sav**

Dose	Participant's Libido	Partner's Libido
Placebo	3	4
	2	1
	5	5
	2	1
	2	2
	2	2
	7	7
	2	4
	4	5
Low Dose	7	5
	5	3
	3	1
	4	2
	4	2
	7	6
	5	4
	4	2
High Dose	9	1
	2	3
	6	5
	3	4
	4	3
	4	3
	4	2
	6	0
	4	1
	6	3
	2	0
	8	1
	5	0

TABLE 11.2 Means (and standard deviations) from **ViagraCovariate.sav**

Dose	Participant's Libido	Partner's Libido
Placebo	3.22 (1.79)	3.44 (2.07)
Low Dose	4.88 (1.46)	3.12 (1.73)
High Dose	4.85 (2.12)	2.00 (1.63)

11.4.2. Initial considerations: testing the independence of the independent variable and covariate ②

In section 11.3.1, I mentioned that before including a covariate in an analysis we should check that it is independent from the experimental manipulation. In this case, the proposed covariate is partner's libido, and we need to check that this variable was roughly equal across levels of our independent variable. In other words, is the mean level of partner's libido roughly equal across our three Viagra groups? We can test this by running an ANOVA with **Partner_Libido** as the outcome and **Dose** as the predictor.

SELF-TEST Conduct an ANOVA to test whether partner's libido (our covariate) is independent of the dose of Viagra (our independent variable).

SPSS Output 11.1 shows the results of such an ANOVA. The main effect of dose is not significant, $F(2, 27) = 1.98$, $p = .16$, which shows that the average level of partner's libido was roughly the same in the three Viagra groups. In other words, the means for partner's libido in Table 11.2 are not significantly different in the placebo, low- and high-dose groups. This result means that it is appropriate to use partner's libido as a covariate in the analysis.

SPSS OUTPUT 11.1

Tests of Between-Subjects Effects

Dependent Variable:Partner's Libido

Source	Type III Sum of Squares	df	Mean Square	F	Sig.
Corrected Model	12.769[a]	2	6.385	1.979	.158
Intercept	234.592	1	234.592	72.723	.000
Dose	12.769	2	6.385	1.979	.158
Error	87.097	27	3.226		
Total	324.000	30			
Corrected Total	99.867	29			

a. R Squared = .128 (Adjusted R Squared = .063)

11.4.3. The main analysis ②

Most of the *General Linear Model* (GLM) procedures in SPSS contain the facility to contain one or more covariates. For designs that don't involve repeated measures it is easiest to conduct ANCOVA via the GLM *Univariate* procedure. To access the main dialog box select Analyze General Linear Model ▸ Univariate... (see Figure 11.3). The main dialog box is similar to that for one-way ANOVA, except that there is a space to specify covariates. Select **Libido** and drag this variable to the box labelled *Dependent Variable* or click on ➔. Select **Dose** and drag it to the box labelled *Fixed Factor(s)* and then select **Partner_Libido** and drag it to the box labelled *Covariate(s)*.

11.4.4. Contrasts and other options ②

There are various dialog boxes that can be accessed from the main dialog box. The first thing to notice is that if a covariate is selected, the *post hoc* tests are disabled (you cannot access this dialog box). *Post hoc* tests are not designed for situations in which a covariate is specified; however, some comparisons can still be done using contrasts.

Click on Contrasts... to access the contrasts dialog box. This dialog box is different to the one we met in Chapter 10 in that you cannot enter codes to specify particular contrasts. Instead,

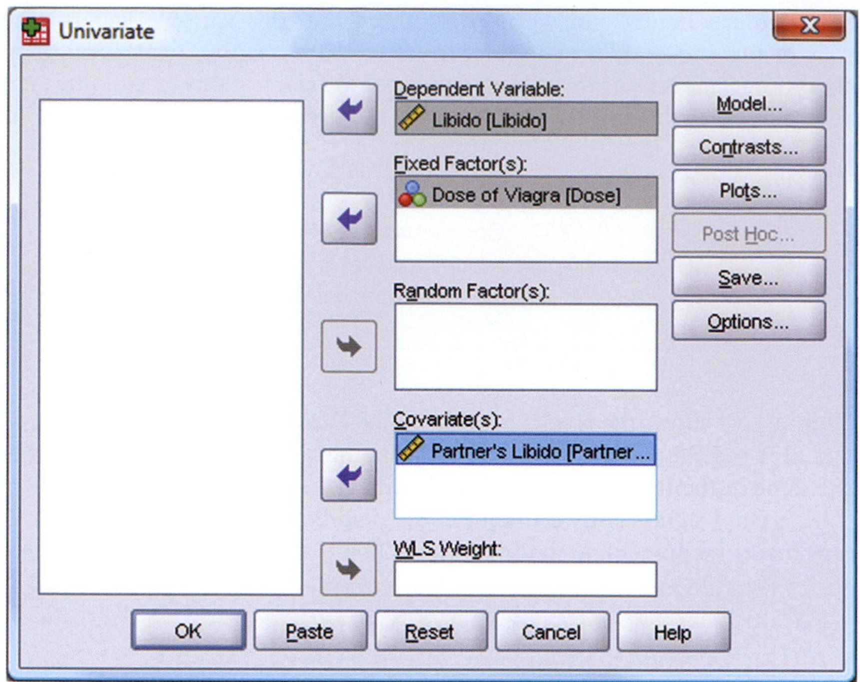

FIGURE 11.3 Main dialog box for GLM univariate

you can specify one of several standard contrasts. These standard contrasts were listed in Table 10.6. In this example, there was a placebo control condition (coded as the first group), so a sensible set of contrasts would be simple contrasts comparing each experimental group with the control. To select a type of contrast click on None to access a drop-down list of possible contrasts. Select a type of contrast (in this case *Simple*) from this list and the list will automatically disappear. For simple contrasts you have the option of specifying a reference category (which is the category against which all other groups are compared). By default the reference category is the last category: because in this case the control group was the first category (assuming that you coded placebo as 1) we need to change this option by selecting First. When you have selected a new contrast option, you must click on Change to register this change. The final dialog box should look like Figure 11.4. Click on Continue to return to the main dialog box.

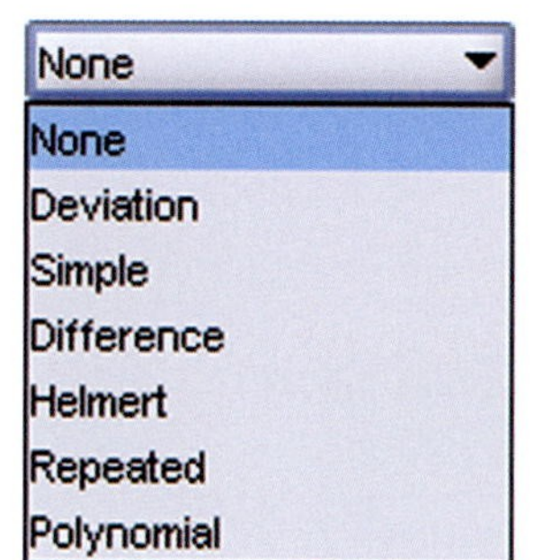

Another way to get *post hoc* tests is by clicking on Options... to access the options dialog box (see Figure 11.5). To specify *post hoc* tests, select the independent variable (in this case **Dose**) from the box labelled *Estimated Marginal Means: Factor(s) and Factor Interactions* and drag it to the box labelled *Display Means for* or click on ➔. Once a variable has been transferred, the box labelled *Compare main effects* becomes active and you should select this option (Compare main effects). If this option is selected, the box labelled *Confidence interval adjustment* becomes active and you can click on LSD (none) to see a choice of three adjustment levels. The default is to have no adjustment and simply perform a Tukey LSD *post hoc* test (this option is not recommended); the second is to ask for a Bonferroni correction (recommended); the final option is to have a **Sidak correction**. The Sidak correction is similar to the Bonferroni correction but is less conservative and so should be selected if you are concerned about the loss of power associated with Bonferroni corrected values. For this example use the Sidak correction (we will use Bonferroni later in the book). As well as producing *post hoc* tests for the **Dose** variable, placing **Dose** in the *Display Means for* box will create a table of estimated marginal means for this variable. These means provide an estimate of the *adjusted* group means (i.e. the means adjusted for the effect of the covariate). When you have selected the options required (see Jane Superbrain Box 11.1),

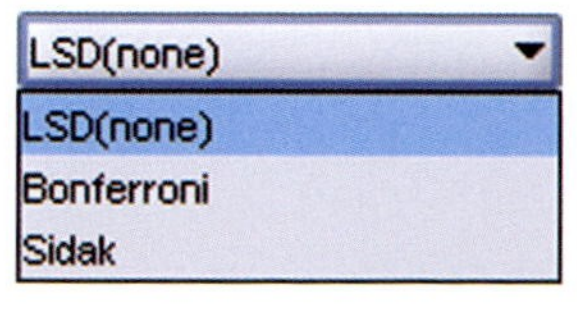

FIGURE 11.4 Options for standard contrasts in GLM univariate

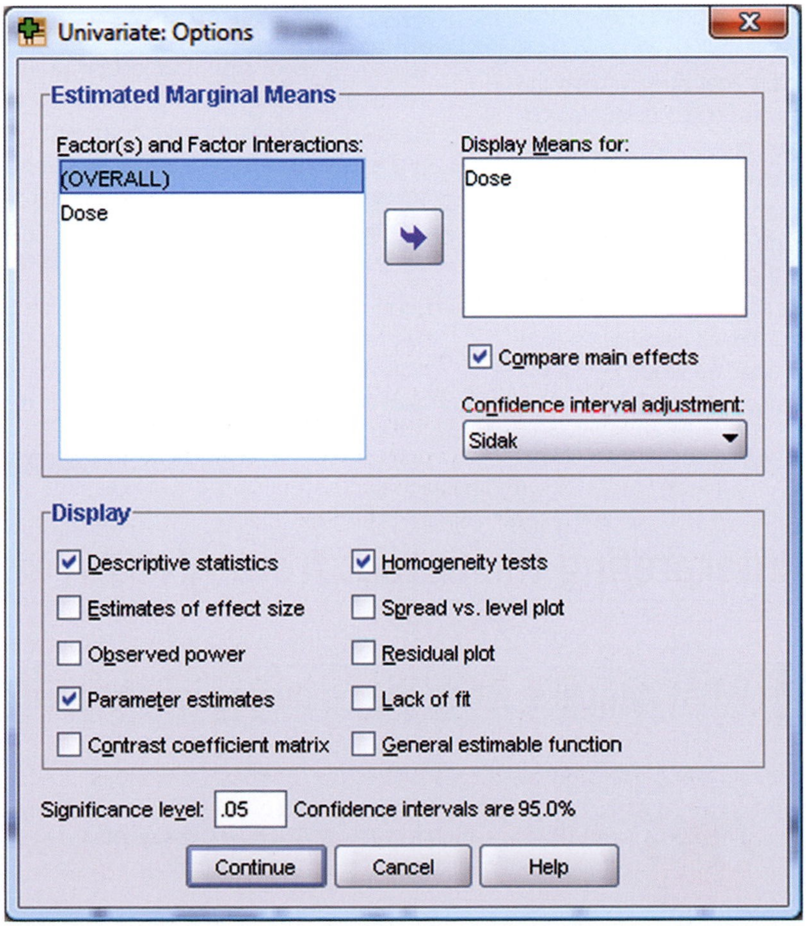

FIGURE 11.5 Options dialog box for GLM univariate

click on Continue to return to the main dialog box. There are other options available from the main dialog box. For example, if you have several independent variables you can plot them against each other (which is useful for interpreting interaction effects – see section 12.3.2). For this analysis, there is only one independent variable and so we can click on OK to run the analysis.

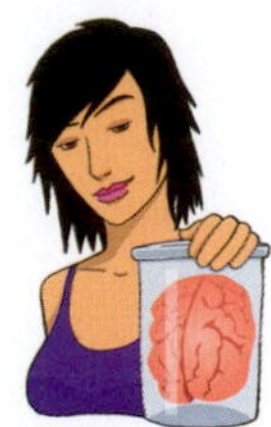

JANE SUPERBRAIN 11.1

Options for ANCOVA ②

The remaining options in this dialog box are as follows:

- ✓ *Descriptive statistics*: This option produces a table of means and standard deviations for each group.
- ✓ *Estimates of effect size*: This option produces the value of **partial eta squared (partial η^2)** – see section 11.8 for a discussion.
- ✓ *Observed power*: This option provides an estimate of the probability that the statistical test could detect the difference between the observed group means (see section 2.6.5). This measure is of little use because if the *F*-test is significant then the probability that the effect was detected will, of course, be high. Likewise, if group differences were small, the observed power would be low. Observed power is of little use and I would advise that power calculations (with regard to sample size) are made before the experiment is conducted (see Cohen, 1988, 1992, and Howell, 2006, for ideas on how to do this by hand, or use the free software G*Power linked from the companion website).

- ✓ *Parameter estimates*: This option produces a table of regression coefficients and their tests of significance for the variables in the regression model (see section 11.5.2).
- ✓ *Contrast coefficient matrix*: This option produces matrices of the coding values used for any contrasts in the analysis. This option is useful only for checking which groups are being compared in which contrast.
- ✓ *Homogeneity tests*: This option produces Levene's test of the homogeneity of variance assumption (see sections 5.6.1 and 10.4.1). In ANCOVA the assumption relates (as in regression) to the homogeneity of *residuals* (see section 7.6).
- ✓ *Spread vs. level plot*: This option produces a chart that plots the mean of each group of a factor (*X*-axis) against the standard deviation of that group (*Y*-axis). This is a useful plot to check that there is no relationship between the mean and standard deviation. If a relationship exists then the data may need to be stabilized using a logarithmic transformation (see Chapter 5).
- ✓ *Residual plot:* This option produces plots of observed-by-predicted-by-standardized residual values. These plots can be used to assess the assumption of equality of variance.

11.5. Interpreting the output from ANCOVA ②

11.5.1. What happens when the covariate is excluded? ②

SELF-TEST Run a one-way ANOVA to see whether the three groups differ in their levels of libido.

SPSS Output 11.2 shows (for illustrative purposes) the ANOVA table for these data when the covariate is not included. It is clear from the significance value, which is greater than .05, that Viagra seems to have no significant effect on libido. It should also be noted that the total amount of variation to be explained (SS_T) was 110.97 (Corrected Total), of which the experimental manipulation accounted for 16.84 units (SS_M), while 94.12 were unexplained (SS_R).

SPSS OUTPUT 11.2

Tests of Between-Subjects Effects

Dependent Variable:Libido

Source	Type III Sum of Squares	df	Mean Square	F	Sig.
Corrected Model	16.844[a]	2	8.422	2.416	.108
Intercept	535.184	1	535.184	153.522	.000
Dose	16.844	2	8.422	2.416	.108
Error	94.123	27	3.486		
Total	683.000	30			
Corrected Total	110.967	29			

a. R Squared = .152 (Adjusted R Squared = .089)

11.5.2. The main analysis ②

SPSS Output 11.3 shows the results of Levene's test (section 5.6.1) and the ANOVA table when partner's libido is included in the model as a covariate. Levene's test is significant, indicating that the group variances are not equal (hence the assumption of homogeneity of variance has been violated). However, as I've mentioned in section 5.6, Levene's test is not necessarily the best way to judge whether variances are unequal enough to cause problems. A good double-check is to look at the highest and lowest variances. For our three groups we have standard deviations of 1.79 (placebo), 1.46 (low dose) and 2.12 (high dose) – see Table 11.1. If we square these values we get variances of 3.20 (placebo), 2.13 (low dose) and 4.49 (high dose). We then take the largest variance and divide it by the smallest: in this case 4.49/2.13 = 2.11. If we look at Figure 5.12 we can get the approximate critical value when comparing three variances and with 10 people per group (we have unequal groups, but this will do as an approximation). The critical value in this situation is approximately 5. Our observed value of 2.11 is less than this critical value of 5 so we probably don't need to worry too much about the differences in variances.

SPSS OUTPUT 11.3

Levene's Test of Equality of Error Variances[a]

Dependent Variable:Libido

F	df1	df2	Sig.
4.618	2	27	.019

Tests the null hypothesis that the error variance of the dependent variable is equal across groups.

a. Design: Intercept + Partner_Libido + Dose

Tests of Between-Subjects Effects

Dependent Variable:Libido

Source	Type III Sum of Squares	df	Mean Square	F	Sig.
Corrected Model	31.920[a]	3	10.640	3.500	.030
Intercept	76.069	1	76.069	25.020	.000
Partner_Libido	15.076	1	15.076	4.959	.035
Dose	25.185	2	12.593	4.142	.027
Error	79.047	26	3.040		
Total	683.000	30			
Corrected Total	110.967	29			

a. R Squared = .288 (Adjusted R Squared = .205)

The format of the ANOVA table is largely the same as without the covariate, except that there is an additional row of information about the covariate (**Partner_Libido**). Looking first at the significance values, it is clear that the covariate significantly predicts the dependent variable, because the significance value is less than .05. Therefore, the person's libido is influenced by their partner's libido. What's more interesting is that when the effect of partner's libido is removed, the effect of Viagra becomes significant (p is .027 which is less than .05). The amount of variation accounted for by the model (SS_M) has increased to 31.92 units (corrected model) of which Viagra accounts for 25.19 units. Most important, the large amount of variation in libido that is accounted for by the covariate has meant that the unexplained variance (SS_R) has been reduced to 79.05 units. Notice that SS_T has not changed; all that has changed is how that total variation is explained.

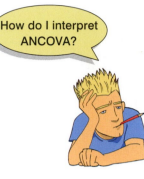

How do I interpret ANCOVA?

This example illustrates how ANCOVA can help us to exert stricter experimental control by taking account of confounding variables to give us a 'purer' measure of effect of the experimental manipulation. Without taking account of the libido of the participants' partners we would have concluded that Viagra had no effect on libido, yet it does. Looking back at the group means from Table 11.1 for the libido data, it seems pretty clear that the significant ANOVA reflects a difference between the placebo group and the two experimental groups (because the low- and high-dose groups have very similar means – 4.88 and 4.85 – whereas the placebo group mean is much lower at 3.22). However, we'll need to check some contrasts to verify this.

SPSS Output 11.4 shows the parameter estimates selected in the options dialog box. These estimates are calculated using a regression analysis with **Dose** split into two dummy coding variables (see section 10.2.3 and section 11.6). SPSS codes the two dummy variables such that the last category (the category coded with the highest value in the data editor – in this case the high-dose group) is the reference category. This reference category (labelled Dose=3 in the output) is coded with 0 for both dummy variables (see section 10.2.3 for a reminder of how dummy coding works). Dose=2, therefore, represents the difference between the group coded as 2 (low dose) and the reference category (high dose), and dose=1 represents the difference between the group coded as 1 (placebo) and the reference category (high dose). The b-values represent the differences between the means of these groups and so the significances of the t-tests tell us whether the group means differ significantly. The degrees of freedom for these t-tests can be calculated as in normal regression (see section 7.2.4) as $N - p - 1$ in which N is the total sample size (in this case 30) and p is the number of predictors (in this case 3, the two dummy variables and the covariate). For these data, $df = 30 - 3 - 1 = 26$. From these estimates we could conclude that the high-dose differs significantly from the placebo group (Dose=1 in the table) but not from the low-dose group (Dose=2 in the table).

The final thing to notice is the value of b for the covariate (0.416). This value tells us that, other things being equal, if a partner's libido increases by one unit, then the person's libido should increase by just under half a unit (although there is nothing to suggest a causal link between the two). The sign of this coefficient tells us the direction of the relationship between the covariate and the outcome. So, in this example, because the coefficient is positive it means that partner's libido has a positive relationship with the participant's libido: as one increases so does the other. A negative coefficient would mean the opposite: as one increases, the other decreases.

SPSS OUTPUT 11.4

Parameter Estimates

Dependent Variable:Libido

					95% Confidence Interval	
Parameter	B	Std. Error	t	Sig.	Lower Bound	Upper Bound
Intercept	4.014	.611	6.568	.000	2.758	5.270
Partner_Libido	.416	.187	2.227	.035	.032	.800
[Dose=1]	-2.225	.803	-2.771	.010	-3.875	-.575
[Dose=2]	-.439	.811	-.541	.593	-2.107	1.228
[Dose=3]	0[a]	.	.	.	.	.

a. This parameter is set to zero because it is redundant.

11.5.3. Contrasts ②

SPSS Output 11.5 shows the result of the contrast analysis specified in Figure 11.4 and compares level 2 (low dose) against level 1 (placebo) as a first comparison, and level 3 (high dose) against level 1 (placebo) as a second comparison. These contrasts are consistent with what was specified: all groups are compared to the first group. The group differences are displayed: a difference value, standard error, significance value and 95% confidence interval. These results show that both the low-dose group (contrast 1, $p = .045$) and high-dose group (contrast 2, $p = .010$) had significantly different libidos than the placebo group. These results are consistent with the regression parameter estimates (in fact, note that contrast 2 is identical to the regression parameters for Dose=1 in the previous section).

SPSS OUTPUT 11.5

Contrast Results (K Matrix)

Dose of Viagra Simple Contrast[a]			Depende... Libido
Level 2 vs. Level 1	Contrast Estimate		1.786
	Hypothesized Value		0
	Difference (Estimate - Hypothesized)		1.786
	Std. Error		.849
	Sig.		.045
	95% Confidence Interval for Difference	Lower Bound	.040
		Upper Bound	3.532
Level 3 vs. Level 1	Contrast Estimate		2.225
	Hypothesized Value		0
	Difference (Estimate - Hypothesized)		2.225
	Std. Error		.803
	Sig.		.010
	95% Confidence Interval for Difference	Lower Bound	.575
		Upper Bound	3.875

a. Reference category = 1

These contrasts and parameter estimates tell us that there were group differences, but to interpret them we need to know the means. We produced the means in Table 11.2 so surely we can just look at these values? Actually we can't because these group means have not been adjusted for the effect of the covariate. These original means tell us nothing about the group differences reflected by the significant ANCOVA. SPSS Output 11.6 gives the adjusted values of the group means and it is these values that should be used for interpretation (this is the main reason for selecting the *Display Means for option*). The **adjusted means** (and our contrasts) show that levels of libido were significantly higher in the low- and high-dose groups compared to the placebo group. The regression parameters also told us that the high- and low-dose groups did not significantly differ ($p = .593$). These conclusions can be verified with the *post hoc* tests specified in the options menu but normally you would do only contrasts or *post hoc* tests, not both.

SPSS OUTPUT 11.6

Estimates

Dependent Variable:Libido

Dose of Viagra	Mean	Std. Error	95% Confidence Interval Lower Bound	Upper Bound
Placebo	2.926[a]	.596	1.701	4.152
Low Dose	4.712[a]	.621	3.436	5.988
High Dose	5.151[a]	.503	4.118	6.184

a. Covariates appearing in the model are evaluated at the following values: Partner's Libido = 2.73.

SPSS OUTPUT 11.7

Pairwise Comparisons

Dependent Variable:Libido

(I) Dose of Viagra	(J) Dose of Viagra	Mean Difference (I-J)	Std. Error	Sig.[a]	95% Confidence Interval for Difference[a]	
					Lower Bound	Upper Bound
Placebo	Low Dose	-1.786	.849	.130	-3.953	.381
	High Dose	-2.225*	.803	.030	-4.273	-.177
Low Dose	Placebo	1.786	.849	.130	-.381	3.953
	High Dose	-.439	.811	.932	-2.509	1.631
High Dose	Placebo	2.225*	.803	.030	.177	4.273
	Low Dose	.439	.811	.932	-1.631	2.509

Based on estimated marginal means

a. Adjustment for multiple comparisons: Sidak.

*. The mean difference is significant at the .05 level.

SPSS Output 11.7 shows the results of the Sidak-corrected *post hoc* comparisons that were requested as part of the options dialog box. The significant difference between the high-dose and placebo groups remains (p = .030), and the high-dose and low-dose groups do not significantly differ (p = .93). However, it is interesting that the significant difference between the low-dose and placebo groups shown by the regression parameters and contrasts (SPSS Output 11.4) is gone (p is only .13).

SELF-TEST Why do you think that the results of the *post hoc* test differ to the contrasts for the comparison of the low-dose and placebo groups?

11.5.4. Interpreting the covariate ②

I've already mentioned that the parameter estimates (SPSS Output 11.4) tell us how to interpret the covariate. If the b-value for the covariate is positive then it means that the covariate and the outcome variable have a positive relationship (as the covariate increases, so does the outcome). If the b-value is negative it means the opposite: that the covariate and the outcome variable have a negative relationship (as the covariate increases, the outcome decreases). For these data the b-value was positive, indicating that as the partner's libido increases, so does the participant's libido. Another way to discover the same thing is simply to draw a scatterplot of the covariate against the outcome. We came across scatterplots in section 4.8 so have a look back there to find out how to produce one. Figure 11.6 shows the resulting scatterplot for these data and confirms what we already know: the effect of the covariate is that as partner's libido increases, so does the participant's libido (as shown by the slope of the regression line).

11.6. ANCOVA run as a multiple regression ②

Although the ANCOVA is essentially done, it is a useful educational exercise to rerun the analysis as a hierarchical multiple regression (it will, I hope, help you to understand what's happening when we do an ANCOVA).

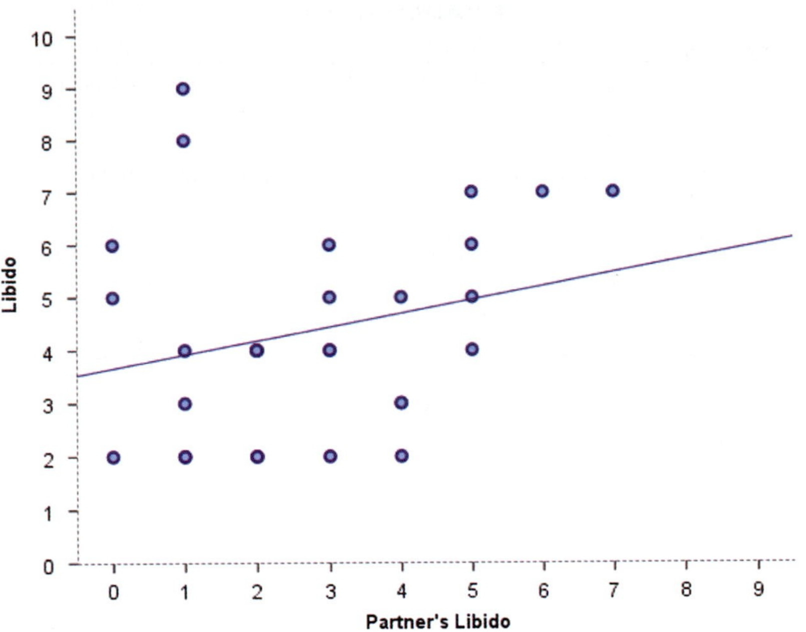

FIGURE 11.6 Scatterplot of libido against partner's libido

LABCOAT LENI'S REAL RESEARCH 11.1

Space invaders ②

Anxious people tend to interpret ambiguous information in a negative way. For example, being highly anxious myself, if I overheard a student saying 'Andy Field's lectures are really *different*' I would assume that 'different' meant rubbish, but it could also mean 'refreshing' or 'innovative'. One current mystery is how these interpretational biases develop in children. Peter Muris and his colleagues addressed this issue in an ingenious study. Children did a task in which they imagined that they were astronauts who had discovered a new planet. Although the planet was similar to Earth, some things were different. They were given some scenarios about their time on the planet (e.g. 'On the street, you encounter a spaceman. He has a sort of toy handgun and he fires at you ...') and the child had to decide which of two outcomes occurred. One outcome was positive ('You laugh: it is a water pistol and the weather is fine anyway') and the other negative ('Oops, this hurts! The pistol produces a red beam which burns your skin!'). After each response the child was told whether their choice was correct. Half of the children were *always* told that the negative interpretation was correct, and the remainder were told that the positive interpretation was correct.

Over 30 scenarios children were trained to interpret their experiences on the planet as negative or positive. Muris et al. then gave children a standard measure of interpretational biases in everyday life to see whether the training had created a bias to interpret things negatively. In doing so, they could ascertain whether children learn interpretational biases through feedback (e.g. from parents) about how to disambiguate ambiguous situations.

The data from this study are in the file **Muris et al (2008).sav**. The independent variable is **Training** (positive or negative) and the outcome was the child's interpretational bias score (**Interpretational_Bias**) – a high score reflects a tendency to interpret situations negatively. It is important to factor in the **Age** and **Gender** of the child and also their natural anxiety level (which they measured with a standard questionnaire of child anxiety called the **SCARED**) because these things affect interpretational biases also. Labcoat Leni wants you to carry out a one-way ANCOVA on these data to see whether **Training** significantly affected children's **Interpretational_Bias** using **Age**, **Gender** and **SCARED** as covariates. What can you conclude?

Answers are in the additional material on the companion website (or look at pages 475–476 in the original article).

MURIS P. ET AL. (2008). *CHILD PSYCHIATRY AND HUMAN DEVELOPMENT*, 39, 469–480.

SELF-TEST Add two dummy variables to the file **ViagraCovariate.sav** that compare the low dose to the placebo (**Low_Placebo**) and the high dose to the placebo (**High_Placebo**) – see section 10.2.3 for help. If you get stuck then download **ViagraCovariateDummy.sav**.

SELF-TEST Run a hierarchical regression analysis with **Libido** as the outcome. In the first block enter partner's libido (**Partner_Libido**) as a predictor, and then in a second block enter both dummy variables (Forced entry) – see section 7.7 for help.

The summary of the regression model resulting from the self-test (SPSS Output 11.8) shows us the goodness of fit of the model first when only the covariate is used in the model, and second when both the covariate and the dummy variables are used.

Can I run ANCOVA using the regression procedure?

Therefore, the difference between the values of R^2 (.288 – .061 = .227) represents the individual contribution of the dose of Viagra. We can say that the dose of Viagra accounted for 22.7% of the variation in libido, whereas partner's libido accounted for only 6.1%. This additional information provides some insight into the substantive importance of Viagra. The next table is the ANOVA table, which is again divided into two sections. The top half represents the effect of the covariate alone, whereas the bottom half represents the whole model (i.e. covariate and dose of Viagra included). Notice at the bottom of the ANOVA table (the bit for Model 2) that the entire model (partner's libido and the dummy variables) accounts for 31.92 units of variance (SS_M), there are 110.97 units in total (SS_T) and the unexplained variance (SS_R) is 79.05. These values and those of F and the significance are the same as in the ANCOVA summary table in SPSS Output 11.3 – see the row labelled 'Corrected Model'.

SPSS OUTPUT 11.8

Model Summary

Model	R	R Square	Adjusted R Square	Std. Error of the Estimate
1	.246[a]	.061	.027	1.929
2	.536[b]	.288	.205	1.744

a. Predictors: (Constant), Partner's Libido

b. Predictors: (Constant), Partner's Libido, Dummy Variable 1 (Placebo vs. Low), Dummy Variable 2 (Placebo vs. High)

ANOVA[c]

Model		Sum of Squares	df	Mean Square	F	Sig.
1	Regression	6.734	1	6.734	1.809	.189[a]
	Residual	104.232	28	3.723		
	Total	110.967	29			
2	Regression	31.920	3	10.640	3.500	.030[b]
	Residual	79.047	26	3.040		
	Total	110.967	29			

a. Predictors: (Constant), Partner's Libido

b. Predictors: (Constant), Partner's Libido, Dummy Variable 1 (Placebo vs. Low), Dummy Variable 2 (Placebo vs. High)

c. Dependent Variable: Libido

SPSS Output 11.9 shows the remainder of the regression analysis. This table of regression coefficients is more interesting. Again, this table is split into two: the top half shows the effect when only the covariate is in the model and the bottom half contains the whole model. When the dose of Viagra is considered with the covariate, the value of b for the covariate is .416, which corresponds to the value in the ANCOVA parameter estimates (SPSS Output 11.4). The *b*-values for the dummy variables represent the difference between the means of the low-dose group and the placebo group (**Low_Placebo**) and the high-dose group and the placebo group (**High_Placebo**) – see section 10.2.3 for an explanation of why. The means of the low- and high-dose groups were 4.88 and 4.85 respectively, and the mean of the placebo group was 3.22. Therefore, the *b*-values for the two dummy variables should be roughly the same (4.88 – 3.22 = 1.66 for **Low_Placebo** and 4.85 – 3.22 = 1.63 for **High_Placebo**). The astute among you might notice from the SPSS output that, in fact, the *b*-values are not only very different from each other (which shouldn't be the case because the high- and low-dose group means are virtually the same), but also different from the values I've just calculated. Does this mean I've been lying to you for the past 50 pages about what the beta values represent? Well, even I'm not that horrible; the reason for this apparent anomaly is because the *b*-values in this regression represent the differences between the *adjusted* means, not the original means; that is, the difference between the means of each group and the placebo when these means have been adjusted for the partner's libido. The adjusted values were given in SPSS Output 11.6 and from this table we can see that:

$$
\begin{aligned}
b_{\text{Dummy1}} &= \overline{X}_{\text{Low(adjusted)}} - \overline{X}_{\text{Placebo(adjusted)}} = 4.71 - 2.93 = 1.78 \\
b_{\text{Dummy2}} &= \overline{X}_{\text{High(adjusted)}} - \overline{X}_{\text{Placebo(adjusted)}} = 5.15 - 2.93 = 2.22
\end{aligned}
\qquad (11.2)
$$

These are the values that you can see in the SPSS table. The *t*-tests conducted on these values show that the significant ANCOVA reflected a significant difference between the high-dose and placebo groups and also between the low-dose and placebo groups.[2] You should also notice that the significances of the *t*-values are the same as we saw in the contrasts table of the original ANCOVA (see SPSS Output 11.5). As a final point, we obviously don't know whether there was a difference between the low-dose and high-dose groups: to find this out we would need to use different dummy coding (perhaps comparing the high and low to the placebo and then comparing high to low like we did for the planned comparisons in Chapter 10 – see Jane Superbrain Box 11.2).

SPSS OUTPUT 11.9

Coefficients[a]

Model		Unstandardized Coefficients		Standardized Coefficients	t	Sig.
		B	Std. Error	Beta		
1	(Constant)	3.657	.634		5.764	.000
	Partner's Libido	.260	.193	.246	1.345	.189
2	(Constant)	1.789	.867		2.063	.049
	Partner's Libido	.416	.187	.395	2.227	.035
	Dummy Variable 1 (Placebo vs. Low)	1.786	.849	.411	2.102	.045
	Dummy Variable 2 (Placebo vs. High)	2.225	.803	.573	2.771	.010

a. Dependent Variable: Libido

To summarize, you don't have to run ANCOVA through the regression menus of SPSS; I have done this merely to illustrate that when we do ANCOVA we are using a regression model. In other words, we could just ignore the ANCOVA menu and run the analysis as regression. However, you wouldn't do both. One instance in which running ANCOVA through the regression menu is helpful is when you want to do contrasts (see Jane Superbrain Box 11.2).

[2] As I mentioned earlier in this chapter, the degrees of freedom for these *t*-tests are $N-p-1$, as in any regression analysis. *N* is the total sample size (in this case 30) and *p* is the number of predictors (in this case 3, the two dummy variables and the covariate). For these data we get $df = 30-3-1 = 26$.

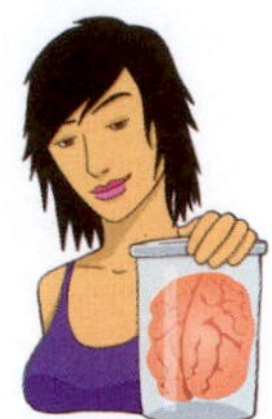

JANE SUPERBRAIN 11.2

Planned contrasts for ANCOVA ③

You may have noticed that although we can ask SPSS to do certain standard contrasts, there is no option for specifying planned contrasts like there was with one-way independent ANOVA (see section 10.3.1). However, these contrasts can be done if we run the ANCOVA through the regression menu. Imagine you chose some planned contrasts as in Chapter 10, in which the first contrast compared the placebo group to all doses of Viagra, and the second contrast then compared the high and low doses (see section 10.2.11). We saw in sections 10.2.11 and 10.3.1 that to do this in SPSS we had to enter certain numbers to code these contrasts. For the first contrast we discovered an appropriate set of codes would be −2 for the placebo group and then 1 for both the high- and low-dose groups. For the second contrast the codes would be 0 for the placebo group, −1 for the low-dose group and 1 for the high-dose group (see Table 10.4). If you want to do these contrasts for ANCOVA, then you enter these values as two dummy variables. So, taking the data in this example, we'd add a column called **Dummy1** and in that column we'd put the value −2 for every person who was in the placebo group, and the value 1 for all other participants. We'd then add a second column called **Dummy2**, in which we'd place the value 0 for everyone in the placebo group, −1 for everyone in the low-dose group and 1 for those in the high-dose group. The completed data would be as in the file **ViagraCovariateContrasts.sav**.

Run the analysis as described in section 11.6. The resulting output will begin with a model summary and ANOVA table that should be identical to those in SPSS Output 11.8 (because we've done the same thing as before, the only difference is how the model variance is subsequently broken down with the contrasts). The regression coefficients for the dummy variables will be different though because we've now specified different codes:

Coefficients[a]

Model		Unstandardized Coefficients		Standardized Coefficients		
		B	Std. Error	Beta	t	Sig.
1	(Constant)	3.657	.634		5.764	.000
	Partner's Libido	.260	.193	.246	1.345	.189
2	(Constant)	3.126	.625		5.002	.000
	Partner's Libido	.416	.187	.395	2.227	.035
	Dummy Variable 1 (Placebo vs. Low & High)	.668	.240	.478	2.785	.010
	Dummy Variable 2 (Low vs. High)	.220	.406	.094	.541	.593

a. Dependent Variable: Libido

The first dummy variable compares the placebo group with the low- and high-dose groups. As such, it compares the adjusted mean of the placebo group (2.93) with the average of the adjusted means for the low- and high-dose groups ((4.71+5.15)/2 = 4.93). The *b*-value for the first dummy variable should therefore be the difference between these values: 4.93 − 2.93 = 2. However, we also discovered in a rather complex and boring bit of section 10.2.11.2 that this value gets divided by the number of groups within the contrast (i.e. 3) and so will be 2/3 = .67 (as it is in the output). The associated *t*-statistic is significant, indicating that the placebo group was significantly different from the combined mean of the Viagra groups.

The second dummy variable compares the low- and high-dose groups, and so the *b*-value should be the difference between the adjusted means of these groups: 5.15 − 4.71 = 0.44. We again discovered in section 10.2.11.2 that this value also gets divided by the number of groups within the contrast (i.e. 2) and so will be 0.44/2 = 0.22 (as in the output). The associated *t*-statistic is not significant (its significance is .59 which is greater than .05), indicating that the high-dose group did not produce a significantly higher libido than the low-dose group.

This illustrates how you can apply the principles from section 10.2.11 to ANCOVA: although SPSS doesn't provide an easy interface to do planned contrasts, they can be done if you use the regression menus rather than the ANCOVA ones!

11.7. Testing the assumption of homogeneity of regression slopes ③

We saw earlier that when we conduct ANCOVA we assume *homogeneity of regression slopes*. This just means that we assume that the relationship between the outcome (dependent variable) and the covariate is the same in each of our treatment groups. Figure 11.7 shows scatterplots that display the relationship between partner's libido (the covariate) and the outcome (participant's libido) for each of the three experimental conditions (different colours and symbols). Each symbol represents the data from a particular participant, and the type of symbol tells us the group (circles = placebo, triangles = low dose, squares = high dose). The lines are the regression slopes for the particular group, they summarize the relationship between libido and partner's libido shown by the dots (blue = placebo group, green = low-dose group, red = high-dose group). It should be clear that there is a positive relationship (the regression line slopes upwards from left to right) between partner's libido and participant's libido in both the placebo and low-dose conditions. In fact, the slopes of the lines for these two groups (blue and green) are very similar, showing that the relationship between libido and partner's libido is very similar in these two groups. This situation is an example of homogeneity of regression slopes (the regression slopes in the two groups are similar). However, in the high-dose condition there appears to be no relationship at all between participant's libido and that of their partner (the squares are fairly randomly scattered and the regression line is very flat and shows a slightly negative relationship). The slope of this line is very different to the other two, and this difference gives us cause to doubt whether there is homogeneity of regression slopes (because the relationship between participant's libido and that of their partner is different in the high-dose group to the other two groups).

To test the assumption of homogeneity of regression slopes we need to rerun the ANCOVA but this time use a customized model. Access the main dialog box as before and place the variables in the same boxes as before (so the finished box should look like Figure 11.3). To customize the model we need to access the model dialog box (Figure 11.8) by clicking on

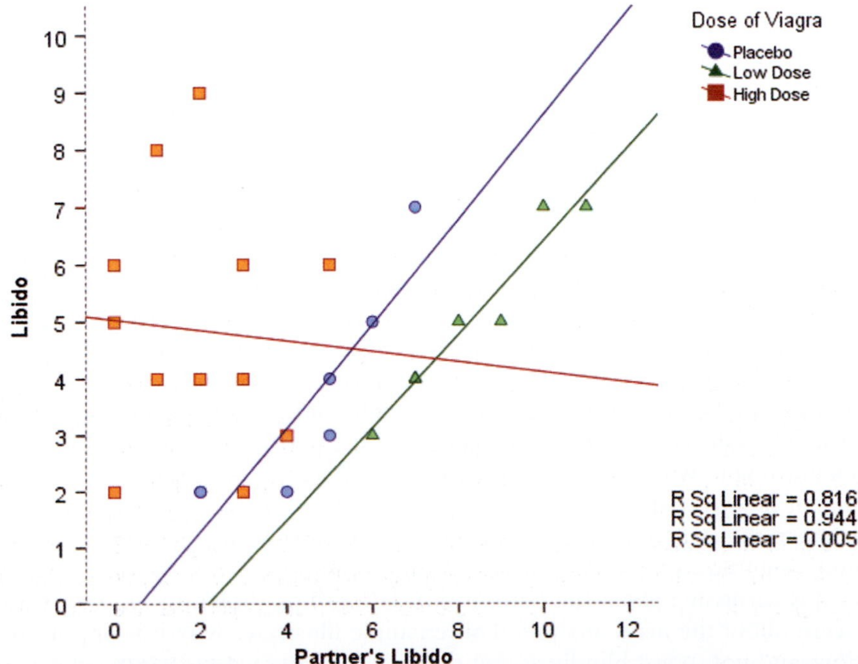

FIGURE 11.7 Scatterplot and regression lines of libido against partner's libido for each of the experimental conditions

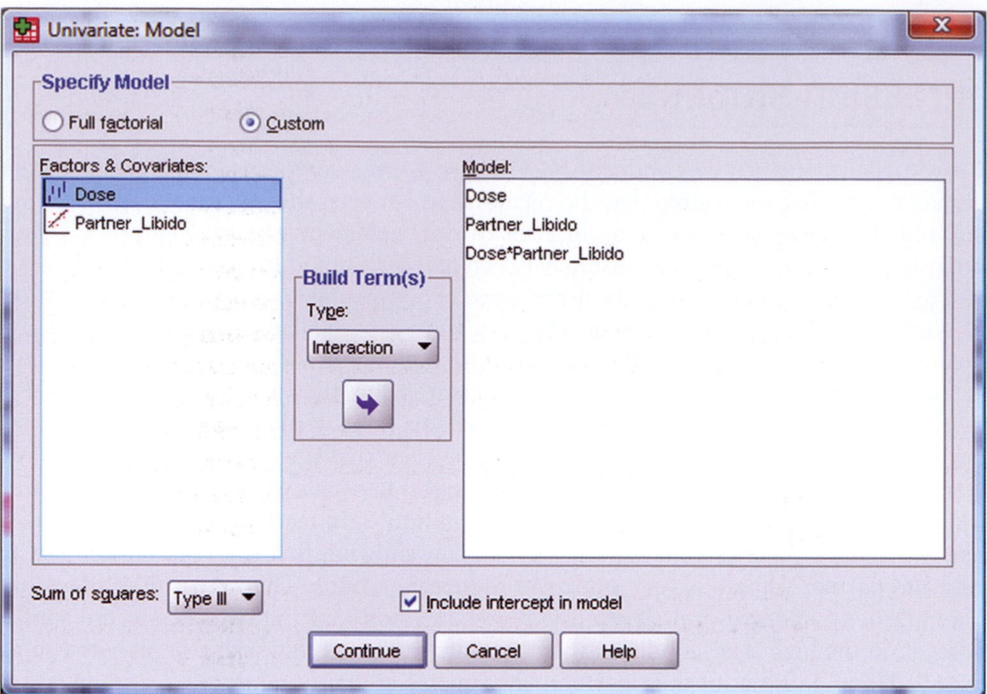

FIGURE 11.8 GLM univariate model dialog box

Model... . To customize your model, select ⦿ Custom to activate the dialog box in Figure 11.8. The variables specified in the main dialog box are listed on the left-hand side. To test the assumption of homogeneity of regression slopes, we need to specify a model that includes the interaction between the covariate and independent variable. Ordinarily, the ANCOVA includes only the main effect of dose and partner's libido and does not include this interaction term. To test this interaction term it's important to still include the main effects of dose and partner so that the interaction term is tested controlling for these main effects. If we don't include the main effects then variance in libido may become attributed to the interaction term that would otherwise be attributed to main effects.

Hence, to begin with you should select **Dose** and **Partner_Libido** (you can select both of them at the same time by holding down *Ctrl*). Then, click on the drop-down menu and change it to Main effects ▾. Having selected this, click on ➜ to move the main effects of **Dose** and **Partner_Libido** to the box labelled *Model*. Next we need to specify the interaction term. To do this, select **Dose** and **Partner_Libido** simultaneously (by holding down the *Ctrl* key while you click on the two variables), then select Interaction ▾ in the drop-down list and click on ➜. This action moves the interaction of **Dose** and **Partner_Libido** to the box labelled *Model*. The finished dialog box should look like Figure 11.8. Having specified our two main effects and the interaction term, click on Continue to return to the main dialog box and then click on OK to run the analysis.

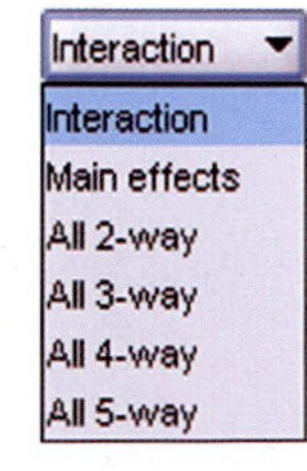

SPSS Output 11.10 shows the main summary table for the ANCOVA including the interaction term. The effects of the dose of Viagra and the partner's libido are still significant, but the main thing in which we're interested is the interaction term, so look at the significance value of the covariate by outcome interaction (**Dose×Partner_Libido**), if this effect is significant then the assumption of homogeneity of regression slopes has been broken. The effect here is significant ($p < .05$); therefore the assumption is not tenable. Although this finding is not surprising given the pattern of relationships shown in Figure 11.7, it does raise concern about the main analysis. This example illustrates why it is important to test assumptions and not to just blindly accept the results of an analysis.

SPSS OUTPUT 11.10

Tests of Between-Subjects Effects

Dependent Variable:Libido

Source	Type III Sum of Squares	df	Mean Square	F	Sig.
Corrected Model	52.346[a]	5	10.469	4.286	.006
Intercept	53.542	1	53.542	21.921	.000
Dose	36.558	2	18.279	7.484	.003
Partner_Libido	17.182	1	17.182	7.035	.014
Dose * Partner_Libido	20.427	2	10.213	4.181	.028
Error	58.621	24	2.443		
Total	683.000	30			
Corrected Total	110.967	29			

a. R Squared = .472 (Adjusted R Squared = .362)

CRAMMING SAM'S TIPS ANCOVA

- Analysis of covariance (ANCOVA) compares several means, but adjusting for the effect of one or more other variables (called *covariates*); for example, if you have several experimental conditions and want to adjust for the age of the participants.
- Before the analysis you should check that the independent variables and covariate(s) are independent. You can do this using ANOVA or a *t*-test to check that levels of the covariate do not differ significantly across groups.
- In the table labelled *Tests of* **Between-Subjects Effects**, look at the column labelled *Sig.* for both the covariate and the independent variable. If the value is less than .05 then for the covariate it means that this variable has a significant relationship to the outcome variable; for the independent variable it means that the means are significantly different across the experimental conditions after partialling out the effect that the covariate has on the outcome.
- As with ANOVA, if you have generated specific hypotheses before the experiment use *planned comparisons*, but if you don't have specific hypotheses use *post hoc* tests. Although SPSS will let you specify certain standard contrasts, other planned comparisons will have to be done by analysing the data using the regression procedure in SPSS.
- For contrasts and post *hoc tests*, again look to the columns labelled *Sig.* to discover if your comparisons are significant (they will be if the significance value is less than .05).
- Test the same assumptions as for ANOVA, but in addition you should test the assumption of *homogeneity of regression slopes*. This has to be done by customizing the ANCOVA model in SPSS to look at the independent variable × covariate interaction.

11.8. Calculating the effect size ②

We saw in the previous chapter that we can use eta squared, η^2, as an effect size measure in ANOVA. This effect size is just r^2 by another name and is calculated by dividing the effect of interest, SS_M, by the total amount of variance in the data, SS_T. As such, it is the proportion of total variance explained by an effect. In ANCOVA (and some of the more complex ANOVAs that we'll encounter in future chapters), we have more than one effect; therefore, we could calculate eta squared for each effect. However, we can also use an effect size measure called partial eta squared (partial η^2). This differs from eta squared in that it looks not at the proportion of total variance that a variable explains, but at the proportion of variance that a variable explains that *is not explained by other variables in the analysis*. Let's look at this with our example; say we want to know the effect size of the dose of Viagra.

Partial eta squared is the proportion of variance in libido that the dose of Viagra shares that is not attributed to partner's libido (the covariate). If you think about the variance that the covariate cannot explain, there are two sources: it cannot explain the variance attributable to the dose of Viagra, SS_{Viagra}, and it cannot explain the error variability, SS_R. Therefore, we use these two sources of variance instead of the total variability, SS_T, in the calculation. The difference between eta squared and partial eta squared is shown as:

$$\eta^2 = \frac{SS_{Effect}}{SS_{Total}} \qquad \text{Partial } \eta^2 = \frac{SS_{Effect}}{SS_{Effect} + SS_{Residual}} \tag{11.3}$$

We can get SPSS to produce partial eta squared for us (see Jane Superbrain Box 11.1). To illustrate its calculation let's look at our Viagra example. We need to use the sums of squares in SPSS Output 11.3 for the effect of dose (25.19), the covariate (15.08) and the error (79.05):

$$\begin{aligned}\text{Partial } \eta^2_{Dose} &= \frac{SS_{Dose}}{SS_{Dose} + SS_{Residual}} \\ &= \frac{25.19}{25.19 + 79.05} \\ &= \frac{25.19}{104.24} \\ &= .24\end{aligned} \qquad \begin{aligned}\text{Partial } \eta^2_{PartnerLibido} &= \frac{SS_{PartnerLibido}}{SS_{PartnerLibido} + SS_{Residual}} \\ &= \frac{15.08}{15.08 + 79.05} \\ &= \frac{15.08}{94.13} \\ &= .16\end{aligned}$$

These values show that **Dose** explained a bigger proportion of the variance not attributable to other variables than **Partner_Libido**.

SELF-TEST Rerun the ANCOVA but select ☑ Estimates of effect size in Figure 11.5. Do the values of partial eta squared match the ones we have just calculated?

As with ANOVA, you can also use omega squared (ω^2). However, as we saw in section 10.5 this measure can be calculated only when we have equal numbers of participants in each group (which is not the case in this example!). So, we're a bit stumped!

However, not all is lost because, as I've said many times already, the overall effect size is not nearly as interesting as the effect size for more focused comparisons. These are easy to calculate because we selected regression parameters (see SPSS Output 11.4) and so we have t-statistics for the covariate and comparisons between the low- and high-dose groups and the placebo and high-dose group. These t-statistics have 26 degrees of freedom (see section 11.5.1). We can use the same equation as in section 9.4.6:[3]

$$r_{contrast} = \sqrt{\frac{t^2}{t^2 + df}}$$

[3] Strictly speaking, we have to use a slightly more elaborate procedure when groups are unequal. It's a bit beyond the scope of this book but Rosnow, Rosenthal, and Rubin (2000) give a very clear account.

Therefore we get (t from SPSS Output 11.4):

$$r_{\text{Covariate}} = \sqrt{\frac{2.23^2}{2.23^2 + 26}} = \sqrt{\frac{4.97}{30.97}} = .40$$

$$r_{\text{High Dose vs. Placebo}} = \sqrt{\frac{-2.77^2}{-2.77^2 + 26}} = \sqrt{\frac{7.67}{33.67}} = .48$$

$$r_{\text{High vs. Low Dose}} = \sqrt{\frac{-0.54^2}{-0.54^2 + 26}} = \sqrt{\frac{0.29}{26.29}} = .11$$

If you think back to our benchmarks for effect sizes, the effect of the covariate and the difference between the high dose and the placebo both represent medium to large effect sizes (they're all between .4 and .5). Therefore, as well as being statistically significant, these effects are substantive findings. The difference between the high- and low-dose groups was a fairly small effect.

11.9. Reporting results ②

Reporting ANCOVA is much the same as reporting ANOVA except we now have to report the effect of the covariate as well. For the covariate and the experimental effect we give details of the F-ratio and the degrees of freedom from which it was calculated. In both cases, the F-ratio was derived from dividing the mean squares for the effect by the mean squares for the residual. Therefore, the degrees of freedom used to assess the F-ratio are the degrees of freedom for the effect of the model (df_M = 1 for the covariate and 2 for the experimental effect) and the degrees of freedom for the residuals of the model (df_R = 26 for both the covariate and the experimental effect) – see SPSS Output 11.3. Therefore, the correct way to report the main findings would be:

✓ The covariate, partner's libido, was significantly related to the participant's libido, $F(1, 26) = 4.96$, $p < .05$, $r = .40$. There was also a significant effect of Viagra on levels of libido after controlling for the effect of partner's libido, $F(2, 26) = 4.14$, $p < .05$, *partial* $\eta^2 = .24$.

We can also report some contrasts (see SPSS Output 11.4):

✓ Planned contrasts revealed that having a high dose of Viagra significantly increased libido compared to having a placebo, $t(26) = -2.77$, $p < .05$, $r = .48$, but not compared to having a low dose, $t(26) = -0.54$, $p > .05$, $r = .11$.

11.10. What to do when assumptions are violated in ANCOVA ③

In previous chapters we have seen that when the assumptions of a test have been violated, there is often a non-parametric test to which we can turn (Chapter 15). However, as we start to discover more complicated procedures, we will also see that the squid of despair squirts its ink of death on our data when they violate assumptions. For complex analyses, there are not non-parametric counterparts that are easily run through SPSS. ANCOVA is the first such example of a test that does not have an SPSS-friendly non-parametric test. As such, if our data violate the assumptions of normality, or homogeneity of variance, the only real solutions are robust methods (see section 5.7.4) such as those described in Wilcox's Chapter 11 and the associated files for the software R (Wilcox, 2005). Use the SPSS R plugin to do these tests directly from within SPSS. You could also use the robust regression procedures in Wilcox's book because, as we have seen, ANCOVA is simply a version of regression. Also, if you have violated the assumption of homogeneity of regression slopes then you can explicitly model this variation using multilevel linear models (see Chapter 19). If you run ANCOVA as a multilevel model you can also bootstrap the parameters to get robust estimates.

What have I discovered about statistics? ②

This chapter has shown you how the general linear model that was described in Chapter 10 can be extended to include additional variables. The advantages of doing so are that we can remove the variance in our outcome that is attributable to factors other than our experimental manipulation. This gives us tighter experimental control, and may also help us to explain some of our error variance, and, therefore, give us a purer measure of the experimental manipulation. We didn't go into too much theory about ANCOVA, just looked conceptually at how the regression model can be expanded to include these additional variables (*covariates*). Instead we jumped straight into an example, which was to look at the effect of Viagra on libido (as in Chapter 10) but including partner's libido as a covariate. I explained how to do the analysis on SPSS and interpret the results but also showed how the same output could be obtained by running the analysis as a regression. This was to try to get the message home that ANOVA and ANCOVA are merely forms of regression! Anyway, we finished off by looking at an additional assumption that has to be considered when doing ANCOVA: the assumption of homogeneity of regression slopes. This just means that the relationship between the covariate and the outcome variable should be the same in all of your experimental groups.

Having seen Iron Maiden in all of their glory, I was inspired. Although I had briefly been deflected from my destiny by the shock of grammar school, I was back on track. I *had* to form a band. There was just one issue: no one else played a musical instrument. The solution was easy: through several months of covert subliminal persuasion I convinced my two best friends (both called Mark oddly enough) that they wanted nothing more than to start learning the drums and bass guitar. A power trio was in the making!

Key terms that I've discovered

Adjusted mean
Analysis of covariance (ANCOVA)
Covariate
Homogeneity of regression slopes
Partial eta squared (partial η^2)
Partial out
Sidak correction

Smart Alex's tasks

- **Task 1**: Stalking is a very disruptive and upsetting (for the person being stalked) experience in which someone (the stalker) constantly harasses or obsesses about another person. It can take many forms, from sending intensely disturbing letters threatening to boil your cat if you don't reciprocate the stalker's undeniable love for you, to literally following you around your local area in a desperate attempt to see which CD you buy on a Saturday (as if it would be anything other than Fugazi!). A psychologist, who'd had enough of being stalked by people, decided to try two different therapies on different groups of stalkers (25 stalkers in each group – this variable is called **Group**). To the first group of stalkers he gave what he termed cruel-to-be-kind therapy. This therapy was based on punishment for stalking behaviours; in short, every time the stalkers followed him around, or sent him a letter, the psychologist attacked them with a cattle prod until they stopped their stalking behaviour. It was hoped that the stalkers would learn an aversive reaction to anything resembling stalking. The second therapy was psychodyshamic therapy, which is a recent development on Freud's psychodynamic therapy that acknowledges what a sham this kind of treatment is (so, you could say it's based on Fraudian theory!). The stalkers were hypnotized and regressed into their childhood; the therapist would also discuss their penis (unless it is a woman, in which case they discussed their lack of penis), the penis of their father, their dog's penis, the penis of the cat down the road and anyone else's penis that sprang to mind. At the end of therapy, the psychologist measured the number of hours in the week that the stalker spent stalking their prey (this variable is called **stalk2**). Now, the therapist believed that the success of therapy might well depend on how bad the problem was to begin with, so before therapy the therapist measured the number of hours that the patient spent stalking as an indicator of how much of a stalker the person was (this variable is called **stalk1**). The data are in the file **Stalker.sav**. Analyse the effect of therapy on stalking behaviour after therapy, controlling for the amount of stalking behaviour before therapy. ②

- **Task 2**: A marketing manager for a certain well-known drinks manufacturer was interested in the therapeutic benefit of certain soft drinks for curing hangovers. He took 15 people out on the town one night and got them drunk. The next morning as they awoke, dehydrated and feeling as though they'd licked a camel's sandy feet clean with their tongue, he gave five of them water to drink, five of them Lucozade (in case this isn't sold outside of the UK, it's a very nice glucose-based drink) and the remaining five a leading brand of cola (this variable is called **drink**). He then measured how well they felt (on a scale from 0 = I feel like death to 10 = I feel really full of beans and healthy) two hours later (this variable is called **well**). He wanted to know which drink produced the greatest level of wellness. However, he realized it was important to control for how drunk the person got the night before, and so he measured this on

a scale of 0 = as sober as a nun to 10 = flapping about like a haddock out of water on the floor in a puddle of their own vomit. The data are in the file **HangoverCure.sav**. Conduct an ANCOVA to see whether people felt better after different drinks when controlling for how drunk they were the night before. ②

The answers are on the companion website and task 1 has a full interpretation in Field & Hole (2003).

Further reading

Howell, D. C. (2006). *Statistical methods for psychology* (6th ed.). Belmont, CA: Thomson. (Or you might prefer his *Fundamental Statistics for the Behavioral Sciences,* also in its 6th edition, 2007.)

Miller, G. A., & Chapman, J. P. (2001). Misunderstanding analysis of covariance. *Journal of Abnormal Psychology*, *110*(1), 40–48.

Rutherford, A. (2000). *Introducing ANOVA and ANCOVA: A GLM approach*. London: Sage.

Wildt, A. R., & Ahtola, O. (1978). *Analysis of covariance*. Sage university paper series on quantitative applications in the social sciences, 07-012. Newbury Park, CA: Sage. (This text is pretty high level but very comprehensive if you want to know the maths behind ANCOVA.)

Online tutorials

The companion website contains the following Flash movie tutorials to accompany this chapter:

- ANCOVA Using SPSS
- The *R* plugin

Interesting real research

Muris, P., Huijding, J., Mayer, B. and Hameetman, M. (2008). A space odyssey: Experimental manipulation of threat perception and anxiety-related interpretation bias in children. *Child Psychiatry and Human Development, 39*(4), 469–480.

Factorial ANOVA (GLM 3)

12

FIGURE 12.1 Andromeda coming to a living room near you in 1988 (L–R: Malcolm, me and the two Marks)

12.1. What will this chapter tell me? ②

After persuading my two friends (Mark and Mark) to learn the bass and drums, I took the rather odd decision to *stop* playing the guitar. I didn't stop, as such, but I focused on singing instead. In retrospect, I'm not sure why because I am *not* a good singer. Mind you, I'm not a good guitarist either. The upshot was that a classmate, Malcolm, ended up as our guitarist. I really can't remember how or why we ended up in this configuration, but we called ourselves Andromeda, we learnt several Queen and Iron Maiden songs and we were truly awful. I have some tapes somewhere to prove just what a cacophony of tuneless drivel we produced, but the chances of their appearing on the companion website are slim at best. Suffice it to say, you'd be hard pushed to recognize *which* Iron Maiden and Queen songs we were trying to play. I try to comfort myself with the fact that we were only 14 or 15 at the time, but even youth does not excuse the depths of ineptitude to which we sank. Still, we garnered a reputation for being too loud in school assembly and we did a successful tour of our friends' houses (much to their parents' amusement I'm sure). We even started to write a few songs (I wrote one called 'Escape From Inside' about the film *The Fly* that contained the wonderful rhyming couplet of 'I am a fly, I want to die': genius!). The only thing that we did that resembled the activities of a 'proper' band was to split up due to 'musical differences', these differences being that Malcolm wanted to write 15-part symphonies about a boy's journey to worship electricity pylons and discover a mythical beast called the cuteasaurus, whereas I wanted to write songs about flies, and dying. When we could not agree on a musical direction the split became inevitable. We could have tested empirically the best musical direction for the band by Malcolm and I both writing a 15-part symphony and a 3-minute song about a fly. If we played these songs to various people and measured their screams of agony then we could ascertain the best musical direction to gain popularity. We have two variables that predict screams: whether I or Malcolm wrote the song (songwriter), and whether the song was a 15-part symphony or a song about a fly (song type). The one-way ANOVA that we encountered in Chapter 10 cannot deal with two predictor variables – this is a job for factorial ANOVA!

12.2. Theory of factorial ANOVA (between-groups) ②

In the previous two chapters we have looked at situations in which we've tried to test for differences between groups when there has been a single independent variable (i.e. one variable has been manipulated). However, at the beginning of Chapter 10 I said that one of the advantages of ANOVA was that we could look at the effects of more than one independent variable (and how these variables interact). This chapter extends what we already know about ANOVA to look at situations where there are two independent variables. We've already seen in the previous chapter that it's very easy to incorporate a second variable into the ANOVA framework when that variable is a continuous variable (i.e. not split into groups), but now we'll move on to situations where there is a second independent variable that has been systematically manipulated by assigning people to different conditions.

12.2.1. Factorial designs ②

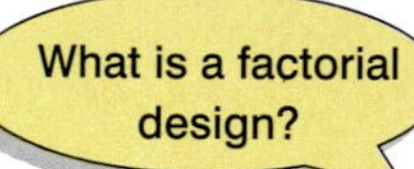
What is a factorial design?

In the previous two chapters we have explored situations in which we have looked at the effects of a single independent variable on some outcome. However, independent variables often get lonely and want to have friends. Scientists are obliging individuals and often put a second (or third) independent variable into their designs to keep the others company. When an experiment has two or more independent variables it is known as a *factorial design* (this is because variables are sometimes referred to as *factors*). There are several types of factorial design:

- **Independent factorial design**: In this type of experiment there are several independent variables or predictors and each has been measured using different participants (between groups). We discuss this design in this chapter.
- **Repeated-measures (related) factorial design**: This is an experiment in which several independent variables or predictors have been measured, but the same participants have been used in all conditions. This design is discussed in Chapter 13.
- **Mixed design**: This is a design in which several independent variables or predictors have been measured; some have been measured with different participants whereas others used the same participants. This design is discussed in Chapter 14.

As you might imagine, it can get quite complicated analysing these types of experiments. Fortunately, we can extend the ANOVA model that we encountered in the previous two chapters to deal with these more complicated situations. When we use ANOVA to analyse a situation in which there are two or more independent variables it is sometimes called **factorial ANOVA**; however, the specific names attached to different ANOVAs reflect the experimental design that they are being used to analyse (see Jane Superbrain Box 12.1). This section extends the one-way ANOVA model to the factorial case (specifically when there are two independent variables). In subsequent chapters we will look at repeated-measures designs, factorial repeated-measures designs and finally mixed designs.

JANE SUPERBRAIN 12.1

Naming ANOVAs ②

ANOVAs can be quite confusing because there appears to be lots of them. When you read research articles you'll quite often come across phrases like 'a two-way independent ANOVA was conducted', or 'a three-way repeated-measures ANOVA was conducted'. These names may look confusing but they are quite easy if you break them down. All ANOVAs have two things in common: they involve some quantity of independent variables and these variables can be measured using either the same or different participants. If the same participants are used we typically use the term *repeated measures* and if different participants are used we use the term *independent*. When there are two or more independent variables, it's possible that some variables use the same participants whereas others use different participants. In this case we use the term *mixed*. When we name an ANOVA, we are simply telling the reader how many independent variables we used and how they were measured. In general terms we could write the name of an ANOVA as:

- A (*number of independent variables*) way of *how these variables were measured* ANOVA.

By remembering this you can understand the name of any ANOVA you come across. Look at these examples and try to work out how many variables were used and how they were measured:

- One-way independent ANOVA
- Two-way repeated-measures ANOVA
- Two-way mixed ANOVA
- Three-way independent ANOVA

The answers you should get are:

- One independent variable measured using different participants.
- Two independent variables both measured using the same participants.
- Two independent variables: one measured using different participants and the other measured using the same participants.
- Three independent variables all of which are measured using different participants.

12.2.2. An example with two independent variables ②

Throughout this chapter we'll use an example that has two independent variables. This is known as a two-way ANOVA (see Jane Superbrain Box 12.1). I'll look at an example with two independent variables because this is the simplest extension of the ANOVAs that we have already encountered.

TABLE 12.1 Data for the beer-goggles effect

Alcohol	None		2 Pints		4 Pints	
Gender	**Female**	**Male**	**Female**	**Male**	**Female**	**Male**
	65	50	70	45	55	30
	70	55	65	60	65	30
	60	80	60	85	70	30
	60	65	70	65	55	55
	60	70	65	70	55	35
	55	75	60	70	60	20
	60	75	60	80	50	45
	55	65	50	60	50	40
Total	485	535	500	535	460	285
Mean	60.625	66.875	62.50	66.875	57.50	35.625
Variance	24.55	106.70	42.86	156.70	50.00	117.41

An anthropologist was interested in the effects of alcohol on mate selection at night-clubs. Her rationale was that after alcohol had been consumed, subjective perceptions of physical attractiveness would become more inaccurate (the well-known '**beer-goggles effect**'). She was also interested in whether this effect was different for men and women. She picked 48 students: 24 male and 24 female. She then took groups of eight participants to a night-club and gave them no alcohol (participants received placebo drinks of alcohol-free lager), 2 pints of strong lager, or 4 pints of strong lager. At the end of the evening she took a photograph of the person that the participant was chatting up. She then got a pool of independent judges to assess the attractiveness of the person in each photograph (out of 100). The data are in Table 12.1 and **Goggles.sav**.

12.2.3. Total sums of squares (SS_T) ②

Two-way ANOVA is conceptually very similar to one-way ANOVA. Basically, we still find the total sum of squared errors (SS_T) and break this variance down into variance that can be explained by the experiment (SS_M) and variance that cannot be explained (SS_R). However, in two-way ANOVA, the variance explained by the experiment is made up of not one experimental manipulation but two. Therefore, we break the model sum of squares down into variance explained by the first independent variable (SS_A), variance explained by the second independent variable (SS_B) and variance explained by the interaction of these two variables ($SS_{A \times B}$) – see Figure 12.2.

We start off in the same way as we did for one-way ANOVA. That is, we calculate how much variability there is between scores when we ignore the experimental condition from which they came. Remember from one-way ANOVA (equation (10.4)) that SS_T is calculated using the following equation:

$$SS_T = s^2_{grand}(N - 1)$$

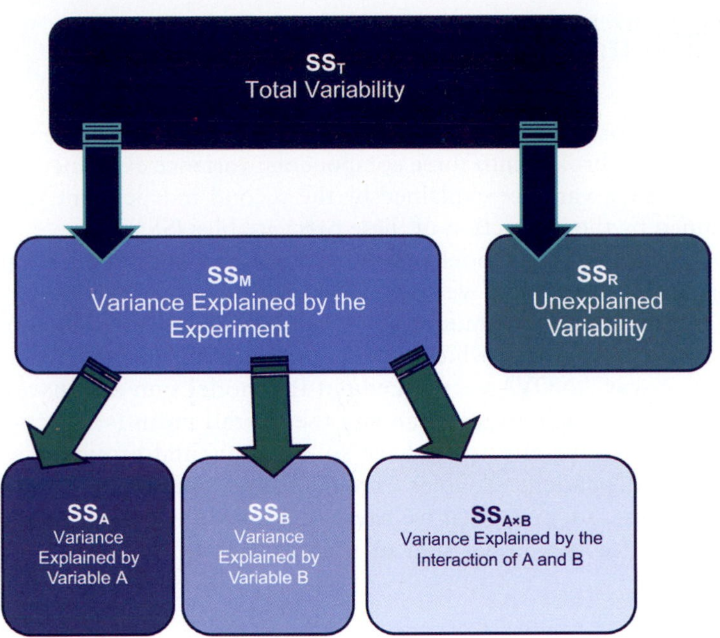

FIGURE 12.2 Breaking down the variance in two-way ANOVA

The grand variance is simply the variance of all scores when we ignore the group to which they belong. So if we treated the data as one big group it would look as follows:

65	50	70	45	55	30
70	55	65	60	65	30
60	80	60	85	70	30
60	65	70	65	55	55
60	70	65	70	55	35
55	75	60	70	60	20
60	75	60	80	50	45
55	65	50	60	50	40
		Grand Mean = 58.33			

If we calculate the variance of all of these scores, we get 190.78 (try this on your calculators if you don't trust me). We used 48 scores to generate this value, and so N is 48. As such the equation becomes:

$$\begin{aligned}\text{SS}_\text{T} &= s^2_\text{grand}(N-1)\\ &= 190.78(48-1)\\ &= 8966.66\end{aligned}$$

The degrees of freedom for this SS will be $N-1$, or 47.

12.2.4. The model sum of squares (SS_M) ②

The next step is to work out the model sum of squares. As I suggested earlier, this sum of squares is then further broken into three components: variance explained by the first independent variable (SS_A), variance explained by the second independent variable (SS_B) and variance explained by the interaction of these two variables ($SS_{A \times B}$).

Before we break down the model sum of squares into its component parts, we must first calculate its value. We know we have 8966.66 units of variance to be explained and our first step is to calculate how much of that variance is explained by our experimental manipulations overall (ignoring which of the two independent variables is responsible). When we did one-way ANOVA we worked out the model sum of squares by looking at the difference between each group mean and the overall mean (see section 10.2.6). We can do the same here. We effectively have six experimental groups if we combine all levels of the two independent variables (three doses for the male participants and three doses for the females). So, given that we have six groups of different people we can then apply the equation for the model sum of squares that we used for one-way ANOVA (equation (10.5)):

$$SS_M = \sum n_k (\bar{x}_k - \bar{x}_{\text{grand}})^2$$

The grand mean is the mean of all scores (we calculated this above as 58.33) and n is the number of scores in each group (i.e. the number of participants in each of the six experimental groups; eight in this case). Therefore, the equation becomes:

$$\begin{aligned} SS_M &= 8(60.625 - 58.33)^2 + 8(66.875 - 58.33)^2 + 8(62.5 - 58.33)^2 + \ldots \\ &\quad + 8(66.875 - 58.33)^2 + 8(57.5 - 58.33)^2 + 8(35.625 - 58.33)^2 \\ &= 8(2.295)^2 + 8(8.545)^2 + 8(4.17)^2 + 8(8.545)^2 + 8(-0.83)^2 + 8(-22.705)^2 \\ &= 42.1362 + 584.1362 + 139.1112 + 584.1362 + 5.5112 + 4124.1362 \\ &= 5479.167 \end{aligned}$$

The degrees of freedom for this SS will be the number of groups used, k, minus 1. We used six groups and so $df = 5$.

At this stage we know that the model (our experimental manipulations) can explain 5479.167 units of variance out of the total of 8966.66 units. The next stage is to further break down this model sum of squares to see how much variance is explained by our independent variables separately.

12.2.4.1. The main effect of gender (SS_A) ②

To work out the variance accounted for by the first independent variable (in this case, gender) we need to group the scores in the data set according to which gender they belong. So, basically we ignore the amount of drink that has been drunk, and we just place all of the male scores into one group and all of the female scores into another. So, the data will look like this (note that the first box contains the three female columns from our original table and the second box contains the male columns):

A_1: Female		
65	70	55
70	65	65
60	60	70
60	70	55
60	65	55
55	60	60
60	60	50
55	50	50
Mean Female = 60.21		

A_2: Male		
50	45	30
55	60	30
80	85	30
65	65	55
70	70	35
75	70	20
75	80	45
65	60	40
Mean Male = 56.46		

We can then apply the equation for the model sum of squares that we used to calculate the overall model sum of squares:

$$\mathrm{SS_A} = \sum n_k \left(\overline{x}_k - \overline{x}_{\text{grand}}\right)^2$$

The grand mean is the mean of all scores (above) and n is the number of scores in each group (i.e. the number of males and females; 24 in this case). Therefore, the equation becomes:

$$\begin{aligned}\mathrm{SS_{Gender}} &= 24(60.21-58.33)^2+24(56.46-58.33)^2\\ &= 24(1.88)^2+24(-1.87)^2\\ &= 84.8256+83.9256\\ &= 168.75\end{aligned}$$

The degrees of freedom for this SS will be the number of groups used, k, minus 1. We used two groups (males and females) and so $df = 1$. To sum up, the main effect of gender compares the mean of all males against the mean of all females (regardless of which alcohol group they were in).

12.2.4.2. The main effect of alcohol (SS_B) ②

To work out the variance accounted for by the second independent variable (in this case, alcohol) we need to group the scores in the data set according to how much alcohol was consumed. So, basically we ignore the gender of the participant, and we just place all of the scores after no drinks in one group, the scores after 2 pints in another group and the scores after 4 pints in a third group. So, the data will look like this:

B_1: None	
65	50
70	55
60	80
60	65
60	70
55	75
60	75
55	65
Mean None = 63.75	

B_2: 2 Pints	
70	45
65	60
60	85
70	65
65	70
60	70
60	80
50	60
Mean 2 Pints = 64.6875	

B_3: 4 Pints	
55	30
65	30
70	30
55	55
55	35
60	20
50	45
50	40
Mean 4 Pints = 46.5625	

We can then apply the same equation for the model sum of squares that we used for the overall model sum of squares and for the main effect of gender:

$$SS_B = \sum n_k \left(\bar{x}_k - \bar{x}_{\text{grand}}\right)^2$$

The grand mean is the mean of all scores (58.33 as before) and n is the number of scores in each group (i.e. the number of scores in each of the boxes above, in this case 16). Therefore, the equation becomes:

$$\begin{aligned} SS_{\text{Alcohol}} &= 16(63.75 - 58.33)^2 + 16(64.6875 - 58.33)^2 + 16(46.5625 - 58.33)^2 \\ &= 16(5.42)^2 + 16(6.3575)^2 + 16(-11.7675)^2 \\ &= 470.0224 + 646.6849 + 2215.5849 \\ &= 3332.292 \end{aligned}$$

The degrees of freedom for this SS will be the number of groups used minus 1 (see section 10.2.6). We used three groups and so $df = 2$. To sum up, the main effect of alcohol compares the means of the no alcohol, 2 pint and 4 pint groups (regardless of whether the scores come from men or women).

12.2.4.3. The interaction effect ($SS_{A\times B}$) ②

The final stage is to calculate how much variance is explained by the **interaction** of the two variables. The simplest way to do this is to remember that the SS_M is made up of three components (SS_A, SS_B and $SS_{A\times B}$). Therefore, given that we know SS_A and SS_B we can calculate the interaction term using subtraction:

$$SS_{A\times B} = SS_M - SS_A - SS_B$$

Therefore, for these data, the value is:

$$\begin{aligned} SS_{A\times B} &= SS_M - SS_A - SS_B \\ &= 5479.167 - 168.75 - 3322.292 \\ &= 1978.125 \end{aligned}$$

The degrees of freedom can be calculated in the same way, but are also the product of the degrees of freedom for the main effects (either method works):

$$\begin{aligned} df_{A\times B} &= df_M - df_A - df_B \\ &= 5 - 1 - 2 \\ &= 2 \end{aligned} \qquad \begin{aligned} df_{A\times B} &= df_A \times df_B \\ &= 1 \times 2 \\ &= 2 \end{aligned}$$

12.2.5. The residual sum of squares (SS_R) ②

The residual sum of squares is calculated in the same way as for one-way ANOVA (see section 10.2.7) and again represents individual differences in performance or the variance that can't be explained by factors that were systematically manipulated. We saw in one-way ANOVA that the value is calculated by taking the squared error between each data point and its corresponding group mean. An alternative way to express this was as (see equation (10.7)):

$$SS_R = s^2_{\text{group }1}(n_1 - 1) + s^2_{\text{group }2}(n_2 - 1) + s^2_{\text{group }3}(n_3 - 1) + \ldots + s^2_{\text{group }n}(n_n - 1)$$

So, we use the individual variances of each group and multiply them by one less than the number of people within the group (n). We have the individual group variances in our original table of data (Table 12.1) and there were eight people in each group (therefore, $n = 8$) and so the equation becomes:

$$\begin{aligned}
SS_R &= s^2_{\text{group }1}(n_1 - 1) + s^2_{\text{group }2}(n_2 - 1) + s^2_{\text{group }3}(n_3 - 1) + s^2_{\text{group }4}(n_4 - 1) + \ldots \\
&\quad + s^2_{\text{group }5}(n_5 - 1) + s^2_{\text{group }6}(n_6 - 1) \\
&= (24.55)(8-1) + (106.7)(8-1) + (42.86)(8-1) + (156.7)(8-1) + \cdots \\
&\quad + (50)(8-1) + (117.41-41)(8-1) \\
&= (24.55 \times 7) + (106.7 \times 7) + (42.86 \times 7) + (156.7 \times 7) + (50 \times 7) + \cdots \\
&\quad + (117.41 \times 7) \\
&= 171.85 + 746.9 + 300 + 1096.9 + 350 + 821.87 \\
&= 3487.52
\end{aligned}$$

The degrees of freedom for each group will be one less than the number of scores per group (i.e. 7). Therefore, if we add the degrees of freedom for each group, we get a total of $6 \times 7 = 42$.

12.2.6. The *F*-ratios ②

Each effect in a two-way ANOVA (the two main effects and the interaction) has its own *F*-ratio. To calculate these we have to first calculate the mean squares for each effect by taking the sum of squares and dividing by the respective degrees of freedom (think back to section 10.2.8). We also need the mean squares for the residual term. So, for this example we'd have four mean squares calculated as follows:

$$\begin{aligned}
MS_A &= \frac{SS_A}{df_A} = \frac{168.75}{1} = 168.75 \\
MS_B &= \frac{SS_B}{df_B} = \frac{3332.292}{2} = 1666.146 \\
MS_{A\times B} &= \frac{SS_{A\times B}}{df_{A\times B}} = \frac{1978.125}{2} = 989.062 \\
MS_R &= \frac{SS_R}{df_R} = \frac{3487.52}{42} = 83.036
\end{aligned}$$

The *F*-ratios for the two independent variables and their interactions are then calculated by dividing their mean squares by the residual mean squares. Again, if you think back to one-way ANOVA this is exactly the same process!

$$F_A = \frac{MS_A}{MS_R} = \frac{168.75}{83.036} = 2.032$$

$$F_B = \frac{MS_B}{MS_R} = \frac{1666.146}{83.036} = 20.065$$

$$F_{A \times B} = \frac{MS_{A \times B}}{MS_R} = \frac{989.062}{83.036} = 11.911$$

Each of these F-ratios can be compared against critical values (based on their degrees of freedom, which can be different for each effect) to tell us whether these effects are likely to reflect data that have arisen by chance, or reflect an effect of our experimental manipulations (these critical values can be found in the Appendix). If an observed F exceeds the corresponding critical values then it is significant. SPSS will calculate these F-ratios and exact significance for each, but what I hope to have shown you in this section is that two-way ANOVA is basically the same as one-way ANOVA except that the model sum of squares is partitioned into three parts: the effect of each of the independent variables and the effect of how these variables interact.

12.3. Factorial ANOVA using SPSS ②

12.3.1. Entering the data and accessing the main dialog box ②

To enter these data into the SPSS Data Editor, remember that *levels of a between-group variable go in a single column*. Applying this rule to these data, we need to create two different coding variables in the data editor to represent gender and alcohol consumption. So, create a variable called **Gender** on the data editor and activate the labels dialog box. You should define value labels to represent the two genders. We have had a lot of experience with coding values, so you should be fairly happy about assigning numerical codes to different groups. I recommend using the code male = 0 and female = 1. Once you have done this, you can enter a code of 0 or 1 in this column indicating to which group the person belonged. Create a second variable called **Alcohol** and assign group codes by using the labels dialog box. I suggest that you code this variable with three values: placebo (no alcohol) = 1, 2 pints = 2 and 4 pints = 3. You can now enter 1, 2 or 3 into this column to represent the amount of alcohol consumed by the participant. Remember that if you turn the *value labels* option on you will see text in the data editor rather than the numerical codes. Now, the way this coding works is as follows:

Gender	Alcohol	Participant was
0	1	Male who consumed no alcohol
0	2	Male who consumed 2 pints
0	3	Male who consumed 4 pints
1	1	Female who consumed no alcohol
1	2	Female who consumed 2 pints
1	3	Female who consumed 4 pints

Once you have created the two coding variables, you can create a third variable in which to place the values of the dependent variable. Call this variable **Attractiveness** and use the *labels* option to give it the fuller name of 'Attractiveness of Date'.

SELF-TEST Use the Chart Builder to plot a line graph (with error bars) of the attractiveness of the date with alcohol consumption on the *x*-axis and different coloured lines to represent males and females.

In this example, there are two independent variables and different participants were used in each condition: hence, we can use the general factorial ANOVA procedure in SPSS. This procedure is designed for analysing between-group factorial designs. To access the main dialog box use the file path Analyze General Linear Model ▸ Univariate.... The resulting dialog box is shown in Figure 12.3. First, select the dependent variable **Attractiveness** from the variables list on the left-hand side of the dialog box and drag it to the space labelled *Dependent Variable* or click on ➜. In the space labelled *Fixed Factor(s)* we need to place any independent variables relevant to the analysis. Select **Alcohol** and **Gender** in the variables list (these variables can be selected simultaneously by holding down *Ctrl* while clicking on the variables) and drag them to the *Fixed Factor(s)* box (or click on ➜). There are various other spaces that are available for conducting more complex analyses such as random factors ANOVA and factorial ANCOVA. Random factors ANOVA is beyond the scope of this book (interested readers should consult Jackson & Brashers, 1994) and factorial ANCOVA simply extends the principles described at the beginning of this chapter to include a covariate (as in the last chapter).

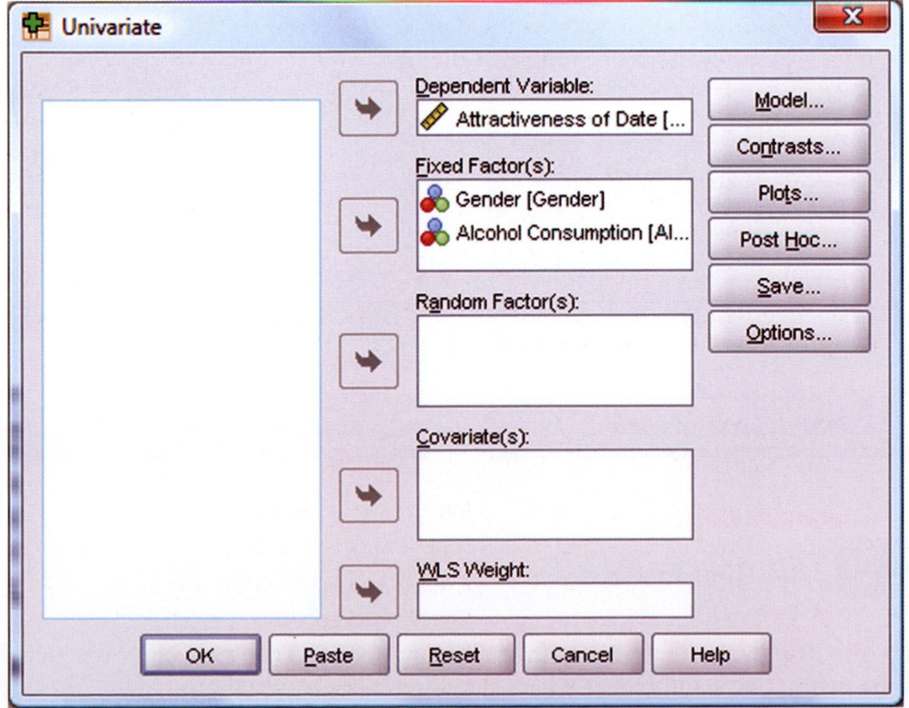

FIGURE 12.3 Main dialog box for univariate ANOVA

OLIVER TWISTED

Please, Sir, can I ... customize my model?

'My friend told me that there are different types of sums of squares,' complains Oliver with an air of impressive authority, 'why haven't you told us about them? Is it because you have a microbe for a brain?' No, it's not Oliver, it's because everyone but you will find this very tedious. If you want to find out more about what the Model... button does, and the different types of sums of squares that can be used in ANOVA, then the additional material on the website will tell you.

12.3.2. Graphing interactions ②

Once the relevant variables have been selected, you can click on Plots... to access the dialog box in Figure 12.4. The plots dialog box allows you to select line graphs of your data and these graphs are very useful for interpreting interaction effects (however, really we should plot graphs of the means before the data are analysed). We have only two independent variables, and the most useful plot is one that shows the interaction between these variables (the plot that displays levels of one independent variable against the other). In this case, the **interaction graph** will help us to interpret the combined effect of gender and alcohol consumption. Select **Alcohol** from the variables list on the left-hand side of the dialog box and drag it to the space labelled *Horizontal Axis* (or click on ➜). In the space labelled *Separate Lines* place the remaining independent variable, **Gender**. It doesn't matter which way round the variables are plotted; you should use your discretion as to which way produces the most sensible graph. When you have moved the two independent variables to the appropriate box, click on Add and this plot will be added to the list at the bottom of the box. You can plot a whole variety of graphs, and if you had a third independent variable, you have the option of plotting different graphs for each level of that third variable by specifying a variable under the heading *Separate Plots*. When you have finished specifying graphs, click on Continue to return to the main dialog box.

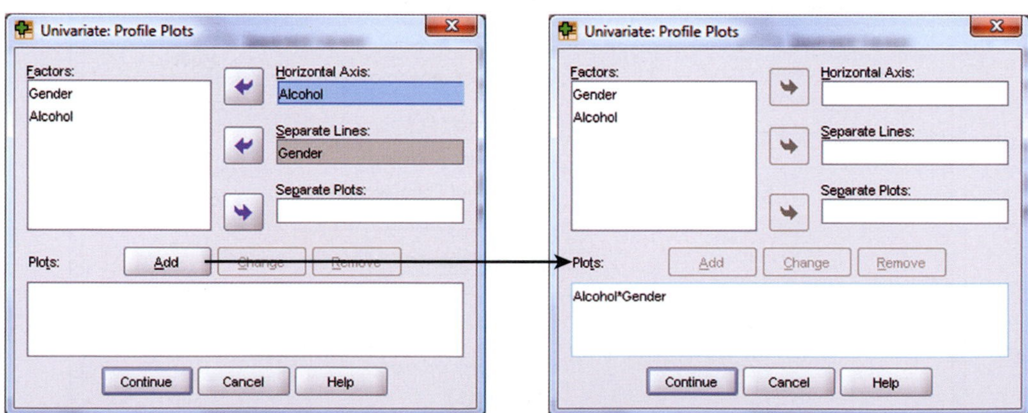

FIGURE 12.4 The plots dialog box

12.3.3. Contrasts ②

We saw in Chapter 10 that it's useful to follow up ANOVA with contrasts that break down the main effects and tell us where the differences between groups lie. For one-way ANOVA, SPSS has a procedure for entering codes that define the contrasts we want to

OLIVER TWISTED

Please, Sir, can I have some more … contrasts?

'I don't want to use standard contrasts,' sulks Oliver as he stamps his feet on the floor, 'they smell of rotting cabbage.' I think actually, Oliver, the stench of rotting cabbage is probably because you stood your Dickensian self under a window when someone emptied their toilet bucket into the street. Nevertheless, I do get asked a fair bit about how to do contrasts with syntax and since I'm a complete masochist I've prepared a fairly detailed guide in the additional material for this chapter. These contrasts are useful to follow up a significant interaction effect.

do. However, for two-way ANOVA no such facility exists and instead we are restricted to doing one of several standard contrasts. These standard contrasts are described in Table 10.6. To be fair, these contrasts will give you what you want in many different situations; however, if they don't and you want to define your own contrasts then this has to be done using syntax (see Oliver Twisted).

We can use standard contrasts for this example. The effect of gender has only two levels, so we don't really need contrasts for this main effect. The effect of alcohol has three levels: none, 2 pints and 4 pints. We could select a simple contrast for this variable, and use the first category as a reference category. This would compare the 2 pint group to the no alcohol group, and then compare the 4 pint category to the no alcohol group. As such, the alcohol groups would get compared to the no alcohol group. We could also select a *repeated* contrast. This would compare the 2 pint group to the no alcohol, and then the 4 pint group to the 2 pint group (so it moves through the groups comparing each group to the one before). Again, this might be useful. We could also do a *Helmert* contrast, which compares each category against all subsequent categories, so in this case would compare the no alcohol group to the remaining categories (that is all of the groups that had some alcohol) and then would move on to the 2 pint category and compare this to the 4 pint category. Any of these would be fine, but they give us contrasts only for the main effects. In reality, most of the time we want contrasts for our interaction term, and they can be obtained only through syntax (oh well, looks like you might have to look at Oliver Twisted after all!).

To get contrasts for the main effect of alcohol click on Contrasts... in the main dialog box. We have used the contrasts dialog box before in section 11.4.4 and so refer back to that section to help you select a Helmert contrast for the alcohol variable. Once the contrasts have been selected (Figure 12.5), click on Continue to return to the main dialog box.

Univariate: Contrasts

Factors:

Gender(None)

Alcohol(Helmert)

Change Contrast

Contrast: Helmert | Change

Reference Category: Last | First

Continue | Cancel | Help

FIGURE 12.5 Defining contrasts in factorial ANOVA

12.3.4. *Post hoc* tests ②

The *post hoc* tests dialog box is obtained by clicking on Post Hoc... in the main dialog box (Figure 12.6). The variable **Gender** has only two levels and so we don't need to select *post hoc* tests for that variable (because any significant effects can reflect only the difference between males and females). However, there were three levels of the **Alcohol** variable (no alcohol, 2 pints and 4 pints); hence we can conduct *post hoc* tests (although remember that normally you would conduct contrasts *or post hoc* tests, not both). First, you should select the variable **Alcohol** from the box labelled *Factors* and transfer it to the box labelled *Post Hoc Tests for*. My recommendations for which *post hoc* procedures to use are in section 10.2.12 (and I don't want to repeat myself). Suffice to say you should select the ones in Figure 12.6! Click on Continue to return to the main dialog box.

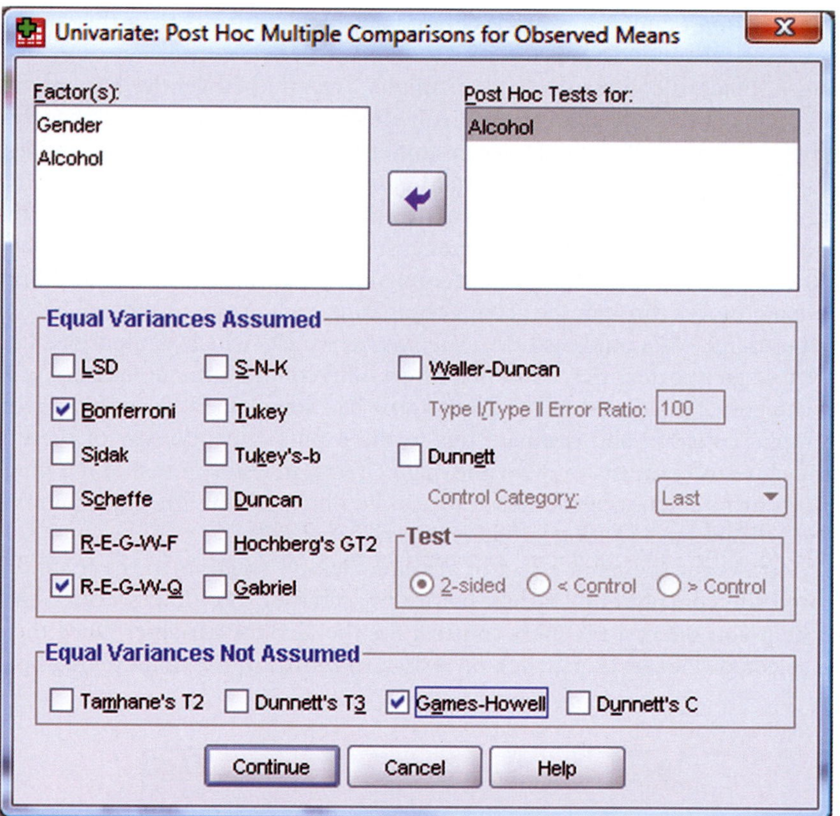

FIGURE 12.6 Dialog box for post hoc tests

12.3.5. Options ②

Click on Options... to activate the dialog box in Figure 12.7. The options for factorial ANOVA are fairly straightforward. First you can ask for some descriptive statistics, which will display a table of the means and standard deviations. This is a useful option to select because it assists in interpreting the final results. A vital option to select is the homogeneity of variance tests, which will produce Levene's test (section 5.6.1) to test the hypothesis

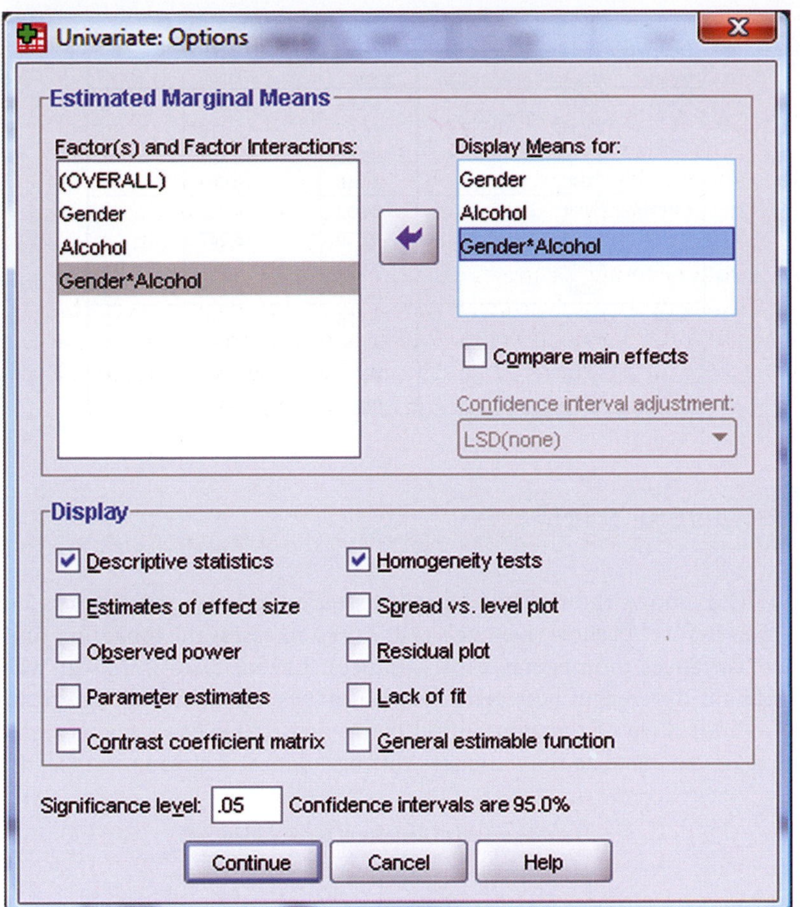

FIGURE 12.7 Dialog box for options

that the variances of each group are equal. You can also select ☑ Estimates of effect size if you want SPSS to calculate partial eta squared for you (see section 11.8). Once these options have been selected click on Continue to return to the main dialog box, then click on OK to run the analysis.

12.4. Output from factorial ANOVA ②

12.4.1. Output for the preliminary analysis ②

SPSS Output 12.1 shows the initial output from factorial ANOVA. This table of descriptive statistics is produced because we asked for descriptives in the options dialog box (see Figure 12.7) and it displays the means, standard deviations and number of participants in all conditions of the experiment. So, for example, we can see that in the no alcohol condition, males typically chatted up a female who was rated at about 67% on the attractiveness scale, whereas females selected a male who was rated as 61% on that scale. These means will be useful in interpreting the direction of any effects that emerge in the analysis.

SPSS OUTPUT 12.1

Descriptive Statistics

Dependent Variable:Attractiveness of Date

Gender	Alcohol Consumption	Mean	Std. Deviation	N
Male	None	66.88	10.329	8
	2 Pints	66.87	12.518	8
	4 Pints	35.62	10.836	8
	Total	56.46	18.503	24
Female	None	60.62	4.955	8
	2 Pints	62.50	6.547	8
	4 Pints	57.50	7.071	8
	Total	60.21	6.338	24
Total	None	63.75	8.466	16
	2 Pints	64.69	9.911	16
	4 Pints	46.56	14.343	16
	Total	58.33	13.812	48

12.4.2. Levene's test ②

SPSS Output 12.2 shows the results of Levene's test. We have come across Levene's test numerous times before! In short, Levene's test is used to assess the tenability of the assumption of equal variances (homogeneity of variance). Levene's test looks at whether there are any significant differences between group variances and so a non-significant result (as found here) is indicative of the assumption being met. If Levene's test is significant then steps can be taken to equalize the variances through data transformation (see Chapter 5).

SPSS OUTPUT 12.2

Levene's Test of Equality of Error Variances[a]

Dependent Variable:Attractiveness of Date

F	df1	df2	Sig.
1.527	5	42	.202

Tests the null hypothesis that the error variance of the dependent variable is equal across groups.

a. Design: Intercept + Gender + Alcohol + Gender * Alcohol

12.4.3. The main ANOVA table ②

SPSS Output 12.3 is the most important part of the output because it tells us whether any of the independent variables have had an effect on the dependent variable. The important things to look at in the table are the significance values of the independent variables. The first thing to notice is that there is a significant main effect of alcohol (because the significance value is less than .05). The *F*-ratio is highly significant, indicating that the amount of alcohol consumed significantly affected who the participant would try to chat up. This means that overall, when we ignore whether the participant was male or female, the amount of alcohol influenced their mate selection. The best way to see what this means is to look at a bar chart of the average attractiveness at each level of alcohol (ignore gender completely). This graph plots the means in SPSS Output 12.1 that we calculated in section 12.2.4.2.

SELF-TEST Plot error bar graphs of the main effects of alcohol and gender.

SPSS OUTPUT 12.3

Tests of Between-Subjects Effects

Dependent Variable:Attractiveness of Date

Source	Type III Sum of Squares	df	Mean Square	F	Sig.
Corrected Model	5479.167[a]	5	1095.833	13.197	.000
Intercept	163333.333	1	163333.333	1967.025	.000
Gender	168.750	1	168.750	2.032	.161
Alcohol	3332.292	2	1666.146	20.065	.000
Gender * Alcohol	1978.125	2	989.062	11.911	.000
Error	3487.500	42	83.036		
Total	172300.000	48			
Corrected Total	8966.667	47			

a. R Squared = .611 (Adjusted R Squared = .565)

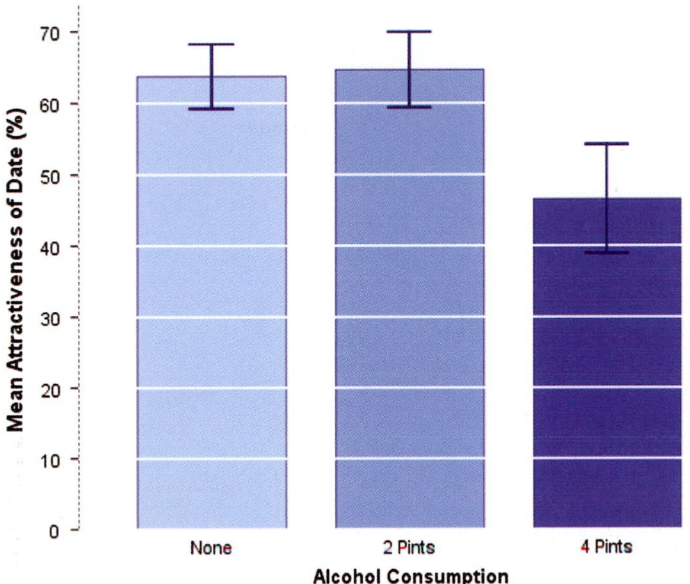

FIGURE 12.8 Graph showing the main effect of alcohol

Figure 12.8 clearly shows that when you ignore gender the overall attractiveness of the selected mate is very similar when no alcohol has been drunk and when 2 pints have been drunk (the means of these groups are approximately equal). Hence, this significant main effect is *likely* to reflect the drop in the attractiveness of the selected mates when 4 pints have been drunk. This finding seems to indicate that a person is willing to accept a less attractive mate after 4 pints.

The next part of SPSS Output 12.3 tells us about the main effect of gender. This time the *F*-ratio is not significant ($p = .161$, which is larger than .05). This effect means that overall, when we ignore how much alcohol had been drunk, the gender of the participant did not influence the attractiveness of the partner that the participant selected. In other words, other things being equal, males and females selected equally attractive mates. The bar chart (that you have hopefully produced from the self-test) of the average attractiveness of mates for men and women (ignoring how much alcohol had been consumed) reveals the meaning of this main effect. Figure 12.9 plots the means in SPSS Output 12.1 that we calculated in section 12.2.4.1. This graph shows that the average attractiveness of the partners of male and female participants was fairly similar (the means are different by only 4%). Therefore, this non-significant effect reflects the fact that the mean attractiveness was similar. We can conclude from this that, *other things being equal*, men and women chose equally attractive partners.

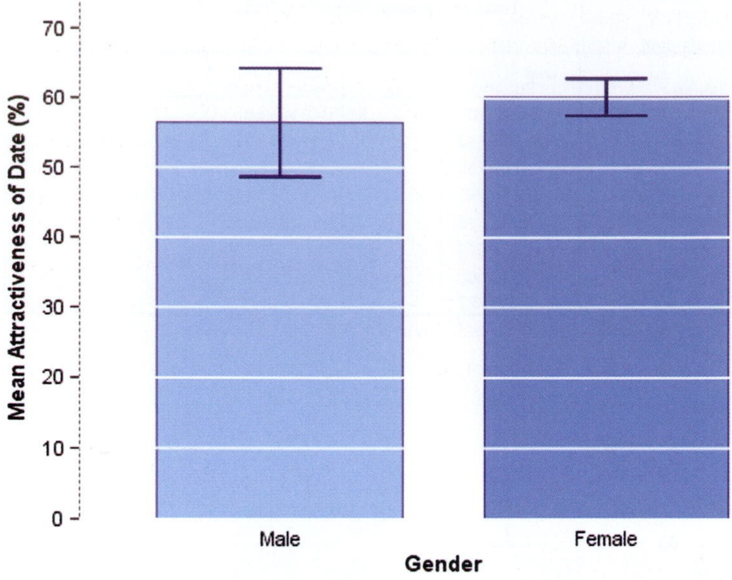

FIGURE 12.9 Graph to show the main effect of gender on mate selection

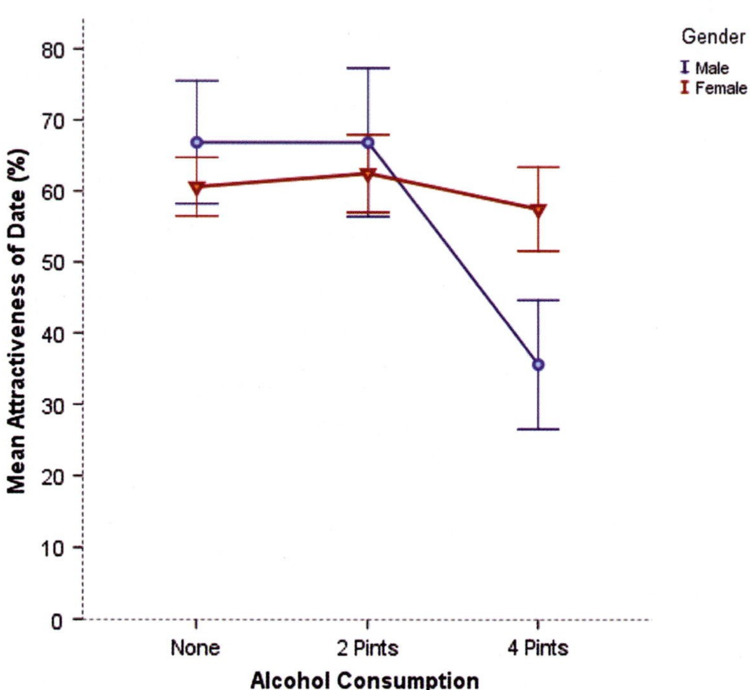

FIGURE 12.10 Graph of the interaction of gender and alcohol consumption in mate selection

Finally, SPSS Output 12.3 tells us about the interaction between the effect of gender and the effect of alcohol. The *F*-value is highly significant (because the *p*-value is less than .05). What this actually means is that the effect of alcohol on mate selection was different for male participants than it was for females. The SPSS output includes a plot that we asked for (see Figure 12.4) which tells us something about the nature of this interaction effect (see Figure 12.10 which is a nicer version of the graph in your output).

Figure 12.10 clearly shows that for women, alcohol has very little effect: the attractiveness of their selected partners is quite stable across the three conditions (as shown by the near-horizontal line). However, for the men, the attractiveness of their partners is stable when only a small amount has been drunk, but rapidly declines when more is drunk. Non-parallel lines usually indicate a significant interaction effect. In this particular graph the lines actually cross, which indicates a fairly large interaction between independent variables. The interaction tells us that alcohol has little effect on mate selection until 4 pints have been drunk and that the effect of alcohol is prevalent only in male participants. In short, the results show that women maintain high standards in their mate selection regardless of alcohol, whereas men have a few beers and then try to get off with anything on legs! One interesting point that these data demonstrate is that we earlier concluded that alcohol significantly affected how attractive a mate was selected (the **Alcohol** main effect); however, the interaction effect tells us that this is true only in males (females appear unaffected). This shows how misleading main effects can be: it is usually the interactions between variables that are most interesting in a factorial design.

How do I interpret interactions?

12.4.4. Contrasts ②

SPSS Output 12.4 shows the results of our Helmert contrast on the effect of alcohol. This helps us to break down the effect of alcohol. The top of the table shows the contrast for *Level 1 vs. Later*, which in this case means the no alcohol group compared to the two alcohol groups. This tests whether the mean of the no alcohol group (63.75) is different to the mean of the 2 pint and 4 pint groups combined ((64.69 + 46.56)/2 = 55.625). This is a difference of 8.125 (63.75 – 55.63), which we're told by both the *Contrast Estimate* and the *Difference* in the table. The important thing to look at is the value of *Sig.*, which tells us if this difference is significant. It is, because *Sig.* is .006, which is smaller than .05. We're also told the confidence interval for this difference and because it doesn't cross zero we can be safe in the knowledge that, assuming this sample is one of the 95 out of 100 that produces a confidence interval containing the true value of the difference, the real difference is more than zero (between 2.49 and 13.76 to be precise). So we could conclude that the effect of alcohol is that any amount of alcohol reduces the attractiveness of the dates selected compared to when no alcohol is drunk. Of course this is misleading because, in fact, the means for the no alcohol and 2 pint groups are fairly similar (63.75 and 64.69), so 2 pints of alcohol don't reduce the attractiveness of selected dates! The reason why the comparison is significant is because it's testing the combined effect of 2 and 4 pints, and because 4 pints has such a drastic effect it drags down the overall mean. This shows why you need to be careful about how you interpret these contrasts: you need to have a look at the remaining contrast as well.

The bottom of the table shows the contrast for *Level 2 vs. Level 3*, which in this case means the 2 pint group compared to the 4 pint group. This tests whether the mean of the 2 pint group (64.69) is different to the mean of the 4 pint groups combined (46.56). This is a difference of 18.13 (64.69 – 46.56), which we're told by both the *Contrast Estimate* and the *Difference* in the table. Again, the important thing to look at is the value of *Sig.*, which tells us if this difference is significant. It is, because *Sig.* is .000, which is smaller than .05. We're also told the confidence interval for this difference and because it doesn't cross zero we can be safe in the knowledge that, assuming this confidence interval is one of the 95 out of 100 that contains the true value of the difference, the real difference is more than zero (between 11.62 and 24.63 to be precise). This tells us that having 4 pints significantly reduced the attractiveness of selected dates compared to having only 2 pints.

SPSS OUTPUT 12.4

Contrast Results (K Matrix)

Alcohol Consumption Helmert Contrast			Dependent Variable Attractiveness of Date
Level 1 vs. Later	Contrast Estimate		8.125
	Hypothesized Value		0
	Difference (Estimate - Hypothesized)		8.125
	Std. Error		2.790
	Sig.		.006
	95% Confidence Interval for Difference	Lower Bound	2.494
		Upper Bound	13.756
Level 2 vs. Level 3	Contrast Estimate		18.125
	Hypothesized Value		0
	Difference (Estimate - Hypothesized)		18.125
	Std. Error		3.222
	Sig.		.000
	95% Confidence Interval for Difference	Lower Bound	11.623
		Upper Bound	24.627

12.4.5. Simple effects analysis ③

One popular way to break down an interaction term is to use a technique called '**simple effects analysis**.' This analysis basically looks at the effect of one independent variable at individual levels of the other independent variable. So, for example, in our beer-goggles data we could do a simple effects analysis looking at the effect of gender at each level of alcohol. This would mean taking the average attractiveness of the date selected by men and comparing it to that for women after no drinks, then making the same comparison for 2 pints and then finally for 4 pints. Another way of looking at this is to say we would compare each triangle to the corresponding circle in Figure 12.10: based on the graph we might expect to find no difference after no alcohol and after 2 pints (in both cases the triangle and circle are located in about the same position) but we would expect a difference after 4 pints (because the circle and triangle are quite far apart). The alternative way to do it would be to compare the mean attractiveness after the none, 2 pints and 4 pints for men and then in a separate analysis do the same but for women. (This would be a bit like doing a one-way ANOVA on the effect of alcohol in men, and then doing a different one-way ANOVA for the effect of alcohol in women.) These analyses can't be run through the usual dialog boxes, but they can be run using syntax – see SPSS Tip 12.1.

OLIVER TWISTED

Please, Sir, can I have some more … simple effects?

'I want to impress my friends by doing a simple effects analysis by hand' boasts Oliver. You don't really need to know how simple effects analyses are calculated to run them, Oliver, but seeing as you asked it is explained in the additional material available from the companion website.

12.4.6. *Post hoc* analysis ②

The *post hoc* tests (SPSS Output 12.5) break down the main effect of alcohol and can be interpreted as if a one-way ANOVA had been conducted on the **Alcohol** variable (i.e. the reported effects for alcohol are collapsed with regard to gender). The Bonferroni and Games–Howell tests show the same pattern of results: when participants had drunk no alcohol or 2 pints of alcohol, they selected equally attractive mates. However, after 4 pints had been consumed, participants selected significantly less attractive mates than after both 2 pints ($p < .001$) and no alcohol ($p < .001$). It is interesting to note that the mean attractiveness of partners after no alcohol and 2 pints were so similar that the probability of the obtained difference between those means is 1 (i.e. completely probable!). The REGWQ test confirms that the means of the placebo and 2 pint conditions were equal, whereas the mean of the 4 pint group was different. It should be noted that these *post hoc* tests ignore the interactive effect of gender and alcohol.

In summary, we should conclude that alcohol has an effect on the attractiveness of selected mates. Overall, after a relatively small dose of alcohol (2 pints) humans are still in control of their judgements and the attractiveness levels of chosen partners are consistent with a control group (no alcohol consumed). However, after a greater dose of alcohol, the attractiveness of chosen mates decreases significantly. This effect is what is referred to as the 'beer-goggles effect'! More interesting, the interaction shows a gender difference in the beer-goggles effect. Specifically, it looks as though men are significantly more likely to pick less attractive mates when drunk. Women, in comparison, manage to maintain their standards despite being drunk. What we still don't know is whether women will become susceptible to the beer-goggles effect at higher doses of alcohol.

SPSS OUTPUT 12.5

Multiple Comparisons

Dependent Variable:Attractiveness of Date

	(I) Alcohol Consumption	(J) Alcohol Consumption	Mean Difference (I-J)	Std. Error	Sig.	95% Confidence Interval Lower Bound	95% Confidence Interval Upper Bound
Bonferroni	None	2 Pints	-.94	3.222	1.000	-8.97	7.10
		4 Pints	17.19*	3.222	.000	9.15	25.22
	2 Pints	None	.94	3.222	1.000	-7.10	8.97
		4 Pints	18.13*	3.222	.000	10.09	26.16
	4 Pints	None	-17.19*	3.222	.000	-25.22	-9.15
		2 Pints	-18.13*	3.222	.000	-26.16	-10.09
Games-Howell	None	2 Pints	-.94	3.259	.955	-8.98	7.11
		4 Pints	17.19*	4.164	.001	6.80	27.58
	2 Pints	None	.94	3.259	.955	-7.11	8.98
		4 Pints	18.13*	4.359	.001	7.31	28.94
	4 Pints	None	-17.19*	4.164	.001	-27.58	-6.80
		2 Pints	-18.13*	4.359	.001	-28.94	-7.31

Based on observed means.
The error term is Mean Square(Error) = 83.036.

*. The mean difference is significant at the .05 level.

Attractiveness of Date

	Alcohol Consumption	N	Subset 1	Subset 2
Ryan-Einot-Gabriel-Welsch Range[a]	4 Pints	16	46.56	
	None	16		63.75
	2 Pints	16		64.69
	Sig.		1.000	.772

Means for groups in homogeneous subsets are displayed.
Based on observed means.
The error term is Mean Square(Error) = 83.036.

a. Alpha = .05.

CRAMMING SAM'S TIPS

- Two-way independent ANOVA compares several means when there are two independent variables and different participants have been used in all experimental conditions. For example, if you wanted to know whether different teaching methods worked better for different subjects, you could take students from four courses (Psychology, Geography, Management and Statistics) and assign them to either lecture-based or book-based teaching. The two variables are course and method of teaching. The outcome might be the end of year mark (as a percentage).
- Test for homogeneity of variance using *Levene's test*. Find the table with this label: if the value in the column labelled *Sig.* is less than .05 then the assumption is violated.
- In the table labelled **Tests of Between-Subjects Effects**, look at the column labelled *Sig.* for all three of your effects: there should be a main effect of each variable and an effect of the interaction between the two variables; if the value is less than .05 then the effect is significant. For main effects consult *post hoc* tests to see which groups differ, and for the interaction look at an interaction graph or conduct simple effects analysis.
- For *post hoc* tests, again look at the columns labelled *Sig.* to discover if your comparisons are significant (they will be if the significance value is less than .05).
- Test the same assumptions as for one way independent ANOVA (see Chapter 10).

SPSS TIP 12.1 Simple effects analysis on SPSS ③

Unfortunately, simple effects analyses can't be done through the dialog boxes and instead you have to use SPSS syntax (see section 3.7. to remind yourself about the syntax window). The syntax you need to use in this example is:

```
GLM Attractiveness by gender alcohol
/emmeans = tables(gender*alcohol)compare(gender).
```

*This initiates the ANOVA by specifying the outcome or dependent variable (**Attractiveness**) and then the BY command is followed by our independent variables (**Gender** and **Alcohol**). The line beginning 'emmeans' specifies the simple effects. For example, 'compare(Gender)' will look at the effect of gender at each level of alcohol. This syntax for looking at the effect of gender at different levels of alcohol is stored in a file called **GogglesSimpleEffects.sps** for you to look at should you not wish to go to the effort of typing the two lines above. Open this file (make sure you also have **Goggles.sav** loaded into the data editor) and run the syntax. The

output you get will be the same as for the main analysis in the chapter but will contain an extra table at the end containing the simple effects:

Univariate Tests

Dependent Variable: Attractiveness of Date

Alcohol Consumption		Sum of Squares	df	Mean Square	F	Sig.
None	Contrast	156.250	1	156.250	1.882	.177
	Error	3487.500	42	83.036		
2 Pints	Contrast	76.562	1	76.562	.922	.342
	Error	3487.500	42	83.036		
4 Pints	Contrast	1914.062	1	1914.062	23.051	.000
	Error	3487.500	42	83.036		

Each F tests the simple effects of Gender within each level combination of the other effects shown. These tests are based on the linearly independent pairwise comparisons among the estimated marginal means.

Looking at the significance values for each simple effect, it appears that there was no significant difference between men and women at level 1 of alcohol (i.e. no alcohol), $p = .18$, or at level 2 of alcohol (2 pints), $p = .34$, but there was a very significant difference ($p = .001$) at level 3 of alcohol (which judging from the graph reflects the fact that the mean for men is considerably lower than for women).

12.5. Interpreting interaction graphs ②

We've already had a look at one interaction graph when we interpreted the analysis in this chapter. However, interactions are very important, and the key to understanding them is being able to interpret interaction graphs. In the example in this chapter we used Figure 12.10 to conclude that the interaction probably reflected the fact that men and women chose equally attractive dates after no alcohol and 2 pints, but that at 4 pints men's standards dropped significantly more than women's. Imagine we'd got the profile of results shown in Figure 12.11; do you think we would've still got a significant interaction effect?

This profile of data probably would also give rise to a significant interaction term because, although the attractiveness of men and women's dates are similar after no alcohol and 4 pints of alcohol, there is a big difference after 2 pints. This reflects a scenario in which the beer-goggles effect is equally big in men and women after 4 pints (and doesn't exist after no alcohol) but kicks in quicker for men: the attractiveness of their dates plummets after 2 pints, whereas women maintain their standards until 4 pints (at which point they'd happily date an unwashed skunk). Let's try another example. Is there a significant interaction in Figure 12.12?

For the data in Figure 12.12 there is unlikely to be a significant interaction because the effect of alcohol is the same for men and women. So, for both men and women, the attractiveness of their dates after no alcohol is quite high, but after 2 pints all types drop by a similar amount (the slope of the male and female lines is about the same). After 4 pints

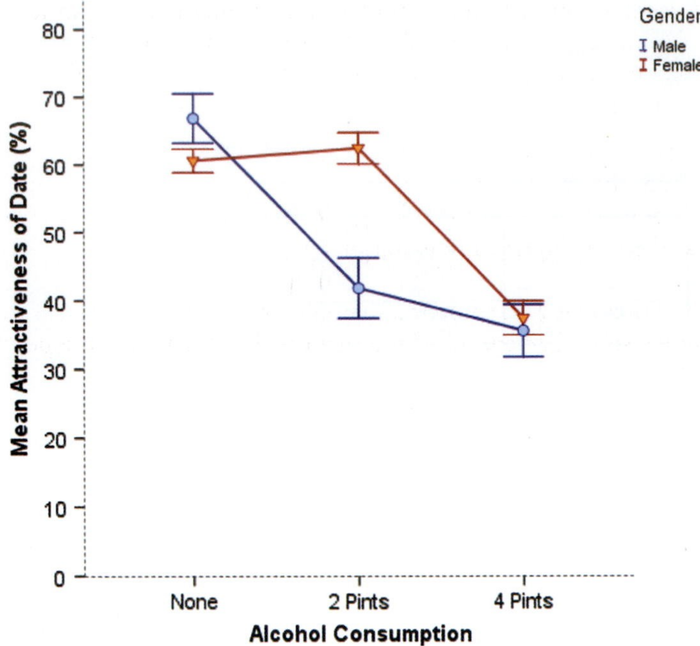

FIGURE 12.11 Another interaction graph

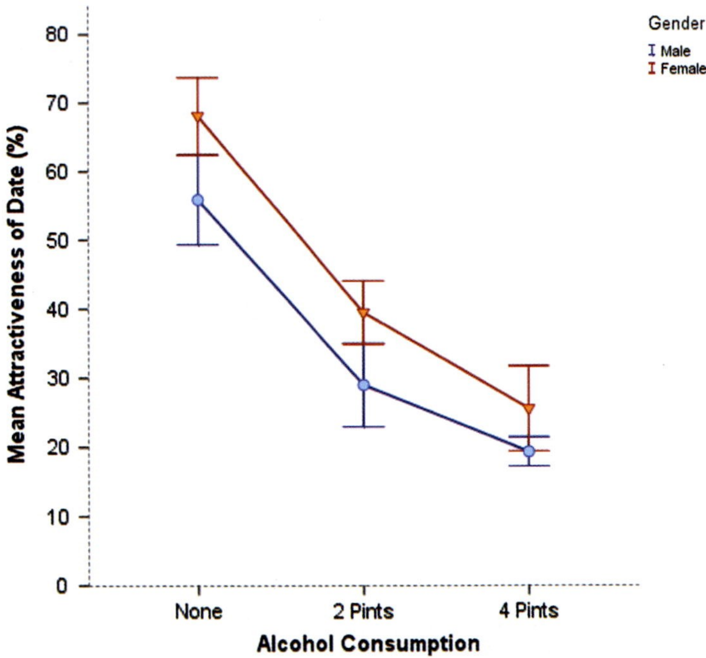

FIGURE 12.12 A 'lack of' interaction graph

there is a further drop and, again, this drop is about the same in men and women (the lines again slope at about the same angle). The fact that the line for males is lower than for females just reflects the fact that across all conditions, men have lower standards than their female counterparts: this reflects a main effect of gender (i.e. males generally chose less attractive dates than females at all levels of alcohol). Two general points that we can make from these examples are that:

- Significant interactions are shown up by non-parallel lines on an interaction graph. However, it's important to remember that this doesn't mean that non-parallel lines automatically mean that the interaction is significant: whether the interaction is significant will depend on the degree to which the lines are not parallel!
- If the lines on an interaction graph cross then obviously they are not parallel and this can be a dead giveaway that you have a possible significant interaction. However, contrary to popular belief it isn't *always* the case that if the lines of the interaction graph cross then the interaction is significant.

A further complication is that sometimes people draw bar charts rather than line charts. Figure 12.13 shows some bar charts of interactions between two independent variables.

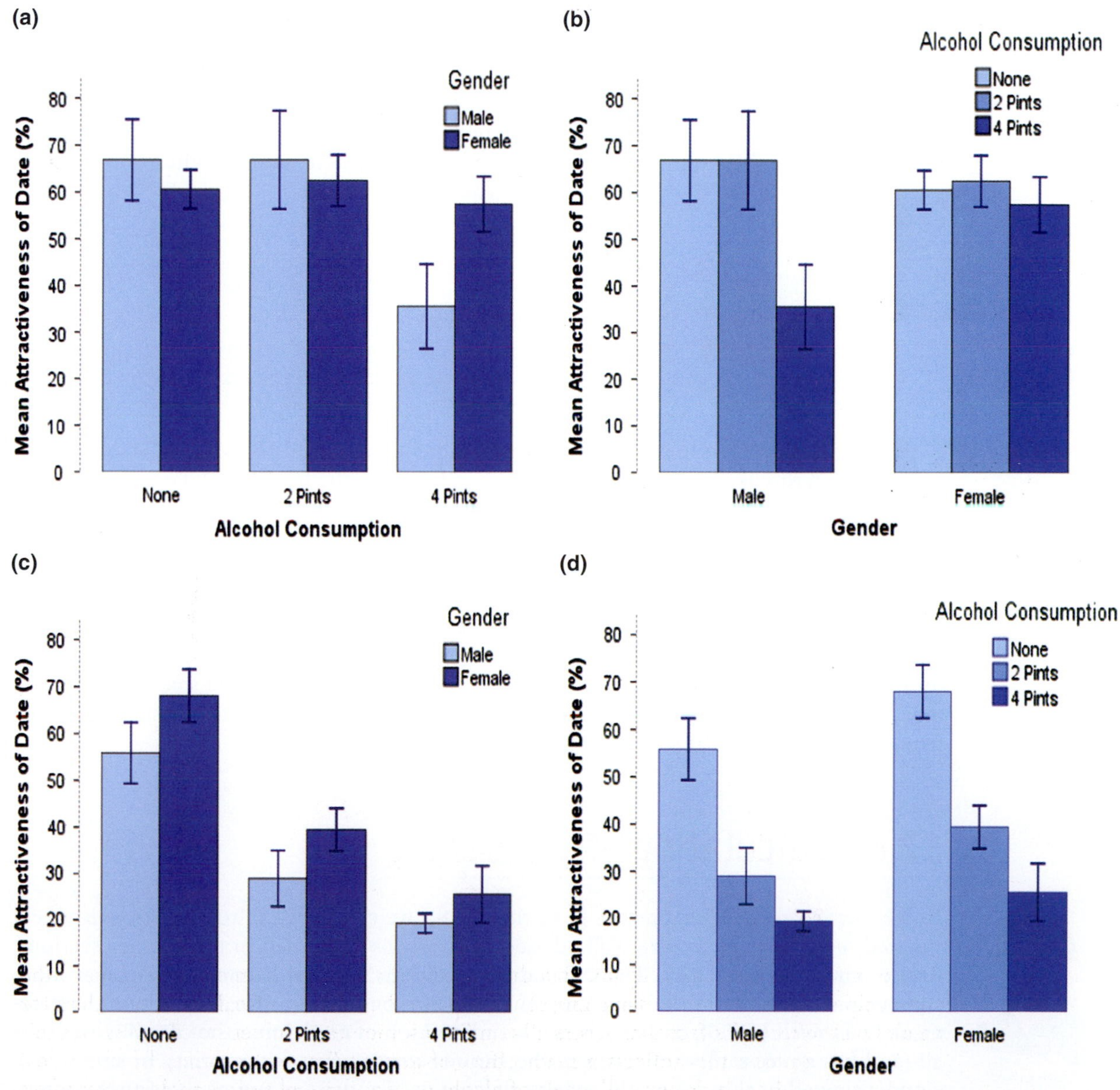

FIGURE 12.13 Bar charts showing interactions between two variables

Panels (a) and (b) actually display the data from the example used in this chapter (in fact, why not have a go at plotting them!). As you can see, there are two ways to present the same data: panel (a) shows the data when levels of alcohol are placed along the x-axis and different-coloured bars are used to show means for males and females, and panel (b) shows the opposite scenario where gender is plotted on the x-axis and different colours distinguish the dose of alcohol. Both of these graphs show an interaction effect. What you're looking for is for the differences between coloured bars to be different at different points along the x-axis. So, for panel (a) you'd look at the difference between the light- and dark-blue bars for no alcohol, and then look to 2 pints and ask, 'Is the difference between the bars different to when I looked at no alcohol?' In this case the dark- and light-blue bars look the same at no alcohol as they do at 2 pints: hence, no interaction. However, we'd then move on to look at 4 pints, and we'd again ask, 'Is the difference between the light- and dark-blue bars different to what it has been in any of the other conditions?' In this case the answer is yes: for no alcohol and 2 pints, the light- and dark-blue bars were about the same height, but at 4 pints the dark-blue bar is much higher than the light one. This shows an interaction: the pattern of responses changes at 4 pints. Panel (b) shows the same thing but plotted the other way around. Again we look at the pattern of responses. So, first we look at the men and see that the pattern is that the first two bars are the same height, but the last bar is much shorter. The interaction effect is shown up by the fact that for the women there is a different pattern: all three bars are about the same height.

SELF-TEST What about panels (c) and (d): do you think there is an interaction?

Again, they display the same data in two different ways, but it's different data to what we've used in this chapter. First let's look at panel (c): for the no alcohol data, the dark bar is a little bit bigger than the light one; moving on to the 2 pint data the dark bar is also a little bit taller than the light bar; and finally for the 4 pint data the dark bar is again higher than the light one. In all conditions the same pattern is shown – the dark-blue bar is a bit higher than the light-blue one (i.e. females pick more attractive dates than men regardless of alcohol consumption) – therefore, there is no interaction. Looking at panel (d) we see a similar result. For men, the pattern is that attractiveness ratings fall as more alcohol is drunk (the bars decrease in height) and then for the women we see the same pattern: ratings fall as more is drunk. This again is indicative of no interaction: the change in attractiveness due to alcohol is similar in men and women.

12.6. Calculating effect sizes ③

SMART ALEX ONLY

As we saw in previous chapters (e.g. section 11.8), we can get SPSS to produce partial eta squared, η^2. However, you're well advised, for reasons explained in these other sections, to use omega squared (ω^2). The calculation of omega squared becomes somewhat more cumbersome in factorial designs ('somewhat' being one of my characteristic understatements!). Howell (2006), as ever, does a wonderful job of explaining the complexities of it all (and has a nice table summarizing the various components for a variety of situations). Condensing all of this down, I'll just say that we need to first compute a variance component for each of the effects (the two main effects and the interaction term) and the error, and then use these to calculate effect sizes for each. If we call the first main effect A, the second main effect B and the interaction effect A × B, then the variance components for each of these is based on the mean squares of each effect and the sample sizes on which they're based:

$$\hat{\sigma}_{\alpha}^2 = \frac{(a-1)(\text{MS}_\text{A} - \text{MS}_\text{R})}{nab}$$

$$\hat{\sigma}_{\beta}^2 = \frac{(b-1)(\text{MS}_\text{B} - \text{MS}_\text{R})}{nab}$$

$$\hat{\sigma}_{\alpha\beta}^2 = \frac{(a-1)(b-1)(\text{MS}_{\text{A}\times\text{B}} - \text{MS}_\text{R})}{nab}$$

In these equations, a is the number of levels of the first independent variable, b is the number of levels of the second independent variable and n is the number of people per condition. Let's calculate these for our data. We need to look at SPSS Output 12.3 to find out the mean squares for each effect, and for the error term. Our first independent variable was alcohol. This had three levels (hence $a = 3$) and had a mean squares of 1666.146. Our second independent variable was gender, which had two levels (hence $b = 2$) and a mean squares of 168.75. The number of people in each group was 8 and the residual mean squares were 83.036. Therefore, our equations become:

$$\hat{\sigma}_{\alpha}^2 = \frac{(3-1)(1666.146 - 83.036)}{8 \times 3 \times 2} = 65.96$$

$$\hat{\sigma}_{\beta}^2 = \frac{(2-1)(168.75 - 83.036)}{8 \times 3 \times 2} = 1.79$$

$$\hat{\sigma}_{\alpha\beta}^2 = \frac{(3-1)(2-1)(989.062 - 83.036)}{8 \times 3 \times 2} = 37.75$$

We also need to estimate the total variability and this is just the sum of these other variables plus the residual mean squares:

$$\begin{aligned}\hat{\sigma}_{\text{total}}^2 &= \hat{\sigma}_{\alpha}^2 + \hat{\sigma}_{\beta}^2 + \hat{\sigma}_{\alpha\beta}^2 + \text{MS}_\text{R} \\ &= 65.96 + 1.79 + 37.75 + 83.04 \\ &= 188.54\end{aligned}$$

The effect size is then simply the variance estimate for the effect in which you're interested divided by the total variance estimate:

$$\omega_{\text{effect}}^2 = \frac{\hat{\sigma}_{\text{effect}}^2}{\hat{\sigma}_{\text{total}}^2}$$

As such, for the main effect of alcohol we get:

$$\omega_{\text{alcohol}}^2 = \frac{\hat{\sigma}_{\text{alcohol}}^2}{\hat{\sigma}_{\text{total}}^2} = \frac{65.96}{188.54} = 0.35$$

For the main effect of gender we get:

$$\omega^2_{\text{gender}} = \frac{\hat{\sigma}^2_{\text{gender}}}{\hat{\sigma}^2_{\text{total}}} = \frac{1.79}{188.54} = .009$$

For the interaction of gender and alcohol we get:

$$\omega^2_{\text{alcohol} \times \text{gender}} = \frac{\hat{\sigma}^2_{\text{alcohol} \times \text{gender}}}{\hat{\sigma}^2_{\text{total}}} = \frac{37.75}{188.54} = .20$$

To make these values comparable to *r* we can take the square root, which gives us effect sizes of .59 for alcohol, .09 for gender and .45 for the interaction term. As such, the effects of alcohol and the interaction are fairly large, but the effect of gender, which was non-significant in the main analysis, is very small indeed (close to zero in fact). It's also possible to calculate effect sizes for our simple effects analysis (if you read section 12.4.5). These effects have 1 degree of freedom for the model (which means they're comparing only two things) and in these situations *F* can be converted to *r* using the following equation (which just uses the *F*-ratio and the residual degrees of freedom):[1]

$$r = \sqrt{\frac{F(1, df_{\text{R}})}{F(1, df_{\text{R}}) + df_{\text{R}}}}$$

Looking at SPSS Tip 12.1, we can see that we got *F*-ratios of 1.88, .92 and 23.05 for the effects of gender at no alcohol, 2 pints and 4 pints respectively. For each of these, the degrees of freedom were 1 for the model and 42 for the residual. Therefore, we get the following effect sizes:

$$r_{\text{Gender(No Alcohol)}} = \sqrt{\frac{1.88}{1.88 + 42}} = 0.21$$

$$r_{\text{Gender(2 Pints)}} = \sqrt{\frac{0.92}{0.92 + 42}} = 0.15$$

$$r_{\text{Gender(4 Pints)}} = \sqrt{\frac{23.05}{23.05 + 42}} = 0.6$$

EVERYBODY

Therefore, the effect of gender is very small at both no alcohol and 2 pints, but becomes large at 4 pints of alcohol.

12.7. Reporting the results of two-way ANOVA ②

As with the other ANOVAs we've encountered, we have to report the details of the *F*-ratio and the degrees of freedom from which it was calculated. For the various effects in these data the *F*-ratio will be based on different degrees of freedom: it was derived from dividing

[1] If your *F* compares more than two things then a different equation is needed (see Rosenthal et al. (2000: 44), but I prefer to try to keep effect sizes to situations in which only two things are being compared because interpretation is easier.

the mean squares for the effect by the mean squares for the residual. For the effects of alcohol and the alcohol × gender interaction, the model degrees of freedom were 2 ($df_M = 2$), but for the effect of gender the degrees of freedom were only 1 ($df_M = 1$). For all effects, the degrees of freedom for the residuals were 42 ($df_R = 42$). We can, therefore, report the three effects from this analysis as follows:

- ✓ There was a significant main effect of the amount of alcohol consumed at the night-club, on the attractiveness of the mate they selected, $F(2, 42) = 20.07, p < .001, \omega^2 = .35$. The Games–Howell *post hoc* test revealed that the attractiveness of selected dates was significantly lower after 4 pints than both after 2 pints and no alcohol (both *ps* < .001). The attractiveness of dates after 2 pints and no alcohol were not significantly different.
- ✓ There was a non-significant main effect of gender on the attractiveness of selected mates, $F(1, 42) = 2.03, p = .161$ $\omega^2 = .009$.
- ✓ There was a significant interaction effect between the amount of alcohol consumed and the gender of the person selecting a mate, on the attractiveness of the partner selected, $F(2, 42) = 11.91, p < .001, \omega^2 = .20$. This indicates that male and female genders were affected differently by alcohol. Specifically, the attractiveness of partners was similar in males ($M = 66.88$, SD = 10.33) and females ($M = 60.63$, SD = 4.96) after no alcohol; the attractiveness of partners was also similar in males ($M = 66.88$, SD = 12.52) and females ($M = 62.50$, SD = 6.55) after 2 pints; however, attractiveness of partners selected by males ($M = 35.63$, SD = 10.84) was significantly lower than those selected by females ($M = 57.50$, SD = 7.07) after 4 pints.

LABCOAT LENI'S REAL RESEARCH 12.1

Don't forget your toothbrush? ②

We have all experienced that feeling after we have left the house of wondering whether we locked the door, or if we remembered to close the window, or if we remembered to remove the bodies from the fridge in case the police turn up. This behaviour is normal; however, people with obsessive compulsive disorder (OCD) tend to check things excessively. They might, for example, check whether they have locked the door so often that it takes them an hour to leave their house. It is a very debilitating problem.

One theory of this checking behaviour in OCD suggests that it is caused by a combination of the mood you are in (positive or negative) interacting with the rules you use to decide when to stop a task (do you continue until you feel like stopping, or until you have done the task as best as you can?). Davey, Startup, Zara, MacDonald & Field (2003) tested this hypothesis by inducing a negative, positive or no mood in different people and then asking them to imagine that they were going on holiday and to generate as many things as they could that they should check before they went away. Within each mood group, half of the participants were instructed to generate as many items as they could (known as an 'as many as can' stop rule), whereas the remainder were asked to generate items for as long as they felt like continuing the task (known as a 'feel like continuing' stop rule). The data are in the file **Davey2003.sav**.

Davey et al. hypothesized that people in negative moods, using an 'as many as can' stop rule would generate more items than those using a 'feel like continuing' stop rule. Conversely, people in a positive mood would generate more items when using a 'feel like continuing' stop rule compared to 'an as many as can' stop rule. Finally, in neutral moods, the stop rule used shouldn't affect the number of items generated. Draw an error bar chart of the data and then conduct the appropriate analysis to test Davey et al.'s hypotheses.

Answers are in the additional material on the companion website (or look at pages 148–149 in the original article).

DAVEY, G. C. L. ET AL. (2003). *JOURNAL OF BEHAVIOR THERAPY & EXPERIMENTAL PSYCHIATRY, 34*, 141–160.

12.8. Factorial ANOVA as regression ③

We saw in section 10.2.3 that one-way ANOVA could be conceptualized as a regression equation (a general linear model). In this section we'll consider how we extend this linear model to incorporate two independent variables. To keep things as simple as possible I just want you to imagine that we have only two levels of the alcohol variable in our example (none and 4 pints). As such, we have two variables, each with two levels. All of the general linear models we've considered in this book take the general form of:

$$\text{outcome}_i = (\text{model}) + \text{error}_i$$

For example, when we encountered multiple regression in Chapter 7 we saw that this model was written as (see equation (7.9)):

$$Y_i = (b_0 + b_1X_{1i} + b_2X_{2i} + \ldots + b_nX_{ni}) + \varepsilon_i$$

Also, when we came across one-way ANOVA, we adapted this regression model to conceptualize our Viagra example, as (see equation (10.2)):

$$\text{Libido}_i = (b_0 + b_2\text{High}_i + b_2\text{Low}_i) + \varepsilon_i$$

In this model, the High and Low variables were dummy variables (i.e. variables that can take only values of 0 or 1). In our current example, we have two variables: gender (male or female) and alcohol (none and 4 pints). We can code each of these with zeros and ones: for example, we could code gender as male = 0, female = 1; and we could code the alcohol variable as 0 = none, 1 = 4 pints. We could then directly copy the model we had in one-way ANOVA:

$$\text{Attractiveness}_i = (b_0 + b_1\text{Gender}_i + b_2\text{Alcohol}_i) + \varepsilon_i$$

Now the astute among you might say, 'Where has the interaction term gone?' Well, of course, we have to include this too, and so the model simply extends to become (first expressed generally and then in terms of this specific example):

$$\begin{aligned}\text{Attractive}_i &= (b_0 + b_1\text{A}_i + b_2\text{B}_i + b_3\text{AB}_i) + \varepsilon_i \\ \text{Attractive}_i &= (b_0 + b_1\text{Gender}_i + b_2\text{Alcohol}_i + b_3\text{Interaction}_i) + \varepsilon_i\end{aligned} \quad (12.1)$$

The question is: how do we code the interaction term? The interaction term represents the combined effect of alcohol and gender and in fact to get any interaction term in regression you simply multiply the variables involved in that interaction term. This is why you see interaction terms written as gender × alcohol, because in regression terms the interaction variable literally is the two variables multiplied by each other. Table 12.2 shows the resulting variables for the regression (note that the interaction variable is simply the value of the gender dummy variable multiplied by the value of the alcohol dummy variable). So, for example, a male receiving 4 pints of alcohol would have a value of 0 for the gender variable, 1 for the alcohol variable and 0 for the interaction variable. The group means for the various combinations of gender and alcohol are also included because they'll come in useful in due course.

To work out what the b-values represent in this model we can do the same as we did for the t-test and one-way ANOVA; that is, look at what happens when we insert values of our predictors (gender and alcohol)! To begin with, let's see what happens when we look at men who had no alcohol. In this case, the value of gender is 0, the value of alcohol is 0 and the value of the interaction is also 0. The outcome we predict (as with one-way ANOVA) is the mean of this group (66.875), so our model becomes:

TABLE 12.2 Coding scheme for factorial ANOVA

Gender	Alcohol	Dummy (Gender)	Dummy (Alcohol)	Interaction	Mean
Male	None	0	0	0	66.875
Male	4 Pints	0	1	0	35.625
Female	None	1	0	0	60.625
Female	4 Pints	1	1	1	57.500

$$\text{Attractive}_i = (b_0 + b_1\text{Gender}_i + b_2\text{Alcohol}_i + b_3\text{Interaction}_i) + \varepsilon_i$$
$$\overline{X}_{\text{Men, None}} = b_0 + (b_1 \times 0) + (b_2 \times 0) + (b_3 \times 0)$$
$$b_0 = \overline{X}_{\text{Men, None}}$$
$$b_0 = 66.875$$

So, the constant b_0 in the model represents the mean of the group for which all variables are coded as 0. As such it's the mean value of the base category (in this case men who had no alcohol). Now, let's see what happens when we look at females who had no alcohol. In this case, the gender variable is 1 and the alcohol and interaction variables are still 0. Also remember that b_0 is the mean of the men who had no alcohol. The outcome is the mean for women who had no alcohol. Therefore, the equation becomes:

$$\overline{X}_{\text{Women, None}} = b_0 + (b_1 \times 1) + (b_2 \times 0) + (b_3 \times 0)$$
$$\overline{X}_{\text{Women, None}} = b_0 + b_1$$
$$\overline{X}_{\text{Women, None}} = \overline{X}_{\text{Men, None}} + b_1$$
$$b_1 = \overline{X}_{\text{Women, None}} - \overline{X}_{\text{Men, None}}$$
$$b_1 = 60.625 - 66.875$$
$$b_1 = -6.25$$

So, b_1 in the model represents the difference between men and women who had no alcohol. More generally we can say it's the effect of gender for the base category of alcohol (the base category being the one coded with 0, in this case no alcohol). Now let's look at males who had 4 pints of alcohol. In this case, the gender variable is 0, the alcohol variable is 1 and the interaction variable is still 0. We can also replace b_0 with the mean of the men who had no alcohol. The outcome is the mean for men who had 4 pints. Therefore, the equation becomes:

$$\overline{X}_{\text{Men, 4 Pints}} = b_0 + (b_1 \times 0) + (b_2 \times 1) + (b_3 \times 0)$$
$$\overline{X}_{\text{Men, 4 Pints}} = b_0 + b_2$$
$$\overline{X}_{\text{Men, 4 Pints}} = \overline{X}_{\text{Men, None}} + b_2$$
$$b_2 = \overline{X}_{\text{Men, 4 Pints}} - \overline{X}_{\text{Men, None}}$$
$$b_2 = 35.625 - 66.875$$
$$b_2 = -31.25$$

So, b_2 in the model represents the difference between having no alcohol and 4 pints in men. Put more generally, it's the effect of alcohol in the base category of gender (i.e. the category of gender that was coded with 0, in this case men). Finally, we can look at females who had 4 pints of alcohol. In this case, the gender variable is 1, the alcohol variable is 1 and the interaction variable is also 1. We can also replace b_0, b_1, and b_2, with what we now know they represent. The outcome is the mean for women who had 4 pints. Therefore, the equation becomes:

$$\overline{X}_{\text{Women, 4 Pints}} = b_0 + (b_1 \times 1) + (b_2 \times 1) + (b_3 \times 1)$$

$$\overline{X}_{\text{Women, 4 Pints}} = b_0 + b_1 + b_2 + b_3$$

$$\overline{X}_{\text{Women, 4 Pints}} = \overline{X}_{\text{Men, None}} + \left(\overline{X}_{\text{Women, None}} - \overline{X}_{\text{Men, None}}\right) + \left(\overline{X}_{\text{Men, 4 Pints}} - \overline{X}_{\text{Men, None}}\right) + b_3$$

$$\overline{X}_{\text{Women, 4 Pints}} = \overline{X}_{\text{Women, None}} + \overline{X}_{\text{Men, 4 Pints}} - \overline{X}_{\text{Men, None}} + b_3$$

$$b_3 = \overline{X}_{\text{Men, None}} - \overline{X}_{\text{Women, None}} + \overline{X}_{\text{Women, 4 Pints}} - \overline{X}_{\text{Men, 4 Pints}}$$

$$b_3 = 66.875 - 60.625 + 57.500 - 35.625$$

$$b_3 = 28.125$$

So, b_3 in the model really compares the difference between men and women in the no alcohol condition to the difference between men and women in the 4 pint condition. Put another way, it compares the effect of gender after no alcohol to the effect of gender after 4 pints.[2] If you think about it in terms of an interaction graph this makes perfect sense. For example, the top left-hand side of Figure 12.14 shows the interaction graph for these data. Now imagine we calculated the difference between men and women for the no alcohol groups. This would be the difference between the lines on the graph for the no alcohol group (the difference between group means, which is 6.25). If we then do the same for the 4 pints group, we find that the difference between men and women is −21.875. If we plotted these two values as a new graph we'd get a line connecting 6.25 to −21.875 (see the bottom left-hand side of Figure 12.14). This reflects the difference between the effect of gender after no alcohol compared to after 4 pints. We know that beta values represent gradients of lines and in fact b_3 in our model is the gradient of this line! (This is 6.25 − (−21.875) = 28.125.) Let's also see what happens if there isn't an interaction effect: the right-hand side of Figure 12.14 shows the same data except that the mean for the females who had 4 pints has been changed to 30. If we calculate the difference between men and women after no alcohol we get the same as before: 6.25. If we calculate the difference between men and women after 4 pints we now get 5.625. If we again plot these differences on a new graph, we find a virtually horizontal line. So, when there's no interaction, the line connecting the effect of gender after no alcohol and after 4 pints is flat and the resulting b_3 in our model would be close to 0 (remember that a zero gradient means a flat line). In fact its actual value would be 6.25 − 5.625 = 0.625.

SELF-TEST The file **GogglesRegression.sav** contains the dummy variables used in this example. Just to prove that all of this works, use this file and run a multiple regression on the data.

[2] In fact, if you re arrange the terms in the equation you'll see that you can also phrase the interaction the opposite way around: it represents the effect of alcohol in men compared to women.

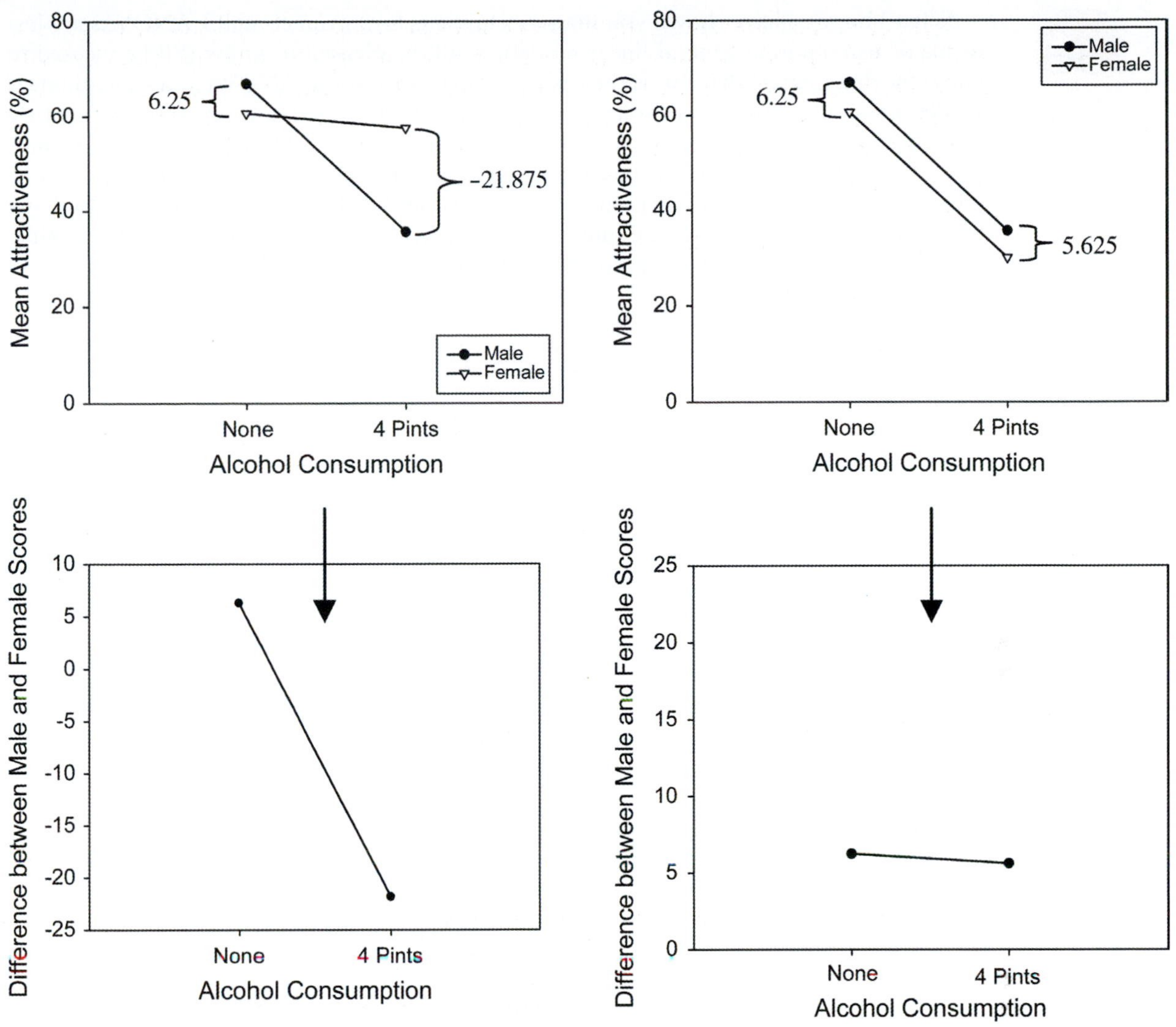

FIGURE 12.14 Breaking down what an interaction represents

The resulting table of coefficients is in SPSS Output 12.6. The important thing to note is that the various beta values are the same as we've just calculated, which should hopefully convince you that factorial ANOVA is, as is everything it would seem, just regression dressed up in a different costume!

SPSS OUTPUT 12.6

Coefficients[a]

Model		Unstandardized Coefficients		Standardized Coefficients		
		B	Std. Error	Beta	t	Sig.
1	(Constant)	66.875	3.055		21.890	.000
	Gender	-6.250	4.320	-.219	-1.447	.159
	Alcohol Consumption	-31.250	4.320	-1.094	-7.233	.000
	Interaction	28.125	6.110	.853	4.603	.000

a. Dependent Variable: Attractiveness of Date

What I hope to have shown you in this example is how even complex ANOVAs are just forms of regression (a general linear model). You'll be pleased to know (I'll be pleased to know for that matter) that this is the last I'm going to say about ANOVA as a general linear model. I hope I've given you enough background so that you get a sense of the fact that we can just keep adding in independent variables into our model. All that happens is these new variables just get added into a multiple regression equation with an associated beta value (just like the regression chapter). Interaction terms can also be added simply by multiplying the variables that interact. These interaction terms will also have an associated beta value. So, any ANOVA (no matter how complex) is just a form of multiple regression.

12.9. What to do when assumptions are violated in factorial ANOVA ③

There is not a simple non-parametric counterpart of factorial ANOVA. As such, if our data violate the assumption of normality the only real solution is robust methods (see section 5.7.4) such as those described in Wilcox's Chapter 7 and the associated files for the software R (Wilcox, 2005), which you can implement direct from SPSS using the R plugin (see the companion website for some demonstration movies of using R from SPSS). Also, if you have violated the assumption of homogeneity of variance then you can try to implement corrections based on the Welch procedure that was described in the previous chapter. However, this is quite technical, SPSS doesn't do it, and if you have anything more complicated than a 2 × 2 design then, really, it would be less painful to cover your body in paper cuts and then bathe in chilli sauce (see Algina & Olejnik, 1984).

What have I discovered about statistics? ②

This chapter has been a whistle-stop tour of factorial ANOVA. In fact we'll come across more factorial ANOVAs in the next two chapters, but for the time being we've just looked at the situation where there are two independent variables, and different people have been used in all experimental conditions. We started off by looking at how to calculate the various sums of squares in this analysis, but most important we saw that we get three effects: two main effects (the effect of each of the independent variables) and an interaction effect. We moved on to see how this analysis is done on SPSS and how the output is interpreted. Much of this was similar to the ANOVAs we've come across in previous chapters, but one big difference was the interaction term. We spent a bit of time exploring interactions (and especially interaction graphs) to see what an interaction looks like and how to spot it! The brave readers also found out how to follow up an interaction with simple effects analysis and also discovered that even complex ANOVAs are simply regression analyses in disguise. Finally we discovered that calculating effect sizes in factorial designs is a complete headache and should be attempted only by the criminally insane. So far we've steered clear of repeated-measures designs, but in the next chapter I have to resign myself to the fact that I can't avoid explaining them for the rest of my life.☹

We also discovered that no sooner had I started my first band than it disintegrated. I went with drummer Mark to sing in a band called the Outlanders, who were much better musically but were not, if the truth be told, metal enough for me. They also sacked me after a very short period of time for not being able to sing like Bono (an insult at the time, but in retrospect ...).

Key terms that I've discovered

Beer-goggles effect
Factorial ANOVA
Independent factorial design
Interaction graph
Mixed design
Related factorial design
Simple effects analysis

Smart Alex's tasks

- **Task 1**: People's musical tastes tend to change as they get older (my parents, for example, after years of listening to relatively cool music when I was a kid in the 1970s, subsequently hit their mid-forties and developed a worrying obsession with country and western music – or maybe it was the stress of having me as a teenage son!). Anyway, this worries me immensely as the future seems incredibly bleak if it is spent listening to Garth Brooks and thinking 'oh boy, did I underestimate Garth's immense talent when I was in my 20s'. So, I thought I'd do some research to find out whether my fate really was sealed, or whether it's possible to be old and like good music too. First, I got myself two groups of people (45 people in each group): one group contained young people (which I arbitrarily decided was under 40 years of age) and the other group contained more mature individuals (above 40 years of age). This is my first independent variable, **age**, and it has two levels (less than or more than 40 years old). I then split each of these groups of 45 into three smaller groups of 15 and assigned them to listen to either Fugazi (who everyone knows are the coolest band on the planet),[3] ABBA or Barf Grooks (who is a lesser known country and western musician not to be confused with anyone who has a similar name and produces music that makes you want to barf). This is my second independent variable, **music**, and has three levels (Fugazi, ABBA or Barf Grooks). There were different participants in all conditions, which means that of the 45 under-forties, 15 listened to Fugazi, 15 listened to ABBA and 15 listened to Barf Grooks; likewise, of the 45 over-forties, 15 listened to Fugazi, 15 listened to ABBA and 15 listened to Barf Grooks. After listening to the music I got each person to rate it on a scale ranging from −100 (I hate this foul music of Satan) through 0 (I am completely indifferent) to +100 (I love this music so much I'm going to explode). This variable is called **liking**. The data are in the file **Fugazi.sav**. Conduct a two-way independent ANOVA on them. ②

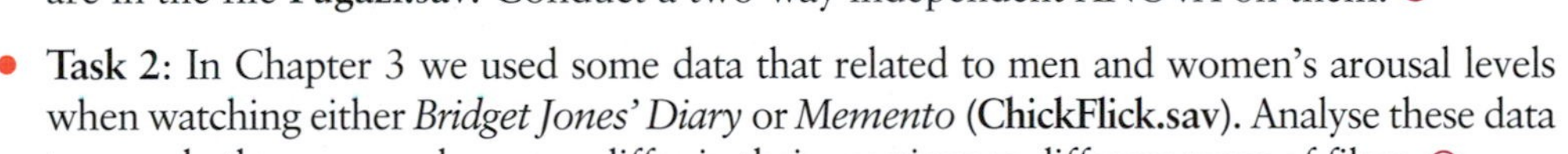

- **Task 2**: In Chapter 3 we used some data that related to men and women's arousal levels when watching either *Bridget Jones' Diary* or *Memento* (**ChickFlick.sav**). Analyse these data to see whether men and women differ in their reactions to different types of films. ②
- **Task 3**: At the start of this chapter I described a way of empirically researching whether I wrote better songs than my old band mate Malcolm, and whether this depended on the type of song (a symphony or song about flies). The outcome variable would be the number of screams elicited by audience members during the songs. These data are in the file **Escape From Inside.sav**. Draw an error bar graph (lines) and analyse and interpret these data. ②

- **Task 4**: Using SPSS Tip 12.1, change the syntax in **GogglesSimpleEffects.sps** to look at the effect of alcohol at different levels of gender. ③

The answers are on the companion website. Task 1 is an example from Field & Hole (2003) and so has a more detailed answer in there if you feel like you want it.

[3] See http://www.dischord.com.

Further reading

Howell, D. C. (2006). *Statistical methods for psychology* (6th ed.). Belmont, CA: Thompson. (Or you might prefer his *Fundamental Statistics for the Behavioral Sciences,* also in its 6th edition, 2007.)

Rosenthal, R., Rosnow, R. L., & Rubin, D. B. (2000). *Contrasts and effect sizes in behavioural research: a correlational approach*. Cambridge: Cambridge University Press. (This is quite advanced but really cannot be bettered for contrasts and effect size estimation.)

Rosnow, R. L., & Rosenthal, R. (2005). *Beginning behavioural research: a conceptual primer* (5th ed.). Englewood, Cliffs, NJ: Pearson/Prentice Hall. (Has some wonderful chapters on ANOVA, with a particular focus on effect size estimation, and some very insightful comments on what interactions actually mean.)

Online tutorials

The companion website contains the following Flash movie tutorials to accompany this chapter:

- The *R* plugin
- Two-Way Independent ANOVA using SPSS

Interesting real research

Davey, G. C. L., Startup, H. M., Zara, A., MacDonald, C. B., & Field, A. P. (2003). Perseveration of checking thoughts and mood-as-input hypothesis. *Journal of Behavior Therapy & Experimental Psychiatry*, *34*, 141–160.

Repeated-measures designs (GLM 4)

13

FIGURE 13.1
Scansion in the early days; I used to stare a lot (L–R: me, Mark and Mark)

13.1. What will this chapter tell me? ②

At the age of 15, I was on holiday with my friend Mark (the drummer) in Cornwall. I had a pretty decent mullet by this stage (nowadays I just wish I had enough hair to grow a mullet … or perhaps not) and had acquired a respectable collection of heavy metal T-shirts from going to various gigs. We were walking along the cliff tops one evening at dusk reminiscing about our times in Andromeda. We came to the conclusion that the only thing we hadn't enjoyed about that band was Malcolm and that maybe we should reform it with a different guitarist.[1] As I was wondering who we could get to play guitar, Mark pointed

[1] I feel bad about saying this because Malcolm was a very nice guy and, to be honest, at that age (and some would argue beyond) I could be a bit of a cock.

out the blindingly obvious: I played guitar. So, when we got home Scansion was born.[2] As the singer, guitarist and songwriter I set about writing some songs. I moved away from writing about flies and set my sights on the pointlessness of existence, death, betrayal and so on. We had the dubious honour of being reviewed in the music magazine *Kerrang!* (in a live review they called us 'twee', which is really not what you want to be called if you're trying to make music so heavy that it ruptures the bowels of Satan himself). Our highlight, however, was playing a gig at the famous Marquee Club in London (this club has closed now, not as a result of us playing there I hasten to add, but in its day it started the careers of people like Jimi Hendrix, the Who, Iron Maiden and Led Zeppelin).[3] This was the biggest gig of our career and it was essential that we played like we never had before. As it turned out, we did: I ran on stage, fell over and in the process de-tuned my guitar beyond recognition and broke the zip on my trousers. I spent the whole gig out of tune and spread-eagle to prevent my trousers falling down. Like I said, I'd never played like *that* before. We used to get quite obsessed with comparing how we played at different gigs. I didn't know about statistics then (happy days) but if I had I would have realized that we could rate ourselves and compare the mean ratings for different gigs; because we would always be the ones doing the rating, this would be a repeated-measures design, so we would need a repeated-measures ANOVA to compare these means. That's what this chapter is about.

13.2. Introduction to repeated-measures designs ②

Over the last three chapters we have looked at a procedure called ANOVA which is used for testing differences between several means. So far we've concentrated on situations in which different people contribute to different means; put another way, different people take part in different experimental conditions. Actually, it doesn't have to be different people (I tend to say people because I'm a psychologist and so spend my life torturing, I mean testing, children in the name of science), it could be different plants, different companies, different plots of land, different viral strains, different goats or even different duck-billed platypuses (or whatever the plural is). Anyway, the point is I've completely ignored situations in which the same people (plants, goats, hamsters, seven-eyed green galactic leaders from space, or whatever) contribute to the different means. I've put it off long enough, and now I'm going to take you through what happens when we do ANOVA on repeated-measures data.

SELF-TEST What is a repeated-measures design? (Clue: it is described in Chapter 1.)

Repeated-measures is a term used when the same participants participate in all conditions of an experiment. For example, you might test the effects of alcohol on enjoyment of a party. Some people can drink a lot of alcohol without really feeling the consequences,

[2] Scansion is a term for the rhythm of poetry. We got the name by searching through a dictionary until we found a word that we liked. Originally we didn't think it was 'metal' enough, and we decided that any self-respecting heavy metal band needed to have a big spiky 'X' in their name. So, for the first couple of years we spelt it 'Scanxion'. Like I said, I could be a bit of a cock back then.

[3] http://www.themarqueeclub.net.

whereas others, like myself, have only to sniff a pint of lager and they start flapping around on the floor waving their arms and legs around shouting 'Look at me, I'm Andy, King of the lost world of the Haddocks.' Therefore, it is important to control for individual differences in tolerance to alcohol and this can be achieved by testing the same people in all conditions of the experiment: participants could be given a questionnaire assessing their enjoyment of the party after they had consumed 1 pint, 2 pints, 3 pints and 4 pints of lager.

We saw in Chapter 1 that this type of design has several advantages; however, there is a big disadvantage. In Chapter 10 we saw that the accuracy of the *F*-test in ANOVA depends upon the assumption that scores in different conditions are independent (see section 10.2.10). When repeated-measures are used this assumption is violated: scores taken under different experimental conditions are likely to be related because they come from the same participants. As such, the conventional *F*-test will lack accuracy. The relationship between scores in different treatment conditions means that an additional assumption has to be made and, put simplistically, we assume that the relationship between pairs of experimental conditions is similar (i.e. the level of dependence between experimental conditions is roughly equal). This assumption is called the assumption of **sphericity**, which, trust me, is a pain in the neck to try to pronounce when you're giving statistics lectures at 9 a.m.

13.2.1. The assumption of sphericity ②

The assumption of sphericity can be likened to the assumption of homogeneity of variance in between-group ANOVA. Sphericity (denoted by ε and sometimes referred to as *circularity*) is a more general condition of **compound symmetry**. Compound symmetry holds true when both the variances across conditions are equal (this is the same as the homogeneity of variance assumption in between-group designs) and the covariances between pairs of conditions are equal. So, we assume that the variation within experimental conditions is fairly similar and that no two conditions are any more dependent than any other two. Although compound symmetry has been shown to be a sufficient condition for ANOVA using repeated-measures data, it is not a necessary condition. Sphericity is a less restrictive form of compound symmetry (in fact much of the early research into repeated-measures ANOVA confused compound symmetry with sphericity). Sphericity refers to the equality of variances of the *differences* between treatment levels. So, if you were to take each pair of treatment levels, and calculate the differences between each pair of scores, then it is necessary that these differences have approximately equal variances. As such, *you need at least three conditions for sphericity to be an issue.*

What is sphericity?

13.2.2. How is sphericity measured? ②

If we were going to check the assumption of sphericity by hand rather than getting SPSS to do it for us then we could start by calculating the differences between pairs of scores in all combinations of the treatment levels. Once this has been done, we could calculate the variance of these differences. Table 13.1 shows data from an experiment with three conditions. The differences between pairs of scores are computed for each participant and the variance for each set of differences is calculated. We saw above that sphericity is met when these variances are roughly equal. For these data, sphericity will hold when:

$$\text{variance}_{\text{A}-\text{B}} \approx \text{variance}_{\text{A}-\text{C}} \approx \text{variance}_{\text{B}-\text{C}}$$

TABLE 13.1 Hypothetical data to illustrate the calculation of the variance of the differences between conditions

Group A	Group B	Group C	A–B	A–C	B–C
10	12	8	–2	2	5
15	15	12	0	3	3
25	30	20	–5	5	10
35	30	28	5	7	2
30	27	20	3	10	7
		Variance:	15.7	10.3	10.3

In these data there is some deviation from sphericity because the variance of the differences between conditions A and B (15.7) is greater than the variance of the differences between A and C and between B and C (10.3). However, these data have *local circularity* (or local sphericity) because two of the variances of differences are identical. Therefore, the sphericity assumption has been met for any multiple comparisons involving these conditions (for a discussion of local circularity see Rouanet and Lépine, 1970). The deviation from sphericity in the data in Table 13.1 does not seem too severe (all variances are *roughly* equal), but can we assess whether a deviation is severe enough to warrant action?

13.2.3. Assessing the severity of departures from sphericity ②

SPSS produces a test known as **Mauchly's test**, which tests the hypothesis that the variances of the differences between conditions are equal. Therefore, if Mauchly's test statistic is significant (i.e. has a probability value less than .05) we should conclude that there are significant differences between the variances of differences and, therefore, the condition of sphericity is not met. If, however, Mauchly's test statistic is non-significant (i.e. $p > .05$) then it is reasonable to conclude that the variances of differences are not significantly different (i.e. they are roughly equal). So, in short, if Mauchly's test is significant then we must be wary of the *F*-ratios produced by the computer. However, like any significance test it is dependent on sample size: in big samples small deviations from sphericity can be significant, and in small samples large violations can be non-significant.

13.2.4. What is the effect of violating the assumption of sphericity? ③

SMART ALEX ONLY

Rouanet and Lépine (1970) provided a detailed account of the validity of the *F*-ratio under violations of the sphericity assumption. They argued that there are two different *F*-ratios that can be used to assess treatment comparisons, labelled F' and F'' respectively. F' refers to an *F*-ratio derived from the mean squares of the comparison in question and the specific error term for the comparison of interest – this is the *F*-ratio normally used. F'' is derived not from the specific error mean square but from the total error mean squares for *all* repeated-measures comparisons. Rouanet and Lépine (1970) showed that for F'' to be valid, *overall* sphericity must hold (i.e. the whole data set must be spherical), but for F' to be valid, sphericity must hold for the *specific comparison in question* (see also Mendoza, Toothaker, & Crain, 1976). F' is the statistic generally used and the effect of violating sphericity is a loss of power (compared to when F'' is used) and a test statistic (F-ratio) that simply cannot be compared to tabulated values of the F-distribution (see Oliver Twisted).

EVERYBODY

OLIVER TWISTED

Please, Sir, can I have some more … sphericity?

'Balls …,' says Oliver, '… are spherical, and I like balls. Maybe I'll like sphericity too if only you could explain it to me in more detail.' Be careful what you wish for, Oliver. In my youth I wrote an article called 'A bluffer's guide to sphericity', which I used to cite in this book, roughly on this page. A few people ask me for it, so I thought I might as well reproduce it in the additional material for this chapter.

13.2.5. What do you do if you violate sphericity? ②

If data violate the sphericity assumption there are several corrections that can be applied to produce a valid F-ratio. SPSS produces three corrections based upon the estimates of sphericity advocated by Greenhouse and Geisser (1959) and Huynh and Feldt (1976). Both of these estimates give rise to a correction factor that is applied to the degrees of freedom used to assess the observed F-ratio. The calculation of these estimates is beyond the scope of this book (interested readers should consult Girden, 1992); we need know only that the three estimates differ. The **Greenhouse–Geisser correction** (usually denoted as $\hat{\varepsilon}$) varies between $1/k - 1$ (where k is the number of repeated-measures conditions) and 1. The closer that $\hat{\varepsilon}$ is to 1, the more homogeneous the variances of differences, and hence the closer the data are to being spherical. For example, in a situation in which there are five conditions the lower limit of $\hat{\varepsilon}$ will be $1/(5 - 1)$, or 0.25 (known as the **lower-bound estimate** of sphericity).

What do I do if sphericity is violated?

Huynh and Feldt (1976) reported that when the Greenhouse–Geisser estimate is greater than 0.75 too many false null hypotheses fail to be rejected (i.e. the correction is too conservative) and Collier, Baker, Mandeville, and Hayes (1967) showed that this was also true when the sphericity estimate was as high as 0.90. Huynh and Feldt, therefore, proposed their own less conservative correction (usually denoted as $\hat{\varepsilon}$). However, Maxwell and Delaney (1990) report that $\hat{\varepsilon}$ overestimates sphericity. Stevens (2002) therefore recommends taking an average of the two and adjusting *df* by this averaged value. Girden (1992) recommends that when estimates of sphericity are greater than 0.75 then the **Huynh–Feldt correction** should be used, but when sphericity estimates are less than 0.75 or nothing is known about sphericity at all, then the Greenhouse–Geisser correction should be used instead. We will see how these values are used in due course.

A final option, when you have data that violate sphericity, is to use multivariate test statistics (MANOVA – see Chapter 16), because they are not dependent upon the assumption of sphericity (see O'Brien & Kaiser, 1985). MANOVA is covered in depth in Chapter 16, but the repeated-measures procedure in SPSS automatically produces multivariate test statistics. However, there may be trade-offs in power between these univariate and multivariate tests (see Jane Superbrain Box 13.1).

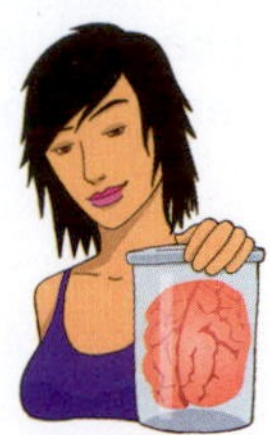

JANE SUPERBRAIN 13.1

Power in ANOVA and MANOVA ③

There is a trade-off in test power between univariate and multivariate approaches (although some authors argue that this can be overcome with suitable mastery of the techniques – O'Brien and Kaiser, 1985). Davidson (1972) compared the power of adjusted univariate techniques with those of Hotelling's T^2 (a MANOVA test statistic) and found that the univariate technique was relatively powerless to detect small reliable changes between highly correlated conditions when other less correlated conditions were also present. Mendoza, Toothaker, and Nicewander (1974) conducted a Monte Carlo study comparing univariate and multivariate techniques under violations of compound symmetry and normality and found that 'as the degree of violation of compound symmetry increased, the empirical power for the multivariate tests also increased. In contrast, the power for the univariate tests generally decreased' (p. 174). Maxwell and Delaney (1990) noted that the univariate test is relatively more powerful than the multivariate test as *n* decreases and proposed that 'the multivariate approach should probably not be used if *n* is less than *a* + 10 (*a* is the number of levels for repeated-measures)' (p. 602). As a rule it seems that when you have a large violation of sphericity ($\varepsilon < 0.7$) and your sample size is greater than (*a* + 10) then multivariate procedures are more powerful, but with small sample sizes or when sphericity holds ($\varepsilon > 0.7$) the univariate approach is preferred (Stevens, 2002). It is also worth noting that the power of MANOVA increases and decreases as a function of the correlations between dependent variables (see Jane Superbrain Box 16.1) and so the relationship between treatment conditions must be considered also.

13.3. Theory of one-way repeated-measures ANOVA ②

In a **repeated-measures ANOVA** the effect of our experiment is shown up in the within-participant variance (rather than in the between-group variance). Remember that in independent ANOVA (section 10.2) the within-participant variance is our residual variance (SS_R); it is the variance created by individual differences in performance. This variance is not contaminated by the experimental effect, because whatever manipulation we've carried out has been done on different people. However, when we carry out our experimental manipulation on the same people then the within-participant variance will be made up of two things: the effect of our manipulation and, as before, individual differences in performance. So, some of the within-subjects variation comes from the effects of our experimental manipulation: we did different things in each experimental condition to the participants, and so variation in an individual's scores will partly be due to these manipulations. For example, if everyone scores higher in one condition than another, it's reasonable to assume that this happened not by chance, but because we did something different to the participants in one of the conditions compared to any other one. *Because* we did the *same* thing to everyone within a particular condition, any variation that cannot be explained by the manipulation we've carried out must be due to random factors outside our control, unrelated to our experimental manipulations (we could call this 'error'). As in independent ANOVA, we use an *F*-ratio that compares the size of the variation due to our experimental manipulations to the size of the variation due to random factors, the only difference being how we calculate these variances. If the variance due to our manipulations is big relative to the variation due to random factors, we get a big value of *F*, and we can conclude that the observed results are unlikely to have occurred if there was no effect in the population.

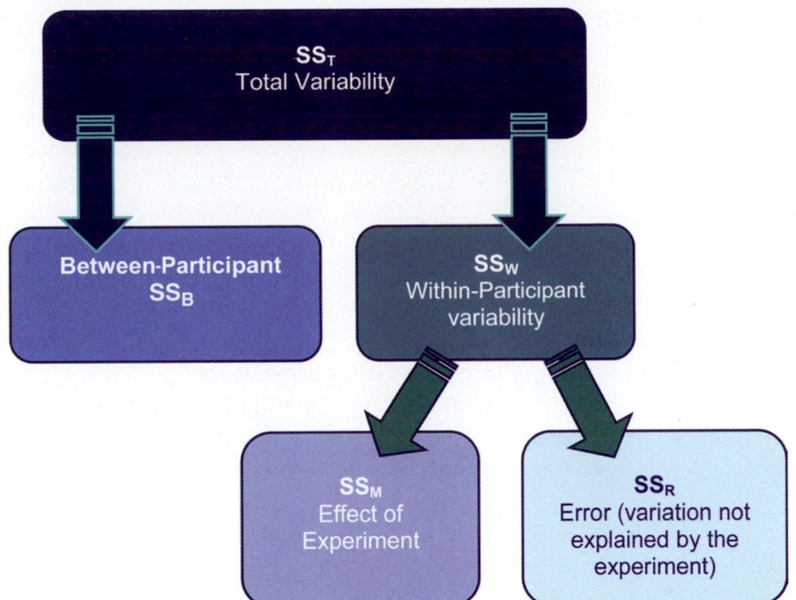

FIGURE 13.2 Partitioning variance for repeated-measures ANOVA

Figure 13.2 shows how the variance is partitioned in a repeated-measures ANOVA. The important thing to note is that we have the same types of variances as in independent ANOVA: we have a total sum of squares (SS_T), a model sum of squares (SS_M) and a residual sum of squares (SS_R). The *only* difference between repeated-measures and independent ANOVA is from where those sums of squares come: in repeated-measures ANOVA the model and residual sums of squares are both part of the within-participant variance. Let's have a look at an example.

I'm a celebrity, get me out of here is a TV show in the UK in which celebrities (well, they're not really celebrities as such, more like ex-celebrities) in a pitiful attempt to salvage their careers (or just have careers in the first place) go and live in the jungle in Australia for a few weeks. During the show these contestants have to do various humiliating and degrading tasks to win food for their camp mates. These tasks invariably involve creepy-crawlies in places where creepy-crawlies shouldn't go; for example, you might be locked in a coffin full of rats, forced to put your head in a bowl of large spiders, or have eels and cockroaches poured onto you. It's cruel, voyeuristic, gratuitous, car crash TV, and I love it. As a vegetarian, a particular favourite task for me is the bushtucker trials in which the celebrities have to eat things like live stick insects, witchetty grubs, fish eyes and kangaroo testicles/penises. Honestly, your mental image of someone is forever scarred by seeing a fish eye exploding in their mouth (here's praying that Angela Gossow never goes on the show, although she'd probably just eat the other contestants which could enhance rather than detract from her appeal). I've often wondered (perhaps a little too much) which of the bushtucker foods is the most revolting. Imagine that I tested this by getting eight celebrities, and forced them to eat four different animals (the aforementioned stick insect, kangaroo testicle, fish eye and witchetty grub) in counterbalanced order. On each occasion I measured the time it took the celebrity to retch, in seconds. This design is repeated-measures because every celebrity eats every food. The independent variable was the type of food eaten and the dependent variable was the time taken to retch.

Table 13.2 shows the data for this example. There were four foods, each eaten by eight different celebrities. Their times taken to retch are shown in the table. In addition, the mean amount of time to retch for each celebrity is shown in the table (and the variance in the time taken to retch), and also the mean time to retch for each animal. The total variance in retching time will, in part, be caused by the fact that different animals are more or less palatable (the manipulation), and will, in part, be caused by the fact that the celebrities themselves will differ in their constitution (individual differences).

TABLE 13.2 Data for the bushtucker example

Celebrity	Stick Insect	Kangaroo Testicle	Fish Eye	Witchetty Grub	Mean	s^2
1	8	7	1	6	5.50	9.67
2	9	5	2	5	5.25	8.25
3	6	2	3	8	4.75	7.58
4	5	3	1	9	4.50	11.67
5	8	4	5	8	6.25	4.25
6	7	5	6	7	6.25	0.92
7	10	2	7	2	5.25	15.58
8	12	6	8	1	6.75	20.92
Mean	**8.13**	**4.25**	**4.13**	**5.75**		

13.3.1. The total sum of squares (SS_T) ②

SMART ALEX ONLY

Remember from one-way independent ANOVA that SS_T is calculated using the following equation (see equation (10.4)):

$$SS_T = s^2_{\text{grand}}(N - 1)$$

Well, in repeated-measures designs the total sum of squares is calculated in exactly the same way. The grand variance in the equation is simply the variance of all scores when we ignore the group to which they belong. So if we treated the data as one big group it would look as follows:

8	7	1	6
9	5	2	5
6	2	3	8
5	3	1	9
8	4	5	8
7	5	6	7
10	2	7	2
12	6	8	1

Grand Mean = 5.56
Grand Variance = 8.19

The variance of these scores is 8.19 (try this on your calculators). We used 32 scores to generate this value, so N is 32. As such the equation becomes:

$$\begin{aligned} SS_T &= s^2_{grand}(N-1) \\ &= 8.19(32-1) \\ &= 253.89 \end{aligned}$$

The degrees of freedom for this sum of squares, as with the independent ANOVA, will be $N-1$, or 31.

13.3.2. The within-participant (SS_W) ②

The crucial difference in this design is that there is a variance component called the within-participant variance (this arises because we've manipulated our independent variable within each participant). This is calculated using a sum of squares. Generally speaking, when we calculate any sum of squares we look at the squared difference between the mean and individual scores. This can be expressed in terms of the variance across scores and the number of scores on which that variance is based. For example, when we calculated the residual sum of squares in independent ANOVA (SS_R) we used the following equation (look back to equation (10.7)):

$$SS_R = \sum_{i=1}^{n}(x_i - \bar{x}_i)^2$$

$$SS_R = s^2(n-1)$$

This equation gave us the variance between individuals within a particular group, and so is an estimate of individual differences within a particular group. Therefore, to get the total value of individual differences we have to calculate the sum of squares within each group and then add them up:

$$SS_R = s^2_{group\ 1}(n_1-1) + s^2_{group\ 2}(n_2-1) + s^2_{group\ 3}(n_3-1)\ldots$$

This is all well and good when we have different people in each group, but in repeated-measures designs we've subjected people to more than one experimental condition, and, therefore, we're interested in the variation not within a group of people (as in independent ANOVA) but within an actual person. That is, how much variability is there within an individual? To find this out we actually use the same equation but we adapt it to look at people rather than groups. So, if we call this sum of squares SS_W (for within-participant SS) we could write it as:

$$SS_W = s^2_{Person\ 1}(n_1-1) + s^2_{Person\ 2}(n_2-1) + s^2_{Person\ 3}(n_3-1) + \ldots + s^2_{Person\ n}(n_n-1)$$

This equation simply means that we are looking at the variation in an individual's scores and then adding these variances for all the people in the study. The *n*s simply represent the number of scores on which the variances are based (i.e. the number of experimental conditions, or in this case the number of foods). All of the variances we need are in Table 13.2, so we can calculate SS_W as:

$$\begin{aligned} SS_W &= s^2_{Celebrity\ 1}(n_1-1) + s^2_{Celebrity\ 2}(n_2-1) + \ldots + s^2_{Celebrity\ n}(n_n-1) \\ &= 9.67(4-1) + 8.25(4-1) + 7.58(4-1) + 11.67(4-1) \\ &\quad + 4.25(4-1) + 0.92(4-1) + 15.58(4-1) + 20.92(4-1) \\ &= 29 + 24.75 + 22.75 + 35 + 12.75 + 2.75 + 46.75 + 62.75 \\ &= 236.50 \end{aligned}$$

The degrees of freedom for each person are $n - 1$ (i.e. the number of conditions minus 1). To get the total degrees of freedom we add the *df*s for all participants. So, with eight participants (celebrities) and four conditions (i.e. $n = 4$) there are three degrees of freedom for each celebrity and $8 \times 3 = 24$ degrees of freedom in total.

13.3.3. The model sum of squares (SS_M) ②

So far, we know that the total amount of variation within the data is 253.58 units. We also know that 236.50 of those units are explained by the variance created by individuals' (celebrities') performances under different conditions. Now some of this variation is the result of our experimental manipulation and some of this variation is simply random fluctuation. The next step is to work out how much variance is explained by our manipulation and how much is not.

In independent ANOVA, we worked out how much variation could be explained by our experiment (the model SS) by looking at the means for each group and comparing these to the overall mean. So, we measured the variance resulting from the differences between group means and the overall mean (see equation (10.5)). We do exactly the same thing with a repeated-measures design. First we calculate the mean for each level of the independent variable (in this case the mean time to retch for each food) and compare these values to the overall mean of all foods.

So, we calculate this SS in the same way as for independent ANOVA:

1. Calculate the difference between the mean of each group and the grand mean.
2. Square each of these differences.
3. Multiply each result by the number of participants that contribute to that mean (n_i).
4. Add the values for each group together:

$$SS_M = \sum_{i=1}^{k} n_i(\bar{x}_i - \bar{x}_{grand})^2$$

Using the means from the bushtucker data (see Table 13.2), we can calculate SS_M as follows:

$$\begin{aligned} SS_M &= 8(8.13 - 5.56)^2 + 8(4.25 - 5.56)^2 + 8(4.13 - 5.56)^2 + 8(5.75 - 5.56)^2 \\ &= 8(2.57)^2 + 8(-1.31)^2 + 8(-1.44)^2 + 8(0.196)^2 \\ &= 83.13 \end{aligned}$$

For SS_M, the degrees of freedom (df_M) are again one less than the number of things used to calculate the sum of squares. For the model sums of squares we calculated the sum of squared errors between the four means and the grand mean. Hence, we used four things to calculate these sums of squares. Therefore, the degrees of freedom will be 3. So, as with independent ANOVA the model degrees of freedom are always the number of conditions (k) minus 1:

$$df_M = k - 1 = 3$$

13.3.4. The residual sum of squares (SS_R) ②

We now know that there are 253.58 units of variation to be explained in our data, and that the variation across our conditions accounts for 236.50 units. Of these 236.50 units, our experimental manipulation can explain 83.13 units. The final sum of squares is the residual sum of squares (SS_R), which tells us how much of the variation cannot be explained by the model. This value is the amount of variation caused by extraneous factors outside of experimental control. Knowing SS_W and SS_M already, the simplest way to calculate SS_R is to subtract SS_M from SS_W ($SS_R = SS_W - SS_M$):

$$SS_R = SS_W - SS_M$$

$$\begin{aligned} SS_R &= 236.50 - 83.13 \\ &= 153.37 \end{aligned}$$

The degrees of freedom are calculated in a similar way:

$$\begin{aligned} df_R &= df_W - df_M \\ &= 24 - 3 \\ &= 21 \end{aligned}$$

13.3.5. The mean squares ②

SS_M tells us how much variation the model (e.g. the experimental manipulation) explains and SS_R tells us how much variation is due to extraneous factors. However, because both of these values are summed values the number of scores that were summed influences them. As with independent ANOVA we eliminate this bias by calculating the average sum of squares (known as the *mean squares*, MS), which is simply the sum of squares divided by the degrees of freedom:

$$MS_M = \frac{SS_M}{df_M} = \frac{83.13}{3} = 27.71$$

$$MS_R = \frac{SS_R}{df_R} = \frac{153.37}{21} = 7.30$$

MS_M represents the average amount of variation explained by the model (e.g. the systematic variation), whereas MS_R is a gauge of the average amount of variation explained by extraneous variables (the unsystematic variation).

13.3.6. The *F*-ratio ②

The *F*-ratio is a measure of the ratio of the variation explained by the model and the variation explained by unsystematic factors. It can be calculated by dividing the model mean squares by the residual mean squares. You should recall that this is exactly the same as for independent ANOVA:

$$F = \frac{MS_M}{MS_R}$$

So, as with the independent ANOVA, the *F*-ratio is still the ratio of systematic variation to unsystematic variation. As such, it is the ratio of the experimental effect to the effect on performance of unexplained factors. For the bushtucker data, the *F*-ratio is:

$$F = \frac{\text{MS}_\text{M}}{\text{MS}_\text{R}} = \frac{27.71}{7.30} = 3.79$$

This value is greater than 1, which indicates that the experimental manipulation had some effect above and beyond the effect of extraneous factors. As with independent ANOVA this value can be compared against a critical value based on its degrees of freedom (df_M and df_R), which are 3 and 21 in this case.

13.3.7. The between-participant sum of squares ②

I mentioned that the total variation is broken down into a within-participant variation and a between-participant variation. We sort of forgot about the between-participant variation because we didn't need it to calculate the *F*-ratio. However, I will just briefly mention what it represents. The easiest way to calculate this term is by subtraction, because we know from Figure 13.2 that:

$$\text{SS}_\text{T} = \text{SS}_\text{B} + \text{SS}_\text{W}$$

Now, we have already calculated SS_W and SS_T so by rearranging the equation and replacing the values of these terms, we get:

$$\begin{aligned}\text{SS}_\text{B} &= \text{SS}_\text{T} - \text{SS}_\text{W}\\ \text{SS}_\text{B} &= 253.58 - 236.89\\ &= 17.39\end{aligned}$$

EVERYBODY

This term represents individual differences between cases. So, in this example, different celebrities will have different tolerances of eating these sorts of food. This is shown by the means for the celebrities in Table 13.2. For example, celebrity 4 (M = 4.50) was, on average, more than 2 seconds quicker to retch than participant 8 (M = 6.75). Celebrity 8 just had a better constitution than celebrity 4. The between-participant sum of squares reflects these differences between individuals. In this case only 17.08 units of variation in the times to retch can be explained by individual differences between our celebrities.

13.4. One-way repeated-measures ANOVA using SPSS ②

13.4.1. The main analysis ②

Sticking with the bushtucker example, we know that *each row of the data editor should represent data from one entity while each column represents a level of a variable* (SPSS

Tip 3.2). Therefore, separate columns represent levels of a repeated-measures variable. As such, there is no need for a coding variable (as with between-group designs). The data are in Table 13.2 and can be entered into the SPSS Data Editor in the same format as this table (you don't need to include the columns labelled *Celebrity*, *Mean* or s^2 as they were included only to clarify that the celebrities ate the same food and to help explain how this ANOVA is calculated). To begin with, create a variable called **stick** and use the labels dialog box to give this variable a full title of 'Stick Insect'. In the next column, create a variable called **testicle**, and give this variable a full title of 'Kangaroo Testicle'. The principle should now be clear: apply it to create the remaining variables called **eye** ('Fish Eye') and **witchetty** ('Witchetty Grub'). These data can also be found in the file **Bushtucker.sav**.

To conduct an ANOVA using a repeated-measures design, activate the define factors dialog box by selecting Analyze General Linear Model ▸ Repeated Measures.... In the define factors dialog box (Figure 13.3), you are asked to supply a name for the within-subject (repeated-measures) variable. In this case the repeated-measures variable was the type of animal eaten in the bushtucker trial, so replace the word *factor1* with the word *Animal*. The name you give to the repeated-measures variable cannot have spaces in it. When you have given the repeated-measures factor a name, you have to tell the computer how many levels there were to that variable (i.e. how many experimental conditions there were). In this case, there were four different animals eaten by each person, so we have to enter the number 4 into the box labelled *Number of Levels*. Click on Add to add this variable to the list of repeated-measures variables. This variable will now appear in the white box at the bottom of the dialog box and appears as *Animal(4)*. If your design has several repeated-measures variables then you can add more factors to the list (see two-way ANOVA example below). When you have entered all of the repeated-measures factors that were measured click on Define to go to the main dialog box.

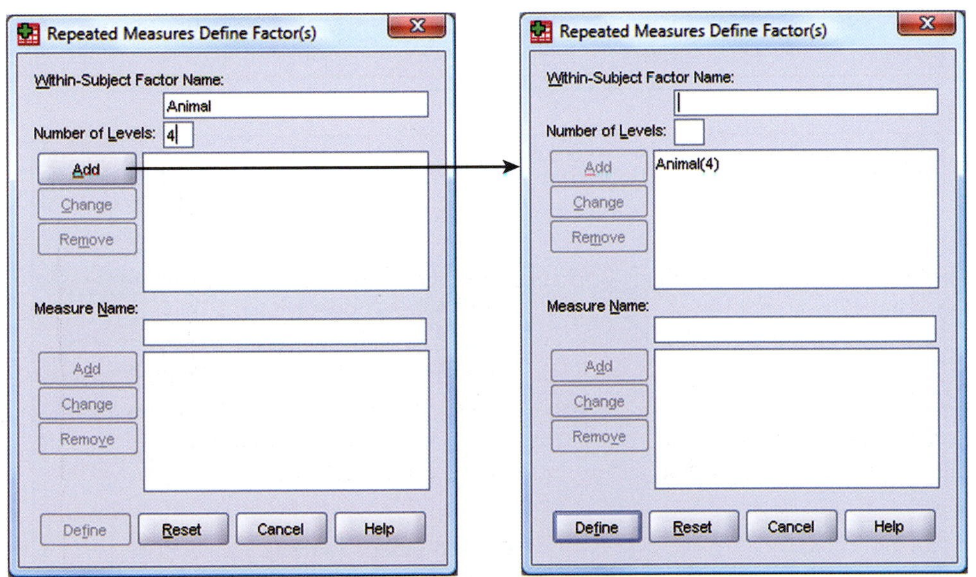

FIGURE 13.3 The define factors dialog box for repeated-measures ANOVA

The main dialog box (Figure 13.4) has a space labelled *Within-Subjects Variables* that contains a list of four question marks followed by a number. These question marks are for the variables representing the four levels of the independent variable. The variables corresponding to these levels should be selected and placed in the appropriate space. We have only four variables in the data editor, so it is possible to select all four variables at once (by clicking on the variable at the top, pressing the *Shift* key and then clicking on the last variable that you want to select). The selected variables can then be dragged to the box labelled *Within-Subjects Variables* (or click on ➜). When all four variables have been transferred, you can select various options for the analysis. There are several options that can be accessed with the buttons at the side of the main dialog box. These options are similar to the ones we have already encountered.

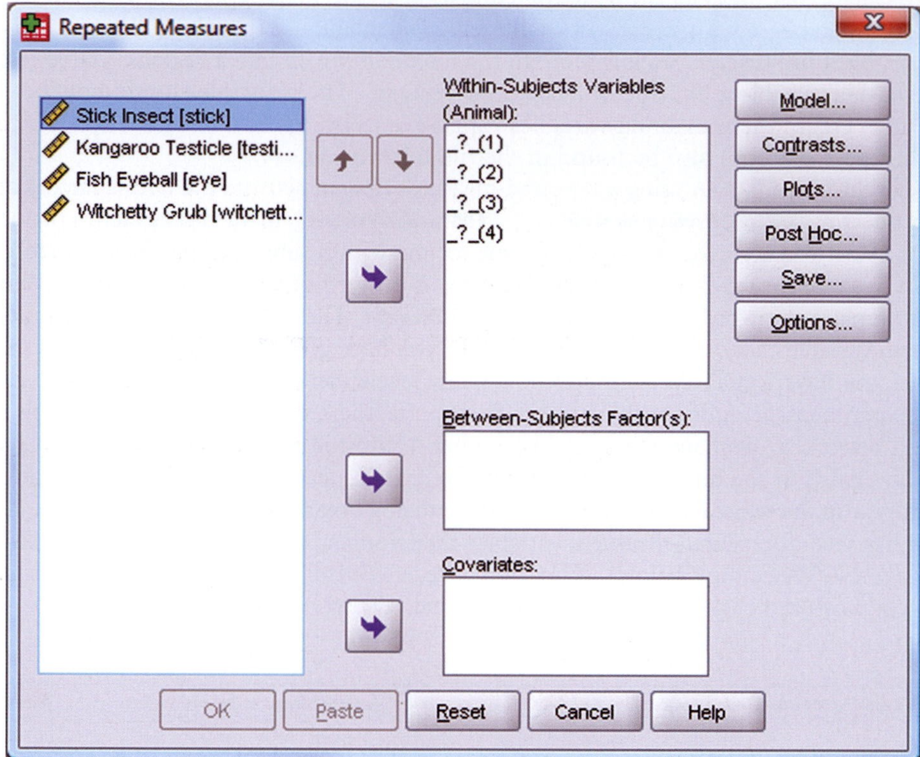

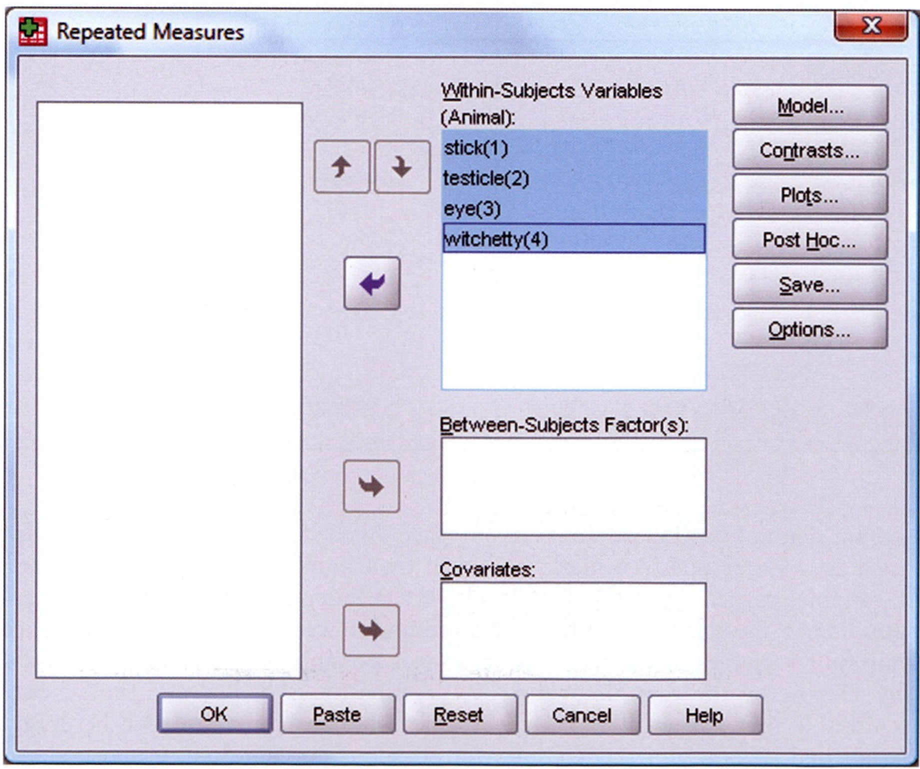

FIGURE 13.4 The main dialog box for repeated-measures ANOVA (before and after completion)

13.4.2. Defining contrasts for repeated-measures ②

It is not possible to specify user-defined planned comparisons for repeated-measures designs in SPSS.[4] However, there is the option to conduct one of the many standard contrasts that we have come across previously (see section 11.4.4 for details of changing contrasts). If you click on Contrasts... in the main dialog box you can access the contrasts dialog box (Figure 13.5). The default contrast is a polynomial contrast, but to change this default select a variable in the box labelled *Factors*, click on None, select a contrast from the list and then click on Change. If you choose to conduct a simple contrast then you can specify whether you would like to compare groups against the first or last category. The first category would be the one entered as (1) in the main dialog box and, for these data, the last category would be the one entered as (4). Therefore, the order in which you enter variables in the main dialog box is important for the contrasts you choose.

There is no particularly good contrast for the data we have (the simple contrast is not very useful because we have no control category) so let's use the *repeated* contrast, which will compare each animal against the previous animal. This contrast can be useful in repeated-measures designs in which the levels of the independent variable have a meaningful order. An example is if you have measured the dependent variable at successive points in time, or administered increasing doses of a drug. When you have selected this contrast, click on Continue to return to the main dialog box.

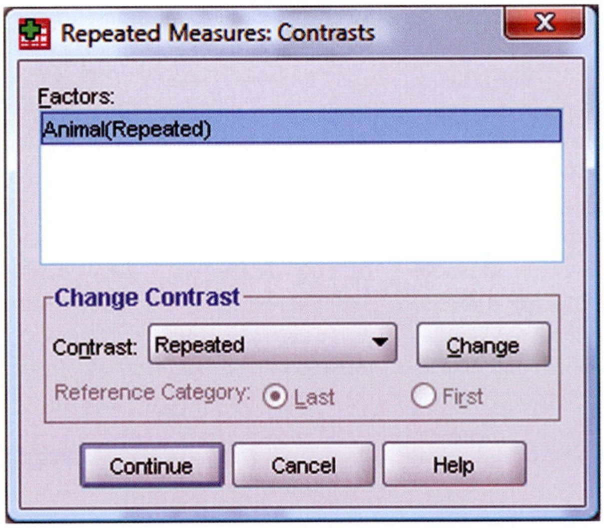

FIGURE 13.5 Repeated-measures contrasts

13.4.3. *Post hoc* tests and additional options ③

Not only does sphericity create problems for the *F* in repeated-measures ANOVA, but also it causes some amusing complications for *post hoc* tests (see Jane Superbrain Box 13.2).[5]

[4] Actually, as I mentioned in the previous chapter, you can, but only using SPSS syntax. Those who are not already feeling like sticking their head in an industrial-sized mincing machine can read the file **ContrastsUsingSyntax.pdf** on the companion website. Those who do feel like sticking their head in the aforementioned mincing machine can read the file as well: it will have much the same effect (at least it did on me)!

[5] David Howell has a good discussion of this issue and suggestions for doing *post hoc* tests in repeated-measures designs on his web page (http://www.uvm.edu/~dhowell/StatPages/More_Stuff/RepMeasMultComp/RepMeasMultComp.html).

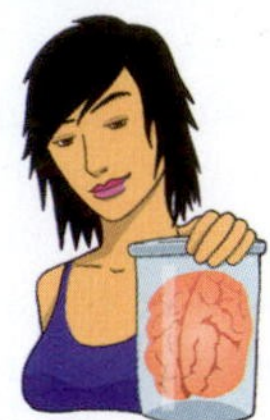

JANE SUPERBRAIN 13.2

Sphericity and post hoc tests ③

The violation of sphericity has implications for multiple comparisons. Boik (1981) provided an estimable account of the effects of non-sphericity on *post hoc* tests in repeated-measures designs, and concluded that even very small departures from sphericity produce large biases in the *F*-test. He recommends against using these tests for repeated-measure contrasts. When experimental error terms are small, the power to detect relatively strong effects can be as low as .05 (when sphericity = .80). Boik argues that the situation for *multiple* comparisons cannot be improved and concludes by recommending a multivariate analogue. Mitzel and Games (1981) found that when sphericity does not hold ($\varepsilon < 1$) the pooled error term conventionally employed in pairwise comparisons resulted in non-significant differences between two means declared significant (i.e. a lenient Type I error rate) or undetected differences (a conservative Type I error rate). Mitzel and Games, therefore, recommended the use of separate error terms for each comparison. Maxwell (1980) systematically tested the power and alpha levels for five *post hoc* tests under repeated-measures conditions. The tests assessed were Tukey's wholly significant difference (WSD) test, which uses a pooled error term; Tukey's procedure but with a separate error term with either $(n - 1)$ *df* (labelled SEP1) or $(n - 1)(k - 1)$ *df* (labelled SEP2); Bonferroni's procedure (BON); and a multivariate approach – the Roy–Bose simultaneous confidence interval (SCI). Maxwell (1980) tested these a priori procedures varying the sample size, number of levels of the repeated factor and departure from sphericity. He found that the multivariate approach was always 'too conservative for practical use' (p. 277) and this was most extreme when *n* (the number of participants) is small relative to *k* (the number of conditions). Tukey's test inflated the alpha rate unacceptably with increasing departures from sphericity even when a separate error term was used (SEP1 and SEP2). The Bonferroni method, however, was extremely robust (although *slightly* conservative) and controlled alpha levels regardless of the manipulation. Therefore, in terms of Type I error rates the Bonferroni method was best.

In terms of test power (the Type II error rate) for a small sample ($n = 8$) Maxwell found WSD to be most powerful under conditions of non-sphericity, but this advantage was severely reduced when $n = 15$. Keselman and Keselman (1988) extended Maxwell's work within unbalanced designs. They too used Tukey's WSD, a modified WSD (with non-pooled error variance), Bonferroni *t*-statistics and a multivariate approach, and found that when unweighted means were used (with unbalanced designs) none of the four tests could control the Type I error rate. When weighted means were used only the multivariate tests could limit alpha rates, although Bonferroni *t*-statistics were considerably better than the two Tukey methods. In terms of power Keselman and Keselman (1988) concluded that 'as the number of repeated treatment levels increases, BON is substantially more powerful than SCI' (p. 223).

If you don't want to worry about what these complications are then the take-home message is that when sphericity is violated, the Bonferroni method seems to be generally the most robust of the univariate techniques, especially in terms of power and control of the Type I error rate. When sphericity is definitely not violated, Tukey's test can be used. In either case, the Games–Howell procedure, which uses a pooled error term, is preferable to Tukey's test.

These sphericity-related complications mean that the standard *post hoc* tests that we have seen for independent designs are not available for repeated-measures analyses (you will find that if you access the *post hoc* test dialog box it will not list any repeated-measured factors). The good news, though, is that you can do some basic *post hoc* procedures through the additional options. These options can be accessed by clicking on Options... in the main dialog box to open the *GLM Repeated Measures: Options* dialog box (see Figure 13.6). To specify *post hoc* tests, select the repeated-measures variable (in this case **Animal**) from the box labelled *Estimated Marginal Means: Factor(s) and Factor Interactions* and drag it to the box labelled *Display Means for* (or click on ➔). Once a variable has been transferred, select ☑ Compare main effects (which will now be active). If this option is selected, the box labelled *Confidence interval adjustment* becomes active and you can click on LSD (none) ▾ to see a choice of three adjustment levels. The default is to have no adjustment and simply perform a Tukey LSD *post hoc* test (this is not recommended). The second option is a Bonferroni correction (recommended for the reasons mentioned above), and the final option is a Sidak correction, which should be selected if you are concerned about the loss of power associated with Bonferroni corrected values.

There are also syntax files available for conducting repeated-measures *post hoc* tests (available at http://www.spss.com/tech/macros/). Of these macros the Dunn–Sidak method is probably best because it is less conservative than Bonferroni corrected comparisons.

The options dialog box (Figure 13.6) has other useful options too. You can ask for descriptive statistics, which will provide the means, standard deviations and number of participants for each level of the independent variable. You can also ask for a transformation matrix, which provides the coding values for any contrast selected in the contrasts dialog box (Figure 13.5) and is very useful for interpreting the contrasts in more complex designs. SPSS can also be asked to print out the hypothesis, error and residual sum of squares and cross-product matrices (SSCPs) and we will learn about the importance of these matrices in Chapter 16. The option for homogeneity of variance tests will be active only when there is a between-group factor as well (mixed designs – see the next chapter). You can also change the level of significance at which to test any *post hoc* tests; generally, the .05 level is acceptable. When you have selected the options of interest, click on Continue to return to the main dialog box, and then click on OK to run the analysis.

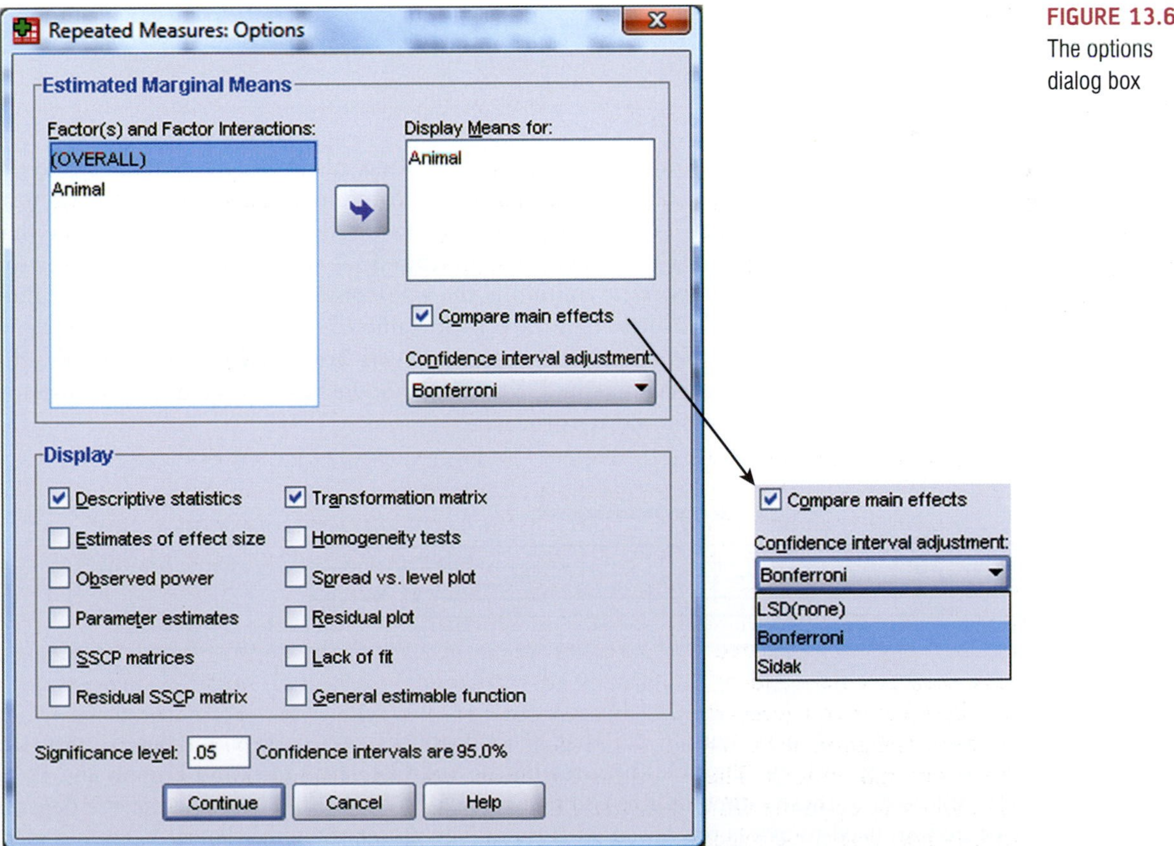

FIGURE 13.6 The options dialog box

13.5. Output for one-way repeated-measures ANOVA ②

13.5.1. Descriptives and other diagnostics ①

SPSS Output 13.1 shows the initial diagnostics statistics. First, we are told the variables that represent each level of the independent variable. This box is useful to check that the variables were entered in the correct order. The next table provides basic descriptive statistics for the four levels of the independent variable. From this table we can see that, on average, the time taken to retch was longest after eating the stick insect, and quickest after eating a testicle or eyeball. These mean values are useful for interpreting any effects that may emerge from the main analysis.

SPSS OUTPUT 13.1

Within-Subjects Factors

Measure:MEASURE_1

Animal	Dependent Variable
1	stick
2	testicle
3	eye
4	witchetty

Descriptive Statistics

	Mean	Std. Deviation	N
Stick Insect	8.12	2.232	8
Kangaroo Testicle	4.25	1.832	8
Fish Eyeball	4.12	2.748	8
Witchetty Grub	5.75	2.915	8

13.5.2. Assessing and correcting for sphericity: Mauchly's test ②

In section 13.2.3 you were told that SPSS produces a test of whether the data violate the assumption of sphericity. The next part of the output contains information about this test. Mauchly's test (see also SPSS Tip 13.1) should be non-significant if we are to assume that the condition of sphericity has been met. SPSS Output 13.2 shows Mauchly's test for the bushtucker data, and the important column is the one containing the significance value. The significance value (.047) is less than the critical value of .05, so we reject the assumption that the variances of the differences between levels are equal. In other words, the assumption of sphericity has been violated. Knowing that we have violated this assumption a pertinent question is: how should we proceed?

SPSS OUTPUT 13.2

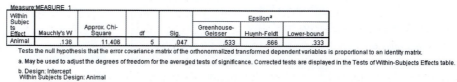

Mauchly's Test of Sphericity[b]

Measure:MEASURE_1

Within Subjects Effect	Mauchly's W	Approx. Chi-Square	df	Sig.	Epsilon[a] Greenhouse-Geisser	Huynh-Feldt	Lower-bound
Animal	.136	11.406	5	.047	.533	.666	.333

Tests the null hypothesis that the error covariance matrix of the orthonormalized transformed dependent variables is proportional to an identity matrix.

a. May be used to adjust the degrees of freedom for the averaged tests of significance. Corrected tests are displayed in the Tests of Within-Subjects Effects table.

b. Design: Intercept
Within Subjects Design: Animal

We discovered in section 13.2.5 that SPSS produces two corrections based upon the estimates of sphericity advocated by Greenhouse and Geisser (1959) and Huynh and Feldt (1976). Both of these estimates give rise to a correction factor that is applied to the degrees of freedom used to assess the observed *F*-ratio. The closer the *Greenhouse–Geisser correction* $\hat{\varepsilon}$ is to 1, the more homogeneous the variances of differences, and hence the closer the

SPSS TIP 13.1 My Mauchly's test looks weird ②

Mauchly's Test of Sphericity[b]

Measure:MEASURE_1

Within Subjects Effect	Mauchly's W	Approx. Chi-Square	df	Sig.	Epsilon[a]		
					Greenhouse-Geisser	Huynh-Feldt	Lower-bound
Animal	1.000	.000	0	.	1.000	1.000	1.000

Tests the null hypothesis that the error covariance matrix of the orthonormalized transformed dependent variables is proportional to an identity matrix.

a. May be used to adjust the degrees of freedom for the averaged tests of significance. Corrected tests are displayed in the Tests of Within-Subjects Effects table.

b. Design: Intercept
Within Subjects Design: Animal

Sometimes the SPSS output for Mauchly's test looks strange. In particular, when you look at the significance, all you see is a dot. There is no significance value. This is the case in the output above, which is from an ANOVA done comparing only the stick insect and kangaroo testicle conditions of our current example. Naturally, you fear that SPSS has gone crazy and is going to break into your bedroom at night and tattoo the equation for the Greenhouse–Geisser correction on your face. The reason that this happens is that (as I mentioned in section 13.2.1) you need at least three conditions for sphericity to be an issue (read that section if you want to know why). Therefore, if you have a repeated-measures variable that has only two levels then sphericity is met. Hence, the estimates computed by SPSS are 1 (perfect sphericity) and the resulting significance test cannot be computed (hence the reason why the table has a value of 0 for the chi-square test and degrees of freedom and a blank space for the significance). It would be a lot easier if SPSS just didn't produce the table, but then I guess we'd all be confused about why the table hadn't appeared; maybe it should just print in big letters 'Hooray! Hooray! Sphericity has gone away!' We can dream.

data are to being spherical. In a situation in which there are four conditions (as with our data) the lower limit of $\hat{\varepsilon}$ will be $1/(4 - 1)$, or 0.33 (the lower-bound estimate in the table). SPSS Output 13.2 shows that the calculated value of $\hat{\varepsilon}$ is 0.533. This is closer to the lower limit of 0.33 than it is to the upper limit of 1 and it therefore represents a substantial deviation from sphericity. We will see how these values are used in the next section.

13.5.3. The main ANOVA ②

SPSS Output 13.3 shows the results of the ANOVA for the within-subject variable. This table can be read much the same as for one-way between-group ANOVA (see Chapter 10). There is a sum of squares for the repeated-measures effect of **Animal**, which tells us how much of the total variability is explained by the experimental effect. Note the value of 83.13, which is the model sum of squares (SS_M) that we calculated in section 13.3.3. There is also an error term, which is the amount of unexplained variation across the conditions of the repeated-measures variable. This is the residual sum of squares (SS_R) that was calculated in section 13.3.4 and note that the value is 153.38 (which is the same value as calculated). As I explained earlier, these sums of squares are converted into mean squares by dividing by the degrees of freedom. As we saw before, the *df* for the effect of **Animal** is simply $k - 1$, where k is the number of levels of the independent variable. The error *df* is $(n - 1)(k - 1)$, where n is the number of participants (or in this case, the number of celebrities) and k is as before. The *F*-ratio is obtained by dividing the mean squares for the experimental effect (27.71) by the error mean squares (7.30). As with between-group ANOVA, this test statistic represents the ratio of systematic variance to unsystematic variance. The value of $F = 3.79$ (the same as we calculated earlier) is then compared against a critical value for 3 and 21 degrees of freedom. SPSS displays the exact significance level for the *F*-ratio. The significance of F is .026, which is significant because it is less than the criterion value of .05. We can, therefore, conclude that there was a significant difference between the four animals in their capacity to induce retching when eaten. However, this main test does not tell us which animals differed from each other.

SPSS OUTPUT 13.3 Repeated measures ANOVA

Tests of Within-Subjects Effects

Measure:MEASURE_1

Source		Type III Sum of Squares	df	Mean Square	F	Sig.
Animal	Sphericity Assumed	83.125	3	27.708	3.794	.026
	Greenhouse-Geisser	83.125	1.599	52.001	3.794	.063
	Huynh-Feldt	83.125	1.997	41.619	3.794	.048
	Lower-bound	83.125	1.000	83.125	3.794	.092
Error(Animal)	Sphericity Assumed	153.375	21	7.304		
	Greenhouse-Geisser	153.375	11.190	13.707		
	Huynh-Feldt	153.375	13.981	10.970		
	Lower-bound	153.375	7.000	21.911		

Although this result seems very plausible, we have learnt that the violation of the sphericity assumption makes the F-test inaccurate. We know from SPSS Output 13.2 that these data were non-spherical and so we need to make allowances for this violation. The table in SPSS Output 13.3 shows the F-ratio and associated degrees of freedom when sphericity is assumed and the significant F-statistic indicated some difference(s) between the mean time to retch after eating the four animals. This table also contains several additional rows giving the corrected values of F for the three different types of adjustment (Greenhouse–Geisser, Huynh–Feldt and lower-bound). Notice that in all cases the F-ratios remain the same; it is the degrees of freedom that change (and hence the critical value against which the obtained F-statistic is compared). The degrees of freedom have been adjusted using the estimates of sphericity calculated in SPSS Output 13.2. The adjustment is made by multiplying the degrees of freedom by the estimate of sphericity (see the previous Oliver Twisted).[6] The new degrees of freedom are then used to ascertain the significance of F. For these data the corrections result in the observed F being non-significant when using the Greenhouse–Geisser correction (because $p > .05$). However, it was noted earlier that this correction is quite conservative, and so can miss effects that genuinely exist. It is, therefore, useful to consult the Huynh–Feldt corrected F-statistic. Using this correction, the F-value is still significant because the probability value of .048 is just below the criterion value of .05. So, by this correction we would accept the hypothesis that the lecturers differed in their marking. However, it was also noted earlier that this correction is quite liberal and so tends to accept values as significant when, in reality, they are not significant. This leaves us with the puzzling dilemma of whether or not to accept this F-statistic as significant (and also illustrates how ridiculous it is to have a fixed criterion like .05 against which to determine significance). I mentioned earlier that Stevens (2002) recommends taking an average of the two estimates, and certainly when the two corrections give different results (as is the case here) this can be useful. If the two corrections give rise to the same conclusion it makes little difference which you choose to report (although if you accept the F-statistic as significant you might as well report the more conservative Greenhouse–Geisser estimate to avoid criticism!). Although it is easy to calculate the average of the two correction factors and to correct the degrees of freedom accordingly, it is not so easy to then calculate an exact probability for those degrees of freedom. Therefore, should you ever be faced with this perplexing situation (and to be honest that's fairly unlikely) I recommend taking an average of the two significance values to give you a rough idea of which correction is giving the most accurate answer. In this case, the average

[6] For example, the Greenhouse–Geisser estimate of sphericity was 0.533. The original degrees of freedom for the model were 3; this value is corrected by multiplying by the estimate of sphericity ($3 \times 0.533 = 1.599$). Likewise the error *df* was 21; this value is corrected in the same way ($21 \times 0.533 = 11.19$). The F-ratio is then tested against a critical value with these new degrees of freedom (1.599, 11.19). The other corrections are applied in the same way.

of the two p-values is (.063 + .048)/2 = .056. Therefore, we should probably go with the Greenhouse–Geisser correction and conclude that the F-ratio is non-significant.

These data illustrate how important it is to use a valid critical value of F: it can potentially mean the difference between making a Type I error and not. However, it also highlights how arbitrary it is that we use a .05 level of significance. These two corrections produce significance values that differ by only .015 and yet they lead to completely opposite conclusions! The decision about 'significance' has, in some ways, become rather arbitrary. The F, and hence the size of effect, is unaffected by these corrections and so whether the p falls slightly above or slightly below .05 is less important than how big the effect is. We might be well advised to look at an effect size to see whether the effect is substantive regardless of its significance.

We also saw earlier that a final option, when you have data that violate sphericity, is to use multivariate test statistics (MANOVA – see Chapter 16), because they do not make this assumption (see O'Brien and Kaiser, 1985). MANOVA is covered in depth in Chapter 16, but the repeated-measures procedure in SPSS automatically produces multivariate test statistics. SPSS Output 13.4 shows the multivariate test statistics for this example (details of these test statistics can be found in section 16.4.4). The column displaying the significance values shows that the multivariate tests are significant (because p is .002, which is less than the criterion value of .05). This result supports a decision to conclude that there are significant differences between the time taken to retch after eating different animals.

SPSS OUTPUT 13.4

Multivariate Tests[b]

Effect		Value	F	Hypothesis df	Error df	Sig.
Animal	Pillai's Trace	.942	26.955[a]	3.000	5.000	.002
	Wilks' Lambda	.058	26.955[a]	3.000	5.000	.002
	Hotelling's Trace	16.173	26.955[a]	3.000	5.000	.002
	Roy's Largest Root	16.173	26.955[a]	3.000	5.000	.002

a. Exact statistic

b. Design: Intercept
Within Subjects Design: Animal

13.5.4. Contrasts ②

The transformation matrix requested in the options is shown in SPSS Output 13.5 and we have to draw on our knowledge of contrast coding (see Chapter 10) to interpret this table. The first thing to remember is that a code of 0 means that the group is not included in a contrast. Therefore, contrast 1 (labelled *Level 1 vs. Level 2* in the table) ignores the fish eyeball and witchetty grub. The next thing to remember is that groups with a negative weight are compared to groups with a positive weight. In this case this means that the first contrast compares the stick insect against the kangaroo testicle. Using the same logic, contrast 2 (labelled *Level 2 vs. Level 3*) ignores the stick insect and witchetty grub and compares the kangaroo testicle with the fish eye.

SELF-TEST What does contrast 3 (*Level 3 vs. Level 4*) compare?

Finally, contrast 3 compares the fish eyeball with the witchetty grub. This pattern of contrasts is consistent with what we expect to get from a repeated contrast (i.e. all groups except the first are compared to the preceding category).

SPSS OUTPUT 13.5

Animal[a]

Measure:MEASURE_1

Dependent Variable	Animal		
	Level 1 vs. Level 2	Level 2 vs. Level 3	Level 3 vs. Level 4
Stick Insect	1	0	0
Kangaroo Testicle	-1	1	0
Fish Eyeball	0	-1	1
Witchetty Grub	0	0	-1

a. The contrasts for the within subjects factors are:
Animal: Repeated contrast

Above the transformation matrix, we should find a summary table of the contrasts (SPSS Output 13.6). Each contrast is listed in turn, and as with between-group contrasts, an *F*-test is performed that compares the two chunks of variation. So, looking at the significance values from the table, we could say that celebrities took significantly longer to retch after eating the stick insect compared to the kangaroo testicle (*Level 1 vs. Level 2*), but that the time to retch was roughly the same after eating the kangaroo testicle and the fish eyeball (*Level 2 vs. Level 3*) and the time taken to retch was not significantly different after eating a fish eyeball compared to eating a witchetty grub (*Level 3 vs. Level 4*).

SPSS OUTPUT 13.6

Tests of Within-Subjects Contrasts

Measure:MEASURE_1

Source	Animal	Type III Sum of Squares	df	Mean Square	F	Sig.
Animal	Level 1 vs. Level 2	120.125	1	120.125	22.803	.002
	Level 2 vs. Level 3	.125	1	.125	.011	.920
	Level 3 vs. Level 4	21.125	1	21.125	.796	.402
Error(Animal)	Level 1 vs. Level 2	36.875	7	5.268		
	Level 2 vs. Level 3	80.875	7	11.554		
	Level 3 vs. Level 4	185.875	7	26.554		

However, it's worth remembering that by some criteria our main effect of the type of animal eaten was not significant, and if this is the case then the significant contrast should be ignored. We have to make some kind of decision about whether we think there really is an effect of eating different animals or not before we can look at further tests. Personally, given the multivariate tests, I would be inclined to conclude that the main effect of animal was significant and proceed with further tests. The important point to note is that the sphericity in our data has raised the issue that statistics is not a recipe book and that sometimes we have to use our own discretion to interpret data (it's comforting to know that the computer does not have all of the answers – but it's alarming to realize that this means that we have to know some of the answers ourselves).

13.5.5. *Post hoc* tests ②

If you selected *post hoc* tests for the repeated-measures variable in the options dialog box (see section 13.4.3), then the table in SPSS Output 13.7 will be produced in the output viewer window.

The arrangement of the table in SPSS Output 13.7 is similar to the table produced for between-group *post hoc* tests: the difference between group means is displayed, the standard error, the significance value and a confidence interval for the difference between means. By looking at the significance values and the means (in SPSS Output 13.1) we can see that the time to retch was significantly longer after eating a stick insect compared to a kangaroo testicle (p = .012) and a fish eye (p = .006) but not compared to a witchetty grub. The time to retch after eating a kangaroo testicle was not significantly different to after eating a fish eyeball or witchetty grub (both *p*s > .05). Finally, the time to retch was not significantly different after eating a fish eyeball compared to a witchetty grub (p > .05).

SPSS OUTPUT 13.7

Pairwise Comparisons

Measure:MEASURE_1

(I) Animal	(J) Animal	Mean Difference (I-J)	Std. Error	Sig.[a]	95% Confidence Interval for Difference[a]	
					Lower Bound	Upper Bound
1	2	3.875*	.811	.012	.925	6.825
	3	4.000*	.732	.006	1.339	6.661
	4	2.375	1.792	1.000	-4.141	8.891
2	1	-3.875*	.811	.012	-6.825	-.925
	3	.125	1.202	1.000	-4.244	4.494
	4	-1.500	1.336	1.000	-6.359	3.359
3	1	-4.000*	.732	.006	-6.661	-1.339
	2	-.125	1.202	1.000	-4.494	4.244
	4	-1.625	1.822	1.000	-8.249	4.999
4	1	-2.375	1.792	1.000	-8.891	4.141
	2	1.500	1.336	1.000	-3.359	6.359
	3	1.625	1.822	1.000	-4.999	8.249

Based on estimated marginal means

*. The mean difference is significant at the .05 level.

a. Adjustment for multiple comparisons: Bonferroni.

CRAMMING SAM'S TIPS

- The one-way repeated-measures ANOVA compares several means, when those means have come from the same participants; for example, if you measured people's statistical ability each month over a year-long course.
- In repeated-measures ANOVA there is an additional assumption: *sphericity*. This assumption needs to be considered only when you have three or more repeated-measures conditions. Test for sphericity using *Mauchly's test*. Find the table with this label: if the value in the column labelled *Sig.* is less than .05 then the assumption is violated. If the significance of Mauchly's test is greater than .05 then the assumption of sphericity has been met.
- The table labelled **Tests of Within-Subjects Effects** shows the main result of your ANOVA. If the assumption of sphericity has been met then look at the row labelled *Sphericity Assumed*. If the assumption was violated then read the row labelled *Greenhouse-Geisser* (you can also look at *Huynh-Feldt* but you'll have to read this chapter to find out the relative merits of the two procedures). Having selected the appropriate row, look at the column labelled *Sig.* if the value is less than .05 then the means of the groups are significantly different.
- For contrasts and *post hoc* tests, again look to the columns labelled *Sig.* to discover if your comparisons are significant (they will be if the significance value is less than .05).

13.6. Effect sizes for repeated-measures ANOVA ③

As with independent ANOVA the best measure of the overall effect size is omega squared (ω^2). However, just to make life even more complicated than it already is, the equations we've previously used for omega squared can't be used for repeated-measures data! If you do use the same equation on repeated-measures data it will slightly overestimate the effect size. For the sake of simplicity some people do use the same equation for one-way independent and repeated-measures ANOVAs (and I'm guilty of this in another book), but I'm afraid that in this book we're going to hit simplicity in the face with stingy the particularly poison-ridden jellyfish, and embrace complexity like a particularly hot date.

In repeated-measures ANOVA, the equation for omega squared is (hang onto your hats):

$$\omega^2 = \frac{\left[\frac{k-1}{nk}(\text{MS}_\text{M} - \text{MS}_\text{R})\right]}{\text{MS}_\text{R} + \frac{\text{MS}_\text{B} - \text{MS}_\text{R}}{k} + \left[\frac{k-1}{nk}(\text{MS}_\text{M} - \text{MS}_\text{R})\right]} \quad (13.1)$$

I know what you're thinking and it's something along the lines of 'are you having a bloody laugh?' Well, no, I'm not, but really the equation isn't too bad if you break it down. First, there are some mean squares that we've come across before (and calculated before). There's the mean square for the model (MS_M) and the residual mean square (MS_R) both of which can be obtained from the ANOVA table that SPSS produces (SPSS Output 13.3). There's also k, the number of conditions in the experiment, which for these data would be 4 (there were four animals), and there's n, the number of people that took part (in this case, the number of celebrities, 8). The main problem is this term MS_B. Back at the beginning of section 13.3 (Figure 13.2) I mentioned that the total variation is broken down into a within-participant variation and a between-participant variation. In section 13.3.7 we saw that we could calculate this term from:

$$\text{SS}_\text{T} = \text{SS}_\text{B} + \text{SS}_\text{W}$$

The problem is that SPSS doesn't give us SS_W in the output, but we know that this is made up of SS_M and SS_R, which we are given. By substituting these terms and rearranging the equation we get:

$$\text{SS}_\text{T} = \text{SS}_\text{B} + \text{SS}_\text{M} + \text{SS}_\text{R}$$

$$\text{SS}_\text{B} = \text{SS}_\text{T} - \text{SS}_\text{M} - \text{SS}_\text{R}$$

The next problem is that SPSS, which is clearly trying to hinder us at every step, doesn't give us SS_T and I'm afraid (unless I've missed something in the output) you're just going to have to calculate it by hand (see section 13.3.1). From the values we calculated earlier, you should get:

$$\begin{aligned}\text{SS}_\text{B} &= 253.89 - 83.13 - 153.38 \\ &= 17.38\end{aligned}$$

The next step is to convert this to a mean squares by dividing by the degrees of freedom, which in this case are the number of people in the sample minus 1 ($N - 1$):

$$\begin{aligned}\text{MS}_\text{B} &= \frac{\text{SS}_\text{B}}{df_\text{B}} = \frac{\text{SS}_\text{B}}{\text{N} - 1} \\ &= \frac{17.38}{8-1} \\ &= 2.48\end{aligned}$$

Having done all this and probably died of boredom in the process, we must now resurrect our corpses with renewed vigour for the effect size equation, which becomes:

$$\begin{aligned} &= \frac{\left[\frac{4-1}{8\times 4}(27.71 - 7.30)\right]}{7.30 + \frac{2.48 - 7.30}{4} + \left[\frac{4-1}{8\times 4}(27.71 - 7.30)\right]} \\ &= \frac{1.91}{8.01} \\ &= .24\end{aligned}$$

So, we get an omega squared of .24.

I've mentioned at various other points that it's actually more useful to have effect size measures for focused comparisons anyway (rather than the main ANOVA), and so a slightly easier approach to calculating effect sizes is to calculate them for the contrasts we did (see SPSS Output 13.6). For these we can use the equation that we've seen before to convert the *F*-values (because they all have 1 degree of freedom for the model) to *r*:

$$r = \sqrt{\frac{F(1, df_R)}{F(1, df_R) + df_R}}$$

For the three comparisons we did, we would get:

$$r_{\text{Stick insect vs. kangaroo testicle}} = \sqrt{\frac{22.80}{22.80 + 7}} = .87$$

$$r_{\text{kangaroo testicle vs. fish eyeball}} = \sqrt{\frac{0.01}{0.01 + 7}} = .04$$

$$r_{\text{Fish eyeball vs. witchetty grub}} = \sqrt{\frac{0.80}{0.80 + 7}} = .32$$

EVERYBODY

The difference between the stick insect and the testicle was a large effect, between the fish eye and witchetty grub a medium effect, but between the testicle and eyeball a very small effect.

13.7. Reporting one-way repeated-measures ANOVA ②

When we report repeated-measures ANOVA, we give the same details as with an independent ANOVA. The only additional thing we should concern ourselves with is reporting the corrected degrees of freedom if sphericity was violated. Personally, I'm also keen on reporting the results of sphericity tests as well. As with the independent ANOVA the degrees of freedom used to assess the *F*-ratio are the degrees of freedom for the effect of the model ($df_M = 1.60$) and the degrees of freedom for the residuals of the model ($df_R = 11.19$). Remember that in this example we corrected both using the Greenhouse–Geisser estimates of sphericity, which is why the degrees of freedom are as they are. Therefore, we could report the main finding as:

✓ The results show that the time to retch was not significantly affected by the type of animal eaten, $F(1.60, 11.19) = 3.79, p > .05$.

However, as I mentioned earlier, because the multivariate tests were significant we should probably be confident that the differences between the animals is significant. We could report these multivariate tests. There are four different test statistics, but in most situations you should probably report Pillai's trace, *V* (see Chapter 16). You should report the value of *V* as well as the associated *F* and its degrees of freedom (all from SPSS Output 13.5). If you choose to report the sphericity test as well, you should report the chi-square approximation, its degrees of freedom and the significance value. It's also nice to report the degree of sphericity by reporting the epsilon value. We'll also report the effect size in this improved version:

✓ Mauchly's test indicated that the assumption of sphericity had been violated, $\chi^2(5) = 11.41$, $p < .05$, therefore multivariate tests are reported ($\varepsilon = .53$). The results show that the time to retch was significantly affected by the type of animal eaten, $V = 0.94$, $F(3, 5) = 26.96$, $p < .01$, $\omega^2 = .24$.

Alternatively, we could report the Huynh–Feldt corrected values:

✓ Mauchly's test indicated that the assumption of sphericity had been violated, $\chi^2(5) = 11.41$, $p < .05$, therefore degrees of freedom were corrected using Huynh–Feldt estimates of sphericity ($\varepsilon = .67$). The results show that the time to retch was significantly affected by the type of animal eaten, $F(2, 13.98) = 3.79$, $p < .05$, $\omega^2 = .24$.

LABCOAT LENI'S REAL RESEARCH 13.1

Who's afraid of the big bad wolf? ②

Field, A. P. (2006). *Journal of Abnormal Psychology, 115*(4), 742–752.

I'm going to let my ego get the better of me and talk about some of my own research. When I'm not scaring my students with statistics, I scare small children with Australian marsupials. There is a good reason for doing this, which is to try to discover how children develop fears (which will help us to prevent them). Most of my research looks at the effect of giving children information about animals or situations that are novel to them (rather like a parent, teacher or TV show would do). In one particular study (Field, 2006), I used three novel animals (the quoll, quokka and cuscus) and children were told negative things about one of the animals, positive things about another, and were given no information about the third (our control). I then asked the children to place their hands in three wooden boxes each of which they believed contained one of the aforementioned animals. My hypothesis was that they would take longer to place their hand in the box containing the animal about which they had heard negative information.

The data from this part of the study are in the file **Field(2006).sav**. Labcoat Leni wants you to carry out a one-way repeated-measures ANOVA on the times taken for children to place their hands in the three boxes (negative information, positive information, no information). First, draw an error bar graph of the means, then do some normality tests on the data, then do a log transformation on the scores, and do the ANOVA on these log-transformed scores (if you read the paper you'll notice that I found that the data were not normally distributed, so I log transformed them before doing the ANOVA). Do children take longer to put their hands in a box that they believe contains an animal about which they have been told nasty things?

Answers are in the additional material on the companion website (or look at page 748 in the original article).

13.8. Repeated-measures with several independent variables ②

We have seen already that simple between-group designs can be extended to incorporate a second (or third) independent variable. It is equally easy to incorporate a second, third or even fourth independent variable into a repeated-measures analysis. As an example, some social scientists were asked to research whether imagery could influence public attitudes towards alcohol. There is evidence that attitudes towards stimuli can be changed using positive and negative imagery (Field, 2005c; Stuart, Shimp, & Engle, 1987) and these researchers were interested in answering two questions. On the one hand, the government had funded them to look at whether negative imagery in advertising could be used to change attitudes towards alcohol. Conversely, an alcohol company had provided funding to see whether positive imagery could be used to improve attitudes towards alcohol. The scientists designed a study to address both issues. Table 13.3 illustrates the experimental design and contains the data for this example (each row represents a single participant).

TABLE 13.3 Data from **Attitude.sav**

Drink	Beer			Wine			Water		
Image	+ve	−ve	Neut	+ve	−ve	Neut	+ve	−ve	Neut
Male	1	6	5	38	−5	4	10	−14	−2
	43	30	8	20	−12	4	9	−10	−13
	15	15	12	20	−15	6	6	−16	1
	40	30	19	28	−4	0	20	−10	2
	8	12	8	11	−2	6	27	5	−5
	17	17	15	17	−6	6	9	−6	−13
	30	21	21	15	−2	16	19	−20	3
	34	23	28	27	−7	7	12	−12	2
	34	20	26	24	−10	12	12	−9	4
	26	27	27	23	−15	14	21	−6	0
Female	1	−19	−10	28	−13	13	33	−2	9
	7	−18	6	26	−16	19	23	−17	5
	22	−8	4	34	−23	14	21	−19	0
	30	−6	3	32	−22	21	17	−11	4
	40	−6	0	24	−9	19	15	−10	2
	15	−9	4	29	−18	7	13	−17	8
	20	−17	9	30	−17	12	16	−4	10
	9	−12	−5	24	−15	18	17	−4	8
	14	−11	7	34	−14	20	19	−1	12
	15	−6	13	23	−15	15	29	−1	10

Participants viewed a total of nine mock adverts over three sessions. In one session, they saw three adverts: (1) a brand of beer (Brain Death) presented with a negative image (a dead body with the slogan 'drinking Brain Death makes your liver explode'); (2) a brand of wine (Dangleberry) presented in the context of a positive image (a sexy naked man or woman – depending on the participant's gender – and the slogan 'drinking Dangleberry wine makes you a horny stud muffin'); and (3) a brand of water (Puritan) presented alongside a neutral image (a person watching television accompanied by the slogan 'drinking Puritan water makes you behave completely normally'). In a second session (a week later), the participants saw the same three brands, but this time Brain Death was accompanied by the positive imagery, Dangleberry by the neutral image and Puritan by the negative. In a third session, the participants saw Brain Death accompanied by the neutral image, Dangleberry by the negative image and Puritan by the positive. After each advert participants were asked to rate the drinks on a scale ranging from −100 (dislike very much) through 0 (neutral) to 100 (like very much). The order of adverts was randomized, as was the order in which people participated in the three sessions. This design is quite complex. There are two independent variables: the type of drink (beer, wine or water) and the type of imagery used (positive, negative or neutral). These two variables completely cross over, producing nine experimental conditions.

13.8.1. The main analysis ②

To enter these data into SPSS we need to remember that each row represents a single participant's data. If a person participates in all experimental conditions (in this case (s)he sees all types of stimuli presented with all types of imagery) then each experimental condition must be represented by a column in the data editor. In this experiment there are nine experimental conditions and so the data need to be entered in nine columns (so, the format is identical to Table 13.3). You should create the following nine variables in the data editor with the names as given. For each one, you should also enter a full variable name (see section 3.4.2) for clarity in the output:

beerpos	Beer	+	Sexy Person
beerneg	Beer	+	Corpse
beerneut	Beer	+	Person in Armchair
winepos	Wine	+	Sexy Person
wineneg	Wine	+	Corpse
wineneut	Wine	+	Person in Armchair
waterpos	Water	+	Sexy Person
waterneg	Water	+	Corpse
waterneut	Water	+	Person in Armchair

SELF-TEST Once these variables have been created, enter the data as in Table 13.3. If you have problems entering the data then use the file **Attitude.sav**.

To access the define factors dialog box select Analyze General Linear Model ▸ Repeated Measures.... In the define factors dialog box you are asked to supply a name for the within-subject (repeated-measures) variable. In this case there are two within-subject factors: **Drink** (beer, wine or water) and **Imagery** (positive, negative and neutral). Replace the word *factor1* with the word *Drink*. When you have given this repeated-measures factor a name, you have to tell the computer how many levels there were to that variable. In this case, there were three types of drink, so we have to enter the number 3 into the box labelled *Number of Levels*. Click on Add to add this variable to the list of repeated-measures variables. This variable will now appear in the white box at the bottom of the dialog box and appears as *Drink(3)*. We now have to repeat this process for the second independent variable. Enter the word *Imagery* into the space labelled *Within-Subject Factor Name* and then, because there were three levels of this variable, enter the number 3 into the space labelled *Number of Levels*. Click on Add to include this variable in the list of factors; it will appear as *Imagery(3)*. The finished dialog box is shown in Figure 13.7. When you have entered both of the within-subject factors click on Define to go to the main dialog box.

The main dialog box is essentially the same as when there is only one independent variable except that there are now nine question marks (Figure 13.8). At the top of the *Within-Subjects Variables* box, SPSS states that there are two factors: **Drink** and **Imagery**. In the box below there is a series of question marks followed by bracketed numbers. The numbers in brackets represent the levels of the factors (independent variables):

_?_(1,1)	variable representing 1st level of drink and 1st level of imagery
_?_(1,2)	variable representing 1st level of drink and 2nd level of imagery
_?_(1,3)	variable representing 1st level of drink and 3rd level of imagery
_?_(2,1)	variable representing 2nd level of drink and 1st level of imagery
_?_(2,2)	variable representing 2nd level of drink and 2nd level of imagery
_?_(2,3)	variable representing 2nd level of drink and 3rd level of imagery
_?_(3,1)	variable representing 3rd level of drink and 1st level of imagery
_?_(3,2)	variable representing 3rd level of drink and 2nd level of imagery
_?_(3,3)	variable representing 3rd level of drink and 3rd level of imagery

In this example, there are two independent variables and so there are two numbers in the brackets. The first number refers to levels of the first factor listed above the box (in this case **Drink**). The second number in the bracket refers to levels of the second factor listed above the box (in this case **Imagery**). As with one-way repeated-measures ANOVA, you are required to replace these question marks with variables from the list on the left-hand side of the dialog box. With between-group designs, in which coding variables are used, the levels of a particular factor are specified by the codes assigned to them in the data editor. However, in repeated-measures designs, no such coding scheme is used and so

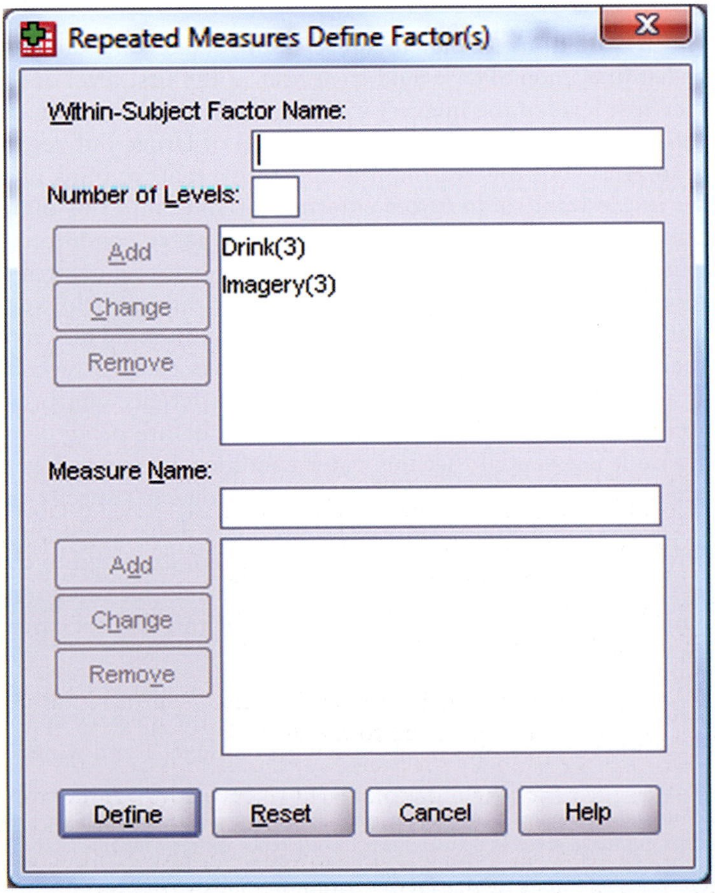

Repeated Measures Define Factor(s)

Within-Subject Factor Name:

Number of Levels:

Add

Change

Remove

Drink(3)

Imagery(3)

Measure Name:

Add

Change

Remove

Define Reset Cancel Help

FIGURE 13.7 The define factors dialog box for factorial repeated-measures ANOVA

FIGURE 13.8

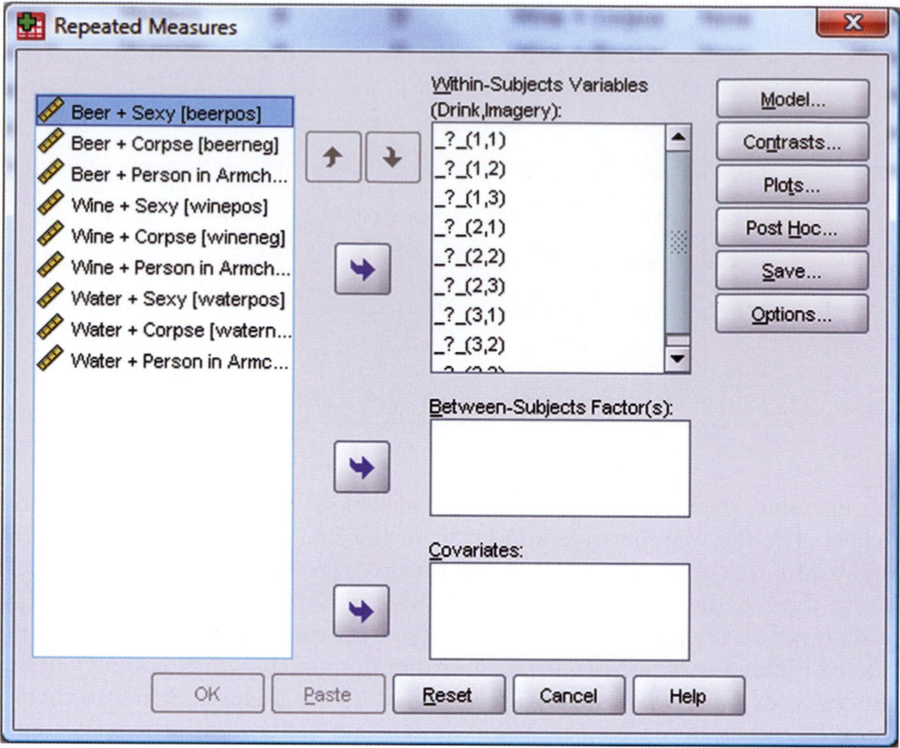

we determine which condition to assign to a level at this stage. For example, if we entered **beerpos** into the list first, then SPSS would treat beer as the first level of **Drink** and positive imagery as the first level of the **Imagery** variable. However, if we entered **wineneg** into the list first, SPSS would consider wine as the first level of **Drink** and negative imagery as the first level of **Imagery**. For this reason, it is imperative that we think about the type of contrasts that we might want to do *before* entering variables into this dialog box. In this design, if we look at the first variable, **Drink**, there were three conditions, two of which involved alcoholic drinks. In a sense, the water condition acts as a control to whether the effects of imagery are specific to alcohol. Therefore, for this variable we might want to compare the beer and wine condition with the water condition. This comparison could be done by either specifying a simple contrast (see Table 10.6) in which the beer and wine conditions are compared to the water, or using a difference contrast in which both alcohol conditions are compared to the water condition before being compared to each other. In either case it is essential that the water condition be entered as either the first or last level of the independent variable **Drink** (because you can't specify the middle level as the reference category in a simple contrast). Now, let's think about the second factor. The imagery factor also has a control category that was not expected to change attitudes (neutral imagery). As before, we might be interested in using this category as a reference category in a simple contrast[7] and so it is important that this neutral category is entered as either the first or last level.

Based on what has been discussed about using contrasts, it makes sense to have water as level 3 of the **Drink** factor and neutral as the third level of the imagery factor. The remaining levels can be decided arbitrarily. I have chosen beer as level 1 and wine as level 2 of the

[7] We expect positive imagery to improve attitudes, whereas negative imagery should make attitudes more negative. Therefore, it does not make sense to do a Helmert or difference contrast for this factor because the effects of the two experimental conditions will cancel each other out.

Drink factor. For the **Imagery** variable I chose positive as level 1 and negative as level 2. These decisions mean that the variables should be entered as follows:

beerpos	→	_?_(1,1)
beerneg	→	_?_(1,2)
beerneut	→	_?_(1,3)
winepos	→	_?_(2,1)
wineneg	→	_?_(2,2)
wineneut	→	_?_(2,3)
waterpos	→	_?_(3,1)
waterneg	→	_?_(3,2)
waterneut	→	_?_(3,3)

Coincidentally, this order is the order in which variables are listed in the data editor; this coincidence occurred simply because I thought ahead about what contrasts would be done, and then entered variables in the appropriate order! When these variables have been transferred, the dialog box should look exactly like Figure 13.9. The buttons at the side of the screen have already been described for the one-independent-variable case and so I will describe only the buttons most relevant to this analysis.

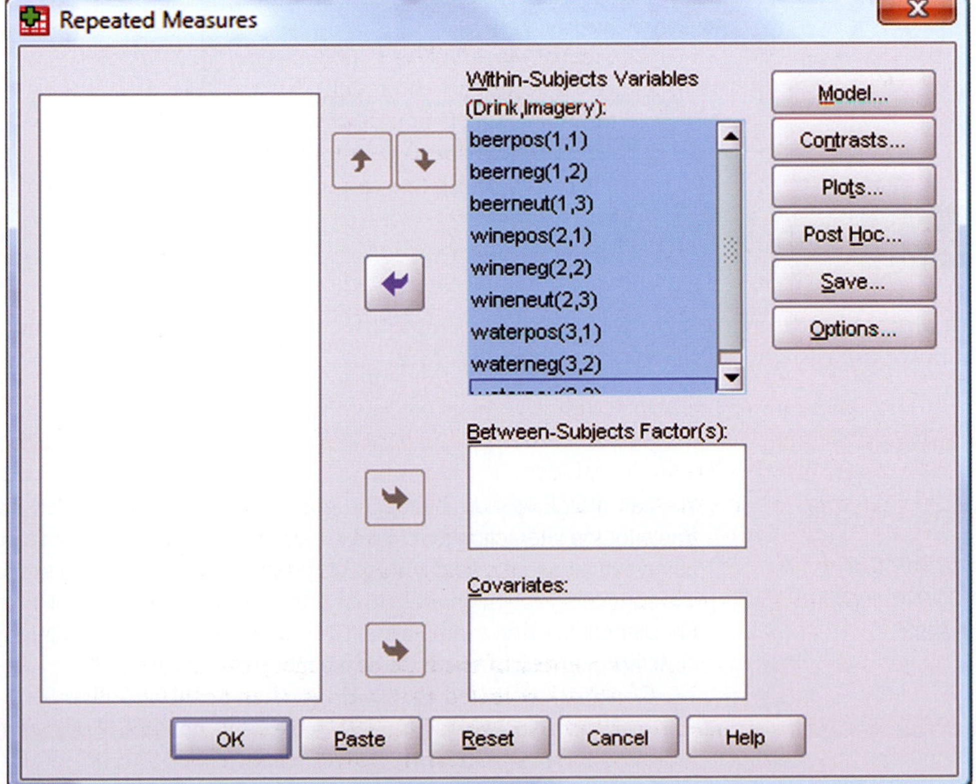

FIGURE 13.9

13.8.2. Contrasts ②

Following the main analysis it is interesting to compare levels of the independent variables to see whether they differ. As we've seen, there's no facility for entering contrast codes (unless you use syntax) so we have to rely on the standard contrasts available (see Table 10.6). Figure 13.10 shows the dialog box for conducting contrasts and is obtained by clicking on Contrasts... in the main dialog box. In the previous section I described why it might be interesting to use the water and neutral conditions as base categories for the drink and imagery factors respectively. We have used the contrasts dialog box before in sections 11.4.4 and 13.4.2 and so all I will say is that you should select a simple contrast for each independent variable. For both independent variables, we entered the variables such that the control category was the last one; therefore, we need not change the reference category for the simple contrast. Once the contrasts have been selected, click on Continue to return to the main dialog box. An alternative to the contrasts available here is to do a simple effects analysis.

FIGURE 13.10

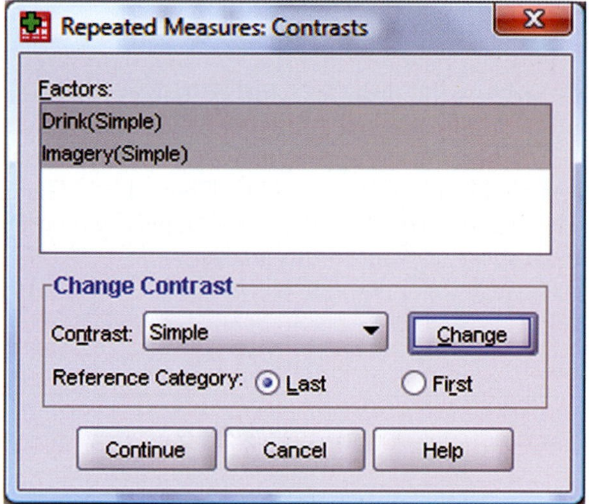

13.8.3. Simple effects analysis ③

With repeated-measures designs we can still do simple effects through SPSS syntax, but the syntax we use is slightly different – see SPSS Tip 13.2.

OLIVER TWISTED

Please, Sir, can I have some more … contrasts?

We can also follow up interaction effects with specially-defined contrasts for the interaction term. Like simple effects this can be done only using syntax and it's a fairly involved process. However, if this sounds like something you might want to do then the additional material for this chapter contains an example that I've prepared that walks you through specifying contrasts across an interaction.

SPSS TIP 13.2 Simple effects analysis on SPSS ③

We saw in the previous chapter than another way to break down an interaction term is to use a technique called 'simple effects' analysis. This analysis looks at the effect of one independent variable at individual levels of the other independent variable. So, for this example, we could look at the effect of drink for positive imagery, then for negative imagery and then for neutral imagery. Alternatively, we could analyse the effect of imagery separately for beer, wine and water. With repeated measures-designs we can still do simple effects through SPSS syntax, but the syntax we use is slightly different. The syntax you need to use in this example is:

```
MANOVA
beerpos beerneg beerneut winepos wineneg wineneut waterpos waterneg waterneut
/WSFACTORS drink(3) imagery(3)
```

*This initiates the ANOVA by specifying the variables in the data editor that relate to the levels of our repeated-measures variables. The /WSFACTORS command then defines the two repeated-measures variables that we have. The order that we list the variables from the data editor is important. So, because we've defined drink(3) imagery(3), SPSS starts at level 1 of drink, and then because we've specified three levels of imagery, it uses the first three variables listed as the levels of imagery at level 1 of drink. It then moves on to level 2 of drink and again looks to the next three variables in the list to be the relevant levels of imagery. Finally it moves to level 3 of drink and uses the next three variables (the last three in this case) to be the levels of imagery. This is hard to explain, but look at the order of variables, and see that the first three relate to beer (and differ according to imagery), then the next three are wine and the three levels of imagery, and the final three are water ordered again according to imagery. Because we ordered them in this way we have to define drink(3) and then imagery(3). (It would be equally valid to write /WSFACTORS imagery(3) drink(3), but only if initially we'd ordered the variables beerpos winepos waterpos beerneg wineneg waterneg beerneut wineneut waterneut.)

```
/WSDESIGN = MWITHIN drink(1) MWITHIN drink(2) MWITHIN drink(3)
```

*This specifies the simple effects. For example, MWITHIN drink(1) asks SPSS to analyse the effect of imagery at level 1 of drink (i.e. when beer was used). If we wanted to compare drink at levels of imagery, then we'd write this the opposite way around: /WSDESIGN = MWITHIN imagery(1) MWITHIN imagery(2) MWITHIN imagery(3)

```
/PRINT
      SIGNIF( UNIV MULT AVERF HF GG ).
```

*These final lines just ask for the main ANOVA to be printed (SIGNIF). The syntax for looking at the effect of imagery at different levels of drink is stored in a file called **SimpleEffectsAttitude.sps** for you to look at. Open this file (make sure you also have **Attitude.sav** loaded into the data editor) and run the syntax. The output you get will be in the form of text (rather than nice tables). Part of it will replicate the main ANOVA results. The simple effects are presented like this:

(Continued)

(Continued)

```
* * * * * * A n a l y s i s  o f  V a r i a n c e -- design  1 * * * * * *

Tests involving 'MWITHIN DRINK(1)' Within-Subject Effect.

Tests of Significance for T1 using UNIQUE sums of squares
Source of Variation          SS      DF       MS         F  Sig of F

WITHIN+RESIDUAL         7829.67      19   412.09
MWITHIN DRINK(1)        8401.67       1  8401.67   20.39       .000

- - - - - - - - - - - - - - - - - - - - - - - - - - - - - - - - - - - - - - -

* * * * * * A n a l y s i s   o f   V a r i a n c e -- design   1 * * * * * *

Tests involving 'MWITHIN DRINK(2)' Within-Subject Effect.

Tests of Significance for T2 using UNIQUE sums of squares
Source of Variation           SS       DF         MS         F  Sig of F

WITHIN+RESIDUAL          376.00      19      19.79
MWITHIN DRINK(2)        4166.67       1    4166.67    210.55      .000

- - - - - - - - - - - - - - - - - - - - - - - - - - - - - - - - - - - - - - -

* * * * * * A n a l y s i s   o f   V a r i a n c e -- design   1 * * * * * *

Tests involving 'MWITHIN DRINK(3)' Within-Subject Effect.

Tests of Significance for T3 using UNIQUE sums of squares
Source of Variation           SS       DF         MS         F  Sig of F

WITHIN+RESIDUAL         1500.32      19      78.96
MWITHIN DRINK(3)         742.02       1     742.02      9.40      .006

- - - - - - - - - - - - - - - - - - - - - - - - - - - - - - - - - - - - - - -
```

The table labelled 'MWITHIN DRINK(1)' gives us an ANOVA of the effect of imagery for beer and the subsequent tables are for wine and water respectively. Looking at the significance values for each simple effect, it appears that there were significant effects of imagery at all levels of drink!

13.8.4. Graphing interactions ②

When we had only one independent variable, we ignored the plots dialog box; however, if there are two or more factors, the plots dialog box is a convenient way to plot the means for each level of the factors (although really you should do some proper graphs before the analysis). To access this dialog box click on Plots... . Select **Drink** from the variables list on the left-hand side of the dialog box and drag it to the space labelled *Horizontal Axis* or click on ➔. In the space labelled *Separate Lines* we need to place the remaining independent variable: **Imagery**. As before, it is down to your discretion which way round the graph is plotted. When you have moved the two independent variables to the appropriate box, click on Add and this interaction graph will be added to the list at the bottom of the box (see Figure 13.11). When you have finished specifying graphs, click on Continue to return to the main dialog box.

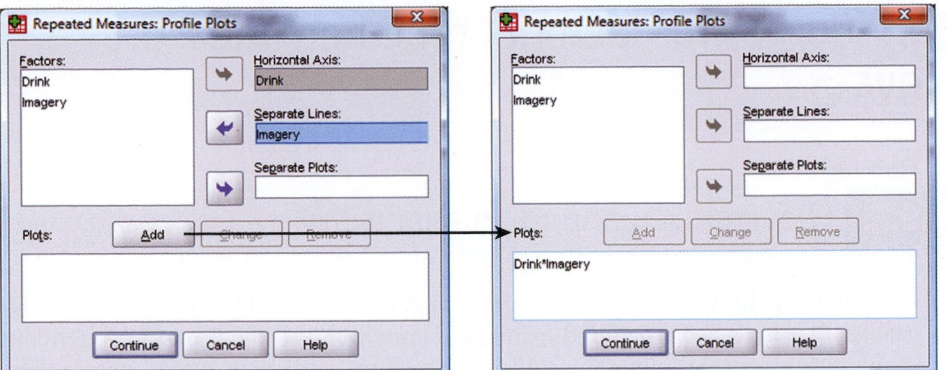

FIGURE 13.11

13.8.5. Other options ②

As for the one-way ANOVA, the *post hoc* tests are disabled because this design has only repeated-measures variables. Therefore, the only remaining options are in the options dialog box, which is accessed by clicking on Options... . The options here are the same as for the one-way ANOVA. I recommend selecting some descriptive statistics and you might also want to select some multiple comparisons by selecting all factors in the box labelled *Factor(s) and Factor Interactions* and dragging them to the box labelled *Display Means for,* or clicking on ➜ (see Figure 13.12). Having selected these variables, you should select ☑ Compare main effects and select an appropriate correction (I chose Bonferroni). The only remaining option of particular interest is to select the *Transformation matrix* option. This option produces a lot of extra output but is important for interpreting the output from the contrasts.

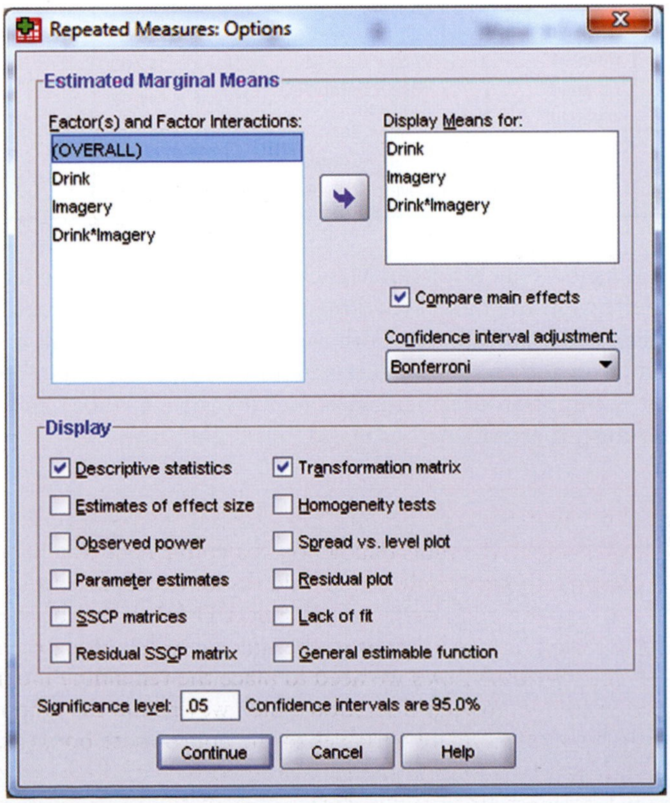

FIGURE 13.12

13.9. Output for factorial repeated-measures ANOVA ②

13.9.1. Descriptives and main analysis ②

SPSS Output 13.8 shows the initial output from this ANOVA. The first table merely lists the variables that have been included from the data editor and the level of each independent variable that they represent. This table is more important than it might seem, because it enables you to verify that you entered the variables in the correct order for the comparisons that you want to do. The second table is a table of descriptives and provides the mean and standard deviation for each of the nine conditions. The names in this table are the names I gave the variables in the data editor (therefore, if you didn't give these variables full names, this table will look slightly different).

The descriptives are interesting in that they tell us that the variability among scores was greatest when beer was used as a product (compare the standard deviations of the beer variables against the others). Also, when a corpse image was used, the ratings given to the products were negative (as expected) for wine and water but not for beer (so, for some reason, negative imagery didn't seem to work when beer was used as a stimulus). The values in this table will help us later to interpret the main effects of the analysis.

SPSS OUTPUT 13.8

Within-Subjects Factors

Measure:MEASURE_1

Drink	Imagery	Dependent Variable
1	1	beerpos
	2	beerneg
	3	beerneut
2	1	winepos
	2	wineneg
	3	wineneut
3	1	waterpos
	2	waterneg
	3	waterneu

Descriptive Statistics

	Mean	Std. Deviation	N
Beer + Sexy	21.05	13.008	20
Beer + Corpse	4.45	17.304	20
Beer + Person in Armchair	10.00	10.296	20
Wine + Sexy	25.35	6.738	20
Wine + Corpse	-12.00	6.181	20
Wine + Person in Armchair	11.65	6.243	20
Water + Sexy	17.40	7.074	20
Water + Corpse	-9.20	6.802	20
Water + Person in Armchair	2.35	6.839	20

SPSS Output 13.9 shows the results of Mauchly's sphericity test (see section 13.2.3) for each of the three effects in the model (two main effects and one interaction). The significance values of these tests indicate that both the main effects of **Drink** and **Imagery** have violated this assumption and so the *F*-values should be corrected (see section 13.5.2). For the interaction the assumption of sphericity is met (because $p > .05$) and so we need not correct the *F*-ratio for this effect.

SPSS OUTPUT 13.9

Mauchly's Test of Sphericity[b]

Measure: MEASURE_1

Within Subjects Effect	Mauchly's W	Approx. Chi-Square	df	Sig.	Epsilon[a] Greenhouse-Geisser	Huynh-Feldt	Lower-bound
DRINK	.267	23.753	2	.000	.577	.591	.500
IMAGERY	.662	7.422	2	.024	.747	.797	.500
DRINK * IMAGERY	.595	9.041	9	.436	.798	.979	.250

Tests the null hypothesis that the error covariance matrix of the orthonormalized transformed dependent variables is proportional to an identity matrix.

a. May be used to adjust the degrees of freedom for the averaged tests of significance. Corrected tests are displayed in the layers (by default) of the Tests of Within Subjects Effects table.

b. Design: Intercept - Within Subjects Design: DRINK+IMAGERY+DRINK*IMAGERY

SPSS Output 13.10 shows the results of the ANOVA (with corrected *F*-values). The output is split into sections that refer to each of the effects in the model and the error terms associated with these effects. By looking at the significance values it is clear that there is a significant effect of the type of drink used as a stimulus, a significant main effect of the type of imagery used and a significant interaction between these two variables. I will examine each of these effects in turn.

SPSS OUTPUT 13.10

Tests of Within-Subjects Effects

Measure:MEASURE_1

Source		Type III Sum of Squares	df	Mean Square	F	Sig.
Drink	Sphericity Assumed	2092.344	2	1046.172	5.106	.011
	Greenhouse-Geisser	2092.344	1.154	1812.764	5.106	.030
	Huynh-Feldt	2092.344	1.181	1770.939	5.106	.029
	Lower-bound	2092.344	1.000	2092.344	5.106	.036
Error(Drink)	Sphericity Assumed	7785.878	38	204.892		
	Greenhouse-Geisser	7785.878	21.930	355.028		
	Huynh-Feldt	7785.878	22.448	346.836		
	Lower-bound	7785.878	19.000	409.783		
Imagery	Sphericity Assumed	21628.678	2	10814.339	122.565	.000
	Greenhouse-Geisser	21628.678	1.495	14468.490	122.565	.000
	Huynh-Feldt	21628.678	1.594	13571.496	122.565	.000
	Lower-bound	21628.678	1.000	21628.678	122.565	.000
Error(Imagery)	Sphericity Assumed	3352.878	38	88.234		
	Greenhouse-Geisser	3352.878	28.403	118.048		
	Huynh-Feldt	3352.878	30.280	110.729		
	Lower-bound	3352.878	19.000	176.467		
Drink * Imagery	Sphericity Assumed	2624.422	4	656.106	17.155	.000
	Greenhouse-Geisser	2624.422	3.194	821.778	17.155	.000
	Huynh-Feldt	2624.422	3.914	670.462	17.155	.000
	Lower-bound	2624.422	1.000	2624.422	17.155	.001
Error(Drink*Imagery)	Sphericity Assumed	2906.689	76	38.246		
	Greenhouse-Geisser	2906.689	60.678	47.903		
	Huynh-Feldt	2906.689	74.373	39.083		
	Lower-bound	2906.689	19.000	152.984		

13.9.2. The effect of drink ②

The first part of SPSS Output 13.10 tells us the effect of the type of drink used in the advert. For this effect we must look at one of the corrected significance values because sphericity was violated (see above). All of the corrected values are significant and so we should report the conservative Greenhouse–Geisser corrected values of the degrees of freedom. This effect tells us that if we ignore the type of imagery that was used, participants still rated some types of drink significantly differently.

In section 13.8.5 we requested that SPSS display means for all of the effects in the model (before conducting *post hoc* tests) and if you scan through your output you should find the table in SPSS Output 13.11 in a section headed *Estimated Marginal Means.*[8] SPSS Output 13.11 is a table of means for the main effect of drink with the associated standard errors. The levels of this variable are labelled 1, 2 and 3 and so we must think back to how we entered

[8] These means are obtained by taking the average of the means in SPSS Output 13.8 for a given condition. For example, the mean for the beer condition (ignoring imagery) is:

$$\begin{aligned}\overline{X}_{\text{Beer}} &= \frac{\overline{X}_{\text{Beer + Sexy}} + \overline{X}_{\text{Beer + Corpse}} + \overline{X}_{\text{Beer + Neutral}}}{3} \\ &= \frac{21.05 + 4.45 + 10.00}{3} \\ &= 11.83\end{aligned}$$

SPSS OUTPUT 13.11

Estimates

Measure:MEASURE_1

Drink	Mean	Std. Error	95% Confidence Interval	
			Lower Bound	Upper Bound
1	11.833	2.621	6.348	17.319
2	8.333	.574	7.131	9.535
3	3.517	1.147	1.116	5.918

FIGURE 13.13

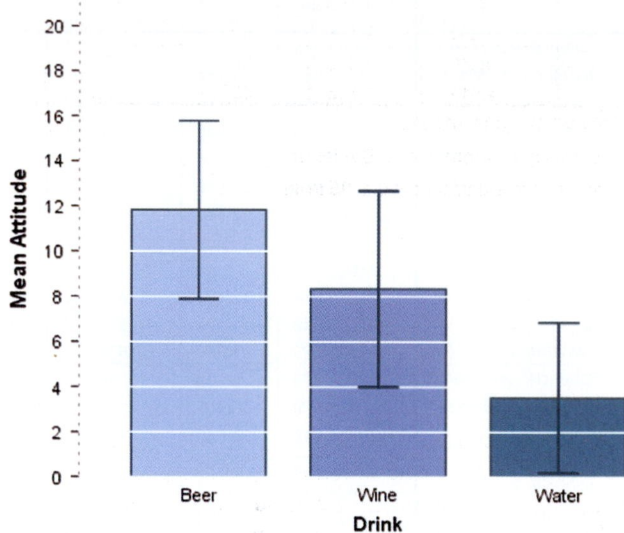

the variable to see which row of the table relates to which condition. We entered this variable with the beer condition first and the water condition last. Figure 13.13 uses this information to display the means for each condition. It is clear from this graph that beer and wine were rated higher than water (with beer being rated most highly). To see the nature of this effect we can look at the *post hoc* tests (see below) and the contrasts (see section 13.9.5).

SPSS Output 13.12 shows the pairwise comparisons for the main effect of drink corrected using a Bonferroni adjustment. This table indicates that the significant main effect reflects a significant difference ($p < .01$) between levels 2 and 3 (wine and water). Curiously, the difference between the beer and water conditions is larger than that for wine and water yet this effect is non-significant ($p > .05$). This inconsistency can be explained by looking at the standard error in the beer condition compared to the wine condition. The standard error for the wine condition is incredibly small and so the difference between means is relatively large (see Chapter 9).

SELF-TEST Try rerunning these *post hoc* tests but select the uncorrected values (LSD) in the options dialog box (see section 13.8.5). You should find that the difference between beer and water is now significant ($p = .02$).

This finding highlights the importance of controlling the error rate by using a Bonferroni correction. Had we not used this correction we could have concluded erroneously that beer was rated significantly more highly than water.

SPSS OUTPUT 13.12

Pairwise Comparisons

Measure:MEASURE_1

(I) Drink	(J) Drink	Mean Difference (I-J)	Std. Error	Sig.[a]	95% Confidence Interval for Difference[a] Lower Bound	Upper Bound
1	2	3.500	2.849	.703	-3.980	10.980
	3	8.317	3.335	.066	-.438	17.072
2	1	-3.500	2.849	.703	-10.980	3.980
	3	4.817*	1.116	.001	1.886	7.747
3	1	-8.317	3.335	.066	-17.072	.438
	2	-4.817*	1.116	.001	-7.747	-1.886

Based on estimated marginal means

a. Adjustment for multiple comparisons: Bonferroni.

*. The mean difference is significant at the .05 level.

13.9.3. The effect of imagery ②

SPSS Output 13.10 also indicates that the effect of the type of imagery used in the advert had a significant influence on participants' ratings of the stimuli. Again, we must look at one of the corrected significance values because sphericity was violated (see above). All of the corrected values are highly significant and so we can again report the Greenhouse–Geisser corrected values of the degrees of freedom. This effect tells us that if we ignore the type of drink that was used, participants' ratings of those drinks were different according to the type of imagery that was used. In section 13.8.5 we requested means for all of the effects in the model and if you scan through your output you should find the table in SPSS Output 13.13 (after the pairwise comparisons for the main effect of drink). SPSS Output 13.13 is a table of means for the main effect of imagery with the associated standard errors. The levels of this variable are labelled 1, 2 and 3 and so we must think back to how we entered the variable to see which row of the table relates to which condition. We entered this variable with the positive condition first and the neutral condition last. Figure 13.14 uses this information to illustrate the means for each condition. It is clear from this graph that positive imagery resulted in very positive ratings (compared to the neutral imagery) and negative imagery resulted in negative ratings (especially compared to the effect of neutral imagery). To see the nature of this effect we can look at the *post hoc* tests (see below) and the contrasts (see section 13.9.5).

SPSS Output 13.14 shows the pairwise comparisons for the main effect of imagery corrected using a Bonferroni adjustment. This table indicates that the significant main effect reflects significant differences (all $p < .01$) between levels 1 and 2 (positive and negative), between levels 1 and 3 (positive and neutral) and between levels 2 and 3 (negative and neutral).

SPSS OUTPUT 13.13

Estimates

Measure:MEASURE_1

Imagery	Mean	Std. Error	95% Confidence Interval Lower Bound	Upper Bound
1	21.267	.977	19.222	23.312
2	-5.583	1.653	-9.043	-2.124
3	8.000	.969	5.972	10.028

FIGURE 13.14

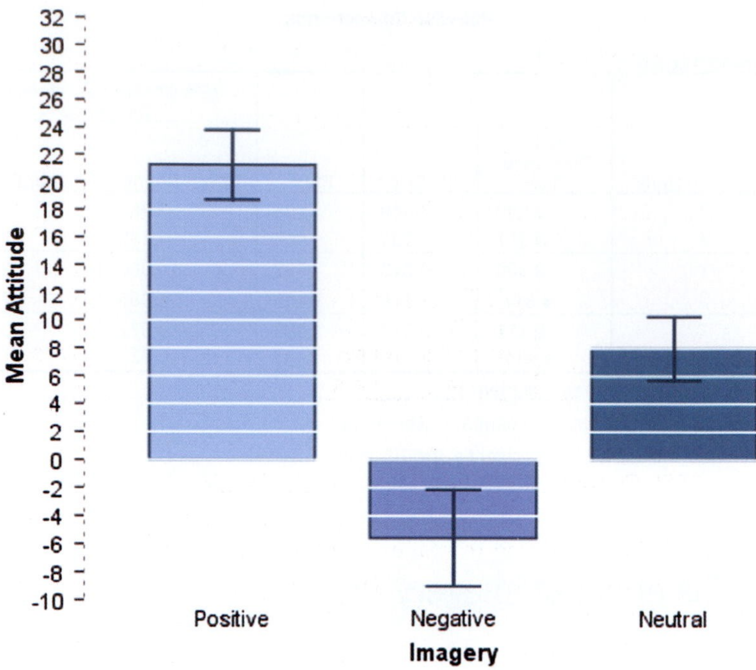

SPSS OUTPUT 13.14

Pairwise Comparisons

Measure:MEASURE_1

(I) Imagery	(J) Imagery	Mean Difference (I-J)	Std. Error	Sig.[a]	95% Confidence Interval for Difference[a] Lower Bound	Upper Bound
1	2	26.850*	1.915	.000	21.824	31.876
	3	13.267*	1.113	.000	10.346	16.187
2	1	-26.850*	1.915	.000	-31.876	-21.824
	3	-13.583*	1.980	.000	-18.781	-8.386
3	1	-13.267*	1.113	.000	-16.187	-10.346
	2	13.583*	1.980	.000	8.386	18.781

Based on estimated marginal means

*. The mean difference is significant at the .05 level.

a. Adjustment for multiple comparisons: Bonferroni.

13.9.4. The interaction effect (drink x imagery) ②

SPSS Output 13.10 indicated that imagery interacted in some way with the type of drink used as a stimulus. From that table we should report that there was a significant interaction between the type of drink used and imagery associated with it, $F(4, 76) = 17.16$, $p < .001$. This effect tells us that the type of imagery used had a different effect depending on which type of drink it was presented alongside. As before, we can use the means that we requested in section 13.8.5 to determine the nature of this interaction (this table should be below the pairwise comparisons for imagery and is shown in SPSS Output 13.15). The table of means in SPSS Output 13.15 is essentially the same as the initial descriptive statistics in SPSS Output 13.8 except that the standard errors are displayed rather than the standard deviations.

SPSS OUTPUT 13.15

Estimates

Measure:MEASURE_1

Drink	Imagery	Mean	Std. Error	95% Confidence Interval Lower Bound	Upper Bound
1	1	21.050	2.909	14.962	27.138
	2	4.450	3.869	-3.648	12.548
	3	10.000	2.302	5.181	14.819
2	1	25.350	1.507	22.197	28.503
	2	-12.000	1.382	-14.893	-9.107
	3	11.650	1.396	8.728	14.572
3	1	17.400	1.582	14.089	20.711
	2	-9.200	1.521	-12.384	-6.016
	3	2.350	1.529	-.851	5.551

The means in SPSS Output 13.15 are used to create the plot that we requested in section 13.8.4 and this graph is essential for interpreting the interaction. Figure 13.15 shows the interaction graph (slightly modified to make it look prettier!) and we are looking for non-parallel lines. The graph shows that the pattern of responding across drinks was similar when positive and neutral imagery were used. That is, ratings were positive for beer, they were slightly higher for wine and then they went down slightly for water. The fact that the line representing positive imagery is higher than the neutral line indicates that positive imagery gave rise to higher ratings than neutral imagery across all drinks. The bottom line (representing negative imagery) shows a different effect: ratings were lower for wine and water but not for beer. Therefore, negative imagery had the desired effect on attitudes towards wine and water, but for some reason attitudes towards beer remained fairly neutral. Therefore, the interaction is likely to reflect the fact that negative imagery has a different effect to both positive and neutral imagery (because it decreases ratings rather than increasing them). This interaction is completely in line with the experimental predictions. To verify the interpretation of the interaction effect, we need to look at the contrasts that we requested in section 13.8.2.

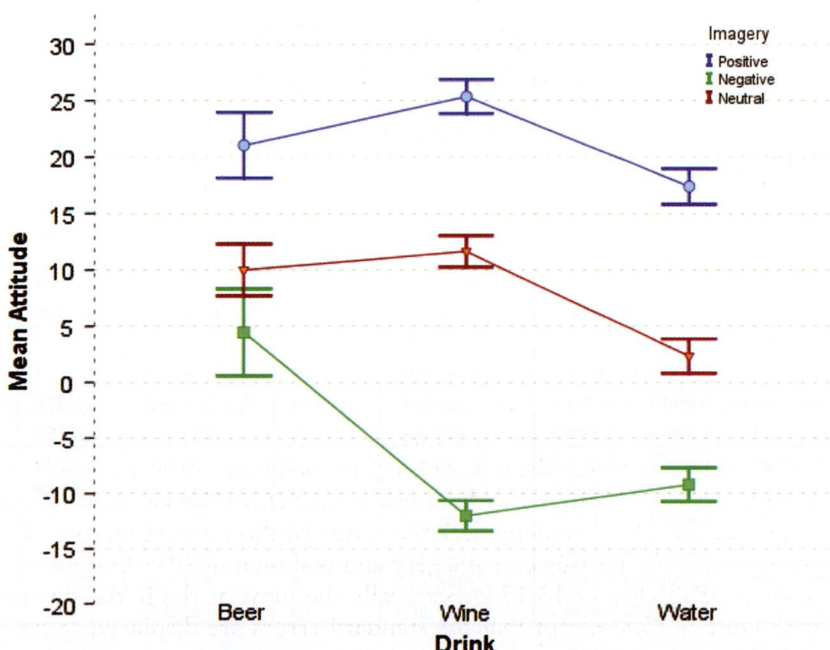

FIGURE 13.15 Interaction graph for **Attitude.sav**. The type of imagery is represented by the three lines: positive imagery (circles), negative imagery (squares) and neutral imagery (triangles)

13.9.5. Contrasts for repeated-measures variables ②

In section 13.8.2 we requested simple contrasts for the **Drink** variable (for which water was used as the control category) and for the **Imagery** category (for which neutral imagery was used as the control category). SPSS Output 13.16 shows the summary results for these contrasts. The table is split up into main effects and interactions, and each effect is split up into components of the contrast. So, for the main effect of drink, the first contrast compares level 1 (beer) against the base category (in this case, the last category: water). If you are confused as to which level is which you are reminded that SPSS Output 13.8 lists them for you. This result is significant, $F(1, 19) = 6.22$, $p < .05$, which contradicts what was found using *post hoc* tests (see SPSS Output 13.12).

SELF-TEST Why do you think that this contradiction has occurred?

The next contrast compares level 2 (wine) with the base category (water) and confirms the significant difference found with the *post hoc* tests, $F(1, 19) = 18.61$, $p < .001$. For the imagery main effect, the first contrast compares level 1 (positive) to the base category (the last category: neutral) and verifies the significant difference found with the *post hoc* tests, $F(1, 19) = 142.19$, $p < .001$. The second contrast confirms the significant difference in ratings found in the negative imagery condition compared to the neutral, $F(1, 19) = 47.07$, $p < .001$. These contrasts are all very well, but they tell us only what we already knew (although note the increased statistical power with these tests shown by the higher significance values). The contrasts become much more interesting when we look at the interaction term.

Tests of Within-Subjects Contrasts

Measure:MEASURE_1

Source	Drink	Imagery	Type III Sum of Squares	df	Mean Square	F	Sig.
Drink	Level 1 vs. Level 3	Imagery	1383.339	1	1383.339	6.218	.022
	Level 2 vs. Level 3	Imagery	464.006	1	464.006	18.613	.000
Error(Drink)	Level 1 vs. Level 3	Imagery	4226.772	19	222.462		
	Level 2 vs. Level 3	Imagery	473.661	19	24.930		
Imagery	Drink * Imagery	Level 1 vs. Level 3	3520.089	1	3520.089	142.194	.000
		Level 2 vs. Level 3	3690.139	1	3690.139	47.070	.000
Error(Imagery)	Drink * Imagery	Level 1 vs. Level 3	470.356	19	24.756		
		Level 2 vs. Level 3	1489.528	19	78.396		
Drink * Imagery	Level 1 vs. Level 3	Level 1 vs. Level 3	320.000	1	320.000	1.576	0.225
		Level 2 vs. Level 3	720.000	1	720.000	6.752	0.018
	Level 2 vs. Level 3	Level 1 vs. Level 3	36.450	1	36.450	.235	0.633
		Level 2 vs. Level 3	2928.200	1	2928.200	26.906	0.000
Error(Drink*Imagery)	Level 1 vs. Level 3	Level 1 vs. Level 3	3858.000	19	203.053		
		Level 2 vs. Level 3	2026.000	19	106.632		
	Level 2 vs. Level 3	Level 1 vs. Level 3	2946.550	19	155.082		
		Level 2 vs. Level 3	2067.800	19	108.832		

SPSS OUTPUT 13.16

13.9.5.1. Beer vs. water, positive vs. neutral imagery ②

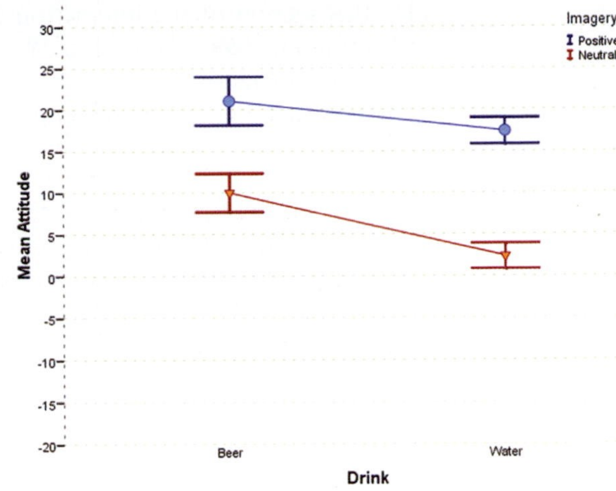

The first interaction term looks at level 1 of drink (beer) compared to level 3 (water), when positive imagery (level 1) is used compared to neutral (level 3). This contrast is non-significant. This result tells us that the increased liking found when positive imagery is used (compared to neutral imagery) is the same for both beer and water. In terms of the interaction graph (Figure 13.15) it means that the distance between the circle and the triangle in the beer condition is the same as the distance between the circle and the triangle in the water condition. If we just plot this section of the interaction graph then it's easy to see that the lines are approximately parallel, indicating no significant interaction. We could conclude that the improvement of ratings due to positive imagery compared to neutral is not affected by whether people are evaluating beer or water.

13.9.5.2. Beer vs. water, negative vs. neutral imagery ②

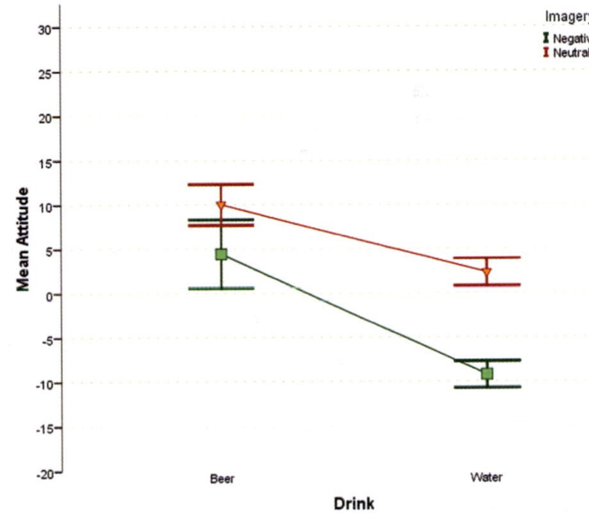

The second interaction term looks at level 1 of drink (beer) compared to level 3 (water), when negative imagery (level 2) is used compared to neutral (level 3). This contrast is significant, $F(1, 19) = 6.75$, $p < .05$. This result tells us that the decreased liking found when negative imagery is used (compared to neutral imagery) is different when beer is used compared to when water is used. In terms of the interaction graph (Figure 13.15) it means that the distance between the square and the triangle in the beer condition (a small difference) is significantly smaller than the distance between the square and the triangle in the water condition (a larger difference). We could conclude that the decrease in ratings due to negative imagery (compared to neutral) found when water is used in the advert is smaller than when beer is used.

13.9.5.3. Wine vs. water, positive vs. neutral imagery ②

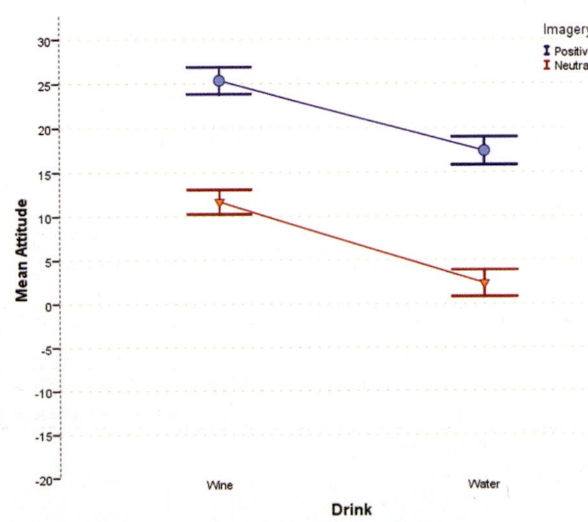

The third interaction term looks at level 2 of drink (wine) compared to level 3 (water), when positive imagery (level 1) is used compared to neutral (level 3). This contrast is non-significant, indicating that the increased liking found when positive imagery is used (compared to neutral imagery) is the same for both wine and water. In terms of the interaction graph (Figure 13.15) it means that the distance between the circle and the triangle in the wine condition is the same as the distance between the circle and the triangle in the water condition. If we just plot this section of the interaction graph then it's easy to see that the lines are parallel, indicating no interaction effect. We could conclude that the improvement of ratings due to positive imagery compared to neutral is not affected by whether people are evaluating wine or water.

13.9.5.4. Wine vs. water, negative vs. neutral imagery ②

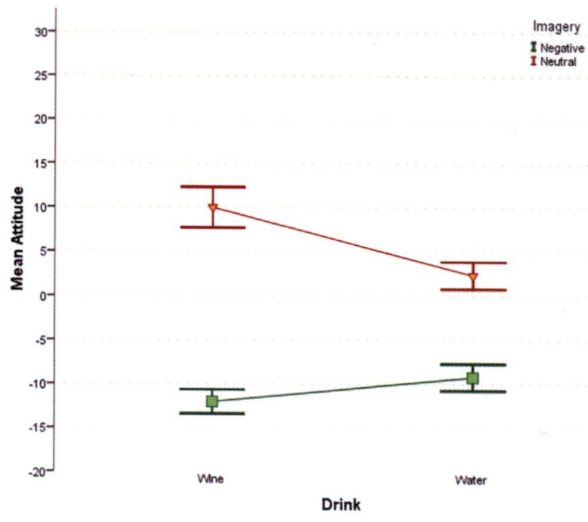

The final interaction term looks at level 2 of drink (wine) compared to level 3 (water), when negative imagery (level 2) is used compared to neutral (level 3). This contrast is significant, $F(1, 19) = 26.91$, $p < .001$. This result tells us that the decreased liking found when negative imagery is used (compared to neutral imagery) is different when wine is used compared to when water is used. In terms of the interaction graph (Figure 13.15) it means that the distance between the square and the triangle in the wine condition (a big difference) is significantly larger than the distance between the square and the triangle in the water condition (a smaller difference). We could conclude that the decrease in ratings due to negative imagery (compared to neutral) is significantly greater when wine is advertised than when water is advertised.

13.9.5.5. Limitations of these contrasts ②

These contrasts, by their nature, tell us nothing about the differences between the beer and wine conditions (or the positive and negative conditions) and different contrasts would have to be run to find out more. However, what is clear so far is that relative to the neutral condition, positive images increased liking for the products more or less regardless of the product; however, negative imagery had a greater effect on wine and a lesser effect on beer. These differences were not predicted. Although it may seem tiresome to spend so long interpreting an analysis so thoroughly, you are well advised to take such a systematic approach if you want to truly understand the effects that you obtain. Interpreting interaction terms is complex, and I can think of a few well-respected researchers who still struggle with them, so don't feel disheartened if you find them hard. Try to be thorough, and break each effect down as much as possible using contrasts and hopefully you will find enlightenment.

CRAMMING SAM'S TIPS

- Two-way repeated-measures ANOVA compares several means when there are two independent variables, and the same participants have been used in all experimental conditions.
- Test the assumption of *sphericity* when you have three or more repeated-measures conditions. Test for sphericity using *Mauchly's test*. Find the table with this label: if the value in the column labelled *Sig.* is less than .05 then the assumption is violated. If the significance of Mauchly's test is greater than .05 then the assumption of sphericity has been met. You should test this assumption for all effects (in a two-way ANOVA this means you test it for the effect of both variables and the interaction term).

- The table labelled **Tests of Within-Subjects Effects** shows the main result of your ANOVA. In a two-way ANOVA you will have three effects: a main effect of each variable and the interaction between the two. For *each* effect, if the assumption of sphericity has been met then look at the row labelled *Sphericity Assumed.* If the assumption was violated then read the row labelled *Greenhouse-Geisser* (you can also look at *Huynh-Feldt* but you'll have to read this chapter to find out the relative merits of the two procedures). Having selected the appropriate row, look at the column labelled *Sig.* If the value is less than .05 then the means of the groups are significantly different.
- Break down the main effects and interaction terms using contrasts. These contrasts appear in the table labelled **Tests of Within-Subjects Contrasts**, again look to the columns labelled *Sig.* to discover if your comparisons are significant (they will be if the significance value is less than .05).

13.10. Effect sizes for factorial repeated-measures ANOVA ③

Calculating omega squared for a one-way repeated-measures ANOVA was hair-raising enough, and as I keep saying, effect sizes are really more useful when they describe a focused effect, so I'd advise calculating effect sizes for your contrasts when you've got a factorial design (and any main effects that compare only two groups). SPSS Output 13.16 shows the values for several contrasts, all of which have 1 degree of freedom for the model (i.e. they represent a focused and interpretable comparison) and have 19 residual degrees of freedom. We can use these *F*-ratios and convert them to an effect size *r*, using a formula we've come across before:

$$r = \sqrt{\frac{F(1, df_{\text{R}})}{F(1, df_{\text{R}}) + df_{\text{R}}}}$$

For the two comparisons we did for the drink variable (SPSS Output 13.16), we would get:

$$r_{\text{Beer vs. Water}} = \sqrt{\frac{6.22}{6.22 + 19}} = 0.50$$

$$r_{\text{Wine vs. Water}} = \sqrt{\frac{18.61}{18.61 + 19}} = 0.70$$

Therefore, both comparisons yielded very large effect sizes. For the two comparisons we did for the imagery variable (SPSS Output 13.16), we would get:

$$r_{\text{Positive vs. Neutral}} = \sqrt{\frac{142.19}{142.19 + 19}} = 0.94$$

$$r_{\text{Negative vs. Neutral}} = \sqrt{\frac{47.07}{47.07 + 19}} = 0.84$$

Again, both comparisons yield very large effect sizes. For the interaction term, we had four contrasts, but again we can convert them to r because they all have 1 degree of freedom for the model (SPSS Output 13.16):

$$r_{\text{Beer vs. Water, Positive vs. Neutral}} = \sqrt{\frac{1.58}{1.58 + 19}} = 0.28$$

$$r_{\text{Beer vs. Water, Negative vs. Neutral}} = \sqrt{\frac{6.75}{6.75 + 19}} = 0.51$$

$$r_{\text{Wine vs. Water, Positive vs. Neutral}} = \sqrt{\frac{0.24}{0.24 + 19}} = 0.11$$

$$r_{\text{Wine vs. Water, Negative vs. Neutral}} = \sqrt{\frac{26.91}{26.91 + 19}} = 0.77$$

EVERYBODY

As such, the two effects that were significant (beer vs. water, negative vs. neutral and wine vs. water, negative vs. neutral) yield large effect sizes. The two effects that were not significant yielded a medium effect size (beer vs. water, positive vs. neutral) and a small effect size (wine vs. water, positive vs. neutral).

13.11. Reporting the results from factorial repeated-measures ANOVA ②

We can report a factorial repeated-measures ANOVA in much the same way as any other ANOVA. Remember that we've got three effects to report, and these effects might have different degrees of freedom. For the main effects of drink and imagery, the assumption of sphericity was violated so we'd have to report the Greenhouse–Geisser corrected degrees of freedom. We can, therefore, begin by reporting the violation of sphericity:

✓ Mauchly's test indicated that the assumption of sphericity had been violated for the main effects of drink, $\chi^2(2) = 23.75$, $p < .001$, and imagery, $\chi^2(2) = 7.42$, $p < .05$. Therefore degrees of freedom were corrected using Greenhouse–Geisser estimates of sphericity ($\varepsilon = .58$ for the main effect of drink and .75 for the main effect of imagery).

We can then report the three effects from this analysis as follows:

✓ All effects are reported as significant at $p < .05$. There was a significant main effect of the type of drink on ratings of the drink, $F(1.15, 21.93) = 5.11$. Contrasts revealed that ratings of beer, $F(1, 19) = 6.22$, $r = .50$, and wine, $F(1, 19) = 18.61$, $r = .70$, were significantly higher than water.

✓ There was also a significant main effect of the type of imagery on ratings of the drinks, $F(1.50, 28.40) = 122.57$. Contrasts revealed that ratings after positive imagery were significantly higher than after neutral imagery, $F(1, 19) = 142.19$, $r = .94$. Conversely, ratings after negative imagery were significantly lower than after neutral imagery, $F(1, 19) = 47.07$, $r = .84$.

✓ There was a significant interaction effect between the type of drink and the type of imagery used, $F(4, 76) = 17.16$. This indicates that imagery had different effects on people's ratings depending on which type of drink was used. To break down this interaction, contrasts were performed comparing all drink types to their baseline (water) and all imagery types to their baseline (neutral imagery). These revealed significant interactions when comparing negative imagery to neutral imagery both for beer compared to water, $F(1, 19) = 6.75$, $r = .51$, and wine compared to water, $F(1, 19) = 26.91$, $r = .77$. Looking at the interaction graph, these effects reflect that negative imagery (compared to neutral) lowered scores significantly more in water than it did for beer, and lowered scores significantly more for wine than it did for water. The remaining contrasts revealed no significant interaction term when comparing positive imagery to neutral imagery both for beer compared to water, $F(1, 19) = 1.58$, $r = .28$, and wine compared to water, $F(1, 19) < 1$, $r = .11$. However, these contrasts did yield small to medium effect sizes.

13.12. What to do when assumptions are violated in repeated-measures ANOVA ③

When you have only one independent variable then you can use a nonparametric test called Friedman's ANOVA (see Chapter 15) if you find that your assumptions are being irksome. However, for factorial repeated measures designs there is not a non-parametric counterpart. At the risk of sounding like a broken record, this means that if our data violate the assumption of normality then the solution is to read Rand Wilcox's book. I know I say this in every chapter, but it really is the definitive source. So, to get a robust method (see section 5.7.4) for factorial repeated-measures ANOVA designs, read Wilcox's Chapter 8, get the associated files for the software R (Wilcox, 2005), use the SPSS R plugin (see the companion website) to implement these files and away you go to statistics oblivion.

What have I discovered about statistics? ②

This chapter has helped us to walk through the murky swamp of repeated-measures designs. We discovered that it was infested with rabid leg-eating crocodiles. The first thing we learnt was that with repeated-measures designs there is yet another assumption to worry about: *sphericity*. Having recovered from this shock revelation, we were fortunate to discover that this assumption, if violated, can be easily remedied. Sorted! We then moved on to look at the theory of repeated-measures ANOVA for one independent variable. Although not essential by any stretch of the imagination, this was a useful exercise to demonstrate that basically it's exactly the same as when we have an independent design (well, there are a few subtle differences but I was trying to emphasize the similarities). We then worked through an example on SPSS, before tackling the particularly foul-tempered, starving hungry, and mad as 'Stabby' the mercury-sniffing hatter, piranha fish of omega squared. That's a road I kind of regretted going down after I'd started, but, stubborn as ever, I persevered. This led us ungracefully on to factorial repeated-measures designs and specifically the situation where we have two independent variables. We learnt that as with other factorial designs we have to worry about interaction terms. But, we also discovered some useful ways to break these terms down using contrasts.

By 16 I had started my first 'serious' band. We actually stayed together for about 7 years (with the same line-up, and we're still friends now) before Mark (drummer) moved to Oxford, I moved to Brighton to do my Ph.D., and rehearsing became a mammoth feat of organization. We had a track on a CD, some radio play and transformed from a thrash metal band to a blend of Fugazi, Nirvana and metal. I never split my trousers during a gig again (although I did once split my head open). Why didn't we make it? Well, Mark was an astonishingly good drummer so it wasn't his fault, the other Mark was an extremely good bassist too (of the three of us he is the one that has always been in a band since we split up), so the weak link was me. This was especially unfortunate given that I had three roles in the band (guitar, singing, songs) – my poor band mates never stood a chance.☺ I stopped playing music for quite a few years after we split. I still wrote songs (for personal consumption) but the three of us were such close friends that I couldn't bear the thought of playing with other people. At least not for a few years ...

Key terms that I've discovered

Compound symmetry
Greenhouse–Geisser correction
Huynh–Feldt correction
Lower bound
Mauchly's test
Repeated-measures ANOVA
Sphericity

Smart Alex's tasks

- **Task 1**: There is often concern among students as to the consistency of marking between lecturers. It is common that lecturers obtain reputations for being 'hard' or 'light' markers (or to use the students' terminology, 'evil manifestations from Beelzebub's bowels' and 'nice people') but there is often little to substantiate these reputations. A group of students investigated the consistency of marking by submitting the same essays to four different lecturers. The mark given by each lecturer was recorded for each of the eight essays. It was important that the same essays were used for all lecturers because this eliminated any individual differences in the standard of work that each lecturer marked. This design is repeated-measures because every lecturer marked every essay. The independent variable was the lecturer who marked the report and the dependent variable was the percentage mark given. The data are in the file **TutorMarks.sav**. Conduct a one-way ANOVA on these data by hand. ②
- **Task 2**: Repeat the analysis above on SPSS and interpret the results. ②
- **Task 3**: Imagine I wanted to look at the effect alcohol has on the roving eye. The 'roving eye' effect is the propensity of people in relationships to 'eye-up' members of the opposite sex. I took 20 men and fitted them with incredibly sophisticated glasses that could track their eye movements and record both the movement and the object being observed (this is the point at which it should be apparent that I'm making it up as I go along). Over four different nights I plied these poor souls with 1, 2, 3 or 4 pints of strong lager in a night-club. Each night I measured how many different women they eyed up (a woman was categorized as having been eyed up if the man's eye moved from her head to her toe and back up again). To validate this measure we also

collected the amount of dribble on the man's chin while looking at a woman. The data are in the file **RovingEye.sav**. Analyse them with a one-way ANOVA. ②

- **Task 4**: In the previous chapter we came across the beer-goggles effect, a severe perceptual distortion after imbibing several pints of alcohol. The specific visual distortion is that previously unattractive people suddenly become the hottest thing since Spicy Gonzalez' extra-hot Tabasco-marinated chillies. In short, one minute you're standing in a zoo admiring the orang-utans, and the next you're wondering why someone would put Angela Gossow in a cage. Anyway, in that chapter, a blatantly fabricated data set demonstrated that the beer-goggles effect was much stronger for men than women, and took effect only after two pints. Imagine we wanted to follow this finding up to look at what factors mediate the beer-goggles effect. Specifically, we thought that the beer-goggles effect might be made worse by the fact that it usually occurs in clubs which have dim lighting. We took a sample of 26 men (because the effect is stronger in men) and gave them various doses of alcohol over four different weeks (0 pints, 2 pints, 4 pints and 6 pints of lager). This is our first independent variable, which we'll call alcohol consumption, and it has four levels. Each week (and, therefore, in each state of drunkenness) participants were asked to select a mate in a normal club (that had dim lighting) and then select a second mate in a specially designed club that had bright lighting. As such, the second independent variable was whether the club had dim or bright lighting. The outcome measure was the attractiveness of each mate as assessed by a panel of independent judges. To recap, all participants took part in all levels of the alcohol consumption variable, and selected mates in both brightly and dimly lit clubs. The data are in the file **BeerGogglesLighting.sav**. Analyse them with a two-way repeated-measures ANOVA. ②
- **Task 5**: Using SPSS Tip 13.2, change the syntax in **SimpleEffectsAttitude.sps** to look at the effect of drink at different levels of imagery. ③

Answers can be found on the companion website.

Further reading

Field, A. P. (1998). A bluffer's guide to sphericity. *Newsletter of the Mathematical, Statistical and Computing section of the British Psychological Society*, 6(1), 13–22. (Available in the additional material for this chapter.)

Howell, D. C. (2006). *Statistical methods for psychology* (6th ed.). Belmont, CA: Thomson. (Or you might prefer his *Fundamental Statistics for the Behavioral Sciences*, also in its 6th edition, 2007.)

Rosenthal, R., Rosnow, R. L., & Rubin, D. B. (2000). *Contrasts and effect sizes in behavioural research: a correlational approach*. Cambridge: Cambridge University Press. (This is quite advanced but really cannot be bettered for contrasts and effect size estimation.)

Online tutorials

The companion website contains the following Flash movie tutorials to accompany this chapter:

- Repeated Measures ANOVA using SPSS
- The *R* Plugin

Interesting real research

Field, A. P. (2006). The behavioral inhibition system and the verbal information pathway to children's fears. *Journal of Abnormal Psychology*, *115*(4), 742–752.

14 Mixed design ANOVA (GLM 5)

FIGURE 14.1
My 18th birthday cake

14.1. What will this chapter tell me? ①

Most teenagers are anxious and depressed, but I probably had more than my fair share. The parasitic leech that was the all boys grammar school that I attended had feasted on my social skills leaving in its wake a terrified husk. Although I had no real problem with playing my guitar and shouting in front of people, speaking to them was another matter entirely. In the band I felt at ease, in the real world I did not. Your 18th birthday is a time of great joy, where (in the UK at any rate) you cast aside the shackles of childhood and embrace the exciting new world of adult life. Your birthday cake might symbolize this happy transition by reflecting one of your great passions. Mine had a picture on it of a long-haired person who looked somewhat like me, slitting his wrists. That pretty much sums it up. Still, you can't lock yourself in your bedroom with your Iron Maiden CDs for ever and soon enough I tried to integrate with society. Between the ages of 16 and 18 this pretty much involved getting drunk. I quickly discovered that getting drunk made it much easier to speak to people, and getting *really* drunk made you unconscious and then the problem of speaking to people went away entirely. This situation was exacerbated by the sudden presence of girls in my social circle. I hadn't seen a girl since Clair Sparks; they were particularly problematic because not only had you to talk to them, but what you said had to be really impressive because then they might become your girlfriend. Also, in 1990, girls didn't like to talk about Iron Maiden – they probably still don't. Speed dating[1] didn't exist back then, but if it had it would have been a sick and twisted manifestation of hell on earth for me. The idea of having a highly pressured social situation where you *have* to think of something witty and amusing to say or be thrown to the baying vultures of eternal loneliness would have had me injecting pure alcohol into my eyeballs; at least that way I could be in a coma and unable to see the disappointment on the faces of those forced to spend 3 minutes in my company. That's what this chapter is all about: speed dating, oh, and mixed ANOVA too, but if I mention that you'll move swiftly on to the next chapter when the bell rings.

14.2. Mixed designs ②

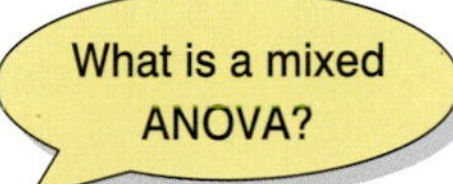

If you thought that the previous chapter was bad, well, I'm about to throw an added complication into the mix. We can combine repeated measures and independent designs, and this chapter looks at this situation. As if this wasn't bad enough, I'm also going to use this as an excuse to show you a design with three independent variables (at this point you should imagine me leaning back in my chair, cross-eyed, dribbling and laughing maniacally). A mixture of between-group and repeated-measures variables is called a **mixed design**. It should be obvious that you need at least two independent variables for this type of design to be possible, but you can have more complex scenarios too (e.g. two between-group and one repeated measures, one between-group and two repeated measures, or even two of each). SPSS allows you to test almost any design you might want to, and of virtually any degree of complexity. However, interaction terms are difficult enough to interpret with only two variables, so imagine how difficult they are if you include four! The best advice I can offer is to stick to three or fewer independent variables if you want to be able to interpret your interaction terms,[2] and certainly don't exceed four unless you want to give yourself a migraine.

This chapter will go through an example of a **mixed ANOVA**. There won't be any theory because really and truly you've probably had enough ANOVA theory by now to have a good idea of what's going on (you can read this as 'it's too complex for me and I'm going to cover up my own incompetence by pretending you don't need to know about it'). So, we look at an example using SPSS and then interpret the output. In the process you'll hopefully develop your understanding of interactions and how to break them down using contrasts.

[1] In case speed dating goes out of fashion and no one knows what I'm going on about, the basic idea is that lots of men and women turn up to a venue (or just men or just women if it's a gay night), one-half of the group sit individually at small tables and the remainder choose a table, get 3 minutes to impress the other person at the table with their tales of heteroscedastic data, then a bell rings and they get up and move to the next table. Having worked around all of the tables, the end of the evening is spent either stalking the person whom you fancied or avoiding the hideous mutant who was going on about hetero…something or other.

[2] Fans of irony will enjoy the four-way ANOVAs that I conducted in Field and Davey (1999) and Field and Moore (2005), to name but two examples!

14.3. What do men and women look for in a partner? ②

The example we're going to use in this chapter stays with the dating theme. It seems that lots of magazines go on all the time about how men and women want different things from relationships (or perhaps it's just my girlfriend's copies of *Marie Clare*, which I don't read – honestly). The big question to which we all want to know the answer is: are looks or personality more important? Imagine you wanted to put this to the test. You devised a cunning plan whereby you'd set up a speed-dating night. Little did the people who came along know that you'd got some of your friends to act as the dates. Each date varied in their attractiveness (attractive, average or ugly) and their charisma (charismatic, average and dull) and by combining these characteristics you get nine different combinations. Each combination was represented by one of your stooge dates. As such, your stooge dates were made up of nine different people. Three were extremely attractive people but differed in their personality: one had tons of charisma,[3] one had some charisma and the other was as dull as this book. Another three people were of average attractiveness, and again differed in their personality: one was highly charismatic, one had some charisma and the third was a dullard. The final three were, not wishing to be unkind in any way, pig-ugly, and again one was charismatic, one had some charisma and the final poor soul was mind-numbingly tedious. Obviously you had two sets of stooge dates: one set was male and the other female so that your participants could match up with dates of the appropriate sex.

The participants themselves were not these nine stooges, but 10 men and 10 women who came to the speed-dating event that you had set up. Over the course of the evening they speed dated all nine members of the opposite sex that you'd set up for them. After their 5 minute date, they rated how much they'd like to have a proper date with the person as a percentage (100% = 'I'd pay large sums of money for your phone number', 0% = 'I'd pay a large sum of money for a plane ticket to get me as far away from you as possible'). As such, each participant rated nine different people who varied in their attractiveness and personality. So, there are two repeated-measures variables: **Looks** (with three levels because the person could be attractive, average or ugly) and **Personality** (again with three levels because the person could have lots of charisma, have some charisma or be a dullard). The people giving the ratings could be male or female, so we should also include the gender of the person making the ratings (male or female), and this, of course, will be a between-group variable. The data are in Table 14.1.

14.4. Mixed ANOVA on SPSS ②

14.4.1. The main analysis ②

To enter these data into SPSS we use the same procedure as the two-way repeated-measures ANOVA that we came across in the previous chapter. Remember that each row in the data editor represents a single participant's data. If a person participates in all experimental conditions (in this case they date all of the people who differ in attractiveness and all of the people who differ in their charisma) then each experimental condition must be represented

[3] The highly attractive people with tons of charisma were, of course, taken to a remote cliff top and shot after the experiment because life is hard enough without having people like that floating around making you feel inadequate.

TABLE 14.1 Data from **LooksOrPersonality.sav** (Att = Attractive, Av = Average, Ug = Ugly)

	High Charisma			Some Charisma			Dullard		
Looks	**Att**	**Av**	**Ugly**	**Att**	**Av**	**Ug**	**Att**	**Av**	**Ug**
Male	86	84	67	88	69	50	97	48	47
	91	83	53	83	74	48	86	50	46
	89	88	48	99	70	48	90	45	48
	89	69	58	86	77	40	87	47	53
	80	81	57	88	71	50	82	50	45
	80	84	51	96	63	42	92	48	43
	89	85	61	87	79	44	86	50	45
	100	94	56	86	71	54	84	54	47
	90	74	54	92	71	58	78	38	45
	89	86	63	80	73	49	91	48	39
Female	89	91	93	88	65	54	55	48	52
	84	90	85	95	70	60	50	44	45
	99	100	89	80	79	53	51	48	44
	86	89	83	86	74	58	52	48	47
	89	87	80	83	74	43	58	50	48
	80	81	79	86	59	47	51	47	40
	82	92	85	81	66	47	50	45	47
	97	69	87	95	72	51	45	48	46
	95	92	90	98	64	53	54	53	45
	95	93	96	79	66	46	52	39	47

by a column in the data editor. In this experiment there are nine experimental conditions and so the data need to be entered in nine columns (the format is identical to Table 14.1). Therefore, create the following nine variables in the data editor with the names as given. For each one, you should also enter a full variable name (see section 3.4.2) for clarity in the output:

att_high	Attractive	+	High Charisma
av_high	Average Looks	+	High Charisma
ug_high	Ugly	+	High Charisma
att_some	Attractive	+	Some Charisma
av_some	Average Looks	+	Some Charisma
ug_some	Ugly	+	Some Charisma
att_none	Attractive	+	Dullard
av_none	Average Looks	+	Dullard
ug_none	Ugly	+	Dullard

SELF-TEST Once these variables have been created, enter the data as in Table 14.1. If you have problems entering the data then use the file **LooksOrPersonality.sav.**

First we have to define our repeated-measures variables, so access the define factors dialog box by selecting Analyze General Linear Model ▸ Repeated Measures... As with two-way repeated-measures ANOVA (see the previous chapter) we need to give names to our repeated-measures variables and specify how many levels they have. In this case there are two within-subject factors: **Looks** (attractive, average or ugly) and **Charisma** (high charisma, some charisma and dullard). In the define factors dialog box replace the word *factor1* with the word *Looks*. When you have given this repeated-measures factor a name, tell the computer that this variable has three levels by typing the number 3 into the box labelled *Number of Levels*. Click on Add to add this variable to the list of repeated-measures variables. This variable will now appear in the white box at the bottom of the dialog box and appears as *Looks(3)*. Now repeat this process for the second independent variable. Enter the word *Charisma* into the space labelled *Within-Subject Factor Name* and then, because there were three levels of this variable, enter the number 3 into the space labelled *Number of Levels*. Click on Add to include this variable in the list of factors; it will appear as *Charisma(3)*. The finished dialog box is shown in Figure 14.2. When you have entered both of the within-subject factors click on Define to go to the main dialog box.

The main dialog box is the same as when we did a factorial repeated-measures ANOVA in the previous chapter (see Figure 14.3). At the top of the *Within-Subjects Variables* box,

FIGURE 14.2 The define factors dialog box for factorial repeated-measures ANOVA

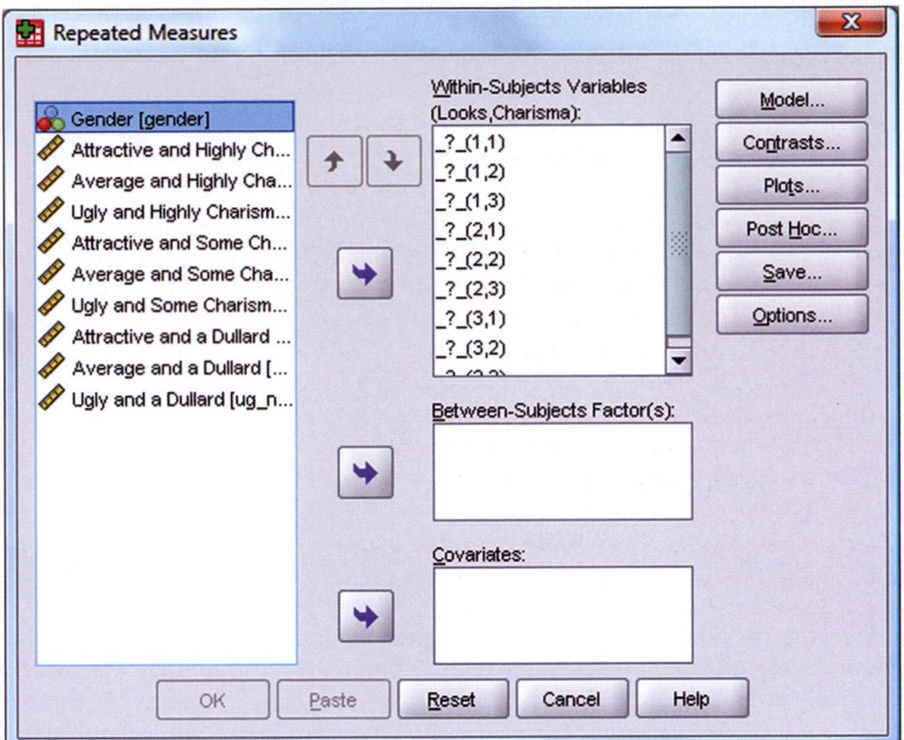

FIGURE 14.3

SPSS states that there are two factors: **Looks** and **Charisma**. In the box below there is a series of question marks followed by bracketed numbers. The numbers in brackets represent the levels of the factors (independent variables) – see the previous chapter for a more detailed explanation.

In this example, there are two independent variables and so there are two numbers in the brackets. The first number refers to levels of the first factor listed above the box (in this case **Looks**). The second number in the brackets refers to levels of the second factor listed above the box (in this case **Charisma**). As with the other repeated-measures ANOVAs we've come across, we have to replace the question marks with variables from the list on the left-hand side of the dialog box. With between-group designs, in which coding variables are used, the levels of a particular factor are specified by the codes assigned to them in the data editor. However, in repeated-measures designs, no such coding scheme is used and so we determine which condition to assign to a level at this stage (again look back to the previous chapter for more about this). For this reason, it is imperative that we think about the type of contrasts that we might want to do *before* entering variables into this dialog box. In this experiment, if we look at the first variable, **Looks**, there were three conditions: attractive, average and ugly. In many ways it makes sense to compare the attractive and ugly conditions to the average because the average person represents the norm (although it wouldn't be wrong to, for example, compare attractive and average to ugly). This comparison could be done by specifying a simple contrast (see Table 10.6) provided that we make sure that average is coded as our first or last category. Now, let's think about the second factor. The **Charisma** factor also has a category that represents the norm, and that is some charisma. Again we could use this as a control against which to compare our two extremes (lots of charisma and none whatsoever). Therefore, we could again conduct a simple contrast comparing everything against some charisma; therefore, this must be entered as either the first or last level.

Based on what has been discussed about using contrasts, it makes sense to have average as level 3 of the **Looks** factor and some charisma as the third level of the **Charisma** factor. The remaining levels can be decided arbitrarily. I have chosen attractive as level 1 and ugly as level 2 of the **Looks** factor. For the **Charisma** variable I chose high charisma as level 1 and none as level 2. These decisions mean that the variables should be entered as follows:

att_high	→	_?_(1,1)
att_none	→	_?_(1,2)
att_some	→	_?_(1,3)
ug_high	→	_?_(2,1)
ug_none	→	_?_(2,2)
ug_some	→	_?_(2,3)
av_high	→	_?_(3,1)
av_none	→	_?_(3,2)
av_some	→	_?_(3,3)

Unlike in the previous chapter, I've deliberately made this order different to how the variables are entered into the data editor. This is simply to illustrate that we can enter the variables in any order we like. So far the procedure has been similar to other factorial repeated-measures designs. However, we have a mixed design here, so we also need to specify our between-group factor as well. We do this by selecting **Gender** in the variables list and dragging it to the box labelled *Between-Subjects Factors* (or click on →). The completed dialog box should look exactly like Figure 14.4. I've already discussed the options for the buttons at the bottom of this dialog box, so I'll talk only about the ones of particular interest for this example.

FIGURE 14.4

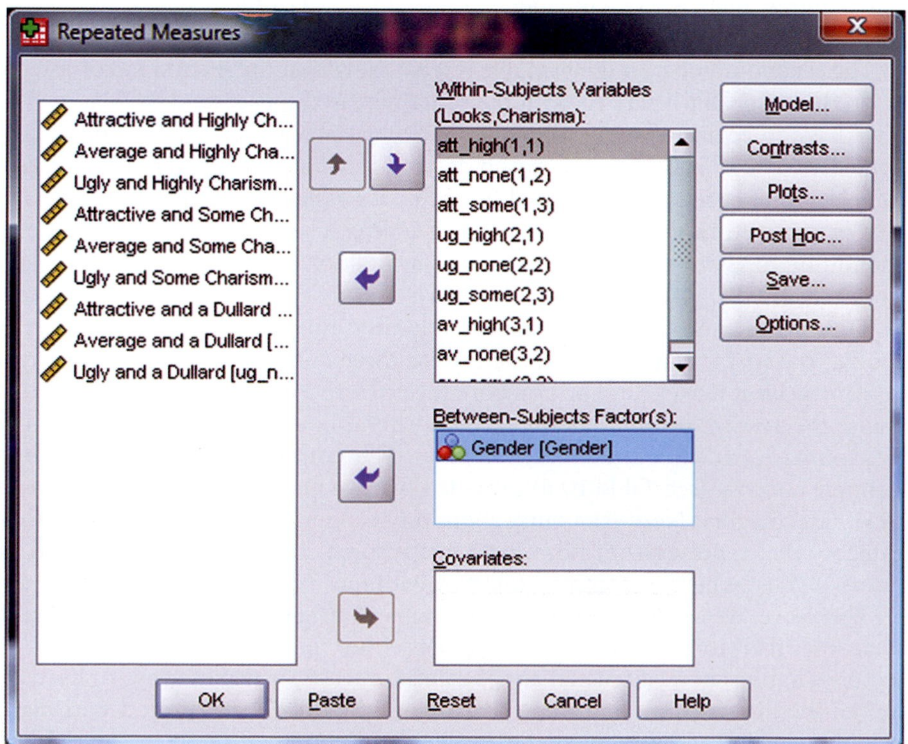

14.4.2. Other options ②

Following the main analysis it is interesting to compare levels of the independent variables to see whether they differ. As we've seen, there's no facility for entering contrast codes (unless you use syntax) so we have to rely on the standard contrasts available (see Table 10.6). Figure 14.5 shows the dialog box for conducting contrasts and is obtained by clicking on Contrasts... in the main dialog box. In the previous section I described why it might be interesting to use the average attractiveness and some charisma conditions as base categories for the looks and charisma factors respectively. We have used the contrasts dialog box before in sections 11.4.4 and 13.4.2 and so all I will say is that you should select a simple contrast for each independent variable. For both independent variables, we entered the variables such that the control category was the last one; therefore, we need not change the reference category for the simple contrast. Once the contrasts have been selected, click on Continue to return to the main dialog box.

Gender has only two levels (male or female) so we don't actually need to specify contrasts for this variable. The addition of a between-group factor also means that we can select *post hoc* tests for this variable by clicking on Post Hoc... . This action brings up the *post hoc* test dialog box (see section 10.3.2), which can be used as previously explained.

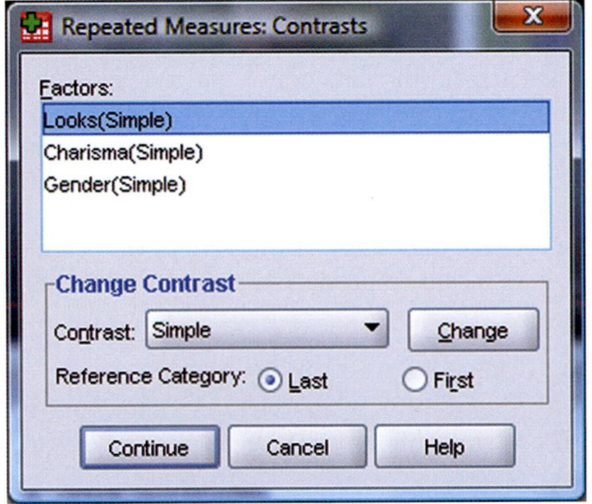

FIGURE 14.5

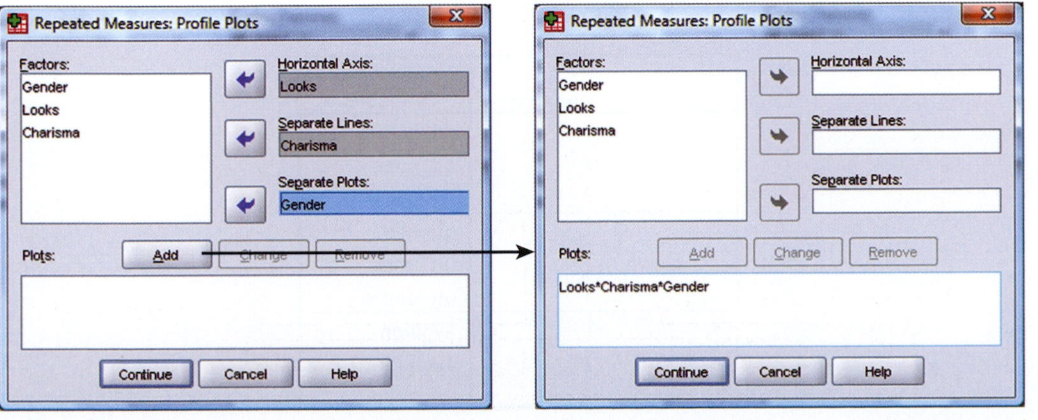

FIGURE 14.6 The plots dialog box for a three-way mixed ANOVA

However, we need not specify any *post hoc* tests here because **Gender** has only two levels.

The addition of a third independent variable makes it necessary to choose a different graph to the one in the previous chapter's example. Click on Plots... to access the dialog box in Figure 14.6. Place **Looks** in the slot labelled *Horizontal Axis* and **Charisma** in slot labelled *Separate Line*, and finally, place **Gender** in the slot labelled *Separate Plots*. When all three variables have been specified, don't forget to click on Add to add this combination to the list of plots. By asking SPSS to plot the looks × charisma × gender interaction, we should get the interaction graph for looks and charisma, but a separate version of this graph will be produced for male and female participants.

As far as other options are concerned, you should select the same ones that were chosen for the previous example (see section 13.8.5). It is worth selecting estimated marginal means for all effects (because these values will help you to understand any significant effects), but to save space I did not ask for confidence intervals for these effects because we have considered this part of the output in some detail already. When all of the appropriate options have been selected, run the analysis.

14.5. Output for mixed factorial ANOVA: main analysis ③

The initial output is the same as the two-way ANOVA example: there is a table listing the repeated-measures variables from the data editor and the level of each independent variable that they represent. The second table contains descriptive statistics (mean and standard deviation) for each of the nine conditions split according to whether participants were male or female (see SPSS Output 14.1). The names in this table are the names I gave the variables in the data editor (therefore, your output may differ slightly). These descriptive statistics are interesting because they show us the pattern of means across all experimental conditions (so, we use these means to produce the graphs of the three-way interaction).

SELF-TEST SPSS Output 14.2 shows the results of Mauchly's sphericity test. Based on what you have already learnt, was sphericity violated?

SPSS OUTPUT 14.1

Within-Subjects Factors

Measure:MEASURE_1

Looks	Charisma	Dependent Variable
1	1	att_high
	2	att_none
	3	att_some
2	1	ug_high
	2	ug_none
	3	ug_some
3	1	av_high
	2	av_none
	3	av_some

Descriptive Statistics

	Gender	Mean	Std. Deviation	N
Attractive and Highly Charismatic	Male	88.30	5.697	10
	Female	89.60	6.637	10
	Total	88.95	6.057	20
Attractive and a Dullard	Male	87.30	5.438	10
	Female	51.80	3.458	10
	Total	69.55	18.743	20
Attractive and Some Charisma	Male	88.50	5.740	10
	Female	87.10	6.806	10
	Total	87.80	6.170	20
Ugly and Highly Charismatic	Male	56.80	5.731	10
	Female	86.70	5.438	10
	Total	71.75	16.274	20
Ugly and a Dullard	Male	45.80	3.584	10
	Female	46.10	3.071	10
	Total	45.95	3.252	20
Ugly and Some Charisma	Male	48.30	5.376	10
	Female	51.20	5.453	10
	Total	49.75	5.476	20
Average and Highly Charismatic	Male	82.80	7.005	10
	Female	88.40	8.329	10
	Total	85.60	8.022	20
Average and a Dullard	Male	47.80	4.185	10
	Female	47.00	3.742	10
	Total	47.40	3.885	20
Average and Some Charisma	Male	71.80	4.417	10
	Female	68.90	5.953	10
	Total	70.35	5.314	20

SPSS OUTPUT 14.2

Mauchly's Test of Sphericity[b]

Measure: MEASURE_1

					Epsilon[a]		
Within Subjects Effect	Mauchly's W	Approx. Chi-Square	df	Sig.	Greenhouse-Geisser	Huynh-Feldt	Lower-bound
LOOKS	.960	.690	2	.708	.962	1.000	.500
CHARISMA	.929	1.246	2	.536	.934	1.000	.500
LOOKS * CHARISMA	.613	8.025	9	.534	.799	1.000	.250

Tests the null hypothesis that the error covariance matrix of the orthonormalized transformed dependent variables is proportional to an identity matrix.

a. May be used to adjust the degrees of freedom for the averaged tests of significance. Corrected tests are displayed in the Tests of Within-Subjects Effects table.

b.
Design: Intercept+GENDER
Within Subjects Design: LOOKS+CHARISMA+LOOKS*CHARISMA

SPSS Output 14.2 shows the results of Mauchly's sphericity test for each of the three repeated-measures effects in the model. None of the effects violate the assumption of sphericity because all of the values in the column labelled *Sig.* are above .05; therefore, we can assume sphericity when we look at our *F*-statistics.

SPSS Output 14.3 shows the summary table of the repeated-measures effects in the ANOVA with corrected *F*-values. As with factorial repeated-measures ANOVA, the output is split into sections for each of the effects in the model and their associated error terms. The only difference is that the interactions between our between-group variable of gender and the repeated-measures effects are included also.

Again, we need to look at the column labelled *Sig.* and if the values in this column are less than .05 for a particular effect then it is statistically significant. Working down from the top of the table we find a significant effect of looks, which means that if we ignore whether the date was charismatic, and whether the rating was from a man or a woman, then the attractiveness of a person significantly affected the ratings they received. The looks × gender interaction is also significant, which means that although the ratings were affected by whether the date was attractive, average or ugly, the way in which ratings were affected by attractiveness was different in male and female raters.

Next, we find a significant effect of charisma, which means that if we ignore whether the date was attractive, and whether the rating was from a man or a woman, then the charisma of a person significantly affected the ratings they received. The charisma × gender interaction is also significant, indicating that this effect of charisma differed in male and female raters.

There is a significant interaction between looks and charisma, which means that if we ignore the gender of the rater, the profile of ratings across different levels of attractiveness was different for highly charismatic dates, charismatic dates and dullards. (It is equally true to say this the opposite way around: the profile of ratings across different levels of charisma was different for attractive, average and ugly dates.) Just to add to the mounting confusion, the looks × charisma × gender interaction is also significant, meaning that the looks × charisma interaction was significantly different in men and women participants!

This is all a lot to take in so we'll look at each of these effects in turn in subsequent sections. First, though, we need to see what has happened to our main effect of gender.

SPSS OUTPUT 14.3

Tests of Within-Subjects Effects

Measure:MEASURE_1

Source		Type III Sum of Squares	df	Mean Square	F	Sig.
Looks	Sphericity Assumed	20779.633	2	10389.817	423.733	.000
	Greenhouse-Geisser	20779.633	1.923	10803.275	423.733	.000
	Huynh-Feldt	20779.633	2.000	10389.817	423.733	.000
	Lower-bound	20779.633	1.000	20779.633	423.733	.000
Looks * Gender	Sphericity Assumed	3944.100	2	1972.050	80.427	.000
	Greenhouse-Geisser	3944.100	1.923	2050.527	80.427	.000
	Huynh-Feldt	3944.100	2.000	1972.050	80.427	.000
	Lower-bound	3944.100	1.000	3944.100	80.427	.000
Error(Looks)	Sphericity Assumed	882.711	36	24.520		
	Greenhouse-Geisser	882.711	34.622	25.496		
	Huynh-Feldt	882.711	36.000	24.520		
	Lower-bound	882.711	18.000	49.040		
Charisma	Sphericity Assumed	23233.600	2	11616.800	328.250	.000
	Greenhouse-Geisser	23233.600	1.868	12437.761	328.250	.000
	Huynh-Feldt	23233.600	2.000	11616.800	328.250	.000
	Lower-bound	23233.600	1.000	23233.600	328.250	.000
Charisma * Gender	Sphericity Assumed	4420.133	2	2210.067	62.449	.000
	Greenhouse-Geisser	4420.133	1.868	2366.252	62.449	.000
	Huynh-Feldt	4420.133	2.000	2210.067	62.449	.000
	Lower-bound	4420.133	1.000	4420.133	62.449	.000
Error(Charisma)	Sphericity Assumed	1274.044	36	35.390		
	Greenhouse-Geisser	1274.044	33.624	37.891		
	Huynh-Feldt	1274.044	36.000	35.390		
	Lower-bound	1274.044	18.000	70.780		
Looks * Charisma	Sphericity Assumed	4055.267	4	1013.817	36.633	.000
	Greenhouse-Geisser	4055.267	3.197	1268.295	36.633	.000
	Huynh-Feldt	4055.267	4.000	1013.817	36.633	.000
	Lower-bound	4055.267	1.000	4055.267	36.633	.000
Looks * Charisma * Gender	Sphericity Assumed	2669.667	4	667.417	24.116	.000
	Greenhouse-Geisser	2669.667	3.197	834.945	24.116	.000
	Huynh-Feldt	2669.667	4.000	667.417	24.116	.000
	Lower-bound	2669.667	1.000	2669.667	24.116	.000
Error(Looks*Charisma)	Sphericity Assumed	1992.622	72	27.675		
	Greenhouse-Geisser	1992.622	57.554	34.622		
	Huynh-Feldt	1992.622	72.000	27.675		
	Lower-bound	1992.622	18.000	110.701		

SELF-TEST What is the difference between a main effect and an interaction?

14.5.1. The main effect of gender ②

The main effect of gender is listed separately from the repeated-measures effects in a table labelled **Tests of Between-Subjects Effects.** Before looking at this table it is important to check the assumption of homogeneity of variance using Levene's test (see section 5.6.1).

SELF-TEST Was the assumption of homogeneity of variance met (SPSS Output 14.4)?

SPSS OUTPUT 14.4

Levene's Test of Equality of Error Variances[a]

	F	df1	df2	Sig.
Attractive and Highly Charismatic	1.131	1	18	.302
Attractive and a Dullard	1.949	1	18	.180
Attractive and Some Charisma	.599	1	18	.449
Ugly and Highly Charismatic	.005	1	18	.945
Ugly and a Dullard	.082	1	18	.778
Ugly and Some Charisma	.124	1	18	.729
Average and Highly Charismatic	.102	1	18	.753
Average and a Dullard	.004	1	18	.950
Average and Some Charisma	1.763	1	18	.201

Tests the null hypothesis that the error variance of the dependent variable is equal across groups.

a.

Design: Intercept+GENDER
Within Subjects Design: LOOKS+CHARISMA+LOOKS*CHARISMA

Tests of Between-Subjects Effects

Measure:MEASURE_1
Transformed Variable:Average

Source	Type III Sum of Squares	df	Mean Square	F	Sig.
Intercept	94027.756	1	94027.756	20036.900	.000
Gender	.022	1	.022	.005	.946
Error	84.469	18	4.693		

SPSS produces a table listing Levene's test for each of the repeated-measures variables in the data editor, and we need to look for any variable that has a significant value. SPSS Output 14.4 shows both tables. The table showing Levene's test indicates that variances are homogeneous for all levels of the repeated-measures variables (because all significance values are greater than .05). If any values were significant, then this would compromise the

SPSS OUTPUT 14.5

Estimates

Measure:MEASURE_1

Gender	Mean	Std. Error	95% Confidence Interval	
			Lower Bound	Upper Bound
Male	68.600	.685	67.161	70.039
Female	68.533	.685	67.094	69.973

FIGURE 14.7

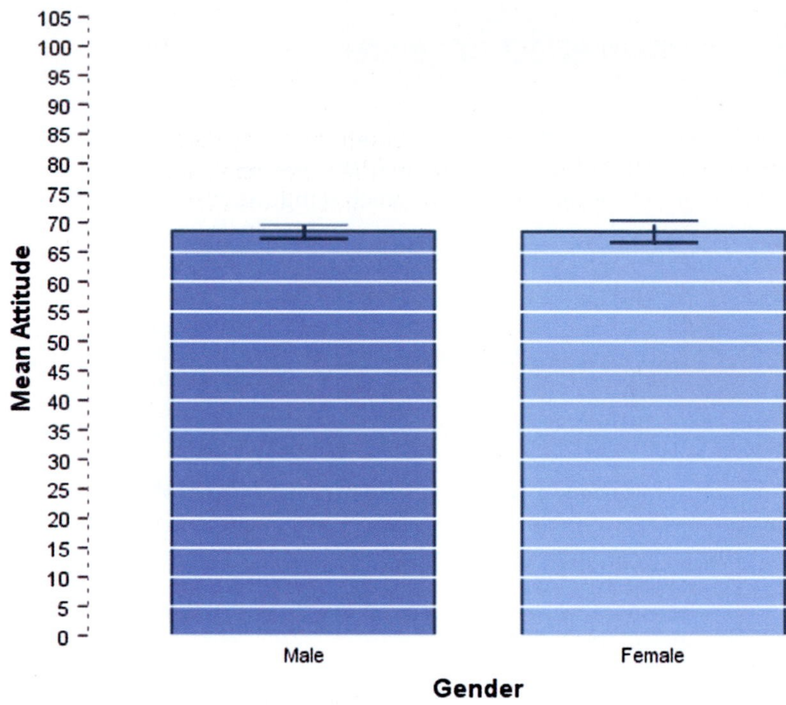

accuracy of the *F*-test for gender, and we would have to consider transforming all of our data to stabilize the variances between groups (see Chapter 5). Fortunately, in this example a transformation is unnecessary. The second table shows the ANOVA summary table for the main effect of gender, and this reveals a non-significant effect (because the significance of .946 is greater than the standard cut-off point of .05).

We can report that there was a non-significant (*ns*) main effect of gender, $F(1, 18) < 1$, *ns*. This effect tells us that if we ignore all other variables, male participants' ratings were basically the same as females'. If you requested that SPSS display means for the gender effect you should scan through your output and find the table in a section headed *Estimated Marginal Means*. SPSS Output 14.5 is a table of means for the main effect of gender with the associated standard errors. This information is plotted in Figure 14.7. It is clear from this graph that men and women's ratings were generally the same.

14.5.2. The main effect of looks ②

SELF-TEST Based on the previous section and what you have learned in previous chapters, can you interpret the main effect of 'looks'? (SPSS Output 14.3)?

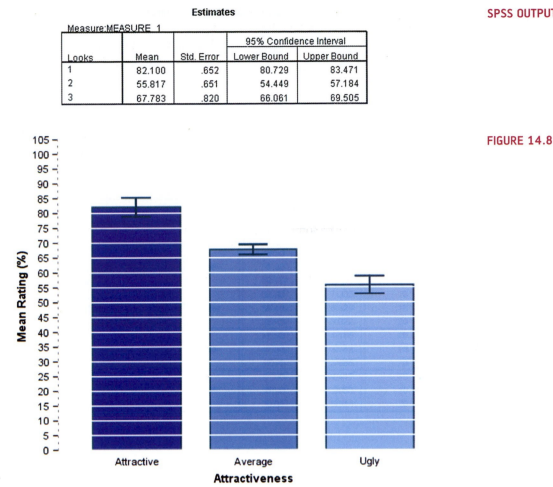

SPSS OUTPUT 14.6

Estimates

Measure:MEASURE_1

Looks	Mean	Std. Error	95% Confidence Interval	
			Lower Bound	Upper Bound
1	82.100	.652	80.729	83.471
2	55.817	.651	54.449	57.184
3	67.783	.820	66.061	69.505

FIGURE 14.8

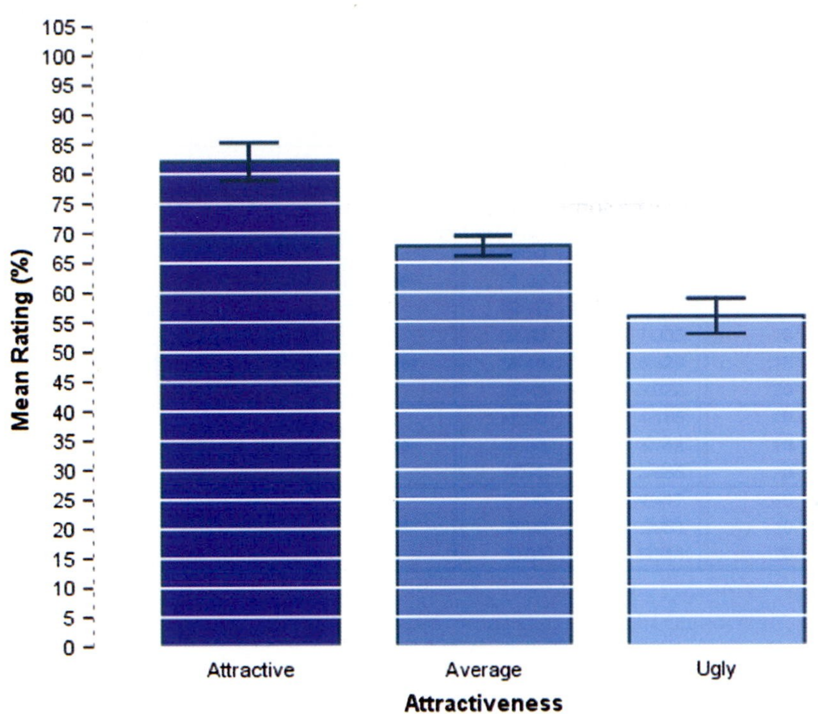

We came across the main effect of looks in SPSS Output 14.3. Now we're going to have a look at what this effect means. We can report that there was a significant main effect of looks, $F(2, 36) = 423.73$, $p < .001$. This effect tells us that if we ignore all other variables, ratings were different for attractive, average and unattractive dates. If you requested that SPSS display means for the looks effect (I'll assume you did from now on) you will find the table in a section headed *Estimated Marginal Means*. SPSS Output 14.6 is a table of means for the main effect of looks with the associated standard errors. The levels of looks are labelled simply 1, 2 and 3, and it's down to you to remember how you entered the variables (or you can look at the summary table that SPSS produces at the beginning of the output – see SPSS Output 14.1). If you followed what I did then level 1 is attractive, level 2 is ugly and level 3 is average. To make things easier, this information is plotted in Figure 14.8. You can see that as attractiveness falls, the mean rating falls too. So this main effect seems to reflect that the raters were more likely to express a greater interest in going out with attractive people than average or ugly people. However, we really need to look at some contrasts to find out exactly what's going on.

SPSS Output 14.7 shows the contrasts that we requested. For the time being, just look at the row labelled *Looks*. Remember that we did a simple contrast, and so we get a contrast comparing level 1 to level 3, and then comparing level 2 to level 3; because of the order in which we entered the variables, these contrasts represent attractive compared to average (level 1 vs. level 3) and ugly compared to average (level 2 vs. level 3). Looking at the values of F for each contrast, and their related significance values, tells us that the effect

SPSS OUTPUT 14.7

Tests of Within-Subjects Contrasts

Measure:MEASURE_1

Source	Looks	Charisma	Type III Sum of Squares	df	Mean Square	F	Sig.
Looks	Level 1 vs. Level 3	Charisma	4099.339	1	4099.339	226.986	.000
	Level 2 vs. Level 3	Charisma	2864.022	1	2864.022	160.067	.000
Looks * Gender	Level 1 vs. Level 3	Charisma	781.250	1	781.250	43.259	.000
	Level 2 vs. Level 3	Charisma	540.800	1	540.800	30.225	.000
Error(Looks)	Level 1 vs. Level 3	Charisma	325.078	18	18.060		
	Level 2 vs. Level 3	Charisma	322.067	18	17.893		
Charisma	Looks * Charisma	Level 1 vs. Level 3	3276.800	1	3276.800	109.937	.000
		Level 2 vs. Level 3	4500.000	1	4500.000	227.941	.000
Charisma * Gender	Looks * Charisma	Level 1 vs. Level 3	810.689	1	810.689	27.199	.000
		Level 2 vs. Level 3	665.089	1	665.089	33.689	.000
Error(Charisma)	Looks * Charisma	Level 1 vs. Level 3	536.511	18	29.806		
		Level 2 vs. Level 3	355.356	18	19.742		
Looks * Charisma	Level 1 vs. Level 3	Level 1 vs. Level 3	3976.200	1	3976.200	21.944	.000
		Level 2 vs. Level 3	441.800	1	441.800	4.091	.058
	Level 2 vs. Level 3	Level 1 vs. Level 3	911.250	1	911.250	6.231	.022
		Level 2 vs. Level 3	7334.450	1	7334.450	88.598	.000
Looks * Charisma * Gender	Level 1 vs. Level 3	Level 1 vs. Level 3	168.200	1	168.200	.928	.348
		Level 2 vs. Level 3	6552.200	1	6552.200	60.669	.000
	Level 2 vs. Level 3	Level 1 vs. Level 3	1711.250	1	1711.250	11.701	.003
		Level 2 vs. Level 3	110.450	1	110.450	1.334	.263
Error(Looks*Charisma)	Level 1 vs. Level 3	Level 1 vs. Level 3	3261.600	18	181.200		
		Level 2 vs. Level 3	1944.000	18	108.000		
	Level 2 vs. Level 3	Level 1 vs. Level 3	2632.500	18	146.250		
		Level 2 vs. Level 3	1490.100	18	82.783		

of attractiveness represented the fact that attractive dates were rated significantly higher than average dates, $F(1, 18) = 226.99$, $p < .001$, and average dates were rated significantly higher than ugly ones, $F(1, 18) = 160.07$, $p < .001$.

14.5.3. The main effect of charisma ②

The main effect of charisma is in SPSS Output 14.3. We can report that there was a significant main effect of charisma, $F(2, 36) = 328.25$, $p < .001$. This effect tells us that if we ignore all other variables, ratings were different for highly charismatic, a bit charismatic and dullard people. The table labelled **Charisma** in the section headed *Estimated Marginal Means* tells us what this effect means (SPSS Output 14.8). Again, the levels of charisma are labelled simply 1, 2 and 3. If you followed what I did then level 1 is high charisma, level 2 is no charisma and level 3 is some charisma. This information is plotted in Figure 14.9: as charisma declines, the mean rating falls too. So this main effect seems to reflect that the raters were more likely to express a greater interest in going out with charismatic people than average people or dullards.

SPSS Output 14.7 shows the contrasts that we requested. Looking at the row labelled *Charisma* and remembering that we requested simple contrasts, we get a contrast comparing level 1 to level 3, and then comparing level 2 to level 3. How we interpret these contrasts depends on the order in which we entered the repeated-measures variables: in this case these contrasts represent high charisma compared to some charisma (level 1 vs. level 3) and no charisma compared to some charisma (level 2 vs. level 3). The contrasts tell us that the effect of charisma represented the fact that highly charismatic dates were rated significantly higher than dates with some charisma, $F(1, 18) = 109.94$, $p < .001$, and dates with some charisma were rated significantly higher than dullards, $F(1, 18) = 227.94$, $p < .001$.

SPSS OUTPUT 14.8

Estimates

Measure:MEASURE_1

Charisma	Mean	Std. Error	95% Confidence Interval	
			Lower Bound	Upper Bound
1	82.100	1.010	79.978	84.222
2	54.300	.573	53.096	55.504
3	69.300	.732	67.763	70.837

FIGURE 14.9

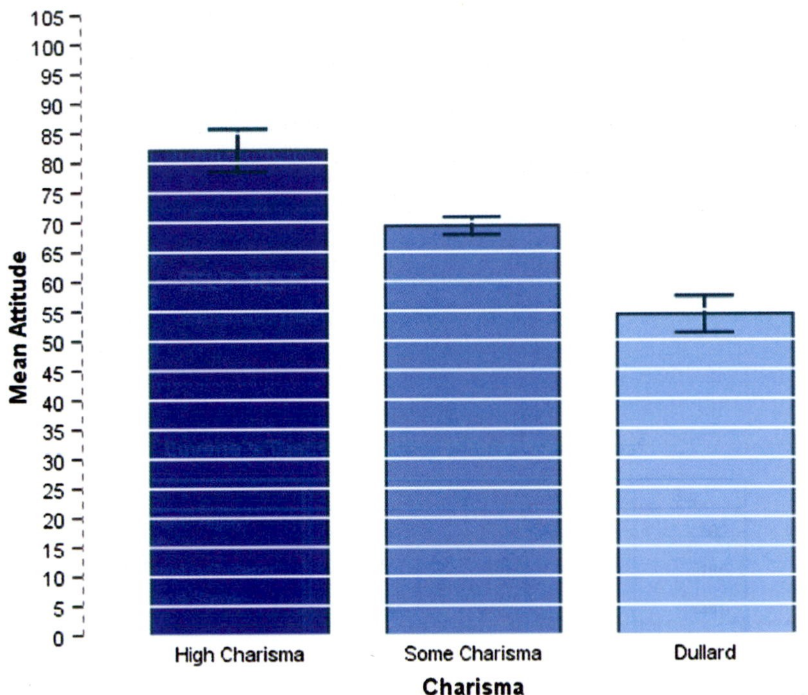

14.5.4. The interaction between gender and looks ②

SPSS Output 14.3 indicated that gender interacted in some way with the attractiveness of the date. From the summary table we can report that there was a significant interaction between the attractiveness of the date and the gender of the participant, $F(2, 36) = 80.43$, $p < .001$. This effect tells us that the profile of ratings across dates of different attractiveness was different for men and women. We can use the estimated marginal means to determine the nature of this interaction (or we could have asked SPSS for a plot of gender × looks using the dialog box in Figure 14.6). The means and interaction graph (Figure 14.10 and SPSS Output 14.9) shows the meaning of this result. The graph shows the average male ratings of dates of different attractiveness ignoring how charismatic the date was (circles). The women's scores are shown as squares. The graph clearly shows that male and female ratings are very similar for average-looking dates, but men give higher ratings (i.e. they're really keen to go out with these people) than women for attractive dates, but women express more interest in going out with ugly people than men. In general this interaction seems to suggest than men's interest in dating a person is more influenced by their looks than for females. Although both male's and female's interest decreases as attractiveness decreases, this decrease is more pronounced for men. This interaction can be clarified using the contrasts in SPSS Output 14.7.

SPSS OUTPUT 14.9

4. Gender * Looks

Measure:MEASURE_1

Gender	Looks	Mean	Std. Error	95% Confidence Interval	
				Lower Bound	Upper Bound
Male	1	88.033	.923	86.095	89.972
	2	50.300	.921	48.366	52.234
	3	67.467	1.159	65.031	69.902
Female	1	76.167	.923	74.228	78.105
	2	61.333	.921	59.399	63.267
	3	68.100	1.159	65.665	70.535

FIGURE 14.10

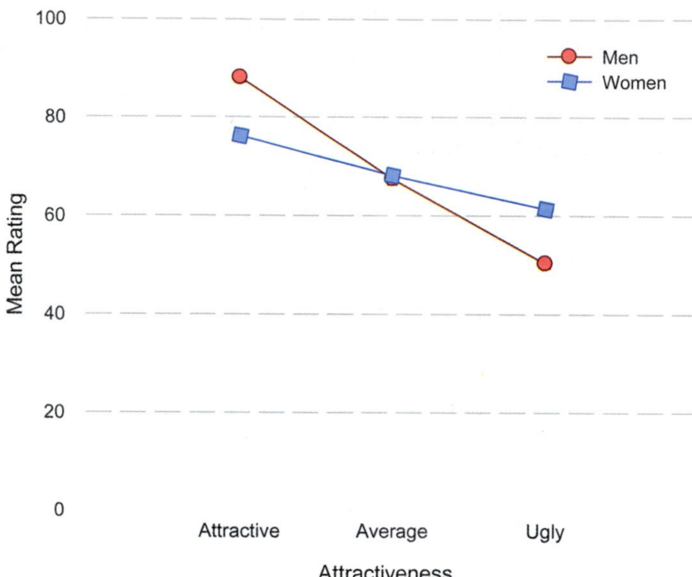

14.5.4.1. Looks x gender interaction 1: attractive vs. average, male vs. female ②

The contrast for the first interaction term looks at level 1 of looks (attractive) compared to level 3 (average), comparing male and female scores. This contrast is highly significant, $F(1, 18) = 43.26$, $p < .001$. This result tells us that the increased interest in attractive dates compared to average-looking dates found for men is significantly more than for women. So, in Figure 14.10 the slope of the line between the circles representing male ratings of attractive dates and average dates is steeper than the line joining the squares representing female ratings of attractive dates and average dates. We can conclude that the preferences for attractive dates, compared to average-looking dates, are greater for males than females.

14.5.4.2. Looks x gender interaction 2: ugly vs. average, male vs. female ②

The second contrast compares level 2 of looks (ugly) to level 3 (average), comparing male and female scores. This contrast is highly significant, $F(1, 18) = 30.23$, $p < .001$. This tells us that the decreased interest in ugly dates compared to average-looking dates found for men is significantly more than for women. So, in Figure 14.10 the slope of the line between the circles representing male ratings of ugly dates and average dates is steeper than the line joining the squares representing female ratings of ugly dates and average dates. We can conclude that the preferences for average-looking dates, compared to ugly dates, are greater for males than females.

14.5.5. The interaction between gender and charisma ②

SPSS Output 14.3 indicated that gender interacted in some way with how charismatic the date was. From the summary table we should report that there was a significant interaction between the attractiveness of the date and the gender of the participant, $F(2, 36) = 62.45$, $p < .001$. This effect tells us that the profile of ratings across dates of different levels of charisma was different for men and women. The estimated marginal means (or a plot of gender × charisma using the dialog box in Figure 14.6) tell us the meaning of this interaction (see Figure 14.11 and SPSS Output 14.10). The graph shows the average male ratings of dates of different levels of charisma ignoring how attractive they were (circles). The women's scores are shown as squares. The graph shows almost the reverse pattern as for the attractiveness data; again male and female ratings are very similar for dates with normal amounts of charisma, but this time men show more interest in dates who are dullards than women do, and women show slightly more interest in very charismatic dates than men do. In general this interaction seems to suggest than women's interest in dating a person is more influenced by their charisma than for men. Although both male's and female's interest decreases as charisma decreases, this decrease is more pronounced for females. This interaction can be clarified using the contrasts in SPSS Output 14.7.

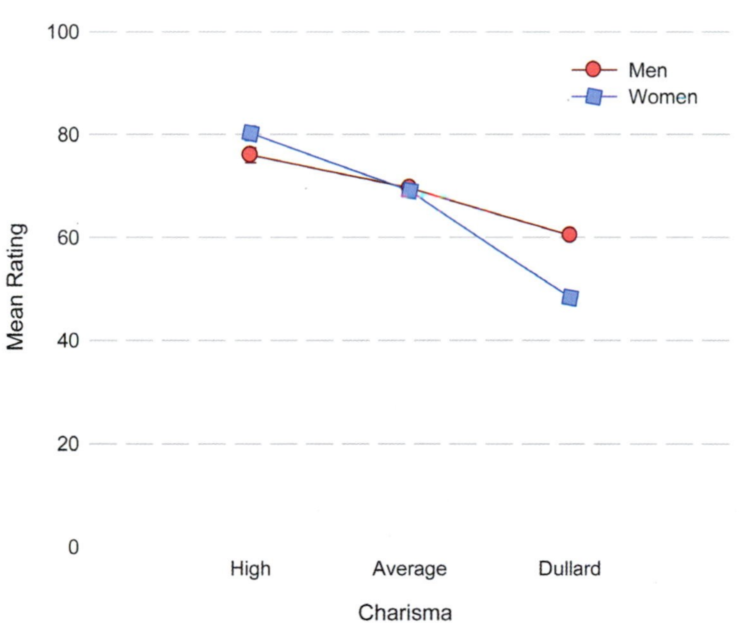

FIGURE 14.11

SPSS OUTPUT 14.10

5. Gender * Charisma

Measure:MEASURE_1

Gender	Charisma	Mean	Std. Error	95% Confidence Interval Lower Bound	Upper Bound
Male	1	75.967	1.428	72.966	78.967
	2	60.300	.810	58.598	62.002
	3	69.533	1.035	67.360	71.707
Female	1	88.233	1.428	85.233	91.234
	2	48.300	.810	46.598	50.002
	3	69.067	1.035	66.893	71.240

14.5.5.1. Charisma × gender interaction 1: high vs. some charisma, male vs. female ②

The first contrast for this interaction term looks at level 1 of charisma (high charisma) compared to level 3 (some charisma), comparing male and female scores. This contrast is highly significant, $F(1, 18) = 27.20$, $p < .001$. This result tells us that the increased interest in highly charismatic dates compared to averagely charismatic dates found for women is significantly more than for men. So, in Figure 14.11 the slope of the line between the squares representing female ratings of very charismatic dates and dates with some charisma is steeper than the line joining the circles representing male ratings of very charismatic dates and dates with some charisma. We can conclude that the preferences for very charismatic dates, compared to averagely charismatic dates, are greater for females than males.

14.5.5.2. Charisma × gender interaction 2: dullard vs. some charisma, male vs. female ②

The second contrast for this interaction term looks at level 2 of charisma (dullard) compared to level 3 (some charisma), comparing male and female scores. This contrast is highly significant, $F(1, 18) = 33.69$, $p < .001$. This result tells us that the decreased interest in dullard dates compared to averagely charismatic dates found for women is significantly more than for men. So, in Figure 14.11 the slope of the line between the squares representing female ratings of dates with some charisma and dullard dates is steeper than the line joining the circles representing male ratings of dates with some charisma and dullard dates. We can conclude that the preferences for dates with some charisma over dullards are greater for females than males.

14.5.6. The interaction between attractiveness and charisma ②

SPSS Output 14.3 indicated that the attractiveness of the date interacted in some way with how charismatic the date was. From the summary table we should report that there was a significant interaction between the attractiveness of the date and the charisma of the date, $F(4, 72) = 36.63$, $p < .001$. This effect tells us that the profile of ratings across dates of different levels of charisma was different for attractive, average and ugly dates. The estimated marginal means (or a plot of looks × charisma using the dialog box in Figure 14.6) tell us the meaning of this interaction (see SPSS Output 14.11 and Figure 14.12).

SPSS OUTPUT 14.11

6. Looks * Charisma

Measure:MEASURE_1

Looks	Charisma	Mean	Std. Error	95% Confidence Interval Lower Bound	95% Confidence Interval Upper Bound
1	1	88.950	1.383	86.045	91.855
	2	69.550	1.019	67.409	71.691
	3	87.800	1.408	84.842	90.758
2	1	71.750	1.249	69.126	74.374
	2	45.950	.746	44.382	47.518
	3	49.750	1.211	47.206	52.294
3	1	85.600	1.721	81.985	89.215
	2	47.400	.888	45.535	49.265
	3	70.350	1.172	67.888	72.812

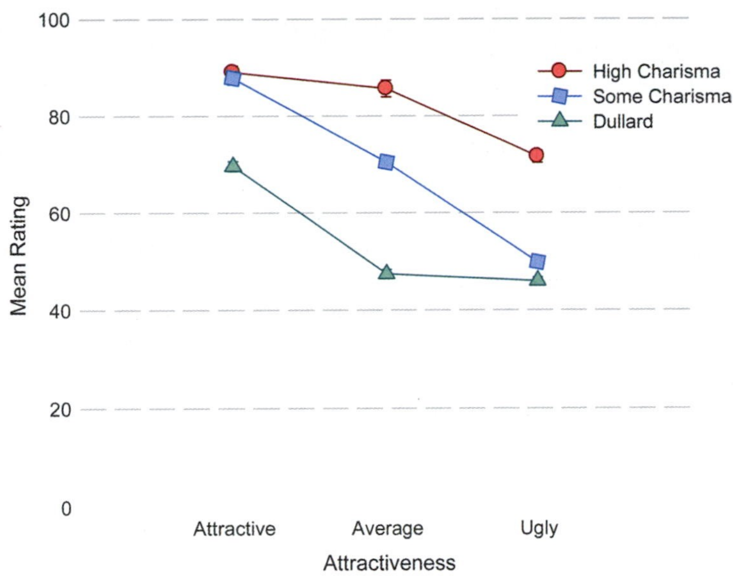

FIGURE 14.12

The graph shows the average ratings of dates of different levels of attractiveness when the date also had high levels of charisma (circles), some charisma (squares) and no charisma (triangles). Look first at the difference between attractive and average-looking dates. The interest in highly charismatic dates doesn't change (the line is more or less flat between these two points), but for dates with some charisma or no charisma interest levels decline. So, if you have lots of charisma you can get away with being average looking and people will still want to date you. Now, if we look at the difference between average-looking and ugly dates, a different pattern is observed. For dates with no charisma (triangles) there is no difference between ugly and average people (so if you're a dullard you have to be really attractive before people want to date you). However, for those with charisma, there is a decline in interest if you're ugly (so, if you're ugly, having charisma won't help you much). This interaction is very complex, but we can break it down using the contrasts in SPSS Output 14.7.

14.5.6.1. Looks × charisma interaction 1: attractive vs. average, high charisma vs. some charisma ②

The first contrast for this interaction term investigates level 1 of looks (attractive) compared to level 3 (average looking), comparing level 1 of charisma (high charisma) to level 3 of charisma (some charisma). This is like asking 'is the difference between high charisma and some charisma the same for attractive people and average-looking people?' The best way to understand what this contrast is testing is to extract the relevant bit of the interaction graph in Figure 14.12. If you look at this you can see that the interest (as indicated by high ratings) in attractive dates was the same

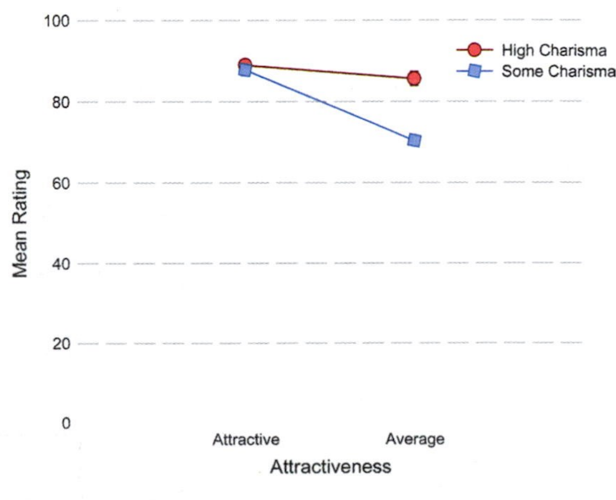

regardless of whether they had high or average charisma. However, for average-looking dates, there was more interest when that person had high charisma rather than average. The contrast is highly significant, $F(1, 18) = 21.94$, $p < .001$, and tells us that as dates become less attractive there is a greater decline in interest when charisma is average compared to when charisma is high.

14.5.6.2. Looks x charisma interaction 2: attractive vs. average, dullard vs. some charisma ②

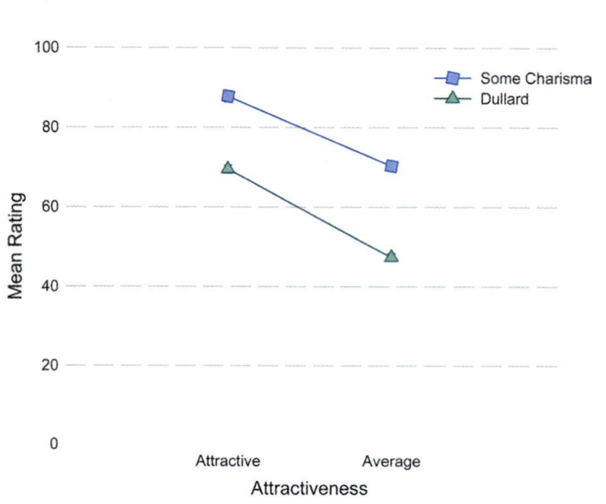

The second contrast for this interaction term investigates level 1 of looks (attractive) compared to level 3 (average looking), when comparing level 2 of charisma (dullard) to level 3 of charisma (some charisma). This is like asking 'is the difference between no charisma and some charisma the same for attractive people and average-looking people?' Again, the best way to understand what this contrast is testing is to extract the relevant bit of the interaction graph in Figure 14.12. If you look at this you can see that the interest (as indicated by high ratings) in attractive dates was higher when they had some charisma than when they were a dullard. The same is also true for average-looking dates. In fact the two lines are fairly parallel. The contrast is not significant, $F(1, 18) = 4.09$, *ns*, and tells us that as dates become less attractive there is a decline in interest both when charisma is low and when there is no charisma at all.

14.5.6.3. Looks x charisma interaction 3: ugly vs. average, high charisma vs. some charisma ②

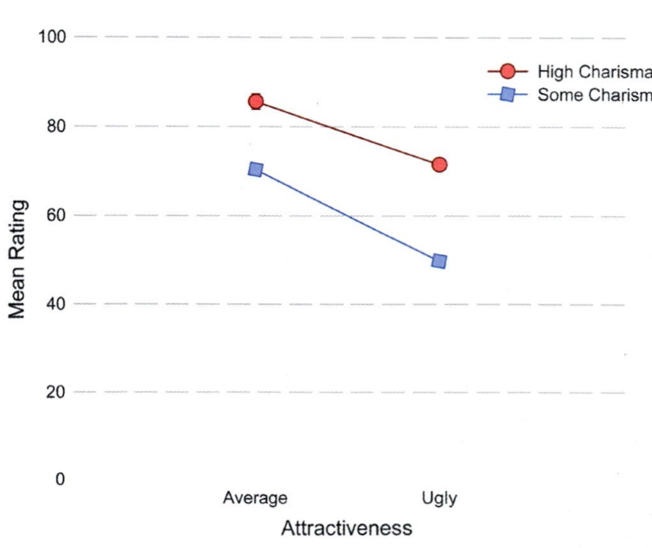

The third contrast for this interaction term investigates level 2 of looks (ugly) compared to level 3 (average looking), comparing level 1 of charisma (high charisma) to level 3 of charisma (some charisma). This is like asking 'is the difference between high charisma and some charisma the same for ugly people and average-looking people?' If we again extract the relevant bit of the interaction graph in Figure 14.12 you can see that the interest (as indicated by high ratings) decreases from average-looking dates to ugly ones in both high- and some-charisma dates; however, this fall is slightly greater in the low-charisma dates (the line connecting the squares is slightly steeper). The contrast is significant, $F(1, 18) = 6.23$, $p < .05$, and tells us that as dates become less attractive there is a greater decline in interest when charisma is low compared to when charisma is high.

14.5.6.4. Looks × charisma interaction 4: ugly vs. average, dullard vs. some charisma ②

The final contrast for this interaction term investigates level 2 of looks (ugly) compared to level 3 (average looking), when comparing level 2 of charisma (dullard) to level 3 of charisma (some charisma). This is like asking 'is the difference between no charisma and some charisma the same for ugly people and average-looking people?' If we extract the relevant bit of the interaction graph in Figure 14.12 you can see that the interest (as indicated by high ratings) in average-looking dates was higher when they had some charisma than when they were a dullard, but for ugly dates the ratings were roughly the same regardless of the level of charisma. This contrast is highly significant, $F(1, 18) = 88.60$, $p < .001$, and tells us that as dates become less attractive the decline in interest in dates with a bit of charisma is significantly greater than for dullards.

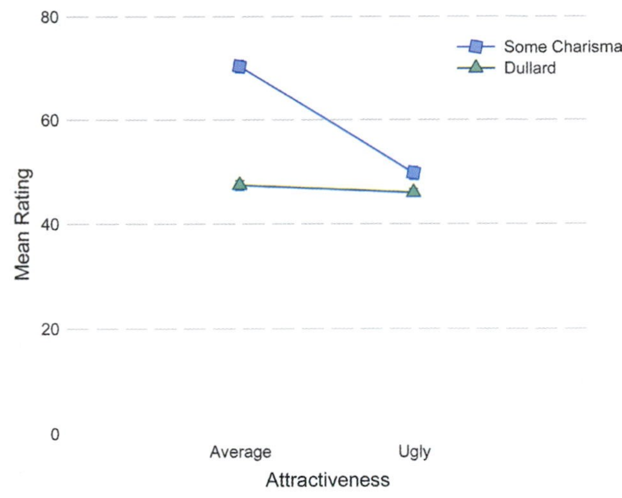

14.5.7. The interaction between looks, charisma and gender ③

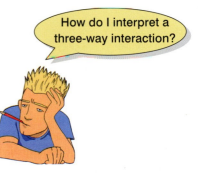

How do I interpret a three-way interaction?

The three-way interaction tells us whether the looks × charisma interaction described above is the same for men and women (i.e. whether the combined effect of attractiveness of the date and their level of charisma is the same for male participants as for female subjects). SPSS Output 14.3 tells us that there is a significant three-way looks × charisma × gender interaction, $F(4, 72) = 24.12$, $p < .001$. The nature of this interaction is revealed in Figure 14.13, which shows the looks × charisma interaction for men and women separately (the means on which this graph is based appear in SPSS Output 14.12). The male graph shows that when dates are attractive, men will express a high interest regardless of charisma levels (the circle, square and dot all overlap). At the opposite end of the attractiveness scale, when a date is ugly, regardless of charisma men will express very little interest (ratings are all low). The only time charisma makes any difference to a man is if the date is average looking, in which case high charisma boosts interest, being a dullard reduces interest, and having a bit of charisma leaves things somewhere in between. The take-home message is that men are superficial cretins who are more interested in physical attributes. The picture for women is very different. If someone has high levels of charisma then it doesn't really matter what they look like, women will express an interest in them (the line of circles is relatively flat). At the other extreme, if the date is a dullard, then they will express no interest in them, regardless of how attractive they are (the line of triangles is relatively flat). The only time attractiveness makes a difference is when someone has an average amount of charisma, in which case being attractive boosts interest, and being ugly reduces it. Put another way, women prioritize charisma over physical appearance. Again, we can look at some contrasts to further break this interaction down (SPSS Output 14.7). These contrasts are similar to those for the looks × charisma interaction, but they now also take into account the effect of gender as well!

SPSS OUTPUT 14.12

7. Gender * Looks * Charisma

Measure:MEASURE_1

Gender	Looks	Charisma	Mean	Std. Error	95% Confidence Interval	
					Lower Bound	Upper Bound
Male	1	1	88.300	1.956	84.191	92.409
		2	87.300	1.441	84.273	90.327
		3	88.500	1.991	84.317	92.683
	2	1	56.800	1.767	53.089	60.511
		2	45.800	1.055	43.583	48.017
		3	48.300	1.712	44.703	51.897
	3	1	82.800	2.434	77.687	87.913
		2	47.800	1.255	45.163	50.437
		3	71.800	1.657	68.318	75.282
Female	1	1	89.600	1.956	85.491	93.709
		2	51.800	1.441	48.773	54.827
		3	87.100	1.991	82.917	91.283
	2	1	86.700	1.767	82.989	90.411
		2	46.100	1.055	43.883	48.317
		3	51.200	1.712	47.603	54.797
	3	1	88.400	2.434	83.287	93.513
		2	47.000	1.255	44.363	49.637
		3	68.900	1.657	65.418	72.382

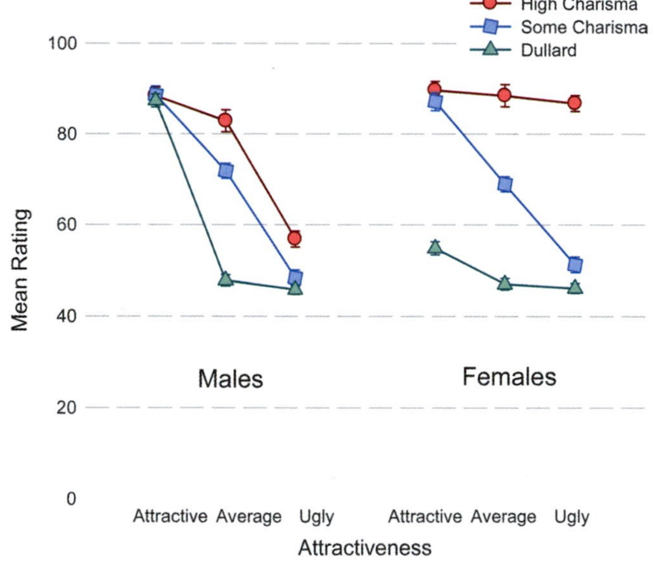

FIGURE 14.13 Graphs showing the looks by charisma interaction for men and women. Lines represent high charisma (circles), some charisma (squares) and no charisma (triangles)

14.5.7.1. Looks × charisma × gender interaction 1: attractive vs. average, high charisma vs. some charisma, male vs. female ③

The first contrast for this interaction term compares level 1 of looks (attractive) to level 3 (average looking), when level 1 of charisma (high charisma) is compared to level 3 of charisma (some charisma) in males compared to females, $F(1, 18) < 1$, *ns*. If we extract the relevant bits of the interaction graph in Figure 14.3, we can see that interest (as indicated by high ratings) in attractive dates was the same regardless of whether they had high or some charisma. However, for average-looking dates, there was more interest when that person had high charisma rather than some charisma. Importantly, this pattern of results is the same in males and females and this is reflected in the non-significance of this contrast.

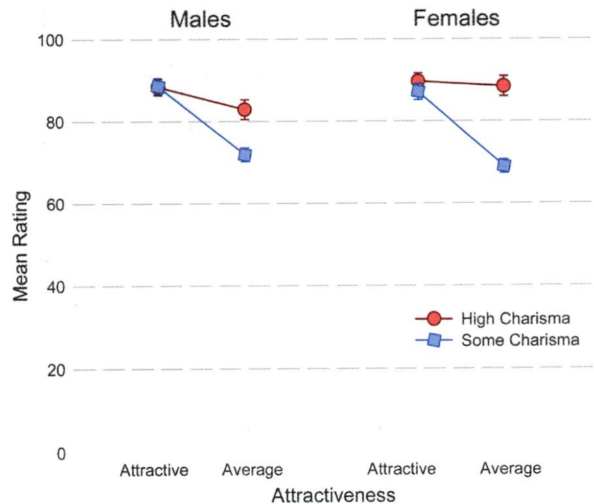

14.5.7.2. Looks x charisma x gender interaction 2: attractive vs. average, dullard vs. some charisma, male vs. female ③

The second contrast for this interaction term compares level 1 of looks (attractive) to level 3 (average looking), when level 2 of charisma (dullard) is compared to level 3 of charisma (some charisma), in men compared to women. Again, we extract the relevant bit of the interaction graph in Figure 14.3 and you can see that the patterns are different for men and women. This is reflected by the fact that the contrast is significant, $F(1, 18) = 60.67$, $p < .001$. To unpick this we need to look at the graph. First, if we look at average-looking dates, for both men and women more interest is expressed when the date has some charisma than when they have none (and the distance between the square and the triangle is about the same). So the difference doesn't appear to be here. If we now look at attractive dates, we see that men are equally interested in their dates regardless of their charisma, but for women, they're much less interested in an attractive person if they are a dullard. Put another way, for attractive dates, the distance between the square and the triangle is much smaller for men than it is for women. Another way to look at it is that for dates with some charisma, the reduction in interest as attractiveness goes down is about the same in men and women (the lines with squares have the same slope). However, for dates who are dullards, the decrease in interest if these dates are average looking rather than attractive is much more dramatic in men than women (the line with triangles is much steeper for men than it is for women).

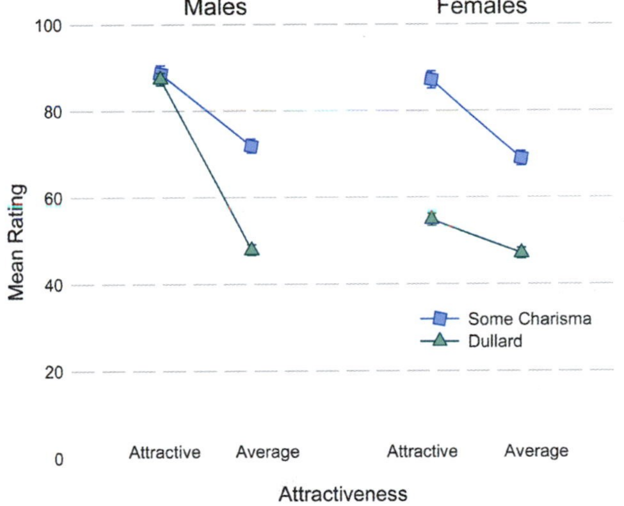

14.5.7.3. Looks x charisma x gender interaction 3: ugly vs. average, high charisma vs. some charisma, males vs. females ③

The third contrast for this interaction term compares level 2 of looks (ugly) to level 3 (average looking), when level 1 of charisma (high charisma) is compared to level 3 of charisma (some charisma), in men compared to women. Again, we extract the relevant bit of

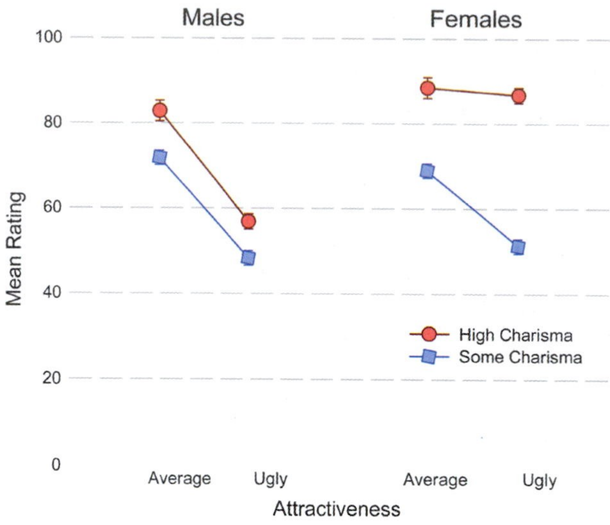

the interaction graph in Figure 14.3 and you can see that the patterns are different for men and women. This is reflected by the fact that the contrast is significant, $F(1, 18) = 11.70$, $p < .01$. To unpick this we need to look at the graph. First, let's look at the men. For men, as attractiveness goes down, so does interest when the date has high charisma and when they have some charisma. In fact the lines are parallel. So, regardless of charisma, there is a similar reduction in interest as attractiveness declines. For women the picture is quite different. When charisma is high, there is no decline in interest as attractiveness falls (the line connecting the circles is flat); however, when charisma is lower, the attractiveness of the date does matter and interest is lower in an ugly date than in an average-looking date. Another way to look at it is that for dates with some charisma, the reduction in interest as attractiveness goes down is about the same in men and women (the lines with squares have the same slope). However, for dates who have high charisma, the decrease in interest if these dates are ugly rather than average looking is much more dramatic in men than women (the line with circles is much steeper for men than it is for women). This is what the significant contrast tells us.

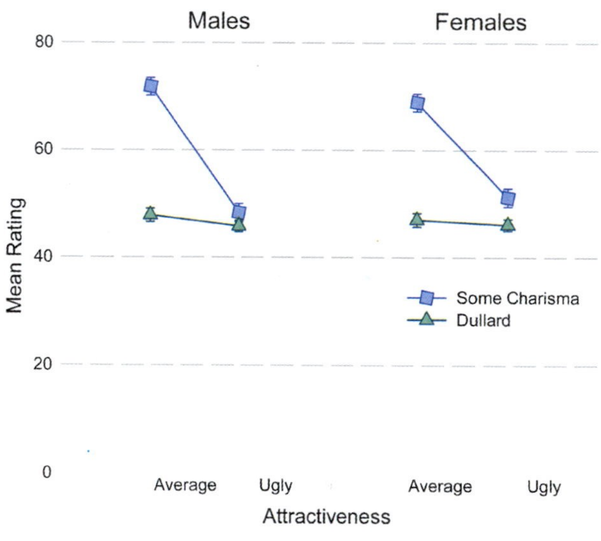

14.5.7.4. Looks x charisma x gender interaction 4: ugly vs. average, dullard vs. some charisma, male vs. female ③

The final contrast for this interaction term compares level 2 of looks (ugly) to level 3 (average looking), when comparing level 2 of charisma (dullard) to level 3 of charisma (some charisma), in men compared to women. If we extract the relevant bits of the interaction graph in Figure 14.3, we can see that interest (as indicated by high ratings) in ugly dates was the same regardless of whether they had some charisma or were a dullard. However, for average-looking dates, there was more interest when that person had some charisma rather than if they were a dullard. Importantly, this pattern of results is the same in males and females and this is reflected in the non-significance of this contrast, $F(1, 18) = 1.33$, *ns*.

14.5.8. Conclusions ③

These contrasts tell us nothing about the differences between the attractive and ugly conditions, or the high-charisma and dullard conditions, because these were never compared. We could rerun the analysis and specify our contrasts differently to get these effects. However, what is clear from our data is that differences exist between men and women in terms of how they're affected by the looks and personality of potential dates. Men appear to be enthusiastic about dating anyone who is attractive regardless of how awful their personality. Women are almost completely the opposite: they are enthusiastic about dating anyone with a lot of charisma, regardless of how they look (and are unenthusiastic about dating people without charisma regardless of how attractive they look). The only consistency between men and women is when there is some charisma (but not lots), in which case for both genders the attractiveness influences how enthusiastic they are about dating the person.

What should be even clearer from this chapter is that when more than two independent variables are used in ANOVA, it yields complex interaction effects that require a great deal of concentration to interpret (imagine interpreting a four-way interaction!). Therefore, it is essential to take a systematic approach to interpretation and plotting graphs is a particularly useful way to proceed. It is also advisable to think carefully about the appropriate contrasts to use to answer the questions you have about your data. It is these contrasts that will help you to interpret interactions, so make sure you select sensible ones!

CRAMMING SAM'S TIPS

- Mixed ANOVA compares several means when there are two or more independent variables, and at least one of them has been measured using the same participants and at least one other has been measured using different participants.
- Test the assumption of *sphericity* for the repeated-measures variable(s) when they have three or more conditions using *Mauchly's test*. If the value in the column labelled *Sig.* is less than .05 then the assumption is violated. You should test this assumption for all effects (if there are two or more repeated-measures variables this means you test the assumption for all variables and the corresponding interaction terms).
- The table labelled **Tests of Within-Subjects Effects** shows the results of your ANOVA for the repeated-measures variables and all of the interaction effects. For *each* effect, if the assumption of sphericity has been met then look at the row labelled *Sphericity Assumed*. Otherwise, read the row labelled *Greenhouse-Geisser* or *Huynh-Feldt* (read the previous chapter to find out the relative merits of the two procedures). Having selected the appropriate row, look at the column labelled *Sig.* If the value is less than .05 then the means are significantly different.
- The table labelled **Tests of Between-Subjects Effects** shows the results of your ANOVA for the between-group variables. Look at the column labelled *Sig.* If the value is less than .05 then the means of the groups are significantly different.
- Break down the main effects and interaction terms using contrasts. These contrasts appear in the table labelled **Tests of Within-Subjects Contrasts**, again look to the columns labelled *Sig.* to discover if your comparisons are significant (they will be if the significance value is less than .05).
- Look at the means, or better still draw graphs, to help you interpret the contrasts.

14.6. Calculating effect sizes ③

I keep emphasizing the fact that effect sizes are really more useful when they summarize a focused effect. This also gives me a useful excuse to circumvent the complexities of omega squared in mixed designs (it's the road to madness I assure you). Therefore, just calculate effect sizes for your contrasts when you've got a factorial design (and any main effects that compare only two groups). SPSS Output 14.7 shows the values for several contrasts, all of which have 1 degree of freedom for the model (i.e. they represent a focused and interpretable comparison) and have 18 residual degrees of freedom. We can use these F-ratios and convert them to an effect size r, using a formula we've come across before:

$$r = \sqrt{\frac{F(1, df_R)}{F(1, df_R) + df_R}}$$

First, we can deal with the main effect of gender because this compares only two groups:

$$r_{\text{Gender}} = \sqrt{\frac{0.005}{0.005 + 18}} = 0.02$$

For the two comparisons we did for the looks variable (SPSS Output 14.7), we would get:

$$r_{\text{Attractive vs. Average}} = \sqrt{\frac{226.99}{226.99 + 18}} = 0.96$$

$$r_{\text{Ugly vs. Average}} = \sqrt{\frac{160.07}{160.07 + 18}} = 0.95$$

Therefore, both comparisons yielded massive effect sizes. For the two comparisons we did for the charisma variable (SPSS Output 14.7), we would get:

$$r_{\text{High vs. Some}} = \sqrt{\frac{109.94}{109.94 + 18}} = 0.93$$

$$r_{\text{Dullard vs. Some}} = \sqrt{\frac{227.94}{227.94 + 18}} = 0.96$$

Again, both comparisons yield massive effect sizes. For the looks × gender interaction, we again had two contrasts:

$$r_{\text{Attractive vs. Average, Male vs. Female}} = \sqrt{\frac{43.26}{43.26 + 18}} = 0.84$$

$$r_{\text{Ugly vs. Average, Male vs. Female}} = \sqrt{\frac{30.23}{30.23 + 18}} = 0.79$$

Again, these are massive effects. For the charisma × gender interaction, the two contrasts give us:

$$r_{\text{High vs. Some, Male vs. Female}} = \sqrt{\frac{27.20}{27.20 + 18}} = 0.78$$

$$r_{\text{Dullard vs. Some, Male vs. Female}} = \sqrt{\frac{33.69}{33.69 + 18}} = 0.81$$

Yet again massive effects (can't you tell the data are fabricated!).

Moving on to the looks × charisma interaction, we get four contrasts:

$$r_{\text{Attractive vs. Average, High vs. Some}} = \sqrt{\frac{21.94}{21.94 + 18}} = 0.74$$

$$r_{\text{Attractive vs. Average, Dullard vs.Some}} = \sqrt{\frac{4.09}{4.09 + 18}} = 0.43$$

$$r_{\text{Ugly vs. Average, High vs. Some}} = \sqrt{\frac{6.23}{6.23 + 18}} = 0.51$$

$$r_{\text{Ugly vs. Average, Dullard vs. Some}} = \sqrt{\frac{88.60}{88.60 + 18}} = 0.91$$

All of these effects are in the medium to massive range. Finally, for the looks × charisma × gender interaction we had four contrasts:

$$r_{\text{Attractive vs. Average, High vs. Some, Male vs. Female}} = \sqrt{\frac{0.93}{0.93 + 18}} = 0.22$$

$$r_{\text{Attractive vs. Average, Dullard vs. Some, Male vs. Female}} = \sqrt{\frac{60.67}{60.67 + 18}} = 0.88$$

$$r_{\text{Ugly vs. Average, High vs. Some, Male vs. Female}} = \sqrt{\frac{11.70}{11.70 + 18}} = 0.63$$

$$r_{\text{Ugly vs. Average, Dullard vs. Some, Male vs. Female}} = \sqrt{\frac{1.33}{1.33 + 18}} = 0.26$$

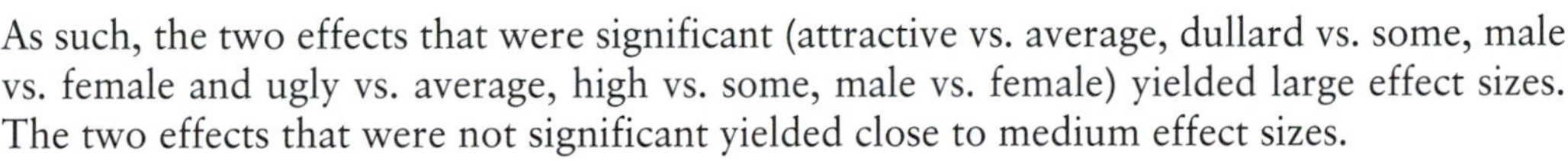

As such, the two effects that were significant (attractive vs. average, dullard vs. some, male vs. female and ugly vs. average, high vs. some, male vs. female) yielded large effect sizes. The two effects that were not significant yielded close to medium effect sizes.

14.7. Reporting the results of mixed ANOVA ②

As you've probably gathered, when you have more than two independent variables there's a hell of a lot of information to report. You have to report all of the main effects, all of the interactions and any contrasts you may have done. This can take up a lot of space and one good tip is: reserve the most detail for the effects that actually matter (e.g. main effects are usually not that interesting if you've got a significant interaction that includes that variable). I'm a big fan of giving brief explanations of results in the results section to really get the message across about what a particular effect is telling us and so I tend to not just report results, but offer some interpretation as well. Having said that, some journal editors are big fans of telling me my results sections are too long. So, you should probably ignore everything I say. Assuming we want to report all of our effects, we could do it something like this (although not as a list!):

- ✓ All effects are reported as significant at $p < .05$. There was a significant main effect of the attractiveness of the date on interest expressed by participant, $F(2, 36) = 423.73$. Contrasts revealed that attractive dates were more desirable than average-looking ones, $F(1, 18) = 226.99$, $r = .96$, and ugly dates were less desirable than average-looking ones $F(1, 18) = 160.07$, $r = .95$.
- ✓ There was also a significant main effect of the amount of charisma the date possessed on the interest expressed in dating them, $F(2, 36) = 328.25$. Contrasts revealed that dates with high charisma were more desirable than dates with some charisma, $F(1, 18) = 109.94$, $r = .93$, and dullards were less desirable than dates with some charisma, $F(1, 18) = 227.94$, $r = .96$.
- ✓ There was no significant effect of gender, indicating that ratings from male and female participants were in general the same, $F(1, 18) < 1$, $r = .02$.
- ✓ There was a significant interaction effect between the attractiveness of the date and the gender of the participant, $F(2, 36) = 80.43$. This indicates that the desirability of dates of different levels of attractiveness differed in men and women. To break down this interaction, contrasts were performed comparing each level of attractiveness to average looking across male and female participants. These revealed significant interactions when comparing male and female scores to attractive dates compared to average-looking dates, $F(1, 18) = 43.26$, $r = .84$, and to ugly dates compared to average dates, $F(1, 18) = 30.23$,

r = .79. Looking at the interaction graph, this suggests that male and female ratings are very similar for average-looking dates, but men rate attractive dates higher than women, whereas women rate ugly dates higher than men. Although both male's and female's interest decreases as attractiveness decreases, this decrease is more pronounced for men, suggesting that when charisma is ignored, men's interest in dating a person is more influenced by their looks than for females.

✓ There was a significant interaction effect between the level of charisma of the date and the gender of the participant, $F(2, 36) = 62.45$. This indicates that the desirability of dates of different levels of charisma differed in men and women. To break down this interaction, contrasts were performed comparing each level of charisma to the middle category of 'some charisma' across male and female participants. These revealed significant interactions when comparing male and female scores to highly charismatic dates compared to dates with some charisma, $F(1, 18) = 27.20$, $r = .78$, and to dullards compared to dates with some charisma, $F(1, 18) = 33.69$, $r = .81$. The interaction graph reveals that men show more interest in dates who are dullards than women do, and women show slightly more interest in very charismatic dates than men do. Both male's and female's interest decrease as charisma decreases, but this decrease is more pronounced for females, suggesting women's interest in dating a person is more influenced by their charisma than for men.

✓ There was a significant interaction effect between the level of charisma of the date and the attractiveness of the date, $F(4, 72) = 36.63$. This indicates that the desirability of dates of different levels of charisma differed according to their attractiveness. To break down this interaction, contrasts were performed comparing each level of charisma to the middle category of 'some charisma' across each level of attractiveness compared to the category of average attractiveness. The first contrast revealed a significant interaction when comparing attractive dates to average-looking dates when the date had high charisma compared to some charisma, $F(1, 18) = 21.94$, $r = .74$, and tells us that as dates become less attractive there is a greater decline in interest when charisma is low compared to when charisma is high. The second contrast compared attractive dates to average-looking dates when the date was a dullard compared to when they had some charisma. This was not significant, $F(1, 18) = 4.09$, $r = .43$, and tells us that as dates become less attractive there is decline in interest both when charisma is low and when there is no charisma at all. The third contrast compared ugly dates to average-looking dates when the date had high charisma compared to some charisma. This was significant, $F(1, 18) = 6.23$, $r = .51$, and tells us that as dates become less attractive there is a greater decline in interest when charisma is low compared to when charisma is high. The final contrast compared ugly dates to average-looking dates when the date was a dullard compared to when they had some charisma. This contrast was highly significant, $F(1, 18) = 88.60$, $r = .91$, and tells us that as dates become less attractive the decline in interest in dates with a bit of charisma is significantly greater than for dullards.

✓ Finally, the looks × charisma × gender interaction was significant $F(4, 72) = 24.12$. This indicates that the looks × charisma interaction described previously was different in male and female participants. Again, contrasts were used to break down this interaction; these contrasts compared male and females scores at each level of charisma compared to the middle category of 'some charisma' across each level of attractiveness compared to the category of average attractiveness. The first contrast revealed a non-significant difference between male and female responses when comparing attractive dates to average-looking dates when the date had high charisma compared to some charisma, $F(1, 18) < 1$, $r = .22$, and tells us that for both males and females, as dates become less attractive there is a greater decline in interest when charisma is low compared to when charisma is high. The second contrast investigated differences between males and females when comparing attractive dates to average-looking dates when the date was a dullard compared to when they had some charisma. This was significant, $F(1, 18) = 60.67$, $r = .88$, and tells us that for dates with some charisma, the reduction in interest as attractiveness goes down is about the same in men and women, but for dates who are dullards, the decrease in interest if these dates are average looking rather than attractive is much more dramatic in men than women. The third contrast looked for differences between males and females when comparing ugly dates to average-looking dates when the date had high charisma compared to some charisma. This was significant, $F(1, 18) = 11.70$, $r = .63$, and tells us that for dates with some charisma, the reduction in interest as attractiveness goes down is about the same in men and women, but for dates who have high charisma, the decrease in interest if these dates are ugly rather than average looking is much more dramatic in men than women. The final contrast looked for differences between men and women when comparing ugly dates to average-looking dates when the date was a dullard compared to when they had some charisma. This contrast was not significant, $F(1, 18) = 1.33$, $r = .26$, and tells us that for both men and women, as dates become less attractive the decline in interest in dates with a bit of charisma is significantly greater than for dullards.

LABCOAT LENI'S REAL RESEARCH 14.1

Keep the faith(ful)? ③

People can be jealous. People can be especially jealous when they think that their partner is being unfaithful. An evolutionary view of jealousy suggests that men and women have evolved distinctive types of jealousy because male and female reproductive success is threatened by different types of infidelity. Specifically, a woman's sexual infidelity deprives her mate of a reproductive opportunity and in some cases burdens him with years investing in a child that is not his. Conversely, a man's sexual infidelity does not burden his mate with unrelated children, but may divert his resources from his mate's progeny. This diversion of resources is signalled by emotional attachment to another female. Consequently, men's jealousy mechanism should have evolved to prevent a mate's *sexual* infidelity, whereas in women it has evolved to prevent emotional infidelity. If this is the case then men and women should divert their attentional resources towards different cues to infidelity: women should be 'on the look out' for emotional infidelity, whereas men should be watching out for sexual infidelity.

Achim Schützwohl put this theory to the test in a unique study in which men and women saw sentences presented on a computer screen (Schützwohl, 2008). On each trial, participants saw a target sentence that was always emotionally neutral (e.g. 'The gas station is at the other side of the street'). However, the trick was that before each of these targets, a distracter sentence was presented that could also be affectively neutral, or could indicate sexual infidelity (e.g. 'Your partner suddenly has difficulty becoming sexually aroused when he and you want to have sex') or emotional infidelity (e.g. 'Your partner doesn't say "I love you" to you anymore'). The idea was that if these distractor sentences grabbed a person's attention then (1) they would remember them, and (2) they would not remember the target sentence that came afterwards (because their attentional resources were still focused on the distractor). These effects should show up only in people currently in a relationship. The outcome was the number of sentences that a participant could remember (out of 6), and the predictors were whether the person had a partner or not (**Relationship**), whether the trial used a neutral distractor, an emotional infidelity distractor or a sexual infidelity distractor, and whether the sentence was a distractor or the target following the distractor. Schützwohl analysed men and women's data separately. The predictions are that women should remember more emotional infidelity sentences (distractors) but fewer of the targets that followed those sentences (target). For men, the same effect should be found but for sexual infidelity sentences.

The data from this study are in the file **Schützwohl (2008).sav**. Labcoat Leni wants you to carry out two three-way mixed ANOVAs (one for men and the other for women) to test these hypotheses. Answers are in the additional material on the companion website (or look at pages 638–642 in the original article).

SCHÜTZWOHL, A. (2008). *PERSONALITY AND INDIVIDUAL DIFFERENCES, 44,* 633–644.

14.8. What to do when assumptions are violated in mixed ANOVA ③

If I had £1 (or $1, 1 euro or whatever currency you fancy) for every time someone had told me with 100% confidence that there was no 'non-parametric' equivalent of mixed ANOVA, then I'd have a nice shiny new drum kit. As with other factorial ANOVAs, there is not a non-parametric counterpart of mixed ANOVA, as such, but there are robust methods that can be used (see section 5.7.4). As in previous chapters, my advice is to read Rand Wilcox's book (Wilcox, 2005), then use the R plugin for SPSS to run the analysis (see Oliver Twisted).

OLIVER TWISTED

Please, Sir, can I have some more … robust tests?

'There is not a non-parametric equivalent of mixed ANOVA', sobs Oliver, 'how will I analyse my data?' That's £1, thanks. In return for your money, Oliver, I've prepared a flash movie on the companion website that shows you how to use the R plugin to run a robust mixed ANOVA. What a bargain.

What have I discovered about statistics? ②

Three-way ANOVA is a confusing nut to crack. I've probably done hundreds of three-way ANOVAs in my life and still I kept getting confused throughout writing this chapter (and so if you're confused after reading it it's not your fault, it's mine). Hopefully, what you should have discovered is that ANOVA is flexible enough that you can mix and match independent variables that are measured using the same or different participants. In addition we've looked at how ANOVA is also flexible enough to go beyond merely including two independent variables. Hopefully, you've also started to realize why there are good reasons to limit the number of independent variables that you include (for the sake of interpretation).

Of course far more interesting than that is that you've discovered that men are superficial creatures who value looks over charisma, and that women are prepared to date the hunchback of Notre Dame provided he has a sufficient amount of charisma. This is why as a 16–18 year old my life was so complicated, because where on earth do you discover your hidden charisma? Luckily for me, some girls find alcoholics appealing. The girl I was particularly keen on at 16 was, as it turned out, keen on me too. I refused to believe this for at least a month. All of our friends were getting bored of us declaring our undying love for each other to them but then not speaking to each other; they eventually intervened. There was a party one evening and all of her friends had spent hours convincing me to ask her on a date, guaranteeing me that she would say 'yes'. I had psyched myself up, I was going to do it, I was actually going to ask a girl out on a date. My whole life had been leading up to this moment and I must not do anything to ruin it. By the time she arrived my nerves had got the better of me and she had to step over my paralytic corpse to get into the house. Later on, my friend Paul Spreckley (see Figure 9.1) physically carried the girl in question from another room and put her next to me and then said something to the effect of 'Andy, I'm going to sit here until you ask her out.' He had a long wait but eventually, miraculously, the words came out of my mouth. What happened next is the topic for another book, not about statistics.

Key terms that I've discovered

Mixed ANOVA

Mixed design

Smart Alex's tasks

- **Task 1**: I am going to extend the example from the previous chapter (advertising and different imagery) by adding a between-group variable into the design.[4] To recap, in case you haven't read the previous chapter, participants viewed a total of nine mock adverts over three sessions. In these adverts there were three products (a brand of beer, Brain Death, a brand of wine, Dangleberry, and a brand of water, Puritan). These could be presented alongside positive, negative or neutral imagery. Over the three sessions and nine adverts, each type of product was paired with each type of imagery (read the previous chapter if you need more detail). After each advert participants rated the drinks on a scale ranging from −100 (dislike very much) through 0 (neutral) to 100 (like very much). The design, thus far, has two independent variables: the type of drink (beer, wine or water) and the type of imagery used (positive, negative or neutral). These two variables completely cross over, producing nine experimental conditions. Now imagine that I also took note of each person's gender. Subsequent to the previous analysis it occurred to me that men and women might respond differently to the products (because, in keeping with stereotypes, men might mostly drink lager whereas women might drink wine). Therefore, I wanted to reanalyse the data taking this additional variable into account. Now, gender is a between-group variable because a participant can be only male or female: they cannot participate as a male and then change into a female and participate again! The data are the same as in the previous chapter (Table 13.3) and can be found in the file **MixedAttitude.sav**. Run a mixed ANOVA on these data. ③

- **Task 2**: Text messaging is very popular among mobile phone owners, to the point that books have been published on how to write in text speak (BTW, hope u kno wat I mean by txt spk). One concern is that children may use this form of communication so much that it will hinder their ability to learn correct written English. One concerned researcher conducted an experiment in which one group of children was encouraged to send text messages on their mobile phones over a six-month period. A second group was forbidden from sending text messages for the same period. To ensure that kids in this latter group didn't use their phones, this group were given armbands that administered painful shocks in the presence of microwaves (like those emitted from phones). There were 50 different participants: 25 were encouraged to send text messages, and 25 were forbidden. The outcome was a score on a grammatical test (as a percentage) that was measured both before and after the experiment. The first independent variable was, therefore, text message use (text messagers versus controls) and the second independent variable was the time at which grammatical ability was assessed (before or after the experiment). The data are in the file **TextMessages.sav**. ③

- **Task 3**: A researcher was interested in the effects on people's mental health of participating in *Big Brother* (see Chapter 1 if you don't know what *Big Brother* is).

[4] Previously the example contained two repeated-measures variables (drink type and imagery type), but now it will include three variables (two repeated-measures and one between-group).

The researcher hypothesized that they start off with personality disorders that are exacerbated by being forced to live with people as attention seeking as themselves. To test this hypothesis, she gave eight contestants a questionnaire measuring personality disorders before they entered the house, and again when they left the house. A second group of eight people acted as a waiting list control. These were people short listed to go into the house, but never actually made it. They too were given the questionnaire at the same points in time as the contestants. The data are in **BigBrother.sav**. Conduct a mixed ANOVA on the data. ②

Answers can be found on the companion website. Some more detailed comments about task 2 can be found in Field and Hole (2003).

Further reading

Field, A. P. (1998). A bluffer's guide to sphericity. *Newsletter of the Mathematical, Statistical and Computing section of the British Psychological Society*, 6(1), 13–22. (Available in the additional material on the companion website.)

Howell, D. C. (2006). *Statistical methods for psychology* (6th ed.). Belmont, CA: Thomson. (Or you might prefer his *Fundamental Statistics for the Behavioral Sciences,* also in its 6th edition, 2007.)

Online tutorials

The companion website contains the following Flash movie tutorials to accompany this chapter:

- Mixed ANOVA using SPSS
- Robust Mixed ANOVA using *R*

Interesting real research

Schützwohl, A. (2008). The disengagement of attentive resources from task-irrelevant cues to sexual and emotional infidelity. *Personality and Individual Differences*, *44*, 633–644.

Non-parametric tests

15

FIGURE 15.1 In my office during my Ph.D., probably preparing some teaching. Yes, as if my life isn't embarrassing enough, I had quite long hair back then

15.1. What will this chapter tell me? ①

After my psychology degree (at City University, London) I went to the University of Sussex to do my Ph.D. (also in psychology) and like many people I had to teach to survive. Much to my dread I was allocated to teach second-year undergraduate statistics. This was possibly the worst combination of events that I could imagine. I was still very shy at the time, and I didn't have a clue about research methods. Standing in front of a room full of strangers and trying to teach them ANOVA was only marginally more appealing than dislocating my knees and running a marathon – with broken glass in my trainers (sneakers). I obsessively prepared for my first session so that it would go well; I created handouts, I invented examples, I rehearsed what I would say. I went in terrified but at least knowing that if preparation was any predictor of success then I would be OK. About half way through the first session as I was mumbling on to a room of bored students, one of them rose majestically from their chair. She walked slowly towards me and I'm convinced that she was surrounded by an aura of bright white light and dry ice. Surely she had been chosen by her peers to impart a message of gratitude for the hours of preparation I had done and the skill with which I was un-clouding their brains of the mysteries of ANOVA. She stopped beside me. We stood inches apart and my eyes raced around the floor looking for the reassurance of my shoelaces: 'No one in this room has a rabbit[1] clue what you're going on about,' she spat before storming out. Scales have not been invented

[1] She didn't say rabbit, but she did say a word that describes what rabbits do a lot; it begins with an 'f' and the publishers think that it will offend you.

yet to measure how much I wished I'd ran the dislocated-knees marathon that morning and then taken the day off. I was absolutely mortified. To this day I have intrusive thoughts about groups of students in my lectures walking zombie-like towards the front of the lecture theatre chanting 'No one knows what you're going on about' before devouring my brain in a rabid feeding frenzy. The point is that sometimes our lives, like data, go horribly, horribly wrong. This chapter is about data that are as wrong as dressing a cat in a pink tutu.

15.2. When to use non-parametric tests ①

What are non-parametric tests?

We've seen in the last few chapters how we can use various techniques to look for differences between means. However, all of these tests rely on parametric assumptions (see Chapter 5). Data are often unfriendly and don't always turn up in nice normally distributed packages! Just to add insult to injury, it's not always possible to correct for problems with the distribution of a data set – so, what do we do in these cases? The answer is that we have to use special kinds of statistical procedures known as **non-parametric tests**. Non-parametric tests are sometimes known as assumption-free tests because they make fewer assumptions about the type of data on which they can be used.[2] Most of these tests work on the principle of **ranking** the data: that is, finding the lowest score and giving it a rank of 1, then finding the next highest score and giving it a rank of 2, and so on. This process results in high scores being represented by large ranks, and low scores being represented by small ranks. The analysis is then carried out on the ranks rather than the actual data. This process is an ingenious way around the problem of using data that break the parametric assumptions. Some people believe that non-parametric tests have less power than their parametric counterparts, but as we will see in Jane Superbrain Box 15.1 below this is not always true. In this chapter we'll look at four of the most common non-parametric procedures: the Mann–Whitney test, the Wilcoxon signed-rank test, Friedman's test and the Kruskal–Wallis test. For each of these we'll discover how to carry out the analysis on SPSS and how to interpret and report the results.

15.3. Comparing two independent conditions: the Wilcoxon rank-sum test and Mann–Whitney test ①

When you want to test differences between two conditions and different participants have been used in each condition then you have two choices: the **Mann–Whitney test** (Mann & Whitney, 1947) and the **Wilcoxon rank-sum test** (Wilcoxon, 1945; Figure 15.2). These tests are the non-parametric equivalent of the independent *t*-test. In fact both tests are equivalent, and there's another, more famous, Wilcoxon test, so it gets extremely confusing for most of us.

For example, a neurologist might collect data to investigate the depressant effects of certain recreational drugs. She tested 20 clubbers in all: 10 were given an ecstasy tablet to take on a Saturday night and 10 were allowed to drink only alcohol. Levels of depression were measured using the Beck Depression Inventory (BDI) the day after and midweek. The data are in Table 15.1.

[2] Non-parametric tests sometimes get referred to as distribution-free tests, with an explanation that they make *no* assumptions about the distribution of the data. Technically, this isn't true: they do make distributional assumptions (e.g. the ones in this chapter all assume a continuous distribution), but they are less restrictive ones than their parametric counterparts.

FIGURE 15.2
Frank Wilcoxon

TABLE 15.1 Data for drug experiment

Participant	Drug	BDI (Sunday)	BDI (Wednesday)
1	Ecstasy	15	28
2	Ecstasy	35	35
3	Ecstasy	16	35
4	Ecstasy	18	24
5	Ecstasy	19	39
6	Ecstasy	17	32
7	Ecstasy	27	27
8	Ecstasy	16	29
9	Ecstasy	13	36
10	Ecstasy	20	35
11	Alcohol	16	5
12	Alcohol	15	6
13	Alcohol	20	30
14	Alcohol	15	8
15	Alcohol	16	9
16	Alcohol	13	7
17	Alcohol	14	6
18	Alcohol	19	17
19	Alcohol	18	3
20	Alcohol	18	10

15.3.1. Theory ②

The logic behind the Wilcoxon rank-sum and Mann–Whitney tests is incredibly elegant. First, let's imagine a scenario in which there is no difference in depression levels between ecstasy and alcohol users. If we were to rank the data *ignoring the group to which a person belonged* from lowest to highest (i.e. give the lowest score a rank of 1 and the next lowest a rank of 2 etc.), then what should you find? Well, if there's no difference between the groups then you expect to find a similar number of high and low ranks in each group; specifically, if you added up the ranks, then you'd expect the summed total of ranks in each group to be about the same. Now think about what would happen if there was a difference between the groups. Let's imagine that the ecstasy group is more depressed than the alcohol group. If you rank the scores as before, then you would expect the higher ranks to be in the ecstasy group and the lower ranks to be in the alcohol group. Again, if we summed the ranks in each group, we'd expect the sum of ranks to be higher in the ecstasy group than in the alcohol group.

The Mann–Whitney and Wilcoxon rank-sum tests both work on this principle. In fact, when the groups have unequal numbers of participants in them then the test statistic (W_s) for the Wilcoxon rank-sum test is simply the sum of ranks in the group that contains the fewer people; when the group sizes are equal it's the value of the smaller summed rank. Let's have a look at how ranking works in practice.

Figure 15.3 shows the ranking process for both the Wednesday and Sunday data. To begin with, let's use our data for Wednesday, because it's more straightforward. First, just arrange the scores in ascending order, attach a label to remind you which group they came from (I've used A for alcohol and E for ecstasy), then starting at the lowest score assign potential ranks starting with 1 and going up to the number of scores you have. The reason why I've called these potential ranks is because sometimes the same score occurs more than once in a data set (e.g. in these data a score of 6 occurs twice, and a score of 35 occurs three times). These are called *tied ranks* and these values need to be given the same rank, so all we do is assign a rank that is the average of the potential ranks for those scores. So, with our two scores of 6, because they would've been ranked as 3 and 4, we take an average of these values (3.5) and use this value as a rank for both occurrences of the score! Likewise, with the three scores of 35, we have potential ranks of 16, 17 and 18; we actually use the average of these three ranks, $(16 + 17 + 18)/3 = 17$. When we've ranked the data, we add up all of the ranks for the two groups. So, add the ranks for the scores that came from the alcohol group (you should find the sum is 59) and then add the ranks for the scores that came from the ecstasy group (this value should be 151). We take the lowest of these sums to be our test statistic, therefore the test statistic for the Wednesday data is $W_s = 59$.

SELF-TEST Based on what you have just learnt, try ranking the Sunday data. (The answers are in Figure 15.3 – there are lots of tied ranks and the data are generally horrible.)

You should find that when you've ranked the data, and added the ranks for the two groups, the sum of ranks for the alcohol group is 90.5 and for the ecstasy group it is 119.5. The lowest of these sums is our test statistic; therefore the test statistic for the Sunday data is $W_s = 90.5$.

Wednesday Data

Score	3	5	6	6	7	8	9	10	17	24	27	28	29	30	32	35	35	35	36	39
Potential Rank	1	2	3	4	5	6	7	8	9	10	11	12	13	14	15	16	17	18	19	20
Actual Rank	1	2	3.5	3.5	5	6	7	8	9	10	11	12	13	14	15	17	17	17	19	20
Group	A	A	A	A	A	A	A	A	A	E	E	E	E	A	E	E	E	E	E	E

Sum of Ranks for Alcohol (A) = 59

Sum of Ranks for Ecstasy (E) = 151

Sunday Data

Score	13	13	14	15	15	15	16	16	16	16	17	18	18	18	19	19	20	20	27	35
Potential Rank	1	2	3	4	5	6	7	8	9	10	11	12	13	14	15	16	17	18	19	20
Actual Rank	1.5	1.5	3	5	5	5	8.5	8.5	8.5	8.5	11	13	13	13	15.5	15.5	17.5	17.5	19	20
Group	A	E	A	A	A	E	A	A	E	E	E	E	A	A	E	A	E	A	E	E

Sum of Ranks for Alcohol (A) = 90.5

Sum of Ranks for Ecstasy (E) = 119.5

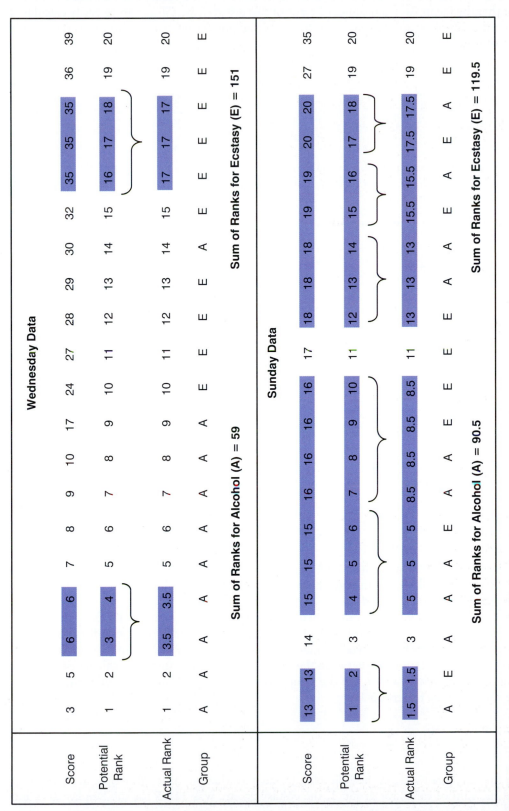

FIGURE 15.3 Ranking the depression scores for Wednesday

The next issue is: how do we determine whether this test statistic is significant? It turns out that the mean ($\overline{W}_s$) and standard error of this test statistic ($SE_{\overline{W}_s}$) can be easily calculated from the sample sizes of each group (n_1 is the sample size of group 1 and n_2 is the sample size of group 2):

$$\overline{W}_s = \frac{n_1(n_1 + n_2 + 1)}{2}$$

$$SE_{\overline{W}_s} = \sqrt{\frac{n_1 n_2 (n_1 + n_2 + 1)}{12}}$$

For our data, we actually have equal-sized groups and there are 10 people in each, so n_1 and n_2 are both 10. Therefore, the mean and standard deviation are:

$$\overline{W}_s = \frac{10(10 + 10 + 1)}{2} = 105$$

$$SE_{\overline{W}_s} = \sqrt{\frac{(10 \times 10)(10 + 10 + 1)}{12}} = 13.23$$

If we know the test statistic, the mean of test statistics and the standard error, then we can easily convert the test statistic to a z-score using the equation that we came across way back in Chapter 1:

$$z = \frac{X - \overline{X}}{s} = \frac{W_s - \overline{W}_s}{SE_{\overline{W}_s}}$$

If we calculate this value for the Sunday and Wednesday depression scores we get:

$$z_{\text{Sunday}} = \frac{W_s - \overline{W}_s}{SE_{\overline{W}_s}} = \frac{90.5 - 105}{13.23} = -1.10$$

$$z_{\text{Wednesday}} = \frac{W_s - \overline{W}_s}{SE_{\overline{W}_s}} = \frac{59 - 105}{13.23} = -3.48$$

If these values are bigger than 1.96 (ignoring the minus sign) then the test is significant at $p < .05$. So, it looks as though there is a significant difference between the groups on Wednesday, but not on Sunday.

The procedure I've actually described is the Wilcoxon rank-sum test. The Mann–Whitney test, with which many of you may be more familiar, is basically the same. It is based on a test statistic U, which is derived in a fairly similar way to the Wilcoxon procedure (in fact there's a direct relationship between the two). If you're interested, U is calculated using an equation in which n_1 and n_2 are the sample sizes of groups 1 and 2 respectively, and R_1 is the sum of ranks for group 1:

$$U = n_1 n_2 + \frac{N_1(N_1 + 1)}{2} - R_1$$

So, for our data we'd get the following (remember that we have 10 people in each group and the sum of ranks for group 1, the ecstasy group, was 119.5 for the Sunday data and 151 for the Wednesday data):

$$U_{\text{Sunday}} = (10 \times 10) + \frac{10(11)}{2} - 119.50 = 35.50$$

$$U_{\text{Wednesday}} = (10 \times 10) + \frac{10(11)}{2} - 151.00 = 4.00$$

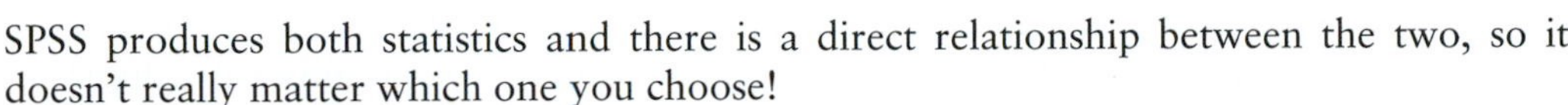

SPSS produces both statistics and there is a direct relationship between the two, so it doesn't really matter which one you choose!

15.3.2. Inputting data and provisional analysis ①

SELF-TEST See whether you can use what you have learnt about data entry to enter the data in Table 15.1 into SPSS.

When the data are collected using different participants in each group, we need to input the data using a coding variable. So, the data editor will have three columns of data. The first column is a coding variable (called something like **Drug**) which, in this case, will have only two codes (for convenience I suggest 1 = ecstasy group and 2 = alcohol group). The second column will have values for the dependent variable (BDI) measured the day after (call this variable **Sunday_BDI**) and the third will have the midweek scores on the same questionnaire (call this variable **Wednesday_BDI**). When you enter the data into SPSS remember to tell the computer that a code of 1 represents the group that was given ecstasy and a code of 2 represents the group that was restricted to alcohol (see section 3.4.2.3).

There were no specific predictions about which drug would have the most effect so the analysis should be two-tailed. First, we would run some exploratory analyses on the data and because we're going to be looking for group differences we need to run these exploratory analyses for each group.

SELF-TEST Carry out some analyses to test for normality and homogeneity of variance in these data (see sections 5.5 and 5.6).

The results of your exploratory analysis are shown in SPSS Output 15.1. These tables show first of all that for the Sunday data the distributions for ecstasy, $D(10) = 0.28$, $p < .05$, appears to be non-normal whereas the alcohol data, $D(10) = 0.17$, *ns*, were normal; we can tell this by whether the significance of the K–S and Shapiro–Wilk tests are less than .05 (and, therefore, significant) or greater than .05 (and, therefore, non-significant, *ns*). For the Wednesday data, although the data for ecstasy were normal, $D(10) = 0.24$, *ns*, the data for alcohol appeared to be significantly non-normal, $D(10) = 0.31$, $p < .01$. This finding would alert us to the fact that the sampling distribution might also be non-normal for the Sunday and Wednesday data and that a non-parametric test should be used. You should note that the Shapiro–Wilk statistic yields exact significance values whereas the K–S test sometimes gives an approximation of .2 for the significance (see the Sunday data for the alcohol group) because SPSS cannot calculate exact significances. This finding highlights an important difference between the K–S test and the Shapiro–Wilk test: in general the Shapiro–Wilk test is more accurate (see Chapter 5). The second table in SPSS Output 15.1 shows the results of Levene's test. For the Sunday data, $F(1, 18) = 3.64$, *ns*, and for Wednesday, $F(1, 18) = 0.51$, *ns*, the variances are not significantly different, indicating that the assumption of homogeneity has been met.

SPSS OUTPUT 15.1

Tests of Normality

	Type of Drug	Kolmogorov-Smirnov[a] Statistic	df	Sig.	Shapiro-Wilk Statistic	df	Sig.
Beck Depression Inventory (Sunday)	Ecstasy	.276	10	.030	.811	10	.020
	Alcohol	.170	10	.200*	.959	10	.780
Beck Depression Inventory (Wednesday)	Ecstasy	.235	10	.126	.941	10	.566
	Alcohol	.305	10	.009	.753	10	.004

a. Lilliefors Significance Correction

*. This is a lower bound of the true significance.

Test of Homogeneity of Variance

		Levene Statistic	df1	df2	Sig.
Beck Depression Inventory (Sunday)	Based on Mean	3.644	1	18	.072
	Based on Median	1.880	1	18	.187
	Based on Median and with adjusted df	1.880	1	10.076	.200
	Based on trimmed mean	2.845	1	18	.109
Beck Depression Inventory (Wednesday)	Based on Mean	.508	1	18	.485
	Based on Median	.091	1	18	.766
	Based on Median and with adjusted df	.091	1	11.888	.768
	Based on trimmed mean	.275	1	18	.606

15.3.3. Running the analysis ①

First, access the main dialog box (see Figure 15.4) by selecting Analyze Nonparametric Tests ▸ 2 Independent Samples... . Once the dialog box is activated, select both dependent variables from the list (click on **Beck Depression Inventory Sunday [Sunday_BDI]** then, holding down *Ctrl*, click on **Beck Depression Inventory Wednesday [Wednesday_BDI]**) and drag them to the box labelled *Test Variable List* (or click on ➡). Next, select the independent variable, in this case **Type of Drug [Drug]**, and transfer it to the box labelled *Grouping Variable*. When the grouping variable has been selected the Define Groups... button becomes active and you should click on it to activate the define groups dialog box. SPSS needs to know what numeric codes you assigned to your two groups, and there is a space for you to type the codes. In this example we coded our ecstasy group as 1 and our alcohol group as 2, and so you should type these two values in the appropriate space. When you have defined the groups, click on Continue to return to the main dialog box. The main dialog box also provides the facility to do tests other than the Mann–Whitney test and these alternatives are explained in SPSS Tip 15.1.

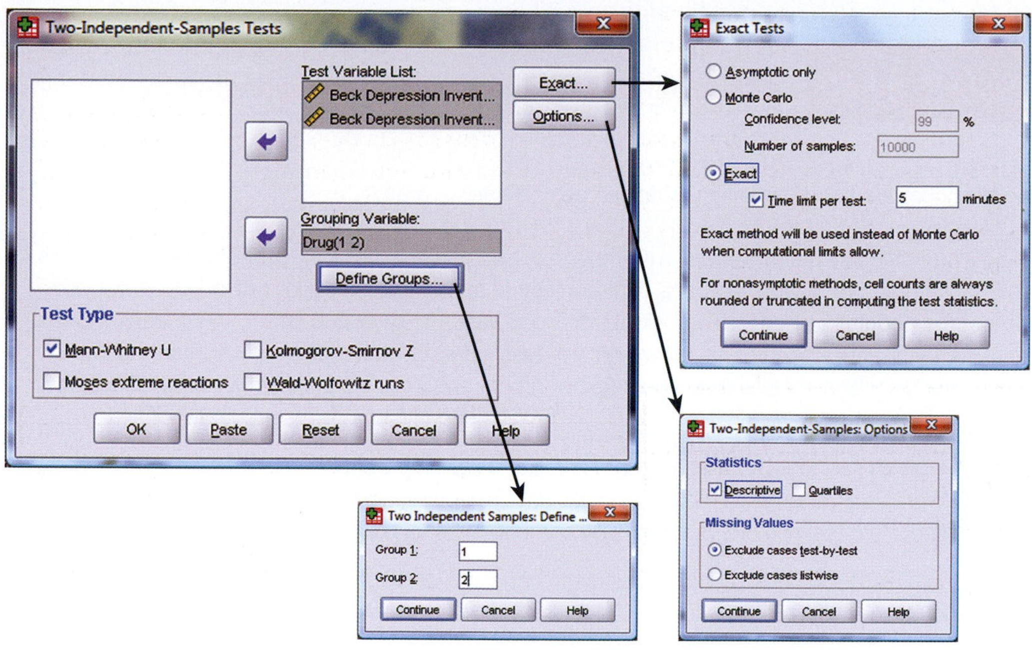

FIGURE 15.4 Dialog boxes for the Mann–Whitney test

If you click on Exact... then another dialog box appears.[3] By default SPSS calculates the significance of the Mann–Whitney test using a method that is accurate with large samples (called the *Asymptotic Method*); however, when samples are smaller, or the data are particularly poorly distributed, then more accurate methods are available. The most accurate method is an *exact* test, which calculates the significance of the Mann–Whitney test exactly. However, to get this precision, there is a price, and because of the complexities of the computation SPSS can take some time to find a solution – especially in large samples. A slightly less labour-intensive method is to estimate the significance using the **Monte Carlo method**.[4] This basically involves creating a distribution similar to that found in the sample and then taking several samples (the default is 10,000) from this distribution and from those samples the mean significance value and the confidence interval around it can be created. If that didn't make any sense to you then, fear not, the rule of thumb is that when samples are large you should probably opt for the Monte Carlo method, and when you have small samples (as we have) it's worth opting for the exact test (as I have done in this example). Finally, clicking on Options... opens another dialog box that gives you options for the analysis. These options are not particularly useful because, for example, the option that provides descriptive statistics does so for the entire data set (so doesn't break down values according to group membership). For this reason, I recommend obtaining descriptive statistics using the methods we learnt about in sections 5.4.3 and 5.5. To run the analyses return to the main dialog box and click on OK.

[3] This button will appear only if you have the *Exact tests* module of SPSS installed. Remember this in future sections too.

[4] If you're wondering why it's called the Monte Carlo method it's because back in the late 1800s when Karl Pearson was trying to simulate data he didn't have a computer to do it for him. So he used to toss coins. A lot. That is, until a friend suggested that roulette wheels, if unbiased, were excellent random number generators. Rather than trying to persuade the Royal Society to fund trips to Monte Carlo casinos to collect data from their roulette wheels, he purchased copies of *Le Monaco*, a weekly Paris periodical that published exactly the data that he required, at the cost of 1 franc (Pearson, 1894; Plackett, 1983). When simulated data are used to test a statistical method, or to estimate a statistic, it is known as the Monte Carlo method even though we use computers now and not roulette wheels.

SPSS TIP 15.1 Other options for the Mann–Whitney test ②

In the main dialog box there are some other tests that can be selected:

- **Kolmogorov-Smirnov Z**: In Chapter 5 we met a Kolmogorov–Smirnov test that tested whether a sample was from a normally distributed population. This is a different test! In fact, it tests whether two groups have been drawn from the same population (regardless of what that population may be). In effect, this means it does much the same as the Mann–Whitney test! However, this test tends to have better power than the Mann–Whitney test when sample sizes are less than about 25 per group, and so is worth selecting if that's the case.
- **Moses Extreme Reactions**: Great name – makes me think of a bearded man standing on Mount Sinai reading a stone tablet and then suddenly bursting into a wild rage, smashing the tablet and screaming 'What do you mean, do not worship any other God?' Sadly, this test isn't as exciting as my mental image. It's a bit like a non-parametric Levene's test (section 5.6.1); it basically compares the variability of scores in the two groups.
- **Wald-Wolfowitz runs**: Despite sounding like a particularly bad case of diarrhoea this test is another variant on the Mann–Whitney test. In this test the scores are rank ordered as in the Mann–Whitney test, but rather than analysing the ranks, this test looks for 'runs' of scores from the same group within the ranked order. Now, if there's no difference between groups then obviously ranks from the two groups should be randomly interspersed. However, if the groups are different then you should see more ranks from one group at the lower end and more ranks from the other group at the higher end. By looking for clusters of scores in this way the test can determine if the groups differ.

15.3.4. Output from the Mann–Whitney test ①

I explained earlier that the Mann–Whitney test works by looking at differences in the ranked positions of scores in different groups. Therefore, the first part of the output summarizes the data after they have been ranked. Specifically, SPSS tells us the average and total ranks in each condition (see SPSS Output 15.2). Remember that the Mann–Whitney test relies on scores being ranked from lowest to highest; therefore, the group with the lowest mean rank is the group with the greatest number of lower scores in it. Similarly, the group that has the highest mean rank should have a greater number of high scores within it. Therefore, this initial table can be used to ascertain which group had the highest scores, which is useful in case we need to interpret a significant result. You should note that the sums of ranks are the same as those calculated in section 15.3.1 (which is something of a relief to me!).

SPSS OUTPUT 15.2

Ranks

	Type of Drug	N	Mean Rank	Sum of Ranks
Beck Depression Inventory (Sunday)	Ecstasy	10	11.95	119.50
	Alcohol	10	9.05	90.50
	Total	20		
Beck Depression Inventory (Wednesday)	Ecstasy	10	15.10	151.00
	Alcohol	10	5.90	59.00
	Total	20		

The second table (SPSS Output 15.3) provides the actual test statistics for the Mann–Whitney test, the Wilcoxon procedure and the corresponding z-score. SPSS Output 15.3

has a column for each variable (one for **Sunday_BDI** and one for **Wednesday_BDI**) and in each column there is the value of Mann–Whitney's *U* statistic, the value of Wilcoxon's statistic and the associated *z* approximation. Note that the values of *U*, W_s and the associated *z*-score are the same as we calculated in section 15.3.1.

The important part of the table is the significance value of the test, which gives the two-tailed probability that a test statistic of at least that magnitude is a chance result. This significance value can be used as it is when no prediction has been made about which group will differ from which. However, if a prediction has been made (e.g. if we said that ecstasy users would be more depressed than alcohol users the day after taking the drug) then we can calculate the one-tailed probability by taking the two-tailed value and dividing it by 2 (SPSS Tip 9.1). For these data, the Mann–Whitney test is non-significant (two-tailed) for the depression scores taken on the Sunday. This finding indicates that ecstasy is no more of a depressant, the day after taking it, than alcohol: both groups report comparable levels of depression. However, for the midweek measures the results are highly significant ($p < .001$). The value of the mean rankings indicates that the ecstasy group had significantly higher levels of depression midweek than the alcohol group. This conclusion is reached by noting that for the Wednesday scores, the average rank is higher in the ecstasy users (15.10) than in the alcohol users (5.90).

SPSS OUTPUT 15.3 (without Monte Carlo exact significance)

Test Statistics[b]

	Beck Depression Inventory (Sunday)	Beck Depression Inventory (Wednesday)
Mann-Whitney U	35.500	4.000
Wilcoxon W	90.500	59.000
Z	-1.105	-3.484
Asymp. Sig. (2-tailed)	.269	.000
Exact Sig. [2*(1-tailed Sig.)]	.280[a]	.000[a]

a. Not corrected for ties.

b. Grouping Variable: Type of Drug

SPSS Output 15.4 shows the output for the Mann–Whitney test when exact significance is selected for. I've just included this to show you that you get some extra lines that give exact significance values (both one- and two-tailed). These don't actually change our conclusions, but be aware that you should probably consult these values in preference to the asymptotic value, especially when sample sizes are small.

SPSS OUTPUT 15.4 (with Monte Carlo exact significance)

Test Statistics[b]

	Beck Depression Inventory (Sunday)	Beck Depression Inventory (Wednesday)
Mann-Whitney U	35.500	4.000
Wilcoxon W	90.500	59.000
Z	-1.105	-3.484
Asymp. Sig. (2-tailed)	.269	.000
Exact Sig. [2*(1-tailed Sig.)]	.280[a]	.000[a]
Exact Sig. (2-tailed)	.288	.000
Exact Sig. (1-tailed)	.144	.000
Point Probability	.013	.000

a. Not corrected for ties.

b. Grouping Variable: Type of Drug

15.3.5. Calculating an effect size ②

As we've seen throughout this book, it's important to report effect sizes so that people have a standardized measure of the size of the effect you observed, which they can compare to other studies. SPSS doesn't calculate an effect size for us, but we can calculate approximate effect sizes really easily thanks to the fact that SPSS converts the test statistics into a *z*-score. The equation to convert a *z*-score into the effect size estimate, *r*, is as follows (from Rosenthal, 1991: 19):

$$r = \frac{Z}{\sqrt{N}}$$

in which *z* is the *z*-score that SPSS produces and *N* is the size of the study (i.e. the number of total observations) on which *z* is based. In this case SPSS Output 15.3 tells us that *z* is −1.11 for the Sunday data and −3.48 for the Wednesday data. In both cases we had 10 ecstasy users and 10 alcohol users and so the total number of observations was 20. The effect sizes are therefore:

$$r_{\text{Sunday}} = \frac{-1.11}{\sqrt{20}} = -0.25$$

$$r_{\text{Wednesday}} = \frac{-3.48}{\sqrt{20}} -0.78$$

This represents a small to medium effect for the Sunday data (it is below the .3 criterion for a medium effect size) and a huge effect for the Wednesday data (the effect size is well above the .5 threshold for a large effect). The Sunday data show how a fairly large effect size can still be non-significant in a small sample!

15.3.6. Writing the results ①

For the Mann–Whitney test, we need to report only the test statistic (which is denoted by *U*) and its significance. Of course, we really ought to include the effect size as well. So, we could report something like:

✓ Depression levels in ecstasy users (Mdn = 17.50) did not differ significantly from alcohol users (Mdn = 16.00) the day after the drugs were taken, $U = 35.50$, $z = -1.11$, *ns*, $r = -.25$. However, by Wednesday, ecstasy users (Mdn = 33.50) were significantly more depressed than alcohol users (Mdn = 7.50), $U = 4.00$, $z = -3.48$, $p < .001$, $r = -.78$.

Note that I've reported the median for each condition – this statistic is more appropriate than the mean for non-parametric tests. We could also choose to report Wilcoxon's test rather than Mann–Whitney's *U* statistic and this would be as follows:

✓ Depression levels in ecstasy users (Mdn = 17.50) did not significantly differ from alcohol users (Mdn = 16.00) the day after the drugs were taken, $W_s = 90.50$, $z = -1.11$, *ns*, $r = -.25$. However, by Wednesday, ecstasy users (Mdn = 33.50) were significantly more depressed than alcohol users (Mdn = 7.50), $W_s = 59.00$, $z = -3.48$, $p < .001$, $r = -.78$.

CRAMMING SAM'S TIPS

- The Mann–Whitney test and Wilcoxon rank-sum test compare two conditions when different participants take part in each condition and the resulting data violate any assumption of the independent *t*-test.
- Look at the row labelled *Asymp. Sig. (2-tailed)*. If the value is less than .05 then the two groups are significantly different. (If you opted for exact tests then look at the row labelled *Exact Sig. (2-tailed)*.)
- The values of the ranks tell you how the groups differ (the group with the highest scores will have the highest ranks).
- SPSS provides only the two-tailed significance value; the one-tailed significance is this value divided by 2. (If you opted for exact tests then look at the row labelled *Exact Sig. (1-tailed)*.)
- Report the *U* statistic (or W_s if you prefer), the corresponding *z*, and the significance value. Also report the medians and their corresponding ranges (or draw a boxplot).
- You should calculate the effect size and report this too!

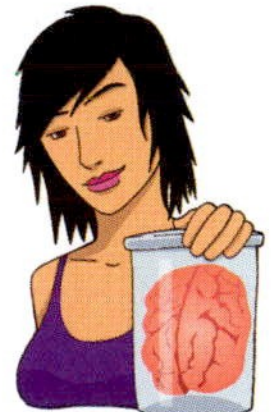

JANE SUPERBRAIN 15.1

Non-parametric tests and statistical power ②

Ranking the data is a useful way around the distributional assumptions of parametric tests but there is a price to pay: by ranking the data we lose some information about the magnitude of differences between scores. Consequently, non-parametric tests can be less powerful than their parametric counterparts. Statistical power (section 2.6.5) refers to the ability of a test to find an effect that genuinely exists. So, by saying that non-parametric tests are less powerful, we mean that if there is a genuine effect in our data then a parametric test is more likely to detect it than a non-parametric one. However, this statement is true only *if the assumptions of the parametric test are met*. So, if we use a parametric test and a non-parametric test on the same data, and those data meet the appropriate assumptions, then the parametric test will have greater power to detect the effect than the non-parametric test.

The problem is that to define the power of a test we need to be sure that it controls the Type I error rate (the number of times a test will find a significant effect when in reality there is no effect to find – see section 2.6.2). We saw in Chapter 2 that this error rate is normally set at 5%. We know that when the sampling distribution is normally distributed then the Type I error rate of tests based on this distribution is indeed 5%, and so we can work out the power. However, when data are not normal the Type I error rate of tests based on this distribution won't be 5% (in fact we don't know what it is for sure as it will depend on the shape of the distribution) and so we have no way of calculating power (because power is linked to the Type I error rate – see section 2.6.5). So, although you often hear (in the first edition of this book for example!) of non-parametric tests having an increased chance of a Type II error (i.e. more chance of accepting that there is no difference between groups when, in reality, a difference exists), this is true only if the sampling distribution is normally distributed.

15.4. Comparing two related conditions: the Wilcoxon signed-rank test ①

The **Wilcoxon signed-rank test** (Wilcoxon, 1945), not to be confused with the rank-sum test in the previous section, is used in situations in which there are two sets of scores to compare, but these scores come from the same participants. As such, think of it as the non-parametric equivalent of the dependent *t*-test (or a Mann–Whitney test for repeated-measures data). Imagine the experimenter in the previous section was now interested in the *change* in depression levels, within people, for each of the two drugs. We now want to compare the BDI scores on Sunday to those on Wednesday. We still have to use a non-parametric test because the distributions of scores for both drugs were non-normal on one of the two days (see SPSS Output 15.1).

15.4.1. Theory of the Wilcoxon signed-rank test ②

SMART ALEX ONLY

The Wilcoxon signed-rank test works in a fairly similar way to the dependent *t*-test (Chapter 9) in that it is based on the differences between scores in the two conditions you're comparing. Once these differences have been calculated they are ranked (just like in section 15.3.1) but the sign of the difference (positive or negative) is assigned to the rank. If we use the same data as before we can compare depression scores on Sunday to those on Wednesday for the two drugs separately.

Table 15.2 shows the ranking for these data. Remember that we're ranking the two drugs separately. First, we calculate the difference between Sunday and Wednesday (that's just Sunday's score subtracted from Wednesday's). If the difference is zero (i.e. the scores are the same on Sunday and Wednesday) then we exclude these data from the ranking. We make a note of the sign of the difference (was it positive or negative?) and then rank the differences (starting with the smallest) ignoring whether they are positive or negative. The ranking is the same as in section 15.3.1, and we deal with tied scores in exactly the same way. Finally we collect together the ranks that came from a positive difference between the conditions, and add them up to get the sum of positive ranks (T_+). We also add up the ranks that came from negative differences between the conditions to get the sum of negative ranks (T_-). So, for ecstasy, $T_+ = 36$ and $T_- = 0$ (in fact there were no negative ranks), and for alcohol, $T_+ = 8$ and $T_- = 47$. The test statistic, T, is the smaller of the two values, and so is 0 for ecstasy and 8 for alcohol.

To calculate the significance of the test statistic (T), we again look at the mean ($\overline{T}$) and standard error ($SE_{\overline{T}}$), which, like the Mann–Whitney and rank-sum test in the previous section, are functions of the sample size, n (because we used the same participants, there is only one sample size):

$$\overline{T} = \frac{n(n+1)}{4}$$

$$SE_{\overline{T}} = \sqrt{\frac{n(n+1)(2n+1)}{24}}$$

In both groups, n is simply 10 (because that's how many participants were used). However, remember that for our ecstasy group we excluded two people because they had differences of zero, therefore the sample size we use is 8, not 10. This gives us:

TABLE 15.2 Ranking data in the Wilcoxon signed-rank test

BDI Sunday	BDI Wednesday	Difference	Sign	Rank	Positive Ranks	Negative Ranks
			Ecstasy			
15	28	13	+	2.5	2.5	
35	35	0	Exclude			
16	35	19	+	6	6	
18	24	6	+	1	1	
19	39	20	+	7	7	
17	32	15	+	4.5	4.5	
27	27	0	Exclude			
16	29	13	+	2.5	2.5	
13	36	23	+	8	8	
20	35	15	+	4.5	4.5	
				Total =	**36**	**0**
			Alcohol			
16	5	−11	−	9		9
15	6	−9	−	7		7
20	30	10	+	8	+8	
15	8	−7	−	3.5		3.5
16	9	−7	−	3.5		3.5
13	7	−6	−	2		2
14	6	−8	−	5.5		5.5
19	17	−2	−	1		1
18	3	−15	−	10		10
18	10	−8	−	5.5		5.5
				Total =	**8**	**47**

$$\overline{T}_{\text{Ecstasy}} = \frac{8(8+1)}{4} = 18$$

$$\text{SE}_{\overline{T}_{\text{Ecstasy}}} = \sqrt{\frac{8(8+1)(16+1)}{24}} = 7.14$$

For the alcohol group there were no exclusions so we get:

$$\overline{T}_{\text{Alcohol}} = \frac{10(10+1)}{4} = 27.50$$

$$\text{SE}_{\overline{T}_{\text{Alcohol}}} = \sqrt{\frac{10(10+1)(20+1)}{24}} = 9.81$$

As before, if we know the test statistic, the mean of test statistics and the standard error, then we can easily convert the test statistic to a z-score using the equation that we came across way back in Chapter 1 and the previous section:

$$z = \frac{X - \overline{X}}{s} = \frac{T - \overline{T}}{SE_{\overline{T}}}$$

If we calculate this value for the ecstasy and alcohol depression scores we get:

$$z_{\text{Ecstasy}} = \frac{T - \overline{T}}{SE_{\overline{T}}} = \frac{0 - 18}{7.14} = -2.52$$

$$z_{\text{Alcohol}} = \frac{T - \overline{T}}{SE_{\overline{T}}} = \frac{8 - 27.5}{9.81} = -1.99$$

If these values are bigger than 1.96 (ignoring the minus sign) then the test is significant at $p < .05$. So, it looks as though there is a significant difference between depression scores on Wednesday and Sunday for both ecstasy and alcohol.

15.4.2. Running the analysis ①

To do the same analysis on SPSS we can use the same data as before, but because we want to look at the change for each drug *separately*, we need to use the *split file* command and ask SPSS to split the file by the variable **Type of Drug [Drug]**. This process ensures that any subsequent analysis is done for the ecstasy group and the alcohol group separately.

SELF-TEST Split the file by **Drug** (see section 5.4.3).

Once the file has been split, select the Wilcoxon test dialog box (Figure 15.5) by selecting Analyze Nonparametric Tests ▸ 2 Related Samples.... This dialog box allows you to select other tests too (see SPSS Tip 15.2).

Once the dialog box is activated, select two variables from the list (click on the first variable with the mouse and then, while holding down the *Ctrl* key, click on the second). You can also select the variables one at a time and transfer them: for example, you could select **Beck Depression Inventory (Sunday) [Sunday_BDI]** and drag it to the column labelled *Variable 1* in the box labelled *Test Pairs* (or click on ➔), and then select **Beck Depression Inventory (Wednesday) [Wednesday_BDI]** and drag it to the column labelled *Variable 2* (or click on ➔). To carry out several Wilcoxon tests select another pair of variables, transfer them to the variables list, and then select another pair and so on. Each pair appears as a new row in the box labelled *Test Pairs*. If you click on Exact... then another dialog box appears that allows you to select for SPSS to compute exact significance values (see section 15.3.3). I won't go into this again, but suffice it to say that when samples are large you should probably opt for the Monte Carlo method, and when you have small samples it's worth opting for the exact test. I haven't opted for either in this example. If you click on

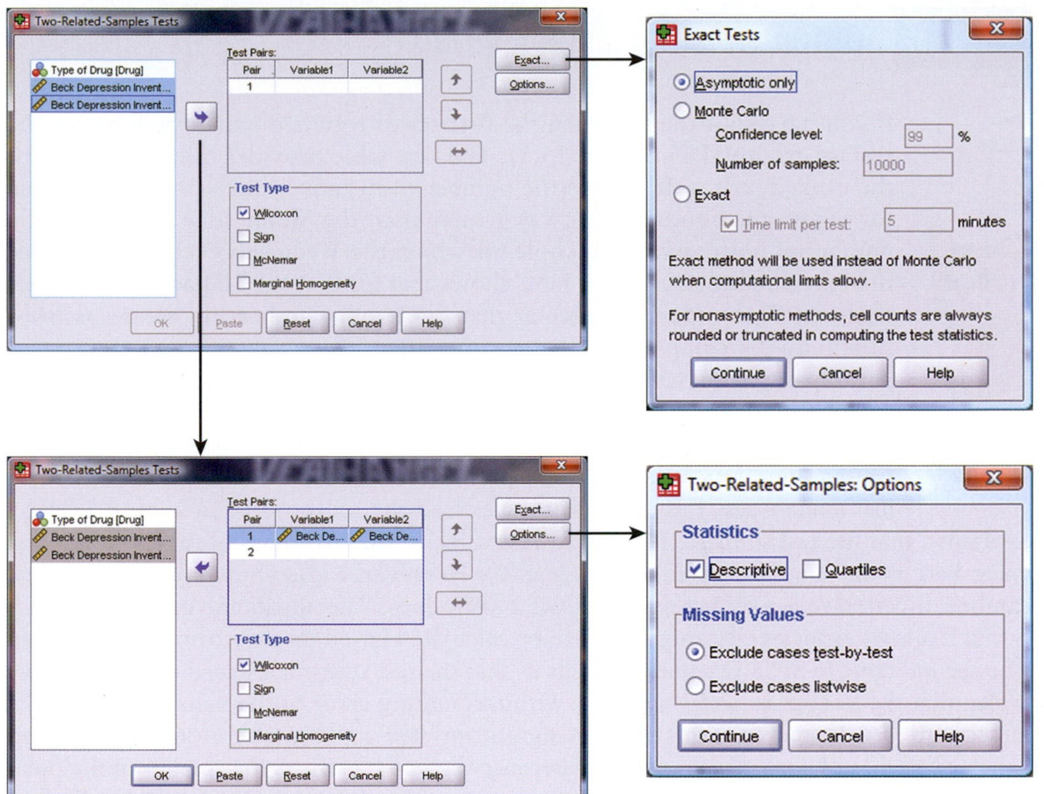

FIGURE 15.5 Dialog boxes for the Wilcoxon signed-rank test

SPSS TIP 15.2 Other options for the Wilcoxon signed-rank test ②

In the main dialog box there are some other tests that can be selected:

- **Sign**: The sign test basically does the same thing as the Wilcoxon signed-rank test, except that it is based only on the direction of difference (positive or negative). The magnitude of change is completely ignored (unlike in the Wilcoxon test where the rank tells us something about the relative magnitude of change). For these reasons the sign test lacks power (it's not very good at detecting effects) unless sample sizes are very small (six or less). So, frankly I don't see the point.
- **McNemar**: McNemar's test is useful when you have nominal rather than ordinal data. It's typically used when you're looking for changes in people's scores and it compares the number of people who changed their response in one direction (i.e. scores increased) to those who changed in the opposite direction (scores decreased). So, this test needs to be used when you've got two related dichotomous variables.
- Marginal Homogeneity: This produces an extension of McNemar's test but for ordinal variables. It does much the same as the Wilcoxon test as far as I can tell.

Options... then a dialog box appears that gives you the chance to select descriptive statistics. Unlike the Mann–Whitney test, the descriptive statistics here are worth having, because it is the change across variables (columns in the data editor) that is relevant. To run the analysis, return to the main dialog box and click on OK.

15.4.3. Output for the ecstasy group ①

What are the effects of ecstasy?

If you have split the file, then the first set of results obtained will be for the ecstasy group (SPSS Output 15.5). The first table provides information about the ranked scores. It tells us the number of negative ranks (these are people for whom the Sunday score was greater than the Wednesday score) and the number of positive ranks (people for whom the Wednesday score was greater than the Sunday score). The table shows that for 8 of the 10 participants, their score on Wednesday was greater than on Sunday, indicating greater depression midweek compared to the morning after. There were two tied ranks (i.e. participants who scored the same on both days). The table also shows the average number of negative and positive ranks and the sum of positive and negative ranks. Below the table are footnotes, which tell us to what the positive and negative ranks relate (so provide the same kind of explanation as I've just made – see, I'm not clever, I just read the footnotes!). In section 15.4.1 I explained that the test statistic, T, is the lowest value of the two types of ranks, so our test value here is the sum of negative ranks (e.g. 0). However, I also showed how this value can be converted to a z-score and this is what SPSS does. The advantage of this approach is that it allows exact significance values to be calculated based on the normal distribution. The second table in SPSS Output 15.5 tells us that the test statistic is based on the negative ranks, that the z-score is −2.53 (which is within rounding error of the value we calculated in section 15.4.1) and that this value is significant at $p = .012$. Therefore, because this value is based on the *negative* ranks (and because the test statistic is the smaller of the positive and negative ranks, the majority of ranks must have been positive), we should conclude that when taking ecstasy there was a significant increase in depression (as measured by the BDI) from the morning after to midweek. If the test statistic had been based on the positive ranks then this would have told us that the results were in the opposite direction (i.e. BDI scores were greater the morning after compared to midweek). Therefore, we can conclude that for ecstasy users there was a significant increase in depression from the next day to midweek ($z = -2.53$, $p < .05$).

SPSS OUTPUT 15.5

Ranks[d]

		N	Mean Rank	Sum of Ranks
Beck Depression Inventory (Wednesday) - Beck Depression Inventory (Sunday)	Negative Ranks	0[a]	.00	.00
	Positive Ranks	8[b]	4.50	36.00
	Ties	2[c]		
	Total	10		

a. Beck Depression Inventory (Wednesday) < Beck Depression Inventory (Sunday)

b. Beck Depression Inventory (Wednesday) > Beck Depression Inventory (Sunday)

c. Beck Depression Inventory (Wednesday) = Beck Depression Inventory (Sunday)

d. Type of Drug = Ecstasy

Test Statistics[b,c]

	Beck Depression Inventory (Wednesday) - Beck Depression Inventory (Sunday)
Z	-2.527[a]
Asymp. Sig. (2-tailed)	.012

a. Based on negative ranks.

b. Wilcoxon Signed Ranks Test

c. Type of Drug = Ecstasy

15.4.4. Output for the alcohol group ①

The remainder of the output should contain the same two tables but for the alcohol group (if it does not, then you probably forgot to split the file). As before, the first table in SPSS Output 15.6 provides information about the ranked scores. It tells us the number of negative ranks (these are people who were more depressed on Sunday than on Wednesday) and the number of positive ranks (people who were more depressed on Wednesday than on Sunday). The table shows that for 9 of the 10 participants, their score on Sunday was greater than on Wednesday, indicating greater depression the morning after compared to midweek. Unlike the ecstasy takers there were no tied ranks. The table also shows the average number of negative and positive ranks and the sum of positive and negative ranks. Below the table are footnotes that tell us to what the positive and negative ranks relate. As before, the lowest value of ranked scores is converted to a z-score (in this case 8). The second table tells us that the test statistic is based on the positive ranks, that the z-score is −1.99 (this is the value we calculated in section 15.4.1; I point this out merely because I'm so amazed that my hand calculations actually worked!) and that this value is significant at $p = .047$. Therefore, we should conclude (based on the fact that *positive* ranks were used) that when taking alcohol there was a significant decline in depression (as measured by the BDI) from the morning after to midweek ($z = -1.99$, $p < .05$).

From the results of the two different groups, we can see that there is an opposite effect when alcohol is taken to when ecstasy is taken. Alcohol makes you slightly depressed the morning after but this depression has dropped by midweek. Ecstasy also causes some depression the morning after consumption; however, this depression increases towards the middle of the week. Of course, to see the true effect of the morning after we would have had to take measures of depression before the drugs were administered! This opposite effect between groups of people is known as an interaction (i.e. you get one effect under certain circumstances and a different effect under other circumstances and we came across these in Chapters 12 to 14.

SPSS OUTPUT 15.6

Ranks[d]

		N	Mean Rank	Sum of Ranks
Beck Depression Inventory (Wednesday) - Beck Depression Inventory (Sunday)	Negative Ranks	9[a]	5.22	47.00
	Positive Ranks	1[b]	8.00	8.00
	Ties	0[c]		
	Total	10		

a. Beck Depression Inventory (Wednesday) < Beck Depression Inventory (Sunday)

b. Beck Depression Inventory (Wednesday) > Beck Depression Inventory (Sunday)

c. Beck Depression Inventory (Wednesday) = Beck Depression Inventory (Sunday)

d. Type of Drug = Alcohol

Test Statistics[b,c]

	Beck Depression Inventory (Wednesday) - Beck Depression Inventory (Sunday)
Z	-1.990[a]
Asymp. Sig. (2-tailed)	.047

a. Based on positive ranks.

b. Wilcoxon Signed Ranks Test

c. Type of Drug = Alcohol

15.4.5. Calculating an effect size ②

The effect size can be calculated in the same way as for the Mann–Whitney test (see the equation in section 15.3.5). In this case SPSS Output 15.6 tells us that for the ecstasy group z is −2.53, and for the alcohol group is −1.99. In both cases we had 20 observations (although we only used 10 people and tested them twice, it is the number of observations, not the number of people, that is important here). The effect size is therefore:

$$r_{\text{Ecstasy}} = \frac{-2.53}{\sqrt{20}} = -0.57$$

$$r_{\text{Alcohol}} = \frac{-1.99}{\sqrt{20}} = -0.44$$

This represents a large change in levels of depression when ecstasy is taken (it is above Cohen's benchmark of .5) and a medium to large change in depression when alcohol is taken (it is between Cohen's criteria of .3 and .5 for a medium and large effect respectively).

15.4.6. Writing the results ①

For the Wilcoxon test, we need to report only the test statistic (which is denoted by the letter T and the smallest of the two sum of ranks), its significance and preferably an effect size. So, we could report something like:

✓ For ecstasy users, depression levels were significantly higher on Wednesday (Mdn = 33.50) than on Sunday (Mdn = 17.50), $T = 0$, $p < .05$, $r = -.57$. However, for alcohol users the opposite was true: depression levels were significantly lower on Wednesday (Mdn = 7.50) than on Sunday (Mdn = 16.0), $T = 8$, $p < .05$, $r = -.44$.

Alternatively, we could report the values of z:

✓ For ecstasy users, depression levels were significantly higher on Wednesday (Mdn = 33.50) than on Sunday (Mdn = 17.50), $z = -2.53$, $p < .05$, $r = -.57$. However, for alcohol users the opposite was true: depression levels were significantly lower on Wednesday (Mdn = 7.50) than on Sunday (Mdn = 16.0), $z = -1.99$, $p < .05$, $r = -.44$.

CRAMMING SAM'S TIPS

- The Wilcoxon signed-rank test compares two conditions when the same participants take part in each condition and the resulting data violate an assumption of the dependent t-test.
- Look at the row labelled *Asymp. Sig. (2-tailed)*. If the value is less than .05 then the two groups are significantly different.
- Look at positive and negative ranks (and the footnotes explaining what they mean) to tell you how the groups differ (the greater number of ranks in a particular direction tells you the direction of the result).
- SPSS provides only the two-tailed significance value. If you want the one-tailed significance just divide the value by 2.
- Report the T statistic, the corresponding z, the significance value, and an effect size if possible. Also report the medians and their corresponding ranges (or draw a boxplot).

LABCOAT LENI'S REAL RESEARCH 15.1

Having a quail of a time? ①

We encountered some research in Chapter 2 in which we discovered that you can influence aspects of male quail's sperm production through 'conditioning'. The basic idea is that the male is granted access to a female for copulation in a certain chamber (e.g. one that is coloured green) but gains no access to a female in a different context (e.g. a chamber with a tilted floor). The male, therefore, learns that when he is in the green chamber his luck is in, but if the floor is tilted then frustration awaits. For other males the chambers will be reversed (i.e. they get sex only when in the chamber with the tilted floor). The human equivalent (well, sort of) would be if you always managed to pull in the Pussycat Club but never in the Honey Club.[5] During the test phase, males get to mate in both chambers; the question is: after the males have learnt that they will get a mating opportunity in a certain context, do they produce more sperm or better-quality sperm when mating in that context compared to the control context? (That is, are you more of a stud in the Pussycat Club? OK, I'm going to stop this analogy now.)

Mike Domjan and his colleagues predicted that if conditioning evolved because it increases reproductive fitness then males who mated in the context that had previously signalled a mating opportunity would fertilize a significantly greater number of eggs than quails that mated in their control context (Matthews, Domjan, Ramsey, & Crews, 2007). They put this hypothesis to the test in an experiment that is utter genius. After training, they allowed 14 females to copulate with two males (counterbalanced): one male copulated with the female in the chamber that had previously signalled a reproductive opportunity (**Signalled**), whereas the second male copulated with the same female but in the chamber that had not previously signalled a mating opportunity (**Control**). Eggs were collected from the females for 10 days after the mating and a genetic analysis was used to determine the father of any fertilized eggs.

The data from this study are in the file **Matthews et al. (2007).sav**. Labcoat Leni wants you to carry out a Wilcoxon signed-rank test to see whether more eggs were fertilized by males mating in their signalled context compared to males in their control context.

Answers are in the additional material on the companion website (or look at page 760 in the original article).

MATTHEWS, R. C. ET AL. (2007). *PSYCHOLOGICAL SCIENCE, 18*(9), 758–762.

15.5. Differences between several independent groups: the Kruskal–Wallis test ①

In Chapter 10 we discovered a technique called one-way independent ANOVA that could be used to test for differences between several independent groups. I mentioned several times in that chapter that the *F* statistic can be robust to violations of its assumptions (section 10.2.10). We also saw that there are measures that can be taken when you have heterogeneity of variance (Jane Superbrain Box 10.2). However, there is another alternative: the one-way independent ANOVA has a non-parametric counterpart called the **Kruskal–Wallis test** (Kruskal & Wallis, 1952; Figure 15.6). If you have data that have violated an assumption then this test can be a useful way around the problem.

I read a story in a newspaper claiming that scientists had discovered that the chemical genistein, which is naturally occurring in soya, was linked to lowered sperm counts in western males. In fact, when you read the actual study, it had been conducted on rats,

[5] These are both clubs in Brighton that I don't go to because although my social skills are marginally better than they used to be, they're not *that* good.

FIGURE 15.6 Joseph Kruskal spotting some more errors in his well-thumbed first edition of *Discovering Statistics...* by that idiot Field

it found no link to lowered sperm counts, but there was evidence of abnormal sexual development in male rats (probably because this chemical acts like oestrogen). The journalist naturally interpreted this as a clear link to apparently declining sperm counts in western males (never trust what you read in the newspaper!). Anyway, as a vegetarian who eats lots of soya products and probably would like to have kids one day, I might want to test this idea in humans rather than rats. I took 80 males and split them into four groups that varied in the number of soya meals they ate per week over a year-long period. The first group was a control group and had no soya meals at all per week (i.e. none in the whole year); the second group had one soya meal per week (that's 52 over the year); the third group had four soya meals per week (that's 208 over the year); and the final group had seven soya meals a week (that's 364 over the year). At the end of the year, all of the participants were sent away to produce some sperm that I could count (when I say 'I', I mean someone else in a laboratory as far away from me as humanly possible).[6]

15.5.1. Theory of the Kruskal–Wallis test ②

SMART ALEX ONLY

The theory for the Kruskal–Wallis test is very similar to that of the Mann–Whitney (and Wilcoxon rank-sum) test, so before reading on look back at section 15.3.1. Like the Mann–Whitney test, the Kruskal–Wallis test is based on ranked data. So, to begin with, you simply order the scores from lowest to highest, ignoring the group to which the score belongs, and then assign the lowest score a rank of 1, the next highest a rank of 2 and so on (see section 15.3.1 for more detail). When you've ranked the data you collect the scores back into their groups and simply add up the ranks for each group. The sum of ranks for each group is denoted by R_i (where i is used to denote the particular group).

Table 15.3 shows the raw data for this example along with the ranks.

SELF-TEST Have a go at ranking the data and see if you get the same results as me!

Once the sum of ranks has been calculated for each group, the test statistic, H, is calculated as:

$$H = \frac{12}{N(N+1)} \sum_{i=1}^{k} \frac{R_i^2}{n_i} - 3(N+1) \tag{15.1}$$

In this equation, R_i is the sum of ranks for each group, N is the total sample size (in this case 80) and n_i is the sample size of a particular group (in this case we have equal sample

[6] In case any medics are reading this chapter, these data are made up and, because I have absolutely no idea what a typical sperm count is, they're probably ridiculous. I apologise and you can laugh at my ignorance!

TABLE 15.3 Data for the soya example with ranks

No Soya		1 Soya Meal		4 Soya Meals		7 Soya Meals	
Sperm (Millions)	Rank	Sperm (Millions)	Rank	Sperm (Millions)	Rank	Sperm (Millions)	Rank
0.35	4	0.33	3	0.40	6	0.31	1
0.58	9	0.36	5	0.60	10	0.32	2
0.88	17	0.63	11	0.96	19	0.56	7
0.92	18	0.64	12	1.20	21	0.57	8
1.22	22	0.77	14	1.31	24	0.71	13
1.51	30	1.53	32	1.35	27	0.81	15
1.52	31	1.62	34	1.68	35	0.87	16
1.57	33	1.71	36	1.83	37	1.18	20
2.43	41	1.94	38	2.10	40	1.25	23
2.79	46	2.48	42	2.93	48	1.33	25
3.40	55	2.71	44	2.96	49	1.34	26
4.52	59	4.12	57	3.00	50	1.49	28
4.72	60	5.65	61	3.09	52	1.50	29
6.90	65	6.76	64	3.36	54	2.09	39
7.58	68	7.08	66	4.34	58	2.70	43
7.78	69	7.26	67	5.81	62	2.75	45
9.62	72	7.92	70	5.94	63	2.83	47
10.05	73	8.04	71	10.16	74	3.07	51
10.32	75	12.10	77	10.98	76	3.28	53
21.08	80	18.47	79	18.21	78	4.11	56
Total (R_i)	**927**		**883**		**883**		**547**

sizes and they are all 20). Therefore, all we really need to do for each group is square the sum of ranks and divide this value by the sample size for that group. We then add up these values. That deals with the middle part of the equation; the rest of it involves calculating various values based on the total sample size. For these data we get:

$$
\begin{aligned}
H &= \frac{12}{80(81)}\left(\frac{972^2}{20} + \frac{883^2}{20} + \frac{833^2}{20} + \frac{547^2}{20}\right) - 3(81) \\
&= \frac{12}{6480}(42{,}966.45 + 38{,}984.45 + 38{,}984.45 + 14{,}960.45) - 243 \\
&= 0.0019(135{,}895.8) - 243 \\
&= 251.66 - 243 \\
&= 8.659
\end{aligned}
$$

This test statistic has a special kind of distribution known as the chi-square distribution (see Chapter 18) and for this distribution there is one value for the degrees of freedom, which is one less than the number of groups ($k - 1$): in this case 3.

15.5.2. Inputting data and provisional analysis ①

SELF-TEST See whether you can enter the data in Table 15.3 into SPSS (you don't need to enter the ranks). Then conduct some exploratory analyses on the data (see sections 5.5 and 5.6).

When the data are collected using different participants in each group, we input the data using a coding variable. So, the data editor will have two columns of data. The first column is a coding variable (called something like **Soya**) which, in this case, will have four codes (for convenience I suggest 1 = no soya, 2 = one soya meal per week, 3 = four soya meals per week and 4 = seven soya meals per week). The second column will have values for the dependent variable (sperm count) measured at the end of the year (call this variable **Sperm**). When you enter the data into SPSS remember to tell the computer which group is represented by which code (see section 3.4.2.3). The data can be found in the file **Soya.sav**.

First, we would run some exploratory analyses on the data and because we're going to be looking for group differences we need to run these exploratory analyses for each group. If you do these analyses you should find the same tables shown in SPSS Output 15.7. The first table shows that the Kolmogorov–Smirnov test (see section 5.5) was not significant for the control group ($D(20) = .181$, $p > .05$) but the Shapiro–Wilk test is significant and this test is actually more accurate (though less widely reported) than the Kolmogorov–Smirnov test (see Chapter 5). Data for the group that ate one soya meal per week were significantly different from normal ($D(20) = .207$, $p < .05$), as were the data for those that ate four ($D(20) = .267$, $p < .01$) and seven ($D(20) = .204$, $p < .05$). The second table shows the results of Levene's test (section 5.6.1). The assumption of homogeneity of variance has been violated, $F(3, 76) = 5.12$, $p < .01$, and this is shown by the fact that the significance of Levene's test is less than .05. As such, these data are not normally distributed, and the groups have heterogeneous variances!

SPSS OUTPUT 15.7

Tests of Normality

	Number of Soya Meals Per Week	Kolmogorov-Smirnov[a]			Shapiro-Wilk		
		Statistic	df	Sig.	Statistic	df	Sig.
Sperm Count (Millions)	No Soya Meals	.181	20	.085	.805	20	.001
	1 Soya Meal Per Week	.207	20	.024	.826	20	.002
	4 Soyal Meals Per Week	.267	20	.001	.743	20	.000
	7 Soya Meals Per Week	.204	20	.028	.912	20	.071

a. Lilliefors Significance Correction

Test of Homogeneity of Variance

		Levene Statistic	df1	df2	Sig.
Sperm Count (Millions)	Based on Mean	5.117	3	76	.003
	Based on Median	2.860	3	76	.042
	Based on Median and with adjusted df	2.860	3	58.107	.045
	Based on trimmed mean	4.070	3	76	.010

15.5.3. Doing the Kruskal–Wallis test on SPSS ①

First, access the main dialog box (see Figure 15.7) by selecting Analyze Nonparametric Tests ▸ K Independent Samples... . Once the dialog box is activated, select the dependent variable from

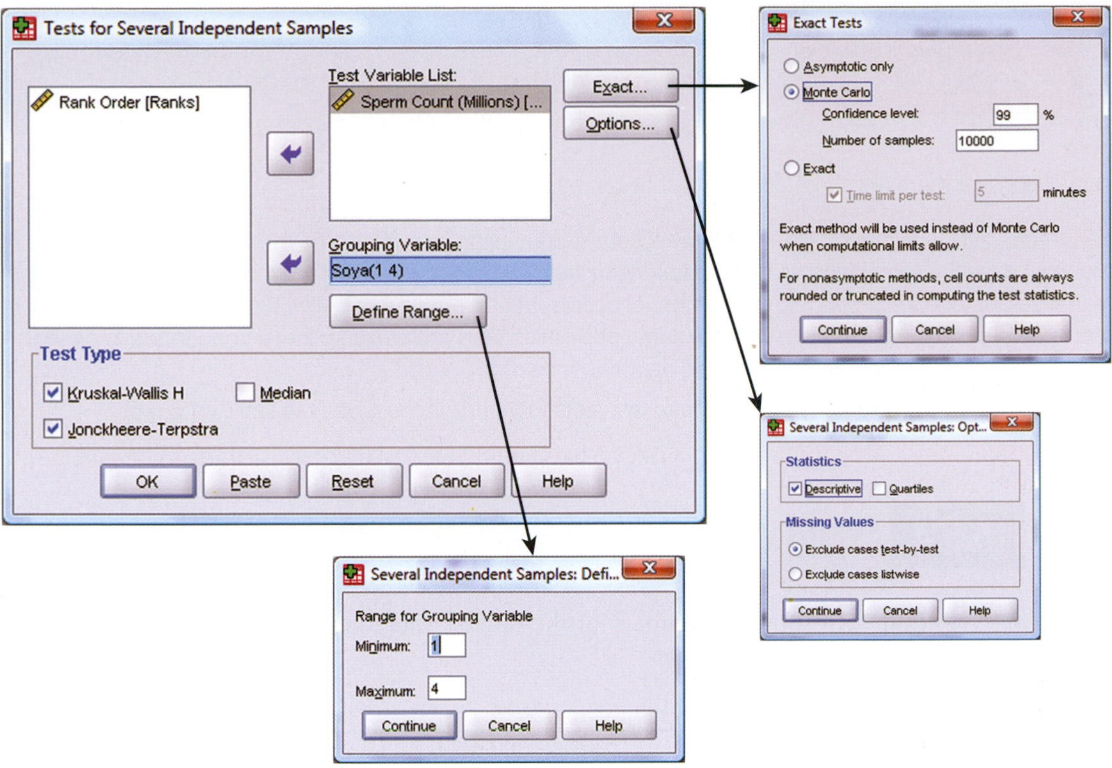

FIGURE 15.7 Dialog boxes for the Kruskal–Wallis test

the list (click on **Sperm Count (Millions)**) and drag it to the box labelled *Test Variable List* (or click on ➜). Next, select the independent variable (the grouping variable), in this case **Soya**, and drag it to the box labelled *Grouping Variable*. When the grouping variable has been selected the Define Range... button becomes active and you should click on it to activate the define range dialog box. SPSS needs to know the range of numeric codes you assigned to your groups, and there is a space for you to type the minimum and maximum code. If you followed my coding scheme, then the minimum code we used was 1, and the maximum was 4, so type these numbers into the appropriate spaces. When you have defined the groups, click on Continue to return to the main dialog box. The main dialog box also provides options to conduct some tests similar to the Kruskal–Wallis test (see SPSS Tip 15.3).

If you click on Exact... then you get a dialog box for selecting exact significance values for the Kruskal–Wallis test. I've explained this option in section 15.3.3, so I won't repeat myself here. I'll just recap by saying that when samples are large you should probably opt for the Monte Carlo method (as I have done in this example) and when you have small samples it's worth opting for the exact test. Finally, if you click on Options... then another dialog box appears that gives you options for the analysis. These options are not particularly useful because, for example, the option that provides descriptive statistics does so for the entire data set (so doesn't break down values according to group membership). For this reason, I recommend obtaining descriptive statistics using the methods we learnt about in sections 5.4.3 and 5.5. The final option you can ask for is for the Jonckheere–Terpstra trend test (select ☑ Jonckheere-Terpstra). This is useful if you want to look for a linear trend in the data (see section 10.2.11.5). To run the analyses return to the main dialog box and click on OK .

SPSS TIP 15.3 Other options for Kruskal–Wallis ②

In the main dialog box there are some other tests that can be selected:

- **Median**: This tests whether samples are drawn from a population with the same median. So, in effect the median test does the same thing as the Kruskal–Wallis test. It works on the basis of producing a contingency table that is split for each group into the number of scores that fall above and below the observed median of the entire data set. If the groups are from the same population then you'd expect these frequencies to be the same in all conditions (about 50% above and about 50% below).
- **Jonckheere-Terpstra**: The Jonckheere–Terpstra test tests for trends in the data (see section 15.5.6).

15.5.4. Output from the Kruskal–Wallis test ①

SPSS Output 15.8 shows a summary of the ranked data in each condition and we'll need these for interpreting any effects.

SPSS OUTPUT 15.8

Ranks

	Number of Soya Meals	N	Mean Rank
Sperm Count (Millions)	No Soya Meals	20	46.35
	1 Soya Meal Per Week	20	44.15
	4 Soyal Meals Per Week	20	44.15
	7 Soya Meals Per Week	20	27.35
	Total	80	

SPSS Output 15.9 shows the test statistic, *H*, for the Kruskal–Wallis test (although SPSS labels it chi-square, because of its distribution, rather than *H*), its associated degrees of freedom (in this case we had four groups so the degrees of freedom are 4–1, or 3) and the significance. The crucial thing to look at is the significance value, which is .034; because this value is less than .05 we could conclude that the amount of soya meals eaten per week does significantly affect sperm counts. Note also the Monte Carlo estimate of significance, which is slightly lower (.031). This is the value we ought to look to rather than the asymptotic value if it yields different results. The confidence interval for significance is also useful: it is .027 to .036 and the fact that the boundary does not cross .05 is important because it means that, assuming this confidence interval is one of the 99 out of 100 that contains the true value of the significance of the test statistic, the true value is less than .05. This gives us a lot of confidence that the significant effect is genuine. Like a one-way ANOVA, though, this test tells us only that a difference exists; it doesn't tell us exactly where the differences lie.

SPSS OUTPUT 15.9

Test Statistics[b,c]

			Sperm Count (Millions)
Chi-Square			8.659
df			3
Asymp. Sig.			.034
Monte Carlo Sig.	Sig.		.031[a]
	99% Confidence Interval	Lower Bound	.027
		Upper Bound	.036

a. Based on 10000 sampled tables with starting seed 2000000.

b. Kruskal Wallis Test

c. Grouping Variable: Number of Soya Meals Per Week

SELF-TEST Use the Chart Builder to draw a boxplot of these data?

One way to see which groups differ is to look at a boxplot (see section 5.4.2) of the groups (see Figure 15.8). The first thing to note is that there are some outliers (note the circles and asterisks that lie above the top whiskers) – these are men who produced a particularly rampant amount of sperm. Using the control as our baseline, the medians of the first three groups seem quite similar; however, the median of the group which ate seven soya meals per week does seem a little lower, so perhaps this is where the difference lies. However, these conclusions are subjective. What we really need are some contrasts or *post hoc* tests like we used in ANOVA (see sections 10.2.11 and 10.2.12).

15.5.5. *Post hoc* tests for the Kruskal–Wallis test ②

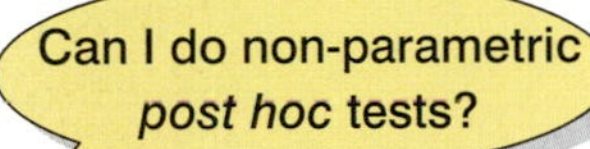

There are two ways to do non-parametric *post hoc* procedures, the first being to use Mann–Whitney tests (section 15.3). However, if we use lots of Mann–Whitney tests we will inflate the Type I error rate (section 10.2.1) and this is precisely why we don't begin by doing lots of Mann–Whitney tests! However, if we want to use lots of Mann–Whitney tests to follow up a Kruskal–Wallis test, we can if we make some kind of adjustment to ensure that the Type I errors don't build up to more than .05. The easiest method is to use a Bonferroni correction, which in its simplest form just means that instead of using .05 as the critical value for significance for each test, you use a critical value of .05 divided by the number of tests you've conducted. If you do this, you'll soon discover that you quickly end up using a critical value for significance that is so small that it is very restrictive. Therefore, it's a good idea to be selective about the comparisons you make. In this

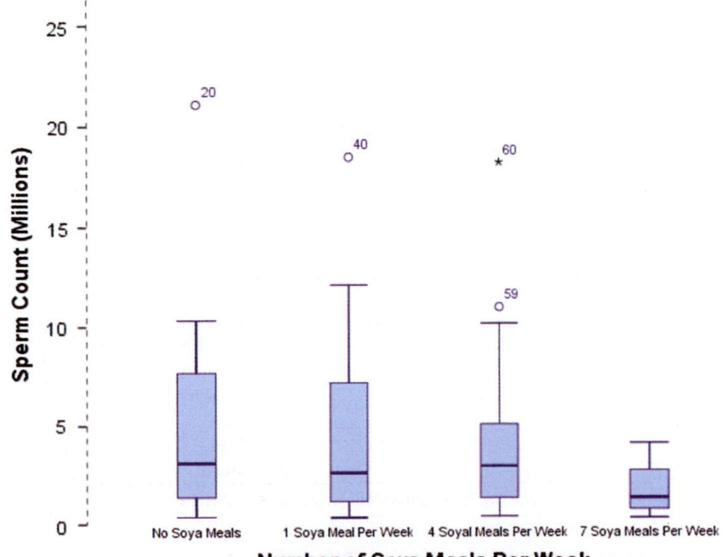

FIGURE 15.8 Boxplot for the sperm counts of individuals eating different numbers of soya meals per week

example, we have a control group which had no soya meals. As such, a nice succinct set of comparisons would be to compare each group against the control:

- Test 1: one soya meal per week compared to no soya meals
- Test 2: four soya meals per week compared to no soya meals
- Test 3: seven soya meals per week compared to no soya meals

This results in three tests, so rather than use .05 as our critical level of significance, we'd use .05/3 = .0167. If we didn't use focused tests and just compared all groups to all other groups we'd end up with six tests rather than three (no soya vs. 1 meal, no soya vs. 4 meals, no soya vs. 7 meals, 1 meal vs. 4 meals, 1 meal vs. 7 meals, 4 meals vs. 7 meals), meaning that our critical value would fall to .05/6 = .0083.

SELF-TEST Carry out the three Mann–Whitney tests suggested above.

SPSS Output 15.10 shows the test statistics from doing Mann–Whitney tests on the three focused comparisons that I suggested. Remember that we are now using a critical value of .0167, so the only comparison that is significant is when comparing those that had seven soya meals a week to those that had none (because the observed significance value of .009 is less than .0167). The other two comparisons produce significance values that are greater than .0167 so we'd have to say they're non-significant. So the effect we got seems to mainly reflect the fact that eating soya seven times per week lowers (I know this from the medians in Figure 15.8) sperm counts compared to eating no soya. However, eating some soya (one meal or four meals) doesn't seem to affect sperm counts significantly.

SPSS OUTPUT 15.10

No Soya vs. 1 Meal per week:

Test Statistics[b]

	Sperm Count (Millions)
Mann-Whitney U	191.000
Wilcoxon W	401.000
Z	-.243
Asymp. Sig. (2-tailed)	.808
Exact Sig. [2*(1-tailed Sig.)]	.820[a]

a. Not corrected for ties.

b. Grouping Variable: Number of Soya Meals Per Week

No Soya vs. 4 Meals per week:

Test Statistics[b]

	Sperm Count (Millions)
Mann-Whitney U	188.000
Wilcoxon W	398.000
Z	-.325
Asymp. Sig. (2-tailed)	.745
Exact Sig. [2*(1-tailed Sig.)]	.758[a]

a. Not corrected for ties.

b. Grouping Variable: Number of Soya Meals Per Week

No Soya vs. 7 Meals per week:

Test Statistics[b]

	Sperm Count (Millions)
Mann-Whitney U	104.000
Wilcoxon W	314.000
Z	-2.597
Asymp. Sig. (2-tailed)	.009
Exact Sig. [2*(1-tailed Sig.)]	.009[a]

a. Not corrected for ties.

b. Grouping Variable: Number of Soya Meals Per Week

The second way to do *post hoc* tests is essentially the same as doing Mann–Whitney tests on all possible comparisons, but for the sake of completeness I'll run you through it! It is described by Siegel and Castellan (1988) and involves taking the difference between the mean ranks of the different groups and comparing this to a value based on the value of z (corrected for the number of comparisons being done) and a constant based on the total sample size and the sample size in the two groups being compared. The inequality is:

$$\left|\overline{R}_u - \overline{R}_v\right| \geq z_{\alpha/k(k-1)}\sqrt{\frac{N(N+1)}{12}\left(\frac{1}{n_u}+\frac{1}{n_v}\right)} \qquad (15.2)$$

The left-hand side of this inequality is just the difference between the mean rank of the two groups being compared, but ignoring the sign of the difference (so the two vertical lines that enclose the difference between mean ranks just indicate that if the difference is negative then we ignore the negative sign and treat it as positive). For the rest of the expression, k is the number of groups (in the soya example, 4), N is the total sample size (in this case 80), n_u is the number of people in the first group that's being compared (we have equal group sizes in the soya example so it will be 20 regardless of which groups we compare), and n_v is the number of people in the second group being compared (again this will be 20 regardless of which groups we compare because we have equal group sizes in the soya example). The only other thing we need to know is $z_{\alpha/k(k-1)}$, and to get this value we need to decide a level for α, which is the level of significance at which we want to work. You should know by now that in the social sciences we traditionally work at a .05 level of significance, so α will be .05. We then calculate $k(k-1)$, which for these data will be $4(4-1) = 12$. Therefore, $\alpha/k(k-1) = .05/12 = .00417$. So, $z_{\alpha/k(k-1)}$ just means 'the value of z for which only $\alpha/k(k-1)$ other values of z are bigger' (or in this case 'the value of z for which only .00417 other values of z are bigger'). In practical terms this means we go to table A.1 in the Appendix, look at the column labelled *Smaller Portion* and find the number .00417 (or the nearest value to this, which if you look at the table is .00415), and we then look in the same row at the column labelled z. In this case, you should find that the value of z is 2.64. The next thing to do is to calculate the right-hand side of inequality (15.2):

$$\begin{aligned}
\text{critical difference} &= z_{\alpha/k(k-1)}\sqrt{\frac{N(N+1)}{12}\left(\frac{1}{n_u}+\frac{1}{n_v}\right)} \\
&= 2.64\sqrt{\frac{80(80+1)}{2}\left(\frac{1}{20}+\frac{1}{20}\right)} \\
&= 2.64\sqrt{540(0.1)} \\
&= 2.64\sqrt{54} \\
&= 19.40
\end{aligned}$$

For this example, because the sample sizes across groups are equal, this critical difference can be used for all comparisons. However, when sample sizes differ across groups, the critical difference will have to be calculated for each comparison individually. The next step is simply to calculate all of the differences between the mean ranks of all of the groups (the mean ranks can be found in SPSS Output 15.8), as in Table 15.4.

Inequality (15.2) basically means that if the differences between mean ranks is bigger than or equal to the critical difference for that comparison, then that difference is significant. In this case, because we have only one critical difference, it means that if any difference is bigger than 19.40, then it is significant. As you can see, all differences are below this value so we would have to conclude that none of the groups were significantly different!

TABLE 15.4 Differences between mean ranks for the soya data

Comparison	$\bar{R}_u$	$\bar{R}_v$	$\bar{R}_u - \bar{R}_v$	$\lvert\bar{R}_u - \bar{R}_v\rvert$
No Meals–1 Meal	46.35	44.15	2.20	2.20
No Meals–4 Meals	46.35	44.15	2.20	2.20
No-Meals–7 Meals	46.35	27.35	19.00	19.00
1 Meal–4 Meals	44.15	44.15	0.00	0.00
1 Meal–7 Meals	44.15	27.35	16.80	16.80
4 Meals–7 Meals	44.15	27.35	16.80	16.80

EVERYBODY

This contradicts our earlier findings where the Mann–Whitney test for the no-meals group compared to the seven-meals group was deemed significant; why do you think that is? Well, for our Mann–Whitney tests, we did only three comparisons and so only corrected the significance value for the three tests we'd done (.05/3 = .0167). Earlier on in this section I said that if we compared all groups against all other groups, that would be six comparisons, so we could accept a difference as being significant only if the significance value was less than .05/6 = .0083. If we go back to our one significant Mann–Whitney test (SPSS Output 15.10) the significance value was .009; therefore, if we had done all six comparisons this would've been non-significant (because .009 is bigger than .0083)! This illustrates what I said earlier about the benefits of choosing selective comparisons.

15.5.6. Testing for trends: the Jonckheere–Terpstra test ②

Back in section 15.5.3 we selected an option for the Jonckheere–Terpstra test, ☑ Jonckheere-Terpstra (Jonckheere, 1954; Terpstra, 1952). This statistic tests for an ordered pattern to the medians of the groups you're comparing. Essentially it does the same thing as the Kruskal–Wallis test (i.e. test for a difference between the medians of the groups) but it incorporates information about whether the order of the groups is meaningful. As such, you should use this test when you expect the groups you're comparing to produce a meaningful order of medians. So, in the current example we expect that the more soya a person eats, the more their sperm count will go down. Therefore, the control group should have the highest sperm count, those having one soya meal per week should have a lower sperm count, the sperm count in the four meals per week group should be smaller still, and the seven meals per week group should have the lowest sperm count. Therefore, there is an order to our medians: they should decrease across the groups. Conversely there might be situations where you expect your medians to increase. For example, there's a phenomenon in psychology known as the 'mere exposure effect' which basically means that the more you're exposed to something, the more you'll like it. Record companies use this to good effect by making sure songs are played on radio for about two months prior to their release, so on the day of release, everyone loves the song and is dying to have it and rushes out to buy it, sending it to number one.[7] Anyway, if you took three groups and exposed them to a song

[7] Although in most cases the mere exposure effect seems to have the reverse effect on me: the more I hear the manufactured rubbish that gets into the charts, the more I want to rid my brain of the mental anguish it creates by making myself deaf by ramming hot irons into my ears.

10 times, 20 times and 30 times respectively and then measured how much people liked the song, you'd expect the medians to increase. Those who heard it 10 times would like it a bit, but those who heard it 20 times would like it more, and those who heard it 30 times would like it the most.

The Jonckheere–Terpstra test (actually referred to more often just as the Jonckheere test) was designed for these situations. In SPSS, it works on the principle that your coding variable (the one that defines the groups) specifies the order in which you expect the medians to change (it doesn't matter whether you expect them to increase or decrease). For our soya example, we coded our groups as 1 = no soya, 2 = one soya meal per week, 3 = four soya meals per week and 4 = seven soya meals per week, so it would test whether the median sperm count increases or decreases across the groups when they're ordered in that way. Obviously we could change the coding scheme and test whether the medians were ordered in a different way. The important thing is that the test determines whether the medians of the groups ascend or descend in the *order specified by the coding variable*.

SPSS OUTPUT 15.11

Jonckheere-Terpstra Test[b]

			Sperm Count (Millions)
Number of Levels in Number of Soya Meals Per Week			4
N			80
Observed J-T Statistic			912.000
Mean J-T Statistic			1200.000
Std. Deviation of J-T Statistic			116.333
Std. J-T Statistic			-2.476
Asymp. Sig. (2-tailed)			.013
Monte Carlo Sig. (2-tailed)	Sig.		.012[a]
	99% Confidence Interval	Lower Bound	.009
		Upper Bound	.015
Monte Carlo Sig. (1-tailed)	Sig.		.006[a]
	99% Confidence Interval	Lower Bound	.004
		Upper Bound	.008

a. Based on 10000 sampled tables with starting seed 2000000.

b. Grouping Variable: Number of Soya Meals Per Week

We saw how to specify the test in section 15.5.3, and SPSS Output 15.11 shows the output from the test for the soya data. This table tells you the number of groups being compared, 4 (just in case you hadn't noticed). It also tells you the value of test-statistic, J, which is 912. In large samples (more than about eight per group) this test statistic has a sampling distribution that is normal, and a mean and standard deviation that are easily defined and calculated (the output tells us that the mean is 1200 and the standard deviation is 116.33). Knowing these things, we can convert to a z-score by taking the test statistic, subtracting the mean of the sampling distribution from it and then dividing the result by the standard deviation ($z = (912–1200)/116.33 = –2.476$). This is much the same as what we did for the Mann–Whitney and Wilcoxon tests. This z-score can then be compared against the values of a normal distribution, and because Jonckheere's test should normally be one-tailed (we specify before the experiment the order of the medians) we're looking for a value greater than 1.65 (when we ignore the sign). A value of 2.47 is, therefore, significant. The sign of the z-value does tell us something useful, though: if it is positive then it indicates a trend of ascending medians (i.e. the medians get bigger as

OLIVER TWISTED

Please, Sir, can I have some more … Jonck?

'I want to know how the Jonckheere–Terpstra test actually works,' complains Oliver. Of course you do, Oliver, sleep is hard to come by these days. I am only too happy to oblige my little syphilitic friend. The additional material for this chapter on the companion website has a complete explanation of the test and how it works. I bet you're glad you asked.

the values of the coding variable get bigger), but if it is negative (as it is here) it indicates a trend of descending medians (the medians get smaller as the value of the coding variable gets bigger). In this example, because we coded the variables as 1 = no soya, 2 = one soya meal per week, 3 = four soya meals per week and 4 = seven soya meals per week, it means that the medians get smaller as we go from no soya, to one soya meal, to four soya meals and on to seven soya meals.

You'll also notice that there are one- and two-tailed significance values estimated using Monte Carlo methods. These have appeared because we chose that option for the Kruskal–Wallis test. They confirm what we have already found.

15.5.7. Calculating an effect size ②

Unfortunately there isn't an easy way to convert a chi-square statistic that has more than 1 degree of freedom to an effect size *r*. You could use the significance value of the Kruskal–Wallis test statistic to find an associated value of *z* from a table of probability values for the normal distribution (like that in the Appendix). From this you could use the conversion to *r* that we used in section 15.3.5. However, this kind of effect size is rarely that useful (because it's summarizing a general effect). In most cases it's more interesting to know the effect size for a focused comparison (such as when comparing two things). For this reason, I'd suggest just calculating effect sizes for the Mann–Whitney tests that we used to follow up the main analysis.

For the first comparison (no soya vs. 1 meal) SPSS Output 15.10 shows us that *z* is −0.243, and because this is based on comparing two groups each containing 20 observations, we had 40 observations in total. The effect size is therefore:

$$\begin{aligned} r_{\text{NoSoya}-1\text{Meal}} &= \frac{-0.243}{\sqrt{40}} \\ &= -.04 \end{aligned}$$

This represents a very small effect because it is close to zero, which tells us that the effect on sperm counts of having one soya meal per week compared to no soya meals was negligible.

For the second comparison (no soya vs. 4 meals) SPSS Output 15.10 shows us that *z* is −0.325, and this was again based on 40 observations. The effect size is therefore:

$$\begin{aligned} r_{\text{NoSoya}-1\text{Meal}} &= \frac{-0.325}{\sqrt{40}} \\ &= -.05 \end{aligned}$$

This again represents a negligible effect because it is close to zero, which tells us that the effect on sperm counts of having four soya meals per week compared to no soya meals was negligible.

For the final comparison (no soya vs. 7 meals) SPSS Output 15.10 shows us that z is −2.597, and this was again based on 40 observations. The effect size is therefore:

$$r_{\text{NoSoya}-1\text{Meal}} = \frac{-2.597}{\sqrt{40}} = -.41$$

This represents a medium effect, which tells us that the effect of seven soya meals a week lowering sperm counts (compared to having none) was a fairly substantive finding.

We can also calculate an effect size for Jonckheere's test if we want to by using the same equation. This test involves all data, so we have to use 80 as our value of N:

$$r_{\text{Jonckheere}} = \frac{-2.476}{\sqrt{80}} = -.28$$

15.5.8. Writing and interpreting the results ①

For the Kruskal–Wallis test, we need only report the test statistic (which we saw earlier is denoted by H), its degrees of freedom and its significance. So, we could report something like:

✓ Sperm counts were significantly affected by eating soya meals, $H(3) = 8.66$, $p < .05$.

However, we need to report the follow-up tests as well (including their effect sizes):

✓ Sperm counts were significantly affected by eating soya meals, $H(3) = 8.66$, $p < .05$. Mann–Whitney tests were used to follow up this finding. A Bonferroni correction was applied and so all effects are reported at a .0167 level of significance. It appeared that sperm counts were no different when one soya meal ($U = 191$, $r = -.04$) or four soya meals ($U = 188$, $r = -.05$) were eaten per week compared to none. However, when seven soya meals were eaten per week sperm counts were significantly lower than when no soya was eaten ($U = 104$, $r = -.41$). We can conclude that if soya is eaten every day it significantly reduces sperm counts compared to eating none; however, eating soya less than every day has no significant effect on sperm counts ('phew!' says the vegetarian man!).

Or, we might want to report our trend:

✓ All effects are reported at $p < .05$. Sperm counts were significantly affected by eating soya meals ($H(3) = 8.66$). Jonckheere's test revealed a significant trend in the data: as more soya was eaten, the median sperm count decreased, $J = 912$, $z = -2.48$, $r = -.28$.

CRAMMING SAM'S TIPS

- The Kruskal–Wallis test compares several conditions when different participants take part in each condition and the resulting data violate an assumption of one-way independent ANOVA.
- Look at the row labelled *Asymp. Sig.* If the value is less than .05 then the groups are significantly different.
- You can follow up the main analysis with Mann–Whitney tests between pairs of conditions, but only accept them as significant if they're significant below .05/number of tests.
- If you predict that the means will increase or decrease across your groups in a certain order then do Jonckheere's trend test.
- Report the *H* statistic, the degrees of freedom and the significance value for the main analysis. For any *post hoc* tests, report the *U* statistic and an effect size if possible (you can also report the corresponding *z* and the significance value). Also report the medians and their corresponding ranges (or draw a boxplot).

LABCOAT LENI'S REAL RESEARCH 15.2

Eggs-traordinary! ①

There seems to be a lot of sperm in this book (not literally I hope) – it's possible that I have a mild obsession. We saw in Labcoat Leni's Real Research 15.1 that male quail fertilized more eggs if they had been trained to be able to predict when a mating opportunity would arise. However, some quail develop fetishes. Really. In the previous example the type of compartment acted as a predictor of an opportunity to mate, but in studies where a terrycloth object acts as a sign that a mate will shortly become available, some quail start to direct their sexual behaviour towards the terrycloth object. (I may regret this anology but in human terms if you imagine that every time you were going to have sex with your boyfriend you gave him a green towel a few moments before seducing him, then after enough seductions he would start rubbing his crotch against any green towel he saw. If you've ever wondered why you boyfriend rubs his crotch on green towels, then I hope this explanation has been enlightening.) In evolutionary terms, this fetishistic behaviour seems counterproductive because sexual behaviour becomes directed towards something that cannot provide reproductive success. However, perhaps this behaviour serves to prepare the organism for the 'real' mating behaviour.

Hakan Çetinkaya and Mike Domjan conducted a brilliant study in which they sexually conditioned male quail (Çetinkaya & Domjan, 2006). All quail experienced the terrycloth stimulus and an opportunity to mate, but for some the terrycloth stimulus immediately preceded the mating opportunity (paired group) whereas for others they experienced it 2 hours after the mating opportunity (this was the control group because the terrycloth stimulus did not predict a mating opportuinity). In the paired group, quail were classified as fetishistic or not depending on whether they engaged in sexual behaviour with the terrycloth object.

During a test trial the quail mated with a female and the researchers measured the percentage of eggs fertilized, the time spent near the terrycloth object, the latency to initiate copulation, and copulatory efficiency. If this fetishistic behaviour provides an evolutionary advantage then we would expect the fetishistic quail to fertilize more eggs, initiate copulation faster and be more efficient in their copulations.

The data from this study are in the file **Çetinkaya & Domjan (2006).sav**. Labcoat Leni wants you to carry out a Kruskal–Wallis test to see whether fetishist quail produced a higher percentage of fertilized eggs and initiated sex more quickly.

Answers are in the additional material on the companion website (or look at pages 429–430 in the original article).

ÇETINKAYA, H., & DOMJAN, M. (2006). *JOURNAL OF COMPARATIVE PSYCHOLOGY, 120*(4), 427–432.

15.6. Differences between several related groups: Friedman's ANOVA ①

In Chapter 13 we discovered a technique called one-way related ANOVA that could be used to test for differences between several related groups. Although, as we've seen, ANOVA can be robust to violations of its assumptions, there is another alternative to the repeated-measures case: **Friedman's ANOVA** (Friedman, 1937). As such, it is used for testing differences between conditions when there are more than two conditions and the same participants have been used in all conditions (each case contributes several scores to the data). If you have violated some assumption of parametric tests then this test can be a useful way around the problem.

Young people (women especially) can become obsessed with body weight and diets, and because the media are insistent on ramming ridiculous images of stick-thin celebrities down our throats (should that be 'into our eyes'?) and brainwashing us into believing that these emaciated corpses are actually attractive, we all end up terribly depressed that we're not perfect (because we don't have a couple of slugs stuck to our faces instead of lips). Then corporate parasites jump on our vulnerability by making loads of money on diets that will help us attain the body beautiful! Well, not wishing to miss out on this great opportunity to exploit people's insecurities I came up with my own diet called the Andikins diet.[8] The principle is that you follow my lifestyle: you eat no meat, drink lots of Darjeeling tea, eat shed-loads of lovely European cheese, lots of fresh crusty bread, pasta, chocolate at every available opportunity (especially when writing books), then enjoy a few beers at the weekend, play football and rugby twice a week and play your drum kit for an hour a day or until your neighbour threatens to saw your arms off and beat you around the head with them for making so much noise. To test the efficacy of my wonderful new diet, I took 10 women who considered themselves to be in need of losing weight and put them on this diet for two months. Their weight was measured in kilograms at the start of the diet and then after one month and two months.

15.6.1. Theory of Friedman's ANOVA ②

SMART ALEX ONLY

The theory for Friedman's ANOVA is much the same as the other tests we've seen in this chapter: it is based on ranked data. To begin with, you simply place your data for different conditions into different columns (in this case there were three conditions so we have three columns). The data for the diet example are in Table 15.5; note that the data are in different columns and so each row represents the weight of a different person. The next thing we have to do is rank the data *for each person*. So, we start with person 1, we look at their scores (in this case person 1 weighed 63.75 kg at the start, 65.38 kg after one month on the diet, and 81.34 kg after two months on the diet), and then we give the lowest one a rank of 1, the next highest a rank of 2 and so on (see section 15.3.1 for more detail). When you've ranked the data for the first person, you move on to the next person and, starting at 1 again, rank their lowest score, then rank the next highest as 2 and so on. You do this for all people from whom you've collected data. You then simply add up the ranks for each condition (R_i, where i is used to denote the particular group).

SELF-TEST Have a go at ranking the data and see if you get the same results as in Table 15.5.

[8] Not to be confused with the Atkins diet obviously.☺

TABLE 15.5 Data for the diet example with ranks

	Weight				Weight		
	Start	Month 1	Month 2		Start (Ranks)	Month 1 (Ranks)	Month 2 (Ranks)
Person 1	63.75	65.38	81.34		1	2	3
Person 2	62.98	66.24	69.31		1	2	3
Person 3	65.98	67.70	77.89		1	2	3
Person 4	107.27	102.72	91.33		3	2	1
Person 5	66.58	69.45	72.87		1	2	3
Person 6	120.46	119.96	114.26		3	2	1
Person 7	62.01	66.09	68.01		1	2	3
Person 8	71.87	73.62	55.43		2	3	1
Person 9	83.01	75.81	71.63		3	2	1
Person 10	76.62	67.66	68.60		3	1	2
				R_i	19	20	21

Once the sum of ranks has been calculated for each group, the test statistic, F_r, is calculated as:

$$F_r = \left[\frac{12}{Nk(k+1)} \sum_{i=1}^{k} R_i^2 \right] - 3N(k+1) \tag{15.3}$$

In this equation, R_i is the sum of ranks for each group, N is the total sample size (in this case 10) and k is the number of conditions (in this case 3). This equation is very similar to that of the Kruskal–Wallis test (compare equations (15.1) and (15.3)). All we need to do for each condition is square the sum of ranks and then add up these values. That deals with the middle part of the equation; the rest of it involves calculating various values based on the total sample size and the number of conditions. For these data we get:

$$\begin{aligned} F_r &= \left[\frac{12}{(10 \times 3)(3+1)} (19^2 + 20^2 + 21^2) \right] - (3 \times 10)(3+1) \\ &= \frac{12}{120}(361 + 400 + 441) - 120 \\ &= 0.1(1202) - 120 \\ &= 120.2 - 120 \\ &= 0.2 \end{aligned}$$

EVERYBODY

When the number of people tested is large (bigger than about 10) this test statistic, like the Kruskal–Wallis test in the previous section, has a chi-square distribution (see Chapter 18) and for this distribution there is one value for the degrees of freedom, which is one less than the number of groups ($k - 1$): in this case 2.

15.6.2. Inputting data and provisional analysis ①

SELF-TEST Using what you know about inputting data, try to enter these data into SPSS and run some exploratory analyses (see Chapter 5)

When the data are collected using the same participants in each condition, the data are entered using different columns. So, the data editor will have three columns of data. The first column is for the data from the start of the diet (called something like **Start**), the second column will have values for the weights after one month (called **Month1**) and the final column will have the weights at the end of the diet (called **Month2**). The data can be found in the file **Diet.sav**.

First, we run some exploratory analyses on the data. With a bit of luck you'll get the same table shown in SPSS Output 15.12, which shows that the Kolmogorov–Smirnov test (see section 5.5) was not significant for the initial weights at the start of the diet ($D(10) = .23$, $p > .05$), but the Shapiro–Wilk test is significant and this test is actually more accurate than the Kolmogorov–Smirnov test. The data one month into the diet were significantly different from normal ($D(10) = .34$, $p < .01$). The data at the end of the diet do appear to be normal, though ($D(10) = .20$, $p > .05$). Some of these data are not normally distributed.

SPSS OUTPUT 15.12

Tests of Normality

	Kolmogorov-Smirnov[a]			Shapiro-Wilk		
	Statistic	df	Sig.	Statistic	df	Sig.
Weight at Start (kg)	.228	10	.149	.784	10	.009
Weight after 1 month (kg)	.335	10	.002	.685	10	.001
Weight after 2 months (kg)	.203	10	.200*	.877	10	.121

a. Lilliefors Significance Correction

*. This is a lower bound of the true significance.

15.6.3. Doing Friedman's ANOVA on SPSS ①

First, access the main dialog box (see Figure 15.9) by selecting Analyze Nonparametric Tests ▸ K Related Samples... . Once the dialog box is activated, select the three variables that represent the dependent variable at the different levels of the independent variable from the list (click on **Start** and then, holding down the *Ctrl* key, click on **Month1** and **Month2**). Drag them to the box labelled *Test Variables* (or click on ➔). If you click on Exact... then you get a dialog box for selecting exact significance values for Friedman's ANOVA (see section 15.3.3). Remember that I've said that when samples are large you should probably opt for the Monte Carlo method but with small samples it's worth opting for the exact test. Well, we've got a relatively small sample here so we can select the exact tests. Finally, click on Statistics... to select some descriptive statistics. To run the analyses return to the main dialog box and click on OK . You can also conduct some related tests using the same dialog box as for Friedman's ANOVA (see SPSS Tip 15.4).

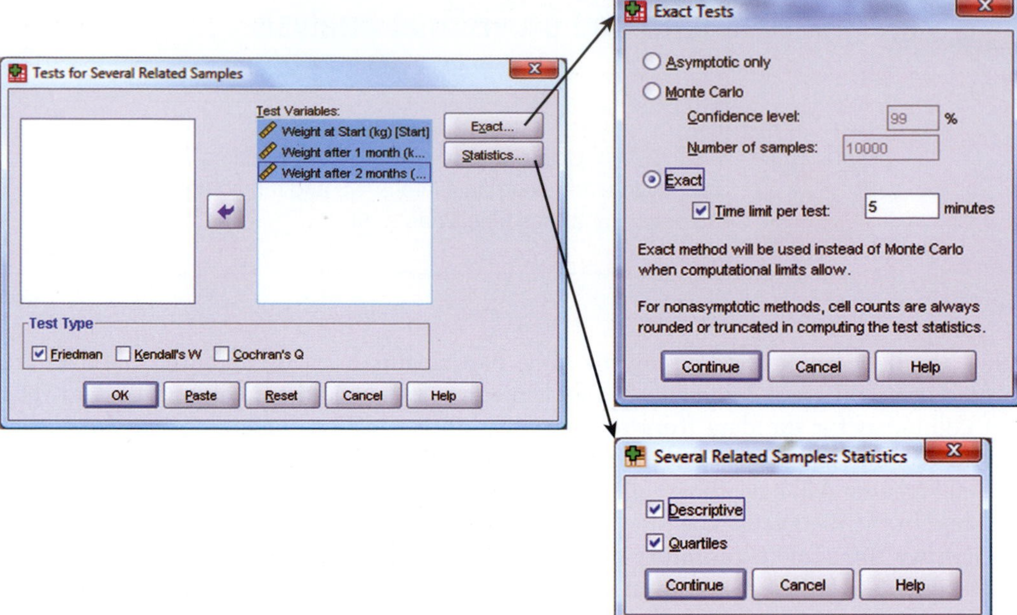

FIGURE 15.9 Dialog boxes for Friedman's ANOVA

SPSS TIP 15.4 Other options for Friedman's ANOVA ②

In the main dialog box there are some other tests that can be selected:

- **Kendall's W**: This is similar to Friedman's ANOVA but is used specifically for looking at the agreement between raters. If, for example, we asked 10 different women to rate the attractiveness of Justin Timberlake, David Beckham and Barack Obama we could use this test to look at the extent to which they agree. This test is particularly useful because, like the correlation coefficient, Kendall's *W* has a limited range: it ranges from 0 (no agreement between judges) to 1 (complete agreement between judges).
- **Cochran's Q**: This test is an extension of McNemar's test (see SPSS Tip 15.3) and is basically a Friedman test for when you have dichotomous data. So imagine you asked 10 people whether they'd like to snog Justin Timberlake, David Beckham and Barack Obama and they could answer only yes or no. If we coded responses as 0 (no) and 1 (yes) we could do the Cochran test on these data).

15.6.4. Output from Friedman's ANOVA ①

As we've seen, Friedman's ANOVA, like all the non-parametric tests in this chapter, is based on the ranks, not the actual scores. SPSS Output 15.13 shows the mean rank in each condition. These mean ranks are important later for interpreting any effects; they show that the ranks were fairly similar across the conditions.

SPSS Output 15.14 shows the test statistic (SPSS calls this *Chi-Square* rather that F_r because F_r has a chi-square distribution). The value of this statistic is 0.2, the same value that we calculated earlier. We're also told the test statistic's degrees of freedom (in this case we had three groups so the degrees of freedom are 3 – 1, or 2), and the significance. If you ask for exact significance then this value is given too. The significance value is .905 (or .974 if you read the exact significance), which is well above .05, therefore we could conclude that the Andikins diet does not have any effect: the weights didn't significantly change over the course of the diet.

SPSS OUTPUT 15.13

Descriptive Statistics

	N	Mean	Std. Deviation	Minimum	Maximum	Percentiles		
						25th	50th (Median)	75th
Weight at Start (kg)	10	78.0543	20.23008	62.01	120.46	63.5549	69.2288	89.0709
Weight after 1 month (kg)	10	77.4635	18.61502	65.38	119.96	66.2065	68.5728	82.5385
Weight after 2 months (kg)	10	77.0668	16.10612	55.43	114.26	68.4525	72.2493	83.8365

Ranks

	Mean Rank
Weight at Start (kg)	1.90
Weight after 1 month (kg)	2.00
Weight after 2 months (kg)	2.10

SPSS OUTPUT 15.14

Test Statistics[a]

N	10.000
Chi-Square	.200
df	2.000
Asymp. Sig.	.905
Exact Sig.	.974
Point Probability	.143

a. Friedman Test

15.6.5. *Post hoc* tests for Friedman's ANOVA ②

In normal circumstances we wouldn't do any follow-up tests because the overall effect from Friedman's ANOVA was not significant. However, in case you get a result that is significant we will have a look at what options you have. As with the Kruskal–Wallis test, there are two ways to do non-parametric *post hoc* procedures, which are in essence the same. The first is to use Wilcoxon signed-rank tests (section 15.4) but correcting for the number of tests we do (see sections 10.2.1 and 15.5.5 for the reasons why). The way we correct for the number of tests is to accept something as significant only if its significance is less than α/number of comparisons (the Bonferroni correction). In the social sciences this usually means .05/number of comparisons. In this example, we have only three groups, so if we compare all of the groups we simply get three comparisons:

- Test 1: Weight at the start of the diet compared to at one month.
- Test 2: Weight at the start of the diet compared to at two months.
- Test 3: Weight at one month compared to at two months.

Therefore, rather than use .05 as our critical level of significance, we'd use .05/3 = .0167. In fact we wouldn't bother with *post hoc* tests at all for this example because the main ANOVA was non-significant, but I'll go through the motions to illustrate what to do.

SELF-TEST Carry out the three Wilcoxon tests suggested above (see Figure 15.5).

SPSS Output 15.15 shows the Wilcoxon signed-rank test statistics from doing the three comparisons. Remember that we are now using a critical value of .0167, and in fact none of the comparisons are significant because they have one-tailed significance values of .500, .423 and .461 (this isn't surprising because the main analysis was non-significant).

SPSS OUTPUT 15.15

Ranks

		N	Mean Rank	Sum of Ranks
Weight after 1 month (kg) - Weight at Start (kg)	Negative Ranks	4[a]	7.00	28.00
	Positive Ranks	6[b]	4.50	27.00
	Ties	0[c]		
	Total	10		
Weight after 2 months (kg) - Weight at Start (kg)	Negative Ranks	5[d]	6.00	30.00
	Positive Ranks	5[e]	5.00	25.00
	Ties	0[f]		
	Total	10		
Weight after 2 months (kg) - Weight after 1 month (kg)	Negative Ranks	4[g]	7.25	29.00
	Positive Ranks	6[h]	4.33	26.00
	Ties	0[i]		
	Total	10		

a. Weight after 1 month (kg) < Weight at Start (kg)

b. Weight after 1 month (kg) > Weight at Start (kg)

c. Weight after 1 month (kg) = Weight at Start (kg)

d. Weight after 2 months (kg) < Weight at Start (kg)

e. Weight after 2 months (kg) > Weight at Start (kg)

f. Weight after 2 months (kg) = Weight at Start (kg)

g. Weight after 2 months (kg) < Weight after 1 month (kg)

h. Weight after 2 months (kg) > Weight after 1 month (kg)

i. Weight after 2 months (kg) = Weight after 1 month (kg)

Test Statistics[b]

	Weight after 1 month (kg) - Weight at Start (kg)	Weight after 2 months (kg) - Weight at Start (kg)	Weight after 2 months (kg) - Weight after 1 month (kg)
Z	-.051[a]	-.255[a]	-.153[a]
Asymp. Sig. (2-tailed)	.959	.799	.878
Exact Sig. (2-tailed)	1.000	.846	.922
Exact Sig. (1-tailed)	.500	.423	.461
Point Probability	.039	.038	.038

a. Based on positive ranks.

b. Wilcoxon Signed Ranks Test

SMART ALEX ONLY

The second way to do *post hoc* tests is very similar to what we did for the Kruskal–Wallis test in section 15.5.5 and is, likewise, described by Siegel and Castellan (1988). Again, we take the difference between the mean ranks of the different groups and compare these differences to a value based on the value of z (corrected for the number of comparisons being done) and a constant based on the total sample size, N (10 in this example) and the number of conditions, k (3 in this case). The inequality is:

$$\left|\bar{R}_u \bar{R}_v\right| \geq z_{\alpha/k(k-1)}\sqrt{\frac{k(k+1)}{6N}} \qquad (15.4)$$

The left-hand side of this inequality is just the difference between the mean rank of the two groups being compared, but ignoring the sign of the difference. As with Kruskal–Wallis, we need to know $z_{\alpha/k(k-1)}$, and if we stick to tradition and use an α level of .05, knowing that k is 3, we get $\alpha/k(k-1) = .05/3(3-1) = .05/6 = .00833$. So, $z_{\alpha/k(k-1)}$ just means 'the value of z for which only $\alpha/k(k-1)$ other values of z are bigger' (or in this case 'the value of z for which only .00833 other values of z are bigger'). Therefore, we go to table A.1 in the Appendix,

TABLE 15.6 Differences between mean ranks for the diet data

Comparison	$\bar{R}_u$	$\bar{R}_v$	$\bar{R}_u - \bar{R}_v$	$\lvert\bar{R}_u - \bar{R}_v\rvert$
Start–1 Month	1.90	2.00	−0.10	0.10
Start–2 Months	1.90	2.10	−0.20	0.20
1 Month–2 Months	2.00	2.10	−0.10	0.10

look at the column labelled *Smaller Portion* and find the number .00833 and then find the value in the same row in the column labelled z. In this case there are values of .00842 and .00820, which give z-values of 2.39 and 2.40 respectively; because .00833 lies about midway between the values we found, we could just take the midpoint of the two z-values, 2.395, or we could err on the side of caution and use 2.40. I'll err on the cautious side and use 2.40. We can now calculate the right-hand side of inequality (15.4):

$$\begin{aligned}\text{critical difference} &= z_{\alpha/k(k-1)}\sqrt{\frac{k(k+1)}{6N}}\\ &= 2.40\sqrt{\frac{3(3+1)}{6(10)}}\\ &= 2.40\sqrt{\frac{12}{60}}\\ &= 2.40\sqrt{0.2}\\ &= 1.07\end{aligned}$$

When the same people have been used, the same critical difference can be used for all comparisons. The next step is simply to calculate all of the differences between the mean ranks of all of the groups (the mean ranks can be found in SPSS Output 15.13), as in Table 15.6.

Inequality (15.4) means that if the differences between mean ranks is bigger than or equal to the critical difference, then that difference is significant. In this case, it means that if any difference is bigger than 1.07, then it is significant. All differences are below this value so we could conclude that none of the groups were significantly different and this is consistent with the non-significance of the initial ANOVA.

EVERYBODY

15.6.6. Calculating an effect size ②

As I mentioned before, there isn't an easy way to convert a chi-square statistic that has more than 1 degree of freedom to an effect size r and, in any case, it's not always that helpful to have an effect size for a general effect like that tested by Friedman's ANOVA.[9] Therefore, it's more sensible (in my opinion at least) to calculate effect sizes for any comparisons you've done after the ANOVA. As we saw in section 15.4.5, it's straightforward to get an effect size r from the Wilcoxon signed-rank test. Alternatively, we could just calculate effect sizes for the Wilcoxon tests that we used to follow up the main analysis. These effect sizes will be very informative in their own right.

[9] If you really want to, though, you can (as with the Kruskal–Wallis test) use the significance value of the chi-square test statistic to find an associated value of z from a table of probability values for the normal distribution (see Appendix) and then use the conversion to r that we've seen throughout this chapter.

For the first comparison (start weight vs. 1 month) SPSS Output 15.15 shows us that z is −0.051, and because this is based on comparing two conditions each containing 10 observations, we had 20 observations in total (remember it isn't important that the observations come from the same people). The effect size is therefore:

$$r_{\text{Start}-1\text{Month}} = \frac{-0.051}{\sqrt{20}} = -0.01$$

For the second comparison (start weight vs. 2 months) SPSS Output 15.15 shows us that z is −0.255, and this was again based on 20 observations. The effect size is therefore:

$$r_{\text{Start}-2\text{Months}} = \frac{-0.255}{\sqrt{20}} = -0.06$$

For the final comparison (1 month vs. 2 months) SPSS Output 15.15 shows us that z is −0.153, and this was again based on 20 observations. The effect size is therefore:

$$r_{\text{Start}-2\text{Months}} = \frac{-0.153}{\sqrt{20}} = -0.03$$

Unsurprisingly, given the lack of significance of the Wilcoxon tests, these all represent virtually non-existent effects: they are all very close to zero.

15.6.7. Writing and interpreting the results ①

For Friedman's ANOVA we need only report the test statistic (which we saw earlier is denoted by χ^2),[10] its degrees of freedom and its significance. So, we could report something like:

✓ The weight of participants did not significantly change over the two months of the diet, $\chi^2(2) = 0.20$, $p > .05$.

Although with no significant initial analysis we wouldn't report *post hoc* tests for these data, in case you need to, you should say something like this (remember that the test statistic T is the smaller of the two sums of ranks for each test and these values are in SPSS Output 15.15):

✓ The weight of participants did not significantly change over the two months of the diet, $\chi^2(2) = 0.20$, $p > .05$. Wilcoxon tests were used to follow up this finding. A Bonferroni correction was applied and so all effects are reported at a .0167 level of significance. It appeared that weight didn't significantly change from the start of the diet to one month, $T = 27$, $r = -0.01$, from the start of the diet to two months, $T = 25$, $r = -0.06$, or from one month to two months, $T = 26$, $r = -0.03$. We can conclude that the Andikins diet, like its creator, is a complete failure.

[10] The test statistic is often denoted as χ^2_F but the official APA style guide doesn't recognize this term.

CRAMMING SAM'S TIPS

- Friedman's ANOVA compares several conditions when the same participants take part in each condition and the resulting data violate an assumption of one-way repeated-measures ANOVA.
- Look at the row labelled *Asymp. Sig*. If the value is less than .05 then the conditions are significantly different.
- You can follow up the main analysis with Wilcoxon signed-rank tests between pairs of conditions, but only accept them as significant if they're significant below .05/number of tests.
- Report the χ^2 statistic, its degrees of freedom and significance. For any *post hoc* tests report the *T* statistic, and an effect size if possible. You can also report the value of *z* and its significance value. Report the medians and their ranges (or draw a boxplot).

What have I discovered about statistics? ①

This chapter has dealt with an alternative approach to violations of parametric assumptions, which is to use tests based on ranking the data. We started with the Wilcoxon rank-sum test and the Mann–Whitney test, which is used for comparing two independent groups. This test allowed us to look in some detail at the process of ranking data. We then moved on to look at the Wilcoxon signed-rank test, which is used to compare two related conditions. We moved on to more complex situations in which there are several conditions (the Kruskal–Wallis test for independent conditions and Friedman's ANOVA for related conditions). For each of these tests we looked at the theory of the test (although these sections could be ignored) and then focused on how to conduct them on SPSS, how to interpret the results and how to report the results of the test. In the process we discovered that drugs make you depressed, soya reduces your sperm count, and my lifestyle is not conducive to losing weight!

We also discovered that my teaching career got off to an inauspicious start. As it turned out, one of the reasons that the class did not have a clue what I was talking about was because I hadn't been shown their course handouts and I was trying to teach them ANOVA using completely different equations to their lecturer (there are many ways to compute an ANOVA). The other reason was that I was a rubbish teacher. This event did change my life, though, because the experience was so awful that I did everything in my power to make sure that it didn't happen again. After years of experimentation I can now pass on the secret of avoiding students telling you how awful your ANOVA classes are: the more penis jokes you tell, the less likely you are to be emotionally crushed by dissatisfied students.

Key terms that I've discovered

Cochran's Q
Friedman's ANOVA
Jonckheere–Terpstra test
Kendall's *W*
Kolmogorov–Smirnov *Z*
Kruskal–Wallis test
Mann–Whitney test
McNemar's test
Median test
Monte Carlo method
Moses extreme reactions
Non-parametric tests
Ranking
Sign test
Wald–Wolfowitz runs
Wilcoxon rank-sum test
Wilcoxon signed-rank test

Smart Alex's tasks

- **Task 1**: A psychologist was interested in the cross-species differences between men and dogs. She observed a group of dogs and a group of men in a naturalistic setting (20 of each). She classified several behaviours as being dog-like (urinating against trees and lamp posts, attempts to copulate with anything that moved, and attempts to lick their own genitals). For each man and dog she counted the number of dog-like behaviours displayed in a 24 hour period. It was hypothesized that dogs would display more dog-like behaviours than men. The data are in the file **MenLikeDogs.sav.** Analyse them with a Mann–Whitney test. ①
- **Task 2**: There's been much speculation over the years about the influence of subliminal messages on records. To name a few cases, both Ozzy Osbourne and Judas Priest have been accused of putting backward masked messages on their albums that subliminally influence poor unsuspecting teenagers into doing things like blowing their heads off with shotguns. A psychologist was interested in whether backward masked messages really did have an effect. He took the master tapes of Britney Spears' 'Baby one more time' and created a second version that had the masked message 'deliver your soul to the dark lord' repeated in the chorus. He took this version, and the original, and played one version (randomly) to a group of 32 people. He took the same group six months later and played them whatever version they hadn't heard the time before. So each person heard both the original and the version with the masked message, but at different points in time. The psychologist measured the number of goats that were sacrificed in the week after listening to each version. It was hypothesized that the backward message would lead to more goats being sacrificed. The data are in the file **DarkLord.sav**. Analyse them with a Wilcoxon signed-rank test. ①
- **Task 3**: A psychologist was interested in the effects of television programmes on domestic life. She hypothesized that through 'learning by watching', certain programmes might actually encourage people to behave like the characters within them. This in turn could affect the viewer's own relationships (depending on whether the programme depicted harmonious or dysfunctional relationships). She took episodes of three popular TV shows and showed them to 54 couples after which the couple were left alone in the room for an hour. The experimenter measured the number of times the couple argued. Each couple viewed all three of the TV programmes at different points in time (a week apart) and the order in which the programmes were viewed was counterbalanced over couples. The TV programmes selected were *Eastenders* (which typically portrays the lives of extremely miserable, argumentative,

London folk who like nothing more than to beat each other up, lie to each other, sleep with each other's wives and generally show no evidence of any consideration to their fellow humans!), *Friends* (which portrays a group of unrealistically considerate and nice people who love each other oh so very much – but for some reason I love it anyway!), and a National Geographic programme about whales (this was supposed to act as a control). The data are in the file **Eastenders.sav**. Access them and conduct Friedman's ANOVA on the data. ①

- **Task 4**: A researcher was interested in trying to prevent coulrophobia (fear of clowns) in children. She decided to do an experiment in which different groups of children (15 in each) were exposed to different forms of positive information about clowns. The first group watched some adverts for McDonald's in which their mascot Ronald McDonald is seen cavorting about with children going on about how they should love their mums. A second group was told a story about a clown who helped some children when they got lost in a forest (although what on earth a clown was doing in a forest remains a mystery). A third group was entertained by a real clown, who came into the classroom and made balloon animals for the children.[11] A final group acted as a control condition and they had nothing done to them at all. The researcher took self-report ratings of how much the children liked clowns resulting in a score for each child that could range from 0 (not scared of clowns at all) to 5 (very scared of clowns). The data are in the file **coulrophobia.sav**. Access them and conduct a Kruskal–Wallis test. ①

Answers can be found on the companion website and because these examples are used in Field and Hole (2003), you could steal this book or photocopy Chapter 7 to get some very detailed answers.

Further reading

Siegel, S., & Castellan, N. J. (1988). *Nonparametric statistics for the behavioral sciences* (2nd ed.). New York: McGraw-Hill. (This has become the definitive text on non-parametric statistics, and is the only book seriously worth recommending as 'further' reading. It is probably not a good book for stats-phobes, but if you've coped with my chapter then this book will be an excellent next step.)

Wilcox, R. R. (2005). *Introduction to robust estimation and hypothesis testing* (2nd ed.). Burlington, MA: Elsevier. (This book is quite technical, compared to this one, but really is a wonderful resource. Wilcox describes how to use an astonishing range of robust tests that can't be done directly in SPSS!)

Online tutorial

The companion website contains the following Flash movie tutorial to accompany this chapter:

- Nonparametric tests using SPSS

Interesting real research

Çetinkaya, H., & Domjan, M. (2006). Sexual fetishism in a quail (*Coturnix japonica*) model system: Test of reproductive success. *Journal of Comparative Psychology*, *120*(4), 427–432.

Matthews, R. C., Domjan, M., Ramsey, M., & Crews, D. (2007). Learning effects on sperm competition and reproductive fitness. *Psychological Science*, *18*(9), 758–762.

[11] Unfortunately, the first time they attempted the study the clown accidentally burst one of the balloons. The noise frightened the children and they associated that fear response with the clown. All 15 children are currently in therapy for coulrophobia!

16 Multivariate analysis of variance (MANOVA)

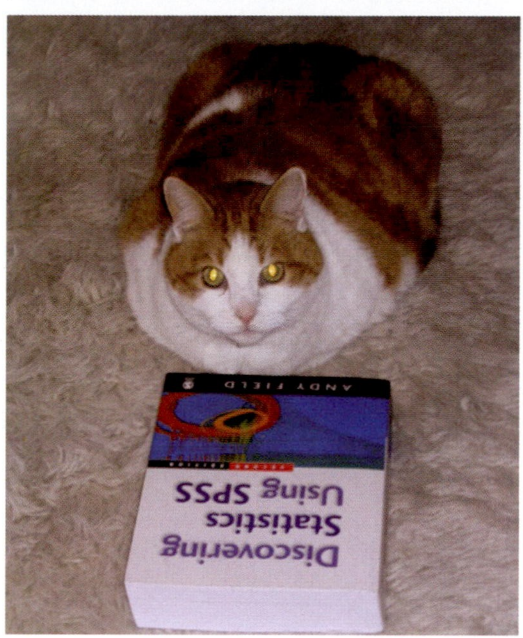

FIGURE 16.1 Fuzzy doing some light reading

16.1. What will this chapter tell me? ②

Having had what little confidence I had squeezed out of me by my formative teaching experiences, I decided that I could either kill myself, or get a cat. I'd wanted to do both for years but when I was introduced to a little four-week old bundle of gingerness the choice was made. Fuzzy (as I named him) was born on 8 April 1996 and has been my right-hand feline ever since. He is like the Cheshire cat in Lewis Carroll's *Alice's adventures in Wonderland*[1] in that he seemingly vanishes and reappears at will: I go to find clothes in my wardrobe and notice

[1] This is one of my favourite books from my childhood. For those that haven't read it, the Cheshire cat is a big fat cat mainly remembered for vanishing and reappearing out of nowhere; on one occasion it vanished leaving only its smile behind.

a ginger face peering out at me, I put my pants in the laundry basket and he looks up at me from a pile of smelly socks, I go to have a bath and he's sitting in it, and I shut the bedroom door yet wake up to find him asleep next to me. His best vanishing act was a few years ago when I moved house. He'd been locked up in his travel basket (which he hates) during the move, so once we were in our new house I thought I'd let him out as soon as possible. I found a quiet room, checked the doors and windows to make sure he couldn't escape, opened the basket, gave him a cuddle and left him to get to know his new house. When I returned five minutes later, he was gone. The door had been shut, the windows closed and the walls were solid (I checked). He had literally vanished into thin air and he didn't even leave behind his smile. Before his dramatic disappearance, Fuzzy had stopped my suicidal tendencies, and there is lots of research showing that having a pet is good for your mental health. If you wanted to test this you could compare people with pets against those without to see if they had better mental health. However, the term *mental health* covers a wide range of concepts including (to name a few) anxiety, depression, general distress and psychosis. As such, we have four outcome measures and all the tests we have encountered allow us to look at one. Fear not, when we want to compare groups on several outcome variables we can extend ANOVA to become a MANOVA. That's what this chapter is all about.

16.2. When to use MANOVA ②

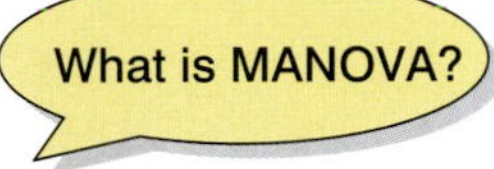

Over Chapters 9–14, we have seen how the general linear model (GLM) can be used to detect group differences on a single dependent variable. However, there may be circumstances in which we are interested in several dependent variables and in these cases the simple ANOVA model is inadequate. Instead, we can use an extension of this technique known as **multivariate analysis of variance** (or **MANOVA**). MANOVA can be thought of as ANOVA for situations in which there are several dependent variables. The principles of ANOVA extend to MANOVA in that we can use MANOVA when there is only one independent variable or when there are several, we can look at interactions between independent variables, and we can even do contrasts to see which groups differ from each other. ANOVA can be used only in situations in which there is one dependent variable (or outcome) and so is known as a **univariate** test (univariate quite obviously means 'one variable'); MANOVA is designed to look at several dependent variables (outcomes) simultaneously and so is a **multivariate** test (multivariate means 'many variables'). This chapter will explain some basics about MANOVA for those of you who want to skip the fairly tedious theory sections and just get on with the test. However, for those who want to know more there is a fairly lengthy theory section to try to explain the workings of MANOVA. We then look at an example using SPSS and see how the output from MANOVA can be interpreted. This leads us to look at another statistical test known as **discriminant function analysis**.

16.3. Introduction: similarities and differences to ANOVA ②

If we have collected data about several dependent variables then we could simply conduct a separate ANOVA for each dependent variable (and if you read research articles you'll find that it is not unusual for researchers to do this!). Think back to Chapter 10 and you should remember that a similar question was posed regarding why ANOVA was used in preference to multiple *t*-tests. The answer to why MANOVA is used instead of multiple ANOVAs is the same: the

Why not do lots of ANOVAs?

more tests we conduct on the same data, the more we inflate the familywise error rate (see section 10.2.1). The more dependent variables that have been measured, the more ANOVAs would need to be conducted and the greater the chance of making a Type I error. However, there are other reasons for preferring MANOVA to several ANOVAs. For one thing, there is important additional information that is gained from a MANOVA. If separate ANOVAs are conducted on each dependent variable, then any relationship between dependent variables is ignored. As such, we lose information about any correlations that might exist between the dependent variables. MANOVA, by including all dependent variables in the same analysis, takes account of the relationship between outcome variables. Related to this point, ANOVA can tell us only whether groups differ along a single dimension whereas MANOVA has the power to detect whether groups differ along a combination of dimensions. For example, ANOVA tells us how scores on a single dependent variable distinguish groups of participants (so, for example, we might be able to distinguish drivers, non-drivers and drunk drivers by the number of pedestrians they kill). MANOVA incorporates information about several outcome measures and, therefore, informs us of whether groups of participants can be distinguished by a combination of scores on several dependent measures. For example, it may not be possible to distinguish drivers, non-drivers and drunk drivers only by the number of pedestrians that they kill, but they might be distinguished by *a combination* of the number of pedestrians they kill, the number of lamp posts they hit, and the number of cars they crash into. So, in this sense MANOVA has greater power to detect an effect, because it can detect whether groups differ along a combination of variables, whereas ANOVA can detect only if groups differ along a single variable (see Jane Superbrain Box 16.1). For these reasons, MANOVA is preferable to conducting several ANOVAs.

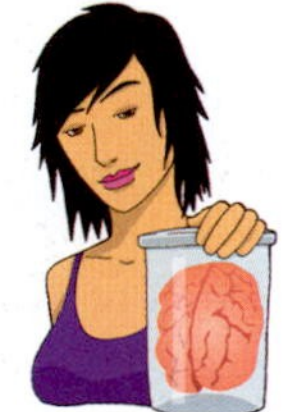

JANE SUPERBRAIN 16.1

The power of MANOVA ③

I mentioned in the previous section that MANOVA had greater power than ANOVA to detect effects because it could take account of the correlations between dependent variables (Huberty & Morris, 1989). However, the issue of power is more complex than alluded to by my simple statement. Ramsey (1982) found that as the correlation between dependent variables increased, the power of MANOVA decreased. This led Tabachnick and Fidell (2007) to recommend that MANOVA 'works best with highly negatively correlated DVs, and acceptably well with moderately correlated DVs in either direction' and that 'MANOVA also is wasteful when DVs are uncorrelated' (p. 268). In contrast, Stevens' (1980) investigation of the effect of dependent variable correlations on test power revealed that 'the power with high intercorrelations is in most cases greater than that for moderate intercorrelations, and in some cases it is dramatically higher' (p. 736). These findings are slightly contradictory, which leaves us with the puzzling conundrum of what, exactly, is the relationship between power and intercorrelation of the dependent variables? Luckily, Cole, Maxwell, Arvey, and Salas (1994) have done a great deal to illuminate this relationship. They found that the power of MANOVA depends on a combination of the correlation between dependent variables and the effect size to be detected. In short, if you are expecting to find a large effect, then MANOVA will have greater power if the measures are somewhat different (even negatively correlated) and if the group differences are in the same direction for each measure. If you have two dependent variables, one of which exhibits a large group difference, and one of which exhibits a small, or no, group difference, then power will be increased if these variables are highly correlated. The take-home message from Cole et al.'s work is that if you are interested in how powerful the MANOVA is likely to be you should consider not just the intercorrelation of dependent variables but also the size and pattern of group differences that you expect to get. However, it should be noted that Cole et al.'s work is limited to the case of where two groups are being compared and power considerations are more complex in multiple group situations.

16.3.1. Words of warning ②

From my description of MANOVA it is probably looking like a pretty groovy little test that allows you to measure hundreds of dependent variables and then just sling them into the analysis. This is not the case. It is not a good idea to lump all of your dependent variables together in a MANOVA unless you have a good theoretical or empirical basis for doing so. I mentioned way back at the beginning of this book that statistical procedures are just a way of number crunching and so even if you put rubbish into an analysis you will still reach conclusions that are statistically meaningful, but are unlikely to be empirically meaningful. In circumstances where there is a good theoretical basis for including some but not all of your dependent variables, you should run separate analyses: one for the variables being tested on a heuristic basis and one for the theoretically meaningful variables. The point to take on board here is not to include lots of dependent variables in a MANOVA just because you have measured them.

16.3.2. The example for this chapter ②

Throughout the rest of this chapter we're going to use a single example to look at how MANOVA works and then how to conduct one on SPSS. Imagine that we were interested in the effects of cognitive behaviour therapy on obsessive compulsive disorder (OCD). OCD is a disorder characterized by intrusive images or thoughts that the sufferer finds abhorrent (in my case this might be the thought of someone carrying out a *t*-test on data that are not normally distributed, or imagining your parents have died). These thoughts lead the sufferer to engage in activities to neutralize the unpleasantness of these thoughts (these activities can be mental, such as doing a MANOVA in my head to make me feel better about the *t*-test thought, or physical, such as touching the floor 23 times so that your parents won't die). Now, we could compare a group of OCD sufferers after cognitive behaviour therapy (CBT) and after behaviour therapy (BT) with a group of OCD sufferers who are still awaiting treatment (a no-treatment condition, NT).[2] Now, most psychopathologies have both behavioural and cognitive elements to them. For example, in OCD if someone had an obsession with germs and contamination, this disorder might manifest itself in obsessive hand-washing and would influence not just how many times they actually wash their hands (behaviour), but also the number of times they think about washing their hands (cognitions). Similarly, someone with an obsession about bags won't just think about bags a lot, but they might carry out bag-related behaviours (such as saying 'bag' repeatedly, or buying lots of bags). If we are interested in seeing how successful a therapy is, it is not enough to look only at behavioural outcomes (such as whether obsessive behaviours are reduced); it is important to establish whether cognitions are being changed also. Hence, in this example two dependent measures were taken: the occurrence of obsession-related behaviours (**Actions**) and the occurrence of obsession-related cognitions (**Thoughts**). These dependent variables were measured on a single day and so represent the number of obsession-related behaviours/thoughts in a normal day.

The data are in Table 16.1 and can be found in the file **OCD.sav**. Participants belonged to group 1 (CBT), group 2 (BT) or group 3 (NT) and within these groups all participants had both actions and thoughts measured.

[2] The non-psychologists out there should note that behaviour therapy works on the basis that if you stop the maladaptive behaviours the disorder will go away, whereas cognitive therapy is based on the idea that treating the maladaptive cognitions will stop the disorder.

TABLE 16.1 Data from **OCD.sav**

	DV 1: Actions			DV 2: Thoughts		
Group:	CBT (1)	BT (2)	NT (3)	CBT (1)	BT (2)	NT (3)
	5	4	4	14	14	13
	5	4	5	11	15	15
	4	1	5	16	13	14
	4	1	4	13	14	14
	5	4	6	12	15	13
	3	6	4	14	19	20
	7	5	7	12	13	13
	6	5	4	15	18	16
	6	2	6	16	14	14
	4	5	5	11	17	18
$\overline{X}$	**4.90**	**3.70**	**5.00**	**13.40**	**15.20**	**15.00**
s	**1.20**	**1.77**	**1.05**	**1.90**	**2.10**	**2.36**
s^2	**1.43**	**3.12**	**1.11**	**3.60**	**4.40**	**5.56**
	$\overline{X}_{\text{grand (Actions)}} = 4.53$			$\overline{X}_{\text{grand (Thoughts)}} = 14.53$		
	$s^2_{\text{grand (Actions)}} = 2.1195$			$s^2_{\text{grand (Thoughts)}} = 4.8780$		

16.4. Theory of MANOVA ③

SMART ALEX ONLY

The theory of MANOVA is very complex to understand without knowing matrix algebra, and frankly matrix algebra is way beyond the scope of this book (those with maths brains can consult Namboodiri, 1984; Stevens, 2002). However, I intend to give a flavour of the conceptual basis of MANOVA, using matrices, without requiring you to understand exactly how those matrices are used. Those interested in the exact underlying theory of MANOVA should read Bray and Maxwell's (1985) superb monograph.

16.4.1. Introduction to matrices ③

A **matrix** is simply a collection of numbers arranged in columns and rows. In fact, throughout this book you have been using a matrix without even realizing it: the SPSS Data Editor. In the data editor we have numbers arranged in columns and rows and this is exactly what a matrix is. A matrix can have many columns and many rows and we usually specify the dimensions of the matrix using numbers. So, a 2 × 3 matrix is a matrix with two rows and three columns, and a 5 × 4 matrix is one with five rows and four columns (examples below):

$$\begin{pmatrix} 2 & 5 & 6 \\ 3 & 5 & 8 \end{pmatrix} \qquad \begin{pmatrix} 2 & 4 & 6 & 8 \\ 3 & 4 & 6 & 7 \\ 4 & 3 & 5 & 8 \\ 2 & 5 & 7 & 9 \\ 4 & 6 & 6 & 9 \end{pmatrix}$$

2 × 3 matrix 5 × 4 matrix

You can think of these matrices in terms of each row representing the data from a single participant and each column as representing data relating to a particular variable. So, for the 5 × 4 matrix we can imagine a situation where five participants were tested on four variables: so, the first participant scored 2 on the first variable and 8 on the fourth variable. The values within a matrix are typically referred to as *components* or *elements*.

A **square matrix** is one in which there are an equal number of columns and rows. In this type of matrix it is sometimes useful to distinguish between the diagonal components (i.e. the values that lie on the diagonal line from the top left component to the bottom right component) and the off-diagonal components (the values that do not lie on the diagonal). In the matrix below, the diagonal components are 5, 12, 2 and 6 because they lie along the diagonal line. The off-diagonal components are all of the other values. A square matrix in which the diagonal elements are equal to 1 and the off-diagonal elements are equal to 0 is known as an **identity matrix**:

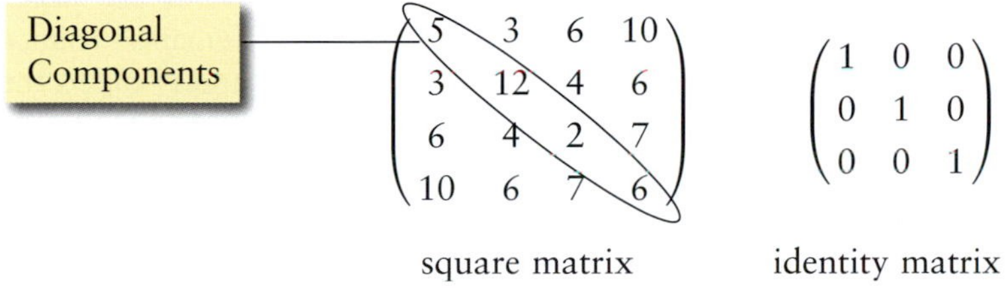

square matrix identity matrix

Hopefully, the concept of a matrix should now be slightly less scary than it was previously: it is not some magical mathematical entity, merely a way of representing a data set – just like a spreadsheet.

Now, there is a special case of a matrix where there are data from only one person, and this is known as a *row vector*. Likewise, if there is only one column in a matrix this is known as a *column vector*. In the examples below, the row vector can be thought of as a single person's score on four different variables, whereas the column vector can be thought of as five participants' scores on one variable:

$$\begin{pmatrix} 2 & 6 & 4 & 8 \end{pmatrix} \qquad \begin{pmatrix} 8 \\ 6 \\ 10 \\ 15 \\ 6 \end{pmatrix}$$

row vector column vector

Armed with this knowledge of what vectors are, we can have a brief look at how they are used to conduct MANOVA.

16.4.2. Some important matrices and their functions ③

As with ANOVA, we are primarily interested in how much variance can be explained by the experimental manipulation (which in real terms means how much variance is explained by the fact that certain scores appear in certain groups). Therefore, we need to know the sum of squares due to the grouping variable (the systematic variation, SS_M), the sum of squares due to natural differences between participants (the residual variation, SS_R) and of course the total amount of variation that needs to be explained (SS_T); for more details about these sources of variation reread Chapters 7 and 10. However, I mentioned that MANOVA also takes into account several dependent variables simultaneously and it does this by using a matrix that contains information about the variance accounted for by each dependent variable. For the univariate *F*-test (e.g. ANOVA) we calculated the ratio of systematic variance to unsystematic variance for a single dependent variable. In MANOVA, the test statistic is derived by comparing the ratio of systematic to unsystematic variance for several dependent variables. This comparison is made by using the ratio of a matrix representing the systematic variance of all dependent variables to a matrix representing the unsystematic variance of all dependent variables. To sum up, the test statistic in both ANOVA and MANOVA represents the ratio of the effect of the systematic variance to the unsystematic variance; in ANOVA these variances are single values, but in MANOVA each is a matrix containing many variances and covariances.

The matrix that represents the systematic variance (or the model sum of squares for all variables) is denoted by the letter ***H*** and is called the **hypothesis sum of squares and cross-products matrix** (or **hypothesis SSCP**). The matrix that represents the unsystematic variance (the residual sums of squares for all variables) is denoted by the letter ***E*** and is called the **error sum of squares and cross-products matrix** (or **error SSCP**). Finally, there is a matrix that represents the total amount of variance present for each dependent variable (the total sums of squares for each dependent variable) and this is denoted by ***T*** and is called the **total sum of squares and cross-products matrix** (or **total SSCP**).

Later, I will show how these matrices are used in exactly the same way as the simple sums of squares (SS_M, SS_R and SS_T) in ANOVA to derive a test statistic representing the ratio of systematic to unsystematic variance in the model. The observant among you may have noticed that the matrices I have described are all called **sum of squares and cross-products (SSCP) matrices**. It should be obvious why these matrices are referred to as sum of squares matrices, but why is there a reference to cross-products in their name?

SELF-TEST Can you remember (from Chapter 6) what a cross-product is?

Cross-products represent a total value for the combined error between two variables (so, in some sense they represent an unstandardized estimate of the total correlation between two variables). As such, whereas the sum of squares of a variable is the total squared difference between the observed values and the mean value, the cross-product is the total combined error between two variables. I mentioned earlier that MANOVA had the power to account for any correlation between dependent variables and it does this by using these cross-products.

16.4.3. Calculating MANOVA by hand: a worked example ③

To begin with let's carry out univariate ANOVAs on each of the two dependent variables in our OCD example (see Table 16.1). A description of the ANOVA model can be found in Chapter 10 and I will draw heavily on the assumption that you have read this chapter; if you are hazy on the details of Chapter 10 then now would be a good time to (re)read sections 10.2.5 to 10.2.9.

16.4.3.1. Univariate ANOVA for DV 1 (actions) ②

There are three sums of squares that need to be calculated. First we need to assess how much variability there is to be explained within the data (SS_T), next we need to see how much of this variability can be explained by the model (SS_M), and finally we have to assess how much error there is in the model (SS_R). From Chapter 10 we can calculate each of these values:

- $SS_{T(Actions)}$: The total sum of squares is obtained by calculating the difference between each of the 20 scores and the mean of those scores, then squaring these differences and adding these squared values up. Alternatively, you can get SPSS to calculate the variance for the action data (regardless of which group the score falls into) and then multiplying this value by the number of scores minus 1:

$$\begin{aligned}SS_T &= s^2_{grand}(n-1)\\ &= 2.1195(30-1)\\ &= 2.1195 \times 29\\ &= 61.47\end{aligned}$$

- $SS_{M(Actions)}$: This value is calculated by taking the difference between each group mean and the grand mean and then squaring them. Multiply these values by the number of scores in the group and then add them together:

$$\begin{aligned}SS_M &= 10(4.90-4.53)^2 + 10(3.70-4.53)^2 + 10(5.00-4.53)^2\\ &= 10(0.37)^2 + 10(-0.83)^2 + 10(0.47)^2\\ &= 1.37 + 6.89 + 2.21\\ &= 10.47\end{aligned}$$

- $SS_{R(Actions)}$: This value is calculated by taking the difference between each score and the mean of the group from which it came. These differences are then squared and then added together. Alternatively we can get SPSS to calculate the variance within each group, multiply each group variance by the number of scores minus 1 and then add them together:

$$\begin{aligned}SS_R &= s^2_{CBT}(n_{CBT}-1) + s^2_{BT}(n_{BT}-1) + s^2_{NT}(n_{NT}-1)\\ &= (1.433)(10-1) + (3.122)(10-1) + (1.111)(10-1)\\ &= (1.433 \times 9) + (3.122 \times 9)(1.111 \times 9)\\ &= 12.9 + 28.1 + 10.0\\ &= 51.00\end{aligned}$$

The next step is to calculate the average sums of squares (the mean square) of each by dividing by the degrees of freedom (see section 10.2.8):

SS	*df*	MS
$SS_{M(Actions)} = 10.47$	2	5.235
$SS_{R(Actions)} = 51.00$	27	1.889

The final stage is calculate F by dividing the mean squares for the model by the mean squares for the error in the model:

$$F = \frac{MS_M}{MS_R} = \frac{5.235}{1.889} = 2.771$$

This value can then be evaluated against critical values of F. The point to take home here is the calculation of the various sums of squares and what each one relates to.

16.4.3.2. Univariate ANOVA for DV 2 (thoughts) ②

As with the data for dependent variable 1, there are three sums of squares that need to be calculated as before:

- $SS_{T(Thoughts)}$:

$$\begin{aligned} SS_T &= s^2_{grand}(n-1) \\ &= 4.878(30-1) \\ &= 4.878 \times 29 \\ &= 141.46 \end{aligned}$$

- $SS_{M(Thoughts)}$:

$$\begin{aligned} SS_M &= 10(13.40 - 14.53)^2 + 10(15.2 - 14.53)^2 + 10(15.0 - 14.53)^2 \\ &= 10(-1.13)^2 + 10(0.67)^2 + 10(0.47)^2 \\ &= 12.77 + 4.49 + 2.21 \\ &= 19.47 \end{aligned}$$

- $SS_{R(Thoughts)}$:

$$\begin{aligned} SS_R &= s^2_{CBT}(n_{CBT} - 1) + s^2_{BT}(n_{BT} - 1) + s^2_{NT}(n_{NT} - 1) \\ &= (3.6)(10-1) + (4.4)(10-1) + (5.56)(10-1) \\ &= (3.6 \times 9) + (4.4 \times 9)(5.56 \times 9) \\ &= 32.4 + 39.6 + 50.0 \\ &= 122 \end{aligned}$$

The next step is to calculate the average sums of squares (the mean square) of each by dividing by the degrees of freedom (see section 10.2.8):

SS	df	MS
$SS_{M(Thoughts)} = 19.47$	2	9.735
$SS_{R(Thoughts)} = 122.00$	27	4.519

The final stage is to calculate *F* by dividing the mean squares for the model by the mean squares for the error in the model:

$$F = \frac{MS_M}{MS_R} = \frac{9.735}{4.519} = 2.154$$

This value can then be evaluated against critical values of *F*. Again, the point to take home here is the calculation of the various sums of squares and what each one relates to.

16.4.3.3. The relationship between DVs: cross-products ②

We know already that MANOVA uses the same sums of squares as ANOVA, and in the next section we will see exactly how it uses these values. However, I have also mentioned that MANOVA takes account of the relationship between dependent variables by using the cross-products. There are three different cross-products that are of interest and these three cross-products relate to the three sums of squares that we calculated for the univariate ANOVAs: that is, there is a total cross-product, a cross-product due to the model and a residual cross-product. Let's look at the total cross-product (CP_T) first.

I mentioned in Chapter 6 that the cross-product was the difference between the scores and the mean in one group multiplied by the difference between the scores and the mean in the other group. In the case of the total cross-product, the mean of interest is the grand mean for each dependent variable (see Table 16.2). Hence, we can adapt the cross-product equation described in Chapter 6 using the two dependent variables. The resulting equation for the total cross-product is described as in equation (16.1). Therefore, for each dependent variable you take each score and subtract from it the grand mean for that variable. This leaves you with two values per participant (one for each dependent variable) which should be multiplied together to get the cross-product for each participant. The total can then be found by adding the cross-products of all participants. Table 16.2 illustrates this process:

$$CP_T = \sum \left(x_{i(\text{Actions})} - \overline{X}_{\text{grand(Actions)}}\right)\left(x_{i(\text{Thoughts})} - \overline{X}_{\text{grand(Thoughts)}}\right) \tag{16.1}$$

The total cross-product is a gauge of the overall relationship between the two variables. However, we are also interested in how the relationship between the dependent variables is influenced by our experimental manipulation and this relationship is measured by the model cross-product (CP_M). The CP_M is calculated in a similar way to the model sum of squares. First, the difference between each group mean and the grand mean is calculated for each dependent variable. The cross-product is calculated by multiplying the differences found for each group. Each product is then multiplied by the number of scores within the group (as was done with the sum of squares). This principle is illustrated in the following equation and Table 16.3:

$$CP_M = \sum n\left[\left(\overline{x}_{\text{group(Actions)}} - \overline{X}_{\text{grand(Actions)}}\right)\left(\overline{x}_{\text{group(Thoughts)}} - \overline{X}_{\text{grand(Thoughts)}}\right)\right] \tag{16.2}$$

TABLE 16.2 Calculation of the total cross-product

Group	Actions	Thoughts	Actions $-\overline{X}_{grand(Actions)}$ (D_1)	Thoughts $-\overline{X}_{grand(Thoughts)}$ (D_2)	$D_1 \times D_2$
CBT	5	14	0.47	−0.53	−0.25
	5	11	0.47	−3.53	−1.66
	4	16	−0.53	1.47	−0.78
	4	13	−0.53	−1.53	0.81
	5	12	0.47	−2.53	−1.19
	3	14	−1.53	−0.53	0.81
	7	12	2.47	−2.53	−6.25
	6	15	1.47	0.47	0.69
	6	16	1.47	1.47	2.16
	4	11	−0.53	−3.53	1.87
BT	4	14	−0.53	−0.53	0.28
	4	15	−0.53	0.47	−0.25
	1	13	−3.53	−1.53	5.40
	1	14	−3.53	−0.53	1.87
	4	15	−0.53	0.47	−0.25
	6	19	1.47	4.47	6.57
	5	13	0.47	−1.53	−0.72
	5	18	0.47	3.47	1.63
	2	14	−2.53	−0.53	1.34
	5	17	0.47	2.47	1.16
NT	4	13	−0.53	−1.53	0.81
	5	15	0.47	0.47	0.22
	5	14	0.47	−0.53	−0.25
	4	14	−0.53	−0.53	0.28
	6	13	1.47	−1.53	−2.25
	4	20	−0.53	5.47	−2.90
	7	13	2.47	−1.53	−3.78
	4	16	−0.53	1.47	−0.78
	6	14	1.47	−0.53	−0.78
	5	18	0.47	3.47	1.63
$\overline{X}_{grand}$	4.53	14.53		$CP_T = \Sigma(D_1 \times D_2) = 5.47$	

Finally, we also need to know how the relationship between the two dependent variables is influenced by individual differences in participants' performances. The residual cross-product (CP_R) tells us about how the relationship between the dependent variables is affected by individual differences, or error in the model. The CP_R is calculated in a similar way to the total cross-product except that the group means are used rather than the grand mean (see equation (16.3)). So, to calculate each of the difference scores, we take each score and subtract from it the mean of the group to which it belongs (see Table 16.4):

$$CP_R = \sum \left(x_{i(\text{Actions})} - \overline{X}_{\text{group(Actions)}}\right)\left(x_{i(\text{Thoughts})} - \overline{X}_{\text{group(Thoughts)}}\right) \quad (16.3)$$

TABLE 16.3 Calculating the model cross-product

	$\bar{X}_{group}$ Actions	$\bar{X}_{group} - \bar{X}_{grand}$ (D_1)	$\bar{X}_{group}$ Thoughts	$\bar{X}_{group} - \bar{X}_{grand}$ (D_2)	$D_1 \times D_2$	$N(D_1 \times D_2)$
CBT	4.9	0.37	13.4	−1.13	−0.418	−4.18
BT	3.7	−0.83	15.2	0.67	−0.556	−5.56
NT	5.0	0.47	15.0	0.47	0.221	2.21
$\bar{X}_{grand}$	4.53		14.53	$CP_M = \Sigma N(D_1 \times D_2) = -7.53$		

The observant among you may notice that the residual cross-product can also be calculated by subtracting the model cross-product from the total cross-product:

$$
\begin{aligned}
CP_R &= CP_T - CP_M \\
&= 5.47 - (-7.53) = 13
\end{aligned}
$$

However, it is useful to calculate the residual cross-product manually in case of mistakes in the calculation of the other two cross-products. The fact that the residual and model cross-products should sum to the value of the total cross-product can be used as a useful double-check.

Each of the different cross-products tells us something important about the relationship between the two dependent variables. Although I have used a simple scenario to keep the maths relatively simple, these principles can be easily extended to more complex scenarios. For example, if we had measured three dependent variables then the cross-products between pairs of dependent variables are calculated (as they were in this example) and entered into the appropriate SSCP matrix (see next section). As the complexity of the situation increases, so does the amount of calculation that needs to be done. At times such as these the benefit of software like SPSS becomes ever more apparent!

16.4.3.4. The total SSCP matrix (*T*) ③

In this example we have only two dependent variables and so all of the SSCP matrices will be 2 × 2 matrices. If there had been three dependent variables then the resulting matrices would all be 3 × 3 matrices. The total SSCP matrix, *T*, contains the total sums of squares for each dependent variable and the total cross-product between the two dependent variables. You can think of the first column and first row as representing one dependent variable and the second column and row as representing the second dependent variable:

	Column 1 Actions	Column 2 Thoughts
Row 1 Actions	$SS_{T(Actions)}$	CP_T
Row 1 Thoughts	CP_T	$SS_{T(Thoughts)}$

TABLE 16.4 Calculation of CP_R

Group	Actions	Actions $-\bar{X}_{group(Actions)}$ (D_1)	Thoughts	Thoughts $-\bar{X}_{group(Thoughts)}$ (D_2)	$D_1 \times D_2$
CBT	5	0.10	14	0.60	0.06
	5	0.10	11	−2.40	−0.24
	4	−0.90	16	2.60	−2.34
	4	−0.90	13	−0.40	0.36
	5	0.10	12	−1.40	−0.14
	3	−1.90	14	0.60	−1.14
	7	2.10	12	−1.40	−2.94
	6	1.10	15	1.60	1.76
	6	1.10	16	2.60	2.86
	4	−0.90	11	−2.40	2.16
$\bar{X}_{CBT}$	**4.9**		**13.4**		**Σ = 0.40**
BT	4	0.30	14	−1.20	−0.36
	4	0.30	15	−0.20	−0.06
	1	−2.70	13	−2.20	5.94
	1	−2.70	14	−1.20	3.24
	4	0.30	15	−0.20	−0.06
	6	2.30	19	3.80	8.74
	5	1.30	13	−2.20	−2.86
	5	1.30	18	2.80	3.64
	2	−1.70	14	−1.20	2.04
	5	1.30	17	1.80	2.34
$\bar{X}_{BT}$	**3.7**		**15.2**		**Σ = 22.60**
NT	4	−1.00	13	−2.00	2.00
	5	0.00	15	0	0.00
	5	0.00	14	−1.00	0.00
	4	−1.00	14	−1.00	1.00
	6	1.00	13	−2.00	−2.00
	4	−1.00	20	5.00	−5.00
	7	2.00	13	−2.00	−4.00
	4	−1.00	16	1.00	−1.00
	6	1.00	14	−1.00	−1.00
	5	0.00	18	3.00	0.00
$\bar{X}_{NT}$	**5**		**15**		**Σ = −10.00**
					$CP_R = \Sigma(D_1 \times D_2) = 13$

We calculated these values in the previous sections and so we can simply place the appropriate values in the appropriate cell of the matrix:

$$T = \begin{pmatrix} 61.47 & 5.47 \\ 5.47 & 141.47 \end{pmatrix}$$

From the values in the matrix (and what they represent) it should be clear that the total SSCP represents both the total amount of variation that exists within the data and the total co-dependence that exists between the dependent variables. You should also note that the off-diagonal components are the same (they are both the total cross-product) because this value is equally important for both of the dependent variables.

16.4.3.5. The residual SSCP matrix (*E*) ③

The residual (or error) sum of squares and cross-product matrix, *E*, contains the residual sums of squares for each dependent variable and the residual cross-product between the two dependent variables. This SSCP matrix is similar to the total SSCP except that the information relates to the error in the model:

	Column 1 Actions	Column 2 Thoughts
Row 1 Actions	$SS_{R(Actions)}$	CP_R
Row 1 Thoughts	CP_R	$SS_{R(Thoughts)}$

We calculated these values in the previous sections and so we can simply place the appropriate values in the appropriate cell of the matrix:

$$E = \begin{pmatrix} 51 & 13 \\ 13 & 122 \end{pmatrix}$$

From the values in the matrix (and what they represent) it should be clear that the residual SSCP represents both the unsystematic variation that exists for each dependent variable and the co-dependence between the dependent variables that is due to chance factors alone. As before, the off-diagonal elements are the same (they are both the residual cross-product).

16.4.3.6. The model SSCP matrix (*H*) ③

The model (or hypothesis) sum of squares and cross-product matrix, *H*, contains the model sums of squares for each dependent variable and the model cross-product between the two dependent variables:

	Column 1 Actions	Column 2 Thoughts
Row 1 Actions	$SS_{M(Actions)}$	CP_M
Row 1 Thoughts	CP_M	$SS_{M(Thoughts)}$

These values were calculated in the previous sections and so we can simply place the appropriate values in the appropriate cell of the matrix (see below). From the values in the matrix (and what they represent) it should be clear that the model SSCP represents both the systematic variation that exists for each dependent variable and the co-dependence between the dependent variables that is due to the model (i.e. is due to the experimental manipulation). As before, the off-diagonal components are the same (they are both the model cross-product):

$$H = \begin{pmatrix} 10.47 & -7.53 \\ -7.53 & 19.47 \end{pmatrix}$$

Matrices are additive, which means that you can add (or subtract) two matrices together by adding (or subtracting) corresponding components. Now, when we calculated univariate ANOVA we saw that the total sum of squares was the sum of the model sum of squares and the residual sum of squares (i.e. $SS_T = SS_M + SS_R$). The same is true in MANOVA except that we are adding matrices rather than single values:

$$
\begin{aligned}
T &= H + E \\
T &= \begin{pmatrix} 10.47 & -7.53 \\ -7.53 & 19.47 \end{pmatrix} + \begin{pmatrix} 51 & 13 \\ 13 & 122 \end{pmatrix} \\
&= \begin{pmatrix} 10.47 + 51 & -7.53 + 13 \\ -7.53 + 13 & 19.47 + 122 \end{pmatrix} \\
&= \begin{pmatrix} 61.47 & 5.47 \\ 5.47 & 141.47 \end{pmatrix}
\end{aligned}
$$

The demonstration that these matrices add up should (hopefully) help you to understand that the MANOVA calculations are conceptually the same as for univariate ANOVA – the difference is that matrices are used rather than single values.

16.4.4. Principle of the MANOVA test statistic ④

In univariate ANOVA we calculate the ratio of the systematic variance to the unsystematic variance (i.e. we divide SS_M by SS_R).[3] The conceptual equivalent would therefore be to divide the matrix *H* by the matrix *E*. There is, however, a problem in that matrices are not divisible by other matrices! However, there is a matrix equivalent to division, which is to multiply by what's known as the inverse of a matrix. So, if we want to divide *H* by *E* we have to multiply *H* by the inverse of *E* (denoted as E^{-1}). So, therefore, the test statistic is based upon the matrix that results from multiplying the model SSCP with the inverse of the residual SSCP. This matrix is called **HE^{-1}**.

Calculating the inverse of a matrix is incredibly difficult and there is no need for you to understand how it is done because SPSS will do it for you. However, the interested reader should consult either Stevens (2002) or Namboodiri (1984) – these texts provide very accessible accounts of how to derive an inverse matrix. For readers who do consult these sources, see Oliver Twisted. For the uninterested reader, you'll have to trust me on the following:

[3] In reality we use the mean squares but these values are merely the sums of squares corrected for the degrees of freedom.

$$E^{-1} = \begin{pmatrix} 0.0202 & -0.0021 \\ -0.0021 & 0.0084 \end{pmatrix}$$

$$HE^{-1} = \begin{pmatrix} 0.2273 & -0.0852 \\ -0.1930 & 0.1794 \end{pmatrix}$$

Remember that HE^{-1} represents the ratio of systematic variance in the model to the unsystematic variance in the model and so the resulting matrix is conceptually the same as the *F*-ratio in univariate ANOVA. There is another problem, though. In ANOVA, when we divide the systematic variance by the unsystematic variance we get a single figure: the *F*-ratio. In MANOVA, when we divide the systematic variance by the unsystematic variance we get a matrix containing several values. In this example, the matrix contains four values, but had there been three dependent variables the matrix would have had nine values. In fact, the resulting matrix will always contain p^2 values, where p is the number of dependent variables. The problem is how to convert these matrix values into a meaningful single value. This is the point at which we have to abandon any hope of understanding the maths behind the test and talk conceptually instead.

16.4.4.1. Discriminant function variates ④

The problem of having several values with which to assess statistical significance can be simplified considerably by converting the dependent variables into underlying dimensions or factors (this process will be discussed in more detail in Chapter 17). In Chapter 7, we saw how multiple regression worked on the principle of fitting a linear model to a set of data to predict an outcome variable (the dependent variable in ANOVA terminology). This linear model was made up of a combination of predictor variables (or independent variables) each of which had a unique contribution to this linear model. We can do a similar thing here, except that we are interested in the opposite problem (i.e. predicting an independent variable from a set of dependent variables). So, it is possible to calculate underlying linear dimensions of the dependent variables. These linear combinations of the dependent variables are known as *variates* (or sometimes called *latent variables* or *factors*). In this context we wish to use these linear variates to predict which group a person belongs to (i.e. whether they were given CBT, BT or no treatment), so we are using them to discriminate groups of people. Therefore, these variates are called *discriminant functions* or **discriminant function variates**. Although I have drawn a parallel between these discriminant functions and the model in multiple regression, there is a difference in that we can extract several discriminant functions from a set of dependent variables, whereas in multiple regression all independent variables are included in a single model.

That's the theory in simplistic terms, but how do we discover these discriminant functions? Well, without going into too much detail, we use a mathematical procedure of maximization, such that the first discriminant function (V_1) is the linear combination of dependent variables that maximizes the differences between groups.

It follows from this that the ratio of systematic to unsystematic variance (SS_M/SS_R) will be maximized for this first variate, but subsequent variates will have smaller values of this ratio. Remember that this ratio is an analogue of what the *F*-ratio represents in univariate ANOVA, and so in effect we obtain the maximum possible value of the *F*-ratio when we look at the first discriminant function. This variate can be described in terms of a linear regression equation (because it is a linear combination of the dependent variables):

$$\begin{aligned} Y &= b_0 + b_1X_1 + b_2X_2 \\ V_1 &= b_0 + b_1\text{DV}_1 + b_2\text{DV}_2 \qquad (16.4) \\ V_1 &= b_0 + b_1\text{Actions} + b_2\text{Thoughts} \end{aligned}$$

Equation (16.4) shows the multiple regression equation for two predictors and then extends this to show how a comparable form of this equation can describe discriminant functions. The *b*-values in the equation are weights (just as in regression) that tell us something about the contribution of each dependent variable to the variate in question. In regression, the values of *b* are obtained by the method of least squares; in discriminant function analysis the values of *b* are obtained from the *eigenvectors* (see Jane Superbrain Box 7.2) of the matrix HE^{-1}. We can actually ignore b_0 as well because this serves only to locate the variate in geometric space, which isn't necessary when we're using it to discriminate groups.

In a situation in which there are only two dependent variables and two groups for the independent variable, there will be only one variate. This makes the scenario very simple: by looking at the discriminant function of the dependent variables, rather than looking at the dependent variables themselves, we can obtain a single value of SS_M/SS_R for the discriminant function, and then assess this value for significance. However, in more complex cases where there are more than two dependent variables or more than three levels of the independent variable (as is the case in our example), there will be more than one variate. The number of variates obtained will be the smaller of *p* (the number of dependent variables) or $k - 1$ (where *k* is the number of levels of the independent variable). In our example, both *p* and $k - 1$ are 2, so we should be able to find two variates. I mentioned earlier that the *b*-values that describe the variates are obtained by calculating the eigenvectors of the matrix HE^{-1} and, in fact, there will be two eigenvectors derived from this matrix: one with the *b*-values for the first variate, and one with the *b*-values of the second variate. Conceptually speaking, eigenvectors are the vectors associated with a given matrix that are unchanged by transformation of that matrix to a diagonal matrix (look back to Jane Superbrain Box 7.2 for a visual explanation of eigenvectors and eigenvalues). A diagonal matrix is simply a matrix in which the off-diagonal elements are zero and by changing HE^{-1} to a diagonal matrix we eliminate all of the off-diagonal elements (thus reducing the number of values that we must consider for significance testing). Therefore, by calculating the eigenvectors and eigenvalues, we still end up with values that represent the ratio of systematic to unsystematic variance (because they are unchanged by the transformation), but there are considerably less of them. The calculation of eigenvectors is extremely complex (insane students can consider reading Namboodiri, 1984), so you can trust me that for the matrix HE^{-1} the eigenvectors obtained are:

$$\text{eigenvector}_1 = \begin{pmatrix} 0.603 \\ -0.335 \end{pmatrix}$$

$$\text{eigenvector}_2 = \begin{pmatrix} 0.425 \\ 0.339 \end{pmatrix}$$

Replacing these values into the two equations for the variates and bearing in mind we can ignore b_0 we obtain the models described in the following equation:

$$\begin{aligned} V_1 &= b_0 + 0.603\text{Actions} - 0.335\text{Thoughts} \\ V_2 &= b_0 + 0.425\text{Actions} - 0.339\text{Thoughts} \end{aligned} \tag{16.5}$$

It is possible to use the equations for each variate to calculate a score for each person on the variate. For example, the first participant in the CBT group carried out 5 obsessive actions and had 14 obsessive thoughts. Therefore, this participant's score on variate 1 would be −1.675:

$$V_1 = (0.603 \times 5) - (0.335 \times 14) = -1.675$$

The score for variate 2 would be 6.871:

$$V_2 = (0.425 \times 5) - (0.339 \times 14) = 6.871$$

If we calculated these variate scores for each participant and then calculated the SSCP matrices (e.g. H, E, T and HE^{-1}) that we used previously, we would find that all of them have cross-products of zero. The reason for this is because the variates extracted from the data are orthogonal, which means that they are uncorrelated. In short, the variates extracted are independent dimensions constructed from a linear combination of the dependent variables that were measured.

This data reduction has a very useful property in that if we look at the matrix HE^{-1} calculated from the variate scores (rather than the dependent variables) we find that all of the off-diagonal elements (the cross-products) are zero. The diagonal elements of this matrix represent the ratio of the systematic variance to the unsystematic variance (i.e. SS_M/SS_R) for each of the underlying variates. So, for the data in this example, this means that instead of having four values representing the ratio of systematic to unsystematic variance, we now have only two. This reduction may not seem a lot. However, in general if we have p dependent variables, then ordinarily we would end up with p^2 values representing the ratio of systematic to unsystematic variance; by looking at discriminant functions, we reduce this number to p. If there were four dependent variables we would end up with four values rather than sixteen (which highlights the benefit of this process).

For the data in our example, the matrix HE^{-1} calculated from the variate scores is:

$$HE^{-1}_{\text{variates}} = \begin{pmatrix} 0.335 & 0.000 \\ 0.000 & 0.073 \end{pmatrix}$$

It is clear from this matrix that we have two values to consider when assessing the significance of the group differences. It probably seems like a complex procedure to reduce the data down in this way: however, it transpires that the values along the diagonal of the matrix for the variates (namely 0.335 and 0.073) are the *eigenvalues* of the original HE^{-1} matrix. Therefore, these values can be calculated directly from the data collected without first forming the eigenvectors. If you have lost all sense of rationality and want to see how these eigenvalues are calculated then see Oliver Twisted. These eigenvalues are conceptually equivalent to the F-ratio in ANOVA and so the final step is to assess how large these values are compared to what we would expect by chance alone. There are four ways in which the values are assessed.

OLIVER TWISTED

Please, Sir, can I have some more … maths?

'You are a bit stupid. I think it would be fun to check your maths so that we can see exactly how much of a village idiot you are,' mocks Oliver. Luckily you can. Never one to shy from public humiliation on a mass scale, I have provided the matrix calculations for this example on the companion website. Find a mistake, go on, you know that you can …

16.4.4.2. Pillai–Bartlett trace (V) ④

The **Pillai–Bartlett trace** (also known as Pillai's trace) is given by equation (16.6) in which λ represents the eigenvalues for each of the discriminant variates and s represents the number of variates. Pillai's trace is the sum of the proportion of explained variance on the discriminant functions. As such, it is similar to the ratio of SS_M/SS_T, which is known as R^2:

$$V = \sum_{i=1}^{s} \frac{\lambda_i}{1+\lambda_i} \tag{16.6}$$

For our data, Pillai's trace turns out to be 0.319, which can be transformed to a value that has an approximate *F*-distribution:

$$V = \frac{0.335}{1+0.335} + \frac{0.073}{1+0.073} = 0.319$$

FIGURE 16.2 Harold Hotelling enjoying my favourite activity of drinking tea

16.4.4.3. Hotelling's T^2 ④

The **Hotelling–Lawley trace** (also known as Hotelling's T^2; Figure 16.2) is simply the sum of the eigenvalues for each variate (see equation (16.7)) and so for these data its value is 0.408 (0.335 + 0.073). This test statistic is the sum of SS_M/SS_R for each of the variates and so it compares directly to the *F*-ratio in ANOVA:

$$T = \sum_{i=1}^{s} \lambda_i \tag{16.7}$$

16.4.4.4. Wilks's lambda (Λ) ④

Wilks's lambda is the product of the *unexplained* variance on each of the variates (see equation (16.8) – the Π symbol is similar to the summation symbol (Σ) that we have encountered already except that it means *multiply* rather than add up). So, Wilks's lambda represents the ratio of error variance to total variance (SS_R/SS_T) for each variate:

$$\Lambda = \prod_{i=1}^{s} \frac{1}{1+\lambda_i} \tag{16.8}$$

For the data in this example the value is 0.698, and it should be clear that large eigenvalues (which in themselves represent a large experimental effect) lead to small values of Wilks's lambda: hence statistical significance is found when Wilks's lambda is small:

$$\Lambda = \left(\frac{1}{1+0.335}\right)\left(\frac{1}{1+0.073}\right) = 0.698$$

16.4.4.5. Roy's largest root ④

Roy's largest root always makes me think of some bearded statistician with a garden spade digging up an enormous parsnip (or similar root vegetable); however, it isn't a parsnip but, as the name suggests, is simply the eigenvalue for the first variate. So, in a sense it is the same as the Hotelling–Lawley trace but for the first variate only, that is:

$$\Theta = \lambda_{\text{Largest}} \tag{16.9}$$

As such, Roy's largest root represents the proportion of explained variance to unexplained variance (SS_M/SS_R) for the first discriminant function.[4] For the data in this example, the value of Roy's largest root is simply 0.335 (the eigenvalue for the first variate). So, this value is conceptually the same as the *F*-ratio in univariate ANOVA. It should be apparent, from what we have learnt about the maximizing properties of these discriminant variates, that Roy's root represents the maximum possible between-group difference given the data collected. Therefore, this statistic should in many cases be the most powerful.

16.5. Practical issues when conducting MANOVA ③

There are three main practical issues to be considered before running MANOVA. First off, as always we have to consider the assumptions of the test. Next, for the main analysis there are four commonly used ways of assessing the overall significance of a MANOVA and debate exists about which method is best in terms of power and sample size considerations. Finally, we also need to think about what analysis to do *after* the MANOVA: like ANOVA, MANOVA is a two-stage test in which an overall (or omnibus) test is first performed before more specific procedures are applied to tease apart group differences. As you will see, there is substantial debate over how best to further analyse and interpret group differences when the overall MANOVA is significant. We will look at these issues in turn.

16.5.1. Assumptions and how to check them ③

MANOVA has similar assumptions to ANOVA but extended to the multivariate case:

- **Independence**: Observations should be statistically independent.
- **Random sampling**: Data should be randomly sampled from the population of interest and measured at an interval level.
- **Multivariate normality**: In ANOVA, we assume that our dependent variable is normally distributed within each group. In the case of MANOVA, we assume that the dependent variables (collectively) have multivariate normality within groups.
- **Homogeneity of covariance matrices**: In ANOVA, it is assumed that the variances in each group are roughly equal (homogeneity of variance). In MANOVA we must assume that this is true for each dependent variable, but also that the correlation between any two dependent variables is the same in all groups. This assumption is examined by testing whether the population **variance–covariance matrices** of the different groups in the analysis are equal.[5]

Most of the assumptions can be checked in the same way as for univariate tests (see Chapter 10); the additional assumptions of multivariate normality and equality of covariance matrices require different procedures. The assumption of multivariate normality cannot be tested

[4] This statistic is sometimes characterized as $\lambda_{\text{largest}}/(1 + \lambda_{\text{largest}})$ but this is not the statistic reported by SPSS.

[5] For those of you who read about SSCP matrices, if you think about the relationship between sums of squares and variance, and cross-products and correlations, it should be clear that a variance-covariance matrix is basically a standardized form of an SSCP matrix.

on SPSS and so the only practical solution is to check the assumption of univariate normality for each dependent variable in turn (see Chapter 5). This solution is practical (because it is easy to implement) and useful (because univariate normality is a necessary condition for multivariate normality), but it does not *guarantee* multivariate normality. So, although this approach is the best we can do, I urge interested readers to consult Stevens (2002) who provides some alternative solutions.

The assumption of equality of covariance matrices is more easily checked. First, for this assumption to be true the univariate tests of equality of variances between groups should be met. This assumption is easily checked using Levene's test (see section 5.6.1). As a preliminary check, Levene's test should not be significant for any of the dependent variables. However, Levene's test does not take account of the covariances and so the variance–covariance matrices should be compared between groups using **Box's test**. This test should be non-significant if the matrices are the same. The effect of violating this assumption is unclear, except that Hotelling's T^2 is robust in the two-group situation when sample sizes are equal (Hakstian, Roed, & Lind, 1979).

Box's test is susceptible to deviations from multivariate normality and so can be non-significant not because the matrices are similar, but because the assumption of multivariate normality is not tenable. Hence, it is vital to have some idea of whether the data meet the multivariate normality assumption before interpreting the result of Box's test. Also, as with any significance test, in large samples Box's test could be significant even when covariance matrices are relatively similar. As a general rule, if sample sizes are equal then disregard Box's test, because (1) it is unstable and (2) in this situation we can assume that Hotelling's and Pillai's statistics are robust (see section 16.5.2). However, if group sizes are different, then robustness cannot be assumed (especially if Box's test is significant at $p < .001$). The more dependent variables you have measured, and the greater the differences in sample sizes, the more distorted the probability values produced by SPSS become. Tabachnick and Fidell (2007) suggest that if the larger samples produce greater variances and covariances then the probability values will be conservative (and so significant findings can be trusted). However, if it is the smaller samples that produce the larger variances and covariances then the probability values will be liberal and so significant differences should be treated with caution (although non-significant effects can be trusted). Therefore, the variance–covariance matrices for samples should be inspected to assess whether the printed probabilities for the multivariate test statistics are likely to be conservative or liberal. In the event that you cannot trust the printed probabilities, there is little you can do except equalize the samples by randomly deleting cases in the larger groups (although with this loss of information comes a loss of power).

16.5.2. Choosing a test statistic ③

Only when there is one underlying variate will the four test statistics necessarily be the same. Therefore, it is important to know which test statistic is best in terms of test power and robustness. A lot of research has investigated the power of the four MANOVA test statistics (Olson, 1974, 1976, 1979; Stevens, 1980). Olson (1974) observed that for small and moderate sample sizes the four statistics differ little in terms of power. If group differences are concentrated on the first variate (as will often be the case in social science research) Roy's statistic should prove most powerful (because it takes account of only that first variate), followed by Hotelling's trace, Wilks's lambda and Pillai's trace. However, when groups differ along more than one variate, the power ordering is the reverse (i.e. Pillai's trace is most powerful and Roy's root is least). One final issue pertinent to test power is that of sample size and the number of dependent variables. Stevens (1980) recommends using fewer than 10 dependent variables unless sample sizes are large.

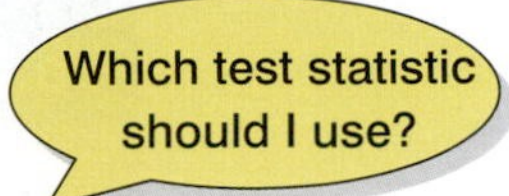
Which test statistic
should I use?

In terms of robustness, all four test statistics are relatively robust to violations of multivariate normality (although Roy's root is affected by platykurtic distributions – see Olson, 1976). Roy's root is also not robust when the homogeneity of covariance matrix assumption is untenable (Stevens, 1979). The work of Olson and Stevens led Bray and Maxwell (1985) to conclude that when sample sizes are equal the Pillai–Bartlett trace is the most robust to violations of assumptions. However, when sample sizes are unequal this statistic is affected by violations of the assumption of equal covariance matrices. As a rule, with unequal group sizes, check the assumption of homogeneity of covariance matrices using Box's test; if this test is non-significant, *and if the assumption of multivariate normality is tenable* (which allows us to assume that Box's test is accurate), then assume that Pillai's trace is accurate.

16.5.3. Follow-up analysis ③

There is some controversy over how best to follow up the main MANOVA. The traditional approach is to follow a significant MANOVA with separate ANOVAs on each of the dependent variables. If this approach is taken, you might well wonder why we bother with the MANOVA in the first place (earlier on I said that multiple ANOVAs were a bad thing to do). Well, the ANOVAs that follow a significant MANOVA are said to be 'protected' by the initial MANOVA (Bock, 1975). The idea is that the overall multivariate test protects against inflated Type I error rates because if that initial test is non-significant (i.e. the null hypothesis is true) then any subsequent tests are ignored (any significance must be a Type I error because the null hypothesis is true). However, the notion of protection is somewhat fallacious because a significant MANOVA, more often than not, reflects a significant difference for one, but not all, of the dependent variables. Subsequent ANOVAs are then carried out on all of the dependent variables, but the MANOVA protects only the dependent variable for which group differences genuinely exist (see Bray and Maxwell, 1985: 40–41). Therefore, you might want to consider applying a Bonferroni correction to the subsequent ANOVAs (Harris, 1975).

By following up a MANOVA with ANOVAs you assume that the significant MANOVA is not due to the dependent variables representing a set of underlying dimensions that differentiate the groups. Therefore, some researchers advocate the use of discriminant analysis, which finds the linear combination(s) of the dependent variables that best *separates* (or discriminates) the groups. This procedure is more in keeping with the ethos of MANOVA because it embraces the relationships that exist between dependent variables and it is certainly useful for illuminating the relationship between the dependent variables and group membership. The major advantage of this approach over multiple ANOVAs is that it reduces and explains the dependent variables in terms of a set of underlying dimensions thought to reflect substantive theoretical dimensions. By default the standard GLM procedure in SPSS provides univariate ANOVAs, but not the discriminant analysis. However, the discriminant analysis can be accessed via different menus.

16.6. MANOVA on SPSS ②

In the remainder of this chapter we will use the OCD data to illustrate how MANOVA is done (those of you who skipped the theory section should refer to Table 16.1).

16.6.1. The main analysis ②

Either load the data in the file **OCD.sav**, or enter the data manually. If you enter the data manually you need three columns: one column must be a coding variable for the **Group** variable (I used the codes CBT = 1, BT = 2, NT = 3), and in the remaining two columns enter the scores for each dependent variable respectively. Once the data have been entered, access the main MANOVA dialog box (see Figure 16.3) by selecting Analyze General Linear Model ▶ Multivariate....

The ANOVAs (and various multiple comparisons) carried out after the main MANOVA are identical to running separate ANOVA procedures in SPSS for each of the dependent variables. Hence, the main dialog box and options for MANOVA are very similar to the factorial ANOVA procedure we met in Chapter 12. The main difference to the main dialog box is that the space labelled *Dependent Variables* has room for several variables. Select the two dependent variables from the variables list (i.e. **Actions** and **Thoughts**) and drag them to the *Dependent Variables* box (or click on ➡). Select **group** from the variables list and drag it (or click on ➡) to the *Fixed Factor(s)* box. There is also a box in which you can place covariates. For this analysis there are no covariates; however, you can apply the principles of ANCOVA to the multivariate case and conduct multivariate analysis of covariance (MANCOVA). Once you have specified the variables in the analysis, you can select any of the other dialog boxes by clicking the buttons on the right-hand side:

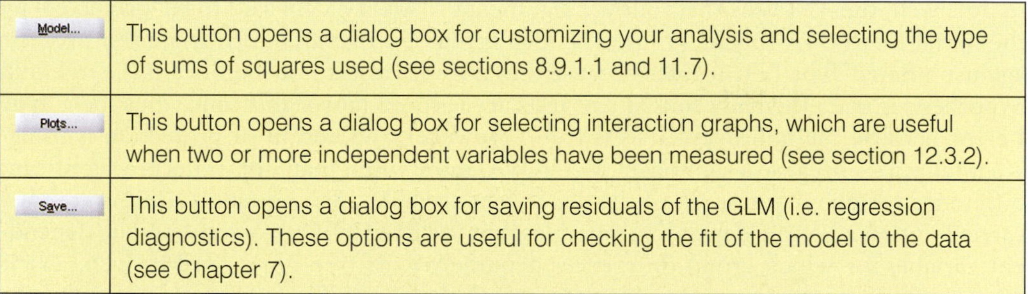

Model...	This button opens a dialog box for customizing your analysis and selecting the type of sums of squares used (see sections 8.9.1.1 and 11.7).
Plots...	This button opens a dialog box for selecting interaction graphs, which are useful when two or more independent variables have been measured (see section 12.3.2).
Save...	This button opens a dialog box for saving residuals of the GLM (i.e. regression diagnostics). These options are useful for checking the fit of the model to the data (see Chapter 7).

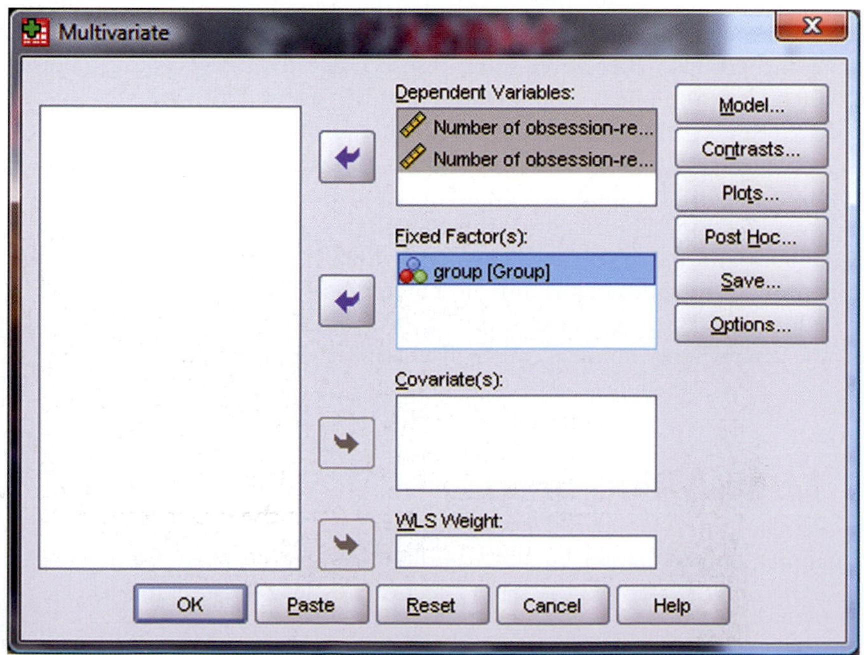

FIGURE 16.3 Main dialog box for MANOVA

16.6.2. Multiple comparisons in MANOVA ②

The default way to follow up a MANOVA is to look at individual univariate ANOVAs for each dependent variable. For these tests, SPSS has the same options as in the univariate ANOVA procedure (see Chapter 10). The Contrasts... button opens a dialog box for specifying one of several standard contrasts for the independent variable(s) in the analysis. Table 10.6 describes what each of these tests compares, but for this example it makes sense to use a *simple* contrast that compares each of the experimental groups to the no-treatment control group. The no-treatment control group was coded as the last category (it had the highest code in the data editor), so we need to select the group variable and change the contrast to a simple contrast using the last category as the reference category (see Figure 16.4). For more details about contrasts see section 10.2.11.

Instead of running a contrast, we could carry out *post hoc* tests on the independent variable to compare each group to all other groups. To access the *post hoc* tests dialog box click on Post Hoc... . The dialog box is the same as that for factorial ANOVA (see Figure 12.6) and the choice of test should be based on the same criteria as outlined in section 10.2.12. For the purposes of this example, I suggest selecting two of my usual recommendations: REGWQ and Games–Howell. Once you have selected *post hoc* tests return to the main dialog box.

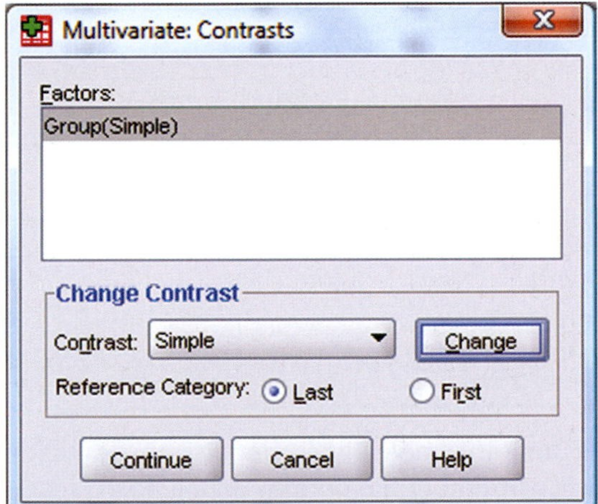

FIGURE 16.4 Contrasts for independent variable(s) in MANOVA

16.6.3. Additional options ③

To access the options dialog box, click on Options... in the main dialog box (see Figure 16.5). The resulting dialog box is fairly similar to that of factorial ANOVA (see section 12.3.5); however, there are a few additional options that are worth mentioning:

- *SSCP Matrices*: If this option is selected, SPSS will produce the model SSCP matrix, the error SSCP matrix and the total SSCP matrix. This option can be useful for understanding the computation of the MANOVA. However, if you didn't read the theory section you might be happy not to select this option and not worry about these matrices!
- *Residual SSCP Matrix*: If this option is selected, SPSS produces the error SSCP matrix, the error variance–covariance matrix and the error correlation matrix. The error variance–covariance matrix is the matrix upon which **Bartlett's test of sphericity** is based. Bartlett's test examines whether this matrix is proportional to an identity matrix (i.e. that the covariances are zero and the variances – the values along the diagonal – are roughly equal).

The remaining options are the same as for factorial ANOVA and so have been described in Chapter 12. I recommend rereading that chapter before deciding which options are useful.

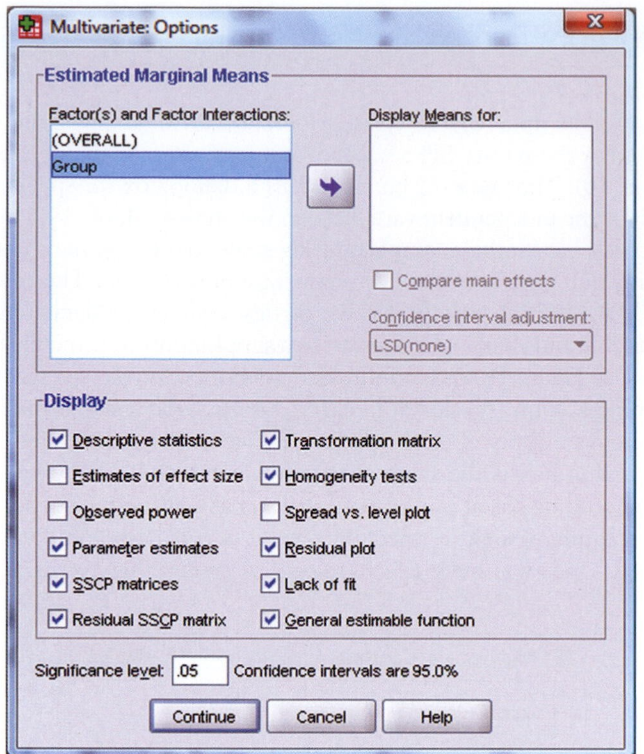

FIGURE 16.5 Additional options in MANOVA

16.7. Output from MANOVA ③

16.7.1. Preliminary analysis and testing assumptions ③

SPSS Output 16.1 shows an initial table of descriptive statistics that is produced by clicking on the descriptive statistics option in the options dialog box (Figure 16.5). This table contains the overall and group means and standard deviations for each dependent variable in turn. These values correspond to those calculated by hand in Table 16.1 and by looking at that table it should be clear what this part of the output tells us. It is clear from the means that participants had many more obsession-related thoughts than behaviours.

SPSS Output 16.2 shows Box's test of the assumption of equality of covariance matrices (see section 16.5.1). This statistic tests the null hypothesis that the variance–covariance matrices are the same in all three groups. Therefore, if the matrices are equal (and therefore the assumption of homogeneity is met) this statistic should be *non-significant*. For these data $p = .18$ (which is greater than .05): hence, the covariance matrices are roughly equal and the assumption is tenable.

If the value of Box's test was significant ($p < .05$) then the covariance matrices are significantly different and so the homogeneity assumption would have been violated. Bartlett's test of sphericity tests whether the assumption of sphericity has been met and is useful only in univariate repeated-measures designs because MANOVA does not require this assumption.

16.7.2. MANOVA test statistics ③

SPSS Output 16.3 shows the main table of results. Test statistics are quoted for the intercept of the model (even MANOVA can be characterized as a regression model, although how this is

SPSS OUTPUT 16.1

Descriptive Statistics

	group	Mean	Std. Deviation	N
Number of obsession-related behaviours	CBT	4.90	1.197	10
	BT	3.70	1.767	10
	No Treatment Control	5.00	1.054	10
	Total	4.53	1.456	30
Number of obsession-related thoughts	CBT	13.40	1.897	10
	BT	15.20	2.098	10
	No Treatment Control	15.00	2.357	10
	Total	14.53	2.209	30

SPSS OUTPUT 16.2

Box's Test of Equality of Covariance Matrices[a]

Box's M	9.959
F	1.482
df1	6.000
df2	18168.923
Sig.	.180

Tests the null hypothesis that the observed covariance matrices of the dependent variables are equal across groups.

a. Design: Intercept + Group

Bartlett's Test of Sphericity[a]

Likelihood Ratio	.042
Approx. Chi-Square	5.511
df	2.000
Sig.	.064

Tests the null hypothesis that the residual covariance matrix is proportional to an identity matrix.

a. Design: Intercept + Group

done is beyond the scope of my brain) and for the group variable. For our purposes, the group effects are of interest because they tell us whether or not the therapies had an effect on the OCD clients. You'll see that SPSS lists the four multivariate test statistics and their values correspond to those calculated in sections 16.4.4.2 to 16.4.4.5. In the next column these values are transformed into an *F*-ratio with 2 degrees of freedom. The column of real interest, however, is the one containing the significance values of these *F*-ratios. For these data, Pillai's trace ($p = .049$), Wilks's lambda ($p = .050$) and Roy's largest root ($p = .020$) all reach the criterion for significance of .05. However, Hotelling's trace ($p = .051$) is non-significant by this criterion. This scenario is interesting, because the test statistic we choose determines whether or not we reject the null hypothesis that there are no between-group differences. However, given what we know about the robustness of Pillai's trace when sample sizes are equal, we might be well advised to trust the result of that test statistic, which indicates a significant difference. This example highlights the additional power associated with Roy's root (you should note how this statistic is considerably more significant than all others) when the test assumptions have been met and when the group differences are focused on one variate (which they are in this example, as we will see later).

From this result we should probably conclude that the type of therapy employed had a significant effect on OCD. The nature of this effect is not clear from the multivariate test statistic. First, it tells us nothing about which groups differed from which, and second it tells us nothing about whether the effect of therapy was on the obsession-related thoughts, the obsession-related behaviours, or a combination of both. To determine the nature of the effect, SPSS provides us with univariate tests.

16.7.3. Univariate test statistics ②

SPSS Output 16.4 initially shows a summary table of Levene's test of equality of variances for each of the dependent variables. These tests are the same as would be found if a one-way ANOVA had been conducted on each dependent variable in turn (see section 12.4.2). Levene's

SPSS OUTPUT 16.3

Multivariate Tests[c]

Effect		Value	F	Hypothesis df	Error df	Sig.
Intercept	Pillai's Trace	.983	745.230[a]	2.000	26.000	.000
	Wilks' Lambda	.017	745.230[a]	2.000	26.000	.000
	Hotelling's Trace	57.325	745.230[a]	2.000	26.000	.000
	Roy's Largest Root	57.325	745.230[a]	2.000	26.000	.000
Group	Pillai's Trace	.318	2.557	4.000	54.000	.049
	Wilks' Lambda	.699	2.555[a]	4.000	52.000	.050
	Hotelling's Trace	.407	2.546	4.000	50.000	.051
	Roy's Largest Root	.335	4.520[b]	2.000	27.000	.020

a. Exact statistic

b. The statistic is an upper bound on F that yields a lower bound on the significance level.

c. Design: Intercept + Group

test should be non-significant for all dependent variables if the assumption of homogeneity of variance has been met. The results for these data clearly show that the assumption has been met. This finding not only gives us confidence in the reliability of the univariate tests to follow, but also strengthens the case for assuming that the multivariate test statistics are robust.

The next part of the output contains the ANOVA summary table for the dependent variables. The row of interest is that labelled *Group* (you'll notice that the values in this row are the same as for the row labelled *Corrected Model*: this is because the model fitted to the data contains only one independent variable: **Group**). The row labelled *Group* contains an ANOVA summary table for each of the dependent variables, and values are given for the sums of squares for both actions and thoughts (these values correspond to the values of SS_M calculated in sections 16.4.3.1 and 16.4.3.2 respectively). The row labelled *Error* contains information about the residual sums of squares and mean squares for each of the dependent variables: these values of SS_R were calculated in sections 16.4.3.1 and 16.4.3.2 and I urge you to look back to these sections to consolidate what these values mean. The row labelled *Corrected Total* contains the values of the total sums of squares for each dependent variable (again, these values of SS_T were calculated in sections 16.4.3.1 and 16.4.3.2). The important parts of this table are the columns labelled *F* and *Sig.* in which the *F*-ratios for each univariate ANOVA and their significance values are listed. What should be clear from SPSS Output 16.4 and the calculations made in sections 16.4.3.1 and 16.4.3.2 is that the values associated with the univariate ANOVAs conducted after the MANOVA are *identical* to those obtained if one-way ANOVA was conducted on each dependent variable. This fact illustrates that MANOVA offers only hypothetical protection of inflated Type I error rates: there is no real-life adjustment made to the values obtained.

The values of p in SPSS Output 16.4 indicate that there was a non-significant difference between therapy groups in terms of both obsession-related thoughts (p = .136) and obsession-related behaviours (p = .080). These two results should lead us to conclude that the type of therapy has had no significant effect on the levels of OCD experienced by clients. Those of you that are still awake may have noticed something odd about this example: the multivariate test statistics led us to conclude that therapy had had a significant impact on OCD, yet the univariate results indicate that therapy has not been successful.

SELF-TEST Why might the univariate tests be non-significant when the multivariate tests were significant?

The reason for the anomaly in these data is simple: the multivariate test takes account of the correlation between dependent variables and so for these data it has more power

SPSS OUTPUT 16.4

Levene's Test of Equality of Error Variances[a]

	F	df1	df2	Sig.
Number of obsession-related behaviours	1.828	2	27	.180
Number of obsession-related thoughts	.076	2	27	.927

Tests the null hypothesis that the error variance of the dependent variable is equal across groups.

a. Design: Intercept + Group

Tests of Between-Subjects Effects

Source	Dependent Variable	Type III Sum of Squares	df	Mean Square	F	Sig.
Corrected Model	Number of obsession-related behaviours	10.467[a]	2	5.233	2.771	.080
	Number of obsession-related thoughts	19.467[b]	2	9.733	2.154	.136
Intercept	Number of obsession-related behaviours	616.533	1	616.533	326.400	.000
	Number of obsession-related thoughts	6336.533	1	6336.533	1402.348	.000
Group	Number of obsession-related behaviours	10.467	2	5.233	2.771	.080
	Number of obsession-related thoughts	19.467	2	9.733	2.154	.136
Error	Number of obsession-related behaviours	51.000	27	1.889		
	Number of obsession-related thoughts	122.000	27	4.519		
Total	Number of obsession-related behaviours	678.000	30			
	Number of obsession-related thoughts	6478.000	30			
Corrected Total	Number of obsession-related behaviours	61.467	29			
	Number of obsession-related thoughts	141.467	29			

a. R Squared = .170 (Adjusted R Squared = .109)

b. R Squared = .138 (Adjusted R Squared = .074)

to detect group differences. With this knowledge in mind, the univariate tests are not particularly useful for interpretation, because the groups differ along a combination of the dependent variables. To see how the dependent variables interact we need to carry out a discriminant function analysis, which will be described in due course.

16.7.4. SSCP matrices ③

If you selected the two options to display SSCP matrices (section 16.6.3), then SPSS will produce the tables in SPSS Output 16.5 and SPSS Output 16.6. The first table (SPSS Output 16.5) displays the model SSCP (H), which is labelled *Hypothesis Group* (I have shaded this matrix blue) and the error SSCP (E) which is labelled *Error* (I have shaded this matrix yellow). The matrix for the intercept is displayed also, but this matrix is not important for our purposes. It should be pretty clear that the values in the model and error matrices displayed in SPSS Output 16.5 correspond to the values we calculated in sections 16.4.3.6 and 16.4.3.5 respectively. These matrices are useful, therefore, for gaining insight into the pattern of the data, and especially in looking at the values of the cross-products to indicate the relationship between dependent variables. In this example, the sums of squares for the error SSCP matrix are substantially bigger than in the model (or group) SSCP matrix, whereas the absolute value of the cross-products is fairly similar. This pattern suggests that if the MANOVA is significant then it might be the relationship between dependent variables that is important rather than the individual dependent variables themselves.

SPSS OUTPUT 16.5

Between-Subjects SSCP Matrix

			Number of obsession-related behaviours	Number of obsession-related thoughts
Hypothesis	Intercept	Number of obsession-related behaviours	616.533	1976.533
		Number of obsession-related thoughts	1976.533	6336.533
	Group	Number of obsession-related behaviours	**10.467**	**-7.533**
		Number of obsession-related thoughts	**-7.533**	**19.467**
	Error	Number of obsession-related behaviours	**51.000**	**13.000**
		Number of obsession-related thoughts	**13.000**	**122.000**

Based on Type III Sum of Squares

SPSS Output 16.6 shows the residual SSCP matrix again, but this time it includes the variance–covariance matrix and the correlation matrix. These matrices are all related. If you look back to Chapter 6, you should remember that the covariance is calculated by dividing the cross-product by the number of observations (i.e. the covariance is the average cross-product). Likewise, the variance is calculated by dividing the sums of squares by the degrees of freedom (and so similarly represents the average sum of squares). Hence, the variance–covariance matrix represents the average form of the SSCP matrix. Finally, we saw in Chapter 6 that the correlation was a standardized version of the covariance (where the standard deviation is also taken into account) and so the correlation matrix represents the standardized form of the variance–covariance matrix. As with the SSCP matrix, these other matrices are useful for assessing the extent of the error in the model. The variance–covariance matrix is especially useful because Bartlett's test of sphericity is based on this matrix. Bartlett's test examines whether this matrix is proportional to an identity matrix. In section 16.4.1 we saw that an identity matrix was one in which the diagonal elements were one and the off-diagonal elements were zero. Therefore, Bartlett's test effectively tests whether the diagonal elements of the variance–covariance matrix are equal (i.e. group variances are the same), and that the off-diagonal elements are approximately zero (i.e. the dependent variables are not correlated). In this case, the variances are quite different (1.89 compared to 4.52) and the covariances slightly different from zero (0.48), so Bartlett's test has come out as nearly significant (see SPSS Output 16.2). Although this discussion is irrelevant to the multivariate tests, I hope that by expanding upon them here you can relate these ideas back to the issues of sphericity raised in Chapter 13, and see more clearly how this assumption is tested.

SPSS OUTPUT 16.6

Residual SSCP Matrix

		Number of obsession-related behaviours	Number of obsession-related thoughts
Sum-of-Squares and Cross-Products	Number of obsession-related behaviours	51.000	13.000
	Number of obsession-related thoughts	13.000	122.000
Covariance	Number of obsession-related behaviours	1.889	.481
	Number of obsession-related thoughts	.481	4.519
Correlation	Number of obsession-related behaviours	1.000	.165
	Number of obsession-related thoughts	.165	1.000

Based on Type III Sum of Squares

16.7.5. Contrasts ③

I need to begin this section by reminding you that because the univariate ANOVAs were both non-significant we should not interpret these contrasts. However, purely to give you an example to follow for when your main analysis is significant, we'll look at this part of the output anyway. In section 16.6.2 I suggested carrying out a *simple* contrast that compares each of the therapy groups to the no-treatment control group. SPSS Output 16.7 shows the results of these contrasts. The table is divided into two sections conveniently labelled *Level 1 vs. Level 3* and *Level 2 vs. Level 3* where the numbers correspond to the coding of the group variable (i.e. 1 represents the lowest code used in the data editor and 3 the highest). If you coded the group variable using the same codes as I did, then these contrasts represent CBT vs. NT and BT vs. NT respectively. Each contrast is performed on both dependent variables separately and so they are identical to the contrasts that would be obtained from a univariate ANOVA. The table provides values for the contrast estimate and the hypothesized value (which will always be zero because we are testing the null hypothesis that the difference between groups is zero). The observed estimated difference is then tested to see whether it is significantly different from zero based on the standard error (it might help to reread Chapter 2 for some theory on this kind of hypothesis testing). A 95% confidence interval is produced for the estimated difference.

SPSS OUTPUT 16.7

Contrast Results (K Matrix)

group Simple Contrast[a]			Dependent Variable: Number of obsession-related behaviours	Dependent Variable: Number of obsession-related thoughts
Level 1 vs. Level 3	Contrast Estimate		-.100	-1.600
	Hypothesized Value		0	0
	Difference (Estimate - Hypothesized)		-.100	-1.600
	Std. Error		.615	.951
	Sig.		.872	.104
	95% Confidence Interval for Difference	Lower Bound	-1.361	-3.551
		Upper Bound	1.161	.351
Level 2 vs. Level 3	Contrast Estimate		-1.300	.200
	Hypothesized Value		0	0
	Difference (Estimate - Hypothesized)		-1.300	.200
	Std. Error		.615	.951
	Sig.		.044	.835
	95% Confidence Interval for Difference	Lower Bound	-2.561	-1.751
		Upper Bound	-.039	2.151

a. Reference category = 3

The first thing that you might notice (from the values of *Sig.*) is that when we compare CBT to NT there are no significant differences in thoughts ($p = .104$) or behaviours ($p = .872$) because both values are above the .05 threshold. However, comparing BT to NT, there is no significant difference in thoughts ($p = .835$) but there is a significant difference in behaviours between the groups ($p = .044$, which is less than .05). The confidence intervals confirm these findings. We have seen before that a 95% confidence interval is an interval that contains the true value of the difference between groups 95% of the time. If these boundaries cross zero (i.e. the lower is a negative number and the upper a positive value), then this tells us that the true value of the group difference could be zero (i.e. there will be no difference between the groups). Therefore, we cannot be confident that the observed group difference is meaningful because the true group difference in the population could be zero. If, however, the confidence interval does not cross zero (i.e. both values are positive or negative), then we can be confident that the true value of the group difference is different from zero. As such, we can be confident that genuine group differences exist. For these data all confidence intervals include zero (the lower bounds are negative whereas the upper bounds are positive) except for the BT vs. NT contrast for behaviours and so only this contrast is significant. This is a little unexpected because the univariate ANOVA for behaviours was non-significant and so we would not expect there to be group differences.

CRAMMING SAM'S TIPS **MANOVA**

- MANOVA is used to test the difference between groups across several dependent variables simultaneously.
- Box's test looks at the assumption of equal covariance matrices. This test can be ignored when sample sizes are equal because when they are some MANOVA test statistics are robust to violations of this assumption. If group sizes differ this test should be inspected. If the value of *Sig.* is less than .001 then the results of the analysis should not be trusted (see section 16.7.1).
- The table labelled **Multivariate Tests** gives us the results of the MANOVA. There are four test statistics (Pillai's Trace, Wilks's Lambda, Hotelling's Trace and Roy's Largest Root). I recommend using Pillai's trace. If the value of *Sig.* for this statistic is less than .05 then the groups differ significantly with respect to the dependent variables.
- ANOVAs can be used to follow up the MANOVA (a different ANOVA for each dependent variable). The results of these are listed in the table entitled *Tests of Between-Subjects Effects*. These ANOVAs can in turn be followed up using contrasts (see Chapters 10 to 14). Personally I don't recommend this approach and suggest conducting a *discriminant function analysis*.

16.8. Reporting results from MANOVA ②

Reporting a MANOVA is much like reporting an ANOVA. As you can see in SPSS Output 16.3, the multivariate tests are converted into approximate *F*s, and people often just report these *F*s just as they would for ANOVA (i.e. they give details of the *F*-ratio and the degrees of freedom from which it was calculated). For our effect of group, we would report the hypothesis *df* and the error *df*. Therefore, we could report these analyses as:

✓ There was a significant effect of therapy on the number of obsessive thoughts and behaviours, $F(4, 54) = 2.56$, $p < .05$.

However, personally, I think the multivariate test statistic should be quoted as well. There are four different multivariate tests reported in SPSS Output 16.3; I'll report each one in turn (note that the degrees of freedom and value of *F* change), but in reality you would just report one of the four:

✓ Using Pillai's trace, there was a significant effect of therapy on the number of obsessive thoughts and behaviours, $V = 0.32$, $F(4, 54) = 2.56$, $p < .05$.

✓ Using Wilks's statistic, there was a significant effect of therapy on the number of obsessive thoughts and behaviours, $\Lambda = 0.70$, $F(4, 52) = 2.56$, $p = .05$.

✓ Using Hotelling's trace statistic, there was not a significant effect of therapy on the number of obsessive thoughts and behaviours, $T = 0.41$, $F(4, 50) = 2.55$, $p > .05$.

✓ Using Roy's largest root, there was a significant effect of therapy on the number of obsessive thoughts and behaviours, $\Theta = 0.35$, $F(2, 27) = 4.52$, $p < .05$.

We can also report the follow-up ANOVAs in the usual way (see SPSS Output 16.4):

✓ Using Pillai's trace, there was a significant effect of therapy on the number of obsessive thoughts and behaviours, $V = 0.32$, $F(4, 54) = 2.56$, $p < .05$. However, separate univariate ANOVAs on the outcome variables revealed non-significant treatment effects on obsessive thoughts, $F(2, 27) = 2.15$, $p > .05$, and behaviours, $F(2, 27) = 2.77$, $p > .05$.

LABCOAT LENI'S REAL RESEARCH 16.1

A lot of hot air! ④

Have you ever wondered what researchers do in their spare time? Well, some of them spend it tracking down the sounds of people burping and farting! It has long been established that anxiety and disgust are linked. Anxious people are, typically, easily disgusted. Throughout this book I have talked about how you cannot infer causality from relationships between variables. This has been a bit of a conundrum for anxiety researchers: does anxiety cause feelings of digust or does a low threshold for being disgusted cause anxiety? Two colleagues of mine at Sussex addressed this in an unusual study in which they induced feelings of anxiety, feelings of disgust, or a neutral mood and they looked at the effect that these induced moods had on feelings of anxiety, sadness, happiness, anger, disgust and contempt. To induce these moods, they used three different types of manipulation: vignettes (e.g. 'you're swimming in a dark lake and something brushes your leg' for anxiety, and 'you go into a public toilet and find it has not been flushed. The bowl of the toilet is full of diarrhoea' for disgust), music (e.g. some scary music for anxiety, and a tape of burps, farts and vomitting for disgust), videos (e.g. a clip from *Silence of the lambs* for anxiety and a scene from *Pink flamingos* in which Divine eats dog faeces) and memory (remembering events from the past that had made the person anxious, disgusted or neutral).

Different people underwent anxious, disgust and neutral mood inductions. Within these groups, the induction was done using either vignettes and music, videos, or memory recall and music for different people. The outcome variables were the change (from before to after the induction) in six moods: anxiety, sadness, happiness, anger, disgust and contempt.

The data are in the file **Marzillier and Davey (2005).sav**. Draw an error bar graph of the changes in moods in the different conditions, then conduct a 3 (Mood: anxiety, disgust, neutral) × 3 (Induction: vignettes + music, videos, memory recall + music) MANOVA on these data.

Whatever you do, don't imagine what their fart tape sounded like while you do the analysis!

Answers are in the additional material on the companion website (or look at page 738 of the original article).

MARZILLIER, S. L., & DAVEY, G. C. L. (2005). *COGNITION AND EMOTION*, 19, 729–750.

16.9. Following up MANOVA with discriminant analysis ③

I mentioned earlier on that a significant MANOVA could be followed up using either univariate ANOVA or **discriminant analysis** (sometimes called discriminant function analysis). In the example in this chapter, the univariate ANOVAs were not a useful way of looking at what the multivariate tests showed because the relationship between dependent variables is obviously having an effect. However, these data were designed especially to illustrate how the univariate ANOVAs should be treated cautiously and in real life a significant MANOVA is likely to be accompanied by at least one significant ANOVA. However, this does not mean that the relationship between dependent variables is not important, and it is still vital to investigate the nature of this relationship. Discriminant analysis is the best way to achieve this, and I strongly recommend that you follow up a MANOVA with both univariate tests and discriminant analysis if you want to fully understand your data.

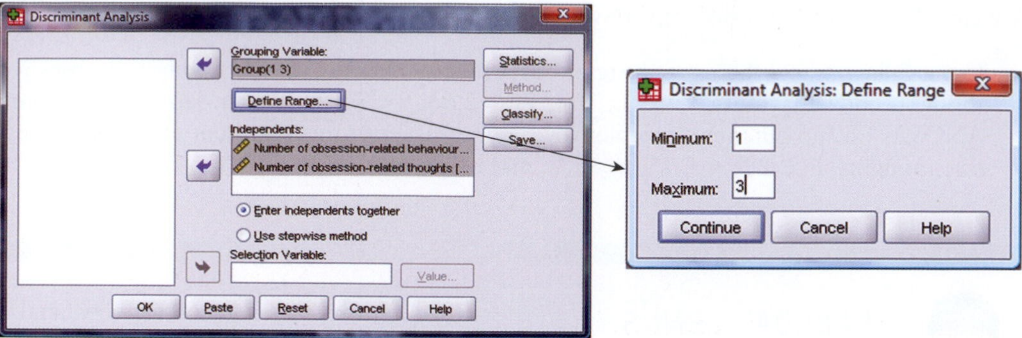

FIGURE 16.6 Main dialog box for discriminant analysis

Discriminant analysis is quite straightforward in SPSS: to access the main dialog box (see Figure 16.6) select Analyze Classify ▸ Discriminant.... The main dialog box will list the variables in the data editor on the left-hand side and provides two spaces on the right: one for the group variable and one for the predictors. In discriminant analysis we look to see how we can best separate (or discriminate) a set of groups using several predictors (so it is a little like logistic regression but where there are several groups rather than two).[6] It might be confusing to think of actions and thoughts as independent variables (after all, they were dependent variables in the MANOVA!) which is why I refer to them as predictors – this is another example of why it is useful not to refer to variables as independent variables and dependent variables in correlational analysis.

To run the analysis, select the variable **Group** and drag it to the box labelled *Grouping Variable* (or click on ➡). Once this variable has been transferred, the Define Range... button will become active and you should click this button to activate a dialog box in which you can specify the value of the highest and lowest coding values (1 and 3 in this case). Once you have specified the codings used for the grouping variable, you should select the variables **Actions** and **Thoughts** (click on them both while holding down the *Ctrl* key) and drag them to the box labelled *Independents* (or click on ➡). There are two options available to determine how the predictors are entered into the model. The default is that both predictors are entered together (Enter independents together) and this is the option we require (because in MANOVA the dependent variables are analysed simultaneously). It is possible to enter the dependent variables in a stepwise manner (Use stepwise method) and if this option is selected the Method... button becomes active, which opens a dialog box for specifying the criteria upon which predictors are entered. For the purpose of following up MANOVA, we need only be concerned with the remaining options.

Click on Statistics... to activate the dialog box in Figure 16.7. This dialog box allows us to request group means, univariate ANOVAs and Box's test of equality of covariance matrices, all of which have already been provided in the MANOVA output (so we need not ask for them again). Furthermore, we can ask for the within-group correlation and covariance matrices, which are the same as the residual correlation and covariance matrices seen in SPSS Output 16.6. There is also an option to display a separate-groups covariance matrix, which can be useful for gaining insight into the relationships between dependent variables for each group (this matrix is something that the MANOVA procedure doesn't display and I recommend selecting it). Finally, we can ask for a total covariance matrix, which displays covariances and variances of the dependent variables overall. Another useful option is to select *Unstandardized* function coefficients. This option will produce the unstandardized *b*s for each variate (see equation (16.5)). When you have finished with this dialog box, click on Continue to return to the main dialog box.

[6] In fact, I could have just as easily described discriminant analysis rather than logistic regression in Chapter 8. Because they are different ways of achieving the same end result. However, logistic regression has far fewer restrictive assumptions and is generally more robust, which is why I have limited the coverage of discriminant analysis to this chapter.

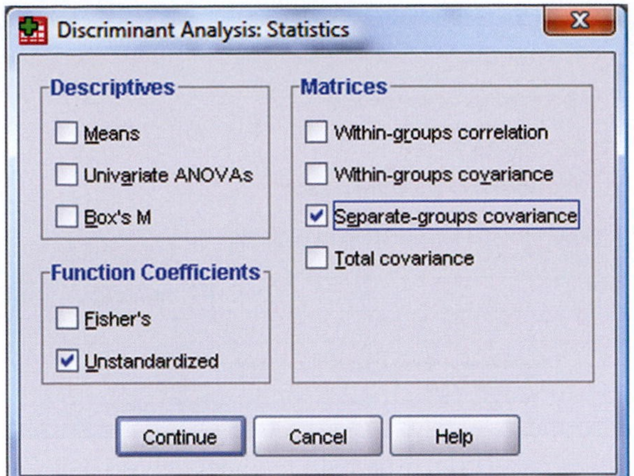

FIGURE 16.7 Statistics options for discriminant analysis

Discriminant Analysis: Classification
Prior Probabilities
All groups equal
Compute from group sizes
Use Covariance Matrix
Within-groups
Separate-groups
Display
Casewise results
Limit cases to first:
Summary table
Leave-one-out classification
Plots
Combined-groups
Separate-groups
Territorial map
Replace missing values with mean
Continue
Cancel
Help

FIGURE 16.8 Discriminant analysis classification options

If you click on Classify... you will access the dialog box in Figure 16.8. In this dialog box there are several options available. First, you can select how prior probabilities are determined: if your group sizes are equal then you should leave the default setting as it is; however, if you have an unbalanced design then it is beneficial to base prior probabilities on the observed group sizes. The default option for basing the analysis on the within-group covariance matrix is fine (because this is the matrix upon which the MANOVA is based). You should also request a combined-groups plot, which will plot the variate scores for each participant grouped according to the therapy they were given. The separate-groups plots show the same thing but using different graphs for each of the groups; when the number of groups is small it is better to select a combined plot because it is easier to interpret. The remaining options are of little interest when using discriminant analysis to follow up MANOVA. The only option that is useful is the summary table, which provides an overall gauge of how well the discriminant variates classify the actual participants. When you have finished with the options click on Continue to return to the main dialog box.

The final options are accessed by clicking on Save... to access the dialog box in Figure 16.9. There are three options available, two of which relate to the predicted group memberships and probabilities of group memberships from the model. These values are comparable to those obtained from a logistic regression analysis (see Chapter 8). The final option is to provide the **discriminant scores**. These are the scores for each person, on each variate, obtained from equation (16.5). These scores can be useful because the variates that the analysis identifies may represent underlying social or psychological constructs. If these constructs are identifiable, then it is useful for interpretation to know what a participant scores on each dimension.

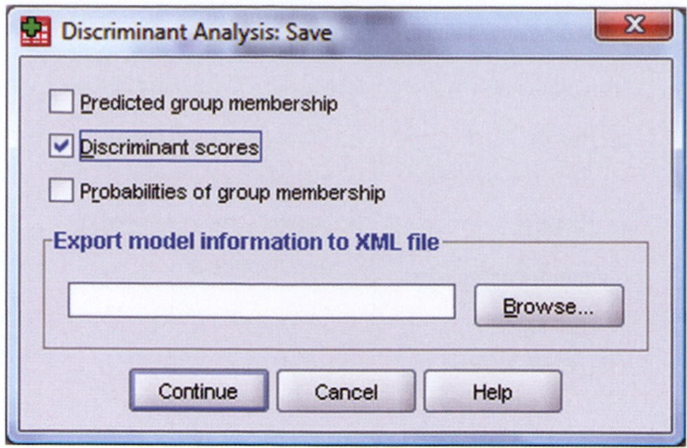

FIGURE 16.9 The save new variables dialog box in discriminant analysis

16.10. Output from the discriminant analysis ④

SPSS Output 16.8 shows the covariance matrices for separate groups (selected in Figure 16.7). These matrices are made up of the variances of each dependent variable for each group (in fact these values are shown in Table 16.1). The covariances are obtained by taking the cross-products between the dependent variables for each group (shown in Table 16.4 as .40, 22.6 and −10) and dividing each by 9, the degrees of freedom, $N-1$ (where N is the number of observations). The values in this table are useful because they give us some idea of how the relationship between dependent variables changes from group to group. For example, in the CBT group behaviours and thoughts have virtually no relationship because the covariance is almost zero. In the BT group thoughts and actions are positively related, so as the number of behaviours decreases, so does the number of thoughts. In the NT condition there is a negative relationship, so if the number of thoughts increases then the number of behaviours decreases. It is important to note that these matrices don't tell us about the substantive importance of the relationships because they are unstandardized (see Chapter 6), they merely give a basic indication.

SPSS OUTPUT 16.8

Covariance Matrices

group		Number of obsession-related behaviours	Number of obsession-related thoughts
CBT	Number of obsession-related behaviours	1.433	.044
	Number of obsession-related thoughts	.044	3.600
BT	Number of obsession-related behaviours	3.122	2.511
	Number of obsession-related thoughts	2.511	4.400
No Treatment Control	Number of obsession-related behaviours	1.111	-1.111
	Number of obsession-related thoughts	-1.111	5.556

SPSS Output 16.9 shows the initial statistics from the discriminant analysis. At first we are told the eigenvalues for each variate and you should note that the values correspond to the values of the diagonal elements of the matrix HE^{-1} (for the calculation see Oliver Twisted). These eigenvalues are converted into percentage of variance accounted for, and the first variate accounts for 82.2% of variance compared to the second variate, which accounts for only 17.8%. This table also shows the canonical correlation, which we can square to use as an effect size (just like R^2 which we have encountered in regression).

SPSS OUTPUT 16.9

Eigenvalues

Function	Eigenvalue	% of Variance	Cumulative %	Canonical Correlation
1	.335[a]	82.2	82.2	.501
2	.073[a]	17.8	100.0	.260

a. First 2 canonical discriminant functions were used in the analy

Wilks's Lambda

Test of Function	Wilks's Lambda	Chi-square	df	Sig.
1 through 2	.699	9.508	4	.050
2	.932	1.856	1	.173

The next part of the output shows the significance tests of the variates. These show the significance of both variates ('1 through 2' in the table), and the significance after the first variate has been removed ('2' in the table). So, effectively we test the model as a whole, and then peel away variates one at a time to see whether what's left is significant. In this case with only two variates we get only two steps: the whole model, and then the model after the first variate is removed (which leaves only the second variate). When both variates are tested in combination Wilks's lambda has the same value (0.699), degrees of freedom (4) and significance value (.05) as in the MANOVA (see SPSS Output 16.3). The important point to note from this table is that the two variates significantly discriminate the groups in combination ($p = .05$), but the second variate alone is non-significant, $p = .173$. Therefore, the group differences shown by the MANOVA can be explained in terms of *two* underlying dimensions in combination.

SPSS OUTPUT 16.10

Standardized Canonical Discriminant Function Coefficients

	Function	
	1	2
Number of obsession-related behaviours	.829	.584
Number of obsession-related thoughts	-.713	.721

Structure Matrix

	Function	
	1	2
Number of obsession-related behaviours	.711*	.703
Number of obsession-related thoughts	-.576	.817*

Pooled within-groups correlations between discriminating variables and standardized canonical discriminant functions
Variables ordered by absolute size of correlation within function.

*. Largest absolute correlation between each variable and any discriminant function

The tables in SPSS Output 16.10 are the most important for interpretation. The first table shows the standardized discriminant function coefficients for the two variates. These values are standardized versions of the values in the eigenvectors calculated in section 16.4.4.1. Recall that if the variates can be expressed in terms of a linear regression equation (see equation (16.4)), the standardized discriminant function coefficients are equivalent to the standardized betas in regression. The structure matrix below shows the same information but in a slightly different form. The values in this matrix are the canonical variate correlation coefficients. These values are comparable to factor loadings and indicate the substantive nature of the variates (see Chapter 17). Bargman (1970) argues that when some dependent variables have high canonical variate correlations while others have low ones, then the ones with high correlations contribute most to group separation. As such they represent the relative contribution of each dependent variable to group separation (see Bray and Maxwell, 1985: 42–45). Hence, the coefficients in these tables tell us the relative contribution of each variable to the variates.

If we look at variate 1 first, thoughts and behaviours have the opposite effect (behaviour has a positive relationship with this variate whereas thoughts have a negative relationship). Given that these values (in both tables) can vary between 1 and −1, we can also see that both relationships are strong (although behaviours have slightly larger contribution to the first variate). The first variate, then, could be seen as one that differentiates thoughts and behaviours (it affects thoughts and behaviours in the opposite way). Both thoughts and behaviours have a strong positive relationship with the second variate. This tells us that this variate represents something that affects thoughts and behaviours in a similar way. Remembering that ultimately these variates are used to differentiate groups, we could say that the first variate differentiates groups by some factor that affects thoughts and behaviours differently, whereas the second variate differentiates groups on some dimension that affects thoughts and behaviours in the same way.

SPSS OUTPUT 16.11

Canonical Discriminant Function Coefficients

	Function	
	1	2
Number of obsession-related behaviours	.603	.425
Number of obsession-related thoughts	-.335	.339
(Constant)	2.139	-6.857

Unstandardized coefficients

Functions at Group Centroids

	Function	
group	1	2
CBT	.601	-.229
BT	-.726	-.128
No Treatment Control	.125	.357

Unstandardized canonical discriminant functions evaluated at group means

SPSS Output 16.11 tells us first the canonical discriminant function coefficients, which are the unstandardized versions of the standardized coefficients described above. These values are the values of b in equation (16.4) and you'll notice that these values correspond to the values in the eigenvectors derived in section 16.4.4.1 and used in equation (16.5). The values are less useful than the standardized versions, but do demonstrate from where the standardized versions come.

The centroids are simply the mean variate scores for each group. For interpretation we should look at the sign of the centroid (positive or negative). We can also use a combined-groups plot (selected using the dialog box in Figure 16.8). This graph plots the variate scores for each person, grouped according to the experimental condition to which that person belonged. In addition, the group centroids from SPSS Output 16.11 are shown as blue squares. The graph (Figure 16.10) and the tabulated values of the centroids (SPSS Output 16.11) tell us that (look at the big squares labelled with the group initials) variate 1 discriminates the BT group from the CBT (look at the horizontal distance between these centroids). The second variate differentiates the no-treatment group from the two interventions (look at the vertical distances), but this difference is not as dramatic as for the first variate. Remember that the variates significantly discriminate the groups in combination (i.e. when both are considered).

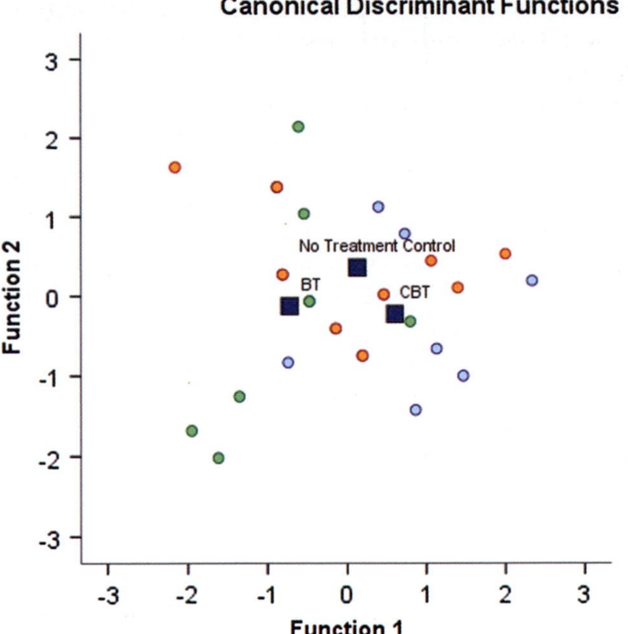

FIGURE 16.10 Combined-groups plot

CRAMMING SAM'S TIPS

- Discriminant function analysis (DFA) can be used after MANOVA to see how the dependent variables discriminate the groups.
- DFA identifies variates (combinations of the dependent variables) and to find out how many variates are significant look at the tables labelled **Wilks's Lambda**: if the value of *Sig.* is less than .05 then the variate is significantly discriminating the groups.
- Once the significant variates have been identified, use the table labelled **Standardized Canonical Discriminant Function Coefficients** to find out how the dependent variables contribute to the variates. High scores indicate that a dependent variable is important for a variate, and variables with positive and negative coefficients are contributing to the variate in opposite ways.
- Finally, to find out which groups are discriminated by a variate look at the table labelled **Functions at Group Centroids**: for a given variate, groups with values opposite in sign are being discriminated by that variate.

16.11. Reporting results from discriminant analysis ②

The guiding principal (for the APA, whose guidelines, as a psychologist, are the ones that I try to follow) in presenting data is to give the readers enough information to be able to judge for themselves what your data mean. The APA does not have specific guidelines for what needs to be reported for discriminant analysis. Personally, I would suggest reporting percentage of variance explained (which gives the reader the same information as the eigenvalue but in a more palatable form) and the squared canonical correlation for each variate (this is the appropriate effect size measure for discriminant analysis). I would also report the chi-square significance tests of the variates. All of these values can be found in SPSS Output 16.9 (although remember to square the canonical correlation). It is probably also useful to quote the values in the structure matrix in SPSS Output 16.10 (which will tell the reader about how the outcome variables relate to the underlying variates). Finally, although I won't reproduce it below, you could consider including a (well-edited) copy of the combined-groups centroid plot (Figure 16.10), which will help readers to determine how the variates contribute to distinguishing your groups. We could, therefore, write something like this:

✓ The MANOVA was followed up with discriminant analysis, which revealed two discriminant functions. The first explained 82.2% of the variance, canonical $R^2 = .25$, whereas the second explained only 17.8%, canonical $R^2 = .07$. In combination these discriminant functions significantly differentiated the treatment groups, $\Lambda = 0.70$, $\chi^2(4) = 9.51$, $p = .05$, but removing the first function indicated that the second function did not significantly differentiate the treatment groups, $\Lambda = 0.93$, $\chi^2(1) = 1.86$, $p > .05$. The correlations between outcomes and the discriminant functions revealed that obsessive behaviours loaded fairly evenly highly onto both functions ($r = .71$ for the first function and $r = .70$ for the second); obsessive thoughts loaded more highly on the second function ($r = .82$) than the first function ($r = -.58$). The discriminant function plot showed that the first function discriminated the BT group from the CBT group, and the second function differentiated the no-treatment group from the two interventions.

16.12. Some final remarks ④

16.12.1. The final interpretation ④

So far we have gathered an awful lot of information about our data, but how can we bring all of it together to answer our research question: can therapy improve OCD and if so which therapy is best? Well, the MANOVA tells us that therapy can have a significant effect on OCD symptoms, but the non-significant univariate ANOVAs suggested that this improvement is not simply in terms of either thoughts or behaviours. The discriminant analysis suggests that the group separation can be best explained in terms of one underlying dimension. In this context the dimension is likely to be OCD itself (which we can realistically presume is made up of both thoughts and behaviours). So, therapy doesn't necessarily change behaviours or thoughts per se, but it does influence the underlying dimension of OCD. So, the answer to the first question seems to be: yes, therapy can influence OCD, but the nature of this influence is unclear.

The next question is more complex: which therapy is best? Figure 16.11 shows graphs of the relationships between the dependent variables and the group means of the original data. The graph of the means shows that for actions, BT reduces the number of obsessive behaviours, whereas CBT and NT do not. For thoughts, CBT reduces the number of obsessive thoughts, whereas BT and NT do not (check the pattern of the bars). Looking now at the relationships between thoughts and actions, in the BT group there is a positive relationship between thoughts and actions, so the more obsessive thoughts a person has, the more obsessive behaviours they carry out. In the CBT group there is no relationship at all (thoughts and actions vary quite independently). In the no-treatment group there is a negative (and non-significant incidentally) relationship between thoughts and actions.

What we have discovered from the discriminant analysis is that BT and CBT can be differentiated from the control group based on variate 2, a variate that has a similar effect on both thoughts and behaviours. We could say then that BT and CBT are both better than a no-treatment group at changing obsessive thoughts and behaviours. We also discovered that BT and CBT could be distinguished by variate 1, a variate that had the opposite effects on thoughts and behaviours. Combining this information with that in Figure 16.11 we could conclude that BT is better at changing behaviours and CBT is better at changing thoughts. So, the NT group can be distinguished from the CBT and BT groups using a

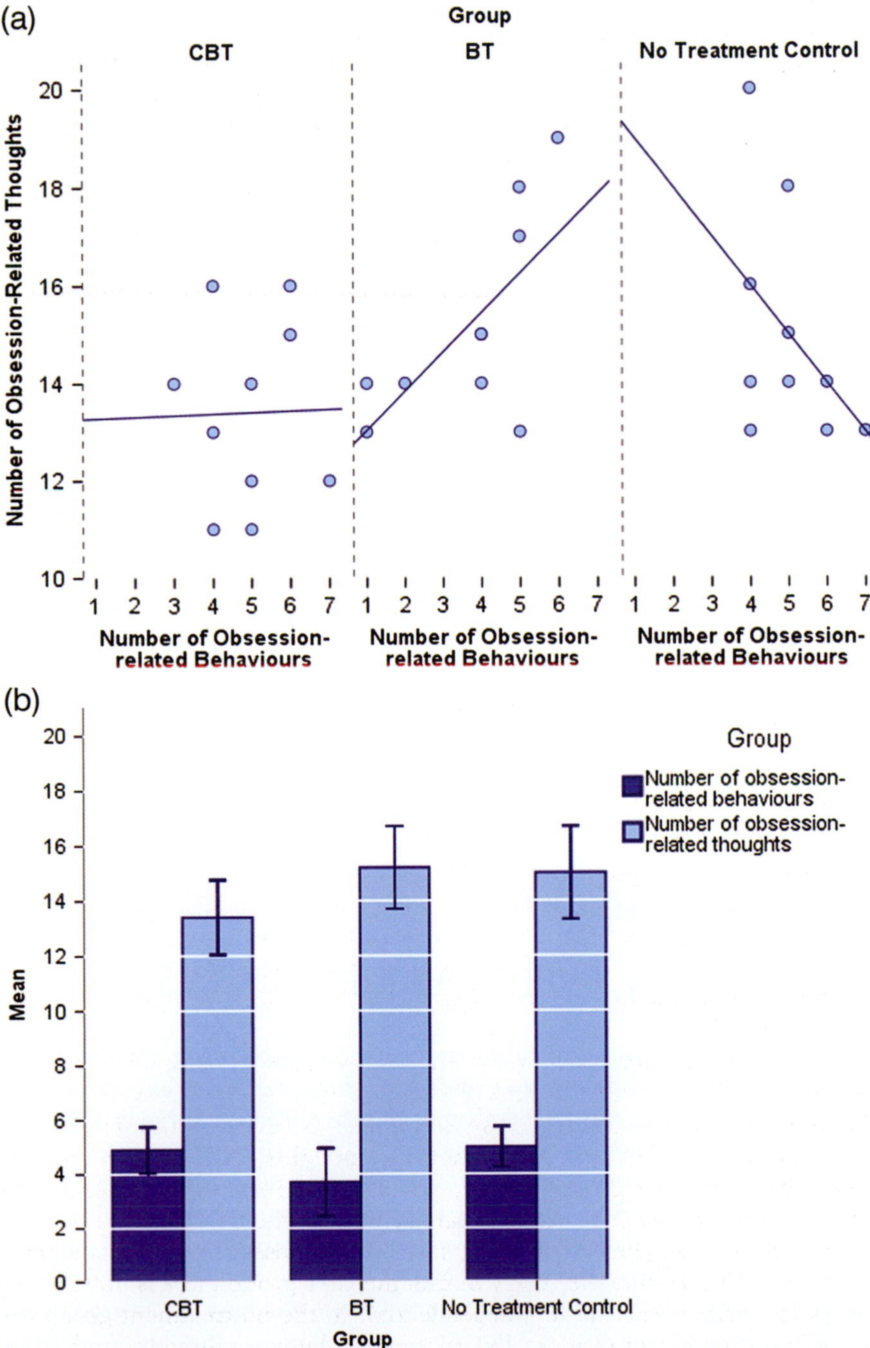

FIGURE 16.11 Graphs showing (a) the relationships and (b) the means (and 95% confidence intervals) between the dependent variables in each therapy group

variable that affects both thoughts and behaviours. Also, the CBT and BT groups can be distinguished by a variate that has opposite effects on thoughts and behaviours. So, some therapy is better than none, but the choice of CBT or BT depends on whether you think it's more important to target thoughts (CBT) or behaviours (BT).

16.12.2. Univariate ANOVA or discriminant analysis?

This example should have made clear that univariate ANOVA and discriminant analysis are ways of answering different questions arising from a significant MANOVA. If univariate ANOVAs are chosen, Bonferroni corrections should be applied to the level at which you accept significance. The truth is that you should run both analyses to get a full picture of what is happening in your data. The advantage of discriminant analysis is that it tells you something about the underlying dimensions within your data (which is especially useful if you have employed several dependent measures in an attempt to capture some social or psychological construct). Even if univariate ANOVAs are significant, the discriminant analysis provides useful insight into your data and should be used. I hope that this chapter will convince you of this recommendation!

16.13. What to do when assumptions are violated in MANOVA ③

SPSS doesn't offer a non-parametric version of MANOVA; however, some ideas have been put forward based on ranked data (much like the non-parametric tests we saw in Chapter 15). Although discussion of these tests is well beyond the scope of this book, there are some techniques that can be beneficial when multivariate normality or homogeneity of covariance matrices cannot be assumed (Zwick, 1985). In addition, there are robust methods (see section 5.7.4) described by Wilcox (Wilcox, 2005) for fairly straightforward designs with multiple outcome variables (for example, the Munzel–Brunner method). The SPSS R plugin can be used to run Wilcox's files for the software R from within SPSS. The companion website has some demonstration movies of how to use the R plugin.

What have I discovered about statistics? ②

In this chapter we've cackled maniacally in the ear of MANOVA, force-fed discriminant function analysis cod-liver oil, and discovered to our horror that Roy has a large root. There are sometimes situations in which several outcomes have been measured in different groups and we discovered that in these situations the ANOVA technique can be extended and is called MANOVA (multivariate analysis of variance). The reasons for using this technique rather than running lots of ANOVAs is that we retain control over the Type I error rate, and we can incorporate the relationships between outcome variables into the analysis. Some of you will have then discovered that MANOVA works in very similar ways to ANOVA, but just with matrices rather than single values. Others will have discovered that it's best

to ignore the theory sections of this book. We had a look at an example of MANOVA on SPSS and discovered that just to make life as confusing as possible you get four test statistics relating to the same effect! Of these, I tried to convince you that Pillai's trace was the safest option. Finally, we had a look at the two options for following up MANOVA: running lots of ANOVAs, or discriminant function analysis. Of these, discriminant function analysis gives us the most information, but can be a bit of nightmare to interpret.

We also discovered that pets can be therapeutic. I left the whereabouts of Fuzzy a mystery. Now admit it, how many of you thought he was dead? He's not: he is lying next to me as I type this sentence. After frantically searching the house I went back to the room that he had vanished from to check again whether there was a hole that he could have wriggled through. As I scuttled around on my hands and knees tapping the walls, a little ginger (and sooty) face popped out from the fireplace with a look as if to say 'have you lost something?' (see the picture). Yep, freaked out by the whole moving experience, he had done the only sensible thing and hidden up the chimney! Cats, you gotta love 'em.

Key terms that I've discovered

- Bartlett's test of sphericity
- Box's test
- Discriminant analysis
- Discriminant function variates
- Discriminant scores
- Error SSCP (E)
- HE^{-1}
- Homogeneity of covariance matrices
- Hotelling–Lawley trace (T^2)
- Hypothesis SSCP (H)
- Identity matrix
- Matrix
- Multivariate
- Multivariate analysis of variance (or MANOVA)
- Multivariate normality
- Pillai–Bartlett trace (V)
- Roy's largest root
- Square matrix
- Sum of squares and cross-products (SSCP) matrix
- Total SSCP (T)
- Univariate
- Variance–covariance matrix
- Wilks's lambda (Λ)

Smart Alex's tasks

- **Task 1**: A clinical psychologist noticed that several of his manic psychotic patients did chicken impersonations in public. He wondered whether this behaviour could be used to diagnose this disorder and so decided to compare his patients against a normal sample. He observed 10 of his patients as they went through a normal day. He also needed to observe 10 of the most normal people he could find: naturally he chose to observe lecturers at the University of Sussex. He measured all participants using two dependent variables: first, how many chicken impersonations they did in the streets of Brighton over the course of a day, and, second, how good their impersonations were (as scored out of 10 by an independent farmyard noise expert). The data are in the file **chicken.sav**. Use MANOVA and DFA to find out whether these variables could be used to distinguish manic psychotic patients from those without the disorder. ③

- **Task 2**: I was interested in whether students' knowledge of different aspects of psychology improved throughout their degree. I took a sample of first years, second years and third years and gave them five tests (scored out of 15) representing different aspects of psychology: **Exper** (experimental psychology such as cognitive and neuropsychology etc.); **Stats** (statistics); **Social** (social psychology); **Develop** (developmental psychology); **Person** (personality). Your task is to: (1) carry out an appropriate general analysis to determine whether there are overall group differences along these five measures; (2) look at the scale-by-scale analyses of group differences produced in the output and interpret the results accordingly; (3) select contrasts that test the hypothesis that second and third years will score higher than first years on all scales; (4) select tests that compare all groups to each other and briefly compare these results with the contrasts; and (5) carry out a separate analysis in which you test whether a combination of the measures can successfully discriminate the groups (comment only briefly on this analysis). Include only those scales that revealed group differences for the contrasts. How do the results help you to explain the findings of your initial analysis? The data are in the file **psychology.sav**. ④

Answers can be found on the companion website.

Further reading

Bray, J. H., & Maxwell, S. E. (1985). *Multivariate analysis of variance*. Sage university paper series on quantitative applications in the social sciences, 07-054. Newbury Park, CA: Sage. (This monograph on MANOVA is superb: I cannot recommend anything better.)

Huberty, C. J., & Morris, J. D. (1989). Multivariate analysis versus multiple univariate analysis. *Psychological Bulletin*, *105*(2), 302–308.

Online tutorials

The companion website contains the following Flash movie tutorials to accompany this chapter:

- MANOVA using SPSS
- The *R* plugin

Interesting real research

Marzillier, S. L., & Davey, G. C. L. (2005). Anxiety and disgust: Evidence for a unidirectional relationship. *Cognition and Emotion*, *19*(5), 729–750.

Exploratory factor analysis 17

FIGURE 17.1 Me at Niagara Falls in 1998. I was in the middle of writing the first edition of this book at the time. Note how fresh faced I look

17.1. What will this chapter tell me? ①

I was a year or so into my Ph.D., and thanks to my initial terrible teaching experiences I had developed a bit of an obsession with over-preparing for classes. I wrote detailed handouts and started using funny examples. Through my girlfriend at the time I met Dan Wright (a psychologist, who was in my department but sadly moved recently to Florida). He had published a statistics book of his own and was helping his publishers to sign up new authors. On the basis that my handouts were quirky and that I was too young to realize that writing a textbook at the age of 23 was academic suicide (really, textbooks take a long time to write and they are not at all valued compared to research articles) I was duly signed up. The commissioning editor was a man constantly on the verge of spontaneously combusting with intellectual energy. He can start a philosophical debate about literally anything: should he ever be trapped in an elevator he will be compelled to attempt to penetrate the occupants' minds with probing arguments that the elevator doesn't exist, that they don't exist, and that their entrapment is an illusory construct generated by their erroneous beliefs in the physical world. Ultimately though, he'd still be a man trapped in an elevator (with several exhausted corpses). A combination of his unfaltering self-confidence and my fear of social interactions with people that I don't know coupled with utter bemusement that anyone would want me to write a book made me incapable of saying anything sensible to him. Ever. He must have thought that he had signed up an imbecile. He was probably right. (I find him less intimidating since thinking up the elevator scenario.) The trouble with agreeing to write books is that you then have to write them.

For the next two years or so I found myself trying to juggle my research, a lectureship at the University of London, and writing a book. Had I been writing a book on heavy metal it would have been fine because all of the information was moshing away in my memory waiting to stage-dive out. Sadly, however, I had agreed to write a book on something that I knew nothing about: statistics. I soon discovered that writing the book was like doing a **factor analysis**: in factor analysis we take a lot of information (variables) and SPSS effortlessly reduces this mass of confusion to a simple message (fewer variables) that is easier to digest. SPSS does this (sort of) by filtering out the bits of the information overload that we don't need to know about. It takes a few seconds. Similarly, my younger self took a mass of information about statistics that I didn't understand and filtered it down into a simple message that I *could* understand: I became a living, breathing factor analysis ... except that, unlike SPSS, it took me two years and some considerable effort.

17.2. When to use factor analysis ②

In the social sciences we are often trying to measure things that cannot directly be measured (so-called **latent variables**). For example, management researchers (or psychologists even) might be interested in measuring 'burnout', which is when someone who has been working very hard on a project (a book, for example) for a prolonged period of time suddenly finds themselves devoid of motivation, inspiration, and wants to repeatedly headbutt their computer screaming 'please Mike, unlock the door, let me out of the basement, I need to feel the soft warmth of sunlight on my skin'. You can't measure burnout directly: it has many facets. However, you can measure different aspects of burnout: you could get some idea of motivation, stress levels, whether the person has any new ideas and so on. Having done this, it would be helpful to know whether these differences really do reflect a single variable. Put another way, are these different variables driven by the same underlying variable? This chapter will look at factor analysis (and principal component analysis) – a technique for identifying groups or clusters of variables. This technique has three main uses: (1) to understand the structure of a set of variables (e.g. pioneers of intelligence such as Spearman and Thurstone used factor analysis to try to understand the structure of the latent variable 'intelligence'); (2) to construct a questionnaire to measure an underlying variable (e.g. you might design a questionnaire to measure burnout); and (3) to reduce a data set to a more manageable size while retaining as much of the original information as possible (e.g. we saw in Chapter 7 that multicollinearity can be a problem in multiple regression, and factor analysis can be used to solve this problem by combining variables that are collinear). Through this chapter we'll discover what factors are, how we find them, and what they tell us (if anything) about the relationship between the variables we've measured.

17.3. Factors ②

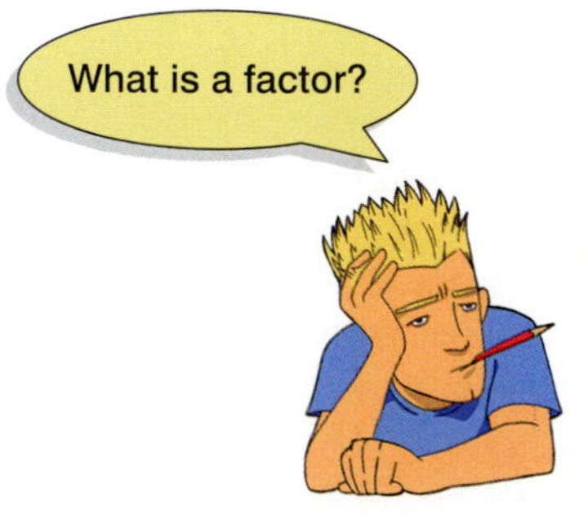

If we measure several variables, or ask someone several questions about themselves, the correlation between each pair of variables (or questions) can be arranged in what's known as an *R-matrix*. An *R*-matrix is just a correlation matrix: a table of correlation coefficients between variables (in fact, we saw small versions of these matrices in Chapter 6). The diagonal elements of an *R*-matrix are all ones because each variable will correlate perfectly with itself. The off-diagonal elements are the correlation coefficients between pairs of variables, or questions.[1] The existence of clusters of large correlation coefficients between subsets of variables suggests that those variables could be measuring aspects of the same underlying dimension. These underlying dimensions are known

[1] This matrix is called an *R*-matrix, or *R*, because it contains correlation coefficients and *r* usually denotes Pearson's correlation (see Chapter 6) – the *r* turns into a capital letter when it denotes a matrix.

as factors (or *latent variables*). By reducing a data set from a group of interrelated variables to a smaller set of factors, factor analysis achieves parsimony by explaining the maximum amount of common variance in a correlation matrix using the smallest number of explanatory constructs.

There are numerous examples of the use of factor analysis in the social sciences. The trait theorists in psychology used factor analysis endlessly to assess personality traits. Most readers will be familiar with the extraversion–introversion and neuroticism traits measured by Eysenck (1953). Most other personality questionnaires are based on factor analysis – notably Cattell's (1966a) 16 personality factors questionnaire – and these inventories are frequently used for recruiting purposes in industry (and even by some religious groups). However, although factor analysis is probably most famous for being adopted by psychologists, its use is by no means restricted to measuring dimensions of personality. Economists, for example, might use factor analysis to see whether productivity, profits and workforce can be reduced down to an underlying dimension of company growth, and Jeremy Miles told me of a biochemist who used it to analyse urine samples!

Let's put some of these ideas into practice by imagining that we wanted to measure different aspects of what might make a person popular. We could administer several measures that we believe tap different aspects of popularity. So, we might measure a person's social skills (Social Skills), their selfishness (Selfish), how interesting others find them (Interest), the proportion of time they spend talking about the other person during a conversation (Talk 1), the proportion of time they spend talking about themselves (Talk 2), and their propensity to lie to people (the Liar scale). We can then calculate the correlation coefficients for each pair of variables and create an *R*-matrix. Table 17.1 shows this matrix. Any significant correlation coefficients are shown in bold type. It is clear that there are two clusters of interrelating variables. Therefore, these variables might be measuring some common underlying dimension. The amount that someone talks about the other person during a conversation seems to correlate highly with both the level of social skills and how interesting the other finds that person. Also, social skills correlates well with how interesting others perceive a person to be. These relationships indicate that the better your social skills, the more interesting and talkative you are likely to be. However, there is a second cluster of variables. The amount that people talk about themselves within a conversation correlates with how selfish they are and how much they lie. Being selfish also correlates with the degree to which a person tells lies. In short, selfish people are likely to lie and talk about themselves.

In factor analysis we strive to reduce this *R*-matrix down to its underlying dimensions by looking at which variables seem to cluster together in a meaningful way. This data reduction is achieved by looking for variables that correlate highly with a group of other variables, but do not correlate with variables outside of that group. In this example, there appear to be two clusters that fit the bill. The first factor seems to relate to general sociability, whereas the second factor seems to relate to the way in which a person treats others socially (we might call it Consideration). It might, therefore, be assumed that popularity depends not only on your ability to socialize, but also on whether you are genuine towards others.

TABLE 17.1 An *R*-matrix

	Talk 1	*Social Skills*	*Interest*	*Talk 2*	*Selfish*	*Liar*
Talk 1	1.000					
Social Skills	**.772**	1.000				
Interest	**.646**	**.879**	1.000			
Talk 2	.074	−.120	.054	1.000		
Selfish	−.131	.031	−.101	**.441**	1.000	
Liar	.068	.012	.110	**.361**	**.277**	1.000

Factor 1

Factor 2

17.3.1. Graphical representation of factors ②

Factors (not to be confused with independent variables in factorial ANOVA) are statistical entities that can be visualized as classification axes along which measurement variables can be plotted. In plain English, this statement means that if you imagine factors as being the axis of a graph, then we can plot variables along these axes. The co-ordinates of variables along each axis represent the strength of relationship between that variable and each factor. Figure 17.2 shows such a plot for the popularity data (in which there were only two factors). The first thing to notice is that for both factors, the axis line ranges from −1 to 1, which are the outer limits of a correlation coefficient. Therefore, the position of a given variable depends on its correlation to the two factors. The circles represent the three variables that correlate highly with factor 1 (Sociability: horizontal axis) but have a low correlation with factor 2 (Consideration: vertical axis). Conversely, the triangles represent variables that correlate highly with consideration to others but have a low correlation to sociability. From this plot, we can tell that selfishness, the amount a person talks about themselves and their propensity to lie all contribute to a factor which could be called consideration of others. Conversely, how much a person takes an interest in other people, how interesting they are and their level of social skills contribute to a second factor, sociability. This diagram therefore supports the structure that was apparent in the *R*-matrix. Of course, if a third factor existed within these data it could be represented by a third axis (creating a 3-D graph). It should also be apparent that if more than three factors exist in a data set, then they cannot all be represented by a 2-D drawing.

If each axis on the graph represents a factor, then the variables that go to make up a factor can be plotted according to the extent to which they relate to a given factor. The co-ordinates of a variable, therefore, represent its relationship to the factors. In an ideal world a variable should have a large co-ordinate for one of the axes and low co-ordinates for any other factors. This scenario would indicate that this particular variable related to only one factor. Variables that have large co-ordinates on the same axis are assumed to measure different aspects of some common underlying dimension. The co-ordinate of a variable along

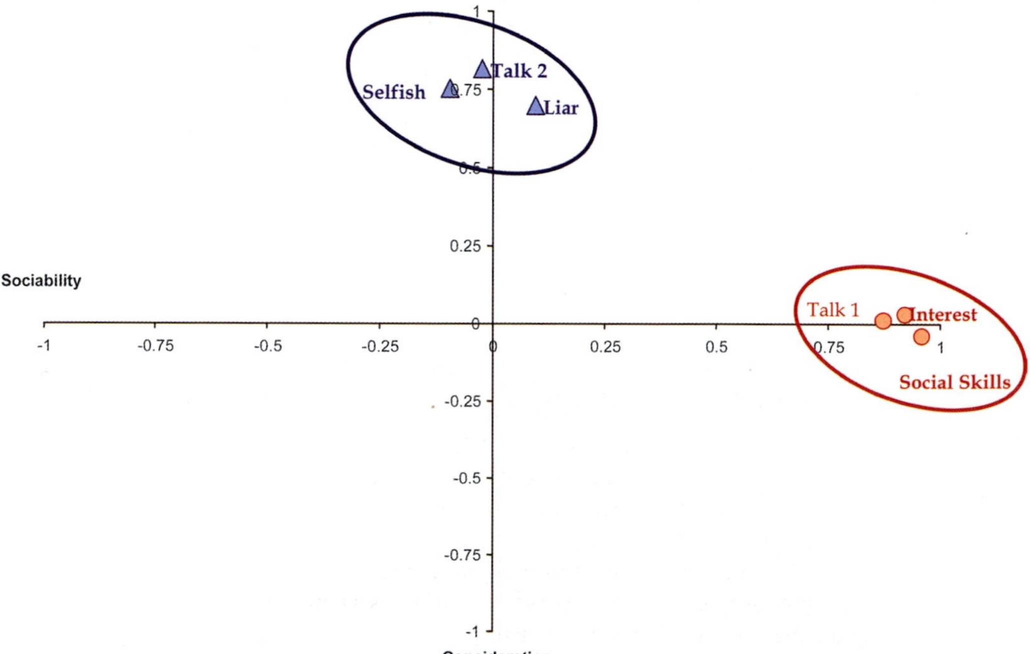

FIGURE 17.2 Example of a factor plot

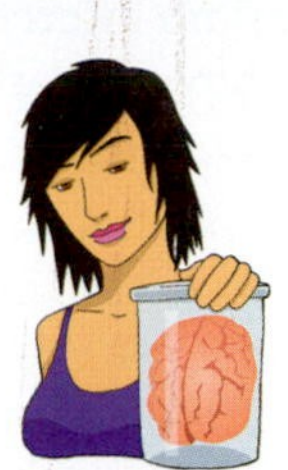

JANE SUPERBRAIN 17.1

What's the difference between a pattern matrix and a structure matrix? ③

Throughout my discussion of factor loadings I've been quite vague. Sometimes I've said that these loadings can be thought of as the correlation between a variable and a given factor, then at other times I've described these loadings in terms of regression coefficients (*b*). Now, it should be obvious from what we discovered in Chapters 6 and 7 that correlation coefficients and regression coefficients are quite different things, so what the hell am I going on about: shouldn't I make up my mind what the factor loadings actually are?

Well, in vague terms (the best terms for my brain) both correlation coefficients and regression coefficients represent the relationship between a variable and linear model in a broad sense, so the key take-home message is that factor loadings tell us about the relative contribution that a variable makes to a factor. As long as you understand that much, you have no problems.

However, the factor loadings in a given analysis can be both correlation coefficients and regression coefficients. In a few section's time we'll discover that the interpretation of factor analysis is helped greatly by a technique known as *rotation*. Without going into details, there are two types: orthogonal and oblique rotation (see section 17.4.6). When orthogonal rotation is used, any underlying factors are assumed to be independent, and the factor loading *is* the correlation between the factor and the variable, but is also the regression coefficient. Put another way, the values of the correlation coefficients are the same as the values of the regression coefficients. However, there are situations in which the underlying factors are assumed to be related or correlated to each other. In these situations, oblique rotation is used and the resulting correlations between variables and factors will differ from the corresponding regression coefficients. In this case, there are, in effect, two different sets of factor loadings: the correlation coefficients between each variable and factor (which are put in the *factor* **structure matrix**) and the regression coefficients for each variable on each factor (which are put in the *factor* **pattern matrix**). These coefficients can have quite different interpretations (see Graham, Guthrie, & Thompson, 2003).

a classification axis is known as a **factor loading**. The factor loading can be thought of as the Pearson correlation between a factor and a variable (see Jane Superbrain Box 17.1). From what we know about interpreting correlation coefficients (see section 6.5.2.3) it should be clear that if we square the factor loading we obtain a measure of the substantive importance of a particular variable to a factor.

17.3.2. Mathematical representation of factors ②

The axes drawn in Figure 17.2 are straight lines and so can be described mathematically by the equation of a straight line. Therefore, factors can also be described in terms of this equation.

SMART ALEX ONLY

SELF-TEST What is the equation of a straight line?

Equation (17.1) reminds us of the equation describing a linear model and then applies this to the scenario of describing a factor. You'll notice that there is no intercept in the equation, the reason being that the lines intersect at zero (hence the intercept is also zero). The *b*s in the equation represent the factor loadings.

$$\begin{aligned} Y_i &= b_1X_{1i} + b_2X_{2i} + \ldots + b_nX_{ni} + \varepsilon_i \\ \text{Factor}_i &= b_1\text{Variable}_{1i} + b_2\text{Variable}_{2i} + \ldots + b_n\text{Variable}_{ni} + \varepsilon_i \end{aligned} \tag{17.1}$$

Sticking with our example of popularity, we found that there were two factors underlying this construct: general sociability and consideration. We can, therefore, construct an equation that describes each factor in terms of the variables that have been measured. The equations are as follows:

$$\begin{aligned} Y_i &= b_1X_{1i} + b_2X_{2i} + \ldots + b_nX_{ni} + \varepsilon_i \\ \text{Sociability}_i &= b_1\text{Talk } 1_i + b_2 \text{ Social Skills}_i + b_3\text{Interest}_i \\ &\quad + b_4\text{Talk } 2_i + b_5\text{Selfish}_i + b_6\text{Liar}_i + \varepsilon_i \\ \text{Consideration}_i &= b_1\text{Talk } 1_i + b_2 \text{ Social Skills}_i + b_3\text{Interest}_i \\ &\quad + b_4\text{Talk } 2_i + b_5\text{Selfish}_i + b_6\text{Liar}_i + \varepsilon_i \end{aligned} \tag{17.2}$$

First, notice that the equations are identical in form: they both include all of the variables that were measures. However, the values of *b* in the two equations will be different (depending on the relative importance of each variable to the particular factor). In fact, we can replace each value of *b* with the co-ordinate of that variable on the graph in Figure 17.2 (i.e. replace the values of *b* with the factor loading). The resulting equations are as follows:

$$\begin{aligned} Y_i &= b_1X_{1i} + b_2X_{2i} + \ldots + b_nX_{ni} + \varepsilon_i \\ \text{Sociability}_i &= 0.87\text{Talk } 1_i + 0.96\text{Social Skills}_i + 0.92\text{Interest}_i \\ &\quad + 0.00\text{Talk } 2_i - 0.10\text{Selfish}_i + 0.09\text{Liar}_i + \varepsilon_i \\ \text{Consideration}_i &= 0.01\text{Talk } 1_i - 0.03\text{Social Skills}_i + 0.04\text{Interest}_i \\ &\quad + 0.82\text{Talk } 2_i + 0.75\text{Selfish}_i + 0.70\text{Liar}_i + \varepsilon_i \end{aligned} \tag{17.3}$$

Notice that, for the sociability factor, the values of *b* are high for Talk 1, Social Skills and Interest. For the remaining variables (Talk 2, Selfish and Liar) the values of *b* are very low (close to 0). This tells us that three of the variables are very important for that factor (the ones with high values of *b*) and three are very unimportant (the ones with low values of *b*). We saw that this point is true because of the way that three variables clustered highly on the factor plot. The point to take on board here is that the factor plot and these equations represent the same thing: the factor loadings in the plot are simply the *b*-values in these equations (but see Jane Superbrain Box 17.1). For the second factor, inconsideration to others, the opposite pattern can be seen in that Talk 2, Selfish and Liar all have high values of *b* whereas the remaining three variables have *b*-values close to 0. In an ideal world, variables would have very high *b*-values for one factor and very low *b*-values for all other factors.

These factor loadings can be placed in a matrix in which the columns represent each factor and the rows represent the loadings of each variable onto each factor. For the popularity data this matrix would have two columns (one for each factor) and six rows (one for each variable). This matrix, usually denoted *A*, can be seen below. To understand what the matrix means, try relating the elements to the loadings in equation (17.3). For example, the top row represents the first variable, Talk 1, which had a loading of .87 for the first factor (Sociability) and a loading of .01 for the second factor (Consideration). This matrix is called the **factor matrix** or **component matrix** (if doing **principal component analysis**) – see Jane Superbrain Box 17.1 to find out about the different forms of this matrix:

$$A = \begin{pmatrix} 0.87 & 0.01 \\ 0.96 & -0.03 \\ 0.92 & 0.04 \\ 0.00 & 0.82 \\ -0.10 & 0.75 \\ 0.09 & 0.70 \end{pmatrix}$$

The major assumption in factor analysis is that these algebraic factors represent real-world dimensions, the nature of which must be *guessed at* by inspecting which variables have high loads on the same factor. So, psychologists might believe that factors represent dimensions of the psyche, education researchers might believe they represent abilities, and sociologists might believe they represent races or social classes. However, it is an extremely contentious point whether this assumption is tenable and some believe that the dimensions derived from factor analysis are real only in the statistical sense – and are real-world fictions.

EVERYBODY

17.3.3. Factor scores ②

A factor can be described in terms of the variables measured and the relative importance of them for that factor (represented by the value of b). Therefore, having discovered which factors exist, and estimated the equation that describes them, it should be possible to also estimate a person's score on a factor, based on their scores for the constituent variables. As such, if we wanted to derive a score of sociability for a particular person, we could place their scores on the various measures into equation (17.3). This method is known as a *weighted average*. In fact, this method is overly simplistic and rarely used, but it is probably the easiest way to explain the principle. For example, imagine the six scales all range from 1 to 10 and that someone scored the following: Talk 1 (4), Social Skills (9), Interest (8), Talk 2 (6), Selfish (8) and Liar (6). We could replace these values into equation (17.3) to get a score for this person's sociability and their consideration to others (see equation (17.4)). The resulting scores of 19.22 and 15.21 reflect the degree to which this person is sociable and their inconsideration to others respectively. This person scores higher on sociability than inconsideration. However, the scales of measurement used will influence the resulting scores, and if different variables use different measurement scales, then **factor scores** for different factors cannot be compared. As such, this method of calculating factor scores is poor and more sophisticated methods are usually used:

$$\begin{aligned} \text{Sociability} &= 0.87\text{Talk 1} + 0.96\text{Social Skills} + 0.92\text{Interest} + 0.00\text{Talk 2} \\ &\quad - 0.10\text{Selfish} + 0.09\text{Liar} \\ &= (0.87 \times 4) + (0.96 \times 9) + (0.92 \times 8) + (0.00 \times 6) - (0.10 \times 8) + (0.09 \times 6) \\ &= 19.22 \\ \text{Consideration} &= 0.01\text{Talk 1} - 0.03\text{Social Skills} + 0.04\text{Interest} + 0.82\text{Talk 2} \\ &\quad + 0.75\text{Selfish} + 0.70\text{Liar} \\ &= (0.01 \times 4) - (0.03 \times 9) + (0.04 \times 8) + (0.82 \times 6) + (0.75 \times 8) + (0.70 \times 6) \\ &= 15.21 \end{aligned} \qquad (17.4)$$

17.3.3.1. The regression method ④

There are several sophisticated techniques for calculating factor scores that use factor score coefficients as weights in equation (17.1) rather than using the factor loadings. The form of the equation remains the same, but the *b*s in the equation are replaced with these factor score coefficients. Factor score coefficients can be calculated in several ways. The simplest way is the regression method. In this method the factor loadings are adjusted to take account of the initial correlations between variables; in doing so, differences in units of measurement and variable variances are stabilized.

To obtain the matrix of factor score coefficients (*B*) we multiply the matrix of factor loadings by the inverse (R^{-1}) of the original correlation or *R*-matrix. You might remember from the previous chapter that matrices cannot be divided (see section 16.4.4.1). Therefore, if we want to divide by a matrix it cannot be done directly and instead we multiply by its inverse. Therefore, by multiplying the matrix of factor loadings by the inverse of the correlation matrix we are, conceptually speaking, dividing the factor loadings by the correlation coefficients. The resulting factor score matrix, therefore, represents the relationship between each variable and each factor taking into account the original relationships between pairs of variables. As such, this matrix represents a purer measure of the *unique* relationship between variables and factors.

The matrices for the popularity data are shown below. The resulting matrix of factor score coefficients, *B*, comes from SPSS. The matrices R^{-1} and *A* can be multiplied by hand to get the matrix *B* and those familiar with matrix algebra (or who have consulted Namboodiri, 1984, or Stevens, 2002) might like to verify the result (see Oliver Twisted). To get the same degree of accuracy as SPSS you should work to at least 5 decimal places:

$$B = R^{-1}A = \begin{pmatrix} 4.76 & -7.46 & 3.91 & -2.35 & 2.42 & -0.49 \\ -7.46 & 18.49 & -12.42 & 5.45 & -5.54 & 1.22 \\ 3.91 & -12.42 & 10.07 & -3.65 & 3.79 & -0.96 \\ -2.35 & -5.45 & -3.65 & -2.97 & -2.16 & -0.02 \\ 2.42 & -5.54 & 3.79 & -2.16 & 2.98 & -0.56 \\ -0.49 & 1.22 & -0.96 & 0.02 & -0.56 & 1.27 \end{pmatrix} \begin{pmatrix} 0.87 & 0.01 \\ 0.96 & -0.03 \\ 0.92 & 0.04 \\ 0.00 & 0.82 \\ -0.10 & 0.75 \\ 0.09 & 0.70 \end{pmatrix}$$

$$= \begin{pmatrix} 0.343 & 0.006 \\ 0.376 & -0.020 \\ 0.362 & 0.020 \\ 0.000 & 0.473 \\ -0.037 & 0.437 \\ 0.039 & 0.405 \end{pmatrix}$$

The pattern of the loadings is the same for the factor score coefficients: that is, the first three variables have high loadings for the first factor and low loadings for the second, whereas the pattern is reversed for the last three variables. The difference is only in the actual value of the weightings, which are smaller because the correlations between variables are now accounted for. These factor score coefficients can be used to replace the *b*-values in equation (17.2):

$$\begin{aligned} \text{Sociability} &= 0.343\text{Talk 1} + 0.376\text{Social Skills} + 0.362\text{Interest} \\ &\quad + 0.000\text{Talk 2} - 0.037\text{Selfish} + 0.039\text{Liar} \\ &= (0.343 \times 4) + (0.376 \times 9) + (0.362 \times 8) + (0.000 \times 6) \\ &\quad - (0.037 \times 8) + (0.039 \times 6) \\ &= 7.59 \end{aligned}$$

$$\begin{aligned}\text{Consideration} &= 0.006\text{Talk 1} - 0.020\text{Social Skills} + 0.020\text{Interest} \\ &\quad + 0.473\text{Talk 2} + 0.437\text{Selfish} + 0.405\text{Liar} \\ &= (0.006 \times 4) + (0.020 \times 9) + (0.020 \times 8) + (0.473 \times 6) \\ &\quad + (0.437 \times 8) + (0.405 \times 6) \\ &= 8.768\end{aligned} \tag{17.5}$$

Equation (17.5) shows how these coefficient scores are used to produce two factor scores for each person. In this case, the participant had the same scores on each variable as were used in equation (17.4.) The resulting scores are much more similar than when the factor loadings were used as weights because the different variances among the six variables have now been controlled for. The fact that the values are very similar reflects the fact that this person not only scores highly on variables relating to sociability, but is also inconsiderate (i.e. they score equally highly on both factors). This technique for producing factor scores ensures that the resulting scores have a mean of 0 and a variance equal to the squared multiple correlation between the estimated factor scores and the true factor values. However, the downside of the regression method is that the scores can correlate not only with factors other than the one on which they are based, but also with other factor scores from a different orthogonal factor.

OLIVER TWISTED

Please, Sir, can I have some more … matrix algebra?

'*The Matrix* …' enthuses Oliver '… that was a good film. I want to dress in black and glide through the air as though time has stood still. Maybe the matrix of factor scores is as cool as the film.' I think you might be disappointed Oliver, but we'll give it a shot. The matrix calculations of factor scores are detailed in the additional material for this chapter on the companion website. Be afraid, be very afraid …

17.3.3.2. Other methods ②

To overcome the problems associated with the regression technique, two adjustments have been proposed: the *Bartlett method* and the **Anderson–Rubin method**. SPSS can produce factor scores based on any of these methods. The Bartlett method produces scores that are unbiased and that correlate only with their own factor. The mean and standard deviation of the scores is the same as for the regression method. However, factor scores can still correlate with each other. The Anderson–Rubin method is a modification of the Bartlett method that produces factor scores that are uncorrelated and standardized (they have a mean of 0 and a standard deviation of 1). Tabachnick and Fidell (2007) conclude that the Anderson–Rubin method is best when uncorrelated scores are required but that the regression method is preferred in other circumstances simply because it is most easily understood. Although it isn't important that you understand the maths behind any of the methods, it is important that you understand what the factor scores represent: namely, a composite score for each individual on a particular factor.

17.3.3.3. Uses of factor scores ②

There are several uses of factor scores. First, if the purpose of the factor analysis is to reduce a large set of data to a smaller subset of measurement variables, then the factor scores tell us an individual's score on this subset of measures. Therefore, any further analysis can be carried out on the factor scores rather than the original data. For example, we could carry out a *t*-test to see whether females are significantly more sociable than males using the factor scores for *sociability*. A second use is in overcoming collinearity problems in regression. If, following a multiple regression analysis, we have identified sources of multicollinearity then the interpretation of the analysis is questioned (see section 7.6.2.3). In this situation, we can simply carry out a factor analysis on the predictor variables to reduce them down to a subset of uncorrelated factors. The variables causing the multicollinearity will combine to form a factor. If we then rerun the regression but using the factor scores as predictor variables then the problem of multicollinearity should vanish (because the variables are now combined into a single factor). There are ways in which we can ensure that the factors are uncorrelated (one way is to use the Anderson–Rubin method – see above). By using uncorrelated factor scores as predictors in the regression we can be confident that there will be no correlation between predictors: hence, no multicollinearity!

17.4. Discovering factors ②

By now, you should have some grasp of the concept of what a factor is, how it is represented graphically, how it is represented algebraically, and how we can calculate composite scores representing an individual's 'performance' on a single factor. I have deliberately restricted the discussion to a conceptual level, without delving into how we actually find these mythical beasts known as factors. This section will look at how we find factors. Specifically we will examine different types of methods, look at the maths behind one method (principal components), investigate the criteria for determining whether factors are important, and discover how to improve the interpretation of a given solution.

17.4.1. Choosing a method ②

The first thing you need to know is that there are several methods for unearthing factors in your data. The method you chose will depend on what you hope to do with the analysis. Tinsley and Tinsley (1987) give an excellent account of the different methods available. There are two things to consider: whether you want to generalize the findings from your sample to a population and whether you are exploring your data or testing a specific hypothesis. This chapter describes techniques for exploring data using factor analysis. Testing hypotheses about the structures of latent variables and their relationships to each other requires considerable complexity and can be done with computer programs such as AMOS (which some of you might find hidden away in the *Analyze* menu of SPSS). Those interested in hypothesis testing techniques (known as **confirmatory factor analysis**) are advised to read Pedhazur and Schmelkin (1991: Chapter 23) for an introduction. Assuming we want to explore our data, we then need to consider whether we want to apply our findings to the sample collected (descriptive method) or to generalize our findings to a population (inferential methods). When factor analysis was originally developed it was assumed that it would be used to explore data to generate future hypotheses. As such, it was assumed that the technique would be applied to the entire population of interest. Therefore, certain techniques assume that the sample used is the population, and so results cannot be extrapolated beyond that particular sample. Principal component analysis is an example of one of these techniques, as is principal factors analysis (*principal axis factoring*) and image covariance analysis (*image factoring*). Of these, principal component analysis and principal factors analysis are the preferred methods and usually result in similar solutions (see section 17.4.3). When these methods are used, conclusions are restricted to the sample collected and generalization of the results can be achieved only if analysis using different samples reveals the same factor structure.

Another approach has been to assume that participants are randomly selected and that the variables measured constitute the population of variables in which we're interested. By assuming this, it is possible to develop techniques from which the results can be generalized from the sample participants to a larger population. However, a constraint is that any findings hold true only for the set of variables measured (because we've assumed this set constitutes the entire population of variables). Techniques in this category include the maximum-likelihood method (see Harman, 1976) and Kaiser's alpha factoring. The choice of method depends largely on what generalizations, if any, you want to make from your data.[2]

17.4.2. Communality ②

Before continuing it is important that you understand some basic things about the variance within an *R*-matrix. It is possible to calculate the variability in scores (the variance) for any given measure (or variable). You should be familiar with the idea of variance by now and comfortable with how it can be calculated (if not see Chapter 2). The total variance for a particular variable will have two components: some of it will be shared with other variables or measures (**common variance**) and some of it will be specific to that measure (**unique variance**). We tend to use the term *unique variance* to refer to variance that can be reliably attributed to only one measure. However, there is also variance that is specific to one measure but not reliably so; this variance is called *error* or **random variance**. The proportion of common variance present in a variable is known as the **communality**. As such, a variable that has no specific variance (or random variance) would have a communality of 1; a variable that shares none of its variance with any other variable would have a communality of 0.

In factor analysis we are interested in finding common underlying dimensions within the data and so we are primarily interested only in the common variance. Therefore, when we run a factor analysis it is fundamental that we know how much of the variance present in our data is common variance. This presents us with a logical impasse: to do the factor analysis we need to know the proportion of common variance present in the data, yet the only way to find out the extent of the common variance is by carrying out a factor analysis! There are two ways to approach this problem. The first is to assume that all of the variance is common variance. As such, we assume that the communality of every variable is 1. By making this assumption we merely transpose our original data into constituent linear components (known as principal component analysis). The second approach is to estimate the amount of common variance by estimating communality values for each variable. There are various methods of estimating communalities but the most widely used (including **alpha factoring**) is to use the squared multiple correlation (SMC) of each variable with all others. So, for the popularity data, imagine you ran a multiple regression using one measure (Selfish) as the outcome and the other five measures as predictors: the resulting multiple R^2 (see section 7.5.2) would be used as an estimate of the communality for the variable Selfish. This second approach is used in factor analysis. These estimates allow the factor analysis to be done. Once the underlying factors have been extracted, new communalities can be calculated that represent the multiple correlation between each variable and the factors extracted. Therefore, the communality is a measure of the proportion of variance explained by the extracted factors.

[2] It's worth noting at this point that principal components analysis is not in fact the same as factor analysis. This doesn't stop idiots like me from discussing them as though they are, but more on that later.

17.4.3. Factor analysis vs. principal component analysis ②

I have just explained that there are two approaches to locating underlying dimensions of a data set: factor analysis and principal component analysis. These techniques differ in the communality estimates that are used. Simplistically, though, factor analysis derives a mathematical model from which factors are estimated, whereas principal component analysis merely decomposes the original data into a set of linear variates (see Dunteman, 1989: Chapter 8, for more detail on the differences between the procedures). As such, only factor analysis can estimate the underlying factors and it relies on various assumptions for these estimates to be accurate. Principal component analysis is concerned only with establishing which linear components exist within the data and how a particular variable might contribute to that component. In terms of theory, this chapter is dedicated to principal component analysis rather than factor analysis. The reasons are that principal component analysis is a psychometrically sound procedure, it is conceptually less complex than factor analysis, and it bears numerous similarities to discriminant analysis (described in the previous chapter).

However, we should consider whether the techniques provide different solutions to the same problem. Based on an extensive literature review, Guadagnoli and Velicer (1988) concluded that the solutions generated from principal component analysis differ little from those derived from factor analytic techniques. In reality, there are some circumstances for which this statement is untrue. Stevens (2002) summarizes the evidence and concludes that with 30 or more variables and communalities greater than 0.7 for all variables, different solutions are unlikely; however, with fewer than 20 variables and any low communalities (< 0.4) differences can occur.

The flip-side of this argument is eloquently described by Cliff (1987) who observed that proponents of factor analysis 'insist that components analysis is at best a common factor analysis with some error added and at worst an unrecognizable hodgepodge of things from which nothing can be determined' (p. 349). Indeed, feeling is strong on this issue with some arguing that when principal component analysis is used it should not be described as a factor analysis and that you should not impute substantive meaning to the resulting components. However, to non-statisticians the difference between a principal component and a factor may be difficult to conceptualize (they are both linear models), and the differences arise largely from the calculation.[3]

17.4.4. Theory behind principal component analysis ③

SMART ALEX ONLY

Principal component analysis works in a very similar way to MANOVA and discriminant function analysis (see previous chapter). Although it isn't necessary to understand the mathematical principles in any detail, readers of the previous chapter may benefit from some comparisons between the two techniques. For those who haven't read that chapter, I suggest you flick through it before moving ahead!

In MANOVA, various sum of squares and cross-product matrices were calculated that contained information about the relationships between dependent variables. I mentioned before that these SSCP matrices could be easily converted to variance–covariance matrices, which represent the same information but in averaged form (i.e. taking account of the number of observations). I also said that by dividing each element by the relevant standard deviation the variance–covariance matrices becomes standardized. The result is a

[3] For this reason I have used the terms *components* and *factors* interchangeably throughout this chapter. Although this use of terms will reduce some statisticians (and psychologists) to tears I'm banking on these people not needing to read this book! I acknowledge the methodological differences, but I think it's easier for students if I dwell on the similarities between the techniques and not the differences.

correlation matrix. In principal component analysis we usually deal with correlation matrices (although it is possible to analyse a variance–covariance matrix too) and the point to note is that this matrix pretty much represents the same information as an SSCP matrix in MANOVA. The difference is just that the correlation matrix is an averaged version of the SSCP that has been standardized.

In MANOVA, we used several SSCP matrices that represented different components of experimental variation (the model variation and the residual variation). In principal component analysis the covariance (or correlation) matrix cannot be broken down in this way (because all data come from the same group of participants). In MANOVA, we ended up looking at the variates or components of the SSCP matrix that represented the ratio of the model variance to the error variance. These variates were linear dimensions that separated the groups tested, and we saw that the dependent variables mapped onto these underlying components. In short, we looked at whether the groups could be separated by some linear combination of the dependent variables. These variates were found by calculating the eigenvectors of the SSCP. The number of variates obtained was the smaller of p (the number of dependent variables) or $k - 1$ (where k is the number of groups). In component analysis we do something similar (I'm simplifying things a little, but it will give you the basic idea). That is, we take a correlation matrix and calculate the variates. There are no groups of observations, and so the number of variates calculated will always equal the number of variables measured (p). The variates are described, as for MANOVA, by the eigenvectors associated with the correlation matrix. The elements of the eigenvectors are the weights of each variable on the variate (see equation (16.5)). These values are the factor loadings described earlier. The largest eigenvalue associated with each of the eigenvectors provides a single indicator of the substantive importance of each variate (or component). The basic idea is that we retain factors with relatively large eigenvalues and ignore those with relatively small eigenvalues.

In summary, component analysis works in a similar way to MANOVA. We begin with a matrix representing the relationships between variables. The linear components (also called variates, or factors) of that matrix are then calculated by determining the eigenvalues of the matrix. These eigenvalues are used to calculate eigenvectors, the elements of which provide the loading of a particular variable on a particular factor (i.e. they are the b-values in equation (17.1)). The eigenvalue is also a measure of the substantive importance of the eigenvector with which it is associated.

EVERYBODY

17.4.5. Factor extraction: eigenvalues and the scree plot ②

How many factors should I extract?

Not all factors are retained in an analysis, and there is debate over the criterion used to decide whether a factor is statistically important. I mentioned above that eigenvalues associated with a variate indicate the substantive importance of that factor. Therefore, it seems logical that we should retain only factors with large eigenvalues. How do we decide whether or not an eigenvalue is large enough to represent a meaningful factor? Well, one technique advocated by Cattell (1966b) is to plot a graph of each eigenvalue (Y-axis) against the factor with which it is associated (X-axis). This graph is known as a **scree plot** (because it looks like a rock face with a pile of debris, or scree, at the bottom). I mentioned earlier that it is possible to obtain as many factors as there are variables and that each has an associated eigenvalue. By graphing the eigenvalues, the relative importance of each factor becomes apparent. Typically there will be a few factors with quite high eigenvalues, and many factors with relatively low eigenvalues, so this graph has a very characteristic shape: there is a sharp descent in the curve followed by a tailing off (see Figure 17.3). Cattell (1966b) argued that the cut-off point for selecting factors should be at the point of inflexion of this curve. The point of inflexion is where the

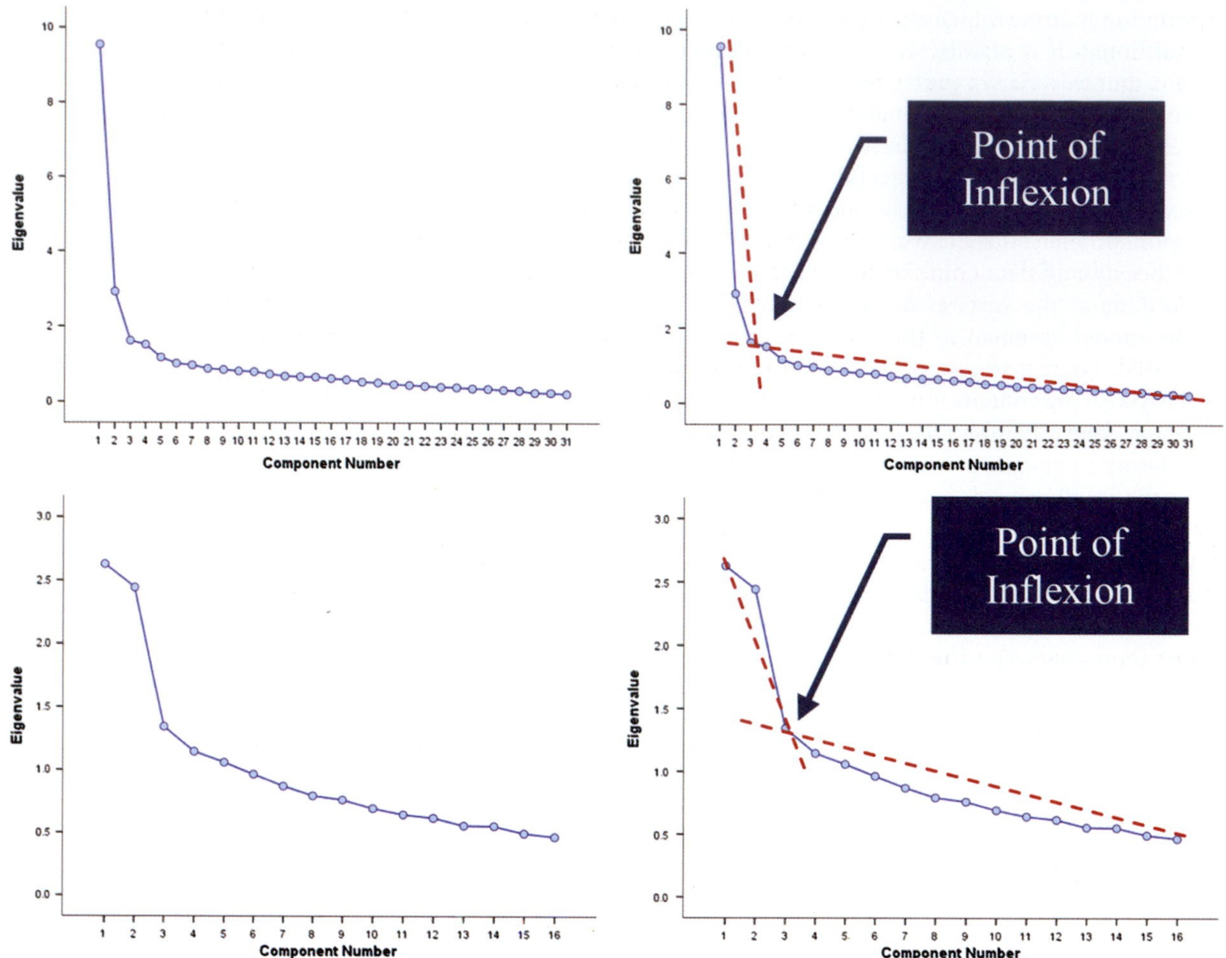

FIGURE 17.3 Examples of scree plots for data that probably have two underlying factors

slope of the line changes dramatically: so, in Figure 17.3, imagine drawing a straight line that summarizes the vertical part of the plot and another that summarizes the horizontal part (the red dashed lines); then the point of inflexion is the data point at which these two lines meet. In both examples in Figure 17.3 the point of inflexion occurs at the third data point (factor); therefore, we would extract two factors. Thus, you retain (or extract) only factors to the left of the point of inflexion (and do not include the factor at the point of inflexion itself).[4] With a sample of more than 200 participants, the scree plot provides a fairly reliable criterion for factor selection (Stevens, 2002).

Although scree plots are very useful, factor selection should not be based on this criterion alone. Kaiser (1960) recommended retaining all factors with eigenvalues greater than 1. This criterion is based on the idea that the eigenvalues represent the amount of variation explained by a factor and that an eigenvalue of 1 represents a substantial amount of variation. Jolliffe (1972, 1986) reports that **Kaiser's criterion** is too strict and suggests the third

[4] Actually if you read Cattell's original paper he advised including the factor at the point of inflexion as well because it is 'desirable to include at least one common error factor as a "garbage can"'. The idea is that the point of inflexion represents an error factor. However, in practice this garbage can factor is rarely retained; also Thurstone argued that it is better to retain too few rather than too many factors so most people do *not* to retain the factor at the point of inflexion.

option of retaining all factors with eigenvalues more than 0.7. The difference between how many factors are retained using Kaiser's methods compared to Jolliffe's can be dramatic.

You might well wonder how the methods compare. Generally speaking, Kaiser's criterion overestimates the number of factors to retain (see Jane Superbrain Box 17.2) but there is some evidence that it is accurate when the number of variables is less than 30 and the resulting communalities (after extraction) are all greater than 0.7. Kaiser's criterion can also be accurate when the sample size exceeds 250 and the average communality is greater than or equal to 0.6. In any other circumstances you are best advised to use a scree plot provided the sample size is greater than 200 (see Stevens, 2002, for more detail). By default, SPSS uses Kaiser's criterion to extract factors. Therefore, if you use the scree plot to determine how many factors are retained you may have to rerun the analysis specifying that SPSS extracts the number of factors you require.

However, as is often the case in statistics, the three criteria often provide different solutions! In these situations the communalities of the factors need to be considered. In principal component analysis we begin with communalities of 1 with all factors retained (because we assume that all variance is common variance). At this stage all we have done is to find the linear variates that exist in the data – so we have just transformed the data without discarding any information. However, to discover what common variance *really* exists between variables we must decide which factors are meaningful and discard any that are too trivial to consider. Therefore, we discard some information. The factors we retain will not explain all of the variance in the data (because we have discarded some information) and so the communalities after extraction will always be less than 1. The factors retained do not map perfectly onto the original variables – they merely reflect the common variance present in the data. If the communalities represent a loss of information then they are important

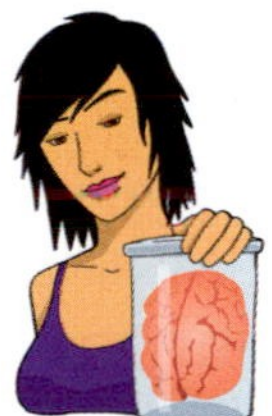

JANE SUPERBRAIN 17.2

How many factors do I retain? ③

The discussion of factor extraction in the text is somewhat simplified. In fact, there are fundamental problems with Kaiser's criterion (Nunnally & Bernstein, 1994). For one thing an eigenvalue of 1 means different things in different analyses: with 100 variables it means that a factor explains 1% of the variance, but with 10 variables it means that a factor explains 10% of the variance. Clearly, these two situations are very different and a single rule that covers both is inappropriate. An eigenvalue of 1 also means only that the factor explains as much variance as a variable, which rather defeats the original intention of the analysis to reduce variables down to 'more substantive' underlying factors (Nunnally & Bernstein, 1994). Consequently, Kaiser's criterion often overestimates the number of factors. On this basis Jolliffe's criterion is even worse (a factor explains less variance than a variable!).

There are more complex ways to determine how many factors to retain, but they are not easy to do on SPSS (which is why I'm discussing them outside of the main text). The best is probably parallel analysis (Horn, 1965). Essentially each eigenvalue (which represents the size of the factor) is compared against an eigenvalue for the corresponding factor in many randomly generated data sets that have the same characteristics as the data being analysed. In doing so, each eigenvalue is being compared to an eigenvalue from a data set that has no underlying factors. This is a bit like asking whether our observed factor is bigger than a non-existing factor. Factors that are bigger than their 'random' counterparts are retained. Of parallel analysis, the scree plot and Kaiser's criterion, Kaiser's criterion is, in general, worst and parallel analysis best (Zwick & Velicer, 1986). If you want to do parallel analysis then SPSS syntax is available (O'Connor, 2000) from http://flash.lakeheadu.ca/~boconno2/nfactors.html.

statistics. The closer the communalities are to 1, the better our factors are at explaining the original data. It is logical that the more factors retained, the greater the communalities will be (because less information is discarded); therefore, the communalities are good indices of whether too few factors have been retained. In fact, with *generalized least-squares factor analysis* and *maximum-likelihood factor analysis* you can get a statistical measure of the goodness of fit of the factor solution (see the next chapter for more on goodness-of-fit tests). This basically measures the proportion of variance that the factor solution explains (so, can be thought of as comparing communalities before and after extraction).

As a final word of advice, your decision on how many factors to extract will depend also on why you're doing the analysis; for example, if you're trying to overcome multicollinearity problems in regression, then it might be better to extract too many factors than too few.

17.4.6. Improving interpretation: factor rotation ③

Once factors have been extracted, it is possible to calculate to what degree variables load onto these factors (i.e. calculate the loading of the variable on each factor). Generally, you will find that most variables have high loadings on the most important factor and small loadings on all other factors. This characteristic makes interpretation difficult, and so a technique called factor rotation is used to discriminate between factors. If a factor is a classification axis along which variables can be plotted, then factor **rotation** effectively rotates these factor axes such that variables are loaded maximally to only one factor. Figure 17.4 demonstrates how this process works using an example in which there are only two factors. Imagine that a sociologist was interested in classifying university lecturers as a demographic group. She discovered that two underlying dimensions best describe this group: alcoholism and achievement (go to any academic conference and you'll see that academics drink heavily!). The first factor, alcoholism, has a cluster of variables associated with it (green circles) and these could be measures such as the number of units drunk in a week, dependency and obsessive personality. The second factor, achievement, also has a cluster of variables associated with it (blue circles) and these could be measures relating to salary, job status and number of research publications. Initially, the full lines represent the factors, and by looking at the co-ordinates it should be clear that the blue circles have high loadings for factor 2 (they are a long way up this axis) and medium loadings on factor 1 (they are not very far up this axis). Conversely, the green circles have high loadings for factor 1 and medium loadings for factor 2. By rotating the axes (dashed lines), we ensure that both clusters of variables are intersected by the factor to which they relate most. So, after rotation, the loadings of the variables are maximized onto one factor (the factor that intersects the cluster) and minimized on the remaining factor(s). If an axis passes through a cluster of variables, then these variables will have a loading of approximately zero on the opposite axis. If this idea is confusing, then look at Figure 17.4 and think about the values of the co-ordinates before and after rotation (this is best achieved by turning the book when you look at the rotated axes).

There are two types of rotation that can be done. The first is **orthogonal rotation**, and the left-hand side of Figure 17.4 represents this method. In Chapter 10 we saw that the term *orthogonal* means unrelated, and in this context it means that we rotate factors while keeping them independent, or unrelated. Before rotation, all factors are independent (i.e. they do not correlate at all) and orthogonal rotation ensures that the factors remain uncorrelated. That is why in Figure 17.4 the axes are turned while remaining perpendicular.[5] The other form of rotation is **oblique rotation**. The difference with oblique rotation is that the factors are allowed to correlate (hence, the axes of the right-hand diagram of Figure 17.4 do not remain perpendicular).

[5] This term means that the axes are at right angles to one another.

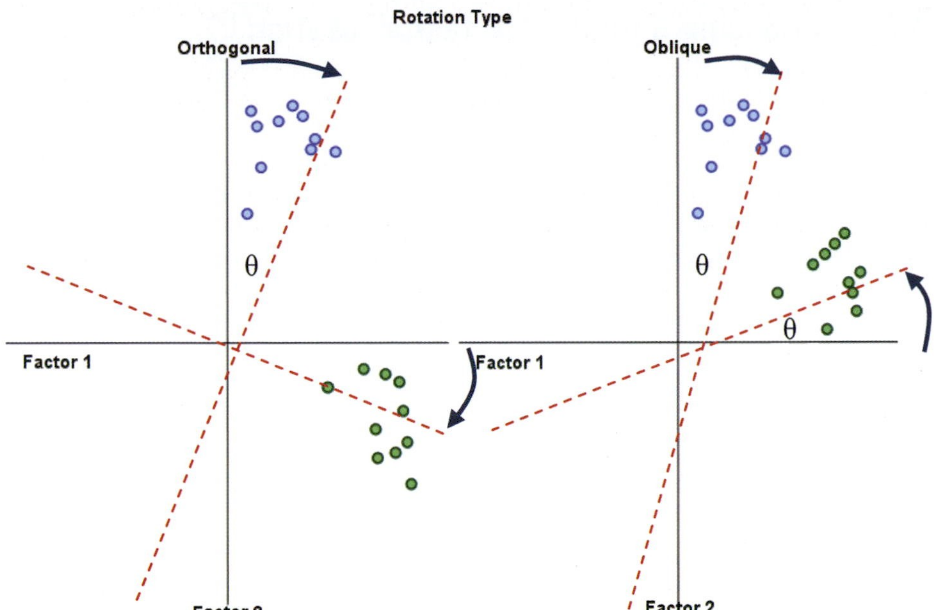

FIGURE 17.4 Schematic representations of factor rotation. The left graph displays orthogonal rotation whereas the right graph displays oblique rotation (see text for more details). θ is the angle through which the axes are rotated

The choice of rotation depends on whether there is a good theoretical reason to suppose that the factors should be related or independent (but see my later comments on this), and also how the variables cluster on the factors before rotation. On the first point, we might not expect alcoholism to be completely independent of achievement (after all, high achievement leads to high stress, which can lead to the drinks cabinet!). Therefore, on theoretical grounds, we might choose oblique rotation. On the second point, Figure 17.4 demonstrates how the positioning of clusters is important in determining how successful the rotation will be (note the position of the green circles). Specifically, if an orthogonal rotation was carried out on the right-hand diagram it would be considerably less successful in maximizing loadings than the oblique rotation that is displayed. One approach is to run the analysis using both types of rotation. Pedhazur and Schmelkin (1991) suggest that if the oblique rotation demonstrates a negligible correlation between the extracted factors then it is reasonable to use the orthogonally rotated solution. If the oblique rotation reveals a correlated factor structure, then the orthogonally rotated solution should be discarded. In any case, an oblique rotation should be used only if there are good reasons to suppose that the underlying factors *could* be related in theoretical terms.

The mathematics behind factor rotation is complex (especially oblique rotation). However, in oblique rotation, because each factor can be rotated by different amounts a **factor transformation matrix, Λ,** is needed. The factor transformation matrix is a square matrix and its size depends on how many factors are extracted from the data. If two factors are extracted then it will be a 2 × 2 matrix, but if four factors are extracted then it becomes a 4 × 4 matrix. The values in the factor transformation matrix consist of sines and cosines of the angle of axis rotation (θ). This matrix is multiplied by the matrix of unrotated factor loadings, **A**, to obtain a matrix of rotated factor loadings.

For the case of two factors the factor transformation matrix would be:

$$\Lambda = \begin{pmatrix} \cos\theta & -\sin\theta \\ \sin\theta & \cos\theta \end{pmatrix}$$

Therefore, you should think of this matrix as representing the angle through which the axes have been rotated, or the degree to which factors have been rotated. The angle of rotation necessary to optimize the factor solution is found in an iterative way (see SPSS Tip 8.1) and different methods can be used.

17.4.6.1. Choosing a method of factor rotation ③

SPSS has three methods of orthogonal rotation (**varimax**, **quartimax** and **equamax**) and two methods of oblique rotation (**direct oblimin** and **promax**). These methods differ in how they rotate the factors and, therefore, the resulting output depends on which method you select. Quartimax rotation attempts to maximize the spread of factor loadings for a variable across all factors. Therefore, interpreting variables becomes easier. However, this often results in lots of variables loading highly onto a single factor. Varimax is the opposite in that it attempts to maximize the dispersion of loadings within factors. Therefore, it tries to load a smaller number of variables highly onto each factor resulting in more interpretable clusters of factors. Equamax is a hybrid of the other two approaches and is reported to behave fairly erratically (see Tabachnick and Fidell, 2007). For a first analysis, you should probably select varimax because it is a good general approach that simplifies the interpretation of factors.

The case with oblique rotations is more complex because correlation between factors is permitted. In the case of direct oblimin, the degree to which factors are allowed to correlate is determined by the value of a constant called delta. The default value in SPSS is 0 and this ensures that high correlation between factors is not allowed (this is known as direct quartimin rotation). If you choose to set delta to greater than 0 (up to 0.8), then you can expect highly correlated factors; if you set delta to less than 0 (down to –0.8) you can expect less correlated factors. The default setting of 0 is sensible for most analyses and I don't recommend changing it unless you know what you are doing (see Pedhazur and Schmelkin, 1991: 620). Promax is a faster procedure designed for very large data sets.

In theory, the exact choice of rotation will depend largely on whether or not you think that the underlying factors should be related. If you expect the factors to be independent then you should choose one of the orthogonal rotations (I recommend varimax). If, however, there are theoretical grounds for supposing that your factors might correlate, then direct oblimin should be selected. In practice, there are strong grounds to believe that orthogonal rotations are a complete nonsense for naturalistic data, and certainly for any data involving humans (can you think of any psychological construct that is not in any way correlated with some other psychological construct?) As such, some argue that orthogonal rotations should never be used.

17.4.6.2. Substantive importance of factor loadings ②

Once a factor structure has been found, it is important to decide which variables make up which factors. Earlier I said that the factor loadings were a gauge of the substantive importance of a given variable to a given factor. Therefore, it makes sense that we use these values to place variables with factors. It is possible to assess the statistical significance of a factor loading (after all, it is simply a correlation coefficient or regression coefficient); however, there are various reasons why this option is not as easy as it seems (see Stevens, 2002: 393). Typically, researchers take a loading of an absolute value of more than 0.3 to be important. However, the significance of a factor loading will depend on the sample size. Stevens (2002) produced a table of critical values against which loadings can be compared. To summarize, he recommends that for a sample size of 50 a loading of 0.722 can be considered significant, for 100 the loading should be greater than 0.512, for 200 it should be greater than 0.364, for 300 it should be greater than 0.298, for 600 it should be greater than 0.21, and for 1000 it should be greater than 0.162. These values are based on an alpha level of .01 (two-tailed), which allows for the fact that several loadings will need to be tested (see Stevens, 2002, for further detail). Therefore, in very large samples, small loadings can be considered statistically meaningful. SPSS does not provide significance tests of factor loadings but by applying Stevens' guidelines you should gain some insight into the structure of variables and factors.

The significance of a loading gives little indication of the substantive importance of a variable to a factor. This value can be found by squaring the factor loading to give an estimate of the amount of variance in a factor accounted for by a variable (like R^2). In this respect Stevens (2002) recommends interpreting only factor loadings with an absolute value greater than 0.4 (which explain around 16% of the variance in the variable).

17.5. Research example ②

One of the uses of factor analysis is to develop questionnaires: after all, if you want to measure an ability or trait, you need to ensure that the questions asked relate to the construct that you intend to measure. I have noticed that a lot of students become very stressed about SPSS. Therefore I wanted to design a questionnaire to measure a trait that I termed 'SPSS anxiety'. I decided to devise a questionnaire to measure various aspects of students' anxiety towards learning SPSS. I generated questions based on interviews with anxious and non-anxious students and came up with 23 possible questions to include. Each question was a statement followed by a five-point Likert scale ranging from 'strongly disagree' through 'neither agree nor disagree' to 'strongly agree'. The questionnaire is printed in Figure 17.5.

The questionnaire was designed to predict how anxious a given individual would be about learning how to use SPSS. What's more, I wanted to know whether anxiety about SPSS could be broken down into specific forms of anxiety. In other words, what latent variables contribute to anxiety about SPSS? With a little help from a few lecturer friends I collected 2571 completed questionnaires (at this point it should become apparent that this example is fictitious!). The data are stored in the file **SAQ.sav**. Load this file into SPSS and have a look at the variables and their properties. The first thing to note is that each question (variable) is represented by a different column. We know that in SPSS, cases (or people's data) are stored in rows and variables are stored in columns and so this layout is consistent with past chapters. The second thing to notice is that there are 23 variables labelled **q1** to **q23** and that each has a label indicating the question. By labelling my variables I can be very clear about what each variable represents (this is the value of giving your variables full titles rather than just using restrictive column headings).

OLIVER TWISTED

Please, Sir, can I have some more ... questionnaires?

'I'm going to design a questionnaire to measure one's propensity to pick a pocket or two,' says Oliver, 'but how would I go about doing it?' You'd read the useful information about the dos and don'ts of questionnaire design in the additional material for this chapter on the companion website, that's how. Rate how useful it is on a Likert scale from 1 = not useful at all, to 5 = very useful.

17.5.1. Before you begin ②

17.5.1.1. Sample size ②

Correlation coefficients fluctuate from sample to sample, much more so in small samples than in large. Therefore, the reliability of factor analysis is also dependent on sample size. Much has been written about the necessary sample size for factor analysis resulting in

FIGURE 17.5 The SPSS anxiety questionnaire (SAQ)

SD = Strongly Disagree, D = Disagree, N = Neither, A = Agree, SA = Strongly Agree

		SD	D	N	A	SA
1	Statistics make me cry	○	○	○	○	○
2	My friends will think I'm stupid for not being able to cope with SPSS	○	○	○	○	○
3	Standard deviations excite me	○	○	○	○	○
4	I dream that Pearson is attacking me with correlation coefficients	○	○	○	○	○
5	I don't understand statistics	○	○	○	○	○
6	I have little experience of computers	○	○	○	○	○
7	All computers hate me	○	○	○	○	○
8	I have never been good at mathematics	○	○	○	○	○
9	My friends are better at statistics than me	○	○	○	○	○
10	Computers are useful only for playing games	○	○	○	○	○
11	I did badly at mathematics at school	○	○	○	○	○
12	People try to tell you that SPSS makes statistics easier to understand but it doesn't	○	○	○	○	○
13	I worry that I will cause irreparable damage because of my incompetence with computers	○	○	○	○	○
14	Computers have minds of their own and deliberately go wrong whenever I use them	○	○	○	○	○
15	Computers are out to get me	○	○	○	○	○
16	I weep openly at the mention of central tendency	○	○	○	○	○
17	I slip into a coma whenever I see an equation	○	○	○	○	○
18	SPSS always crashes when I try to use it	○	○	○	○	○
19	Everybody looks at me when I use SPSS	○	○	○	○	○
20	I can't sleep for thoughts of eigenvectors	○	○	○	○	○
21	I wake up under my duvet thinking that I am trapped under a normal distribution	○	○	○	○	○
22	My friends are better at SPSS than I am	○	○	○	○	○
23	If I am good at statistics people will think I am a nerd	○	○	○	○	○

many 'rules of thumb'. The common rule is to suggest that a researcher has at least 10–15 participants per variable. Although I've heard this rule bandied about on numerous occasions its empirical basis is unclear (however, Nunnally, 1978, did recommend having 10 times as many participants as variables). Kass and Tinsley (1979) recommended having between 5 and 10 participants per variable up to a total of 300 (beyond which test parameters tend to be stable regardless of the participant to variable ratio). Indeed, Tabachnick and Fidell (2007) agree that 'it is comforting to have at least 300 cases for factor analysis' (p. 613) and Comrey and Lee (1992) class 300 as a good sample size, 100 as poor and 1000 as excellent.

Fortunately, recent years have seen empirical research done in the form of experiments using simulated data (so-called Monte Carlo studies). Arrindell and van der Ende (1985) used real-life data to investigate the effect of different participant to variable ratios. They concluded that changes in this ratio made little difference to the stability of factor solutions. Guadagnoli and Velicer (1988) found that the most important factors in determining reliable factor solutions was the absolute sample size and the absolute magnitude of factor loadings. In short, they argue that if a factor has four or more loadings greater than 0.6 then it is reliable regardless of sample size. Furthermore, factors with 10 or more loadings greater than 0.40 are reliable if the sample size is greater than 150. Finally, factors with a few low loadings should not be interpreted unless the sample size is 300 or more. MacCallum, Widaman, Zhang, and Hong (1999) have shown that the minimum sample size or sample to variable ratio depends on other aspects of the design of the study. In short, their study indicated that as communalities become lower the importance of sample size increases. With all communalities above 0.6, relatively small samples (less than 100) may be perfectly adequate. With communalities in the 0.5 range, samples between 100 and 200 can be good enough provided there are relatively few factors each with only a small number of indicator variables. In the worst scenario of low communalities (well below 0.5) and a larger number of underlying factors they recommend samples above 500.

What's clear from this work is that a sample of 300 or more will probably provide a stable factor solution but that a wise researcher will measure enough variables to adequately measure all of the factors that theoretically they would expect to find.

Another alternative is to use the **Kaiser–Meyer–Olkin measure of sampling adequacy (KMO)** (Kaiser, 1970). The KMO can be calculated for individual and multiple variables and represents the ratio of the squared correlation between variables to the squared partial correlation between variables. The KMO statistic varies between 0 and 1. A value of 0 indicates that the sum of partial correlations is large relative to the sum of correlations, indicating diffusion in the pattern of correlations (hence, factor analysis is likely to be inappropriate). A value close to 1 indicates that patterns of correlations are relatively compact and so factor analysis should yield distinct and reliable factors. Kaiser (1974) recommends accepting values greater than 0.5 as barely acceptable (values below this should lead you to either collect more data or rethink which variables to include). Furthermore, values between 0.5 and 0.7 are mediocre, values between 0.7 and 0.8 are good, values between 0.8 and 0.9 are great and values above 0.9 are superb (Hutcheson & Sofroniou, 1999).

17.5.1.2. Correlations between variables ③

When I was an undergraduate my statistics lecturer always used to say 'if you put garbage in, you get garbage out'. This saying applies particularly to factor analysis because SPSS will always find a factor solution to a set of variables. However, the solution is unlikely to have any real meaning if the variables analysed are not sensible. The first thing to do when conducting a factor analysis or principal component analysis is to look at the intercorrelation between variables. There are essentially two potential problems: (1) correlations that are not high enough; and (2) correlations that are too high. The correlations between variables can be checked using the *correlate* procedure (see Chapter 6) to create a correlation matrix of all variables. This matrix can also be created as part of the main factor analysis. In both cases the remedy is to remove variables from the analysis. We will look at each problem in turn.

If our test questions measure the same underlying dimension (or dimensions) then we would expect them to correlate with each other (because they are measuring the same thing). Even if questions measure different aspects of the same things (e.g. we could measure overall anxiety in terms of sub-components such as worry, intrusive thoughts and physiological arousal), there should still be high intercorrelations between the variables relating to these sub-traits. We can test for this problem first by visually scanning the correlation matrix and looking for correlations below about .3 (you could use the significance of correlations but, given the large sample sizes normally used with factor analysis, this approach isn't helpful because even very small correlations will be significant in large samples). If any variables have lots of correlations below .3 then consider excluding them. It should be immediately clear that this approach is very subjective: I've used fuzzy terms such as 'about .3' and 'lots of', but I have to because every data set is different. Analysing data really is a skill, not a recipe book!

If you want an objective test of whether correlations (overall) are too small then we can test for a very extreme scenario. If the variables in our correlation matrix did not correlate at all, then our correlation matrix would be an identity matrix (i.e. the off-diagonal components are zero – see section 16.4.2). In Chapter 16 we came across a test that examines whether the population correlation matrix resembles an identity matrix: *Bartlett's test*. If the population correlation matrix resembles an identity matrix then it means that every variable correlates very badly with all other variables (i.e. all correlation coefficients are close to zero). If it *were* an identity matrix then it would mean that all variables are perfectly independent from one another (all correlation coefficients are zero). Given that we are looking for clusters of variables that measure similar things, it should be obvious why this scenario is problematic: if no variables correlate then there are no clusters to find. Bartlett's test tells us whether our correlation matrix is significantly different from an identity matrix. Therefore, if it is significant then it means that the correlations between variables are (overall) significantly different from zero. So, if Bartlett's test is significant then it is good news. However, as with any significance test it depends on sample sizes and in factor analysis we typically use very large samples. Therefore, although a non-significant Bartlett's test is certainly cause for concern, a significant test does not necessarily mean that correlations are big enough to make the analysis meaningful. If you do identify any variables, that seem to have very low correlations with lots of other variables, then exclude them from the factor analysis.

The opposite problem is when variables correlate too highly. Although mild multicollinearity is not a problem for factor analysis it is important to avoid extreme multicollinearity (i.e. variables that are very highly correlated) and **singularity** (variables that are perfectly correlated). As with regression, multicollinearity causes problems in factor analysis because it becomes impossible to determine the unique contribution to a factor of the variables that are highly correlated (as was the case for multiple regression). Multicollinearity does not cause a problem for principal component analysis. Therefore, as well as scanning the correlation matrix for low correlations, we could also look out for very high correlations ($r > .8$). The problem with a heuristic such as this is that the effect of two variables correlating with $r = .9$ might be less than the effect of, say, three variables that all correlate at $r = .6$. In other words, eliminating such highly correlating variables might not be getting at the cause of the multicollinearity (Rockwell, 1975).

Multicollinearity can be detected by looking at the determinant of the *R*-matrix, denoted $|R|$ (see Jane Superbrain Box 17.3). One simple heuristic is that the determinant of the *R*-matrix should be greater than 0.00001. However, Haitovsky (1969) proposed a significance test of whether the determinant is zero (i.e. the matrix is singular). If this test is significant it tells us that the correlation matrix is significantly different from a singular matrix, which implies that there is no severe multicollinearity. Simple eh? Well, not quite

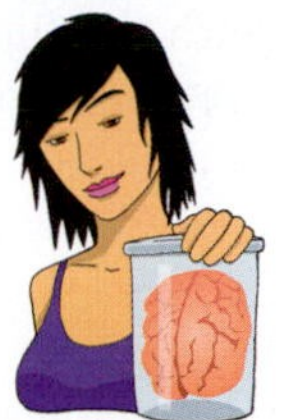

JANE SUPERBRAIN 17.3

What is the determinant? ③

The determinant of a matrix is an important diagnostic tool in factor analysis, but the question of what it is is not easy to answer because it has a mathematical definition and I'm not a mathematician. Rather than pretending that I understand the maths, all I'll say is that a good explanation of how the determinant is derived can be found at mathworld.wolfram.com. However, we can bypass the maths and think about the determinant conceptually. The way that I think of the determinant is as describing the 'area' of the data. In Jane Superbrain Box 7.2 we saw these two diagrams:

At the time I used these to describe eigenvectors and eigenvalues (which describe the shape of the data). The determinant is related to eigenvalues and eigenvectors but instead of describing the height and width of the data it describes the overall area. So, in the left diagram, the determinant of those data would represent the area inside the red dashed elipse. These variables have a low correlation so the determinant (area) is big; the biggest value it can be is 1. In the right diagram, the variables are perfectly correlated or singular, and the elipse (red dashed line) has been squashed down to a straight line. In other words, the opposite sides of the ellipse have actually met each other and there is no distance between them at all. Put another way, the area, or determinant, is zero. Therefore, the determinant tells us whether the correlation matrix is singular (determinant is 0), or if all variables are completely unrelated (determinant is 1), or somewhere in between!

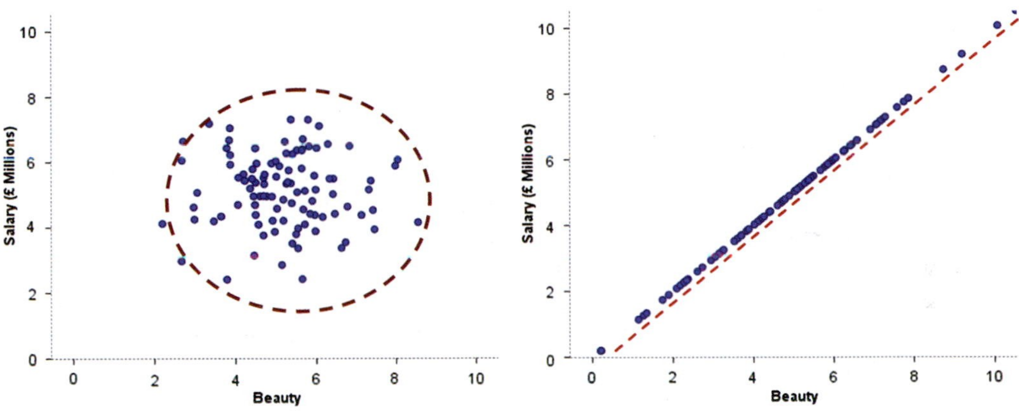

so simple because SPSS doesn't do this test! However, it can be done manually without too much trauma:

$$\text{Haitovsky's } \chi^2_{\text{H}} = \left[1 + \frac{(2p+5)}{6} - N\right] ln(1 - |R|) \tag{17.6}$$

in which p is the number of variables in the correlation matrix, N is the total sample size, $|R|$ is the determinant of the correlation matrix and ln is the natural logarithm (this is a standard mathematical function that we came across in Chapter 8 and you can find it on your calculator usually labelled as ln or log_e). The resulting test statistic has a chi-square distribution with $p(p-1)/2$ degrees of freedom. We'll see how to use this equation in due course.

If you have reason to believe that the correlation matrix has multicollinearity then you could look through the correlation matrix for variables that correlate very highly

($R > 0.8$) and consider eliminating one of the variables (or more depending on the extent of the problem) before proceeding. You may have to try some trial and error to work out which variables are creating the problem (it's not always the two with the highest correlation, it could be a larger number of variables with correlations that are not obviously too large).

17.5.1.3. The distribution of data ②

As well as looking for interrelations, you should ensure that variables have roughly normal distributions and are measured at an interval level (which Likert scales are, perhaps wrongly, assumed to be!). The assumption of normality is most important if you wish to generalize the results of your analysis beyond the sample collected. You can do factor analysis on non-continuous data; for example, if you had dichotomous variables, it's possible (using syntax) to do the factor analysis direct from the correlation matrix, but you should construct the correlation matrix from tetrachoric correlation coefficients (http://ourworld.compuserve.com/homepages/jsuebersax/tetra.htm). This keeps the analysis close to the familiar framework of factor analysis, and the only hassle is computing the correlations (but see the website for software options). Alternatively you can use software other than SPSS, such as MPLUS (http://www.statmodel.com/), to analyse the raw data.

17.6. Running the analysis ②

Access the main dialog box (Figure 17.6) by selecting Analyze Dimension Reduction ▸ Factor... . Simply select the variables you want to include in the analysis (remember to exclude any variables that were identified as problematic during the data screening) and transfer them to the box labelled *Variables* by clicking on ➔.

There are several options available, the first of which can be accessed by clicking on Descriptives... to access the dialog box in Figure 17.7. The *Univariate descriptives* option provides means and standard deviations for each variable. Most of the other options relate to the correlation matrix of variables (the *R*-matrix described earlier). The *Coefficients* option produces the *R*-matrix, and the *Significance levels* option will produce a matrix indicating the significance value of each correlation in the *R*-matrix. You can also ask for

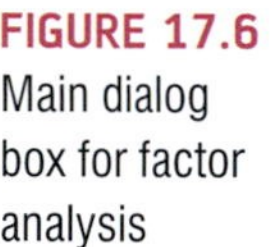

FIGURE 17.6 Main dialog box for factor analysis

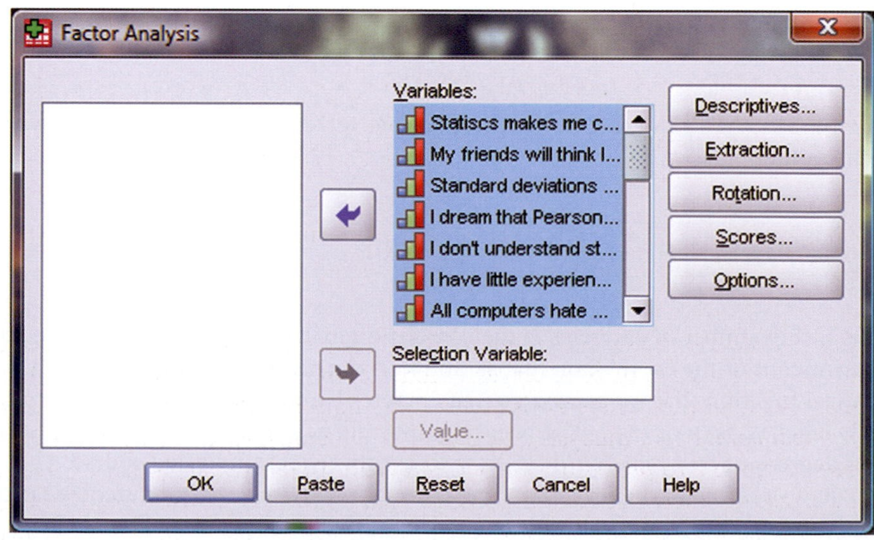

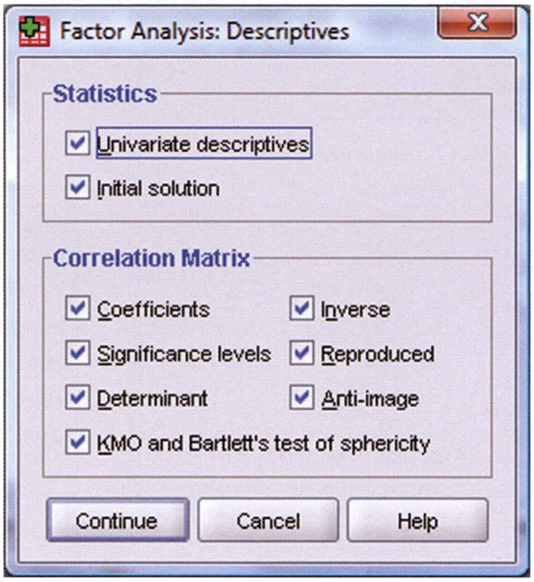

FIGURE 17.7
Descriptives in factor analysis

the *Determinant* of this matrix and this option is vital for testing for multicollinearity or singularity (see section 17.5.1.2).

KMO and Bartlett's test of sphericity produces the Kaiser–Meyer–Olkin measure of sampling adequacy and Bartlett's test. With a sample of 2571 we shouldn't have cause to worry about the sample size (see section 17.5.1.1). We have already stumbled across KMO (see section 17.5.1.1) and Bartlett's test (see section 17.5.1.2) and have seen the various criteria for adequacy.

The *Reproduced* option produces a correlation matrix based on the model (rather than the real data). Differences between the matrix based on the model and the matrix based on the observed data indicate the residuals of the model (i.e. differences). SPSS produces these residuals in the lower table of the reproduced matrix and we want relatively few of these values to be greater than .05. Luckily, to save us scanning this matrix, SPSS produces a summary of how many residuals lie above .05. The *Reproduced* option should be selected to obtain this summary. The *Anti-image* option produces an anti-image matrix of covariances and correlations. These matrices contain measures of sampling adequacy for each variable along the diagonal and the negatives of the partial correlation/covariances on the off-diagonals. The diagonal elements, like the KMO measure, should all be greater than 0.5 at a bare minimum if the sample is adequate for a given pair of variables. If any pair of variables has a value less than this, consider dropping one of them from the analysis. The off-diagonal elements should all be very small (close to zero) in a good model. When you have finished with this dialog box click on Continue to return to the main dialog box.

17.6.1. Factor extraction on SPSS ②

To access the extraction dialog box (Figure 17.8), click on Extraction... in the main dialog box. There are several ways of conducting a factor analysis (see section 17.4.1) and when and where you use the various methods depend on numerous things. For our purposes we will use *principal component* analysis (Principal components ▾) which strictly speaking isn't factor analysis; however, the two procedures may often yield similar results (see section 17.4.3).

In the *Analyze* box there are two options: to analyse the *Correlation matrix* or to analyse the *Covariance matrix*. You should be happy with the idea that these two matrices

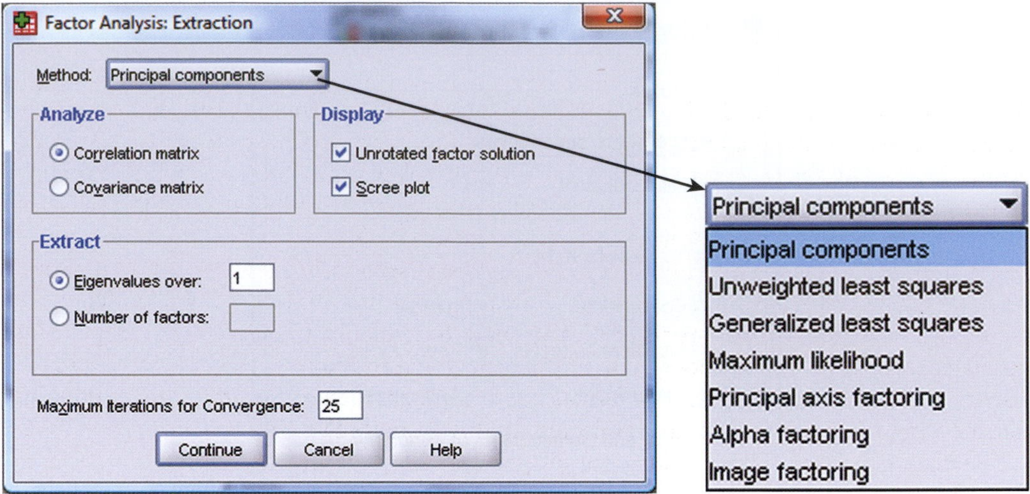

FIGURE 17.8 Dialog box for factor extraction

are actually different versions of the same thing: the correlation matrix is the standardized version of the covariance matrix. Analysing the correlation matrix is a useful default method because it takes the standardized form of the matrix; therefore, if variables have been measured using different scales this will not affect the analysis. In this example, all variables have been measured using the same measurement scale (a five-point Likert scale), but often you will want to analyse variables that use different measurement scales. Analysing the correlation matrix ensures that differences in measurement scales are accounted for. In addition, even variables measured using the same scale can have very different variances and this too creates problems for principal component analysis. Using the correlation matrix eliminates this problem also. There are statistical reasons for preferring to analyse the covariance matrix[6] and generally the results will differ from analysis on the correlation matrix. However, the covariance matrix should be analysed only when your variables are commensurable.

The *Display* box has two options within it: to display the *Unrotated factor solution* and a *Scree plot*. The scree plot was described earlier and is a useful way of establishing how many factors should be retained in an analysis. The unrotated factor solution is useful in assessing the improvement of interpretation due to rotation. If the rotated solution is little better than the unrotated solution then it is possible that an inappropriate (or less optimal) rotation method has been used.

The *Extract* box provides options pertaining to the retention of factors. You have the choice of either selecting factors with eigenvalues greater than a user-specified value or retaining a fixed number of factors. For the *Eigenvalues over* option the default is Kaiser's recommendation of eigenvalues over 1, but you could change this to Jolliffe's recommendation of 0.7 or any other value you want. It is probably best to run a primary analysis with the *Eigenvalues over* 1 option selected, select a scree plot and compare the results. If looking at the scree plot and the eigenvalues over 1 lead you to retain the same number of factors then continue with the analysis and be happy. If the two criteria give different results then examine the communalities and decide for yourself which of the two criteria to believe. If you decide to use the scree plot then you may want to redo the analysis specifying the number of factors to extract. The number of factors to be extracted can be specified by selecting *Number of factors* and then typing the appropriate number in the space provided (e.g. 4).

[6] The reason being that correlation coefficients are insensitive to variations in the dispersion of data whereas covariance is and so produces better-defined factor structures (Tinsley & Tinsley, 1987)

17.6.2. Rotation ②

We have already seen that the interpretability of factors can be improved through rotation. Rotation maximizes the loading of each variable on one of the extracted factors while minimizing the loading on all other factors. This process makes it much clearer which variables relate to which factors. Rotation works through changing the absolute values of the variables while keeping their differential values constant. Click on Rotation... to access the dialog box in Figure 17.9. I've discussed the various rotation options in section 17.4.6.1, but, to summarize, the exact choice of rotation will depend on whether or not you think that the underlying factors should be related. If there are theoretical grounds to think that the factors are independent (unrelated) then you should choose one of the orthogonal rotations (I recommend varimax). However, if theory suggests that your factors might correlate then one of the oblique rotations (direct oblimin or promax) should be selected. In this example I've selected varimax.

The dialog box also has options for displaying the *Rotated solution* and a *Loading plot*. The rotated solution is displayed by default and is essential for interpreting the final rotated analysis. The loading plot will provide a graphical display of each variable plotted against the extracted factors up to a maximum of three factors (unfortunately SPSS cannot produce four- or five-dimensional graphs!). This plot is basically similar to Figure 17.2 and it uses the factor loading of each variable for each factor. With two factors these plots are fairly interpretable, and you should hope to see one group of variables clustered close to the *X*-axis and a different group of variables clustered around the *Y*-axis. If all variables are clustered between the axes, then the rotation has been relatively unsuccessful in maximizing the loading of a variable onto a single factor. With three factors these plots can become quite messy and certainly put considerable strain on the visual system! However, they can still be a useful way to determine the underlying structures within the data.

A final option is to set the *Maximum Iterations for Convergence*, which specifies the number of times that the computer will search for an optimal solution. In most circumstances the default of 25 is more than adequate for SPSS to find a solution for a given data set. However, if you have a large data set (like we have here) then the computer might have difficulty finding a solution (especially for oblique rotation). To allow for the large data set we are using, change the value to 30.

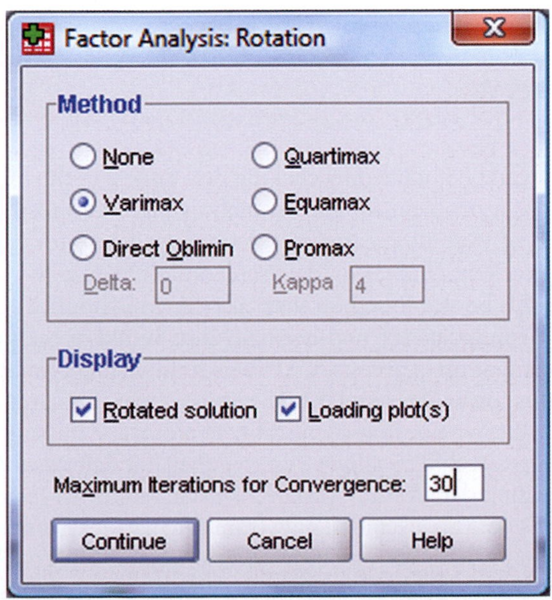

FIGURE 17.9 *Factor Analysis: Rotation* dialog box

17.6.3. Scores ②

The factor scores dialog box (Figure 17.10) can be accessed by clicking on Scores... in the main dialog box. This option allows you to save factor scores (see section 17.3.3) for each case in the data editor. SPSS creates a new column for each factor extracted and then places the factor score for each case within that column. These scores can then be used for further analysis, or simply to identify groups of participants who score highly on particular factors. There are three methods of obtaining these scores, all of which were described in sections 17.3.3.1 and 17.3.3.2. If you want to ensure that factor scores are uncorrelated then select the *Anderson-Rubin* method; if correlations between factor scores are acceptable then choose the *Regression* method.

As a final option, you can ask SPSS to produce the factor score coefficient matrix. This matrix was the matrix *B* described in section 17.3.3.1. This matrix can be useful if, for whatever reason, you wish to construct factor equations such as those in equation (17.5), because it provides you with the values of *b* for each of the variables.

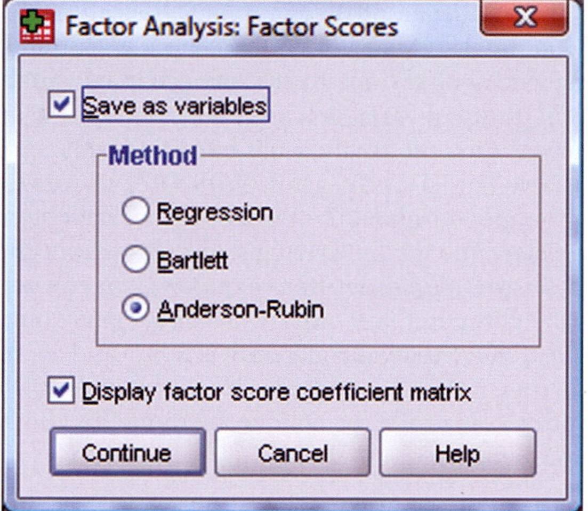

FIGURE 17.10 *Factor Analysis: Factor Scores* dialog box

17.6.4. Options ②

The options dialog box can be obtained by clicking on Options... in the main dialog box (Figure 17.11). Missing data are a problem for factor analysis just like most other procedures and SPSS provides a choice of excluding cases or estimating a value for a case. Tabachnick and Fidell (2007) have an excellent chapter on data screening (Chapter 4 – see also the rather less excellent Chapter 5 of this book). Based on their advice, you should consider the distribution of missing data. If the missing data are non-normally distributed or the sample size after exclusion is too small then estimation is necessary. SPSS uses the mean as an estimate (*Replace with mean*). These procedures lower the standard deviation of variables and so can lead to significant results that would otherwise be non-significant. Therefore, if missing data are random, you might consider excluding cases. SPSS allows you to either *Exclude cases listwise*, in which case any participant with missing data for any variable is excluded, or to *Exclude cases pairwise*, in which case a participant's data are excluded only from calculations for which a datum is missing (see SPSS Tip 6.1). If you exclude cases pairwise your estimates can go all over the place so it's probably safest to opt to exclude cases listwise unless this results in a massive loss of data.

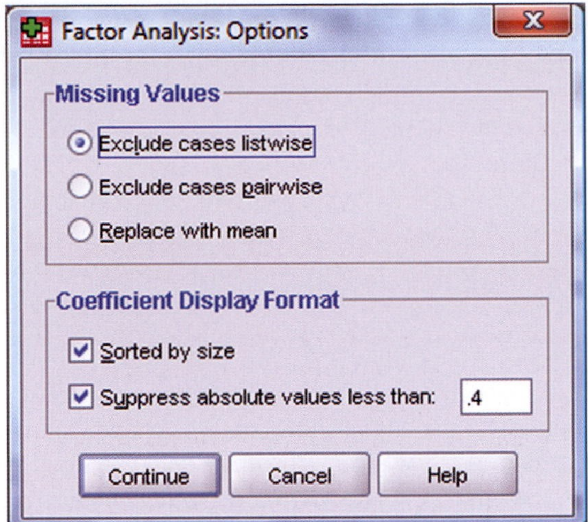

FIGURE 17.11 *Factor Analysis: Options* dialog box

The final two options relate to how coefficients are displayed. By default SPSS will list variables in the order in which they are entered into the data editor. Usually, this format is most convenient. However, when interpreting factors it is sometimes useful to list variables by size. By selecting *Sorted by size*, SPSS will order the variables by their factor loadings. In fact, it does this sorting fairly intelligently so that all of the variables that load highly onto the same factor are displayed together. The second option is to *Suppress absolute values less than* a specified value (by default 0.1). This option ensures that factor loadings within ±0.1 are not displayed in the output. Again, this option is useful for assisting in interpretation. The default value is probably sensible, but on your first analysis I recommend changing it either to 0.4 (for interpretation purposes) or to a value reflecting the expected value of a significant factor loading given the sample size (see section 17.4.6.2). This will make interpretation simpler. You can, if you like, rerun the analysis and set this value lower just to check you haven't missed anything (like a loading of .39). For this example set the value at 0.4.

17.7. Interpreting output from SPSS ②

Select the same options as I have in the screen diagrams and run a factor analysis with orthogonal rotation.

SELF-TEST Having done this, select the *Direct Oblimin* option in Figure 17.9 and repeat the analysis. You should obtain two outputs identical in all respects except that one used an orthogonal rotation and the other an oblique.

For the purposes of saving space in this section I set the default SPSS options such that each variable is referred to only by its label on the data editor (e.g. Q12). On the output *you* obtain, you should find that SPSS uses the value label (the question itself) in all of the output. When using the output in this chapter just remember that Q1 represents question 1, Q2 represents question 2 and Q17 represents question 17. By referring back to Figure 17.5 and matching the question number to the variable name you can identify each question.

Sometimes the factor analysis doesn't work, the KMO test and determinant are nowhere to be found and SPSS spits out an error message about a 'non positive definite matrix' (see SPSS Tip 17.1).

SPSS TIP 17.1 **Error messages about a 'non positive definite matrix'** ④

What is a non-positive definite matrix?: As we have seen, factor analysis works by looking at your correlation matrix. This matrix has to be 'positive definite' for the analysis to work. What does that mean in plain English? It means lots of horrible things mathematically (e.g. the eigenvalues and determinant of the matrix have to be positive) and about the best explanation I've seen is at http://www2.gsu.edu/~mkteer/npdmatri.html. In more basic terms, factors are like lines floating in space, and eigenvalues measure the length of those lines. If your eigenvalue is negative then it means that the length of your line/factor is negative too. It's a bit like me asking you how tall you are, and you responding 'I'm minus 175 cm tall'. That would be nonsense. By analogy, if a factor has negative length, then that too is nonsense. When SPSS decomposes the correlation matrix to look for factors, if it comes across a negative eigenvalue it starts thinking 'oh dear, I've entered some weird parallel universe where the usual rules of maths no longer apply and things can have negative lengths, and this probably means that time runs backwards, my mum is my dad, my sister is a dog, my head is a fish, and my toe is a frog called Gerald'. As you might well imagine, it does the sensible thing and decides not to proceed.

Things like the KMO test and the determinant rely on a positive definite matrix; if you don't have one they can't be computed.

Why have I got a non-positive definite matrix?: The most likely answer is that you have too many variables and too few cases of data, which makes the correlation matrix a bit unstable. It could also be that you have too many highly correlated items in your matrix (singularity, for example, tends to mess things up). In any case it means that your data are bad, naughty data, and not to be trusted; if you let them loose then you have only yourself to blame for the consequences.

What can I do?: Other than cry, there's not that much you can do. You could try to limit your items, or selectively remove items (especially highly correlated ones) to see if that helps. Collecting more data can help too. There are some mathematical fudges you can do, but they're not as tasty as vanilla fudge and they are hard to implement easily.

17.7.1. Preliminary analysis ②

The first body of output concerns data screening, assumption testing and sampling adequacy. You'll find several large tables (or matrices) that tell us interesting things about our data. If you selected the *Univariate descriptives* option in Figure 17.7 then the first table will contain descriptive statistics for each variable (the mean, standard deviation and number of cases). This table is not included here, but you should have enough experience to be able to interpret it. The table also includes the number of missing cases; this summary is a useful way to determine the extent of missing data.

SPSS Output 17.1 shows the *R*-matrix (or correlation matrix)[7] produced using the *Coefficients* and *Significance levels* options in Figure 17.7. The top half of this table contains

[7] To save space I have edited out several columns of data from the large tables: only data for the first and last five questions in the questionnaire are included.

SPSS OUTPUT 17.1

Correlation Matrix[a]

		Q01	Q02	Q03	Q04	Q05	Q19	Q20	Q21	Q22	Q23
Correlation	Q01	1.000	-.099	-.337	.436	.402	-.189	.214	.329	-.104	-.004
	Q02	-.099	1.000	.318	-.112	-.119	.203	-.202	-.205	.231	.100
	Q03	-.337	.318	1.000	-.380	-.310	.342	-.325	-.417	.204	.150
	Q04	.436	-.112	-.380	1.000	.401	-.186	.243	.410	-.098	-.034
	Q05	.402	-.119	-.310	.401	1.000	-.165	.200	.335	-.133	-.042
	Q06	.217	-.074	-.227	.278	.257	-.167	.101	.272	-.165	-.069
	Q07	.305	-.159	-.382	.409	.339	-.269	.221	.483	-.168	-.070
	Q08	.331	-.050	-.259	.349	.269	-.159	.175	.296	-.079	-.050
	Q09	-.092	.315	.300	-.125	-.096	.249	-.159	-.136	.257	.171
	Q10	.214	-.084	-.193	.216	.258	-.127	.084	.193	-.131	-.062
	Q11	.357	-.144	-.351	.369	.298	-.200	.255	.346	-.162	-.086
	Q12	.345	-.195	-.410	.442	.347	-.267	.298	.441	-.167	-.046
	Q13	.355	-.143	-.318	.344	.302	-.227	.204	.374	-.195	-.053
	Q14	.338	-.165	-.371	.351	.315	-.254	.226	.399	-.170	-.048
	Q15	.246	-.165	-.312	.334	.261	-.210	.206	.300	-.168	-.062
	Q16	.499	-.168	-.419	.416	.395	-.267	.265	.421	-.156	-.082
	Q17	.371	-.087	-.327	.383	.310	-.163	.205	.363	-.126	-.092
	Q18	.347	-.164	-.375	.382	.322	-.257	.235	.430	-.160	-.080
	Q19	-.189	.203	.342	-.186	-.165	1.000	-.249	-.275	.234	.122
	Q20	.214	-.202	-.325	.243	.200	-.249	1.000	.468	-.100	-.035
	Q21	.329	-.205	-.417	.410	.335	-.275	.468	1.000	-.129	-.068
	Q22	-.104	.231	.204	-.098	-.133	.234	-.100	-.129	1.000	.230
	Q23	-.004	.100	.150	-.034	-.042	.122	-.035	-.068	.230	1.000
Sig. (1-tailed)	Q01		.000	.000	.000	.000	.000	.000	.000	.000	.410
	Q02	.000		.000	.000	.000	.000	.000	.000	.000	.000
	Q03	.000	.000		.000	.000	.000	.000	.000	.000	.000
	Q04	.000	.000	.000		.000	.000	.000	.000	.000	.043
	Q05	.000	.000	.000	.000		.000	.000	.000	.000	.017
	Q06	.000	.000	.000	.000	.000	.000	.000	.000	.000	.000
	Q07	.000	.000	.000	.000	.000	.000	.000	.000	.000	.000
	Q08	.000	.006	.000	.000	.000	.000	.000	.000	.000	.005
	Q09	.000	.000	.000	.000	.000	.000	.000	.000	.000	.000
	Q10	.000	.000	.000	.000	.000	.000	.000	.000	.000	.001
	Q11	.000	.000	.000	.000	.000	.000	.000	.000	.000	.000
	Q12	.000	.000	.000	.000	.000	.000	.000	.000	.000	.009
	Q13	.000	.000	.000	.000	.000	.000	.000	.000	.000	.004
	Q14	.000	.000	.000	.000	.000	.000	.000	.000	.000	.007
	Q15	.000	.000	.000	.000	.000	.000	.000	.000	.000	.001
	Q16	.000	.000	.000	.000	.000	.000	.000	.000	.000	.000
	Q17	.000	.000	.000	.000	.000	.000	.000	.000	.000	.000
	Q18	.000	.000	.000	.000	.000	.000	.000	.000	.000	.000
	Q19	.000	.000	.000	.000	.000		.000	.000	.000	.000
	Q20	.000	.000	.000	.000	.000	.000		.000	.000	.039
	Q21	.000	.000	.000	.000	.000	.000	.000		.000	.000
	Q22	.000	.000	.000	.000	.000	.000	.000	.000		.000
	Q23	.410	.000	.000	.043	.017	.000	.039	.000	.000	

a. Determinant = 5.271E-04

the Pearson correlation coefficient between all pairs of questions whereas the bottom half contains the one-tailed significance of these coefficients. You should be comfortable with the idea that to do a factor analysis we need to have variables that correlate fairly well, but not perfectly. Also, any variables that correlate with no others should be eliminated. Therefore, we can use this correlation matrix to check the pattern of relationships. First, scan the matrix for correlations greater than .3, then look for variables that only have a small number of correlations greater than this value. Then scan the correlation coefficients themselves and look for any greater than 0.9. If any are found then you should be aware that a problem could arise because of multicollinearity in the data.

You can also check the determinant of the correlation matrix and, if necessary, eliminate variables that you think are causing the problem. The determinant is listed at the bottom of the matrix (blink and you'll miss it). For these data its value is 5.271E-04 (which is 0.0005271) which is greater than the necessary value of 0.00001 (see section 17.6).[8] We can also compute Haitovsky's (1969) test of whether the determinant is 0 using equation

[8] SPSS 16 appears to report the determinant as = .001 for these data, which is wrong and I'm not sure why it does this. Hopefully SPSS will sort out this problem in the future.

(17.6). We have a sample of 2571 (*N*), 23 variables (*p*) and a determinant of 0.0005271, which gives us:

$$\begin{aligned}\text{Haitovsky's } \chi^2_H &= \left[1 + \frac{(2 \times 23 + 5)}{6} - 2571\right] ln(1 - 0.0005271) \\ &= [1 + 8.5 - 2571] ln(0.99947) \\ &= 1.35\end{aligned}$$

This test statistic has $p(p - 1)/2$ degrees of freedom, which is equal to $23(23 - 1)/2 = 253$. We can look in Appendix A.4, and for $df = 253$ the critical values are 233.99 ($df = 200$) and 341.40 ($df = 300$) and in both cases the observed chi-square is much smaller than these values indicating non-significance. As such, our determinant is not significantly different from zero.

Therefore, we have rather contradictory evidence about whether multicollinearity is a problem for these data. If we were doing factor analysis then we might want to explore this further and try to eliminate the problem (although it's not entirely obvious from the correlation matrix where the problem lies). However, because we are performing principal component analysis, we don't need to worry about multicollinearity.

In summary, all questions in the SAQ correlate reasonably well with all others and none of the correlation coefficients are excessively large; therefore, we won't eliminate any questions at this stage.

SPSS Output 17.2 shows the inverse of the correlation matrix (R^{-1}), which is used in various calculations (including factor scores – see section 17.3.3.1). This matrix is produced using the *Inverse* option in Figure 17.7 but in all honesty is useful only if you want some insight into the calculations that go on in a factor analysis. Most of us have more interesting things to do than gain insight into the workings of factor analysis and the practical use of this matrix is minimal – so ignore it.

SPSS Output 17.3 shows several very important parts of the output: the Kaiser–Meyer–Olkin measure of sampling adequacy, Bartlett's test of sphericity and the anti-image

SPSS OUTPUT 17.2

Inverse of Correlation Matrix

	Q01	Q02	Q03	Q04	Q05	Q19	Q20	Q21	Q22	Q23
Q01	1.595	-.028	.087	-.268	-.233	.017	-.024	.011	.002	-.078
Q02	-.028	1.232	-.224	-.057	.013	-.037	.076	.062	-.148	-.003
Q03	.087	-.224	1.661	.138	.057	-.175	.118	.122	-.009	-.103
Q04	-.268	-.057	.138	1.626	-.203	-.049	-.006	-.149	-.045	-.023
Q05	-.233	.013	.057	-.203	1.410	-.024	-.016	-.074	.045	-.006
Q06	.034	-.078	-.072	-.011	-.055	-.023	.080	.069	.058	.025
Q07	.039	.025	.127	-.152	-.072	.105	.077	-.386	.019	-.012
Q08	-.087	-.051	-.013	-.134	-.045	.074	.034	-.039	-.035	.003
Q09	-.023	-.242	-.208	.043	-.027	-.141	.050	-.047	-.156	-.110
Q10	-.017	-.015	-.023	.009	-.124	-.012	.056	.026	.023	.017
Q11	-.075	.061	.121	-.041	.000	-.010	-.140	-.009	.055	.015
Q12	-.011	.046	.147	-.259	-.091	.060	-.100	-.141	.026	-.038
Q13	-.145	-.011	-.055	.040	.007	.014	.028	-.061	.077	-.042
Q14	-.064	.033	.115	-.007	-.040	.063	.002	-.110	.041	-.034
Q15	.138	.050	.013	-.098	.021	.013	-.054	.058	.034	-.030
Q16	-.454	-.017	.142	-.063	-.155	.071	-.008	-.158	-.005	.033
Q17	-.084	-.045	.063	-.064	-.030	-.074	.025	-.077	.015	.080
Q18	-.041	.028	.070	-.044	.004	.047	-.004	-.136	-.037	.033
Q19	.017	-.037	-.175	-.049	-.024	1.264	.120	.048	-.141	-.045
Q20	-.024	.076	.118	-.006	-.016	.120	1.370	-.511	-.014	-.034
Q21	.011	.062	.122	-.149	-.074	.048	-.511	1.830	-.036	.018
Q22	.002	-.148	-.009	-.045	.045	-.141	-.014	-.036	1.200	-.202
Q23	-.078	-.003	-.103	-.023	-.006	-.045	-.034	.018	-.202	1.094

SPSS OUTPUT 17.3

KMO and Bartlett's Test

Kaiser-Meyer-Olkin Measure of Sampling Adequacy.		.930
Bartlett's Test of Sphericity	Approx. Chi-Square	19334.492
	df	253
	Sig.	.000

Anti-image Matrices

Statistics=Anti-image Correlation

Variables	Question_01	Question_02	Question_03	Question_04	Question_05	Question_19	Question_20	Question_21	Question_22	Question_23
Question_01	**.930**[a]	-.020	.053	-.167	-.156	.012	-.016	.006	.001	-.059
Question_02	-.020	**.875**[a]	-.157	-.041	.010	-.029	.059	.041	-.121	-.002
Question_03	.053	-.157	**.951**[a]	.084	.037	-.121	.078	.070	-.007	-.076
Question_04	-.167	-.041	.084	**.955**[a]	-.134	-.034	-.004	-.086	-.033	-.017
Question_05	-.156	.010	.037	-.134	**.960**[a]	-.018	-.011	-.046	.035	-.005
Question_06	.020	-.053	-.042	-.007	-.035	-.015	.051	.039	.040	.018
Question_07	.023	.016	.072	-.087	-.044	.068	.048	-.208	.013	-.008
Question_08	-.049	-.033	-.007	-.075	-.027	.047	.021	-.020	-.023	.002
Question_09	-.016	-.193	-.142	.030	-.020	-.111	.038	-.031	-.126	-.092
Question_10	-.012	-.012	-.016	.006	-.093	-.009	.043	.017	.019	.015
Question_11	-.041	.038	.064	-.022	-3.269E-5	-.006	-.082	-.005	.034	.010
Question_12	-.007	.031	.087	-.154	-.058	.040	-.065	-.079	.018	-.028
Question_13	-.085	-.008	-.032	.023	.004	.009	.018	-.033	.052	-.030
Question_14	-.040	.023	.069	-.004	-.026	.044	.001	-.063	.029	-.026
Question_15	.089	.037	.008	-.062	.014	.009	-.037	.035	.025	-.024
Question_16	-.264	-.011	.081	-.036	-.096	.047	-.005	-.085	-.003	.023
Question_17	-.047	-.029	.035	-.035	-.018	-.047	.015	-.041	.010	.055
Question_18	-.023	.018	.039	-.025	.002	.030	-.003	-.072	-.024	.023
Question_19	.012	-.029	-.121	-.034	-.018	**.941**[a]	.091	.031	-.115	-.038
Question_20	-.016	.059	.078	-.004	-.011	.091	**.889**[a]	-.323	-.011	-.028
Question_21	.006	.041	.070	-.086	-.046	.031	-.323	**.929**[a]	-.024	.013
Question_22	.001	-.121	-.007	-.033	.035	-.115	-.011	-.024	**.878**[a]	-.176
Question_23	-.059	-.002	-.076	-.017	-.005	-.038	-.028	.013	-.176	**.766**[a]

a. Measures of Sampling Adequacy(MSA)

correlation and covariance matrices (note that these matrices have been edited down to contain only the first and last five variables). The anti-image correlation and covariance matrices provide similar information (remember the relationship between covariance and correlation) and so only the anti-image correlation matrix need be studied in detail as it is the most informative. These tables are obtained using the *KMO and Bartlett's test of sphericity* and the *Anti-image* options in Figure 17.7.

We came across the KMO statistic in section 17.5.1.1 and saw that Kaiser (1974) recommends a bare minimum of 0.5 and that values between 0.5 and 0.7 are mediocre, values between 0.7 and 0.8 are good, values between 0.8 and 0.9 are great and values above 0.9 are superb (Hutcheson & Sofroniou, 1999). For these data the value is 0.93, which falls into the range of being superb, so we should be confident that the sample size is adequate for factor analysis.

I mentioned that KMO can be calculated for multiple and individual variables. The KMO values for individual variables are produced on the diagonal of the anti-image correlation matrix (I have highlighted these cells). These values make the anti-image correlation matrix an extremely important part of the output (although the anti-image covariance matrix can be ignored). As well as checking the overall KMO statistic, it is important to examine the diagonal elements of the anti-image correlation matrix: the value should be above the bare minimum of 0.5 for all variables (and preferably higher). For these data all values are well above 0.5, which is good news! If you find any variables with values below 0.5 then you should consider excluding them from the analysis (or run the analysis with and without that variable and note the difference). Removal of a variable affects the KMO statistics, so if you do remove a variable be sure to re-examine the new anti-image correlation matrix. As for the rest of the anti-image correlation matrix, the off-diagonal elements represent the partial correlations between variables. For a good factor analysis we want these correlations to be very small (the smaller, the better). So, as a final check you can just look through to see that the off-diagonal elements are small (they should be for these data).

Bartlett's measure tests the null hypothesis that the original correlation matrix is an identity matrix. For factor analysis to work we need some relationships between variables and if the *R*-matrix were an identity matrix then all correlation coefficients would be zero. Therefore, we want this test to be *significant* (i.e. have a significance value less than .05). A significant test tells us that the *R*-matrix is not an identity matrix; therefore, there are some relationships between the variables we hope to include in the analysis. For these data, Bartlett's test is highly significant ($p < .001$), and therefore factor analysis is appropriate.

CRAMMING SAM'S TIPS **Preliminary analysis**

- Scan the **Correlation Matrix**: look for variables that don't correlate with any other variables, or correlate very highly ($r = .9$) with one or more other variable. In factor analysis, check that the determinant of this matrix is bigger than 0.00001; if it is then multicollinearity isn't a problem.
- In the table labelled **KMO and Bartlett's Test** the KMO statistic should be greater than 0.5 as a bare minimum; if it isn't collect more data. Bartlett's test of sphericity should be significant (the value of *Sig*. should be less than .05). You can also check the KMO statistic for individual variables by looking at the diagonal of the **Anti-Image Matrices**: again, these values should be above 0.5 (this is useful for identifying problematic variables if the overall KMO is unsatisfactory).

17.7.2. Factor extraction ②

The first part of the factor extraction process is to determine the linear components within the data set (the eigenvectors) by calculating the eigenvalues of the *R*-matrix (see section 17.4.4). We know that there are as many components (eigenvectors) in the *R*-matrix as there are variables, but most will be unimportant. To determine the importance of a particular vector we look at the magnitude of the associated eigenvalue. We can then apply criteria to determine which factors to retain and which to discard. By default SPSS uses Kaiser's criterion of retaining factors with eigenvalues greater than 1 (see Figure 17.8).

SPSS Output 17.4 lists the eigenvalues associated with each linear component (factor) before extraction, after extraction and after rotation. Before extraction, SPSS has identified 23 linear components within the data set (we know that there should be as many eigenvectors as there are variables and so there will be as many factors as variables – see section 17.4.4). The eigenvalues associated with each factor represent the variance explained by that particular linear component and SPSS also displays the eigenvalue in terms of the percentage of variance explained (so, factor 1 explains 31.696% of total variance). It should be clear that the first few factors explain relatively large amounts of variance (especially factor 1) whereas subsequent factors explain only small amounts of variance. SPSS then extracts all factors with eigenvalues greater than 1, which leaves us with four factors. The eigenvalues associated with these factors are again displayed (and the percentage of variance explained) in the columns labelled *Extraction Sums of Squared Loadings*. The values in this part of the table are the same as the values before extraction, except that the values for the discarded factors are ignored (hence, the table is blank after the fourth factor). In the final part of the table (labelled *Rotation Sums of Squared Loadings*), the eigenvalues of the factors after rotation are displayed. Rotation has the effect of optimizing the factor structure and one consequence for these data is that the relative importance of the four factors is equalized. Before rotation, factor 1 accounted for considerably more variance than the remaining three (31.696%

SPSS OUTPUT 17.4

Total Variance Explained

Component	Initial Eigenvalues			Extraction Sums of Squared Loadings			Rotation Sums of Squared Loadings		
	Total	% of Variance	Cumulative %	Total	% of Variance	Cumulative %	Total	% of Variance	Cumulative %
1	7.290	31.696	31.696	7.290	31.696	31.696	3.730	16.219	16.219
2	1.739	7.560	39.256	1.739	7.560	39.256	3.340	14.523	30.742
3	1.317	5.725	44.981	1.317	5.725	44.981	2.553	11.099	41.842
4	1.227	5.336	50.317	1.227	5.336	50.317	1.949	8.475	50.317
5	.988	4.295	54.612						
6	.895	3.893	58.504						
7	.806	3.502	62.007						
8	.783	3.404	65.410						
9	.751	3.265	68.676						
10	.717	3.117	71.793						
11	.684	2.972	74.765						
12	.670	2.911	77.676						
13	.612	2.661	80.337						
14	.578	2.512	82.849						
15	.549	2.388	85.236						
16	.523	2.275	87.511						
17	.508	2.210	89.721						
18	.456	1.982	91.704						
19	.424	1.843	93.546						
20	.408	1.773	95.319						
21	.379	1.650	96.969						
22	.364	1.583	98.552						
23	.333	1.448	100.000						

Extraction Method: Principal Component Analysis.

compared to 7.560, 5.725 and 5.336%), but after extraction it accounts for only 16.219% of variance (compared to 14.523, 11.099 and 8.475% respectively).

SPSS Output 17.5 shows the table of communalities before and after extraction. Remember that the communality is the proportion of common variance within a variable (see section 17.4.1). Principal component analysis works on the initial assumption that all variance is common; therefore, before extraction the communalities are all 1 (see column labelled *Initial*). In effect, all of the variance associated with a variable is assumed to be common variance. Once factors have been extracted, we have a better idea of how much variance is, in reality, common. The communalities in the column labelled *Extraction* reflect this common variance. So, for example, we can say that 43.5% of the variance associated with question 1 is common, or shared, variance. Another way to look at these communalities is in terms of the proportion of variance explained by the underlying factors. Before extraction, there are as many factors as there are variables, so all variance is explained by the factors and communalities are all 1. However, after extraction some of the factors are discarded and so some information is lost. The retained factors cannot explain all of the variance present in the data, but they can explain some. The amount of variance in each variable that can be explained by the retained factors is represented by the communalities after extraction.

SPSS Output 17.5 also shows the component matrix before rotation. This matrix contains the loadings of each variable onto each factor. By default SPSS displays all loadings; however, we requested that all loadings less than 0.4 be suppressed in the output (see Figure 17.11) and so there are blank spaces for many of the loadings. This matrix is not particularly important for interpretation, but it is interesting to note that before rotation most variables load highly onto the first factor (that is why this factor accounts for most of the variance in SPSS Output 17.4).

At this stage SPSS has extracted four factors. Factor analysis is an exploratory tool and so it should be used to guide the researcher to make various decisions: you shouldn't leave the computer to make them. One important decision is the number of factors to extract. In section 17.4.5 we saw various criteria for assessing the importance of factors. By Kaiser's criterion we should extract four factors and this is what SPSS has done. However,

SPSS OUTPUT 17.5

Communalities

	Initial	Extraction
Q01	1.000	.435
Q02	1.000	.414
Q03	1.000	.530
Q04	1.000	.469
Q05	1.000	.343
Q06	1.000	.654
Q07	1.000	.545
Q08	1.000	.739
Q09	1.000	.484
Q10	1.000	.335
Q11	1.000	.690
Q12	1.000	.513
Q13	1.000	.536
Q14	1.000	.488
Q15	1.000	.378
Q16	1.000	.487
Q17	1.000	.683
Q18	1.000	.597
Q19	1.000	.343
Q20	1.000	.484
Q21	1.000	.550
Q22	1.000	.464
Q23	1.000	.412

Extraction Method: Principal Component

Component Matrix[a]

	Component			
	1	2	3	4
Q18	.701			
Q07	.685			
Q16	.679			
Q13	.673			
Q12	.669			
Q21	.658			
Q14	.656			
Q11	.652			-.400
Q17	.643			
Q04	.634			
Q03	-.629			
Q15	.593			
Q01	.586			
Q05	.556			
Q08	.549	.401		-.417
Q18	.437			
Q20	.436		-.404	
Q19	-.427			
Q09		.627		
Q02		.548		
Q22		.465		
Q06	.562		.571	
Q23				.507

Extraction Method: Principal Component Analysis.
a. 4 components extracted.

this criterion is accurate when there are less than 30 variables and communalities after extraction are greater than 0.7 or when the sample size exceeds 250 and the average communality is greater than 0.6. The communalities are shown in SPSS Output 17.5, and only one exceeds 0.7. The average of the communalities can be found by adding them up and dividing by the number of communalities (11.573/23 = 0.503). So, on both grounds Kaiser's rule may not be accurate. However, you should consider the huge sample that we have, because the research into Kaiser's criterion gives recommendations for much smaller samples. By Jolliffe's criterion (retain factors with eigenvalues greater than 0.7) we should retain 10 factors (see SPSS Output 17.4), but there is little to recommend this criterion over Kaiser's. As a final guide we can use the scree plot which we asked SPSS to produce by using the option in Figure 17.8. The scree plot is shown in SPSS Output 17.6. This curve is difficult to interpret because it begins to tail off after three factors, but there is another drop after four factors before a stable plateau is reached. Therefore, we could probably justify retaining either two or four factors. Given the large sample, it is probably safe to assume Kaiser's criterion; however, you might like to rerun the analysis specifying that SPSS extract only two factors (see Figure 17.8) and compare the results.

SPSS Output 17.7 shows an edited version of the reproduced correlation matrix that was requested using the option in Figure 17.7. The top half of this matrix (labelled **Reproduced Correlations)** contains the correlation coefficients between all of the questions based on the factor model. The diagonal of this matrix contains the communalities after extraction for each variable (you can check the values against SPSS Output 17.5).

The correlations in the reproduced matrix differ from those in the *R*-matrix because they stem from the model rather than the observed data. If the model were a perfect fit of the data then we would expect the reproduced correlation coefficients to be the same as the original correlation coefficients. Therefore, to assess the fit of the model we can look at

SPSS OUTPUT 17.6

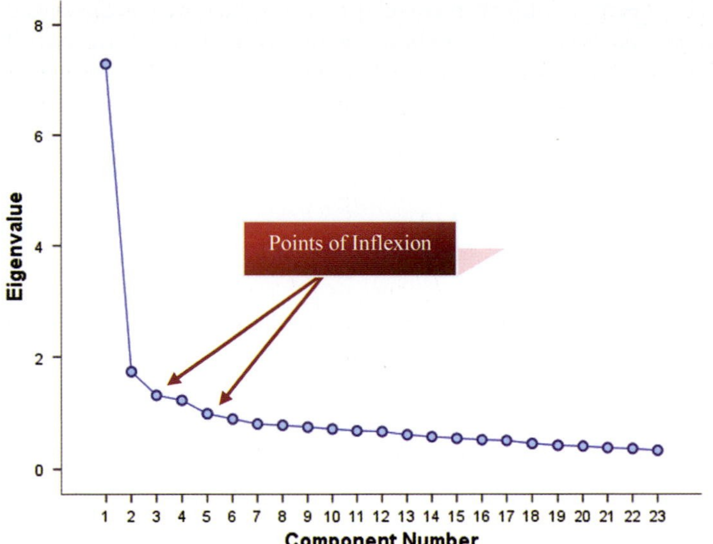

SPSS OUTPUT 17.7

Reproduced Correlations

		Question_01	Question_02	Question_03	Question_04	Question_05	Question_19	Question_20	Question_21	Question_22	Question_23
Reproduced Correlation	Question_01	**.435**[a]	-.112	-.372	.447	.376	-.204	.342	.449	-.025	.045
	Question_02	-.112	**.414**[a]	.380	-.134	-.122	.357	-.301	-.254	.333	.246
	Question_03	-.372	.380	**.530**[a]	-.399	-.345	.403	-.440	-.488	.275	.158
	Question_04	.447	-.134	-.399	**.469**[a]	.399	-.231	.353	.480	-.050	.042
	Question_05	.376	-.122	-.345	.399	**.343**[a]	-.207	.292	.412	-.060	.028
	Question_06	.218	-.033	-.200	.278	.273	-.147	-.021	.244	-.209	-.082
	Question_07	.366	-.148	-.373	.419	.380	-.254	.219	.430	-.179	-.037
	Question_08	.412	.002	-.270	.390	.312	-.104	.164	.282	-.099	-.136
	Question_09	-.042	.430	.352	-.073	-.080	.363	-.218	-.191	.417	.323
	Question_10	.172	-.061	-.181	.212	.205	-.137	.006	.188	-.197	-.110
	Question_11	.423	-.097	-.357	.419	.348	-.198	.200	.342	-.209	-.202
	Question_12	.402	-.219	-.440	.448	.397	-.302	.354	.503	-.136	.011
	Question_13	.347	-.122	-.342	.395	.360	-.231	.163	.384	-.203	-.079
	Question_14	.362	-.155	-.373	.411	.370	-.254	.241	.431	-.159	-.022
	Question_15	.311	-.158	-.337	.343	.306	-.236	.175	.336	-.230	-.141
	Question_16	.440	-.217	-.458	.466	.400	-.299	.373	.494	-.152	-.049
	Question_17	.439	-.048	-.331	.434	.359	-.162	.196	.347	-.145	-.140
	Question_18	.368	-.149	-.376	.424	.388	-.259	.215	.439	-.183	-.032
	Question_19	-.204	.357	.403	-.231	-.207	**.343**[a]	-.308	-.324	.294	.196
	Question_20	.342	-.301	-.440	.353	.292	-.308	**.484**[a]	.457	-.068	.021
	Question_21	.449	-.254	-.488	.480	.412	-.324	.457	**.550**[a]	-.096	.032
	Question_22	-.025	.333	.275	-.050	-.060	.294	-.068	-.096	**.464**[a]	.408
	Question_23	.045	.246	.158	.042	.028	.196	.021	.032	.408	**.412**[a]
Residual[b]	Question_01		.013	.035	-.011	.027	.015	-.128	-.120	-.079	-.049
	Question_02	.013		-.062	.022	.003	-.153	.099	.049	-.102	-.147
	Question_03	.035	-.062		.019	.035	-.061	.115	.071	-.071	-.008
	Question_04	-.011	.022	.019		.002	.045	-.110	-.070	-.049	-.076
	Question_05	.027	.003	.035	.002		.041	-.092	-.078	-.072	-.070
	Question_06	.000	-.041	-.027	.000	-.016	-.020	.122	.029	.043	.013
	Question_07	-.061	-.011	-.009	-.010	-.041	-.015	.002	.053	.010	-.033
	Question_08	-.081	-.052	.011	-.041	-.044	-.056	.011	.014	.020	.086
	Question_09	-.050	-.115	-.052	-.051	-.016	-.114	.060	.055	-.161	-.152
	Question_10	.042	-.023	-.013	.003	.053	.010	.078	.005	.066	.048
	Question_11	-.066	-.046	.006	-.051	-.050	-.002	.056	.005	.047	.116
	Question_12	-.057	.024	.030	-.006	-.050	.036	-.056	-.062	-.031	-.057
	Question_13	.008	-.021	.024	-.051	-.058	.004	.041	-.010	.007	.026
	Question_14	-.024	-.009	.002	-.060	-.055	.000	-.015	-.032	-.011	-.027
	Question_15	-.065	-.007	.025	-.009	-.045	.027	.031	-.036	.062	.079
	Question_16	.059	.050	.039	-.050	-.005	.032	-.108	-.074	-.004	-.032
	Question_17	-.069	-.039	.003	-.052	-.049	-.001	.009	.016	.019	.049
	Question_18	-.020	-.015	.001	-.042	-.066	.003	.020	-.009	.023	-.049
	Question_19	.015	-.153	-.061	.045	.041		.060	.049	-.060	-.073
	Question_20	-.128	.099	.115	-.110	-.092	.060		.010	-.032	-.056
	Question_21	-.120	.049	.071	-.070	-.078	.049	.010		-.033	-.100
	Question_22	-.079	-.102	-.071	-.049	-.072	-.060	-.032	-.033		-.177
	Question_23	-.049	-.147	-.008	-.076	-.070	-.073	-.056	-.100	-.177	

Extraction Method: Principal Component Analysis.

a. Reproduced communalities

b. Residuals are computed between observed and reproduced correlations. There are 91 (35.0%) nonredundant residuals with absolute values greater than 0.05.

the differences between the observed correlations and the correlations based on the model. For example, if we take the correlation between questions 1 and 2, the correlation based on the observed data is –0.099 (taken from SPSS Output 17.1). The correlation based on the model is –0.112, which is slightly higher. We can calculate the difference as follows:

$$\begin{aligned} \text{residual} &= r_{\text{observed}} - r_{\text{from model}} \\ \text{residual}_{Q_1 Q_2} &= (-0.099) - (-0.112) \\ &= 0.013 \end{aligned}$$

You should notice that this difference is the value quoted in the lower half of the reproduced matrix (labelled *Residual*) for questions 1 and 2. Therefore, the lower half of the reproduced matrix contains the differences between the observed correlation coefficients and the ones predicted from the model. For a good model these values will all be small. In fact, we want most values to be less than 0.05. Rather than scan this huge matrix, SPSS provides a footnote summary, which states how many residuals have an absolute value greater than 0.05. For these data there are 91 residuals (35%) that are greater than 0.05. There are no hard and fast rules about what proportion of residuals should be below 0.05; however, if more than 50% are greater than 0.05 you probably have grounds for concern.

CRAMMING SAM'S TIPS **Factor extraction**

- To decide how many factors to extract look at the table labelled **Communalities** and the column labelled *Extraction*. If these values are all 0.7 or above and you have less than 30 variables then the SPSS default option for extracting factors is fine (Kaiser's criterion of retaining factors with eigenvalues greater than 1). Likewise, if your sample size exceeds 250 and the average of the communalities is 0.6 or greater then the default option is fine. Alternatively, with 200 or more participants the scree plot can be used.
- Check the bottom of the table labelled **Reproduced Correlations** for the percentage of 'nonredundant residuals with absolute values > 0.05'. This percentage should be less than 50% and the smaller it is, the better.

17.7.3. Factor rotation ②

The first analysis I asked you to run was using an orthogonal rotation. However, you were asked to rerun the analysis using oblique rotation too. In this section the results of both analyses will be reported so as to highlight the differences between the outputs. This comparison will also be a useful way to show the circumstances in which one type of rotation might be preferable to another.

17.7.3.1. Orthogonal rotation (varimax) ②

SPSS Output 17.8 shows the rotated component matrix (also called the rotated factor matrix in factor analysis) which is a matrix of the factor loadings for each variable onto each factor. This matrix contains the same information as the component matrix in SPSS

SPSS OUTPUT 17.8

Rotated Component Matrix[a]

	Component			
	1	2	3	4
I have little experience of computers	.800			
SPSS always crashes when I try to use it	.684			
I worry that I will cause irreparable damage because of my incompetenece with computers	.647			
All computers hate me	.638			
Computers have minds of their own and deliberately go wrong whenever I use them	.579			
Computers are useful only for playing games	.550			
Computers are out to get me	.459			
I can't sleep for thoughts of eigen vectors		.677		
I wake up under my duvet thinking that I am trapped under a normal distribtion		.661		
Standard deviations excite me		-.567		
People try to tell you that SPSS makes statistics easier to understand but it doesn't	.473	.523		
I dream that Pearson is attacking me with correlation coefficients		.516		
I weep openly at the mention of central tendency		.514		
Statiscs makes me cry		.496		
I don't understand statistics		.429		
I have never been good at mathematics			.833	
I slip into a coma whenever I see an equation			.747	
I did badly at mathematics at school			.747	
My friends are better at statistics than me				.648
My friends are better at SPSS than I am				.645
If I'm good at statistics my friends will think I'm a nerd				.586
My friends will think I'm stupid for not being able to cope with SPSS				.543
Everybody looks at me when I use SPSS				.427

Extraction Method: Principal Component Analysis.
Rotation Method: Varimax with Kaiser Normalization.

a. Rotation converged in 9 iterations.

Component Transformation Matrix

Component	1	2	3	4
1	.635	.585	.443	-.242
2	.137	-.168	.488	.846
3	.758	-.513	-.403	.008
4	.067	.605	-.635	.476

Extraction Method: Principal Component Analysis.
Rotation Method: Varimax with Kaiser Normalization.

Output 17.5 except that it is calculated *after* rotation. There are several things to consider about the format of this matrix. First, factor loadings less than 0.4 have not been displayed because we asked for these loadings to be suppressed using the option in Figure 17.11. If you didn't select this option, or didn't adjust the criterion value to 0.4, then your output will differ. Second, the variables are listed in the order of size of their factor loadings. By default, SPSS orders the variables as they are in the data editor; however, we asked for the output to be *Sorted by size* using the option in Figure 17.11. If this option was not selected your output will look different. Finally, for all other parts of the output I suppressed the variable labels (for reasons of space) but for this matrix I have allowed the variable labels to be printed to aid interpretation.

The original logic behind suppressing loadings less than 0.4 was based on Stevens' (2002) suggestion that this cut-off point was appropriate for interpretative purposes (i.e. loadings greater than 0.4 represent substantive values). However, this means that we have suppressed several loadings that are undoubtedly significant (see section 17.4.6.2). However, significance itself is not important.

Compare this matrix to the unrotated solution (SPSS Output 17.5). Before rotation, most variables loaded highly onto the first factor and the remaining factors didn't really get a look in. However, the rotation of the factor structure has clarified things considerably: there are four factors and variables load very highly onto only one factor (with the exception of one question). The suppression of loadings less than 0.4 and ordering variables by loading size also make interpretation considerably easier (because you don't have to scan the matrix to identify substantive loadings).

The next step is to look at the content of questions that load onto the same factor to try to identify common themes. If the mathematical factor produced by the analysis represents some real-world construct then common themes among highly loading questions can help us identify what the construct might be. The questions that load highly on factor 1 seem to all relate to using computers or SPSS. Therefore we might label this factor *fear of computers*. The questions that load highly on factor 2 all seem to relate to different aspects of statistics; therefore, we might label this factor *fear of statistics*. The three questions that load highly on factor 3 all seem to relate to mathematics; therefore, we might label this factor *fear of mathematics*. Finally, the questions that load highly on factor 4 all contain some component of social evaluation from friends; therefore, we might label this factor *peer evaluation*. This analysis seems to reveal that the initial questionnaire, in reality, is composed of four subscales: fear of computers, fear of statistics, fear of maths and fear of negative peer evaluation. There are two possibilities here. The first is that the SAQ failed to measure what it set out to (namely, SPSS anxiety) but does measure some related constructs. The second is that these four constructs are sub-components of SPSS anxiety; however, the factor analysis does not indicate which of these possibilities is true.

The final part of the output is the factor transformation matrix (see section 17.4.6). This matrix provides information about the degree to which the factors were rotated to obtain a solution. If no rotation were necessary, this matrix would be an identity matrix. If orthogonal rotation were completely appropriate then we would expect a symmetrical matrix (same values above and below the diagonal). However, in reality the matrix is not easy to interpret although very unsymmetrical matrices might be taken as a reason to try oblique rotation. For the inexperienced factor analyst you are probably best advised to ignore the factor transformation matrix.

17.7.3.2. Oblique rotation ②

When an oblique rotation is conducted the factor matrix is split into two matrices: the *pattern matrix* and the *structure matrix* (see Jane Superbrain Box 17.1). For orthogonal rotation these matrices are the same. The pattern matrix contains the factor loadings and is comparable to the factor matrix that we interpreted for the orthogonal rotation. The structure matrix takes into account the relationship between factors (in fact it is a product of the pattern matrix and the matrix containing the correlation coefficients between factors). Most researchers interpret the pattern matrix, because it is usually simpler; however, there are situations in which values in the pattern matrix are suppressed because of relationships between the factors. Therefore, the structure matrix is a useful

SPSS OUTPUT 17.9

Pattern Matrix[a]

	Component			
	1	2	3	4
I can't sleep for thoughts of eigen vectors	.706			
I wake up under my duvet thinking that I am trapped under a normal distribtion	.591			
Standard deviations excite me	-.511			
I dream that Pearson is attacking me with correlation coefficients	.405			
I weep openly at the mention of central tendency	.400			
Statiscs makes me cry				
I don't understand statistics				
My friends are better at SPSS than I am		.643		
My friends are better at statistics than me		.621		
If I'm good at statistics my friends will think I'm a nerd		.615		
My friends will think I'm stupid for not being able to cope with SPSS		.507		
Everybody looks at me when I use SPSS				
I have little experience of computers			.885	
SPSS always crashes when I try to use it			.713	
All computers hate me			.653	
I worry that I will cause irreparable damage because of my incompetenece with computers			.650	
Computers have minds of their own and deliberately go wrong whenever I use them			.588	
Computers are useful only for playing games			.585	
People try to tell you that SPSS makes statistics easier to understand but it doesn't	.412		.462	
Computers are out to get me			.411	
I have never been good at mathematics				-.902
I slip into a coma whenever I see an equation				-.774
I did badly at mathematics at school				-.774

Extraction Method: Principal Component Analysis.
Rotation Method: Oblimin with Kaiser Normalization.
a. Rotation converged in 29 iterations.

double-check and Graham et al. (2003) recommend reporting both (with some useful examples of why this can be important).

For the pattern matrix for these data (SPSS Output 17.9) the same four factors seem to have emerged (although for some variables the factor loadings are too small to be displayed). Factor 1 seems to represent fear of statistics, factor 2 represents fear of peer evaluation, factor 3 represents fear of computers and factor 4 represents fear of mathematics. The structure matrix (SPSS Output 17.10) differs in that shared variance is not ignored. The picture becomes more complicated because with the exception of factor 2, several variables load highly onto more than one factor. This has occurred because of the relationship between factors 1 and 3 and factors 3 and 4. This example should highlight why the pattern matrix is preferable for interpretative reasons: because it contains information about the *unique* contribution of a variable to a factor.

The final part of the output is a correlation matrix between the factors (SPSS Output 17.11). This matrix contains the correlation coefficients between factors. As predicted from the structure matrix, factor 2 has little or no relationship with any other factors (correlation coefficients are low), but all other factors are interrelated to some degree (notably factors 1 and 3 and factors 3 and 4). The fact that these correlations exist tell us that the constructs measured can be interrelated. If the constructs were independent then we would expect oblique rotation to provide an identical solution to an orthogonal rotation and the

SPSS OUTPUT 17.10

Structure Matrix

	Component			
	1	2	3	4
I wake up under my duvet thinking that I am trapped under a normal distribtion	.695		.477	
I can't sleep for thoughts of eigen vectors	.685			
Standard deviations excite me	-.632		-.407	
I weep openly at the mention of central tendency	.567		.516	-.491
I dream that Pearson is attacking me with correlation coefficients	.548		.487	-.485
Statiscs makes me cry	.520		.413	-.501
I don't understand statistics	.462		.453	
My friends are better at SPSS than I am		.660		
My friends are better at statistics than me		.653		
If I'm good at statistics my friends will think I'm a nerd		.588		
My friends will think I'm stupid for not being able to cope with SPSS		.546		
Everybody looks at me when I use SPSS	-.435	.446		
I have little experience of computers			.777	
SPSS always crashes when I try to use it	.404		.761	
All computers hate me	.401		.723	
I worry that I will cause irreparable damage because of my incompetenece with computers			.723	-.429
Computers have minds of their own and deliberately go wrong whenever I use them	.426		.671	
People try to tell you that SPSS makes statistics easier to understand but it doesn't	.576		.606	
Computers are out to get me			.561	-.441
Computers are useful only for playing games			.556	
I have never been good at mathematics				-.855
I slip into a coma whenever I see an equation			.453	-.822
I did badly at mathematics at school			.451	-.818

Extraction Method: Principal Component Analysis.
Rotation Method: Oblimin with Kaiser Normalization.

SPSS OUTPUT 17.11

Component Correlation Matrix

Component	1	2	3	4
1	1.000	-.154	.364	-.279
2	-.154	1.000	-.185	8.155E-02
3	.364	-.185	1.000	-.464
4	-.279	8.155E-02	-.464	1.000

Extraction Method: Principal Component Analysis.
Rotation Method: Oblimin with Kaiser Normalization.

component correlation matrix should be an identity matrix (i.e. all factors have correlation coefficients of 0). Therefore, this final matrix gives us a guide to whether it is reasonable to assume independence between factors: for these data it appears that we cannot assume independence. Therefore, the results of the orthogonal rotation should not be trusted: the obliquely rotated solution is probably more meaningful.

On a theoretical level the dependence between our factors does not cause concern; we might expect a fairly strong relationship between fear of maths, fear of statistics and fear of computers. Generally, the less mathematically and technically minded people struggle with statistics. However, we would not expect these constructs to correlate with fear of peer evaluation (because this construct is more socially based). In fact, this factor is the one that correlates fairly badly with all others – so on a theoretical level, things have turned out rather well!

CRAMMING SAM'S TIPS Interpretation

- If you've conduced orthogonal rotation then look at the table labelled **Rotated Component Matrix**. For each variable, note the component for which the variable has the highest loading. Also, for each component, note the variables that load highly onto it (by high I'd say loadings should be above 0.4 when you ignore the plus or minus sign). Try to make sense of what the factors represent by looking for common themes in the items that load onto them.
- If you've conducted oblique rotation then look at the table labelled **Pattern Matrix**. For each variable, note the component for which the variable has the highest loading. Also, for each component, note the variables that load highly onto it (by high I'd say loadings should be above 0.4 when you ignore the plus or minus sign). Double check what you find by doing the same thing for the **Structure Matrix**. Try to make sense of what the factors represents by looking for common themes in the items that load onto them.

17.7.4. Factor scores ②

Having reached a suitable solution and rotated that solution we can look at the factor scores. SPSS Output 17.12 shows the component score matrix *B* (see section 17.3.3.1) from which the factor scores are calculated and the covariance matrix of factor scores. The component score matrix is not particularly useful in itself. It can be useful in understanding how the factor scores have been computed, but with large data sets like this one you are unlikely to want to delve into the mathematics behind the factor scores. However, the covariance matrix of scores is useful. This matrix in effect tells us the relationship between factor scores (it is an unstandardized correlation matrix). If factor scores are uncorrelated then this matrix should be an identity matrix (i.e. diagonal elements will be 1 but all other elements are 0). For these data the covariances are all zero indicating that the resulting scores are uncorrelated.

In the original analysis we asked for scores to be calculated based on the Anderson–Rubin method (hence why they are uncorrelated). You will find these scores in the data editor. There should be four new columns of data (one for each factor) labelled *FAC1_1*, *FAC2_1*, *FAC3_1* and *FAC4_1* respectively. If you asked for factor scores in the oblique rotation then these scores will appear in the data editor in four other columns labelled *FAC2_1* and so on.

SELF-TEST Using what you learnt in section 7.8.6 use the *Case Summaries* command to list the factor scores for these data (given that there are over 2500 cases, you might like to restrict the output to the first 10 or 20).

SPSS Output 17.13 shows the factor scores for the first 10 participants. It should be pretty clear that participant 9 scored highly on all four factors and so this person is very anxious about statistics, computing and maths, but less so about peer evaluation (factor 4).

SPSS OUTPUT 17.12

Component Score Coefficient Matrix

	Component			
	1	2	3	4
Question_01	-.053	.173	.089	.110
Question_02	.102	-.129	.086	.282
Question_03	.087	-.195	.013	.137
Question_04	-.011	.170	.045	.107
Question_05	.021	.131	.014	.083
Question_06	.383	-.211	-.088	.014
Question_07	.213	.004	-.078	.038
Question_08	-.129	-.074	.460	.013
Question_09	.025	-.029	.108	.354
Question_10	.244	-.161	-.021	-.036
Question_11	-.066	-.087	.379	-.059
Question_12	.097	.161	-.116	.051
Question_13	.224	-.065	-.019	.013
Question_14	.180	.040	-.084	.043
Question_15	.114	-.055	.061	-.058
Question_16	-.015	.146	.046	.014
Question_17	-.057	-.067	.372	.005
Question_18	.242	-.001	-.104	.043
Question_19	.048	-.115	.061	.199
Question_20	-.195	.359	-.061	-.002
Question_21	-.039	.270	-.064	.059
Question_22	-.036	.162	-.048	.382
Question_23	.032	.211	-.162	.379

Extraction Method: Principal Component Analysis.
Rotation Method: Varimax with Kaiser Normalization.

Component Score Covariance Matrix

Component	1	2	3	4
1	1.000	.000	.000	.000
2	.000	1.000	.000	.000
3	.000	.000	1.000	.000
4	.000	.000	.000	1.000

Extraction Method: Principal Component Analysis.
Rotation Method: Varimax with Kaiser Normalization.

SPSS OUTPUT 17.13

Case Summaries[a]

	FAC1_1	FAC2_1	FAC3_1	FAC4_1
1	.10587	-.92816	-1.82760	-.45953
2	-.58279	-.18932	-.04137	.29332
3	-.54761	.02953	.19918	-.97142
4	.74661	.72268	-.68654	-.18547
5	.25169	-.51489	-.63354	.68301
6	1.91614	-.27356	-.68201	-.52709
7	-.26051	-1.40947	-.00431	.91008
8	-.28572	-.91761	-.08947	1.03746
9	1.71750	1.15102	3.15659	.81082
10	-.69151	-.73318	.18376	1.49541
Total N	10	10	10	10

a. Limited to first 10 cases.

Factor scores can be used in this way to assess the relative fear of one person compared to another, or we could add the scores up to obtain a single score for each participant (that we might assume represents SPSS anxiety as a whole). We can also use factor scores in regression when groups of predictors correlate so highly that there is multicollinearity.

17.7.5. Summary ②

To sum up, the analyses revealed four underlying scales in our questionnaire that may, or may not, relate to genuine sub-components of SPSS anxiety. It also seems as though an obliquely rotated solution was preferred due to the interrelationships between factors. The use of factor analysis is purely exploratory; it should be used only to guide future hypotheses, or to inform researchers about patterns within data sets. A great many decisions are left to the researcher using factor analysis and I urge you to make informed decisions, rather than basing decisions on the outcomes you would like to get. The next question is whether or not our scale is reliable.

17.8. How to report factor analysis ①

As with any analysis, when reporting factor analysis we need to provide our readers with enough information to make an informed opinion about our data. As a bare minimum we should be very clear about our criteria for extracting factors and the method of rotation used. We must also produce a table of the rotated factor loadings of all items and flag (in bold) values above a criterion level (I would personally choose .40 but I discussed the various criteria you could use in section 17.4.6.2). You should also report the percentage of variance that each factor explains and possibly the eigenvalue too. Table 17.2 shows an example of such a table for the SAQ data; note that I have also reported the sample size in the title.

In my opinion, a table of factor loadings and a description of the analysis are a bare minimum, though. You could consider (if it's not too large) including the table of correlations from which someone could reproduce your analysis (should they want to). You could also consider including some information on sample size adequacy. For this example we might write something like this (although obviously you don't have to cite this book as much as I have):

✓ A principal component analysis (PCA) was conducted on the 23 items with orthogonal rotation (varimax). The Kaiser–Meyer–Olkin measure verified the sampling adequacy for the analysis, KMO = .93 ('superb' according to Field, 2009), and all KMO values for individual items were > .77, which is well above the acceptable limit of .5 (Field, 2009). Bartlett's test of sphericity χ^2 (253) = 19334.49, $p < .001$, indicated that correlations between items were sufficiently large for PCA. An initial analysis was run to obtain eigenvalues for each component in the data. Four components had eigenvalues over Kaiser's criterion of 1 and in combination explained 50.32% of the variance. The scree plot was slightly ambiguous and showed inflexions that would justify retaining both components 2 and 4. Given the large sample size, and the convergence of the scree plot and Kaiser's criterion on four components, this is the number of components that were retained in the final analysis. Table 17.2 shows the factor loadings after rotation. The items that cluster on the same components suggest that component 1 represents a fear of computers, component 2 a fear of statistics, component 3 a fear of maths and component 4 peer evaluation concerns.

Finally, if you have used oblique rotation you should consider reporting a table of both the structure and pattern matrix because the loadings in these tables have different interpretations (see Jane Superbrain Box 17.1).

[illegible] factor analysis results for the SPSS anxiety questionnaire (N = 2571)

	Rotated Factor Loadings			
	Fear of Computers	**Fear of Statistics**	**Fear of Maths**	**Peer Evaluation**
[illegible]s	**.80**	−.01	.10	−.07
[illegible] use it	**.68**	.33	.13	−.08
I worry that I will cause irreparable damage because of my incompetence with computers	**.65**	.23	.23	−.10
All computers hate me	**.64**	.33	.16	−.08
Computers have minds of their own and deliberately go wrong whenever I use them	**.58**	.36	.14	−.07
Computers are useful only for playing games	**.55**	.00	.13	−.12
Computers are out to get me	**.46**	.22	.29	−.19
I can't sleep for thoughts of eigenvectors	-.04	**.68**	.08	−.14
I wake up under my duvet thinking that I am trapped under a normal distribution	.29	**.66**	.16	−.07
Standard deviations excite me	−.20	**−.57**	−.18	.37
People try to tell you that SPSS makes statistics easier to understand but it doesn't	**.47**	**.52**	.10	−.08
I dream that Pearson is attacking me with correlation coefficients	.32	**.52**	.31	.04
I weep openly at the mention of central tendency	.33	**.51**	.31	−.12
Statistics makes me cry	.24	**.50**	.36	.06
I don't understand statistics	.32	**.43**	.24	.02
I have never been good at mathematics	.13	.17	**.83**	.01
I slip into a coma whenever I see an equation	.27	.22	**.75**	−.04
I did badly at mathematics at school	.26	.21	**.75**	−.14
My friends are better at statistics than me	−.09	−.20	.12	**.65**
My friends are better at SPSS than I am	−.19	.03	−.10	**.65**
If I'm good at statistics my friends will think I'm a nerd	−.02	.17	−.20	**.59**
My friends will think I'm stupid for not being able to cope with SPSS	−.01	−.34	.07	**.54**
Everybody looks at me when I use SPSS	−.15	−.37	−.03	**.43**
Eigenvalues	3.73	3.34	2.55	1.95
% of variance	16.22	14.52	11.10	8.48
α	.82	.82	.82	.57

Note: Factor loadings over .40 appear in bold.

LABCOAT LENI'S REAL RESEARCH 17.1

World wide addiction? ②

The Internet is now a houshold tool. In 2007 it was estimated that around 179 million people worldwide used the Internet (over 100 million of those were in the USA and Canada). From the increasing populatrity (and usefulness) of the Internet has emerged a new phenomenon: Internet addiction. This is now a serious and recognized problem, but until very recently it was very difficult to research this topic because there was not a psychometrically sound measure of Internet addition. That is, until Laura Nichols and Richard Nicki developed the Internet Addiction Scale, IAS (Nichols & Nicki, 2004). (Incidentally, while doing some research on this topic I encountered an Internet addiction recovery website that I won't name but offered a whole host of resources that would keep you online for ages, such as questionnaires, an online support group, videos, articles, a recovery blog and podcasts. It struck me that this was a bit like having a recovery centre for heroin addiction where the addict arrives to be greeted by a nice-looking counsellor who says 'there's a huge pile of heroin in the corner over there, just help yourself').

Anyway, Nichols and Nicki developed a 36-item questionnaire to measure Internet addiction. It contained items such as 'I have stayed on the Internet longer than I intended to' and 'My grades/work have suffered because of my Internet use' which could be responded to on a 5-point scale (Never, Rarely, Sometimes, Frequently, Always). They collected data from 207 people to validate this measure.

The data from this study are in the file **Nichols & Nicki (2004).sav**. The authors dropped two items because they had low means and variances, and dropped three others because of relatively low correlations with other items. They performed a principal component analysis on the remaining 31 items. Labcoat Leni wants you to run some descriptive statistics to work out which two items were dropped for having low means/variances, then inspect a correlation matrix to find the three items that were dropped for having low correlations. Finally, he wants you to run a principal component analysis on the data.

Answers are in the additional material on the companion website (or look at the original article).

NICHOLS, L.A., & NICKI, R. (2004). *PSYCHOLOGY OF ADDICTIVE BEHAVIORS, 18*(4), 381–384.

17.9. Reliability analysis ②

17.9.1. Measures of reliability ③

If you're using factor analysis to validate a questionnaire, it is useful to check the reliability of your scale.

SELF-TEST Thinking back to Chapter 1, what are reliability and test–retest reliability?

Reliability means that a measure (or in this case questionnaire) should consistently reflect the construct that it is measuring. One way to think of this is that, other things being equal, a person should get the same score on a questionnaire if they complete it at two different points in time (we have already discovered that this is called test–retest reliability). So,

How do I tell if my questionnaire is reliable?

someone who is terrified of statistics and who scores highly on our SAQ should score similarly highly if we tested them a month later (assuming they hadn't gone into some kind of statistics-anxiety therapy in that month). Another way to look at reliability is to say that two people who are the same in terms of the construct being measured should get the same score. So, if we took two people who were equally statistics-phobic, then they should get more or less identical scores on the SAQ. Likewise, if we took two people who loved statistics, they should both get equally low scores. It should be apparent that if we took someone who loved statistics and someone who was terrified of it, and they got the same score on our questionnaire, then it wouldn't be an accurate measure of statistical anxiety! In statistical terms, the usual way to look at reliability is based on the idea that individual items (or sets of items) should produce results consistent with the overall questionnaire. So, if we take someone scared of statistics, then their overall score on the SAQ will be high; if the SAQ is reliable then if we randomly select some items from it the person's score on those items should also be high.

The simplest way to do this in practice is to use **split-half reliability**. This method randomly splits the data set into two. A score for each participant is then calculated based on each half of the scale. If a scale is very reliable a person's score on one half of the scale should be the same (or similar) to their score on the other half: therefore, across several participants, scores from the two halves of the questionnaire should correlate perfectly (well, very highly). The correlation between the two halves is the statistic computed in the split-half method, with large correlations being a sign of reliability. The problem with this method is that there are several ways in which a set of data can be split into two and so the results could be a product of the way in which the data were split. To overcome this problem, Cronbach (1951) came up with a measure that is loosely equivalent to splitting data in two in every possible way and computing the correlation coefficient for each split. The average of these values is equivalent to **Cronbach's alpha, α**, which is the most common measure of scale reliability.[9]

Cronbach's α is:

$$\alpha = \frac{N^2\overline{\text{Cov}}}{\sum s^2_{\text{item}} + \sum \text{Cov}_{\text{item}}} \tag{17.7}$$

which may look complicated, but actually isn't. The first thing to note is that for each item on our scale we can calculate two things: the variance within the item, and the covariance between a particular item and any other item on the scale. Put another way, we can construct a variance–covariance matrix of all items. In this matrix the diagonal elements will be the variance within a particular item, and the off-diagonal elements will be covariances between pairs of items. The top half of the equation is simply the number of items (*N*) squared multiplied by the average covariance between items (the average of the off-diagonal elements in the aforementioned variance–covariance matrix). The bottom half is just the sum of all the item variances and item covariances (i.e. the sum of everything in the variance–covariance matrix).

[9] Although this is the easiest way to conceptualize Cronbach's α, whether or not it is exactly equal to the average of all possible split-half reliabilities depends on exactly how you calculate the split-half reliability (see the glossary for computational details). If you use the Spearman–Brown formula, which takes no account of item standard deviations, then Cronbach's α will be equal to the average split half-reliability only when the item standard deviations are equal; otherwise α will be smaller than the average. However, if you use a formula for split-half reliability that does account for item standard deviations (such as Flanagen, 1937; Rulon, 1939) then α will always equal the average split-half reliability (see Cortina, 1993).

There is a standardized version of the coefficient too, which essentially uses the same equation except that correlations are used rather than covariances, and the bottom half of the equation uses the sum of the elements in the correlation matrix of items (including the ones that appear on the diagonal of that matrix). The normal alpha is appropriate when items on a scale are summed to produce a single score for that scale (the standardized alpha is not appropriate in these cases). The standardized alpha is useful, though, when items on a scale are standardized before being summed.

17.9.2. Interpreting Cronbach's α (some cautionary tales ...) ②

You'll often see in books, journal articles, or be told by people that a value of .7 to .8 is an acceptable value for Cronbach's α; values substantially lower indicate an unreliable scale. Kline (1999) notes that although the generally accepted value of .8 is appropriate for cognitive tests such as intelligence tests, for ability tests a cut-off point of .7 is more suitable. He goes on to say that when dealing with psychological constructs values below even .7 can, realistically, be expected because of the diversity of the constructs being measured.

However, Cortina (1993) notes that such general guidelines need to be used with caution because the value of α depends on the number of items on the scale. You'll notice that the top half of the equation for α includes the number of items squared. Therefore, as the number of items on the scale increases, α will increase. Therefore, it's possible to get a large value of α because you have a lot of items on the scale, and not because your scale is reliable! For example, Cortina reports data from two scales, both of which have $\alpha = .8$. The first scale has only three items, and the average correlation between items was a respectable .57; however, the second scale had 10 items with an average correlation between these items of a less respectable .28. Clearly the internal consistency of these scales differs enormously, yet according to Cronbach's α they are both equally reliable!

A second common interpretation of alpha is that it measures 'unidimensionality', or the extent to which the scale measures one underlying factor or construct. This interpretation stems from the fact that when there is one factor underlying the data, α is a measure of the strength of that factor (see Cortina, 1993). However, Grayson (2004) demonstrates that data sets with the same α can have very different structures. He showed that $\alpha = .8$ can be achieved in a scale with one underlying factor, with two moderately correlated factors and with two uncorrelated factors. Cortina (1993) has also shown that with more than 12 items, and fairly high correlations between items ($r > .5$), α can reach values around and above .7 (.65 to .84). These results compellingly show that α should not be used as a measure of 'unidimensionality'. Indeed, Cronbach (1951) suggested that if several factors exist then the formula should be applied separately to items relating to different factors. In other words, if your questionnaire has subscales, α should be applied separately to these subscales.

Eek! My alpha is negative! What do I do?

The final warning is about items that have a reverse phrasing. For example, in our SAQ that we used in the factor analysis part of this chapter, we had one item (question 3) that was phrased the opposite way around to all other items. The item was 'standard deviations excite me'. Compare this to any other item and you'll see it requires the opposite response. For example, item 1 is 'statistics make me cry'. Now, if you don't like statistics then you'll strongly agree with this statement and so will get a score of 5 on our scale. For item 3, if you hate statistics then standard deviations are unlikely to excite you so you'll strongly disagree and get a score of 1 on the scale. These reverse-phrased items are important for reducing response bias; participants will actually have to read the items in case they are phrased the other way around. For factor analysis, this reverse phrasing doesn't matter, all that happens is you get a negative factor loading for any reversed items (in fact, look at SPSS Output 17.8 and you'll see that item 3 has a negative factor loading). However, in reliability analysis these reverse-scored items do make a difference. To see why, think about the equation for Cronbach's α. In this equation, the top half incorporates the *average* covariance between items. If an item is reverse phrased then it will have a negative relationship with other items, hence the covariances between this item and other items will be negative. The average covariance is obviously the sum of covariances divided by the number of covariances, and by including a bunch of negative values we reduce the sum of covariances, and hence we also reduce Cronbach's α, because the top half of the equation gets smaller. In extreme cases, it is even possible to get a negative value for Cronbach's α, simply because the magnitude of negative covariances is bigger than the magnitude of positive ones! A negative Cronbach's α doesn't make much sense, but it does happen, and if it does, ask yourself whether you included any reverse-phrased items!

If you have reverse-phrased items then you have to also reverse the way in which they're scored before you conduct reliability analysis. This is quite easy. To take our SAQ data, we have one item which is currently scored as 1 = strongly disagree, 2 = disagree, 3 = neither, 4 = agree and 5 = strongly agree. This is fine for items phrased in such a way that agreement indicates statistics anxiety, but for item 3 (standard deviations excite me), disagreement indicates statistics anxiety. To reflect this numerically, we need to reverse the scale such that 1 = strongly agree, 2 = agree, 3 = neither, 4 = disagree and 5 = strongly disagree. This way, an anxious person still gets 5 on this item (because they'd strongly disagree with it).

To reverse the scoring find the maximum value of your response scale (in this case 5) and add 1 to it (so you get 6 in this case). Then for each person, you take this value and subtract from it the score they actually got. Therefore, someone who scored 5 originally now scores 6 – 5 = 1, and someone who scored 1 originally now gets 6 – 1 = 5. Someone in the middle of the scale with a score of 3 will still get 6 – 3 = 3! Obviously it would take a long time to do this for each person, but we can get SPSS to do it for us

SELF-TEST Using what you learnt in Chapter 5, use the *Compute* command to reverse-score item 3. (Clue: Remember that you are simply changing the variable to 6 minus its original value.)

17.9.3. Reliability analysis on SPSS ②

Let's test the reliability of the SAQ using the data in **SAQ.sav**. Now, you should have reverse scored item 3 (see above), but if you can't be bothered then load up the file **SAQ (Item 3 Reversed).sav** instead. Remember also that I said we should conduct reliability analysis on any subscales individually. If we use the results from our orthogonal rotation (look back at SPSS Output 17.8), then we have four subscales:

1 Subscale 1 (*Fear of computers*): items 6, 7, 10, 13, 14, 15, 18
2 Subscale 2 (*Fear of statistics*): items 1, 3, 4, 5, 12, 16, 20, 21
3 Subscale 3 (*Fear of mathematics*): items 8, 11, 17
4 Subscale 4 (*Peer evaluation*): items 2, 9, 19, 22, 23

To conduct each reliability analysis on these data you need to select Analyze Scale ▸ Reliability Analysis... to display the dialog box in Figure 17.12. Select any

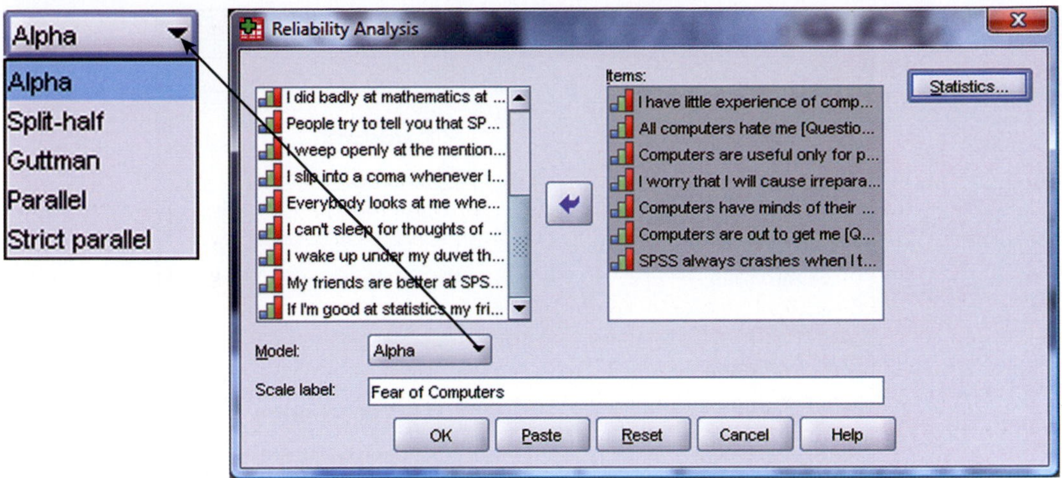

FIGURE 17.12 Main dialog box for reliability analysis

items from the list that you want to analyse (to begin with, let's do the items from the fear of computers subscale) on the left-hand side of the dialog box and drag them to the box labelled Items (or click on ➡). Remember that you can select several items at the same time if you hold down the *Ctrl* key while you click on the variables.

There are several reliability analyses you can run, but the default option is Cronbach's α. You can change the method (e.g. to the split-half method) by clicking on Alpha ▾ to reveal a drop-down list of possibilities, but the default method is a good one to select. Also, it's a good idea to type the name of the scale (in this case 'Fear of Computers') into the box labelled *Scale Label* because this will add a header to the SPSS output with whatever you type in this box: typing a sensible name here will make your output easier to follow.

If you click on Statistics... you can access the dialog box in Figure 17.13. In the statistics dialog box you can select several things, but the one most important for questionnaire reliability is: *Scale if item deleted*. This option provides a value of Cronbach's α for each item on your scale. It tells us what the value of α would be if that item were deleted. If our questionnaire is reliable then we would not expect any one item to greatly affect the overall reliability. In other words, no item should cause a substantial decrease in α. If it does then we have serious cause for concern and you should consider dropping that item from the questionnaire. As .8 is seen as a good value for α, we would hope that all values of *alpha if item deleted* should be around .8 or higher.

The inter-item correlations and covariances (and summaries) provide us with correlation coefficients and averages for items on our scale. We should already have these values from our factor analysis so there is little point in selecting these options. However, if you haven't already done a factor analysis then it's useful to ask for inter-item correlations because the overall α is affected by the number of items being analysed, and so you might want to check back to see whether the items seem to interrelate well. Options like the *ANOVA Table* will simply compare the central tendency of different items on the questionnaire using an *F test* (it conducts a one-way repeated-measures ANOVA on the items on the questionnaire), a *Friedman chi-square* (if your data are ranked), or a *Cochran chi-square* (if your data are dichotomous, e.g. if items on the questionnaire had yes/no responses). The *Hotelling's T-square* does much the same but produces the multivariate equivalent of the *F test*. These tests are useful if you want to check that items have similar distributional properties (i.e. the same average value), but given the large sample sizes you ought to be using for factor analysis, they will inevitably produce significant results even when only small differences exist between the means of questionnaire items.

You can also use this dialog box to get an **intraclass correlation coefficient (ICC)**. The correlation coefficients that we encountered earlier in this book measure the relation between variables that measure different things. For example, the correlation between listening to Deathspell Omega and Satanism represents two classes of measures: the type of music a

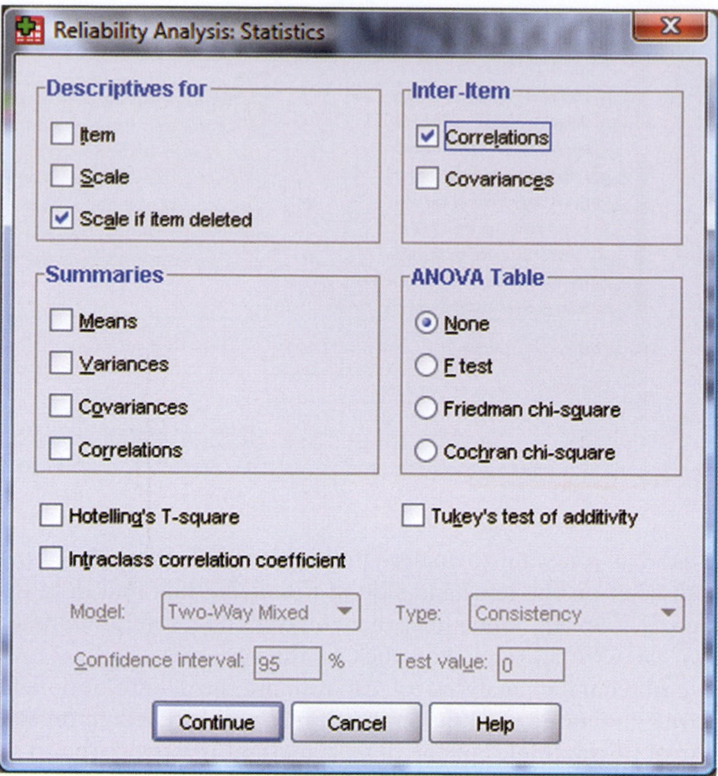

FIGURE 17.13 Statistics for reliability analysis

person likes and their religious beliefs. Intraclass correlations measure the relationship between two variables that measure the same thing (i.e. variables within the same class). Two common uses are in comparing paired data (such as twins) on the same measure, and assessing the consistency between judges' ratings of a set of objects (hence the reason why it is found in the reliability statistics in SPSS). If you'd like to know more see section 19.2.1.

Use the simple set of options in Figure 17.13 to run a basic reliability analysis. Click on Continue to return to the main dialog box and then click on OK to run the analysis.

17.9.4. Interpreting the output ②

SPSS Output 17.14 shows the results of this basic reliability analysis for the fear of computing subscale. The values in the column labelled *Corrected Item-Total Correlation* are the correlations between each item and the total score from the questionnaire. In a reliable scale all items should correlate with the total. So, we're looking for items that don't correlate with the overall score from the scale: if any of these values are less than about .3 then we've got problems, because it means that a particular item does not correlate very well with the scale overall. Items with low correlations may have to be dropped. For these data, all data have item-total correlations above 0.3, which is encouraging.

The values in the column labelled *Cronbach's Alpha if Item is Deleted* are the values of the overall α if that item isn't included in the calculation. As such, they reflect the change in Cronbach's α that would be seen if a particular item were deleted. The overall α is .823, and so all values in this column should be around that same value. What we're actually looking for is values of α greater than the overall α. If you think about it, if the deletion of an item increases Cronbach's α then this means that the deletion of that item improves reliability. Therefore, any items that have values of α in the column labelled *Cronbach's Alpha if item is Deleted* that are substantially

SPSS OUTPUT 17.14

Inter-Item Correlation Matrix

	I have little experience of computers	All computers hate me	Computers are useful only for playing games	I worry that I will cause irreparable damage because of my incompetenece with computers	Computers have minds of their own and deliberately go wrong whenever I use them	Computers are out to get me	SPSS always crashes when I try to use it
I have little experience of computers	1.000	.514	.322	.466	.402	.360	.513
All computers hate me	.514	1.000	.284	.442	.441	.391	.501
Computers are useful only for playing games	.322	.284	1.000	.302	.255	.295	.293
I worry that I will cause irreparable damage because of my incompetenece with computers	.466	.442	.302	1.000	.450	.342	.533
Computers have minds of their own and deliberately go wrong whenever I use them	.402	.441	.255	.450	1.000	.380	.498
Computers are out to get me	.360	.391	.295	.342	.380	1.000	.343
SPSS always crashes when I try to use it	.513	.501	.293	.533	.498	.343	1.000

Item-Total Statistics

	Scale Mean if Item Deleted	Scale Variance if Item Deleted	Corrected Item-Total Correlation	Squared Multiple Correlation	Cronbach's Alpha if Item Deleted
I have little experience of computers	15.87	17.614	.619	.398	.791
All computers hate me	15.17	17.737	.619	.395	.790
Computers are useful only for playing games	15.81	20.736	.400	.167	.824
I worry that I will cause irreparable damage because of my incompetenece with computers	15.64	18.809	.607	.384	.794
Computers have minds of their own and deliberately go wrong whenever I use them	15.22	18.719	.577	.350	.798
Computers are out to get me	15.33	19.322	.491	.250	.812
SPSS always crashes when I try to use it	15.52	17.832	.647	.447	.786

Reliability Statistics

Cronbach's Alpha	Cronbach's Alpha Based on Standardized Items	N of Items
.823	.821	7

greater than the overall α may need to be deleted from the scale to improve its reliability. None of the items here would substantially affect reliability if they were deleted. The worst offender is question 10: deleting this question would increase the α from .823 to .824. Nevertheless this increase is negligible and both values reflect a good degree of reliability.

Finally, and perhaps most important, the value of *Alpha* at the very bottom is Cronbach's α: the overall reliability of the scale. To reiterate, we're looking for values in the range of .7 to .8 (or thereabouts) bearing in mind what we've already noted about effects from the number of items. In this case α is slightly above .8, and is certainly in the region indicated by Kline (1999), so this probably indicates good reliability. As a final point, it's worth noting that if items do need to be removed at this stage then you should rerun your factor analysis as well to make sure that the deletion of the item has not affected the factor structure!

OK, let's move on to do the fear of statistics subscale (items 1, 3, 4, 5, 12, 16, 20 and 21). I won't go through the SPSS again, but SPSS Output 17.15 shows the output from the analysis (to save space I've omitted the inter-item correlations). The values in the column labelled *Corrected Item-Total Correlation* are again all above .3, which is good. The values in the column labelled *Cronbach's Alpha if Item is Deleted* are the values of the overall α if that item isn't included in the calculation. The overall α is .821, and none of the items here would increase the reliability if they were deleted. This indicates that all items are positively contributing to the overall reliability. The overall α is also excellent (.821) because it is above .8, and indicates good reliability.

Just to illustrate the importance of reverse scoring items before running reliability analysis, SPSS Output 17.16 shows the reliability analysis for the fear of statistics subscale but done on the original data (i.e. without item 3 being reverse scored). Note that the overall α is considerably lower (.605 rather than .821). Also, note that this item has a negative item-total correlation (which is a good way to spot if you have a potential reverse-scored item in the data that hasn't been reverse scored). Finally, note that for item 3, the α if item deleted is .8. That is, if this item were deleted then the reliability would improve from about .6 to about .8. This, I hope, illustrates that failing to reverse-score items that have been phrased oppositely to other items on the scale will mess up your reliability analysis.

Moving swiftly on to the fear of maths subscale (items 8, 11 and 17), SPSS Output 17.17 shows the output from the analysis. The values in the column labelled *Corrected Item-Total Correlation* are again all above .3, which is good, and the values in the column labelled *Cronbach's Alpha*

SPSS OUTPUT 17.15

Item-Total Statistics

	Scale Mean if Item Deleted	Scale Variance if Item Deleted	Corrected Item-Total Correlation	Squared Multiple Correlation	Cronbach's Alpha if Item Deleted
Statiscs makes me cry	21.76	21.442	.536	.343	.802
Standard deviations excite me	20.72	19.825	.549	.309	.800
I dream that Pearson is attacking me with correlation coefficients	21.35	20.410	.575	.355	.796
I don't understand statistics	21.41	20.942	.494	.272	.807
People try to tell you that SPSS makes statistics easier to understand but it doesn't	20.97	20.639	.572	.337	.796
I weep openly at the mention of central tendency	21.25	20.451	.597	.389	.793
I can't sleep for thoughts of eigen vectors	20.51	21.176	.419	.244	.818
I wake up under my duvet thinking that I am trapped under a normal distribtion	20.96	19.939	.606	.399	.791

Reliability Statistics

Cronbach's Alpha	Cronbach's Alpha Based on Standardized Items	N of Items
.821	.823	8

SPSS OUTPUT 17.16

Item-Total Statistics

	Scale Mean if Item Deleted	Scale Variance if Item Deleted	Corrected Item-Total Correlation	Squared Multiple Correlation	Cronbach's Alpha if Item Deleted
Statiscs makes me cry	20.93	12.125	.505	.343	.521
Standard deviations excite me	20.72	19.825	-.549	.309	.800
I dream that Pearson is attacking me with correlation coefficients	20.52	11.447	.526	.355	.505
I don't understand statistics	20.58	11.714	.466	.272	.523
People try to tell you that SPSS makes statistics easier to understand but it doesn't	20.14	11.739	.501	.337	.515
I weep openly at the mention of central tendency	20.42	11.584	.529	.389	.507
I can't sleep for thoughts of eigen vectors	19.68	12.107	.353	.244	.558
I wake up under my duvet thinking that I am trapped under a normal distribtion	20.13	11.189	.541	.399	.497

Reliability Statistics

Cronbach's Alpha	Cronbach's Alpha Based on Standardized Items	N of Items
.605	.641	8

SPSS OUTPUT 17.17

Item-Total Statistics

	Scale Mean if Item Deleted	Scale Variance if Item Deleted	Corrected Item-Total Correlation	Squared Multiple Correlation	Cronbach's Alpha if Item Deleted
I have never been good at mathematics	4.72	2.470	.684	.470	.740
I did badly at mathematics at school	4.70	2.453	.682	.467	.742
I slip into a coma whenever I see an equation	4.49	2.504	.652	.425	.772

Reliability Statistics

Cronbach's Alpha	Cronbach's Alpha Based on Standardized Items	N of Items
.819	.819	3

if Item is Deleted indicate that none of the items here would increase the reliability if they were deleted because all values in this column are less than the overall reliability of .819. As with the previous two subscales, the overall α is around .8, which indicates good reliability.

Finally, if you run the analysis for the final subscale of peer evaluation, you should get the output in SPSS Output 17.18. The values in the column labelled *Corrected Item-Total Correlation* are all around .3, and in fact for item 23 the value is below .3. This indicates fairly bad internal consistency and identifies item 23 as a potential problem. The values in the column labelled *Cronbach's Alpha if Item is Deleted* indicate that none of the items here would increase the reliability if they were deleted because all values in this column are less than the overall reliability of .57. Unlike the previous subscales, the overall α is quite low and although this is in keeping with what Kline (1999) says we should expect for this kind of social science data, it is well below the other scales. The scale has five items, compared to seven, eight and

Item-Total Statistics

	Scale Mean if Item Deleted	Scale Variance if Item Deleted	Corrected Item-Total Correlation	Squared Multiple Correlation	Cr Al
My friends will think I'm stupid for not being able to cope with SPSS	11.46	8.119	.339	.134	
My friends are better at statistics than me	10.24	6.395	.391	.167	
Everybody looks at me when I use SPSS	10.79	7.381	.316	.106	
My friends are better at SPSS than I am	10.20	7.282	.378	.144	
If I'm good at statistics my friends will think I'm a nerd	9.65	7.988	.239	.069	

Reliability Statistics

Cronbach's Alpha	Cronbach's Alpha Based on Standardized Items	N of Items
.570	.572	5

CRAMMING SAM'S TIPS Reliability

- Reliability is really the consistency of a measure.
- Reliability analysis can be used to measure the consistency of a questionnaire.
- Remember to reverse-score any items that were reverse phrased on the original questionnaire before you run the analysis.
- Run separate reliability analyses for all subscales of your questionnaire.
- Cronbach's α indicates the overall reliability of a questionnaire and values around 0.8 are good (or 0.7 for ability tests and such like).
- The *Cronbach's Alpha if Item Deleted* tells you whether removing an item will improve the overall reliability: values greater than the overall reliability indicate that removing that item will improve the overall reliability of the scale. Look for items that dramatically increase the value of α.
- If you do remove items, rerun your factor analysis to check that the factor structure still holds!

three on the other scales, so its reduced reliability is not going to be dramatically affected by the number of items (in fact, it has more items than the fear of maths subscale). If you look at the items on this subscale, they cover quite diverse themes of peer evaluation, and this might explain the relative lack of consistency. This might lead us to rethink this subscale.

17.10. How to report reliability analysis ②

You can report the reliabilities in the text using the symbol α and remembering that because Cronbach's α can't be larger than 1 then we drop the zero before the decimal place (if we are following APA format):

✓ The fear of computers, fear of statistics and fear of maths subscales of the SAQ all had high reliabilities, all Cronbach's $\alpha = .82$. However, the fear of negative peer evaluation subscale had relatively low reliability, Cronbach's $\alpha = .57$.

However, the most common way to report reliability analysis when it follows a factor analysis is to report the values of Cronbach's α as part of the table of factor loadings. For example, in Table 17.2 notice that in the last row of the table I have quoted the value of Cronbach's α for each subscale in turn.

What have I discovered about statistics? ②

This chapter has made us tiptoe along the craggy rockface that is factor analysis. This is a technique for identifying clusters of variables that relate to each other. One of the difficult things with statistics is realizing that they are subjective: many books (this one included I suspect) create the impression that statistics are like a cook book and if you follow the instructions you'll get a nice tasty chocolate cake (yum!). Factor analysis perhaps more than any other test in this book illustrates how incorrect this is. The world of statistics is full of arbitrary rules that we probably shouldn't follow (.05 being the classic example) and nearly all of the time, whether you realize it or not, we should act upon our own discretion. So, if nothing else I hope you've discovered enough to give you sufficient discretion about factor analysis to act upon! We saw that the first stage of factor analysis is to scan your variables to check that they relate to each other to some degree but not too strongly. The factor analysis itself has several stages: check some initial issues (e.g. sample size adequacy), decide how many factors to retain, and finally decide which items load onto which factors (and try to make sense of the meaning of the factors). Having done all that you can consider whether the items you have are reliable measures of what you're trying to measure.

We also discovered that at the age of 23 I took it upon myself to become a living homage to the digestive system. I furiously devoured articles and books on statistics (some of them I even understood), I mentally chewed over them, I broke them down with the stomach acid of my intellect, I stripped them of their goodness and nutrients, I compacted them down, and after about two years I forced the smelly brown remnants of those intellectual meals out of me in the form of a book. I was mentally exhausted at the end of it; 'It's a good job I'll never have to do that again' I thought.

Key terms that I've discovered

Alpha factoring
Anderson–Rubin method
Common variance
Communality
Component matrix
Confirmatory factor analysis CFA)
Cronbach's α
Direct oblimin
Equamax
Extraction
Factor analysis
Factor loading
Factor matrix
Factor scores
Factor transformation matrix, Λ
Intraclass correlation coefficient (ICC)
Kaiser–Meyer–Olkin (KMO) measure of sampling adequacy
Kaiser's criterion
Latent variable
Oblique rotation
Orthogonal rotation
Pattern matrix
Principal component analysis (PCA)
Promax
Quartimax
Random variance
Rotation
Scree plot
Singularity
Split-half reliability
Structure matrix
Unique variance
Varimax

Smart Alex's tasks

- **Task 1**: The University of Sussex is constantly seeking to employ the best people possible as lecturers (no, really, it is). Anyway, they wanted to revise a questionnaire based on Bland's theory of research methods lecturers. This theory predicts that good research methods lecturers should have four characteristics: (1) a profound love of statistics; (2) an enthusiasm for experimental design; (3) a love of teaching; and (4) a complete absence of normal interpersonal skills. These characteristics should be related (i.e. correlated). The 'Teaching of Statistics for Scientific Experiments' (TOSSE) already existed, but the university revised this questionnaire and it became the 'Teaching of Statistics for Scientific Experiments – Revised' (TOSSE-R). They gave this questionnaire to 239 research methods lecturers around the world to see if it supported Bland's theory. The questionnaire is in Figure 17.14, and the data are in **TOSSE-R.sav**. Conduct a factor analysis (with appropriate rotation) to see the factor structure of the data. ②

SD = Strongly Disagree, D = Disagree, N = Neither, A = Agree, SA = Strongly Agree						
		SD	**D**	**N**	**A**	**SA**
1	I once woke up in the middle of a vegetable patch hugging a turnip that I'd mistakenly dug up thinking it was Roy's largest root	O	O	O	O	O
2	If I had a big gun I'd shoot all the students I have to teach	O	O	O	O	O
3	I memorize probability values for the *F*-distribution	O	O	O	O	O
4	I worship at the shrine of Pearson	O	O	O	O	O
5	I still live with my mother and have little personal hygiene	O	O	O	O	O
6	Teaching others makes me want to swallow a large bottle of bleach because the pain of my burning oesophagus would be light relief in comparison	O	O	O	O	O
7	Helping others to understand sums of squares is a great feeling	O	O	O	O	O
8	I like control conditions	O	O	O	O	O
9	I calculate 3 ANOVAs in my head before getting out of bed every morning	O	O	O	O	O
10	I could spend all day explaining statistics to people	O	O	O	O	O
11	I like it when people tell me I've helped them to understand factor rotation	O	O	O	O	O
12	People fall asleep as soon as I open my mouth to speak	O	O	O	O	O
13	Designing experiments is fun	O	O	O	O	O
14	I'd rather think about appropriate dependent variables than go to the pub	O	O	O	O	O
15	I soil my pants with excitement at the mere mention of factor analysis	O	O	O	O	O
16	Thinking about whether to use repeated or independent measures thrills me	O	O	O	O	O
17	I enjoy sitting in the park contemplating whether to use participant observation in my next experiment	O	O	O	O	O
18	Standing in front of 300 people in no way makes me lose control of my bowels	O	O	O	O	O
19	I like to help students	O	O	O	O	O

20	Passing on knowledge is the greatest gift you can bestow on an individual	O	O	O	O	O
21	Thinking about Bonferroni corrections gives me a tingly feeling in my groin	O	O	O	O	O
22	I quiver with excitement when thinking about designing my next experiment	O	O	O	O	O
23	I often spend my spare time talking to the pigeons ... and even they die of boredom	O	O	O	O	O
24	I tried to build myself a time machine so that I could go back to the 1930s and follow Fisher around on my hands and knees licking the floor on which he'd just trodden	O	O	O	O	O
25	I love teaching	O	O	O	O	O
26	I spend lots of time helping students	O	O	O	O	O
27	I love teaching because students have to pretend to like me or they'll get bad marks	O	O	O	O	O
28	My cat is my only friend	O	O	O	O	O

FIGURE 17.14 The Teaching of Statistics for Scientific Experiments – Revised (TOSSE-R)

- **Task 2:** Dr Sian Williams (University of Brighton) devised a questionnaire to measure organizational ability. She predicted five factors to do with organizational ability: (1) preference for organization; (2) goal achievement; (3) planning approach; (4) acceptance of delays; and (5) preference for routine. These dimensions are *theoretically independent*. Williams' questionnaire contains 28 items using a 7-point Likert scale (1 = strongly disagree, 4 = neither, 7 = strongly agree). She gave it to 239 people. Run a principal component analysis on the data in **Williams.sav**. ②

1	I like to have a plan to work to in everyday life
2	I feel frustrated when things don't go to plan
3	I get most things done in a day that I want to
4	I stick to a plan once I have made it
5	I enjoy spontaneity and uncertainty
6	I feel frustrated if I can't find something I need
7	I find it difficult to follow a plan through
8	I am an organized person
9	I like to know what I have to do in a day
10	Disorganized people annoy me
11	I leave things to the last minute
12	I have many different plans relating to the same goal
13	I like to have my documents filed and in order
14	I find it easy to work in a disorganized environment

15	I make 'to do' lists and achieve most of the things on it
16	My workspace is messy and disorganized
17	I like to be organized
18	Interruptions to my daily routine annoy me
19	I feel that I am wasting my time
20	I forget the plans I have made
21	I prioritize the things I have to do
22	I like to work in an organized environment
23	I feel relaxed when I don't have a routine
24	I set deadlines for myself and achieve them
25	I change rather aimlessly from one activity to another during the day
26	I have trouble organizing the things I have to do
27	I put tasks off to another day
28	I feel restricted by schedules and plans

Answers can be found on the companion website.

Further reading

Cortina, J. M. (1993). What is coefficient alpha? An examination of theory and applications. *Journal of Applied Psychology*, *78*, 98–104. (A very readable paper on Cronbach's α.)

Dunteman, G. E. (1989). *Principal components analysis*. Sage university paper series on quantitative applications in the social sciences, 07-069. Newbury Park, CA: Sage. (This monograph is quite high level but comprehensive.)

Pedhazur, E., & Schmelkin, L. (1991). *Measurement, design and analysis: an integrated approach*. Hillsdale, NJ: Erlbaum. (Chapter 22 is an excellent introduction to the theory of factor analysis.)

Tabachnick, B. G., & Fidell, L. S. (2007). *Using multivariate statistics* (5th ed.). Boston: Allyn & Bacon. (Chapter 13 is a technical but wonderful overview of factor analysis.)

Online tutorial

The companion website contains the following Flash movie tutorial to accompany this chapter:

- Principal Component Analysis (PCA) using SPSS

Interesting real research

Nichols, L. A., & Nicki, R. (2004). Development of a psychometrically sound internet addiction scale: a preliminary step. *Psychology of Addictive Behaviors*, *18*(4), 381–384.

18 Categorical data

FIGURE 18.1 Midway through writing the second edition of this book, things had gone a little strange

18.1. What will this chapter tell me? ①

We discovered in the previous chapter that I wrote a book. This book. There are a lot of good things about writing books. The main benefit is that your parents are impressed. Well, they're not *that* impressed actually because they think that a good book sells as many copies as *Harry Potter* and that people should queue outside bookshops for the latest enthralling instalment of *Discovering Statistics* … . My parents are, consequently, quite baffled about how this book is seen as successful, yet I don't get invited to dinner by the Queen. Nevertheless, given that my family don't really understand what I do, books are tangible proof that I do *something*. The size of this book and the fact it has equations in it is an added bonus because it makes me look cleverer than I actually am. However, there is a price to pay, which is immeasurable mental anguish. In England we don't talk about our emotions, because we fear that if they get out into the open, civilization as we know it will collapse, so I definitely will not mention that the writing process for the second edition was so stressful that I came within one of Fuzzy's whiskers of a total meltdown. It took me two years to recover, just in time to start thinking about this third edition. Still, it was worth it because the feedback suggests that some people found the book vaguely useful. Of course, the publishers don't care about helping people, they care only about raking in as much cash as possible to feed their cocaine habits and champagne addictions. Therefore, they are obsessed with sales figures and comparisons with other books. They have databases that have sales figures of this book and its competitors in different 'markets' (you are not a person, you are a 'consumer' and you don't live in a country, you live in a 'market') and they gibber and twitch at their consoles creating frequency distributions (with 3-D effects) of these values. The data they get are frequency data (the number of books sold in a certain timeframe). Therefore, if they wanted to compare sales of this book to its competitors, in different countries, they would need to read this chapter because it's all about analysing data, for which we know only the frequency with which events occur. Of course, they won't read this chapter, but they should ...

18.2. Analysing categorical data ①

Sometimes, we are interested not in test scores, or continuous measures, but in *categorical variables*. These are not variables involving cats (although the examples in this chapter might convince you otherwise), but are what we have mainly used as grouping variables. They are variables that describe categories of entities (see section 1.5.1.2). We've come across these types of variables in virtually every chapter of this book. There are different types of categorical variable (see section 6.5.5), but in theory a person, or case, should fall into only one category. Good examples of categorical variables are gender (with few exceptions people can be only biologically male or biologically female),[1] pregnancy (a woman can be only pregnant or not pregnant) and voting in an election (as a general rule you are allowed to vote for only one candidate). In all cases (except logistic regression) so far, we've used such categorical variables to predict some kind of continuous outcome, but there are times when we want to look at relationships between lots of categorical variables. This chapter looks at two techniques for doing this. We begin with the simple case of two categorical variables and discover the chi-square statistic (which we're not really discovering because we've unwittingly come across it countless times before). We then extend this model to look at relationships between several categorical variables.

18.3. Theory of analysing categorical data ①

We will begin by looking at the simplest situation that you could encounter; that is, analysing two categorical variables. If we want to look at the relationship between two categorical variables then we can't use the mean or any similar statistic because we don't have any variables that have been measured continuously. Trying to calculate the mean of a categorical variable is completely meaningless because the numeric values you attach to different categories are arbitrary, and the mean of those numeric values will depend on how many members each category has. Therefore, when we've measured only categorical variables, we analyse frequencies. That is, we analyse the number of things that fall into each combination

[1] Before anyone rips my arms from their sockets and beats me around the head with them, I am aware that numerous chromosomal and hormonal conditions exist that complicate the matter. Also, people can have a different gender identity to their biological gender.

TABLE 18.1 Contingency table showing how many cats will line dance after being trained with different rewards

		Training		
		Food as Reward	**Affection as Reward**	**Total**
Could They Dance?	Yes	28	48	76
	No	10	114	124
	Total	38	162	200

of categories. If we take an example, a researcher was interested in whether animals could be trained to line dance. He took 200 cats and tried to train them to line dance by giving them either food or affection as a reward for dance-like behaviour. At the end of the week they counted how many animals could line dance and how many could not. There are two categorical variables here: **training** (the animal was trained using either food or affection, not both) and **dance** (the animal either learnt to line dance or it did not). By combining categories, we end up with four different categories. All we then need to do is to count how many cats fall into each category. We can tabulate these frequencies as in Table 18.1 (which shows the data for this example) and this is known as a **contingency table**.

18.3.1. Pearson's chi-square test ①

If we want to see whether there's a relationship between two categorical variables (i.e. does the amount of cats that line dance relate to the type of training used?) we can use the Pearson's **chi-square test** (Fisher, 1922; Pearson, 1900). This is an extremely elegant statistic based on the simple idea of comparing the frequencies you observe in certain categories to the frequencies you might expect to get in those categories by chance. All the way back in Chapters 2, 7 and 10 we saw that if we fit a model to any set of data we can evaluate that model using a very simple equation (or some variant of it):

$$\text{deviation} = \sum(\text{observed} - \text{model})^2$$

This equation was the basis of our sums of squares in regression and ANOVA. Now, when we have categorical data we can use the same equation. There is a slight variation in that we divide by the model scores as well, which is actually much the same process as dividing the sum of squares by the degrees of freedom in ANOVA. So, basically, what we're doing is standardizing the deviation for each observation. If we add all of these standardized deviations together the resulting statistic is Pearson's chi-square (χ^2) given by:

$$\chi^2 = \sum \frac{(\text{observed}_{ij} - \text{model}_{ij})^2}{\text{model}_{ij}}, \qquad (18.1)$$

in which i represents the rows in the contingency table and j represents the columns. The observed data are, obviously, the frequencies in Table 18.1, but we need to work out what the model is. In ANOVA the model we use is group means, but as I've mentioned we can't work with means when we have only categorical variables so we work with frequencies instead. Therefore, we use 'expected frequencies'. One way to estimate the expected frequencies would be to say 'well, we've got 200 cats in total, and four categories, so the expected value is simply 200/4 = 50'. This would be fine if, for example, we had the same number of cats that had affection as a reward and food as a reward; however, we didn't: 38 got food and 162 got affection as a reward. Likewise there are not equal numbers that could and couldn't dance. To take account of this, we calculate expected frequencies for each of the cells in the table (in this case there are four cells) and we use the column and row totals for a particular cell to calculate the expected value:

$$\text{model}_{ij} = E_{ij} = \frac{\text{row total}_i \times \text{column total}_j}{n}$$

n is simply the total number of observations (in this case 200). We can calculate these expected frequencies for the four cells within our table (row total and column total are abbreviated to RT and CT respectively):

$$\text{model}_{\text{Food, Yes}} = \frac{\text{RT}_{\text{Yes}} \times \text{CT}_{\text{Food}}}{n} = \frac{76 \times 38}{200} = 14.44$$

$$\text{model}_{\text{Food, No}} = \frac{\text{RT}_{\text{No}} \times \text{CT}_{\text{Food}}}{n} = \frac{124 \times 38}{200} = 23.56$$

$$\text{model}_{\text{Affection, Yes}} = \frac{\text{RT}_{\text{Yes}} \times \text{CT}_{\text{Affection}}}{n} = \frac{76 \times 162}{200} = 61.56$$

$$\text{model}_{\text{Affection, No}} = \frac{\text{RT}_{\text{No}} \times \text{CT}_{\text{Affection}}}{n} = \frac{124 \times 162}{200} = 100.44$$

Given that we now have these model values, all we need to do is take each value in each cell of our data table, subtract from it the corresponding model value, square the result, and then divide by the corresponding model value. Once we've done this for each cell in the table, we just add them up!

$$\begin{aligned}\chi^2 &= \frac{(28-14.44)^2}{14.44} + \frac{(10-23.56)^2}{23.56} + \frac{(48-61.56)^2}{61.56} + \frac{(114-100.44)^2}{100.44}\\ &= \frac{(13.56)^2}{14.44} + \frac{(-13.56)^2}{23.56} + \frac{(-13.568)^2}{61.56} + \frac{(13.56)^2}{100.44}\\ &= 12.73 + 7.80 + 2.99 + 1.83\\ &= 25.35\end{aligned}$$

This statistic can then be checked against a distribution with known properties. All we need to know is the degrees of freedom and these are calculated as $(r - 1)(c - 1)$ in which r is the number of rows and c is the number of columns. Another way to think of it is the number of levels of each variable minus one multiplied. In this case we get $df = (2 - 1)(2 - 1) = 1$. If you were doing the test by hand, you would find a critical value for the chi-square distribution with $df = 1$ and if the observed value was bigger than this critical value you would say that there was a significant relationship between the two variables. These critical values are produced in Appendix A.4, and for $df = 1$ the critical values are 3.84 ($p = .05$) and 6.63 ($p = .01$) and so because the observed chi-square is bigger than these values it is significant at $p < .01$. However, if you use SPSS, it will simply produce an estimate of the precise probability of obtaining a chi-square statistic at least as big as (in this case) 25.35 if there were no association in the population between the variables.

18.3.2. Fisher's exact test ①

There is one problem with the chi-square test, which is that the sampling distribution of the test statistic has an *approximate* chi-square distribution. The larger the sample is, the better this approximation becomes and in large samples the approximation is good enough to not worry about the fact that it is an approximation. However, in small samples, the approximation is not good enough, making significance tests of the chi-square distribution inaccurate. This is why you often read that to use the chi-square test the expected frequencies in each cell must be greater than 5 (see section 18.4). When the expected frequencies are greater than 5, the sampling distribution is probably close enough to a perfect chi-square distribution for us not to worry. However, when the expected frequencies are too low, it probably means that the sample size is too small and that the sampling distribution of the test statistic is too deviant from a chi-square distribution to be of any use.

Fisher came up with a method for computing the exact probability of the chi-square statistic that is accurate when sample sizes are small. This method is called **Fisher's exact test** (Fisher, 1922) even though it's not so much of a test as a way of computing the exact probability of the chi-square statistic. This procedure is normally used on 2 × 2 contingency tables (i.e. two variables each with two options) and with small samples. However, it can be used on larger contingency tables and with large samples, but on larger contingency tables it becomes computationally intensive and you might find SPSS taking a long time to give you an answer. In large samples there is really no point because it was designed to overcome the problem of small samples, so you don't need to use it when samples are large.

18.3.3. The likelihood ratio ②

An alternative to Pearson's chi-square is the likelihood ratio statistic, which is based on maximum-likelihood theory. The general idea behind this theory is that you collect some data and create a model for which the probability of obtaining the observed set of data is maximized, then you compare this model to the probability of obtaining those data under the null hypothesis. The resulting statistic is, therefore, based on comparing observed frequencies with those predicted by the model:

$$L\chi^2 = 2\sum \text{observed}_{ij} \ln\left(\frac{\text{observed}_{ij}}{\text{model}_{ij}}\right) \tag{18.2}$$

in which i and j are the rows and columns of the contingency table and ln is the natural logarithm (this is the standard mathematical function that we came across in Chapter 8 and you can find it on your calculator usually labelled as ln or $\log_e$). Using the same model and observed values as in the previous section, this would give us:

$$\begin{aligned} L\chi^2 &= 2\left[28 \times \ln\left(\frac{28}{14.44}\right) + 10 \times \ln\left(\frac{10}{23.56}\right) + 48 \times \ln\left(\frac{48}{61.56}\right) + 114 \times \ln\left(\frac{114}{100.44}\right)\right] \\ &= 2[(28 \times 0.662) + (10 \times -0.857) + (48 \times -0.249) + (114 \times 0.0.127)] \\ &= 2[18.54 - 8.57 - 11.94 + 14.44] \\ &= 24.94 \end{aligned}$$

As with Pearson's chi-square, this statistic has a chi-square distribution with the same degrees of freedom (in this case 1). As such, it is tested in the same way: we could look up the critical value of chi-square for the number of degrees of freedom that we have. As before, the value we have here will be significant because it is bigger than the critical values of 3.84 ($p = .05$) and 6.63 ($p = .01$). For large samples this statistic will be roughly the same as Pearson's chi-square, but is preferred when samples are small.

18.3.4. Yates's correction ②

When you have a 2 × 2 contingency table (i.e. two categorical variables each with two categories) then Pearson's chi-square tends to produce significance values that are too small (in other words, it tends to make a Type I error). Therefore, Yates suggested a correction to the Pearson formula (usually referred to as **Yates's continuity correction**). The basic idea is that when you calculate the deviation from the model (the observed$_{ij}$ – Model$_{ij}$ in equation (18.1)) you subtract 0.5 from the absolute value of this deviation before you square it. In plain English this means you calculate the deviation, ignore whether it is positive or negative, subtract 0.5 from the value and then square it. Pearson's equation then becomes:

$$\chi^2 = \sum \frac{\left(|\text{observed}_{ij} - \text{model}_{ij}| - 0.5\right)^2}{\text{model}_{ij}}$$

For the data in our example this just translates into:

$$\begin{aligned}\chi^2 &= \frac{(13.56 - 0.5)^2}{14.44} + \frac{(13.56 - 0.5)^2}{23.56} + \frac{(13.56 - 0.5)^2}{61.56} + \frac{(13.56 - 0.5)^2}{100.44} \\ &= 11.81 + 7.24 + 2.77 + 1.70 \\ &= 23.52\end{aligned}$$

The key thing to note is that it lowers the value of the chi-square statistic and, therefore, makes it less significant. Although this seems like a nice solution to the problem, there is a fair bit of evidence that this overcorrects and produces chi-square values that are too small! Howell (2006) provides an excellent discussion of the problem with Yates's correction for continuity if you're interested; all I will say is that although it's worth knowing about, it's probably best ignored!

18.4. Assumptions of the chi-square test ①

It should be obvious that the chi-square test does not rely on assumptions such as having continuous normally distributed data like most of the other tests in this book (categorical data cannot be normally distributed because they aren't continuous). However, the chi-square test still has two important assumptions:

1. Pretty much all of the tests we have encountered in this book have made an assumption about the independence of data and the chi-square test is no exception. For the chi-square test to be meaningful it is imperative that each person, item or entity contributes to only one cell of the contingency table. Therefore, you cannot use a chi-square test on a repeated-measures design (e.g. if we had trained some cats with food to see if they would dance and then trained the same cats with affection to see if they would dance we couldn't analyse the resulting data with Pearson's chi-square test).

2 The expected frequencies should be greater than 5. Although it is acceptable in larger contingency tables to have up to 20% of expected frequencies below 5, the result is a loss of statistical power (so, the test may fail to detect a genuine effect). Even in larger contingency tables no expected frequencies should be below 1. Howell (2006) gives a nice explanation of why violating this assumption creates problems. If you find yourself in this situation consider using Fisher's exact test (section 18.3.2).

Finally, although it's not an assumption, it seems fitting to mention in a section in which a gloomy and foreboding tone is being used that proportionately small differences in cell frequencies can result in statistically significant associations between variables if the sample is large enough (although it might need to be very large indeed). Therefore, we must look at row and column percentages to interpret any effects we get. These percentages will reflect the patterns of data far better than the frequencies themselves (because these frequencies will be dependent on the sample sizes in different categories).

18.5. Doing chi-square on SPSS ①

There are two ways in which categorical data can be entered: enter the raw scores, or enter weighted cases. We'll look at both in turn.

18.5.1. Entering data: raw scores ①

If we input the raw scores, it means that every row of the data editor represents each entity about which we have data (in this example, each row represents a cat). So, you would create two coding variables (**Training** and **Dance**) and specify appropriate numeric codes for each. The **Training** could be coded with 0 to represent a food reward and 1 to represent affection, and **Dance** could be coded with 0 to represent an animal that danced and 1 to represent one that did not. For each animal, you put the appropriate numeric code into each column. So a cat that was trained with food that did not dance would have 0 in the training column and 1 in the dance column. The data in the file **Cats.sav** are entered in this way and you should be able to identify the variables described. There were 200 cats in all and so there are 200 rows of data.

18.5.2. Entering data: weight cases ①

An alternative method of data entry is to create the same coding variables as before, but to have a third variable that represents the number of animals that fell into each combination of categories. In other words, we input the frequency data (the number of cases that fall into a particular category). We could call this variable **Frequency**. Figure 18.2 shows the data editor with this third variable added. Now, instead of having 200 rows, each one representing a different animal, we have one row representing each combination of categories and a variable telling us how many animals fell into this category combination. So, the first row represents cats that had food as a reward and then danced. The variable **Frequency** tells us that there were 28 cats that had food as a reward and then danced. This information was previously represented by 28 different rows in the file **Cats.sav** and so you can see how this method of data entry saves you a lot of time! Extending this principle, we can see that when affection was used as a reward 114 cats did not dance.

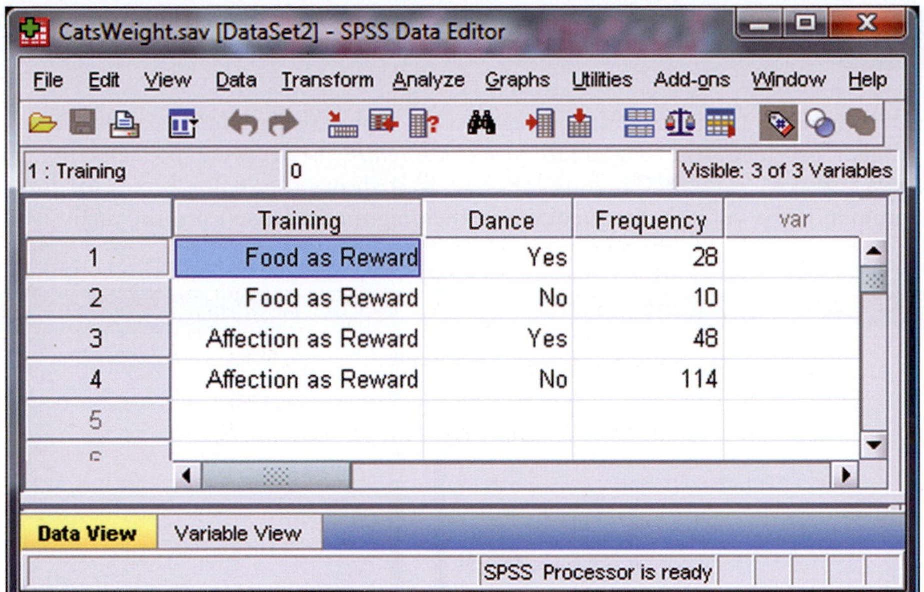

FIGURE 18.2 Data entry using weighted cases

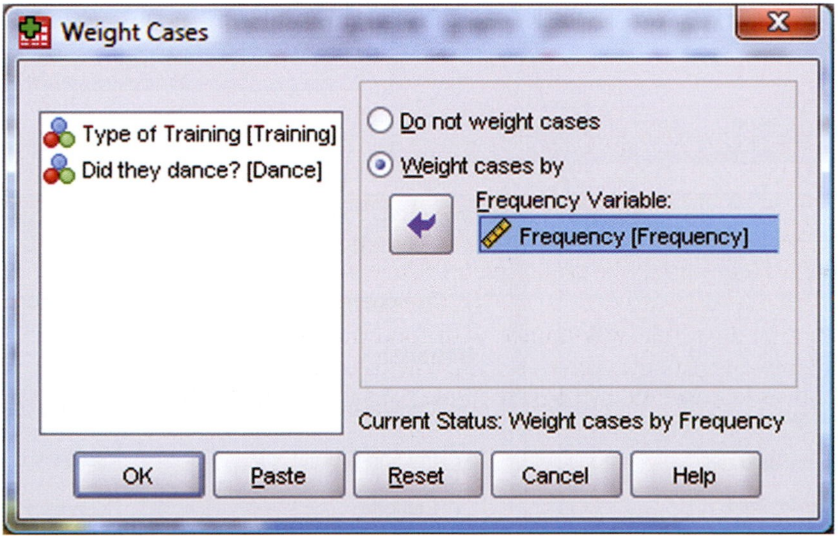

FIGURE 18.3 The dialog box for the *weight cases* command

Entering data using a variable representing the number of cases that fall into a combination of categories can be quite labour saving. However, to analyse data entered in this way we must tell the computer that the variable **Frequency** represents the number of cases that fell into a particular combination of categories. To do this, access the *Weight Cases* dialog box in Figure 18.3 by selecting Data ▸ Weight Cases.... Select ⦿ Weight cases by and then select the variable in which the number of cases is specified (in this case **Frequency**) and drag it to the box labelled *Frequency variable* (or click on ➜). This process tells the computer that it should weight each category combination by the number in the column labelled **Frequency**. Therefore, the computer will pretend, for example, that there are 28 rows of data that have the category combination 0, 0 (representing cats trained with food and that danced). Data entered in this way are in the file **CatsWeight.sav** and if you use this file you must remember to weight the cases as described.

18.5.3. Running the analysis ①

Summarizing data that fall into categories is done using the *crosstabs* command (which also produces the chi-square test). *Crosstabs* is accessed by selecting Analyze Descriptive Statistics ▸ Crosstabs.... Figure 18.4 shows the dialog boxes for the *crosstabs* command (the variable **Frequency** is in the diagram because I ran the analysis on the

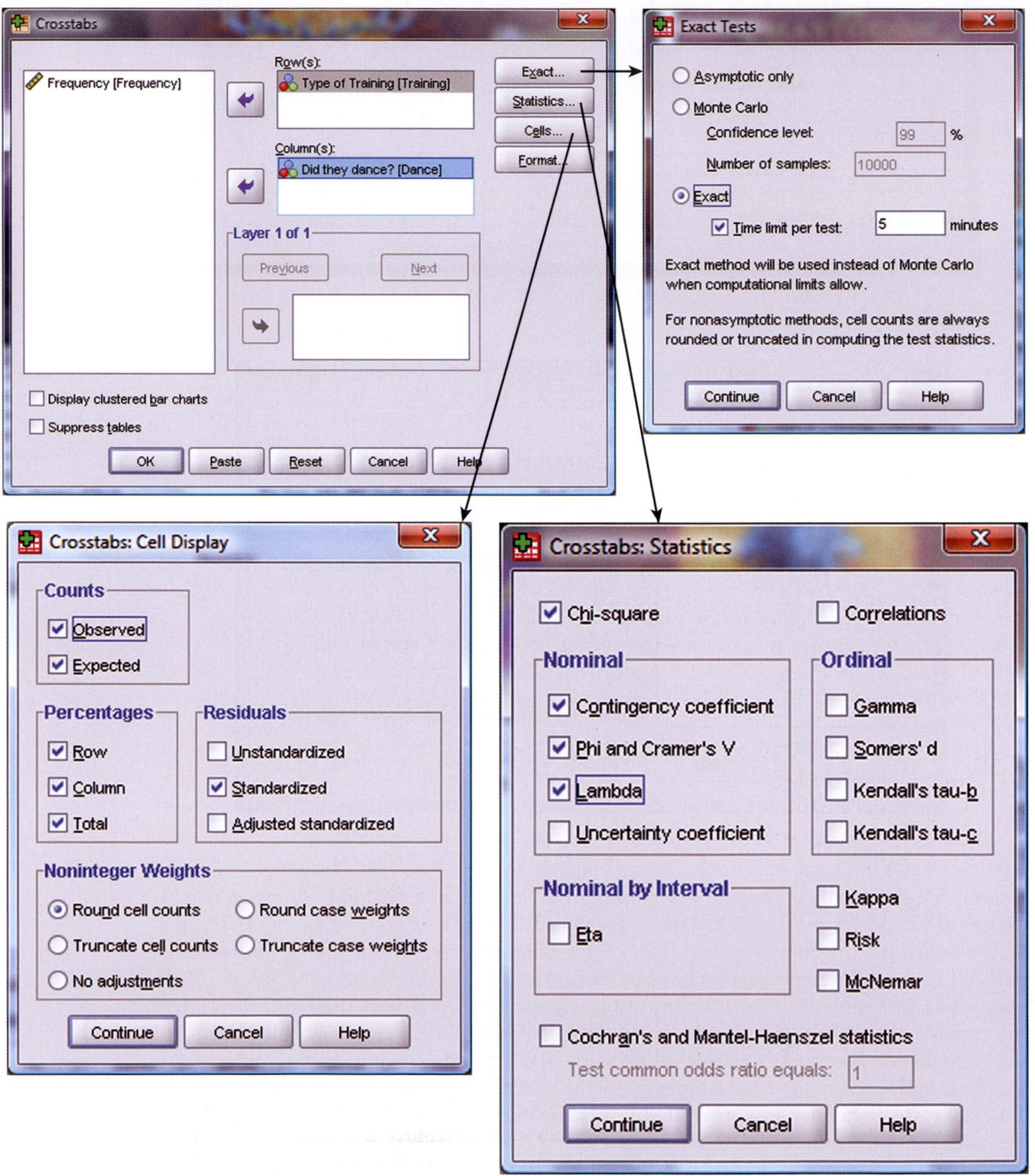

FIGURE 18.4 Dialog boxes for the *crosstabs* command

CatsWeight.sav data). First, select one of the variables of interest in the variable list and drag it into the box labelled *Row(s)* (or click on ➔). For this example, I selected **Training** to be the rows of the table. Next, select the other variable of interest (**Dance**) and drag it to the box labelled *Column(s)* (or click on ➔). In addition, it is possible to select a layer variable (i.e. you can split the rows of the table into further categories). If you had a third categorical variable (as we will later in this chapter) you could split the contingency table by this variable (so layers of the table represent different categories of this third variable).

If you click on Statistics... a dialog box appears in which you can specify various statistical tests. The most important options under the statistics menu for categorical data are described in SPSS Tip 18.1.

SPSS TIP 18.1 Statistical options for *crosstabs* ②

- **Chi-square**: This performs the basic Pearson chi-square test (section 18.3.1).
- **Phi** and **Cramer's V**: These are measures of the strength of association between two categorical variables. Phi is used with 2 × 2 contingency tables (tables in which you have two categorical variables and each variable has only two categories). Phi is calculated by taking the chi-square value and dividing it by the sample size and then taking the square root of this value. If one of the two categorical variables contains more than two categories then Cramer's *V* is preferred to phi because phi fails to reach its minimum value of 0 (indicating no association) in these circumstances.
- **Goodman and Kruskal's lambda (λ)**: This measures the proportional reduction in error that is achieved when membership of a category of one variable is used to predict category membership of the other variable. A value of 1 means that one variable perfectly predicts the other, whereas a value of 0 indicates that one variable in no way predicts the other.
- **Kendall's statistic**: This statistic is discussed in section 6.5.4.

Select the chi-square test, the continuity correction, phi and lambda, and then click on Continue. If you click on Cells... a dialog box appears in which you can specify the type of data displayed in the crosstabulation table. It is important that you ask for expected counts because for chi-square to be accurate these expected counts must exceed certain values. The basic rule of thumb is that with 2 × 2 contingency tables no expected values should be below 5. In larger tables the rule is that all expected counts should be greater than 1 and no more than 20% of expected counts should be less than 5. It is also useful to have a look at the row, column and total percentages because these values are usually more easily interpreted than the actual frequencies and provide some idea of the origin of any significant effects. The second important option is to select some standardized residuals. These values are important for breaking down a significant chi-square test (should we get one). Once these options have been selected, click on Continue to return to the main dialog box.

From here you can click on Exact... (if you have exact tests installed) to compute Fisher's exact test (section 18.3.2). You can use this option if your sample is small or if your expected frequencies are too low (see 18.4). Select the *Exact* test option; we don't really need it for these data but it will be a useful way to see how it is used. Click on Continue to return to the main dialog box and then click on OK to run the analysis.

18.5.4. Output for the chi-square test ①

The crosstabulation table produced by SPSS (SPSS Output 18.1) contains the number of cases that fall into each combination of categories and is rather like our original contingency table. We can see that in total 76 cats danced (38% of the total) and of these 28 were trained using food (36.8% of the total that danced) and 48 were trained with affection (63.2% of the total that danced). Further, 124 cats didn't dance at all (62% of the total) and of those that didn't dance, 10 were trained using food as a reward (8.1% of the total that didn't dance) and a massive 114 were trained using affection (91.9% of the total that didn't dance). The numbers of cats can be read from the rows labelled *Count* and the percentages are read from the rows labelled *% within Did they dance?* We can also look at the percentages within the training categories by looking at the rows labelled *% within Type of Training*. This tells us, for example, that of those trained with food as a reward, 73.7% danced and 26.3% did not. Similarly, for those trained with affection only 29.6% danced compared to 70.4% that didn't. In summary, when food was used as a reward most cats would dance, but when affection was used most cats refused to dance.

SPSS OUTPUT 18.1

Type of Training * Did they dance? Crosstabulation

			Did they dance?		
			Yes	No	Total
Type of Training	Food as Reward	Count	28	10	38
		Expected Count	14.4	23.6	38.0
		% within Type of Training	73.7%	26.3%	100.0%
		% within Did they dance?	36.8%	8.1%	19.0%
		% of Total	14.0%	5.0%	19.0%
		Std. Residual	3.6	-2.8	
	Affection as Reward	Count	48	114	162
		Expected Count	61.6	100.4	162.0
		% within Type of Training	29.6%	70.4%	100.0%
		% within Did they dance?	63.2%	91.9%	81.0%
		% of Total	24.0%	57.0%	81.0%
		Std. Residual	-1.7	1.4	
	Total	Count	76	124	200
		Expected Count	76.0	124.0	200.0
		% within Type of Training	38.0%	62.0%	100.0%
		% within Did they dance?	100.0%	100.0%	100.0%
		% of Total	38.0%	62.0%	100.0%

Before moving on to look at the test statistics itself it is vital that we check that the assumption for chi-square has been met. The assumption is that in 2 × 2 tables (which is what we have here), all expected frequencies should be greater than 5. If you look at the expected counts in the crosstabulation table (which incidentally are the same as we calculated earlier), it should be clear that the smallest expected count is 14.4 (for cats that were trained with food and did dance). This value exceeds 5 and so the assumption has been met. If you found an expected count lower than 5 the best remedy is to collect more data to try to boost the proportion of cases falling into each category.

As we saw earlier, Pearson's chi-square test examines whether there is an association between two categorical variables (in this case the type of training and whether the animal danced or not). As part of the *crosstabs* procedure SPSS produces a table that includes the chi-square statistic and its significance value (SPSS Output 18.2). The Pearson chi-square

Chi-Square Tests

SPSS OUTPUT 18.2

	Value	df	Asymp. Sig. (2-sided)	Exact Sig. (2-sided)	Exact Sig. (1-sided)	Point Probability
Pearson Chi-Square	25.356[a]	1	.000	.000	.000	
Continuity Correction[b]	23.520	1	.000			
Likelihood Ratio	24.932	1	.000	.000	.000	
Fisher's Exact Test				.000	.000	
Linear-by-Linear Association	25.229[c]	1	.000	.000	.000	.000
N of Valid Cases	200					

a. 0 cells (.0%) have expected count less than 5. The minimum expected count is 14.44.

b. Computed only for a 2x2 table

c. The standardized statistic is 5.023.

statistic tests whether the two variables are independent. If the significance value is small enough (conventionally *Sig.* must be less than .05) then we reject the hypothesis that the variables are independent and gain confidence in the hypothesis that they are in some way related. The value of the chi-square statistic is given in the table (and the degrees of freedom) as is the significance value. The value of the chi-square statistic is 25.356, which is within rounding error of what we calculated in section 18.3.1. This value is highly significant ($p < .001$), indicating that the type of training used had a significant effect on whether an animal would dance.

A series of other statistics are also included in the table (many of which have to be specifically requested using the options in the dialog box in Figure 18.4). *Continuity Correction* is Yates's continuity corrected chi-square (see section 18.3.4) and its value is the same as the value we calculated earlier (23.52). As I mentioned earlier, this test is probably best ignored anyway, but it does confirm the result from the main chi-square test. The *Likelihood Ratio* is the statistic we encountered in section 18.3.3 (and is again within rounding error of the value we calculated: 24.93). Again this confirms the main chi-square result, but this statistic would be preferred in smaller samples.

Underneath the chi-square table there are several footnotes relating to the assumption that expected counts should be greater than 5. If you forgot to check this assumption yourself, SPSS kindly gives a summary of the number of expected counts below 5. In this case, there were no expected frequencies less than 5 so we know that the chi-square statistic should be accurate.

The highly significant result indicates that there is an association between the type of training and whether the cat danced or not. What we mean by an association is that the pattern of responses (i.e. the proportion of cats that danced to the proportion that did not) in the two training conditions is significantly different. This significant finding reflects the fact that when food is used as a reward, about 74% of cats learn to dance and 26% do not, whereas when affection is used, the opposite is true (about 70% refuse to dance and 30% do dance). Therefore, we can conclude that the type of training used significantly influences the cats: they will dance for food but not for love! Having lived with a lovely cat for many years now, this supports my cynical view that they will do nothing unless there is a bowl of cat-food waiting for them at the end of it!

How do I interpret chi-square?

If requested, SPSS will produce another table of output containing some additional statistical tests. Most of these tests are measures of the strength of association. These measures are based on modifying the chi-square statistic to take account of sample size and degrees of freedom and they try to restrict the range of the test statistic from 0 to 1 (to make them similar to the correlation coefficient described in Chapter 6). These are shown in SPSS Output 18.3.

- *Phi*: This statistic is accurate for 2 × 2 contingency tables. However, for tables with greater than two dimensions the value of phi may not lie between 0 and 1 because the chi-square value can exceed the sample size. Therefore, Pearson suggested the use of the coefficient of contingency.
- *Contingency Coefficient*: This coefficient ensures a value between 0 and 1 but, unfortunately, it seldom reaches its upper limit of 1 and for this reason Cramer devised Cramer's *V*.
- *Cramer's V*: When both variables have only two categories, phi and Cramer's *V* are identical. However, when variables have more than two categories Cramer's statistic can attain its maximum of one – unlike the other two – and so it is the most useful.

For these data, Cramer's statistic is 0.36 out of a possible maximum value of 1. This represents a medium association between the type of training and whether the cats danced or not (if you think of it like a correlation coefficient then this represents a medium effect size). This value is highly significant ($p < .001$) indicating that a value of the test statistic that is this big is unlikely to have happened by chance, and therefore the strength of the relationship is significant. These results confirm what the chi-square test already told us but also give us some idea of the size of effect.

SPSS OUTPUT 18.3

Symmetric Measures

		Value	Approx. Sig.	Exact Sig.
Nominal by Nominal	Phi	.356	.000	.000
	Cramer's V	.356	.000	.000
	Contingency Coefficient	.335	.000	.000
	N of Valid Cases	200		

18.5.5. Breaking down a significant chi-square test with standardized residuals ②

Although in a 2 × 2 contingency table like the one we have in this example the nature of the association can be quite clear from just the cell percentages or counts, in larger contingency tables it can be useful to do a finer-grained investigation of the table. In a way, you can think of a significant chi-square test in much the same way as a significant interaction in ANOVA: it is an effect that needs to be broken down further. One very easy way to break down a significant chi-square test is to use data that we already have – the standardized residual.

Just like regression, the residual is simply the error between what the model predicts (the expected frequency) and the data actually observed (the observed frequency):

$$\text{residual}_{ij} = \text{observed}_{ij} - \text{model}_{ij}$$

in which *i* and *j* represent the two variables (i.e. the rows and columns in the contingency table). This is the same as every other residual or deviation that we have encountered in this book (compare this equation to, for example, equation (2.4)). To standardize this equation, we simply divide by the square root of the expected frequency:

$$\text{standardized residual} = \frac{\text{observed}_{ij} - \text{model}_{ij}}{\sqrt{\text{model}_{ij}}}$$

Does this equation look familiar? Well, it's basically part of equation (18.1). The only difference is that rather than looking at squared deviations, we're looking at the pure deviation. Remember that the rationale for squaring deviations in the first place is simply to make them positive so that they don't cancel out when we add them. The chi-square statistic is based on adding together values, so it is important that the deviations are squared so that they don't cancel out. However, if we're not planning to add up the deviations or residuals then we can inspect them in their unsquared form. There are two important things about these standardized residuals:

1 Given that the chi-square statistic is the sum of these standardized residuals (sort of), then if we want to decompose what contributes to the overall association that the chi-square statistic measures, then looking at the individual standardized residuals is a good idea because they have a direct relationship with the test statistic.

2 These standardized residuals behave like any other (see section 7.6.1.1) in the sense that each one is a z-score. This is very useful because it means that just by looking at a standardized residual we can assess its significance (see section 1.7.4). As we have learnt many times before, if the value lies outside of ±1.96 then it is significant at $p < .05$, if it lies outside ±2.58 then it is significant at $p < .01$ and if it lies outside ±3.29 then it is significant at $p < .001$.

Fortunately, when we selected ☑ Standardized in Figure 18.4, SPSS produced these standardized residuals and we can see them in SPSS Output 18.1. There are four residuals: one for each combination of the type of training and whether the cats danced. When food was used as a reward the standardized residual was significant for both those that danced ($z = 3.6$) and those that didn't dance ($z = -2.8$) because both values are bigger than 1.96 (when you ignore the minus sign). The plus or minus sign (and the counts and expected counts within the cells) tells us that when food was used as a reward significantly more cats than expected danced, and significantly less cats than expected did not dance. When affection was used as a reward the standardized residual was not significant for both those that danced ($z = -1.7$) and those that didn't dance ($z = 1.4$) because they are both smaller than 1.96 (when you ignore the minus sign). This tells us that when affection was used as a reward as many cats as expected danced and did not dance. In a nutshell, the cells for when food was used as a reward both significantly contribute to the overall chi-square statistic. Put another way, the association between the type of reward and dancing is mainly driven by when food is a reward.

18.5.6. Calculating an effect size ②

Although Cramer's *V* is an adequate effect size (in the sense that it is constrained to fall between 0 and 1 and is, therefore, easily interpretable), a more common and possibly more useful measure of effect size for categorical data is the **odds ratio**, which we encountered in Chapter 8 (it was called *Exp(B)* in that chapter). Odds ratios are most interpretable in 2 × 2 contingency tables and are probably not useful for larger contingency tables. However, this isn't as restrictive as you might think because as I've said more times than I care to recall in the GLM chapters, effect sizes are only ever useful when they summarize a focused comparison. A 2 × 2 contingency table is the categorical data equivalent of a focused comparison!

The odds ratio is simple enough to calculate. If we look at our example, we can first calculate the odds that a cat danced given that they had food as a reward. This is simply the number of cats that were given food and danced, divided by the number of cats given food that didn't dance:

$$\begin{aligned}\text{odds}_{\text{dancing after food}} &= \frac{\text{number that had food and danced}}{\text{number that had food but didn't dance}}\\ &= \frac{28}{10}\\ &= 2.8\end{aligned}$$

Next we calculate the odds that a cat danced given that they had affection as a reward. This is simply the number of cats that were given affection and danced, divided by the number of cats given affection that didn't dance:

$$\begin{aligned}\text{odds}_{\text{dancing after affection}} &= \frac{\text{number that had affection and danced}}{\text{number that had affection but didn't dance}}\\ &= \frac{48}{114}\\ &= 0.421\end{aligned}$$

The odds ratio is simply the odds of dancing after food divided by the odds of dancing after affection:

$$\begin{aligned}\text{odds ratio} &= \frac{\text{odds}_{\text{dancing after food}}}{\text{odds}_{\text{dancing after affection}}}\\ &= \frac{2.8}{0.421}\\ &= 6.65\end{aligned}$$

What this tells us is that if a cat was trained with food the odds of their dancing were 6.65 times higher than if they had been trained with affection. As you can see, this is an extremely elegant and easily understood metric for expressing the effect you've got!

18.5.7. Reporting the results of chi-square ①

When reporting Pearson's chi-square we simply report the value of the test statistic with its associated degrees of freedom and the significance value. The test statistic, as we've seen, is denoted by χ^2. The SPSS output tells us that the value of χ^2 was 25.36, that the degrees of freedom on which this was based were 1, and that it was significant at $p < .001$. It's also useful to reproduce the contingency table and my vote would go to quoting the odds ratio too. As such, we could report:

✓ There was a significant association between the type of training and whether or not cats would dance χ^2 (1) = 25.36, $p < .001$. This seems to represent the fact that, based on the odds ratio, the odds of cats dancing were 6.65 times higher if they were trained with food than if trained with affection.

CRAMMING SAM'S TIPS

- If you want to test the relationship between two categorical variables you can do this with *Pearson's chi-square test* or the *likelihood ratio statistic*.
- Look at the table labelled **Chi-Square tests**; if the *Exact Sig.* value is less than .05 for the row labelled *Pearson chi-square* then there is a significant relationship between your two variables.
- Check underneath this table to make sure that no expected frequencies are less than 5.
- Look at the crosstabulation table to work out what the relationship between the variables is. Better still, look out for significant standardized residuals (values outside of ±1.96), and calculate the *odds ratio*.
- Report the χ^2 statistic, the degrees of freedom, and the significance value. Also report the contingency table.

LABCOAT LENI'S REAL RESEARCH 18.1

Is the black American happy? ①

When I was doing my psychology degree I spent a lot of time reading about the civil rights movement in the USA. Although I was supposed to be reading psychology, I became more interested in Malcolm X and Martin Luther King Jr. This is why I find Beckham's 1929 study of black Americans such an interesting piece of research. Beckham was a black American academic who founded the psychology laboratory at Howard University, Washington, DC, and his wife Ruth was the first black woman ever to be awarded a Ph.D. (also in psychology) at the University of Minnesota. The article needs to be placed within the era in which it was published. To put some context on the study, it was published 36 years before the Jim Crow laws were finally overthrown by the Civil Rights Act of 1964, and in a time when black Americans were segregated, openly discriminated against and were victims of the most abominable violations of civil liberties and human rights. For a richer context I suggest reading James Baldwin's superb novel *The fire next time*. Even the language of the study and the data from it are an uncomfortable reminder of the era in which it was conducted.

Beckham sought to measure the psychological state of black Americans with three questions put to 3443 black Americans from different walks of life. He asked them whether they thought black Americans were happy, whether they personally were happy as a black American, and whether black Americans *should* be happy. They could answer only *yes* or *no* to each question. By today's standards the study is quite simple, and he did no formal statistical analysis of his data (Fisher's article containing the popularized version of the chi-square test was published only seven years earlier in a statistics journal that would not have been read by psychologists). I love this study, though, because it demonstrates that you do not need elaborate methods to answer important and far-reaching questions; with just three questions, Beckham told the world an enormous amount about very real and important psychological and sociological phenomena.

The frequency data (number of yes and no responses within each employment category) from this study are in the file **Beckham(1929).sav**. Labcoat Leni wants you to carry out three chi-square tests (one for each question that was asked). What conclusions can you draw?

Answers are in the additional material on the companion website.

BECKHAM, A. S. (1929). *JOURNAL OF ABNORMAL AND SOCIAL PSYCHOLOGY, 24*, 186–190.

18.6. Several categorical variables: loglinear analysis ③

So far we've looked at situations in which there are only two categorical variables. However, often we want to analyse more complex contingency tables in which there are three or more variables. For example, what about if we took the example we've just used but also collected data from a sample of 70 dogs? We might want to compare the behaviour in dogs to that in cats. We would now have three variables: **Animal** (dog or cat), **Training** (food as reward or affection as reward) and **Dance** (did they dance or not?). This couldn't be analysed with the Pearson chi-square and instead has to be analysed with a technique called **loglinear analysis**.

18.6.1. Chi-square as regression ④

SMART ALEX ONLY

To begin with, let's have a look at how our simple chi-square example can be expressed as a regression model. Although we already know about as much as we need to about the chi-square test, if we want to understand more complex situations life becomes considerably easier if we consider our model as a general linear model (i.e. regression). All of the general linear models we've considered in this book take the general form of:

$$\text{outcome}_i = (\text{model}) + \text{error}_i$$

For example, when we encountered multiple regression in Chapter 7 we saw that this model was written as (see equation (7.9)):

$$Y_i = (b_0 + b_1X_{1i} + b_2X_{2i} + \ldots + b_nX_{ni}) + \varepsilon_i$$

Also, when we came across one-way ANOVA, we adapted this regression model to conceptualize our Viagra example, as (see equation (10.2)):

$$\text{Libido}_i = b_0 + b_2\text{High}_i + b_1\text{Low}_i + \varepsilon_i$$

The *t*-test was conceptualized in a similar way. In all cases the same basic equation is used; it's just the complexity of the model that changes. With categorical data we can use the same model in much the same way as with regression to produce a linear model. In our current example we have two categorical variables: training (food or affection) and dance (yes they did dance or no they didn't dance). Both variables have two categories and so we can represent each one with a single dummy variable (see section 7.11.1) in which one category is coded as 0 and the other as 1. So for training, we could code 'food' as 0 and 'affection' as 1, and we could code the dancing variable as 0 for 'yes' and 1 for 'no' (see Table 18.2).

TABLE 18.2 Coding scheme for dancing cats

Training	Dance	Dummy (Training)	Dummy (Dance)	Interaction	Frequency
Food	Yes	0	0	0	28
Food	No	0	1	0	10
Affection	Yes	1	0	0	48
Affection	No	1	1	1	114

This situation might be familiar if you think back to factorial ANOVA (section 12.8) in which we also had two variables as predictors. In that situation we saw that when there are two variables the general linear model became (think back to equation (12.1)):

$$\text{outcome}_i = (b_0 + b_1 A_i + b_2 B_i + b_3 AB_i) + \varepsilon_i$$

in which A represents the first variable, B represents the second and AB represents the interaction between the two variables. Therefore, we can construct a linear model using these dummy variables that is exactly the same as the one we used for factorial ANOVA (above). The interaction term will simply be the training variable multiplied by the dance variable (look at Table 18.2 and if it doesn't make sense look back to section 12.8 because the coding is exactly the same as this example):

$$\begin{aligned} \text{outcome}_i &= (\text{model}) + \text{Error}_i \\ \text{outcome}_{ij} &= (b_0 + b_1 \text{Training}_i + b_2 \text{Dance}_j + b_3 \text{Interaction}_{ij}) \end{aligned} \tag{18.3}$$

However, because we're using categorical data, to make this model linear we have to actually use log values (see Chapter 8) and so the actual model becomes.[2]

$$\begin{aligned} \ln(O_i) &= \ln(model) + \ln(\varepsilon_i) \\ \ln(O_{ij}) &= (b_0 + b_1 \text{Training}_i + b_2 \text{Dance}_j + b_3 \text{Interaction}_{ij}) + \ln(\varepsilon_i) \end{aligned} \tag{18.4}$$

The training and dance variables and the interaction can take the values 0 and 1, depending on which combination of categories we're looking at (Table 18.2). Therefore, to work out what the *b*-values represent in this model we can do the same as we did for the *t*-test and ANOVA and look at what happens when we replace training and dance with values of 0 and 1. To begin with, let's see what happens when we look at when training and dance are both zero. This represents the category of cats that got food reward and did line dance. When we used this sort of model for the *t*-test and ANOVA the outcomes we used were taken from the observed data: we used the group means (e.g. see sections 9.7 and 10.2.3). However, with categorical variables, means are rather meaningless because we haven't measured anything on an ordinal or interval scale, instead we merely have frequency data. Therefore, we use the observed frequencies (rather than observed means) as our outcome instead. In Table 18.1 we saw that there were 28 cats that had food for a reward and did line dance. If we use this as the observed outcome then the model can be written as (if we ignore the error term for the time being):

$$\ln(O_{ij}) = b_0 + b_1 \text{Training}_i + b_2 \text{Dance}_j + b_3 \text{Interaction}_{ij}$$

For cats that had food reward and did dance, the training and dance variables and the interaction will all be 0 and so the equation reduces down to:

$$\begin{aligned} \ln(O_{\text{Food, Yes}}) &= b_0 + (b_1 \times 0) + (b_2 \times 0) + (b_3 \times 0) \\ \ln(O_{\text{Food, Yes}}) &= b_0 \\ \ln(28) &= b_0 \\ b_0 &= 3.332 \end{aligned}$$

[2] Actually, the convention is to denote b_0 as θ and the b values as λ, but I think these notational changes serve only to confuse people so I'm sticking with b because I want to emphasize the similarities to regression and ANOVA.

Therefore, b_0 in the model represents the log of the observed value when all of the categories are zero. As such it's the log of the observed value of the base category (in this case cats that got food and danced). Now, let's see what happens when we look at cats that had affection as a reward and danced. In this case, the training variable is 1 and the dance variable and the interaction are still 0. Also, our outcome now changes to be the observed value for cats that received affection and danced (from Table 18.1 we can see the value is 48). Therefore, the equation becomes:

$$\ln(O_{\text{Affection, Yes}}) = b_0 + (b_1 \times 1) + (b_2 \times 0) + (b_3 \times 0)$$
$$\ln(O_{\text{Affection, Yes}}) = b_0 + b_1$$
$$b_1 = \ln(O_{\text{Affection, Yes}}) - b_0$$

Remembering that b_0 is the expected value for cats that had food and danced, we get:

$$\begin{aligned} b_1 &= \ln(O_{\text{Affection, Yes}}) - \ln(O_{\text{Food, Yes}}) \\ &= \ln(48) - \ln(28) \\ &= 3.871 - 3.332 \\ &= 0.539 \end{aligned}$$

The important thing is that b_1 is the difference between the log of the observed frequency for cats that received affection and danced, and the log of the observed values for cats that received food and danced. Put another way, within the group of cats that danced it represents the difference between those trained using food and those trained using affection.

Now, let's see what happens when we look at cats that had food as a reward and did not dance. In this case, the training variable is 0, the dance variable is 1 and the interaction is again 0. Our outcome now changes to be the observed frequency for cats that received food but did not dance (from Table 18.1 we can see the value is 10). Therefore, the equation becomes:

$$\ln(O_{\text{Food, No}}) = b_0 + (b_1 \times 0) + (b_2 \times 1) + (b_3 \times 0)$$
$$\ln(O_{\text{Food, No}}) = b_0 + b_2$$
$$b_2 = \ln(O_{\text{Food, No}}) - b_0$$

Remembering that b_0 is the expected value for cats that had food and danced, we get:

$$\begin{aligned} b_2 &= \ln(O_{\text{Food, No}}) - \ln(O_{\text{Food, Yes}}) \\ &= \ln(10) - \ln(28) \\ &= 2.303 - 3.332 \\ &= -1.029 \end{aligned}$$

The important thing is that b_2 is the difference between the log of the observed frequency for cats that received food and danced, and the log of the observed frequency for cats that received food and didn't dance. Put another way, within the group of cats that received food as a reward it represents the difference between cats that didn't dance and those that did. Finally, we can look at cats that had affection and danced. In this case, the training and dance variables are both 1 and the interaction (which is the value of training multiplied by the value of dance) is also 1. We can also replace b_0, b_1 and b_2 with what we now know they represent. The outcome is the log of the observed frequency for cats that received affection but didn't dance (this expected value is 114 – see Table 18.1). Therefore, the equation becomes (I've used the shorthand of A for affection, F for food, Y for yes, and N for No):

$$\begin{aligned}
\ln(O_{A,N}) &= b_0 + (b_1 \times 1) + (b_2 \times 1) + (b_3 \times 1) \\
\ln(O_{A,N}) &= b_0 + b_1 + b_2 + b_3 \\
\ln(O_{A,N}) &= \ln(O_{F,Y}) + (\ln(O_{A,Y}) - \ln(O_{F,Y})) + (\ln(O_{F,N}) - \ln(O_{F,Y})) + b_3 \\
\ln(O_{A,N}) &= \ln(O_{A,Y}) + \ln(O_{F,N}) - \ln(O_{F,Y}) + b_3 \\
b_3 &= \ln(O_{A,N}) - \ln(O_{F,N}) + \ln(O_{F,Y}) - \ln(O_{A,Y}) \\
&= \ln(114) - \ln(10) + \ln(28) - \ln(48) \\
&= 1.895
\end{aligned}$$

So, b_3 in the model really compares the difference between affection and food when the cats didn't dance to the difference between food and affection when the cats did dance. Put another way, it compares the effect of training when cats didn't dance to the effect of training when they did dance.

The final model is therefore:

$$\ln(O_{ij}) = 3.332 + 0.539\text{Training} - 1.029\text{Dance} + 1.895\text{Interaction} + \ln(\varepsilon_{ij})$$

The important thing to note here is that everything is exactly the same as factorial ANOVA except that we dealt with log-transformed values (in fact compare this section to section 12.8 to see just how similar everything is). In case you still don't believe me that this works as a general linear model, I've prepared a file called **CatRegression.sav** which contains the two variables **Dance** and **Training** (both dummy coded with 0 and 1 as described above) and the interaction (**Interaction**). There is also a variable called **Observed** which contains the observed frequencies in Table 18.1 for each combination of **Dance** and **Training**. Finally, there is a variable called **LnObserved**, which is the natural logarithm of these observed frequencies (remember that throughout this section we've dealt with the log observed values).

SELF-TEST Run a multiple regression analysis using **CatsRegression.sav** with **LnObserved** as the outcome, and **Training**, **Dance** and **Interaction** as your three predictors.

SPSS Output 18.4 shows the resulting coefficients table from this regression. The important thing to note is that the constant, b_0, is 3.332 as calculated above, the beta value for type of training, b_1, is 0.539 and for dance, b_2, is –1.030, both of which are within rounding error of what was calculated above. Also the coefficient for the interaction, b_3, is 1.895 as predicted. There is one interesting point, though: all of the standard errors are zero, or put differently there is *no* error at all in this model (which is also why there are no significance tests). This is because the various combinations of coding variables completely explain the observed values. This is known as a **saturated model** and I will return to this point later, so bear it in mind. For the time being, I hope this convinces you that chi-square can be conceptualized as a linear model.

SPSS OUTPUT 18.4

Coefficients[a]

Model		Unstandardized Coefficients		Standardized Coefficients	t	Sig.
		B	Std. Error	Beta		
1	(Constant)	3.332	.000		.	.
	Type of Training	.539	.000	.307	.	.
	Did they dance?	-1.030	.000	-.725	.	.
	Interaction	1.895	.000	1.361	.	.

a. Dependent Variable: LN (Observed values)

OK, this is all very well, but the heading of this section did rather imply that I would show you how the chi-square test can be conceptualized as a linear model. Well, basically, the chi-square test looks at whether two variables are independent; therefore, it has no interest in the combined effect of the two variables, only their unique effect. Thus, we can conceptualize chi-square in much the same way as the saturated model, except that we don't include the interaction term. If we remove the interaction term, our model becomes:

$$\ln(\text{model}_{ij}) = b_0 + b_1\text{Training}_i + b_2\text{Dance}_j$$

With this new model, we cannot predict the observed values like we did for the saturated model because we've lost some information (namely, the interaction term). Therefore, the outcome from the model changes, and therefore the beta values change too. We saw earlier that the chi-square test is based on 'expected frequencies'. Therefore, if we're conceptualizing the chi-square test as a linear model, our outcomes will be these expected values. If you look back to the beginning of this chapter you'll see we already have the expected frequencies based on this model. We can recalculate the beta values based on these expected values:

$$\ln(E_{ij}) = b_0 + b_1\text{Training}_i + b_2\text{Dance}_j$$

For cats that had food reward and did dance, the training and dance variables will be 0 and so the equation reduces down to:

$$\begin{aligned}
\ln(E_{\text{Food, Yes}}) &= b_0 + (b_1 \times 0) + (b_2 \times 0) \\
\ln(E_{\text{Food, Yes}}) &= b_0 \\
b_0 &= \ln(14.44) \\
&= 2.67
\end{aligned}$$

Therefore, b_0 in the model represents the log of the expected value when all of the categories are zero.

When we look at cats that had affection as a reward and danced, the training variable is 1 and the dance variable is still 0. Also, our outcome now changes to be the expected value for cats that received affection and danced:

$$\begin{aligned}
\ln(E_{\text{Affection, Yes}}) &= b_0 + (b_1 \times 1) + (b_2 \times 0) \\
\ln(E_{\text{Affection, Yes}}) &= b_0 + b_1 \\
b_1 &= \ln(E_{\text{Affection, Yes}}) - b_0 \\
&= \ln(E_{\text{Affection, Yes}}) - \ln(E_{\text{Food, Yes}}) \\
&= \ln(61.56) - \ln(14.44) \\
&= 1.45
\end{aligned}$$

The important thing is that b_1 is the difference between the log of the expected frequency for cats that received affection and did dance and the log of the expected values for cats that received food and danced. In fact, the value is the same as the column marginal, that is the difference between the total number of cats getting affection and the total number of cats getting food: ln(162) – ln(38) = 1.45. Put simply, it represents the main effect of the type of training.

When we look at cats that had food as a reward and did not dance, the training variable is 0 and the dance variable is 1. Our outcome now changes to be the expected frequency for cats that received food but did not dance:

$$
\begin{aligned}
\ln(E_{\text{Food, No}}) &= b_0 + (b_1 \times 0) + (b_2 \times 1)\\
\ln(E_{\text{Food, No}}) &= b_0 + b_2\\
b_2 &= \ln(O_{\text{Food, No}}) - b_0\\
&= \ln(O_{\text{Food, No}}) - \ln(O_{\text{Food, yes}})\\
&= \ln(23.56) - \ln(14.44)\\
&= 0.49
\end{aligned}
$$

Therefore, b_2 is the difference between the log of the expected frequencies for cats that received food and didn't or did dance. In fact, the value is the same as the row marginal, that is the difference between the total number of cats that did and didn't dance: ln(124) – ln (76) = 0.49. In simpler terms, it is the main effect of whether or not the cat danced.

We can double-check all of this by looking at the final cell:

$$
\begin{aligned}
\ln(E_{\text{Affection, No}}) &= b_0 + (b_1 \times 1) + (b_2 \times 1)\\
\ln(E_{\text{Affection, No}}) &= b_0 + b_1 + b_2\\
\ln(100.44) &= 2.67 + 1.45 + 0.49\\
4.61 &= 4.61
\end{aligned}
$$

The final chi-square model is therefore:

$$
\begin{aligned}
\ln(O_i) &= \text{model}_{ij} + ln(\varepsilon_i)\\
\ln(O_i) &= 2.67 + 1.45\text{Training} + 0.49\text{Dance} + \ln(\varepsilon_i)
\end{aligned}
$$

We can rearrange this to get some residuals (the error term):

$$\ln(\varepsilon_i) = \ln(O_i) - (\text{model})$$

In this case, the model is merely the expected frequencies that were calculated for the chi-square test, so the residuals are the differences between the observed and expected frequencies.

SELF-TEST To show that this all actually works, run another multiple regression analysis using **CatsRegression.sav**. This time the outcome is the log of expected frequencies (**LnExpected**) and **Training** and **Dance** are the predictors (the interaction is not included).

This demonstrates how chi-square can work as a linear model, just like regression and ANOVA, in which the beta values tell us something about the relative differences in frequencies across categories of our two variables. If nothing else made sense I want you to leave this section aware that chi-square (and analysis of categorical data generally) can be expressed as a linear model (although we have to use log values). We can express categories of a variable using dummy variables, just as we did with regression and ANOVA, and the resulting beta values can be calculated in exactly the same way as for regression and ANOVA. In ANOVA, these beta values represented differences between the means of a particular category compared against a baseline category. With categorical data, the beta values represent the same thing, the only difference being that rather than dealing with means, we're dealing with expected values. Grasping this idea (that regression, *t*-tests, ANOVAs and categorical data analysis are basically the same) will help (me) considerably in the next section.

18.6.2. Loglinear analysis ③

In the previous section, after nearly reducing my brain to even more of a rotting vegetable than it already is trying to explain how categorical data analysis is just another form of regression, I ran the data through an ordinary regression on SPSS to prove that I wasn't talking complete gibberish. At the time I rather glibly said 'oh, by the way, there's no error in the model, that's odd isn't it?' and sort of passed this off by telling you that it was a 'saturated' model and not to worry too much about it because I'd explain it all later just as soon as I'd worked out what the hell was going on. That seemed like a good avoidance tactic at the time but unfortunately I now have to explain what I was going on about.

To begin with, I hope you're now happy with the idea that categorical data can be expressed in the form of a linear model provided that we use log values (this, incidentally, is why the technique we're discussing is called loglinear analysis). From what you hopefully already know about ANOVA and linear models generally, you should also be cosily tucked up in bed with the idea that we can extend any linear model to include any amount of predictors and any resulting interaction terms between predictors. Therefore, if we can represent a simple two-variable categorical analysis in terms of a linear model, then it shouldn't amaze you to discover that if we have more than two variables this is no problem: we can extend the simple model by adding whatever variables and the resulting interaction terms. This is all you really need to know. So, just as in multiple regression and ANOVA, if we think of things in terms of a linear model, then conceptually it becomes very easy to understand how the model expands to incorporate new variables. So, for example, if we have three predictors (A, B and C) in ANOVA (think back to section 14.4) we end up with three two-way interactions (AB, AC, BC) and one three-way interaction (ABC). Therefore, the resulting linear model of this is just:

$$\text{outcome}_i = (b_0 + b_1\text{A} + b_2\text{B} + b_3\text{C} + b_4\text{AB} + b_5\text{AC} + b_6\text{BC} + b_7\text{ABC}) + \varepsilon_i$$

In exactly the same way, if we have three variables in a categorical data analysis we get an identical model, but with an outcome in terms of logs:

$$\ln(\text{O}_{ijk}) = (b_0 + b_1\text{A}_i + b_2\text{B}_j + b_3\text{C}_k + b_4\text{AB}_{ij} + b_5\text{AC}_{ik} + b_6\text{BC}_{jk} + b_7\text{ABC}_{ijk}) + \ln(\varepsilon_{ijk})$$

Obviously the calculation of beta values and expected values from the model becomes considerably more cumbersome and confusing, but that's why we invented computers – so that we don't have to worry about it! Loglinear analysis works on these principles. However, as we've seen in the two-variable case, when our data are categorical and we include all of the available terms (main effects and interactions) we get no error: our predictors can perfectly predict our outcome (the expected values). So, if we start with the most complex model possible, we will get no error. The job of loglinear analysis is to try to fit a simpler model to the data without any substantial loss of predictive power. Therefore, loglinear analysis typically works on a principle of backward elimination (yes, the same kind of backward elimination that we can use in multiple regression – see section 7.5.3.3). So we begin with the saturated model, and then we remove a predictor from the model and using this new model we predict our data (calculate expected frequencies, just like the chi-square test) and then see how well the model fits the data (i.e. are the expected frequencies close to the observed frequencies?). If the fit of the new model is not very different from the more complex model, then we abandon the complex model in favour of the new one. Put another way, we assume the term we removed was not having a significant impact on the ability of our model to predict the observed data.

However, the analysis doesn't just remove terms randomly, it does it hierarchically. So, we start with the saturated model and then remove the highest-order interaction, and assess the effect that this has. If removing the interaction term has no effect on the model then it's obviously not having much of an effect; therefore, we get rid of it and move on to remove any lower-order interactions. If removing these interactions has no effect then we carry on to any main effects until we find an effect that does affect the fit of the model if it is removed.

To put this in more concrete terms, at the beginning of the section on loglinear analysis I asked you to imagine we'd extended our training and line-dancing example to incorporate a sample of dogs. So, we now have three variables: animal (dog or cat), training (food or affection) and dance (did they dance or not?). Just as in ANOVA this results in three main effects:

- Animal
- Training
- Dance

three interactions involving two variables:

- Animal × Training
- Animal × Dance
- Training × Dance

and one interaction involving all three variables:

- Animal × Training × Dance

When I talk about backward elimination all I mean is that loglinear analysis starts by including all of these effects; it then takes the highest-order interaction (in this case the three-way interaction of animal × training × dance) and removes it. It constructs a new model without this interaction, and from the model calculates expected frequencies. It then compares these expected frequencies (or model frequencies) to the observed frequencies using the standard equation for the likelihood ratio statistic (see 18.3.3). If the new model significantly changes the likelihood ratio statistic, then removing this interaction term has a significant effect on the fit of the model and we know that this effect is statistically important. If this is the case then SPSS will stop there and tell you that you have a significant three-way interaction! It won't test any other effects because with categorical data all lower-order effects are consumed within higher-order effects. If, however, removing the three-way interaction doesn't significantly affect the fit of the model then SPSS moves on to lower-order interactions. Therefore, it looks at the animal × training, animal × dance and training × dance interactions in turn and constructs models in which these terms are not present. For each model it again calculates expected values and compares them to the observed data using a likelihood ratio statistic.[3] Again, if any one of these models does result in a significant change in the likelihood ratio then the term is retained and SPSS won't move on to look at any main effects involved in that interaction (so, if the animal × training interaction is significant it won't look at the main effects of animal or training). However, if the likelihood ratio is unchanged then the analysis removes the offending interaction term and moves on to look at main effects.

I mentioned that the likelihood ratio statistic (see section 18.3.3) is used to assess each model. From the equation it should be clear how this equation can be adapted to fit any model: the observed values are the same throughout, and the model frequencies are simply the expected frequencies from the model being tested. For the saturated model, this statistic will always be 0 (because the observed and model frequencies are the same so the ratio of observed to model frequencies will be 1, and ln(1) = 0), but as we've seen, in other cases it will provide a measure of how well the model fits the observed frequencies. To test whether a new model has changed the likelihood ratio, all we need do is to take the likelihood ratio for a model and subtract from it the likelihood statistic for the previous model (provided the models are hierarchically structured):

$$L\chi^2_{\text{Change}} = L\chi^2_{\text{Current Model}} - L\chi^2_{\text{Previous Model}} \qquad (18.5)$$

I've tried in this section to give you a flavour of how loglinear analysis works, without actually getting too much into the nitty-gritty of the calculations. I've tried to show you how we can conceptualize a chi-square analysis as a linear model and then relied on what I've previously told you about ANOVA to hope that you can extrapolate these conceptual ideas to understand roughly what's going on. The curious among you might want to know exactly how everything is calculated and to these people I have two things to say: 'I don't know' and 'I know a really good place where you can buy a straitjacket'. If you're that interested then Tabachnick and Fidell (2007) have, as ever, written a wonderfully detailed and lucid chapter on the subject which frankly puts this feeble attempt to shame. Still, assuming you're happy to live in relative ignorance, we'll now have a look at how to do a loglinear analysis.

18.7. Assumptions in loglinear analysis ②

Loglinear analysis is an extension of the chi-square test and so has similar assumptions; that is, an entity should fall into only one cell of the contingency table (i.e. cells of the table must be independent) and the expected frequencies should be large enough for a reliable analysis. In loglinear analysis with more than two variables it's all right to have up to 20% of cells with expected frequencies less than 5; however, all cells must have expected frequencies greater than 1. If this assumption is broken the result is a radical reduction in test power – so dramatic in fact that it may not be worth bothering with the analysis at all. Remedies for problems with expected frequencies are: (1) collapse the data across one of the variables (preferably the one you least expect to have an effect!); (2) collapse levels of one of the variables; (3) collect more data; or (4) accept the loss of power.

If you want to collapse data across one of the variables then certain things have to be considered:

[3] It's worth mentioning that for every model, the computation of expected values differs, and as the designs get more complex, the computation gets increasingly tedious and incomprehensible (at least to me); however, you don't need to know the calculations to get a feel for what is going on.

1 The highest-order interaction should be non-significant.

2 At least one of the lower-order interaction terms involving the variable to be deleted should be non-significant.

Let's take the example we've been using. Say we wanted to delete the animal variable; then for this to be valid, the animal × training × dance variable should be non-significant, and either the animal × training or the animal × dance interaction should also be non-significant. You can also collapse categories within a variable. So, if you had a variable of 'season' relating to spring, summer, autumn and winter, and you had very few observations in winter, you could consider reducing the variable to three categories: spring, summer, autumn/winter perhaps. However, you should really only combine categories that it makes theoretical sense to combine.

Finally, some people overcome the problem by simply adding a constant to all cells of the table, but there really is no point in doing this because it doesn't address the issue of power.

18.8. Loglinear analysis using SPSS ②

18.8.1. Initial considerations ②

Data are entered for loglinear analysis in the same way as for the chi-square test (see sections 18.5.1 and 18.5.2). The data for the cat and dog example are in the file **CatsandDogs.sav**; open this file. Notice that it has three variables (**Animal**, **Training** and **Dance**) and each one contains codes representing the different categories of these variables. To begin with, we should use the *crosstabs* command to produce a contingency table of the data.

SELF-TEST Use section 18.5.3 to help you to create a contingency table of these data with dance as the columns, the type of training as rows and the type of animal as a layer

The crosstabulation table produced by SPSS (SPSS Output 18.5) contains the number of cases that fall into each combination of categories. The top half of this table is the same as SPSS Output 18.1 because the data are the same (we've just added some dogs) and if you look back in this chapter there's a summary of what this tells us. For the dogs we can summarize the data in a similar way. In total 49 dogs danced (70% of the total) and of these 20 were trained using food (40.8% of the total that danced) and 29 were trained with affection (59.2% of the total that danced). Further, 21 dogs didn't dance at all (30% of the total) and of those that didn't dance, 14 were trained using food as a reward (66.7% of the total that didn't dance) and 7 were trained using affection (33.3% of the total that didn't dance). The numbers of dogs can be read from the rows labelled *Count* and the percentages are read from the rows labelled *% within Did they dance?* In summary, a lot more dogs danced (70%) than didn't (30%). About half of those that danced were trained with affection and about half with food as a reward. In short, dogs seem more willing to dance than cats (70% compared to 38%), and they're not too worried what training method is used.

SPSS OUTPUT 18.5

Type of Training * Did they dance? * Animal Crosstabulation

Animal				Did they dance? Yes	No	Total
Cat	Type of Training	Food as Reward	Count	28	10	38
			Expected Count	14.4	23.6	38.0
			% within Type of Training	73.7%	26.3%	100.0%
			% within Did they dance?	36.8%	8.1%	19.0%
			% of Total	14.0%	5.0%	19.0%
			Std. Residual	3.6	-2.8	
		Affection as Reward	Count	48	114	162
			Expected Count	61.6	100.4	162.0
			% within Type of Training	29.6%	70.4%	100.0%
			% within Did they dance?	63.2%	91.9%	81.0%
			% of Total	24.0%	57.0%	81.0%
			Std. Residual	-1.7	1.4	
		Total	Count	76	124	200
			Expected Count	76.0	124.0	200.0
			% within Type of Training	38.0%	62.0%	100.0%
			% within Did they dance?	100.0%	100.0%	100.0%
			% of Total	38.0%	62.0%	100.0%
Dog	Type of Training	Food as Reward	Count	20	14	34
			Expected Count	23.8	10.2	34.0
			% within Type of Training	58.8%	41.2%	100.0%
			% within Did they dance?	40.8%	66.7%	48.6%
			% of Total	28.6%	20.0%	48.6%
			Std. Residual	-.8	1.2	
		Affection as Reward	Count	29	7	36
			Expected Count	25.2	10.8	36.0
			% within Type of Training	80.6%	19.4%	100.0%
			% within Did they dance?	59.2%	33.3%	51.4%
			% of Total	41.4%	10.0%	51.4%
			Std. Residual	.8	-1.2	
		Total	Count	49	21	70
			Expected Count	49.0	21.0	70.0
			% within Type of Training	70.0%	30.0%	100.0%
			% within Did they dance?	100.0%	100.0%	100.0%
			% of Total	70.0%	30.0%	100.0%

Before moving on to look at the test statistics it is vital that we check that the assumptions of loglinear analysis have been met: specifically, there should be no expected counts less than 1, and no more than 20% less than 5. If you look at the expected counts in the crosstabulation table, it should be clear that the smallest expected count is 10.2 (for dogs that were trained with food but didn't dance). This value still exceeds 5 and so the assumption has been met.

18.8.2. The loglinear analysis ②

Having established that the assumptions have been met we can move on to the main analysis. The way to run loglinear analysis that is consistent with my section on the theory of the analysis is to select Analyze Loglinear ▸ Model Selection... to access the dialog box in Figure 18.5. Select any variables that you want to include in the analysis by selecting them with the mouse (remember that you can select several at the same time by holding down the *Ctrl* key) and then dragging them to the box labelled *Factor(s)* (or click on ➜). When there

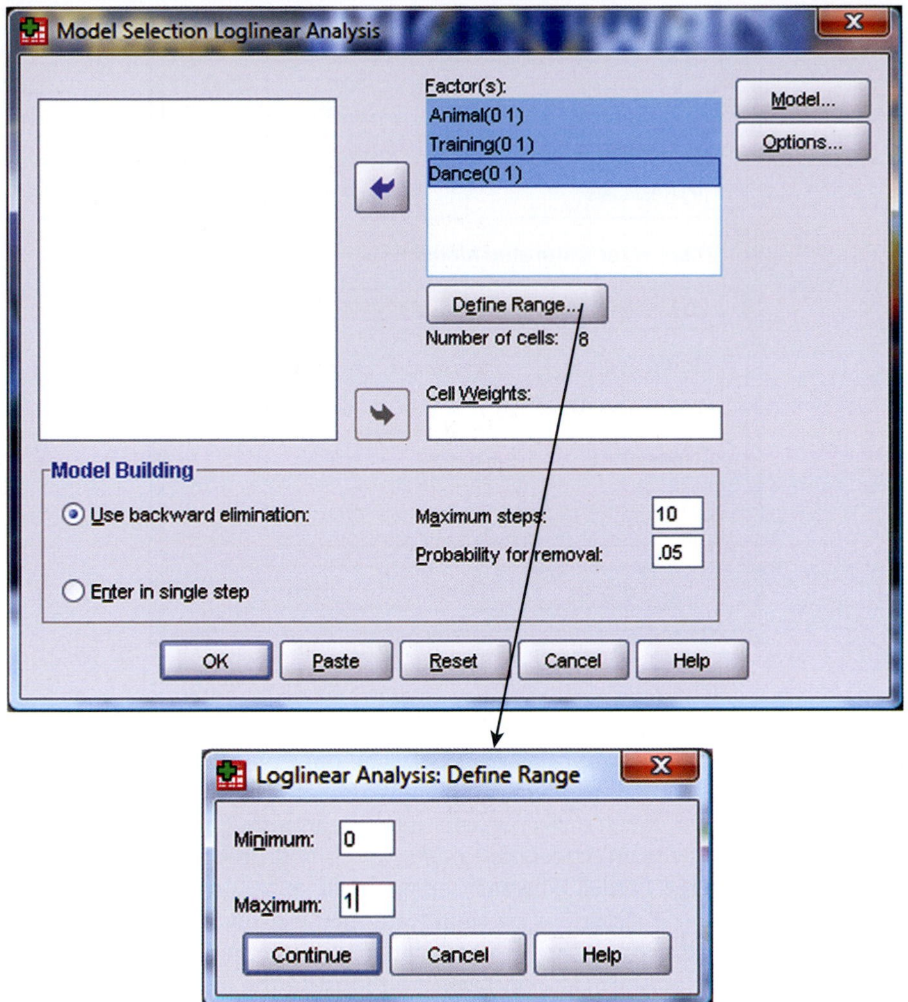

FIGURE 18.5 Main dialog box for loglinear analysis

is a variable in this box the Define Range... button becomes active. Just like the *t*-test and several of the non-parametric tests we encountered in Chapter 15, we have to tell SPSS the codes that we've used to define our categorical variables. Select a variable in the *Factor(s)* box and then click on Define Range... to activate a dialog box that allows you to specify the value of the minimum and maximum code that you've used for that variable. In fact all three variables in this example have the same codes (they all have two categories and I coded them all with 0 and 1) so we can select all three, then click on Define Range... and type 0 in the *Minimum* box and 1 in the *Maximum* box. When you've done this click on Continue to return to the main dialog box.

The default options in this main box are fine; the main thing to note is that by default SPSS uses backward elimination (as I've described elsewhere). You can actually select *Enter in a single step*, which is a non-hierarchical method (in which all effects are entered and evaluated, like forced entry in multiple regression). In loglinear analysis the combined effects take precedence over lower-order effects and so there is little to recommend non-hierarchical methods.

If you click on Model... then this will open a dialog box very similar to those we saw in ANCOVA (e.g. see Figure 11.8). By default SPSS fits the saturated model, and this is what we should be fitting. However, you can define your own model if you like by specifying individual main effects and interaction terms. Unless you have a very good reason for not fitting the saturated model, then leave well alone!

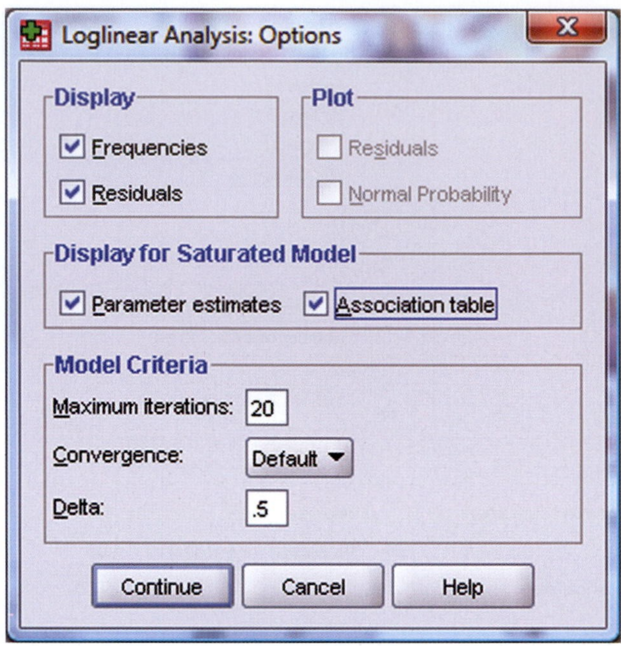

FIGURE 18.6 Options for loglinear analysis

Clicking on Options... opens the dialog box in Figure 18.6. There are few options to play around with really (the default options are fine). The only two things you can select are *Parameter estimates*, which will produce a table of parameter estimates for each effect (by parameter estimates I just mean a *z*-score and associated confidence interval). The other thing is that you can ask for an *Association table*, which will produce chi-square statistics for all of the effects in the model. This may be useful in some situations, but as I've said before, if the higher-order interactions are significant then we shouldn't really be interested in the lower-order effects because they're confounded with the higher-order effects. When you've finished with the options click on Continue to return to the main dialog box and then click on OK to run the analysis.

18.9. Output from loglinear analysis ③

SPSS Output 18.6 shows the initial output from the loglinear analysis. The first table tells us that we have 270 cases (remember that we had 200 cats and 70 dogs and this is a useful check that no cats or dogs have been lost – they do tend to wander off). SPSS then lists all of the factors in the model and the number of levels they have (in this case all have two levels). To begin with, SPSS fits the saturated model (all terms are in the model including the highest-order interaction, in this case the animal × training × dance interaction). The second table gives us the observed and expected counts for each of the combinations of categories in our model. These values should be the same as the original contingency table except that each cell has 0.5 added to it (this value is the default and is fine, but if you want to change it you can do so by changing *Delta* in Figure 18.6).

The final bit of this initial output gives us two goodness-of-fit statistics (Pearson's chi-square and the likelihood ratio statistic, both of which we came across at the beginning of this chapter). In this context these tests are testing the hypothesis that the frequencies predicted by the model (the expected frequencies) are significantly different from the actual frequencies in our data (the observed frequencies). Now, obviously, if our model is a good fit of the data then the observed and expected frequencies should be very similar

SPSS OUTPUT 18.6

Data Information

		N
Cases	Valid	270
	Out of Range[a]	0
	Missing	0
	Weighted Valid	270
Categories	Animal	2
	Type of Training	2
	Did they dance?	2

a. Cases rejected because of out of range factor values.

Cell Counts and Residuals

Animal	Type of Training	Did they dance?	Observed Count[a]	Observed %	Expected Count	Expected %	Residuals	Std. Residuals
Cat	Food as Reward	Yes	28.500	10.6%	28.500	10.6%	.000	.000
		No	10.500	3.9%	10.500	3.9%	.000	.000
	Affection as Reward	Yes	48.500	18.0%	48.500	18.0%	.000	.000
		No	114.500	42.4%	114.500	42.4%	.000	.000
Dog	Food as Reward	Yes	20.500	7.6%	20.500	7.6%	.000	.000
		No	14.500	5.4%	14.500	5.4%	.000	.000
	Affection as Reward	Yes	29.500	10.9%	29.500	10.9%	.000	.000
		No	7.500	2.8%	7.500	2.8%	.000	.000

a. For saturated models, .500 has been added to all observed cells.

Goodness-of-Fit Tests

	Chi-Square	df	Sig.
Likelihood Ratio	.000	0	.
Pearson	.000	0	.

(i.e. not significantly different). Therefore, we want these statistics to be non-significant. A significant result would mean that our model was significantly different from our data (i.e. the model is a bad fit of the data). In large samples these statistics should give the same results but the likelihood ratio statistic is preferred in small samples. In this example, both statistics are 0 and yield a probability value, *p*, of '.', which is a rather confusing way of saying that the probability cannot be computed. The reason why it cannot be computed is because at this stage the model *perfectly* predicts the data. If you read the theory section this shouldn't surprise you because I showed there that the saturated model is a perfect fit of the data and I also mentioned that the resulting likelihood ratio would be zero. What's interesting in loglinear analysis is what bits of the model we can then remove without significantly affecting the fit of the model.

The next part of the output (SPSS Output 18.7) tells us something about which components of the model can be removed. The first bit of the output is labelled **K-Way and Higher-Order Effects** and there are rows showing likelihood ratio and Pearson chi-square statistics when $K = 1$, 2 and 3 (as we go down the rows of the table). The first row ($K = 1$) tells us whether removing the one-way effects (i.e. the main effects of animal, training and dance) and any higher-order effects will significantly affect the fit of the model. There are lots of higher-order effects here – there are the two-way interactions and the three-way interaction – and so this is basically testing whether if we remove everything from the model there will be a significant effect on the fit of the model. This effect is highly significant. If this test was non-significant (if the values of *Sig.* were above .05) then this would tell you that removing everything from your model would not affect the fit of the model (in other words, overall the combined effect of your variables and interactions is not significant). The next row of the table ($K = 2$) tells us whether removing the two-way interactions (i.e. the animal × training, animal × dance and training × dance interactions)

SPSS OUTPUT 18.7

K-Way and Higher-Order Effects

	K	df	Likelihood Ratio Chi-Square	Likelihood Ratio Sig.	Pearson Chi-Square	Pearson Sig.	Number of Iterations
K-way and Higher Order Effects[a]	1	7	200.163	.000	253.556	.000	0
	2	4	72.267	.000	67.174	.000	2
	3	1	20.305	.000	20.778	.000	4
K-way Effects[b]	1	3	127.896	.000	186.382	.000	0
	2	3	51.962	.000	46.396	.000	0
	3	1	20.305	.000	20.778	.000	0

a. Tests that k-way and higher order effects are zero.

b. Tests that k-way effects are zero.

and any higher-order effects will affect the model. In this case there is a higher-order effect (the three-way interaction) so this is testing whether removing the two-way interactions *and* the three-way interaction would affect the fit of the model. This is also highly significant indicating that if we removed the two-way interactions and the three-way interaction then this would have a significant detrimental effect on the model. The final row ($K = 3$) is testing whether removing the three-way effect *and* higher-order effects will significantly affect the fit of the model. Now of course, the three-way interaction is the highest-order effect that we have, so this is simply testing whether removal of three-way interaction (i.e. the animal × training × dance interaction) will significantly affect the fit of the model. If you look at the two columns labelled *Sig.* then you can see that both chi-square and likelihood ratio tests agree that removing this interaction will significantly affect the fit of the model (because the probability value is less than .05).

The next part of the table expresses the same thing but without including the higher-order effects. It's labelled *K-way Effects* and it lists tests for when $K = 1$, 2 and 3. The first row ($K = 1$), therefore, tests whether removing the main effects (the one-way effects) has a significant detrimental effect on the model. The probability values are smaller than .05 indicating that if we removed the main effects of animal, training and dance from our model it would significantly affect the fit of the model (in other words, these effects in combination are significant predictors of the data). The second row ($K = 2$) tests whether removing the two-way interactions has a significant detrimental effect on the model. The probability values are less than .05 indicating that if we removed the animal × training, animal × dance and training × dance interactions then this would significantly reduce how well the model fits the data. In other words, one or more of these two-way interactions is a significant predictor of the data. The final row ($K = 3$) tests whether removing the three-way interaction has a significant detrimental effect on the model. The probability values are less than .05 indicating that if we removed the animal × training × dance interaction then this would significantly reduce how well the model fits the data. In other words, this three-way interaction is a significant predictor of the data. This row should be identical to the final row of the top of the table (the *K-way and Higher Order Effects*) because it is the highest-order effect and so in the previous table there were no higher-order effects to include in the test (look at the output and you'll see the results are identical).

What this is actually telling us is that the three-way interaction is significant: removing it from the model has a significant effect on how well the model fits the data. We also know that removing all two-way interactions has a significant effect on the model, but you have to remember that loglinear analysis should be done hierarchically and so these two-way interactions aren't of interest to us because the three-way interaction is significant (we'd look only at these effects if the three-way interaction were non-significant).

If you selected an *Association table* in Figure 18.6 then you'll get the table in SPSS Output 18.8. This simply breaks down the table that we've just looked at into its component parts. So, for example, although we know from the previous output that removing all of the two-way interactions significantly affects the model, we don't know which of the

SPSS OUTPUT 18.8

Partial Associations

Effect	df	Partial Chi-Square	Sig.	Number of Iterations
Animal*Training	1	13.760	.000	2
Animal*Dance	1	13.748	.000	2
Training*Dance	1	8.611	.003	2
Animal	1	65.268	.000	2
Training	1	61.145	.000	2
Dance	1	1.483	.223	2

SPSS OUTPUT 18.9

Parameter Estimates

Effect	Parameter	Estimate	Std. Error	Z	Sig.	95% Confidence Interval	
						Lower Bound	Upper Bound
Animal*Training*Dance	1	.360	.083	4.320	.000	.197	.523
Animal*Training	1	-.402	.083	-4.823	.000	-.565	-.239
Animal*Dance	1	-.197	.083	-2.364	.018	-.360	-.034
Training*Dance	1	.104	.083	1.251	.211	-.059	.268
Animal	1	.404	.083	4.843	.000	.240	.567
Training	1	-.328	.083	-3.937	.000	-.492	-.165
Dance	1	.232	.083	2.782	.005	.069	.395

two-way interactions is having the effect. This table tells us. We get a Pearson chi-square test for each of the two-way interactions and the main effects and the column labelled *Sig.* tells us which of these effects is significant (values less than .05 are significant). We can tell from this that the animal × dance, training × dance and the animal × training interactions are all significant. Likewise, we saw in the previous output that removing the one-way effects (the main effects of animal, training and dance) also significantly affected the fit of the model, and these findings are confirmed here because the main effects of animal and training are both significant. However, the main effect of dance is not (the probability value is greater than .05). Interesting as these findings are, we should ignore them because of the hierarchical nature of loglinear analysis: these effects are all confounded with the higher-order interaction of animal × training × dance.

If you selected the *Parameter estimates* in Figure 18.6 then you'll get the table in SPSS Output 18.9. This simply tells us the same thing as the previous table (i.e. it provides individual estimates for each effect) but it does so using a z-score rather than a chi-square test. This can be useful because we get confidence intervals, and also because the value of z gives us a useful comparison between effects (if you ignore the plus or minus sign, the bigger the z, the more significant the effect). So, if you look at the z-values you can see that the main effect of animal is the most important effect in the model ($z = 4.84$) followed by the animal × training interaction ($z = -4.82$) and then the animal × training × dance interaction ($z = 4.32$) and so on. However, it's worth reiterating that in this case we don't need to concern ourselves with anything other than the three-way interaction.

The final bit of output (SPSS Output 18.10) deals with the backward elimination. SPSS will begin with the highest-order effect (in this case the animal × training × dance interaction); it removes it from the model, sees what effect this has, and if it doesn't have a significant effect then it moves on to the next highest effects (in this case the two-way interactions). However, we've already seen that removing the three-way interaction will have a significant effect and this is confirmed at this stage by the table labelled **Step Summary**, which confirms that removing the three-way interaction has a significant effect on the model. Therefore, the analysis stops here: the three-way interaction is not removed and SPSS evaluates this final model.

SPSS OUTPUT 18.10

Step Summary

Step[a]		Effects	Chi-Square[c]	df	Sig.	Number of Iterations
0	Generating Class[b]	Animal*Training*Dance	.000	0	.	
	Deleted Effect 1	Animal*Training*Dance	20.305	1	.000	4
1	Generating Class[b]	Animal*Training*Dance	.000	0	.	

a. At each step, the effect with the largest significance level for the Likelihood Ratio Change is deleted, provided the significance level is larger than .050.

b. Statistics are displayed for the best model at each step after step 0.

c. For 'Deleted Effect', this is the change in the Chi-Square after the effect is deleted from the model.

Cell Counts and Residuals

Animal	Type of Training	Did they dance?	Observed		Expected		Residuals	Std. Residuals
			Count	%	Count	%		
Cat	Food as Reward	Yes	28.000	10.4%	28.000	10.4%	.000	.000
		No	10.000	3.7%	10.000	3.7%	.000	.000
	Affection as Reward	Yes	48.000	17.8%	48.000	17.8%	.000	.000
		No	114.000	42.2%	114.000	42.2%	.000	.000
Dog	Food as Reward	Yes	20.000	7.4%	20.000	7.4%	.000	.000
		No	14.000	5.2%	14.000	5.2%	.000	.000
	Affection as Reward	Yes	29.000	10.7%	29.000	10.7%	.000	.000
		No	7.000	2.6%	7.000	2.6%	.000	.000

Goodness-of-Fit Tests

	Chi-Square	df	Sig.
Likelihood Ratio	.000	0	.
Pearson	.000	0	.

I don't need a loglinear analysis to tell me that cats are vastly superior to dogs!

Finally SPSS evaluates this final model with the likelihood ratio statistic and we're looking for a non-significant test statistic which indicates that the expected values generated by the model are not significantly different from the observed data (put another way, the model is a good fit of the data). In this case the result is very non-significant indicating that the model is a good fit of the data.[4]

The next step is to try to interpret this interaction. The first useful thing we can do is to plot the frequencies across all of the different categories. You should plot the frequencies in terms of the percentage of the total (these values can be found in the crosstabulation table in SPSS Output 18.5 in the rows labelled *% of total*). The resulting graph is shown in Figure 18.7 and this shows what we already know about cats: they will dance (or do anything else for that matter) when there is food involved but if you train them with affection they're not interested. Dogs on the other hand will dance when there's affection involved (actually more dogs danced than didn't dance regardless of the type of reward, but the effect is more pronounced when affection was the training method). In fact, both animals show similar responses to food training, it's just that cats won't do anything for affection. So cats are sensible creatures that only do stupid stuff when there's something in it for them (i.e. food), whereas dogs are just plain stupid!

SELF-TEST Can you use the Chart Builder to replicate the graph in Figure 18.7?

[4] The fact that the analysis has stopped here is unhelpful because I can't show you how it would proceed in the event of a non-significant three-way interaction. However, it does keep things simple and if you're interested in exploring loglinear analysis further, the task at the end of the chapter shows you what happens when the highest order interaction is not significant.

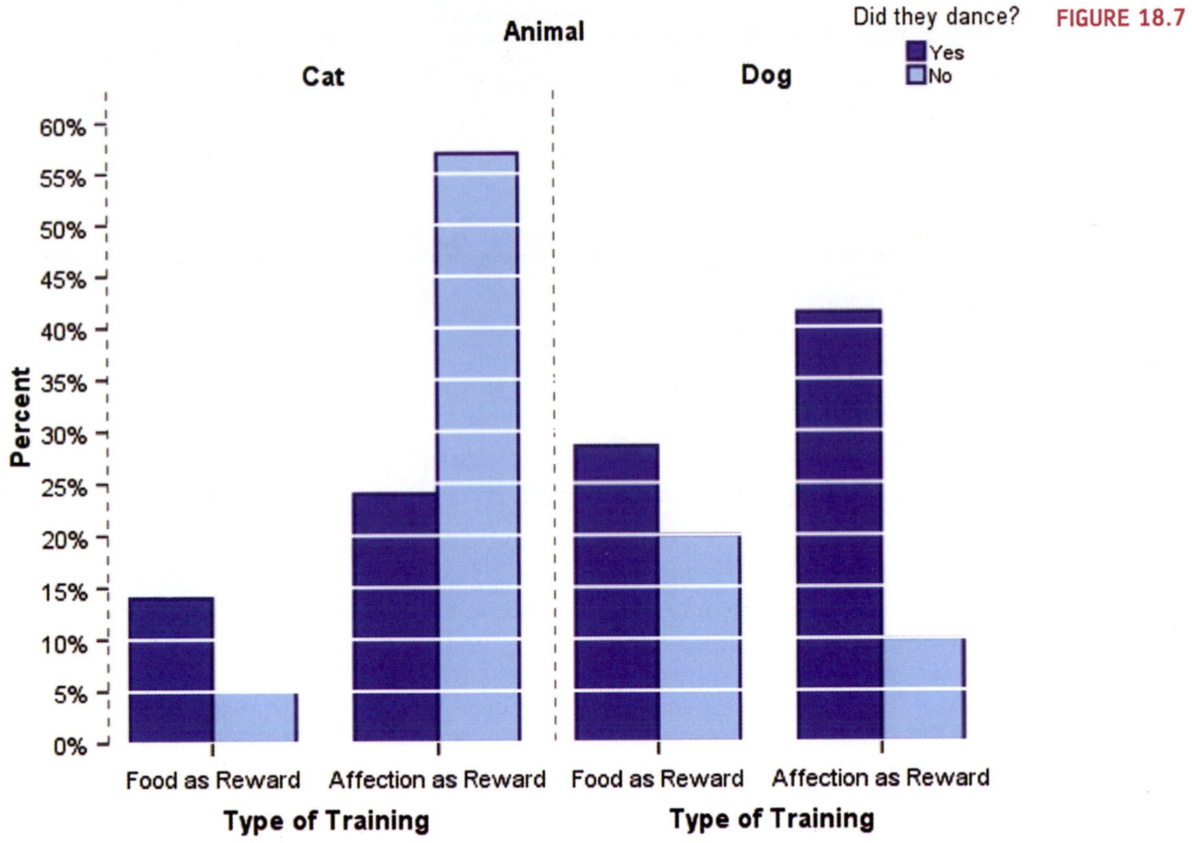

FIGURE 18.7

18.10. Following up loglinear analysis ②

An alternative way to interpret a three-way interaction is to conduct chi-square analysis at different levels of one of your variables. For example, to interpret our animal × training × dance interaction, we could perform a chi-square test on training and dance but do this separately for dogs and cats (in fact the analysis for cats will be the same as the example we used for chi-square). You can then compare the results in the different animals.

SELF-TEST Use the split file command (see section 5.4.3) to run a chi-square test on **Dance** and **Training** for dogs and cats.

The results and interpretation for cats are in SPSS Output 18.2 and for dogs the output is shown in SPSS Output 18.11. For dogs there is still a significant relationship between the types of training and whether they danced but it is weaker (the chi-square is 3.93 compared to 25.2 in the cats).[5] This reflects the fact that dogs are more likely to dance if given affection than if given food, the opposite of cats!

[5] The chi-square statistic depends on the sample size, so really you need to calculate effect sizes and compare them to make this kind of statement (unless you had equal numbers of dogs and cats!).

SPSS OUTPUT 18.11

Chi-Square Tests[d]

	Value	df	Asymp. Sig. (2-sided)	Exact Sig. (2-sided)	Exact Sig. (1-sided)	Point Probability
Pearson Chi-Square	3.932[a]	1	.047	.068	.042	
Continuity Correction[b]	2.966	1	.085			
Likelihood Ratio	3.984	1	.046	.068	.042	
Fisher's Exact Test				.068	.042	
Linear-by-Linear Association	3.876[c]	1	.049	.068	.042	.030
N of Valid Cases	70					

a. 0 cells (.0%) have expected count less than 5. The minimum expected count is 10.20.

b. Computed only for a 2x2 table

c. The standardized statistic is -1.969.

d. Animal = Dog

18.11. Effect sizes in loglinear analysis ②

As with Pearson's chi-square, one of the most elegant ways to report your effects is in terms of odds ratios. Odds ratios are easiest to understand for 2 × 2 contingency tables and so if you have significant higher-order interactions, or your variables have more than two categories, it is worth trying to break these effects down into logical 2 × 2 tables and calculating odds ratios that reflect the nature of the interaction. So, for example, in this example we could calculate odds ratios for dogs and cats separately. We have the odds ratios for cats already (section 18.5.5), and for dogs we would get:

$$\begin{aligned}\text{odds}_{\text{dancing after food}} &= \frac{\text{number that had food and danced}}{\text{number that had food but didn't dance}}\\ &= \frac{20}{14}\\ &= 1.43\end{aligned}$$

$$\begin{aligned}\text{odds}_{\text{dancing after affection}} &= \frac{\text{number that had affection and danced}}{\text{number that had affection but didn't dance}}\\ &= \frac{29}{7}\\ &= 4.14\end{aligned}$$

$$\begin{aligned}\text{odds ratio} &= \frac{\text{odds}_{\text{dancing after food}}}{\text{odds}_{\text{dancing after affection}}}\\ &= \frac{1.43}{4.14}\\ &= 0.35\end{aligned}$$

This tells us that if a dog was trained with food the odds of their dancing were 0.35 times the odds if they were rewarded with affection (i.e. they were less likely to dance). Another way to say this is that the odds of their dancing were 1/0.35 = 2.90 times lower if they were trained with food instead of affection. Compare this to cats where the odds of dancing were 6.65 higher if they were trained with food rather than affection. As you can see, comparing the odds ratios for dogs and cats is an extremely elegant way to present the three-way interaction term in the model.

18.12. Reporting the results of loglinear analysis ②

When reporting loglinear analysis you need to report the likelihood ratio statistic for the final model, usually denoted just by χ^2. For any terms that are significant you should report the chi-square change, or you could consider reporting the z-score for the effect and its associated confidence interval. If you break down any higher-order interactions in subsequent analyses then obviously you need to report the relevant chi-square statistics (and odds ratios). For this example we could report:

✓ The three-way loglinear analysis produced a final model that retained all effects. The likelihood ratio of this model was χ^2 (0) = 0, p = 1. This indicated that the highest-order interaction (the animal × training × dance interaction) was significant, χ^2 (1) = 20.31, $p < .001$. To break down this effect, separate chi-square tests on the training and dance variables were performed separately for dogs and cats. For cats, there was a significant association between the type of training and whether or not cats would dance, χ^2 (1) = 25.36, $p < .001$; this was true in dogs also, χ^2 (1) = 3.93, $p < .05$. Odds ratios indicated that the odds of dancing were 6.65 higher after food than affection in cats, but only 0.35 in dogs (i.e. in dogs, the odds of dancing were 2.90 times lower if trained with food compared to affection). Therefore, the analysis seems to reveal a fundamental difference between dogs and cats: cats are more likely to dance for food rather than affection, whereas dogs are more likely to dance for affection than food.

CRAMMING SAM'S TIPS

- If you want to test the relationship between more than two categorical variables you can do this with *loglinear analysis*.
- Loglinear analysis is hierarchical: the initial model contains all main effects and interactions. Starting with the highest-order interaction, terms are removed to see whether their removal significantly affects the fit of the model. If it does then this term is not removed and all lower-order effects are ignored.
- Look at the table labelled **K-Way and Higher Order Effects** to see which effects have been retained in the final model. Then look at the table labelled **Partial Associations** to see the individual significance of the retained effects (look at the column labelled *Prob*: values less than .05 indicate significance).
- Look at the **Goodness-of-fit test statistics** for the final model: if this model is a good fit of the data then this statistic should be non-significant (*Sig.* should be bigger than .05).
- Look at the crosstabulation table to interpret any significant effects (% of total for cells is the best thing to look at).

What have I discovered about statistics? ①

When I wrote the first edition of this book I had always intended to do a chapter on loglinear analysis, but by the time I got to that chapter I had already written 300 pages more than I was contracted to do, and had put so much effort into the rest of it that, well, the thought of that extra chapter was making me think of large cliffs and jumping. When the second edition needed to be written, I wanted to make sure that at the very least I did a loglinear chapter. However, when I came to it, I'd already written 200 pages more than I was supposed to for this new edition, and with deadlines fading into the distance, history was repeating itself. It won't surprise you to know then that I was really happy to have written the damn thing! This chapter has taken a very brief look at analysing categorical data. What I've tried to do is to show you how really we approach categorical data in much the same way as any other kind of data: we fit a model, we calculate the deviation between our model and the observed data, and we use that to evaluate the model we've fitted. I've also tried to show that the model we fit is the same one that we've come across throughout this book: it's a linear model (regression). When we have only two variables we can use Pearson's chi-square test or the likelihood ratio test to look at whether those two variables are associated. In more complex situations, we simply extend these models into something known as a loglinear model. This is a bit like ANOVA for categorical data: for every variable we have, we get a main effect but we also get interactions between variables. Loglinear analysis simply evaluates all of these effects hierarchically to tell us which ones best predict our outcome.

Fortunately the experience of this loglinear chapter taught me a valuable lesson, which is never to agree to write a chapter about something that you know very little about, and if you do then definitely don't leave it until the very end of the writing process when you're under pressure and mentally exhausted. It's lucky that we learn from our mistakes isn't it …?

Key terms that I've discovered

Chi-square test
Contingency table
Cramer's V
Fisher's exact test
Goodman and Kruskal's λ
Loglinear analysis
Odds ratio
Phi
Saturated model
Yates's continuity correction

Smart Alex's tasks

- **Task 1**: Certain editors at Sage like to think they're a bit of a whiz at football (soccer if you prefer). To see whether they are better than Sussex lecturers and postgraduates we invited various employees of Sage to join in our football matches (oh, sorry,

I mean we invited them down for important meetings about books). Every player was only allowed to play in one match. Over many matches, we counted the number of players that scored goals. The data are in the file **SageEditorsCan'tPlayFootball.sav**. Do a chi-square test to see whether more publishers or academics scored goals. We predict that Sussex people will score more than Sage people. ③

- **Task 2**: I wrote much of this update while on sabbatical in The Netherlands (I have a real soft spot for Holland). However, living there for three months did enable me to notice certain cultural differences to England. The Dutch are famous for travelling by bike; they do it much more than the English. However, I noticed that many more Dutch people cycle while steering with only one hand. I pointed this out to one of my friends, Birgit Mayer, and she said that I was being a crazy English fool and that Dutch people did not cycle one-handed. Several weeks of my pointing at one-handed cyclists and her pointing at two-handed cyclists ensued. To put it to the test I counted the number of Dutch and English cyclists who ride with one or two hands on the handlebars (**Handlebars.sav**). Can you work out whether Birgit or I am right? ①

- **Task 3**: I was interested in whether horoscopes are just a figment of people's minds. Therefore, I got 2201 people, made a note of their star sign (this variable, obviously, has 12 categories: Capricorn, Aquarius, Pisces, Aries, Taurus, Gemini, Cancer, Leo, Virgo, Libra, Scorpio and Sagittarius) and whether they believed in horoscopes (this variable has two categories: believer or unbeliever). I then sent them a horoscope in the post of what would happen over the next month: everybody, regardless of their star sign, received the same horoscope, which read 'August is an exciting month for you. You will make friends with a tramp in the first week of the month and cook him a cheese omelette. Curiosity is your greatest virtue, and in the second week, you'll discover knowledge of a subject that you previously thought was boring, statistics perhaps. You might purchase a book around this time that guides you towards this knowledge. Your new wisdom leads to a change in career around the third week, when you ditch your current job and become an accountant. By the final week you find yourself free from the constraints of having friends, your boy/girlfriend has left you for a Russian ballet dancer with a glass eye, and you now spend your weekends doing loglinear analysis by hand with a pigeon called Hephzibah for company.' At the end of August I interviewed all of these people and I classified the horoscope as having come true, or not, based on how closely their lives had matched the fictitious horoscope. The data are in the file **Horoscope.sav**. Conduct a loglinear analysis to see whether there is a relationship between the person's star sign, whether they believe in horoscopes and whether the horoscope came true. ③

- **Task 4**: On my statistics course students have weekly SPSS classes in a computer laboratory. These classes are run by postgraduate tutors but I often pop in to help out. I've noticed in these sessions that many students are studying Facebook rather more than they are studying the very interesting statistics assignments that I have set them. I wanted to see the impact that this behaviour had on their exam performance. I collected data from all 260 students on my course. First I checked their **Attendance** and classified them as having attended either more or less than 50% of their lab classes. Next, I classified them as being either someone who looked at **Facebook** during their lab class, or someone who never did. Lastly, after the Research Methods in Psychology (RMiP) exam, I classified them as having either passed or failed (**Exam**). The data are in **Facebook.sav**. Do a loglinear analysis on the data to see if there is an association between studying Facebook and failing your exam. ③

Answers can be found on the companion website.

Further reading

Hutcheson, G., & Sofroniou, N. (1999). *The multivariate social scientist*. London: Sage.
Tabachnick, B. G., & Fidell, L. S. (2007). *Using multivariate statistics* (5th ed.). Boston: Allyn & Bacon. (Chapter 16 is a fantastic account of loglinear analysis.)

Online tutorial

The companion website contains the following Flash movie tutorial to accompany this chapter:

- Chi-Square test using SPSS

Interesting real research

Beckham, A. S. (1929). Is the Negro happy? A psychological analysis. *Journal of Abnormal and Social Psychology*, *24*, 186–190.

Multilevel linear models 19

FIGURE 19.1
Having a therapy session in 2007

19.1. What will this chapter tell me? ①

Over the last couple of chapters we saw that I had gone from a child having dreams and aspirations of being a rock star, to becoming a living (barely) statistical test. A more dramatic demonstration of my complete failure to achieve my life's ambitions I can scarcely imagine. Having devoted far too much of my life to statistics it was time to unlock the latent rock star once more. The second edition of the book had left me in desperate need for some therapy and, therefore, at the age of 29 I decided to start playing the drums (there's a joke in there somewhere about it being the perfect instrument for a failed musician, but really they're much harder to play than people think). A couple of years later I had a call from an old friend of mine, Doug, who used to be in a band that my old band Scansion used to play with a lot: 'Remember the last time I saw you we talked about you coming and having a jam with us?' I had absolutely no recollection whatsoever of him saying this so I responded

'Yes'. 'Well, how about it then?' he said. 'OK,' I said, 'you arrange it and I'll bring my guitar.' 'No, you whelk,' he said, 'we want you to drum and maybe you could learn some of the songs on the CD I gave you last year?' I'd played his band's CD and I liked it, but there was no way on this earth that I could play the drums as well as their drummer. 'Sure, no problem,' I lied. I spent the next two weeks playing along to this CD as if my life depended on it and when the rehearsal came, much as I'd love to report that I drummed like a lord, I didn't. I did, however, nearly have a heart attack and herniate everything in my body that it's possible to herniate (really, the music is pretty fast!). Still, we had another rehearsal, and then another and, well, three years down the line we're still having them. The only difference is that now I can play the songs at a speed that makes their old recordings seem as though a sedated snail was on the drums (www.myspace.com/fracturepattern). The point is that it's never too late to learn something new. This is just as well because, as a man who clearly doesn't learn from his mistakes, I agreed to write a chapter on multilevel linear models, a subject about which I know absolutely nothing. I'm writing it last, when I feel mentally exhausted and stressed. Hopefully at some point between now and the end of writing the chapter I will learn something. With a bit of luck you will too.

19.2. Hierarchical data ②

What are hierarchical data?

In all of the analyses in this book so far we have treated data as though they are organized at a single level. However, in the real world, data are often hierarchical. This just means that some variables are clustered or *nested* within other variables. For example, when I'm not writing statistics books I spend most of my time researching how anxiety develops in children below the age of 10. This typically involves my running experiments in schools. When I run research in a school, I test children who have been assigned to different classes, and who are taught by different teachers. The classroom that a child is in could conceivably affect my results. Let's imagine I test in two different classrooms. The first class is taught by Mr Nervous. Mr Nervous is very anxious and often when he supervises children he tells them to be careful, or that things that they do are dangerous, or that they might hurt themselves. The second class is taught by Little Miss Daredevil.[1] She is very carefree and she believes that children in her class should have the freedom to explore new experiences. Therefore, she is always telling them not to be scared of things and to explore new situations. One day I go into the school to test the children. I take in a big animal carrier, which I tell them has an animal inside. I measure whether they will put their hand in the carrier to stroke the animal. Children taught by Mr Nervous have grown up in an environment where their teacher reinforces caution, whereas children taught by Miss Daredevil have been encouraged to embrace new experiences. Therefore, we might expect Mr Nervous's children to be more reluctant to put their hand in the box because of the classroom experiences that they have had. The classroom is, therefore, known as a contextual variable. In reality, as an experimenter I would be interested in a much more complicated situation. For example, I might tell some of the children that the animal is a bloodthirsty beast, whereas I tell others that the animal is friendly. Now obviously I'm expecting the information I give the children to affect their enthusiasm for stroking the animal. However, it's also possible that their classroom has an effect. Therefore, my manipulation of the information that I give the children also has to be placed within the context of the classroom to which the

[1] Those of you who don't spot the Mr Men references here, check out http://www.mrmen.com. Mr Nervous used to be called Mr Jelly and was a pink jelly-shaped blob, which in my humble opinion was better than his current incarnation.

child belongs. My threat information is likely to have more impact on Mr Nervous's children than it will on Miss Daredevil's children. One consequence of this is that children within Mr Nervous's class will be more similar to each other than they are to children in Miss Daredevil's class and vice versa.

Figure 19.2 illustrates this scenario more generally. In a big data set, we might have collected data from lots of children. This is the bottom of the hierarchy and is known as a *level 1* variable. So, children (or cases) are our level 1 variable. However, these children are organized by classroom (children are said to be *nested* within classes). The class to which a child belongs is a level up from the participant in the hierarchy and is said to be a *level 2* variable.

The situation that I have just described is the simplest hierarchy that you can have because there are just two levels. However, you can have other layers to your hierarchy. The easiest way to explain this is to stick with our example of my testing children in different classes and then to point out the obvious fact that classrooms are themselves nested within schools. Therefore, if I ran a study incorporating lots of different schools, as well as different classrooms within those schools, then I would have to add another level to the hierarchy. We can apply the same logic as before, in that children in particular schools will be more similar to each other than to children in different schools. This is because schools tend to reflect their social demographic (which can differ from school to school) and they may differ in their policies also. Figure 19.3 shows this scenario. There are now three levels in the hierarchy: the child (level 1), the class to which the child belongs (level 2) and the school within which that class exists (level 3). In this situation we have two contextual variables: school and classroom.

Hierarchical data structures need not apply only to between-participant situations. We can also think of data as being nested within people. In this situation the case, or person, is not at the bottom of the hierarchy (level 1), but is further up. A good example is memory. Imagine that after giving children threat information about my caged animal I asked them a week later to recall everything they could about the animal. For each child there are many facts that they could recall. Let's say that I originally gave them 15 pieces of information; some children might recall all 15 pieces of information, but others might remember only 2 or 3 bits of information. The bits of information, or memories, are nested within the person and their recall depends on the person. The probability of a given memory being recalled depends on what other memories are available, and the recall of one memory may have knock-on effects for what other memories are recalled. Therefore, memories are not independent units. As such, the person acts as a context within which memories are recalled (Wright, 1998).

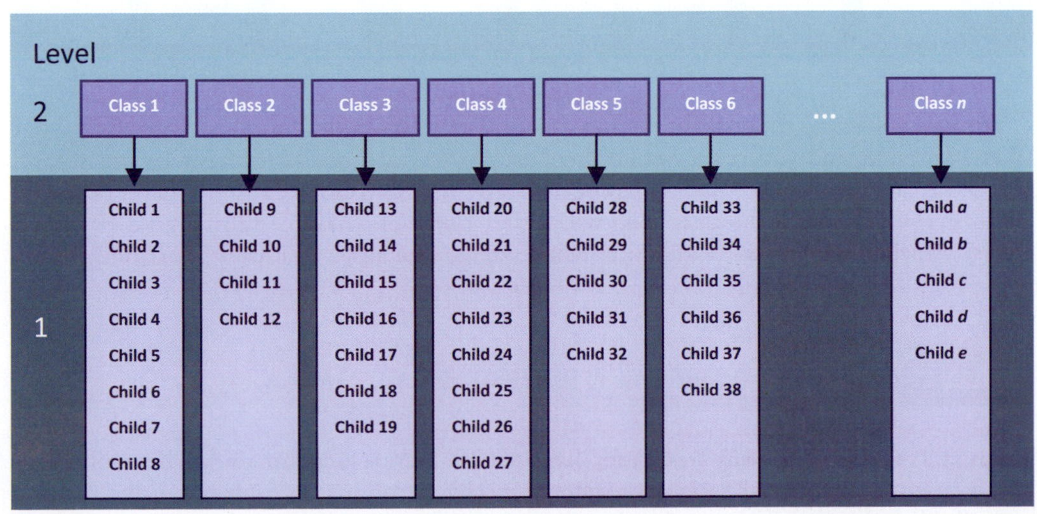

FIGURE 19.2 An example of a two-level hierarchical data structure. Children (level 1) are organized within classrooms (level 2)

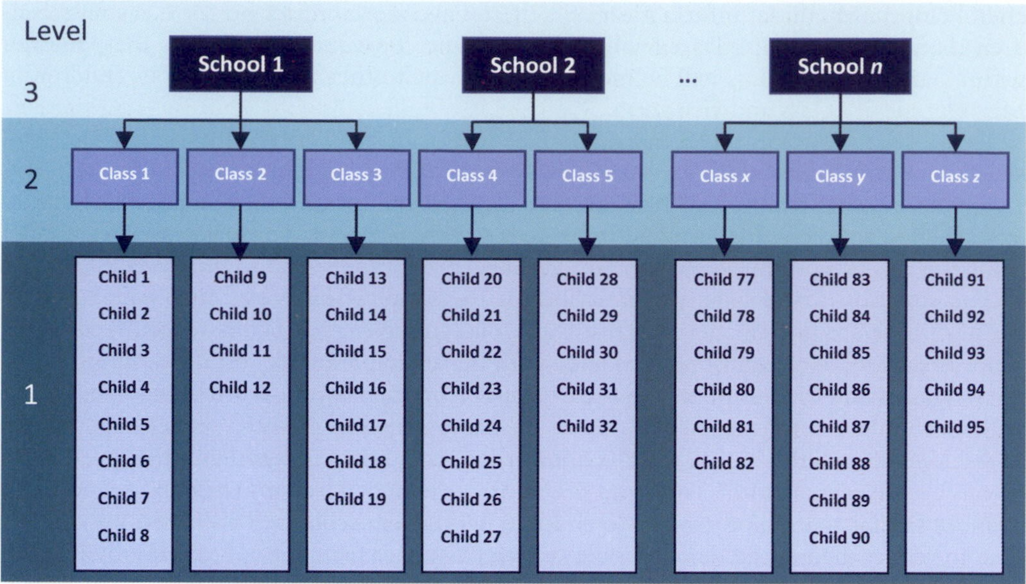

FIGURE 19.3 An example of a three-level hierarchical data structure

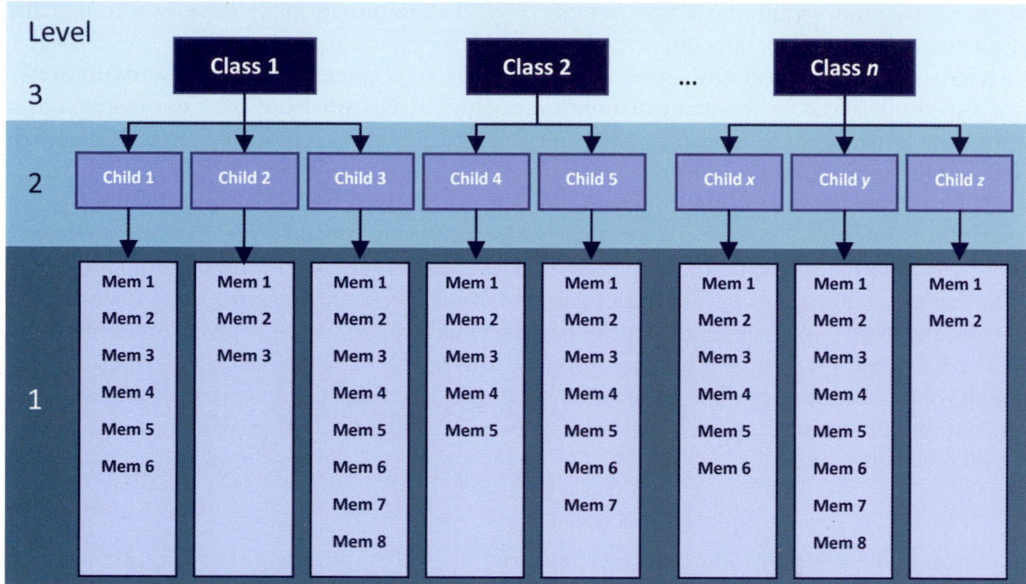

FIGURE 19.4 An example of a three-level hierarchical data structure, where the level 1 variable is a repeated measure (memories recalled)

Figure 19.4 shows the structure of the situation that I have just described. The child is our level 2 variable, and within each child there are several memories (our level 1 variable). Of course we can also have levels of the hierarchy above the child. So, we could still, for example, factor in the context of the class from which they came (as I have done in Figure 19.4) as a level 3 variable. Indeed, we could even include the school again as a level 4 variable!

19.2.1. The intraclass correlation ②

You might well wonder why it matters that data are hierarchical (or not). The main problem is that the contextual variables in the hierarchy introduce dependency in the data. In plain English this means that residuals will be correlated. I have alluded to this fact already when I noted that children in Mr Nervous's class would be more similar to each other than to children in Miss Daredevil's class. In some sense, having the same teacher makes children more similar to each other. This similarity is a problem because in nearly every test we have covered in this book we assume that cases are independent. In other words, there is absolutely no correlation between residual scores of one child and another. However, when people are sampled from similar contexts, this independence is unlikely to be true. For example, Charlotte and Emily's responses to the animal in the carrier have both been influenced by Mr Nervous's cautious manner, so their behaviour will be similar. Likewise, Kiki and Jip's responses to the animal in the box have both been influenced by Miss Daredevil's carefree manner, so their behaviour will be similar too. We have seen before that in ANOVA, for example, a lack of independence between cases is a huge problem that really affects the resulting test statistic – and not in a good way! (See section 10.2.10.)

By thinking about contextual variables and factoring them into the analysis we can overcome this problem of non-independent observations. One way that we can do this is to use the intraclass correlation (ICC). We came across this measure in section 17.9.3 as a measure of inter-rater reliability, but it can also be used as a measure of dependency between scores. We'll skip the formalities of calculating the ICC (but see Oliver Twisted if you're keen to know), and we'll just give a conceptual grasp of what it represents. In our two-level example of children within classes, the ICC represents the proportion of the total variability in the outcome that is attributable to the classes. It follows that if a class has had a big effect on the children within it then the variability within the class will be small (the children will behave similarly). As such, variability in the outcome within classes is minimized, and variability in the outcome between classes is maximized; therefore, the ICC is large. Conversely, if the class has little effect on the children then the outcome will vary a lot within classes, which will make differences between classes relatively small. Therefore, the ICC is small too. Thus, the ICC tells us that variability within levels of a contextual variable (in this case the class to which a child belongs) is small, but between levels of a contextual variable (comparing classes) is large. As such the ICC is a good gauge of whether a contextual variable has an effect on the outcome.

OLIVER TWISTED

Please, Sir, can I have some more ... ICC?

'I have a dependency on gruel,' whines Oliver. 'Maybe I could measure this dependency if I knew more about the ICC.' We'll you're so high on gruel Oliver that you have rather missed the point. Still, I did write an article on the ICC once upon a time (Field, 2005a) and it's reproduced in the additional web material for your delight and amusement.

19.2.2. Benefits of multilevel models ②

Multilevel linear models have numerous uses. To convince you that trawling through this chapter is going to reward you with statistical possibilities beyond your wildest dreams, here are just a few (slightly overstated) benefits of multilevel models:

- **Cast aside the assumption of homogeneity of regression slopes**: We saw in Chapter 11 that when we use analysis of covariance we have to assume that the relationship between our covariate and our outcome is the same across the different groups that make up our predictor variable. However, this doesn't always happen. Luckily, in multilevel models we can explicitly model this variability in regression slopes, thus overcoming this inconvenient problem.

- **Say 'bye bye' to the assumption of independence**: We saw in Chapter 10 that when we use independent ANOVA we have to assume that the different cases of data are independent. If this is not true, little lizards climb out of your mattress while you're asleep and eat you. Again, multilevel models are specifically designed to allow you to model these relationships between cases. Also, in Chapter 7 we saw that multiple regression relies on having independent observations. However, there are situations in which you might want to measure someone on more than one occasion (i.e. over time). Ordinary regression turns itself into cheese and hides in the fridge at the prospect of cases of data that are related. Multilevel models eat these data for breakfast, with a piece of regression-flavoured cheese.
- **Laugh in the face of missing data**: I've spent a lot of this book extolling the virtues of balanced designs and not having missing data. Regression, ANOVA, ANCOVA and most of the other tests we have covered do strange things when data are missing or the design is not balanced. This can be a real pain. Multilevel models open the door to missing data, invite them to sit by the fire and make them a cup of tea. Multilevel models expect missing data, they love them in fact. So, if you have some kind of ANOVA or regression (of any variety) for which you have missing data, fear not, just do a multilevel model.

I think you'll agree that multilevel models are pretty funky. 'Is there anything they can't do?' I hear you cry. Well, no, not really.

19.3. Theory of multilevel linear models ③

The underlying theory of multilevel models is very complicated indeed – far too complicated for my little peanut of a brain to comprehend. Fortunately, the advent of computers and software like SPSS makes it possible for feeble-minded individuals such as myself to take advantage of this wonderful tool without actually needing to know the maths. Better still, this means I can get away with not explaining the maths (and really, I'm not kidding, I don't understand any of it). What I will do though is try to give you a flavour of what multilevel models are and what they do by describing the key concepts within the framework of linear models that has permeated this whole book.

19.3.1. An example ②

Throughout the first part of the chapter we will use an example to illustrate some of the concepts in multilevel models. Cosmetic surgery is on the increase at the moment. In the USA, there was a 1600% increase in cosmetic surgical and non-surgical treatments between 1992 and 2002, and in 2004, 65,000 people in the UK underwent privately and publicly funded operations (Kellett, Clarke, & McGill, 2008). With the increasing popularity of this surgery, many people are starting to question the motives of those who want to go under the knife. There are two main reasons to have cosmetic surgery: (1) to help a physical problem such as having breast reduction surgery to relieve back ache; and (2) to change your external appearance, for example by having a face lift. Related to this second point, there is even some case for arguing that cosmetic surgery could be performed as a psychological intervention: to improve self-esteem (Cook, Rosser, & Salmon, 2006; Kellett et al., 2008). The main example for this chapter looks at the effects of cosmetic surgery on quality of life. The variables in the data file **Cosmetic Surgery.sav**, are:

- **Post_QoL**: This is a measure of quality of life after the cosmetic surgery. This is our outcome variable.
- **Base_QoL**: We need to adjust our outcome for quality of life before the surgery.
- **Surgery**: This variable is a dummy variable that specifies whether the person has undergone cosmetic surgery (1) or whether they are on the waiting list (0), which acts as our control group.
- **Clinic**: This variable specifies which of 10 clinics the person attended to have their surgery.
- **Age**: This variable tells us the person's age in years.
- **BDI**: It is becoming increasingly apparent that people volunteering for cosmetic surgery (especially when the surgery is purely for vanity) might have very different personality profiles than the general public (Cook, Rossera, Toone, James, & Salmon, 2006). In particular, these people might have low self-esteem or be depressed. When looking at quality of life it is important to assess natural levels of depression and this variable used the Beck Depression Inventory (BDI) to do just that.
- **Reason**: This dummy variable specifies whether the person had/is waiting to have surgery purely to change their appearance (1), or because of a physical reason (0).
- **Gender**: This variable simply specifies whether the person was a man (1) or a woman (0).

When conducting hierarchical models we generally work up from a very simple model to more complicated models and we will take that approach in this chapter. In doing so I hope to illustrate multilevel modelling by attaching it to frameworks that you already understand, such as ANOVA and ANCOVA.

Figure 19.5 shows the hierarchical structure of the data. Essentially, people being treated in the same surgeries are not independent of each other because they will have had surgery from the same surgeon. Surgeons will vary in how good they are, and quality of life will to some extent depend on how well the surgery went (if they did a nice neat job then quality of life should be higher than if they left you with unpleasant scars). Therefore, people within clinics will be more similar to each other than people in different clinics. As such, the person undergoing surgery is the level 1 variable, but there is a level 2 variable, a variable higher in the hierarchy, which is the clinic attended.

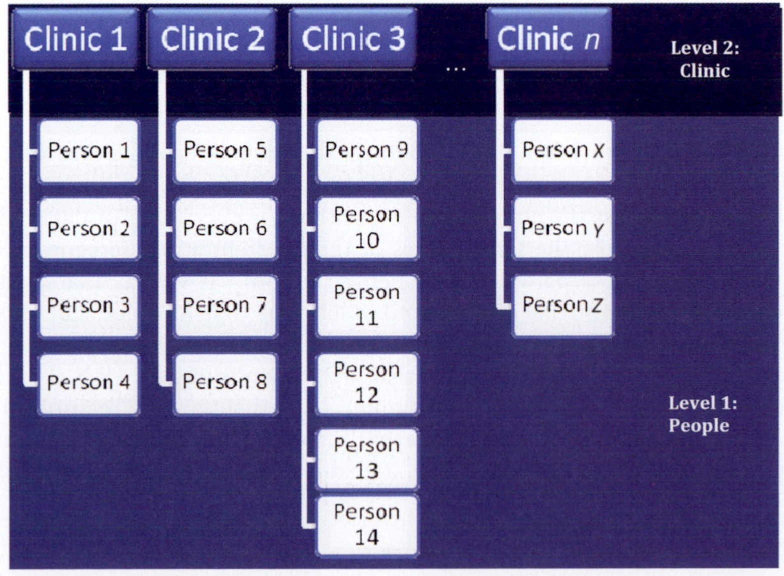

FIGURE 19.5 Diagram to show the hierarchical structure of the cosmetic surgery data set. People are clustered within clinics. Note that for each person there would be a series of variables measured: surgery, BDI, age, gender, reason and pre-surgery quality of life

19.3.2. Fixed and random coefficients ③

Throughout this book we have discussed effects and variables and these concepts should be very familiar to you by now. However, we have viewed these effects and variables in a relatively simple way: we have not distinguished between whether something is fixed or random.

What we mean by 'fixed' and 'random' can be a bit confusing because the terms are used in a variety of contexts. You hear people talk about **fixed effects** and **random effects**. An effect in an experiment is said to be a fixed effect if all possible treatment conditions that a researcher is interested in are present in the experiment. An effect is said to be random if the experiment contains only a random sample of possible treatment conditions. This distinction is important because fixed effects can be generalized only to the situations in your experiment, whereas random effects can be generalized beyond the treatment conditions in the experiment (provided that the treatment conditions are representative). For example, in our Viagra example from Chapter 10, the effect is fixed if we say that we are interested only in the three conditions that we had (placebo, low dose and high dose) and we can generalize our findings only to the situation of a placebo, low dose and high dose. However, if we were to say that the three doses were only a sample of possible doses (maybe we could have tried a very high dose), then it is a random effect and we can generalize beyond just placebos, low doses and high doses. All of the effects in this book so far we have treated as fixed effects. The vast majority of academic research that you read will treat variables as fixed effects.

People also talk about **fixed variables** and **random variables**. A fixed variable is one that is not supposed to change over time (e.g. for most people their gender is a fixed variable – it never changes), whereas a random one varies over time (e.g. your weight is likely to fluctuate over time).

In the context of multilevel models we need to make a distinction between **fixed coefficients** and **random coefficients**. In the regressions, ANOVAs and ANCOVAs throughout this book we have assumed that the regression parameters are fixed. We have seen numerous times that a linear model is characterized by two things: the intercept, b_0, and the slope, b_1:

$$Y_i = b_0 + b_1 X_{1i} + \varepsilon_i$$

Note that the outcome (*Y*), the predictor (*X*) and the error (ε) all vary as a function of *i*, which normally represents a particular case of data. In other words, it represents the level 1 variable. If, for example, we wanted to predict Sam's score, we could replace the *i*s with her name:

$$Y_{\text{Sam}} = b_0 + b_1 X_{1\text{Sam}} + \varepsilon_{\text{Sam}}$$

This is just some basic revision. Now, when we do a regression like this we assume that the *b*s are fixed and we estimate them from the data. In other words, we're assuming that the model holds true across the entire sample and that for every case of data in the sample we can predict a score using the same values of the gradient and intercept. However, we can also conceptualize these parameters as being random.[2] If we say that a parameter is random then we assume not that it is a fixed value, but that its value can vary. Up until now we have thought of regression models as having fixed intercepts and fixed slopes, but this opens up three new possibilities for us that are shown in Figure 19.6. This figure uses the data from our ANCOVA example in Chapter 11 and shows the relationship between a person's libido and that of their partner overall (the dashed line) and separately for the three groups in the study (a placebo group, a group that had a low dose of Viagra and a group that had a high dose).

[2] In a sense random isn't an intuitive term for us non-statisticians because it implies that values are plucked out of thin air (randomly selected). However, this is not the case, they are carefully estimated just as fixed parameters are.

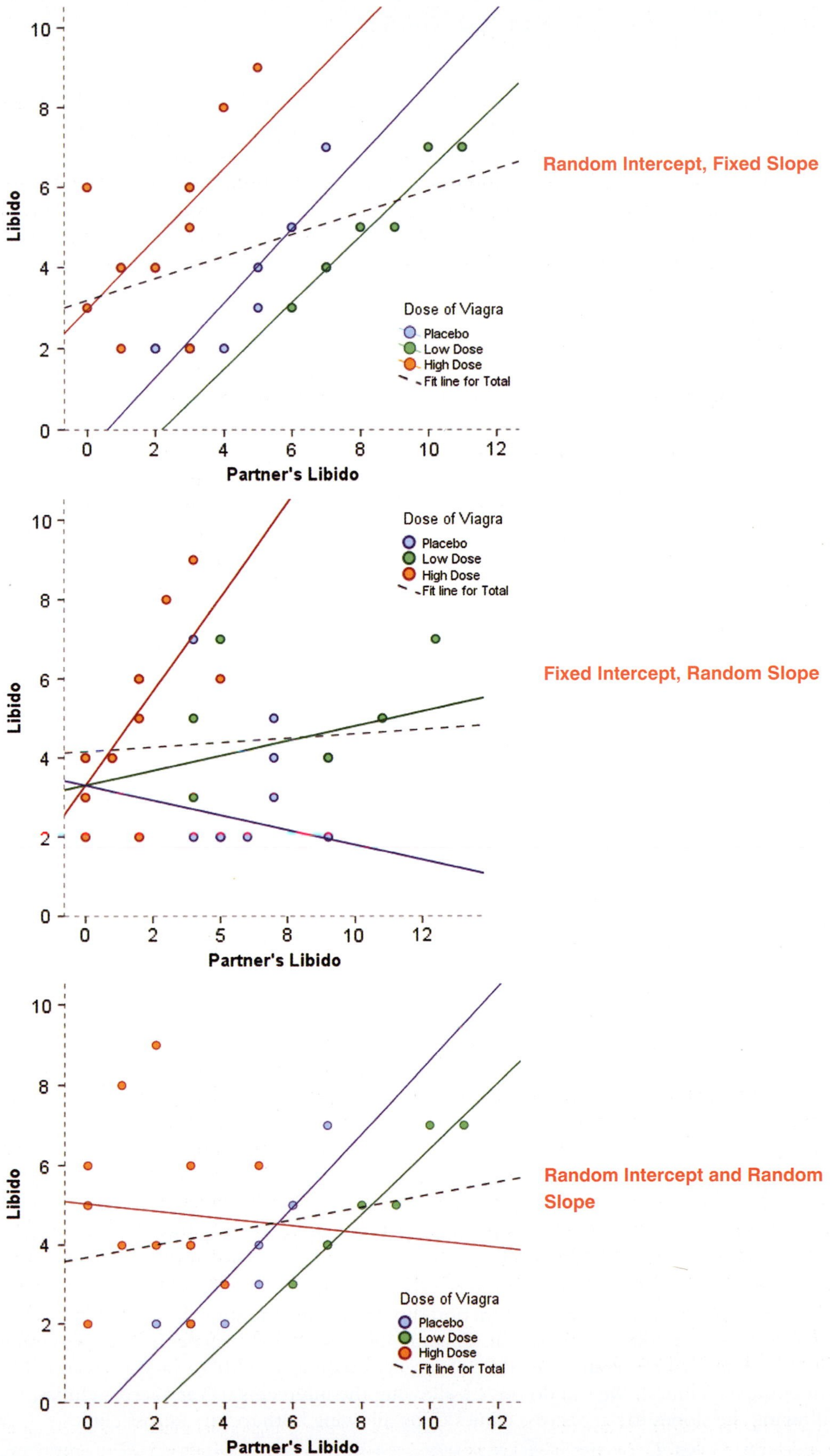

FIGURE 19.6 Data sets showing an overall model (dashed line) and the models for separate contexts within the data (i.e. groups of cases)

19.3.2.1. The random intercept model ③

The simplest way to introduce random parameters into the model is to assume that the intercepts vary across contexts (or groups) – because the intercepts vary, we call them random intercepts. For our libido data this is like assuming that the relationship between libido and partner's libido is the same in the placebo, low- and high-dose groups (i.e. the slope is the same), but that the models for each group are in different locations (i.e. the intercepts are different). This is shown in the diagram in which the models within the different contexts (colours) have the same shape (slope) but are located in different geometric space (they have different intercepts – top panel of Figure 19.6).

19.3.2.2. Random slope model ③

We can also assume that the slopes vary across contexts – i.e. we assume random slopes. For our libido data this is like assuming that the relationship between libido and partner's libido is different in the placebo, low- and high-dose groups (i.e. the slopes are different), but that the models for each group are fixed at the same geometric location (i.e. the intercepts are the same). This is what happens when we violate the assumption of homogeneity of regression slopes in ANCOVA. Homogeneity of regression slopes is the assumption that regression slopes are the same across contexts. If this assumption is not tenable than we can use a multilevel model to explicitly estimate that variability in slopes. This is shown in the diagram in which the models within the different contexts (colours) converge on a single intercept but have different slopes (middle panel of Figure 19.6).

19.3.2.3. The random intercept and slope model ③

The most realistic situation is to assume that both intercepts and slopes vary around the overall model. This is shown in the diagram in which the models within the different contexts (colours) have different slopes but are also located in different geometric space and so have different intercepts (bottom panel of Figure 19.6).

19.4. The multilevel model ④

We have seen conceptually what a random intercept, random slope and random intercept and slope model looks like. Now let's look at how we actually represent the models. To keep things concrete, let's use our example. For the sake of simplicity, let's imagine first that we wanted to predict someone's quality of life (QoL) after cosmetic surgery. We can represent this as a linear model as follows:

$$\text{QoL After Surgery}_i = b_0 + b_1\text{Surgery}_i + \varepsilon_i \tag{19.1}$$

We have seen equations like this many times and it represents a linear model: regression, a *t*-test (in this case) and ANOVA. In this example, we had a contextual variable, which was the clinic in which the cosmetic surgery was conducted. We might expect the effect of surgery on quality of life to vary as a function of which clinic the surgery was conducted at because surgeons will differ in their skill. This variable is a level 2 variable. As such we could allow the model that represents the effect of surgery on quality of life to vary across the different contexts (clinics). We can do this by allowing the intercepts to vary across clinics, or by allowing the slopes to vary across clinics or by allowing both to vary across clinics.

To begin with, let's say we want to include a random intercept for quality of life. All we do is add a component to the intercept that measures the variability in intercepts, u_{0j}. Therefore, the intercept changes from b_0 to become $b_0 + u_{0j}$. This term estimates the intercept of the overall model fitted to the data, b_0, and the variability of intercepts around that overall model, u_{0j}. The overall model becomes.[3]

$$Y_{ij} = (b_0 + u_{0j}) + b_1X_{ij} + \varepsilon_{ij} \tag{19.2}$$

The *j*s in the equation reflect levels of the variable over which the intercept varies (in this case the clinic) – the level 2 variable. Another way that we could write this is to take out the error terms so that it looks like an ordinary regression equation except that the intercept has changed from a fixed, b_0, to a random one, b_{0j}, which is defined in a separate equation:

$$\begin{aligned} Y_{ij} &= b_{0j} + b_1X_{ij} + \varepsilon_{ij} \\ b_{0j} &= b_0 + u_{0j} \end{aligned} \tag{19.3}$$

Therefore, if we want to know the estimated intercept for Clinic 7, we simply replace the *j* with 'clinic 7' in the second equation:

$$b_{0\text{Clinic7}} = b_0 + u_{0\text{Clinic7}}$$

If we want to include random slopes for the effect of surgery on quality of life, then all we do is add a component to the slope of the overall model that measures the variability in slopes, u_{1j}. Therefore, the gradient changes from b_1 to become $(b_1 + u_{1j})$. This term estimates the slope of the overall model fitted to the data, b_1, and the variability of slopes in different contexts around that overall model, u_{1j}. The overall model becomes (compare to the random intercept model above):

$$Y_{ij} = b_0 + (b_1 + u_{1j})X_{ij} + \varepsilon_{ij} \tag{19.4}$$

Again we can take the error terms out into a separate equation to make the link to a familiar linear model even clearer. It now looks like an ordinary regression equation except that the slope has changed from a fixed, b_1, to a random one, b_{1j}, which is defined in a separate equation:

$$\begin{aligned} Y_{ij} &= b_{0i} + b_{1j}X_{ij} + \varepsilon_{ij} \\ b_{1j} &= b_1 + u_{1j} \end{aligned} \tag{19.5}$$

If we want to model a situation with random slopes *and* intercepts, then we combine the two models above. We still estimate the intercept and slope of the overall model (b_0 and b_1) but we also include the two terms that estimate the variability in intercepts, u_{0j}, and slopes, u_{1j}. The overall model becomes (compare to the two models above):

$$Y_{ij} = (b_0 + u_{0j}) + (b_1 + u_{1j})X_{ij} + \varepsilon_{ij} \tag{19.6}$$

We can link this more directly to a simple linear model if we take some of these extra terms out into separate equations. We could write this model as a basic linear model, except

[3] Some people use gamma (γ), not b, to represent the parameters, but I prefer b because it makes the link to the other linear models that we have used in this book clearer.

we've replaced our fixed intercept and slope (b_0 and b_1) with their random counterparts (b_{0j} and b_{1j}):

$$\begin{aligned} Y_{ij} &= b_{0j} + b_{1j}X_{ij} + \varepsilon_{ij} \\ b_{0j} &= b_0 + u_{0j} \\ b_{1j} &= b_1 + u_{1j} \end{aligned} \tag{19.7}$$

The take-home point is that we're not doing anything terribly different from the rest of the book: it's basically just a posh regression.

Now imagine we wanted to add in another predictor, for example quality of life before surgery. Knowing what we do about multiple regression we shouldn't be invading the personal space of the idea that we can simply add this variable in with an associated beta:

$$\text{QoL After Surgery}_i = b_0 + b_1\text{Surgery}_i + b_2\text{QoL Before Surgery}_i + \varepsilon_i \tag{19.8}$$

This is all just revision of ideas from earlier in the book. Remember also that the i represents the level 1 variable, in this case the people we tested. Therefore, we can predict a given person's quality of life after surgery by replacing the i with their name:

$$\text{QoL After}_{\text{Sam}} = b_0 + b_1\text{Surgery}_{\text{Sam}} + b_2\text{QoL Before}_{\text{Sam}} + \varepsilon_{\text{Sam}}$$

Now, if we want to allow the intercept of the effect of surgery on quality of life after surgery to vary across contexts then we simply replace b_0 with b_{0j}. If we want to allow the slope of the effect of surgery on quality of life after surgery to vary across contexts then we replace b_1 with b_{1j}. So, even with a random intercept and slope, our model stays much the same:

$$\begin{aligned} \text{QoL After}_{ij} &= b_{0j} + b_{1j}\text{Surgery}_{ij} + b_2\text{QoL Before}_{ij} + \varepsilon_{ij} \\ b_{0j} &= b_0 + u_{0j} \\ b_{1j} &= b_1 + u_{1j} \end{aligned} \tag{19.9}$$

Remember that the j in the equation relates to the level 2 contextual variable (clinic in this case). So, if we wanted to predict someone's score we wouldn't just do it from their name, but also from the clinic they attended. Imagine our guinea pig Sam had her surgery done at Clinic 7; then we could replace the is and js as follows:

$$\begin{aligned} \text{QoL After Surgery}_{\text{Sam, Clinic7}} = {} & b_{0\text{Clinic7}} + b_{1\text{Clinic7}}\text{Surgery}_{\text{Sam, Clinic7}} \\ & + b_2\text{QoL Before Surgery}_{\text{Sam, Clinic7}} + \varepsilon_{\text{Sam, Clinic7}} \end{aligned}$$

I want to sum up by just reiterating that all we're really doing in a multilevel model is a fancy regression in which we allow either the intercepts or slopes, or both, to vary across different contexts. All that really changes is that for every parameter that we allow to be random, we get an estimate of the variability of that parameter as well as the parameter itself. So, there isn't anything terribly complicated; we can add new predictors to the model and for each one decide whether its regression parameter is fixed or random.

19.4.1. Assessing the fit and comparing multilevel models ④

As in logistic regression (Chapter 8) the overall fit of a multilevel model is tested using a chi-square likelihood ratio test (see section 18.3.3) and just as in logistic regression, SPSS reports the −2 log-likelihood (see section 8.3.1). Essentially, the smaller the value of the log-likelihood, the better. SPSS also produces four adjusted versions of the log-likelihood value. All of these can be interpreted in the same way as the log-likelihood, but they have been corrected for various things:

- *Akaike's information criterion* (**AIC**): This is basically a goodness-of-fit measure that is corrected for model complexity. That just means that it takes into account how many parameters have been estimated.
- *Hurvich and Tsai's criterion* (**AICC**): This is the same as AIC but is designed for small samples.
- *Bozdogan's criterion* (**CAIC**): Again this can be interpreted in the same way as the AIC, but this version corrects not just for model complexity but for sample size too.
- *Schwarz's Bayesian criterion* (**BIC**): This statistic is again comparable to the AIC, although it is slightly more conservative (it corrects more harshly for the number of parameters being estimated). It should be used when sample sizes are large and the number of parameters is small.

All of these measures are similar but the AIC and BIC are the most commonly used. None of them are intrinsically interpretable (it's not meaningful to talk about their values being large or small per se); however, they are all useful as a way of comparing models. The value of AIC, AICC, CAIC and BIC can all be compared to their equivalent values in other models. In all cases smaller values mean better-fitting models.

Many writers recommend building up multilevel models starting with a 'basic' model in which all parameters are fixed and then adding in random coefficients as appropriate and exploring confounding variables (Raudenbush & Bryk, 2002; Twisk, 2006). One advantage of doing this is that you can compare the fit of the model as you make parameters random, or as you add in variables. To compare models we simply subtract the log-likelihood of the new model from the value for the old:

$$\begin{aligned} \chi^2_{\text{Change}} &= (-2\text{Log} - \text{Likelihood}_{\text{Old}}) | (-2\text{Log} - \text{Likelihood}_{\text{New}}) \\ df_{\text{Change}} &= \text{Number of Parameters}_{\text{Old}} - \text{Number of Parameters}_{\text{New}} \end{aligned} \qquad (19.10)$$

This equation is the same as equations (18.5) and (8.6), but written in a way that uses the names of the actual values that SPSS produces. There are two caveats to this equation: (1) it works only if full maximum-likelihood estimation is used (and not restricted maximal likelihood, see SPSS Tip 19.1); and (2) the new model contains all of the effects of the older model.

19.4.2. Types of covariance structures ④

If you have any random effects or repeated measures in your multilevel model then you have to decide upon the *covariance structure* of your data. If you have random effects and repeated measures then you can specify different covariance structures for each. The covariance structure simply specifies the form of the variance–covariance matrix (a matrix in which the diagonal elements are variances and the off-diagonal elements are covariances). There are various forms that this matrix could take and we have to tell SPSS what form we think it *does* take. Of course we might not know what form it takes (most of the time we'll be taking an educated guess), so it is sometimes useful to run the model with different covariance structures defined and use the goodness-of-fit indices (the AIC, AICC, CAIC and BIC) to see whether changing the covariance structure improves the fit of the model (remember that a smaller value of these statistics means a better-fitting model).

The covariance structure is important because SPSS uses it as a starting point to estimate the model parameters. As such, you will get different results depending on which covariance structure you choose. If you specify a covariance structure that is too simple then you are more likely to make a Type I error (finding a parameter is significant when in reality it is not), but if you specify one that is too complex then you run the risk of a Type II error (finding parameters to be non-significant when in reality they are). SPSS has 17 different covariance structures that you can use. We will look at four of the commonest covariance structures to give you a feel for what they are and when they should be used. In each case I use a representation of the variance–covariance matrix to illustrate. With all of these matrices you could imagine that the rows and columns represents four different clinics in our cosmetic surgery data:

$\begin{pmatrix} 1 & 0 & 0 & 0 \\ 0 & 1 & 0 & 0 \\ 0 & 0 & 1 & 0 \\ 0 & 0 & 0 & 1 \end{pmatrix}$	**Variance components**: This covariance structure is very simple and assumes that all random effects are independent (this is why all of the covariances in the matrix are 0). Variances of random effects are assumed to be the same (hence why they are 1 in the matrix) and sum to the variance of the outcome variable. In SPSS this is the default covariance structure for random effects and is sometimes called the independence model.
$\begin{pmatrix} \sigma_1^2 & 0 & 0 & 0 \\ 0 & \sigma_1^2 & 0 & 0 \\ 0 & 0 & \sigma_1^2 & 0 \\ 0 & 0 & 0 & \sigma_1^2 \end{pmatrix}$	**Diagonal**: This variance structure is like variance components except that variances are assumed to be heterogeneous (this is why the diagonal of the matrix is made up of different variance terms). This structure again assumes that variances are independent and, therefore, that all of the covariances are 0. In SPSS this is the default covariance structure for repeated measures.
$\begin{pmatrix} 1 & \rho & \rho^2 & \rho^2 \\ \rho & 1 & \rho & \rho^2 \\ \rho^2 & \rho & 1 & \rho \\ \rho^2 & \rho^2 & \rho & 1 \end{pmatrix}$	**AR(1)**: This stands for first-order autoregressive structure. In layman's terms this means that the relationship between variances changes in a systematic way. If you imagine the rows and columns of the matrix to be points in time, then it assumes that the correlations between repeated measurements is highest at adjacent time points. So, in the first column, the correlation between time points 1 and 2 is ρ; let's assume that this value is .3. As we move to time point 3, the correlation between time point 1 and 3 is ρ^2, or .09. In other words, it has decreased: scores at time point 1 are more related to scores at time 2 than they are to scores at time 3. At time 4, the correlation goes down again to ρ^3 or .027. So, the correlations between time points next to each other are assumed to be ρ, scores two intervals apart are assumed to have correlations of ρ^2, and scores three intervals apart are assumed to have correlations of ρ^3. So the correlation between scores gets smaller over time. Variances are assumed to be homogeneous but there is a version of this covariance structure where variance can be heterogeneous. This structure is often used for repeated-measures data (especially when measurements are taken over time such as in growth models).
$\begin{pmatrix} \sigma_1^2 & \sigma_{21} & \sigma_{31} & \sigma_{41} \\ \sigma_{21} & \sigma_2^2 & \sigma_{32} & \sigma_{42} \\ \sigma_{31} & \sigma_{32} & \sigma_3^2 & \sigma_{43} \\ \sigma_{41} & \sigma_{42} & \sigma_{43} & \sigma_4^2 \end{pmatrix}$	**Unstructured**: This covariance structure is completely general and is, therefore, the default option for random effects in SPSS. Covariances are assumed to be completely unpredictable: they do not conform to a systematic pattern.

CRAMMING SAM'S TIPS Multilevel models

- Multilevel models should be used to analyse data that have a hierarchical structure. For example, you might measure depression after psychotherapy. In your sample, patients will see different therapists within different clinics. This is a three-level hierarchy with depression scores from patients (level 1), nested within therapists (level 2) who are themselves nested within clinics (level 3).
- Hierarchical models are just like regression, except that you can allow parameters to vary (this is called a random effect). In ordinary regression, parameters generally are a fixed value estimated from the sample (a fixed effect).
- If we estimate a linear model within each context (e.g. the therapist or clinic to use the example above) rather than the sample as a whole, then we can assume that the intercepts of these models vary (a random intercepts model), or that the slopes of these models differ (a random slopes model) or that both vary.
- We can compare different models (assuming that they differ in only one additional parameter) by looking at the difference in the −2 log-likelihood. Usually we would do this when we have changed only one parameter (added one new thing to the model).
- For any model we have to assume a covariance structure. For random intercepts models the default of *variance components* is fine, but when slopes are random an *unstructured* covariance structure is often assumed. When data are measured over time an autoregressive structure (AR1) is often assumed.

19.5. Some practical issues ③

19.5.1. Assumptions ③

Multilevel linear models are an extension of regression so all of the assumptions for regression apply to multilevel models (see section 7.6.2). There is a caveat, though, which is that the assumptions of independence and independent errors can sometimes be solved by a multilevel model because the purpose of this model is to factor in the correlations between cases caused by higher-level variables. As such, if a lack of independence is being caused by a level 2 or level 3 variable then a multilevel model should make this problem go away (although not always). As such, try to check the usual assumptions in the usual way.

There are two additional assumptions in multilevel models that relate to the random coefficients. These coefficients are assumed to be normally distributed around the overall model. So, in a random intercepts model the intercepts in the different contexts are assumed to be normally distributed around the overall model. Similarly, in a random slopes model, the slopes of the models in different contexts are assumed to be normally distributed.

Also it's worth mentioning that multicollinearity can be a particular problem in multilevel models if you have interactions that cross levels in the data hierarchy (cross-level interactions). However, centring predictors can help matters enormously (Kreft & de Leeuw, 1998), and we will see how to centre predictors in section 19.5.3.

19.5.2. Sample size and power ③

As you might well imagine, the situation with power and sample size is very complex indeed. One complexity is that we are trying to make decisions about our power to detect both fixed and random effects coefficients. Kreft and de Leeuw (1998) do a tremendous job of making sense of things for us. Essentially, the take-home message is the more data, the better. As more levels are introduced into the model, more parameters need to be estimated and the larger the sample sizes need to be. Kreft and de Leeuw conclude that if you are looking for cross-level interactions then you should aim to have more than 20 contexts (groups) in the higher-level variable, and that group sizes 'should not be too small'. They conclude by saying that there are so many factors involved in multilevel analysis that it is impossible to produce any meaningful rules of thumb.

Twisk (2006) agrees that the number of contexts relative to individuals within those contexts is important. He also points out that standard sample size and power calculations can be used but then 'corrected' for the multilevel component of the analysis (by factoring, among other things, the intraclass correlation). However, there are two corrections that he discusses that yield very different sample sizes! He recommends using sample size calculations with caution.

The easiest option is to get a computer to do it for you. HLM (http://www.ssicentral.com/hlm/index.html) will do power calculations for multilevel models, and for two-level models you could try Tom Snijders' PinT program (http://stat.gamma.rug.nl/multilevel.htm).

19.5.3. Centring variables ④

What is centring and do I need to do it?

Centring refers to the process of transforming a variable into deviations around a fixed point. This fixed point can be any value that you choose, but typically we use the grand mean. We have already come across a form of centring way back in Chapter 1, when we discovered how to compute *z*-scores. When we calculate a *z*-score we take each score and subtract from it the mean of all scores (this centres the values at 0), and then divide by the standard deviation (this changes the units of measurement to standard deviations). When we centre a variable around the mean we simply subtract the mean from all of the scores: this centres the variables around 0.

There are two forms of centring that are typically used in multilevel modelling: **grand mean centring** and **group mean centring**. Grand mean centring means that for a given variable we take each score and subtract from it the mean of all scores (for that variable). Group mean centring means that for a given variable we take each score and subtract from it the mean of the scores (for that variable) within a given group. In both cases it is usually only level 1 predictors that are centred (in our cosmetic surgery example this would be predictors such as age, BDI and pre-surgery quality of life). If group mean centring is used then a level 1 variable is typically centred around means of a level 2 variable (in our cosmetic surgery data this would mean that, for example, the age of a person would be centred around the mean of age for the clinic at which the person had their surgery).

Centring can be used in ordinary multiple regression too, and because this form of regression is already familiar to you I'd like to begin by looking at the effects of centring in regression. In multiple regression the intercept represents the value of the outcome when all of the predictors take a value of 0. There are some predictors for which a value of 0 makes little sense. For example, if you were using heart rate as a predictor variable then a value of 0 would be meaningless (no one will have a heart rate of 0 unless they are dead). As such, the intercept in this case has no real-world use: why would you want to know the value of the outcome when heart rate was 0 given than no alive person would even have a heart rate that low? Centring heart rate around its mean changes the meaning of the intercept. The intercept becomes the value of the outcome when heart rate is its average value. In more general terms, if all predictors are centred around their mean then the intercept is the value of the outcome when all predictors are the value of their mean. Centring can, therefore, be a useful tool for interpretation when a value of 0 for the predictor is meaningless.

The effect of centring in multilevel models, however, is much more complicated. There are some excellent reviews that look in detail at the effects of centering on multilevel models (Kreft & de Leeuw, 1998; Kreft, de Leew, & Aiken, 1995), and here I will just give a very basic précis of what they say. Essentially if you fit a multilevel model using the raw score predictors and then fit the same model but with grand mean centred predictors then the resulting models are equivalent. By this, I mean that they will fit the data equally well, have the same predicted values, and the residuals will be the same. The parameters themselves (the *b*s) will, of course, be different but there will be a direct relationship between the parameters from the two models (i.e. they can be directly transformed into each other). Therefore, grand mean centring doesn't change the model, but it would change your interpretation of the parameters (you can't interpret them as though they are raw scores). When group mean centring is used the picture is much more complicated. In this situation the raw score model is not equivalent to the centred model in either the fixed part or the random part. One exception is when only the intercept is random (which arguably is an unusual situation), and the group means are reintroduced into the model as level 2 variables (Kreft & de Leeuw, 1998).

The decision about whether to centre or not is quite complicated and you really need to make the decision yourself in a given analysis. Centring can be a useful way to combat multicollinearity between predictor variables. It's also helpful when predictors do not have a meaningful zero point. Finally, multilevel models with centred predictors tend to be more stable, and estimates from these models can be treated as more or less independent of each other, which might be desirable. If group mean centring is used then the group means should be reintroduced as a level 2 variable unless you want to look at the effect of your 'group' or level 2 variable uncorrected for the mean effect of the centred level 1 predictor, such as when fitting a model when time is your main explanatory variable (Kreft & de Leeuw, 1998).

OLIVER TWISTED

Please, Sir, can I have some more … centring?

'Recentgin' babbles Oliver as he stumbles drunk out of Mrs Moonshine's alcohol emporium. 'I need some more recent gin.' I think you mean *centring* Oliver, not *recentgin*. If you want to know how to centre your variables using SPSS, then the additional material for this chapter on the companion website will tell you.

19.6. Multilevel modelling on SPSS ④

SPSS is not the best program in the world for multilevel modelling. Most people who do serious multilevel modelling tend to use specialist software such as MLwiN, HLM, SAS and R. There are several excellent books that compare the various packages and SPSS tends to fare pretty badly in all of them (Tabachnick & Fidell, 2001; Twisk, 2006). The main area where SPSS is behind its competitors is that it cannot do multilevel modelling when the outcome variable is categorical, yet this is bread and butter (albeit staggeringly complicated bread and butter) for the other packages mentioned. The second problem is that SPSS cannot produce bootstrap estimates of the model parameters, and these can be a very useful way to circumvent pesky distributional assumptions (see section 5.7.4). Other packages have these facilities. SPSS also has (and it's not just me that says this) a completely indecipherable windows interface for doing multilevel models (it is much easier to do using syntax).

We saw in section 19.4.1 that it is useful to build up models starting with a 'basic' model in which all parameters are fixed and then add random coefficients as appropriate before exploring confounding variables. We will take this approach to look at an example of conducting a multilevel model on SPSS.

19.6.1. Entering the data ②

Data entry depends a bit on the type of multilevel model that you wish to run: the data layout is slightly different when the same variables are measured at several points in time. However, we will look at the case of repeated-measures data in a second example. In this first example, the situation we have is very much like multiple regression in that data from each person who had surgery are not measured over multiple time points. Figure 19.7 shows the data layout. Each row represents a case of data (in this case a person who had surgery). Their scores on the various variables are simply entered in different columns. So, for example, the first person was 31 years old, had a BDI score of 12, they were in the waiting list control group at clinic 1, were female and were waiting for surgery for a physical reason.

19.6.2. Ignoring the data structure: ANOVA ②

First of all, let's ground the example in something very familiar to us: ANOVA. Let's say for the time being that we were interested only in the effect that surgery has on post-operative quality of life. We could analyse this with a simple one-way independent ANOVA (or indeed a *t*-test), and the model is described by equation (19.1).

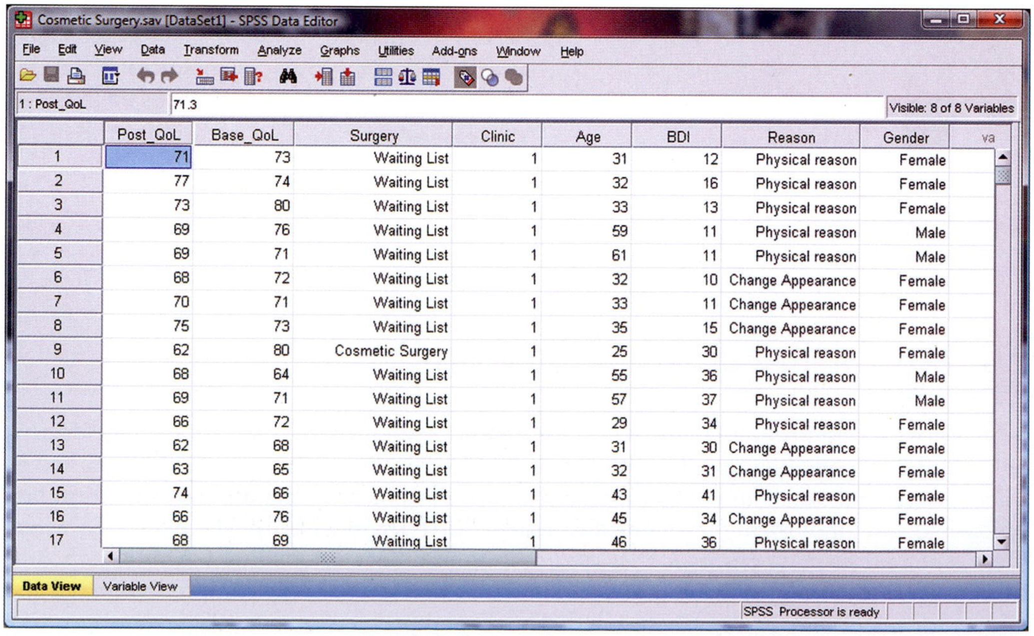

FIGURE 19.7 Data layout for multilevel modelling with no repeated measure

Cosmetic Surgery.sav [DataSet1] - SPSS Data Editor

1 : Post_QoL 71.3 — Visible: 8 of 8 Variables

	Post_QoL	Base_QoL	Surgery	Clinic	Age	BDI	Reason	Gender
1	71	73	Waiting List	1	31	12	Physical reason	Female
2	77	74	Waiting List	1	32	16	Physical reason	Female
3	73	80	Waiting List	1	33	13	Physical reason	Female
4	69	76	Waiting List	1	59	11	Physical reason	Male
5	69	71	Waiting List	1	61	11	Physical reason	Male
6	68	72	Waiting List	1	32	10	Change Appearance	Female
7	70	71	Waiting List	1	33	11	Change Appearance	Female
8	75	73	Waiting List	1	35	15	Change Appearance	Female
9	62	80	Cosmetic Surgery	1	25	30	Physical reason	Female
10	68	64	Waiting List	1	55	36	Physical reason	Male
11	69	71	Waiting List	1	57	37	Physical reason	Male
12	66	72	Waiting List	1	29	34	Physical reason	Female
13	62	68	Waiting List	1	31	30	Change Appearance	Female
14	63	65	Waiting List	1	32	31	Change Appearance	Female
15	74	66	Waiting List	1	43	41	Physical reason	Female
16	66	76	Waiting List	1	45	34	Change Appearance	Female
17	68	69	Waiting List	1	46	36	Physical reason	Female

Data View | Variable View — SPSS Processor is ready

SELF-TEST Using what you know about ANOVA, conduct a one-way ANOVA using **Surgery** as the predictor and **Post_QoL** as the outcome.

In reality we wouldn't do an ANOVA, I'm just using it as a way of showing you that multilevel models are not big and scary, but are simply extensions of what we have done before. SPSS Output 19.1 shows the results of the ANOVA that you should get if you did the self-test. We find a non-significant effect of surgery on quality of life, $F(1, 274) = 0.33$, $p > .05$.

SPSS OUTPUT 19.1

ANOVA

Quality of Life After Cosmetic Surgery

	Sum of Squares	df	Mean Square	F	Sig.
Between Groups	28.620	1	28.620	.330	.566
Within Groups	23747.883	274	86.671		
Total	23776.504	275			

To run a multilevel model we use the *Mixed Models* command. To access this command select Analyze Mixed Models ▸ Linear..., which will bring up the dialog box in Figure 19.8. This dialog box is for specifying the hierarchical nature of the data and because for the time being we are ignoring the hierarchical structure of our data, we will ignore this dialog box for now.

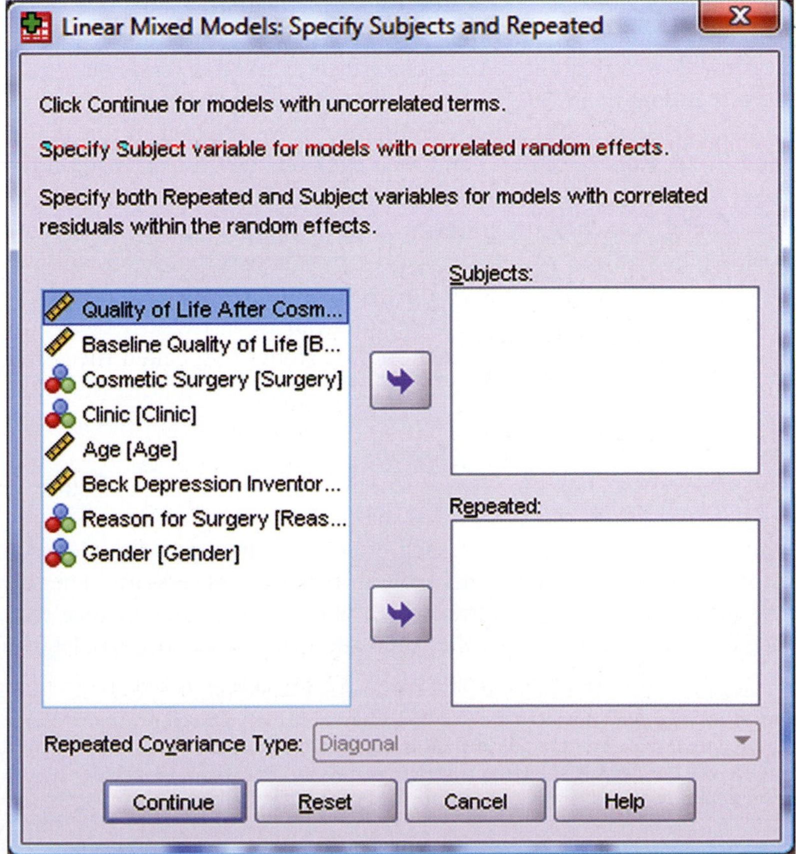

FIGURE 19.8 The initial mixed models dialog box

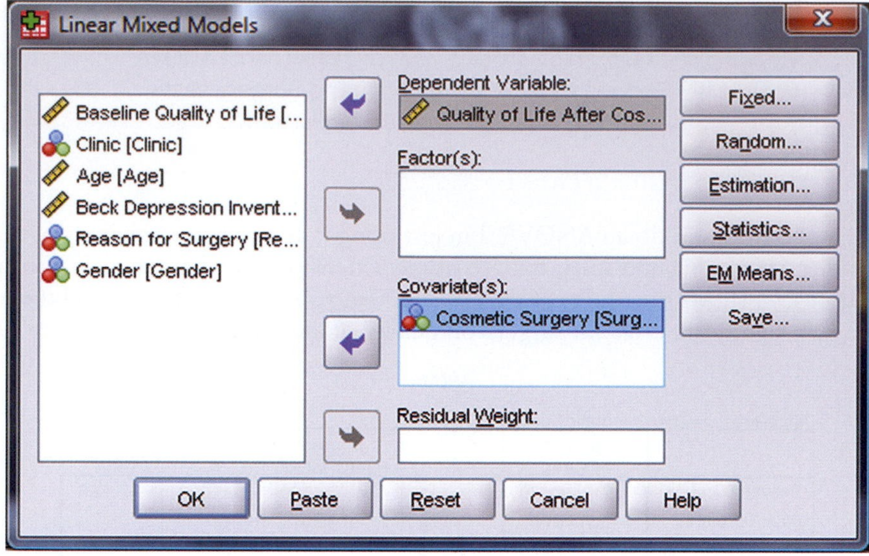

FIGURE 19.9 The main mixed models dialog box

Click on Continue to move to the main dialog box (Figure 19.9), which should look very familiar to many other dialog boxes that we have seen before. First we must specify our outcome variable, which is quality of life (QoL) after surgery, so select **Post_QoL** and drag it to the space labelled *Dependent Variable* (or click on ➜). Next we need to specify our predictor, which is whether or not the person has had surgery. Therefore, select **Surgery** and drag it to the space labelled *Covariate(s)* (or click on ➜).[4]

You'll notice several buttons at the side of the main dialog box. We use Fixed... to specify fixed effects in our model, and Random... to specify, yes, you've guessed it, random effects. To begin with we are going to treat our effects as fixed, so click on Fixed... to bring up the dialog box in Figure 19.10. We have only one variable specified as a predictor, and we want this to be treated as a fixed effect; therefore, we select it in this dialog box from the list labelled *Factors and Covariates* and then click on Add to transfer it to the *Model*. Click on Continue to return to the main dialog box.

In the main dialog box click on Estimation... to open the dialog box in Figure 19.11 (left panel). This dialog box allows you to change the parameters that SPSS will use when estimating the model. For example, if you don't get a solution then you could increase the number of iterations (SPSS Tip 8.1). The defaults can be left alone, but you do need to decide whether to use the maximum likelihood, or something called the restricted maximum-likelihood estimation method. There are pros and cons to both (see SPSS Tip 19.1) but because we want to compare models as we build them up, we will select ⦿ Maximum Likelihood (ML). Click on Continue to return to the main dialog box.

In the main dialog box click on Statistics... to open the dialog box in Figure 19.11 (right panel). There are two useful options in this dialog box. The first is to request *Parameter estimates*. This will give us *b*-values for each effect and their significance (so, it will give us similar information to the coefficients table in multiple regression). The second useful option is *Tests for covariance parameters*, which will give us a significance test of each of the covariance estimates in the model (i.e. the values of *u* in equations (19.3), (19.5) and

[4] You might wonder why we don't drag it to the *Factors* box given that it is a categorical variable. I wondered that too, but when I did drag it there the resulting analysis is wrong. Given this variable is coded 0 and 1 it shouldn't make any difference whether we specify it as a covariate of a factor, but when we include the hierarchical data structure it does. I don't know why, maybe you can email SPSS and then tell me.

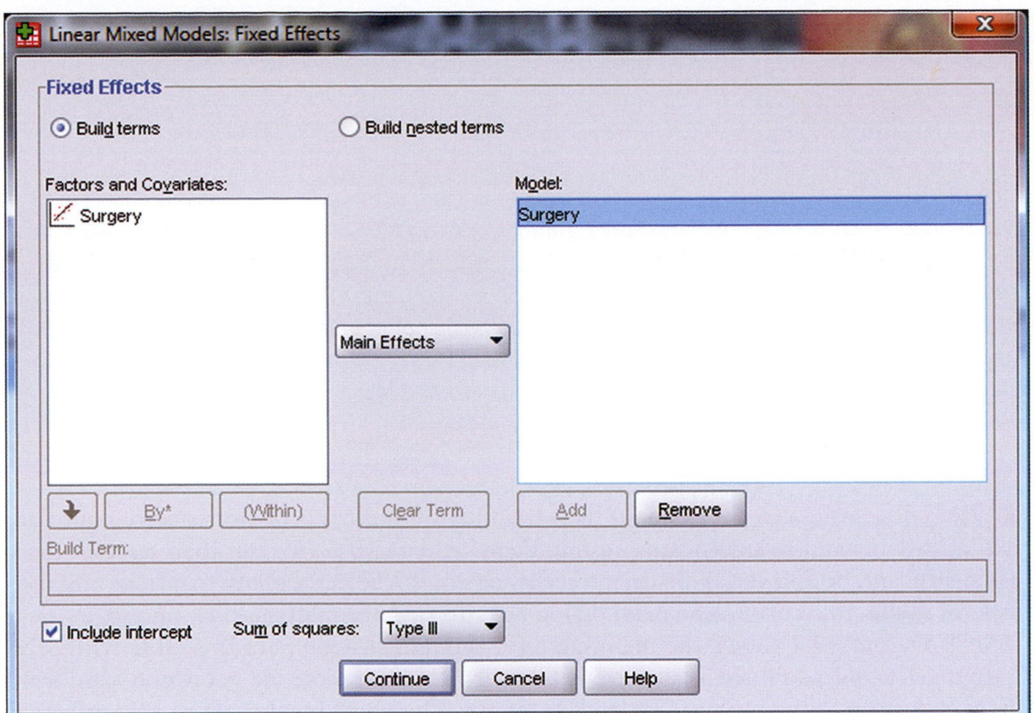

FIGURE 19.10 The dialog box for specifying fixed effects in mixed models

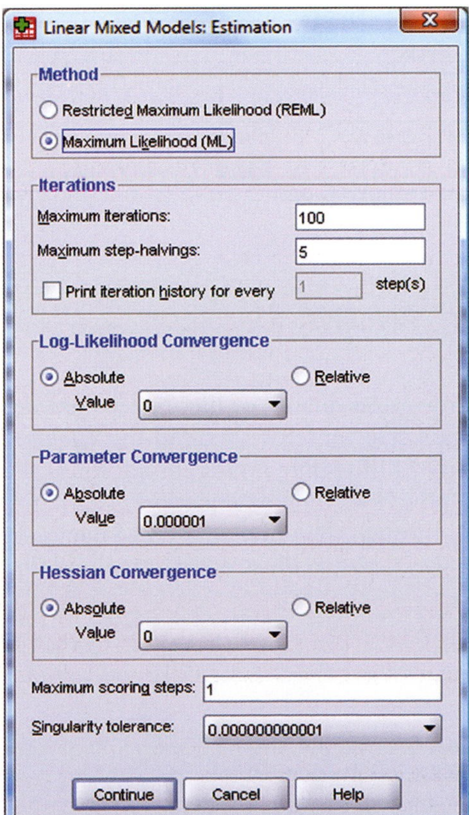

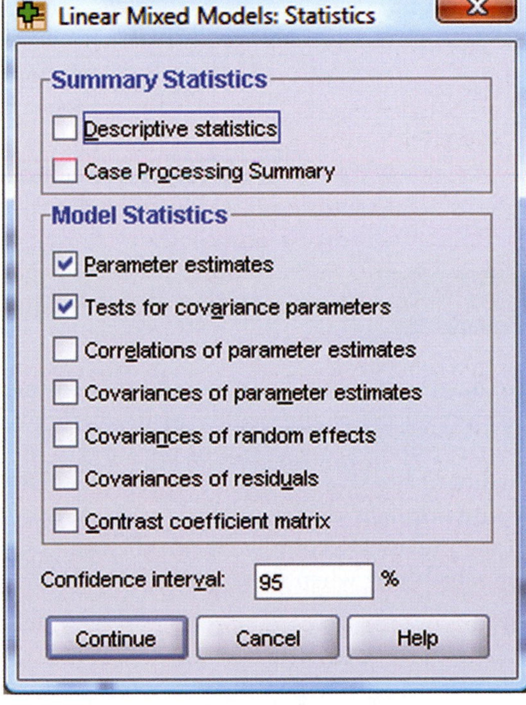

FIGURE 19.11 The estimation and statistics options for mixed models

SPSS TIP 19.1 Estimation ③

SPSS gives you the choice of two methods for estimating the parameters in the analysis: maximum likelihood (ML), which we have encountered before, and restricted maximum likelihood (REML). The conventional wisdom seems to be that ML produces more accurate estimates of fixed regression parameters, whereas REML produces more accurate estimates of random variances (Twisk, 2006). As such, the choice of estimation procedure depends on whether your hypotheses are focused on the fixed regression parameters or on estimating variances of the random effects. However, in many situations the choice of ML or REML will make only small differences to the parameter estimates. Also, if you want to compare models you must use ML.

(19.7)). These estimates tell us about the variability of intercepts or slopes across our contextual variable and so significance testing them can be useful (we can then say that there was significant, or not, variability in intercepts or slopes). Select these two options and then click on Continue to return to the main dialog box. To run the analysis, click on OK.

SPSS Output 19.2 shows the main table for the model. Compare this table with SPSS Output 19.1 and you'll see that there is basically no difference: we get a non-significant effect of surgery with an *F* of 0.33, and a *p* of .56. The point I want you to absorb here is that if we ignore the hierarchical structure of the data then what we are left with is something very familiar: an ANOVA/regression. The numbers are more or less exactly the same; all that has changed is that we have used different menus to get to the same end point.

SPSS OUTPUT 19.2

Type III Tests of Fixed Effects[a]

Source	Numerator df	Denominator df	F	Sig.
Intercept	1	276	6049.727	.000
Surgery	1	276	.333	.565

a. Dependent Variable: Quality of Life After Cosmetic Surgery.

19.6.3. Ignoring the data structure: ANCOVA ②

We have seen that there is no effect of cosmetic surgery on quality of life, but we did not take into account the quality of life before surgery. Let's, therefore, extend the example a little to look at the effect of the surgery on quality of life while taking into account the quality of life scores before surgery. Our model is now described by equation (19.8). You would normally do this analysis with an ANCOVA, through the univariate GLM menu. As in the previous section we'll run the analysis both ways, just to illustrate that we're doing the same thing when we run a hierarchical model.

SELF-TEST Using what you know about ANCOVA, conduct a one-way ANCOVA using **Surgery** as the predictor, **Post_QoL** as the outcome and **Base_QoL** as the covariate.

As before, we probably wouldn't do an ANCOVA using the mixed model command, but it's a useful illustration. SPSS Output 19.3 shows the results of the ANCOVA that you should get if you did the self-test. With baseline quality of life included we find a significant effect of surgery on quality of life, $F(1, 273) = 4.04$, $p < .05$. Baseline quality of life also predicted quality of life after surgery, $F(1, 273) = 214.89$, $p < .001$.

SPSS OUTPUT 19.3

Tests of Between-Subjects Effects

Dependent Variable:Quality of Life After Cosmetic Surgery

Source	Type III Sum of Squares	df	Mean Square	F	Sig.
Corrected Model	10488.253[a]	2	5244.127	107.738	.000
Intercept	1713.257	1	1713.257	35.198	.000
Base_QoL	10459.633	1	10459.633	214.888	.000
Surgery	196.816	1	196.816	4.043	.045
Error	13288.250	273	48.675		
Total	1004494.530	276			
Corrected Total	23776.504	275			

a. R Squared = .441 (Adjusted R Squared = .437)

Select Analyze Mixed Models ▸ Linear... again, and just like last time ignore the first dialog box because for the time being we are ignoring the hierarchical structure of our data. We can leave the main dialog box (Figure 19.12) as it was in the last analysis except that we now need to add the baseline quality of life as another predictor. To do this, select **Base_QoL** and drag it to the space labelled *Covariate(s)* (or click on ➔).

We need to add this new variable to our model as a fixed effect, so click on Fixed... to bring up the dialog box in Figure 19.13. Select **Base_QoL** in the list labelled *Factors and Covariates* and then click on Add to transfer it to the *Model*. Click on Continue to return to the main dialog box and click on OK to run the analysis.

SPSS Output 19.4 shows the main table for the model. Compare this table with SPSS Output 19.3 and you'll see that again there is no difference: we get a significant effect of surgery with an F of 4.08, $p < .05$, and a significant effect of baseline quality of life with an F of 217.25, $p < .001$. We can also see that the regression coefficient for surgery is −1.70. Again, the results are pretty similar to when we ran the analysis as ANCOVA (the values are

FIGURE 19.12
The main mixed models dialog box

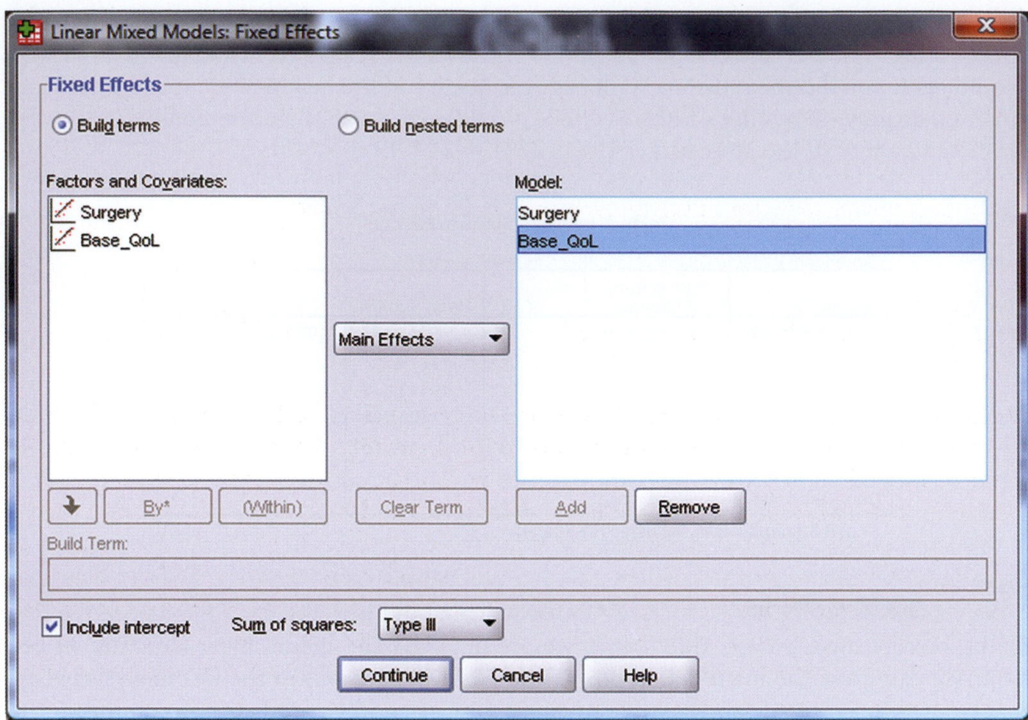

FIGURE 19.13 The dialog box for specifying fixed effects in mixed models

SPSS OUTPUT 19.4

Model Dimension[a]

		Number of Levels	Number of Parameters
Fixed Effects	Intercept	1	1
	Surgery	1	1
	Base_QoL	1	1
	Residual		1
	Total	3	4

a. Dependent Variable: Quality of Life After Cosmetic Surgery.

Information Criteria[a]

-2 Log Likelihood	1852.543
Akaike's Information Criterion (AIC)	1860.543
Hurvich and Tsai's Criterion (AICC)	1860.690
Bozdogan's Criterion (CAIC)	1879.024
Schwarz's Bayesian Criterion (BIC)	1875.024

The information criteria are displayed in smaller-is-better forms.

a. Dependent Variable: Quality of Life After Cosmetic Surgery.

Estimates of Fixed Effects[a]

Parameter	Estimate	Std. Error	df	t	Sig.	95% Confidence Interval Lower Bound	95% Confidence Interval Upper Bound
Intercept	18.147025	2.891820	276.000	6.275	.000	12.454198	23.839851
Surgery	-1.697233	.839442	276	-2.022	.044	-3.349756	-.044710
Base_QoL	.665036	.045120	276.000	14.739	.000	.576213	.753858

a. Dependent Variable: Quality of Life After Cosmetic Surgery.

Type III Tests of Fixed Effects[a]

Source	Numerator df	Denominator df	F	Sig.
Intercept	1	276.000	39.379	.000
Surgery	1	276	4.088	.044
Base_QoL	1	276.000	217.249	.000

a. Dependent Variable: Quality of Life After Cosmetic Surgery.

slightly different because here we're using maximum likelihood methods to estimate the parameters of the model but in ANCOVA we use ordinary least squares methods). Hopefully this has convinced you that we're just doing a regression here, something you have been doing throughout this book. This technique isn't radically different, and if you think about it as just an extension of what you already know, then it's relatively easy to understand. So, having shown you that we can do basic analyses through the mixed models command, let's now use its power to factor in the hierarchical structure of the data.

19.6.4. Factoring in the data structure: random intercepts ③

We have seen that when we factor in the pre-surgery quality of life scores, which themselves significantly predict post-surgery quality of life scores, surgery seems to positively affect quality of life. However, at this stage we have ignored the fact that our data have a hierarchical structure. Essentially we have violated the independence assumption because scores from people who had their surgery at the same clinic are likely to be related to each other (and certainly more related than with people at different clinics). We have seen that violating the assumption of independence can have some quite drastic consequences (see section 10.2.10). However, rather than just panic and gibber about our *F*-ratio being inaccurate, we can model this covariation within clinics explicitly by including the hierarchical data structure in our analysis.

To begin with, we will include the hierarchy in a fairly crude way by assuming simply that intercepts vary across clinics. Our model is now described by:

$$\text{QoL After Surgery}_{ij} = b_{0j} + b_1\text{Surgery}_{ij} + b_2\text{QoL Before Surgery}_{ij} + \varepsilon_{ij}$$
$$b_{0j} = b_0 + u_{0j}$$

To run a multilevel model we use the *Mixed Models* option by selecting Analyze Mixed Models ▸ Linear..., which will bring up the dialog box in Figure 19.8. This time we don't want to ignore this dialog box, but instead want to specify our level 2 variable (**Clinic**). We specify contextual variables that group participants (or subjects) in the box labelled *Subjects*. Select **Clinic** from the list of variables and drag it to the box labelled *Subjects* (or click on ➔). The completed dialog box is shown in Figure 19.14.

Click on Continue to access the main dialog box. We don't need to change this because all we are doing in this model is changing the intercept from being fixed to random. Therefore, the main dialog box should still look like Figure 19.12. We also don't need to re-specify our fixed effects so there is no need to click on Fixed... unless you want to check that the dialog box still looks like Figure 19.13. However, we do need to specify a random effect for the first time, so click on Random... in the main dialog box to access the dialog box in Figure 19.15. The first thing we need to do is to specify our contextual variable. We do this by selecting it from the list of contextual variables that we have told SPSS about in Figure 19.14. These appear in the section labelled *Subjects* and because we only specified one variable, there is only one variable in the list, **Clinic**. Select this variable and drag it to the area labelled *Combinations* (or click on ➔). We want to specify only that the intercept is random, and we do this by selecting ☑ Include intercept. Notice in this dialog box that there is a drop-down list to specify the type of covariance (Variance Components ▾). For a random intercept model this default option is fine. Click on Continue to return to the main dialog box and then click on OK to run the analysis.

The output of this analysis is shown in SPSS Output 19.5. The first issue is whether allowing the intercepts to vary has made a difference to the model. We can test this from the change in the −2 log-likelihood (equation (19.10)). In our new model the −2LL is

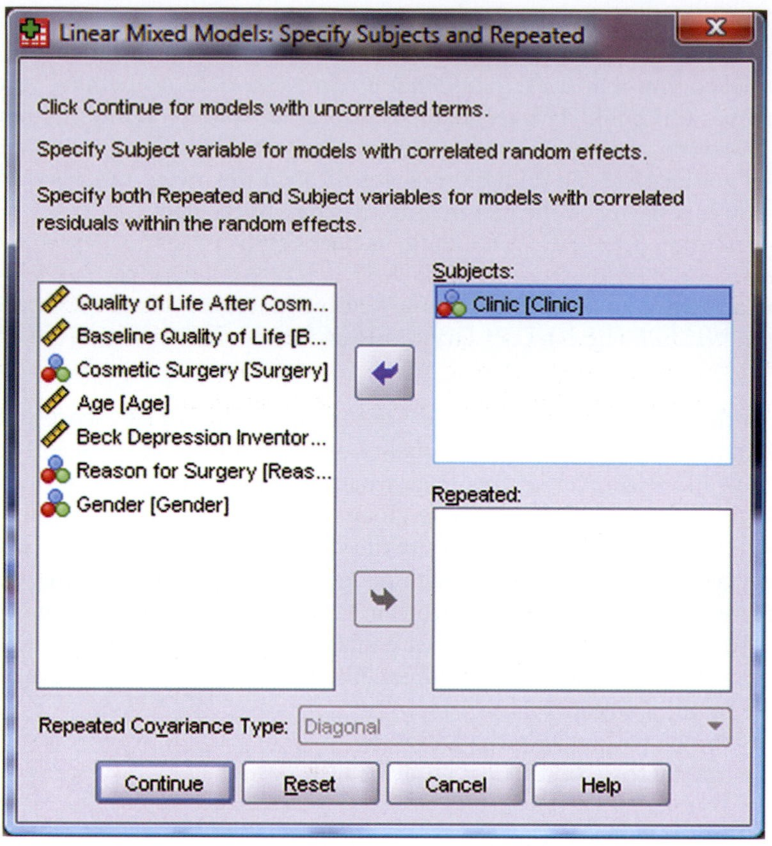

FIGURE 19.14 Specifying a level 2 variable in a hierarchical linear model

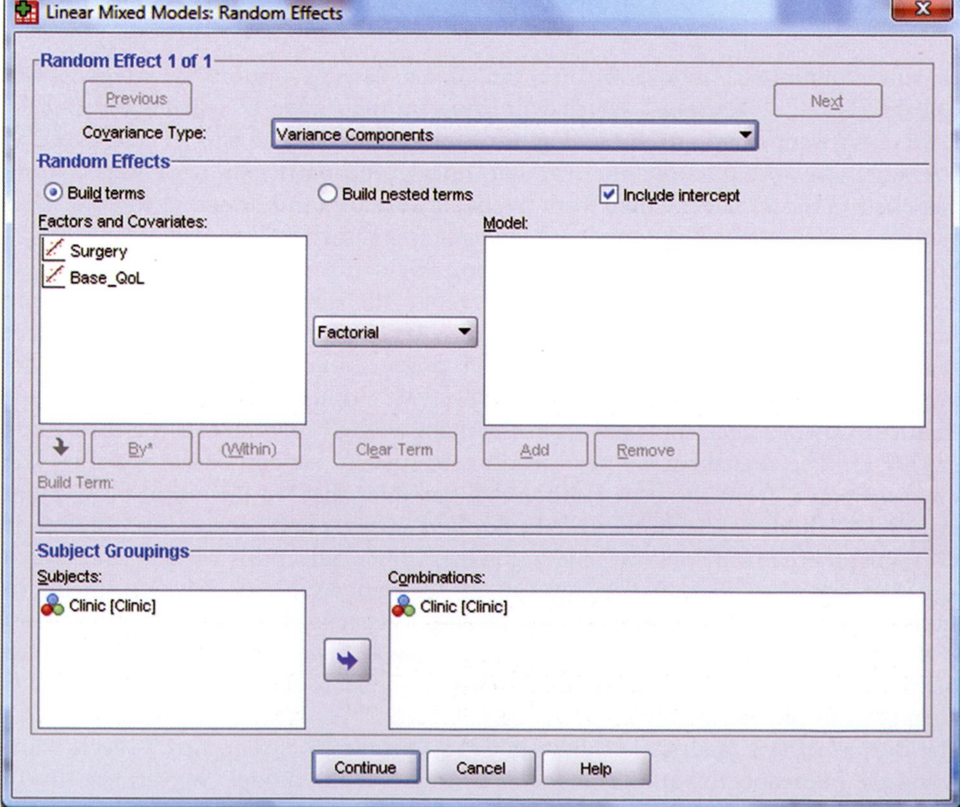

FIGURE 19.15 The dialog box for specifying random effects in mixed models

1837.49 (SPSS Output 19.5) based on a total of five parameters. In the old model (SPSS Output 19.4) the −2LL was 1852.54, based on four parameters. Therefore:

$$\chi^2_{\text{Change}} = 1852.54 - 1837.49 = 15.05$$

$$df_{\text{Change}} = 5 - 4 = 1$$

If we look at the critical values for the chi-square statistic with 1 degree of freedom in Appendix A4, they are 3.84 ($p < .05$) and 6.63 ($p < .01$); therefore, this change is highly significant. Put another way, it is important that we modelled this variability in intercepts because when we do our model is significantly improved. We can conclude then that the intercepts for the relationship between surgery and quality of life (when controlling for baseline quality of life) vary significantly across the different clinics.

SPSS OUTPUT 19.5

Model Dimension[b]

		Number of Levels	Covariance Structure	Number of Parameters	Subject Variables
Fixed Effects	Intercept	1		1	
	Surgery	1		1	
	Base_QoL	1		1	
Random Effects	Intercept[a]	1	Variance Components	1	Clinic
Residual				1	
Total		4		5	

a. As of version 11.5, the syntax rules for the RANDOM subcommand have changed. Your command syntax may yield results that differ from those produced by prior versions. If you are using SPSS 11 syntax, please consult the current syntax reference guide for more information.

b. Dependent Variable: Quality of Life After Cosmetic Surgery.

df for – 2LL

Information Criteria[a]

-2 Log Likelihood	1837.490
Akaike's Information Criterion (AIC)	1847.490
Hurvich and Tsai's Criterion (AICC)	1847.712
Bozdogan's Criterion (CAIC)	1870.592
Schwarz's Bayesian Criterion (BIC)	1865.592

The information criteria are displayed in smaller-is-better forms.

a. Dependent Variable: Quality of Life After Cosmetic Surgery.

– 2LL

Type III Tests of Fixed Effects[a]

Source	Numerator df	Denominator df	F	Sig.
Intercept	1	163.879	73.305	.000
Surgery	1	275.631	.139	.709
Base_QoL	1	245.020	83.159	.000

a. Dependent Variable: Quality of Life After Cosmetic Surgery.

bs

Estimates of Fixed Effects[a]

Parameter	Estimate	Std. Error	df	t	Sig.	95% Confidence Interval Lower Bound	Upper Bound
Intercept	29.563601	3.452958	163.879	8.562	.000	22.745578	36.381624
Surgery	-.312999	.838551	275.631	-.373	.709	-1.963776	1.337779
Base_QoL	.478630	.052486	245.020	9.119	.000	.375248	.582012

a. Dependent Variable: Quality of Life After Cosmetic Surgery.

Var(ε_{ij})

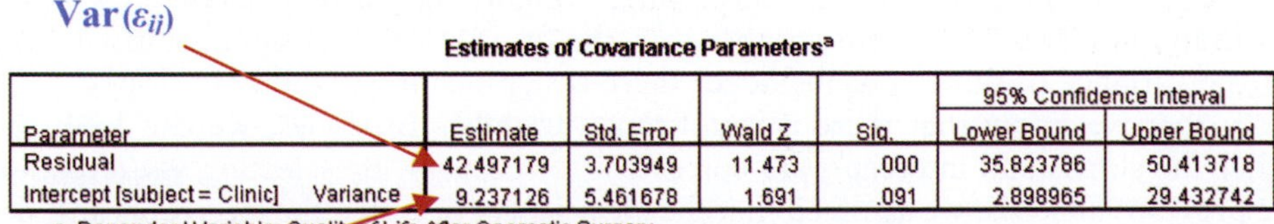

Estimates of Covariance Parameters[a]

Parameter		Estimate	Std. Error	Wald Z	Sig.	95% Confidence Interval Lower Bound	Upper Bound
Residual		42.497179	3.703949	11.473	.000	35.823786	50.413718
Intercept [subject = Clinic]	Variance	9.237126	5.461678	1.691	.091	2.898965	29.432742

a. Dependent Variable: Quality of Life After Cosmetic Surgery.

Var(u_{0j})

You will also notice that the significance of the variance estimate for the intercept (9.24) is tested using the Wald statistic, which is a standard z-score in this case ($z = 1.69$). You should be cautious in interpreting the Wald statistic because, for random parameters especially, it can be quite unpredictable (for fixed effects it should be OK). The change in the −2LL is much more reliable, and you should use this to assess the significance of changes to the model – just like with logistic regression (Chapter 8).

By allowing the intercept to vary we also have a new regression parameter for the effect of surgery, which is −.31 compared to −1.70 when the intercept was fixed (SPSS Output 19.4). In other words, by allowing the intercepts to vary over clinics, the effect of surgery has decreased dramatically. In fact, it is not significant any more, $F(1, 275.63) = 0.14$, $p > .05$. This shows how, had we ignored the hierarchical structure in our data, we would have reached very different conclusions to what we have found here.

19.6.5. Factoring in the data structure: random intercepts and slopes ④

We have seen that including a random intercept is important for this model (it changes the log-likelihood significantly). However, we could now look at whether adding a random slope will also be beneficial by adding this term to the model. The model is now described by equation (19.9), which we saw earlier on; it can be specified in SPSS with only minor modifications to the dialog boxes. All we are doing is adding another random term to the model; therefore, the only changes we need to make are in the dialog box accessed by clicking on Random... . (If you are starting from scratch then follow the instructions for setting up the dialog box in the previous section.) We need to select the predictor (**Surgery**) from the list of *Factors and covariates* and add it to the model by clicking on Add (see Figure 19.16). Click on Continue to return to the main dialog box and then click on OK to run the analysis

All we're interested in at this stage is estimating the effect of including the variance in slopes. SPSS Output 19.6 gives us the −2LL for the new model and the value of the variance in slopes (29.63). To find the significance of the variance in slopes, we subtract this value from the −2LL for the previous model. This gives us a chi-square statistic with df = 1 (because we have added only one new parameter to the model: the variance in slopes). In our new model the −2LL is 1816 (SPSS Output 19.6) based on a total of six parameters. In the old model (SPSS Output 19.5) the −2LL was 1837.49, based on five parameters. Therefore:

$$\chi^2_{\text{Change}} = 1837.5 - 1816 = 21.5$$
$$df_{\text{Change}} = 6 - 5 = 1$$

Comparing this value to the same critical values as before for the chi-square statistic with $df = 1$ (i.e. 3.84 and 6.63) shows that this change is highly significant because 21.49 is much larger than these two values. Put another way, the fit of our model significantly improved when the variance of slopes was included: there is significant variability in slopes.

Now that we know that there is significant variability in slopes, we can look to see whether the slopes and intercepts are correlated (or covary). By selecting Variance Components ▾ in the previous analysis, we assumed that the covariances between the intercepts and slopes were zero. Therefore, SPSS estimated only the variance of slopes. This was a useful thing

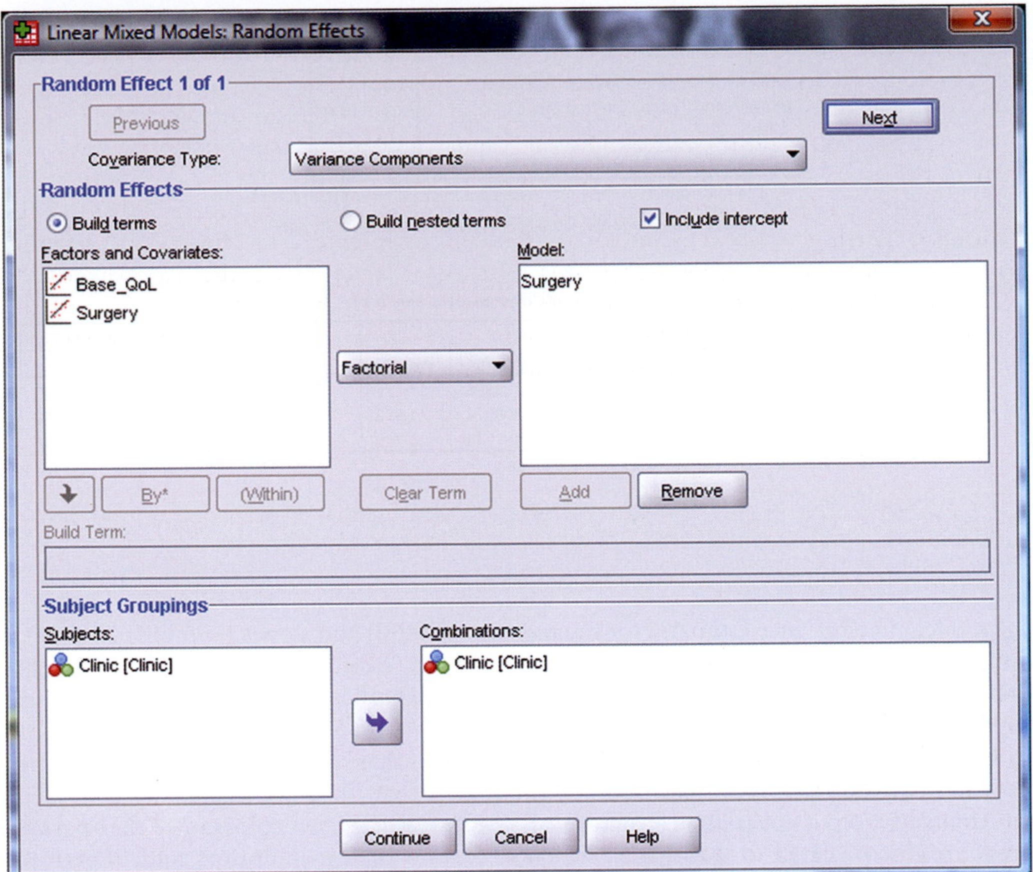

FIGURE 19.16 The dialog box for specifying random effects in mixed models

to do because it allowed us to look at the effect of the variance of slopes in isolation. If we now want to include the covariance between random slopes and random intercepts we do this by clicking on Variance Components in Figure 19.16 to access the drop-down list, and selecting Unstructured instead. By changing to Unstructured, we remove the assumption that the covariances between slopes and intercepts are zero, and so SPSS will estimate this covariance. As such, by changing to Unstructured, we add a new term to the model that estimates the covariance between random slopes and intercepts. Redo the analysis but change Variance Components to Unstructured in Figure 19.16.

The output of this analysis is shown in SPSS Output 19.7. The first issue is whether adding the covariance between slopes and intercepts has made a difference to the model using the change in the −2LL (equation (19.10)). In our new model the −2LL is 1798.62 (SPSS Output 19.7) based on a total of seven parameters. In the old model (SPSS Output 19.6) the −2LL was 1816, based on six parameters. Therefore:

$$\chi^2_{\text{Change}} = 1816 - 1798.62 = 17.38$$
$$df_{\text{Change}} = 7 - 6 = 1$$

This change is highly significant at $p < .01$ because 17.38 is bigger than the critical value of 6.63 for the chi-square statistic with 1 degree of freedom (see Appendix A4). Put another way, our model is significantly improved when the covariance term is included in

SPSS OUTPUT 19.6

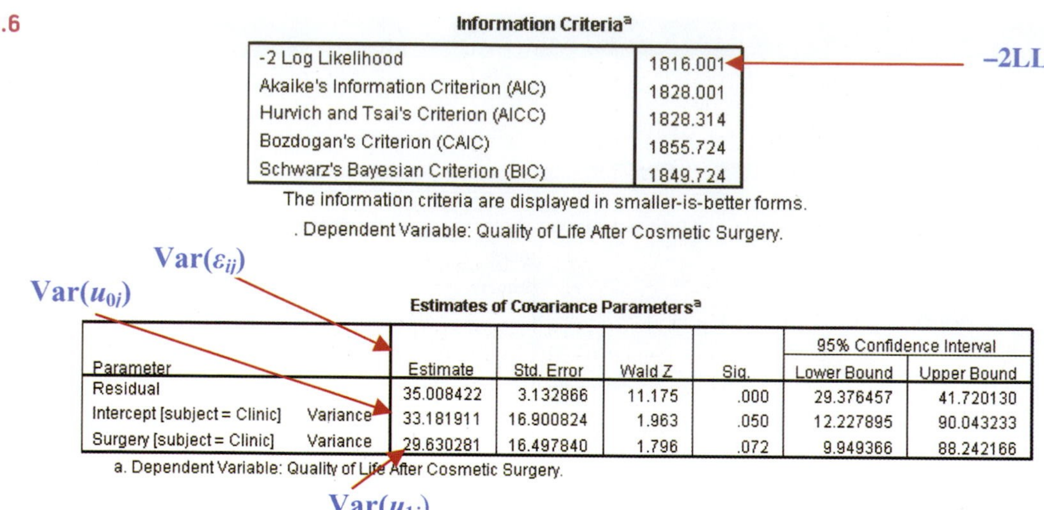

Information Criteria[a]

-2 Log Likelihood	1816.001
Akaike's Information Criterion (AIC)	1828.001
Hurvich and Tsai's Criterion (AICC)	1828.314
Bozdogan's Criterion (CAIC)	1855.724
Schwarz's Bayesian Criterion (BIC)	1849.724

The information criteria are displayed in smaller-is-better forms.

. Dependent Variable: Quality of Life After Cosmetic Surgery.

Estimates of Covariance Parameters[a]

Parameter		Estimate	Std. Error	Wald Z	Sig.	95% Confidence Interval	
						Lower Bound	Upper Bound
Residual		35.008422	3.132866	11.175	.000	29.376457	41.720130
Intercept [subject = Clinic]	Variance	33.181911	16.900824	1.963	.050	12.227895	90.043233
Surgery [subject = Clinic]	Variance	29.630281	16.497840	1.796	.072	9.949366	88.242166

a. Dependent Variable: Quality of Life After Cosmetic Surgery.

the model. The variance estimates for the intercept (37.60) and slopes (–36.68 and 38.41), and their associated significance based on the Wald test, confirm this because all three estimates are close to significance (although I reiterate my earlier point that the Wald statistic should be interpreted with caution).

One important part of the output to take note of is that the random part of the slopes now has two values (–36.68 and 38.41). The reason that there are two values is because we changed from a covariance structure of Variance Components, which assumes that parameters are uncorrelated to Unstructured, which makes no such assumption, and, therefore, the covariance is estimated too. The first of these values is the covariance between the random slope and random intercept, and the second is the variance of the random slopes. We encountered covariance in Chapter 6 and saw that it is an unstandardized measure of the relationship between variables. In other words, it's like a correlation. Therefore, the covariance term tells us whether there is a relationship or interaction between the random slope and the random intercept within the model. The actual size of this value is not terribly important because it is unstandardized (so we can't compare the size of covariances measured across different variables), but the direction of it is. In this case the covariance is negative (–36.68) indicating a negative relationship between the intercepts and the slopes. Remember that we are looking at the effect of surgery on quality of life in 10 different clinics, so this means that, across these clinics, as the intercept for the relationship between surgery and quality of life increases, the value of the slope decreases. This is best understood using a diagram and Figure 19.17 shows the observed values of quality of life after surgery plotted against those predicted by our model. In this diagram each line represents a different clinic. We can see that the 10 clinics differ: those with low intercepts (low values on the *y*-axis) have quite steep positive slopes. However, as the intercept increases (as we go from the line that crosses the *y*-axis at the lowest point up to the line that hits the *y*-axis at the highest point) the slopes of the lines get flatter (the slope decreases). The negative covariance between slope and intercept reflects this relationship. Had it been positive it would mean the opposite: as intercepts increase, the slopes increase also.

The second term that we get with the random slope is its variance (in this case 38.41). This tells us how much the slopes vary around a single slope fitted to the entire data set (i.e. ignoring the clinic from which the data came). This confirms what our chi-square test showed us: that the slopes across clinics are significantly different.

SPSS OUTPUT 19.7

Model Dimension[b]

		Number of Levels	Covariance Structure	Number of Parameters	Subject Variables
Fixed Effects	Intercept	1		1	
	Base_QoL	1		1	
	Surgery	1		1	
Random Effects	Intercept + Surgery[a]	2	Unstructured	3	Clinic
Residual				1	
Total		5		7	

a. As of version 11.5, the syntax rules for the RANDOM subcommand have changed. Your command syntax may yield results that differ from those produced by prior versions. If you are using SPSS 11 syntax, please consult the current syntax reference guide for more information.

b. Dependent Variable: Quality of Life After Cosmetic Surgery.

df for –2LL

Information Criteria[a]

-2 Log Likelihood	1798.624
Akaike's Information Criterion (AIC)	1812.624
Hurvich and Tsai's Criterion (AICC)	1813.042
Bozdogan's Criterion (CAIC)	1844.967
Schwarz's Bayesian Criterion (BIC)	1837.967

The information criteria are displayed in smaller-is-better forms.

a. Dependent Variable: Quality of Life After Cosmetic Surgery.

–2LL

Type III Tests of Fixed Effects[a]

Source	Numerator df	Denominator df	F	Sig.
Intercept	1	84.954	107.284	.000
Surgery	1	9.518	.097	.762
Base_QoL	1	265.933	33.984	.000

a. Dependent Variable: Quality of Life After Cosmetic Surgery.

bs

Estimates of Fixed Effects[a]

Parameter	Estimate	Std. Error	df	t	Sig.	95% Confidence Interval Lower Bound	95% Confidence Interval Upper Bound
Intercept	40.102525	3.871729	84.954	10.358	.000	32.404430	47.800620
Surgery	-.654530	2.099413	9.518	-.312	.762	-5.364643	4.055583
Base_QoL	.310218	.053214	265.933	5.830	.000	.205443	.414993

a. Dependent Variable: Quality of Life After Cosmetic Surgery.

$Var(u_{0j})$ = Variance of intercepts

$Var(\varepsilon_{ij})$ = Variance of residuals

Estimates of Covariance Parameters[a]

Parameter		Estimate	Std. Error	Wald Z	Sig.	95% Confidence Interval Lower Bound	95% Confidence Interval Upper Bound
Residual		34.955705	3.116670	11.216	.000	29.351106	41.630504
Intercept + Surgery [subject = Clinic]	UN (1,1)	37.609439	18.726052	2.008	.045	14.173482	99.796926
	UN (2,1)	-36.680707	18.763953	-1.955	.051	-73.457378	.095965
	UN (2,2)	38.408857	20.209811	1.901	.057	13.694612	107.724141

a. Dependent Variable: Quality of Life After Cosmetic Surgery.

$Cov(u_{0j}, u_{1j})$ = Covariance between intercepts and Slopes

$Var(u_{1j})$ = Variance of Slopes

We can conclude then that the intercepts and slopes for the relationship between surgery and quality of life (when controlling for baseline quality of life) vary significantly across the different clinics. By allowing the intercept and slopes to vary we also have a new regression parameter for the effect of surgery, which is –.65 compared to –0.31 when the slopes were fixed (SPSS Output 19.5). In other words, by allowing the intercepts to vary over clinics, the effect of surgery has increased slightly, although it is still nowhere near significant, $F(1, 9.518) = 0.10$, $p > .05$. This shows how, had we ignored the hierarchical structure in our data, we would have reached very different conclusions to what we have found here.

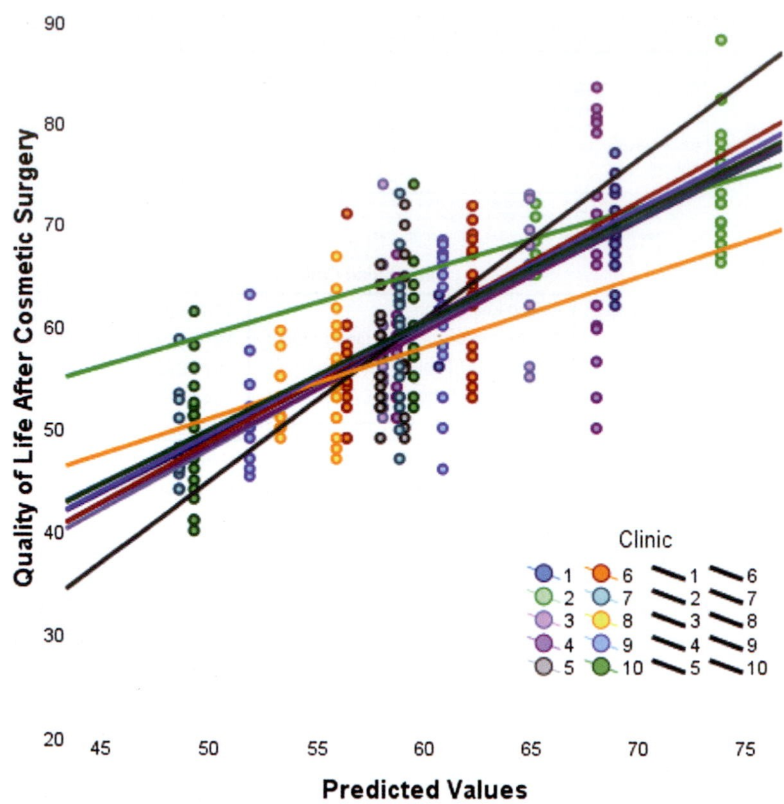

FIGURE 19.17 Predicted values from the model (surgery predicting quality of life after controlling for baseline quality of life) plotted against the observed values

19.6.6. Adding an interaction to the model ④

We can now build up the model by adding in another variable. One of the variables we measured was the reason for the person having cosmetic surgery: was it to resolve a physical problem or was it purely for vanity? We can add this variable to the model, and also look at whether it interacts with surgery in predicting quality of life.[5] Our model will simply expand to incorporate these new terms, and each term will have a regression coefficient (which we select to be fixed). Therefore, our new model can be described as in the equation below (note that all that has changed is that there are two new predictors):

$$\begin{aligned} \text{QoL After}_{ij} &= b_{0j} + b_{1j}\text{Surgery}_{ij} + b_2\text{QoL Before Surgery}_{ij} + b_3\text{Reason}_{ij} \\ &\quad + b_4\ (\text{Reason} \times \text{Surgery})_{ij} + \varepsilon_{ij} \\ b_{0j} &= b_0 + u_{0j} \\ b_{1j} &= b_1 + u_{1j} \end{aligned} \tag{19.11}$$

To set up this model in SPSS is very easy to do and just requires some minor changes to the dialog boxes that we have already used. First, select Analyze Mixed Models ▸ Linear...; this initial dialog box should be set up as for the previous analysis, but if you're running this analysis without running the prior one first then set up the level 2 variable of

[5] In reality, because we would use the change in the −2LL to see whether effects are significant, we would build this new model up a term at a time. Therefore, we would first include only **Reason** in the model, then in a separate analysis we would add the interaction. By doing so we can calculate the change in −2LL for each effect. To save space I'm going to put both into the model in a single step.

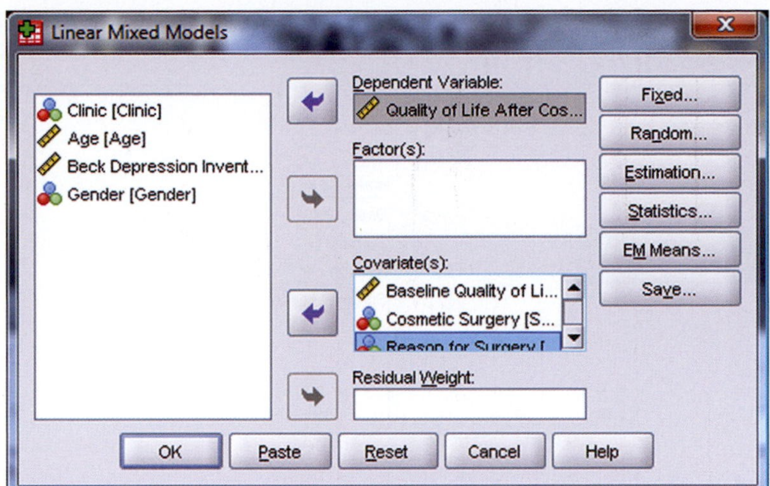

FIGURE 19.18 The main mixed models dialog box

Clinic as in the previous sections (the completed dialog box is shown in Figure 19.14). Click on Continue to access the main dialog box. If you're continuing the previous analysis then this dialog box will already be set up with the previous model. If you've jumped straight into this analysis then set this dialog box up as with our previous model (Figure 19.12). We have two new covariates to add to the model: the effect of the reason for the surgery (**Reason**) and the interaction of **Reason** and **Surgery**. At this stage we simply need to add **Reason** as a covariate, so select this variable and drag it to the space labelled *Covariate(s)* (or click on ➜). The completed dialog box is in Figure 19.18.

We need to add these fixed effects to our model, so click on Fixed... to bring up the dialog box in Figure 19.19. First let's specify the main effect of **Reason**; to do this, select this variable in the list labelled *Factors and Covariates* and then click on Add to transfer it to *Model*. To specify the interaction term, first click on Factorial ▾ and change it to Interaction ▾. Next, select **Surgery** from *Factors and Covariates* and then while holding down the *Ctrl* key select **Reason**. With both variables selected click on Add to transfer them to *Model* as an interaction effect. The dialog box should now look like Figure 19.19. Click on Continue to return to the main dialog box. We don't need to specify any extra random coefficients so we can leave the dialog box accessed through Random... as it is in Figure 19.16, and we can leave the other options as they are in previous analyses. In the main dialog box click on OK to run the analysis.

SPSS Output 19.8 shows the resulting output, which is similar to the previous output except that we now have two new fixed effects. The first issue is whether these new effects make a difference to the model. We can use the log-likelihood statistics again:

$$\chi^2_{\text{Change}} = 1798.62 - 1789.05 = 9.57$$
$$df_{\text{Change}} = 9 - 7 = 2$$

If we look at the critical values for the chi-square statistic in the Appendix, it is 5.99 ($p < .05$, $df = 2$); therefore, this change is significant. We can look at the effects individually in the table of fixed effects. This tells us that quality of life before surgery significantly predicted quality of life after surgery, $F(1, 268.92) = 33.65$, $p < .001$, surgery still did not significantly predict quality of life, $F(1, 15.86) = 2.17$, $p = .161$, but the reason for surgery, $F(1, 259.89) = 9.67$, $p < .01$, and the interaction of the reason for surgery and surgery, $F(1, 217.09) = 6.28$, $p < .05$, both did significantly predict quality of life. The

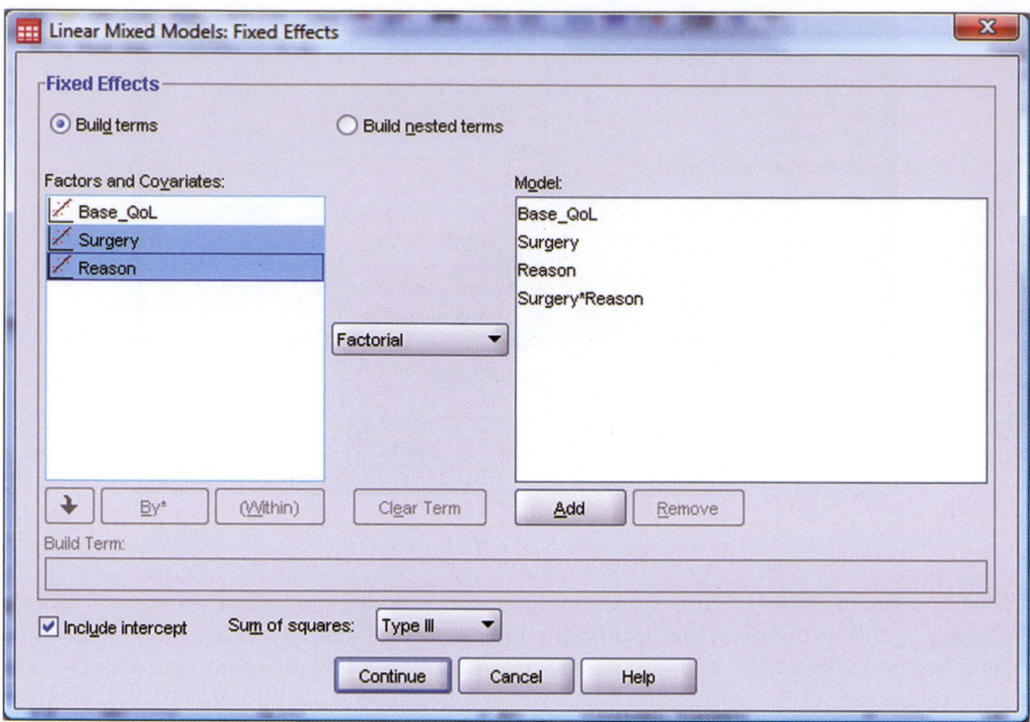

FIGURE 19.19 Specifying a fixed effect interaction in mixed models

table of estimates of fixed effects tells us much the same thing except it also gives us the regression coefficients and their confidence intervals.

The values of the variance for the intercept (30.06) and the slope (29.35) are lower than the previous model but still significant (one-tailed). Also the covariance between the slopes and intercepts is still negative (–28.08). As such our conclusions about our random parameters stay much the same as in the previous model.

The effect of the reason for surgery is easy to interpret. Given that we coded this predictor as 0 = physical reason and 1 = change appearance, the negative coefficient tells us that as reason increases (i.e. as a person goes from having surgery for a physical reason to having surgery to change their appearance) quality of life decreases. However, this effect in isolation isn't that interesting because it includes both people who had surgery and the waiting list controls. More interesting is the interaction term, because this takes account of whether or not the person had surgery. To break down this interaction we could rerun the analysis separately for the two 'reason groups'. Obviously we would remove the interaction term and the main effect of **Reason** from this analysis (because we are analysing the physical reason group separately from the group that wanted to change their appearance). As such, you need to fit the model in the previous section, but first split the file by **Reason**.

SELF-TEST Split the file by **Reason** and then run a multilevel model predicting **Post_QoL** with a random intercept, and random slopes for **Surgery**, and including **Base_QoL** and **Surgery** as predictors.

SPSS OUTPUT 19.8

Model Dimension[b]

		Number of Levels	Covariance Structure	Number of Parameters	Subject Variables
Fixed Effects	Intercept	1		1	
	Base_QoL	1		1	
	Surgery	1		1	
	Reason	1		1	
	Surgery * Reason	1		1	
Random Effects	Intercept + Surgery[a]	2	Unstructured	3	Clinic
Residual				1	
Total		7		9	

a. As of version 11.5, the syntax rules for the RANDOM subcommand have changed. Your command syntax may yield results that differ from those produced by prior versions. If you are using SPSS 11 syntax, please consult the current syntax reference guide for more information.

b. Dependent Variable: Quality of Life After Cosmetic Surgery.

df for –2LL

Information Criteria[a]

-2 Log Likelihood	1789.045
Akaike's Information Criterion (AIC)	1807.045
Hurvich and Tsai's Criterion (AICC)	1807.722
Bozdogan's Criterion (CAIC)	1848.629
Schwarz's Bayesian Criterion (BIC)	1839.629

The information criteria are displayed in smaller-is-better forms.

a. Dependent Variable: Quality of Life After Cosmetic Surgery.

–2LL

Type III Tests of Fixed Effects[a]

Source	Numerator df	Denominator df	F	Sig.
Intercept	1	108.853	122.593	.000
Base_QoL	1	268.920	33.647	.000
Surgery	1	15.863	2.167	.161
Reason	1	259.894	9.667	.002
Surgery * Reason	1	217.087	6.278	.013

a. Dependent Variable: Quality of Life After Cosmetic Surgery.

bs

Estimates of Fixed Effects[a]

Parameter	Estimate	Std. Error	df	t	Sig.	95% Confidence Interval Lower Bound	Upper Bound
Intercept	42.517820	3.840055	108.853	11.072	.000	34.906839	50.128800
Base_QoL	.305356	.052642	268.920	5.801	.000	.201713	.408999
Surgery	-3.187677	2.165484	15.863	-1.472	.161	-7.781510	1.406157
Reason	-3.515148	1.130552	259.894	-3.109	.002	-5.741357	-1.288939
Surgery * Reason	4.221288	1.684798	217.087	2.506	.013	.900633	7.541944

a. Dependent Variable: Quality of Life After Cosmetic Surgery.

$Var(u_{0j})$ = Variance of intercepts

$Var(\varepsilon_{ij})$ = Variance of residuals

Estimates of Covariance Parameters[a]

Parameter		Estimate	Std. Error	Wald Z	Sig.	95% Confidence Interval Lower Bound	Upper Bound
Residual		33.859719	3.024395	11.196	.000	28.421886	40.337948
Intercept + Surgery [subject = Clinic]	UN (1,1)	30.056340	15.444593	1.946	.052	10.978478	82.286775
	UN (2,1)	-28.083657	15.195713	-1.848	.065	-57.866706	1.699393
	UN (2,2)	29.349323	16.404492	1.789	.074	9.813593	87.774453

a. Dependent Variable: Quality of Life After Cosmetic Surgery.

$Cov(u_{0j}, u_{1j})$ = Covariance between intercepts and Slopes

$Var(u_{1j})$ = Variance of Slopes

SPSS Output 19.9 shows the parameter estimates from these analyses. It shows that for those operated on only to change their appearance, surgery almost significantly predicted quality of life after surgery, $b = -4.31$, $t(7.72) = -1.92$, $p = .09$. The negative gradient shows that in these people, quality of life was lower after surgery compared to the control group. However, for those that had surgery to solve a physical problem surgery did not significantly predict quality of life, $b = 1.20$, $t(7.61) = 0.58$, $p = .58$. However, the slope was positive indicating that people who had surgery scored higher on quality of life than

SPSS OUTPUT 19.9 Surgery to Change Appearance:

Estimates of Fixed Effects[a,b]

Parameter	Estimate	Std. Error	df	t	Sig.	95% Confidence Interval Lower Bound	Upper Bound
Intercept	41.786055	5.487873	77.331	7.614	.000	30.859052	52.713059
Base_QoL	.338492	.079035	88.619	4.283	.000	.181440	.495543
Surgery	-4.307014	2.239912	7.719	-1.923	.092	-9.505157	.891130

a. Reason for Surgery = Change Appearance

b. Dependent Variable: Quality of Life After Cosmetic Surgery.

Surgery for a Physical Problem:

Estimates of Fixed Effects[a,b]

Parameter	Estimate	Std. Error	df	t	Sig.	95% Confidence Interval Lower Bound	Upper Bound
Intercept	38.020790	4.666154	93.558	8.148	.000	28.755460	47.286119
Base_QoL	.317710	.068883	172.816	4.612	.000	.181749	.453670
Surgery	1.196550	2.081999	7.614	.575	.582	-3.647282	6.040382

a. Reason for Surgery = Physical reason

b. Dependent Variable: Quality of Life After Cosmetic Surgery.

those on the waiting list (although not significantly so!). The interaction effect, therefore, reflects the difference in slopes for surgery as a predictor of quality of life in those that had surgery for physical problems (slight positive slope) and those that had surgery purely for vanity (a negative slope).

We could sum up these results by saying that quality of life after surgery, after controlling for quality of life before surgery, was lower for those that had surgery to change their appearance than those that had surgery for a physical reason. This makes sense because for those having surgery to correct a physical problem, the surgery has probably bought relief and so their quality of life will improve. However, for those having surgery for vanity they might well discover that having a different appearance wasn't actually at the root of their unhappiness, so their quality of life is lower.

CRAMMING SAM'S TIPS Multilevel models SPSS output

- The **Information Criteria** table can be used to assess the overall fit of the model. The −2LL can be significance tested with *df* = the number of parameters being estimated. It is mainly used, though, to compare models that are the same in all but one parameter by testing the difference in −2LL in the two models against df = 1 (if only one parameter has been changed). The AIC, AICC, CAIC and BIC can also be compared across models (but not significance tested).
- The table of **Type III Tests of Fixed Effects** tells you whether your predictors significantly predict the outcome: look in the column labelled *Sig.* If the value is less than .05 then the effect is significant.
- The table of **Estimates of Fixed Effects** gives us the regression coefficient for each effect and its confidence interval. The direction of these coefficients tells us whether the relationship between each predictor and the outcome is positive or negative.
- The table labelled **Estimates of Covariance Parameters** tells us about any random effects in the model. These values can tell us how much intercepts and slopes varied over our level 1 variable. The significance of these estimates should be treated cautiously. The exact labelling of these effects depends on which covariance structure you selected for the analysis.

19.7. Growth models ④

Growth models are extremely important in many areas of science including psychology, medicine, physics, chemistry or economics. In a growth model the aim is to look at the rate of change of a variable over time: for example, we could look at white blood cell counts, attitudes, radioactive decay or profits. In all cases we're trying to see which model best describes the change over time.

19.7.1. Growth curves (polynomials) ④

Figure 19.20 gives some examples of possible **growth curves**. This diagram shows three **polynomials** representing a linear trend (the blue line) otherwise known as a first-order polynomial, a quadratic trend (the green line) otherwise known as a second-order polynomial, and a cubic trend (the red line) otherwise known as a third-order polynomial. Notice first that the linear trend is a straight line, but as the polynomials increase they get more and more curved, indicating more rapid growth over time. Also, as polynomials increase, the change in the curve is quite dramatic (so dramatic that I adjusted the scale of the graph to fit all three curves on the same diagram). This observation highlights the fact that any growth curve higher than a quadratic (or possibly cubic) trend is very unrealistic in real data. By fitting a growth model to the data we can see which trend best describes the growth of an outcome variable over time (although, no one will believe that a significant fifth-order polynomial is telling us anything meaningful about the real world!).

The growth curves that we have described might seem familiar to you: they are the same as the trends that we described for ordered means in section 10.2.11.5. What we are discussing now is really no different. There are just two important things to remember when fitting growth curves: (1) you can fit polynomials up to one less than the number of time points that you have; and (2) a polynomial is defined by a simple power function. On the first point, this means that with three time points you can fit a linear and quadratic growth curve (or a first- and second-order polynomial), but you cannot fit any higher-order growth curves. Similarly, if you have six time points you can fit up to a fifth-order polynomial. This is the same basic idea as having one less contrast than the number of groups in ANOVA (see section 10.2.11).

On the second point, we have to define growth curves manually in multilevel models in SPSS: there is not a convenient option that we can select to do it for us. However, this is quite easy to do. If *time* is our predictor variable, then a linear trend is tested by including this variable alone. A quadratic or second-order polynomial is tested by including a predictor that is $time^2$, a cubic or third-order polynomial is tested by including a predictor that is $time^3$ and so on. So any polynomial is tested by including a variable that is the predictor to the power of the order of polynomial that you want to test: for a fifth-order polynomial we need a predictor of $time^5$ and for an n-order polynomial we would have to include $time^n$ as a predictor. Hopefully you get the general idea.

19.7.2. An example: the honeymoon period ②

I recently heard a brilliant talk given by Professor Daniel Kahneman, who won the 2002 Nobel Prize for Economics. In this talk Kahneman brought together an enormous amount of research on life satisfaction (he explored questions such as whether people are happier if they are richer). There was one graph in this talk that particularly grabbed my attention. It showed

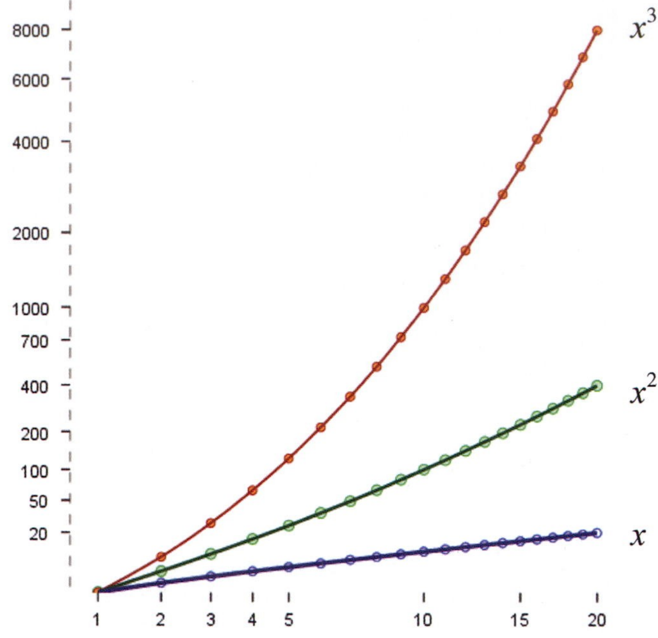

FIGURE 19.20 Illustration of a first-order (linear, blue), second-order (quadratic, green) and third-order (cubic, red) polynomial

that leading up to marriage people reported greater life satisfaction, but by about two years after marriage this life satisfaction decreased back to its baseline level. This graph perfectly illustrated what people talk about as the 'honeymoon period': a new relationship/marriage is great at first (no matter how ill suited you may be) but after six months or so the cracks start to appear and everything turns to elephant dung. Kahneman argued that people adapt to marriage; it does not make them happier in the long run (Kahneman & Krueger, 2006).[6] This got me thinking about relationships not involving marriage (is it marriage that makes you happy, or just being in a long-term relationship?). Therefore, in a completely fictitious parallel world where I don't research child anxiety, but instead concern myself with people's life satisfaction, I collected some data. I organized a massive speed-dating event (see Chapter 14). At the start of the night I measured everyone's life satisfaction (**Satisfaction_Baseline**) on a 10-point scale (0 = completely dissatisfied, 10 = completely satisfied) and their gender (**Gender**). After the speed dating I noted all of the people who had found dates. If they ended up in a relationship with the person that they met on the speed-dating night then I stalked these people over the next 18 months of that relationship. As such, I had measures of their life satisfaction at 6 months (**Satisfaction_6_Months**), 12 months (**Satisfaction_12_Months**) and 18 months (**Satisfaction_18_Months**), after they entered the relationship. None of the people measured were in the same relationship (i.e. I measured only life satisfaction from one of the people in the couple).[7] Also, as is often the case with longitudinal data, I didn't have scores for all people at all time points because not everyone was available at the follow-up sessions. One of the benefits of a multilevel approach is that these missing data do not pose a particular problem. The data are in the file **Honeymoon Period.sav**.

Figure 19.21 shows the data. Each circle is a data point and the line shows the average life satisfaction over time. Basically, from baseline, life satisfaction rises slightly at time 2 (6 months) but then starts to decrease over the next 12 months. There are two things to note about the data. First, time 0 is before the people enter into their new relationship yet

[6] The romantics among you might be relieved to know that others have used the same data to argue the complete opposite: that married people are happier than non-married people in the long term (Easterlin, 2003).

[7] However, I could have measured both people in the couple because using a multilevel model I could have treated people as being nested within 'couples' to take account of the dependency in their data.

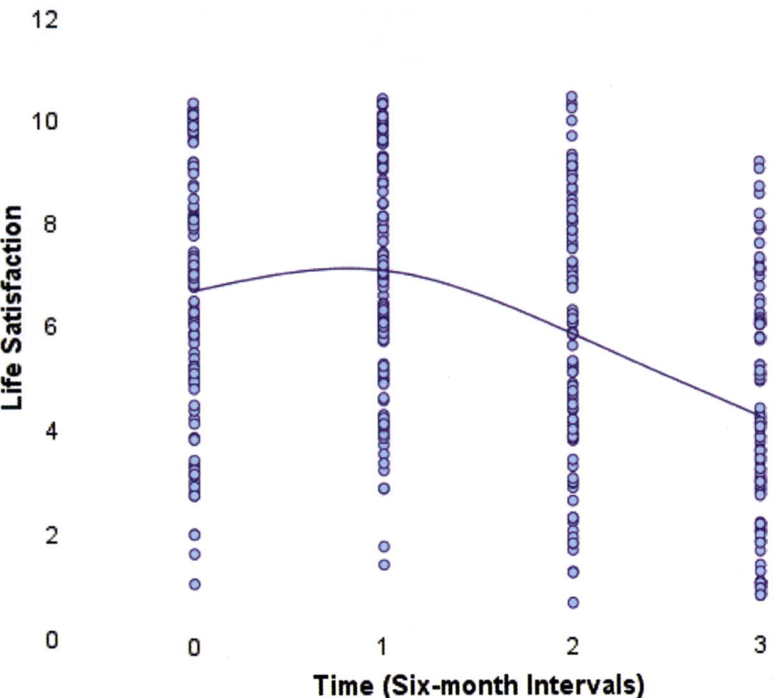

FIGURE 19.21 Life satisfaction over time

already there is a lot of variability in their responses (reflecting the fact that people will vary in their satisfaction due to other reasons such as finances, personality and so on). This suggests that intercepts for life satisfaction differ across people. Second, there is also a lot of variability in life satisfaction after the relationship has started (time 1) and at all subsequent time points, which suggests that the slope of the relationship between time and life satisfaction might vary across people also. If we think of the time points as a level 1 variable that is nested with people (a level 2 variable) then we can easily model this variability in intercepts and slopes within people. We have a situation similar to Figure 19.4 (except with two levels instead of three, although we could add in the location of the speed dating event as a level 3 variable if we had that information!).

19.7.3. Restructuring the data ③

The first problem with having data measured over time is that to do a multilevel model the data need to be in a different format to what we are used to. Figure 19.22 shows how we would normally set up the data editor for a repeated-measures design: each row represents a person, and notice that the repeated-measures variable of time is represented by four different columns. If we were going to run an ordinary repeated-measures ANOVA this data layout would be fine; however, for a multilevel model we need the variable **Time** to be represented by a single column. We could enter all of the data again, but that would be a pain; luckily we don't have to do this because SPSS has a *restructure* command, which is also a pain, but not as much as retyping the data. This command enables you to take your data set and create a new data set that is organized differently.

To access the restructure wizard select Data Restructure.... The steps in the wizard are shown in Figure 19.23. In the first dialog box you need to say whether you are converting variables to

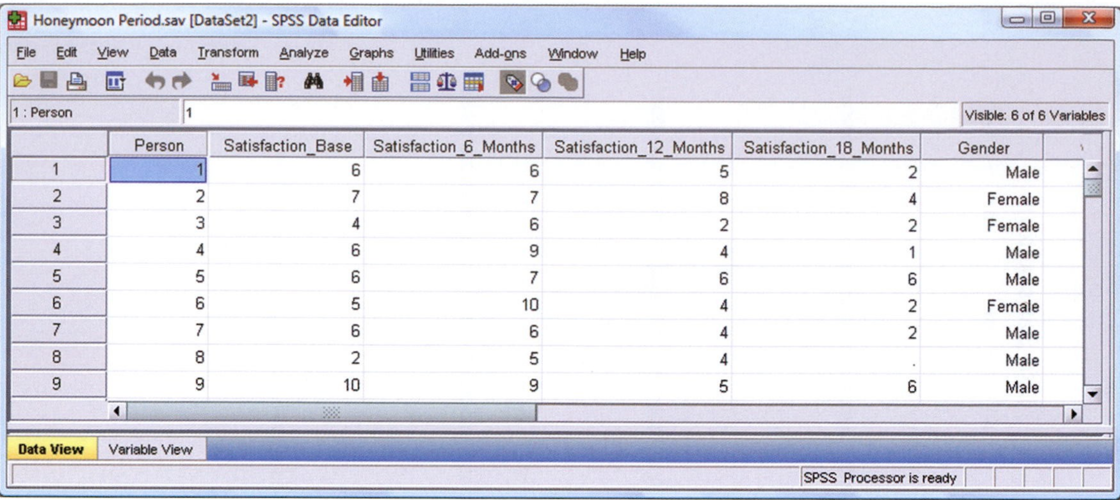

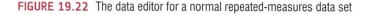

FIGURE 19.22 The data editor for a normal repeated-measures data set

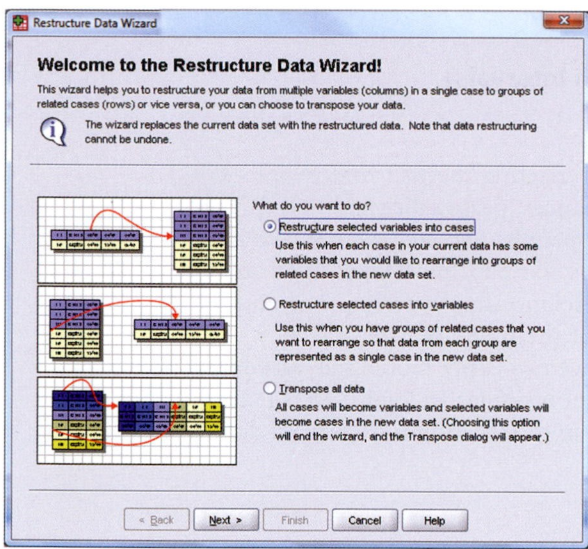

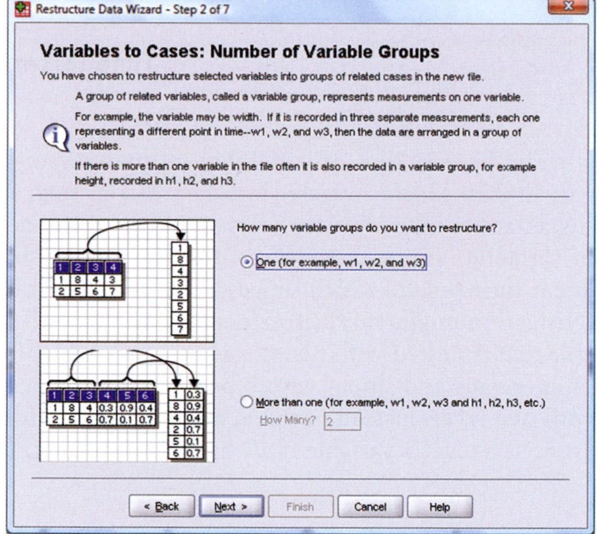

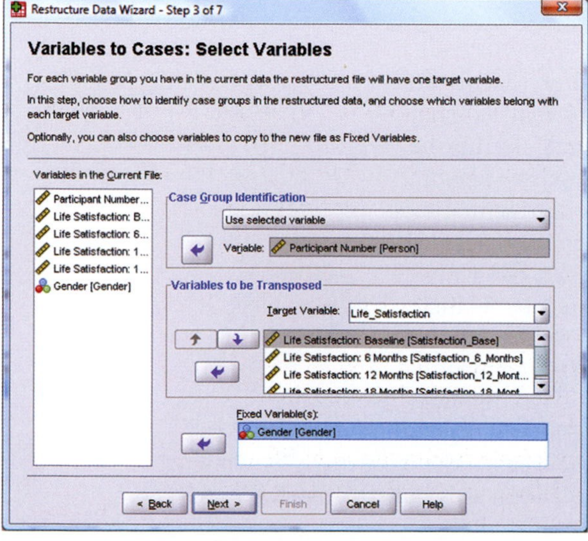

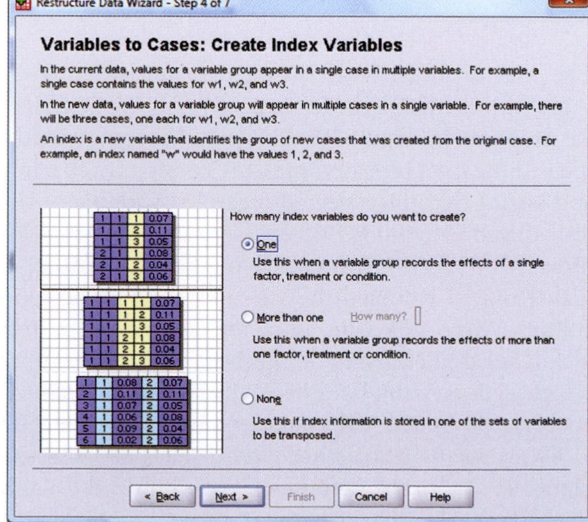

FIGURE 19.23 *Continued*

Restructure Data Wizard - Step 5 of 7

Variables to Cases: Create One Index Variable

You have chosen to create one index variable. The variable's values can be sequential numbers or the names of variables in a group.

In the table you can specify the name and label for the index variable.

What kind of index values?

Sequential numbers
Index Values: 1, 2, 3, 4

Variable names
Index Values: Satisfaction_Base, Satisfaction_6_Months, Satisfaction_12_Months, Satisfa...

Edit the Index Variable Name and Label:

	Name	Label	Levels	Index Values
1	Time	Time	4	1, 2, 3, 4

< Back | Next > | Finish | Cancel | Help

Restructure Data Wizard - Step 6 of 7

Variables to Cases: Options

In this step you can set options that will be applied to the restructured data file.

Handling of Variables not Selected

Drop variable(s) from the new data file

Keep and treat as fixed variable(s)

System Missing or Blank Values in all Transposed Variables

Create a case in the new file

Discard the data

Case Count Variable

Count the number of new cases created by the case in the current data

Name:

Label:

< Back | Next > | Finish | Cancel | Help

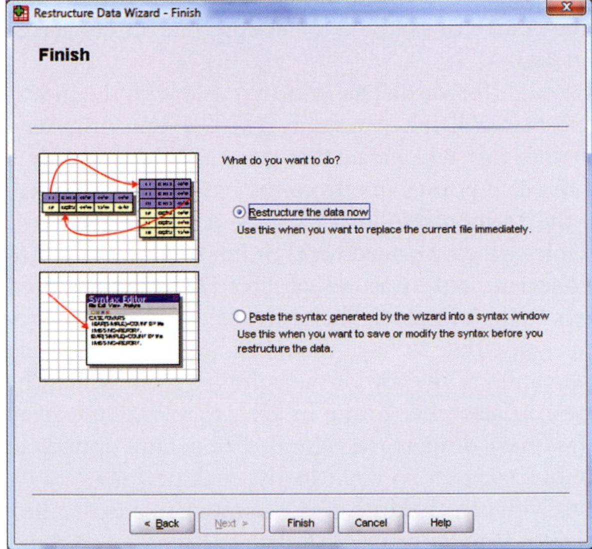

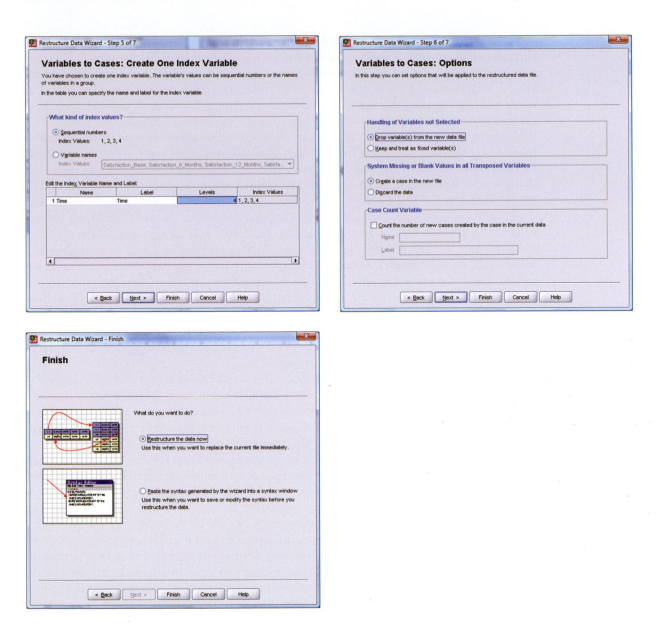

FIGURE 19.23 The data restructure wizard

cases, or cases to variables. We have different levels of time in different columns (variables), and we want them to be in different rows (cases), so we need to select Restructure selected variables into cases. Click on Next > to move to the next dialog box. This dialog box asks you whether you are creating just one new variable in your new data file from different columns in the old data file, or whether you want to create more than one new variable. In our case we are going to create one variable representing life satisfaction; therefore, select One (for example, w1, w2, and w3). When you have done this click on Next >. The next dialog box is crucial because it's where you set up the new data file. By default, SPSS creates a variable in your new data file called **id** which tells you from which person the data came (i.e. which row of the original data file). It does this by using the case numbers in the original data file. This default is fine, but if you want to change it (or the name **id**) then go to the section labelled *Case Group Identification* and change Use case number to be Use selected variable and then select a variable from your data file to act as a label in the new data file. For example, in the diagram I have chosen the variable **Person** from the original data set to identify participants in the new data file.

In the section labelled *Variables to be Transposed* there is a drop-down list labelled *Target Variable* which should contain an item labelled (rather unimaginatively) *trans1*. There is one item because we specified that we wanted one new variable in the previous dialog box (if we had asked for more than one new variable this drop-down list would contain as many items as variables that we requested). We can change the name of *trans1* by selecting the variables in the drop-down list and then editing their names. I suggest that you rename the variable *Life_Satisfaction*. We then need to tell SPSS which columns are associated with these two variables. Select from the list labelled *Variables in the Current File* the four variables that represent the different time points at which life satisfaction was measured (**Satisfaction_Baseline, Satisfaction_6_Months, Satisfaction_12_Months, Satisfaction_18_ Months**). If you hold down the *Ctrl* key then you can select all four variables and either drag them across or click on ➡. It's important that you select the variables in the correct order: SPSS assumes that the first variable that it encounters is the first level of the repeated measure, and the second variable is the second level and so on. Once the variables are transferred, you can reorder them by using ⬆ or ⬇ to move selected variables up or down the list. Finally, there is a space to select *Fixed Variable(s)*. Drag variables here that do not vary at level 1 of your hierarchy. In this example, this means that we can select variables that are different in different people (they vary at level 2) but have the same value at the different time points (they do not vary at level 1). The only variable that we have like this is **Gender**, which did not change over the course of the data collection, but differs across people. When you have finished click on Next ».

The remaining dialog boxes are fairly straightforward. The next two deal with the indexing variable. SPSS creates a new variable that will tell you from which column the data originate. In our case with four time points this will mean that the variable is simply a sequence of numbers from 1 to 4. So, if the data point came from the baseline phase it will be assigned a 1, whereas it if came from the 18-month follow-up phase it will be given a 4. You can select to have a single index variable (which we need here) or not to have one, or to have several. We have restructured only one repeated-measures variable (**Time**), so we need only one index variable to represent levels of this variable. Therefore, select ⦿ One and click on Next ». In the next dialog box you can opt either to have the index variable containing numbers (such as 1, 2, 3, 4) or to use the names at the top of the columns from which the data came. The choice is up to you. You can also change the index variables name from *Index* to something useful such as Time (as I have done in the figure). The default options in the remaining dialog boxes are fine for most purposes so you can just click on Finish to do the restructuring. However, if you want to, you can click on Next » to move to a dialog box that enables you to opt to keep any variables from your original data file that you haven't explicitly specified in the previous dialog boxes. The default is to drop them, and that's fine (any variables from the original data file that you want in the new data file should probably be specified earlier on in the wizard). You can also choose to keep or discard missing data. Again, the default option to keep missing data is advisable here because multilevel models can deal with missing values. Click on Next » to move on to the final dialog box. This box gives you the option to restructure the data (the default) or to paste the syntax into a syntax window so that you can save it as a syntax file. If you're likely to do similar data restructuring then saving the syntax might be useful, but once you have got used to the windows it doesn't take long to restructure new data anyway. Click on Finish to restructure the data.

The restructured data are shown in Figure 19.24; it's useful to compare the restructured data with the old data file in Figure 19.22. Notice that each person is now represented by four rows (one for each time point) and that variables such as gender that are invariant over the time points have the same value within each person. However, our outcome variable (life satisfaction) does change over the four time points (the four rows for each person).

There is only one other thing left to do. In your data set you'll notice that the time points have values from 1 to 4. However, it's useful to centre this variable at 0 because our initial life satisfaction was measured before the new relationship. Therefore, an intercept of 0 is

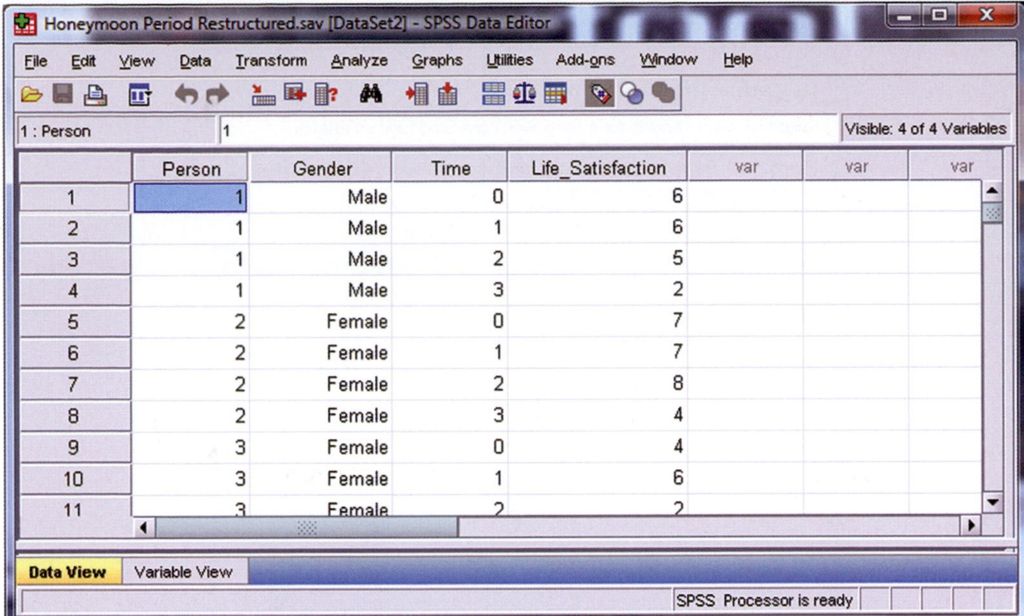

FIGURE 19.24 Data entry for a repeated-measures multilevel model

meaningful for these data: it is the value of life satisfaction when not in a relationship. By centring the scores around a baseline value of 0 we can interpret the intercept much more easily and intuitively. The easiest way to change the values is using the *compute* command to recompute **Time** to be **Time** –1. This will change the values from 1–4 to 0–3.

SELF-TEST Use the *compute* command to transform **Time** into **Time** minus 1.

19.7.4. Running a growth model on SPSS ④

Now that we have our data set up, we can run the analysis. Essentially, we can set up this analysis in a very similar way to the previous example. First, select Analyze Mixed Models ▸ Linear... and in the initial dialog box set up the level 2 variable. In this example, life satisfaction at multiple time points is nested within people. Therefore, the level 2 variable is the person and this variable is represented by the variable labelled **Person**. Select this variable and drag it to the box labelled *Subjects* (or click on ➔), see Figure 19.25. Click on Continue to access the main dialog box.

In the main dialog box we need to set up our predictors and outcome. The outcome was life satisfaction, so select **Life_Satisfaction** and drag it to the box labelled *Dependent Variable* (or click on ➔). Our predictor, or growth variable, is **Time** so select this variable and drag it to the box labelled *Covariate(s)*, or click on ➔, see Figure 19.26.

We need to add the potential growth curves that we want to test as fixed effects to our model, so click on Fixed... to bring up the fixed effects dialog box (Figure 19.27). In section 19.7.1 we discussed different growth curves. With four time points we can fit up to a third-order polynomial. One way to do this would be to start with just the linear effect (**Time**), then run a new

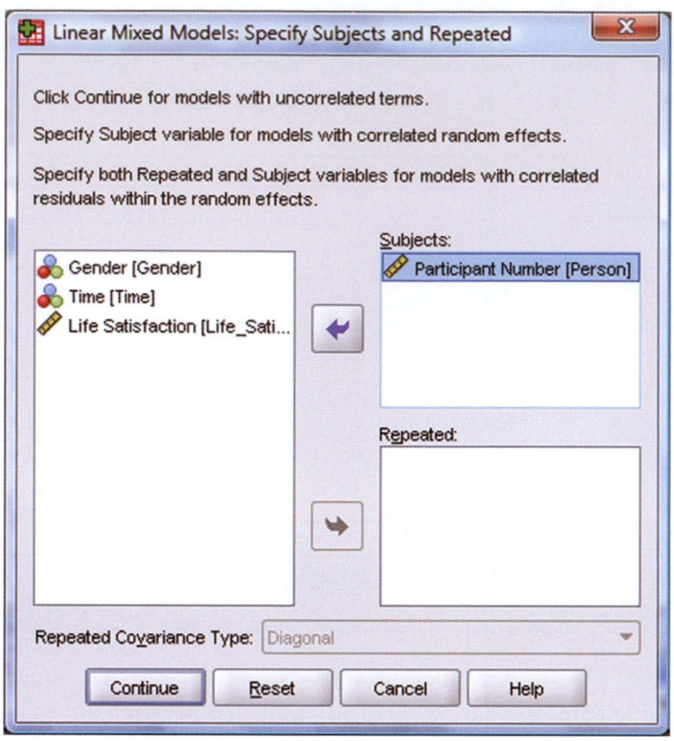

FIGURE 19.25 Setting up the level 2 variable in a growth model

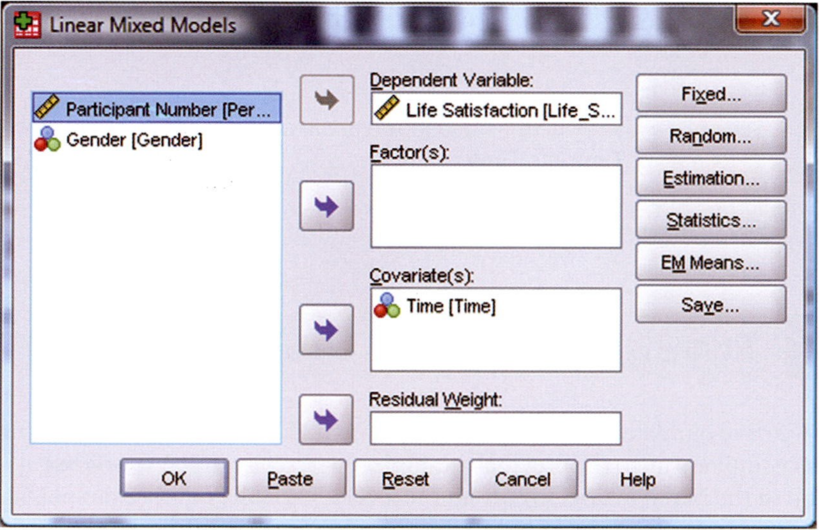

FIGURE 19.26 Setting up the outcome variable and predictor in a multilevel growth model

model with the linear and quadratic (**Time**²) polynomials to see if the quadratic trend improves the model. Finally, run a third model with the linear, quadratic and cubic (**Time**³) polynomial in, and see if the cubic trend adds to the model. So, basically, we add in polynomials one at a time and assess the change in the −2LL. To specify the linear polynomial click on **Time** and then click Add to add it into the model. Click on Continue to return to the main dialog box.

I mentioned earlier on that we expected the relationship between time and life satisfaction to have both a random intercept and a random slope. We need to define these parameters now by clicking on Random... in the main dialog box to access the dialog box in Figure 19.28. The first thing we need to do is to specify our contextual variable. We do this by selecting it from the list

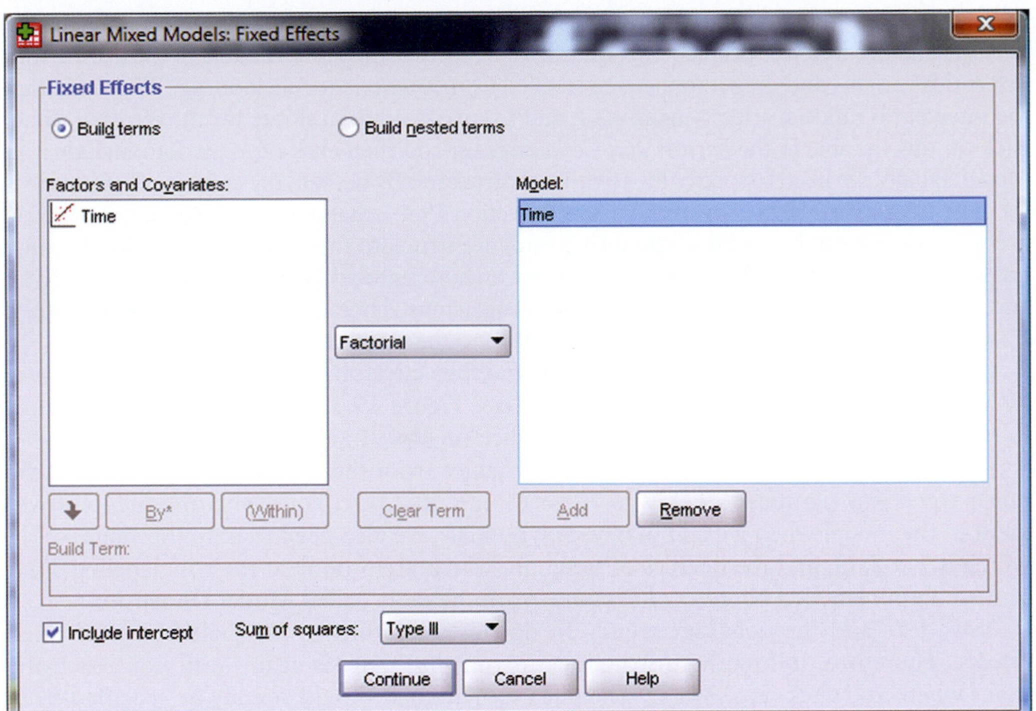

FIGURE 19.27 Setting up the linear polynomial

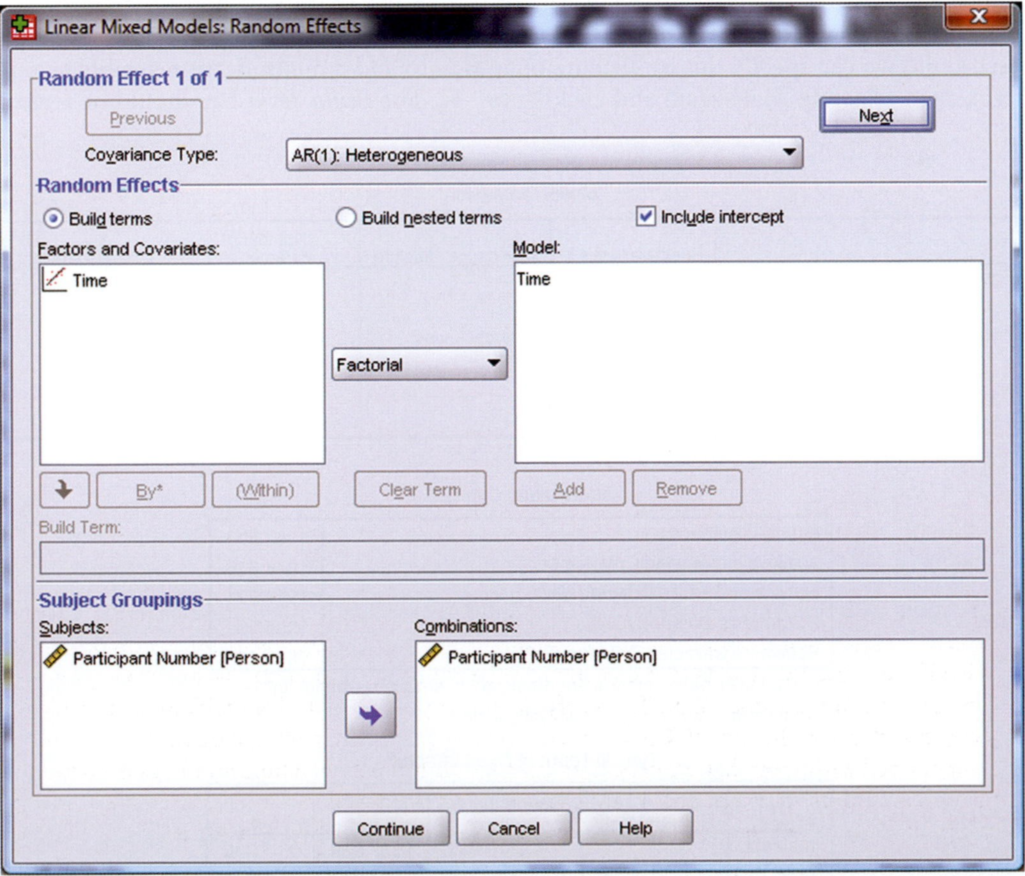

FIGURE 19.28 Defining a random intercept and random slopes in a growth model

of contextual variables that we have told SPSS about already. These appear in the section labelled *Subjects* and because we specified only one variable, there is only one variable in the list, **Person.** Select this variable and drag it to the area labelled *Combinations* (or click on ➜). To specify that the intercept is random select ☑ Include intercept, and to specify random slopes for the effect of **Time,** click on this variable in the *Factors and Covariates* list and then click on Add to include it in *Model*. Finally, we need to specify the covariance structure. By default, the covariance structure is set to be Variance Components ▾. However, we saw in section 19.4.2 that when we have repeated measures over time it can be useful to specify a covariance structure that assumes that scores become less correlated over time. Therefore, let's choose an autoregressive covariance structure, AR(1), and let's also assume that variances will be heterogeneous. Therefore, select AR(1): Heterogeneous ▾ from the drop-down list. Click on Continue to return to the main dialog box.

Click on Estimation... and select ⊙ Maximum Likelihood (ML) and then click on Statistics... and select *Parameter estimates* and *Tests for covariance parameters* (see Figure 19.11). Click on Continue to return to the main dialog box. To run the analysis, click on OK.

SPSS Output 19.10 shows the preliminary tables from the output. We can see that the linear trend was significant, $F(1, 106.72) = 134.26$, $p < .001$. For evaluating the improvement in the model when we add in new polynomials, we also need to note the value −2LL, which is 1862.63, and the degrees of freedom, which are 6 (look at the row labelled *Total* in the column labelled *Number of Parameters*, in the table called **Model Dimension**).

Now, let's add the quadratic trend. To do this we return to the dialog box for fixed effects. Therefore, follow the instructions to run this analysis again until you reach the point where you click on Fixed... . The linear polynomial should already be specified from before (if not, then click on **Time** and then click on Add to add it into the model) and the dialog box will look like Figure 19.27. To add the higher-order polynomials we need to select ⊙ Build nested terms. Select **Time** in the *Factors and Covariates* list and ⬇ will become active; click on this button and **Time** will appear in the space labelled *Build Term*. For the quadratic or second-order polynomial we need to define **Time**^{2} (**Time** multiplied by itself), and we can specify this by clicking on By* to add a multiplication symbol to our term, then selecting **Time** again and clicking on ⬇. The *Build Term* bar should now read

SPSS OUTPUT 19.10

Model Dimension[a]

		Number of Levels	Covariance Structure	Number of Parameters	Subject Variables
Fixed Effects	Intercept	1		1	
	Time	1		1	
Random Effects	Intercept + Time	2	Heterogeneous First-Order Autoregressive	3	Person
Residual				1	
Total		4		6	

a. Dependent Variable: Life Satisfaction.

Information Criteria[a]

-2 Log Likelihood	1862.626
Akaike's Information Criterion (AIC)	1874.626
Hurvich and Tsai's Criterion (AICC)	1874.821
Bozdogan's Criterion (CAIC)	1905.119
Schwarz's Bayesian Criterion (BIC)	1899.119

The information criteria are displayed in smaller-is-better forms.

a. Dependent Variable: Life Satisfaction.

Type III Tests of Fixed Effects[a]

Source	Numerator df	Denominator df	F	Sig.
Intercept	1	113.653	1137.088	.000
Time	1	106.715	134.264	.000

a. Dependent Variable: Life Satisfaction.

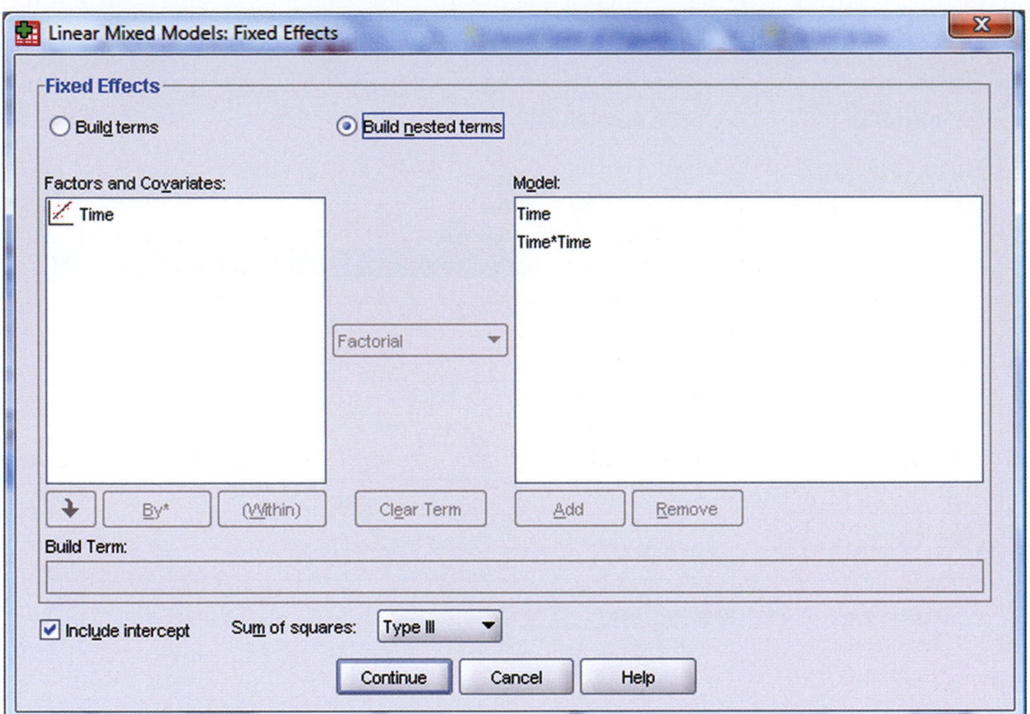

FIGURE 19.29 Specifying a linear trend (*Time*) and a quadratic trend (*Time*Time*)

*Time*Time* (or, put another way, **Time²**). This term is the second-order polynomial, and we click on Add to put it into the model. Click on Continue to return to the main dialog box and click on OK to rerun the analysis.

The output will now include the quadratic polynomial. To see whether this quadratic trend has improved the model we need to compare the −2LL for this new model, to the value when only the linear polynomial was included. The value of −2LL is shown in SPSS Output 19.11, and it is 1802.03. We have added only one term to the model so the new degrees of freedom will have risen by 1, from 6 to 7 (you can check that the new degrees of freedom are 7 in the row labelled *Total* in the column labelled *Number of Parameters*, in the table called **Model Dimension**). We can compute the change in −2LL as a result of the quadratic term by subtracting the −2LL for this model from the −2LL for the model with only the linear trend:

$$\chi^2_{\text{Change}} = 1862.63 - 1802.03 = 60.60$$
$$df_{\text{Change}} = 7 - 6 = 1$$

If we look at the critical values for the chi-square statistic for $df = 1$ in Appendix A4, they are 3.84 ($p < .05$) and 6.63 ($p < .01$); therefore, this change is highly significant because 60.60 is bigger than these values.

SPSS OUTPUT 19.11

Information Criteria[a]

-2 Log Likelihood	1802.026
Akaike's Information Criterion (AIC)	1816.026
Hurvich and Tsai's Criterion (AICC)	1816.287
Bozdogan's Criterion (CAIC)	1851.602
Schwarz's Bayesian Criterion (BIC)	1844.602

The information criteria are displayed in smaller-is-better forms.

a. Dependent Variable: Life Satisfaction.

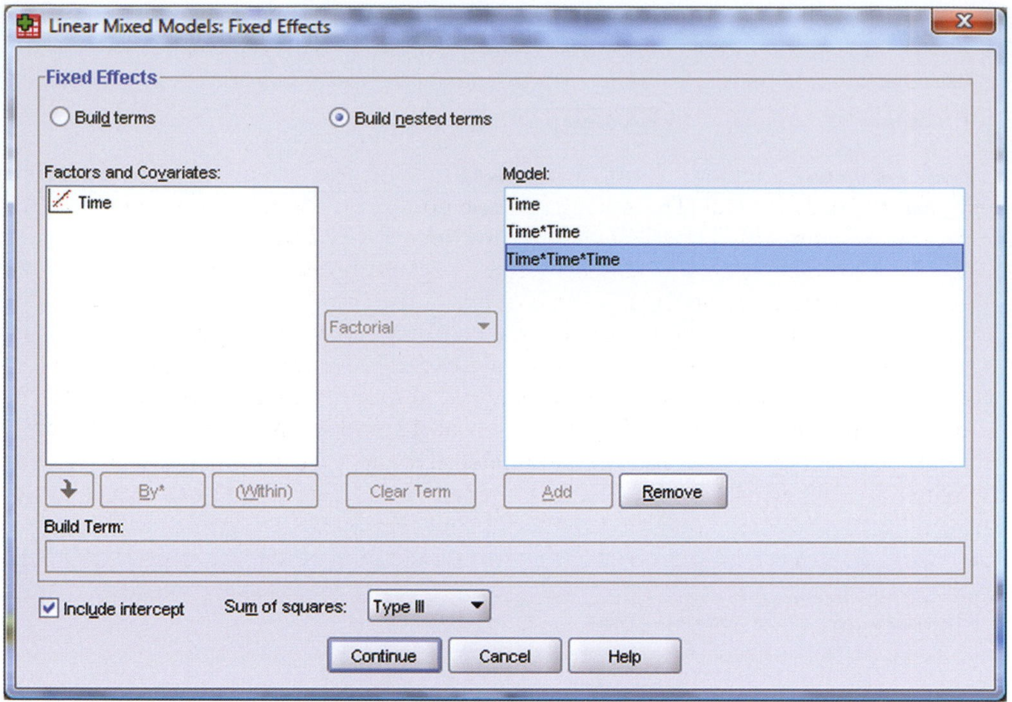

FIGURE 19.30 Specifying linear (*Time*), quadratic (*Time*Time*) and cubic (*Time*Time*Time*) trends

Finally, let's add the cubic trend. To do this we return to the dialog box for fixed effects. As for the quadratic trend, follow the instructions to run this analysis until you reach the point where you click on Fixed... . The linear and quadratic polynomials should already be specified from before and the dialog box will look like Figure 19.29. As for the quadratic trend, make sure Build nested terms is selected. Then, select **Time** in the *Factors and Covariates* list and ↓ will become active; click on this button and **Time** will appear in the space labelled *Build Term*. For the cubic or third-order polynomial we need to define **Time**^{3} (or *Time*Time*Time*). We build this term up in the same way as for the quadratic polynomial. Select **Time**, click on ↓, click on By* , select **Time** again, click on ↓, click on By* again, select **Time** for a third time, click on ↓ and finally click on Add . This should add the third-order polynomial (or *Time*Time*Time*) to the model,[8] see Figure 19.30. Click on Continue to return to the main dialog box and click on OK to rerun the analysis.

The output will now include the cubic polynomial. To see whether this cubic trend has improved the model we again compare the −2LL for this new model to the value in the previous model. The value of −2LL is shown in SPSS Output 19.12, and it is 1798.86. We have added only one term to the model so the new degrees of freedom will have risen by 1, from 7 to 8 (again you can find the value of 8 in the row labelled *Total* in the column labelled *Number of Parameters*, in the table called **Model Dimension**). We can compute the change in −2LL as a result of the cubic-term by subtracting the −2LL for this model from the −2LL for the model with only the linear trend:

$$\chi^2_{\text{Change}} = 1802.03 - 1798.86 = 3.17$$

$$df_{\text{Change}} = 8 - 7 = 1$$

[8] Should you ever want even high-order polynomials (notwithstanding my remark about them having little real-world relevance) then you can extrapolate from what I have told you about the other polynomials; for example, for a fourth-order polynomial you go through the whole process again, but this time creating **Time**^{4} (or *Time*Time*Time*Time*), and for the fifth-order polynomial you create **Time**^{5} (or *Time*Time*Time*Time*Time*).

Using the same critical values for the chi-square statistic as before, we can conclude that this change is not significant because 3.17 is less than the critical value of 3.84.

We will look at the SPSS output for this final model in a little more detail (SPSS Output 19.12). First, we are given the fit indices (the −2LL, AIC, AICC, CAIC and BIC). As we have seen, these are useful mainly for comparing models, so we have used the log-likelihood, for example, to test whether the addition of a polynomial significantly affects the fit of the model. The main part of the output is the table of fixed effects and the parameter estimates. These tell us that the linear, $F(1, 221.39) = 10.01$, $p < .01$, and quadratic, $F(1, 212.49) = 9.41$, $p < .01$, trends both significantly described the pattern of the data over time; however, the cubic trend, $F(1, 214.37) = 3.19$, $p > .05$, does not. This confirms what we already know from comparing the fit of successive models. The trend in the data is best described by a second-order polynomial, or a quadratic trend. This reflects the initial increase in life satisfaction 6 months after finding a new partner but a subsequent reduction in life satisfaction at 12 and 18 months after the start of the relationship (Figure 19.21). The parameter estimates tell us much the same thing. It's worth remembering that this quadratic trend is only an *approximation*: if it were completely accurate then we would predict from the model that couples who had been together for 10 years would have negative life satisfaction, which is impossible given the scale we used to measure it.

SPSS OUTPUT 19.12

Information Criteria[a]

-2 Log Likelihood	1798.857
Akaike's Information Criterion (AIC)	1814.857
Hurvich and Tsai's Criterion (AICC)	1815.193
Bozdogan's Criterion (CAIC)	1855.515
Schwarz's Bayesian Criterion (BIC)	1847.515

The information criteria are displayed in smaller-is-better forms.

a. Dependent Variable: Life Satisfaction.

Type III Tests of Fixed Effects[a]

Source	Numerator df	Denominator df	F	Sig.
Intercept	1	137.405	884.469	.000
Time	1	221.392	10.009	.002
Time * Time	1	212.488	9.408	.002
Time * Time * Time	1	214.371	3.187	.076

a. Dependent Variable: Life Satisfaction.

Estimates of Fixed Effects[a]

Parameter	Estimate	Std. Error	df	t	Sig.	95% Confidence Interval Lower Bound	95% Confidence Interval Upper Bound
Intercept	6.634783	.223093	137.405	29.740	.000	6.193644	7.075921
Time	1.544663	.488236	221.392	3.164	.002	.582478	2.506849
Time * Time	-1.323625	.431531	212.488	-3.067	.002	-2.174256	-.472995
Time * Time * Time	.170266	.095374	214.371	1.785	.076	-.017724	.358257

a. Dependent Variable: Life Satisfaction.

Estimates of Covariance Parameters[a]

Parameter		Estimate	Std. Error	Wald Z	Sig.	95% Confidence Interval Lower Bound	95% Confidence Interval Upper Bound
Residual		1.834247	.178865	10.255	.000	1.515144	2.220556
Intercept + Time [subject = Person]	Var: Intercept	3.889341	.700207	5.555	.000	2.732954	5.535028
	Var: Time	.244426	.096858	2.524	.012	.112420	.531437
	ARH1 rho	-.382572	.151351	-2.528	.011	-.635491	-.055507

a. Dependent Variable: Life Satisfaction.

The final part of the output tells us about the random parameters in the model. First of all, the variance of the random intercepts was Var(u_{0j}) = 3.89. This suggests that we were correct to assume that life satisfaction at baseline varied significantly across people. Also, the variance of the people's slopes varied significantly Var(u_{1j}) = 0.24. This suggests also that the change in life satisfaction over time varied significantly across people too. Finally, the covariance between the slopes and intercepts (–0.38) suggests that as intercepts increased, the slope decreased. (Ideally, all of these terms should have been added in individually so that we could calculate the chi-square statistic for the change in the –2LL for each of them.)

19.7.5. Further analysis ④

It's worth pointing out that I've kept this growth curve analysis simple to give you the basic tools. In the example I allowed only the linear term to have a random intercept and slopes, but given that we discovered that a second-order polynomial described the change in responses, we could redo the analysis and allow random intercepts and slopes for the second-order polynomial also. To do these we would just have to specify these terms in Figure 19.28 in much the same way as we set them up as fixed effects in Figure 19.29. If we were to do this it would make sense to add the random components one at a time and test whether they have a significant impact on the model by comparing the log-likelihood values or other fit indices.

Also, the polynomials I have described are not the only ones that can be used. You could test for a logarithmic trend over time, or even an exponential one.

CRAMMING SAM'S TIPS Growth models

- Growth models are multilevel models in which changes in an outcome over time are modelled using potential growth patterns.
- These growth patterns can be linear, quadratic, cubic, logarithmic, exponential, or anything you like really.
- The hierarchy in the data is that time points are nested within people (or other entities). As such, it's a way of analysing repeated-measures data that have a hierarchical structure.
- The **Information Criteria** table can be used to assess the overall fit of the model. The –2LL can be significance tested with *df* = the number of parameters being estimated. It is mainly used, though, to compare models that are the same in all but one parameter by testing the difference in –2LL in the two models against *df* = 1 (if only one parameter has been changed). The AIC, AICC, CAIC and BIC can also be compared across models (but not significance tested).
- The table of **Type III Tests of Fixed Effects** tells you whether the growth functions that you have entered into the model significantly predict the outcome: look in the column labelled *Sig.* If the value is less than .05 then the effect is significant.
- The table labelled **Estimates of Covariance Parameters** tells us about any random effects in the model. These values can tell us how much intercepts and slopes varied over our level 1 variable. The significance of these estimates should be treated cautiously. The exact labelling of these effects depends on which covariance structure you selected for the analysis.
- An autoregressive covariance structure, AR(1), is often assumed in time course data such as that in growth models.

LABCOAT LENI'S REAL RESEARCH 19.1

A fertile gesture ③

MILLER, G. TYBUR, J.M. & JORDAN, B.D. (2007). *EVOLUTION AND HUMAN BEHAVIOR, 28*, 375–381.

Most female mammals experience a phase of 'estrus' during which they are more sexually receptive, proceptive, selective and attractive. As such, the evolutionary benefit to this phase is believed to be to attract mates of superior genetic stock. However, some people have argued that this important phase became uniquely lost or hidden in human females. Testing these evolutionary ideas is exceptionally difficult but Geoffrey Miller and his colleagues came up with an incredibly elegant piece of research that did just that. They reasoned that if the 'hidden-estrus' theory is incorrect then men should find women most attractive during the fertile phase of their menstrual cycle compared to the pre-fertile (menstrual) and post-fertile (luteal) phase.

To measure how attractive men found women in an ecologically valid way, they came up with the ingeneous idea of collecting data from women working at lap-dancing clubs. These women maximize their tips from male visitors by attracting more dances. In effect the men 'try out' several dancers before choosing a dancer for a prolonged dance. For each dance the male pays a 'tip', therefore the more men that choose a particular woman, the more her earnings will be. As such, each dancer's earnings are a good index of how attractive the male customers have found her. Miller and his colleagues argued, therefore, that if women do have an estrus phase then they will be more attractive during this phase and therefore earn more money. This study is a brilliant example of using a real-world phenomenon to address an important scientific question in an ecologically valid way.

The data for this study are in the file **Miller et al. (2007).sav**. The researcher collected data via a website from several dancers (**ID**), who provided data for multiple lap-dancing shifts (so for each person there are several rows of data). They also measured what phase of the menstrual cycle the women were in at a given shift (**Cyclephase**), and whether they were using hormonal contraceptives (**Contraceptive**) because this would affect their cycle. The outcome was their earnings on a given shift in dollars (**Tips**).

A multilevel model can be used here because the data are unbalanced: each woman differed in the number of shifts they provided data for (the range was 9 to 29 shifts); multilevel models can handle this problem.

Labcoat Leni wants you to carry out a multilevel model to see whether **Tips** can be predicted from **Cyclephase**, **Contraceptive** and their interaction. Is the 'estrus-hidden' hypothesis supported? Answers are in the additional material on the companion website (or look at page 378 in the original article).

19.8. How to report a multilevel model ③

Specific advice on reporting multilevel models is hard to come by. Also, the models themselves can take on so many forms that giving standard advice is hard. If you have built up your model from one with only fixed parameters to one with a random intercept, and then random slope, it is advisable to report all stages of this process (or at the very least report the fixed-effects-only model and the final model). For any model you need to say something about the random effects. For the final model of the cosmetic surgery example you could write something like:

✓ The relationship between surgery and quality of life showed significant variance in intercepts across participants, $\text{var}(u_{0j}) = 30.06$, $\chi^2(1) = 15.05$, $p < .01$. In addition, the slopes varied across participants, $\text{var}(u_{1j}) = 29.35$, $\chi^2(1) = 21.49$, $p < .01$, and the slopes and intercepts negatively and significantly covaried, $\text{cov}(u_{0j}, u_{1j}) = -28.08$, $\chi^2(1) = 17.38$, $p < .01$.

For the model itself, you have two choices. The first is to report the results rather like an ANOVA, with the *F*s and degrees of freedom for the fixed effects, and then report the parameters for the random effects in the text as well. The second is to produce a table of parameters as you would for regression. For example, we might report our cosmetic surgery example as follows:

✓ Quality of life before surgery significantly predicted quality of life after surgery, $F(1, 268.92) = 33.65$, $p < .001$, surgery did not significantly predict quality of life, $F(1, 15.86) = 2.17$, $p = .161$, but the reason for surgery, $F(1, 259.89) = 9.67$, $p < .01$, and the interaction of the reason for surgery and surgery, $F(1, 217.09) = 6.28$, $p < .05$, both did significantly predict quality of life. This interaction was broken down by conducting separate multilevel models on the 'physical reason' and 'attractiveness reason'. The models specified were the same as the main model but excluded the main effect and interaction term involving the reason for surgery. These analyses showed that for those operated on only to change their appearance, surgery almost significantly predicted quality of life after surgery, $b = -4.31$, $t(7.72) = -1.92$, $p = .09$: quality of life was lower after surgery compared to the control group. However, for those that had surgery to solve a physical problem, surgery did not significantly predict quality of life, $b = 1.20$, $t(7.61) = 0.58$, $p = .58$. The interaction effect, therefore, reflects the difference in slopes for surgery as a predictor of quality of life in those that had surgery for physical problems (slight positive slope) and those that had surgery purely for vanity (a negative slope).

Alternatively we could present parameter information in a table:

	b	*SE* b	*95% CI*
Baseline QoL	0.31	0.05	0.20, 0.41
Surgery	−3.19	2.17	−7.78, 1.41
Reason	−3.51	1.13	−5.74, −1.29
Surgery × Reason	4.22	1.68	0.90, 7.54

What have I discovered about statistics? ②

Writing this chapter was quite a steep learning curve for me. I've been meaning to learn about multilevel modelling for ages, and now I finally feel like I know something. This is pretty amazing considering that the bulk of the reading and writing was done between 11 p.m. and 3 a.m. over many nights. However, despite now feeling as though I understand them, I don't, and if you feel like you now understand them then you're wrong. This sounds harsh, but sadly multilevel modelling is very complicated and we have scratched only the surface of what there is to know. Multilevel models often fail to converge with no apology or explanation, and trying to fathom out what's happening can feel like hammering nails into your head.

Needless to say I didn't mention any of this at the start of the chapter because I wanted you to read it. Instead, I lulled you into a false sense of security by looking gently at how data can be hierarchical and how this hierarchical structure can be important. Most of the tests in this book simply ignore the hierarchy. We also saw that hierarchical models are just basically a fancy regression in which you can estimate the variability in the slopes and intercepts within entities. We saw that you should start with a model that ignores the hierarchy and then add in random intercepts and slopes to see if they improve the fit of the model. Having submerged ourselves in the warm bath of standard multilevel models we moved on to the icy lake of growth curves. We saw that there are ways to model trends in the data over time (and that these trends can also have variable intercepts and slopes). We also discovered that these trends have long confusing names like fourth-order polynomial. We asked ourselves why they couldn't have a sensible name, like Kate. In fact, we decided to ourselves that we'd secretly call a linear trend Kate, a quadratic trend Benjamin, a cubic trend Zoë and a fourth-order trend Doug. 'That will show the statisticians' we thought to ourselves, and felt a little bit self-satisfied too.

We also saw that after years of denial, my love of making a racket got the better of me. This brings my life story up to date. Admittedly I left out some of the more colourful bits, but only because I couldn't find an extremely tenuous way to link them to statistics. We saw that over my life I managed to completely fail to achieve any of my childhood dreams. It's OK, I have other ambitions now (a bit smaller scale than 'rock star') and I'm looking forward to failing to achieve them too. The question that remains is whether there is life after *Discovering Statistics*. What effect does writing a statistics book have on your life?

Key terms that I've discovered

AIC
AICC
AR(1)
BIC
CAIC
Centring
Diagonal
Fixed coefficient
Fixed effect
Fixed intercept
Fixed slope
Fixed variable
Grand mean centring
Group mean centring
Growth curve
Multilevel linear model
Polynomial
Random coefficient
Random effect
Random intercept
Random slope
Random variable
Unstructured
Variance components

Smart Alex's tasks

- **Task 1**: Using the cosmetic surgery example, run the analysis described in section 19.6.5 but also including BDI, age and gender as fixed effect predictors. What differences does including these predictors make? ④
- **Task 2**: Using our growth model example in this chapter, analyse the data but include **Gender** as an additional covariate. Does this change your conclusions? ④
- **Task 3: Getting kids to exercise (Hill, Abraham, & Wright, 2007)**: The purpose of this research was to examine whether providing children with a leaflet based on the 'theory of planned behaviour' increases children's exercise. There were four different interventions (**Intervention**): a control group, a leaflet, a leaflet and quiz, and a leaflet and plan. A total of 503 children from 22 different classrooms were sampled (**Classroom**). It was not practical to have children in the same classrooms in different conditions, therefore the 22 classrooms were randomly assigned to the four

different conditions. Children were asked 'On average over the last three weeks, I have exercised energetically for at least 30 minutes ______ times per week' after the intervention (**Post_Exercise**). Run a multilevel model analysis on these data (**Hill et al. (2007).sav**) to see whether the intervention affected the children's exercise levels (the hierarchy in the data is: children within classrooms within interventions). ④

- **Task 4**: Repeat the above analysis but include the pre-intervention exercise scores (**Pre_Exercise**) as a covariate. What difference does this make to the results? ④

Answers can be found on the companion website.

Further reading

Kreft, I., & De Leeuw, J. (1998). *Introducing multilevel modeling*. London: Sage. (This is a fantastic book that is easy to get into but has a lot of depth too.)

Tabachnick, B. G., & Fidell, L. S. (2001). *Using multivariate statistics* (4th ed.). Boston: Allyn & Bacon. (Chapter 15 is a fantastic account of multilevel linear models that goes a bit more in depth than I do.)

Twisk, J. W. R. (2006). *Applied multilevel analysis: a practical guide*. Cambridge: Cambridge University Press. (An absolutely superb introduction to multilevel modelling. This book is exceptionally clearly written and is aimed at novices. Without question, this is the best beginner's guide that I have read.)

Online tutorial

The companion website contains the following Flash movie tutorial to accompany this chapter:

- Mixed Models using SPSS

Interesting real research

Cook, S. A., Rosser, R., & Salmon, P. (2006). Is cosmetic surgery an effective psychotherapeutic intervention? A systematic review of the evidence. *Journal of Plastic, Reconstructive & Aesthetic Surgery*, *59*, 1133–1151.

Miller, G., Tybur, J. M., & Jordan, B. D. (2007). Ovulatory cycle effects on tip earnings by lap dancers: economic evidence for human estrus? *Evolution and Human Behavior*, *28*, 375–381.

EPILOGUE: LIFE AFTER DISCOVERING STATISTICS

Here's some questions that the writer sent
Can an observer be a participant?
Have I seen too much?
Does it count if it doesn't touch?
If the view is all I can ascertain,
Pure understanding is out of range.

(Fugazi, 2001)

When I wrote the first edition of this book my main ambition was to write a statistics book that I would enjoy reading. Pretty selfish I know. I thought that if I had a reference book that had a few examples that amused me then it would make life a lot easier when I needed to look something up. I honestly didn't think anyone would buy the thing (well, apart from my mum and dad) and I anticipated a glut of feedback along the lines of 'the whole of Chapter X is completely wrong and you're an arrant fool', or 'you should be ashamed of how many trees have died in the name of this rubbish, you brainless idiot'. In fact, even the publishers didn't think it would sell (they have only revealed this subsequently I might add). There are several other things that I didn't expect to happen:

1 *Nice emails*: I didn't expect to receive hundreds of extremely nice emails from people who liked the book. To this day it still absolutely amazes me that anyone reads it, let alone takes the time to write me a nice email, and knowing that the book has helped people always puts a huge smile on my face. When the nice comments are followed by four pages of statistics questions the smile fades a bit …
2 *Everybody thinks that I'm a statistician*: I should have seen this one coming really, but since writing a statistics textbook everyone assumes that I'm a statistician. I'm not, I'm a psychologist. Consequently, I constantly disappoint people by not being able to answer their statistics questions. In fact, this book is the sum total of my knowledge about statistics; there is nothing else (statistics-wise) in my brain that isn't in this book. Actually, that's a lie: there is *more* in this book about statistics than in my brain. For example, in the logistic regression chapter there is a new example on multinomial logistic regression. To write this new section I read a lot about multinomial logistic regression because I'd never used it. I wrote that new section about four months ago, and I've now forgotten everything that I wrote. Should I ever need to do a multinomial logistic regression I will read the chapter in this book and think to myself 'wow, it really sounds as though I know what I'm talking about'.
3 *Craziness on a grand scale*: The nicest thing about life after discovering statistics is the effort that people make to demonstrate that they are even stranger than me. All of these people have made life after *Discovering Statistics* … a profoundly enjoyable experience.

- *Catistics*: I've had quite a few photographs of people's cats (and dogs) reading my book (check out my 'discovering catistics' website at http://www.statisticshell.com/catistics.html). There has been many a week where one of these in my inbox has turned what was going to be a steaming turd of a day into a fragrant romp through fields of tulips. How can you not get a big stupid grin on your face when you see these?
- *Facebook*: Two particularly strange people from Exeter (UK) whom I have never met set up an 'Andy Field appreciation society' on Facebook. I don't go there much because it scares me a bit. But secretly I think it is quite cool. It's almost like being the rock star that I always wanted to be, except that when people join a rock star's appreciation society they mean it, but people join mine because it's funny. Nevertheless, beggars can't be choosers and I'm happy to overlook a technicality such as the truth if it means that I can believe that I'm popular.
- *Films*: Possibly the strangest thing to have happened is Julie-Renée Kabriel and her bonkers friends from Washburn University producing a video homage to 'Discovering Stats' (http://www.youtube.com/watch?v=oLsrt594Xxc). I was in equal parts crippled with laughter and utterly bemused watching this video. My parents liked it too. (Oddly enough, it's to the tune of 'Sweet Home Alabama' by Lynyrd Skynyrd; I once gave a talk at Aberdeen University (Scotland) after which I got taken to a bar and ended up (quite unexpectedly) playing drums to that song with a makeshift band of complete strangers.)
- *Invitation to an autopsy*: I got invited to an autopsy. Really! Some (very nice) forensic scientists in Leicester loved this book so much that they felt that I needed to be rewarded for my efforts. They felt that the most appropriate reward would be to offer to take me to see a dead body being carved up (or to spend a day visiting crime scenes). In a strange way, I can see their logic. I haven't been because I'm slightly scared that it's a cruel trick and that it will turn out to be my body on the slab. However, in the interests of having a good story for the next edition I might just go …
- *Befriended by Satan*: I got an email from the manager of a black metal band from London who, while using my book for her studies, was impressed to see that I like black metal bands. My band was playing the next week in London and never one to miss an opportunity, I invited her to come along. She not only turned up, but brought some of the band and some free CDs. They're called Abgott, they rock, and they renamed me 'The Evil Statistic'. I've subsequently spent many a happy night in London listening to deafening music and drinking too much with them. Buy their albums, buy their albums, buy their albums …

Life after *Discovering Statistics* … never ceases to amuse me. I never dreamed for a second that I'd be writing a third edition, or that this book would have become such a huge part of my life. I would recommend writing a statistics book to anyone: it changes your life. You get a constant warm fuzzy feeling from being told that you've helped people, strangers send you photos of their pets, they make films about you, they give you CDs, you get an appreciation society, you can go to see corpses being cut up, join a black metal band (well, maybe not, but if my drumming improves and their drummer's arms and legs fall off, who knows?) and have people constantly overestimate your intelligence. It's a great life and long may the craziness continue.

GLOSSARY

0: the amount of a clue that Sage has about how much effort I put into writing this book.

−2LL: the *log-likelihood* multiplied by minus 2. This version of the likelihood is used in *logistic regression*.

α-level: the probability of making a *Type I error* (usually this value is .05).

A Life: what you don't have when writing statistics textbooks.

Adjusted mean: in the context of *analysis of covariance* this is the value of the group mean adjusted for the effect of the *covariate*.

Adjusted predicted value: a measure of the influence of a particular case of data. It is the predicted value of a case from a model estimated without that case included in the data. The value is calculated by re-estimating the model without the case in question, then using this new model to predict the value of the excluded case. If a case does not exert a large influence over the model then its predicted value should be similar regardless of whether the model was estimated including or excluding that case. The difference between the predicted value of a case from the model when that case was included and the predicted value from the model when it was excluded is the *DFFit*.

Adjusted R^2: a measure of the loss of predictive power or *shrinkage* in regression. The adjusted R^2 tells us how much variance in the outcome would be accounted for if the model had been derived from the population from which the sample was taken.

AIC (Akaike's Information Criterion): a *goodness-of-fit* measure that is corrected for model complexity. That just means that it takes into account how many parameters have been estimated. It is not intrinsically interpretable, but can be compared in different models to see how changing the model affects the fit. A small value represents a better fit of the data.

AICC (Hurvich and Tsai's Criterion): a *goodness-of-fit* measure that is similar to AIC but is designed for small samples. It is not intrinsically interpretable, but can be compared in different models to see how changing the model affects the fit. A small value represents a better fit of the data.

Alpha factoring: a method of *factor analysis*.

Alternative hypothesis: the prediction that there will be an effect (i.e. that your experimental manipulation will have some effect or that certain variables will relate to each other).

Analysis of covariance: a statistical procedure that uses the *F*-ratio to test the overall fit of a linear model controlling for the effect that one or more *covariates* have on the *outcome variable*. In experimental research this linear model tends to be defined in terms of group means and the resulting ANOVA is therefore an overall test of whether group means differ after the variance in the outcome variable explained by any *covariates* has been removed.

Analysis of variance: a statistical procedure that uses the *F*-ratio to test the overall fit of a linear model. In experimental research this linear model tends to be defined in terms of group means and the resulting ANOVA is therefore an overall test of whether group means differ.

ANCOVA: acronym for *analysis of covariance*.

Anderson–Rubin method: a way of calculating *factor scores* which produces scores that are uncorrelated and *standardized* with a mean of 0 and a standard deviation of 1.

ANOVA: acronym for *analysis of variance*.

AR(1): this stands for first-order autoregressive structure. It is a covariance structure used in *multilevel models* in which the relationship between scores changes in a systematic way. It is assumed that the correlation between scores gets smaller over time and variances are assumed to be homogeneous. This structure is often used for repeated-measures data (especially when measurements are taken over time such as in growth models).

Autocorrelation: when the *residuals* of two observations in a regression model are correlated.

b_i: unstandardized regression coefficient. Indicates the strength of relationship between a given predictor, *i*, and an outcome in the units of measurement of the predictor. It is the change in the outcome associated with a unit change in the predictor.

β_i: standardized regression coefficient. Indicates the strength of relationship between a given predictor, *i*, and an outcome in a *standardized* form. It is the change in the outcome (in standard deviations) associated with a one standard deviation change in the predictor.

β-level: the probability of making a *Type II error* (Cohen, 1992, suggests a maximum value of 0.2).

Bar chart: a graph in which a summary statistic (usually the mean) is plotted on the *y*-axis against a categorical variable on the *x*-axis (this categorical variable could represent, for example, groups of people, different times or different experimental conditions). The value of the mean for each category is shown by a bar. Different-coloured bars may be used to represent levels of a second categorical variable.

Bartlett's test of sphericity: unsurprisingly this is a test of the assumption of *sphericity*. This test examines whether a *variance–covariance matrix* is proportional to an *identity matrix*. Therefore, it effectively tests whether the diagonal elements of the variance–covariance matrix are equal (i.e. group variances are the same), and that the off-diagonal elements are approximately zero (i.e. the *dependent variables* are not *correlated*). Jeremy Miles, who does a lot of multivariate stuff, claims he's never ever seen a matrix that reached non-significance using this test and, come to think of it, I've never seen one either (although I do less multivariate stuff) so you've got to wonder about its practical utility.

Beer-goggles effect: the phenomenon that people of the opposite gender (or the same depending on your sexual orientation) appear much more attractive after a few alcoholic drinks.

Between-group design: another name for *independent design*.

Between-subject design: another name for *independent design*.

BIC (Schwarz's Bayesian Criterion): a *goodness-of-fit* statistic comparable to the AIC, although it is slightly more conservative (it corrects more harshly for the number of parameters being estimated). It should be used when sample sizes are large and the number of parameters is small. It is not intrinsically interpretable, but can be compared in different models to see how changing the model affects the fit. A small value represents a better fit of the data.

Bimodal: a description of a distribution of observations that has two *modes*.

Binary logistic regression: *logistic regression* in which the outcome variable has exactly two categories.

Binary variable: a *categorical variable* that has only two mutually exclusive categories (e.g. being dead or alive).

Biserial correlation: a standardized measure of the strength of relationship between two variables when one of the two variables is *dichotomous*. The biserial correlation coefficient is used when one variable is a continuous dichotomy (e.g. has an underlying continuum between the categories).

Bivariate correlation: a correlation between two variables.

Blockwise regression: another name for *hierarchical regression*.

Bonferroni correction: a correction applied to the *α-level* to control the overall *Type I error rate* when multiple significance tests are carried out. Each test conducted should use a criterion of significance of the α-level (normally .05) divided by the number of tests conducted. This is a simple but effective correction, but tends to be too strict when lots of tests are performed.

Bootstrap: a technique from which the sampling distribution of a statistic is estimated by taking repeated samples (with replacement) from the data set (so, in effect, treating the data as a population from which smaller samples are taken). The statistic of interest (e.g. the *mean*, or *b* coefficient) is calculated for each sample, from which the sampling distribution of the statistic is estimated. The standard error of the statistic is estimated as the standard deviation of the sampling distribution created from the bootstrap samples. From this, confidence intervals and significance tests can be computed.

Boredom effect: refers to the possibility that performance in tasks may be influenced (the assumption is a negative influence) by boredom/lack of concentration if there are many tasks, or the task goes on for a long period of time. In short, what you are experiencing reading this glossary is a boredom effect.

Box's test: a test of the assumption of *homogeneity of covariance matrices*. This test should be non-significant if the matrices are roughly the same. Box's test is very susceptible to deviations from *multivariate normality* and so can be non-significant, not because the *variance–covariance matrices* are similar across groups, but because the assumption of multivariate normality is not tenable. Hence, it is vital to have some idea of whether the data meet the multivariate normality assumption (which is extremely difficult) before interpreting the result of Box's test.

Boxplot (a.k.a. box–whisker diagram): a graphical representation of some important characteristics of a set of observations. At the centre of the plot is the *median*, which is surrounded by a box the top and bottom of which are the limits within which the middle 50% of observations fall (the *interquartile range*). Sticking out of the top and bottom of the box are two whiskers which extend to the most and least extreme scores respectively.

Box–whisker plot: see *Boxplot*.

Brown–Forsythe *F*: a version of the *F*-ratio designed to be accurate when the assumption of *homogeneity of variance* has been violated.

CAIC (Bozdogan's criterion): a *goodness-of-fit* measure similar to the *AIC*, but corrects for model complexity and sample size. It is not intrinsically interpretable, but can be compared in different models to see how changing the model affects the fit. A small value represents a better fit of the data.

Categorical variable: any variable made up of categories of objects/entities. The UK degree classifications are a good example because degrees are classified as a 1, 2:1, 2:2, 3, pass or fail. Therefore, graduates form a categorical variable because they will fall into only one of these categories (hopefully the category of students receiving a first).

Central limit theorem: this theorem states that when samples are large (above about 30) the *sampling distribution* will take the shape of a *normal distribution* regardless of the shape of the population from which the sample was drawn. For small samples the *t*-distribution better approximates the shape of the sampling distribution. We also know from this theorem that the *standard deviation* of the sampling distribution (i.e. the *standard error* of the sample *mean*) will be equal to the standard deviation of the sample (*s*) divided by the square root of the sample size (*N*).

Central tendency: a generic term describing the centre of a *frequency distribution* of observations as measured by the *mean*, *mode* and *median*.

Centring: the process of transforming a variable into deviations around a fixed point. This fixed point can be any value that is chosen, but typically a mean is used. To centre a variable the mean is subtracted from each score. See *Grand mean centring*, *Group mean centring*.

Chart Builder: facility in SPSS for drawing graphs that is accessed via the *Graphs* menu.

Chart Editor: a window in SPSS in which graphs from the *SPSS Viewer* can be edited. To access the chart editor window double-click on a graph in the viewer.

Chartjunk: superfluous material that distracts from the data being displayed on a graph.

Chi-square distribution: a *probability distribution* of the sum of squares of several normally distributed variables. It tends to be used to (1) test hypotheses about categorical data, and (2) test the fit of models to the observed data.

Chi-square test: although this term can apply to any *test statistic* having a *chi-square distribution*, it generally refers to Pearson's chi-square test of the independence of two categorical variables. Essentially it tests whether two categorical variables forming a *contingency table* are associated.

Cochran's Q: this test is an extension of *McNemar's test* and is basically a *Friedman's ANOVA* for *dichotomous* data. So imagine you asked 10 people whether they'd like to shoot Justin Timberlake, David Beckham and Simon Cowell and they could answer only yes or no. If we coded responses as 0 (no) and 1 (yes) we could do Cochran's test on these data.

Coefficient of determination: the proportion of variance in one variable explained by a second variable. It is the *Pearson correlation coefficient* squared.

Common variance: variance shared by two or more variables.

Communality: the proportion of a variable's variance that is *common variance*. This term is used primarily in *factor analysis*. A variable that has no *unique variance* (or *random variance*) would have a communality of 1, whereas a variable that shares none of its variance with any other variable would have a communality of 0.

Complete separation: a situation in *logistic regression* when the outcome variable can be perfectly predicted by one predictor or a combination of predictors. Suffice it to say this situation makes your computer have the equivalent of a nervous breakdown: it'll start gibbering, weeping and saying it doesn't know what to do.

Component matrix: general term for the *structure matrix* in SPSS *principal component analysis*.

Compound symmetry: a condition that holds true when both the variances across conditions are equal (this is the same as the *homogeneity of variance* assumption) and the *covariances* between pairs of conditions are also equal.

Confidence interval: for a given statistic calculated for a sample of observations (e.g. the mean), the confidence interval is a range of values around that statistic that are believed to contain, with a certain probability (e.g. 95%), the true value of that statistic (i.e. the population value).

Confirmatory factor analysis (CFA): a version of *factor analysis* in which specific hypotheses about structure and relations between the *latent variables* that underlie the data are tested.

Confounding variable: a variable (that we may or may not have measured) other than the *predictor variables* in which we're interested that potentially affects an *outcome variable*.

Content validity: evidence that the content of a test corresponds to the content of the construct it was designed to cover.

Contingency table: a table representing the cross-classification of two or more *categorical variables*. The levels of each variable are arranged in a grid, and the number of observations falling into each category is noted in the cells of the table. For example, if we took the categorical variables of glossary (with two categories: whether an author was made to write a glossary or not) and mental state (with three categories: normal, sobbing uncontrollably and utterly psychotic), we could construct a table as below. This instantly tells us that 127 authors who were made to write a glossary ended up as utterly psychotic, compared to only 2 who did not write a glossary.

		Glossary		
		Author made to write glossary	No glossary	Total
Mental state	Normal	5	423	428
	Sobbing uncontrollably	23	46	69
	Utterly psychotic	127	2	129
	Total	155	471	626

Continuous variable: a variable that can be measured to any level of precision. (Time is a continuous variable, because there is in principle no limit on how finely it could be measured.)

Cook's distance: a measure of the overall influence of a case on a model. Cook and Weisberg (1982) have suggested that values greater than 1 may be cause for concern.

Correlation coefficient: a measure of the strength of association or relationship between two variables. See *Pearson's correlation coefficient*, *Spearman's correlation coefficient*, *Kendall's tau*.

Correlational research: a form of research in which you observe what naturally goes on in the world without directly interfering with it. This term implies that data will be analysed so as to look at relationships between naturally-occurring variables rather than making statements about cause and effect. Compare with *cross-sectional research* and *experimental research*.

Counterbalancing: a process of systematically varying the order in which experimental conditions are conducted. In the simplest case of there being two conditions (A and B), counterbalancing simply implies that half of the participants complete condition A followed by condition B, whereas the remainder do condition B followed by condition A. The aim is to remove systematic bias caused by *practice effects* or *boredom effects*.

Covariance: a measure of the 'average' relationship between two variables. It is the average *cross-product deviation* (i.e. the cross-product divided by one less than the number of observations).

Covariance ratio (CVR): a measure of whether a case influences the variance of the parameters in a *regression model*. When this ratio is close to 1 the case is having very little influence on the variances of the model parameters. Belsey et al. (1980) recommend the following: if the CVR of a case is greater than 1 + [3(*k* + 1)/*n*] then deleting that case will damage the precision of some of the model's parameters, but if it is less than 1 − [3(*k* + 1)/*n*] then deleting the case will improve the precision of some of the model's parameters (*k* is the number of predictors and *n* is the sample size).

Covariate: a variable that has a relationship with (in terms of *covariance*), or has the potential to be related to, the *outcome variable* we've measured.

Cox and Snell's R^2_{CS}: a version of the *coefficient of determination* for logistic regression. It is based on the log-likelihood of a model (*LL(new)*) and the log-likelihood of the original model (*LL(baseline)*), and the sample size, *n*. However, it is notorious for not reaching its maximum value of 1 (see *Nagelkerke's* R^2_N).

Cramer's V: a measure of the strength of association between two *categorical variables* used when one of these variables has more than two categories. It is a variant of *phi* used because when one or both of the categorical variables contain more than two categories, phi fails to reach its minimum value of 0 (indicating no association).

Criterion validity: evidence that scores from an instrument correspond with or predict concurrent external measures conceptually related to the measured construct.

Cronbach's α: a measure of the reliability of a scale defined by:

$$\alpha = \frac{N^2\overline{\text{Cov}}}{\sum s^2_{\text{item}} + \sum \text{Cov}_{\text{item}}}$$

in which the top half of the equation is simply the number of items (*N*) squared multiplied by the average covariance between items (the average of the off-diagonal elements in the *variance–covariance matrix*). The bottom half is the sum of all the elements in the *variance–covariance matrix*.

Cross-product deviations: a measure of the 'total' relationship between two variables. It is the deviation of one variable from its mean multiplied by the other variable's deviation from its mean.

Cross-sectional research: a form of research in which you observe what naturally goes on in the world without directly interfering with it. This term specifically implies that data come from people at different age points with different people representing each age point. See also *correlational research*.

Cross-validation: assessing the accuracy of a model across different samples. This is an important step in *generalization*. In a *regression model* there are two main methods of cross-validation: *adjusted* R^2 or data splitting, in which the data are split randomly into two halves, and a regression model is estimated for each half and then compared.

Crying: what you feel like doing after writing statistics textbooks.

Cubic trend: if you connected the means in ordered conditions with a line then a cubic trend is shown by two changes in the direction of this line. You must have at least four ordered conditions.

Currency variable: a variable containing values of money.

Data editor: the main window in SPSS in which you enter data and carry out statistical functions.

Data view: there are two ways to view the contents of the *data editor* window. The data view shows you a spreadsheet and can be used for entering raw data. See also *variable view*.

Date variable: variables made up of dates. The data can take forms such as dd-mmm-yyyy (e.g. 21-Jun-1973), dd-mmm-yy (e.g. 21-Jun-73), mm/dd/yy (e.g. 06/21/73), dd.mm.yyyy (e.g. 21.06.1973).

Degrees of freedom: an impossible thing to define in a few pages let alone a few lines. Essentially it is the number of 'entities' that are free to vary when estimating some kind of statistical parameter. In a more practical sense, it has a bearing on significance tests for many commonly used *test statistics* (such as the *F-ratio*, *t-test*, *chi-square statistic*) and determines the exact form of the *probability distribution* for these *test statistics*. The explanation involving rugby players in Chapter 8 is far more interesting…

Deleted residual: a measure of the influence of a particular case of data. It is the difference between the *adjusted predicted value* for a case and the original observed value for that case.

Density plot: similar to a *histogram* except that rather than having a summary bar representing the frequency of scores, it shows each individual score as a dot. They can be useful for looking at the shape of a distribution of scores.

Dependent *t*-test: a test using the *t-statistic* that establishes whether two means collected from the same sample (or related observations) differ significantly.

Dependent variable: another name for *outcome variable*. This name is usually associated with experimental methodology (which is the only time it really makes sense) and is so called because it is the variable that is not manipulated by the experimenter and so its value depends on the variables that have been manipulated. To be honest I just use *outcome variable* all the time – it makes more sense (to me) and is less confusing.

Deviance: the difference between the observed value of a variable and the value of that variable predicted by a statistical model.

Deviation contrast: a non-orthogonal *planned contrast* that compares the mean of each group (except first or last depending on how the contrast is specified) to the overall mean.

DFA: acronym for discriminant function analysis (see *discriminant analysis*).

DFBeta: a measure of the influence of a case on the values of b_i in a *regression model*. If we estimated a regression parameter b_i and then deleted a particular case and re-estimated the same regression parameter b_i, then the difference between these two estimates would be the DFBeta for the case that was deleted. By looking at the values of the DFBetas, it is possible to identify cases that have a large influence on the parameters of the regression model; however, the size of DFBeta will depend on the units of measurement of the regression parameter.

DFFit: a measure of the influence of a case. It is the difference between the *adjusted predicted value* and the original predicted value of a particular case. If a case is not influential then its DFFit should be zero – hence, we expect non-influential cases to have small DFFit values. However, we have the problem that this statistic depends on the units of measurement of the outcome and so a DFFit of 0.5 will be very small if the outcome ranges from 1 to 100, but very large if the outcome varies from 0 to 1.

Diagonal: a covariance structure used in *multilevel models*. This variance structure in which variances are assumed to be heterogeneous and that all of the covariances are 0. In SPSS this is the default covariance structure for repeated measures designs.

Dichotomous: description of a variable that consists of only two categories (e.g. the variable gender is dichotomous because it consists of only two categories: male and female).

Difference contrast: a non-orthogonal *planned contrast* that compares the mean of each condition (except the first) to the overall mean of all previous conditions combined.

Direct oblimin: a method of *oblique rotation*.

Discrete variable: a variable that can only take on certain values (usually whole numbers) on the scale.

Discriminant analysis: also known as discriminant function analysis. This analysis identifies and describes the *discriminant function variates* of a set of variables and is useful as a follow-up test to *MANOVA* as a means of seeing how these variates allow groups of cases to be discriminated.

Discriminant function variate: a linear combination of variables created such that the differences between group means on the transformed variable are maximized. It takes the general form:

$$Variate_1 = b_1X_1 + b_2X_2 + \ldots + b_nX_n.$$

Discriminant score: a score for an individual case on a particular *discriminant function variate* obtained by replacing that case's scores on the measured variables into the equation that defines the variate in question.

Dummy variables: a way of recoding a categorical variable with more than two categories into a series of variables all of which are *dichotomous* and can take on values of only 0 or 1. There are seven basic steps to create such variables: (1) count the number of groups you want to recode and subtract 1; (2) create as many new variables as the value you calculated in step 1 (these are your dummy variables); (3) choose one of your groups as a baseline (i.e. a group against which all other groups should be compared, such as a control group); (4) assign that baseline group values of 0 for all of your dummy variables; (5) for your first dummy variable, assign the value 1 to the first group that you want to compare against the baseline group (assign all other groups 0 for this variable); (6) for the second dummy variable assign the value 1 to the second group that you want to compare against the baseline group (assign all other groups 0 for this variable); (7) repeat this process until you run out of dummy variables.

Durbin–Watson test: tests for serial correlations between errors in *regression models*. Specifically, it tests whether adjacent residuals are correlated, which is useful in assessing the assumption of *independent errors*. The test statistic can vary between 0 and 4 with a value of 2 meaning that the residuals are uncorrelated. A value greater than 2 indicates a negative correlation between adjacent residuals, whereas a value below 2 indicates a positive correlation. The size of the Durbin–Watson statistic depends upon the number of predictors in the model and the number of observations. For accuracy, look up the exact acceptable values in Durbin and Watson's (1951) original paper. As a very conservative rule of thumb, values less than 1 or greater than 3 are definitely cause for concern; however, values closer to 2 may still be problematic depending on the sample and model.

Ecological validity: evidence that the results of a study, experiment or test can be applied, and allow inferences, to real-world conditions.

Eel: long, snakelike, scaleless fishes that lack pelvic fins. From the order Anguilliformes or Apodes, they should probably not be inserted into your anus to cure constipation (or for any other reason).

Effect size: an objective and (usually) standardized measure of the magnitude of an observed effect. Measures include Cohen's *d*, Glass' *g* and Pearson's correlations coefficient, *r*.

Equamax: a method of *orthogonal rotation* that is a hybrid of *quartimax* and *varimax*. It is reported to behave fairly erratically (see Tabachnick and Fidell, 2001) and so is probably best avoided.

Error bar chart: a graphical representation of the mean of a set of observations that includes the 95% confidence interval of the mean. The mean is usually represented as a circle, square or rectangle at the value of the mean (or a bar extending to the value of the mean). The confidence interval is represented by a line protruding from the mean (upwards, downwards or both) to a short horizontal line representing the limits of the confidence interval. Error bars can be drawn using the standard error or standard deviation instead of the 95% confidence interval.

Error SSCP (*E*): the error sum of squares and cross-product matrix. This is a *sum of squares and cross-product matrix* for the error in a predictive *linear model* fitted to *multivariate* data. It represents the *unsystematic variance* and is the multivariate equivalent of the *residual sum of squares*.

Eta squared (η^2): an *effect size* measure that is the ratio of the *model sum of squares* to the *total sum of squares*. So, in essence, *the coefficient of determination* by another name. It doesn't have an awful lot going for it: not only is it biased, but it typically measures the overall effect of an ANOVA and effect sizes are more easily interpreted when they reflect specific comparisons (e.g. the difference between two means).

***Exp(B)*:** the label that SPSS applies to the *odds ratio*. It is an indicator of the change in *odds* resulting from a unit change in the predictor in *logistic regression*. If the value is greater than 1 then it indicates that as the predictor increases, the odds of the outcome occurring increase. Conversely, a value less than 1 indicates that as the predictor increases, the odds of the outcome occurring decrease.

Experimental hypothesis: synonym for *alternative hypothesis*.

Experimental research: a form of research in which one or more variable is systematically manipulated to see their effect (alone or in combination) on an *outcome variable*. This term implies that data will be able to be used to make statements about cause and effect. Compare with *cross-sectional research* and *correlational research*.

Experimentwise error rate: the probability of making a *Type I error* in an experiment involving one or more statistical comparisons when the null hypothesis is true in each case.

Extraction: a term used for the process of deciding whether a *factor* in *factor analysis* is statistically important enough to 'extract' from the data and interpret. The decision is based on the magnitude of the eigenvalue associated with the factor. See *Kaiser's criterion*, *scree plot*.

F_{Max}: see *Hartley's* F_{Max}.

***F*-ratio:** a test statistic with a known *probability distribution* (the

F-distribution). It is the ratio of the average variability in the data that a given model can explain to the average variability unexplained by that same model. It is used to test the overall fit of the model in *simple regression* and *multiple regression*, and to test for overall differences between group means in experiments.

Factor: another name for an *independent variable* or *predictor* that's typically used when describing experimental designs. However, to add to the confusion, it is also used synonymously with *latent variable* in factor analysis.

Factor analysis: a *multivariate* technique for identifying whether the correlations between a set of observed variables stem from their relationship to one or more *latent variables* in the data, each of which takes the form of a *linear model*.

Factor loading: the *regression coefficient* of a variable for the *linear model* that describes a *latent variable* or *factor* in *factor analysis*.

Factor matrix: general term for the *structure matrix* in SPSS *factor analysis*.

Factor scores: a single score from an individual entity representing their performance on some *latent variable*. The score can be crudely conceptualized as follows: take an entity's score on each of the variables that make up the factor and multiply it by the corresponding *factor loading* for the variable, then add these values up (or average them).

Factor transformation matrix, Λ: a matrix used in *factor analysis*. It can be thought of as containing the angles through which factors are rotated in factor *rotation*.

Factorial ANOVA: an analysis of variance involving two or more *independent variables* or *predictors*.

Falsification: the act of disproving a hypothesis or theory.

Familywise error rate: the probability of making a *Type I error* in any family of tests when the null hypothesis is true in each case. The 'family of tests' can be loosely defined as a set of tests conducted on the same data set and addressing the same empirical question.

Fisher's exact test: Fisher's exact test (Fisher, 1922) is not so much of a test as a way of computing the exact probability of a statistic. It was designed originally to overcome the problem that with small samples the sampling distribution of the chi-square statistic deviates substantially from a chi-square distribution. It should be used with small samples.

Fit: how sexually attractive you find a statistical test. Alternatively, it's the degree to which a statistical model is an accurate representation of some observed data. (Incidentally, it's just plain *wrong* to find statistical tests sexually attractive.)

Fixed coefficient: a coefficient or model parameter that is fixed; that is, it cannot vary over situations or contexts (cf. *Random coefficient*).

Fixed effect: an effect in an experiment is said to be a fixed effect if all possible treatment conditions that a researcher is interested in are present in the experiment. Fixed effects can be generalized only to the situations in the experiment. For example, the effect is fixed if we say that we are interested only in the conditions that we had in our experiment (e.g. placebo, low dose and high dose) and we can generalize our findings only to the situation of a placebo, low dose and high dose.

Fixed intercept: a term used in *multilevel modelling* to denote when the intercept in the model is fixed. That is, it is not free to vary across different groups or contexts (cf. *Random intercept*).

Fixed slope: a term used in *multilevel modelling* to denote when the slope of the model is fixed. That is, it is not free to vary across different groups or contexts (cf. *Random slope*).

Fixed variable: a fixed variable is one that is not supposed to change over time (e.g. for most people their gender is a fixed variable – it never changes).

Frequency distribution: a graph plotting values of observations on the horizontal axis, and the frequency with which each value occurs in the data set on the vertical axis (a.k.a. *histogram*).

Friedman's ANOVA: a non-parametric test of whether more than two related groups differ. It is the non-parametric version of one-way *repeated-measures ANOVA*.

Generalization: the ability of a statistical model to say something beyond the set of observations that spawned it. If a model generalizes it is assumed that predictions from that model can be applied not just to the sample on which it is based, but to a wider population from which the sample came.

Glossary: a collection of grossly inaccurate definitions (written late at night when you really ought to be asleep) of things that you thought you understood until some evil book publisher forced you to try to define them.

Goodman and Kruskal's λ: measures the proportional reduction in error that is achieved when membership of a category of one variable is used to predict category membership of the other variable. A value of 1 means that one variable perfectly predicts the other, whereas a value of 0 indicates that one variable in no way predicts the other.

Goodness of fit: an index of how well a model fits the data from which it was generated. It's usually based on how well the data predicted by the model correspond to the data that were actually collected.

Grand mean: the *mean* of an entire set of observations.

Grand mean centring: grand mean *centring* means the transformation of a variable by taking each score and subtracting the mean of all scores (for that variable) from it (cf. *Group mean centring*).

Grand variance: the *variance* within an entire set of observations.

Greenhouse–Geisser correction: an estimate of the departure from *sphericity*. The maximum value is 1 (the data completely meet the assumption of sphericity) and minimum is the *lower bound*. Values below 1 indicate departures from sphericity and are used to correct the *degrees of freedom* associated with the corresponding *F-ratios* by multiplying them by the value of the estimate. Some say the Greenhouse–Geisser correction is too conservative (strict) and recommend the *Huynh–Feldt correction* instead.

Group mean centring: group mean *centering* is to transform a variable by taking each score and subtracting from it the mean of the scores (for that variable) for the group to which that score belongs (cf. *Grand mean centring*).

Growth curve: a curve that summarizes the change in some outcome over time. See *Polynomial*.

Harmonic mean: a weighted version of the *mean* that takes account of the relationship between variance and sample size. It is calculated by summing the reciprocal of all observations, then dividing by the number of observations. The reciprocal of the end product is the harmonic mean:

$$H = \frac{1}{\frac{1}{n}\sum_{i=1}^{n}\frac{1}{x_i}}$$

Hartley's F_{Max}: also known as the *variance ratio*, this is the ratio of the variances between the group with the biggest variance and the group with the smallest variance. This ratio is compared to critical values in a table published by Hartley as a test of *homogeneity of variance*. Some general rules are that with sample sizes (*n*) of 10 per group, an F_{Max} less than 10 is more or less always going to be non-significant, with 15–20 per group the ratio needs to be less than about 5, and with samples of 30–60 the ratio should be below about 2 or 3.

Hat values: another name for *leverage*.

HE^{-1}: this is a matrix that is functionally equivalent to the *hypothesis SSCP* divided by the *error SSCP* in *MANOVA*. Conceptually it represents the ratio of *systematic* to *unsystematic variance*, so is a *multivariate* analogue of the *F-ratio*.

Helmert contrast: a non-orthogonal *planned contrast* that compares the mean of each condition (except the last) to the overall mean all subsequent conditions combined.

Heterogeneity of variance: the opposite of *homogeneity of variance*. This term means that the variance of one variable varies (i.e. is different) across levels of another variable.

Heteroscedasticity: the opposite of *homoscedasticity*. This occurs when the residuals at each level of the predictor variables(s) have unequal variances. Put another way, at each point along any predictor variable, the spread of residuals is different.

Hierarchical regression: a method of *multiple regression* in which the order in which predictors are entered into the regression model is determined by the researcher based on previous research: variables already known to be predictors are entered first, new variables are entered subsequently.

Histogram: a *frequency distribution*.

Homogeneity of covariance matrices: an assumption of some *multivariate* tests such as *MANOVA*. It is an extension of the *homogeneity of variance assumption* in *univariate* analyses. However, as well as assuming that *variances* for each *dependent variable* are the same across groups, it also assumes that relationships (*covariances*) between these dependent variables are roughly equal. It is tested by comparing the population *variance–covariance matrices* of the different groups in the analysis.

Homogeneity of regression slopes: an assumption of *analysis of covariance*. This is the assumption that the relationship between the *covariate* and *outcome variable* is constant across different treatment levels. So, if we had three treatment conditions, if there's a positive relationship between the covariate and the outcome in one group, we assume that there is a similar-sized positive relationship between the covariate and outcome in the other two groups too.

Homogeneity of variance: the assumption that the variance of one variable is stable (i.e. relatively similar) at all levels of another variable.

Homoscedasticity: an assumption in regression analysis that the residuals at each level of the predictor variables(s) have similar variances. Put another way, at each point along any predictor variable, the spread of residuals should be fairly constant.

Hosmer and Lemeshow's R^2_L: a version of the *coefficient of determination* for logistic regression. It is a fairly literal translation in that it is the −2LL for the model divided by the original −2LL; in other words, it's the ratio of what the model can explain compared to what there was to explain in the first place!

Hotelling–Lawley trace (T^2): a *test statistic* in *MANOVA*. It is the sum of the eigenvalues for each *discriminant function variate* of the data and so is conceptually the same as the *F-ratio* in *ANOVA*: it is the sum of the ratio of *systematic* and *unsystematic variance* (SS_M/SS_R) for each of the variates.

Huynh–Feldt correction: an estimate of the departure from *sphericity*. The maximum value is 1 (the data completely meet the assumption of sphericity). Values below this indicate departures from sphericity and are used to correct the *degrees of freedom* associated with the corresponding *F-ratios* by multiplying them by the value of the estimate. It is less conservative than the *Greenhouse–Geisser estimate*, but some say it is too liberal.

Hypothesis: a prediction about the state of the world (see *experimental hypothesis* and *null hypothesis*).

Hypothesis SSCP (*H*): the hypothesis sum of squares and cross-product matrix. This is a *sum of squares and cross-product matrix* for a predictive *linear model* fitted to *multivariate* data. It represents the *systematic variance* and is the multivariate equivalent of the *model sum of squares*.

Identity matrix: a square matrix (i.e. has the same number of rows and columns) in which the diagonal elements are equal to 1, and the off-diagonal elements are equal to 0. The following are all examples:

$$\begin{pmatrix}1 & 0\\0 & 1\end{pmatrix}\begin{pmatrix}1 & 0 & 0\\0 & 1 & 0\\0 & 0 & 1\end{pmatrix}\begin{pmatrix}1 & 0 & 0 & 0\\0 & 1 & 0 & 0\\0 & 0 & 1 & 0\\0 & 0 & 0 & 1\end{pmatrix}$$

Independence: the assumption that one data point does not influence another. When data come from people, it basically means that the behaviour of one person does not influence the behaviour of another.

Independent ANOVA: *analysis of variance* conducted on any design in which all *independent variables* or *predictors* have been manipulated using different participants (i.e. all data come from different entities).

Independent design: an experimental design in which different treatment conditions utilize different organisms (e.g. in psychology, this would mean using different people in different treatment conditions) and so the resulting data are independent (a.k.a. between-group or between-subject designs).

Independent errors: for any two observations in regression the *residuals* should be uncorrelated (or independent).

Independent factorial design: an experimental design incorporating two or more *predictors* (or *independent variables*) all of which have been manipulated using different participants (or whatever entities are being tested).

Independent *t*-test: a test using the *t-statistic* that establishes whether two means collected from independent samples differ significantly.

Independent variable: another name for a *predictor variable*. This name is usually associated with experimental methodology (which is the only time it makes sense) and is so called because it is the variable that is manipulated by the experimenter and so its value does not depend on any other variables (just on the experimenter). I just use the term *predictor variable* all the time because the meaning of the term is not constrained to a particular methodology.

Interaction effect: the combined effect of two or more *predictor variables* on an *outcome variable*.

Interaction graph: a graph showing the means of two or more *independent variables* in which means of one variable are shown at different levels of the other variable. Unusually the means are connected with lines, or are displayed as bars. These graphs are used to help understand *interaction effects*.

Interquartile range: the limits within which the middle 50% of an ordered set of observations falls. It is the difference between the value of the *upper quartile* and *lower quartile*.

Interval variable: data measured on a scale along the whole of which intervals are equal. For example, people's ratings of this book on Amazon.com can range from 1 to 5; for these data to be interval it should be true that the increase in appreciation for this book represented by a change from 3 to 4 along the scale should be the same as the change in appreciation represented by a change from 1 to 2, or 4 to 5.

Intraclass correlation: a *correlation coefficient* that assess the consistency between measures of the same class (i.e. measures of the same thing). (cf. *Pearson product moment correlation* which measures the relationship between variables of a different class.) Two common uses are in comparing paired data (such as twins) on the same measure, and assessing the consistency between judges' ratings of a set of objects. The calculations of these correlations depends on whether a measure of consistency (in which the order of scores from a source is considered but not the actual value around which the scores are anchored) or absolute agreement (in which both the order of scores and the relative values are considered), and whether the scores represent averages of many measures or just a single measure, is required. This measure is also used in *multilevel linear models* to measure the dependency in data within the same context.

Jonckheere–Terpstra test: this statistic tests for an ordered pattern of medians across independent groups. Essentially it does the same thing as the *Kruskal–Wallis test* (i.e. test for a difference between the medians of the groups) but it incorporates information about whether the order of the groups is meaningful. As such, you should use this test when you expect the groups you're comparing to produce a meaningful order of medians.

Kaiser–Meyer–Olkin measure of sampling adequacy (KMO): the KMO can be calculated for individual and multiple variables and represents the ratio of the squared correlation between variables to the squared *partial correlation* between variables. It varies between 0 and 1: a value of 0 indicates that the sum of partial correlations is large relative to the sum of correlations, indicating diffusion in the pattern of correlations (hence, *factor analysis* is likely to be inappropriate); a value close to 1 indicates that patterns of correlations are relatively compact and so factor analysis should yield distinct and reliable factors. Values between .5 and .7 are mediocre, values between .7 and .8 are good, values between .8 and .9 are great and values above .9 are superb (see Hutcheson and Sofroniou, 1999).

Kaiser's criterion: a method of *extraction* in *factor analysis* based on the idea of retaining factors with associated eigenvalues greater than 1. This method appears to be accurate when the number of variables in the analysis is less than 30 and the resulting *communalities* (after *extraction*) are all greater than 0.7, or when the sample size exceeds 250 and the average communality is greater than or equal to 0.6.

Kendall's tau: a non-parametric correlation coefficient similar to *Spearman's correlation coefficient*, but should be used in preference for a small data set with a large number of tied ranks.

Kendall's *W*: this is much the same as *Friedman's ANOVA* but is used specifically for looking at the agreement between raters. So, if, for example, we asked 10 different women to rate the attractiveness of Justin Timberlake, David Beckham and Brad Pitt we could use this test to look at the extent to which they agree. Kendall's *W* ranges from 0 (no agreement between judges) to 1 (complete agreement between judges).

Kolmogorov–Smirnov test: a test of whether a distribution of scores is significantly different from a *normal distribution*. A significant value indicates a deviation from normality, but this test is notoriously affected by large samples in which small deviations from normality yield significant results.

Kolmogorov–Smirnov *Z*: not to be confused with the *Kolmogorov–Smirnov test* that tests whether a sample comes from a normally distributed population. This tests whether two groups have been drawn from the same population (regardless of what that population may be). It does much the same as the *Mann–Whitney test* and *Wilcoxon rank-sum test*! This test tends to have better power than the Mann–Whitney test when sample sizes are less than about 25 per group.

Kruskal–Wallis test: non-parametric test of whether more than two independent groups differ. It is the non-parametric version of one-way *independent ANOVA*.

Kurtosis: this measures the degree to which scores cluster in the tails of a frequency distribution. A distribution with positive Kurtosis (*leptokurtic*, kurtosis > 0) has too many scores in the tails and is too peaked, whereas a distribution with negative kurtosis (*platykurtic*, kurtosis < 0) has too few scores in the tails and is quite flat.

Latent variable: a variable that cannot be directly measured, but is assumed to be related to several variables that can be measured.

Leptokurtic: see *Kurtosis*.

Levels of measurement: the relationship between what is being measured and the numbers obtained on a scale.

Levene's test: tests the hypothesis that the variances in different groups are equal (i.e. the difference between the variances is zero). It basically does a one-way ANOVA on the *deviations* (i.e. the absolute value of the difference between each score and the mean of its group). A significant result indicates that the variances are significantly different – therefore, the assumption of *homogeneity of variances* has been violated. When samples sizes are large, small differences in group variances can produce a significant Levene's test and so the *variance ratio* is a useful double-check.

Leverage: leverage statistics (or hat values) gauge the influence of the observed value of the outcome variable over the predicted values.

The average leverage value is $(k + 1)/n$ in which k is the number of predictors in the model and n is the number of participants. Leverage values can lie between 0 (the case has no influence whatsoever) and 1 (the case has complete influence over prediction). If no cases exert undue influence over the model then we would expect all of the leverage value to be close to the average value. Hoaglin and Welsch (1978) recommend investigating cases with values greater than twice the average $(2(k + 1)/n)$ and Stevens (2002) recommends using three times the average $(3(k + 1)/n)$ as a cut-off point for identifying cases having undue influence.

Likelihood: the probability of obtaining a set of observations given the parameters of a model fitted to those observations.

Linear model: a model that is based upon a straight line.

Line chart: a graph in which a summary statistic (usually the mean) is plotted on the *y*-axis against a categorical variable on the *x*-axis (this categorical variable could represent, for example, groups of people, different times or different experimental conditions). The value of the mean for each category is shown by a symbol and means across categories are connected by a line. Different-coloured lines may be used to represent levels of a second categorical variable.

Logistic regression: a version of *multiple regression* in which the outcome is a *categorical variable*. If the categorical variable has exactly two categories the analysis is called binary logistic regression, and when the outcome has more than two categories it is called multinomial logistic regression.

Log-likelihood: a measure of error, or unexplained variation, in categorical models. It is based on summing the probabilities associated with the predicted and actual outcomes and is analogous to the *residual sum of squares* in multiple regression in that it is an indicator of how much unexplained information there is after the model has been fitted. Large values of the log-likelihood statistic indicate poorly fitting statistical models, because the larger the value of the log-likelihood, the more unexplained observations there are. The log-likelihood is the logarithm of the *likelihood*.

Loglinear analysis: a procedure used as an extension of the *chi-square test* to analyse situations in which we have more than two *categorical variables* and we want to test for relationships between these variables. Essentially, a *linear model* is fit to the data that predicts expected frequencies (i.e. the number of cases expected in a given category). In this respect it is much the same as *analysis of variance* but for entirely categorical data.

Lower bound: the name given to the lowest possible value of the *Greenhouse–Geisser estimate* of *sphericity*. Its value is $1/k-1$, in which k is the number of treatment conditions.

Lower quartile: the value that cuts off the lowest 25% of the data. If the data are ordered and then divided into two halves at the median, then the lower quartile is the median of the lower half of the scores.

Mahalanobis distances: these measure the influence of a case by examining the distance of cases from the mean(s) of the predictor variable(s). One needs to look for the cases with the highest values. It is not easy to establish a cut-off point at which to worry, although Barnett and Lewis (1978) have produced a table of critical values dependent on the number of predictors and the sample size. From their work it is clear that even with large samples ($N = 500$) and five predictors, values above 25 are cause for concern. In smaller samples ($N = 100$) and with fewer predictors (namely, three) values greater than 15 are problematic, and in very small samples ($N = 30$) with only two predictors values greater than 11 should be examined. However, for more specific advice, refer to Barnett and Lewis's (1978) table.

Main effect: the unique effect of a *predictor variable* (or *independent variable*) on an *outcome variable*. The term is usually used in the context of *ANOVA*.

Mann–Whitney test: a *non-parametric test* that looks for differences between two independent samples. That is, it tests whether the populations from which two samples are drawn have the same location. It is functionally the same as *Wilcoxon's rank-sum test*, and both tests are non-parametric equivalents of the *independent t-test*.

MANOVA: acronym for *multivariate analysis of variance*.

Matrix: a collection of numbers arranged in columns and rows. The values within a matrix are typically referred to as *components* or *elements*.

Mauchly's test: a test of the assumption of *sphericity*. If this test is significant then the assumption of *sphericity* has not been met and an appropriate correction must be applied to the *degrees of freedom* of the *F-ratio* in *repeated-measures ANOVA*. The test works by comparing the *variance–covariance matrix* of the data to an *identity matrix;* if the variance–covariance matrix is a scalar multiple of an *identity matrix* then sphericity is met.

Maximum-likelihood estimation: a way of estimating statistical parameters by choosing the parameters that make the data most likely to have happened. Imagine for a set of parameters that we calculated the probability (or likelihood) of getting the observed data; if this probability was high then these particular parameters yield a good fit of the data, but conversely if the probability was low, these parameters are a bad fit of our data. Maximum-likelihood estimation chooses the parameters that maximize the probability.

McNemar's test: this tests differences between two related groups (see *Wilcoxon signed-rank test* and *sign test*), when *nominal data* have been used. It's typically used when we're looking for changes in people's scores and it compares the proportion of people who changed their response in one direction (i.e. scores increased) to those who changed in the opposite direction (scores decreased). So, this test needs to be used when we've got two related dichotomous variables.

Mean: a simple statistical model of the centre of a distribution of scores. A hypothetical estimate of the 'typical' score.

Mean squares: a measure of average variability. For every *sum of squares* (which measure the total variability) it is possible to create mean squares by dividing by the number of things used to calculate the sum of squares (or some function of it).

Measurement error: the discrepancy between the numbers used to represent the thing that we're measuring and the actual value of the thing we're measuring (i.e. the value we would get if we could measure it directly).

Median: the middle score of a set of ordered observations. When there is an even number of observations the median is the average of the two scores that fall either side of what would be the middle value.

Median test: a non-parametric test of whether samples are drawn from a population with the same median. So, in effect it does the same thing as the *Kruskal–Wallis test*. It works on the basis of producing a contingency table that is split for each group into the number of scores that fall above and below the observed median of the entire data set. If the groups are from the same population then these frequencies would be expected to be the same in all conditions (about 50% above and about 50% below).

Meta-analysis: this is a statistical procedure for assimilating research findings. It is based on the simple idea that we can take effect sizes from individual studies that research the same question, quantify the observed effect in a standard way (using *effect sizes*) and then combine these effects to get a more accurate idea of the true effect in the population.

Mixed ANOVA: *analysis of variance* used for a *mixed design*.

Mixed design: an experimental design incorporating two or more *predictors* (or *independent variables*) at least one of which has been manipulated using different participants (or whatever entities are being tested) and at least one of which has been manipulated using the same participants (or entities). Also known as a split-plot design because Fisher developed ANOVA for analysing agricultural data involving 'plots' of land containing crops.

Mode: the most frequently occurring score in a set of data.

Model sum of squares: a measure of the total amount of variability for which a model can account. It is the difference between the *total sum of squares* and the *residual sum of squares*.

Monte Carlo method: a term applied to the process of using data simulations to solve statistical problems. Its name comes from the use of Monte Carlo roulette tables to generate 'random' numbers in the pre-computer age. Karl Pearson, for example, purchased copies of *Le Monaco*, a weekly Paris periodical that published data from the Monte Carlo casinos' roulette wheels. He used these data as pseudo-random numbers in his statistical research.

Moses extreme reactions: a non-parametric test that compares the variability of scores in two groups, so it's a bit like a non-parametric *Levene's test*.

Multicollinearity: a situation in which two or more variables are very closely linearly related.

Multilevel linear model (MLM): a linear model (just like regression, ANCOVA, ANOVA, etc.) in which the hierarchical structure of the data is explicitly considered. In this analysis regression parameters can be fixed (as in regression and ANOVA) but also random (i.e. free to vary across different contexts at a higher level of the hierarchy). This means that for each regression parameter there is a fixed component but also an estimate of how much the parameter varies across contexts (see *Fixed coefficient, Random coefficient*).

Multimodal: description of a distribution of observations that has more than two *modes*.

Multinomial logistic regression: *logistic regression* in which the outcome variable has more than two categories.

Multiple *R*: the multiple correlation coefficient. It is the correlation between the observed values of an outcome and the values of the outcome predicted by a multiple regression model.

Multiple regression: an extension of *simple regression* in which an outcome is predicted by a linear combination of two or more predictor variables. The form of the model is $Y_i = (b_0 + b_1X_{1i} + b_2X_{2i} + \ldots + b_nX_{ni}) + \varepsilon_i$ in which the outcome is denoted as Y and each predictor is denoted as X. Each predictor has a regression coefficient b_i associated with it, and b_0 is the value of the outcome when all predictors are zero.

Multivariate: means 'many variables' and is usually used when referring to analyses in which there is more than one *outcome variable* (e.g. *MANOVA*, *principal component analysis*, etc.).

Multivariate analysis of variance: family of tests that extend the basic *analysis of variance* to situations in which more than one *outcome variable* has been measured.

Multivariate normality: an extension of a normal distribution to multiple variables. It is a *probability distribution* of a set of variables $v' = [v_1, v_2, \ldots, v_n]$ given by:

$$f(v_1, v_2, \ldots, v_n) = 2\pi^{n/2}|\Sigma|^{1/2} \exp\left[-\frac{1}{2}(v-\mu)'\Sigma^{-1}(v-\mu)\right]$$

in which μ is the vector of means of the variables, and Σ is the *variance–covariance* matrix. If that made any sense to you then you're cleverer than I am.

Nagelkerke's R^2_N: a version of the *coefficient of determination* for logistic regression. It is a variation on *Cox and Snell's* R^2_{CS} which overcomes the problem that this statistic has of not being able to reach its maximum value.

Negative skew: see *Skew*.

Nominal variable: where numbers merely represent names. For example, the numbers on sports players shirts: a player with the number 1 on her back is not necessarily worse than a player with a 2 on her back. The numbers have no meaning other than denoting the type of player (i.e. full back, centre forward, etc.).

Noniles: a type of *quantile*; they are values that split the data into nine equal parts. They are commonly used in educational research.

Non-parametric tests: a family of statistical procedures that do not rely on the restrictive assumptions of parametric tests. In particular they do not assume that the sampling distribution is normally distributed.

Normal distribution: a *probability distribution* of a random variable that is known to have certain properties. It is perfectly symmetrical (has a *skew* of 0), and has a *kurtosis* of 0.

Null hypothesis: reverse of the *experimental hypothesis* that your prediction is wrong and the predicted effect doesn't exist.

Numeric variables: variables involving numbers.

Oblique rotation: a method of *rotation* in *factor analysis* that allows the underlying factors to be correlated.

Odds: the probability of an event occurring divided by the probability of that event not occurring.

Odds ratio: the ratio of the *odds* of an event occurring in one group compared to another. So, for example, if the odds of dying after writing a glossary are 4, and the odds of dying after not writing a glossary are 0.25, then the odds ratio is 4/0.25 = 16. This means that the *odds* of dying if you write a glossary are 16 times higher than if you dont. An odds ratio of 1 would indicate that the *odds* of a particular outcome are equal in both groups.

Omega squared: an *effect size* measure associated with ANOVA that is less biased than *eta squared*. It is a (sometimes hideous) function of the *model sum of squares* and the *residual sum of squares* and isn't actually much use because it measures the overall effect of the ANOVA and so can't be interpreted in a meaningful way. In all other respects it's great, though.

One-tailed test: a test of a directional hypothesis. For example, the hypothesis 'the longer I write this glossary, the more I want to place my editor's genitals in a starved crocodile's mouth' requires a one-tailed test because I've stated the direction of the relationship (see also *two-tailed test*).

Ordinal variable: data that tell us not only that things have occurred, but also the order in which they occurred. These data tell us nothing about the differences between values. For example, gold, silver and bronze medals are ordinal: they tell us that the gold medallist was better than the silver medallist, but they don't tell us how much better (was gold a lot better than silver, or were gold and silver very closely competed?).

Orthogonal: means perpendicular (at right angles) to something. It tends to be equated to *independence* in statistics because of the connotation that perpendicular *linear models* in geometric space are completely independent (one is not influenced by the other).

Orthogonal rotation: a method of *rotation* in *factor analysis* that keeps the underlying factors independent (i.e. not correlated).

Outcome variable: a variable whose values we are trying to predict from one or more *predictor variables*.

Outlier: an observation very different from most others. Outliers can bias statistics such as the mean.

Pairwise comparisons: comparisons of pairs of means.

Parametric test: a test that requires data from one of the large catalogue of distributions that statisticians have described. Normally this term is used for parametric tests based on the *normal distribution*, which require four basic assumptions that must be met for the test to be accurate: a normally distributed sampling distribution (see *Normal distribution*), *homogeneity of variance*, *interval* or *ratio data*, and *independence*.

Part correlation: another name for a *semi-partial correlation*.

Partial correlation: a measure of the relationship between two variables while 'controlling' the effect of one or more additional variables has on both.

Partial eta squared (partial η^2): a version of *eta squared* that is the proportion of variance that a variable explains when excluding other variables in the analysis. Eta squared is the proportion of total variance explained by a variable, whereas partial eta squared is the proportion of variance that a variable explains that is not explained by other variables.

Partial out: to partial out the effect of a variable is to remove the variance that the variable shares with other variables in the analysis before looking at their relationships (see *partial correlation*).

Pattern matrix: a matrix in *factor analysis* containing the *regression coefficients* for each variable on each *factor* in the data. See also *Structure matrix*.

Pearson's correlation coefficient: or Pearson's product-moment correlation coefficient to give it its full name, is a *standardized* measure of the strength of relationship between two variables. It can take any value from −1 (as one variable changes, the other changes in the opposite direction by the same amount), through 0 (as one variable changes the other doesn't change at all), to +1 (as one variable changes, the other changes in the same direction by the same amount).

Percentiles: are a type of *quantile*; they are values that split the data into 100 equal parts.

Perfect collinearity: exists when at least one predictor in a *regression model* is a perfect linear combination of the others (the simplest example being two predictors that are perfectly correlated – they have a correlation coefficient of 1).

Phi: a measure of the strength of association between two *categorical variables*. Phi is used with 2 × 2 *contingency tables* (tables which have two categorical variables and each variable has only two categories). Phi is a variant of the *chi square test*, χ^2: $\phi = \sqrt{\frac{\chi^2}{n}}$,

in which n is the total number of observations.

Pillai–Bartlett trace (*V*): a *test statistic* in *MANOVA*. It is the sum of the proportion of explained variance on the *discriminant function variates* of the data. As such, it is similar to the ratio of SS_M/SS_T.

Planned comparisons: another name for *planned contrasts*.

Planned contrasts: a set of comparisons between group means that are constructed before any data are collected. These are theory-led comparisons and are based on the idea of partitioning the variance created by the overall effect of group differences into gradually smaller portions of variance. These tests have more power than *post hoc tests*.

Platykurtic: see *Kurtosis*.

Point–biserial correlation: a standardized measure of the strength of relationship between two variables when one of the two variables is *dichotomous*. The point–biserial correlation coefficient is used when the dichotomy is discrete, or true, dichotomy (i.e. one for which there is no underlying continuum between the categories). An example of this is pregnancy: you can be either pregnant or not, there is no in between.

Polychotomous logistic regression: another name for *multinomial logistic regression*.

Polynomial: a posh name for a *growth curve* or trend over time. If *time* is our predictor variable, then any polynomial is tested by including a variable that is the predictor to the power of the order of polynomial that we want to test: a linear trend is tested by *time* alone, a quadratic or second-order polynomial is tested by including a predictor that is $time^2$, for a fifth-order polynomial we need a predictor of $time^5$ and for an nth-order polynomial we would have to include $time^n$ as a predictor.

Polynomial contrast: a contrast that tests for trends in the data. In its most basic form it looks for a linear trend (i.e. that the group means increase proportionately).

Population: in statistical terms this usually refers to the collection of units (be they people, plankton, plants, cities, suicidal authors, etc.) to which we want to generalize a set of findings or a statistical model.

Positive skew: see *Skew*.

***Post hoc* tests:** a set of comparisons between group means that were not thought of before data were collected. Typically these tests involve comparing the means of all combinations of pairs of groups. To compensate for the number of tests conducted, each test uses a strict criterion for significance. As such, they tend to have less power than *planned contrasts*. They are usually used for exploratory work for which no firm hypotheses were available on which to base planned contrasts.

Power: the ability of a test to detect an effect of a particular size (a value of .8 is a good level to aim for).

P–P plot: short for a probability–probability plot. A graph plotting the cumulative probability of a variable against the cumulative probability of a particular distribution (often a normal distribution). Like a *Q–Q plot*, if values fall on the diagonal of the plot then the variable shares the same distribution as the one specified. Deviations from the diagonal show deviations from the distribution of interest.

Practice effect: refers to the possibility that participants' performance in a task may be influenced (positively or negatively) if they repeat the task because of familiarity with the experimental situation and/or the measures being used.

Predictor variable: a variable that is used to try to predict values of another variable known as an *outcome variable*.

Principal component analysis (PCA): a *multivariate* technique for identifying the linear components of a set of variables.

Probability distribution: a curve describing an idealized *frequency distribution* of a particular variable from which it is possible to ascertain the probability with which specific values of that variable will occur. For categorical variables it is simply a formula yielding the probability with which each category occurs.

Promax: a method of *oblique rotation* that is computationally faster than *direct oblimin* and so useful for large data sets.

Q–Q plot: short for a quantile–quantile plot. A graph plotting the *quantiles* of a variable against the quantiles of a particular distribution (often a normal distribution). Like a *P–P plot*, if values fall on the diagonal of the plot then the variable shares the same distribution as the one specified. Deviations from the diagonal show deviations from the distribution of interest.

Quadratic trend: if the means in ordered conditions are connected with a line then a quadratic trend is shown by one change in the direction of this line (e.g. the line is curved in one place); the line is, therefore, U-shaped. There must be at least three ordered conditions.

Qualitative methods: extrapolating evidence for a theory from what people say or write (contrast with *quantitative methods*).

Quantiles: values that split a data set into equal portions. *Quartiles*, for example, are a special case of quantiles that split the data into four equal parts. Similarly, *percentiles* are points that split the data into 100 equal parts and *noniles* are points that split the data into 9 equal parts (you get the general idea).

Quantitative methods: inferring evidence for a theory through measurement of variables that produce numeric outcomes (contrast with *qualitative methods*).

Quartic trend: if the means in ordered conditions are connected with a line then a quartic trend is shown by three changes in the direction of this line. There must be at least five ordered conditions.

Quartiles: generic term for the three values that cut an ordered data set into four equal parts. The three quartiles are known as the *lower quartile*, the second quartile (or *median*) and the *upper quartile*.

Quartimax: a method of *orthogonal rotation*. It attempts to maximize the spread of factor loadings for a variable across all *factors*. This often results in lots of variables loading highly onto a single *factor*.

Random coefficient: a coefficient or model parameter that is free to vary over situations or contexts (cf. *Fixed coefficient*).

Random effect: an effect is said to be random if the experiment contains only a sample of possible treatment conditions. Random effects can be generalized beyond the treatment conditions in the experiment. For example, the effect is random if we say that the conditions in our experiment (e.g. placebo, low dose and high dose) are only a sample of possible conditions (maybe we could have tried a very high dose). We can generalize this random effect beyond just placebos, low doses and high doses.

Random intercept: a term used in *multilevel modelling* to denote when the intercept in the model is free to vary across different groups or contexts (cf. *Fixed intercept*).

Random slope: a term used in *multilevel modelling* to denote when the slope of the model is free to vary across different groups or contexts (cf. *Fixed slope*).

Random variable: a random variable is one that varies over time (e.g. your weight is likely to fluctuate over time).

Random variance: variance that is unique to a particular variable but not reliably so.

Randomization: the process of doing things in an unsystematic or random way. In the context of experimental research the word usually applies to the random assignment of participants to different treatment conditions.

Range: the range of scores is value of the smallest score subtracted from the highest score. It is a measure of the dispersion of a set of scores. See also *variance*, *standard deviation*, and *interquartile range*.

Ranking: the process of transforming raw scores into numbers that represent their position in an ordered list of those scores. i.e. the raw scores are ordered from lowest to highest and the lowest score is assigned a rank of 1, the next highest score is assigned a rank of 2, and so on.

Ratio variable: an *interval variable* but with the additional property that ratios are meaningful. For example, people's ratings of this book on Amazon.com can range from 1 to 5; for these data to be ratio not only must they have the properties of *interval variables*, but in addition a rating of 4 should genuinely represent someone who enjoyed this book twice as much as someone who rated it as 2. Likewise, someone who rated it as 1 should be half as impressed as someone who rated it as 2.

Regression coefficient: see b_i and β_i.

Regression line: a line on a scatterplot representing the *regression model* of the relationship between the two variables plotted.

Regression model: see *Multiple regression* and *Simple regression*.

Related design: another name for a *repeated-measures design*.

Related factorial design: an experimental design incorporating two or more *predictors* (or *independent variables*) all of which have been manipulated using the same participants (or whatever entities are being tested).

Reliability: the ability of a measure to produce consistent results when the same entities are measured under different conditions.

Repeated contrast: a non-orthogonal *planned contrast* that compares the mean in each condition (except the first) to the mean of the preceding condition.

Repeated-measures ANOVA: an *analysis of variance* conducted on any design in which the *independent variable* (*predictor*) or *variables* (*predictors*) have all been measured using the same participants in all conditions.

Repeated-measures design: an experimental design in which different treatment conditions utilize the same organisms (i.e. in psychology, this would mean the same people take part in all experimental conditions) and so the resulting data are related (a.k.a. related design or within-subject designs).

Residual: the difference between the value a model predicts and the value observed in the data on which the model is based. When the residual is calculated for each observation in a data set the resulting collection is referred to as the *residuals*.

Residuals: see *Residual*.

Residual sum of squares: a measure of the variability that cannot be explained by the model fitted to the data. It is the total squared *deviance* between the observations, and the value of those observations predicted by whatever model is fitted to the data.

Reverse Helmert contrast: another name for a *difference contrast*.

Roa's efficient score statistic: a statistic measuring the same thing as the *Wald statistic* but which is computationally easier to calculate.

Robust test: a term applied to a family of procedures to estimate statistics that are reliable even when the normal assumptions of the statistic are not met.

Rotation: a process in *factor analysis* for improving the interpretability of factors. In essence, an attempt is made to transform the *factors* that emerge from the analysis in such a way as to maximize *factor loadings* that are already large, and minimize factor loadings that are already small. There are two general approaches: *orthogonal rotation* and *oblique rotation*.

Roy's largest root: a *test statistic* in *MANOVA*. It is the eigenvalue for the first *discriminant function variate* of a set of observations. So, it is the same as the *Hotelling–Lawley trace* but for the first variate only. It represents the proportion of explained variance to unexplained variance (SS_M/SS_R) for the first discriminant function.

Sample: a smaller (but hopefully representative) collection of units from a *population* used to determine truths about that population (e.g. how a given population behaves in certain conditions).

Sampling distribution: the *probability distribution* of a statistic. We can think of this as follows: if we take a *sample* from a *population* and calculate some statistic (e.g. the *mean*), the value of this statistic will depend somewhat on the sample we took. As such the statistic will vary slightly from sample to sample. If, hypothetically, we took lots and lots of samples from the population and calculated the statistic of interest we could create a frequency distribution of the values we get. The resulting distribution is what the sampling distribution represents: the distribution of possible values of a given statistic that we could expect to get from a given population.

Sampling variation: the extent to which a statistic (e.g. the mean, median, *t*, *F*, etc.) varies in samples taken from the same population.

Saturated model: a model that perfectly fits the data and, therefore, has no error. It contains all possible *main effects* and *interactions* between variables.

Scatterplot: a graph that plots values of one variable against the corresponding value of another variable (and the corresponding value of a third variable can also be included on a 3-D scatterplot).

Scree plot: a graph plotting each *factor* in a *factor analysis* (*X*-axis) against its associated eigenvalue (*Y*-axis). It shows the relative importance of each factor. This graph has a very characteristic shape (there is a sharp descent in the curve followed by a tailing off) and the point of inflexion of this curve is often used as a means of *extraction*. With a sample of more than 200 participants, this provides a fairly reliable criterion for *extraction* (Stevens, 2002)

Second quartile: another name for the *median*.

Semi-partial correlation: a measure of the relationship between two variables while 'controlling' the effect that one or more additional variables has on one of those variables. If we call our variables *x* and *y*, it gives us a measure of the variance in *y* that *x* alone shares.

Shapiro–Wilk test: a test of whether a distribution of scores is significantly different from a *normal distribution*. A significant value indicates a deviation from normality, but this test is notoriously affected by large samples in which small deviations from normality yield significant results.

Shrinkage: the loss of predictive power of a regression model if the model had been derived from the population from which the sample was taken, rather than the sample itself.

Sidak correction: slightly less conservative variant of a *Bonferroni correction*.

Sign test: tests whether two related samples are different. It does the same thing as the *Wilcoxon signed-rank test*. Differences between the conditions are calculated and the sign of this difference (positive or negative) is analysed because it indicates the direction of differences. The magnitude of change is completely ignored (unlike in Wilcoxon's test where the rank tells us something about the relative magnitude of change), and for this reason it lacks *power*. However, its computational simplicity makes it a nice party trick if ever anyone drunkenly accosts you needing some data quickly analysed without the aid of a computer … doing a sign test in your head really impresses people. Actually it doesn't, they just think you're a sad gimboid.

Simple contrast: a non-orthogonal *planned contrast* that compares the mean in each condition to the mean of either the first or last condition depending on how the contrast is specified.

Simple effects analysis: this analysis looks at the effect of one *independent variable* (categorical *predictor variable*) at individual levels of another independent variable.

Simple regression: a *linear model* in which one variable or outcome is predicted from a single predictor variable. The model takes the form $Y_i = (b_0 + b_1X_i) + \varepsilon_i$ in which *Y* is the outcome variable, *X* is the predictor, b_1 is the regression coefficient associated with the predictor and b_0 is the value of the outcome when the predictor is zero.

Singularity: a term used to describe variables that are perfectly correlated (i.e. the *correlation coefficient* is 1 or −1).

Skew: a measure of the symmetry of a *frequency distribution*. Symmetrical distributions have a skew of 0. When the frequent scores are clustered at the lower end of the distribution and the tail points towards the higher or more positive scores, the value of skew is positive. Conversely, when the frequent scores are clustered at the higher end of the distribution and the tail points towards the lower more negative scores, the value of skew is negative.

SmartViewer: a program that accompanies SPSS that enables output files (.spo files) to be viewed from pre-version 16 editions of SPSS.

Spearman's correlation coefficient: a standardized measure of the strength of relationship between two variables that does not rely on the assumptions of a *parametric test*. It is *Pearson's correlation coefficient* performed on data that have been converted into ranked scores.

Sphericity: a less restrictive form of *compound symmetry* which assumes that the variances of the differences between data taken from the same participant (or other entity being tested) are equal. This assumption is most commonly found in *repeated-measures ANOVA* but applies only where there are more than two points of data from the same participant. (see also *Greenhouse–Geisser correction*, *Huynh–Feldt correction*).

Split-half reliability: a measure of *reliability* obtained by splitting items on a measure into two halves (in some random fashion) and obtaining a score from each half of the scale. The correlation between the two scores, corrected to take account of the fact the correlations are based on only half of the items, is used as a measure of reliability. There are two popular ways to do this. Spearman (1910) and Brown (1910) developed a formula that takes no account of the standard deviation of items:

$$r_{sh} = \frac{2r_{12}}{1 + r_{12}}$$

in which r_{12} is the correlation between the two halves of the scale. Flanagan (1937) and Rulon (1939), however, proposed a measure that does account for item variance:

$$r_{sh} = \frac{4r_{12} \times s_1 \times s_2}{s_T^2}$$

in which s_1 and s_2 are the standard deviations of each half of the scale, and s_T^2 is the variance of the whole test. See Cortina (1993) for more detail.

Square matrix: a *matrix* that has an equal number of columns and rows.

Standard deviation: an estimate of the average variability (spread) of a set of data measured in the same units of measurement as the original data. It is the square root of the variance.

Standard error: the standard deviation of the *sampling distribution* of a statistic. For a given statistic (e.g. the *mean*) it tells us how much variability there is in this statistic across *samples* from the same *population*. Large values, therefore, indicate that a statistic from a given sample may not be an accurate reflection of the population from which the sample came.

Standard error of differences: if we were to take several pairs of samples from a population and calculate their means, then we could also calculate the difference between their means. If we plotted these differences between sample means as a *frequency distribution*, we would have the *sampling distribution* of differences. The standard deviation of this sampling distribution is the *standard error of differences*. As such it is a measure of the variability of differences between sample means.

Standard error of the mean (SE): the full name of the *standard error*.

Standardization: the process of converting a variable into a standard unit of measurement. The unit of measurement typically used is *standard deviation* units (see also *z-scores*). Standardization allows us to compare data when different units of measurement have been used (we could compare weight measured in kilograms to height measured in inches).

Standardized: see *Standardization*.

Standardized DFBeta: a *standardized* version of *DFBeta*. These standardized values are easier to use than DFBeta because universal cut-off points can be applied. Stevens (2002) suggests looking at cases with absolute values greater than 2.

Standardized DFFit: a *standardized* version of *DFFit*.

Standardized residuals: the *residuals* of a model expressed in standard deviation units. Standardized residuals with an absolute value greater than 3.29 (actually we usually just use 3) are cause for concern because in an average sample a value this high is unlikely to happen by chance; if more than 1% of our observations have standardized residuals with an absolute value greater than 2.58 (we usually just say 2.5) there is evidence that the level of error within our model is unacceptable (the model is a fairly poor fit of the sample data); and if more than 5% of observations have standardized residuals with an absolute value greater than 1.96 (or 2 for convenience) then there is also evidence that the model is a poor representation of the actual data.

Stepwise regression: a method of multiple regression in which variables are entered into the model based on a statistical criterion (the semi-partial correlation with the outcome variable). Once a new variable is entered into the model, all variables in the model are assessed to see whether they should be removed.

String variables: variables involving words (i.e. letter strings). Such variables could include responses to open-ended questions such as 'how much do you like writing glossary entries?'; the response might be 'about as much as I like placing my gonads on hot coals'.

Structure matrix: a matrix in *factor analysis* containing the *correlation coefficients* for each variable on each *factor* in the data. When *orthogonal rotation* is used this is the same as the *pattern matrix*, but when oblique rotation is used these matrices are different.

Studentized deleted residual: a measure of the influence of a particular case of data. This is a standardized version of the *deleted residual*.

Studentized residuals: a variation on *standardized residuals*. Studentized residuals are the *unstandardized residual* divided by an estimate of its standard deviation that varies point by point. These residuals have the same properties as the *standardized residuals* but usually provide a more precise estimate of the error variance of a specific case.

Sum of squared errors: another name for the *sum of squares*.

Sum of squares (SS): an estimate of total variability (spread) of a set of data. First the *deviance* for each score is calculated, and then this value is squared. The SS is the sum of these squared deviances.

Sum of squares and cross-products matrix (SSCP matrix): a *square matrix* in which the diagonal elements represent the *sum of squares* for a particular variable, and the off-diagonal elements represent the *cross-products* between pairs of variables. The SSCP matrix is basically the same as the *variance–covariance matrix*, except the SSCP matrix expresses variability and between-variable relationships as total values, whereas the variance–covariance matrix expresses them as average values.

Suppressor effects: when a predictor has a significant effect but only when another variable is held constant.

Syntax: predefined written commands that instruct SPSS what you would like it to do (writing 'bugger off and leave me alone' doesn't seem to work ...).

Syntax Editor: a window in SPSS for writing and editing *syntax*.

Systematic variation: variation due to some genuine effect (be that the effect of an experimenter doing something to all of the participants in one sample but not in other samples, or natural variation between sets of variables). We can think of this as variation that can be explained by the model that we've fitted to the data.

***T*-statistic:** Student's *t* is a *test statistic* with a known *probability distribution* (the *t*-distribution). In the context of regression it is used to test whether a regression coefficient *b* is significantly different from zero; in the context of experimental work it is used to test whether the differences between two means are significantly different from zero. See also *Dependent t-test* and *Independent t-test*.

Tertium quid: the possibility that an apparent relationship between two variables is actually caused by the effect of a third variable on them both (often called *the third-variable problem*).

Test–retest reliability: the ability of a measure to produce consistent results when the same entities are tested at two different points in time.

Test statistic: a statistic for which we know how frequently different values occur. The observed value of such a statistic is typically used to test *hypotheses*.

Theory: although it can be defined more formally, a theory is a hypothesized general principle or set of principles that explain known findings about a topic and from which new hypotheses can be generated.

Tolerance: tolerance statistics measure *multicollinearity* and are simply the reciprocal of the *variance inflation factor* (1/VIF). Values below 0.1 indicate serious problems, although Menard (1995) suggests that values below 0.2 are worthy of concern.

Total SSCP (*T*): the total sum of squares and cross-product matrix. This is a *sum of squares and cross-product matrix* for an entire set of observations. It is the *multivariate* equivalent of the *total sum of squares*.

Total sum of squares: a measure of the total variability within a set of observations. It is the total squared *deviance* between each observation and the overall mean of all observations.

Transformation: the process of applying a mathematical function to all observations in a data set, usually to correct some distributional abnormality such as *skew* or *kurtosis*.

Trimmed mean: a statistic used in many *robust tests*. Imagine we had 20 scores representing the annual income of students (in thousands, rounded to the nearest thousand: 2, 2, 2, 2, 3, 3, 3, 3, 3, 4, 4, 4, 4, 4, 4, 4, 4, 4, 6, 35. The mean income is 5 (£5000). This value is biased by an outlier. A trimmed mean is simply a mean based on the distribution of scores after some percentage of scores has been removed from each extreme of the distribution. So, a 10% trimmed mean will remove 10% of scores from the top and bottom of ordered scores before the mean is calculated. With 20 scores, removing 10% of scores involves removing the top and bottom 2 scores. This gives us: 2, 2, 3, 3, 3, 3, 3, 4, 4, 4, 4, 4, 4, 4, 4, 4, the mean of which is 3.44. The mean depends on a symmetrical distribution to be accurate, but a trimmed mean produces accurate results even when the distribution is not symmetrical. There are more complex examples of robust methods such as the *bootstrap*.

Two-tailed test: a test of a non-directional hypothesis. For example, the hypothesis 'writing this glossary has some effect on what I want to do with my editor's genitals' requires a two-tailed test because it doesn't suggest the direction of the relationship. (See also *One-tailed test.*)

Type I error: occurs when we believe that there is a genuine effect in our population, when in fact there isn't.

Type II error: occurs when we believe that there is no effect in the population when, in reality, there is.

Unique variance: variance that is specific to a particular variable (i.e. is not shared with other variables). We tend to use the term 'unique variance' to refer to variance that can be reliably attributed to only one measure, otherwise it is called *random variance*.

Univariate: means 'one variable' and is usually used to refer to situations in which only one *outcome variable* has been measured (i.e. *ANOVA*, *t-tests*, *Mann–Whitney tests,* etc.).

Unstructured: a covariance structure used in *multilevel models*. This covariance structure is completely general and is, therefore, the default option used in random effects in SPSS. Covariances are assumed to be completely unpredictable: they do not conform to a systematic pattern.

Unstandardized residuals: the *residuals* of a model expressed in the units in which the original outcome variable was measured.

Unsystematic variation: this is variation that isn't due to the effect in which we're interested (so could be due to natural differences between people in different samples such as differences in intelligence or motivation). We can think of this as variation that can't be explained by whatever model we've fitted to the data.

Upper quartile: the value that cuts off the highest 25% of ordered scores. If the scores are ordered and then divided into two halves at the median, then the upper quartile is the median of the top half of the scores.

Validity: evidence that a study allows correct inferences about the question it was aimed to answer or that a test measures what it set out to measure conceptually (see also *Content validity*, *Criterion validity*).

Variables: anything that can be measured and can differ across entities or across time.

Variable view: there are two ways to view the contents of the *data editor* window. The variable view allows you to define properties of the variables for which you wish to enter data. See also *data view*.

Variance: an estimate of average variability (spread) of a set of data. It is the sum of squares divided by the number of values on which the sum of squares is based minus 1.

Variance components: a covariance structure used in *multilevel models*. This covariance structure is very simple and assumes that all random effects are independent and variances of random effects are assumed to be the same and sum to the variance of the outcome variable. In SPSS this is the default covariance structure for random effects.

Variance–covariance matrix: a square matrix (i.e. same number of columns and rows) representing the variables measured. The diagonals represent the *variances* within each variable, whereas the off-diagonals represent the *covariances* between pairs of variables.

Variance inflation factor (VIF): a measure of *multicollinearity*. The VIF indicates whether a predictor has a strong linear relationship with the other predictor(s). Myers (1990) suggests that a value of 10 is a good value at which to worry. Bowerman and O'Connell (1990) suggest that if the average VIF is greater than 1, then multicollinearity may be biasing the regression model.

Variance ratio: see *Hartley's* F_{max}.

Variance sum law: states that the variance of a difference between two independent variables is equal to the sum of their variances.

Varimax: a method of *orthogonal rotation*. It attempts to maximize the dispersion of *factor loadings* within *factors*. Therefore, it tries to load a smaller number of variables highly onto each factor resulting in more interpretable clusters of factors.

Viewer: SPSS window in which output of any analysis is displayed.

VIF: see *Variance inflation factor*.

Wald statistic: a *test statistic* with a known *probability distribution* (a *chi-square distribution*) that is used to test whether the *b* coefficient for a predictor in a *logistic* regression model is significantly different from zero. It is analogous to the *t-statistic* in a *regression model* in that it is simply the *b* coefficient divided by its standard error. The Wald statistic is inaccurate when the regression coefficient (*b*) is large, because the standard error tends to become inflated, resulting in the Wald statistic being underestimated.

Wald–Wolfowitz runs: another variant on the *Mann–Whitney test*. Scores are rank ordered as in the Mann–Whitney test, but rather than analysing the ranks, this test looks for 'runs' of scores from the same group within the ranked order. Now, if there's no difference between groups then obviously ranks from the two groups should be randomly interspersed. However, if the groups are different then one should see more ranks from one group at the lower end, and more ranks from the other group at the higher end. By looking for clusters of scores in this way the test can determine if the groups differ.

Weights: a number by which something (usually a variable in statistics) is multiplied. The weight assigned to a variable determines the influence that variable has within a mathematical equation: large weights give the variable a lot of influence.

Welch's *F*: a version of the *F*-ratio designed to be accurate when the assumption of *homogeneity of variance* has been violated. Not to be confused with the squelch test which is where you shake your head around after writing statistics books to see if you still have a brain.

Wilcoxon's rank-sum test: a *non-parametric test* that looks for differences between two independent samples. That is, it tests whether the populations from which two samples are drawn have the same location. It is functionally the same as the *Mann–Whitney test*, and both tests are non-parametric equivalents of the *independent t-test*.

Wilcoxon signed-rank test: a *non-parametric test* that looks for differences between two related samples. It is the non-parametric equivalent of the *related t-test*.

Wilks's lambda (Λ): a *test statistic* in *MANOVA*. It is the product of the unexplained variance on each of the *discriminant function variates* so it represents the ratio of error variance to total variance (SS_R/SS_T) for each variate.

Within-subject design: another name for a *repeated-measures design*.

Writer's block: something I suffered from a lot while writing this edition. It's when you can't think of any decent examples and so end up talking about sperm the whole time. Seriously, look at this book, it's all sperm this, sperm that, quail sperm, human sperm. Frankly, I'm amazed donkey sperm didn't get in there somewhere. Oh, it just did.

Yates's continuity correction: an adjustment made to the *chi-square test* when the *contingency table* is 2 rows by 2 columns (i.e. there are two categorical variables both of which consist of only two categories). In large samples the adjustment makes little difference and is slightly dubious anyway (see Howell, 2006).

z-score: the value of an observation expressed in standard deviation units. It is calculated by taking the observation, subtracting from it the mean of all observations, and dividing the result by the standard deviation of all observations. By converting a distribution of observations into *z*-scores a new distribution is created that has a mean of 0 and a standard deviation of 1.

APPENDIX

A.1. Table of the standard normal distribution

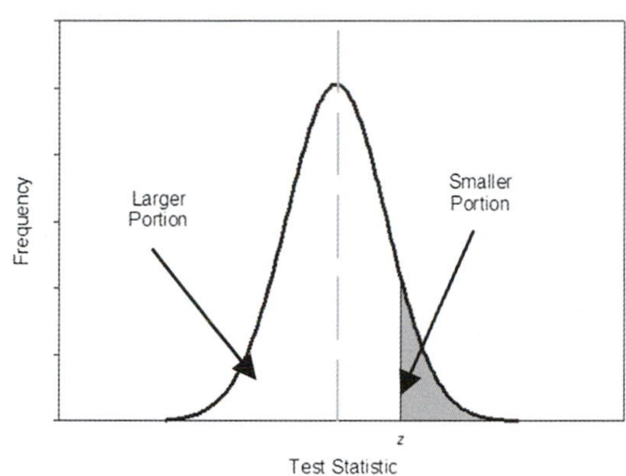

z	Larger Portion	Smaller Portion	y
.00	.50000	.50000	.3989
.01	.50399	.49601	.3989
.02	.50798	.49202	.3989
.03	.51197	.48803	.3988
.04	.51595	.48405	.3986
.05	.51994	.48006	.3984
.06	.52392	.47608	.3982
.07	.52790	.47210	.3980
.08	.53188	.46812	.3977
.09	.53586	.46414	.3973
.10	.53983	.46017	.3970
.11	.54380	.45620	.3965
.12	.54776	.45224	.3961
.13	.55172	.44828	.3956
.14	.55567	.44433	.3951
.15	.55962	.44038	.3945
.16	.56356	.43644	.3939
.17	.56749	.43251	.3932
.18	.57142	.42858	.3925
.19	.57535	.42465	.3918
.20	.57926	.42074	.3910
.21	.58317	.41683	.3902
.22	.58706	.41294	.3894
.23	.59095	.40905	.3885

(Continued)

(Continued)

z	Larger Portion	Smaller Portion	y
.24	.59483	.40517	.3876
.25	.59871	.40129	.3867
.26	.60257	.39743	.3857
.27	.60642	.39358	.3847
.28	.61026	.38974	.3836
.29	.61409	.38591	.3825
.30	.61791	.38209	.3814
.31	.62172	.37828	.3802
.32	.62552	.37448	.3790
.33	.62930	.37070	.3778
.34	.63307	.36693	.3765
.35	.63683	.36317	.3752
.36	.64058	.35942	.3739
.37	.64431	.35569	.3725
.38	.64803	.35197	.3712
.39	.65173	.34827	.3697
.40	.65542	.34458	.3683
.41	.65910	.34090	.3668
.42	.66276	.33724	.3653
.43	.66640	.33360	.3637
.44	.67003	.32997	.3621
.45	.67364	.32636	.3605
.46	.67724	.32276	.3589
.47	.68082	.31918	.3572
.48	.68439	.31561	.3555
.49	.68793	.31207	.3538
.50	.69146	.30854	.3521
.51	.69497	.30503	.3503
.52	.69847	.30153	.3485
.53	.70194	.29806	.3467

z	Larger Portion	Smaller Portion	y
.54	.70540	.29460	.3448
.55	.70884	.29116	.3429
.56	.71226	.28774	.3410
.57	.71566	.28434	.3391
.58	.71904	.28096	.3372
.59	.72240	.27760	.3352
.60	.72575	.27425	.3332
.61	.72907	.27093	.3312
.62	.73237	.26763	.3292
.63	.73565	.26435	.3271
.64	.73891	.26109	.3251
.65	.74215	.25785	.3230
.66	.74537	.25463	.3209
.67	.74857	.25143	.3187
.68	.75175	.24825	.3166
.69	.75490	.24510	.3144
.70	.75804	.24196	.3123
.71	.76115	.23885	.3101
.72	.76424	.23576	.3079
.73	.76730	.23270	.3056
.74	.77035	.22965	.3034
.75	.77337	.22663	.3011
.76	.77637	.22363	.2989
.77	.77935	.22065	.2966
.78	.78230	.21770	.2943
.79	.78524	.21476	.2920
.80	.78814	.21186	.2897
.81	.79103	.20897	.2874
.82	.79389	.20611	.2850
.83	.79673	.20327	.2827

z	Larger Portion	Smaller Portion	y
.84	.79955	.20045	.2803
.85	.80234	.19766	.2780
.86	.80511	.19489	.2756
.87	.80785	.19215	.2732
.88	.81057	.18943	.2709
.89	.81327	.18673	.2685
.90	.81594	.18406	.2661
.91	.81859	.18141	.2637
.92	.82121	.17879	.2613
.93	.82381	.17619	.2589
.94	.82639	.17361	.2565
.95	.82894	.17106	.2541
.96	.83147	.16853	.2516
.97	.83398	.16602	.2492
.98	.83646	.16354	.2468
.99	.83891	.16109	.2444
1.00	.84134	.15866	.2420
1.01	.84375	.15625	.2396
1.02	.84614	.15386	.2371
1.03	.84849	.15151	.2347
1.04	.85083	.14917	.2323
1.05	.85314	.14686	.2299
1.06	.85543	.14457	.2275
1.07	.85769	.14231	.2251
1.08	.85993	.14007	.2227
1.09	.86214	.13786	.2203
1.10	.86433	.13567	.2179
1.11	.86650	.13350	.2155
1.12	.86864	.13136	.2131
1.13	.87076	.12924	.2107

z	Larger Portion	Smaller Portion	y
1.14	.87286	.12714	.2083
1.15	.87493	.12507	.2059
1.16	.87698	.12302	.2036
1.17	.87900	.12100	.2012
1.18	.88100	.11900	.1989
1.19	.88298	.11702	.1965
1.20	.88493	.11507	.1942
1.21	.88686	.11314	.1919
1.22	.88877	.11123	.1895
1.23	.89065	.10935	.1872
1.24	.89251	.10749	.1849
1.25	.89435	.10565	.1826
1.26	.89617	.10383	.1804
1.27	.89796	.10204	.1781
1.28	.89973	.10027	.1758
1.29	.90147	.09853	.1736
1.30	.90320	.09680	.1714
1.31	.90490	.09510	.1691
1.32	.90658	.09342	.1669
1.33	.90824	.09176	.1647
1.34	.90988	.09012	.1626
1.35	.91149	.08851	.1604
1.36	.91309	.08691	.1582
1.37	.91466	.08534	.1561
1.38	.91621	.08379	.1539
1.39	.91774	.08226	.1518
1.40	.91924	.08076	.1497
1.41	.92073	.07927	.1476
1.42	.92220	.07780	.1456
1.43	.92364	.07636	.1435

(Continued)

(Continued)

z	Larger Portion	Smaller Portion	y
1.44	.92507	.07493	.1415
1.45	.92647	.07353	.1394
1.46	.92785	.07215	.1374
1.47	.92922	.07078	.1354
1.48	.93056	.06944	.1334
1.49	.93189	.06811	.1315
1.50	.93319	.06681	.1295
1.51	.93448	.06552	.1276
1.52	.93574	.06426	.1257
1.53	.93699	.06301	.1238
1.54	.93822	.06178	.1219
1.55	.93943	.06057	.1200
1.56	.94062	.05938	.1182
1.57	.94179	.05821	.1163
1.58	.94295	.05705	.1145
1.59	.94408	.05592	.1127
1.60	.94520	.05480	.1109
1.61	.94630	.05370	.1092
1.62	.94738	.05262	.1074
1.63	.94845	.05155	.1057
1.64	.94950	.05050	.1040
1.65	.95053	.04947	.1023
1.66	.95154	.04846	.1006
1.67	.95254	.04746	.0989
1.68	.95352	.04648	.0973
1.69	.95449	.04551	.0957
1.70	.95543	.04457	.0940
1.71	.95637	.04363	.0925
1.72	.95728	.04272	.0909
1.73	.95818	.04182	.0893

z	Larger Portion	Smaller Portion	y
1.74	.95907	.04093	.0878
1.75	.95994	.04006	.0863
1.76	.96080	.03920	.0848
1.77	.96164	.03836	.0833
1.78	.96246	.03754	.0818
1.79	.96327	.03673	.0804
1.80	.96407	.03593	.0790
1.81	.96485	.03515	.0775
1.82	.96562	.03438	.0761
1.83	.96638	.03362	.0748
1.84	.96712	.03288	.0734
1.85	.96784	.03216	.0721
1.86	.96856	.03144	.0707
1.87	.96926	.03074	.0694
1.88	.96995	.03005	.0681
1.89	.97062	.02938	.0669
1.90	.97128	.02872	.0656
1.91	.97193	.02807	.0644
1.92	.97257	.02743	.0632
1.93	.97320	.02680	.0620
1.94	.97381	.02619	.0608
1.95	.97441	.02559	.0596
1.96	.97500	.02500	.0584
1.97	.97558	.02442	.0573
1.98	.97615	.02385	.0562
1.99	.97670	.02330	.0551
2.00	.97725	.02275	.0540
2.01	.97778	.02222	.0529
2.02	.97831	.02169	.0519
2.03	.97882	.02118	.0508

z	Larger Portion	Smaller Portion	y
2.04	.97932	.02068	.0498
2.05	.97982	.02018	.0488
2.06	.98030	.01970	.0478
2.07	.98077	.01923	.0468
2.08	.98124	.01876	.0459
2.09	.98169	.01831	.0449
2.10	.98214	.01786	.0440
2.11	.98257	.01743	.0431
2.12	.98300	.01700	.0422
2.13	.98341	.01659	.0413
2.14	.98382	.01618	.0404
2.15	.98422	.01578	.0396
2.16	.98461	.01539	.0387
2.17	.98500	.01500	.0379
2.18	.98537	.01463	.0371
2.19	.98574	.01426	.0363
2.20	.98610	.01390	.0355
2.21	.98645	.01355	.0347
2.22	.98679	.01321	.0339
2.23	.98713	.01287	.0332
2.24	.98745	.01255	0325
2.25	.98778	.01222	.0317
2.26	.98809	.01191	.0310
2.27	.98840	.01160	.0303
2.28	.98870	.01130	.0297
2.29	.98899	.01101	.0290
2.30	.98928	.01072	.0283
2.31	.98956	.01044	.0277
2.32	.98983	.01017	.0270
2.33	.99010	.00990	.0264

z	Larger Portion	Smaller Portion	y
2.34	.99036	.00964	.0258
2.35	.99061	.00939	.0252
2.36	.99086	.00914	.0246
2.37	.99111	.00889	.0241
2.38	.99134	.00866	.0235
2.39	.99158	.00842	.0229
2.40	.99180	.00820	.0224
2.41	.99202	.00798	.0219
2.42	.99224	.00776	.0213
2.43	.99245	.00755	.0208
2.44	.99266	.00734	.0203
2.45	.99286	.00714	.0198
2.46	.99305	.00695	.0194
2.47	.99324	.00676	.0189
2.48	.99343	.00657	.0184
2.49	.99361	.00639	.0180
2.50	.99379	.00621	.0175
2.51	.99396	.00604	.0171
2.52	.99413	.00587	.0167
2.53	.99430	.00570	.0163
2.54	.99446	.00554	.0158
2.55	.99461	.00539	.0154
2.56	.99477	.00523	.0151
2.57	.99492	.00508	.0147
2.58	.99506	.00494	.0143
2.59	.99520	.00480	.0139
2.60	.99534	.00466	.0136
2.61	.99547	.00453	.0132
2.62	.99560	.00440	.0129
2.63	.99573	.00427	.0126

(Continued)

(Continued)

z	Larger Portion	Smaller Portion	y
2.64	.99585	.00415	.0122
2.65	.99598	.00402	.0119
2.66	.99609	.00391	.0116
2.67	.99621	.00379	.0113
2.68	.99632	.00368	.0110
2.69	.99643	.00357	.0107
2.70	.99653	.00347	.0104
2.71	.99664	.00336	.0101
2.72	.99674	.00326	.0099
2.73	.99683	.00317	.0096
2.74	.99693	.00307	.0093
2.75	.99702	.00298	.0091
2.76	.99711	.00289	.0088
2.77	.99720	.00280	.0086
2.78	.99728	.00272	.0084
2.79	.99736	.00264	.0081
2.80	.99744	.00256	.0079
2.81	.99752	.00248	.0077
2.82	.99760	.00240	.0075
2.83	.99767	.00233	.0073
2.84	.99774	.00226	.0071
2.85	.99781	.00219	.0069

z	Larger Portion	Smaller Portion	y
2.86	.99788	.00212	.0067
2.87	.99795	.00205	.0065
2.88	.99801	.00199	.0063
2.89	.99807	.00193	.0061
2.90	.99813	.00187	.0060
2.91	.99819	.00181	.0058
2.92	.99825	.00175	.0056
2.93	.99831	.00169	.0055
2.94	.99836	.00164	.0053
2.95	.99841	.00159	.0051
2.96	.99846	.00154	.0050
2.97	.99851	.00149	.0048
2.98	.99856	.00144	.0047
2.99	.99861	.00139	.0046
3.00	.99865	.00135	.0044
⋮	⋮	⋮	⋮
3.25	.99942	.00058	.0020
⋮	⋮	⋮	⋮
3.50	.99977	.00023	.0009
⋮	⋮	⋮	⋮
4.00	.99997	.00003	.0001

All values calculated by the author using SPSS.

A.2. Critical values of the *t*-distribution

	Two-Tailed Test		*One-Tailed Test*	
df	***0.05***	***0.01***	***0.05***	***0.01***
1	12.71	63.66	6.31	31.82
2	4.30	9.92	2.92	6.96
3	3.18	5.84	2.35	4.54
4	2.78	4.60	2.13	3.75
5	2.57	4.03	2.02	3.36
6	2.45	3.71	1.94	3.14
7	2.36	3.50	1.89	3.00
8	2.31	3.36	1.86	2.90
9	2.26	3.25	1.83	2.82
10	2.23	3.17	1.81	2.76
11	2.20	3.11	1.80	2.72
12	2.18	3.05	1.78	2.68
13	2.16	3.01	1.77	2.65
14	2.14	2.98	1.76	2.62
15	2.13	2.95	1.75	2.60
16	2.12	2.92	1.75	2.58
17	2.11	2.90	1.74	2.57
18	2.10	2.88	1.73	2.55
19	2.09	2.86	1.73	2.54
20	2.09	2.85	1.72	2.53
21	2.08	2.83	1.72	2.52
22	2.07	2.82	1.72	2.51
23	2.07	2.81	1.71	2.50
24	2.06	2.80	1.71	2.49
25	2.06	2.79	1.71	2.49
26	2.06	2.78	1.71	2.48
27	2.05	2.77	1.70	2.47
28	2.05	2.76	1.70	2.47
29	2.05	2.76	1.70	2.46
30	2.04	2.75	1.70	2.46
35	2.03	2.72	1.69	2.44
40	2.02	2.70	1.68	2.42
45	2.01	2.69	1.68	2.41
50	2.01	2.68	1.68	2.40
60	2.00	2.66	1.67	2.39
70	1.99	2.65	1.67	2.38
80	1.99	2.64	1.66	2.37
90	1.99	2.63	1.66	2.37
100	1.98	2.63	1.66	2.36
∞ (z)	1.96	2.58	1.64	2.33

All values computed by the author using SPSS.

A.3. Critical values of the *F*-distribution

df (denominator)		df (numerator)									
	p	1	2	3	4	5	6	7	8	9	10
1	.05	161.45	199.50	215.71	224.58	230.16	233.99	236.77	238.88	240.54	241.88
	.01	4052.18	4999.50	5403.35	5624.58	5763.65	5858.99	5928.36	5981.07	6022.47	6055.85
2	.05	18.51	19.00	19.16	19.25	19.30	19.33	19.35	19.37	19.38	19.40
	.01	98.50	99.00	99.17	99.25	99.30	99.33	99.36	99.37	99.39	99.40
3	.05	10.13	9.55	9.28	9.12	9.01	8.94	8.89	8.85	8.81	8.79
	.01	34.12	30.82	29.46	28.71	28.24	27.91	27.67	27.49	27.35	27.23
4	.05	7.71	6.94	6.59	6.39	6.26	6.16	6.09	6.04	6.00	5.96
	.01	21.20	18.00	16.69	15.98	15.52	15.21	14.98	14.80	14.66	14.55
5	.05	6.61	5.79	5.41	5.19	5.05	4.95	4.88	4.82	4.77	4.74
	.01	16.26	13.27	12.06	11.39	10.97	10.67	10.46	10.29	10.16	10.05
6	.05	5.99	5.14	4.76	4.53	4.39	4.28	4.21	4.15	4.10	4.06
	.01	13.75	10.92	9.78	9.15	8.75	8.47	8.26	8.10	7.98	7.87
7	.05	5.59	4.74	4.35	4.12	3.97	3.87	3.79	3.73	3.68	3.64
	.01	12.25	9.55	8.45	7.85	7.46	7.19	6.99	6.84	6.72	6.62
8	.05	5.32	4.46	4.07	3.84	3.69	3.58	3.50	3.44	3.39	3.35
	.01	11.26	8.65	7.59	7.01	6.63	6.37	6.18	6.03	5.91	5.81
9	.05	5.12	4.26	3.86	3.63	3.48	3.37	3.29	3.23	3.18	3.14
	.01	10.56	8.02	6.99	6.42	6.06	5.80	5.61	5.47	5.35	5.26
10	.05	4.96	4.10	3.71	3.48	3.33	3.22	3.14	3.07	3.02	2.98
	.01	10.04	7.56	6.55	5.99	5.64	5.39	5.20	5.06	4.94	4.85
11	.05	4.84	3.98	3.59	3.36	3.20	3.09	3.01	2.95	2.90	2.85
	.01	9.65	7.21	6.22	5.67	5.32	5.07	4.89	4.74	4.63	4.54
12	.05	4.75	3.89	3.49	3.26	3.11	3.00	2.91	2.85	2.80	2.75
	.01	9.33	6.93	5.95	5.41	5.06	4.82	4.64	4.50	4.39	4.30
13	.05	4.67	3.81	3.41	3.18	3.03	2.92	2.83	2.77	2.71	2.67
	.01	9.07	6.70	5.74	5.21	4.86	4.62	4.44	4.30	4.19	4.10
14	.05	4.60	3.74	3.34	3.11	2.96	2.85	2.76	2.70	2.65	2.60
	.01	8.86	6.51	5.56	5.04	4.69	4.46	4.28	4.14	4.03	3.94
15	.05	4.54	3.68	3.29	3.06	2.90	2.79	2.71	2.64	2.59	2.54
	.01	8.68	6.36	5.42	4.89	4.56	4.32	4.14	4.00	3.89	3.80
16	.05	4.49	3.63	3.24	3.01	2.85	2.74	2.66	2.59	2.54	2.49
	.01	8.53	6.23	5.29	4.77	4.44	4.20	4.03	3.89	3.78	3.69
17	.05	4.45	3.59	3.20	2.96	2.81	2.70	2.61	2.55	2.49	2.45
	.01	8.40	6.11	5.18	4.67	4.34	4.10	3.93	3.79	3.68	3.59
18	.05	4.41	3.55	3.16	2.93	2.77	2.66	2.58	2.51	2.46	2.41
	.01	8.29	6.01	5.09	4.58	4.25	4.01	3.84	3.71	3.60	3.51

df (denominator)	p	1	2	3	4	5	6	7	8	9	10
		df (numerator)									
19	.05	4.38	3.52	3.13	2.90	2.74	2.63	2.54	2.48	2.42	2.38
	.01	8.18	5.93	5.01	4.50	4.17	3.94	3.77	3.63	3.52	3.43
20	.05	4.35	3.49	3.10	2.87	2.71	2.60	2.51	2.45	2.39	2.35
	.01	8.10	5.85	4.94	4.43	4.10	3.87	3.70	3.56	3.46	3.37
22	.05	4.30	3.44	3.05	2.82	2.66	2.55	2.46	2.40	2.34	2.30
	.01	7.95	5.72	4.82	4.31	3.99	3.76	3.59	3.45	3.35	3.26
24	.05	4.26	3.40	3.01	2.78	2.62	2.51	2.42	2.36	2.30	2.25
	.01	7.82	5.61	4.72	4.22	3.90	3.67	3.50	3.36	3.26	3.17
26	.05	4.23	3.37	2.98	2.74	2.59	2.47	2.39	2.32	2.27	2.22
	.01	7.72	5.53	4.64	4.14	3.82	3.59	3.42	3.29	3.18	3.09
28	.05	4.20	3.34	2.95	2.71	2.56	2.45	2.36	2.29	2.24	2.19
	.01	7.64	5.45	4.57	4.07	3.75	3.53	3.36	3.23	3.12	3.03
30	.05	4.17	3.32	2.92	2.69	2.53	2.42	2.33	2.27	2.21	2.16
	.01	7.56	5.39	4.51	4.02	3.70	3.47	3.30	3.17	3.07	2.98
35	.05	4.12	3.27	2.87	2.64	2.49	2.37	2.29	2.22	2.16	2.11
	.01	7.42	5.27	4.40	3.91	3.59	3.37	3.20	3.07	2.96	2.88
40	.05	4.08	3.23	2.84	2.61	2.45	2.34	2.25	2.18	2.12	2.08
	.01	7.31	5.18	4.31	3.83	3.51	3.29	3.12	2.99	2.89	2.80
45	.05	4.06	3.20	2.81	2.58	2.42	2.31	2.22	2.15	2.10	2.05
	.01	7.23	5.11	4.25	3.77	3.45	3.23	3.07	2.94	2.83	2.74
50	.05	4.03	3.18	2.79	2.56	2.40	2.29	2.20	2.13	2.07	2.03
	.01	7.17	5.06	4.20	3.72	3.41	3.19	3.02	2.89	2.78	2.70
60	.05	4.00	3.15	2.76	2.53	2.37	2.25	2.17	2.10	2.04	1.99
	.01	7.08	4.98	4.13	3.65	3.34	3.12	2.95	2.82	2.72	2.63
80	.05	3.96	3.11	2.72	2.49	2.33	2.21	2.13	2.06	2.00	1.95
	.01	6.96	4.88	4.04	3.56	3.26	3.04	2.87	2.74	2.64	2.55
100	.05	3.94	3.09	2.70	2.46	2.31	2.19	2.10	2.03	1.97	1.93
	.01	6.90	4.82	3.98	3.51	3.21	2.99	2.82	2.69	2.59	2.50
150	.05	3.90	3.06	2.66	2.43	2.27	2.16	2.07	2.00	1.94	1.89
	.01	6.81	4.75	3.91	3.45	3.14	2.92	2.76	2.63	2.53	2.44
300	.05	3.87	3.03	2.63	2.40	2.24	2.13	2.04	1.97	1.91	1.86
	.01	6.72	4.68	3.85	3.38	3.08	2.86	2.70	2.57	2.47	2.38
500	.05	3.86	3.01	2.62	2.39	2.23	2.12	2.03	1.96	1.90	1.85
	.01	6.69	4.65	3.82	3.36	3.05	2.84	2.68	2.55	2.44	2.36
1000	.05	3.85	3.00	2.61	2.38	2.22	2.11	2.02	1.95	1.89	1.84
	.01	6.66	4.63	3.80	3.34	3.04	2.82	2.66	2.53	2.43	2.34

(Continued)

(Continued)

df (denominator)	p	15	20	25	30	40	50	1000
		df (numerator)						
1	.05	245.95	248.01	249.26	250.10	251.14	251.77	254.19
	.01	6157.31	6208.74	6239.83	6260.65	6286.79	6302.52	6362.70
2	.05	19.43	19.45	19.46	19.46	19.47	19.48	19.49
	.01	99.43	99.45	99.46	99.47	99.47	99.48	99.50
3	.05	8.70	8.66	8.63	8.62	8.59	8.58	8.53
	.01	26.87	26.69	26.58	26.50	26.41	26.35	26.14
4	.05	5.86	5.80	5.77	5.75	5.72	5.70	5.63
	.01	14.20	14.02	13.91	13.84	13.75	13.69	13.47
5	05	4.62	4.56	4.52	4.50	4.46	4.44	4.37
	.01	9.72	9.55	9.45	9.38	9.29	9.24	9.03
6	.05	3.94	3.87	3.83	3.81	3.77	3.75	3.67
	.01	7.56	7.40	7.30	7.23	7.14	7.09	6.89
7	.05	3.51	3.44	3.40	3.38	3.34	3.32	3.23
	.01	6.31	6.16	6.06	5.99	5.91	5.86	5.66
8	.05	3.22	3.15	3.11	3.08	3.04	3.02	2.93
	.01	5.52	5.36	5.26	5.20	5.12	5.07	4.87
9	.05	3.01	2.94	2.89	2.86	2.83	2.80	2.71
	.01	4.96	4.81	4.71	4.65	4.57	4.52	4.32
10	.05	2.85	2.77	2.73	2.70	2.66	2.64	2.54
	.01	4.56	4.41	4.31	4.25	4.17	4.12	3.92
11	.05	2.72	2.65	2.60	2.57	2.53	2.51	2.41
	.01	4.25	4.10	4.01	3.94	3.86	3.81	3.61
12	.05	2.62	2.54	2.50	2.47	2.43	2.40	2.30
	.01	4.01	3.86	3.76	3.70	3.62	3.57	3.37
13	.05	2.53	2.46	2.41	2.38	2.34	2.31	2.21
	.01	3.82	3.66	3.57	3.51	3.43	3.38	3.18
14	.05	2.46	2.39	2.34	2.31	2.27	2.24	2.14
	.01	3.66	3.51	3.41	3.35	3.27	3.22	3.02
15	.05	2.40	2.33	2.28	2.25	2.20	2.18	2.07
	.01	3.52	3.37	3.28	3.21	3.13	3.08	2.88
16	.05	2.35	2.28	2.23	2.19	2.15	2.12	2.02
	.01	3.41	3.26	3.16	3.10	3.02	2.97	2.76
17	.05	2.31	2.23	2.18	2.15	2.10	2.08	1.97
	.01	3.31	3.16	3.07	3.00	2.92	2.87	2.66
18	.05	2.27	2.19	2.14	2.11	2.06	2.04	1.92
	.01	3.23	3.08	2.98	2.92	2.84	2.78	2.58

df (denominator)	p	15	20	25	30	40	50	1000
		df (numerator)						
19	0.05	2.23	2.16	2.11	2.07	2.03	2.00	1.88
	0.01	3.15	3.00	2.91	2.84	2.76	2.71	2.50
20	0.05	2.20	2.12	2.07	2.04	1.99	1.97	1.85
	0.01	3.09	2.94	2.84	2.78	2.69	2.64	2.43
22	0.05	2.15	2.07	2.02	1.98	1.94	1.91	1.79
	0.01	2.98	2.83	2.73	2.67	2.58	2.53	2.32
24	0.05	2.11	2.03	1.97	1.94	1.89	1.86	1.74
	0.01	2.89	2.74	2.64	2.58	2.49	2.44	2.22
26	0.05	2.07	1.99	1.94	1.90	1.85	1.82	1.70
	0.01	2.81	2.66	2.57	2.50	2.42	2.36	2.14
28	0.05	2.04	1.96	1.91	1.87	1.82	1.79	1.66
	0.01	2.75	2.60	2.51	2.44	2.35	2.30	2.08
30	0.05	2.01	1.93	1.88	1.84	1.79	1.76	1.63
	0.01	2.70	2.55	2.45	2.39	2.30	2.25	2.02
35	0.05	1.96	1.88	1.82	1.79	1.74	1.70	1.57
	0.01	2.60	2.44	2.35	2.28	2.19	2.14	1.90
40	0.05	1.92	1.84	1.78	1.74	1.69	1.66	1.52
	0.01	2.52	2.37	2.27	2.20	2.11	2.06	1.82
45	0.05	1.89	1.81	1.75	1.71	1.66	1.63	1.48
	0.01	2.46	2.31	2.21	2.14	2.05	2.00	1.75
50	0.05	1.87	1.78	1.73	1.69	1.63	1.60	1.45
	0.01	2.42	2.27	2.17	2.10	2.01	1.95	1.70
60	0.05	1.84	1.75	1.69	1.65	1.59	1.56	1.40
	0.01	2.35	2.20	2.10	2.03	1.94	1.88	1.62
80	0.05	1.79	1.70	1.64	1.60	1.54	1.51	1.34
	0.01	2.27	2.12	2.01	1.94	1.85	1.79	1.51
100	0.05	1.77	1.68	1.62	1.57	1.52	1.48	1.30
	0.01	2.22	2.07	1.97	1.89	1.80	1.74	1.45
150	0.05	1.73	1.64	1.58	1.54	1.48	1.44	1.24
	0.01	2.16	2.00	1.90	1.83	1.73	1.66	1.35
300	0.05	1.70	1.61	1.54	1.50	1.43	1.39	1.17
	.01	2.10	1.94	1.84	1.76	1.66	1.59	1.25
500	.05	1.69	1.59	1.53	1.48	1.42	1.38	1.14
	.01	2.07	1.92	1.81	1.74	1.63	1.57	1.20
1000	.05	1.68	1.58	1.52	1.47	1.41	1.36	1.11
	.01	2.06	1.90	1.79	1.72	1.61	1.54	1.16

All values computed by the author using SPSS.

A.4. Critical values of the chi-square distribution

df	p 0.05	p 0.01
1	3.84	6.63
2	5.99	9.21
3	7.81	11.34
4	9.49	13.28
5	11.07	15.09
6	12.59	16.81
7	14.07	18.48
8	15.51	20.09
9	16.92	21.67
10	18.31	23.21
11	19.68	24.72
12	21.03	26.22
13	22.36	27.69
14	23.68	29.14
15	25.00	30.58
16	26.30	32.00
17	27.59	33.41
18	28.87	34.81
19	30.14	36.19
20	31.41	37.57
21	32.67	38.93
22	33.92	40.29
23	35.17	41.64
24	36.42	42.98

df	p 0.05	p 0.01
25	37.65	44.31
26	38.89	45.64
27	40.11	46.96
28	41.34	48.28
29	42.56	49.59
30	43.77	50.89
35	49.80	57.34
40	55.76	63.69
45	61.66	69.96
50	67.50	76.15
60	79.08	88.38
70	90.53	100.43
80	101.88	112.33
90	113.15	124.12
100	124.34	135.81
200	233.99	249.45
300	341.40	359.91
400	447.63	468.72
500	553.13	576.49
600	658.09	683.52
700	762.66	789.97
800	866.91	895.98
900	970.90	1001.63
1000	1074.68	1106.97

All values computed by the author using SPSS.

REFERENCES

Agresti, A., & Finlay, B. (1986). *Statistical methods for the social sciences* (2nd ed.). San Francisco: Dellen.

Algina, J., & Olejnik, S. F. (1984). Implementing the Welch-James procedure with factorial designs. *Educational and Psychological Measurement, 44*, 39–48.

Arrindell, W. A., & van der Ende, J. (1985). An empirical test of the utility of the observer-to-variables ratio in factor and components analysis. *Applied Psychological Measurement, 9*, 165–178.

Baguley, T. (2004). Understanding statistical power in the context of applied research. *Applied Ergonomics, 35*(2), 73–80.

Bale, C., Morrison, R., & Caryl, P. G. (2006). Chat-up lines as male sexual displays. *Personality and Individual Differences, 40*(4), 655–664.

Bargman, R. E. (1970). *Interpretation and use of a generalized discriminant function*. In R. C. Bose et al. (eds.), *Essays in probability and statistics*. Chapel Hill: University of North Carolina Press.

Barnard, G. A. (1963). Ronald Aylmer Fisher, 1890–1962: Fisher's contributions to mathematical statistics. *Journal of the Royal Statistical Society, Series A (General), 126*, 162–166.

Barnett, V., & Lewis, T. (1978). *Outliers in statistical data*. New York: Wiley.

Beckham, A. S. (1929). Is the Negro happy? A psychological analysis. *Journal of Abnormal and Social Psychology, 24*, 186–190.

Belsey, D. A., Kuh, E., & Welsch, R. (1980). *Regression diagnostics: Identifying influential data and sources of collinearity*. New York: Wiley.

Bemelman, M., & Hammacher, E. R. (2005). Rectal impalement by pirate ship: A case report. *Injury Extra, 36*, 508–510.

Berger, J. O. (2003). Could Fisher, Jeffreys and Neyman have agreed on testing? *Statistical Science, 18*(1), 1–12.

Berry, W. D. (1993). *Understanding regression assumptions*. Sage university paper series on quantitative applications in the social sciences, 07-092. Newbury Park, CA: Sage.

Berry, W. D., & Feldman, S. (1985). *Multiple regression in practice*. Sage university paper series on quantitative applications in the social sciences, 07-050. Beverly Hills, CA: Sage.

Board, B. J., & Fritzon, K. (2005). Disordered personalities at work. *Psychology, Crime & Law, 11*(1), 17–32.

Bock, R. D. (1975). *Multivariate statistical methods in behavioural research*. New York: McGraw-Hill.

Boik, R. J. (1981). A priori tests in repeated measures designs: Effects of nonsphericity. *Psychometrika, 46*(3), 241–255.

Bowerman, B. L., & O'Connell, R. T. (1990). *Linear statistical models: An applied approach* (2nd ed.). Belmont, CA: Duxbury.

Bray, J. H., & Maxwell, S. E. (1985). *Multivariate analysis of variance*. Sage university paper series on quantitative applications in the social sciences, 07-054. Newbury Park, CA: Sage.

Brown, M. B., & Forsythe, A. B. (1974). The small sample behaviour of some statistics which test the equality of several means. *Technometrics, 16*, 129–132.

Brown, W. (1910). Some experimental results in the correlation of mental abilities. *British Journal of Psychology, 3*, 296–322.

Budescu, D. V. (1982). The power of the F test in normal populations with heterogeneous variances. *Educational and Psychological Measurement, 42*, 609–616.

Budescu, D. V., & Appelbaum, M. I. (1981). Variance stabilizing transformations and the power of the F test. *Journal of Educational Statistics, 6*(1), 55–74.

Cattell, R. B. (1966a). *The scientific analysis of personality*. Chicago: Aldine.

Cattell, R. B. (1966b). The scree test for the number of factors. *Multivariate Behavioral Research, 1*, 245–276.

Çetinkaya, H., & Domjan, M. (2006). Sexual fetishism in a quail (*Coturnix japonica*) model system: Test of reproductive success. *Journal of Comparative Psychology, 120*(4), 427–432.

Chamorro-Premuzic, T., Furnham, A., Christopher, A. N., Garwood, J., & Martin, N. (2008). Birds of a feather: Students' preferences for lecturers' personalities as predicted by their own personality and learning approaches. *Personality and Individual Differences, 44*, 965–976.

Chen, P. Y., & Popovich, P. M. (2002). *Correlation: Parametric*

and nonparametric measures. Thousand Oaks, CA: Sage.
Clarke, D. L., Buccimazza, I., Anderson, F. A., & Thomson, S. R. (2005). Colorectal foreign bodies. *Colorectal Disease, 7*(1), 98–103.
Cliff, N. (1987). *Analyzing multivariate data*. New York: Harcourt Brace Jovanovich.
Cohen, J. (1968). Multiple regression as a general data-analytic system. *Psychological Bulletin, 70*(6), 426–443.
Cohen, J. (1988). *Statistical power analysis for the behavioural sciences* (2nd ed.). New York: Academic Press.
Cohen, J. (1990). Things I have learned (so far). *American Psychologist, 45*(12), 1304–1312.
Cohen, J. (1992). A power primer. *Psychological Bulletin, 112*(1), 155–159.
Cohen, J. (1994). The Earth is round (p < .05). *American Psychologist, 49*(12), 997–1003.
Cole, D. A., Maxwell, S. E., Arvey, R., & Salas, E. (1994). How the power of MANOVA can both increase and decrease as a function of the intercorrelations among the dependent variables. *Psychological Bulletin, 115*(3), 465–474.
Collier, R. O., Baker, F. B., Mandeville, G. K., & Hayes, T. F. (1967). Estimates of test size for several test procedures based on conventional variance ratios in the repeated measures design. *Psychometrika, 32*(2), 339–352.
Comrey, A. L., & Lee, H. B. (1992). *A first course in factor analysis* (2nd ed.). Hillsdale, NJ: Erlbaum.
Cook, R. D., & Weisberg, S. (1982). *Residuals and influence in regression*. New York: Chapman & Hall.
Cook, S. A., Rosser, R., & Salmon, P. (2006). Is cosmetic surgery an effective psychotherapeutic intervention? A systematic review of the evidence. *Journal of Plastic, Reconstructive & Aesthetic Surgery, 59*, 1133–1151.
Cook, S. A., Rossera, R., Toone, H., James, M. I., & Salmon, P. (2006). The psychological and social characteristics of patients referred for NHS cosmetic surgery: Quantifying clinical need. *Journal of Plastic, Reconstructive & Aesthetic Surgery 59*, 54–64.
Cooper, C. L., Sloan, S. J., & Williams, S. (1988) *Occupational Stress Indicator Management Guide*. Windsor, UK: NFER-Nelson.
Cooper, M., O'Donnell, D., Caryl, P. G., Morrison, R., & Bale, C. (2007). Chat-up lines as male displays: Effects of content, sex, and personality. *Personality and Individual Differences, 43*(5), 1075–1085.
Cortina, J. M. (1993). What is coefficient alpha? An examination of theory and applications. *Journal of Applied Psychology, 78*, 98–104.
Cox, D. R., & Snell, D. J. (1989). *The analysis of binary data* (2nd ed.). London: Chapman & Hall.
Cronbach, L. J. (1951). Coefficient alpha and the internal structure of tests. *Psychometrika, 16*, 297–334.
Cronbach, L. J. (1957). The two disciplines of scientific psychology. *American Psychologist, 12*, 671–684.
Dalgaard, P. (2002). *Introductory Statistics with R*. New York: Springer.
Davey, G. C. L., Startup, H. M., Zara, A., MacDonald, C. B., & Field, A. P. (2003). Perseveration of checking thoughts and mood-as-input hypothesis. *Journal of Behavior Therapy & Experimental Psychiatry, 34*, 141–160.
Davidson, M. L. (1972). Univariate versus multivariate tests in repeated-measures experiments. *Psychological Bulletin, 77*, 446–452.
DeCarlo, L. T. (1997). On the meaning and use of kurtosis. *Psychological Methods, 2*(3), 292–307.
Domjan, M., Blesbois, E., & Williams, J. (1998). The adaptive significance of sexual conditioning: Pavlovian control of sperm release. *Psychological Science, 9*(5), 411–415.
Donaldson, T. S. (1968). Robustness of the *F*-test to errors of both kinds and the correlation between the numerator and denominator of the *F*-ratio. *Journal of the American Statistical Association, 63*, 660–676.
Dunlap, W. P., Cortina, J. M., Vaslow, J. B., & Burke, M. J. (1996). Meta-analysis of experiments with matched groups or repeated measures designs. *Psychological Methods, 1*(2), 170–177.
Dunteman, G. E. (1989). *Principal components analysis*. Sage university paper series on quantitative applications in the social sciences, 07-069. Newbury Park, CA: Sage.
Durbin, J., & Watson, G. S. (1951). Testing for serial correlation in least squares regression, II. *Biometrika, 30*, 159–178.
Easterlin, R. A. (2003). Explaining Happiness. *Proceedings of the National Academy of Sciences., 100*(19), 11176–11183.
Efron, B., & Tibshirani, R. (1993). *An introduction to the bootstrap*: Chapman and Hall.
Eriksson, S.-G., Beckham, D., & Vassell, D. (2004). Why are the English so shit at penalties? A review. *Journal of Sporting Ineptitude, 31*, 231–1072.
Erlebacher, A. (1977). Design and analysis of experiments contrasting the within- and between-subjects manipulations of the independent variable. *Psychological Bulletin, 84*, 212–219.
Eysenck, H. J. (1953). *The structure of human personality*. New York: Wiley.
Fesmire, F. M. (1988). Termination of intractable hiccups with digital rectal massage. *Annals of Emergency Medicine, 17*(8), 872.
Field, A. P. (1998). A bluffer's guide to sphericity. *Newsletter of the mathematical, statistical and computing section of the British Psychological Society*, 6(1)13–22.
Field, A. P. (2000). *Discovering statistics using SPSS for Windows: Advanced techniques for the beginner*. London: Sage.

Field, A. P. (2001). Meta-analysis of correlation coefficients: A Monte Carlo comparison of fixed- and random-effects methods. *Psychological Methods, 6*(2), 161–180.

Field, A. P. (2005a). Intraclass correlation. In B. Everitt & D. C. Howell (eds.), *Encyclopedia of Statistics in Behavioral Science* (Vol. 2, pp. 948–954). New York: Wiley.

Field, A. P. (2005b). Is the meta-analysis of correlation coefficients accurate when population correlations vary? *Psychological Methods, 10*(4), 444–467.

Field, A. P. (2005c). Learning to like (and dislike): Associative learning of preferences. In A. J. Wills (ed.), *New Directions in Human Associative Learning* (pp. 221–252). Mahwah, NJ: LEA.

Field, A. P. (2005d). Sir Ronald Aylmer Fisher. In B. S. Everitt & D. C. Howell (eds.), *Encyclopedia of Statistics in Behavioral Science* (Vol. 2, pp. 658–659). Chichester: Wiley.

Field, A. P. (2006). The behavioral inhibition system and the verbal information pathway to children's fears. *Journal of Abnormal Psychology, 115*(4), 742–752.

Field, A. P. (2009). *Discovering statistics using SPSS (and sex and drugs and rock' n' roll)* (3rd ed.). London: Sage.

Field, A. P., & Davey, G. C. L. (1999). Reevaluating evaluative conditioning: A nonassociative explanation of conditioning effects in the visual evaluative conditioning paradigm. *Journal of Experimental Psychology – Animal Behavior Processes, 25*(2), 211–224.

Field, A. P., & Hole, G. J. (2003). *How to design and report experiments*. London: Sage.

Field, A. P., & Moore, A. C. (2005). Dissociating the effects of attention and contingency awareness on evaluative conditioning effects in the visual paradigm. *Cognition and Emotion, 19*(2), 217–243.

Fisher, R. A. (1921). On the probable error of a coefficient of correlation deduced from a small sample. *Metron, 1*, 3–32.

Fisher, R. A. (1922). On the interpretation of chi square from contingency tables, and the calculation of P. *Journal of the Royal Statistical Society, 85*, 87–94.

Fisher, R. A. (1925). *Statistical methods for research workers*. Edinburgh: Oliver & Boyd.

Fisher, R. A. (1925/1991). *Statistical methods, experimental design, and scientific inference*. Oxford: Oxford University Press. (This reference is for the 1991 reprint.)

Fisher, R. A. (1956). *Statistical methods and scientific inference*. New York: Hafner.

Flanagen, J. C. (1937). A proposed procedure for increasing the efficiency of objective tests. *Journal of Educational Psychology, 28*, 17–21.

Friedman, M. (1937). The use of ranks to avoid the assumption of normality implicit in the analysis of variance. *Journal of the American Statistical Association, 32*, 675–701.

Gallup, G. G. J., Burch, R. L., Zappieri, M. L., Parvez, R., Stockwell, M., & Davis, J. A. (2003). The human penis as a semen displacement device. *Evolution and Human Behavior, 24*, 277–289.

Games, P. A. (1983). Curvilinear transformations of the dependent variable. *Psychological Bulletin, 93*(2), 382–387.

Games, P. A. (1984). Data transformations, power, and skew: A rebuttal to Levine and Dunlap. *Psychological Bulletin, 95*(2), 345–347.

Games, P. A., & Lucas, P. A. (1966). Power of the analysis of variance of independent groups on non-normal and normally transformed data. *Educational and Psychological Measurement, 26*, 311–327.

Girden, E. R. (1992). *ANOVA: Repeated measures*. Sage university paper series on quantitative applications in the social sciences, 07-084. Newbury Park, CA: Sage.

Glass, G. V. (1966). Testing homogeneity of variances. *American Educational Research Journal, 3*(3), 187–190.

Glass, G. V., Peckham, P. D., & Sanders, J. R. (1972). Consequences of failure to meet assumptions underlying the fixed effects analyses of variance and covariance. *Review of Educational Research, 42*(3), 237–288.

Graham, J. M., Guthrie, A. C., & Thompson, B. (2003). Consequences of not interpreting structure coefficients in published CFA research: A reminder. *Structural Equation Modeling, 10*(1), 142–153.

Grayson, D. (2004). Some myths and legends in quantitative psychology. *Understanding Statistics, 3*(1), 101–134.

Green, S. B. (1991). How many subjects does it take to do a regression analysis? *Multivariate Behavioral Research, 26*, 499–510.

Greenhouse, S. W., & Geisser, S. (1959). On methods in the analysis of profile data. *Psychometrika, 24*, 95–112.

Guadagnoli, E., & Velicer, W. F. (1988). Relation of sample size to the stability of component patterns. *Psychological Bulletin, 103*(2), 265–275.

Haitovsky, Y. (1969). Multicollinearity in regression analysis: A comment. *Review of Economics and Statistics, 51*(4), 486–489.

Hakstian, A. R., Roed, J. C., & Lind, J. C. (1979). Two-sample T2 procedure and the assumption of homogeneous covariance matrices. *Psychological Bulletin, 86*, 1255–1263.

Halekoh, U., & Højsgaard, S. (2007). Overdispersion. Retrieved March 18, 2007 from http://gbi.agrsci.dk/statistics/courses/phd07/material/Day7/overdispersion-handout.pdf.

Hardy, M. A. (1993). *Regression with dummy variables*. Sage university paper series on quantitative applications in the social sciences, 07-093. Newbury Park, CA: Sage.

Harman, B. H. (1976). *Modern factor analysis* (3rd ed., rev.). Chicago: University of Chicago Press.

Harris, R. J. (1975). *A primer of multivariate statistics*. New York: Academic Press.

Hill, C., Abraham, C., & Wright, D. B. (2007). Can theory-based messages in combination with cognitive prompts promote exercise in classroom settings? *Social Science & Medicine, 65*, 1049–1058.

Hoaglin, D., & Welsch, R. (1978). The hat matrix in regression and ANOVA. *American Statistician, 32*, 17–22.

Hoddle, G., Batty, D., & Ince, P. (1998). How not to take penalties in important soccer matches. *Journal of Cretinous Behaviour, 1*, 1–2.

Horn, J. L. (1965). A rationale and test for the number of factors in factor analysis. *Psychometrika, 30*, 179–185.

Hosmer, D. W., & Lemeshow, S. (1989). *Applied logistic regression*. New York: Wiley.

Howell, D. C. (1997). *Statistical methods for psychology* (4th ed.). Belmont, CA: Duxbury.

Howell, D. C. (2006). *Statistical methods for psychology* (6th ed.). Belmont, CA: Thomson.

Huberty, C. J., & Morris, J. D. (1989). Multivariate analysis versus multiple univariate analysis. *Psychological Bulletin, 105*(2), 302–308.

Hughes, J. P., Marice, H. P., & Gathright, J. B. (1976). Method of removing a hollow object from the rectum. *Diseases of the Colon & Rectum, 19*(1), 44–45.

Hume, D. (1739–40). *A treatise of human nature* (ed. L. A. Selby-Bigge). Oxford: Clarendon Press, 1965.

Hume, D. (1748). *An enquiry concerning human understanding*. Chicago: Open Court, 1927.

Hutcheson, G., & Sofroniou, N. (1999). *The multivariate social scientist*. London: Sage.

Huynh, H., & Feldt, L. S. (1976). Estimation of the Box correction for degrees of freedom from sample data in randomised block and split-plot designs. *Journal of Educational Statistics, 1*(1), 69–82.

Iverson, G. R., & Norpoth, H. (1987). *ANOVA* (2nd ed.). Sage University series on quantitative applications in the social sciences, 07-001. Newbury Park, CA: Sage.

Jackson, S., & Brashers, D. E. (1994). *Random factors in ANOVA*. Sage university paper series on quantitative applications in the social sciences, 07-098. Thousand Oaks, CA: Sage.

Jolliffe, I. T. (1972). Discarding variables in a principal component analysis, I: Artificial data. *Applied Statistics, 21*, 160–173.

Jolliffe, I. T. (1986). *Principal component analysis*. New York: Springer.

Jonckheere, A. R. (1954). A distribution-free k-sample test against ordered alternatives. *Biometrika, 41*, 133–145.

Kahneman, D., & Krueger, A. B. (2006). Developments in the measurement of subjective well-being. *Journal of Economic Perspectives, 20*(1), 3–24.

Kaiser, H. F. (1960). The application of electronic computers to factor analysis. *Educational and Psychological Measurement, 20*, 141–151.

Kaiser, H. F. (1970). A second-generation little jiffy. *Psychometrika, 35*, 401–415.

Kaiser, H. F. (1974). An index of factorial simplicity. *Psychometrika, 39*, 31–36.

Kass, R. A., & Tinsley, H. E. A. (1979). Factor analysis. *Journal of Leisure Research, 11*, 120–138.

Kellett, S., Clarke, S., & McGill, P. (2008). Outcomes from psychological assessment regarding recommendations for cosmetic surgery. *Journal of Plastic, Reconstructive & Aesthetic Surgery, 61*, 512–517.

Keselman, H. J., & Keselman, J. C. (1988). Repeated measures multiple comparison procedures: Effects of violating multisample sphericity in unbalanced designs. *Journal of Educational Statistics, 13*(3), 215–226.

Kirk, R. E. (1996). Practical significance: A concept whose time has come. *Educational and Psychological Measurement, 56*(5), 746–759.

Kline, P. (1999). *The handbook of psychological testing* (2nd ed.). London: Routledge.

Klockars, A. J., & Sax, G. (1986). *Multiple comparisons*. Sage university paper series on quantitative applications in the social sciences, 07-061. Newbury Park, CA: Sage.

Koot, V. C. M., Peeters, P. H. M., Granath, F., Grobbee, D. E., & Nyren, O. (2003). Total and cause specific mortality among Swedish women with cosmetic breast implants: Prospective study. *British Medical Journal, 326*(7388), 527–528.

Kreft, I. G. G., & De Leeuw, J. (1998). *Introducing multilevel modeling*. London: Sage.

Kreft, I. G. G., De Leew, J., & Aiken, L. S. (1995). The effect of different forms of centering in hierarchical linear models. *Multivariate Behavioral Research, 30*, 1–21.

Kruskal, W. H., & Wallis, W. A. (1952). Use of ranks in one-criterion variance analysis. *Journal of the American Statistical Association, 47*, 583–621.

Lacourse, E., Claes, M., & Villeneuve, M. (2001). Heavy metal music and adolescent suicidal risk. *Journal of Youth and Adolescence, 30*(3), 321–332.

Lehmann, E. L. (1993). The Fisher, Neyman–Pearson theories of testing hypotheses: One theory or two? *Journal of the American Statistical Association, 88*, 1242–1249.

Lenth, R. V. (2001). Some practical guidelines for effective sample size determination. *American Statistician, 55*(3), 187–193.

Levene, H. (1960). Robust tests for equality of variances. In I. Olkin,

S. G. Ghurye, W. Hoeffding, W. G. Madow, & H. B. Mann (eds.), *Contributions to Probability and Statistics: Essays in Honor of Harold Hotelling* (pp. 278–292). Stanford, CA: Stanford University Press.

Levine, D. W., & Dunlap, W. P. (1982). Power of the F test with skewed data: Should one transform or not? *Psychological Bulletin, 92*(1), 272–280.

Levine, D. W., & Dunlap, W. P. (1983). Data transformation, power, and skew: A rejoinder to Games. *Psychological Bulletin, 93*(3), 596–599.

Lo, S. F., Wong, S. H., Leung, L. S., Law, I. C., & Yip, A. W. C. (2004). Traumatic rectal perforation by an eel. *Surgery, 135*(1), 110–111.

Loftus, G. R., & Masson, M. E. J. (1994). Using confidence intervals in within-subject designs. *Psychonomic Bulletin and Review, 1*(4), 476–490.

Lord, F. M. (1967). A paradox in the interpretation of group comparisons. *Psychological Bulletin, 68*(5), 304–305.

Lord, F. M. (1969). Statistical adjustments when comparing preexisting groups. *Psychological Bulletin, 72*(5), 336–337.

Lunney, G. H. (1970). Using analysis of variance with a dichotomous dependent variable: An empirical study. *Journal of Educational Measurement, 7*(4), 263–269.

MacCallum, R. C., Widaman, K. F., Zhang, S., & Hong, S. (1999). Sample size in factor analysis. *Psychological Methods, 4*(1), 84–99

MacCallum, R. C., Zhang, S., Preacher, K. J., & Rucker, D. D. (2002). On the practice of dichotomization of quantitative variables. *Psychological Methods, 7*(1), 19–40.

Mann, H. B., & Whitney, D. R. (1947). On a test of whether one of two random variables is stochastically larger than the other. *Annals of Mathematical Statistics, 18*, 50–60.

Marzillier, S. L., & Davey, G. C. L. (2005). Anxiety and disgust: Evidence for a unidirectional relationship. *Cognition and Emotion, 19*(5), 729–75.

Mather, K. (1951). R. A. Fisher's *Statistical Methods for Research Workers*: An appreciation. *Journal of the American Statistical Association, 46*, 51–54.

Matthews, R. C., Domjan, M., Ramsey, M., & Crews, D. (2007). Learning effects on sperm competition and reproductive fitness. *Psychological Science, 18*(9), 758–762.

Maxwell, S. E. (1980). Pairwise multiple comparisons in repeated measures designs. *Journal of Educational Statistics, 5*(3), 269–287.

Maxwell, S. E., & Delaney, H. D. (1990). *Designing experiments and analyzing data*. Belmont, CA: Wadsworth.

McDonald, P. T., & Rosenthal, D. (1977). An unusual foreign body in the rectum – A baseball: Report of a case. *Diseases of the Colon & Rectum, 20*(1), 56–57.

McGrath, R. E., & Meyer, G. J. (2006). When effect sizes disagree: The case of r and d. *Psychological Methods, 11*(4), 386–401.

Menard, S. (1995). *Applied logistic regression analysis*. Sage university paper series on quantitative applications in the social sciences, 07-106. Thousand Oaks, CA: Sage.

Mendoza, J. L., Toothaker, L. E., & Crain, B. R. (1976). Necessary and sufficient conditions for F ratios in the L * J * K factorial design with two repeated factors. *Journal of the American Statistical Association, 71*, 992–993.

Mendoza, J. L., Toothaker, L. E., & Nicewander, W. A. (1974). A Monte Carlo comparison of the univariate and multivariate methods for the groups by trials repeated measures design. *Multivariate Behavioural Research, 9*, 165–177.

Miles, J. N. V., & Banyard, P. (2007). *Understanding and using statistics in psychology: a practical introduction*. London: Sage.

Miles, J. N. V. & Shevlin, M. (2001). *Applying regression and correlation: a guide for students and researchers*. London: Sage.

Mill, J. S. (1865). *A system of logic: ratiocinative and inductive*. London: Longmans, Green.

Miller, G. A., & Chapman, J. P. (2001). Misunderstanding analysis of covariance. *Journal of Abnormal Psychology, 110*(1), 40–48.

Miller, G., Tybur, J. M., & Jordan, B. D. (2007). Ovulatory cycle effects on tip earnings by lap dancers: economic evidence for human estrus? *Evolution and Human Behavior, 28*, 375–381.

Mitzel, H. C., & Games, P. A. (1981). Circularity and multiple comparisons in repeated measures designs. *British Journal of Mathematical and Statistical Psychology, 34*, 253–259.

Muris, P., Huijding, J., Mayer, B., & Hameetman, M. (2008). A space odyssey: Experimental manipulation of threat perception and anxiety-related interpretation bias in children. *Child Psychiatry and Human Development 39*(4), 469–480.

Myers, R. (1990). *Classical and modern regression with applications* (2nd ed.). Boston, MA: Duxbury.

Nagelkerke, N. J. D. (1991). A note on a general definition of the coefficient of determination. *Biometrika, 78*, 691–692.

Namboodiri, K. (1984). *Matrix algebra: an introduction*. Sage university paper series on quantitative applications in the social sciences, 07-38. Beverly Hills, CA: Sage.

Nichols, L. A., & Nicki, R. (2004). Development of a psychometrically sound internet addiction scale: A preliminary step. *Psychology of Addictive Behaviors, 18*(4), 381–384.

Nunnally, J. C. (1978). *Psychometric theory*. New York: McGraw-Hill.

Nunnally, J. C., & Bernstein, I. H. (1994). *Psychometric Theory* (3rd ed.). New York: McGraw-Hill.

O'Brien, M. G., & Kaiser, M. K. (1985). MANOVA method for analyzing repeated measures designs: An extensive primer. *Psychological Bulletin, 97*(2), 316–333.

O'Connor, B. P. (2000). SPSS and SAS programs for determining the number of components using parallel analysis and Velicer's MAP test. *Behavior Research Methods, Instrumentation, and Computers, 32*, 396–402.

Olson, C. L. (1974). Comparative robustness of six tests in multivariate analysis of variance. *Journal of the American Statistical Association, 69*, 894–908.

Olson, C. L. (1976). On choosing a test statistic in multivariate analysis of variance. *Psychological Bulletin, 83*, 579–586.

Olson, C. L. (1979). Practical considerations in choosing a MANOVA test statistic: A rejoinder to Stevens. *Psychological Bulletin, 86*, 1350–1352.

Pearson, E. S., & Hartley, H. O. (1954). *Biometrika tables for statisticians, volume I*. New York: Cambridge University Press.

Pearson, K. (1894). Science and Monte Carlo. *The Fortnightly Review, 55*, 183–193.

Pearson, K. (1900). On the criterion that a given system of deviations from the probable in the case of a correlated system of variables is such that it can be reasonably supposed to have arisen from random sampling. *Philosophical Magazine, 50*(5), 157–175.

Pedhazur, E., & Schmelkin, L. (1991). *Measurement, design and analysis: an integrated approach*. Hillsdale, NJ: Erlbaum.

Plackett, R. L. (1983). Karl Pearson and the chi-squared test. *International Statistical Review, 51*(1), 59–72.

Ramsey, P. H. (1982). Empirical power of procedures for comparing two groups on p variables. *Journal of Educational Statistics, 7*, 139–156.

Raudenbush, S. W., & Bryk, A. S. (2002). *Hierarchical linear models* (2nd ed.). Thousand Oaks, CA: Sage.

Rockwell, R. C. (1975). Assessment of multicollinearity: The Haitovsky test of the determinant. *Sociological Methods and Research, 3*(4), 308–320.

Rosenthal, R. (1991). *Meta-analytic procedures for social research* (2nd ed.). Newbury Park, CA: Sage.

Rosenthal, R., Rosnow, R. L., & Rubin, D. B. (2000). *Contrasts and effect sizes in behavioural research: a correlational approach*. Cambridge: Cambridge University Press.

Rosnow, R. L., Rosenthal, R., & Rubin, D. B. (2000). Contrasts and correlations in effect-size estimation. *Psychological Science, 11*, 446–453.

Rosnow, R. L., & Rosenthal, R. (2005). *Beginning behavioural research: a conceptual primer* (5th ed.). Englewood Cliffs, NJ: Pearson/Prentice Hall.

Rouanet, H., & Lépine, D. (1970). Comparison between treatments in a repeated-measurement design: ANOVA and multivariate methods. *British Journal of Mathematical and Statistical Psychology, 23*, 147–163.

Rulon, P. J. (1939). A simplified procedure for determining the reliability of a test by split-halves. *Harvard Educational Review, 9*, 99–103.

Rutherford, A. (2000). *Introducing ANOVA and ANCOVA: A GLM* approach. London: Sage.

Sacco, W. P., Levine, B., Reed, D., & Thompson, K. (1991). Attitudes about condom use as an AIDS-relevant behavior: Their factor structure and relation to condom use. *Psychological Assessment: A Journal of Consulting and Clinical Psychology, 3(2)*, 265–272.

Sacco, W. P., Rickman, R. L., Thompson, K., Levine, B., & Reed, D. L. (1993). Gender differences in aids-relevant condom attitudes and condom use. *AIDS Education and Prevention, 5*(4), 311–326.

Sachdev, Y. V. (1967). An unusual foreign body in the rectum. *Diseases of the Colon & Rectum, 10*(3), 220–221.

Salsburg, D. (2002). *The lady tasting tea: How statistics revolutionized science in the twentieth century*. New York: Owl Books.

Savage, L. J. (1976). On re-reading R. A. Fisher. *Annals of Statistics, 4*, 441–500.

Scariano, S. M., & Davenport, J. M. (1987). The effects of violations of independence in the one-way ANOVA. *American Statistician, 41*(2), 123–129.

Schützwohl, A. (2008). The disengagement of attentive resources from task-irrelevant cues to sexual and emotional infidelity. *Personality and Individual Differences, 44*, 633–644.

Shackelford, T. K., LeBlanc, G. J., & Drass, E. (2000). Emotional reactions to infidelity. *Cognition & Emotion, 14*(5), 643–659.

Shee, J. C. (1964). Pargyline and the cheese reaction. *British Medical Journal, 1*(539), 1441.

Siegel, S., & Castellan, N. J. (1988). *Nonparametric statistics for the behavioral sciences* (2nd ed.). New York: McGraw-Hill.

Spearman, C. (1910). Correlation calculated with faulty data. *British Journal of Psychology, 3*, 271–295.

Stevens, J. P. (1979). Comment on Olson: Choosing a test statistic in multivariate analysis of variance. *Psychological Bulletin, 86*, 365–360.

Stevens, J. P. (1980). Power of the multivariate analysis of variance tests. *Psychological Bulletin, 88*, 728–737.

Stevens, J. P. (2002). *Applied multivariate statistics for the social sciences* (4th ed.). Hillsdale, NJ: Erlbaum.

Strahan, R. F. (1982). Assessing magnitude of effect from rank-order correlation coeffients. *Educational and Psychological Measurement, 42*, 763–765.

Stuart, E. W., Shimp, T. A., & Engle, R. W. (1987). Classical-conditioning of consumer attitudes – 4. Experiments in an advertising context. *Journal of Consumer Research, 14*(3), 334–349.

Studenmund, A. H., & Cassidy, H. J. (1987). *Using econometrics: a practical guide*. Boston: Little Brown.

Tabachnick, B. G., & Fidell, L. S. (2001). *Using multivariate statistics* (4th ed.). Boston: Allyn & Bacon.

Tabachnick, B. G., & Fidell, L. S. (2007). *Using multivariate statistics* (5th ed.). Boston: Allyn & Bacon.

Terpstra, T. J. (1952). The asymptotic normality and consistency of Kendall's test against trend, when ties are present in one ranking. *Indagationes Mathematicae, 14*, 327–333.

Terrell, C. D. (1982a). Significance tables for the biserial and the point biserial. *Educational and Psychological Measurement, 42*, 975–981.

Terrell, C. D. (1982b). Table for converting the point biserial to the biserial. *Educational and Psychological Measurement, 42*, 983–986.

Tinsley, H. E. A., & Tinsley, D. J. (1987). Uses of factor analysis in counseling psychology research. *Journal of Counseling Psychology, 34*, 414–424.

Tomarken, A. J., & Serlin, R. C. (1986). Comparison of ANOVA alternatives under variance heterogeneity and specific noncentrality structures. *Psychological Bulletin, 99*, 90–99.

Toothaker, L. E. (1993). *Multiple comparison procedures*. Sage university paper series on quantitative applications in the social sciences, 07–089. Newbury Park, CA: Sage.

Tufte, E. R. (2001). *The visual display of quantitative information* (2nd ed.). Cheshire, CT: Graphics Press.

Twisk, J. W. R. (2006). *Applied multilevel analysis: a practical guide*. Cambridge: Cambridge University Press.

Umpierre, S. A., Hill, J. A., & Anderson, D. J. (1985). Effect of Coke on sperm motility. *New England Journal of Medicine, 313*(21), 1351–1351.

Wainer, H. (1984). How to display data badly. *American Statistician, 38*(2), 137–147.

Welch, B. L. (1951). On the comparison of several mean values: An alternative approach. *Biometrika, 38*, 330–336.

Wilcox, R. R. (2005). *Introduction to robust estimation and hypothesis testing* (2nd ed.). Burlington, MA: Elsevier.

Wilcoxon, F. (1945). Individual comparisons by ranking methods. *Biometrics, 1*, 80–83.

Wildt, A. R., & Ahtola, O. (1978). *Analysis of covariance*. Sage university paper series on quantitative applications in the social sciences, 07-012. Newbury Park, CA: Sage.

Williams, J. M. G. (2001). *Suicide and attempted suicide*. London: Penguin.

Wright, D. B. (1998). Modeling clustered data in autobiographical memory research: The multilevel approach. *Applied Cognitive Psychology, 12*, 339–357.

Wright, D. B. (2003). Making friends with your data: Improving how statistics are conducted and reported. *British Journal of Educational Psychology, 73*, 123–136.

Wright, D. B., & London, K. (2009). *First steps in statistics* (2nd ed.). London: Sage.

Wright, D. B., & Williams, S. (2003). Producing bad results sections. *The Psychologist, 16*, 646–648.

Yates, F. (1951). The influence of statistical methods for research workers on the development of the science of statistics. *Journal of the American Statistical Association, 46*, 19–34.

Zabell, S. L. (1992). R. A. Fisher and fiducial argument. *Statistical Science, 7*(3), 369–387.

Zwick, R. (1985). Nonparametric one-way multivariate analysis of variance: A computational approach based on the Pillai–Bartlett trace. *Psychological Bulletin, 97*(1), 148–152.

Zwick, W. R., & Velicer, W. F. (1986). Comparison of five rules for determining the number of components to retain. *Psychological Bulletin, 99*(3), 432–442.

INDEX

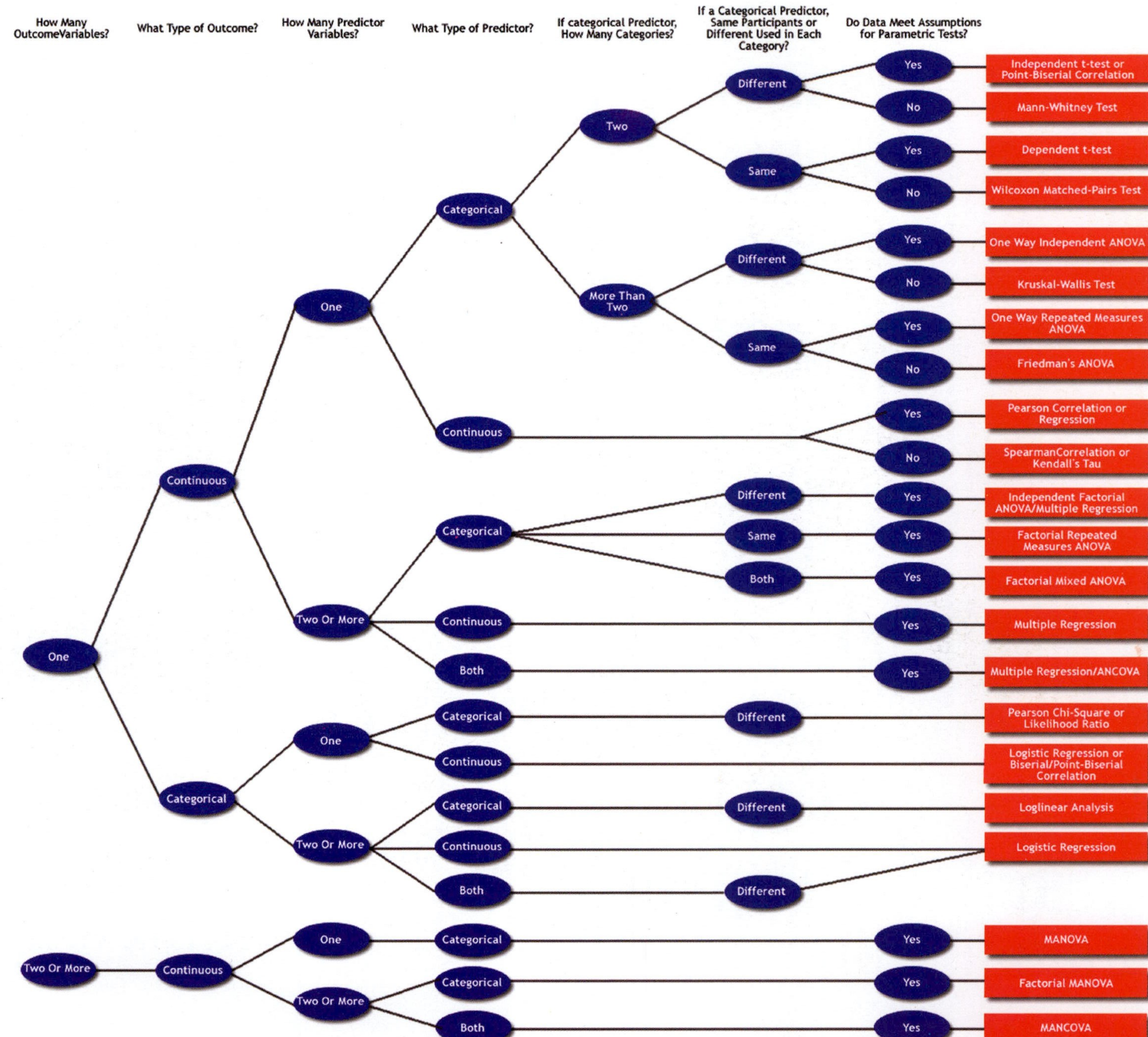
How Many OutcomeVariables?
What Type of Outcome?
How Many Predictor Variables?
What Type of Predictor?
If categorical Predictor, How Many Categories?
If a Categorical Predictor, Same Participants or Different Used in Each Category?
Do Data Meet Assumptions for Parametric Tests?
One
Continuous
One
Categorical
Two
Different
Yes
Independent t-test or Point-Biserial Correlation
No
Mann-Whitney Test
Same
Yes
Dependent t-test
No
Wilcoxon Matched-Pairs Test
More Than Two
Different
Yes
One Way Independent ANOVA
No
Kruskal-Wallis Test
Same
Yes
One Way Repeated Measures ANOVA
No
Friedman's ANOVA
Continuous
Yes
Pearson Correlation or Regression
No
SpearmanCorrelation or Kendall's Tau
Two Or More
Categorical
Different
Yes
Independent Factorial ANOVA/Multiple Regression
Same
Yes
Factorial Repeated Measures ANOVA
Both
Yes
Factorial Mixed ANOVA
Continuous
Yes
Multiple Regression
Both
Yes
Multiple Regression/ANCOVA
Categorical
One
Categorical
Different
Pearson Chi-Square or Likelihood Ratio
Continuous
Logistic Regression or Biserial/Point-Biserial Correlation
Two Or More
Categorical
Different
Loglinear Analysis
Continuous
Logistic Regression
Both
Different
Two Or More
Continuous
One
Categorical
Yes
MANOVA
Two Or More
Categorical
Yes
Factorial MANOVA
Both
Yes
MANCOVA